THE NORTON BOOK OF
NATURE WRITING

THE NORTON BOOK OF
NATURE WRITING

College Edition

Edited by ROBERT FINCH

and JOHN ELDER
Middlebury College

W · W · NORTON & COMPANY · NEW YORK · LONDON

The text of this book is composed in Electra with the display set in
Bernhard Modern
Composition by Tom Ernst
Manufacturing by the Haddon Craftsmen, Inc.
Production manager: Julia Druskin

ISBN 0-393-97816-8 (pbk.)

W. W. Norton & Company, Inc., 500 Fifth Avenue, New York, N.Y. 10110
www.wwnorton.com

W. W. Norton & Company Ltd., Castle House, 75/76 Wells Street, London
W1T 3QT

9 0

CONTENTS

PREFACE

We are pleased to introduce a College Edition of *The Norton Book of Nature Writing*. It offers, for the first time, a paperback edition of this ground-breaking anthology. In addition to its significantly expanded table of contents, the new version responds to numerous requests from instructors wanting to use our book but desiring some study apparatus in order to do so most effectively. This edition has thus been augmented by a *Field Guide* for teachers and students, written by Lilace Mellin Guignard, a writer and teacher at the Center for Environmental Literature of the University of Nevada at Reno. (There continues to be a cloth edition for the trade market, now entitled *Nature Writing: The Tradition in English*, available without any of the study materials.)

The *Field Guide* includes a thematic table of contents with useful groupings of the readings and a set of three study questions for each selection. These questions call attention to literary techniques and themes, suggest connections among various authors in the collection, and invite response to the personal or political challenges posed by particular selections. For students, these questions may offer a further context for their initial encounters with these readings and help to identify rich areas to explore further in their own writing. For teachers, they may serve as prompts for student journals or as topics for discussion in class. It is our hope that these questions will thus enhance the book's usefulness in a wide range of courses, including composition classes focused on the natural environment, introductions to nature writing, and many of the interdisciplinary courses now offered in environmental studies.

We should explain a few matters of format. Ellipses at the beginning or end of a selection or paragraph indicate that the text originally began before our starting point here or continues after our ending. Ellipses centered in a blank line within a selection show that material has been deleted by us at this point, while a blank line without ellipses is a break in the text created by the author. Brackets around a title are used when the title has been supplied by the editors for unnamed excerpts; otherwise, titles for essays or chapters are the authors' own.

Finally, we were fortunate to have the thoughtful comments and suggestions of many colleagues who've taught with our book. In particular, we thank Robert DeMott, Ohio University; Fred Strebeigh, Yale University; David Vanderwerken, Texas Christian University; and Ann Woodlief, Virginia Commonwealth University.

John Elder
Robert Finch

INTRODUCTION

In our introduction to the 1990 *Norton Book of Nature Writing*, we wrote that any such collection might be looked at as "a landscape at a given moment. It may have a pleasing shape, an authority, and a seeming stability. But it remains subject to such processes as erosion and uplift, deposition and depression, even sudden, wholesale shifts of ground. Some new features are brought into being with each new age, only perhaps to disappear in the next."

A dozen years have passed since that volume appeared. It has been a period in which interest in naturist literature has soared, in part because of the ever-increasing prominence of environmental concerns but equally because of the variety and excellence of the writing being produced. Highly gifted new writers have emerged, taking the genre in new directions; established authors have continued to produce important works; older writers whose work had been long neglected have been rediscovered; and new voices from other cultural traditions have been added to the dialogue.

The term *nature writing* came into wide currency several decades before the publication of *The Norton Book of Nature Writing*. One of our guiding impulses behind that particular selection of readings was to trace a literary lineage in English that ran back through Thoreau to Gilbert White. We wanted to evoke the living tradition that included both the eighteenth century's Golden Age of Natural History and the renaissance of nature writing in our own day. Our hope was that the project would thus, in its small way, help both to bring more attention to the voices of contemporary writers and to gain wider recognition for the reflective natural history essay as a legitimate subject for literary study.

During the subsequent years, a number of influences have contributed to the growing visibility of nature writing. The continuing fertility of the writing has spawned numerous anthologies of nature writing, representing both more narrowly defined and more inclusive approaches to the genre.

New essays continue to appear regularly in magazines and journals such as *Orion, Audubon, The Georgia Review,* and *The Hungry Mind Review,* some of which have published special issues devoted to this field. In addition, John Murray's series, *American Nature Writing,* inaugurated in 1994, has provided a yearly showcase for some of the best new work in the genre, and Houghton Mifflin's *Best American Science and Nature Writing* series has recently joined its prestigious annual anthologies of essays and short stories.

Scholarship in the field has flourished, too. The founding of the Association for the Study of Literature and the Environment in 1992 has done much to foster research into nature writing as a genre, as well as to promote dialogue about how its definition might be reconsidered or broadened. Since the appearance of *The Norton Book of Nature Writing,* hundreds of new college courses have been introduced that connect literature and nature, and a number of conferences and writers' workshops are now held annually on related topics. During the same period, writers like John Muir, Mary Austin, Aldo Leopold, Rachel Carson, Edward Abbey, Annie Dillard, and Barry Lopez have begun to be taught regularly alongside writers in more established genres. Such developments have also contributed significantly to the growth of environmental studies programs attempting to integrate the humanities with the sciences.

We believe that these rapid and significant changes do indeed represent wholesale shifts and mark a new epoch in the literary landscape. Our introduction to the earlier collection (still appearing after this one) discussed the evolution of nature writing over a span of more than two centuries. Such a broad perspective informs the present collection as a whole, too. But we want in introducing this revised volume to reflect more specifically on the form's more recent evolution and to share some of the considerations that have shaped our revised table of contents.

As the genre has achieved higher resolution and greater prominence, it has received sharper scrutiny as well. Some readers have noted the extent to which white male writers—as represented both by our own anthology and by those that preceded it—have dominated the tradition of nature writing. We have taken this observation seriously and have been motivated to bring out the present revision in part by a desire to address it. The historical gender imbalance has been increasingly corrected by the large number of women among today's distinguished nature writers, twenty-four of whom we are happy to be adding here. We have also benefited from the work of scholars and anthologists calling attention to a group of women authors who had achieved considerable prominence in the nineteenth and early twentieth centuries but who had largely been forgotten by the middle of the twentieth century. The presence here of such writers as Susan Fenimore Cooper, Mabel Osgood Wright, and Gene Stratton Porter reflects this exciting process of literary recovery.

We were aware, of course, even when compiling *The Norton Book of Nature Writing*, that this particular form of the personal essay has not been the primary genre of choice for many outstanding authors of African American, Native American, Hispanic, Asian American, or other ethnic backgrounds. But a number of writers who are known more for their poetry or fiction have in fact also produced occasional pieces of memorable nature writing. Selections here by writers such as Alice Walker, Jamaica Kincaid, Maxine Hong Kingston, Joseph Bruchac, Richard Wright, Louise Erdrich, and Evelyn White speak in their different ways to the growing consciousness that there can be no fundamental distinction between environmental preservation and social justice, that human compassion and environmental sustainability are branches of the same tree, and that cultural diversity is one of the primary resources we have for ensuring biological diversity.

Such a broadening of the table of contents reflects our recognition that any tradition, if it is to remain vital, must continue to change and incorporate new elements. On the other hand, it also registers the process through which the Thoreauvian inheritance continues to leaven and transform literature of nature in English. A related effort in the current collection has been to find a more international perspective on this genre. Accordingly, we have included more British, Canadian, Australian, and Caribbean authors than in the previous mix. In short, we believe that the list of new authors included in this book demonstrates that nature writing, like other forms of literature and the arts, has been opening its doors to an increasingly diverse group of writers.

Still, despite our belief that *The Norton Book of Nature Writing* reflects a wider variety of cultural perspectives and locales than the previous collection, it is important to emphasize that nature writing as it is exemplified here is far from encompassing the full variety of literature about nature—far even from representing the range of such literature that has been published in English prose. Nor does it seek to do so. There is an important place for anthologies that are unbounded by genre or language or that seek to emphasize other themes, points of view, or cultural traditions. The house of environmental literature is capacious enough to include many green mansions. But our own goal remains to celebrate a distinct and limited, yet enormously rich, literary tradition in English—one whose principal expression has been the personal narrative essay and whose flowering, for various cultural, historical, and geographical reasons, has been especially remarkable in the United States. In this respect, the basic character of *The Norton Book of Nature Writing* is similar to that of the original collection.[1]

Beyond these important matters of definition and inclusion, other features of the table of contents reflect our sense that this is an exciting new moment in the landscape of nature writing. There are, for instance, quite a few pieces by distinguished contemporary authors who have become much more prominent over the past decade. We are grateful for this opportunity to include such fine writers as Rick Bass, Bill McKibben, William Kittredge, Linda Hasselstrom, and Scott Russell Sanders in the new edition; to replace or augment previous selections for authors like Wendell Berry and David Quammen, who were included last time around but who have written significant new works since then; and to have examples of new (or newly published) prose pieces by such major figures as Gary Snyder, Barry Lopez, and Vladimir Nabokov. In addition, we are pleased to include in this book the full text of Thoreau's seminal essay, "Walking."

Three other new features in the landscape of this book should be mentioned, namely, the increased prominence of selections on gardening and agriculture, of animal portraits, and of personal narratives. The added garden and farming essays complement the existing selections on wilderness, drawing on the work of earlier authors like Mabel Osgood Wright, established writers like Wendell Berry and Jamaica Kincaid, and newer voices like Jane Brox, Michael Pollan, and David Mas Masumoto. Such writers have stimulated a new appreciation for cultivated nature as a meeting ground, sometimes an arena, for human desire and natural forces.

An interest in animals as individuals and a depiction of their lives from their own point of view has characterized the work of a number of the early- and mid-twentieth-century nature writers, several of whom we have included in this volume. Prominent among them is Ernest Thompson Seton, a Canadian writer whose true-life animal tales made him one of the most popular writers of his day (and also the target of John Burrough's scathing attack on what he called "nature fakers"). In fact, many of the creators of the modern animal story are Canadian, something that Rebecca Raglan has suggested may reflect a peculiarly Canadian experience with the landscape, namely, that "[I]n the process

---

[1] It should, however, be noted that, in order to accommodate an expanded range of writers, we have decided to be more consistent in restricting our selections to nonfiction in English. Thus, for instance, the excerpts from *Moby-Dick* and Fabre's *Life of the Grasshopper* have been reluctantly deleted.

of confronting the fierceness and loneliness of the northern wilderness, Canadian nature writers have discovered the common bonds they share with other creatures making their home in a difficult land."

Finally, we would note that most of our new choices strengthen the tradition of first-person narrative essays that formed the core of the 1990 *Norton Book of Nature Writing*. While we have tried to make room for a variety of ways of speaking about nature in such modes as pure reflection, philosophy, history, environmental advocacy, traditional myths, testimony and witness, social criticism, and apocalyptic warning, the continued prominence of the *story* reflects the predominately literary intent of this anthology. If there is a clear bias in our choices, it is for the exploratory over the prescriptive, for storytelling over argument or exhortation, for speculation over certainty, and for imagination over analysis. We would argue that nature writing needs to be, as Thomas Babington Macaulay asserted for the writing of history, "not merely traced on the mind but branded into it" through vivid and compelling chronicles of engagement with the perpetual mystery of human existence in a physical and biological universe. The range of this central narrative tradition continues to be augmented by such newer writers as David James Duncan, Trudy Dittmar, Jan Zita Grover, Alison Deming, Ray Gonzalez, Franklin Burroughs, Ellen Meloy, Emily Hiestand, and Sharman Apt Russell.

In the new millennium the genre will undoubtedly take on new shapes, undergo new transformations, and explore new directions. Still, even with all the shifts of topography, population, and climate that have occurred in the past decade, the landscape of nature writing remains in important ways unchanged. It still offers a consoling largeness, into which one may escape from the overheated rooms of human self-absorption and academic specialization. It allows for a vision of wholeness, informed by the natural sciences but attuned to the human and spiritual meanings of our earth and the living communities that it sustains. And it gives access to the unforgettable voices of men and women who, in paying attention to the vivid, dynamic outer world, have also clarified our own sense of identity and purpose.

We have been helped enormously in compiling *The Norton Book of Nature Writing* by the advice, criticism and work of many specialists in literature and the environment. We want especially to acknowledge the suggestions of Scott Slovic, whose knowledge of contemporary environmental literature is unparalleled. Anthologies and bibliographies assembled by the following people have done much to steer us toward new or previously neglected writers and in several cases to particular selections we have chosen for this book: Lorraine Anderson, David

Landis Barnhill, Sean O'Grady, Joseph Barbato, Lisa Weiner-man, Terrell Dixon, Patrick Murphy, John Murray, Daniel Philippon, Sydney Landon Plum, Rochelle Johnson, Daniel Patterson, and Hertha Wong. In addition, these friends and colleagues have inspired and instructed us at many points: Betsy Hilbert, Parker Huber, John Tallmadge, Lawrence Buell, Scott Russell Sanders, Robert Michael Pyle, Ann Zwinger, Cheryll Glotfelty, Michael Branch, John McWilliams, Annie Dillard, John Hanson Mitchell, Dan Peck, Emily Hiestand, Don Mitchell, Janisse Ray, and Lisa New. Even the second time around, we have almost certainly been guilty of glaring omissions. It should go without saying that none of these teachers and guides bears any responsibility for such lapses in knowledge or judgment on our part.

What also often goes without saying is that anthologies like this can be shaped in part by factors other than knowledge and personal taste. One of these is simply space. Once again, despite the generous dimensions allowed us by W. W. Norton, we inevitably had to leave out many fine contemporary nature writers. There are also a few major figures whose work we had hoped to include but whose permission fees exceeded our budgetary limits. We only mention this because it is seldom acknowledged that literary anthologies, like other enterprises, are conditioned to some degree by the realities of the marketplace. On the other hand, we are extremely grateful to those friends, colleagues, editors, and publishers who have donated selections or accepted fees less than they might have commanded in order to help us make this the most comprehensive and representative collection that we could.

Finally, we would like to express our appreciation to John Barstow, our genial and forbearing editor at Norton, to Robyn Renahan, our tireless permissions editor, and to Kirsten George and Anne Majusiak for their valued help with permissions.

> John Elder, Bristol, Vermont
> Robert Finch, Wellfleet, Massachusetts

INTRODUCTION

to the 1990 *Norton Book of Nature Writing*

Nature writing, as a recognizable and distinct tradition in English prose, has existed for over two hundred years. During this time it has engaged the imagination and talents of major literary figures on both sides of the Atlantic, produced works of worldwide influence, and achieved a broad and enthusiastic readership. Since World War II the genre has become an increasingly significant and popular one, producing some of the finest nonfiction prose of our time. Yet too often it continues to be identified with or limited to landscape descriptions, accounts of animal behavior, or treatises on "environmental issues," rather than being acknowledged as a literary form in its own right. This anthology is intended to suggest the broad range of nature writing in English and to indicate the rich tradition within which contemporary writers are working.

Nature writers are the children of Linnaeus. Beginning with his *Systema Naturae* (1735) and elaborating on that general system in such volumes as *Species Plantarum* (1753), Linnaeus introduced a framework within which all living things could be classified and identified. Throughout the second half of the eighteenth century and into the nineteenth, parsons, poets, ladies and gentlemen of leisure in England, explorers and collectors in the wilds of America, all carried their copies of Linnaeus.

We open this collection of nature writing with Gilbert White, one of Linnaeus's early English disciples. A *Natural History of Selborne*, first published in 1789, has exerted a profound and lasting influence on nature writers. Darwin may have taken along the first volume of Charles Lyell's *Principles of Geology* when he shipped on the HMS *Beagle*, but he also found space in his luggage for White. Thoreau's shelf of books at Walden was a short one, but there too A *Natural History of Selborne* was included.

While nature has been the subject of imaginative literature in every country and in every age, nature writing in prose has achieved a unique fullness and continuity within the Anglo-American context. Certain historical facts have contributed to this flourishing tradition. When the gospel according to Linnaeus began to be propagated worldwide, there was already a rich English literature of naturalist theology. Writers like John Ray, in such books as *The Wisdom of God manifested in the Works of Creation* (1691), found in nature a confirmation of their faith. The eighteenth century also saw English empire and industry beginning to produce a large class with the education and leisure to pursue their "natural curiosity." Similarly, the early prominence of nature writing in America is associated with the exploration of a diverse, abundant continent. Naturalists were commissioned, first by amateurs and institutions in Britain and then by the fledgling American government, to travel, draw maps, keep lists, and ship specimens. The cultural climate and the natural environment were thus both auspicious for the burgeoning of a genre of nature writing in English. But in this literary form, as in all others, books are also propagated by previous books. And in ways much more characteristic and specific than the broad trends just mentioned, the tradition represented in this anthology grows out of the entrancing letters of Gilbert White.

In his book *Nature's Economy: The Roots of Ecology,* Donald Worster points to one constant element in nature writing from Gilbert White to the present: "The rise of the natural history essay in the latter half of the nineteenth century was an essential legacy of the Selborne Cult. It was more than a scientific-literary genre modeled after White's pioneering achievement. A constant theme of the nature essayists was the search for a lost pastoral haven, for a home in an inhospitable and threatening world." In Worster's view, subsequent writers like John Burroughs, John Muir, W. H. Hudson, and Richard Jefferies all develop, as White did, models of human life integrated into a beloved landscape. They provide their readers with an antidote to industrialism and urbanization and an alternative to "cold science—not by a retreat into unexamined dogmatism, but by restoring to scientific inquiry some of the warmth, breadth and piety which had been infused into it by the departed parson-naturalist."

The world "pastoral" indicates one useful angle to follow into *A Natural History of Selborne,* as it does into nature writing in general. White was the pastor of Selborne, that Southhampton parish of seven hundred souls into which he had been born. When, in his book's "advertisement," he refers to it as a work of "parochial history," he conveys a proprietary tone that often colors nature writers' descriptions of their chosen or ultimate landscapes—from Thoreau's Concord to Muir's Sierra to Abbey's Arches National Monument. Beyond this parochial dimension, however, "pastoral" relates nature writing to the literary tradition stemming from Theocritus's *Idylls* and Virgil's *Eclogues*. White frequently quotes from Virgil, never more charmingly than when, in his Letter XXXVIII to Daines Barrington, he tests an echo with the *Eclogue's* first line: "Tityre, tu patulae recubans. . ." Reclining under a shady tree, piping and singing love songs, the figure of Tityrus defines the pastoral posture of *otium*, or leisure. But the First Eclogue as a whole makes it clear that such an experience of "the golden age" is always individual and momentary, and that it is always ironically enfolded in a world of cities, labor, and war. It is, explicitly, a fortunate, temporary escape.

Nature writing, on the model of *Selborne*, often depicts a sort of golden age. White's book evokes days of hunting for echoes, nights of testing the hoots of owls with pitch pipes. This English leisure is more active than Roman *otium*, though. White is always walking around looking and poking into things, less like Tityrus under his beech tree than like Winnie the Pooh strolling into the morning: "Well, he was humming this hum to himself, and walking along gaily, wondering what everybody else was doing, and what it felt like being someone else, when suddenly he came to a sandy bank, and in the bank was a large hole. 'Aha!' said Pooh. (*Rum-tum-tiddle-um-tum*)." There is, in fact, an important element of play in much nature writing. One thinks, for instance, of Emerson playing his "ice harp" at Walden Pond or Farley Mowat investigating the diet of wolves by devising his own recipes for Creamed Arctic Mouse. It is as if playing in a landscape were as important as exploring it, or rather, as if the two become one activity in which we rediscover our wholeness as beings in nature.

We enjoy an escapist impulse in certain nature writing, as we do in pastoral poetry and in English children's books by authors like A. A. Milne, Kenneth Grahame, and Beatrix Potter. Escape, first and foremost, from the world of specialized and alienating work. In identifying the period from 1770 to 1880 as "The Golden Age of English Natural History," E. D. H. Johnson emphasized that it was a time when original and important scientific work could be accomplished by classically educated *amateurs*. Gilbert White's letters show us an amateur scientist carrying out original research, in the open air, that is indistinguishable from recreation. We look back at such a life with nostalgia from the divided vantage of C. P. Snow's "two cultures."

If the idyllic flavor of adventures in nature lingers in some writers today, however, the confident and self-contained nature of White's science no longer remains available. Charles Lyell, Darwin, and Alfred Russel Wallace gave a dizzying spin to the static hierarchies of Linnaeus. The dynamism and vastly magnified time frame of the new biology and geology left behind the earlier system, with its tidy, fixed categories. A new era of worldwide biological exploration had also begun, inspired by the South American voyages of Alexander von Humboldt from 1799 to 1804. European scientists were especially staggered by the enormous array of species in the tropics, and the Linnaean system began to look increasingly like an antique stamp album, charming but inadequate.

Darwin himself shared White's classical background, as is reflected in his allusive prose style. But the 1859 publication of *On The Origin of Species* stimulated research in the life-sciences so powerfully that a much higher degree of professionalism and specialization soon developed. As researchers tried to cope with the flood of knowledge, a new protocol of expression developed which was quite different from Darwin's own humane, accessible style. A white-coated, passive, impersonal style became the established voice of "objective science."

In the post-Darwinian world, the older term "natural history" has often been defiantly embraced by nature writers as they respond to the physical creation in ways that, while scientifically informed, are also marked by a personal voice and a concern for literary values. Even if nature writers have often resisted the model of impersonal and specialized science, though, it is also important to note how many of them have been influenced by its concepts, from genetics to molecular biology, from plate tectonics to quantum physics, from population ecology to cognitive theory. No idea has been more influential and inspiring than the grand spectacle of evolution. The selections here by Loren Eiseley, Annie Dillard, and David Rains Wallace, for instance, illustrate how contemporary writers have responded thoughtfully and lyrically to the metaphysical and mythic implications of an evolutionary vision of creation.

Gilbert White established the pastoral dimension of nature writing in the late eighteenth century and remains the patron saint of English nature writing. Henry David Thoreau was an equally crucial figure in mid-nineteenth century America and occupies a similar niche in our country's literary hagiography. The fact that Thoreau is more fully represented in this collection than any other writer reflects our sense that he both touches the genre's roots and anticipates its flowering in this century. In his journals and books he evokes an outwardly quiet life, like White's, of gathered attentiveness to the earth. Walden Pond, no less than Selborne, has been a site of pilgrimage by readers wanting to find an antidote to their own "lives of quiet desperation." Like White, too, Thoreau is deliberately and fiercely parochial and an extraordinarily gifted observer of natural details.

On several important levels, though, this American solitary diverges from his English forerunner. Whereas White conveys a sense of being unconsciously a part of the natural order he beheld, Thoreau brings an ironic awareness to his nature writing, continually recognizing in his wry style that by focusing on nonhuman nature we objectify and abstract it. He acknowledges the advantages of a woodcutter's unlettered response to the woods, and dramatizes the conflicts between his own reverence for "higher laws" and his desire to eat a woodchuck raw. Thoreau's supremely self-conscious style has kept him continuously available to readers who no longer draw a confident distinction between humanity and the rest of the world, and who would find a simpler worship of nature both archaic and incredible.

Another significant way in which Thoreau anticipates contemporary nature writing is in his recognition that the natural environment must be protected. Readers of White have remarked that, though he lived through the American and French Revolutions, as well as the first wave of industrialization in England, none of those events comes into his letters. Moreover, as Mark Daniel noted in his 1983 selection, *The Essential Gilbert White of Selborne*: "White had no sense of responsibility to something called 'nature.'" By contrast, Thoreau constantly reflects about such modern incursions into the landscape as the railroad that ran near his cabin at Walden Pond, and in *The Maine Woods* he calls for actions to preserve our great forests from devastation. Because he knows personally and experimentally that "in wildness is the preservation of the world," he searches single-mindedly for ways in which both literature and society might become more integrally "wild."

The second half of the nineteenth century saw the origins of what we today call the environmental movement. Two of its most influential American voices were John Muir and John Burroughs, literary sons of Thoreau, though hardly twins. Muir led the fight to preserve wilderness with his *Century* articles about Yosemite and Hetch Hetchy, as well as with his founding of the Sierra Club. Burroughs popularized the study of local nature with his many volumes of "ramble" essays, and brought political and economic muscle into the conservation movement by befriending such influential figures as Theodore Roosevelt and Harvey Firestone. W. H. Hudson, writing in England, was a major force in the protection of endangered bird populations, and the rural essays of Richard Jefferies helped create an aesthetic foundation for the landscape preservation work of the National Trust.

In the early twentieth century the activist voice and prophetic anger of nature writers who saw, in Muir's words, that "the money changers were in the temple" continued to grow. Building upon the principles of scientific ecology that were being developed in the 1930s and 1940s, Rachel Carson and Aldo Leopold sought to create a literature in which appreciation of nature's wholeness would lead to ethical principles and social programs.

Today, nature writing in America flourishes as never before. Nonfiction may well be the most vital form of current American literature, and a notable proportion of the best writers of nonfiction practice nature writing. The generous selection of contemporary writers in this anthology reflects not only the impressive flowering of the genre in recent decades but also its extraordinary range of voice and achievement. It has attracted poets such as John Hay, Wendell Berry, and John Haines, novelists like John Updike, Peter Matthiessen, John Fowles, and Ursula K. Le Guin, as well as such preeminent contemporary essayists as Annie Dillard, Edward Hoagland, John McPhee, and Barry Lopez. On the other hand some of its most notable practitioners have been professional scientists in their own right. Writers like Rachel Carson, Loren Eiseley, Lewis Thomas, Chet Raymo, and E. O. Wilson have, in their literary essays, imbued their respective fields of marine biology, anthropology, cellular biology, astronomy, and sociobiology with humanistic concerns and poetic resonance. Writers from such diverse regions of the humanities and the sciences find common ground in the older and more-generalized field of natural history, which, as John Fowles puts it, assumed "that it [was] being presented by an entire human being, with all his complexities, to an audience of other entire human beings."

One common element in so diverse a company of writers is suggested by Thoreau's term "excursions." Contemporary nature writers characteristically take walks through landscapes of associations. Beginning with a closely observed phenomenon, they reflect upon its personal meaning for them. Or, beginning with an argumentative point of view, they venture out into a natural setting that has no vested interest in their opinions and that contradicts or distracts as often as it confirms. In an age that has learned that any theory is subject to almost constant revision, a hallmark of the modern nature essay is its insistent open-endedness.

This process of association often reminds nature writers of other texts as well as other landscapes. Like poets and novelists, they are frequently both inspired by and critical of their predecessors and contemporaries. W. H. Hudson's respectful rebuttal to Melville's "The Whiteness of the Whale" or Henry Beston's homage to "the obstinate and unique genius of Thoreau" or Edward Abbey's remarks on virtually any other writer all provide examples of the ambivalent esteem nature writers have inspired in one another.

To a distinctive degree, nature writing fulfills the essay's purpose of *connection*. It fuses literature's attention to style, form, and the inevitable ironies of expression with a scientific concern for palpable fact. In a time when the natural context of fiction has been attenuated and when much literary theory discovers nothing to read but constructs of self-reflexive language, nature writing asserts both the humane value of literature and the importance to a mature individual's relationship with the world of understanding fundamental physical and biological processes. Alfred North Whitehead called for just such a balance in this eloquent passage from *Science and the Modern World:* "What is wanted is an appreciation of the infinite variety of vivid values achieved by an organism in its proper environment. When you understand all about the sun and all about the atmosphere and all about the rotation of the earth, you may still miss the radiance of the sunset. There is no substitute for the direct perception of the concrete achievement of a thing in its actuality. We want concrete fact with a high light thrown upon what is relevant to its preciousness."

We should include a few remarks about how we chose the selections in this anthology. Our primary aim was to represent, as fully as possible, the range of nature writing in English over the past two centuries. It was particularly important, we felt, to place before American readers the rich and continuing tradition of British nature writing. This tradition often reflects a significantly different sense of landscape and the role of people in it than does its American counterpart, but it also feeds into and enriches the latter.

For the sake of more clearly defining the genre, we have as a rule restricted our selections to nonfiction prose, recognizing, however, that genres do not exist in isolation anymore than writers do. For that matter, in its imaginative cast, nature writing often seems to bear more resemblance to lyric poetry than it does to many other forms of essay writing. It is no accident that the prose works of Coleridge, John Clare, Whitman, Hopkins, and Edward Thomas find a place here.

Landscape and natural settings have also been central elements in the great tradition of both British and American novelists. With writers like Emily Brontë, Thomas Hardy, D. H. Lawrence, Mark Twain, Willa Cather, and William Faulkner, the natural presence often assumes the importance of a major character. Moreover, the work of these and other novelists has strongly influenced many nature writers. Still, the purposes of fiction differ sufficiently from those of nonfiction that, with a few unabashed exceptions, we have chosen to exclude excerpts from novels. The major exception, of course, is *Moby-Dick*. One could argue that much of this archetypal American novel qualifies as nonfiction, and no less a nature writer than Annie Dillard has called it "the best book ever written about nature." But even without arguing its generic status, it seemed clear to us that no representation of nature writing in America would be complete without taking into account Melville and his leviathan.

By restricting our selections to original works in English, we have inevitably excluded such notable literary naturalists as Maurice Maeterlinck, Karl von Frisch, Mikhail Prishvin, and Konrad Lorenz. Here again, however, we have made one important exception: the famous description of the mating habits of the praying mantis by French entomologist Jean Henri Fabre. Not only has Fabre's work, and this essay in particular, had a strong influence on contemporary American nature writers, but the English translation by Alexander Teixeira de Mattos has achieved a classic status of its own.

One thing that struck us in gathering readings for this book was that, although the figure and experience of the Native American come into much early American nature writing, examples of nature writing by Indians in English have not often been available until our day. This is obviously in part a linguistic matter: we have mainly the Victorian translations of a few formal speeches. But it may also reflect the fact that a certain kind of intense and self-conscious awareness of nature follows from a loss of integration between society and nature. Today, through Native American writers like Scott Momaday and Leslie Marmon Silko, as well as through non-Indian writers like Richard Nelson and Gary Nabhan who are seeking to understand the Indian way of knowing the earth, this voice flows into and amplifies the tradition.

The personal element—that is, the filtering of experience through an individual sensibility—is central to what we view as the nature writing tradition. For this reason we have not included selections from a whole line of fine scientific essayists from Thomas Huxley to Stephen Jay Gould, though the ties between the two traditions are strong and significant. On the other hand, we have tried to convey a broader notion of what constitutes "nature" and nature writing by bringing in such nonmainstream examples as Thomas Merton's essay on the teleological connotations of an autumn rainstorm, a chapter from Vicki Hearne's book on the philosophical implications of training animals, and Richard Selzer's treatment of the human body as a landscape.

Though we hope to illuminate a tradition and showcase a genre, we have included no selection simply because it is representative of a particular current or period of nature writing. What finally matters is the vitality of the material itself. Is it a good read? Does each piece, whether old or new, convey a literary power and a character of its own for the contemporary reader? Reflecting our belief that our own time is seeing a remarkable richness in this field, nearly half of the selections are from works published after 1945.

Despite the generous space afforded us by our publisher, we were forced to leave out much we would have liked to include. One unavoidable fact is that some good writers simply do not write, or cannot be adequately represented, in anthologizable dimensions (as Dickens and George Eliot, say, rarely get into English literature anthologies). It also goes without saying that the selections inevitably represent the editors' personal tastes and predilections. But more than that, each age finds its own tastes and needs. Readers will find some familiar names missing; others, once immensely popular, seem to speak to us less strongly today. The anthologist builds in part on previous anthologies, revising, adding, weeding out, recognizing that these new choices, even if right for his or her time, will likewise be revised in turn. To borrow an analogy from the natural sciences: an anthology is somewhat like a landscape at a given moment. It may have a pleasing shape, an authority and a seeming stability. But it remains subject to such processes as erosion and uplift, deposition and depression—even sudden, wholesale shifts of ground. Some new features are brought into being with each new age, only perhaps to disappear in the next. Meanwhile the process as a whole remains vital and dynamic. What is strong endures; what is good shines.

After many months of comparing and sorting our enthusiasms, we sifted through our shopping bags of Xeroxed material to see what we had. What emerged was not so much the prominence of certain recurring ideas and themes, or even patterns of emotions and attitudes evoked by a contemplation of what Thoreau calls "the unhandseled globe." Rather, we were reminded of the English biologist J. S. B. Haldane's remark when asked what he could say about the personality of God on the basis of his evolutionary studies. He replied that the Creator seems to have had "an inordinate fondness for beetles."

Reading this anthology, one might reach the conclusion that nature writers have an inordinate fondness for snakes, turtles, rivers, farm fields, the moon, night, rain, spring, sex, ants, and the nasty habits of insects in general. Birds, on the other hand, though well-represented, are not as prevalent as one might have expected, and fish are virtually ignored. Some geographical areas—the Rocky Mountains, the deserts of the American Southwest, East Africa, the south of England, Cape Cod—receive much attention. Others, such as the southeastern states (with the important exception of William Bartram's *Travels*) are inexplicably underrepresented, and still others, Antarctica for instance, have yet to find their celebrants. Such disparities may, of course, be simply a matter of editorial bias and of historical and cultural circumstance. On the other hand, one is tempted to ask if there may be organisms and locales for which the naturalist imagination has innate affinities. As Barry Lopez speculates when observing the lack of narwhals in Eskimo mythology, are there some animals that are "good to imagine" and others that are not? Or as E. O. Wilson asks in his book *Biophilia*, do human beings instinctively gravitate toward a certain kind of landscape—what he calls "the right place"?

There are certain subjects and ideas which seem to recur persistently, attracting writers of very different temperaments. The ethics of hunting, for instance, are examined in some of the very earliest encounters of Europeans with the enormous herds of mammals on the Great Plains, and the moral implications of taking or exploiting the life of other creatures remains the central topic in many contemporary essays. Evolution has become a nearly universally accepted intellectual principle; yet, as many of these selections illustrate, it contains philosophical, ethical, and social implications with which we have not yet come to terms. Nature's ability to provide relief or solace from human grief has been a constant theme of nature writers. Yet that ability has been called increasingly into question in an age when human activity seems to threaten the planet's basic life-support systems and every landscape is, in Terrence Des Pres's phrase, part of a "nuclear grid." Finally, the recognition of valid alternative conceptions of nature in other cultures has led to a series of provocative inquiries: How does the human mind make sense of nature? How does human meaning in general derive from natural phenomena? And what, ultimately, constitutes the perennial attraction of and need for things natural in our lives?

In the end, however, all of these selections seem to share a larger fundamental intent. In the two centuries since the publication of *A Natural History of Selborne*, nature writers have undertaken excursions away from the dominant literary and scientific models, returning with their testimony about how human beings respond to what is nonhuman, and how individuals and society may achieve more significant and rewarding integration with the earth that sustains them. All literature, by illuminating the full nature of human existence, asks a single question: how shall we live? In our age that question has taken its most urgent form in relation to the natural environment. Because it has never been more necessary, the voice of nature writing has never been stronger than it is today.

John Elder
Robert Finch

Shall I not have intelligence with the earth? Am I not partly leaves and vegetable mould myself?

—Henry David Thoreau, *Walden*

It is interesting to contemplate a tangled bank, clothed with many plants of many kinds, with birds singing on the bushes, with various insects flitting about, and with worms crawling through the damp earth, and to reflect that these elaborately constructed forms, so different from each other and dependent upon each other in so complex a manner, have all been produced by laws acting around us.

—Charles Darwin, *On the Origin of Species*

GILBERT WHITE
1720–1793

The Natural History and Antiquities of Selborne (1789) *presents the letters written by a country curate to two fellow naturalists in other parts of England. Gilbert White's reports, speculations, and queries add up to an engaging self-portrait of a man who delighted in the natural surroundings of the village where he was born, where he served as a priest, and where he eventually died. White's duties in the parish, where his father had served as a priest before him, seem to have left him plenty of time to observe martins, take long daily walks, and correspond with Thomas Pennant and Daines Barrington about "natural curiosities." His remarkable powers of observation and analysis lend vividness to White's letters. As he contemplated the role of earthworms in working the soil, observed the daily habits of his resident tortoise Timothy, or described with precision the aerial mating of swifts, he drew attention to patterns in nature which no one before him had discerned so sharply. He delightfully typified the ideal of amateur science—approaching the physical creation in a playful spirit and making original, substantial contributions to knowledge.*

From THE NATURAL HISTORY AND ANTIQUITIES OF SELBORNE

LETTERS TO THOMAS PENNANT, ESQUIRE

LETTER II

[The Raven-tree]

*   *   * In the centre of this grove there stood an oak, which, though shapely and tall on the whole, bulged out into a large excrescence about the middle of the stern. On this a pair of ravens had fixed their residence for such a series of years, that the oak was distinguished by the title of the Raven-tree. Many were the attempts of the neighbouring youths to get at this eyry: the difficulty whetted their inclinations, and each was ambitious of surmounting the arduous task. But, when they arrived at the swelling, it jutted out so in their way, and was so far beyond their grasp, that the most daring lads were awed, and acknowledged the undertaking to be too hazardous. So the ravens built on, nest upon nest, in perfect security, till the fatal day arrived in which the wood was to be levelled. It was in the month of February, when those birds usually sit. The saw was applied to the butt, the wedges were inserted into the opening, the woods echoed to the heavy blows of the beetle or mallet, the tree nodded to its fall; but still the dam sat on. At last, when it gave way, the bird was flung from her nest; and, though her parental affection deserved a better fate, was whipped down by the twigs, which brought her dead to the ground.

LETTER VIII

[Natural Economy]

*   *   * A circumstance respecting these ponds, though by no means peculiar to them, I cannot pass over in silence; and that is, that instinct by which in summer all the kine, whether oxen, cows, calves, or heifers, retire constantly to the water during the hotter hours; where, being more exempt from flies, and inhaling the coolness of that element, some belly deep, and some only to mid-leg, they ruminate and solace themselves from about ten in the morning till four in the afternoon, and then return to their feeding. During this great proportion of

The Natural History and Antiquities of Selborne (London: B. White and Son, 1789).

the day they drop much dung, in which insects nestle; and so supply food for the fish, which would be poorly subsisted but from this contingency. Thus nature, who is a great economist, converts the recreation of one animal to the support of another!

LETTER XXVII

[Hedge-hogs]

Selborne, Feb. 22, 1770.

Dear Sir,

Hedge-hogs abound in my gardens and fields. The manner in which they eat their roots of the plantain in my grass-walks is very curious: with their upper mandible, which is much longer than their lower, they bore under the plant, and so eat the root off upwards, leaving the tuft of leaves untouched. In this respect they are serviceable, as they destroy a very troublesome weed; but they deface the walks in some measure by digging little round holes. It appears, by the dung that they drop upon the turf, that beetles are no inconsiderable part of their food. In June last I procured a litter of four or five young hedge-hogs, which appeared to be about five or six days old; they, I find, like puppies, are born blind, and could not see when they came to my hands. No doubt their spines are soft and flexible at the time of their birth, or else the poor dam would have but a bad time of it in the critical moment of parturition: but it is plain that they soon harden; for these little pigs had such stiff prickles on their backs and sides as would easily have fetched blood, had they not been handled with caution. Their spines are quite white at this age; and they have little hanging ears, which I do not remember to be discernible in the old ones. They can, in part, at this age draw their skin down over their faces; but are not able to contract themselves into a ball as they do, for the sake of defence, when full grown. The reason, I suppose, is, because the curious muscle that enables the creature to roll itself up into a ball was not then arrived at its full tone and firmness. Hedge-hogs make a deep and warm *hybernaculum* with leaves and moss, in which they conceal themselves for the winter: but I never could find that they stored in any winter provision, as some quadrupeds certainly do.

* * *

I have somewhat to inform you of concerning the moose-deer; but in general foreign animals fall seldom in my way; my little intelligence is confined to the narrow sphere of my own observations at home.

LETTERS TO THE HONORABLE DAINES BARRINGTON

LETTER VII

[Timothy the Tortoise]

Ringmer, near Lewes, Oct. 8, 1770.

* * * A land tortoise, which has been kept for thirty years in a little walled court belonging to the house where I now am visiting, retires under ground about the middle of November, and comes forth again about the middle of April. When it first appears in the spring it discovers very little inclination towards food; but in the height of summer grows voracious: and then as the summer declines it's appetite declines; so that for the last six weeks in autumn it hardly eats at all. Milky plants, such as lettuces, dandelions, sowthistles, are it's favourite dish. In a neighbouring village one was kept till by tradition it was supposed to be an hundred years old. An instance of vast longevity in such a poor reptile!

LETTER XIII

[Timothy Digs In]

April 12, 1772.

Dear Sir,

While I was in Sussex last autumn my residence was at the village near Lewes, from whence I had formerly the pleasure of writing to you. On the first of November I remarked that the old tortoise, formerly mentioned, began first to dig the ground in order to the forming it's hybernaculum, which it had fixed on just beside a great tuft of hepaticas. It scrapes out the ground with it's fore-feet, and throws it up over it's back with it's hind; but the motion of it's legs is ridiculously slow, little exceeding the hour-hand of a clock; and suitable to the composure of an animal said to be a whole month in performing one feat of copulation. Nothing can be more assiduous than this creature night and day in scooping the earth, and forcing it's great body into the cavity; but, as the noons of that season proved unusually warm and sunny, it was continually interrupted, and called forth by the heat in the middle of the day; and though I continued there till the thirteenth of November, yet the work remained unfinished. Harsher weather, and frosty mornings, would have quickened it's operations. No part of it's behaviour ever struck me more than the extreme timidity it always expresses with regard to rain; for though it has a shell that would secure it against the wheel of a loaded cart, yet does it discover as much solicitude about rain as a lady dressed in all her best attire, shuffling away on the first sprinklings, and running its head up in a corner. If attended to, it becomes an excellent weatherglass; for as sure as it walks elate, and as it were on tiptoe, feeding with great earnestness in a morning, so sure will it rain before night. It is totally a diurnal animal, and never pretends to stir after it becomes dark. The tortoise, like other reptiles, has an arbitrary stomach as well as lungs; and can refrain from eating as well as breathing for a great part of the year. When first awakened it eats nothing; nor again in the autumn before it retires: through the height of the summer it feeds voraciously, devouring all the food that comes in it's way. I was much taken with it's sagacity in discerning those that do it kind offices: for, as soon as the good old lady comes in sight who has waited on it for more than thirty years, it hobbles towards it's benefactress with aukward alacrity; but remains inattentive to strangers. Thus not only *"the ox knoweth his owner, and the ass his master's crib,"*[1] but the most abject reptile and torpid of beings distinguishes the hand that feeds it, and is touched with the feelings of gratitude!

I am, &c. &c.

P.S. In about three days after I left Sussex the tortoise retired into the ground under the hepatica.

LETTER XIV

[Parental Instincts in Birds]

Selborne, March 26, 1773.

Dear Sir,

The more I reflect on the στοργη [parental love] of animals, the more I am astonished at it's effects. Nor is the violence of this affection more wonderful than the shortness of it's duration. Thus every hen is in her turn the virago of the yard, in proportion to the helplessness of her brood; and will fly in the face of a dog or a sow in defence of those chickens, which in a few weeks she will drive before her with relentless cruelty.

✻ ✻ ✻

The flycatcher of the *Zoology* (the *stoparola* of Ray) builds every year in the vines that grow on the walls of my house. A pair of these little birds had one year inadvertently placed their nest on a naked bough, perhaps in a shady time, not being aware of the inconvenience that fol-

---

[1] *Isaiah i, 3.* [Gilbert White's note]

lowed. But an hot sunny season coming on before the brood was half fledged, the reflection of the wall became insupportable, and must inevitably have destroyed the tender young, had not affection suggested an expedient, and prompted the parent-birds to hover over the nest all the hotter hours, while with wings expanded, and mouths gaping for breath, they screened off the heat from their suffering offspring.

A farther instance I once saw of notable sagacity in a willowwren, which had built in a bank in my fields. This bird a friend and myself had observed as she sat in her nest; but were particularly careful not to disturb her, though we saw she eyed us with some degree of jealousy. Some days after as we passed that way we were desirous of remarking how this brood went on; but no nest could be found, till I happened to take up a large bundle of long green moss, as it were, carelessly thrown over the nest in order to dodge the eye of any impertinent intruder.

A still more remarkable mixture of sagacity and instinct occurred to me one day as my people were pulling off the lining of an hotbed, in order to add some fresh dung. From out of the side of this bed leaped an animal with great agility that made a most grotesque figure; nor was it without great difficulty that it could be taken; when it proved to be a large white-bellied field-mouse with three or four young clinging to her teats by their mouths and feet. It was amazing that the desultory and rapid motions of this dam should not oblige her litter to quit their hold, especially when it appeared that they were so young as to be both naked and blind!

To these instances of tender attachment, many more of which might be daily discovered by those that are studious of nature, may be opposed that rage of affection, that monstrous perversion of the στοργη [parental love], which induces some females of the brute creation to devour their young because their owners have handled them too freely, or removed them from place to place! Swine, and sometimes the more gentle race of dogs and cats, are guilty of this horrid and preposterous murder. When I hear now and then of an abandoned mother that destroys her offspring, I am not so much amazed; since reason perverted, and the bad passions let loose, are capable of any enormity: but why the parental feelings of brutes, that usually flow in one most uniform tenor, should sometimes be so extravagantly diverted, I leave to abler philosophers than myself to determine.

I am, etc.

LETTER XXI

[Swifts]

Selborne, Sept. 28, 1774.

Dear Sir,

As the swift or black-martin is the largest of the British *hirundines*, so is it undoubtedly the latest comer. For I remember but one instance of its appearing before the last week in April: and in some of our late frosty, harsh springs, it has not been seen till the beginning of May. This species usually arrives in pairs.

The swift, like the sand-martin, is very defective in architecture, making no crust, or shell, for its nest; but forming it of dry grasses and feathers, very rudely and inartificially put together. With all my attention to these birds, I have never been able once to discover one in the act of collecting or carrying in materials: so that I have suspected (since their nests are exactly the same) that they sometimes usurp upon the house-sparrows, and expel them, as sparrows do the house and sand-martin; well remembering that I have seen them squabbling together at the entrance of their holes; and the sparrows up in arms, and much disconcerted at these intruders. And yet I am assured, by a nice observer in such matters, that they do collect feathers for their nests in Andalusia; and that he has shot them with such materials in their mouths.

Swifts, like sand-martins, carry on the business of nidification quite in the dark, in crannies of castles, and towers, and steeples, and upon the tops of the walls of churches under the roof; and therefore cannot be so narrowly watched as those species that build more openly: but, from what I could ever observe, they begin nesting about the middle of May; and I have remarked, from eggs taken, that they have sat hard by the ninth of June. In general they haunt tall buildings, churches, and steeples, and breed only in such: yet in this village some pairs frequent the lowest and meanest cottages, and educate their young under those thatched roofs. We remember but one instance where they breed out of buildings; and that is in the sides of a deep chalk-pit near the town of Odiham, in this county, where we have seen many pairs entering the crevices, and skimming and squeaking round the precipices.

As I have regarded these amusive birds with no small attention, if I should advance something new and peculiar with respect to them, and different from all other birds, I might perhaps be credited; especially as my assertion is the result of many years' exact observation. The fact that I would advance is, that swifts tread, or copulate, on the wing: and I would wish any nice observer, that is startled at this supposition, to use his own eyes, and I think he will soon be convinced. In another class of animals, viz., the insect, nothing is so common as to see the different species of many genera in conjunction as they fly. The swift is almost continually on the wing; and as it never settles on the ground, on trees, or roofs, would seldom find opportunity for amorous rites, was it not enabled to indulge them in the air. If any person would watch these birds of a fine morning in May, as they are sailing round at a great height from the ground, he would see, every now and then, one drop on the back of another, and both of them sink down together for many fathoms with a loud piercing shriek. This I take to be the juncture when the business of generation is carrying on.

As the swift eats, drinks, collects materials for its nest, and, as it seems, propagates on the wing; it appears to live more in the air than any other bird, and to perform all functions there save those of sleeping and incubation.

* * *

Swifts are very anomalous in many particulars, dissenting from all their congeners not only in the number of their young, but in breeding but once in a summer; whereas all the other British *hirundines* breed invariably twice. It is past all doubt that swifts can breed but once, since they withdraw in a short time after the flight of their young, and some time before their congeners bring out their second brood. We may here remark, that, as swifts breed but once in a summer, and only two at a time, and the other *hirundines* twice, the latter, who lay from four to six eggs, increase at an average five times as fast as the former.

But in nothing are swifts more singular than in their early retreat. They retire, as to the main body of them, by the tenth of August, and sometimes a few days sooner: and every straggler invariably withdraws by the twentieth, while their congeners, all of them, stay till the beginning of October; many of them all through that month, and some occasionally to the beginning of November. This early retreat is mysterious and wonderful, since that time is often the sweetest season in the year. But, what is more extraordinary, they begin to retire still earlier in the most southerly parts of Andalusia, where they can be no ways influenced by any defect of heat; or, as one might suppose, defect of food. Are they regulated in their motions with us by a failure of food, or by a propensity to moulting, or by a disposition to rest after so rapid a life, or by what? This is one of those incidents in natural history that not only baffles our searches, but almost eludes our guesses!

* * *

In London a party of swifts frequents the Tower, playing and feeding over the river just below the bridge; others haunt some of the churches of the Borough next the fields; but do not venture, like the house-martin, into the close crowded part of the town.

The Swedes have bestowed a very pertinent name on this swallow, calling it *ring swala*, from the perpetual rings or circles that it takes round the scene of its nidification.

Swifts feed on *coleoptera*, or small beetles with hard cases over their wings, as well as on the softer insects; but it does not appear how they can procure gravel to grind their food, as swallows do, since they never settle on the ground. Young ones, over-run with *hippoboscœ*, are sometimes found, under their nests, fallen to the ground: the number of vermin rendering their abode insupportable any longer. They frequent in this village several abject cottages: yet a succession still haunts the same unlikely roofs: a good proof this that the same birds return to the same spots. As they must stoop very low to get up under these humble eaves, cats lie in wait, and sometimes catch them on the wing.

On the fifth of July, 1775, I again untiled part of a roof over the nest of a swift. The dam sat in the nest; but so strongly was she affected by natural στοργη [parental love] for her brood, which she supposed to be in danger, that, regardless of her own safety, she would not stir, but lay sullenly by them, permitting herself to be taken in hand. The squab young we brought down and placed on the grass-plot, where they tumbled about, and were as helpless as a new-born child. While we contemplated their naked bodies, their unwieldy disproportioned *abdomina*, and their heads, too heavy for their necks to support, we could not but wonder when we reflected that these shiftless beings in a little more than a fortnight would be able to dash through the air almost with the inconceivable swiftness of a meteor; and perhaps, in their emigration must traverse vast continents and oceans as distant as the equator. So soon does nature advance small birds to their ἡλικια [full development], or state of perfection; while the progressive growth of men and large quadrupeds is slow and tedious!

I am, etc.

LETTER XXVII

[The Bee-boy]

Selborne, Dec. 12, 1775.

Dear Sir,

We had in this village more than twenty years ago an idiot-boy, whom I well remember, who, from a child, shewed a strong propensity to bees; they were his food, his amusement, his sole object. And as people of this cast have seldom more than one point in view, so this lad exerted all his few faculties on this one pursuit. In the winter he dosed away his time, within his father's house, by the fire side, in a kind of torpid state, seldom departing from the chimney-corner; but in the summer he was all alert, and in quest of his game in the fields, and on sunny banks. Honey-bees, humble-bees, and wasps, were his prey wherever he found them: he had no apprehensions from their stings, but would seize them *nudis manibus*, and at once disarm them of their weapons, and suck their bodies for the sake of their honey-bags. Sometimes he would fill his bosom between his shirt and his skin with a number of these captives; and sometimes would confine them in bottles. He was a very *merops apiaster, or bee-bird*; and very injurious to men that kept bees; for he would slide into their bee-gardens, and, sitting down before the stools, would rap with his fingers on the hives, and so take the bees as they came out. He has been known to overturn hives for the sake of honey, of which he was passionately fond. Where metheglin was making he would linger round the tubs and vessels, begging a draught of what he called *bee-wine*. As he ran about he used to make a humming noise with his lips, resembling the buzzing of bees. This lad was lean and sallow, and of a cadaverous complexion; and, except in his favourite pursuit, in which he was wonderfully adroit, discovered no manner of understanding. Had his capacity been better, and directed to the same object, he had perhaps abated much of our wonder at the feats of a more modern exhibiter of bees: and we may justly say of him now,

> ————————Thou,
> Had thy presiding star propitious shone,
> Should'st Wildman be ————.

When a tall youth he was removed from hence to a distant village, where he died, as I understand, before he arrived at manhood.

I am, &c.

LETTER XXXI

[A Pregnant Viper]

Selborne, April 29, 1776.

Dear Sir,

On August the 4th, 1775, we surprised a large viper, which seemed very heavy and bloated, as it lay in the grass basking in the sun. When we came to cut it up, we found that the abdomen was crowded with young, fifteen in number; the shortest of which measured full seven inches, and were about the size of full-grown earth-worms. This little fry issued into the world with the true viper-spirit about them, shewing great alertness as soon as disengaged from the belly of the dam: they twisted and wriggled about, and set themselves up, and gaped very wide when touched with a stick, shewing manifest tokens of menace and defiance, though as yet they had no manner of fangs that we could find, even with the help of our glasses.

To a thinking mind nothing is more wonderful than that early instinct which impresses young animals with the notion of the situation of their natural weapons, and of using them properly in their own defence, even before those weapons subsist or are formed. Thus a young cock will spar at his adversary before his spurs are grown; and a calf or a lamb will push with their heads before their horns are sprouted. In the same manner did these young adders attempt to bite before their fangs were in being. The dam however was furnished with very formidable ones, which we lifted up (for they fold down when not used) and cut them off with the point of our scissars.

There was little room to suppose that this brood had ever been in the open air before; and that they were taken in for refuge, at the mouth of the dam, when she perceived that danger was approaching; because then probably we should have found them somewhere in the neck, and not in the abdomen.

LETTER XXXV

[Earthworms]

Selborne, May 20, 1777.

Dear Sir,

Lands that are subject to frequent inundations are always poor; and probably the reason may be because the worms are drowned. The most insignificant insects and reptiles are of much more consequence, and have much more influence in the economy of nature, than the incurious are aware of; and are mighty in their effect, from their minuteness, which renders them less an object of attention; and from their numbers and fecundity. Earth-worms, though in appearance a small and despicable link in the chain of nature, yet, if lost, would make a lamentable chasm. For, to say nothing of half the birds, and some quadrupeds, which are almost entirely supported by them, worms seem to be the great promoters of vegetation, which would proceed but lamely without them, by boring, perforating, and loosening the soil, and rendering it pervious to rains and the fibres of plants, by drawing straws and stalks of leaves and twigs into it; and, most of all, by throwing up such infinite numbers of lumps of earth called worm-casts, which, being their excrement, is a fine manure for grain and grass. Worms probably provide new soil for hills and slopes where the rain washes the earth away; and they affect slopes, probably to avoid being flooded. Gardeners and farmers express their detestation of worms; the former because they render their walks unsightly, and make them much work: and the latter because, as they think, worms eat their green corn. But these men would find that the earth without worms would soon become cold, hard-bound, and void of fermentation; and consequently sterile: and besides, in favour of worms, it should be hinted that green corn, plants, and flowers, are not so much injured by them as by many species of *coleoptera* (scarabs), and *tipulœ* (long-legs), in their larva, or grub-state; and by unnoticed myriads of small shell-less snails, called slugs, which silently and imperceptibly make amazing havoc in the field and garden.

These hints we think proper to throw out in order to set the inquisitive and discerning to work.

A good monography of worms would afford much entertainment and information at the same time, and would open a large and new field in natural history. Worms work most in the spring; but by no means lie torpid in the dead months; are out every mild night in the winter, as any person may be convinced that will take the pains to examine his grassplots with a candle; are hermaphrodites, and much addicted to venery, and consequently very prolific.

I am, etc.

LETTER XXXVIII

[Echoes]

Fortè puer, comitum seductus ab agmine fido,
Dixerat, ecquis adest? et, adest, responderat echo.
Hic stupet; utque aciem partes divisit in omnes;
Voce, veni, clamat magna. Vocat illa vocantem
[Perhaps a youngster, strayed from his friends, had cried,
"Who's there?" "He's there," the echo had then replied.
Baffled, he searches round, and with mighty shout,
"Come out!" he calls, and the voice returns, "Come out!"]
Ovid, *Metamorphoses*

Selborne, Feb. 12, 1778.

Dear Sir,

In a district so diversified as this, so full of hollow vales and hanging woods, it is no wonder that echoes should abound. Many we have discovered that return the cry of a pack of dogs, the notes of a hunting-horn, a tunable ring of bells, or the melody of birds, very agreeably: but we were still at a loss for a polysyllabical, articulate echo, till a young gentleman, who had parted from his company in a summer evening walk, and was calling after them, stumbled upon a very curious one in a spot where it might least be expected. At first he was much surprised, and could not be persuaded but that he was mocked by some boy; but, repeating his trials in several languages, and finding his respondent to be a very adroit polyglot, he then discerned the deception.

This echo in an evening, before rural noises cease, would repeat ten syllables most articulately and distinctly, especially if quick dactyls were chosen. The last syllables of

> Tityre, tu patulæ recubans . . .
> [Tityrus, under the wide-spreading beech . . .]

were as audibly and intelligibly returned as the first: and there is no doubt, could trial have been made, but that at midnight, when the air is very elastic, and a dead stillness prevails, one or two syllables more might have been obtained; but the distance rendered so late an experiment very inconvenient.

Quick dactyls, we observed, succeeded best; for when we came to try it's powers in slow, heavy, embarrassed spondees of the same number of syllables,

> Monstrum horrendum, informe, ingens . . .
> [Horrible monster, void of form and vast . . .]

we could perceive a return but of four or five.

All echoes have some one place to which they are returned stronger and more distinct than to any other; and that is always the place that lies at right angles with the object of repercussion, and is not too near, nor too far off. Buildings, or naked rocks, re-echo much more articulately than hanging wood or vales; because in the latter the voice is as it were entangled, and embarrassed in the covert, and weakened in the rebound.

The true object of this echo, as we found by various experiments, is the stone-built, tiled hop-kiln in Gally-lane, which measures in front 40 feet, and from the ground to the eaves 12 feet. The true *centrum phonicum*, or just distance, is one particular spot in the King's-field, in the path to Nore-hill, on the very brink of the steep balk above the hollow cart-way. In this case there is no choice of distance; but the path, by mere contingency, happens to be the lucky, the identical spot, because the ground rises or falls so immediately, if the speaker either retires or advances, that his mouth would at once be above or below the object.

We measured this polysyllabical echo with great exactness, and found the distance to fall very short of Dr. Plot's rule for distinct articulation: for the Doctor, in his history of Oxfordshire, allows 120 feet for the return of each syllable distinctly: hence this echo, which gives ten distinct syllables, ought to measure 400 yards, or 120 feet to each syllable; whereas our distance is only 258 yards, or near 75 feet, to each syllable. Thus our measure falls short of the Doctor's, as five to eight: but then it must be acknowledged that this candid philosopher was convinced afterwards, that some latitude must be admitted of in the distance of echoes according to time and place.

When experiments of this sort are making, it should always be remembered that weather and the time of day have a vast influence on an echo; for a dull, heavy, moist air deadens and clogs the sound; and hot sunshine renders the air thin and weak, and deprives it of all it's springiness; and a ruffling wind quite defeats the whole. In a still, clear, dewy evening the air is most elastic; and perhaps the later the hour the more so.

Echo has always been so amusing to the imagination, that the poets have personified her; and in their hands she has been the occasion of many a beautiful fiction. Nor need the gravest man be ashamed to appear taken with such a phenomenon, since it may become the subject of philosophical or mathematical inquiries.

One should have imagined that echoes, if not entertaining, must at least have been harmless and inoffensive; yet Virgil advances a strange notion, that they are injurious to bees. After enumerating some probable and reasonable annoyances, such as prudent owners would wish far removed from their bee-gardens, he adds

> ——————aut ubi concava pulsu
> Saxa sonant, vocisque offensa resultant imago.
> [. . . or where the smitten hollow rock resounds,
> And where the voice's echo strikes and, answering, rebounds.]

This wild and fanciful assertion will hardly be admitted by the philosophers of these days; especially as they all now seem agreed that insects are not furnished with any organs of hearing at all. But if it should be urged, that though they cannot *hear* yet perhaps they may *feel* the repercussions of sounds, I grant it is possible they may. Yet that these impressions are distasteful or hurtful, I deny, because bees, in good summers, thrive well in my outlet, where the echoes are very strong: for this village is another Anathoth, a place of *responses* or *echoes*. Besides, it does not appear from experiment that bees are in any way capable of being affected by sounds: for I have often tried my own with a large speaking-trumpet held close to their hives, and with such an exertion of voice as would have haled a ship at the distance of a mile, and still these insects pursued their various employments undisturbed, and without shewing the least sensibility or resentment.

Some time since it's discovery this echo is become totally silent, though the object, or hop-kiln, remains: nor is there any mystery in this defect; for the field between is planted as an hop-garden, and the voice of the speaker is totally absorbed and lost among the poles and entangled foliage of the hops. And when the poles are removed in autumn the disappointment is the same; because a tall quick-set hedge, nurtured up for the purpose of shelter to the hop ground, entirely interrupts the impulse and repercussion of the voice: so that till those obstructions are removed no more of it's garrulity can be expected.

Should any gentleman of fortune think an echo in his park or outlet a pleasing incident, he might *build* one at little or no expense. For whenever he had occasion for a new barn, stable, dog-kennel, or the like structure, it would be only needful to erect this building on the gentle declivity of an hill, with a like rising opposite to it, at a few hundred yards distance; and perhaps success might be the easier ensured could some canal, lake, or stream, intervene. From a seat at the *centrum phonicum* he and his friends might amuse themselves sometimes of an evening with the prattle of this loquacious nymph, of whose complacency and decent reserve more may be said than can with truth of every individual of her sex; since she is —————

—————— quæ nec *reticere* loquenti,
Nec *prior* ipsa *loqui* didicit resonabilis echo.
[. . . answering echo, who has yet to learn
To *keep her peace* when spoken to, or *speak* the *first* in turn.]

I am, &c.

LETTER XLIII

[Revenge of the Hens]

Selborne, Sept. 9, 1778.

Dear Sir,

*   *   *

No inhabitants of a yard seem possessed of such a variety of expression and so copious a language as common poultry. Take a chicken of four or five days old, and hold it up to a window where there are flies, and it will immediately seize its prey, with little twitterings of complacency; but if you tender it a wasp or a bee, at once its note becomes harsh, and expressive of disapprobation and a sense of danger. When a pullet is ready to lay she intimates the event by a joyous and easy soft note. Of all the occurrences of their life that of laying seems to be the most important; for no sooner has a hen disburdened herself, than she rushes forth with a clamorous kind of joy, which the cock and the rest of his mistresses immediately adopt. The tumult is not confined to the family concerned, but catches from yard to yard, and spreads to every homestead within hearing, till at last the whole village is in an uproar. As soon as a hen becomes a mother her new relation demands a new language; she then runs clocking and screaming about, and seems agitated as if possessed. The father of the flock has also a considerable vocabulary; if he finds food, he calls a favourite concubine to partake; and if a bird of prey passes over, with a warning voice he bids his family beware. The gallant chanticleer has, at command, his amorous phrases, and his terms of defiance. But the sound by which he is best known is his crowing: by this he has been distinguished in all ages as the countryman's clock or larum, as the watchman that proclaims the divisions of the night. Thus the poet elegantly styles him:

> . . . the created cock, whose clarion sounds
> The silent hours.

A neighbouring gentleman one summer had lost most of his chickens by a sparrow-hawk, that came gliding down between a faggot-pile and the end of his house to the place where the coops stood. The owner, inwardly vexed to see his flock thus diminishing, hung a setting net adroitly between the pile and the house, into which the caitiff dashed and was entangled. Resentment suggested the law of retaliation: he therefore clipped the hawk's wings, cut off his talons, and, fixing a cork on his bill, threw him down among the brood-hens. Imagination cannot paint the scene that ensued; the expressions that fear, rage, and revenge inspired, were new, or at least such as had been unnoticed before: the exasperated matrons upbraided, they execrated, they insulted, they triumphed. In a word, they never desisted from buffeting their adversary till they had torn him in an hundred pieces.

LETTER L

[Timothy Observed]

Selbourne, April 21, 1780.

Dear Sir,

The old Sussex tortoise, that I have mentioned to you so often, is become my property. I dug it out of it's winter dormitory in March last, when it was enough awakened to express it's resentments by hissing; and, packing it in a box with earth, carried it eighty miles in post-chaises. The rattle and hurry of the journey so perfectly roused it that, when I turned it out on a border, it walked twice down to the bottom of my garden; however, in the evening, the weather being cold, it buried itself in the loose mould, and continues still concealed.

As it will be under my eye, I shall now have an opportunity of enlarging my observations on it's mode of life, and propensities; and perceive already that, towards the time of coming forth, it opens a breathing place in the ground near it's head, requiring, I conclude, a freer respiration, as it becomes more alive. This creature not only goes under the earth from the middle of November to the middle of April, but sleeps great part of the summer; for it goes to bed in the longest days at four in the afternoon, and often does not stir in the morning till late. Besides, it retires to rest for every shower; and does not move at all in wet days.

When one reflects on the state of this strange being, it is a matter of wonder to find that Providence should bestow such a profusion of days, such a seeming waste of longevity, on a reptile that appears to relish it so little as to squander more than two thirds of it's existence in a joyless stupor, and be lost to all sensation for months together in the profoundest of slumbers.

Because we call this creature an abject reptile, we are too apt to undervalue his abilities, and depreciate his powers of instinct. Yet he is, as Mr. Pope says of his lord,

————————Much too wise to walk into a well:

and has so much discernment as not to fall down an haha; but to stop and withdraw from the brink with the readiest precaution.

Though he loves warm weather he avoids the hot sun; because his thick shell, when once heated, would, as the poet says of solid armour—"scald with safety." He therefore spends the more sultry hours under the umbrella of a large cabbage-leaf, or amidst the waving forests of an asparagus-bed.

But as he avoids heat in the summer, so, in the decline of the year, he improves the faint autumnal beams, by getting within the reflection of a fruit-wall; and, though he never has read that planes inclining to the horizon receive a greater share of warmth,[1] he inclines his shell, by tilting it against the wall, to collect and admit every feeble ray.

Pitiable seems the condition of this poor embarrassed reptile: to be cased in a suit of ponderous armour, which he cannot lay aside; to be imprisoned, as it were, within his own shell, must preclude, we should suppose, all activity and disposition for enterprise. Yet there is a season of the year (usually the beginning of June) when his exertions are remarkable. He then walks on tiptoe, and is stirring by five in the morning; and, traversing the garden, examines every wicket and interstice in the fences, through which he will escape if possible: and often has eluded the care of the gardener, and wandered to some distant field. The motives that impel him to undertake these rambles seem to be of the amorous kind: his fancy then becomes intent on sexual attachments, which transport him beyond his usual gravity, and induce him to forget for a time his ordinary solemn deportment.

While I was writing this letter, a moist and warm afternoon, with the thermometer at 50, brought forth troops of *shell-snails*; and, at the same juncture, the *tortoise* heaved up the mould and put out it's head; and the next morning came forth, as it were raised from the dead; and walked about till four in the afternoon. This was a curious coincidence! a very amusing occurrence! to see such a similarity of feelings between the two φερεοικοι! [house-carriers] for so the Greeks called both the *shell-snail* and the *tortoise*.

Summer birds are, this cold and backward spring, unusually late: I have seen but one swallow yet. This conformity with the weather convinces me more and more that they sleep in the winter.

---

[1] *Several years ago a book was written entitled* Fruit-walls improved by inclining them to the horizon: *in which the author has shewn, by calculation, that a much greater number of the rays of the sun will fall on such walls than on those which are perpendicular.* [Gilbert White's note]

HECTOR ST. JOHN de CRÈVECOEUR
1735–1813

The *"American Farmer" of Crèvecoeur's classic account of early American rural life was in reality a French-born aristocrat who went to school in England, fought with Montcalm in the French and Indian Wars, wandered the frontiers of the Colonies, and was forced out of the young Republic in 1780 because of his Loyalist sympathies. Nevertheless, the idyllic years he and his American wife spent on a farm in New York's Orange County before the Revolution provided the basis for* Letters from an American Farmer. *Published in London in 1782, this book became a best-seller in England and Europe and was embraced by the Romantics, along with the works of William Bartram and John James Audubon, as depictions of a wild and unsullied New World paradise.*

Crèvecoeur was an idealist in the best Rousseauian tradition, seeing nature as a system devised by the Creator for humanity's benefit and moral instruction but as one needing selective interference on occasion. He was also a writer of considerable power, and Letters *represents one of the first genuinely imaginative works of nature writing in America. In addition, it marks the beginning of an agrarian literary tradition that includes writers like Aldo Leopold and Wendell Berry. Crèvecoeur's celebrations of farm life are full of rich detailed vignettes of the terms of existence in the wild. Though these accounts often have a strong element of embellishment in them, they also possess an emotional veracity, a compassion for animals, and an appreciation of nature's primitive vitality unmatched by any writer until Thoreau.*

From LETTERS FROM AN AMERICAN FARMER

ON THE SITUATION, FEELINGS, AND
PLEASURES OF AN AMERICAN FARMER

*   *   * Pray do not laugh in thus seeing an artless countryman tracing himself through the simple modifications of his life; remember that you have required it; therefore, with candour, though with diffidence, I endeavour to follow the thread of my feelings, but I cannot tell you all. Often when I plough my low ground, I place my little boy on a chair which screws to the beam of the plough—its motion and that of the horses please him; he is perfectly happy and begins to chat, As I lean over the handle, various are the thoughts which crowd into my mind. I am now doing for him, I say, what my father formerly did for me; may God enable him to live that he may perform the same operations for the same purposes when I am worn out and old! I relieve his mother of some trouble while I have him with me; the odoriferous furrow exhilarates his spirits and seems to do the child a great deal of good, for he looks more blooming since I have adopted that practice; can more pleasure, more dignity, be added to that primary occupation? The father thus ploughing with his child, and to feed his family, is inferior only to the emperor of China ploughing as an example to his kingdom. In the evening, when I return home through my low grounds, I am astonished at the myriads of insects which I perceive dancing in the beams of the setting sun. I was before scarcely acquainted with their existence; they are so small that it is difficult to distinguish them; they are carefully improving this short evening space, not daring to expose themselves to the blaze of our meridian sun. I never see an egg brought on my table but I feel penetrated with the wonderful change it would have undergone but for my gluttony; it might have been a gentle, useful hen leading her chicken with a care and vigilance which speaks shame to many women. A cock perhaps, arrayed with the most majestic plumes, tender to its mate, bold, courageous, endowed with an astonishing instinct, with thoughts, with memory, and every distinguishing characteristic of the reason of man. I never see my trees drop their leaves and their fruit in the autumn, and bud again in the spring, without wonder; the sagacity of those animals which have long been the tenants of my farm astonish me; some of them seem to surpass even men in memory and sagacity. I could tell you singular instances of that kind. What, then, is this instinct which we so debase,

Letters from an American Farmer (London: T. Davies, 1782).

and of which we are taught to entertain so diminutive an idea? My bees, above any other tenants of my farm, attract my attention and respect; I am astonished to see that nothing exists but what has its enemy; one species pursues and lives upon the other: unfortunately, our king-birds are the destroyers of those industrious insects, but on the other hand, these birds preserve our fields from the depredation of crows, which they pursue on the wing with great vigilance and astonishing dexterity.

Thus divided by two interested motives, I have long resisted the desire I had to kill them until last year, when I thought they increased too much, and my indulgence had been carried too far; it was at the time of swarming, when they all came and fixed themselves on the neighbouring trees whence they caught those that returned loaded from the fields. This made me resolve to kill as many as I could, and was just ready to fire when a bunch of bees as big as my fist issued from one of the hives, rushed on one of these birds, and probably stung him, for he instantly screamed and flew, not as before, in an irregular manner, but in a direct line. He was followed by the same bold phalanx, at a considerable distance, which unfortunately, becoming too sure of victory, quitted their military array and disbanded themselves. By this inconsiderate step, they lost all that aggregate of force which had made the bird fly off. Perceiving their disorder, he immediately returned and snapped as many as he wanted; nay, he had even the impudence to alight on the very twig from which the bees had driven him. I killed him and immediately opened his craw, from which I took 171 bees; I laid them all on a blanket in the sun, and to my great surprise, 54 returned to life, licked themselves clean, and joyfully went back to the hive, where they probably informed their companions of such an adventure and escape as I believe had never happened before to American bees! I draw a great fund of pleasure from the quails which inhabit my farm; they abundantly repay me, by their various notes and peculiar tameness, for the inviolable hospitality I constantly show them in the winter. Instead of perfidiously taking advantage of their great and affecting distress when nature offers nothing but a barren universal bed of snow, when irresistible necessity forces them to my barn doors, I permit them to feed unmolested; and it is not the least agreeable spectacle which that dreary season presents, when I see those beautiful birds, tamed by hunger, intermingling with all my cattle and sheep, seeking in security for the poor, scanty grain which but for them would be useless and lost. Often in the angles of the fences where the motion of the wind prevents the snow from settling, I carry them both chaff and grain, the one to feed them, the other to prevent their tender feet from freezing fast to the earth as I have frequently observed them to do.

I do not know an instance in which the singular barbarity of man is so strongly delineated as in the catching and murthering those harmless birds, at that cruel season of the year. Mr.——, one of the most famous and extraordinary farmers that has ever done honour to the province of Connecticut, by his timely and humane assistance in a hard winter, saved this species from being entirely destroyed. They perished all over the country; none of their delightful whistlings were heard the next spring but upon this gentleman's farm; and to his humanity we owe the continuation of their music. When the severities of that season have dispirited all my cattle, no farmer ever attends them with more pleasure than I do; it is one of those duties which is sweetened with the most rational satisfaction. I amuse myself in beholding their different tempers, actions, and the various effects of their instinct now powerfully impelled by the force of hunger. I trace their various inclinations and the different effects of their passions, which are exactly the same as among men; the law is to us precisely what I am in my barnyard, a bridle and check to prevent the strong and greedy from oppressing the timid and weak. Conscious of superiority, they always strive to encroach on their neighbours; unsatisfied with their portion, they eagerly swallow it in order to have an opportunity of taking what is given to others, except they are prevented. Some I chide; others, unmindful of my admonitions, receive some blows. Could victuals thus be given to men without the assistance of any language, I am sure they would not behave better to one another, nor more philosophically than my cattle do. *   *   *

ON SNAKES; AND ON THE HUMMING-BIRD

Why would you prescribe this task; you know that what we take up ourselves seems always lighter than what is imposed on us by others. You insist on my saying something about our snakes; and in relating what I know concerning them, were it not for two singularities, the one of which I saw and the other I received from an eyewitness, I should have but very little to observe. The southern provinces are the countries where Nature has formed the greatest variety of alligators, snakes, serpents, and scorpions from the smallest size up to the pine barren, the largest species known here. We have but two, whose stings are mortal, which deserve to be mentioned; as for the black one, it is remarkable for nothing but its industry, agility, beauty, and the art of enticing birds by the power of its eyes. I admire it much and never kill it, though its formidable length and appearance often get the better of the philosophy of some people, particularly Europeans. The most dangerous one is the pilot, or copperhead, for the poison of which no remedy has yet been discovered. It bears the first name because it always precedes the rattlesnake, that is, quits its state of torpidity in the spring a week before the other. It bears the second name on account of its head being adorned with many copper-coloured spots. It lurks in rocks near the water and is extremely active and dangerous. Let man beware of it! I have heard only of one person who was stung by a copperhead in this country. The poor wretch instantly swelled in a most dreadful manner; a multitude of spots of different hues alternately appeared and vanished on different parts of his body; his eyes were filled with madness and rage; he cast them on all present with the most vindictive looks; he thrust out his tongue as the snakes do; he hissed through his teeth with inconceivable strength and became an object of terror to all bystanders. To the lividness of a corpse he united the desperate force of a maniac; they hardly were able to fasten him so as to guard themselves from his attacks, when in the space of two hours death relieved the poor wretch from his struggles and the spectators from their apprehensions. The poison of the rattlesnake is not mortal in so short a space, and hence there is more time to procure relief; we are acquainted with several antidotes with which almost every family is provided. They are extremely inactive, and if not touched, are perfectly inoffensive. I once saw, as I was travelling, a great cliff which was full of them; I handled several, and they appeared to be dead; they were all entwined together, and thus they remain until the return of the sun. I found them out by following the track of some wild hogs which had fed on them; and even the Indians often regale on them. When they find them asleep, they put a small forked stick over their necks, which they keep immovably fixed on the ground, giving the snake a piece of leather to bite; and this they pull back several times with great force until they observe their two poisonous fangs torn out. Then they cut off the head, skin the body, and cook it as we do eels; and their flesh is extremely sweet and white. I once saw a *tamed one*, as gentle as you can possibly conceive a reptile to be; it took to the water and swam whenever it pleased; and when the boys to whom it belonged called it back, their summons was readily obeyed. It had been deprived of its fangs by the preceding method; they often stroked it with a soft brush, and this friction seemed to cause the most pleasing sensations, for it would turn on its back to enjoy it, as a cat does before the fire. One of this species was the cause, some years ago, of a most deplorable accident, which I shall relate to you as I had it from the widow and mother of the victims. A Dutch farmer of the Minisink went to mowing, with his Negroes, in his boots, a precaution used to prevent being stung. Inadvertently he trod on a snake, which immediately flew at his legs; and as it drew back in order to renew its blow, one of his Negroes cut it in two with his scythe. They prosecuted their work and returned home; at night the farmer pulled off his boots and went to bed, and was soon after attacked with a strange sickness at his stomach; he swelled, and before a physician could be sent for, died. The sudden death of this man did not cause much inquiry; the neighbourhood wondered, as is usual in such cases, and without any further examination, the corpse was buried. A few days after, the son put on his father's boots and went to the meadow; at night he pulled them off, went to bed, and was attacked with the same symptoms about the same time, and died in the morning. A little before he expired, the doctor came, but was not able to assign what could be the cause of so singular a disorder; however, rather than appear wholly at a loss before the country people, he pronounced both father and son to have been bewitched. Some weeks after, the widow sold all the movables for the benefit of the younger children, and the farm was leased. One of the neighbours, who bought the boots, presently put them on, and was attacked in the same manner as the other two had been; but this man's wife, being alarmed by what had happened in the former family, despatched one of her Negroes for an eminent physician, who, fortunately having heard something of the dreadful affair, guessed at the cause, applied oil, etc., and recovered the man. The boots which had been so fatal were then carefully examined, and he found that the two fangs of the snake had been left in the leather after being wrenched out of their sockets by the strength with which the snake had drawn back its head. The bladders which contained the poison and several of the small nerves were still fresh and adhered to the boot. The unfortunate father and son had been poisoned by pulling off these boots, in which action they imperceptibly scratched their legs with the points of the fangs, through the hollow of which some of this astonishing poison was conveyed. You have no doubt heard of their rattles if you have not seen them; the only observation I wish to make is that the rattling is loud and distinct when they are angry and, on the contrary, when pleased, it sounds like a distant trepidation, in which nothing distinct is heard. In the thick settlements, they are now become very scarce, for wherever they are met with, open war is declared against them, so that in a few years there will be none left but on our mountains. The black snake, on the contrary, always diverts me because it excites no idea of danger. Their swiftness is astonishing; they will sometimes equal that of a horse; at other times they will climb up trees in quest of our tree toads or glide on the ground at full length. On some occasions they present themselves half in the reptile state, half erect; their eyes and their heads in the erect posture appear to great advantage; the former display a fire which I have often admired, and it is by these they are enabled to fascinate birds and squirrels. When they have fixed their eyes on an animal, they become immovable, only turning their head sometimes to the right and sometimes to the left, but still with their sight invariably directed to the object. The distracted victim, instead of flying its enemy, seems to be arrested by some invincible power; it screams; now approaches and then recedes; and after skipping about with unaccountable agitation, finally rushes into the jaws of the snake and is swallowed, as soon as it is covered with a slime or glue to make it slide easily down the throat of the devourer.

One anecdote I must relate, the circumstances of which are as true as they are singular. One of my constant walks when I am at leisure is in my lowlands, where I have the pleasure of seeing my cattle, horses, and colts. Exuberant grass replenishes all my fields, the best representative of our wealth; in the middle of that track I have cut a ditch eight feet wide, the banks of which Nature adorns every spring with the wild salendine and other flowering weeds, which on these luxuriant grounds shoot up to a great height. Over this ditch I have erected a bridge, capable of bearing a loaded waggon; on each side I carefully sow every year some grains of hemp, which rise to the height of fifteen feet, so strong and so full of limbs as to resemble young trees; I once ascended one of them four feet above the ground. These produce natural arbours, rendered often still more compact by the assistance of an annual creeping plant, which we call a vine, that never fails to entwine itself among their branches and always produces a very desirable shade. From this simple grove I have amused myself a hundred times in observing the great number of humming-birds with which our country abounds: the wild blossoms everywhere attract the attention of these birds, which like bees subsist by suction. From this retreat I distinctly watch them in all their various attitudes, but their flight is so rapid that you cannot distinguish the motion of their wings. On this little bird Nature has profusely lavished her most splendid colours; the most perfect azure, the most beautiful gold, the most dazzling red, are forever in contrast and help to embellish the plumes of his majestic head. The richest palette of the most luxuriant painter could never invent anything to be compared to the variegated tints with which this insect bird is arrayed. Its bill is as long and as sharp as a coarse sewing needle; like the bee, Nature has taught it to find out in the calyx of flowers and blossoms those mellifluous particles that serve it for sufficient food; and yet it seems to leave them untouched, undeprived of anything that our eyes can possibly distinguish. When it feeds, it appears as if immovable, though continually on the wing; and sometimes, from what motives I know not, it will tear and lacerate flowers into a hundred pieces, for, strange to tell, they are the most irascible of the feathered tribe. Where do passions find room in so diminutive a body? They often fight with the fury of lions until one of the combatants falls a sacrifice and dies. When fatigued, it has often perched within a few feet of me, and on such favourable opportunities I have surveyed it with the most minute attention. Its little eyes appear like diamonds, reflecting light on every side; most elegantly finished in all parts, it is a miniature work of our Great Parent, who seems to have formed it the smallest, and at the same time the most beautiful of the winged species.

As I was one day sitting solitary and pensive in my primitive arbour, my attention was engaged by a strange sort of rustling noise at some paces distant. I looked all around without distinguishing anything, until I climbed one of my great hemp stalks, when to my astonishment I beheld two snakes of considerable length, the one pursuing the other with great celerity through a hemp-stubble field. The aggressor was of the black kind, six feet long; the fugitive was a water snake, nearly of equal dimensions. They soon met, and in the fury of their first encounter, they appeared in an instant firmly twisted together; and whilst their united tails beat the ground, they mutually tried with open jaws to lacerate each other. What a fell aspect did they present! Their heads were compressed to a very small size, their eyes flashed fire; and after this conflict had lasted about five minutes, the second found means to disengage itself from the first and hurried toward the ditch. Its antagonist instantly assumed a new posture, and half creeping and half erect, with a majestic mien, overtook and attacked the other again, which placed itself in the same attitude and prepared to resist. The scene was uncommon and beautiful; for thus opposed, they fought with their jaws, biting each other with the utmost rage; but notwithstanding this appearance of mutual courage and fury, the water snake still seemed desirous of retreating toward the ditch, its natural element. This was no sooner perceived by the keen-eyed black one, than twisting its tail twice round a stalk of hemp and seizing its adversary by the throat, not by means of its jaws but by twisting its own neck twice round that of the water snake, pulled it back from the ditch. To prevent a defeat, the latter took hold likewise of a stalk on the bank, and by the acquisition of that point of resistance, became a match for its fierce antagonist. Strange was this to behold; two great snakes strongly adhering to the ground, mutually fastened together by means of the writhings which lashed them to each other, and stretched at their full length, they pulled but pulled in vain; and in the moments of greatest exertions, that part of their bodies which was entwined seemed extremely small, while the rest appeared inflated and now and then convulsed with strong undulations, rapidly following each other. Their eyes seemed on fire and ready to start out of their heads; at one time the conflict seemed decided; the water snake bent itself into two great folds and by that operation rendered the other more than commonly outstretched; the next minute the new struggles of the black one gained an unexpected superiority; it acquired two great folds likewise, which necessarily extended the body of its adversary in proportion as it had contracted its own. These efforts were alternate; victory seemed doubtful, inclining sometimes to the one side and sometimes to the other, until at last the stalk to which the black snake fastened suddenly gave way, and in consequence of this accident they both plunged into the ditch. The water did not extinguish their vindictive rage; for by their agitations I could trace, though not distinguish, their mutual attacks. They soon reappeared on the surface twisted together, as in their first onset; but the black snake seemed to retain its wonted superiority, for its head was exactly fixed above that of the other, which it incessantly pressed down under the water, until it was stifled and sunk. The victor no sooner perceived its enemy incapable of farther resistance than, abandoning it to the current, it returned on shore and disappeared.

From SKETCHES OF EIGHTEENTH CENTURY AMERICA

ENEMIES OF THE FARMER

* * * Nature has placed a certain degree of antipathy between some species of animals and birds. Often one lives on the other; at other times they only attack each other. The king-bird is the most skilful on the wing of any we have here. Every spring he declares war against all the kites, hawks, and crows which pass within the bounds of his precincts. If they build anywhere on your farm, rest assured that none of those great tyrants of the air will fly over it with impunity. Nor is it an unpleasant sight to behold the contest. Like the Indians, they scream aloud when they go to the attack. They fly at first with an apparent trepidation. They err here and there, and then dart with immense impetuosity and with consummate skill, always getting to the windward of their enemy, let it be even the great bald eagle of the Blue

Sketches of Eighteenth century America (New Haven, Conn.: Yale University Press, 1925).

Mountains. By repeatedly falling on him and striking him with their bills and sometimes by attacking him under the wings, they will make the largest bird accelerate its motion, and describe the most beautiful curves in those rapid descents which they compel him to delineate. This amuses me much, on two accounts: first, because they are doing my work; second, because I have an excellent opportunity of viewing the art of flying carried to a great degree of perfection and varied in a multiplicity of appearances. What a pity that I am very often obliged to shoot these little kings of the air! Hunting crows serves only to make them more hungry, nor will they live on grain, but on bees. These precious insects, these daughters of heaven, serve them as food whenever they can catch them.

The blackbird, which you say resembles your starling, visits us every spring in great numbers. They build their nests in our most inaccessible swamps. Their rough notes are delightful enough at a distance. As soon as the corn sprouts, they come to dig it up; nothing but the gun can possibly prevent them. Hanging of their dead companions has no kind of effect. They are birds that show the greatest degree of temerity of any we have. Sometimes we poison corn, which we strew on the ground with the juice of hitch-root. Sometimes, after soaking it, we pass a horsehair through each grain, which we cut about an inch long. These expedients will destroy a few of them, but either by the effect of inspection or by means of some language unknown, like a great many other phenomena, to Man, the rest will become acquainted with the danger, and in two days the survivors will not touch it, but take the utmost pains to eradicate that which lies three inches under the ground. I have often poisoned the very grains I have planted, but in two days it grows as sweet as ever. Thus while our corn is young it requires a great deal of watching. To prevent these depredations, some counties have raised money and given a small bounty of two pence per head. If they are greatly disturbed while they are hatching, they will soon quit that district.

But after all the efforts of our selfishness, are they not the children of the great Creator as well as we? They are entitled to live and to get their food wherever they can get it. We can better afford to lose a little corn than any other grain because it yields above seventy for one. But Man is a huge monster who devours everything and will suffer nothing to live in peace in his neighbourhood. The easiest, best, and the most philanthropic method is to break up either our summer fallow or our buckwheat ground while our corn is young. They will immediately cease to do us mischief and go to prey on the worms and caterpillars which the plough raises. Their depredations proceeded from hunger, not from premeditated malice. As soon as their young ones are able to fly, they bid us farewell. They retire to some other countries which produce what they want, for as they neither sow nor plant, it is necessary that either Man or Nature should feed them. Towards the autumn, they return in astonishing flocks. If our corn-fields are not then well guarded, they will in a few hours make great havoc. They disappear in about a week.

At this season another animal comes out of our woods and demands of Man his portion. It is the squirrel, of which there are three sorts, the grey, the black, both of the same size, and the little ground one, which harbours under rocks and stones. The two former are the most beautiful inhabitants of our forests. They live in hollow trees which they fill with koka toma nuts, chestnuts, and corn when they can get it. Like man they know the approach of the winter and as wisely know how to prepare against its wants. Some years their numbers are very great. They will travel over our fences with the utmost agility, descend into our fields, cut down an ear perhaps eighteen inches long and heavier than themselves. They will strip it of its husk and, properly balancing it in their mouths, return thus loaded to their trees. For my part, I cannot blame them, but I should blame myself were I peaceably to look on and let them carry all. As we pay no tithes in this country, I think we should be a little more generous than we are to the brute creation. If there are but few, a gun and a dog are sufficient. If they openly declare war in great armies, men collect themselves and go to attack them in their native woods. The county assembles and forms itself into companies to which a captain is appointed. Different districts of woods are assigned them; the rendezvous is agreed on. They march, and that company which kills the most is treated by the rest; thus the day is spent. The meat of these squirrels is an excellent food; they make excellent soup or pies. Their skins are exceedingly tough; they are stronger than eels' skins; we use them to tie our flails with.

Mirth, jollity, coarse jokes, the exhilarating cup, and dancing are always the concomitant circumstances which enliven and accompany this kind of meeting, the only festivals that we simple people are acquainted with in this young country. Religion, which in so many parts of the world affords processions and a variety of other exercises and becomes a source of temporal pleasures to the people, here gives us none. What few it yields are all of the spiritual kind. A few years ago I was invited to one of these parties. At my first entrance into the woods, whilst the affrighted echoes were resounding with the noise of the men and dogs, before I had joined the company, I found a bee-tree, which is my favourite talent. But, behold, it contained also the habitation of a squirrel family. The bees were lodged in one of its principal limbs; the others occupied the body of the tree. For the sake of the former I saved the latter. While I was busy in marking it, I perceived a great number of ants (those busy-bodies) travelling up three deep in a continual succession and returning in the same way. Both these columns were perfectly straight in the ascent as well as in the descent and but a small distance apart. I killed a few which I smelled. I found them all replete with honey. I therefore concluded that these were a set of thieves living on the industrious labours of the others. This intrusion gave me a bad opinion of the vigour and vigilance of the latter. However, as the honey season was not come, I resolved to let them alone and to deliver them from the rapacity of an enemy which they could not repel.

Next day, accompanied by my little boy, I brought a kettle, kindled a fire, boiled some water, and scalded the whole host as it ascended and descended. I did not leave one stirring. The lad asked me a great many questions respecting what I was doing, and the answers I made him afforded me the means of conveying to his mind the first moral ideas I had as yet given him. On my return home I composed a little fable on the subject, which I made him learn by heart. God grant that this trifling incident may serve as the basis of a future moral education.

Thus, sir, do we save our corn. But when it is raised on our lowlands, it is subject to transitory frosts, an accident which I have not mentioned to you yet, but as we can foresee them, it is in our power to avoid the mischief they cause. If in any of our summer months the wind blows north-west two days, on the second night the frost is inevitable. The only means we have to preserve our grain from its bad effects on our low grounds—for it seldom reaches the upland—is to kindle a few fires to the windward as soon as the sun goes down. No sooner does that luminary disappear from our sight than the wind ceases. This is the most favourable moment. The smoke will not rise, but, on the contrary, lie on the ground, mixing with the vapours of the evening. The whole will form a body four feet deep which will cover the face of the earth until the power of the sun dissipates it next morning. Whatever is covered with it is perfectly safe. I had once some hops and pole-beans, about twenty feet high. Whatever grew above the body of the smoke was entirely killed; the rest was saved. I had at the same time upwards of three acres of buckwheat which I had sown early for my bees. I lost not a grain. Some of my neighbours have by this simple method saved their tobacco.

These low grounds are exposed, besides, to the ravages of grasshoppers, an intolerable nuisance. While young and deprived of wings, they may be kept off by means of that admirable contrivance which a Negro found out in South Carolina: a few pots filled with brimstone and tar are kindled at nightfall to the windward of the field; the powerful smell of these two ingredients either kills them or drives them away. But when they have wings, they easily avoid it and transport themselves wherever they please. The damage they cause in our hemp grounds as well as in our meadows is inconceivable. They will eat the leaves of the former to the bare stalks and consume the best of our grasses. The only remedy for the latter is to go to mowing as soon as possible. The former devastation is unavoidable. Some years a certain worm, which I cannot describe, insinuates itself into the heart of the corn-stalk while it is young; and if not killed by squeezing the plant, it will eat the embryo of the great stem, which contains the imperceptible rudiments of our future hopes.

Sometimes our rye is attacked by a small animalcula of the worm kind which lodges itself in the stem just below the first joint. There it lives on the sap as it ascends. The ear becomes white and grainless, the perfect symbol of sterility.

I should have never done, were I to recount to you all the inconveniences and accidents which the grains of our fields, the trees of our orchards, as well as those of the woods, are exposed to. If bountiful Nature is kind to us on the one hand, on the other she wills that we shall purchase her kindness not only with sweats and labour but with vigilance and care. These calamities remind us of our precarious situation. The field and meadow-mice come in also for their share, and sometimes take more from Man than he can well spare. The rats are so multiplied that no one can imagine the great quantities of grain they destroy every year. Some farmers, more unfortunate than the others, have lost half of their crops after they were safely lodged in their barns. I'd forgive Nature all the rest if she would rid us of these cunning, devouring thieves which no art can subdue. When the floods rise on our low grounds, the mice quit their burrows and come to our stacks of grain or to our heaps of turnips, which are buried under the earth out of the reach of the frost. There, secured from danger, they find a habitation replenished with all they want. I must not, however, be murmuring and ungrateful. If Nature has formed mice, she has created also the fox and the owl. They both prey on these. Were it not for their kind assistance, the mice would drive us out of our farms.

Thus one species of evil is balanced by another; thus the fury of one element is repressed by the power of the other. In the midst of this great, this astonishing equipoise, Man struggles and lives. *   *   *

WILLIAM BARTRAM
1739–1823

William Bartram's Travels Through North & South Carolina, Georgia, East & West Florida, the Cherokee Country, the Extensive Territories of the Muscogulges, or Creek Confederacy, and the Country of the Chactaws *(1791) has entranced readers from its first publication to the present day. Mangrove swamps and alligators, poisonous snakes and wolves seem to have been as delightful to Bartram, by the very fact of their existence, as they would have been terrifying to one more concerned with protecting his own existence. Joy was inseparable from the study of nature for Bartram, and he frequently weaves Linnaean nomenclature into his rhapsodic descriptions in ways that express a litany of praise. Such a blending of precision with adoration marks his account of the "crystal bason," which is included here. This passage so impressed Coleridge that he worked a number of details from it into his description of the enchanted landscape in "Kubla Khan."*

As the full title of Bartram's book indicates, he was also interested, in a particular and discriminating sense, in the native peoples of the Southeast. We can be grateful for the presence in our early literature of nature of a few writers like Bartram and Thoreau, who combined a vivid and informed love of the land with respect for the cultures that had evolved within it before the coming of white settlers.

From TRAVELS THROUGH NORTH & SOUTH CAROLINA, GEORGIA, EAST & WEST FLORIDA, . . .

[Ephemera]

* * * Early in the evening, after a pleasant day's voyage, I made a convenient and safe harbour, in a little lagoon, under an elevated bank, on the West shore of the river; where I shall entreat the reader's patience, whilst we behold the closing scene of the short-lived Ephemera, and communicate to each other the reflections which so singular an exhibition might rationally suggest to an inquisitive mind. Our place of observation is happily situated under the protecting shade of majestic Live Oaks, glorious Magnolias, and the fragrant Orange, open to the view of the great river and still waters of the lagoon just before us.

At the cool eve's approach, the sweet enchanting melody of the feathered songsters gradually ceases, and they betake themselves to their leafy coverts for security and repose.

Solemnly and slowly move onward, to the river's shore, the rustling clouds of the Ephemera. How awful the procession! innumerable millions of winged beings, voluntarily verging on to destruction, to the brink of the grave, where they behold bands of their enemies with wide open jaws, ready to receive them. But as if insensible of their danger, gay and tranquil each meets his beloved mate in the still air, inimitably bedecked in their new nuptial robes. What eye can trace them, in their varied wanton amorous chaces, bounding and fluttering on the odoriferous air! With what peace, love, and joy, do they end the last moments of their existence!

I think we may assert, without any fear of exaggeration, that there are annually of these beautiful winged beings, which rise into existence, and for a few moments take a transient view of the glory of the Creator's works, a number greater than the whole race of mankind that have ever existed since the creation; and that, only from the shores of this river. How many then must have been produced since the creation, when we consider the number of large rivers in America, in comparison with which, this river is but a brook or rivulet.

The importance of the existence of these beautiful and delicately

Travels Through North & South Carolina, Georgia, East & West Florida, the Cherokee Country, the Extensive Territories of the Muscogulges, or Creek Confederacy, and the County of the Chactaws (Philadelphia: James and Johnson, 1791).

formed little creatures, whose frame and organization are equally wonderful, more delicate, and perhaps as complicated as those of the most perfect human being, is well worth a few moments contemplation; I mean
particularly when they appear in the fly state. And if we consider the very
short period of that stage of existence, which we may reasonably suppose
to be the only space of their life that admits of pleasure and enjoyment,
what a lesson doth it not afford us of the vanity of our own pursuits!

Their whole existence in this world is but one complete year: and at
least three hundred and sixty days of that time they are in the form of an
ugly grub, buried in mud, eighteen inches under water, and in this condition scarcely locomotive, as each larva or grub has but its own narrow
solitary cell, from which it never travels or moves, but in a perpendicular
progression of a few inches, up and down, from the bottom to the surface
of the mud, in order to intercept the passing atoms for its food, and get a
momentary respiration of fresh air; and even here it must be perpetually
on its guard, in order to escape the troops of fish and shrimps watching to
catch it, and from whom it has no escape, but by instantly retreating back
into its cell. One would be apt almost to imagine them created merely
for the food of fish and other animals. * * *

[Encounters with Alligators]

* * * The evening was temperately cool and calm. The crocodiles
began to roar and appear in uncommon numbers along the shores and
in the river. I fixed my camp in an open plain, near the utmost projection of the promontory, under the shelter of a large live oak, which
stood on the highest part of the ground, and but a few yards from my
boat. From this open, high situation, I had a free prospect of the river,
which was a matter of no trivial consideration to me, having good reason to dread the subtle attacks of the alligators, who were crowding
about my harbour. Having collected a good quantity of wood for the
purpose of keeping up a light and smoke during the night, I began to
think of preparing my supper, when, upon examining my stores, I
found but a scanty provision. I thereupon determined, as the most
expeditious way of supplying my necessities, to take my bob and try for
some trout. About one hundred yards above my harbour began a cove
or bay of the river, out of which opened a large lagoon. The mouth or
entrance from the river to it was narrow, but the waters soon after
spread and formed a little lake, extending into the marshes: its entrance
and shores within I observed to be verged with floating lawns of the pistia and nymphea and other aquatic plants; these I knew were excellent
haunts for trout.

The verges and islets of the lagoon were elegantly embellished with flowering plants and shrubs; the laughing coots with wings half spread were tripping over the little coves, and hiding themselves in the tufts of grass; young broods of the painted summer teal, skimming the still surface of the waters, and following the watchful parent unconscious of danger, were frequently surprised by the voracious trout; and he, in turn, as often by the subtle greedy alligator. Behold him rushing forth from the flags and reeds. His enormous body swells. His plaited tail brandished high, floats upon the lake. The waters like a cataract descend from his opening jaws. Clouds of smoke issue from his dilated nostrils. The earth trembles with his thunder. When immediately from the opposite coast of the lagoon, emerges from the deep his rival champion. They suddenly dart upon each other. The boiling surface of the lake marks their rapid course, and a terrific conflict commences. They now sink to the bottom folded together in horrid wreaths. The water becomes thick and discoloured. Again they rise, their jaws clap together, re-echoing through the deep surrounding forests. Again they sink, when the contest ends at the muddy bottom of the lake, and the vanquished makes a hazardous escape, hiding himself in the muddy turbulent waters and sedge on a distant shore. The proud victor exulting returns to the place of action. The shores and forests resound his dreadful roar, together with the triumphing shouts of the plaited tribes around, witnesses of the horrid combat.

My apprehensions were highly alarmed after being a spectator of so dreadful a battle. It was obvious that every delay would but tend to increase my dangers and difficulties, as the sun was near setting, and the alligators gathered around my harbour from all quarters. From these considerations I concluded to be expeditious in my trip to the lagoon, in order to take some fish. Not thinking it prudent to take my fusee with me, lest I might lose it overboard in case of a battle, which I had every reason to dread before my return, I therefore furnished myself with a club for my defence, went on board, and penetrating the first line of those which surrounded my harbour, they gave way; but being pursued by several very large ones, I kept strictly on the watch, and paddled with all my might towards the entrance of the lagoon, hoping to be sheltered there from the multitude of my assailants; but ere I had half-way reached the place, I was attacked on all sides, several endeavouring to overset the canoe. My situation now became precarious to the last degree: two very large ones attacked me closely, at the same instant, rushing up with their heads and part of their bodies above the water, roaring terribly and belching floods of water over me. They struck their jaws together so close to my ears, as almost to stun me, and

I expected every moment to be dragged out of the boat and instantly devoured. But I applied my weapons so effectually about me, though at random, that I was so successful as to beat them off a little; when, finding that they designed to renew the battle, I made for the shore, as the only means left me for my preservation; for, by keeping close to it, I should have my enemies on one side of me only, whereas I was before surrounded by them; and there was a probability, if pushed to the last extremity, of saving myself, by jumping out of the canoe on shore, as it is easy to outwalk them on land, although comparatively as swift as lightning in the water. I found this last expedient alone could fully answer my expectations, for as soon as I gained the shore, they drew off and kept aloof. This was a happy relief, as my confidence was, in some degree, recovered by it. On recollecting myself, I discovered that I had almost reached the entrance of the lagoon, and determined to venture in, if possible, to take a few fish, and then return to my harbour, while day-light continued; for I could now, with caution and resolution, make my way with safety along shore; and indeed there was no other way to regain my camp, without leaving my boat and making my retreat through the marshes and reeds, which, if I could even effect, would have been in a manner throwing myself away, for then there would have been no hopes of ever recovering my bark, and returning in safety to any settlements of men. I accordingly proceeded, and made good my entrance into the lagoon, though not without opposition from the alligators, who formed a line across the entrance, but did not pursue me into it, nor was I molested by any there, though there were some very large ones in a cove at the upper end. I soon caught more trout than I had present occasion for, and the air was too hot and sultry to admit of their being kept for many hours, even though salted or barbecued. I now prepared for my return to camp, which I succeeded in with but little trouble, by keeping close to the shore; yet I was opposed upon re-entering the river out of the lagoon, and pursued near to my landing (though not closely attacked), particularly by an old daring one, about twelve feet in length, who kept close after me; and when I stepped on shore and turned about, in order to draw up my canoe, he rushed up near my feet, and lay there for some time, looking me in the face, his head and shoulders out of water. I resolved he should pay for his temerity, and having a heavy load in my fusee, I ran to my camp, and returning with my piece, found him with his foot on the gunwale of the boat, in search of fish. On my coming up he withdrew sullenly and slowly into the water, but soon returned and placed himself in his former position, looking at me, and seeming neither fearful nor any way disturbed. I soon dispatched him by lodging the contents of my gun in his head, and then proceeded to cleanse and prepare my fish for supper; and accordingly took them out of the boat, laid them down on the sand close to the water, and began to scale them; when, raising my head, I saw before me, through the clear water, the head and shoulders of a very large alligator, moving slowly towards me. I instantly stepped back, when, with a sweep of his tail, he brushed off several of my fish. It was certainly most providential that I looked up at that instant, as the monster would probably, in less than a minute, have seized and dragged me into the river. This incredible boldness of the animal disturbed me greatly, supposing there could now be no reasonable safety for me during the night, but by keeping continually on the watch: I therefore, as soon as I had prepared the fish, proceeded to secure myself and effects in the best manner I could. In the first place, I hauled my bark upon the shore, almost clear out of the water, to prevent their oversetting or sinking her; after this, every moveable was taken out and carried to my camp, which was but a few yards off; then ranging some dry wood in such order as was the most convenient, I cleared the ground round about it, that there might be no impediment in my way, in case of an attack in the night, either from the water or the land; for I discovered by this time, that this small isthmus, from its remote situation and fruitfulness, was resorted to by bears and wolves. Having prepared myself in the best manner I could, I charged my gun, and proceeded to reconnoitre my camp and the adjacent grounds; when I discovered that the peninsula and grove, at the distance of about two hundred yards from my encampment, on the land side, were invested by a cypress swamp, covered with water, which below was joined to the shore of the little lake, and above to the marshes surrounding the lagoon; so that I was confined to an islet exceedingly circumscribed, and I found there was no other retreat for me, in case of an attack, but by either ascending one of the large oaks, or pushing off with my boat.

It was by this time dusk, and the alligators had nearly ceased their roar, when I was again alarmed by a tumultuous noise that seemed to be in my harbour, and therefore engaged my immediate attention. Returning to my camp, I found it undisturbed, and then continued on to the extreme point of the promontory, where I saw a scene, new and surprising, which at first threw my senses into such a tumult, that it was some time before I could comprehend what was the matter; however, I soon accounted for the prodigious assemblage of crocodiles at this place, which exceeded every thing of the kind I had ever heard of.

How shall I express myself so as to convey an adequate idea of it to the reader, and at the same time avoid raising suspicions of my verac-

ity? Should I say, that the river (in this place) from shore to shore, and
perhaps near half a mile above and below me, appeared to be one solid
bank of fish, of various kinds, pushing through this narrow pass of St.
Juan's into the little lake, on their return down the river, and that the
alligators were in such incredible numbers, and so close together from
shore to shore, that it would have been easy to have walked across on
their heads, had the animals been harmless? What expressions can sufficiently declare the shocking scene that for some minutes continued,
whilst this mighty army of fish were forcing the pass? During this
attempt, thousands, I may say hundreds of thousands, of them were
caught and swallowed by the devouring alligators. I have seen an alligator take up out of the water several great fish at a time, and just squeeze
them betwixt his jaws, while the tails of the great trout flapped about
his eyes and lips, ere he had swallowed them. The horrid noise of their
closing jaws, their plunging amidst the broken banks of fish, and rising
with their prey some feet upright above the water, the floods of water
and blood rushing out of their mouths, and the clouds of vapour issu-
ing from their wide nostrils, were truly frightful. This scene continued
at intervals during the night, as the fish came to the pass. After this
sight, shocking and tremendous as it was, I found myself somewhat easier and more reconciled to my situation; being convinced that their
extraordinary assemblage here was owing to the annual feast of fish;
and that they were so well employed in their own element, that I had
little occasion to fear their paying me a visit.

It being now almost night, I returned to my camp, where I had left
my fish broiling, and my kettle of rice stewing; and having with me oil,
pepper, and salt, and excellent oranges hanging in abundance over my
head (a valuable substitute for vinegar) I sat down and regaled myself
cheerfully. Having finished my repast, I rekindled my fire for light, and
whilst I was revising the notes of my past day's journey, I was suddenly
roused with a noise behind me toward the main land. I sprang up on
my feet, and listening, I distinctly heard some creature wading the
water of the isthmus. I seized my gun and went cautiously from my
camp, directing my steps towards the noise: when I had advanced
about thirty yards, I halted behind a coppice of orange trees, and soon
perceived two very large bears, which had made their way through the
water, and had landed in the grove, about one hundred yards distance
from me, and were advancing towards me. I waited until they were
within thirty yards of me: they there began to snuff and look towards my
camp: I snapped my piece, but it flashed, on which they both turned
about and galloped off, plunging through the water and swamp, never
halting, as I suppose, until they reached fast land, as I could hear them

leaping and plunging a long time. They did not presume to return again, nor was I molested by any other creature, except being occasionally awakened by the whooping of owls, screaming of bitterns, or the wood-rats running amongst the leaves.

The wood-rat is a very curious animal. It is not half the size of the domestic rat; of a dark brown or black colour; its tail slender and shorter in proportion, and covered thinly with short hair. It is singular with respect to its ingenuity and great labour in the construction of its habitation, which is a conical pyramid about three or four feet high, constructed with dry branches, which it collects with great labour and perseverance, and piles up without any apparent order; yet they are so interwoven with one another, that it would take a bear or wild-cat some time to pull one of these castles to pieces, and allow the animals sufficient time to secure a retreat with their young.

The noise of the crocodiles kept me awake the greater part of the night; but when I arose in the morning, contrary to my expectations, there was perfect peace; very few of them to be seen, and those were asleep on the shore. Yet I was not able to suppress my fears and apprehensions of being attacked by them in future; and indeed yesterday's combat with them, notwithstanding I came off in a manner victorious, or at least made a safe retreat, had left sufficient impression on my mind to damp my courage; and it seemed too much for one of my strength, being alone in a very small boat, to encounter such collected danger. To pursue my voyage up the river, and be obliged every evening to pass such dangerous defiles, appeared to me as perilous as running the gauntlet betwixt two rows of Indians armed with knives and firebrands. I however resolved to continue my voyage one day longer, if I possibly could with safety, and then return down the river, should I find the like difficulties to oppose. Accordingly I got every thing on board, charged my gun, and set sail, cautiously, along shore. As I passed by Battle lagoon, I began to tremble and keep a good look-out; when suddenly a huge alligator rushed out of the reeds, and with a tremendous roar came up, and darted as swift as an arrow under my boat, emerging upright on my lee quarter, with open jaws, and belching water and smoke that fell upon me like rain in a hurricane. I laid soundly about his head with my club, and beat him off; and after plunging and darting about my boat, he went off on a straight line through the water, seemingly with the rapidity of lightning, and entered the cape of the lagoon. I now employed my time to the very best advantage in paddling close along shore, but could not forbear looking now and then behind me, and presently perceived one of them coming up again. The water of the river hereabouts was shoal and very clear; the monster came up with the usual roar and menaces, and passed close by the side of my boat, when I could distinctly see a young brood of alligators, to the number of one hundred or more, following after her in a long train. They kept close together in a column, without straggling off to the one side or the other; the young appeared to be of an equal size, about fifteen inches in length, almost black, with pale yellow transverse waved clouds or blotches, much like rattlesnakes in colour. I now lost sight of my enemy again.

Still keeping close along shore, on turning a point or projection of the river bank, at once I beheld a great number of hillocks or small pyramids, resembling hay-cocks, ranged like an encampment along the banks. They stood fifteen or twenty yards distant from the water, on a high marsh, about four feet perpendicular above the water. I knew them to be the nests of the crocodile, having had a description of them before; and now expected a furious and general attack, as I saw several large crocodiles swimming abreast of these buildings. These nests being so great a curiosity to me, I was determined at all events immediately to land and examine them. Accordingly, I ran my bark on shore at one of their landing-places, which was a sort of nick or little dock, from which ascended a sloping path or road up to the edge of the meadow, where their nests were; most of them were deserted, and the great thick whitish egg-shells lay broken and scattered upon the ground round about them.

The nests or hillocks are of the form of an obtuse cone, four feet high and four or five feet in diameter at their bases; they are constructed with mud, grass and herbage. At first they lay a floor of this kind of tempered mortar on the ground, upon which they deposit a layer of eggs, and upon this a stratum of mortar, seven or eight inches in thickness, and then another layer of eggs; and in this manner one stratum upon another, nearly to the top. I believe they commonly lay from one to two hundred eggs in a nest: these are hatched, I suppose, by the heat of the sun; and perhaps the vegetable substances mixed with the earth, being acted upon by the sun, may cause a small degree of fermentation, and so increase the heat in those hillocks. The ground for several acres about these nests shewed evident marks of a continual resort of alligators; the grass was every where beaten down, hardly a blade or straw was left standing; whereas, all about, at a distance, it was five or six feet high, and as thick as it could grow together. The female, as I imagine, carefully watches her own nest of eggs until they are all hatched; or perhaps while she is attending her own brood, she takes under her care and protection as many as she can get at one time, either from her own particular nest or others; but certain it is, that the

young are not left to shift for themselves; for I have had frequent opportunities of seeing the female alligator leading about the shores her train
of young ones, just as a hen does her brood of chickens; and she is
equally assiduous and courageous in defending the young, which are
under her care, and providing for their subsistence; and when she is
basking upon the warm banks, with her brood around her, you may
hear the young ones continually whining and barking like young puppies. I believe but few of a brood live to the years of full growth and
magnitude, as the old feed on the young as long as they can make prey
of them.

The alligator when full grown is a very large and terrible creature,
and of prodigious strength, activity and swiftness in the water. I have
seen them twenty feet in length, and some are supposed to be
twenty-two or twenty-three feet. Their body is as large as that of a horse;
their shape exactly resembles that of a lizard, except their tail, which is
flat or cuneiform, being compressed on each side, and gradually diminishing from the abdomen to the extremity, which, with the whole body
is covered with horny plates or squammæ, impenetrable when on the
body of the live animal, even to a rifle ball, except about their head and
just behind their fore-legs or arms, where it is said they are only vulnerable. The head of a full grown one is about three feet, and the mouth
opens nearly the same length; their eyes are small in proportion, and
seem sunk deep in the head, by means of the prominency of the brows;
the nostrils are large, inflated and prominent on the top, so that the
head in the water resembles, at a distance, a great chunk of wood floating about. Only the upper jaw moves, which they raise almost perpendicular, so as to form a right angle with the lower one. In the fore-part
of the upper jaw, on each side, just under the nostrils, are two very
large, thick, strong teeth or tusks, not very sharp, but rather the shape of
a cone: these are as white as the finest polished ivory, and are not covered by any skin or lips, and always in sight, which gives the creature a
frightful appearance: in the lower jaw are holes opposite to these teeth,
to receive them: when they clap their jaws together it causes a surprising noise, like that which is made by forcing a heavy plank with violence upon the ground, and may be heard at a great distance.

But what is yet more surprising to a stranger, is the incredible loud
and terrifying roar, which they are capable of making, especially in the
spring season, their breeding time. It most resembles very heavy distant
thunder, not only shaking the air and waters, but causing the earth to
tremble; and when hundreds and thousands are roaring at the same
time, you can scarcely be persuaded, but that the whole globe is violently and dangerously agitated.

An old champion, who is perhaps absolute sovereign of a little lake or lagoon (when fifty less than himself are obliged to content themselves with swelling and roaring in little coves round about) darts forth from the reedy coverts all at once, on the surface of the waters, in a right line; at first seemingly as rapid as lightning, but gradually more slowly until he arrives at the centre of the lake, when he stops. He now swells himself by drawing in wind and water through his mouth, which causes a loud sonorous rattling in the throat for near a minute, but it is immediately forced out again through his mouth and nostrils, with a loud noise, brandishing his tail in the air, and the vapour ascending from his nostrils like smoke. At other times, when swollen to an extent ready to burst, his head and tail lifted up, he spins or twirls round on the surface of the water. He acts his part like an Indian chief when rehearsing his feats of war; and then retiring, the exhibition is continued by others who dare to step forth, and strive to excel each other, to gain the attention of the favourite female.

Having gratified my curiosity at this general breeding-place and nursery of crocodiles, I continued my voyage up the river without being greatly disturbed by them. In my way I observed islets or floating fields of the bright green Pistia, decorated with other amphibious plants, as Senecio Jacobea, Persicaria amphibia, Coreopsis bidens, Hydrocotyle fluitans, and many others of less note. * * *

["The Crystal Bason"]

* * * I now directed my steps towards my encampment, in a different direction. I seated myself upon a swelling green knoll, at the head of the crystal bason. Near me, on the left, was a point or projection of an entire grove of the aromatic Illicium Floridanum; on my right, and all around behind me, was a fruitful Orange grove, with Palms and Magnolias interspersed; in front, just under my feet, was the inchanting and amazing crystal fountain, which incessantly threw up, from dark, rocky caverns below, tons of water every minute, forming a bason, capacious enough for large shallops to ride in, and a creek of four or five feet depth of water, and near twenty yards over, which meanders six miles through green meadows, pouring its limpid waters into the great Lake George, where they seem to remain pure and unmixed. About twenty yards from the upper edge of the bason, and directly opposite to the mouth or outlet of the creek, is a continual and amazing ebullition, where the waters are thrown up in such abundance and amazing force, as to jet and swell up two or three feet above the common surface: white sand and small particles of shells are thrown up with the waters,

near to the top, when they diverge from the centre, subside with the expanding flood, and gently sink again, forming a large rim or funnel round about the aperture or mouth of the fountain, which is a vast perforation through a bed of rocks, the ragged points of which are projected out on every side. Thus far I know to be matter of real fact, and I have related it as near as I could conceive or express myself. But there are yet remaining scenes inexpressibly admirable and pleasing.

Behold, for instance, a vast circular expanse before you, the waters of which are so extremely clear as to be absolutely diaphanous or transparent as the ether; the margin of the bason ornamented with a great variety of fruitful and floriferous trees, shrubs, and plants, the pendant golden Orange dancing on the surface of the pellucid waters, the balmy air vibrating with the melody of the merry birds, tenants of the encircling aromatic grove.

At the same instant innumerable bands of fish are seen, some clothed in the most brilliant colours; the voracious crocodile stretched along at full length, as the great trunk of a tree in size; the devouring garfish, inimical trout, and all the varieties of gilded painted bream; the barbed catfish, dreaded sting-ray, skate, and flounder, spotted bass, sheeps head and ominous drum; all in their separate bands and communities, with free and unsuspicious intercourse performing their evolutions: there are no signs of enmity, no attempt to devour each other; the different bands seem peaceably and complaisantly to move a little aside, as it were to make room for others to pass by.

But behold yet something far more admirable, see whole armies descending into an abyss, into the mouth of the bubbling fountain: they disappear! are they gone for ever? I raise my eyes with terror and astonishment; I look down again to the fountain with anxiety, when behold them as it were emerging from the blue ether of another world, apparently at a vast distance; at their first appearance, no bigger than flies or minnows; now gradually enlarging, their brilliant colours begin to paint the fluid.

Now they come forward rapidly, and instantly emerge, with the elastic expanding column of crystalline waters, into the circular bason or funnel: see now how gently they rise, some upright, others obliquely, or seem to lie as it were on their sides, suffering themselves to be gently lifted or borne up by the expanding fluid towards the surface, sailing or floating like butterflies in the cerulean ether: then again they as gently descend, diverge and move off; when they rally, form again, and rejoin their kindred tribes.

This amazing and delightful scene, though real, appears at first but as a piece of excellent painting; there seems no medium; you imagine the picture to be within a few inches of your eyes, and that you may without the least difficulty touch any one of the fish, or put your finger upon the crocodile's eye, when it really is twenty or thirty feet under water.

And although this paradise of fish may seem to exhibit a just representation of the peaceable and happy state of nature which existed before the fall, yet in reality it is a mere representation; for the nature of the fish is the same as if they were in Lake George or the river; but here the water or element in which they live and move, is so perfectly clear and transparent, it places them all on an equality with regard to their ability to injure or escape from one another; (as all river fish of prey, or such as feed upon each other, as well as the unwieldy crocodile, take their prey by surprise; secreting themselves under covert or in ambush, until an opportunity offers, when they rush suddenly upon them:) but here is no covert, no ambush; here the trout freely passes by the very nose of the alligator, and laughs in his face, and the bream by the trout.

But what is really surprising is, that the consciousness of each other's safety, or some other latent cause, should so absolutely alter their conduct, for here is not the least attempt made to injure or disturb one another. * * *

ALEXANDER WILSON
1766–1813

Emigrating from Scotland as an adult, Wilson fell in love with the wildlife of his adopted country and was soon roaming the woods of Pennyslvania studying and painting birds. With the publication of his seven-volume American Ornithology *between 1808 and 1814, he gained an authority in the field that caused even Jefferson to write to him with his naturalist questions. Nonetheless, Wilson was often in conflict with other ornithologists of his day and criticized them publicly. The love of nature, like other forms of love, can turn to jealousy. Yet within Wilson's proprietary feelings about American birds, there may also have been the more positive element of protective loyalty to old relationships. Joseph Kastner describes an incident, just a few years after Wilson's arrival in America, when he bitterly complained about some birds shot by Charles Willson Peale and his son Rubens in their own naturalist pursuits. "One of the victims was a cardinal he knew well: it would come and sing at his window every morning."*

From AMERICAN ORNITHOLOGY; OR, THE NATURAL HISTORY OF THE BIRDS OF THE UNITED STATES

IVORY-BILLED WOODPECKER

This majestic and formidable species, in strength and magnitude, stands at the head of the whole class of Woodpeckers hitherto discovered. He may be called the king or chief of his tribe; and Nature seems to have designed him a distinguished characteristic, in the superb carmine crest, and bill of polished ivory, with which she has ornamented him. His eye is brilliant and daring; and his whole frame so admirably adapted for his mode of life, and method of procuring subsistence, as to impress on the mind of the examiner the most reverential ideas of the Creator. His manners have also a dignity in them superior to the common herd of Woodpeckers. Trees, shrubbery, orchards, rails, fenceposts, and old prostrate logs, are alike interesting to those, in their humble and indefatigable search for prey; but the royal hunter now before us, scorns the humility of such situations, and seeks the most towering trees of the forest; seeming particularly attached to those prodigious cypress swamps, whose crowded giant sons stretch their bare and blasted, or moss-hung, arms midway to the skies. In these almost inaccessible recesses, amid ruinous piles of impending timber, his trumpet-like note, and loud strokes, resound through the solitary, savage wilds, of which he seems the sole lord and inhabitant. Wherever he frequents, he leaves numerous monuments of his industry behind him. We there see enormous pine-trees, with cart-loads of bark lying around their roots, and chips of the trunk itself in such quantities, as to suggest the idea that half a dozen of axemen had been at work for the whole morning. The body of the tree is also disfigured with such numerous and so large excavations, that one can hardly conceive it pos-

American Ornithology; or, The Natural History of the Birds of the United States, 9 vols. (Philadelphia: Bradford and Inskeep, 1808–1814)).

sible for the whole to be the work of a Woodpecker. With such strength, and an apparatus so powerful, what havoc might he not commit, if numerous, on the most useful of our forest trees; and yet with all these appearances, and much of vulgar prejudice against him, it may fairly be questioned whether he is at all injurious; or, at least, whether his exertions do not contribute most powerfully to the protection of our timber. Examine closely the tree where he has been at work, and you will soon perceive, that it is neither from motives of mischief nor amusement that he slices off the bark, or digs his way into the trunk. For the second and healthy tree is not in the least the object of his attention. The diseased, infested with insects, and hastening to putrefaction, are *his* favorites; there the deadly crawling enemy have formed a lodgment, between the bark and tender wood, to drink up the very vital part of the tree. It is the ravages of these vermin which the intelligent proprietor of the forest deplores, as the sole perpetrators of the destruction of his timber. Would it be believed that the larvæ of an insect, or fly, no larger than a grain of rice, should silently, and in one season, destroy some thousand acres of pine trees, many of them from two to three feet in diameter, and a hundred and fifty feet high! Yet whoever passes along the high road from Georgetown to Charleston, in South Carolina, about twenty miles from the former place, can have striking and melancholy proofs of this fact. In some places the whole woods, as far as you can see around you, are dead, stripped of the bark, their wintry-looking arms and bare trunks bleaching in the sun, and tumbling in ruins before every blast, presenting a frightful picture of desolation. And yet ignorance and prejudice stubbornly persist in directing their indignation against the bird now before us, the constant and mortal enemy of these very vermin, as if the hand that probed the wound, to extract its cause, should be equally detested with that which inflicted it; or as if the thief-catcher should be confounded with the thief. Until some effectual preventive, or more complete mode of destruction, can be devised against these insects, and their larvæ, I would humbly suggest the propriety of protecting, and receiving with proper feelings of gratitude, the services of this and the whole tribe of Woodpeckers, letting the odium of guilt fall to its proper owners.

In looking over the accounts given of the Ivory-billed Woodpecker by the naturalists of Europe, I find it asserted, that it inhabits from New Jersey to Mexico. I believe, however, that few of them are ever seen to the north of Virginia, and very few of them even in that state. The first place I observed this bird at, when on my way to the south, was about twelve miles north of Wilmington, in North Carolina. There I found the bird from which the drawing of the figure in the plate was taken.

This bird was only wounded slightly in the wing, and on being caught, uttered a loudly-reiterated, and most piteous note, exactly resembling the violent crying of a young child; which terrified my horse so, as nearly to have cost me my life. It was distressing to hear it. I carried it with me in the chair, under cover, to Wilmington. In passing through the streets, its affecting cries surprised every one within hearing, particularly the females, who hurried to the doors and windows, with looks of alarm and anxiety. I drove on, and on arriving at the piazza of the hotel, where I intended to put up, the landlord came forward, and a number of other persons who happened to be there, all equally alarmed at what they heard; this was greatly increased by my asking whether he could furnish me with accommodations for myself and my baby. The man looked blank, and foolish, while the others stared with still greater astonishment. After diverting myself for a minute or two at their expense, I drew my Woodpecker from under the cover, and a general laugh took place. I took him up stairs, and locked him up in my room, while I went to see my horse taken care of. In less than an hour I returned, and on opening the door he set up the same distressing shout, which now appeared to proceed from the grief that he had been discovered in his attempts at escape. He had mounted along the side of the window, nearly as high as the ceiling, a little below which he had begun to break through. The bed was covered with large pieces of plaster; the lath was exposed for at least fifteen inches square, and a hole, large enough to admit the fist, opened to the weather-boards; so that in less than another hour he would certainly have succeeded in making his way through. I now tied a string round his leg, and fastening it to the table, again left him. I wished to preserve his life, and had gone off in search of suitable food for him. As I reascended the stairs, I heard him again hard at work, and on entering had the mortification to perceive that he had almost entirely ruined the mahogany table to which he was fastened, and on which he had wreaked his whole vengeance. While engaged in taking the drawing, he cut me severely in several places, and on the whole, displayed such a noble and unconquerable spirit, that I was frequently tempted to restore him to his native woods. He lived with me nearly three days, but refused all sustenance, and I witnessed his death with regret.

The head and bill of this bird is in great esteem among the southern Indians, who wear them by way of amulet or charm, as well as ornament; and, it is said, dispose of them to the northern tribes at considerable prices. An Indian believes that the head, skin, or even feathers of certain birds, confer on the wearer all the virtues or excellencies of those birds. Thus I have seen a coat made of the skins, heads and claws of the Raven; caps stuck round with heads of Butcher-birds, Hawks and Eagles; and as the disposition and courage of the Ivory-billed Woodpecker are well known to the savages, no wonder they should attach great value to it, having both beauty, and, in their estimation, distinguished merit to recommend it.

This bird is not migratory, but resident in the countries where it inhabits. In the low counties of the Carolinas, it usually prefers the large-timbered cypress swamps for breeding in. In the trunk of one of these trees, at a considerable height, the male and female alternately, and in conjunction, dig out a large and capacious cavity for their eggs and young. Trees thus dug out have frequently been cut down, with sometimes the eggs and young in them. This hole according to information, for I have never seen one myself, is generally a little winding, the better to keep out the weather, and from two to five feet deep. The eggs are said to be generally four, sometimes five, as large as a pullet's, pure white, and equally thick at both ends; a description that, except in size, very nearly agrees with all the rest of our Woodpeckers. The young begin to be seen abroad about the middle of June. Whether they breed more than once in the same season is uncertain.

So little attention do the people of the countries where these birds inhabit, pay to the minutiæ of natural history, that, generally speaking, they make no distinction between the Ivory-billed and Pileated Woodpecker, represented in the same plate; and it was not till I showed them the two birds together, that they knew of any difference. The more intelligent and observing part of the natives, however, distinguish them by the name of the large and lesser *Logcocks*. They seldom examine them but at a distance, gunpowder being considered too precious to be thrown away on Woodpeckers; nothing less than a Turkey being thought worth the value of a load.

The food of this bird consists, I believe, entirely of insects and their larvæ. The Pileated Woodpecker is suspected of sometimes tasting the Indian corn; the Ivory-billed never. His common note, repeated every three or four seconds, very much resembles the tone of a trumpet, or the high note of a clarionet, and can plainly be distinguished at the distance of more than half a mile; seeming to be immediately at hand, though perhaps more than one hundred yards off. This it utters while mounting along the trunk, or digging into it. At these times it has a stately and novel appearance; and the note instantly attracts the notice of a stranger. Along the borders of the Savannah river, between Savannah and Augusta, I found them very frequently; but my horse no sooner heard their trumpet-like note, than remembering his former alarm, he became almost ungovernable.

The Ivory-billed Woodpecker is twenty inches long, and thirty inches in extent; the general color is black, with a considerable gloss of green when exposed to a good light; iris of the eye vivid yellow; nostrils covered with recumbent white hairs; fore part of the head black, rest of the crest of a most splendid red, spotted at the bottom with white, which is only seen when the crest is erected, as represented in the plate; this long red plumage being ash-colored at its base, above that white, and ending in brilliant red; a stripe of white proceeds from a point, about half an inch below each eye, passes down each side of the neck, and along the back, where they are about an inch apart, nearly to the rump; the first five primaries are wholly black, on the next five the white spreads from the tip higher and higher to the secondaries, which are wholly white from their coverts downwards: these markings, when the wings are shut, make the bird appear as if his back were white, hence he has been called, by some of our naturalists, the large White-backed Woodpecker; the neck is long; the beak an inch broad at the base, of the color and consistence of ivory, prodigiously strong, and elegantly fluted; the tail is black, tapering from the two exterior feathers, which are three inches shorter than the middle ones, and each feather has the singularity of being greatly concave below; the wing is lined with yellowish white; the legs are about an inch and a quarter long, the exterior toe about the same length, the claws exactly semicircular and remarkably powerful, the whole of a light blue or lead color. The female is about half an inch shorter, the bill rather less, and the whole plumage of the head black, glossed with green; in the other parts of the plumage she exactly resembles the male. In the stomachs of three which I opened, I found large quantities of a species of worm called *borers*, two or three inches long, of a dirty cream-color, with a black head; the stomach was an oblong pouch, not muscular like the gizzards of some others. The tongue was worm-shaped, and for half an inch at the tip as hard as horn, flat, pointed, of the same white color as the bill, and thickly barbed on each side.

JOHN LEONARD KNAPP
1767–1845

When reading Gilbert White as an undergraduate, the young Charles Darwin is supposed to have exclaimed, "Why does not every gentleman become a naturalist?" While Darwin's explorations and researches eventually led him outward to an epochal transformation of the life sciences, there were also gentlemen who stuck closer to White's own Linnaean model as a way to organize days of rambling and observing out-of-doors. John Leonard Knapp's Journal of a Naturalist (1829) acknowledges The Natural History of Selborne in the first sentence of its Preface, but then goes on to say, "The works do not, I apprehend, interfere with each other. The meditations of separate naturalists in fields, in wilds, in woods, may yield a similarity of ideas; yet the different aspects under which the same things are viewed, and characters considered, afford infinite variety of description and narrative . . . ." Much of the charm of this book, written by Knapp on his estate near Bristol, comes from the sense that, despite the systematic, detailed treatment of soils, birds, and other game and the skillful drawings of details from nature, he is presenting, and means to present, only an aspect: a personal, familiar view of nature such as is available to, and distinct for, every new observer. Writing and nature, in such an approach, renew one another.

From THE JOURNAL OF A NATURALIST

THE HEDGEHOG

Notwithstanding all the persecutions from prejudice and wantonness to which the hedgehog (erinaceus europæus) is exposed, it is yet common with us; sleeping by day in a bed of leaves and moss, under the cover of a very thick bramble or furze-bush, and at times in some

The Journal of a Naturalist (London: John Murray, 1829).

hollow stump of a tree. It creeps out in the summer evenings; and, running about with more agility than its dull appearance promises, feeds
on dew-worms and beetles, which it finds among the herbage, but
retires with trepidation at the approach of man. In the autumn, crabs,
haws, and the common fruits of the hedge, constitute its diet. In the
winter, covering itself deeply in moss and leaves, it sleeps during the
severe weather; and, when drawn out from its bed, scarcely anything of
the creature is to be observed, it exhibiting only a ball of leaves, which
it seems to attach to its spines by repeatedly rolling itself round in its
nest. Thus comfortably invested, it suffers little from the season. Some
strong smell must proceed from this animal, as we find it frequently,
with our sporting dogs, even in this state; and every village boy with his
cur detects the haunts of the poor hedgehog, and as assuredly worries
and kills him. Killing everything, and cruelty, are the common vices of
the ignorant; and unresisting innocence becomes a ready victim to
prejudice or power. The snake, the blindworm, and the toad, are all
indiscriminately destroyed as venomous animals whenever found; and
it is well for the last-mentioned poor animal, which, Boyle says, "lives
on poison, and is all venom," if prolonged sufferings do not finish its
being: but even we, who should know better, yet give rewards for the
wretched urchin's head! that very ancient prejudice of its drawing milk
from the udders of resting cows being still entertained, without any
consideration of its impracticability from the smallness of the hedgehog's mouth; and so deeply is this character associated with its name,
that we believe no argument would persuade to the contrary, or remon-
strance avail with our idle boys, to spare the life of this most harmless
and least obtrusive creature in existence.

   If we were to detail the worst propensities of man, disgusting as they
might be, yet the one most eminently offensive would be, cruelty—a
compound of tyranny, ingratitude, and pride; tyranny, because there is
the power—ingratitude, for the most harmless and serviceable are usually the object—pride, to manifest a contempt of the weakness of
humanity. There is no one creature, whose services Providence has
assigned to man, that contributes more to his wants, is more conducive
to his comforts, than the horse; nor is there one which is subjected to
more afflictions than this his faithful servant. The ass, probably and
happily, is not a very sensitive animal, but the poor horse no sooner
becomes the property of man in the lower walks of life, than he commonly has his ears shorn off; his knees are broken, his wind is broken,
his body is starved, and his eyes——!! I fear, in these grades of society,
mercy is only known by the name of cowardice, and compassion designated simplicity and effeminacy; and so we become cruel, and con-

sider it as valiance and manliness. Cruelty is a vice repeatedly marked in Scripture as repugnant to the primest attributes of our Maker, "because he delighteth in mercy." One of the three requisites necessary for man to obtain the favour of Heaven, and which was of more avail than sacrifice and oblation, was that of "showing mercy," and He, who has left us so many examples in a life of compassion and pity, hath most strongly enforced this virtue, by assuring us, that the "merciful are blessed, for they will obtain mercy."

Hedgehogs were formerly an article of food; but this diet was pronounced to be dry, and not nutritive, "because he putteth forth so many prickles." All plants producing thorns, or tending to any roughness, were considered to be of a drying nature; and, upon this foundation, the ashes of the hedgehog were administered as a "great desiccative of fistulas."

The spines of the hedgehog are moveable, not fixed and resisting, but loose in the skin, and when dry, fall backward and forward upon being moved; yet, from the peculiar manner in which they are inserted, it requires more force to draw them out than may be at first sight expected. The hair of most creatures seems to arise from a bulbous root fixed in the skin; but the spines of the hedgehog have their lower ends fined down to a thin neck or thread, which, passing through a small orifice in the skin, is secured on the under side by a round head like that of a pin, or are riveted, as it were, by the termination being enlarged and rounded; and these heads are all visible when the skin becomes dry, as if studded by small pins thrust through. Hence they are moveable in all directions, and, resting upon the muscle of the creature, must be the medium of a very sensible perception to the animal, and more so than hair could be, which does not seem to penetrate so far as the muscular fibre. Now, this little quadruped, upon suspicion of harm, rolls itself up in a ball, hiding his nose and eyes in the hollow of his stomach, and thus the common organs of perception, hearing, seeing, smelling, are precluded from action; but by the sensibility of the spines, he seems fully acquainted with every danger that may threaten him; and upon any attempt to uncoil himself, if these spines be touched, he immediately retracts, assuming his globular form again, awaiting a more secure period for retreat.

THE WATER SHREW

That little animal the water shrew (*sorex fodiens*) appears to be but partially known, but is probably more generally diffused than we imagine. The common shrew in particular seasons gambols through our hedgerows, squeaking and rustling about the dry foliage, and is observed by every one; but the water shrew inhabits places that secrete it from general notice, and appears to move only in the evenings, which occasions it being so seldom observed. That this creature was an occasional resident in our neighbourhood, was manifest from the dead bodies of two or three having occurred in my walks; but it was some time before I discovered a little colony of them quietly settled in one of my ponds, overshadowed with bushes and foliage. It is very amusing to observe the actions of these creatures, all life and animation in an element they could not be thought any way calculated for enjoying; but they swim admirably, frolicking over the floating leaves of the pondweed, and up the foliage of the flags, which, bending with their weight, will at times souse them in the pool, and away they scramble to another, searching apparently for the insects that frequent such places, and feeding on drowned moths (*phalæna potamogeta*) and similar insects. They run along the margin of the water, rooting amid the leaves and mud with their long noses for food, like little ducks, with great earnestness and perseverance. Their power of vision seems limited to a confined circumference. The smallness of their eyes, and the growth of the fur about them, are convenient for the habits of the animal, but impediments to extended vision; so that, with caution, we can approach them in their gambols, and observe all their actions. The general blackness of the body, and the triangular spot beneath the tail, as mentioned by Pennant, afford the best ready distinction of this mouse from the common shrew. Both our species of sorex seem to feed by preference on insects and worms; and thus, like the mole, their flesh is rank and offensive to most creatures, which reject them as food. The common shrew, in spring and summer, is ordinarily in motion even during the day, from sexual attachment, which occasions the destruction of numbers by cats, and other prowling animals; and thus we find them strewed in our paths, by gateways, and in our garden walks, dropped by these animals in their progress. It was once thought that some periodical disease occasioned this mortality of the species; but I think we may now conclude that violence alone is the cause of their destruction in these instances. The bite of this creature was considered by the ancients as peculiarly noxious, even to horses and large cattle; and variety of the most extraordinary remedies for the wound, and preventives against it, are mentioned by Pliny and others. The prejudices of antiquity, long as they usually are in keeping possession of the mind, have not been remembered by us; and we only know the hardy shrew now as a perfectly harmless animal, though we still retain a name for it expressive of something malignant and spiteful.

DAVID THOMPSON
1770–1857

During the nineteenth century, David Thompson was best known as an English explorer and cartographer whose travels through western Canada and the northwestern United States formed the basis for most subsequent maps and explorations. He explored the upper Missouri River almost a decade before Lewis and Clark and was the first white man to trace the entire course of the Columbia River. His journals were not published until 1916, but they have subsequently been recognized as a classic of early Canadian literature. In particular, his vivid and sympathetic portraits of individual wild animals anticipate Thoreau as well as such developers of the modern "animal story" as Ernest Thomas Seton.

From DAVID THOMPSON'S NARRATIVE OF HIS EXPLORATIONS IN WESTERN AMERICA, 1784–1812

[The Ermine]

* * * All the Animals of this Region are known to the civilized world, I shall therefore only give those traits of them which naturalists do not, or have not noticed in their descriptions. There are two species of the Mouse, the common, and the field Mouse with a short tail; they appear to be numerous, and build a House where we will, as soon as it is inhabited they make their appearance; but the country is clear of the plague of the Norway Rat, which, although he comes from England, part owner of the cargo, as yet has not travelled beyond the Factories at the sea side. The Ermine, this active little animal is an Ermine only in winter, in summer of a light brown color, he is most indefatigable after

David Thompson's Narrative of His Explorations in Western America, 1784-1812 (Toronto: The Champlain Society, 1916).

mice and small birds, and in the season, a plunderer of eggs; wherever we build, some of them soon make their burrows, and sometimes become too familiar. Having in June purchased from a Native about three dozen of Gull eggs, I put them in a room, up stairs, a plain flight of about eight feet. The Ermine soon found them, and having made a meal of one egg, was determined to carry the rest to his burrow for his young; I watched to see how he would take the eggs down stairs; holding an egg between his throat and two fore paws, he came to the head of the stairs; there he made a long stop, at a loss how to get the egg down without breaking it, his resolution was taken, and holding fast to the egg dropped down to the next stair on his neck and back; and thus to the floor, and carried it to his nest: he returned and brought two more eggs in the same manner; while he was gone for the fourth, I took the three eggs away; laying down the egg he brought, he looked all around for the others, standing on his hind legs and chattering, he was evidently in a fighting humour; at length he set off and brought another, these two I took away, and he arrived with the sixth egg, which I allowed him to keep; he was too fatigued to go for another. The next morning he returned, but the eggs were in a basket out of his reach, he knew where they were but could not get at them, and after chattering awhile, had to look for other prey. In winter we take the Ermine in small traps for the skin, which is valued to ornament dresses. *   *   *

[The Marten]

I have often tried to tame the Marten, but could never trust him beyond his chain: to one which I kept some time, I brought a small hawk slightly wounded, and placed it near him, he seemed willing to get away; and did not like it; two days after I winged a middle sized owl, and brought it to him, he appeared afraid of it, and would willingly have run away, but did not dare to cease watching it. Shortly after I found a Hare in one of the snares just taken. I brought it alive to near the Marten, he became much agitated, the skin of his head distorted to a ferocious aspect; he chattered, sprung to the Hare, as if with mortal hatred; this appeared to me strangely unaccountable, all this state of excitement against a weak animal it's common prey. Walking quickly through the Forest to visit the snares and traps, I have several times been amused with the Marten trying to steal the Hare, suspended by a snare from a pole; the Marten is very active, but the soft snow does not allow him to spring more than his own height above the surface; the Hare is suspended full five feet above the surface; determined to get the Hare, he finds the pole to which the Hare is hanging, and running along the pole, when near the small end, his weight over balances the other end, and the Marten is precipitated into the snow with the hare, before he recovers, the pole has risen with the Hare out of his reach; he would stand on his hind feet, chatter at the hare with vexation; return to the Pole, to try to get the hare, to be again plunged in the snow; how long he would have continued, I do not know, the cold did not allow me to remain long; seeing me, he ran away. ＊ ＊ ＊

[The Wolverine]

The Wolverene, is an animal unknown to other parts of the world, and we would willingly dispense with his being round here. It is a strong, well made, powerful animal; his legs short, armed with long sharp claws, he climbs trees with ease and nothing is safe that he can get at; by nature a plunderer, and mischievous, he is the plague of the country.

A party of six men were sent to square timber for the Factory, and as usual left their heavy axes where they were working, when they went to the tent for the night. One morning the six axes were not to be found, and as they knew there was no person within many miles of them they were utterly at a loss what to think or do. They were all from the very north of Scotland, and staunch believers in ghosts, fairies and such like folk, except one; at length one of them who thought himself wiser than the rest, addressed his unbelieving companion, "Now Jamie, you infidel, this comes of your laughing at ghosts and fairies, I told you that they would make us suffer for it, here now all our axes are gone and if a ghost has not taken them, what has?" Jamie was sadly puzzled what to say, for the axes were gone; fortunately the Indian lad who was tenting with them, to supply them with grouse came to them; they told him all their axes were taken away, upon looking about he perceived the footmarks of a Wolverene, and told them who the thief was, which they could not believe until tracking the Wolverene, he found one of the axes hid under the snow: in like manner three more were found, the others were carried to some distance and took two hours to find them, they were all hidden separately, and to secure their axes they had to shoulder them every evening to their tent. During the winter hunt, the feathers of the birds are the property of the hunters; and those of the white Grouse sell for six pence a pound to the Officer's of the ship, we gave our share to Robert Tennant, whom we called Old Scot. He had collected the feathers of about 300 grouse in a canvas bag, and to take it to the Factory, tied it on the Dog's sled, but some snow having fallen in the night, the hauling was heavy; and after going a short distance the bag of feathers had to be left, which was suspended to the branch of a tree; On our return we were surprized to see feathers on the snow, on coming to the tree on which we had hung the bag we found a wolverene had cut it down, torn the bag to pieces, and scattered the feathers so as hardly to leave two together. He was too knowing for a trap but [was] killed by a set Gun. In trapping of Martens, ranges of traps sometimes extend forty miles, or more. An old trapper always begins with a Wolverene trap, and at the end of every twenty traps makes one for the Wolverene, this is a work of some labor, as the trap must be strongly made and well loaded, for this strong animal, his weight is about that of an english Mastiff, but more firmly made; his skin is thick, the hair coarse, of a dark brown color, value about ten shillings, but to encourage the natives to kill it, [it] is valued at two beavers, being four times it's real value. *   *   *

[The Aurora Borealis]

Hitherto I have said little on the Aurora Borealis of the northern countries; at Hudson's Bay they are north westward, and only occasionally brilliant. I have passed four winters between the Bay and the Rein Deer's Lake, the more to the westward, the higher and brighter is this electric fluid, but always westward; but at this, the Rein Deer's Lake, as the winter came on, especially in the months of February and March, the whole heavens were in a bright glow. We seemed to be in the centre of it's action, from the horizon in every direction from north to south, from east to west, the Aurora was equally bright, sometimes, indeed often, with a tremulous motion in immense sheets, slightly tinged with the colors of the Rainbow, would roll, from horizon to horizon. Sometimes there would be a stillness of two minutes; the Dogs howled with fear, and their brightness was often such that with only their light I could see to shoot an owl at twenty yards; in the rapid motions of the Aurora we were all perswaded we heard them, reason told me I did not, but it was cool reason against sense. My men were positive they did hear the rapid motions of the Aurora, this was the eye deceiving the ear; I had my men blindfolded by turns, and then enquired of them, if they heard the rapid motions of the Aurora. They soon became sensible they did not, and yet so powerful was the Illusion of the eye on the ear, that they still believed they heard the Aurora. What is the cause that this place seems to be in the centre of the most vivid brightness and extension of the Aurora: from whence this immense extent of electric fluid, how is it formed, whither does it go. Questions without an answer. I am well acquainted with all the countries to the westward. The farther west the less is this Aurora. At the Mountains it is not seen.

DOROTHY WORDSWORTH
1771–1855

For most of the century after her death, only a few excerpts from Dorothy Wordsworth's journals were widely known. Her place in literary history rested mainly on her relationships with her brother William and with Samuel Taylor Coleridge at the period of their greatest creativity. But since the fuller publication of her Alfoxden Journal, covering the year 1798, and her Grasmere Journals, of 1800–1803, Dorothy Wordsworth has increasingly become valued as an artist in her own right. Her responses to flowers and weather, as well as to the rural poor, are startlingly vivid. It is not hard to understand why William so prized the company of his sister when he was composing his poetry and why he asked her permission to read and draw from her journals. She achieves a lyrical immediacy that makes a reader see what she has seen, feel what she has felt.

From JOURNALS OF DOROTHY WORDSWORTH

THE ALFOXDEN JOURNAL 1798

Alfoxden, 20th January 1798.

The green paths down the hillsides are channels for streams. The young wheat is streaked by silver lines of water running between the ridges, the sheep are gathered together on the slopes. After the wet dark days, the country seems more populous. It peoples itself in the sunbeams. The garden, mimic of spring, is gay with flowers. The purple-starred hepatica spreads itself in the sun, and the clustering snow-drops put forth their white heads, at first upright, ribbed with green, and like a rosebud; when completely opened, hanging their heads downwards, but slowly lengthening their slender stems. The slanting woods of an unvarying brown, showing the light through the thin net-work of their upper boughs. Upon the highest ridge of that

Journals of Dorothy Wordsworth (London: Oxford University Press, 1971).

round hill covered with planted oaks, the shafts of the trees show in the light like the columns of a ruin.

21st. Walked on the hill-tops—a warm day. Sate under the firs in the park. The tops of the beeches of a brown-red, or crimson. Those oaks, fanned by the sea breeze, thick with feathery sea-green moss, as a grove not stripped of its leaves. Moss cups more proper than acorns for fairy goblets.

22nd. Walked through the wood to Holford. The ivy twisting round the oaks like bristled serpents. The day cold—a warm shelter in the hollies, capriciously bearing berries. Query: Are the male and female flowers on separate trees?

23rd. Bright sunshine, went out at 3 o'clock. The sea perfectly calm blue, streaked with deeper colour by the clouds, and tongues or points of sand; on our return of a gloomy red. The sun gone down. The crescent moon, Jupiter, and Venus. The sound of the sea distinctly heard on the tops of the hills, which we could never hear in summer. We attribute this partly to the bareness of the trees, but chiefly to the absence of the singing of birds, the hum of insects, that noiseless noise which lives in the summer air. The villages marked out by beautiful beds of smoke. The turf fading into the mountain road. The scarlet flowers of the moss.

24th. Walked between half-past three and half-past five. The evening cold and clear. The sea of a sober grey, streaked by the deeper grey clouds. The half dead sound of the near sheep-bell, in the hollow of the sloping coombe, exquisitely soothing.

25th. Went to Poole's after tea. The sky spread over with one continuous cloud, whitened by the light of the moon, which, though her dim shape was seen, did not throw forth so strong a light as to chequer the earth with shadows. At once the clouds seemed to cleave asunder, and left her in the centre of a black-blue vault. She sailed along, followed by multitudes of stars, small, and bright, and sharp. Their brightness seemed concentrated, (half-moon).

26th. Walked upon the hill-tops; followed the sheep tracks till we overlooked the larger coombe. Sat in the sunshine. The distant sheep-bells, the sound of the stream; the woodman winding along the half-marked road with his laden pony; locks of wool still spangled with the dewdrops; the blue-grey sea, shaded with immense masses of cloud, not streaked; the sheep glittering in the sunshine. Returned through the wood. The trees skirting the wood, being exposed more directly to the action of the sea breeze, stripped of the net-work of their upper boughs, which are stiff and erect, like black skeletons; the ground strewed with the red berries of the holly. Set forward before two o'clock. Returned a little after four.

27th. Walked from seven o'clock till half-past eight. Upon the whole an uninteresting evening. Only once while we were in the wood the moon burst through the invisible veil which enveloped her, the shadows of the oaks blackened, and their lines became more strongly marked. The withered leaves were coloured with a deeper yellow, a brighter gloss spotted the hollies; again her form became dimmer; the sky flat, unmarked by distances, a white thin cloud. The manufacturer's dog makes a strange, uncouth howl, which it continues many minutes after there is no noise near it but that of the brook. It howls at the murmur of the village stream.

28th. Walked only to the mill.

29th. A very stormy day. William walked to the top of the hill to see the sea. Nothing distinguishable but a heavy blackness. An immense bough riven from one of the fir trees.

30th. William called me into the garden to observe a singular appearance about the moon. A perfect rainbow, within the bow one star, only of colours more vivid. The semi-circle soon became a complete circle, and in the course of three or four minutes the whole faded away. Walked to the blacksmith's and the baker's; an uninteresting evening.

31st. Set forward to Stowey at half-past five. A violent storm in the wood; sheltered under the hollies. When we left home the moon immensely large, the sky scattered over with clouds. These soon closed in, contracting the dimensions of the moon without concealing her. The sound of the pattering shower, and the gusts of wind, very grand. Left the wood when nothing remained of the storm but the driving wind, and a few scattering drops of rain. Presently all clear, Venus first showing herself between the struggling clouds; afterwards Jupiter appeared. The hawthorn hedges, black and pointed, glittering with millions of diamond drops; the hollies shining with broader patches of light. The road to the village of Holford glittered like another stream. On our return, the wind high—a violent storm of hail and rain at the Castle of Comfort. All the Heavens seemed in one perpetual motion when the rain ceased; the moon appearing, now half veiled, and now retired behind heavy clouds, the stars still moving, the roads very dirty.

1st February. About two hours before dinner, set forward towards Mr Bartholomew's. The wind blew so keen in our faces that we felt ourselves inclined to seek the covert of the wood. There we had a warm shelter, gathered a burthen of large rotten boughs blown down by the wind of the preceding night. The sun shone clear, but all at once a heavy blackness hung over the sea. The trees almost *roared*, and the ground seemed in motion with the multitudes of dancing leaves, which made a rustling sound, distinct from that of the trees. Still the asses pastured in quietness under the hollies, undisturbed by these forerunners of the storm. The wind beat furiously against us as we returned. Full moon. She rose in uncommon majesty over the sea, slowly ascending through the clouds. Sat with the window open an hour in the moonlight.

2nd. Walked through the wood, and on to the Downs before dinner; a warm pleasant air. The sun shone, but was often obscured by straggling clouds. The redbreasts made a ceaseless song in the woods. The wind rose very high in the evening. The room smoked so that we were obliged to quit it. Young lambs in a green pasture in the Coombe, thick legs, large heads, black staring eyes.

3rd. A mild morning, the windows open at breakfast, the redbreasts singing in the garden. Walked with Coleridge over the hills. The sea at first obscured by vapour; that vapour afterwards slid in one mighty mass along the sea-shore; the islands and one point of land clear beyond it. The distant country (which was purple in the clear dull air), overhung by straggling clouds that sailed over it, appeared like the darker clouds, which are often seen at a great distance apparently motionless, while the nearer ones pass quickly over them, driven by the lower winds. I never saw such a union of earth, sky, and sea. The clouds beneath our feet spread themselves to the water, and the clouds of the sky almost joined them. Gathered sticks in the wood; a perfect stillness. The redbreasts sang upon the leafless boughs. Of a great number of sheep in the field, only one standing. Returned to dinner at five o'clock. The moonlight still and warm as a summer's night at nine o'clock.

4th. Walked a great part of the way to Stowey with Coleridge. The morning warm and sunny. The young lasses seen on the hill-tops, in the villages and roads, in their summer holiday clothes—pink petticoats and blue. Mothers with their children in arms, and the little ones that could just walk, tottering by their side. Midges or small flies spinning in the sunshine; the songs of the lark and redbreast; daisies upon the turf; the hazels in blossom; honeysuckles budding. I saw one solitary strawberry flower under a hedge. The furze gay with blossom. The moss rubbed from the pailings by the sheep, that leave locks of wool, and the red marks with which they are spotted, upon the wood.

THE GRASMERE JOURNALS 1800–1803

[April 1802]

Thursday 29th. A beautiful morning. The sun shone and all was pleasant. We sent off our parcel to Coleridge by the waggon. Mr Simpson heard the Cuckow today. Before we went out after I had written down the Tinker (which William finished this morning) Luff called. He was very lame, limped into the kitchen—he came on a little

Pony. We then went to John's Grove, sate a while at first. Afterwards William lay, and I lay in the trench under the fence—he with his eyes shut and listening to the waterfalls and the Birds. There was no one waterfall above another—it was a sound of waters in the air—the voice of the air. William heard me breathing and rustling now and then but we both lay still, and unseen by one another. He thought that it would be as sweet thus to lie so in the grave, to hear the *peaceful* sounds of the earth and just to know that our dear friends were near. The Lake was still. There was a Boat out. Silver How reflected with delicate purple and yellowish hues as I have seen Spar. Lambs on the island and running races together by the half dozen in the round field near us. The copses green*ish*, hawthorn green.—Came home to dinner then went to Mr Simpson. We rested a long time under a wall. Sheep and lambs were in the field—cottages smoking. As I lay down on the grass, I observed the glittering silver line on the ridges of the Backs of the sheep, owing to their situation respecting the Sun—which made them look beautiful but with something of strangeness, like animals of another kind—as if belonging to a more splendid world. Met old Mr S. at the door—Mrs S. poorly. I got mullens and pansies. I was sick and ill and obliged to come home soon. We went to bed immediately—I slept up stairs. The air coldish where it was felt somewhat frosty.

Friday April 30th. We came into the orchard directly after Breakfast, and sate there. The lake was calm—the sky cloudy. We saw two fishermen by the lake side. William began to write the poem of the Celandine. I wrote to Mary H. sitting on the fur gown. Walked backwards and forwards with William—he repeated his poem to me. Then he got to work again and could not give over—he had not finished his dinner till 5 o'clock. After dinner we took up the fur gown into The Hollins above. We found a sweet seat and thither we will often go. We spread the gown put on each a cloak and there we lay. William fell asleep—he had a bad head ache owing to his having been disturbed the night before with reading C.'s letter which Fletcher had brought to the door. I did not sleep but I lay with half shut eyes, looking at the prospect as in a vision almost I was so resigned to it. Loughrigg Fell was the most distant hill, then came the Lake slipping in between the copses and above the copse the round swelling field, nearer to me a wild intermixture of rocks trees, and slacks of grassy ground.—When we turned the corner of our little shelter we saw the Church and the whole vale. It is a blessed place. The Birds were about us on all sides—Skobbys Robins Bullfinches. Crows now and then flew over our heads as we were warned by the sound of the beating of the air above. We stayed till the light of day was going and the little Birds had begun to settle their singing. But there was a thrush not far off that seemed to sing louder and clearer than the thrushes had sung when it was quite day. We came in at 8 o'clock, got tea. Wrote to Coleridge, and I wrote to Mrs Clarkson part of a letter. We went to bed at 20 minutes past 11 with prayers that Wm might sleep well.

Saturday May 1st. Rose not till $\frac{1}{2}$ past 8. A heavenly morning. As soon as Breakfast was over we went into the garden and sowed the scarlet beans about the house. It was a clear sky a heavenly morning. I sowed the flowers William helped me. We then went and sate in the orchard till dinner time. It was very hot. William wrote the Celandine. We planned a shed for the sun was too much for us. After dinner we went again to our old resting place in the Hollins under the Rock. We first lay under a holly where we saw nothing but the holly tree and a budding elm mossed with [?] and the sky above our heads. But that holly tree had a beauty about it more than its own, knowing as we did where we were. When the sun had got low enough we went to the Rock shade. Oh the overwhelming beauty of the vale below—greener than green. Two Ravens flew high high in the sky and the sun shone upon their bellys and their wings long after there was none of his light to be seen but a little space on the top of Loughrigg Fell. We went down to tea at 8 o'clock—had lost the poem and returned after tea. The Landscape was fading, sheep and lambs quiet among the Rocks. We walked towards Kings and backwards and forwards. The sky was perfectly cloudless. N.B. Is it often so? 3 solitary stars in the middle of the blue vault one or two on the points of the high hills. Wm wrote the Celandine 2nd part tonight. Heard the cuckow today this first of May.

MERIWETHER LEWIS
1774–1809

The 8,000-mile transcontinental expedition of Meriwether Lewis and William Clark from 1804 to 1806 to find a navigable inland passage from the Mississippi to the Pacific Ocean was the first great exploration of a young nation and opened the way for U.S. expansionism in the nineteenth century. The original combined journals of their trip were not published until 1904, but in their energy, their dramatic and candid accounts of the expedition's trials and adventures, their rich natural history, their descriptions of and interactions with Native Americans, and their linguistic inventiveness, The Journals of Lewis and Clark have come to be regarded by many critics as our national poem or epic, a mythic prefiguration of a people's movement westward. The landscape they traveled through seems Edenic, not only in its pristine abundance but also because language itself seems stretched as the two men named and described hundreds of new plants and animals for the first time, including Ursa horribilis, variously known as the brown, white, and "Grizly" bear. A protegé of Thomas Jefferson, Meriwether Lewis was the more-polished writer and more-accomplished naturalist, and the following passages are generally ascribed to him.

From THE JOURNALS OF LEWIS AND CLARK

Wednesday April 10th 1805.

The country on both sides of the missouri from the tops of the river hills, is one continued level fertile plain as far as the eye can reach, in which there is not even a solitary tree or shrub to be seen, except such as from their moist situations or the steep declivities of hills are sheltered from the ravages of the fire. about 1$\frac{1}{2}$ miles down this bluff from this point, the bluff is now on fire and throws out considerable quantities of smoke which has a strong sulphurious smell. the appearance of the coal in the blufs continues as yesterday ＊　　＊　　＊

Saturday April 13th.

＊　　＊　　＊ we found a number of carcases of the Buffaloe lying along shore, which had been drowned by falling through the ice in winter and lodged on shore by the high water when the river broke up about the first of this month. we saw also many tracks of the white bear of enormous size, along the river shore and about the carcases of the Buffaloe, on which I presume they feed. we have not as yet seen one of these anamals, tho' their tracks are so abundant and recent. the men as well as ourselves are anxious to meet with some of these bear. the Indians give a very formidable account of the streng[t]h and ferocity of this anamal, which they never dare to attack but in parties of six eight or ten persons; and are even then frequently defeated with the loss of one or more of their party. the savages attack this anamal with their

The Journals of Lewis and Clark (New York: Viking Penguin, 1989).

bows and arrows and the indifferent guns with which the traders furnish them, with these they shoot with such uncertainty and at so short a distance, that (*unless shot thro' head or heart wound not mortal*) they frequently mis their aim & fall a sacrefice to the bear. two Minetaries were killed during the last winter in an attack on a white bear. this anamall is said more frequently to attack a man on meeting with him, than to flee from him. When the Indians are about to go in quest of the white bear, previous to their departure, they paint themselves and perform all those supersticious rights commonly observed when they are about to make war uppon a neighbouring nation. * * *

Wednesday April 17th. 1805.

we saw immence quantities of game in every direction around us as we passed up the river; consisting of herds of Buffaloe, Elk, and Antelopes with some deer and woolves. tho' we continue to see many tracks of the bear we have seen but very few of them, and those are at a great distance generally runing from us; I the[re]fore presume that they are extreemly wary and shy; the Indian account of them dose not corrispond with our experience so far. one black bear passed near the perogues on the 16th. and was seen by myself and the party but he so quickly disappeared that we did not shoot at him. at the place we halted to dine on the Lard. side we met with a herd of buffaloe of which I killed the fatest as I conceived among them, however on examining it I found it so poar that I thought it unfit for uce and only took the tongue; the party killed another which was still more lean. just before we encamped this evening we saw some tracks of Indians who had passed about 24 hours; they left four rafts of tim[ber] on the Stard. side, on which they had passed. we supposed them to have been a party of the Assinniboins who had been to war against the rocky Mountain Indians, and then on their return. Capt. Clark saw a Curlou to-day. there were three beaver taken this morning by the party. the men prefer the flesh of this anamal, to that of any other which we have, or are able to procure at this moment. I eat very heartily of the beaver myself, and think it excellent; particularly the tale, and liver

* * *

Thursday April 25th. 1805.

* * * the whol face of the country was covered with herds of Buffaloe, Elk & Antelopes; deer are also abundant, but keep themselves more concealed in the woodland. the buffaloe Elk and Antelope are so gentle that we pass near them while feeding, without apearing to excite any alarm among them; and when we attract their attention, they frequently

approach us more nearly to discover what we are; and in some instances pursue us a considerable distance apparently with that view. * * *

Friday April 26th. 1805.

* * * after I had completed my observations in the evening I walked down and joined the party at their encampment on the point of land formed by the junction of the rivers; found them all in good health, and much pleased at having arrived at this long wished for spot, and in order to add in some measure to the general pleasure which semed to pervade our little community, we ordered a dram to be issued to each person; this soon produced the fiddle, and they spent the evening with much hilarity, singing & dancing, and seemed as perfectly to forget their past toils, as they appeared regardless of those to come. * * *

Saturday May 4th. 1805.

* * * I saw immense quantities of buffaloe in every direction, also some Elk deer and goats; having an abundance of meat on hand I passed them without firing on them; they are extreemly gentle the bull buffaloe particularly will scarcely give way to you. I passed several in the open plain within fifty paces, they viewed me for a moment as something novel and then very unconcernedly continued to feed.

Sunday May 5th. 1805.

* * * Capt. Clark and Drewyer killed the largest brown bear this evening which we have yet seen. it was a most tremendious looking anamal, and extreemly hard to kill notwithstanding he had five balls through his lungs and five others in various parts he swam more than half the distance acoss the river to a sandbar, & it was at least twenty minutes before he died; he did not attempt to attack, but fled and made the most tremendous roaring from the moment he was shot. We had no means of weighing this monster; Capt. Clark thought he would weigh 500 lbs. for my own part I think the estimate too small by 100 lbs. he measured 8. Feet $7^{1}/_{2}$ Inches from the nose to the extremety of the hind feet, 5 F. $10^{1}/_{2}$ Ins. arround the breast, 1 F. 11. I. arround the middle of the arm, & 3.F. 11.I. arround the neck; his tallons which were five in number on each foot were $4^{3}/_{8}$ Inches in length. he was in good order, we therefore divided him among the party and made them boil the oil and put it in a cask for future uce; the oil is as hard as hogs lard when cool, much more so than that of the black bear. this bear differs

from the common black bear in several respects; it's tallons are much longer and more blont, it's tale shorter, it's hair which is of a redish or bey brown, is longer thicker and finer than that of the black bear; his liver lungs and heart are much larger even in proportion with his size; the heart particularly was as large as that of a large Ox. his maw was also ten times the size of black bear, and was filled with flesh and fish. his testicles were pendant from the belly and placed four inches assunder in separate bags or pouches. this animal also feeds on roots and almost every species of wild fruit.

✳ ✳ ✳

Tuesday May 14th. 1805.

In the evening the men in two of the rear canoes discovered a large brown bear lying in the open grounds about 300 paces from the river, and six of them went out to attack him, all good hunters; they took the advantage of a small eminence which concealed them and got within 40 paces of him unperceived, two of them reserved their fires as had been previously conscerted, the four others fired nearly at the same time and put each his bullet through him, two of the balls passed through the bulk of both lobes of his lungs, in an instant this monster ran at them with open mouth, the two who had reserved their fir[e]s discharged their pieces at him as he came towards them, boath of them struck him, one only slightly and the other fortunately broke his shoulder, this however only retarded his motion for a moment only, the men unable to reload their guns took to flight, the bear pursued and had very nearly overtaken them before they reached the river; two of the party betook themselves to a canoe and the others separated an[d] concealed themselves among the willows, reloaded their pieces, each discharged his piece at him as they had an opportunity they struck him several times again but the guns served only to direct the bear to them, in this manner he pursued two of them seperately so close that they were obliged to throw aside their guns and pouches and throw themselves into the river altho' the bank was nearly twenty feet perpendicular; so enraged was this anamal that he plunged into the river only a few feet behind the second man he had compelled [to] take refuge in the water, when one of those who still remained on shore shot him through the head and finally killed him; they then took him on shore and butch[er]ed him when they found eight balls had passed through him in different directions; the bear being old the flesh was indifferent, they therefore only took the skin and fleece, the latter made us several gallons of oil ✳ ✳ ✳

Wednesday May 29th. 1805.

* * * Today we passed on the Stard. side the remains of a vast
many mangled carcases of Buffalow which had been driven over a
precipice of 120 feet by the Indians and perished; the water appeared to
have washed away a part of this immence pile of slaughter and still
their remained the fragments of at least a hundred carcases they created a most horrid stench. in this manner the Indians of the Missouri
distroy vast herds of buffaloe at a stroke; for this purpose one of the
most active and fleet young men is scelected and disguised in a robe of
buffaloe skin, having also the skin of the buffaloe's head with the years
and horns fastened on his head in form of a cap, thus caparisoned he
places himself at a convenient distance between a herd of buffaloe and
a precipice proper for the purpose, which happens in many places on
this river for miles together; the other indians now surround the herd
on the back and flanks and at a signal agreed on all shew themselves at
the same time moving forward towards the buffaloe; the disguised
indian or decoy has taken care to place himself sufficiently nigh the
buffaloe to be noticed by them when they take to flight and runing
before them they follow him in full speede to the precipice, the cattle
behind driving those in front over and seeing them go do not look or
hesitate about following untill the whole are precipitated down the pre-
cepice forming one common mass of dead an[d] mangled carcases: the
decoy in the mean time has taken care to secure himself in some cran-
ney or crivice of the clift which he had previously prepared for that
purpose. the part of the decoy I am informed is extreamly dangerous, if
they are not very fleet runers the buffaloe tread them under foot and
crush them to death, and sometimes drive them over the precipice
also, where they perish in common with the buffaloe. we saw a great
many wolves in the neighbourhood of these mangled carcases they
were fat and extreemly gentle, Capt. C. who was on shore killed one of
them with his espontoon. just above this place we came too for dinner
opposite the entrance of a bold runing river 40 Yds. wide which falls in
on Lard. side. this stream we called Slaughter river. * * *

Friday May 31st. 1805. —

* * * The hills and river Clifts which we passed today exhibit a
most romantic appearance. The bluffs of the river rise to the hight of
from 2 to 300 feet and in most places nearly perpendicular; they are
formed of remarkable white sandstone which is sufficiently soft to give
way readily to the impression of water; two or thre thin horizontal stratas
of white freestone, on which the rains or water make no impression, lie

imbeded in these clifts of soft stone near the upper part of them; the earth on the top of these Clifts is a dark rich loam, which forming a graduly ascending plain extends back from $\frac{1}{2}$ a mile to a mile where the hills commence and rise abruptly to a hight of about 300 feet more. The water in the course of time in decending from those hills and plains on either side of the river has trickled down the soft sand clifts and woarn it into a thousand grotesque figures, which with the help of a little immagination and an oblique view, at a distance are made to represent eligant ranges of lofty freestone buildings, having their parapets well stocked with statuary; collumns of various sculpture both grooved and plain, are also seen supporting long galleries in front of those buildings; in other places on a much nearer approach and with the help of less immagination we see the remains or ruins of eligant buildings; some collumns standing and almost entire with their pedestals and capitals; others retaining their pedestals but deprived by time or accident of their capitals, some lying prostrate an broken other[r]s in the form of vast pyramids of conic structure bearing a serees of other pyramids on their tops becoming less as they ascend and finally terminating in a sharp point. nitches and alcoves of various forms and sizes are seen at different hights as we pass. a number of the small martin which build their nests with clay in a globular form attatched to the wall within those nitches, and which were seen hovering about the tops of the collumns did not the less remind us of some of those large stone buildings in the U. States. the thin stratas of hard freestone intermixed with the soft sandstone seems to have aided the water in forming this curious scenery. As we passed on it seemed as if those seens of visionary inchantment would never have and [an] end; for here it is too that nature presents to the view of the traveler vast ranges of walls of tolerable workmanship, so perfect indeed are those walls that I should have thought that nature had attempted here to rival the human art of masonry had I not recollected that she had first began her work.

＊   ＊   ＊

Thursday June 13th. 1805.

I had proceded about two miles with Goodrich at some distance behind me whin my ears were saluted with the agreeable sound of a fall of water and advancing a little further I saw the spray arrise above the plain like a collumn of smoke which would frequently dispear again in an instant caused I presume by the wind which blew pretty hard from the S.W. I did not however loose my direction to this point which soon began to make a roaring too tremendious to be mistaken for any cause short of the great falls of the Missouri.

I hurryed down the hill which was about 200 feet high and difficult of access, to gaze on this sublimely grand specticle.

immediately at the cascade the river is about 300 yrds. wide; about ninty or a hundred yards of this next the Lard. bluff is a smoth even sheet of water falling over a precipice of at least eighty feet, the remaining part of about 200 yards on my right formes the grandest sight I ever beheld, the hight of the fall is the same of the other but the irregular and somewhat projecting rocks below receives the water in it's passage down and brakes it into a perfect white foam which assumes a thousand forms in a moment sometimes flying up in jets of sparkling foam to the hight of fifteen or twenty feet and are scarcely formed before large roling bodies of the same beaten and foaming water is thrown over and conceals them. in short the rocks seem to be most happily fixed to present a sheet of the whitest beaten froath for 200 yards in length and about 80 feet perpendicular. the water after decending strikes against the butment before mentioned or that on which I stand and seems to reverberate and being met by the more impetuous courant they roll and swell into half formed billows of great hight which rise and again disappear in an instant.

the buffaloe have a large beaten road to the water, for it is but in very few places that these anamals can obtain water near this place owing to the steep and inaccessible banks. I see several skelletons of the buffaloe lying in the edge of the water near the Stard. bluff which I presume have been swept down by the current and precipitated over this tremendious fall.

from the reflection of the sun on the sprey or mist which arrises from these falls there is a beatifull rainbow produced which adds not a little to the beauty of this majestically grand senery. after wrighting this imperfect discription I again viewed the falls and was so much disgusted with the imperfect idea which it conveyed of the scene that I determined to draw my pen across it and begin agin, but then reflected that I could not perhaps succeed better than pening the first impressions of the mind; I wished for the pencil of Salvator Rosa or the pen of Thompson, that I might be enabled to give to the enlightened world some just idea of this truly magnifficent and sublimely grand object, which has from the commencement of time been concealed from the view of civilized man; but this was fruitless and vain. I most sincerely regreted that I had not brought a crimee [camera] obscura with me by the assistance of which even I could have hoped to have done better but alas this was also out of my reach; I therefore with the assistance of my pen only indeavoured to trace some of the stronger features of this

seen by the assistance of which and my recollection aided by some able
pencil I hope still to give to the world some faint idea of an object
which at this moment fills me with such pleasure and astonishment;
and which of it's kind I will venture to ascert is second to but one in the
known world. I retired to the shade of a tree where I determined to fix
my camp for the present and dispatch a man in the morning to inform
Capt. C. and the party of my success in finding the falls and settle in
their minds all further doubts as to the Missouri.

*   *   *

Friday June 14th. 1805.

*   *   * I decended the hill and directed my course to the bend of the
Missouri near which there was a herd of at least a thousand buffaloe;
here I thought it would be well to kill a buffaloe and leave him untill
my return from the river and if I then found that I had not time to get
back to camp this evening to remain all night here there being a few
sticks of drift wood lying along shore which would answer for my fire,
and a few s[c]attering cottonwood trees a few hundred yards below
which would afford me at least the semblance of a shelter. under this
impression I selected a fat buffaloe and shot him very well, through the
lungs; while I was gazeing attentively on the poor anamal discharging
blood in streams from his mouth and nostrils, expecting him to fall
every instant, and having entirely forgotten to reload my rifle, a large
white, or reather brown bear, had perceived and crept on me within 20
steps before I discovered him; in the first moment I drew up my gun to
shoot, but at the same instant recolected that she was not loaded and
that he was too near for me to hope to perform this opperation before he
reached me, as he was then briskly advancing on me; it was an open
level plain, not a bush within miles nor a tree within less than three
hundred yards of me; the river bank was sloping and not more than
three feet above the level of the water; in short there was no place by
means of which I could conceal myself from this monster untill I could
charge my rifle; in this situation I thought of retreating in a brisk walk as
fast as he was advancing until I could reach a tree about 300 yards below
me, but I had no sooner terned myself about but he pitched at me, open
mouthed and full speed, I ran about 80 yards and found he gained on
me fast, I then run into the water the idea struk me to get into the water
to such debth that I could stand and he would be obliged to swim, and
that I could in that situation defend myself with my espontoon; accordingly I ran haistily into the water about waist deep, and faced about and
presented the point of my espontoon, at this instant he arrived at the
edge of the water within about 20 feet of me; the moment I put myself

in this attitude of defence he sudonly wheeled about as if frightened, declined the combat on such unequal grounds, and retreated with quite as great precipitation as he had just before pursued me. as soon as I saw him run of[f] in that manner I returned to the shore and charged my gun, which I had still retained in my hand throughout this curious adventure. I saw him run through the level open plain about three miles, till he disappeared in the woods on medecine river; during the whole of this distance he ran at full speed, sometimes appearing to look behind him as if he expected pursuit.

I now began to reflect on this novil occurrence and indeavoured to account for this sudden retreat of the bear. I at first thought that perhaps he had not smelt me before he arrived at the waters edge so near me, but I then reflected that he had pursued me for about 80 or 90 yards before I took [to] the water and on examination saw the grownd toarn with his tallons immediately on the imp[r]ession of my steps; and the cause of his allarm still remains with me misterious and unaccountable. so it was and I felt myself not a little gratifyed that he had declined the combat. my gun reloaded I felt confidence once more in my strength; and determined not to be thwarted in my design of visiting medecine river, but determined never again to suffer my peice to be longer empty than the time she necessarily required to charge her.

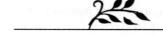

CHARLES WATERTON
1782–1865

Scion of an ancient and distinguished Yorkshire family of landed gentry, Charles Waterton was sent to British Guiana at the age of twenty-two to manage his family's sugar and coffee plantations. His true avocations, however, were natural history and exploration, and in 1813 he made the first of four expeditions into the Amazonian wilderness. On returning to England, he published an account of his travels, Wanderings in South America *(1825), whose lively style, vivid descriptions, and penchant for dramatic and somewhat exaggerated adventures made it one of the most popular nineteenth-century books about that continent. The following account of his wrestling match with a caiman—a South American*

cousin of the alligator—is quintessential Waterton. Its "tall tale" flavor
anticipates the style of such later writers as Mark Twain, Farley Mowat,
and Edward Abbey. Despite the characteristic macho tone of his adventures, Waterton was an early and ardent conservationist, and when he
succeeded to the family seat, Walton Hall, in 1806, he turned its grounds
into the first protected bird sanctuary in the British Isles.

From WANDERINGS IN SOUTH AMERICA, THE NORTH-WEST
OF THE UNITED STATES, AND THE ANTILLES

* * * I had long wished to examine the native haunts of the
Cayman; but as the river Demerara did not afford a specimen of the
large kind, I was obliged to go to the river Essequibo to look for one.

I got the canoe ready, and went down in it to George-town; where, having put in the necessary articles for the expedition, not forgetting a couple
of large shark-hooks, with chains attached to them, and a coil of strong
new rope, I hoisted a little sail, which I had got made on purpose, and at
six o'clock in the morning shaped our course for the river Essequibo. I
had put a pair of shoes on to prevent the tar at the bottom of the canoe
from sticking to my feet. The sun was flaming hot, and from eleven
o'clock till two beat perpendicularly upon the top of my feet, betwixt the
shoes and the trowsers. Not feeling it disagreeable, or being in the least
aware of painful consequences, as I had been barefoot for months, I neglected to put on a pair of short stockings which I had with me. I did not
reflect, that sitting still in one place, with your feet exposed to the sun, was
very different from being exposed to the sun while in motion.

We went ashore, in the Essequibo, about three o'clock in the afternoon, to choose a place for the night's residence, to collect firewood,
and to set the fish-hooks. It was then that I first began to find my legs
very painful: they soon became much inflamed, and red, and blistered;
and it required considerable caution not to burst the blisters, otherwise
sores would have ensued. I immediately got into the hammock, and
there passed a painful and sleepless night, and for two days after, I was
disabled from walking.

About midnight, as I was lying awake, and in great pain, I heard the
Indian say, 'Massa, massa, you no hear Tiger?' I listened attentively,
and heard the softly sounding tread of his feet as he approached us.
The moon had gone down; but every now and then we could get a

Wanderings in South America, the North-West of the United States, and the Antilles
(London: J. Mawman, 1825).

glance of him by the light of our fire; he was the Jaguar, for I could see the spots on his body. Had I wished to have fired at him I was not able to take a sure aim, for I was in such pain that I could not turn myself in my hammock. The Indian would have fired, but I would not allow him to do so, as I wanted to see a little more of our new visitor; for it is not every day or night that the traveller is favoured with an undisturbed sight of the Jaguar in his own forests.

Whenever the fire got low, the Jaguar came a little nearer, and when the Indian renewed it, he retired abruptly; sometimes he would come within twenty yards, and then we had a view of him, sitting on his hind legs like a dog; sometimes he moved slowly to and fro, and at other times we could hear him mend his pace, as if impatient. At last, the Indian not relishing the idea of having such company in the neighbourhood, could contain himself no longer, and set up a most tremendous yell. The Jaguar bounded off like a race-horse, and returned no more; it appeared by the print of his feet the next morning, that he was a full-grown Jaguar.

In two days after this we got to the first falls in the Essequibo. There was a superb barrier of rocks quite across the river. In the rainy season these rocks are for the most part under water; but it being now dry weather, we had a fine view of them, while the water from the river above them rushed through the different openings in majestic grandeur. Here, on a little hill, jutting out into the river, stands the house of Mrs Peterson, the last house of people of colour up this river; I hired a negro from her, and a coloured man, who pretended that they knew the haunts of the Cayman, and understood every thing about taking him. We were a day in passing these falls and rapids, celebrated for the Pacou, the richest and most delicious fish in Guiana. The coloured man was now in his element; he stood in the head of the canoe, and with his bow and arrow shot the Pacou as they were swimming in the stream. The arrow had scarcely left the bow before he had plunged headlong into the river, and seized the fish as it was struggling with it. He dived and swam like an otter, and rarely missed the fish he aimed at.

Did my pen, gentle reader, possess descriptive powers, I would here give thee an idea of the enchanting scenery of the Essequibo; but that not being the case, thou must be contented with a moderate and well-intended attempt.

Nothing could be more lovely than the appearance of the forest on each side of this noble river. Hills rose on hills in fine gradation, all covered with trees of gigantic height and size. Here their leaves were of a lively purple, and there of the deepest green. Sometimes the Carcara extended its scarlet blossoms from branch to branch, and gave the tree the appearance as though it had been hung with garlands.

This delightful scenery of the Essequibo made the soul overflow with joy, and caused you to rove in fancy through fairy-land; till, on turning an angle of the river, you were recalled to more sober reflections on seeing the once grand and towering Mora, now dead and ragged in its topmost branches, while its aged trunk, undermined by the rushing torrent, hung as though in sorrow over the river, which, ere long, would receive it, and sweep it away for ever.

During the day, the trade-wind blew a gentle and refreshing breeze, which died away as the night set in, and then the river was as smooth as glass.

The moon was within three days of being full, so that we did not regret the loss of the sun, which set in all its splendour. Scarce had he sunk behind the western hills, when the goatsuckers sent forth their soft and plaintive cries; some often repeating, 'Who are you—who, who, who are you?' and others, 'Willy, Willy, Willy come go.'

The Indian and Daddy Quashi often shook their head at this, and said they were bringing talk from Yabahou, who is the evil spirit of the Essequibo. It was delightful to sit on the branch of a fallen tree, near the water's edge, and listen to these harmless birds as they repeated their evening song; and watch the owls and vampires as they every now and then passed up and down the river.

The next day, about noon, as we were proceeding onwards, we heard the Campanero tolling in the depth of the forest. Though I should not then have stopped to dissect even a rare bird, having a greater object in view, still I could not resist the opportunity offered of acquiring the Campanero. The place where he was tolling was low and swampy, and my legs not having quite recovered from the effects of the sun, I sent the Indian to shoot the Campanero. He got up to the tree, which he described as very high, with a naked top, and situated in a swamp. He fired at the bird, but either missed it, or did not wound it sufficiently to bring it down. This was the only opportunity I had of getting a Campanero during this expedition. We had never heard one toll before this morning, and never heard one after.

About an hour before sunset, we reached the place which the two men, who had joined us at the falls, pointed out as a proper one to find a Cayman. There was a large creek close by, and a sandbank gently sloping to the water. Just within the forest on this bank, we cleared a place of brushwood, suspended the hammocks from the trees, and then picked up enough of decayed wood for fuel.

The Indian found a large land tortoise, and this, with plenty of fresh fish which we had in the canoe, afforded a supper not to be despised.

The tigers had kept up a continual roaring every night since we had entered the Essequibo. The sound was awfully fine. Sometimes it was in the immediate neighbourhood; at other times it was far off, and echoed amongst the hills like distant thunder.

It may, perhaps, not be amiss to observe here, that when the word tiger is used, it does not mean the Bengal tiger. It means the Jaguar, whose skin is beautifully spotted, and not striped like that of the tiger in the East. It is, in fact, the tiger of the new world, and receiving the name of tiger from the discoverers of South America, it has kept it ever since. It is a cruel, strong, and dangerous beast, but not so courageous as the Bengal tiger.

We now baited a shark-hook with a large fish, and put it upon a board about a yard long, and one foot broad, which we had brought on purpose. This board was carried out in the canoe, about forty yards into the river. By means of a string, long enough to reach the bottom of the river, and at the end of which string was fastened a stone, the board was kept, as it were, at anchor. One end of the new rope I had bought in town, was reeved through the chain of the shark-hook, and the other end fastened to a tree on the sand-bank.

It was now an hour after sunset. The sky was cloudless, and the moon shone beautifully bright. There was not a breath of wind in the heavens, and the river seemed like a large plain of quicksilver. Every now and then a huge fish would strike and plunge in the water; then the owls and goatsuckers would continue their lamentations, and the sound of these was lost in the prowling tiger's growl. Then all was still again and silent as midnight.

The Caymen were now upon the stir, and at intervals their noise could be distinguished amid that of the Jaguar, the owls, the goat-suckers, and frogs. It was a singular and awful sound. It was like a suppressed sigh, bursting forth all of a sudden, and so loud that you might hear it above a mile off. First one emitted this horrible noise, and then another answered him; and on looking at the countenances of the people round me, I could plainly see that they expected to have a Cayman that night.

We were at supper, when the Indian, who seemed to have had one eye on the turtle-pot, and the other on the bait in the river, said he saw the Cayman coming.

Upon looking towards the place, there appeared something on the water like a black log of wood. It was so unlike any thing alive, that I doubted if it were a Cayman; but the Indian smiled, and said he was sure it was one, for he remembered seeing a Cayman, some years ago, when he was in the Essequibo.

At last it gradually approached the bait, and the board began to move. The moon shone so bright, that we could distinctly see him open his huge jaws, and take in the bait. We pulled the rope. He immediately let drop the bait; and then we saw his black head retreating from the board, to the distance of a few yards; and there it remained quite motionless.

He did not seem inclined to advance again; and so we finished our supper. In about an hour's time he again put himself in motion, and took hold of the bait. But, probably, suspecting that he had to deal with knaves and cheats, he held it in his mouth, but did not swallow it. We pulled the rope again, but with no better success than the first time.

He retreated as usual, and came back again in about an hour. We paid him every attention till three o'clock in the morning; when, worn out with disappointment, we went to the hammocks, turned in, and fell asleep.

When day broke, we found that he had contrived to get the bait from the hook, though we had tied it on with string. We had now no more hopes of taking a Cayman, till the return of night. The Indian took off into the woods, and brought back a noble supply of game. The rest of us went into the canoe, and proceeded up the river to shoot fish. We got even more than we could use.

As we approached the shallows, we could see the large Sting-rays moving at the bottom. The coloured man never failed to hit them with his arrow. The weather was delightful. There was scarcely a cloud to intercept the sun's rays.

I saw several scarlet Aras, Anhingas, and ducks, but could not get a shot at them. The parrots crossed the river in innumerable quantities, always flying in pairs. Here, too, I saw the Sun-bird, called Tirana by the Spaniards in the Oroonoque, and shot one of them. The black and white scarlet-headed Finch was very common here. I could never see this bird in the Demerara, nor hear of its being there.

We at last came to a large sand-bank, probably two miles in circumference. As we approached it we could see two or three hundred fresh-water turtle on the edge of the bank. Ere we could get near enough to let fly an arrow at them, they had all sunk into the river and appeared no more.

We went on the sand-bank to look for their nests, as this was the breeding season. The coloured man showed us how to find them. Wherever a portion of the sand seemed smoother than the rest, there was sure to be a turtle's nest. On digging down with our hands, about nine inches deep, we found from twenty to thirty white eggs; in less than an hour we got above two hundred. Those which had a little black spot or two on the shell we ate the same day, as it was a sign that they were not fresh, and of course would not keep: those which had no speck were put into dry sand, and were good some weeks after.

At midnight, two of our people went to this sand-bank, while the rest staid to watch the Cayman. The turtle had advanced on to the sand to lay their eggs, and the men got betwixt them and the water; they brought off half a dozen very fine and well-fed turtle. The eggshell of the fresh-water turtle is not hard like that of the land tortoise, but appears like white parchment, and gives way to the pressure of the fingers; but it is very tough, and does not break. On this sandbank, close to the forest, we found several Guana's nests; but they had never more than fourteen eggs a-piece. Thus passed the day in exercise and knowledge, till the sun's declining orb reminded us it was time to return to the place from whence we had set out.

The second night's attempt upon the Cayman was a repetition of the first, quite unsuccessful. We went a fishing the day after, had excellent sport, and returned to experience a third night's disappointment. On the fourth evening, about four o'clock, we began to erect a stage amongst the trees, close to the water's edge. From this we intended to shoot an arrow into the Cayman: at the end of this arrow was to be attached a string, which would be tied to the rope, and as soon as the Cayman was struck, we were to have the canoe ready, and pursue him in the river.

While we were busy in preparing the stage, a tiger began to roar. We judged by the sound that he was not above a quarter of a mile from us, and that he was close to the side of the river. Unfortunately, the Indian said it was not a Jaguar that was roaring, but a Couguar. The Couguar is of a pale, brownish red colour, and not as large as the Jaguar. As there was nothing particular in this animal, I thought it better to attend to the apparatus for catching the Cayman than to go in quest of the Couguar. The people, however, went in the canoe to the place where the Couguar was roaring. On arriving near the spot, they saw it was not a Couguar but an immense Jaguar, standing on the trunk of an aged Mora-tree, which bended over the river; he growled, and showed his teeth as they approached; the coloured man fired at him with a ball, but probably missed him, and the tiger instantly descended, and took off into the woods. I went to the place before dark, and we searched the forest for about half a mile in the direction he had fled, but we could see no traces of him, or any marks of blood; so I concluded that fear had prevented the man from taking steady aim.

We spent the best part of the fourth night in trying for the Cayman, but all to no purpose. I was now convinced that something was materially wrong. We ought to have been successful, considering our vigilance and attention, and that we had repeatedly seen the Cayman. It was useless to tarry here any longer; moreover, the coloured man began to take airs, and fancied that I could not do without him. I never admit of this in any expedition where I am commander; and so I convinced the man, to his sorrow, that I could do without him; for I paid him what I had agreed to give him, which amounted to eight dollars, and ordered him back in his own curial to Mrs. Peterson's, on the hill at the first falls. I then asked the negro if there were any Indian settlements in the neighbourhood; he said he knew of one, a day and a half off. We went in quest of it, and about one o'clock the next day, the negro showed us the creek where it was.

The entrance was so concealed by thick bushes that a stranger would have passed it without knowing it to be a creek. In going up it we found it dark, winding, and intricate beyond any creek that I had ever seen before. When Orpheus came back with his young wife from Styx, his path must have been similar to this, for Ovid says it was

'Arduus, obliquus, caligine densus opaca,'
['Steep, twisting, and obscured with a dense fog.']

and this creek was exactly so.

When we had got about two-thirds up it, we met the Indians going a fishing. I saw, by the way their things were packed in the curial, that they did not intend to return for some days. However, on telling them what we wanted, and by promising handsome presents of powder, shot, and hooks, they dropped their expedition, and invited us up to the settlement they had just left, and where we laid in a provision of Cassava.

They gave us for dinner boiled ant-bear and red monkey; two dishes unknown even at Beauvilliers in Paris, or at a London city feast. The monkey was very good indeed, but the ant-bear had been kept beyond its time; it stunk like our venison does in England; and so, after tasting it, I preferred dining entirely on monkey. After resting here, we went back to the river. The Indians, three in number, accompanied us in their own curial, and, on entering the river, pointed to a place a little way above, well calculated to harbour a Cayman. The water was deep and still, and flanked by an immense sand-bank; there was also a little shallow creek close by.

On this sand-bank, near the forest, the people made a shelter for the night. My own was already made; for I always take with me a painted sheet, about twelve feet by ten. This, thrown over a pole, supported betwixt two trees, makes you a capital roof with very little trouble.

We showed one of the Indians the shark-hook. He shook his head and laughed at it, and said it would not do. When he was a boy, he had seen his father catch the Caymen, and on the morrow he would make something that would answer.

In the mean time, we set the shark-hook, but it availed us naught; a Cayman came and took it, but would not swallow it.

Seeing it was useless to attend the shark-hook any longer, we left it for the night, and returned to our hammocks.

Ere I fell asleep, a reflection or two broke in upon me. I considered, that as far as the judgment of civilized man went, every thing had been procured and done to ensure success. We had hooks, and lines, and baits, and patience; we had spent nights in watching, had seen the Cayman come and take the bait, and after our expectations had been wound up to the highest pitch, all ended in disappointment. Probably this poor wild man of the woods would succeed by means of a very simple process; and thus prove to his more civilized brother, that notwithstanding books and schools, there is a vast deal of knowledge to be picked up at every step, whichever way we turn ourselves.

In the morning, as usual, we found the bait gone from the shark-hook. The Indians went into the forest to hunt, and we took the canoe to shoot fish, and get another supply of turtle's eggs, which we found in great abundance on this large sand-bank.

We went to the little shallow creek, and shot some young Caymen, about two feet long. It was astonishing to see what spite and rage these little things showed when the arrow struck them; they turned round and bit it, and snapped at us when we went into the water to take them up. Daddy Quashi boiled one of them for his dinner, and found it very sweet and tender. I do not see why it should not be as good as frog or veal.

The day was now declining apace, and the Indian had made his instrument to take the Cayman. It was very simple. There were four pieces of tough hard wood, a foot long, and about as thick as your little finger, and barbed at both ends; they were tied round the end of the rope, in such a manner, that if you conceive the rope to be an arrow, these four sticks would form the arrow's head; so that one end of the four united sticks answered to the point of the arrow-head, while the other end of the sticks expanded at equal distances round the rope, thus—Now it is evident, that if the Cayman swallowed this, (the other end of the rope, which was thirty yards long, being fastened to a tree,) the more he pulled, the faster the barbs would stick into his stomach. This wooden hook, if you may so call it, was well-baited with the flesh of the Acouri, and the entrails were twisted round the rope for about a foot above it.

Nearly a mile from where we had our hammocks, the sand-bank was steep and abrupt, and the river very still and deep; there the Indian pricked a stick into the sand. It was two feet long, and on its extremity was fixed the machine; it hung suspended about a foot from the water, and the end of the rope was made fast to a stake driven well into the sand.

The Indian then took the empty shell of a land tortoise, and gave it some heavy blows with an axe. I asked why he did that. He said, it was to let the Cayman hear that something was going on. In fact, the Indian meant it as the Cayman's dinner-bell.

Having done this, we went back to the hammocks, not intending to visit it again till morning. During the night, the Jaguars roared and grumbled in the forest, as though the world was going wrong with them, and at intervals we could hear the distant Cayman. The roaring of the Jaguars was awful; but it was music to the dismal noise of these hideous and malicious reptiles.

About half past five in the morning, the Indian stole off silently to take a look at the bait. On arriving at the place, he set up a tremendous shout. We all jumped out of our hammocks, and ran to him. The Indians got there before me, for they had no clothes to put on, and I lost two minutes in looking for my trowsers and in slipping into them.

We found a Cayman, ten feet and a half long, fast to the end of the rope. Nothing now remained to do, but to get him out of the water without injuring his scales, 'hoc opus, hic labor' ['this the work, this the labor']. We mustered strong: there were three Indians from the creek, there was my own Indian Yan, Daddy Quashi, the negro from Mrs. Peterson's, James, Mr. R. Edmonstone's man, whom I was instructing to preserve birds, and, lastly, myself.

I informed the Indians that it was my intention to draw him quietly out of the water, and then secure him. They looked and stared at each other, and said, I might do it myself; but they would have no hand in it; the Cayman would worry some of us. On saying this, 'consedere duces,' [the leaders sat down together,'] they squatted on their hams with the most perfect indifference.

The Indians of these wilds have never been subject to the least restraint; and I knew enough of them to be aware, that if I tried to force them against their will, they would take off, and leave me and my presents unheeded, and never return.

Daddy Quashi was for applying to our guns, as usual, considering them our best and safest friends. I immediately offered to knock him down for his cowardice, and he shrunk back, begging that I would be cautious, and not get myself worried; and apologizing for his own want of resolution. My Indian was now in conversation with the others, and they asked if I would allow them to shoot a dozen arrows into him, and thus disable him. This would have ruined all. I had come above three hundred miles on purpose to get a Cayman uninjured, and not to carry back a mutilated specimen. I rejected their proposition with firmness, and darted a disdainful eye upon the Indians.

Daddy Quashi was again beginning to remonstrate, and I chased him on the sand-bank for a quarter of a mile. He told me afterwards, he thought he should have dropped down dead with fright, for he was firmly persuaded, if I had caught him, I should have bundled him into the Cayman's jaws. Here then we stood, in silence, like a calm before a thunder-storm. 'Hoc res summa loco. Scinditur in contraria vulgus.' ['This is the chief matter in the place. The people is torn in contrary directions.'] They wanted to kill him, and I wanted to take him alive.

I now walked up and down the sand, revolving a dozen projects in my head. The canoe was at a considerable distance, and I ordered the people to bring it round to the place where we were. The mast was eight feet long, and not much thicker than my wrist. I took it out of the canoe, and wrapped the sail round the end of it. Now it appeared clear to me, that if I went down upon one knee, and held the mast in the same position as the soldier holds his bayonet when rushing to the charge, I could force it down the Cayman's throat, should he come open-mouthed at me. When this was told to the Indians, they brightened up, and said they would help me to pull him out of the river.

. 'Brave squad!' said I to myself, '"Audax omnia perpeti," ["Bold to endure everything,"] now that you have got me betwixt yourselves and danger.' I then mustered all hands for the last time before the battle. We were, four South American savages, two negroes from Africa, a Creole from Trinidad, and myself a white man from Yorkshire. In fact, a little tower of Babel group, in dress, no dress, address, and language.

Daddy Quashi hung in the rear: I showed him a large Spanish knife, which I always carried in the waistband of my trowsers: it spoke volumes to him, and he shrugged up his shoulders in absolute despair. The sun was just peeping over the high forests on the eastern hills, as if coming to look on, and bid us act with becoming fortitude. I placed all the people at the end of the rope, and ordered them to pull till the Cayman appeared on the surface of the water; and then, should he plunge, to slacken the rope and let him go again into the deep.

I now took the mast of the canoe in my hand (the sail being tied round the end of the mast) and sunk down upon one knee, about four yards from the water's edge, determining to thrust it down his throat, in case he gave me an opportunity. I certainly felt somewhat uncomfortable in this situation, and I thought of Cerberus on the other side of the Styx ferry. The people pulled the Cayman to the surface; he plunged furiously as soon as he arrived in these upper regions, and immediately went below again on their slackening the rope. I saw enough not to fall in love at first sight. I now told them we would run all risks, and have him on land immediately. They pulled again, and out he came, — 'monstrum, horrendum, informe,' ['Horrible monster, void of form']. This was an interesting moment. I kept my position firmly, with my eye fixed steadfast on him.

By the time the Cayman was within two yards of me, I saw he was in a state of fear and perturbation; I instantly dropped the mast, sprung up, and jumped on his back, turning half round as I vaulted, so that I gained my seat with my face in a right position. I immediately seized his fore legs, and, by main force, twisted them on his back; thus they served me for a bridle.

He now seemed to have recovered from his surprise, and probably fancying himself in hostile company, he began to plunge furiously, and lashed the sand with his long and powerful tail. I was out of reach of the strokes of it, by being near his head. He continued to plunge and strike, and made my seat very uncomfortable. It must have been a fine sight for an unoccupied spectator.

The people roared out in triumph, and were so vociferous, that it was some time before they heard me tell them to pull me and my beast of burden farther in land. I was apprehensive the rope might break, and then there would have been every chance of going down to the regions under water with the Cayman. That would have been more perilous than Arion's marine morning ride: —

'Delphini insidens vada cærula sulcat Arion.'
['Astride a dolphin Arion ploughed the sky-blue wave.']

The people now dragged us above forty yards on the sand: it was the first and last time I was ever on a Cayman's back. Should it be asked, how I managed to keep my seat, I would answer, — I hunted some years with Lord Darlington's fox hounds.

After repeated attempts to regain his liberty, the Cayman gave in, and became tranquil through exhaustion. I now managed to tie up his jaws, and firmly secured his fore feet in the position I had held them. We had now another severe struggle for superiority, but he was soon overcome, and again remained quiet. While some of the people were pressing upon his head and shoulders, I threw myself on his tail, and by keeping it down to the sand, prevented him from kicking up another dust. He was finally conveyed to the canoe, and then to the place where we had suspended our hammocks. There I cut his throat; and after breakfast was over, commenced the dissection.

Now that the affray had ceased, Daddy Quashi played a good finger and thumb at breakfast; he said he found himself much revived, and became very talkative and useful, as there was no longer any danger.

He was a faithful, honest negro. His master, my worthy friend Mr. Edmonstone, had been so obliging as to send out particular orders to the colony, that the Daddy should attend me all the time I was in the forest. He had lived in the wilds of Demerara with Mr. Edmonstone for many years; and often amused me with the account of the frays his master had had in the woods with snakes, wild beasts, and runaway negroes. Old age was now coming fast upon him; he had been an able fellow in his younger days, and a gallant one too, for he had a large scar over his eyebrow, caused by the stroke of a cutlass, from another negro, while the Daddy was engaged in an intrigue.

The back of the Cayman may be said to be almost impenetrable to a musket ball; but his sides are not near so strong, and are easily pierced with an arrow; indeed, were they as strong as the back and the belly, there would be no part of the Cayman's body soft and elastic enough to admit of expansion after taking in a supply of food.

The Cayman has no grinders; his teeth are entirely made for snatch and swallow; there are thirty-two in each jaw. Perhaps no animal in existence bears more decided marks in his countenance of cruelty and malice than the Cayman. He is the scourge and terror of all the large rivers in South America near the line.

One Sunday evening, some years ago, as I was walking with Don Felipe de Ynciarte, governor of Angustura, on the bank of the Oroonoque, 'Stop here a minute or two, Don Carlos,' said he to me, 'while I recount a sad accident. One fine evening last year, as the people of Angustura were sauntering up and down here, in the Alameda, I was within twenty yards of this place, when I saw a large Cayman rush out of the river, seize a man, and carry him down, before any body had it in his power to assist him. The screams of the poor fellow were terrible as the Cayman was running off with him. He plunged into the river with his prey; we instantly lost sight of him, and never saw or heard him more.'

I was a day and a half in dissecting our Cayman, and then we got all ready to return to Demerara. * * *

JOHN JAMES AUDUBON
1785–1851

Audubon's elephant folio of The Birds of America *(1827–1838) won him international fame while boosting patriotic pride in America. His compositions emphasized the spectacular drama of wildlife in the New World—showing birds fighting for their lives or hurtling down toward their prey. In his accompanying text* Ornithological Biography, *written in collaboration with the Scottish naturalist William McGillivray, Audubon both told the stories of the birds he had depicted and described his own dangers and triumphs in the wilderness.*

From ORNITHOLOGICAL BIOGRAPHY

PITTING OF THE WOLVES

There seems to be a universal feeling of hostility among men against the Wolf, whose strength, agility, and cunning, which latter is scarcely inferior to that of his relative master Reynard, tend to render him an object of hatred, especially to the husbandman, on whose flocks he is ever apt to commit depredations. In America, where this animal was formerly abundant, and in many parts of which it still occurs in considerable numbers, it is more mercifully dealt with than in other parts of the world. Traps and snares of all sorts are set for catching it , while dogs and horses are trained for hunting the Fox. The Wolf, however, unless in some way injured, being more powerful and perhaps better winded than the Fox, is rarely pursued with hounds or any other dogs in the open chase; but as his depredations are at times extensive and highly injurious to the farmer, the greatest exertions have been used to exterminate his race. Few instances have occurred among us of any attack made by Wolves on man, and only one has come under my own notice.

Ornithological Biography, 5 vols. (Edinburgh, 1831–1839).

Two young Negroes who resided near the banks of the Ohio, in the lower part of the State of Kentucky, about twenty-three years ago, had sweethearts living on a plantation ten miles distant. After the labours of the day were over, they frequently visited the fair ladies of their choice, the nearest way to whose dwelling lay directly across a great cane brake. As to the lover every moment is precious, they usually took this route, to save time. Winter had commenced, cold, dark, and forbidding, and after sunset scarcely a glimpse of light or glow of warmth, one might imagine, could be found in that dreary swamp, excepting in the eyes and bosoms of the ardent youths, or the hungry Wolves that prowled about. The snow covered the earth, and rendered them more easy to be scented from a distance by the famished beasts. Prudent in a certain degree, the young lovers carried their axes on their shoulders, and walked as briskly as the narrow path would allow. Some transient glimpses of light now and then met their eyes, but so faint were they that they believed them to be caused by their faces coming in contact with the slender reeds covered with snow. Suddenly, however, a long and frightful howl burst upon them, and they instantly knew that it proceeded from a troop of hungry, perhaps desperate Wolves. They stopped, and putting themselves in an attitude of defence, awaited the result. All around was dark, save a few feet of snow, and the silence of night was dismal. Nothing could be done to better their situation, and after standing a few minutes in expectation of an attack, they judged it best to resume their march; but no sooner had they replaced their axes on their shoulders, and begun to move, than the foremost found himself assailed by several foes. His legs were held fast as if pressed by a powerful screw, and the torture inflicted by the fangs of the ravenous animal was for a moment excruciating. Several Wolves in the mean time sprung upon the breast of the other Negro, and dragged him to the ground. Both struggled manfully against their foes; but in a short time one of them ceased to move, and the other, reduced in strength, and perhaps despairing of maintaining his ground, still more of aiding his unfortunate companion, sprung to the branch of a tree, and speedily gained a place of safety near the top. The next morning, the mangled remains of his comrade lay scattered around on the snow, which was stained with blood. Three dead wolves lay around, but the rest of the pack had disappeared, and Scipio, sliding to the ground, took up the axes, and made the best of his way home, to relate the sad adventure.

About two years after this occurrence, as I was travelling between Henderson and Vincennes, I chanced to stop for the night at a farmer's house by the side of a road. After putting up my horse and refreshing myself, I entered into conversation with mine host, who asked if I should like to pay a visit to the wolf-pits, which were about half a mile distant.

Glad of the opportunity I accompanied him across the fields to the neighbourhood of a deep wood, and soon saw the engines of destruction. He had three pits, within a few hundred yards of each other. They were about eight feet deep, and broader at bottom, so as to render it impossible for the most active animal to escape from them. The aperture was covered with a revolving platform of twigs, attached to a central axis. On either surface of the platform was fastened a large piece of putrid venison, with other matters by no means pleasant to my olfactory nerves, although no doubt attractive to the wolves. My companion wished to visit them that evening, merely as he was in the habit of doing so daily, for the purpose of seeing that all was right. He said that Wolves were very abundant that autumn, and had killed nearly the whole of his sheep and one of his colts, but that he was now "paying them off in full;" and added that if I would tarry a few hours with him next morning, he would beyond a doubt shew me some sport rarely seen in those parts. We retired to rest in due time, and were up with the dawn.

"I think," said my host, "that all's right, for I see the dogs are anxious to get away to the pits, and although they are nothing but curs, their noses are none the worse for that." As he took up his gun, an axe and a large knife, the dogs began to howl and bark, and whisked around us, as if full of joy. When we reached the first pit, we found the bait all gone, and the platform much injured; but the animal that had been entrapped had scraped a subterranean passage for himself and so scaped. On peeping into the next, he assured me that "three famous fellows were safe enough" in it. I also peeped in and saw the Wolves, two black, and the other brindled, all of goodly size, sure enough. They lay flat on the earth, their ears laid close over the head, their eyes indicating fear more than anger. "But how are we to get them out?"—"How sir," said the farmer, "why by going down to be sure, and ham-stringing them." Being a novice in these matters, I begged to be merely a looker-on. "With all my heart," quoth the farmer, "stand here, and look at me through the brush." Whereupon he glided down, taking with him his axe and knife, and leaving his rifle to my care. I was not a little surprised to see the cowardice of the Wolves. He pulled out successively their hind legs, and with a side stroke of the knife cut the principal tendon above the joint, exhibiting as little fear as if he had been marking lambs.

"Lo!" exclaimed the farmer, when he had got out, "we have forgot the rope; I'll go after it." Off he went accordingly, with as much alacrity as any youngster could shew. In a short time he returned out of breath, and wiping his forehead with the back of his hand—"Now for it." I was desired to raise and hold the platform on its central balance, whilst he, with all the dexterity of an Indian, threw a noose over the neck of one of the Wolves. We hauled it up motionless with fright, as if dead, its disabled legs swinging to and fro, its jaws wide open, and the gurgle in its throat alone indicating that it was alive. Letting him drop on the ground, the farmer loosened the rope by means of a stick, and left him to the dogs, all of which set upon him with great fury and soon worried him to death. The second was dealt with in the same manner; but the third, which was probably the oldest, as it was the blackest, shewed some spirit, the moment it was left loose to the mercy of the curs. This Wolf, which we afterwards found to be a female, scuffled along on its fore legs at a surprising rate, giving a snap every now and then to the nearest dog, which went off howling dismally with a mouthful of skin torn from its side. And so well did the furious beast defend itself, that apprehensive of its escape, the farmer levelled his rifle at it, and shot it through the heart, on which the curs rushed upon it, and satiated their vengeance on the destroyer of their master's flock.

WHITE-HEADED EAGLE

(BALD EAGLE, *HALIAEETUS LEUCOCEPHALUS*)

The figure of this noble bird is well known throughout the civilized world, emblazoned as it is on our national standard, which waves in the breeze of every clime, bearing to distant lands the remembrance of a great people living in a state of peaceful freedom. May that peaceful freedom last forever!

The great strength, daring, and cool courage of the White-headed Eagle, joined to his unequalled power of flight, render him highly conspicuous among his brethren. To these qualities did he add a generous disposition towards others, he might be looked up to as a model of nobility. The ferocious, overbearing, and tyrannical temper which is ever and anon displaying itself in his actions, is, nevertheless, best adapted to his state, and was wisely given him by the Creator to enable him to perform the office assigned to him.

To give you, kind reader, some idea of the nature of this bird, permit me to place you on the Mississippi, on which you may float gently along, while approaching winter brings millions of water-fowel on whistling wings, from the countries of the north, to seek a milder climate in which to sojourn for a season. The Eagle is seen perched, in an erect attitude, on the highest summit of the tallest tree by the margin of the broad stream. His glistening but stern eye looks over the vast expanse. He listens attentively to every sound that comes to his quick ear from afar, glancing now and then on the earth beneath, lest even the light tread of the fawn may pass unheard. His mate is perched on the opposite side, and should all be tranquil and silent, warns him by a cry to continue patient. At this well known call, the male partly opens his broad wings, inclines his body a little downwards, and answers to her voice in tones not unlike the laugh of a maniac. The next moment, he resumes his erect attitude, and again all around is silent. Ducks of many species, the Teal, the Wigeon, the Mallard and others, are seen passing with great rapidity, and following the course of the current; but the Eagle heeds them not: they are at that time beneath his attention. The next moment, however, the wild trumpet-like sound of a yet distant but approaching Swan is heard. A shriek from the female Eagle comes across the stream—for, kind reader, she is fully as alert as her mate. The latter suddenly shakes the whole of his body, and with a few touches of his bill, aided by the action of his cuticular muscles, arranges his plumage in an instant. The snow-white bird is now in sight: her long neck is stretched forward, her eye is on the watch, vigilant as that of her enemy; her large wings seem with difficulty to support the weight of her body, although they flap incessantly. So irksome do her exertions seem, that her very legs are spread beneath her tail, to aid her in her flight. She approaches, however. The Eagle has marked her for his prey. As the Swan is passing the dreaded pair, starts from his perch, in full preparation for the chase, the male bird, with an awful scream, that to the Swan's ear brings more terror than the report of the large duck-gun.

Now is the moment to witness the display of the Eagle's powers. He glides through the air like a falling star; and, like a flash of lightning, comes upon the timorous quarry, which now, in agony and despair, seeks, by various manœuvers, to elude the grasp of his cruel talons. It mounts, doubles, and willingly would plunge into the stream, were it not prevented by the Eagle, which, long possessed of the knowledge that by such a stratagem the Swan might escape him, forces it to remain in the air by attempting to strike it with his talons from beneath. The hope of escape is soon given up by the Swan. It has already become much weakened, and its strength fails at the sight of the courage and swiftness of its antagonist. Its last gasp is about to escape, when the ferocious Eagle strikes with his talons the under side of its wing, and with unresisted power forces the bird to fall in a slanting direction upon the nearest shore.

It is then, reader, that you may see the cruel spirit of this dreaded enemy of the feathered race, whilst, exulting over his prey, he for the first time breathes at ease. He presses down his powerful feet, and drives his sharp claws deeper than ever into the heart of the dying Swan. He shrieks with delight, as he feels the last convulsions of his prey, which has now sunk under his unceasing efforts to render death as painfully felt as it can possibly be. The female has watched every movement of her mate; and if she did not assist him in capturing the Swan, it was not from want of will, but merely that she felt full assurance that the power and courage of her lord were quite sufficient for the deed. She now sails to the spot where he eagerly awaits her, and when she has arrived, they together turn the breast of the luckless Swan upwards, and gorge themselves with gore. *   *   *

JOHN CLARE
1793–1864

"The Northamptonshire peasant boy" was the son of a farm laborer and began working in the fields and with animals at the age of seven. Like many readers in his day, he was inspired by James Thomson's The Seasons *and produced poetry that applied those literary forms and conventions to his own much more direct experience of rural life. When his* Poems Descriptive of Rural Life and Scenery *was published in 1820, Clare became celebrated in ways that destroyed his own economy and stability without giving him a secure new identity. Poverty and disorientation followed, and from 1837 to the end of his life he was an inmate in a lunatic asylum. For many readers today Clare's most exciting writing is found in his notebooks—prose descriptions of the life of the fields that are highly emotional and precise and that are conveyed in a direct, authentic voice.*

THE NATURAL WORLD

I often pulled my hat over my eyes to watch the rising of the lark, or to see the hawk hang in the summer sky and the kite take its circles round the wood. I often lingered a minute on the woodland stile to

The Poet's Eye, ed. Geoffrey Grigson (London: Frederick Muller, 1944).

hear the woodpigeons clapping their wings among the dark oaks. I hunted curious flowers in rapture and muttered thoughts in their praise. I loved the pasture with its rushes and thistles and sheep-tracks. I adored the wild, marshy fen with its solitary heronshaw sweeping along in its melancholy sky. I wandered the heath in raptures among the rabbit burrows and golden-blossomed furze. I dropt down on a thymy molehill or mossy eminence to survey the summer landscape . . . I marked the various colours in flat, spreading fields, checkered into closes of different-tinctured grain like the colours of a map; the copper-tinted clover in blossom; the sun-tanned green of the ripening hay; the lighter hues of wheat and barley intermixed with the sunset glare of yellow charlock and the sunset imitation of the scarlet headaches; the blue corn-bottles crowding their splendid colours in large sheets over the land and troubling the cornfields with destroying beauty; the different greens of the woodland trees, the dark oak, the paler ash, the mellow lime, the white poplars peeping above the rest like leafy steeples, the grey willow shining chilly in the sun, as if the morning mist still lingered on its cool green. I loved the meadow lake with its flags and long purples crowding the water's edge. I listened with delight to hear the wind whisper among the feather-topt reeds, to see the taper bulrush nodding in gentle curves to the rippling water; and I watched with delight on haymaking evenings the setting sun drop behind the Brigs and peep again through the half-circle of the arches as if he longed to stay . . . I observed all this with the same rapture as I have done since. But I knew nothing of poetry. It was felt and not uttered.

From THE NATURAL HISTORY PROSE WRITINGS OF JOHN CLARE

Feb. 7 [1825]

I always think that this month the prophet of spring brings many beautys to the landscape tho a carless observer woud laugh at me for saying so who believes that it brings nothing because he does not give himself the trouble to seek them—I always admire the kindling freshness that the bark of the different sorts of trees & underwood asume in the for-

The Natural History Prose Writings of John Clare, ed. Margaret Grainger (Oxford: Clarendon, 1983).

est—the foulroyce twigs kindle into a vivid color at their tops as red as woodpiegons claws the ash with its grey bark & black swelling buds the Birch with it 'paper rind' & the darker mottled sorts of hazle & black alder with the greener hues of sallow willows & the bramble that still wears its leaves with the privet of a purple hue while the straggling wood briar shines in a brighter & more beautiful green even then leaves can boast at this season too odd forward branches in the new laid hedges of whitethorn begin to freshen into green before the arum dare peep out of its hood or the primrose & violet shoot up a new leaf thro the warm moss & ivy that shelters their spring dwellings the furze too on the common wear a fairer green & here & there an odd branch is covered with golden flowers & the ling or heath nestling among the long grass below (covered with the withered flowers of last year) is sprouting up into fresh hopes of spring the fairey rings on the pastures are getting deeper dyes & the water weeds with long silver green blades of grass are mantling the stagnant ponds in their summer liverys I find more beautys in this month then I can find room to talk about in a letter . . .

I forgot to say in my last that the Nightingale sung as common by day as night & as often tho its a fact that is not generaly known your Londoners are very fond of talking about the bird & I believe fancy every bird they hear after sunset a Nightingale I remember while I was there last while walking with a friend in the fields of Shacklwell we saw a gentleman & lady listening very attentive by the side of a shrubbery & when we came up we heard them lavishing praises on the beautiful song of the nightingale which happened to be a thrush but it did for them & they listend & repeated their praise with heartfelt satisfaction while the bird seemed to know the grand distinction that its song had gaind for it & strove exultingly to keep up the deception by attempting a varied & more louder song the dews was ready to fall but the lady was heedless of the wet grass tho the setting sun as a traveller glad to rest was leaning his enlarged rim on the earth like a table of fire & lessening by degrees out of sight leaving night & a few gilt clouds behind him such is the ignorance of Nature in large Citys that are nothing less than overgrown prisons that shut out the world and all its beautys

The nightingale as I said before is a shoy bird if any one approaches too near her secret haunts its song ceases till they pass when it is resumd as loud as before but I must repeat your quotation from Chaucer to illustrate this

> The new abashèd nightingale
> That stinteth first when she beginneth sing
> When that she heareth any herde's tale

Or in the hedges any [wight] stirring
& after siker doth her voice outring

As soon as they have young their song ceases & is heard no more till
the returning may after they cease singing they make a sort of gurring
guttural noise as if calling the young to their food I know not what its
for else but they make this noise continually & doubtless before the
young leave the nest I have said all I can say about the Nightingale—In
a thicket of blackthorns near our village called 'bushy close' we have
great numbers of them every year but not so many as we used to have
like the Martins & Swallows & other birds of passage they seem to
diminish but for what cause I know not

As to the cuckoo I can give you no further tidings that what I have
given in my last Artis has one in his collection of stuffed birds (but I
have not sufficient scientific curiosity about me to go & take the exact
description of its head rump & wings the length of its tail & the
breadth from the tips of the extended wings these old bookish descriptions you may find in any natural history if they are of any gratification

for my part I love to look on nature with a poetic feeling which mag-
nifys the pleasure I love to see the nightingale in its hazel retreat & the
cuckoo hiding in its solitudes of oaken foliage & not to examine their
carcasses in glass cases yet naturalists & botanists seem to have no taste
for this practical feeling they merely make collections of dryd specimens classing them after Linnaeus into tribes & familys & there they
delight to show them as a sort of ambitious fame with them 'a bird in
the hand is worth two in the bush' well everyone to his hobby

I have none of this curiosity about me tho I feel as happy as they can
about finding a new species of field flower or butterflye which I have
not seen before yet I have no desire further to dry the plant or torture
the butterflye by sticking it on a cork board with a pin—I have no wish
to do this if my feelings woud let me I only crop the blossom of the
flower or take the root from its solitudes if it would grace my garden &
wish the fluttering butterflye to settle till I can come up with it to examine the powdered colours on its wings & then it may dance off again
from fancyd dangers & welcome) I think your feelings are on the side
of Poetry for I have no specimens to send you so be as it may you must
be content with my descriptions & observations I always feel delighted
when an object in nature brings up in ones mind an image of poetry
that describes it from some favourite author you have a better opportunity of consulting books than I have therefore I will set down a list of
favourite Poems & Poets who went to nature for their images so that
you may consult them & share the feelings & pleasures which I
describe—your favourite Chaucer is one Passages in Spenser Cowley's

grasshopper & Swallow Passages in Shakespear Milton's Allegro & Penseroso & Parts of Comus the Elizabethan Poets of glorious memory Gay's Shepherds Week Green's Spleen Thomson's Seasons Collins Ode to Evening Dyer's Grongar Hill & Fleece Shenstone's Schoolmistress Gray's Ode to Spring T. Warton's April Summer Hamlet & Ode to a friend Cowper's Task Wordsworth Logans Ode to the Cuckoo Langhorne's Fables of Flora Jago's Blackbirds Bloomfield Witchwood Forest Shooters hill &c with Hurdis's Evening Walk in the village Curate & many others that may have slipped my memory

it might seem impertinent in me to advise you what to read if you misunderstood my meaning for I dont only do it for your pleasure but I wish you to make extracts from your readings in your letters to me so that I may feel some of my old gratifications agen—a clown may say that he loves the morning but a man of taste feels it in a higher degree by bringing up in his mind that beautiful line of Thomson's 'The meek eyd morn appears mother of dews' The rustic sings beneath the evening moon but it brings no associations he knowns nothing about Miltons description of it 'Now comes still evening on & twilight grey hath in her sober livery all things clad' nor of Collins Ode to Evening

the man of taste looks on the little Celandine in Spring & mutters in his mind some favourite lines from Wordsworths address to that flower he never sees the daisy without thinking of Burns & who sees the taller buttercup carpeting the closes in golden fringe without a remembrance of Chatterton's beautiful mention of it if he knows it 'The kingcup brasted with the morning dew' other flowers crowd my imagination with their poetic associations but I have no room for them the clown knows nothing of these pleasures he knows they are flowers & just turns an eye on them & plods bye therefore as I said before to look on nature with a poetic eye magnifies the pleasure she herself being the very essence & soul of Poesy if I had the means to consult & the health to indulge it I should crowd these letters on Natural History with lucious scraps of Poesy from my favourite Minstrels & make them less barren of amusement & more profitable of perusal In my catalogue of poets I forgot Charlotte Smith whose poetry is full of pleasing images from nature—Does Mr. Whites account of the Cuckoo & Nightingale agree with mine look & tell me in your next

P.S. I can scarcely believe the account which you mention at the end of your letter respecting the mans 'puzzling himself with doubts about the Nightingales singing by day & about the expression of his notes whether they are grave or gay'—you may well exclaim 'what solemn trifling' it betrays such ignorance that I can scarcely believe it—if the man does but go into any village solitude a few miles from

London next may their varied music will soon put away his doubts of its singing by day—nay he may get rid of them now by asking any country clown the question for its such a common fact that all know of it—& as to the 'expression of its notes' if he has any knowledge of nature let him ask himself whether Nature is in the habit of making such happy seeming songs for sorrow as that of the Nightingales—the poets indulgd in fancys but they did not wish that those matter of fact men the Naturalists shoud take them for facts upon their credit—What absurditys for a world that is said to get wiser and wiser every day—

*   *   *

ON ANTS

It has been a commonly believd notion among such naturalists that trusts to books & repeats the old error that ants hurd up & feed on the curnels of grain such as wheat & barley but every common observer knows this to be a falsehood I have noticd them minutely & often & never saw one with such food in its mouth they feed on flyes & caterpillars which I have often seen them tugging home with & for which they climb trees & the stems of flowers—when they first appear in the spring they may be seen carrying out ants in their mouths of a smaller size which they will continue to do a long time transporting them away from home perhaps to form new colonys—they always make a track & keep it & will go for furlongs away from their homes fetching bits of bents & others lugging away with flyes or green maggots which they pick off of flowers & leaves some when overloaded are joind by others till they get a sufficient quantity to master it home—I have often minded that two while passing each other woud pause like old friends longparted see & as if they suddenly reccolected each other they went & put their heads together as if they shook hands or saluted each other when a shower comes on an unusual bustle ensues round their nest some set out & suddenly turn round again without fetching anything home others will hasten to help those that are loaded & when the rain begins to fall the others will leave their loads & make the best of their way as fast as they can their general employment is the gathering of bents &c to cover their habitation which they generally make round an old root which they cut into holes like an honeycomb these holes lead & communicate into each other for a long way in the ground in winter they lye dormant but quickly revive if exposd to the sun there is nothing to be seen of food in their habitations then I have observd stragglers that crawl about the grass seemingly without a purpose & if they accidentally fall into the track of those at labour that quickens their pace & sudden retreat I have fancied these to be the idle & discontented sort of radicals to the government.

The smaller ants calld pismires seem to be under a different sort of government at least there is not that regularity observd among them in their labours as there is among the large ones they do not keep one track as the others do but creep about the grass were they please they are uncommonly fond of bread (which the larger ants will not touch) & when the shepherd litters his crumble from his dinner bag on their hill as he often will to observe them it instantly creates a great bustle among the little colony & they hasten away with it as fast as they can till every morsel is cleand up when they pause about as if looking for more—it is commonly believd by carless observers that every hillock on greens & commons has been first rooted up & afterwards occupied by these little tenants but on the contrary most of the hills they occupy are formed by themselves which they increase every year by bringing up a portion of mold on the surface finely powdered on which they lay their eggs to receive the warmth of the sun & the shepherd by observing their wisdom in this labour judges correctly of the changes of the weather in fact he finds it an infallible almanack when fine weather sets in their eggs are brought nearly to the top of the new addition to their hill & as soon as ever a change is about to take place nay at the approach of a shower they are observed carrying them deeper down to safer situations & if much wet is coming they entirely disappear with them into the bottom of the castle were no rain can reach them for they generally use a composition of clay in making the hills that forces off the wet & keeps it from penetrating into their cells if the crown of one of the hills be taken off with a spade it will appear pierced with holes like a honeycomb—these little things are armd with stings that blister & torture the skin with a pain worse than the keen nettle There is a smaller sort still of a deep black color that like the large ones have no sting I once when sitting at my dinner hour in the fields seeing a colony of the red pismires near one of these black ones tryd the experiment to see wether they woud associate with each other & as soon as I put a black one among them they began to fight with the latter after wounding his antagonist (seeming to be of inferior strength) curld up & dyd at his feet I then put a red one to the colony of the blacks which they instantly seized & tho he generaly contrivd to escape he appeard to be terribly wounded & no doubt was a cripple for life—these little creatures will raise a large tower of earth as thick as a mans arm in the form of a sugar loaf to a foot or a foot & a half high in the grain & long grass for in such places they cannot meet the sun on the ground so they raise these towers on the top of which they lay their eggs & as the grass or grain keeps growing they keep raising their towers till I have met with them as tall as ones knee.

GEORGE CATLIN
1796–1872

Catlin spent decades living among the native peoples of America, paint-ing their portraits and depicting scenes of tribal life. Letters and Notes on the Manners, Customs, and Conditions of the North American Indians . . . (1841) included 300 engravings of his work, as well as many of his observations on the cultures which had developed on the Great Plains. His description of the "buffalo country" through which he trav-eled includes one of the earliest recognitions that, with the coming of the white man, the vast herds would soon be destroyed by "profligate waste."

From LETTERS AND NOTES ON THE MANNERS, CUSTOMS, AND CONDITIONS OF THE NORTH AMERICAN INDIANS . . .

LETTER—NO. 31

MOUTH OF TETON RIVER, UPPER MISSOURI

In former Letters I have given some account of the *Bisons,* or (as they are more familiarly denominated in this country) *Buffaloes,* which inhabit these regions in numerous herds; and of which I must say yet a little more.

These noble animals of the ox species, and which have been so well described in our books on Natural History, are a subject of curious interest and great importance in this vast wilderness; rendered pecu-liarly so at this time, like the history of the poor savage; and from the same consideration, that they are rapidly wasting away at the approach of civilized man—and like him and his character, in a very few years, to live only in books or on canvass.

Letters and Notes on the Manners, Customs, and Conditions of the North American Indians, Written during Eight Years' Travel Amongst the Wildest Tribes of Indians in North America (New York: Wiley and Putnam, 1841).

The word buffalo is undoubtedly most incorrectly applied to these animals, and I can scarcely tell why they have been so called; for they bear just about as much resemblance to the Eastern buffalo, as they do to a zebra or to a common ox. How nearly they may approach to the bison of Europe, which I never have had an opportunity to see, and which, I am inclined to think, is now nearly extinct, I am unable to say; yet if I were to judge from the numerous engravings I have seen of those animals, and descriptions I have read of them, I should be inclined to think, there was yet a wide difference between the bison of the American prairies, and those in the North of Europe and Asia. The American bison, or (as I shall hereafter call it) buffalo, is the largest of the ruminating animals that is now living in America; and seems to have been spread over the plains of this vast country, by the Great Spirit, for the use and subsistence of the red men, who live almost exclusively on their flesh, and clothe themselves with their skins. The reader, by referring back to PLATES 7 and 8, in the beginning of this Work, will see faithful traces of the male and female of this huge animal, in their proud and free state of nature, grazing on the plains of the country to which they appropriately belong. Their colour is a dark brown, but changing very much as the season varies from warm to cold; their hair or fur, from its great length in the winter and spring, and exposure to the weather, turning quite light, and almost to a jet black, when the winter coat is shed off, and a new growth is shooting out.

The buffalo bull often grows to the enormous weight of 2000 pounds, and shakes a long and shaggy black mane, that falls in great profusion and *confusion*, over his head and shoulders; and oftentimes falling down quite to the ground. The horns are short, but very large, and have but one turn, *i.e.* they are a simple arch, without the least approach to a spiral form, like those of the common ox, or of the goat species.

The female is much smaller than the male, and always distinguishable by the peculiar shape of the horns, which are much smaller and more crooked, turning their points more in towards the centre of the forehead.

One of the most remarkable characteristics of the buffalo, is the peculiar formation and expression of the eye, the ball of which is very large and white, and the iris jet black. The lids of the eye seem always to be strained quite open, and the ball rolling forward and down; so that a considerable part of the iris is hidden behind the lower lid, while the pure white of the eyeball glares out over it in an arch, in the shape of a moon at the end of its first quarter.

These animals are, truly speaking, gregarious, but not migratory—they graze in immense and almost incredible numbers at times, and roam about and over vast tracts of country, from East to West, and from West to East, as often as from North to South; which has often been supposed they naturally and habitually did to accommodate themselves to the temperature of the climate in the different latitudes. The limits within which they are found in America, are from the 30th to the 55th degrees of North latitude; and their extent from East to West, which is from the border of our extreme Western frontier limits, to the Western verge of the Rocky Mountains, is defined by quite different causes, than those which the degrees of temperature have prescribed to them on the North and the South. Within these 25 degrees of latitude, the buffaloes seem to flourish, and get their living without the necessity of evading the rigour of the climate, for which Nature seems most wisely to have prepared them by the greater or less profusion of fur, with which she has clothed them.

It is very evident that, as high North as Lake Winnepeg, seven or eight hundred miles North of this, the buffalo subsists itself through the severest winters; getting its food chiefly by browsing amongst the timber, and by pawing through the snow, for a bite at the grass, which in those regions is frozen up very suddenly in the beginning of the winter, with all its juices in it, and consequently furnishes very nutritious and efficient food; and often, if not generally, supporting the animal in better flesh during these difficult seasons of their lives, than they are found to be in, in the 30th degree of latitude, upon the borders of Mexico, where the severity of winter is not known, but during a long and tedious autumn, the herbage, under the influence of a burning sun, is gradually dried away to a mere husk, and its nutriment gone, leaving these poor creatures, even in the dead of winter, to bask in the warmth of a genial sun, without the benefit of a green or juicy thing to bite at.

The place from which I am now writing, may be said to be the very heart or nucleus of the buffalo country, about equi-distant between the two extremes; and of course, the most congenial temperature for them to flourish in. The finest animals that gaze on the prairies are to be found in this latitude; and I am sure I never could send from a better source, some further account of the death and destruction that is dealt among these noble animals, and hurrying on their final extinction.

The Sioux are a bold and desperate set of horsemen, and great hunters; and in the heart of their country is one of the most extensive assortments of goods, of whiskey, and other saleable commodities, as well as a party of the most indefatigable men, who are constantly calling for every robe that can be stripped from these animals' backs.

These are the causes which lead so directly to their rapid destruction; and which open to the view of the traveller so freshly, so vividly, and so familiarly, the scenes of archery—of lancing, and of death-dealing, that belong peculiarly to this wild and shorn country.

The almost countless herds of these animals that are sometimes met with on these prairies, have been often spoken of by other writers, and may yet be seen by any traveller who will take the pains to visit these regions. The "*running seasons*," which is in August and September, is the time when they congregate into such masses in some places, as literally to blacken the prairies for miles together. It is no uncommon thing at this season, at these gatherings, to see several thousands in a mass, eddying and wheeling about under a cloud of dust, which is raised by the bulls as they are pawing in the dirt, or engaged in desperate combats, as they constantly are, plunging and butting at each other in the most furious manner. In these scenes, the males are continually following the females, and the whole mass are in constant motion; and all bellowing (or "roaring") in deep and hollow sounds; which, mingled altogether, appear, at the distance of a mile or two, like the sound of distant thunder.

During the season whilst they are congregated together in these dense and confused masses, the remainder of the country around for many miles, becomes entirely vacated; and the traveller may spend many a toilsome day, and many a hungry night, without being cheered by the sight of one; where, if he retraces his steps a few weeks after, he will find them dispersed, and grazing quietly in little families and flocks, and equally stocking the whole country. Of these quiet little herds, a fair representation will be seen in PLATE 106, where some are grazing, others at play, or lying down, and others indulging in their "wallows." "A bull in his wallow" is a frequent saying in this country; and has a very significant meaning with those who have ever seen a buffalo bull performing *ablution*, or rather endeavouring to cool his heated sides, by tumbling about in a mud puddle.

In the heat of summer, these huge animals, which, no doubt, suffer very much with the great profusion of their long and shaggy hair or fur, often graze on the low grounds in the prairies, where there is a little stagnant water lying amongst the grass, and the ground underneath being saturated with it, is soft, into which the enormous bull, lowered down upon one knee, will plunge his horns, and at last his head, driving up the earth, and soon making an excavation in the ground, into which the water filters from amongst the grass, forming for him in a few moments, a cool and comfortable bath, into which he plunges like a hog in his mire.

In this *delectable* laver, he throws himself flat upon his side, and forcing himself violently around, with his horns and his huge hump on his shoulders presented to the sides, he ploughs up the ground by his rotary motion, sinking himself deeper and deeper in the ground, continually enlarging his pool, in which he at length becomes nearly immersed; and the water and mud about him mixed into a complete mortar, which changes his colour, and drips in streams from every part of him as he rises up upon his feet, a hideous monster of mud and ugliness, too frightful and too eccentric to be described!

It is generally the leader of the herd that takes upon him to make this excavation; and if not (but another one opens the ground), the leader (who is conqueror) marches forward, and driving the other from it plunges himself into it; and having cooled his sides, and changed his colour to a walking mass of mud and mortar; he stands in the pool until inclination induces him to step out, and give place to the next in command, who stands ready; and another, and another, who advance forward in their turns, to enjoy the luxury of the wallow; until the whole band (sometimes an hundred or more) will pass through it in turn; each one throwing his body around in a similar manner; and each one adding a little to the dimensions of the pool, while he carries away in his hair an equal share of the clay, which dries to a grey or whitish colour, and gradually falls off. By this operation, which is done, perhaps, in the space of half an hour, a circular excavation of fifteen or twenty feet in diameter, and two feet in depth, is completed, and left for the water to run into, which soon fills it to the level of the ground.

To these sinks, the waters lying on the surface of the prairies, are continually draining, and in them lodging their vegetable deposits; which, after a lapse of years, fill them up to the surface with a rich soil, which throws up an unusual growth of grass and herbage; forming conspicuous circles which arrest the eye of the traveller, and are calculated to excite his surprise for ages to come.

Many travellers who have penetrated not quite far enough into the Western country to see the habits of these animals, and the manner in which these *mysterious* circles are made; but who have seen the prairies strewed with their bleached bones, and have beheld these strange circles, which often occur in groups, and of different sizes—have come home with beautiful and ingenious theories (which *must needs be made*), for the origin of these singular and unaccountable appearances, which, for want of a rational theory, have generally been attributed to *fairy feet*, and gained the appellation of *"fairy circles."*

Many travellers, again, have supposed that these rings were produced by the dances of the Indians, which are oftentimes (and in fact most generally) performed in a circle; yet a moment's consideration disproves such a probability, inasmuch as the Indians always select the ground for their dancing near the sites of their villages, and that always on a dry and hard foundation; when these "fairy circles" are uniformly found to be on low and wet ground.

* * *

The poor buffaloes have their enemy *man*, besetting and beseiging them at all times of the year, and in all the modes that man in his superior wisdom has been able to devise for their destruction. They struggle in vain to evade his deadly shafts, when he dashes amongst them over the plains on his wild horse—they plunge into the snow-drifts where they yield themselves an easy prey to their destroyers, and they also stand unwittingly and behold him, unsuspected under the skin of a white wolf, insinuating himself and his fatal weapons into close company, when they are peaceably grazing on the level prairies, and shot down before they are aware of their danger.

There are several varieties of the wolf species in this country, the most formidable and most numerous of which are white, often sneaking about in gangs or families of fifty or sixty in numbers, appearing in distance, on the green prairies like nothing but a flock of sheep. Many of these animals grow to a very great size, being I should think, quite a match for the largest Newfoundland dog. At present, whilst the buffaloes are so abundant, and these ferocious animals are glutted with the buffalo's flesh, they are harmless, and everywhere sneak away from man's presence; which I scarcely think will be the case after the buffaloes are all gone, and they are left, as they must be, with scarcely anything to eat. They always are seen following about in the vicinity of herds of buffaloes and stand ready to pick the bones of those that the hunters leave on the ground, or to overtake and devour those that are wounded, which fall an easy prey to them. While the herd of buffaloes are together, they seem to have little dread of the wolf, and allow them to come in close company with them. The Indian then has taken advantage of this fact, and often places himself under the skin of this animal, and crawls for half a mile or more on his hands and knees, until he approaches within a few rods of the unsuspecting group, and easily shoots down the fattest of the throng.

The buffalo is a very timid animal, and shuns the vicinity of man with the keenest sagacity; yet, when overtaken, and harassed or wounded, turns upon its assailants with the utmost fury, who have only to seek safety in flight. In their desperate resistance the finest horses are often destroyed; but the Indian, with his superior sagacity and dexterity, generally finds some effective mode of escape.

During the season of the year whilst the calves are young, the male seems to stroll about by the side of the dam, as if for the purpose of protecting the young, at which time it is exceedingly hazardous to attack them, as they are sure to turn upon their pursuers, who have often to fly to each others assistance. The buffalo calf, during the first six months is red, and has so much the appearance of a red calf in cultivated fields, that it could easily be mingled and mistaken amongst them. In the fall, when it changes its hair it takes a brown coat for the winter, which it always retains. In pursuing a large herd of buffaloes at the season when their calves are but a few weeks old, I have often been exceedingly amused with the curious manœuvres of these shy little things. Amidst the thundering confusion of a throng of several hundreds or several thousands of these animals, there will be many of the calves that lose sight of their dams; and being left behind by the throng, and the swift passing hunters, they endeavour to secrete themselves, when they are exceedingly put to it on a level prairie, where nought can be seen but the short grass of six or eight inches in height, save an occasional bunch of wild sage, a few inches higher, to which the poor affrighted things will run, and dropping on their knees, will push their noses under it, and into the grass, where they will stand for hours, with their eyes shut, imagining themselves securely hid, whilst they are standing up quite straight upon their hind feet and can easily be seen at several miles distance. It is a familiar amusement for us accustomed to these scenes, to retreat back over the ground where we have just escorted the herd, and approach these little trembling things, which stubbornly maintain their positions, with their noses pushed under the grass, and their eyes strained upon us, as we dismount from our horses and are passing around them. From this fixed position they are sure not to move, until hands are laid upon them, and then for the shins of a novice, we can extend our sympathy; or if he can preserve the skin on his bones from the furious buttings of its head, we know how to congratulate him on his signal success and good luck. In these desperate struggles, for a moment, the little thing is conquered, and makes no further resistance. And I have often, in concurrence with a known custom of the country, held my hands over the eyes of the calf, and breathed a few strong breaths into its nostrils; after which I have, with my hunting companions, rode several miles into our encampment, with the little prisoner busily following the heels of my horse the whole way, as closely and as affectionately as its instinct would attach it to the company of its dam!

This is one of the most extraordinary things that I have met with in the habits of this wild country, and although I had often heard of it, and felt unable exactly to believe it, I am now willing to bear testimony to the fact, from the numerous instances which I have witnessed since I came into the country. During the time that I resided at this post, in the spring of the year, on my way up the river, I assisted (in numerous hunts of the buffalo, with the Fur Company's men,) in bringing in, in the above manner, several of these little prisoners, which sometimes followed for five or six miles close to our horses' heels, and even into the Fur Company's Fort, and into the stable where our horses were led. In this way, before I left for the head waters of the Missouri, I think we had collected about a dozen, which Mr. Laidlaw was successfully raising with the aid of a good milch cow, and which were to be committed to the care of Mr. Chouteau to be transported by the return of the steamer, to his extensive plantation in the vicinity of St. Louis.[1]

It is truly a melancholy contemplation for the traveller in this country, to anticipate the period which is not far distant, when the last of these noble animals, at the hands of white and red men, will fall victims to their cruel and improvident rapacity; leaving these beautiful green fields, a vast and idle waste, unstocked and unpeopled for ages to come, until the bones of the one and the traditions of the other will have vanished, and left scarce an intelligible trace behind.

That the reader should not think me visionary in these contemplations, or romancing in making such assertions, I will hand him the following item of the extravagancies which are practiced in these regions, and rapidly leading to the results which I have just named.

When I first arrived at this place, on my way up the river, which was in the month of May, in 1832, and had taken up my lodgings in the Fur Company's Fort, Mr. Laidlaw, of whom I have before spoken, and also his chief clerk, Mr. Halsey, and many of their men, as well as the chiefs of the Sioux, told me, that only a few days before I arrived, (when an immense herd of buffaloes had showed themselves on the opposite side of the river, almost blackening the plains for a great distance,) a party of five or six hundred Sioux Indians on horseback, forded the river about mid-day, and spending a few hours amongst them, recrossed the river at sun-down and came into the Fort with *fourteen hundred fresh buffalo tongues*, which were thrown down in a mass, and for which they required but a few gallons of whiskey, which was soon demolished, indulging them in a little, and harmless carouse.

This profligate waste of the lives of these noble and useful animals, when, from all that I could learn, not a skin or a pound of the meat

---

[1] The fate of these poor little prisoners, I was informed on my return to St. Louis a year afterwards, was a very disastrous one. The steamer having a distance of 1600 miles to perform, and lying a week or two on sand bars, in a country where milk could not be procured, they all perished but one, which is now flourishing in the extensive fields of this gentleman. [Catlin's note]

(except the tongues), was brought in, fully supports me in the seemingly extravagant predictions that I have made as to their extinction, which I am certain is near at hand. In the above extravagant instance, at a season when their skins were without fur and not worth taking off, and their camp was so well stocked with fresh and dried meat, that they had no occasion for using the flesh, there is a fair exhibition of the improvident character of the savage, and also of his recklessness in catering for his appetite, so long as the present inducements are held out to him in his country, for its gratification.

In this singular country, where the poor Indians have no laws or regulations of society, making it a vice or an impropriety to drink to excess, they think it no harm to indulge in the delicious beverage, as long as they are able to buy whiskey to drink. They look to white men as wiser than themselves, and able to set them examples—they see none of these in their country but sellers of whiskey, who are constantly tendering it to them, and most of them setting the example by using it themselves; and they easily acquire a taste, that to be catered for, where whiskey is sold at sixteen dollars per gallon, soon impoverishes them, and must soon strip the skin from the last buffalo's back that lives in their country, to "be dressed by their squaws" and vended to the Traders for a pint of diluted alcohol.

From the above remarks it will be seen, that not only the red men, but red men and white, have aimed destruction at the race of these animals; and with them, *beasts* have turned hunters of buffaloes in this country, slaying them, however, in less numbers, and for far more laudable purpose than that of selling their skins. The white wolves, of which I have spoken in a former epistle, follow the herds of buffaloes as I have said, from one season to another, glutting themselves on the carcasses of those that fall by the deadly shafts of their enemies, or linger with disease or old age to be dispatched by these sneaking cormorants, who are ready at all times kindly to relieve them from the pangs of a lingering death.

Whilst the herd is together, the wolves never attack them, as they instantly gather for combined resistance, which they effectually make. But when the herds are travelling, it often happens that an aged or wounded one, lingers at a distance behind, and when fairly out of sight of the herd, is set upon by these voracious hunters, which often gather to the number of fifty or more, and are sure at last to torture him to death, and use him up at a meal. The buffalo, however, is a huge and furious animal, and when his retreat is cut off, makes desperate and deadly resistance, contending to the last moment for the right of life—and oftentimes deals death by wholesale, to his canine assailants, which he is tossing into the air or stamping to death under his feet.

During my travels in these regions, I have several times come across such a gang of these animals surrounding an old or a wounded bull, where it would seem, from appearances, that they had been for several days in attendance, and at intervals desperately engaged in the effort to take his life. But a short time since, as one of my hunting companions and myself were returning to our encampment with our horses loaded with meat, we discovered at a distance, a huge bull, encircled with a gang of white wolves; we rode up as near as we could without driving them away, and being within pistol shot, we had a remarkably good view, where I sat for a few moments and made a sketch in my note-book; after which, we rode up and gave the signal for them to disperse, which they instantly did, withdrawing themselves to the distance of fifty or sixty rods, when we found, to our great surprise, that the animal had made desperate resistance, until his eyes were entirely eaten out of his head — the grizzle of his nose was mostly gone — his tongue was half eaten off, and the skin and flesh of his legs torn almost literally into strings. In this tattered and torn condition, the poor old veteran stood bracing up in the midst of his devourers, who had ceased hostilities for a few minutes, to enjoy a sort of parley, recovering strength and preparing to resume the attack in a few moments again. In this group, some were reclining, to gain breath, whilst others were sneaking about and licking their chaps in anxiety for a renewal of the attack; and others, less lucky, had been crushed to death by the feet or the horns of the bull. I rode nearer to the pitiable object as he stood bleeding and trembling before me, and said to him, "Now is your time, old fellow, and you had better be off." Though blind and nearly destroyed, there seemed evidently to be a recognition of a friend in me, as he straightened up, and, trembling with excitement, dashed off at full speed upon the prairie, in a straight line. We turned our horses and resumed our march, and when we had advanced a mile or more, we looked back, and on our left, where we saw again the ill-fated animal surrounded by his tormentors, to whose insatiable voracity he unquestionably soon fell a victim.

Thus much I wrote of the buffaloes, and of the accidents that befall them, as well as of the fate that awaits them; and before I closed my book, I strolled out one day to the shade of a plum-tree, where I laid in the grass on a favourite bluff, and wrote thus: —

"It is generally supposed, and familiarly said, that a man 'falls' into a rêverie; but I seated myself in the shade a few minutes since, resolved to *force* myself into one; and for this purpose I laid open a small pocket-map of North America, and excluding my thoughts from every other object in the world, I soon succeeded in producing the desired illusion. This little chart, over which I bent, was seen in all its parts, as nothing but the green and vivid reality. I was lifted up upon an imaginary pair of wings, which easily raised and held me floating in the open air, from whence I could behold beneath me the Pacific and the Atlantic Oceans—the great cities of the East, and the mighty rivers. I could see the blue chain of the great lakes at the North—the Rocky Mountains, and beneath them and near their base, the vast, and almost boundless plains of grass, which were speckled with the bands of grazing buffaloes!

"The world turned gently around, and I examined its surface; continent after continent passed under my eye, and yet amidst them all, I saw not the vast and vivid green, that is spread like a carpet over the Western wilds of my own country. I saw not elsewhere in the world, the myriad herds of buffaloes—my eyes scanned in vain, for they were not. And when I turned again to the wilds of my native land, I beheld them all in motion! For the distance of several hundreds of miles from North to South, they were wheeling about in vast columns and herds—some were scattered, and ran with furious wildness—some lay dead, and others were pawing the earth for a hiding-place—some were sinking down and dying, gushing out their life's blood in deep-drawn sighs—and others were contending in furious battle for the life they possessed, and the ground that they stood upon. They had long since assembled from the thickets, and secret haunts of the deep forest, into the midst of the treeless and bushless plains, as the place for their safety. I could see in an hundred places, amid the wheeling bands, and on their skirts and flanks, the leaping wild horse darting among them. I saw not the arrows, nor heard the twang of the sinewy bows that sent them; but I saw their victims fall!—on other steeds that rushed along their sides, I saw the glistening lances, which seemed to lay across them; their blades were blazing in the sun, till dipped in blood, and then I lost them! In other parts (and there were many), the vivid flash of *fire-arms* was seen—*their* victims fell too, and over their dead bodies hung suspended in air, little clouds of whitened smoke, from under which the flying horsemen had darted forward to mingle again with, and deal death to, the trampling throng.

"So strange were men mixed (both red and white) with the countless herds that wheeled and eddyed about, that all below seemed one vast extended field of battle—whole armies, in some places, seemed to blacken the earth's surface;—in other parts, regiments, battalions, wings, platoons, rank and file, and *"Indian-file"*—all were in motion; and death and destruction seemed to be the watch-word amongst them. In their turmoil, they sent up great clouds of dust, and with them came the mingled din of groans and trampling hoofs, that seemed like the rumbling of a dreadful cataract, or the roaring of distant thunder. Alternate pity and admiration harrowed up in my bosom and my brain, many a hidden thought; and amongst them a few of the beautiful notes that were once sung, and exactly in point: '*Quadrupedante putrem sonitu quatit ungula campum.*' Even such was the din amidst the quadrupeds of these vast plains. And from the craggy cliffs of the Rocky Mountains also were seen descending into the valley, the myriad Tartars, who had not horses to ride, but before their well-drawn bows the fattest of the herds were falling. Hundreds and thousands were strewed upon the plains—they were flayed, and their reddened carcasses left; and about them bands of wolves, and dogs, and buzzards were seen devouring them. Contiguous, and in sight, were the distant and feeble smokes of wigwams and villages, where the skins were dragged, and dressed for white man's luxury! where they were all sold for *whiskey,* and the poor Indians laid drunk, and were crying. I cast my eyes into the towns and cities of the East, and there I beheld buffalo robes hanging at almost every door for traffic; and I saw also the curling smokes of a thousand *Stills*—and I said, 'Oh insatiable man, is thy avarice such! wouldst thou tear the skin from the back of the last animal of this noble race, *and rob thy fellow-man of his meat, and for it give him poison!*'"   *   *   *

RALPH WALDO EMERSON
1803–1882

In the development of American nature writing, Emerson is generally given credit for having inspired such major figures as Thoreau, Burroughs, Whitman, and Muir. There is no question that his first book, Nature, *published in 1835, had enormous influence on the course of the genre, and the opening sections remain powerful statements of the American imperative to "enjoy an original relation to the universe." The prevailing notion of Emerson, though, is that he was primarily an abstract philosopher and armchair naturalist who preferred his nature warmed-over, and except for some of his poems he is rarely anthologized as a nature writer. In his extensive* Journals, *however, he reveals himself as an original observer and vivid recorder of local natural phenomena. His frequent, affectionate visits to Walden Pond remind us that Emerson was no mere absentee landlord of that famous piece of waterfront.*

From NATURE

Nature is but an image or imitation of wisdom,
the last thing of the soul; nature being a thing
which doth only do, but not know. — PLOTINUS

(Motto of 1836)

A subtle chain of countless rings
The next unto the farthest brings;
The eye reads omens where it goes,
And speaks all languages the rose;
And, striving to be man, the worm
Mounts through all the spires of form.

(Motto of 1849)

INTRODUCTION

Our age is retrospective. It builds the sepulchres of the fathers. It writes biographies, histories, and criticism. The foregoing generations beheld God and nature face to face; we, through their eyes. Why should not we also enjoy an original relation to the universe? Why should not we have a poetry and philosophy of insight and not of tradition, and a religion by revelation to us, and not the history of theirs? Embosomed for a season in nature, whose floods of life stream around and through us, and invite us, by the powers they supply, to action proportioned to nature, why should we grope among the dry bones of the past, or put the living generation into masquerade out of its faded wardrobe? The sun shines today also. There is more wool and flax in the fields. There are new lands, new men, new thoughts. Let us demand our own works and laws and worship.

Undoubtedly we have no questions to ask which are unanswerable. We must trust the perfection of the creation so far as to believe that whatever curiosity the order of things has awakened in our minds, the order of things can satisfy. Every man's condition is a solution in hieroglyphic to those inquiries he would put. He acts it as life, before he

Nature (Boston: J. Monroe, 1836).

apprehends it as truth. In like manner, nature is already, in its forms and tendencies, describing its own design. Let us interrogate the great apparition that shines so peacefully around us. Let us inquire, to what end is nature?

All science has one aim, namely, to find a theory of nature. We have theories of races and of functions, but scarcely yet a remote approach to an idea of creation. We are now so far from the road to truth, that religious teachers dispute and hate each other, and speculative men are esteemed unsound and frivolous. But to a sound judgment, the most abstract truth is the most practical. Whenever a true theory appears, it will be its own evidence. Its test is, that it will explain all phenomena. Now many are thought not only unexplained but inexplicable; as language, sleep, madness, dreams, beasts, sex.

Philosophically considered, the universe is composed of Nature and the Soul. Strictly speaking, therefore, all that is separate from us, all which Philosophy distinguishes as the NOT ME, that is, both nature and art, all other men and my own body, must be ranked under this name, NATURE. In enumerating the values of nature and casting up their sum, I shall use the word in both senses; — in its common and in its philosophical import. In inquiries so general as our present one, the inaccuracy is not material; no confusion of thought will occur. *Nature*, in the common sense, refers to essences unchanged by man; space, the air, the river, the leaf. *Art* is applied to the mixture of his will with the same things, as in a house, a canal, a statue, a picture. But his operations taken together are so insignificant, a little chipping, baking, patching, and washing, that in an impression so grand as that of the world on the human mind, they do not vary the result.

I. NATURE

To go into solitude, a man needs to retire as much from his chamber as from society. I am not solitary whilst I read and write, though nobody is with me. But if a man would be alone, let him look at the stars. The rays that come from those heavenly worlds will separate between him and what he touches. One might think the atmosphere was made transparent with this design, to give man, in the heavenly bodies, the perpetual presence of the sublime. Seen in the streets of cities, how great they are! If the stars should appear one night in a thousand years, how would men believe and adore; and preserve for many generations the remembrance of the city of God which had been shown! But every night come out these envoys of beauty, and light the universe with the admonishing smile.

The stars awaken a certain reverence, because though always present, they are inaccessible; but all natural objects make a kindred impression, when the mind is open to their influence. Nature never wears a mean appearance. Neither does the wisest man extort her secret, and lose his curiosity by finding out all her perfection. Nature never became a toy to a wise spirit. The flowers, the animals, the mountains, reflected the wisdom of his best hour, as much as they had delighted the simplicity of his childhood.

When we speak of nature in this manner, we have a distinct but most poetical sense in the mind. We mean the integrity of impression made by manifold natural objects. It is this which distinguishes the stick of timber of the wood-cutter from the tree of the poet. The charming landscape which I saw this morning is indubitably made up of some twenty or thirty farms. Miller owns this field, Locke that, and Manning the woodland beyond. But none of them owns the landscape. There is a property in the horizon which no man has but he whose eye can integrate all the parts, that is, the poet. This is the best part of these men's farms, yet to this their warranty-deeds give no title.

To speak truly, few adult persons can see nature. Most persons do not see the sun. At least they have a very superficial seeing. The sun illuminates only the eye of the man, but shines into the eye and the heart of the child. The lover of nature is he whose inward and outward senses are still truly adjusted to each other; who has retained the spirit of infancy even into the era of manhood. His intercourse with heaven and earth becomes part of his daily food. In the presence of nature a wild delight runs through the man, in spite of real sorrows. Nature says,—he is my creature, and maugre all his impertinent griefs, he shall be glad with me. Not the sun or the summer alone, but every hour and season yields its tribute of delight; for every hour and change corresponds to and authorizes a different state of the mind, from breathless noon to grimmest midnight. Nature is a setting that fits equally well a comic or a mourning piece. In good health, the air is a cordial of incredible virtue. Crossing a bare common, in snow puddles, at twilight, under a clouded sky, without having in my thoughts any occurrence of special good fortune, I have enjoyed a perfect exhilaration. I am glad to the brink of fear. In the woods, too, a man casts off his years, as the snake his slough, and at what period soever of life is always a child. In the woods is perpetual youth. Within these plantations of God, a decorum and sanctity reign, a perennial festival is dressed, and the guest sees not how he should tire of them in a thousand years. In the woods, we return to reason and faith. There I feel that nothing can befall me in life,—no disgrace, no calamity (leaving me my eyes), which nature cannot repair. Standing on the bare ground,—my head bathed by the blithe air and uplifted into infinite space,—all mean egotism vanishes. I become a transparent eyeball; I am nothing; I see all; the currents of the Universal Being circulate through me; I am part or parcel of God. The name of the nearest friend sounds then foreign and accidental: to be brothers, to be acquaintances, master or servant, is then a trifle and a disturbance. I am the lover of uncontained and immortal beauty. In the wilderness, I find something more dear and connate than in streets or villages. In the tranquil landscape, and especially in the distant line of the horizon, man beholds somewhat as beautiful as his own nature.

The greatest delight which the fields and woods minister is the suggestion of an occult relation between man and the vegetable. I am not alone and unacknowledged. They nod to me, and I to them. The waving of the boughs in the storm is new to me and old. It takes me by surprise, and yet is not unknown. Its effect is like that of a higher thought or a better emotion coming over me, when I deemed I was thinking justly or doing right.

Yet it is certain that the power to produce this delight does not reside in nature, but in man, or in a harmony of both. It is necessary to use these pleasures with great temperance. For nature is not always tricked in holiday attire, but the same scene which yesterday breathed perfume and glittered as for the frolic of the nymphs is overspread with melancholy today. Nature always wears the colors of the spirit. To a man laboring under calamity, the heat of his own fire hath sadness in it. Then there is a kind of contempt of the landscape felt by him who has just lost by death a dear friend. The sky is less grand as it shuts down over less worth in the population. * * *

From THE JOURNALS OF
RALPH WALDO EMERSON

July 13. [1833]

I carried my ticket from Mr. Warden to the Cabinet of Natural History in the Garden of Plants. How much finer things are in composition than alone. "'Tis wise in man to make cabinets. When I was come into the Ornithological Chambers I wished I had come only there. The

The Journals of Ralph Waldo Emerson, 10 vols. (Boston: Houghton Mifflin, 1909–1914).

fancy-coloured vests of these elegant beings make me as pensive as the hues and forms of a cabinet of shells, formerly. It is a beautiful collection and makes the visitor as calm and genial as a bridegroom. The limits of the possible are enlarged, and the real is stranger than the imaginary. Some of the birds have a fabulous beauty. One parrot of a fellow called *Psittacus erythropterus* from New Holland deserves as special mention as a picture of Raphael in a gallery. He is the beau of all birds. Then the humming birds, little and gay. Least of all is the *Trochilus Niger*. I have seen beetles larger. The *Trochilus pella* hath such a neck of gold and silver and fire! *Trochilus Delalandi* from Brazil is a glorious little tot, *la mouche magnifique*. Among the birds of Paradise I remarked the *Manucode* or *Paradisea regia* from New Guinea, the *Paradisea Apoda*, and *Paradisea rubra*. Forget not the *Veuve à epaulettes*, or *Emberiza longicauda*, black with fine shoulder-knots; nor the *Ampelis cotinga;* nor the *Phasianus Argus*, a peacock-looking pheasant; nor the *Trogon pavoninus*, called also *Couroncou pavonin*.

I saw black swans and white peacocks; the ibis, the sacred and the rosy; the flamingo, with a neck like a snake; the toucan rightly called *rhinoceros;* and a vulture whom to meet in the wilderness would make your flesh quiver, so like an executioner he looked.

In the other rooms I saw amber containing perfect musquitoes, grand blocks of quartz, native gold in all its forms of crystallization,—threads, plates, crystals, dust; and silver, black as from fire. Ah! said I, this is philanthropy, wisdom, taste,—to form a cabinet of natural history. Many students were there with grammar and note-book, and a class of boys with their tutor from some school.

Here we are impressed with the inexhaustible riches of nature. The universe is a more amazing puzzle than ever, as you glance along this bewildering series of animated forms,—the hazy butterflies, the carved shells, the birds, beasts, fishes, insects, snakes, and the upheaving principle of life everywhere incipient, in the very rock aping organized forms. Not a form so grotesque, so savage, nor so beautiful but is an expression of some property inherent in man the observer,—an occult relation between the very scorpions and man. I feel the centipede in me,—cayman, carp, eagle, and fox. I am moved by strange sympathies; I say continually "I will be a naturalist."

*   *   *

April 11. [1834]

Went yesterday to Cambridge and spent most of the day at Mount Auburn; got my luncheon at Fresh Pond, and went back again to the woods. After much wandering and seeing many things, four snakes gliding up and down a hollow for no purpose that I could see—not to eat, not for love, but only gliding; then a whole bed of *Hepatica triloba,* cousins of the Anemone, all blue and beautiful, but constrained by niggard nature to wear their last year's faded jacket of leaves; then a black-capped titmouse, who came upon a tree, and when I would know his name, sang *chick-a-dee-dee;* then a far-off tree full of clamorous birds, I know not what, but you might hear them half a mile; I forsook the tombs, and found a sunny hollow where the east wind would not blow, and lay down against the side of a tree to most happy beholdings. At least I opened my eyes and let what would pass through them into the soul. I saw no more my relation, how near and petty, to Cambridge or Boston; I heeded no more what minute or hour our Massachusetts clocks might indicate—I saw only the noble earth on which I was born, with the great Star which warms and enlightens it. I saw the clouds that hang their significant drapery over us. It was Day—that was all Heaven said. The pines glittered with their innumerable green needles in the light, and seemed to challenge me to read their riddle. The drab oak-leaves of the last year turned their little somersets and lay still again. And the wind bustled high overhead in the forest top. This gay and grand architecture, from the vault to the moss and lichen on which I lay,—who shall explain to me the laws of its proportions and adornments?

*   *   *

December 10. [1836]

Pleasant walk yesterday, the most pleasant of days. At Walden Pond I found a new musical instrument which I call the ice-harp. A thin coat of ice covered a part of the pond, but melted around the edge of the shore. I threw a stone upon the ice which rebounded with a shrill sound, and falling again and again, repeated the note with pleasing modulation. I thought at first it was the "peep, peep" of a bird I had scared. I was so taken with the music that I threw down my stick and spent twenty minutes in throwing stones single or in handfuls on this crystal drum.

*   *   *

August 4. [1837]

The grass is mown; the corn is ripe; autumnal stars arise. After raffling all day in Plutarch's *Morals,* or shall I say angling there for such fish as I might find, I sallied out this fine afternoon through the woods to Walden Water. The woods were too full of mosquitoes to offer any hospitality to the muse, and when I came to the blackberry vines, the plucking the crude berries at the risk of splintering my hand and with a mosquito mounting guard over every particular berry seemed a little too emblematical of general life whose shining and glossy fruits are very hard beset with thorns and very sour and good for nothing when gathered. But the pond was all blue and beautiful in the bosom of the woods and under the amber sky—like a sapphire lying in the moss. I sat down a long time on the shore to see the show. The variety and density of the foliage at the eastern end of the pond is worth seeing, then the extreme softness and holiday beauty of the summer clouds floating feathery overhead, enjoying, as I fancied, their height and privilege of motion and yet not seeming so much the drapery of this place and hour as forelooking to some pavilions and gardens of festivity beyond. I rejected this fancy with a becoming spirit and insisted that clouds, woods and waters were all there for me. The waterflies were full of happiness. The frogs that shoot from the land as fast as you walk along, a yard ahead of you, are a meritorious beastie. For their cowardice is only greater than their curiosity and desire of acquaintance with you. Three strokes from the shore the little swimmer turns short round, spreads his webbed paddles, and hangs at the surface, looks you in the face and so continues as long as you do not assault him.

August 12. [1837]

If you gather apples in the sunshine or make hay or hoe corn and then retire within doors and strain your body or squeeze your eyes *six hours after,* you shall still see apples hanging in the bright light with leaves and boughs thereto. There lie the impressions still on the retentive organ though I knew it not. So lies the whole series of natural images with which my life has made me acquainted in my memory, though I know it not, and a thrill of passion, a sudden emotion flashes light upon their dark chamber and the Active power seizes instantly the fit image as the word of his momentary thought. So lies all the life I have lived as my dictionary from which to extract the word which I want to dress the new perception of this moment. This is the way to learn Grammar. God never meant that we should learn Language by Colleges or Books. That only can we say which we have lived.

April 26. [1838]

Yesterday afternoon I went to the Cliff with Henry Thoreau. Warm, pleasant, misty weather, which the great mountain amphitheatre seemed to drink in with gladness. A crow's voice filled all the miles of air with sound. A bird's voice, even a piping frog, enlivens a solitude and makes world enough for us. At night I went out into the dark and

saw a glimmering star and heard a frog, and Nature seemed to say, Well
do not these suffice? Here is a new scene, a new experience. Ponder it,
Emerson, and not like the foolish world, hanker after thunders and
multitudes and vast landscapes, the sea or Niagara.

*   *   *

May 14. [1838]

A Bird-while. In a natural chronometer, a Bird-while may be admitted as one of the metres, since the space most of the wild birds will
allow you to make your observations on them when they alight near
you in the woods, is a pretty equal and familiar measure.

*   *   *

April 9. [1840]

We walked this afternoon to Edmund Hosmer's and Walden Pond.
The South wind blew and filled with bland and warm light the dry sunny
woods. The last year's leaves flew like birds through the air. As I sat on the
bank of the Drop, or God's Pond, and saw the amplitude of the little
water, what space, what verge, the little scudding fleets of ripples found to
scatter and spread from side to side and take so much time to cross the
pond, and saw how the water seemed made for the wind, and the wind for
the water, dear playfellows for each other,—I said to my companion, I
declare this world is so beautiful that I can hardly believe it exists. At
Walden Pond the waves were larger and the whole lake in pretty uproar.
Jones Very said, 'See how each wave rises from the midst with an original
force, at the same time that it partakes the general movement!'

September 8. [1840]

I went into the woods. I found myself not wholly present there. If I
looked at a pine-tree or an aster, *that* did not seem to be Nature. Nature
was still elsewhere: this, or this was but outskirt and far-off reflection and
echo of the triumph that had passed by and was now at its glancing
splendor and heyday,—perchance in the neighboring fields, or, if I stood
in the field, then in the adjacent woods. Always the present object gave
me this sense of the stillness that follows a pageant that has just gone by.

*   *   *

[December ? 1840]

Nature ever flows; stands never still. Motion or change is her mode
of existence. The poetic eye sees in Man the Brother of the River, and
in Woman the Sister of the River. Their life is always transition. Hard
blockheads only drive nails all the time; forever remember; which is

fixing. Heroes do not fix, but flow, bend forward ever and invent a resource for every moment. A man is a compendium of nature, an indomitable savage; . . . as long as he has a temperament of his own, and a hair growing on his skin, a pulse beating in his veins, he has a physique which disdains all intrusion, all despotism; it lives, wakes, alters, by omnipotent modes, and is directly related there, amid essences and *billets doux*, to Himmaleh mountain chains, wild cedar swamps, and the interior fires, the molten core of the globe.

*   *   *

June 6. [1841]

I am sometimes discontented with my house because it lies on a dusty road, and with its sills and cellar almost in the water of the meadow. But when I creep out of it into the Night or the Morning and see what majestic and what tender beauties daily wrap me in their bosom, how near to me is every transcendent secret of Nature's love and religion, I see how indifferent it is where I eat and sleep. This very street of hucksters and taverns the moon will transform to a Palmyra, for she is the apologist of all apologists, and will kiss the elm trees alone and hides every meanness in a silver-edged darkness.

Summer, 1841

The metamorphosis of Nature shows itself in nothing more than this, that there is no word in our language that cannot become typical to us of Nature by giving it emphasis. The world is a Dancer; it is a Rosary; it is a Torrent; it is a Boat; a Mist; a Spider's Snare; it is what you will; and the metaphor will hold, and it will give the imagination keen pleasure. Swifter than light the world converts itself into that thing you name, and all things find their right place under this new and capricious classification. There is nothing small or mean to the soul. It derives as grand a joy from symbolizing the Godhead or his universe under the form of a moth or a gnat as of a Lord of Hosts. Must I call the heaven and the earth a maypole and country fair with booths, or an anthill, or an old coat, in order to give you the shock of pleasure which the imagination loves and the sense of spiritual greatness? Call it a blossom, a rod, a wreath of parsley, a tamarisk-crown, a cock, a sparrow, the ear instantly hears and the spirit leaps to the trope. . . .

September, undated. [1843]

The only straight line in Nature that I remember is the spider swinging down from a twig.

February ?, 1844

That bread which we ask of Nature is that she should entrance us, but amidst her beautiful or her grandest pictures I cannot escape the *second thought*. I walked this P.M. in the woods, but there too the snow-banks were sprinkled with tobacco-juice. We have the wish to forget night and day, father and mother, food and ambition, but we never lose our dualism. Blessed, wonderful Nature, nevertheless! without depth, but with immeasurable lateral spaces. If we look before us, if we compute our path, it is very short. Nature has only the thickness of a shingle or a slate: we come straight to the extremes; but sidewise, and at unawares, the present moment opens into other moods and moments, rich, prolific, leading onward without end. . . .

[July, 1844]

Geology has initiated us into the secularity of Nature, and taught us to disuse our dame-school measures, and exchange our Mosaic and Ptolemaic schemes for her large style. We knew nothing rightly, for want of perspective. Now we learn what patient periods must round themselves before the rock is formed; then before the rock is broken, and the first lichen race has disintegrated the thinnest external plate into soil, and opened the door for the remote Flora, Fauna, Ceres, and Pomona to come in. How far off yet is the trilobite! how far the quadruped! how inconceivably remote is man! All duly arrive, and then race after race of men. It is a long way from granite to the oyster; farther yet to Plato and the preaching of the immortality of the soul. Yet all must come, as surely as the first atom has two sides.

August, undated. [1848]

Henry Thoreau is like the wood-god who solicits the wandering poet and draws him into antres vast and desarts idle, and bereaves him of his memory, and leaves him naked, plaiting vines and with twigs in his hand. . . .

I spoke of friendship, but my friends and I are fishes in our habit. As for taking Thoreau's arm, I should as soon take the arm of an elm tree.

[Last days of September.] [1848]

I go twice a week over Concord with Ellery, and, as we sit on the steep park at Conantum, we still have the same regret as oft before. Is all this beauty to perish? Shall none remake this sun and wind, the sky-blue river, the river-blue sky; the yellow meadow spotted with sacks and sheets of cranberry-pickers; the red bushes; the iron-gray house with just the color of the granite rock; the paths of the thicket, in which the only engineers are the cattle grazing on yonder hill; the wide, straggling wild orchard in which Nature has deposited every possible flavor in the apples of different trees? Whole zones and climates she has concentrated into apples. We think of the old benefactors who have conquered these fields; of the old man Moore, who is just dying in these days, who has absorbed such volumes of sunshine like a huge melon or pumpkin in the sun,—who has owned in every part of Concord a woodlot, until he could not find the boundaries of these, and never saw their interiors. But we say, where is he who is to save the present moment, and cause that this beauty be not lost? Shakespeare saw no better heaven or earth, but had the power and need to sing, and seized the dull ugly England, ugly to this, and made it amicable and enviable to all reading men, and now we are fooled into likening this to that; whilst, if one of us had the chanting constitution, that land would no more be heard of.

September 5. [1855]

All the thoughts of a turtle are turtle.

CHARLES DARWIN
1809–1882

On the Origin of Species *(1859) introduced a comprehensive new paradigm for the life sciences and, in doing so, stimulated an enormous increase in research and in the accumulation of data. Darwin's work laid the foundation for the professional and specialized biological disciplines we know today, but his own career was shaped by very different forces. He found his formal schooling uninspiring. His naturalist collections as a boy, his long walks with the botanist James Henslow during his days at Cambridge, and, above all, his great adventure on HMS Beagle's voyage of exploration contributed much more to his eventual vision of life's evolutionary pattern.*

The Beagle's mission was to map the South American coastline, which necessitated lengthy shore visits and exposed Darwin to a wide variety of spectacular landscapes, from the Brazilian rain forest to the Argentine pampas to the fjords of Tierra del Fuego. When the ship touched at the Galapagos Islands, off the coast of Ecuador, he entered a laboratory of natural selection. Although Darwin did not publish his theory of evolution for almost three decades, he was keenly aware as a young naturalist that the Galapagos finches, tortoises, and iguanas differed from corresponding species on the mainland and that they varied intriguingly from island to island of the group.

The transitional character of Darwin, as at once a scientist in the modern sense and a more old fashioned naturalist, might be illustrated by the fact that in setting out to sea, he packed along both the initial volume of Charles Lyell's ground-breaking Principles of Geology and Gilbert White's Natural History and Antiquities of Selborne. The first opened his imagination to science's vast new temporal perspective, while the second grounded his sensibility and his style in an amateur tradition of English natural history.

FROM VOYAGE OF H.M.S. BEAGLE

[Galapagos Tortoises]

*   *   * I will first describe the habits of the tortoise (Testudo nigra, formerly called Indica), which has been so frequently alluded to. These animals are found, I believe, on all the islands of the Archipelago; certainly on the greater number. They frequent in preference the high damp parts, but they likewise live in the lower and arid districts. I have already shown, from the numbers which have been caught in a single day, how very numerous they must be. Some grow to an immense size: Mr. Lawson, an Englishman, and vice-governor of the colony, told us that he had seen several so large, that it required six or eight men to lift them from the ground; and that some had afforded as much as two hundred pounds of meat. The old males are the largest, the females rarely growing to so great a size: the male can readily be distinguished from the female by the greater length of its tail. The tortoises which live on those islands where there is no water, or in the lower and arid

Journal of Researches into the Geology and Natural History of Various Countries Visited by H.M.S. Beagle (London: H. Colburn, 1836).

parts of the others, feed chiefly on the succulent cactus. Those which frequent the higher and damp regions, eat the leaves of various trees, a kind of berry (called guayavita) which is acid and austere, and likewise a pale green filamentous lichen (Usnera plicata), that hangs in tresses from the boughs of the trees.

The tortoise is very fond of water, drinking large quantities, and wallowing in the mud. The larger islands alone possess springs, and these are always situated towards the central parts, and at a considerable height. The tortoises, therefore, which frequent the lower districts, when thirsty, are obliged to travel from a long distance. Hence broad and well-beaten paths branch off in every direction from the wells down to the sea-coast; and the Spaniards by following them up, first discovered the watering-places. When I landed at Chatham Island, I could not imagine what animal travelled so methodically along well-chosen tracks. Near the springs it was a curious spectacle to behold many of these huge creatures, one set eagerly travelling onwards with outstretched necks, and another set returning, after having drunk their fill. When the tortoise arrives at the spring, quite regardless of any spectator, he buries his head in the water above his eyes, and greedily swallows great mouthfuls, at the rate of about ten in a minute. The inhabitants say each animal stays three or four days in the neighbourhood of the water, and then returns to the lower country; but they differed respecting the frequency of these visits. The animal probably regulates them according to the nature of the food on which it has lived. It is, however, certain, that tortoises can subsist even on those islands, where there is no other water than what falls during a few rainy days in the year.

I believe it is well ascertained, that the bladder of the frog acts as a reservoir for the moisture necessary to its existence: such seems to be the case with the tortoise. For some time after a visit to the springs, their urinary bladders are distended with fluid, which is said gradually to decrease in volume, and to become less pure. The inhabitants, when walking in the lower district, and overcome with thirst, often take advantage of this circumstance, and drink the contents of the bladder if full: in one I saw killed, the fluid was quite limpid, and had only a very slightly bitter taste. The inhabitants, however, always first drink the water in the pericardium, which is described as being best.

The tortoises, when purposely moving towards any point, travel by night and day, and arrive at their journey's end much sooner than would be expected. The inhabitants, from observing marked individuals, consider that they travel a distance of about eight miles in two or three days. One large tortoise, which I watched, walked at the rate of sixty yards in ten minutes, that is 360 yards in the hour, or four miles a day,—allowing a little time for it to eat on the road. During the breeding season, when the male and female are together, the male utters a hoarse roar or bellowing, which, it is said, can be heard at the distance of more than a hundred yards. The female never uses her voice, and the male only at these times; so that when the people hear this noise, they know that the two are together. They were at this time (October) laying their eggs. The female, where the soil is sandy, deposits them together, and covers them up with sand; but where the ground is rocky she drops them indiscriminately in any hole: Mr. Bynoe found seven placed in a fissure. The egg is white and spherical; one which I measured was seven inches and three-eighths in circumference, and therefore larger than a hen's egg. The young tortoises, as soon as they are hatched, fall a prey in great numbers to the carrion-feeding buzzard. The old ones seem generally to die from accidents, as from falling down precipices: at least, several of the inhabitants told me, that they had never found one dead without some evident cause.

The inhabitants believe that these animals are absolutely deaf; certainly they do not overhear a person walking close behind them. I was always amused when overtaking one of these great monsters, as it was quietly pacing along, to see how suddenly, the instant I passed, it would draw in its head and legs, and uttering a deep hiss fall to the ground with a heavy sound, as if struck dead. I frequently got on their backs, and then giving a few raps on the hinder part of their shells, they would rise up and walk away;—but I found it very difficult to keep my balance. The flesh of this animal is largely employed, both fresh and salted; and a beautifully clear oil is prepared from the fat. When a tortoise is caught, the man makes a slit in the skin near its tail, so as to see inside its body, whether the fat under the dorsal plate is thick. If it is not, the animal is liberated; and it is said to recover soon from this strange operation. In order to secure the tortoises, it is not sufficient to turn them like turtle, for they are often able to get on their legs again.

There can be little doubt that this tortoise is an aboriginal inhabitant of the Galapagos; for it is found on all, or nearly all, the islands, even on some of the smaller ones where there is no water; had it been an imported species, this would hardly have been the case in a group which has been so little frequented. Moreover, the old Bucaniers found this tortoise in greater numbers even than at present: Wood and Rogers also, in 1708, say that it is the opinion of the Spaniards, that it is found nowhere else in this quarter of the world. It is now widely distributed; but it may be questioned whether it is in any other place an aboriginal. The bones of a tortoise at Mauritius, associated with those of the extinct

Dodo, have generally been considered as belonging to this tortoise: if this had been so, undoubtedly it must have been there indigenous; but M. Bibron informs me that he believes that it was distinct, as the species now living there certainly is. * * *

[Retrospect on Our Voyage]

* * * Our Voyage having come to an end, I will take a short retrospect of the advantages and disadvantages, the pains and pleasures, of our circumnavigation of the world. If a person asked my advice, before undertaking a long voyage, my answer would depend upon his possessing a decided taste for some branch of knowledge, which could by this means be advanced. No doubt it is a high satisfaction to behold various countries and the many races of mankind, but the pleasures gained at the time do not counterbalance the evils. It is necessary to look forward to a harvest, however distant that may be, when some fruit will be reaped, some good effected.

Many of the losses which must be experienced are obvious; such as that of the society of every old friend, and of the sight of those places with which every dearest remembrance is so intimately connected. These losses, however, are at the time partly relieved by the exhaustless delight of anticipating the long wished-for day of return. If, as poets say, life is a dream, I am sure in a voyage these are the visions which best serve to pass away the long night. Other losses, although not at first felt, tell heavily after a period: these are the want of room, of seclusion, of rest; the jading feeling of constant hurry; the privation of small luxuries, the loss of domestic society, and even of music and the other pleasures of imagination. When such trifles are mentioned, it is evident that the real grievances, excepting from accidents, of a sea-life are at an end. The short space of sixty years has made an astonishing difference in the facility of distant navigation. Even in the time of Cook, a man who left his fireside for such expeditions underwent severe privations. A yacht now, with every luxury of life, can circumnavigate the globe. Besides the vast improvements in ships and naval resources, the whole western shores of America are thrown open, and Australia has become the capital of a rising continent. How different are the circumstances to a man ship-wrecked at the present day in the Pacific, to what they were in the time of Cook! Since his voyage a hemisphere has been added to the civilized world.

If a person suffer much from sea-sickness, let him weigh it heavily in the balance. I speak from experience: it is no trifling evil, cured in a week. If, on the other hand, he take pleasure in naval tactics, he will assuredly have full scope for his taste. But it must be borne in mind, how large a proportion of the time, during a long voyage, is spent on the water, as compared with the days in harbour. And what are the boasted glories of the illimitable ocean? A tedious waste, a desert of water, as the Arabian calls it. No doubt there are some delightful scenes. A moonlight night, with the clear heavens and the dark glittering sea, and the white sails filled by the soft air of a gently-blowing trade-wind; a dead calm, with the heaving surface polished like a mirror, and all still except the occasional flapping of the canvas. It is well once to behold a squall with its rising arch and coming fury, or the heavy gale of wind and mountainous waves. I confess, however, my imagination had painted something more grand, more terrific in the full-grown storm. It is an incomparably finer spectacle when beheld on shore, where the waving trees, the wild flight of the birds, the dark shadows and bright lights, the rushing of the torrents, all proclaim the strife of the unloosed elements. At sea the albatross and little petrel fly as if the storm were their proper sphere, the water rises and sinks as if fulfilling its usual task, the ship alone and its inhabitants seem the objects of wrath. On a forlorn and weather-beaten coast, the scene is indeed different, but the feelings partake more of horror than of wild delight.

Let us now look at the brighter side of the past time. The pleasure derived from beholding the scenery and the general aspect of the various countries we have visited, has decidedly been the most constant and highest source of enjoyment. It is probable that the picturesque beauty of many parts of Europe exceeds anything which we beheld. But there is a growing pleasure in comparing the character of the scenery in different countries, which to a certain degree is distinct from merely admiring its beauty. It depends chiefly on an acquaintance with the individual parts of each view: I am strongly induced to believe that, as in music, the person who understands every note will, if he also possesses a proper taste, more thoroughly enjoy the whole, so he who examines each part of a fine view, may also thoroughly comprehend the full and combined effect. Hence, a traveller should be a botanist, for in all views plants form the chief embellishment. Group masses of naked rock even in the wildest forms, and they may for a time afford a sublime spectacle, but they will soon grow monotonous. Paint them with bright and varied colours, as in Northern Chile, they will become fantastic; clothe them with vegetation, they must form a decent, if not a beautiful picture.

When I say that the scenery of parts of Europe is probably superior to anything which we beheld, I except, as a class by itself, that of the intertropical zones. The two classes cannot be compared together; but I have already often enlarged on the grandeur of those regions. As the force of impressions generally depends on preconceived ideas, I may add, that mine were taken from the vivid descriptions in the Personal Narrative of Humboldt, which far exceed in merit anything else which I have read. Yet with these high-wrought ideas, my feelings were far from partaking of a tinge of disappointment on my first and final landing on the shores of Brazil.

Among the scenes which are deeply impressed on my mind, none exceed in sublimity the primeval forests undefaced by the hand of man; whether those of Brazil, where the powers of Life are predominant, or those of Tierra del Fuego, where Death and Decay prevail. Both are temples filled with the varied productions of the God of Nature:—no one can stand in these solitudes unmoved, and not feel that there is more in man than the mere breath of his body. In calling up images of the past, I find that the plains of Patagonia frequently cross before my eyes; yet these plains are pronounced by all wretched and useless. They can be described only by negative characters; without habitations, without water, without trees, without mountains, they support merely a few dwarf plants. Why then, and the case is not peculiar to myself, have these arid wastes taken so firm a hold on my memory? Why have not the still more level, the greener and more fertile Pampas, which are serviceable to mankind, produced an equal impression? I can scarcely analyze these feelings: but it must be partly owing to the free scope given to the imagination. The plains of Patagonia are boundless, for they are scarcely passable, and hence unknown: they bear the stamp of having lasted, as they are now, for ages, and there appears no limit to their duration through future time. If, as the ancients supposed, the flat earth was surrounded by an impassable breadth of water, or by deserts heated to an intolerable excess, who would not look at these last boundaries to man's knowledge with deep but ill-defined sensations?

Lastly, of natural scenery, the views from lofty mountains, though certainly in one sense not beautiful, are very memorable. When looking down from the highest crest of the Cordillera, the mind, undisturbed by minute details, was filled with the stupendous dimensions of the surrounding masses.

Of individual objects, perhaps nothing is more certain to create astonishment than the first sight in his native haunt of a barbarian,—of man in his lowest and most savage state. One's mind hurries back over past centuries, and then asks, could our progenitors have been men like these?—men, whose very signs and expressions are less intelligible to us than those of the domesticated animals; men, who do not possess the instinct of those animals, nor yet appear to boast of human reason, or at least of arts consequent on that reason. I do not believe it is possible to describe or paint the difference between a wild and tame animal: and part of the interest in beholding a savage, is the same which would lead every one to desire to see the lion in his desert, the tiger tearing his prey in the jungle, or the rhinoceros wandering over the wild plains of Africa.

Among the other most remarkable spectacles which we have beheld, may be ranked the Southern Cross, the cloud of Magellan, and the other constellations of the southern hemisphere — the water-spout — the glacier leading its blue stream of ice, overhanging the sea in a bold precipice — a lagoon-island raised by the reef-building corals — an active volcano — and the overwhelming effects of a violent earthquake. These latter phenomena, perhaps, possess for me a peculiar interest, from their intimate connexion with the geological structure of the world. The earthquake, however, must be to every one a most impressive event: the earth, considered from our earliest childhood as the type of solidity, has oscillated like a thin crust beneath our feet; and in seeing the laboured works of man in a moment overthrown, we feel the insignificance of his boasted power.

It has been said, that the love of the chase is an inherent delight in man — a relic of an instinctive passion. If so, I am sure the pleasure of living in the open air, with the sky for a roof and the ground for a table, is part of the same feeling; it is the savage returning to his wild and native habits. I always look back to our boat cruises, and my land journeys, when through unfrequented countries, with an extreme delight, which no scenes of civilization could have created. I do not doubt that every traveller must remember the glowing sense of happiness which he experienced, when he first breathed in a foreign clime, where the civilized man had seldom or never trod.

There are several other sources of enjoyment in a long voyage, which are of a more reasonable nature. The map of the world ceases to be a blank; it becomes a picture full of the most varied and animated figures. Each part assumes its proper dimensions: continents are not looked at in the light of islands, or islands considered as mere specks, which are, in truth, larger than many kingdoms of Europe. Africa, or North and South America, are well-sounding names, and easily pronounced; but it is not until having sailed for weeks along small portions of their shores, that one is thoroughly convinced what vast spaces on our immense world these names imply.

From seeing the present state, it is impossible not to look forward with high expectations to the future progress of nearly an entire hemisphere. The march of improvement, consequent on the introduction of Christianity throughout the South Sea, probably stands by itself in the records of history. It is the more striking when we remember that only sixty years since, Cook, whose excellent judgment none will dispute, could foresee no prospect of a change. Yet these changes have now been effected by the philanthropic spirit of the British nation.

In the same quarter of the globe Australia is rising, or indeed may be said to have risen, into a grand centre of civilization, which, at some not very remote period, will rule as empress over the southern hemisphere. It is impossible for an Englishman to behold these distant colonies, without a high pride and satisfaction. To hoist the British flag, seems to draw with it as a certain consequence, wealth, prosperity, and civilization.

In conclusion, it appears to me that nothing can be more improving to a young naturalist, than a journey in distant countries. It both sharpens, and partly allays that want and craving, which, as Sir J. Herschel remarks, a man experiences although every corporeal sense be fully satisfied. The excitement from the novelty of objects, and the chance of success, stimulate him to increased activity. Moreover, as a number of isolated facts soon become uninteresting, the habit of comparison leads to generalization. On the other hand, as the traveller stays but a short time in each place, his descriptions must generally consist of mere sketches, instead of detailed observations. Hence arises, as I have found to my cost, a constant tendency to fill up the wide gaps of knowledge, by inaccurate and superficial hypotheses.

But I have too deeply enjoyed the voyage, not to recommend any naturalist, although he must not expect to be so fortunate in his companions as I have been, to take all chances, and to start, on travels by land if possible, if otherwise on a long voyage. He may feel assured, he will meet with no difficulties or dangers, excepting in rare cases, nearly so bad as he beforehand anticipates. In a moral point of view, the effect ought to be, to teach him good-humored patience, freedom from selfishness, the habit of acting for himself, and of making the best of every occurence. In short, he ought to partake of the characteristic qualities of most sailors. Travelling ought also to teach him distrust; but at the same time he will discover, how many truly kind-hearted people there are, with whom he never before had, or ever again will have any further communication, who yet are ready to offer him the most disinterested assistance.   *   *   *

From ON THE ORIGIN OF SPECIES
BY MEANS OF NATURAL SELECTION

[Conclusion—"the tangled bank"]

*   *   * Authors of the highest eminence seem to be fully satisfied with the view that each species has been independently created. To my mind it accords better with what we know of the laws impressed on matter by the Creator, that the production and extinction of the past and present inhabitants of the world should have been due to second-ary causes, like those determining the birth and death of the individual. When I view all beings not as special creations, but as the lineal descendants of some few beings which lived long before the first bed of the Cambrian system was deposited, they seem to me to become enno-bled. Judging from the past, we may safely infer that not one living species will transmit its unaltered likeness to a distant futurity. And of the species now living very few will transmit progeny of any kind to a far distant futurity; for the manner in which all organic beings are grouped, shows that the greater number of species in each genus, and all the species in many genera, have left no descendants, but have become utterly extinct. We can so far take a prophetic glance into futu-rity as to foretell that it will be the common and widely-spread species, belonging to the larger and dominant groups within each class, which will ultimately prevail and procreate new and dominant species. As all the living forms of life are the lineal descendants of those which lived long before the Cambrian epoch, we may feel certain that the ordinary succession by generation has never once been broken, and that no cat-aclysm has desolated the whole world. Hence we may look with some confidence to a secure future of great length. And as natural selection works solely by and for the good of each being, all corporeal and men-tal endowments will tend to progress towards perfection.

It is interesting to contemplate a tangled bank, clothed with many plants of many kinds, with birds singing on the bushes, with various insects flitting about, and with worms crawling through the damp earth, and to reflect that these elaborately constructed forms, so different from each other, and dependent upon each other in so complex a manner, have all been produced by laws acting around us. These laws, taken in the largest sense, being Growth with Reproduction; Inheritance which is

On the Origin of Species by Means of Natural Selection (London: J. Murray, 1859).

almost implied by reproduction; Variability from the indirect and direct action of the conditions of life, and from use and disuse: a Ratio of Increase so high as to lead to a Struggle for Life, and as a consequence to Natural Selection, entailing Divergence of Character and the Extinction of less-improved forms. Thus, from the war of nature, from famine and death, the most exalted object which we are capable of conceiving, namely, the production of the higher animals, directly follows. There is grandeur in this view of life, with its several powers, having been originally breathed by the Creator into a few forms or into one; and that, whilst this planet has gone cycling on according to the fixed law of gravity, from so simple a beginning endless forms most beautiful and most wonderful have been, and are being evolved.

From THE DESCENT OF MAN,
AND SELECTION IN RELATION TO SEX

[*Conclusion* — *"the indelible stamp of his lowly origin"*]

*   *   * Man scans with scrupulous care the character and pedigree of his horses, cattle, and dogs before he matches them; but when he comes to his own marriage he rarely, or never, takes any such care. He is impelled by nearly the same motives as the lower animals, when they are left to their own free choice, though he is in so far superior to them that he highly values mental charms and virtues. On the other hand he is strongly attracted by mere wealth or rank. Yet he might by selection do something not only for the bodily constitution and frame of his offspring, but for their intellectual and moral qualities. Both sexes ought to refrain from marriage if they are in any marked degree inferior in body or mind; but such hopes are Utopian and will never be even partially realised until the laws of inheritance are thoroughly known. Everyone does good service, who aids toward this end. When the principles of breeding and inheritance are better understood, we shall not hear ignorant members of our legislature rejecting with scorn a plan for ascertaining whether or not consanguineous marriages are injurious to man.

The advancement of the welfare of mankind is a most intricate problem: all ought to refrain from marriage who cannot avoid abject poverty

The Descent of Man, and Selection in Relation to Sex (London: J. Murray, 1871).

for their children; for poverty is not only a great evil, but tends to its own increase by leading to recklessness in marriage. On the other hand, as Mr. Galton has remarked, if the prudent avoid marriage, whilst the reckless marry, the inferior members tend to supplant the better members of society. Man, like every other animal, has no doubt advanced to his present high condition through a struggle for existence consequent on his rapid multiplication; and if he is to advance still higher, it is to be feared that he must remain subject to a severe struggle. Otherwise he would sink into indolence, and the more gifted men would not be more successful in the battle of life than the less gifted. Hence our natural rate of increase, though leading to many and obvious evils, must not be greatly diminished by any means. There should be open competition for all men; and the most able should not be prevented by laws or customs from succeeding best and rearing the largest number of offspring. Important as the struggle for existence has been and even still is, yet as far as the highest part of man's nature is concerned there are other agencies more important. For the moral qualities are advanced, either directly or indirectly, much more through the effects of habit, the reasoning powers, instruction, religion, &c., than through natural selection; though to this latter agency may be safely attributed the social instincts, which afforded the basis for the development of the moral sense.

The main conclusion arrived at in this work, namely, that man is descended from some lowly organised form, will, I regret to think, be highly distasteful to many. But there can hardly be a doubt that we are descended from barbarians. The astonishment which I felt on first seeing a party of Fuegians on a wild and broken shore will never be forgotten by me, for the reflection at once rushed into my mind—such were our ancestors. These men were absolutely naked and bedaubed with paint, their long hair was tangled, their mouths, frothed with excitement, and their expression was wild, startled, and distrustful. They possessed hardly any arts, and like wild animals lived on what they could catch; they had no government, and were merciless to every one not of their own small tribe. He who has seen a savage in his native land will not feel much shame, if forced to acknowledge that the blood of some more humble creature flows in his veins. For my own part I would as soon be descended from that heroic little monkey, who braved his dreaded enemy in order to save the life of his keeper, or from that old baboon, who descending from the mountains, carried away in triumph his young comrade from a crowd of astonished dogs—as from a savage who delights to torture his enemies, offers up bloody sacrifices, practises infanticide without remorse, treats his wives like slaves, knows no decency, and is haunted by the grossest superstitions.

Man may be excused for feeling some pride at having risen, though not through his own exertions, to the very summit of the organic scale; and the fact of his having thus risen, instead of having been aboriginally placed there, may give him hope for a still higher destiny in the distant future. But we are not here concerned with hopes or fears, only with the truth as far as our reason permits us to discover it; and I have given the evidence to the best of my ability. We must, however, acknowledge, as it seems to me, that man with all his noble qualities, with sympathy which feels for the most debased, with benevolence which extends not only to other men but to the humblest living creature, with his god-like intellect which has penetrated into the movements and constitution of the solar system—with all these exalted powers—Man still bears in his bodily frame the indelible stamp of his lowly origin.

SUSAN FENIMORE COOPER
1813–1894

Scholarship in the field of American nature writing has, over the past decade, brought renewed attention to a number of forgotten or neglected authors. Without doubt the most dramatic of these instances has been the rediscovery of Susan Fenimore Cooper as a pioneer in the genre. Her Rural Hours *(1850) quickly went through several editions. It was a book consulted by Thoreau as he was completing* Walden, *and an important link between his work and the natural-history tradition represented by Gilbert White in England. But until Rochelle Johnson and Daniel Patterson brought out their 1998 edition of the original text, there had not been an unabridged printing of it since 1876. An increasing number of scholars and critics now celebrate Cooper not only as a significant early figure but also as an author whose sustained focus on one small community and on the interaction between its human and nonhuman inhabitants anticipates our contemporary bioregional movement. Loyalty, civic responsibility, a comprehensive perspective, and a good memory are the hallmarks of Cooper's writing, and sources of inspiration for her successors whose goal is to write not only about their place on Earth but on behalf of it, as well.*

From RURAL HOURS

Wednesday, 27th. — Charming day; thermometer 80. Towards sunset strolled in the lane.

The fields which border this quiet bit of road are among the oldest in our neighborhood, belonging to one of the first farms cleared near the village; they are in fine order, and to look at them, one might readily believe these lands had been under cultivation for ages. But such is already very much the character of the whole valley; a stranger moving along the highway looks in vain for any striking signs of a new country; as he passes from farm to farm in unbroken succession, the aspect of the whole region is smiling and fruitful. Probably there is no part of the earth, within the limits of a temperate climate, which has taken the aspect of an old country so soon as our native land; very much is due, in this respect, to the advanced state of civilization in the present age, much to the active, intelligent character of the people, and something, also, to the natural features of the country itself. There are no barren tracts in our midst, no deserts which defy cultivation; even our mountains are easily tilled — arable, many of them, to their very summits — while the most sterile among them are more or less clothed with vegetation in their natural state. Altogether, circumstances have been very much in our favor.

While observing, this afternoon, the smooth fields about us, it was easy, within the few miles of country in sight at the moment, to pick out parcels of land in widely different conditions, and we amused ourselves by following upon the hill-sides the steps of the husbandman, from the first rude clearing, through every successive stage of tillage, all within range of the eye at the same instant. Yonder, for instance, appeared an opening in the forest, marking a new clearing still in the rudest state, black with charred stumps and rubbish; it was only last winter that the timber was felled on that spot, and the soil was first opened to the sunshine, after having been shaded by the old woods for more ages than one can tell. Here, again, on a nearer ridge, lay a spot not only cleared, but fenced, preparatory to being tilled; the decayed

Rural Hours (New York: Putnam, 1850).

trunks and scattered rubbish having been collected in heaps and burnt. Probably that spot will soon be ploughed, but it frequently happens that land is cleared of the wood, and then left in a rude state, as wild pasture-ground; an indifferent sort of husbandry this, in which neither the soil nor the wood receives any attention; but there is more land about us in this condition than one would suppose. The broad hill-side, facing the lane in which we were walking, though cleared perhaps thirty years since, has continued untilled to the present hour. In another direction, again, lies a field of new land, ploughed and seeded for the first time within the last few weeks; the young maize plants, just shooting out their glossy leaves, are the first crop ever raised there, and when harvested, the grain will prove the first fruits the earth has ever yielded to man from that soil, after lying fallow for thousands of seasons. Many other fields in sight have just gone through the usual rotation of crops, showing what the soil can do in various ways; while the farm before us has been under cultivation from the earliest history of the village, yielding every season, for the last half century, its share of grass and grain. To one familiar with the country, there is a certain pleasure in thus beholding the agricultural history of the neighborhood unfolding before one, following upon the farms in sight these progressive steps in cultivation.

The pine stumps are probably the only mark of a new country which would be observed by a stranger. With us, they take the place of rocks, which are not common; they keep possession of the ground a long while—some of those about us are known to have stood more than sixty years, or from the first settlement of the country, and how much longer they will last, time alone can tell. In the first years of cultivation, they are a very great blemish, but after a while, when most of them have been burnt or uprooted, a gray stump here and there, among the grass of a smooth field, does not look so very much amiss, reminding one, as it does, of the brief history of the country. Possibly there may be something of partiality in this opinion, just as some lovers have been found to admire a freckled face, because the rosy cheek of their sweetheart was mottled with brown freckles; people generally may not take the same view of the matter, they may think that even the single stump had better be uprooted. Several ingenious machines have been invented for getting rid of these enemies, and they have already done good service in the county. Some of them work by levers, others by wheels; they usually require three or four men and a yoke of oxen, or a horse, to work them, and it is really surprising what large stumps are drawn out of the earth by these contrivances, the strongest roots cracking and snapping like threads. Some digging about the stump is often necessary as a preliminary step, to enable the chain to be fastened securely, and occasionally the axe is used to relieve the machine; still, they work so expeditiously, that contracts are taken to clear lands in this way, at the rate of twenty or thirty cents a stump, when, according to the old method, working by hand, it would cost, perhaps, two or three dollars to uproot a large one thoroughly. In the course of a day, these machines will tear up from twenty to fifty stumps, according to their size. Those of the pine, hemlock, and chestnut are the most difficult to manage, and these last longer than those of other trees. When uprooted, the stumps are drawn together in heaps and burnt, or frequently they are turned to account as fences, being placed on end, side by side, their roots interlocking, and a more wild and formidable barrier about a quiet field cannot well be imagined. These rude fences are quite common in our neighborhood, and being peculiar, one rather likes them; it is said that they last much longer than other wooden fences, remaining in good condition for sixty years.

But although the stumps remaining here and there may appear to a stranger the only sign of a new country to be found here, yet closer observation will show others of the same character. Those wild pastures upon hill-sides, where the soil has never been ploughed, look very differently from other fallows. Here you observe a little hillock rounding over a decayed stump, there a petty hollow where some large tree has been uprooted by the storm; fern and brake also are seen in patches, instead of the thistle and the mullein. Such open hill-sides, even when rich and grassy, and entirely free from wood or bushes, bear a kind of heaving, billowy character, which, in certain lights, becomes very distinct; these ridges are formed by the roots of old trees, and remain long after the wood has entirely decayed. Even on level ground there is always an elevation about the root of an old tree and upon a hill-side, these petty knolls show more clearly as they are thrown into relief by the light; they become much bolder, also, from the washing of the soil, which accumulates above, and is carried away from the lower side of the trunk, leaving, often, a portion of the root bare in that direction. Of course, the older a wood and the larger its trees, the more clearly will this billowy character be marked. The tracks of the cattle also make the formation more ridge-like, uniting one little knoll with another, for when feeding, they generally follow one another, their heads often turned in one direction, and upon a hill-side they naturally take a horizontal course, as the most convenient. Altogether, the billowy face of these rude hill-sides is quite striking and peculiar, when seen in a favorable light.

But there are softer touches also, telling the same story of recent cultivation. It frequently happens, that walking about our farms, among rich fields, smooth and well worked, one comes to a low bank, or some little nook, a strip of land never yet cultivated, though surrounded on all sides by ripening crops of eastern grains and grasses. One always knows such places by the pretty native plants growing there. It was but the other day we paused to observe a spot of this kind in a fine meadow, near the village, neat and smooth, as though worked from the days of Adam. A path made by the workmen and cattle crosses the field, and one treads at every step upon plantain, that regular path-weed of the Old World; following this track, we come to a little runnel, which is dry and grassy now, though doubtless at one time the bed of a considerable spring; the banks are several feet high, and it is filled with native plants; on one side stands a thorn-tree, whose morning shadow falls upon grasses and clovers brought from beyond the seas, while in the afternoon, it lies on gyromias and moose-flowers, sarsaparillas and cahoshes, which bloomed here for ages, when the eye of the red man alone beheld them. Even within the limits of the village spots may still be found on the bank of the river, which are yet unbroken by the plough, where the trailing arbutus, and squirrel-cups, and May-wings tell us so every spring; in older regions, these children of the forest would long since have vanished from all the meadows and villages, for the plough would have passed a thousand times over every rood of such ground.

The forest flowers, the gray stumps in our fields, and the heaving surface of our wild hill-sides, are not, however, the only waymarks to tell the brief course of cultivation about us. These speak of the fallen forest; but here, as elsewhere, the waters have also left their impression on the face of the earth, and in these new lands the marks of their passage are seen more clearly than in older countries. They are still, in many places, sharp and distinct, as though fresh from the workman's hand. Our valleys are filled with these traces of water-work; the most careless observer must often be struck with their peculiar features, and it appears remarkable that here, at an elevation so much above the great western lakes, upon this dividing ridge, at the very fountain head of a stream, running several hundred miles to the sea, these lines are as frequent and as boldly marked as though they lay in a low country subject to floods. Large mounds rise like islands from the fields, their banks still sharply cut; in other spots a depressed meadow is found below the level of the surrounding country, looking like a drained lake, enclosed within banks as plainly marked as the works of a fortification; a shrunken brook, perhaps, running to-day where a river flowed at some period of past time. Quite near the village, from the lane where we were walking this evening, one may observe a very bold formation of this kind; the bank of the river is high and abrupt at this spot, and it is scooped out into two adjoining basins, not unlike the amphitheatres of ancient times. The central horn, as it were, which divides the two semicircles, stretches out quite a distance into a long, sharp point, very abrupt on both sides. The farther basin is the most regular, and it is also marked by successive ledges like the tiers of seats in those ancient theatres. This spot has long been cleared of wood, and used as a wild pasture; but the soil has never yet been broken by the plough, and we have often paused here to note the singular formation, and the surprising sharpness of the lines. Quite recently they have begun to dig here for sand; and if they continue the work, the character of the place must necessarily be changed. But now, as we note the bold outline of the basin, and watch the lines worked by the waters ages and ages since, still as distinct as though made last year, we see with our own eyes fresh proofs that we are in a new country, that the meadows about us, cleared by our fathers, are the first that have lain on the lap of the old earth, at this point, since yonder bank was shaped by the floods.

HENRY DAVID THOREAU
1817–1862

Thoreau not only traveled a good deal in Concord, Massachusetts, he also lived many lives in it. As schoolteacher, pencil maker, botanist, editor, surveyor, gardener, poet, lecturer, essayist, moral philosopher, political protester, travel writer, devoted brother and son, woodland hermit and village character, vegetarian and luster-after-raw-woodchuck—this most cocksure of American writers consistently refuses to be pigeonholed. The attempts of so many readers to do so spring from his propensity "to make an extreme statement, if so I may make an emphatic one." Thoreau's seeming paradoxes and contradictions are a result not of any conscious attempt at iconoclasm but of the inherent complexity of his perceptions coupled with a passion to "drive life into a corner, and reduce it to its lowest terms." The result, among other things, is one of the most condensed and energetic styles in English. Though his books sold little in his lifetime, Walden *has long been recognized as one of the classics of*

American literature, and Thoreau's ideas and writings have had world-wide influence.

The primary purpose of the following selections is to suggest the range of Thoreau's personality in all its complex and ambivalent guises. His first book, A Week on the Concord and Merrimack Rivers (1849), was based on a rowboat voyage he took with his older brother John in 1839 and was written largely during his stay at Walden Pond. One of the first in a long tradition of American river journey books, it contains superb sketches of the early New England landscape and some of Thoreau's most graceful and congenial writing. Contemplating the rivers' "healthy natural tumult" in the company of his beloved brother, he was never happier. From his masterpiece, Walden: or Life in the Woods (1854), we have included the chapter on "Brute Neighbors." Its descriptions of wild animals (including the famous battle of the ants) remain among the finest of their kind, groundbreaking in their consciously metaphoric overtones and Thoreau's deliberate exploration of the role of the self-aware narrator as both observer and participant.

"Walking" (1862), with its ringing defense and celebration of "wildness," has become one of the gospels of the conservation movement. The passage from The Maine Woods (1864), on the other hand, describing Thoreau's ascent to the summit of Mt. Katahdin, presents a startlingly different view of nature from that afforded by the domesticated fields and ponds of Concord. Here, actual rather than metaphoric wilderness shakes the very foundations of self and of human pretensions to understanding nature. In "the presence of a force not bound to be kind to man," the very structures of language seem to break down.

Finally, we have chosen some selections from the Journals (1906), begun when Thoreau was twenty and not published in his lifetime. The Journals are not a finished literary work by any means, but they contain some of his most brilliant passages. Intended as "a meteorological journal of the mind," they served, not only as a reservoir of material for his lectures and books, but also as a kind of literary laboratory. In these journal entries Thoreau is less studiedly candid and more genuinely personal—revealing, for instance, a poignant attachment to his native town or taking delight in the antics of a kitten a few months before his own death at the age of forty-four. He is also less attitudinizing and more experimental toward his material, exploring radically different approaches to the same subject—sometimes on the same page. Thoreau was a keen observer of natural process, and many of the journal selections seem to intuit modern ecological principles. Of special relevance to contemporary nature writing are his lively discussions of the relative virtues of poetic and scientific approaches in natural history.

From A WEEK ON THE CONCORD AND MERRIMACK RIVERS

CONCORD RIVER

"Beneath low hills, in the broad interval
Through which at will our Indian rivulet
Winds mindful still of sannup and of squaw,
Whose pipe and arrow oft the plough unburies,
Here, in pine houses, built of new-fallen trees,
Supplanters of the tribe, the farmers dwell."

Emerson

The Musketaquid, or Grass-ground River, though probably as old as the
Nile or Euphrates, did not begin to have a place in civilized history, until
the fame of its grassy meadows and its fish attracted settlers out of England
in 1635, when it received the other but kindred name of CONCORD from
the first plantation on its banks, which appears to have been commenced
in a spirit of peace and harmony. It will be Grass-ground River as long as
grass grows and water runs here; it will be Concord River only while men
lead peaceable lives on its banks. To an extinct race it was grass-ground,
where they hunted and fished, and it is still perennial grass-ground to
Concord farmers, who own the Great Meadows, and get the hay from
year to year. "One branch of it," according to the historian of Concord,
for I love to quote so good authority, "rises in the south part of
Hopkinton, and another from a pond and a large cedar-swamp in
Westborough," and flowing between Hopkinton and Southborough,
through Framingham, and between Sudbury and Wayland, where it is
sometimes called Sudbury River, it enters Concord at the south part of
the town, and after receiving the North or Assabeth River, which has its
source a little farther to the north and west, goes out at the northeast
angle, and flowing between Bedford and Carlisle, and through Billerica,
empties into the Merrimack at Lowell. In Concord it is, in summer,
from four to fifteen feet deep, and from one hundred to three hundred
feet wide, but in the spring freshets, when it overflows its banks, it is in
some places nearly a mile wide. Between Sudbury and Wayland the
meadows acquire their greatest breadth, and when covered with water,
they form a handsome chain of shallow vernal lakes, resorted to by
numerous gulls and ducks. Just above Sherman's Bridge, between these

A Week on the Concord and Merrimack Rivers (Boston: Ticknor and Fields, 1849).

towns, is the largest expanse, and when the wind blows freshly in a raw March day, heaving up the surface into dark and sober billows or regular swells, skirted as it is in the distance with alder-swamps and smoke-like maples, it looks like a smaller Lake Huron, and is very pleasant and exciting for a landsman to row or sail over. The farm-houses along the Sudbury shore, which rises gently to a considerable height, command fine water prospects at this season. The shore is more flat on the Wayland side, and this town is the greatest loser by the flood. Its farmers tell me that thousands of acres are flooded now, since the dams have been erected, where they remember to have seen the white honeysuckle or clover growing once, and they could go dry with shoes only in summer. Now there is nothing but blue-joint and sedge and cut-grass there, standing in water all the year round. For a long time, they made the most of the driest season to get their hay, working sometimes till nine o'clock at night, sedulously paring with their scythes in the twilight round the hummocks left by the ice; but now it is not worth the getting when they can come at it, and they look sadly round to their wood-lots and upland as a last resource.

It is worth the while to make a voyage up this stream, if you go no farther than Sudbury, only to see how much country there is in the rear of us; great hills, and a hundred brooks, and farm-houses, and barns, and haystacks, you never saw before, and men everywhere, Sudbury, that is *Southborough* men, and Wayland, and Nine-Acre-Corner men, and Bound Rock, where four towns bound on a rock in the river, Lincoln, Wayland, Sudbury, Concord. Many waves are there agitated by the wind, keeping nature fresh, the spray blowing in your face, reeds and rushes waving; ducks by the hundred, all uneasy in the surf, in the raw wind, just ready to rise, and now going off with a clatter and a whistling like riggers straight for Labrador, flying against the stiff gale with reefed wings, or else circling round first, with all their paddles briskly moving, just over the surf, to reconnoitre you before they leave these parts; gulls wheeling overhead, muskrats swimming for dear life, wet and cold, with no fire to warm them by that you know of; their labored homes rising here and there like haystacks; and countless mice and moles and winged titmice along the sunny windy shore; cranberries tossed on the waves and heaving up on the beach, their little red skiffs beating about among the alders;—such healthy natural tumult as proves the last day is not yet at hand. And there stand all around the alders, and birches, and oaks, and maples full of glee and sap, holding in their buds until the waters subside. You shall perhaps run aground on Cranberry Island, only some spires of last year's pipe-grass above water, to show where the danger is, and get as good a freezing there as anywhere on the Northwest Coast. I never voyaged so far in all my life. You shall see men you never heard of before, whose names you don't know, going away down through the meadows with long ducking-guns, with water-tight boots wading through the fowl-meadow grass, on bleak, wintry, distant shores, with guns at half-cock, and they shall see teal, blue-winged, green-winged, shelldrakes, whistlers, black ducks, ospreys, and many other wild and noble sights before night, such as they who sit in parlors never dream of. You shall see rude and sturdy, experienced and wise men, keeping their castles, or teaming up their summer's wood, or chopping alone in the woods, men fuller of talk and rare adventure in the sun and wind and rain, than a chestnut is of meat; who were out not only in '75 and 1812, but have been out every day of their lives; greater men than Homer, or Chaucer, or Shakespeare, only they never got time to say so; they never took to the way of writing. Look at their fields, and imagine what they might write, if ever they should put pen to paper. Or what have they not written on the face of the earth already, clearing, and burning, and scratching, and harrowing, and ploughing, and subsoiling, in and in, and out and out, and over and over, again and again, erasing what they had already written for want of parchment. *   *   *

From WALDEN: OR, LIFE IN THE WOODS

BRUTE NEIGHBORS

*   *   * Why do precisely these objects which we behold make a world? Why has man just these species of animals for his neighbors; as if nothing but a mouse could have filled this crevice? I suspect that Pilpay & Co. have put animals to their best use, for they are all beasts of burden, in a sense, made to carry some portion of our thoughts.

The mice which haunted my house were not the common ones, which are said to have been introduced into the country, but a wild native kind not found in the village. I sent one to a distinguished naturalist, and it interested him much. When I was building, one of these had its nest underneath the house, and before I had laid the second floor, and swept out the shavings, would come out regularly at lunch time and pick up the crumbs at my feet. It probably had never seen a man before; and it soon became quite familiar, and would run over my

Walden: or, Life in the Woods (Boston: Ticknor and Fields, 1854).

shoes and up my clothes. It could readily ascend the sides of the room by short impulses, like a squirrel, which it resembled in its motions. At length, as I leaned with my elbow on the bench one day, it ran up my clothes, and along my sleeve, and round and round the paper which held my dinner, while I kept the latter close, and dodged and played at bopeep with it; and when at last I held still a piece of cheese between my thumb and finger, it came and nibbled it, sitting in my hand, and afterward cleaned its face and paws, like a fly, and walked away.

A phœbe soon built in my shed, and a robin for protection in a pine which grew against the house. In June the partridge *(Tetrao umbellus)*, which is so shy a bird, led her brood past my windows, from the woods in the rear to the front of my house, clucking and calling to them like a hen, and in all her behavior proving herself the hen of the woods. The young suddenly disperse on your approach, at a signal from the mother, as if a whirlwind had swept them away, and they so exactly resemble the dried leaves and twigs that many a traveller has placed his foot in the midst of a brood, and heard the whir of the old bird as she flew off, and her anxious calls and mewing, or seen her trail her wings to attract his attention, without suspecting their neighborhood. The parent will sometimes roll and spin round before you in such a dishabille, that you cannot, for a few moments, detect what kind of creature it is. The young squat still and flat, often running their heads under a leaf, and mind only their mother's directions given from a distance, nor will your approach make them run again and betray themselves. You may even tread on them, or have your eyes on them for a minute, without discovering them. I have held them in my open hand at such a time, and still their only care, obedient to their mother and their instinct, was to squat there without fear or trembling. So perfect is this instinct, that once, when I had laid them on the leaves again, and one accidentally fell on its side, it was found with the rest in exactly the same position ten minutes afterward. They are not callow like the young of most birds, but more perfectly developed and precocious even than chickens. The remarkably adult yet innocent expression of their open and serene eyes is very memorable. All intelligence seems reflected in them. They suggest not merely the purity of infancy, but a wisdom clarified by experience. Such an eye was not born when the bird was, but is coeval with the sky it reflects. The woods do not yield another such a gem. The traveller does not often look into such a limpid well. The ignorant or reckless sportsman often shoots the parent at such a time, and leaves these innocents to fall a prey to some prowling beast or bird, or gradually mingle with the decaying leaves which they so much resemble. It is said that when hatched by a hen they will directly disperse on some alarm, and so are lost, for they never hear the mother's call which gathers them again. These were my hens and chickens.

It is remarkable how many creatures live wild and free though secret in the woods, and still sustain themselves in the neighborhood of towns, suspected by hunters only. How retired the otter manages to live here! He grows to be four feet long, as big as a small boy, perhaps without any human being getting a glimpse of him. I formerly saw the raccoon in the woods behind where my house is built, and probably still heard their whinnering at night. Commonly I rested an hour or two in the shade at noon, after planting, and ate my lunch, and read a little by a spring which was the source of a swamp and of a brook, oozing from under Brister's Hill, half a mile from my field. The approach to this was through a succession of descending grassy hollows, full of young pitch pines, into a larger wood about the swamp. There, in a very secluded and shaded spot, under a spreading white pine, there was yet a clean, firm sward to sit on. I had dug out the spring and made a well of clear gray water, where I could dip up a pailful without roiling it, and thither I went for this purpose almost every day in midsummer, when the pond was warmest. Thither, too, the woodcock led her brood, to probe the mud for worms, flying but a foot above them down the bank, while they ran in a troop beneath; but at last, spying me, she would leave her young and circle round and round me, nearer and nearer till within four or five feet, pretending broken wings and legs, to attract my attention, and get off her young, who would already have taken up their march, with faint, wiry peep, single file through the swamp, as she directed. Or I heard the peep of the young when I could not see the parent bird. There too the turtle doves sat over the spring, or fluttered from bough to bough of the soft white pines over my head; or the red squirrel, coursing down the nearest bough, was particularly familiar and inquisitive. You only need sit still long enough in some attractive spot in the woods that all its inhabitants may exhibit themselves to you by turns.

I was witness to events of a less peaceful character. One day when I went out to my wood-pile, or rather my pile of stumps, I observed two large ants, the one red, the other much larger, nearly half an inch long, and black, fiercely contending with one another. Having once got hold they never let go, but struggled and wrestled and rolled on the chips incessantly. Looking farther, I was surprised to find that the chips were covered with such combatants, that it was not a *duellum*, but a *bellum*, a war between two races of ants, the red always pitted against the black, and frequently two red ones to one black. The legions of these Myrmidons covered all the hills and vales in my wood-yard, and the ground was already strewn with the dead and dying, both red and black.

It was the only battle which I have ever witnessed, the only battle-field I ever trod while the battle was raging; internecine war; the red republicans on the one hand, and the black imperialists on the other. On every side they were engaged in deadly combat, yet without any noise that I could hear, and human soldiers never fought so resolutely. I watched a couple that were fast locked in each other's embraces, in a little sunny valley amid the chips, now at noonday prepared to fight till the sun went down, or life went out. The smaller red champion had fastened himself like a vice to his adversary's front, and through all the tumblings on that field never for an instant ceased to gnaw at one of his feelers near the root, having already caused the other to go by the board; while the stronger black one dashed him from side to side, and, as I saw on looking nearer, had already divested him of several of his members. They fought with more pertinacity than bulldogs. Neither manifested the least disposition to retreat. It was evident that their battle-cry was "Conquer or die." In the meanwhile there came along a single red ant on the hillside of this valley, evidently full of excitement, who either had despatched his foe, or had not yet taken part in the battle; probably the latter, for he had lost none of his limbs; whose mother had charged him to return with his shield or upon it. Or perchance he was some Achilles, who had nourished his wrath apart, and had now come to avenge or rescue his Patroclus. He saw this unequal combat from afar,—for the blacks were nearly twice the size of the red,—he drew near with rapid pace till he stood on his guard within half an inch of the combatants; then, watching his opportunity, he sprang upon the black warrior, and commenced his operations near the root of his right fore leg, leaving the foe to select among his own members; and so there were three united for life, as if a new kind of attraction had been invented which put all other locks and cements to shame. I should not have wondered by this time to find that they had their respective musical bands stationed on some eminent chip, and playing their national airs the while, to excite the slow and cheer the dying combatants. I was myself excited somewhat even as if they had been men. The more you think of it, the less the difference. And certainly there is not the fight recorded in Concord history, at least, if in the history of America, that will bear a moment's comparison with this, whether for the numbers engaged in it, or for the patriotism and heroism displayed. For numbers and for carnage it was an Austerlitz or Dresden. Concord Fight! Two killed on the patriots' side, and Luther Blanchard wounded! Why here every ant was a Buttrick,—"Fire! for God's sake fire!"—and thousands shared the fate of Davis and Hosmer. There was not one hireling there. I have no doubt that it was a principle they fought for, as much as our ancestors, and not to avoid a three-penny tax on their tea; and the results of this battle will be as important and memorable to those whom it concerns as those of the battle of Bunker Hill, at least.

I took up the chip on which the three I have particularly described were struggling, carried into my house, and placed it under a tumbler on my window-sill, in order to see the issue. Holding a microscope to the first-mentioned red ant, I saw that, though he was assiduously gnawing at the near fore leg of his enemy, having severed his remaining feeler, his own breast was all torn away, exposing what vitals he had there to the jaws of the black warrior, whose breastplate was apparently too thick for him to pierce; and the dark carbuncles of the sufferer's eyes shone with ferocity such as war only could excite. They struggled half an hour longer under the tumbler, and when I looked again the black soldier had severed the heads of his foes from their bodies, and the still living heads were hanging on either side of him likely ghastly trophies at his saddlebow, still apparently as firmly fastened as ever, and he was endeavoring with feeble struggles, being without feelers and with only the remnant of a leg, and I know not how many other wounds, to divest himself of them; which at length, after half an hour more, he accomplished. I raised the glass, and he went off over the window-sill in that crippled state. Whether he finally survived that combat, and spent the remainder of his days in some Hôtel des Invalides, I do not know; but I thought that his industry would not be worth much thereafter. I never learned which party was victorious, nor the cause of the war; but I felt for the rest of that day as if I had had my feelings excited and harrowed by witnessing the struggle, the ferocity and carnage, of a human battle before my door.

Kirby and Spence tell us that the battles of ants have long been celebrated and the date of them recorded, though they say that Huber is the only modern author who appears to have witnessed them. "Aeneas Sylvius," say they, "after giving a very circumstantial account of one contested with great obstinacy by a great and small species on the trunk of a pear tree," adds that "'this action was fought in the pontificate of Eugenius the Fourth, in the presence of Nicholas Pistoriensis, an eminent lawyer, who related the whole history of the battle with the greatest fidelity.' A similar engagement between great and small ants is recorded by Olaus Magnus, in which the small ones, being victorious, are said to have buried the bodies of their own soldiers, but left those of their giant enemies a prey to the birds. This event happened previous to the expulsion of the tyrant Christiern the Second from Sweden." The battle which I witnessed took place in the Presidency of Polk, five years before the passage of Webster's Fugitive-Slave Bill.

Many a village Bose, fit only to course a mud-turtle in a victualling cellar, sported his heavy quarters in the woods, without the knowledge of his master, and ineffectually smelled at old fox burrows and wood-chucks' holes; led perchance by some slight cur which nimbly threaded the wood, and might still inspire a natural terror in its denizens; — now far behind his guide, barking like a canine bull toward some small squirrel which had treed itself for scrutiny, then, cantering off, bending the bushes with his weight, imagining that he is on the track of some stray member of the jerbilla family. Once I was surprised to see a cat walking along the stony shore of the pond, for they rarely wander so far from home. The surprise was mutual. Nevertheless the most domestic cat, which has lain on a rug all her days, appears quite at home in the woods, and, by her sly and stealthy behavior, proves herself more native there than the regular inhabitants. Once, when berrying, I met with a cat with young kittens in the woods, quite wild, and they all, like their mother, had their backs up and were fiercely spitting at me. A few years before I lived in the woods there was what was called a "winged cat" in one of the farm-houses in Lincoln nearest the pond, Mr. Gilian Baker's. When I called to see her in June, 1842, she was gone a-hunting in the woods, as was her wont (I am not sure whether it was a male or female, and so use the more common pronoun), but her mistress told me that she came into the neighborhood a little more than a year before, in April, and was finally taken into their house; that she was of a dark brownish-gray color, with a white spot on her throat, and white feet, and had a large bushy tail like a fox; that in the winter the fur grew thick and flatted out along her sides, forming strips ten or twelve inches long by two and a half wide, and under her chin like a muff, the upper side loose, the under matted like felt, and in the spring these appendages dropped off. They gave me a pair of her "wings," which I keep still. There is no appearance of a membrane about them. Some thought it was part flying squirrel or some other wild animal, which is not impossible, for, according to naturalists, prolific hybrids have been produced by the union of the marten and domestic cat. This would have been the right kind of cat for me to keep, if I had kept any; for why should not a poet's cat be winged as well as his horse?

In the fall the loon (*Colymbus glacialis*) came, as usual, to moult and bathe in the pond, making the woods ring with his wild laughter before I had risen. At rumor of his arrival all the Mill-dam sportsmen are on the alert, in gigs and on foot, two by two and three by three, with patent rifles and conical balls and spy-glasses. They come rustling through the woods like autumn leaves, at least ten men to one loon. Some station themselves on this side of the pond, some on that, for the poor bird can-

not be omnipresent; if he dive here he must come up there. But now the
kind October wind rises, rustling the leaves and rippling the surface of
the water, so that no loon can be heard or seen, though his foes sweep
the pond with spy-glasses, and make the woods resound with their dis-
charges. The waves generously rise and dash angrily, taking sides with
all water-fowl, and our sportsmen must beat a retreat to town and shop
and unfinished jobs. But they were too often successful. When I went to
get a pail of water early in the morning I frequently saw this stately bird
sailing out of my cove within a few rods. If I endeavored to overtake him
in a boat, in order to see how he would manœuvre, he would dive and
be completely lost, so that I did not discover him again, sometimes, till
the latter part of the day. But I was more than a match for him on the
surface. He commonly went off in a rain.

As I was paddling along the north shore one very calm afternoon, for
such days especially they settle on to the lakes, like the milkweed down,
having looked in vain over the pond for a loon, suddenly one, sailing
out from the shore toward the middle a few rods in front of me, set up
his wild laugh and betrayed himself. I pursued with a paddle and he
dived, but when he came up I was nearer than before. He dived again,
but I miscalculated the direction he would take, and we were fifty rods
apart when he came to the surface this time, for I had helped to widen
the interval; and again he laughed long and loud, and with more reason than before. He manœuvred so cunningly that I could not get
within half a dozen rods of him. Each time, when he came to the surface, turning his head this way and that, he coolly surveyed the water
and the land, and apparently chose his course so that he might come
up where there was the widest expanse of water and at the greatest distance from the boat. It was surprising how quickly he made up his
mind and put his resolve into execution. He led me at once to the
widest part of the pond, and could not be driven from it. While he was
thinking one thing in his brain, I was endeavoring to divine his thought
in mine. It was a pretty game, played on the smooth surface of the
pond, a man against a loon. Suddenly your adversary's checker disappears beneath the board, and the problem is to place yours nearest to
where his will appear again. Sometimes he would come up unexpectedly on the opposite side of me, having apparently passed directly
under the boat. So long-winded was he and so unweariable, that when
he had swum farthest he would immediately plunge again, nevertheless; and then no wit could divine where in the deep pond, beneath the
smooth surface, he might be speeding his way like a fish, for he had
time and ability to visit the bottom of the pond in its deepest part. It is
said that loons have been caught in the New York lakes eighty feet

beneath the surface, with hooks set for trout,—though Walden is deeper than that. How surprised must the fishes be to see this ungainly visitor from another sphere speeding his way amid their schools! Yet he appeared to know his course as surely under water as on the surface, and swam much faster there. Once or twice I saw a ripple where he approached the surface, just put his head out to reconnoitre, and instantly dived again. I found that it was as well for me to rest on my oars and wait his reappearing as to endeavor to calculate where he would rise; for again and again, when I was straining my eyes over the surface one way, I would suddenly be startled by his unearthly laugh behind me. But why, after displaying so much cunning, did he invariably betray himself the moment he came up by that loud laugh? Did not his white breast enough betray him? He was indeed a silly loon, I thought. I could commonly hear the plash of the water when he came up, and so also detected him. But after an hour he seemed as fresh as ever, dived as willingly, and swam yet farther than at first. It was surprising to see how serenely he sailed off with unruffled breast when he came to the surface, doing all the work with his webbed feet beneath. His usual note was this demoniac laughter, yet somewhat like that of a water-fowl; but occasionally, when he had balked me most successfully and come up a long way off, he uttered a long-drawn unearthly howl, probably more like that of a wolf than any bird; as when a beast puts his muzzle to the ground and deliberately howls. This was his looning,—perhaps the wildest sound that is ever heard here, making the woods ring far and wide. I concluded that he laughed in derision of my efforts, confident of his own resources. Though the sky was by this time overcast, the pond was so smooth that I could see where he broke the surface when I did not hear him. His white breast, the stillness of the air, and the smoothness of the water were all against him. At length, having come up fifty rods off, he uttered one of those prolonged howls, as if calling on the god of loons to aid him, and immediately there came a wind from the east and rippled the surface, and filled the whole air with misty rain, and I was impressed as if it were the prayer of the loon answered, and his god was angry with me; and so I left him disappearing far away on the tumultuous surface.

For hours, in fall days, I watched the ducks cunningly tack and veer and hold the middle of the pond, far from the sportsman; tricks which they will have less need to practise in Louisiana bayous. When compelled to rise they would sometimes circle round and round and over the pond at a considerable height, from which they could easily see to other ponds and the river, like black motes in the sky; and, when I thought they had gone off thither long since, they would settle down by a slanting flight of a quarter of a mile on to a distant part which was left free; but what beside safety they got by sailing in the middle of Walden I do not know, unless they love its water for the same reason that I do.

WALKING

I wish to speak a word for Nature, for absolute freedom and wildness, as contrasted with a freedom and culture merely civil,—to regard man as an inhabitant, or a part and parcel of Nature, rather than a member of society. I wish to make an extreme statement, if so I may make an emphatic one, for there are enough champions of civilization: the minister and the school-committee and every one of you will take care of that.

I have met with but one or two persons in the course of my life who understood the art of Walking, that is, of taking walks,—who had a genius, so to speak, for *sauntering*: which word is beautifully derived "from idle people who roved about the country, in the Middle Ages, and asked charity, under pretense of going *à la Sainte Terre*," to the Holy Land, till the children exclaimed, "There goes a *Sainte-Terrer*," a Saunterer, a Holy-Lander. They who never go to the Holy Land in their walks, as they pretend, are indeed mere idlers and vagabonds; but they who do go there are saunterers in the good sense, such as I mean. Some, however, would derive the word from *sans terre*, without land or home, which, therefore, in the good sense, will mean, having no particular home, but equally at home everywhere. For this is the secret of successful sauntering. He who sits still in a house all the time may be the greatest vagrant of all; but the saunterer, in the good sense, is no more vagrant than the meandering river, which is all the while sedulously seeking the shortest course to the sea. But I prefer the first, which, indeed, is the most probable derivation. For every walk is a sort of crusade, preached by some Peter the Hermit in us, to go forth and reconquer this Holy Land from the hands of the Infidels.

It is true, we are but faint-hearted crusaders, even the walkers, nowadays, who undertake no persevering, never-ending enterprises. Our expeditions are but tours, and come round again at evening to the old hearth-side from which we set out. Half the walk is but retracing our steps. We should go forth on the shortest walk, perchance, in the spirit

Excursions (Boston: Ticknor and Fields, 1863).

of undying adventure, never to return,—prepared to send back our embalmed hearts only as relics to our desolate kingdoms. If you are ready to leave father and mother, and brother and sister, and wife and child and friends, and never see them again,—if you have paid your debts, and made your will, and settled all your affairs, and are a free man, then you are ready for a walk.

To come down to my own experience, my companion and I, for I sometimes have a companion, take pleasure in fancying ourselves knights of a new, or rather an old, order,—not Equestrians or Chevaliers, not Ritters or Riders, but Walkers, a still more ancient and honorable class, I trust. The chivalric and heroic spirit which once belonged to the Rider seems now to reside in, or perchance to have subsided into, the Walker,—not the Knight, but Walker, Errant. He is a sort of fourth estate, outside of Church and State and People.

We have felt that we almost alone hereabouts practiced this noble art; though, to tell the truth, at least, if their own assertions are to be received, most of my townsmen would fain walk sometimes, as I do, but they cannot. No wealth can buy the requisite leisure, freedom, and independence which are the capital in this profession. It comes only by the grace of God. It requires a direct dispensation from Heaven to become a walker. You must be born into the family of the Walkers. *Ambulator nascitur, non fit.* Some of my townsmen, it is true, can remember and have described to me some walks which they took ten years ago, in which they were so blessed as to lose themselves for half an hour in the woods; but I know very well that they have confined themselves to the highway ever since, whatever pretensions they may make to belong to this select class. No doubt they were elevated for a moment as by the reminiscence of a previous state of existence, when even they were foresters and outlaws.

> "When he came to grene wode,
>   In a mery mornynge,
> There he herde the notes small
>   Of byrdes mery syngynge.
>
> "It is ferre gone, sayd Robyn,
>   That I was last here;
> Me lyste a lytell for to shote
>   At the donne dere."

I think that I cannot preserve my health and spirits, unless I spend four hours a day at least,—and it is commonly more than that,—sauntering through the woods and over the hills and fields, absolutely free from all worldly engagements. You may safely say, A penny for your thoughts, or a thousand pounds. When sometimes I am reminded that the mechanics and shopkeepers stay in their shops not only all the forenoon, but all the afternoon too, sitting with crossed legs, so many of them,—as if the legs were made to sit upon, and not to stand or walk upon,—I think that they deserve some credit for not having all committed suicide long ago.

I, who cannot stay in my chamber for a single day without acquiring some rust, and when sometimes I have stolen forth for a walk at the eleventh hour or four o'clock in the afternoon, too late to redeem the day, when the shades of night were already beginning to be mingled with the daylight, have felt as if I had committed some sin to be atoned for,—I confess that I am astonished at the power of endurance, to say nothing of the moral insensibility, of my neighbors who confine themselves to shops and offices the whole day for weeks and months, aye, and years almost together. I know not what manner of stuff they are of,—sitting there now at three o'clock in the afternoon, as if it were three o'clock in the morning. Bonaparte may talk of the three-o'clock-in-the-morning courage, but it is nothing to the courage which can sit down cheerfully at this hour in the afternoon over against one's self whom you have known all the morning, to starve out a garrison to whom you are bound by such strong ties of sympathy. I wonder that about this time, or say between four and five o'clock in the afternoon, too late for the morning papers and too early for the evening ones, there is not a general explosion heard up and down the street, scattering a legion of antiquated and house-bred notions and whims to the four winds for an airing,—and so the evil cure itself.

How womankind, who are confined to the house still more than men, stand it I do not know; but I have ground to suspect that most of them do not *stand* it at all. When, early in a summer afternoon, we have been shaking the dust of the village from the skirts of our garments, making haste past those houses with purely Doric or Gothic fronts, which have such an air of repose about them, my companion whispers that probably about these times their occupants are all gone to bed. Then it is that I appreciate the beauty and the glory of architecture, which itself never turns in, but forever stands out and erect, keeping watch over the slumberers.

No doubt temperament, and, above all, age, have a good deal to do with it. As a man grows older, his ability to sit still and follow indoor occupations increases. He grows vespertinal in his habits as the evening of life approaches, till at last he comes forth only just before sundown, and gets all the walk that he requires in half an hour.

But the walking of which I speak has nothing in it akin to taking exercise, as it is called, as the sick take medicine at stated hours,—as the swinging of dumb-bells or chairs; but is itself the enterprise and adventure of the day. If you would get exercise, go in search of the springs of life. Think of a man's swinging dumb-bells for his health, when those springs are bubbling up in far-off pastures unsought by him!

Moreover, you must walk like a camel, which is said to be the only beast which ruminates when walking. When a traveler asked Wordsworth's servant to show him her master's study, she answered, "Here is his library, but his study is out of doors."

Living much out of doors, in the sun and wind, will no doubt produce a certain roughness of character,—will cause a thicker cuticle to grow over some of the finer qualities of our nature, as on the face and hands, or as severe manual labor robs the hands of some of their delicacy of touch. So staying in the house, on the other hand, may produce a softness and smoothness, not to say thinness of skin, accompanied by an increased sensibility to certain impressions. Perhaps we should be more susceptible to some influences important to our intellectual and moral growth, if the sun had shone and the wind blown on us a little less; and no doubt it is a nice matter to proportion rightly the thick and thin skin. But methinks that is a scurf that will fall off fast enough,—that the natural remedy is to be found in the proportion which the night bears to the day, the winter to the summer, thought to experience. There will be so much the more air and sunshine in our thoughts. The callous palms of the laborer are conversant with finer tissues of self-respect and heroism, whose touch thrills the heart, than the languid fingers of idleness. That is mere sentimentality that lies abed by day and thinks itself white, far from the tan and callus of experience.

When we walk, we naturally go to the fields and woods: what would become of us, if we walked only in a garden or a mall? Even some sects of philosophers have felt the necessity of importing the woods to themselves, since they did not go to the woods. "They planted groves and walks of Platanes," where they took *subdiales ambulationes* in porticos open to the air. Of course it is of no use to direct our steps to the woods, if they do not carry us thither. I am alarmed when it happens that I have walked a mile into the woods bodily, without getting there in spirit. In my afternoon walk I would fain forget all my morning occupations and my obligations to society. But it sometimes happens that I cannot easily shake off the village. The thought of some work will run in my head and I am not where my body is,—I am out of my senses. In my walks I would fain return to my senses. What business have I in the woods, if I am thinking of something out of the woods? I suspect myself, and cannot help a shudder, when I find myself so implicated even in what are called good works, — for this may sometimes happen.

My vicinity affords many good walks; and though for so many years I have walked almost every day, and sometimes for several days together, I have not yet exhausted them. An absolutely new prospect is a great happiness, and I can still get this any afternoon. Two or three hours' walking will carry me to as strange a country as I expect ever to see. A single farm-house which I had not seen before is sometimes as good as the dominions of the King of Dahomey. There is in fact a sort of harmony discoverable between the capabilities of the landscape within a circle of ten miles' radius, or the limits of an afternoon walk, and the threescore years and ten of human life. It will never become quite familiar to you.

Nowadays almost all man's improvements, so called, as the building of houses, and the cutting down of the forest and of all large trees, simply deform the landscape, and make it more and more tame and cheap. A people who would begin by burning the fences and let the forest stand! I saw the fences half consumed, their ends lost in the middle of the prairie, and some worldly miser with a surveyor looking after his bounds, while heaven had taken place around him, and he did not see the angels going to and fro, but was looking for an old post-hole in the midst of paradise. I looked again, and saw him standing in the middle of a boggy stygian fen, surrounded by devils, and he had found his bounds without a doubt, three little stones, where a stake had been driven, and looking nearer, I saw that the Prince of Darkness was his surveyor.

I can easily walk ten, fifteen, twenty, any number of miles, commencing at my own door, without going by any house, without crossing a road except where the fox and the mink do: first along by the river, and then the brook, and then the meadow and the woodside. There are square miles in my vicinity which have no inhabitant. From many a hill I can see civilization and the abodes of man afar. The farmers and their works are scarcely more obvious than woodchucks and their burrows. Man and his affairs, church and state and school, trade and commerce, and manufactures and agriculture, even politics, the most alarming of them all, — I am pleased to see how little space they occupy in the landscape. Politics is but a narrow field, and that still narrower highway yonder leads to it. I sometimes direct the traveler thither. If you would go to the political world, follow the great road, — follow that market-man, keep his dust in your eyes, and it will lead you straight to it; for it, too, has its place merely, and does not occupy all space. I pass from it as from a bean-field into the forest, and it is forgotten. In one half-hour I can walk off to some portion of the earth's surface where a man does not stand from one year's end to another, and there, consequently, politics are not, for they are but as the cigar-smoke of a man.

The village is the place to which the roads tend, a sort of expansion of the highway, as a lake of a river. It is the body of which roads are the arms and legs,—a trivial or quadrivial place, the thoroughfare and ordinary of travelers. The word is from the Latin *villa*, which together with *via*, a way, or more anciently *ved* and *vella*, Varro derives from *veho*, to carry, because the villa is the place to and from which things are carried. They who got their living by teaming were said *vellaturam facere*. Hence, too, the Latin word *vilis* and our vile; also *villain*. This suggests what kind of degeneracy villagers are liable to. They are wayworn by the travel that goes by and over them, without traveling themselves.

Some do not walk at all; others walk in the highways; a few walk across lots. Roads are made for horses and men of business. I do not travel in them much, comparatively, because I am not in a hurry to get to any tavern or grocery or livery-stable or depot to which they lead. I am a good horse to travel, but not from choice a roadster. The landscape-painter uses the figures of men to mark a road. He would not make that use of my figure. I walk out into a Nature such as the old prophets and poets, Menu, Moses, Homer, Chaucer, walked in. You may name it America, but it is not America; neither Americus Vespucius, nor Columbus, nor the rest were the discoverers of it. There is a truer account of it in mythology than in any history of America, so called, that I have seen.

However, there are a few old roads that may be trodden with profit, as if they led somewhere now that they are nearly discontinued. There is the Old Marlborough Road, which does not go to Marlborough now, methinks, unless that is Marlborough where it carries me. I am the bolder to speak of it here, because I presume that there are one or two such roads in every town.

THE OLD MARLBOROUGH ROAD

Where they once dug for money,
But never found any;
Where sometimes Martial Miles
Singly files,
And Elijah Wood,
I fear for no good:
No other man,
Save Elisha Dugan,—
O man of wild habits,

Partridges and rabbits,
Who hast no cares
Only to set snares,
Who liv'st all alone,
Close to the bone,
And where life is sweetest
Constantly eatest.
When the spring stirs my blood
With the instinct to travel
I can get enough gravel
On the Old Marlborough Road.
Nobody repairs it,
For nobody wears it;
It is a living way,
As the Christians say.
Not many there be
Who enter therein,
Only the guests of the
Irishman Quin.
What is it, what is it,
But a direction out there,
And the bare possibility
Of going somewhere?
Great guide-boards of stone,
But travelers none;
Cenotaphs of the towns
Named on their crowns.
It is worth going to see
Where you *might* be.
What king
Did the thing,
I am still wondering;
Set up how or when,
By what selectmen,
Gourgas or Lee,
Clark or Darby?
They're a great endeavor
To be something forever;
Blank tablets of stone,
Where a traveler might groan,
And in one sentence
Grave all that is known;
Which another might read,
In his extreme need.
I know one or two
Lines that would do,

Literature that might stand
All over the land,
Which a man could remember
Till next December,
And read again in the Spring,
After the thawing.
If with fancy unfurled
You leave your abode,
You may go round the world
By the Old Marlborough Road.

At present, in this vicinity, the best part of the land is not private property; the landscape is not owned, and the walker enjoys comparative freedom. But possibly the day will come when it will be partitioned off into so-called pleasure-grounds, in which a few will take a narrow and exclusive pleasure only,—when fences shall be multipled, and man-traps and other engines invented to confine men to the *public* road, and walking over the surface of God's earth shall be construed to mean trespassing on some gentleman's grounds. To enjoy a thing exclusively is commonly to exclude yourself from the true enjoyment of it. Let us improve our opportunities, then, before the evil days come.

What is it that makes it so hard sometimes to determine whither we will walk? I believe that there is a subtle magnetism in Nature, which, if we unconsciously yield to it, will direct us aright. It is not indifferent to us which way we walk. There is a right way; but we are very liable from heedlessness and stupidity to take the wrong one. We would fain take that walk, never yet taken by us through this actual world, which is perfectly symbolical of the path which we love to travel in the interior and ideal world; and sometimes, no doubt, we find it difficult to choose our direction, because it does not yet exist distinctly in our idea.

When I go out of the house for a walk, uncertain as yet whither I will bend my steps, and submit myself to my instinct to decide for me, I find, strange and whimsical as it may seem, that I finally and inevitably settle southwest, toward some particular wood or meadow or deserted pasture or hill in that direction. My needle is slow to settle,—varies a few degrees, and does not always point due southwest, it is true, and it has good authority for this variation, but it always settles between west and south-southwest. The future lies that way to me, and the earth seems more unexhausted and richer on that side. The outline which would bound my walks would be, not a circle, but a parabola, or rather like one of those cometary orbits which have been thought to be non-returning curves, in this case opening westward, in which my house occupies the place of the sun. I turn round and round irresolute sometimes for a quarter of an hour, until I decide, for a thousandth time, that I will walk into the southwest or west. Eastward I go only by force; but westward I go free. Thither no business leads me. It is hard for me to believe that I shall find fair landscapes or sufficient wildness and freedom behind the eastern horizon. I am not excited by the prospect of a walk thither; but I believe that the forest which I see in the western horizon stretches uninterruptedly toward the setting sun, and there are no towns nor cities in it of enough consequence to disturb me. Let me live where I will, on this side is the city, on that the wilderness, and ever I am leaving the city more and more, and withdrawing into the wilderness. I should not lay so much stress on this fact, if I did not believe that something like this is the prevailing tendency of my countrymen. I must walk toward Oregon, and not toward Europe. And that way the nation is moving, and I may say that mankind progress from east to west. Within a few years we have witnessed the phenomenon of a southeastward migration, in the settlement of Australia; but this affects us as a retrograde movement, and, judging from the moral and physical character of the first generation of Australians, has not yet proved a successful experiment. The eastern Tartars think that there is nothing west beyond Thibet. "The world ends there," say they; "beyond there is nothing but a shoreless sea." It is unmitigated East where they live.

We go eastward to realize history and study the works of art and literature, retracing the steps of the race; we go westward as into the future, with a spirit of enterprise and adventure. The Atlantic is a Lethean stream, in our passage over which we have had an opportunity to forget the Old World and its institutions. If we do not succeed this time, there is perhaps one more chance for the race left before it arrives on the banks of the Styx; and that is in the Lethe of the Pacific, which is three times as wide.

I know not how significant it is, or how far it is an evidence of singularity, that an individual should thus consent in his pettiest walk with the general movement of the race; but I know that something akin to the migratory instinct in birds and quadrupeds,—which, in some instances, is known to have affected the squirrel tribe, impelling them to a general and mysterious movement, in which they were seen, say some, crossing the broadest rivers, each on its particular chip, with its tail raised for a sail, and bridging narrower streams with their dead,—that something like the *furor* which affects the domestic cattle in the spring, and which is referred to a worm in their tails,—affects both nations and individuals, either perennially or from time to time.

Not a flock of wild geese cackles over our town, but it to some extent unsettles the value of real estate here, and, if I were a broker, I should probably take that disturbance into account.

> "Than longen folk to gon on pilgrimages,
> And palmeres for to seken strange strondes."

Every sunset which I witness inspires me with the desire to go to a West as distant and as fair as that into which the sun goes down. He appears to migrate westward daily, and tempt us to follow him. He is the Great Western Pioneer whom the nations follow. We dream all night of those mountain-ridges in the horizon, though they may be of vapor only, which were last gilded by his rays. The island of Atlantis, and the islands and gardens of the Hesperides, a sort of terrestrial paradise, appear to have been the Great West of the ancients, enveloped in mystery and poetry. Who has not seen in imagination, when looking into the sunset sky, the gardens of the Hesperides, and the foundation of all those fables?

Columbus felt the westward tendency more strongly than any before. He obeyed it, and found a New World for Castile and Leon. The herd of men in those days scented fresh pastures from afar.

> "And now the sun had stretched out all the hills,
> And now was dropped into the western bay;
> At last *he* rose, and twitched his mantle blue;
> To-morrow to fresh woods and pastures new."

Where on the globe can there be found an area of equal extent with that occupied by the bulk of our States, so fertile and so rich and varied in its productions, and at the same time so habitable by the European, as this is? Michaux, who knew but part of them, says that "the species of large trees are much more numerous in North America than in Europe; in the United States there are more than one hundred and forty species that exceed thirty feet in height; in France there are but thirty that attain this size." Later botanists more than confirm his observations. Humboldt came to America to realize his youthful dreams of a tropical vegetation, and he beheld it in its greatest perfection in the primitive forests of the Amazon, the most gigantic wilderness on the earth, which he has so eloquently described. The geographer Guyot, himself a European, goes farther,—farther than I am ready to follow him; yet not when he says: "As the plant is made for the animal, as the vegetable world is made for the animal world, America is made for the man of the Old World. . . . The man of the Old World sets out upon his way. Leaving the highlands of Asia, he descends from station to station towards Europe. Each of his steps is marked by a new civilization superior to the preceding, by a greater power of development. Arrived at the Atlantic, he pauses on the shore of this unknown ocean, the bounds of which he knows not, and turns upon his footprints for an instant." When he has exhausted the rich soil of Europe, and reinvigorated himself, "then recommences his adventurous career westward as in the earliest ages." So far Guyot.

From this western impulse coming in contact with the barrier of the Atlantic sprang the commerce and enterprise of modern times. The younger Michaux, in his "Travels West of the Alleghanies in 1802," says that the common inquiry in the newly settled West was, "'From what part of the world have you come?' As if these vast and fertile regions would naturally be the place of meeting and common country of all the inhabitants of the globe."

To use an obsolete Latin word, I might say, *Ex Oriente lux; ex Occidente* FRUX. From the East light; from the West fruit.

Sir Francis Head, an English traveler and a Governor-General of Canada, tells us that "in both the northern and southern hemispheres of the New World, Nature has not only outlined her works on a larger scale, but has painted the whole picture with brighter and more costly colors than she used in delineating and in beautifying the Old World. . . . The heavens of America appear infinitely higher, the sky is bluer, the air is fresher, the cold is intenser, the moon looks larger, the stars are brighter, the thunder is louder, the lightning is vivider, the wind is stronger, the rain is heavier, the mountains are higher, the rivers longer, the forest bigger, the plains broader." This statement will do at least to set against Buffon's account of this part of the world and its productions.

Linnæus said long ago, "Nescio quæ facies *læta, glabra* plantis Americanis: I know not what there is of joyous and smooth in the aspect of American plants;" and I think that in this country there are no, or at most very few, *Africanæ bestiæ,* African beasts, as the Romans called them, and that in this respect also it is peculiarly fitted for the habitation of man. We are told that within three miles of the centre of the East-Indian city of Singapore, some of the inhabitants are annually carried off by tigers; but the traveler can lie down in the woods at night almost anywhere in North America without fear of wild beasts.

These are encouraging testimonies. If the moon looks larger here than in Europe, probably the sun looks larger also. If the heavens of America appear infinitely higher, and the stars brighter, I trust that these facts are symbolical of the height to which the philosophy and poetry and religion of her inhabitants may one day soar. At length, perchance, the immaterial heaven will appear as much higher to the American mind, and the intimations that star it as much brighter. For I believe that climate does thus react on man,—as there is something in the mountain-air that feeds the spirit and inspires. Will not man grow to greater perfection intellectually as well as physically under these influences? Or is it unimportant how many foggy days there are in his life? I trust that we shall be more imaginative, that our thoughts will be clearer, fresher, and more ethereal, as our sky,—our understanding more comprehensive and broader, like our plains,—our intellect generally on a grander scale, like our thunder and lightning, our rivers and mountains and forests,—and our hearts shall even correspond in breadth and depth and grandeur to our inland seas. Perchance there will appear to the traveler something, he knows not what, of *læta* and *glabra*, of joyous and serene, in our very faces. Else to what end does the world go on, and why was America discovered?

To Americans I hardly need to say,—

"Westward the star of empire takes its way."

As a true patriot, I should be ashamed to think that Adam in paradise was more favorably situated on the whole than the backwoodsman in this country.

Our sympathies in Massachusetts are not confined to New England; though we may be estranged from the South, we sympathize with the West. There is the home of the younger sons, as among the Scandinavians they took to the sea for their inheritance. It is too late to be studying Hebrew; it is more important to understand even the slang of to-day.

Some months ago I went to see a panorama of the Rhine. It was like a dream of the Middle Ages. I floated down its historic stream in something more than imagination, under bridges built by the Romans, and repaired by later heroes, past cities and castles whose very names were music to my ears, and each of which was the subject of a legend. There were Ehrenbreitstein and Rolandseck and Coblentz, which I knew only in history. They were ruins that interested me chiefly. There seemed to come up from its waters and its vine-clad hills and valleys a hushed music as of Crusaders departing for the Holy Land. I floated along under the spell of enchantment, as if I had been transported to an heroic age, and breathed an atmosphere of chivalry.

Soon after, I went to see a panorama of the Mississippi, and as I worked my way up the river in the light of to-day, and saw the steam-boats wooding up, counted the rising cities, gazed on the fresh ruins of Nauvoo, beheld the Indians moving west across the stream, and, as before I had looked up the Moselle, now looked up the Ohio and the Missouri and heard the legends of Dubuque and of Wenona's Cliff,—still thinking more of the future than of the past or present,—I saw that this was a Rhine stream of a different kind; that the founda-tions of castles were yet to be laid, and the famous bridges were yet to be thrown over the river; and I felt that *this was the heroic age itself*, though we know it not, for the hero is commonly the simplest and obscurest of men.

The West of which I speak is but another name for the Wild; and what I have been preparing to say is, that in Wildness is the preserva-tion of the World. Every tree sends its fibres forth in search of the Wild. The cities import it at any price. Men plough and sail for it. From the forest and wilderness come the tonics and barks which brace mankind. Our ancestors were savages. The story of Romulus and Remus being suckled by a wolf is not a meaningless fable. The founders of every state which has risen to eminence have drawn their nourishment and vigor from a similar wild source. It was because the children of the Empire were not suckled by the wolf that they were conquered and dis-placed by the children of the northern forests who were.

I believe in the forest, and in the meadow, and in the night in which the corn grows. We require an infusion of hemlock-spruce or arbor-vitæ in our tea. There is a difference between eating and drinking for strength and from mere gluttony. The Hottentots eagerly devour the marrow of the koodoo and other antelopes raw, as a matter of course. Some of our Northern Indians eat raw the marrow of the Arctic rein-deer, as well as the various other parts, including the summits of the antlers, as long as they are soft. And herein, perchance, they have stolen a march on the cooks of Paris. They get what usually goes to feed the fire. This is probably better than stall-fed beef and slaughter-house pork to make a man of. Give me a wildness whose glance no civilization can endure,—as if we lived on the marrow of koodoos devoured raw.

There are some intervals which border the strain of the wood-thrush, to which I would migrate,—wild lands where no settler has squatted; to which, methinks, I am already acclimated.

The African hunter Cummings tells us that the skin of the eland, as well as that of most other antelopes just killed, emits the most delicious perfume of trees and grass. I would have every man so much like a wild antelope, so much a part and parcel of Nature, that his very person should thus sweetly advertise our senses of his presence, and remind us of those parts of Nature which he most haunts. I feel no disposition to be satirical, when the trapper's coat emits the odor of musquash even; it is a sweeter scent to me than that which commonly exhales from the merchant's or the scholar's garments. When I go into their wardrobes and handle their vestments, I am reminded of no grassy plains and flowery meads which they have frequented, but of dusty merchants' exchanges and libraries rather.

A tanned skin is something more than respectable, and perhaps olive is a fitter color than white for a man,—a denizen of the woods. "The pale white man!" I do not wonder that the African pitied him. Darwin the naturalist says, "A white man bathing by the side of a Tahitian was like a plant bleached by the gardener's art, compared with a fine, dark green one, growing vigorously in the open fields."

Ben Jonson exclaims,—

"How near to good is what is fair!"

So I would say,—

How near to good is what is *wild*!

Life consists with wildness. The most alive is the wildest. Not yet subdued to man, its presence refreshes him. One who pressed forward incessantly and never rested from his labors, who grew fast and made infinite demands on life, would always find himself in a new country or wilderness, and surrounded by the raw material of life. He would be climbing over the prostrate stems of primitive forest-trees.

Hope and the future for me are not in lawns and cultivated fields, not in towns and cities, but in the impervious and quaking swamps. When, formerly, I have analyzed my partiality for some farm which I had contemplated purchasing, I have frequently found that I was attracted solely by a few square rods of impermeable and unfathomable bog,—a natural sink in one corner of it. That was the jewel which dazzled me. I derive more of my subsistence from the swamps which surround my native town than from the cultivated gardens in the village. There are no richer parterres to my eyes than the dense beds of dwarf andromeda (*Casandra calyculata*) which cover these tender places on the earth's surface. Botany cannot go farther than tell me the names of the shrubs which grow there,—the high-blueberry, panicled andromeda, lamb-kill, azalea, and rhodora,—all standing in the quaking sphagnum. I often think that I should like to have my house front on this

mass of dull red bushes, omitting other flower plots and borders, transplanted spruce and trim box, even graveled walks, — to have this fertile
spot under my windows, not a few imported barrow-fulls of soil only to
cover the sand which was thrown out in digging the cellar. Why not
put my house, my parlor, behind this plot, instead of behind that meagre assemblage of curiosities, that poor apology for a Nature and Art,
which I call my front yard? It is an effort to clear up and make a decent
appearance when the carpenter and mason have departed, though
done as much for the passer-by as the dweller within. The most tasteful
front-yard fence was never an agreeable object of study to me; the most
elaborate ornaments, acorn-tops, or what not, soon wearied and disgusted me. Bring your sills up to the very edge of the swamp, then
(though it may not be the best place for a dry cellar), so that there be
no access on that side to citizens. Front yards are not made to walk in,
but, at most, through, and you could go in the back way.

Yes, though you may think me perverse, if it were proposed to me to
dwell in the neighborhood of the most beautiful garden that ever human
art contrived, or else of a Dismal Swamp, I should certainly decide for
the swamp. How vain, then, have been all your labors, citizens, for me!

My spirits infallibly rise in proportion to the outward dreariness.
Give me the ocean, the desert, or the wilderness! In the desert, pure air
and solitude compensate for want of moisture and fertility. The traveler
Burton says of it: "Your *morale* improves; you become frank and cordial, hospitable and single-minded. . . . In the desert, spirituous liquors
excite only disgust. There is a keen enjoyment in a mere animal existence." They who have been traveling long on the steppes of Tartary
say: "On reëntering cultivated lands, the agitation, perplexity, and turmoil of civilization oppressed and suffocated us; the air seemed to fail
us, and we felt every moment as if about to die of asphyxia." When I
would recreate myself, I seek the darkest wood, the thickest and most
interminable and, to the citizen, most dismal swamp. I enter a swamp
as a sacred place, — a *sanctum sanctorum*. There is the strength, the
marrow of Nature. The wild-wood covers the virgin-mould, — and the
same soil is good for men and for trees. A man's health requires as
many acres of meadow to his prospect as his farm does loads of muck.
There are the strong meats on which he feeds. A town is saved, not
more by the righteous men in it than by the woods and swamps that
surround it. A township where one primitive forest waves above while
another primitive forest rots below, — such a town is fitted to raise not
only corn and potatoes, but poets and philosophers for the coming
ages. In such a soil grew Homer and Confucius and the rest, and out of
such a wilderness comes the Reformer eating locusts and wild honey.

To preserve wild animals implies generally the creation of a forest for them to dwell in or resort to. So it is with man. A hundred years ago they sold bark in our streets peeled from our own woods. In the very aspect of those primitive and rugged trees there was, methinks, a tanning principle which hardened and consolidated the fibres of men's thoughts. Ah! already I shudder for these comparatively degenerate days of my native village, when you cannot collect a load of bark of good thickness,—and we no longer produce tar and turpentine.

The civilized nations—Greece, Rome, England—have been sustained by the primitive forests which anciently rotted where they stand. They survive as long as the soil is not exhausted. Alas for human culture! little is to be expected of a nation, when the vegetable mould is exhausted, and it is compelled to make manure of the bones of its fathers. There the poet sustains himself merely by his own superfluous fat, and the philosopher comes down on his marrow-bones.

It is said to be the task of the American "to work the virgin soil," and that "agriculture here already assumes proportions unknown everywhere else." I think that the farmer displaces the Indian even because he redeems the meadow, and so makes himself stronger and in some respects more natural. I was surveying for a man the other day a single straight line one hundred and thirty-two rods long, through a swamp, at whose entrance might have been written the words which Dante read over the entrance to the infernal regions,—"Leave all hope, ye that enter,"—that is, of ever getting out again; where at one time I saw my employer actually up to his neck and swimming for his life in his property, though it was still winter. He had another similar swamp which I could not survey at all, because it was completely under water, and nevertheless, with regard to a third swamp, which I did *survey* from a distance, he remarked to me, true to his instincts, that he would not part with it for any consideration, on account of the mud which it contained. And that man intends to put a girdling ditch round the whole in the course of forty months, and so redeem it by the magic of his spade. I refer to him only as the type of a class.

The weapons with which we have gained our most important victories, which should be handed down as heirlooms from father to son, are not the sword and the lance, but the bushwhack, the turf-cutter, the spade, and the boghoe, rusted with the blood of many a meadow, and begrimed with the dust of many a hard-fought field. The very winds blew the Indian's corn-field into the meadow, and pointed out the way which he had not the skill to follow. He had no better implement with which to intrench himself in the land than a clam-shell. But the farmer is armed with plough and spade.

In literature it is only the wild that attracts us. Dullness is but another name for tameness. It is the uncivilized free and wild thinking in "Hamlet" and the "Iliad," in all the Scriptures and Mythologies, not learned in the schools, that delights us. As the wild duck is more swift and beautiful than the tame, so is the wild—the mallard—thought, which 'mid falling dews wings its way above the fens. A truly good book is something as natural, and as unexpectedly and unaccountably fair and perfect, as a wild flower discovered on the prairies of the West or in the jungles of the East. Genius is a light which makes the darkness visible, like the lightning's flash, which perchance shatters the temple of knowledge itself,—and not a taper lighted at the hearthstone of the race, which pales before the light of common day.

English literature, from the days of the minstrels to the Lake Poets,—Chaucer and Spenser and Milton, and even Shakespeare, included,—breathes no quite fresh and, in this sense, wild strain. It is an essentially tame and civilized literature, reflecting Greece and Rome. Her wilderness is a greenwood, her wild man a Robin Hood. There is plenty of genial love of Nature, but not so much of Nature herself. Her chronicles inform us when her wild animals, but not when the wild man in her, became extinct.

The science of Humboldt is one thing, poetry is another thing. The poet to-day, notwithstanding all the discoveries of science, and the accumulated learning of mankind, enjoys no advantage over Homer.

Where is the literature which gives expression to Nature? He would be a poet who could impress the winds and streams into his service, to speak for him; who nailed words to their primitive senses, as farmers drive down stakes in the spring, which the frost has heaved; who derived his words as often as he used them,—transplanted them to his page with earth adhering to their roots; whose words were so true and fresh and natural that they would appear to expand like the buds at the approach of spring, though they lay half-smothered between two musty leaves in a library,—ay, to bloom and bear fruit there, after their kind, annually, for the faithful reader, in sympathy with surrounding Nature.

I do not know of any poetry to quote which adequately expresses this yearning for the Wild. Approached from this side, the best poetry is tame. I do not know where to find in any literature, ancient or modern, any account which contents me of that Nature with which even I am acquainted. You will perceive that I demand something which no Augustan nor Elizabethan age, which no *culture*, in short, can give. Mythology comes nearer to it than anything. How much more fertile a Nature, at least, has Grecian mythology its root in than English literature! Mythology is the crop which the Old World bore before its soil was exhausted, before the fancy and imagination were affected with blight; and which it still bears, wherever its pristine vigor is unabated. All other literatures endure only as the elms which overshadow our houses; but this is like the great dragon-tree of the Western Isles, as old as mankind, and, whether that does or not, will endure as long; for the decay of other literatures makes the soil in which it thrives.

The West is preparing to add its fables to those of the East. The valleys of the Ganges, the Nile, and the Rhine having yielded their crop, it remains to be seen what the valleys of the Amazon, the Plate, the Orinoco, the St. Lawrence, and the Mississippi will produce. Perchance, when, in the course of ages, American liberty has become a fiction of the past, — as it is to some extent a fiction of the present, — the poets of the world will be inspired by American mythology.

The wildest dreams of wild men, even, are not the less true, though they may not recommend themselves to the sense which is most common among Englishmen and Americans to-day. It is not every truth that recommends itself to the common sense. Nature has a place for the wild clematis as well as for the cabbage. Some expressions of truth are reminiscent, — others merely *sensible*, as the phrase is, — others prophetic. Some forms of disease, even, may prophesy forms of health. The geologist has discovered that the figures of serpents, griffins, flying dragons, and other fanciful embellishments of heraldry, have their prototypes in the forms of fossil species which were extinct before man was created, and hence "indicate a faint and shadowy knowledge of a previous state of organic existence." The Hindoos dreamed that the earth rested on an elephant, and the elephant on a tortoise, and the tortoise on a serpent; and though it may be an unimportant coincidence, it will not be out of place here to state, that a fossile tortoise has lately been discovered in Asia large enough to support an elephant. I confess that I am partial to these wild fancies, which transcend the order of time and development. They are the sublimest recreation of the intellect. The partridge loves peas, but not those that go with her into the pot.

In short, all good things are wild and free. There is something in a strain of music, whether produced by an instrument or by the human voice, — take the sound of a bugle in a summer night, for instance, — which by its wildness, to speak without satire, reminds me of the cries emitted by wild beasts in their native forests. It is so much of their wildness as I can understand. Give me for my friends and neighbors wild men, not tame ones. The wildness of the savage is but a faint symbol of the awful ferity with which good men and lovers meet.

I love even to see the domestic animals reassert their native rights,—any evidence that they have not wholly lost their original wild habits and vigor; as when my neighbor's cow breaks out of her pasture early in the spring and boldly swims the river, a cold, gray tide, twenty-five or thirty rods wide, swollen by the melted snow. It is the buffalo crossing the Mississippi. This exploit confers some dignity on the herd in my eyes,—already dignified. The seeds of instinct are preserved under the thick hides of cattle and horses, like seeds in the bowels of the earth, an indefinite period.

Any sportiveness in cattle is unexpected. I saw one day a herd of a dozen bullocks and cows running about and frisking in unwieldy sport, like huge rats, even like kittens. They shook their heads, raised their tails, and rushed up and down a hill, and I perceived by their horns, as well as by their activity, their relation to the deer tribe. But, alas! a sudden loud *Whoa!* would have damped their ardor at once, reduced them from venison to beef, and stiffened their sides and sinews like the locomotive. Who but the Evil One has cried, "Whoa!" to mankind? Indeed, the life of cattle, like that of many men, is but a sort of locomotiveness; they move a side at a time, and man, by his machinery, is meeting the horse and the ox half-way. Whatever part the whip has touched is thenceforth palsied. Who would ever think of a *side* of any of the supple cat tribe, as we speak of a *side* of beef?

I rejoice that horses and steers have to be broken before they can be made the slaves of men, and that men themselves have some wild oats still left to sow before they become submissive members of society. Undoubtedly, all men are not equally fit subjects for civilization; and because the majority, like dogs and sheep, are tame by inherited disposition, this is no reason why the others should have their natures broken that they may be reduced to the same level. Men are in the main alike, but they were made several in order that they might be various. If a low use is to be served, one man will do nearly or quite as well as another; if a high one, individual excellence is to be regarded. Any man can stop a hole to keep the wind away, but no other man could serve so rare a use as the author of this illustration did. Confucius says, "The skins of the tiger and the leopard, when they are tanned, are as the skins of the dog and the sheep tanned." But it is not the part of a true culture to tame tigers, any more than it is to make sheep ferocious; and tanning their skins for shoes is not the best use to which they can be put.

When looking over a list of men's names in a foreign language, as of military officers, or of authors who have written on a particular subject, I am reminded once more that there is nothing in a name. The name Menschikoff, for instance, has nothing in it to my ears more human than a whisker, and it may belong to a rat. As the names of the Poles and Russians are to us, so are ours to them. It is as if they had been named by the child's rigmarole,—*Iery wiery ichery van, tittle-tol-tan.* I see in my mind a herd of wild creatures swarming over the earth, and to each the herdsman has affixed some barbarous sound in his own dialect. The names of men are of course as cheap and meaningless as *Bose* and *Tray,* the names of dogs.

Methinks it would be some advantage to philosophy, if men were named merely in the gross, as they are known. It would be necessary only to know the genus and perhaps the race or variety, to know the individual. We are not prepared to believe that every private soldier in a Roman army had a name of his own,—because we have not supposed that he had a character of his own.

At present our only true names are nicknames. I knew a boy who, from his peculiar energy, was called "Buster" by his playmates, and this rightly supplanted his Christian name. Some travelers tell us that an Indian had no name given him at first, but earned it, and his name was his fame; and among some tribes he acquired a new name with every new exploit. It is pitiful when a man bears a name for convenience merely, who has earned neither name nor fame.

I will not allow mere names to make distinctions for me, but still see men in herds for all them. A familiar name cannot make a man less strange to me. It may be given to a savage who retains in secret his own wild title earned in the woods. We have a wild savage in us, and a savage name is perchance somewhere recorded as ours. I see that my neighbor, who bears the familiar epithet William, or Edwin, takes it off with his jacket. It does not adhere to him when asleep or in anger, or aroused by any passion or inspiration. I seem to hear pronounced by some of his kin at such a time his original wild name in some jaw-breaking or else melodious tongue.

Here is this vast, savage, howling mother of ours, Nature, lying all around, with such beauty, and such affection for her children, as the leopard; and yet we are so early weaned from her breast to society, to that culture which is exclusively an interaction of man on man,—a sort of breeding in and in, which produces at most a merely English nobility, a civilization destined to have a speedy limit.

In society, in the best institutions of men, it is easy to detect a certain precocity. When we should still be growing children, we are already little men. Give me a culture which imports much muck from the meadows, and deepens the soil,—not that which trusts to heating manures, and improved implements and modes of culture only!

Many a poor sore-eyed student that I have heard of would grow faster, both intellectually and physically, if, instead of sitting up so very late, he honestly slumbered a fool's allowance.

There may be an excess even of informing light. Niepce, a Frenchman, discovered "actinism," that power in the sun's rays which produces a chemical effect; that granite rocks, and stone structures, and statues of metal, "are all alike destructively acted upon during the hours of sunshine, and, but for provisions of Nature no less wonderful, would soon perish under the delicate touch of the most subtile of the agencies of the universe." But he observed that "those bodies which underwent this change during the daylight possessed the power of restoring themselves to their original conditions during the hours of night, when this excitement was no longer influencing them." Hence it has been inferred that "the hours of darkness are as necessary to the inorganic creation as we know night and sleep are to the organic kingdom." Not even does the moon shine every night, but gives place to darkness.

I would not have every man nor every part of a man cultivated, any more than I would have every acre of earth cultivated: part will be tillage, but the greater part will be meadow and forest, not only serving an immediate use, but preparing a mould against a distant future, by the annual decay of the vegetation which it supports.

There are other letters for the child to learn than those which Cadmus invented. The Spaniards have a good term to express this wild and dusky knowledge, *Gramática parda*, tawny grammar, a kind of mother-wit derived from that same leopard to which I have referred.

We have heard of a Society for the Diffusion of Useful Knowledge. It is said that knowledge is power; and the like. Methinks there is equal need of a Society for the Diffusion of Useful Ignorance, what we will call Beautiful Knowledge, a knowledge useful in a higher sense: for what is most of our boasted so-called knowledge but a conceit that we know something, which robs us of the advantage of our actual ignorance? What we call knowledge is often our positive ignorance; ignorance our negative knowledge. By long years of patient industry and reading of the newspapers, — for what are the libraries of science but files of newspapers? — a man accumulates a myriad facts, lays them up in his memory, and then when in some spring of his life he saunters into the Great Fields of thought, he, as it were, goes to grass like a horse and leaves all his harness behind in the stable. I would say to the Society for the Diffusion of Useful Knowledge, sometimes, — Go to grass. You have eaten hay long enough. The spring has come with its green crop. The very cows are driven to their country pastures before the end of May; though I have heard of one unnatural farmer who kept his cow in the barn and fed her on hay all the year round. So, frequently, the Society for the Diffusion of Useful Knowledge treats its cattle.

A man's ignorance sometimes is not only useful, but beautiful,
—while his knowledge, so called, is oftentimes worse than useless,
besides being ugly. Which is the best man to deal with,—he who
knows nothing about a subject, and, what is extremely rare, knows that
he knows nothing, or he who really knows something about it, but
thinks that he knows all?

My desire for knowledge is intermittent; but my desire to bathe my
head in atmospheres unknown to my feet is perennial and constant.
The highest that we can attain to is not Knowledge, but Sympathy with
Intelligence. I do not know that this higher knowledge amounts to anything more definite than a novel and grand surprise on a sudden revelation of the insufficiency of all that we called Knowledge before,—a
discovery that there are more things in heaven and earth than are
dreamed of in our philosophy. It is the lighting up of the mist by the
sun. Man cannot *know* in any higher sense than this, any more than he
can look serenely and with impunity in the face of the sun: Ὡς τί νοῶν,
οὐ χεῖνον νοήσεις, — "You will not perceive that, as perceiving a particular thing," say the Chaldean Oracles.

There is something servile in the habit of seeking after a law which
we may obey. We may study the laws of matter at and for our convenience, but a successful life knows no law. It is an unfortunate discovery
certainly, that of a law which binds us where we did not know before
that we were bound. Live free, child of the mist,—and with respect to
knowledge we are all children of the mist. The man who takes the liberty to live is superior to all the laws, by virtue of his relation to the
law-maker. "That is active duty," says the Vishnu Purana, "which is not
for our bondage; that is knowledge which is for our liberation: all other
duty is good only unto weariness; all other knowledge is only the cleverness of an artist."

It is remarkable how few events or crises there are in our histories;
how little exercised we have been in our minds; how few experiences we
have had. I would fain be assured that I am growing apace and rankly,
though my very growth disturb this dull equanimity,—though it be with
struggle through long, dark, muggy nights or seasons of gloom. It would
be well, if all our lives were a divine tragedy even, instead of this trivial
comedy or farce. Dante, Bunyan, and others appear to have been exercised in their minds more than we: they were subjected to a kind of culture such as our district schools and colleges do not contemplate. Even
Mahomet, though many may scream at his name, had a good deal more
to live for, aye, and to die for, than they have commonly.

When, at rare intervals, some thought visits one, as perchance he is
walking on a railroad, then indeed the cars go by without his hearing

them. But soon, by some inexorable law, our life goes by and the cars
return.

> "Gentle breeze, that wanderest unseen,
> And bendest the thistles round Loira of storms,
> Traveler of the windy glens,
> Why hast thou left my ear so soon?"

While almost all men feel an attraction drawing them to society, few
are attracted strongly to Nature. In their reaction to Nature men appear
to me for the most part, notwithstanding their arts, lower than the animals. It is not often a beautiful relation, as in the case of the animals.
How little appreciation of the beauty of the landscape there is among
us! We have to be told that the Greeks called the world Κόσμος,
Beauty, or Order, but we do not see clearly why they did so, and we
esteem it at best only a curious philological fact.

For my part, I feel that with regard to Nature I live a sort of border
life, on the confines of a world into which I make occasional and transient forays only, and my patriotism and allegiance to the State into
whose territories I seem to retreat are those of a moss-trooper. Unto a
life which I call natural I would gladly follow even a will-o'-the-wisp
through bogs and sloughs unimaginable, but no moon nor firefly has
shown me the causeway to it. Nature is a personality so vast and universal that we have never seen one of her features. The walker in the
familiar fields which stretch around my native town sometimes finds
himself in another land than is described in their owners' deeds, as it
were in some far-away field on the confines of the actual Concord,
where her jurisdiction ceases, and the idea which the word Concord
suggests ceases to be suggested. These farms which I have myself surveyed, these bounds which I have set up, appear dimly still as through a
mist; but they have no chemistry to fix them; they fade from the surface
of the glass; and the picture which the painter painted stands out dimly
from beneath. The world with which we are commonly acquainted
leaves no trace, and it will have no anniversary.

I took a walk on Spaulding's Farm the other afternoon. I saw the setting sun lighting up the opposite side of a stately pine wood. Its golden
rays straggled into the aisles of the wood as into some noble hall. I was
impressed as if some ancient and altogether admirable and shining
family had settled there in that part of the land called Concord,
unknown to me,—to whom the sun was servant,—who had not gone
into society in the village,—who had not been called on. I saw their
park, their pleasure-ground, beyond through the wood, in Spaulding's
cranberry-meadow. The pines furnished them with gables as they grew.

Their house was not obvious to vision; the trees grew through it. I do not know whether I heard the sounds of a suppressed hilarity or not. They seemed to recline on the sunbeams. They have sons and daughters. They are quite well. The farmer's cart-path, which leads directly through their hall, does not in the least put them out, as the muddy bottom of a pool is sometimes seen through the reflected skies. They never heard of Spaulding, and do not know that he is their neighbor,—notwithstanding I heard him whistle as he drove his team through the house. Nothing can equal the serenity of their lives. Their coat of arms is simply a lichen. I saw it painted on the pines and oaks. Their attics were in the tops of the trees. They are of no politics. There was no noise of labor. I did not perceive that they were weaving or spinning. Yet I did detect, when the wind lulled and hearing was done away, the finest imaginable sweet musical hum,—as of a distant hive in May, which perchance was the sound of their thinking. They had no idle thoughts, and no one without could see their work, for their industry was not as in knots and excrescences embayed.

But I find it difficult to remember them. They fade irrevocably out of my mind even now while I speak, and endeavor to recall them and recollect myself. It is only after a long and serious effort to recollect my best thoughts that I become again aware of their cohabitancy. If it were not for such families as this, I think I should move out of Concord.

We are accustomed to say in New England that few and fewer pigeons visit us every year. Our forests furnish no mast for them. So, it would seem, few and fewer thoughts visit each growing man from year to year, for the grove in our minds is laid waste,—sold to feed unnecessary fires of ambition, or sent to mill, and there is scarcely a twig left for them to perch on. They no longer build nor breed with us. In some more genial season, perchance, a faint shadow flits across the landscape of the mind, cast by the *wings* of some thought in its vernal or autumnal migration, but, looking up, we are unable to detect the substance of the thought itself. Our winged thoughts are turned to poultry. They no longer soar, and they attain only to a Shanghai and Cochin-China grandeur. Those *gra-a-ate thoughts*, those *gra-a-ate men* you hear of!

We hug the earth,—how rarely we mount! Methinks we might elevate ourselves a little more. We might climb a tree, at least. I found my account in climbing a tree once. It was a tall white-pine, on the top of a hill; and though I got well pitched, I was well paid for it, for I discovered new mountains in the horizon which I had never seen before,—so much more of the earth and the heavens. I might have walked about the foot of the tree for three-score years and ten, and yet I certainly should never have seen them. But, above all, I discovered around me, — it was near the end of June, — on the ends of the topmost branches only, a few minute and delicate red cone-like blossoms, the fertile flower of the white pine looking heavenward. I carried straight-way to the village the topmost spire, and showed it to stranger jurymen who walked the streets, — for it was court-week, — and the farmers and lumber-dealers and wood-choppers and hunters, and not one had ever seen the like before, but they wondered as at a star dropped down. Tell of ancient architects finishing their works on the tops of columns as perfectly as on the lower and more visible parts! Nature has from the first expanded the minute blossoms of the forest only toward the heav-ens, above men's heads and unobserved by them. We see only the flow-ers that are under our feet in the meadows. The pines have developed their delicate blossoms on the highest twigs of the wood every summer for ages, as well over the heads of Nature's red children as of her white ones; yet scarcely a farmer or hunter in the land has ever seen them.

Above all, we cannot afford not to live in the present. He is blessed over all mortals who loses no moment of the passing life in remember-ing the past. Unless our philosophy hears the cock crow in every barn-yard within our horizon, it is belated. That sound commonly reminds us that we are growing rusty and antique in our employments and habits of thought. His philosophy comes down to a more recent time than ours. There is something suggested by it that is a newer testa-ment, — the gospel according to this moment. He has not fallen astern; he has got up early and kept up early, and to be where he is is to be in season, in the foremost rank of time. It is an expression of the health and soundness of Nature, a brag for all the world, — healthiness as of a spring burst forth, a new fountain of the Muses, to celebrate this last instant of time. Where he lives no fugitive slave laws are passed. Who has not betrayed his master many times since last he heard that note?

The merit of this bird's strain is in its freedom from all plaintiveness. The singer can easily move us to tears or to laughter, but where is he who can excite in us a pure morning joy? When, in doleful dumps, breaking the awful stillness of our wooden sidewalk on a Sunday, or, perchance, a watcher in the house of mourning, I hear a cockerel crow far or near, I think to myself, "There is one of us well, at any rate," — and with a sudden gush return to my senses.

We had a remarkable sunset one day last November. I was walking in a meadow, the source of a small brook, when the sun at last, just before setting, after a cold gray day, reached a clear stratum in the horizon, and the softest, brightest morning sunlight fell on the dry grass and on the stems of the trees in the opposite horizon and on the leaves of the shrub-oaks on the hillside, while our shadows stretched long over the meadow eastward, as if we were the only motes in its beams. It was such a light as we could not have imagined a moment before, and the air also was so warm and serene that nothing was wanting to make a paradise of that meadow. When we reflected that this was not a solitary phenomenon, never to happen again, but that it would happen forever and ever an infinite number of evenings, and cheer and reassure the latest child that walked there, it was more glorious still.

The sun sets on some retired meadow, where no house is visible, with all the glory and splendor that it lavishes on cities, and perchance as it has never set before,—where there is but a solitary marsh-hawk to have his wings gilded by it, or only a musquash looks out from his cabin, and there is some little black-veined brook in the midst of the marsh, just beginning to meander, winding slowly round a decaying stump. We walked in so pure and bright a light, gilding the withered grass and leaves, so softly and serenely bright, I thought I had never bathed in such a golden flood, without a ripple or a murmur to it. The west side of every wood and rising ground gleamed like the boundary of Elysium, and the sun on our backs seemed like a gentle herdsman driving us home at evening.

So we saunter toward the Holy Land, till one day the sun shall shine more brightly than ever he has done, shall perchance shine into our minds and hearts, and light up our whole lives with a great awakening light, as warm and serene and golden as on a bankside in autumn.

From THE MAINE WOODS

KTAADN

* * * In the morning, after whetting our appetite on some raw pork, a wafer of hard bread, and a dipper of condensed cloud or water-spout, we all together began to make our way up the falls, which I have described; this time choosing the right hand, or highest peak, which was not the one I had approached before. But soon my companions were lost to my sight behind the mountain ridge in my rear, which still seemed ever retreating before me, and I climbed alone over huge

The Maine Woods (Boston: Ticknor and Fields, 1864).

rocks, loosely poised, a mile or more, still edging toward the clouds; for though the day was clear elsewhere, the summit was concealed by mist. The mountain seemed a vast aggregation of loose rocks, as if some time it had rained rocks, and they lay as they fell on the mountain sides, nowhere fairly at rest, but leaning on each other, all rocking-stones, with cavities between, but scarcely any soil or smoother shelf. They were the raw materials of a planet dropped from an unseen quarry, which the vast chemistry of nature would anon work up, or work down, into the smiling and verdant plains and valleys of earth. This was an undone extremity of the globe; as in lignite, we see coal in the process of formation.

At length I entered within the skirts of the cloud which seemed forever drifting over the summit, and yet would never be gone, but was generated out of that pure air as fast as it flowed away; and when, a quarter of a mile farther, I reached the summit of the ridge, which those who have seen in clearer weather say is about five miles long, and contains a thousand acres of table-land, I was deep within the hostile ranks of clouds, and all objects were obscured by them. Now the wind would blow me out a yard of clear sunlight, wherein I stood; then a gray, dawning light was all it could accomplish, the cloud-line ever rising and falling with the wind's intensity. Sometimes it seemed as if the summit would be cleared in a few moments, and smile in sunshine; but what was gained on one side was lost on another. It was like sitting in a chimney and waiting for the smoke to blow away. It was, in fact, a cloud factory,—these were the cloud-works, and the wind turned them off done from the cool, bare rocks. Occasionally, when the windy columns broke in to me, I caught sight of a dark, damp crag to the right or left; the mist driving ceaselessly between it and me. It reminded me of the creations of the old epic and dramatic poets, of Atlas, Vulcan, the Cyclops, and Prometheus. Such was Caucasus and the rock where Prometheus was bound. Aeschylus had no doubt visited such scenery as this. It was vast, Titanic, and such as man never inhabits. Some part of the beholder, even some vital part, seems to escape through the loose grating of his ribs as he ascends. He is more lone than you can imagine. There is less of substantial thought and fair understanding in him than in the plains where men inhabit. His reason is dispersed and shadowy, more thin and subtile, like the air. Vast, Titanic, inhuman Nature has got him at disadvantage, caught him alone, and pilfers him of some of his divine faculty. She does not smile on him as in the plains. She seems to say sternly, Why came ye here before your time. This ground is not prepared for you. Is it not enough that I smile in the valleys? I have never made this soil for thy feet, this air for thy breathing, these rocks for thy neighbors. I cannot pity nor fondle thee here, but forever relentlessly drive thee hence to where I *am* kind. Why seek me where I have not called thee, and then complain because you find me but a stepmother? Shouldst thou freeze or starve, or shudder thy life away, here is no shrine, nor altar, nor any access to my ear.

> "Chaos and ancient Night, I come no spy
> With purpose to explore or to disturb
> The secrets of your realm, but . . .
> . . . . . as my way
> Lies through your spacious empire up to light."

The tops of mountains are among the unfinished parts of the globe, whither it is a slight insult to the gods to climb and pry into their secrets, and try their effect on our humanity. Only daring and insolent men, perchance, go there. Simple races, as savages, do not climb mountains, — their tops are sacred and mysterious tracts never visited by them. Pomola is always angry with those who climb to the summit of Ktaadn.

According to Jackson, who, in his capacity of geological surveyor of the State, has accurately measured it, — the altitude of Ktaadn is 5300 feet, or a little more than one mile above the level of the sea, — and he adds, "It is then evidently the highest point in the State of Maine, and is the most abrupt granite mountain in New England." The peculiarities of that spacious table-land on which I was standing, as well as the remarkable semi-circular precipice or basin on the eastern side, were all concealed by the mist. I had brought my whole pack to the top, not knowing but I should have to make my descent to the river, and possibly to the settled portion of the State alone, and by some other route, and wishing to have a complete outfit with me. But at length, fearing that my companions would be anxious to reach the river before night, and knowing that the clouds might rest on the mountain for days, I was compelled to descend. Occasionally, as I came down, the wind would blow me a vista open, through which I could see the country eastward, boundless forests, and lakes, and streams, gleaming in the sun, some of them emptying into the East Branch. There were also new mountains in sight in that direction. Now and then some small bird of the sparrow family would flit away before me, unable to command its course, like a fragment of the gray rock blown off by the wind.

I found my companions where I had left them, on the side of the peak, gathering the mountain-cranberries, which filled every crevice between the rocks, together with blueberries, which had a spicier flavor the higher up they grew, but were not the less agreeable to our palates. When the country is settled, and roads are made, these cranberries will perhaps become an article of commerce. From this elevation, just on the skirts of the clouds, we could overlook the country, west and south, for a hundred miles. There it was, the State of Maine, which we had seen on the map, but not much like that, — immeasurable forest for the sun to shine on, that eastern *stuff* we hear of in Massachusetts. No clearing, no house. It did not look as if a solitary traveler had cut so much as a walking-stick there. Countless lakes, — Moosehead in the southwest, forty miles long by ten wide, like a gleaming silver platter at the end of the table; Chesuncook, eighteen long by three wide, without an island; Millinocket, on the south, with its hundred islands; and a hundred others without a name; and mountains, also, whose names, for the most part, are known only to the Indians. The forest looked like a firm grass sward, and the effect of these lakes in its midst has been well compared, by one who has since visited this same spot, to that of a "mirror broken into a thousand fragments, and wildly scattered over the grass, reflecting the full blaze of the sun." It was a large farm for somebody, when cleared. According to the Gazetteer, which was printed before the boundary question was settled, this single Penobscot county, in which we were, was larger than the whole State of Vermont, with its fourteen counties; and this was only a part of the wild lands of Maine. We are concerned now, however, about natural, not political limits. We were about eighty miles, as the bird flies, from Bangor, or one hundred and fifteen, as we had ridden, and walked, and paddled. We had to console ourselves with the reflection that this view was probably as good as that from the peak, as far as it went; and what were a mountain without its attendant clouds and mists? Like ourselves, neither Bailey nor Jackson had obtained a clear view from the summit.

Setting out on our return to the river, still at an early hour in the day, we decided to follow the course of the torrent, which we supposed to be Murch Brook, as long as it would not lead us too far out of our way. We thus traveled about four miles in the very torrent itself, continually crossing and recrossing it, leaping from rock to rock, and jumping with the stream down falls of seven or eight feet, or sometimes sliding down on our backs in a thin sheet of water. This ravine had been the scene of an extraordinary freshet in the spring, apparently accompanied by a slide from the mountain. It must have been filled with a stream of stones and water, at least twenty feet above the present level of the torrent. For a rod or two, on either side of its channel, the trees were barked and splintered up to their tops, the birches bent over, twisted, and sometimes finely split, like a stable-broom; some, a foot in diameter, snapped off, and whole clumps of trees bent over with the weight of rocks piled on them. In one place we noticed a rock, two or three feet in diameter, lodged nearly twenty feet high in the crotch of a tree. For the whole four miles, we saw but one rill emptying in, and the volume of water did not seem to be increased from the first. We traveled thus very rapidly with a downward impetus, and grew remarkably expert at leaping from rock to rock, for leap we must, and leap we did, whether there was any rock at the right distance or not. It was a pleasant picture when the foremost turned about and looked up the winding ravine, walled in with rocks and the green forest, to see, at intervals of a rod or two, a red-shirted or green-jacketed mountaineer against the white torrent, leaping down the channel with his pack on his back, or pausing upon a convenient rock in the midst of the torrent to mend a rent in his clothes, or unstrap the dipper at his belt to take a draught of the water. At one place we were startled by seeing, on a little sandy shelf by the side of the stream, the fresh print of a man's foot, and for a moment realized how Robinson Crusoe felt in a similar case; but at last we remembered that we had struck this stream on our way up, though we could not have told where, and one had descended into the ravine for a drink. The cool air above and the continual bathing of our bodies in mountain water, alternate foot, sitz, douche, and plunge baths, made this walk exceedingly refreshing, and we had traveled only a mile or two, after leaving the torrent, before every thread of our clothes was as dry as usual, owing perhaps to a peculiar quality in the atmosphere.

After leaving the torrent, being in doubt about our course, Tom threw down his pack at the foot of the loftiest spruce-tree at hand, and shinned up the bare trunk some twenty feet, and then climbed through the green tower, lost to our sight, until he held the topmost spray in his hand. McCauslin, in his younger days, had marched through the wilderness with a body of troops, under General Somebody, and with one other man did all the scouting and spying service. The General's word was, "Throw down the top of that tree," and there was no tree in the Maine woods so high that it did not lose its top in such a case. I have heard a story of two men being lost once in these woods, nearer to the settlements than this, who climbed the loftiest pine they could find, some six feet in diameter at the ground, from whose top they discovered a solitary clearing and its smoke. When at this height, some two hundred feet from the ground, one of them became dizzy, and fainted in his companion's arms, and the latter had to accomplish the descent with him, alternately fainting and reviving, as best he could. To Tom we cried, Where away does the summit bear? where the burnt lands? The last he could only conjecture; he descried, however, a little meadow and pond, lying probably in our course, which we concluded to steer for. On reaching this secluded meadow, we found fresh tracks of moose on the shore of the pond, and the water was still unsettled as if they had fled before us. A little farther, in a dense thicket, we seemed to be still on their trail. It was a small meadow, of a few acres, on the mountain side, concealed by the forest, and perhaps never seen by a white man before, where one would think that the moose might browse and bathe, and rest in peace. Pursuing this course, we soon reached the open land, which went sloping down some miles toward the Penobscot.

Perhaps I most fully realized that this was primeval, untamed, and forever untamable *Nature*, or whatever else men call it, while coming down this part of the mountain. We were passing over "Burnt Lands," burnt by lightning, perchance, though they showed no recent marks of fire, hardly so much as a charred stump, but looked rather like a natural pasture for the moose and deer, exceedingly wild and desolate, with occasional strips of timber crossing them, and low poplars springing up, and patches of blueberries here and there. I found myself traversing them familiarly, like some pasture run to waste, or partially reclaimed by man; but when I reflected what man, what brother or sister or kinsman of our race made it and claimed it, I expected the proprietor to rise up and dispute my passage. It is difficult to conceive of a region uninhabited by man. We habitually presume his presence and influence everywhere. And yet we have not seen pure Nature, unless we have seen her thus vast and drear and inhuman, though in the midst of cities. Nature was here something savage and awful, though beautiful. I looked with awe at the ground I trod on, to see what the Powers had made there, the form and fashion and material of their work. This was that Earth of which we have heard, made out of Chaos and Old Night. Here was no man's garden, but the unhandseled globe. It was not lawn, nor pasture, nor mead, nor woodland, nor lea, nor arable, nor waste land. It was the fresh and natural surface of the planet Earth, as it was made forever and ever, — to be the dwelling of man, we say, — so Nature made it, and man may use it if he can. Man was not to be associated with it. It was Matter, vast, terrific, — not his Mother Earth that we have heard of, not for him to tread on, or be buried in, — no, it were being too familiar even to let his bones lie there, — the home, this, of Necessity and Fate. There was clearly felt the presence of a force not bound to be kind to man. It was a place for heathenism and superstitious rites, — to be inhabited by men nearer of kin to the rocks and to wild animals than we. We walked over it with a certain awe, stopping, from time to time, to pick the blueberries which grew there, and had a smart and spicy taste. Perchance where *our* wild pines stand, and leaves lie on their forest floor, in Concord, there were once reapers, and husbandmen planted grain; but here not even the surface had been scarred by man, but it was a specimen of what God saw fit to make this world. What is it to be admitted to a museum, to see a myriad of particular things, compared with being shown some star's surface, some hard matter in its home! I stand in awe of my body, this matter to which I am bound has become so strange to me. I fear not spirits, ghosts, of which I am one,—*that* my body might,—but I fear bodies, I tremble to meet them. What is this Titan that has possession of me? Talk of mysteries! Think of our life in nature,—daily to be shown matter, to come in contact with it,—rocks, trees, wind on our cheeks! the *solid* earth! the *actual* world! the *common sense! Contact! Contact! Who* are we? *where* are we?

Erelong we recognized some rocks and other features in the landscape which we had purposely impressed on our memories, and, quickening our pace, by two o'clock we reached the batteau.[1] * * *

[1] The bears had not touched things on our possessions. They sometimes tear a batteau to pieces for the sake of the tar with which it is besmeared. [Thoreau's note]

From JOURNALS

OCT. 16, 1856

[An Offensive Fungus]

Found amid the sphagnum on the dry bank on the south side of the Turnpike, just below Everett's meadow, a rare and remarkable fungus, such as I have heard of but never seen before. The whole height six and three quarters inches, two thirds of it being buried in the sphagnum. It may be divided into three parts, pileus, stem, and base,—or scrotum, for it is a perfect phallus. One of those fungi named *impudicus*, I think.[1] In all respects a most disgusting object, yet very suggestive. It is hollow from top to bottom, the form of the hollow answering to that of the outside. The color of the outside white excepting the pileus, which is olive-colored and somewhat coarsely corrugated, with an

Journals (Boston: Houghton Mifflin, 1906).

[1] This is very similar to if not the same with that represented in Loudon's *Encyclopædia* and called "*Phallus impudicus*, Stinking Morel, very fetid." [Thoreau's note]

oblong mouth at tip about one eighth of an inch long, or, measuring
the white lips, half an inch. This cap is thin and white within, about
one and three eighths inches high by one and a half wide. The stem
(bare portion) is three inches long (tapering more rapidly than in the
drawing), horizontally viewed of an oval form. Longest diameter at base
one and a half inches, at top (on edge of pileus) fifteen sixteenths of an
inch. Short diameters in both cases about two thirds as much. It is a
delicate white cylinder of a finely honeycombed and crispy material
about three sixteenths of an inch thick, or more, the whole very straight
and regular. The base, or scrotum, is of an irregular bag form, about
one inch by two in the extremes, consisting of a thick trembling gelati-
nous mass surrounding the bottom of the stem and covered with a
tough white skin of a darker tint than the stem. The whole plant rather
frail and trembling. There was at first a very thin delicate white collar
(or *volva?*) about the base of the stem above the scrotum. It was as
offensive to the eye as to the scent, the cap rapidly melting and defiling
what it touched with a fetid, olivaceous, semiliquid matter. In an hour
or two the plant scented the whole house wherever placed, so that it
could not be endured. I was afraid to sleep in my chamber where it had
lain until the room had been well ventilated. It smelled like a dead
rat in the ceiling, in all the ceilings of the house. Pray, what was
Nature thinking of when she made this? She almost puts herself on a
level with those who draw in privies. The cap had at first a smooth
and almost dry surface, of a sort of olive slate-color, but the next day
this colored surface all melted out, leaving deep corrugations or
gills — rather honeycomb-like-cells — with a white bottom.

NOV. 1, 1858

[Here]

As the afternoons grow shorter, and the early evening drives us home
to complete our chores, we are reminded of the shortness of life, and
become more pensive, at least in this twilight of the year. We are
prompted to make haste and finish our work before the night comes. I
leaned over a rail in the twilight on the Walden road, waiting for the
evening mail to be distributed, when such thoughts visited me. I seemed
to recognize the November evening as a familiar thing come round
again, and yet I could hardly tell whether I had ever known it or only
divined it. The November twilights just begun! It appeared like a part of
a panorama at which I sat spectator, a part with which I was perfectly
familiar just coming into view, and I foresaw how it would look and roll

along, and prepared to be pleased. Just such a piece of art merely, though infinitely sweet and grand, did it appear to me, and just as little were any active duties required of me. We are independent on all that we see. The hangman whom I have *seen* cannot hang me. The earth which I have *seen* cannot bury me. Such doubleness and distance does sight prove. Only the rich and such as are troubled with ennui are implicated in the maze of phenomena. You cannot see anything until you are clear of it. The long railroad causeway through the meadows west of me, the still twilight in which hardly a cricket was heard,[1] the dark bank of clouds in the horizon long after sunset, the villagers crowding to the post-office, and the hastening home to supper by candle-light, had I not seen all this before! What new sweet was I to extract from it? Truly they mean that we shall learn our lesson well. Nature gets thumbed like an old spelling-book. The almshouse and Frederick were still as last November. I was no nearer, methinks, nor further off from my friends. Yet I sat the bench with perfect contentment, unwilling to exchange the familiar vision that was to be unrolled for any treasure or heaven that could be imagined. Sure to keep just so far apart in our orbits still, in obedience to the laws of attraction and repulsion, affording each other only steady but indispensable starlight. It was as if I was promised the greatest novelty the world has ever seen or shall see, though the utmost possible novelty would be the difference between me and myself a year ago. This alone encouraged me, and was my fuel for the approaching winter. That we may behold the panorama with this slight improvement or change, this is what we sustain life for with so much effort from year to year.

And yet there is no more tempting novelty than this new November. No going to Europe or another world is to be named with it. Give me the old familiar walk, post-office and all, with this ever new self, with this infinite expectation and faith, which does not know when it is beaten. We'll go nutting once more. We'll pluck the nut of the world, and crack it in the winter evenings. Theatres and all other sightseeing are puppet-shows in comparison. I will take another walk to the Cliff, another row on the river, another skate on the meadow, be out in the first snow, and associate with the winter birds. Here I am at home. In the bare and bleached crust of the earth I recognize my friend.

One actual Frederick that you know is worth a million only read of. Pray, am I altogether a bachelor, or am I a widower, that I should go away and leave my bride? This Morrow that is ever knocking with irresistible force at our door, there is no such guest as that. I will stay at home and receive company.

---

[1] Probably too cool for any these evenings; only in the afternoon. [Thoreau's note]

I want nothing new, if I can have but a tithe of the old secured to
me. I will spurn all wealth beside. Think of the consummate folly of
attempting to go away from *here!* When the constant endeavor should
be to get nearer and nearer *here.* Here are all the friends I ever had or
shall have, and as friendly as ever. Why, I never had any quarrel with a
friend but it was just as sweet as unanimity could be. I do not think we
budge an inch forward or backward in relation to our friends. How
many things can you go away from? They see the comet from the
northwest coast just as plainly as we do, and the same stars through its
tail. Take the shortest way round and stay at home. A man dwells in his
native valley like a corolla in its calyx, like an acorn in its cup. *Here,* of
course, is all that you love, all that you expect, all that you are. Here is
your bride elect, as close to you as she can be got. Here is all the best
and all the worst you can imagine. What more do you want? Bear here
away then! Foolish people imagine that what they imagine is somewhere else. That stuff is not made in any factory but their own.

NOV. 4, 1858

[Seeing]

If, about the last of October, you ascend any hill in the outskirts of
the town and look over the forest, you will see, amid the brown of other
oaks, which are now withered, and the green of the pines, the
bright-red tops or crescent of the scarlet oaks, very equally and thickly
distributed on all sides, even to the horizon. Complete trees standing
exposed on the edges of the forest, where you have never suspected
them, or their tops only in the recesses of the forest surface, or perhaps
towering above the surrounding trees, or reflecting a warm rose red
from the very edge of the horizon in favorable lights. All this you will
see, and much more, if you are prepared to see it,—if you *look* for it.
Otherwise, regular and universal as this phenomenon is, you will think
for threescore years and ten that all the wood is at this season sere and
brown. Objects are concealed from our view not so much because they
are out of the course of our visual ray (continued) as because there is
no intention of the mind and eye toward them. We do not realize how
far and widely, or how near and narrowly, we are to look. The greater
part of the phenomena of nature are for this reason concealed to us all
our lives. Here, too, as in political economy, the supply answers to the
demand. Nature does not cast pearls before swine. There is just as
much beauty visible to us in the landscape as we are prepared to appreciate,—not a grain more. The actual objects which one person will see

from a particular hilltop are just as different from those which another will see as the persons are different. The scarlet oak must, in a sense, be in your eye when you go forth. We cannot see anything until we are possessed with the idea of it, and then we can hardly see anything else. In my botanical rambles I find that first the idea, or image, of a plant occupies my thoughts, though it may at first seem very foreign to this locality, and for some weeks or months I go thinking of it and expecting it unconsciously, and at length I surely see it, and it is henceforth an actual neighbor of mine. This is the history of my finding a score or more of rare plants which I could name.

Take one of our selectmen and put him on the highest hill in the township, and tell him to look! What probably, would he see? What would he *select* to look at? Sharpening his sight to the utmost, and putting on the glasses that suited him best, aye, using a spy-glass if he liked, straining his optic nerve to its utmost, and making a full report. Of course, he would see a Brocken spectre of himself. Now take Julius Cæsar, or Emanuel Swedenborg, or a Fiji-Islander, and set him up there! Let them compare notes afterward. Would it appear that they had enjoyed the same prospect? For aught we know, as strange a man as any of these is always at our elbows. It does not appear that anybody saw Shakespeare when he was about in England looking off, but only some of his raiment.

Why, it takes a sharpshooter to bring down even such trivial game as snipes and woodcocks; he must take very particular aim, and know what he is aiming at. He would stand a very small chance if he fired at random into the sky, being told that snipes were flying there. And so it is with him that shoots at beauty. Not till the sky falls will he catch larks, unless he is a trained sportsman. He will not bag any if he does not already know its seasons and haunts and the color of its wing,—if he has not dreamed of it, so that he can *anticipate* it; then, indeed, he flushes it at every step, shoots double and on the wing, with both barrels, even in corn-fields. The sportsman trains himself, dresses, and watches unweariedly, and loads and primes for his particular game. He prays for it, and so he gets it. After due and long preparation, schooling his eye and hand, dreaming awake and asleep, with gun and paddle and boat, he goes out after meadow-hens,—which most of his townsmen never saw nor dreamed of,—paddles for miles against a head wind, and therefore he gets them. He had them half-way into his bag when he started, and has only to shove them down. The fisherman, too, dreams of fish, till he can almost catch them in his sink-spout. The hen scratches, and finds her food right under where she stands; but such is not the way with the hawk.

The true sportsman can shoot you almost any of his game from his windows. It comes and perches at last on the barrel of his gun; but the rest of the world never see it, with the feathers on. He will keep himself supplied by firing up his chimney. The geese fly exactly under his zenith, and honk when they get there. Twenty musquash have the refusal of each one of his traps before it is empty.

*   *   *

FEB 16, 1860

[Topsell's Gesner]

We cannot spare the very lively and lifelike descriptions of some of the old naturalists. They sympathize with the creatures which they describe. Edward Topsell in his translation of Conrad Gesner, in 1607, called "The History of Four-footed Beasts," says of the antelopes that "they are bred in India and Syria, near the river Euphrates," and then—which enables you to realize the living creature and its habitat—he adds, "and delight much to drink of the cold water thereof." The beasts which most modern naturalists describe do not *delight* in anything, and their water is neither hot nor cold. Reading the above makes you want to go and drink of the Euphrates yourself, if it is warm weather. I do not know how much of his spirit he owes to Gesner, but he proceeds in his translation to say that "they have horns growing forth of the crown of their head, which are very long and sharp; so that Alexander affirmed they pierced through the shields of his soldiers, and fought with them very irefully: at which time his company slew as he travelled to India, eight thousand five hundred and fifty, which great slaughter may be the occasion why they are so rare and seldom seen to this day."

Now here *something* is described at any rate; it is a real account, whether of a real animal or not. You can plainly see the horns which "grew forth" from their crowns, and how well that word "irefully" describes a beast's fighting! And then for the number which Alexander's men slew "as he travelled to India,"—and what a travelling was that, my hearers!—eight thousand five hundred and fifty, just the number you would have guessed after the thousands were given, and [an] easy one to remember too. He goes on to say that "their horns are great and made like a saw, and they with them can cut asunder the branches of osier or small trees, whereby it cometh to pass that many times their necks are taken in the twists of the falling boughs, whereat the beast with repining cry, bewrayeth himself to the hunters, and so is taken." The artist too has done his part equally well, for you are presented with a drawing of the beast with serrated horns, the tail of a lion, a cheek tooth (canine?) as big as a boar's, a stout front, and an exceedingly "ireful" look, as if he were facing all Alexander's army.

Though some beasts are described in this book which have no existence as I can learn but in the imagination of the writers, they really have an existence there, which is saying not a little, for most of our modern authors have not imagined the actual beasts which they presume to describe. The very frontispiece is a figure of "the gorgon," which looks sufficiently like a hungry beast covered with scales, which you may have dreamed of, apparently just fallen on the track of you, the reader, and snuffing the odor with greediness.

These men had an adequate idea of a beast, or what a beast should be, a very *bellua* (the translator makes the word *bestia* to be "*a vastando*"); and they will describe and will draw you a cat with four strokes, more beastly or beast-like to look at than Mr. Ruskin's favorite artist draws a tiger. They had an adequate idea of the wildness of beasts and of men, and in their descriptions and drawings they did not always fail when they *surpassed* nature.

I think that the most important requisite in describing an animal, is to be sure and give its character and spirit, for in that you have, without error, the sum and effect of all its parts, known and unknown. You must tell what it is to man. Surely the most important part of an animal is its *anima*, its vital spirit, on which is based its character and all the peculiarities by which it most concerns us. Yet most scientific books which treat of animals leave this out altogether, and what they describe are as it were phenomena of dead matter. What is most interesting in a dog, for example, is his attachment to his master, his intelligence, courage, and the like, and not his anatomical structure or even many habits which affect us less.

If you have undertaken to write the biography of an animal, you will have to present to us the living creature, *i.e.*, a result which no man can understand, but only in his degree report the impression made on him.

Science in many departments of natural history does not pretend to go beyond the shell; *i.e.*, it does not get to animated nature at all. A history of animated nature must itself be animated.

The ancients, one would say, with their gorgons, sphinxes, satyrs, mantichora, etc., could imagine more than existed, while the moderns cannot imagine so much as exists.

In describing brutes, as in describing men, we shall naturally dwell most on those particulars in which they are most like ourselves, — in which we have most sympathy with them.

We are as often injured as benefited by our systems, for, to speak the truth, no human system is a true one, and a name is at most a mere convenience and carries no information with it. As soon as I begin to be aware of the life of any creature, I at once forget its name. To know the names of creatures is only a convenience to us at first, but so soon as we have learned to distinguish them, the sooner we forget their names the better, so far as any true appreciation of them is concerned. I think, therefore, that the best and most harmless names are those which are an imitation of the voice or note of an animal, or the most poetic ones. But the name adheres only to the accepted and conventional bird or quadruped, never an instant to the real one. There is always something ridiculous in the name of a great man, — as if he were named John Smith. The name is convenient in communicating with others, but it is not to be remembered when I communicate with myself.

If you look over a list of medicinal recipes in vogue in the last century, how foolish and useless they are seen to be! And yet we use equally absurd ones with faith to-day.

When the ancients had not found an animal wild and strange enough to suit them, they created one by the mingled [traits] of the most savage already known, — as hyenas, lionesses, pards, panthers, etc., — one with another. Their beasts were thus of wildness and savageness all compact, and more *ferine* and *terrible* than any of an unmixed breed could be. They allowed nature great license in these directions. The most strange and fearful beasts were by them supposed to be the offspring of two different savage kinds. So fertile were their imaginations, and such fertility did they assign to nature. In the modern account the fabulous part will be omitted, it is true, but the portrait of the real and living creature also.

The old writers have left a more lively and lifelike account of the gorgon than modern writers give us of real animals.

* * *

MARCH 22, 1861

[Tenacity of Life]

When we consider how soon some plants which spread rapidly, by seeds or roots, would cover an area equal to the surface of the globe, how soon some species of trees, as the white willow, for instance, would equal in mass the earth itself; if all their seeds became full-grown trees, how soon some fishes would fill the ocean if all their ova became full-grown fishes, we are tempted to say that every organism, whether animal or vegetable, is contending for the possession of the planet, and, if any one were sufficiently favored, supposing it still possible to grow, as at first, it would at length convert the entire mass of the globe into its own substance. Nature opposes to this many obstacles, as climate, myriads of brute and also human foes, an of competitors which may preoccupy the ground. Each suggests an immense and wonderful greediness and tenactivity of life (I speak of the species, not individual), as if bent on taking entire possession of the globe wherever the climate and soil will permit. And each prevails as much as it does, because of the ample preparations it has made for the contest, — it has secured a myriad chances, — because it never depends on spontaneous generation to save it.

1861

[A Young Kitten]

The kitten can already spit at a fortnight old, and it can mew from the first, though it often makes the motion of mewing without uttering any sound.

The cat about to bring forth seeks out some dark and secret place for the purpose, not frequented by other cats.

The kitten's ears are at first nearly concealed in the fur, and at a fortnight old they are mere broad-based triangles with a side foremost. But the old cat is ears for them at present, and comes running hastily to their aid when she hears them mew and licks them into contentment again. Even at three weeks the kitten cannot fairly walk, but only creeps feebly with outspread legs. But thenceforth its ears visibly though gradually lift and sharpen themselves.

At three weeks old the kitten begins to walk in a staggering and creeping manner and even to play a little with its mother, and, if you put your ear close, you may hear it purr. It is remarkable that it will not wander far from the dark corner where the cat has left it, but will instinctively find its way back to it, probably by the sense of touch, and will rest nowhere else. Also it is careful not to venture too near the edge of a precipice, and its claws are ever extended to save itself in such places. It washes itself somewhat, and assumes many of the attitudes of an old cat at this age. By the disproportionate size of its feet and head and legs now it reminds you [of] a lion.

I saw it scratch its ear to-day, probably for the first time; yet it lifted one of its hind legs and scratched its ear as effectually as an old cat does. So this is instinctive, and you may say that, when a kitten's ear first itches, Providence comes to the rescue and lifts its hind leg for it. You would say that this little creature was as perfectly protected by its instinct in its infancy as an old man can be by his wisdom. I observed when she first noticed the figures on the carpet, and also put up her paws to touch or play with surfaces a foot off. By the same instinct that they find the mother's teat before they can see they scratch their ears and guard against falling.

CHARLES KINGSLEY
1819–1875

Charles Kingsley was one of those Victorians who achieved distinction in several careers while also writing prodigiously. Among his important positions were those of chaplain to Queen Victoria and professor of modern history at Cambridge. For most of his life, however, Kingsley was a parish priest and canon of the Church of England. Natural history was his recreation, as it was for so many middleclass Englishmen of his day. Glaucus; or, The Wonders of the Shore (1855) reflects the influence of the marine biologist Philip Henry Gosse, as well as Kingsley's own residence at the Devon seaport of Torquay. Prose Idylls: New and Old (1873) collected his naturalist sketches, a number of which gravitated to the fen country of eastern England where he grew up. Among his novels, Westward Ho! (1855), Hypatia (1853), Two Years (1857), and the children's classic The Water Babies (1863) were especially noted for their vivid landscape descriptions, ranging from South America to the Egyptian desert to north Devon.

From Glaucus; or, The Wonders of the Shore

[A Repellent Worm]

*   *   * And now, worshipper of final causes and the mere useful in nature, answer but one question,—Why this prodigal variety? All these Nudibranchs live in much the same way: why would not the same

Glaucus; or, The Wonders of the Shore (Boston: Ticknor and Fields, 1855).

mould have done for them all? And why, again, (for we must push the argument a little further,) why have not all the butterflies, at least all who feed on the same plant, the same markings? Of all unfathomable triumphs of design, (we can only express ourselves thus, for honest induction, as Paley so well teaches, allows us to ascribe such results only to the design of some personal will and mind,) what surpasses that by which the scales on a butterfly's wing are arranged to produce a certain pattern of artistic beauty beyond all painter's skill? What a waste of power, on any utilitarian theory of nature! And once more, why are those strange microscopic atomies, the Diatomaceæ and Infusoria, which fill every stagnant pool; which fringe every branch of sea-weed; which form banks hundreds of miles long on the Arctic sea-floor, and the strata of whole moorlands; which pervade in millions the mass of every iceberg, and float aloft in countless swarms amid the clouds of the volcanic dust;—why are their tiny shells of flint as fantastically various in their quaint mathematical symmetry, as they are countless beyond the wildest dreams of the Poet? Mystery inexplicable on the conceited notion which, making man forsooth the centre of the universe, dares to believe that this variety of forms has existed for countless ages in abysmal sea-depths and untrodden forests, only that some few individuals of the Western races might, in these latter days, at last discover and admire a corner here and there of the boundless realms of beauty. Inexplicable, truly, if man be the centre and the object of their existence; explicable enough to him who believes that God has created all things for Himself, and rejoices in His own handiwork, and that the material universe is, as the wise man says, "A platform whereon His Eternal Spirit sports and makes melody." Of all the blessings which the study of nature brings to the patient observer, let none, perhaps, be classed higher than this: that the further he enters into those fairy gardens of life and birth, which Spenser saw and described in his great poem, the more he learns the awful and yet most comfortable truth, that they do not belong to him, but to One greater, wiser, lovelier than he; and as he stands, silent with awe, amid the pomp of Nature's ever-busy rest, hears, as of old. "The Word of the Lord God walking among the trees of the garden in the cool of the day."

One sight more, and we have done. I had something to say, had time permitted, on the ludicrous element which appears here and there in nature. There are animals, like monkeys and crabs, which seem made to be laughed at; by those at least who possess that most indefinable of faculties, the sense of the ridiculous. As long as man possesses muscles especially formed to enable him to laugh, we have no right to suppose (with some) that laughter is an accident of our fallen nature; or to find (with others) the primary cause of the ridiculous in the perception of unfitness or disharmony. And yet we shrink (whether rightly or wrongly, we can hardly tell) from attributing a sense of the ludicrous to the Creator of these forms. It may be a weakness on my part; at least I will hope it is a reverent one: but till we can find something corresponding to what we conceive of the Divine Mind in any class of phenomena, it is perhaps better not to talk about them at all, but observe a stoic "epoché," waiting for more light, and yet confessing that our own laughter is uncontrollable, and therefore we hope not unworthy of us, at many a strange creature and strange doing which we meet, from the highest ape to the lowest polype.

But, in the meanwhile, there are animals in which results so strange, fantastic, even seemingly horrible, are produced, that fallen man may be pardoned if he shrinks from them in disgust. That, at least, must be a consequence of our own wrong state; for everything is beautiful and perfect in its place. It may be answered, 'Yes, in its place; but its place is not yours. You had no business to look at it, and must pay the penalty for intermeddling.' I doubt that answer: for surely, if man have liberty to do anything, he has liberty to search out freely his Heavenly Father's works; and yet every one seems to have his antipathic animal, and I know one bred from his childhood to zoology by land and sea, and bold in asserting, and honest in feeling, that all without exception is beautiful, who yet cannot, after handling, and petting, and admiring all day long every uncouth and venomous beast, avoid a paroxysm of horror at the sight of the common house-spider. At all events, whether we were intruding or not, in turning this stone, we must pay a fine for having done so; for there lies an animal, as foul and monstrous to the eye as 'hydra, gorgon, or chimera dire,' and yet so wondrously fitted for its work, that we must needs endure for our own instruction to handle and look at it. Its name I know not (though it lurks here under every stone), and should be glad to know. It seems some very 'low' Ascarid or Planarian worm. You see it? That black, slimy, knotted lump among the gravel, small enough to be taken up in a dessert-spoon. Look now, as it is raised and its coils drawn out. Three feet! Six—nine at least, with a capability of seemingly endless expansion; a slimy tape of living caoutchouc, some eighth of an inch in diameter, a dark chocolate-black, with paler longitudinal lines. It is alive? It hangs helpless and motionless, a mere velvet string across the hand. Ask the neighbouring Annelids and the fry of the rock fishes, or put it into a vase at home, and see. It lies motionless, trailing itself among the gravel; you cannot tell where it begins or ends; it may be a strip of dead sea-weed, *Himanthalia lorea*, perhaps, or *Chorda filum*; or even a tarred string. So thinks the little fish who plays over and over it, till he touches at last what is too surely a head. In an instant a bell-shaped sucker mouth has fastened to its side. In another instant, from one lip, a concave double proboscis, just like a tapir's (another instance of the repetition of forms), has clasped him like a finger, and now begins the struggle; but in vain. He is being 'played,' with such a fishing-rod as the skill of a Wilson or a Stoddart never could invent; a living line, with elasticity beyond that of the most delicate fly-rod, which follows every lunge, shortening and lengthening, slipping and twining round every piece of gravel and stem of sea-weed, with a tiring drag such as no Highland wrist or step could ever bring to bear on salmon or trout. The victim is tired now; and slowly, yet dexterously, his blind assailant is feeling and shifting along his side, till he reaches one end of him; and then the black lips expand, and slowly and surely the curved finger begins packing him end foremost down into the gullet, where he sinks, inch by inch, till the swelling which marks his place is lost among the coils, and he is probably macerated into a pulp long before he has reached the opposite extremity of his cave of doom. Once safe down, the black murderer contracts again into a knotted heap, and lies like a boa with a stag inside him, motionless and blest.

WALT WHITMAN
1819–1892

In Leaves of Grass *(1855) Whitman threw open the windows of American poetry. His inclusiveness, registered both in an expansive sense of poetic line and in his enthusiastic cataloguing of persons and sensations, also served the nature writers' purpose, by bringing a wider range of life into our literature, into our consciousness. John Burroughs, one of the most influential nature writers of his generation, considered himself Whitman's protégé and disciple during their years together in Washington, D.C. The two often went on bird-watching and flower-viewing walks together. In the collection of Whitman's prose writings called* Specimen Days *(1882), he shows his alertness to the weather and the life of sky and fields, as well as to the varieties of human experience.*

From SPECIMEN DAYS AND COLLECT

THE WHITE HOUSE BY MOONLIGHT

FEBRUARY 24TH, 1863

A spell of fine soft weather. I wander about a good deal, sometimes at night under the moon. Tonight took a long look at the President's house. The white portico—the palacelike, tall, round columns, spotless as snow—the walls also—the tender and soft moonlight, flooding the pale marble, and making peculiar faint languishing shades, not shadows—everywhere a soft, transparent, hazy, thin, blue moon lace, hanging in the air—the brilliant and extra-plentiful clusters of gas, on and around the façade, columns, portico, etc.—everything so white, so marbly pure and dazzling, yet soft—the White House of future poems, and of dreams and dramas, there in the soft and copious moon—the gorgeous front, in the trees, under the lustrous flooding moon, full of reality, full of illusion—the forms of the trees, leafless, silent, in trunk and myriad angles of branches, under the stars and sky—the White House of the land, and of beauty and night—sentries at the gates, and by the portico, silent, pacing there in blue overcoats—stopping you not at all, but eyeing you with sharp eyes, whichever way you move.

* * *

BUMBLEBEES

May-month—month of swarming, singing, mating birds—the bumblebee month—month of the flowering lilac (and then my own birth month). As I jot this paragraph, I am out just after sunrise, and down toward the creek. The lights, perfumes, melodies—the bluebirds, grass-birds, and robins, in every direction—the noisy, vocal, natural concert. For undertones, a neighboring woodpecker tapping his tree, and the distant clarion of chanticleer. Then the fresh earth smells—the colors, the delicate drabs and thin blues of the perspective. The bright green of the grass has received an added tinge from the last two days' mildness and moisture. How the sun silently mounts in the broad clear sky, on his day's journey! How the warm beams bathe all, and come streaming kissingly and almost hot on my face. A while since the croaking of the pond frogs and the first white of the dogwood blossoms. Now the golden dandelions in endless profusion, spotting the ground every-

Specimen Days and Collect (Philadelphia: Rees Welsh, 1882).

where. The white cherry and pear blows—the wild violets, with their blue eyes looking up and saluting my feet, as I saunter the wood edge—the rosy blush of budding apple trees—the light-clear emerald hue of the wheat-fields—the darker green of the rye—a warm elasticity pervading the air—the cedar bushes profusely decked with their little brown apples—the summer fully awakening—the convocation of black birds, garrulous flocks of them, gathering on some tree, and making the hour and place noisy as I sit near.

Later. Nature marches in procession, in sections, like the corps of an army. All have done much for me, and still do. But for the last two days it has been the great wild bee, the humblebee, or "bumble," as the children call him. As I walk, or hobble, from the farmhouse down to the creek, I traverse the before-mentioned lane, fenced by old rails, with many splits, splinters, breaks, holes, etc., the choice habitat of those crooning, hairy insects. Up and down and by and between these rails, they swarm and dart and fly in countless myriads. As I wend slowly along, I am often accompanied with a moving cloud of them. They play a leading part in my morning, mid-day, or sunset rambles, and often dominate the landscape in a way I never before thought of—fill the long lane, not by scores or hundreds only, but by thousands. Large and vivacious and swift, with wonderful momentum and a loud swelling perpetual hum, varied now and then by something almost like a shriek, they dart to and fro, in rapid flashes, chasing each other, and (little things as they are) conveying to me a new and pronounced sense of strength, beauty, vitality, and movement. Are they in their mating season? Or what is the meaning of this plentitude, swiftness, eagerness, display? As I walked, I thought I was followed by a particular swarm, but upon observation I saw that it was a rapid succession of changing swarms, one after another. As I write, I am seated under a big wild cherry tree—the warm day tempered by partial clouds and a fresh breeze, neither too heavy nor light—and here I sit long and long, enveloped in the deep musical drone of these bees, flitting, balancing, darting to and fro about me by hundreds—big fellows with light yellow jackets, great glistening swelling bodies, stumpy heads and gauzy wings—humming their perpetual rich mellow boom. (Is there not a hint in it for a musical composition, of which it should be the background? Some bumblebee symphony?) How it all nourishes, lulls me, in the way most needed; the open air, the rye fields, the apple orchards. The last two days have been faultless in sun, breeze, temperature, and everything; never two more perfect days, and I have enjoyed them wonderfully. My health is somewhat better, and my spirit at peace. (Yet the anniversary of the saddest loss and sorrow of my life is close at hand.) Another jotting, another perfect day: forenoon, from 7 to 9, two hours enveloped in sound of bumble-bees and bird music. Down in the apple trees and in a neighboring cedar were three or four russet-backed thrushes, each singing his best, and roulading in ways I never heard surpassed. Two hours I abandon myself to hearing them, and indolently absorbing the scene. Almost every bird I notice has a special time in the year—sometimes limited to a few days—when it sings its best; and now is the period of these russet-backs. Meanwhile, up and down the lane, the darting, droning, musical bum-blebees. A great swarm again for my entourage as I return home, mov-ing along with me as before. As I write this, two or three weeks later, I am sitting near the brook under a tulip tree, 70 feet high, thick with the fresh verdure of its young maturity—a beautiful object—every branch, every leaf perfect. From top to bottom, seeking the sweet juice in the blossoms, it swarms with myriads of these wild bees, whose loud and steady humming makes an undertone to the whole, and to my mood and the hour. *   *   *

CEDAR APPLES

As I journeyed today in a light wagon ten or twelve miles through the country, nothing pleased me more, in their homely beauty and novelty (I had either never seen the little things to such advantage, or had never noticed them before) than that peculiar fruit, with its profuse clear-yellow dangles of inch-long silk or yarn, in boundless profusion spotting the dark-green cedar bushes—contrasting well with their bronze tufts—the flossy shreds covering the knobs all over, like a shock of wild hair on elfin pates. On my ramble afterward down by the creek I plucked one from its bush, and shall keep it. These cedar apples last only a little while, however, and soon crumble and fade.

*   *   *

SEASHORE FANCIES

Even as a boy, I had the fancy, the wish, to write a piece, perhaps a poem, about the seashore—that suggesting, dividing line, contact, junction, the solid marrying the liquid—that curious, lurking some-thing (as doubtless every objective form finally becomes to the subjec-tive spirit) which means far more than its mere first sight, grand as that is—blending the real and ideal, and each made portion of the other. Hours, days, in my Long Island youth and early manhood, I haunted the shores of Rockaway or Coney Island, or away east to the Hamptons or Montauk. Once, at the latter place (by the old lighthouse, nothing but sea-tossings in sight in every direction as far as the eye could reach), I remember well, I felt that I must one day write a book expressing this liquid, mystic theme. Afterward, I recollect, how it came to me that instead of any special lyrical or epical or literary attempt, the seashore should be an invisible *influence*, a pervading gauge and tally for me, in my composition. (Let me give a hint here to young writers. I am not sure but I have unwittingly followed out the same rule with other powers besides sea and shores—avoiding them, in the way of any dead set at poetizing them, as too big for formal handling—quite satisfied if I could indirectly show that we have met and fused, even if only once, but enough—that we have really absorbed each other and understand each other.) There is a dream, a picture, that for years at intervals (sometimes quite long ones, but surely again, in time) has come noiselessly up before me, and I really believe, fiction as it is, has entered largely into my practical life—certainly into my writings, and shaped and colored them. It is nothing more or less than a stretch of interminable white-brown sand, hard and smooth and broad, with the ocean perpetually, grandly, rolling in upon it, with slow-measured sweep, with rustle and hiss and foam, and many a thump as of low bass drums. This scene, this picture, I say, has risen before me at times for years. Sometimes I wake at night and can hear and see it plainly.

THOUGHTS UNDER AN OAK—A DREAM

JUNE 2, 1878

This is the fourth day of a dark northeast storm, wind and rain. Day before yesterday was my birthday. I have now entered on my 60th year. Every day of the storm, protected by overshoes and a waterproof blanket, I regularly come down to the pond, and ensconce myself under the lee of the great oak; I am here now writing these lines. The dark smoke-colored clouds roll in furious silence athwart the sky; the soft green leaves dangle all round me; the wind steadily keeps up its hoarse, soothing music over my head—Nature's mighty whisper. Seated here in solitude I have been musing over my life—connecting events, dates, as links of a chain, neither sadly nor cheerily, but somehow, today here under the oak, in the rain, in an unusually matter-of-fact spirit. But my great oak—sturdy, vital, green—five feet thick at the butt: I sit a great deal near or under him. Then the tulip tree near by—the Apollo of the woods—tall and graceful, yet robust and sinewy, inimitable in hang of foliage and throwing-out of limb; as if the beauteous, vital, leafy creature could walk, if it only would. (I had a sort of dream-trance the other day, in which I saw my favorite trees step out and promenade up, down and around, very curiously—with a whisper from one, leaning down as he passed me, *"We do all this on the present occasion, exceptionally, just for you."*)

* * *

AMERICA'S CHARACTERISTIC LANDSCAPE

Speaking generally as to the capacity and sure future destiny of that plain and prairie area (larger than any European kingdom) it is the inexhaustible land of wheat, maize, wool, flax, coal, iron, beef and pork, butter and cheese, apples and grapes—land of ten million virgin farms—to the eye at present wild and unproductive—yet experts say that upon it when irrigated may easily be grown enough wheat to feed the world. Then as to scenery (giving my own thought and feeling), while I know the standard claim is that Yosemite, Niagara Falls, the upper Yellowstone and the like, afford the greatest natural shows, I am not so sure but the prairies and plains, while less stunning at first sight, last longer, fill the esthetic sense fuller, precede all the rest, and make North America's characteristic landscape. Indeed through the whole of this journey, with all its shows and varieties, what most impressed me, and will longest remain with me, are these same prairies. Day after day, and night after night, to my eyes, to all my senses—the esthetic one most of all—they silently and broadly unfolded. Even their simplest statistics are sublime.

BEETHOVEN'S SEPTET

FEB. 11, 1880

At a good concert tonight in the foyer of the Opera House, Philadelphia—the band a small but first-rate one. Never did music more sink into and soothe and fill me—never so prove its soul-rousing power, its impossibility of statement. Especially in the rendering of one of Beethoven's master septets by the well-chosen and perfectly combined instruments (violins, viola, clarionet, horn, 'cello, and contrabass) was I carried away, seeing, absorbing many wonders. Dainty abandon, sometimes as if Nature laughing on a hillside in the sunshine; serious and firm monotonies, as of winds; a horn sounding through the tangle of the forest, and the dying echoes; soothing floating of waves, but presently rising in surges, angrily lashing, muttering, heavy; piercing peals of laughter, for interstices; now and then weird, as Nature herself is in certain moods—but mainly spontaneous, easy, careless—often the sentiment of the postures of naked children playing or sleeping. It did me good even to watch the violinists drawing their bows so masterly—every motion a study. I allowed myself, as I sometimes do, to wander out of myself. The conceit came to me of a copious grove of singing birds, and in their midst a simple harmonic duo, two human souls, steadily asserting their own pensiveness, joyousness.

BIRDS—AND A CAUTION

MAY 14, 1888

Home again; down temporarily in the Jersey woods. Between 8 and 9 A.M. a full concert of birds, from different quarters, in keeping with the fresh scent, the peace, the naturalness all around me. I am lately noticing the russet-back, size of the robin or a trifle less, light breast and shoulders, with irregular dark stripes—tail long—sits hunched up by the hour these days, top of a tall bush, or some tree, singing blithely. I often get near and listen as he seems tame; I like to watch the working of his bill and throat, the quaint sidle of his body, and flex of his long tail. I hear the woodpecker, and night and early morning the shuttle of the whippoorwill—noons, the gurgle of thrush delicious, and *meo-o-ow* of the catbird. Many I cannot name; but I do not very particularly seek information. (You must not know too much, or be too precise or scientific about birds and trees and flowers and water craft; a certain free margin, and even vagueness—perhaps ignorance, credulity—helps your enjoyment of these things, and of the sentiment of feathered, wooded, river, or marine Nature generally. I repeat it—don't want to know too exactly, or the reasons why. My own notes have been written offhand in the latitude of middle New Jersey. Though they describe what I saw—what appeared to me—I dare say the expert ornithologist, botanist, or entomologist will detect more than one slip in them.)

JOHN WESLEY POWELL
1834–1902

*Like Audubon and Muir, Powell lives in American literature as a charac-
ter as well as an author. The one-armed Civil War hero, standing in the
prow of a boat plunging through the Colorado's rapids, represents the
bravery of explorers all along the frontier and the ambition of scientists
eager to map and analyze the Republic's new territories. His* Exploration
of The Colorado River of the West and Its Tributaries *(1875) grew out of
a series of articles for* Scribner's Magazine, *in which Powell synthesized
his expeditions of 1869 and 1871. These articles both introduced many
readers to the marvels of the Grand Canyon and addressed the question
of America's destiny, now that the War was past. What country was it
that had been held together at such cost? As had also occurred after the
Revolution, Americans turned to description of our sublime landscape in
defining the national pride available to us as a new people in a new
place. In addition to becoming director of the U.S. Geological Survey,
Major Powell was also to serve as America's director of the Bureau of
Ethnology. His sympathetic interest in native peoples of the Colorado
region is often apparent in his* Exploration.

From EXPLORATION OF THE COLORADO RIVER
OF THE WEST AND ITS TRIBUTARIES

FROM THE LITTLE COLORADO TO THE FOOT OF
THE GRAND CANYON

AUGUST 13.

We are now ready to start on our way down the Great Unknown. Our
boats, tied to a common stake, chafe each other as they are tossed by

Exploration of the Colorado River of the West and Its Tributaries (Washington, D.C.: U.S.
Government, 1875).

the fretful river. They ride high and buoyant, for their loads are lighter than we could desire. We have but a month's rations remaining. The flour has been resifted through the mosquito-net sieve; the spoiled bacon has been dried and the worst of it boiled; the few pounds of dried apples have been spread in the sun and reshrunken to their normal bulk. The sugar has all melted and gone on its way down the river. But we have a large sack of coffee. The lightening of the boats has this advantage: they will ride the waves better and we shall have but little to carry when we make a portage.

We are three quarters of a mile in the depths of the earth, and the great river shrinks into insignificance as it dashes its angry waves against the walls and cliffs that rise to the world above; the waves are but puny ripples, and we but pigmies, running up and down the sands or lost among the boulders.

We have an unknown distance yet to run, an unknown river to explore. What falls there are, we know not; what rocks beset the channel, we know not; what walls rise over the river, we know not. Ah, well! we may conjecture many things. The men talk as cheerfully as ever; jests are bandied about freely this morning; but to me the cheer is somber and the jests are ghastly.

With some eagerness and some anxiety and some misgiving we enter the canyon below and are carried along by the swift water through walls which rise form its very edge. They have the same structure that we noticed yesterday—tiers of irregular shelves below, and, above these, steep slopes to the foot of marble cliffs. We run six miles in a little more than half an hour and emerge into a more open portion of the canyon, where high hills and ledges of rock intervene between the river and the distant walls. Just at the head of this open place the river runs across a dike; that is, a fissure in the rocks, open to depths below, was filled with eruptive matter, and this on cooling was harder than the rocks through which the crevice was made, and when these were washed away the harder volcanic matter remained as a wall, and the river has cut a gateway through it several hundred feet high and as many wide. As it crosses the wall, there is a fall below and a bad rapid, filled with boulders of trap; so we stop to make a portage. Then on we go, gliding by hills and ledges, with distant walls in view; sweeping past sharp angles of rock; stopping at a few points to examine rapids, which we find can be run, until we have made another five miles, when we land for dinner.

Then we let down with lines over a long rapid and start again. Once more the walls close in, and we find ourselves in a narrow gorge, the water again filling the channel and being very swift. With great care and constant watchfulness we proceed, making about four miles this afternoon, and camp in a cave.

AUGUST 14.

At daybreak we walk down the bank of the river, on a little sandy beach, to take a view of a new feature in the canyon. Heretofore hard rocks have given us bad river; soft rocks, smooth water; and a series of rocks harder than any we have experienced sets in. The river enters the gneiss! We can see but a little way into the granite gorge, but it looks threatening.

After breakfast we enter on the waves. At the very introduction it inspires awe. The canyon is narrower than we have ever before seen it; the water is swifter; there are but few broken rocks in the channel; but the walls are set, on either side, with pinnacles and crags; and sharp, angular buttresses, bristling with wind- and wave-polished spires, extend far out into the river.

Ledges of rock jut into the stream, their tops sometimes just below the surface, sometimes rising a few or many feet above; and island ledges and island pinnacles and island towers break the swift course of the stream into chutes and eddies and whirlpools. We soon reach a place where a creek comes in from the left, and just below, the channel is choked with boulders, which have washed down this lateral canyon and formed a dam, over which there is a fall of 30 or 40 feet; but on the boulders foothold can be had, and we make a portage. Three more such dams are found. Over one we make a portage; at the other two are chutes through which we can run.

As we proceed the granite rises higher, until nearly a thousand feet of the lower part of the walls are composed of this rock.

About eleven o'clock we hear a great roar ahead, and approach it very cautiously. The sound grows louder and louder as we run, and at last we find ourselves above a long, broken fall, with ledges and pinnacles of rock obstructing the river. There is a descent of perhaps 75 or 80 feet in a third of a mile, and the rushing waters break into great waves on the rocks, and lash themselves into a mad, white foam. We can land just above, but there is no foothold on either side by which we can make a portage. It is nearly a thousand feet to the top of the granite; so it will be impossible to carry our boats around, though we can climb to the summit up a side gulch and, passing along a mile or two, descend to the river. This we find on examination; but such a portage would be impracticable for us, and we must run the rapid or abandon the river. There is no hesitation. We step into our boats, push off, and away we go, first on smooth but swift water, then we strike a glassy wave and ride to its top, down again into the trough, up again on a higher wave, and down and up on waves higher and still higher until we strike one just as it curls back, and a breaker rolls over our little boat. Still on we speed, shooting past projecting rocks, till the little boat is caught in a whirlpool and spun round several times. At last we pull out again into the stream. And now the other boats have passed us. The open compartment of the "Emma Dean" is filled with water and every breaker rolls over us. Hurled back from a rock, now on this side, now on that, we are carried into an eddy, in which we struggle for a few minutes, and are then out again, the breakers still rolling over us. Our boat is unmanageable, but she cannot sink, and we drift down another hundred yards through breakers—how, we scarcely know. We find the other boats have turned into an eddy at the foot of the fall and are waiting to catch us as we come, for the men have seen that our boat is swamped. They push out as we come near and pull us in against the wall. Our boat bailed, on we go again.

The walls now are more than a mile in height—a vertical distance difficult to appreciate. Stand on the south steps of the Treasury building in Washington and look down Pennsylvania Avenue to the Capitol; measure this distance overhead, and imagine cliffs to extend to that altitude, and you will understand what is meant; or stand at Canal Street in New York and look up broadway to Grace Church, and you have about the distance; or stand at Lake Street bridge in Chicago and look down to the Central Depot, and you have it again.

A thousand feet of this is up through granite crags; then steep slopes and perpendicular cliffs rise one above another to the summit. The gorge is black and narrow below, red and gray and flaring above, with crags and angular projections on the walls, which, cut in many places by side canyons, seem to be a vast wilderness of rocks. Down in these grand, gloomy depths we glide, ever listening, for the mad waters keep up their roar; ever watching, ever peering ahead, for the narrow canyon is winding and the river is closed in so that we can see but a few hundred yards, and what there may be below we know not; so we listen for falls and watch for rocks, stopping now and then in the bay of a recess to admire the gigantic scenery; and ever as we go there is some new pinnacle or tower, some crag or peak, some distant view of the upper plateau, some strangely shaped rock, or some deep, narrow side canyon.

Then we come to another broken fall, which appears more difficult than the one we ran this morning. A small creek comes in on the right, and the first fall of the water is over boulders, which have been carried down by this lateral stream. We land at its mouth and stop for an hour or two to examine the fall. It seems possible to let down with lines, at least a part of the way, from point to point, along the right-hand wall.

So we make a portage over the first rocks and find footing on some boulders below. Then we let down one of the boats to the end of her line, when she reaches a corner of the projecting rock, to which one of the men clings and steadies her while I examine an eddy below. I think we can pass the other boats down by us and catch them in the eddy. This is soon done, and the men in the boats in the eddy pull us to their side. On the shore of this little eddy there is about two feet of gravel beach above the water. Standing on this beach, some of the men take the line of the little boat and let it drift down against another projecting angle. Here is a little shelf, on which a man from my boat climbs, and a shorter line is passed to him, and he fastens the boat to the side of the cliff; then the second one is let down, bringing the line of the third. When the second boat is tied up, the two men standing on the beach above spring into the last boat, which is pulled up alongside of ours; then we let down the boats for 25 or 30 yards by walking along the shelf, landing them again in the mouth of a side canyon. Just below this there is another pile of boulders, over which we make another portage. From the foot of these rocks we can climb to another shelf, 40 or 50 feet above the water.

On this bench we camp for the night. It is raining hard, and we have no shelter, but find a few sticks which have lodged in the rocks, and kindle a fire and have supper. We sit on the rocks all night, wrapped in our *ponchos*, getting what sleep we can.

AUGUST 15.

This morning we find we can let down for 300 or 400 yards, and it is managed in this way: we pass along the wall by climbing from project-ing point to point, sometimes near the water's edge, at other places 50 or 60 feet above, and hold the boat with a line while two men remain aboard and prevent her from being dashed against the rocks and keep the line from getting caught on the wall. In two hours we have brought them all down, as far as it is possible, in this way. A few yards below, the river strikes with great violence against a projecting rock and our boats are pulled up in a little bay above. We must now manage to pull out of this and clear the point below. The little boat is held by the bow obliquely up the stream. We jump in and pull out only a few strokes, and sweep clear of the dangerous rock. The other boats follow in the same manner and the rapid is passed.

It is not easy to describe the labor of such navigation. We must pre-vent the waves from dashing the boats against the cliffs. Sometimes, where the river is swift, we must put a bight of rope about a rock, to prevent the boat from being snatched from us by a wave; but where the plunge is too great or the chute too swift, we must let her leap and catch her below or the undertow will drag her under the falling water and sink her. Where we wish to run her out a little way from shore through a channel between rocks, we first throw in little sticks of drift-wood and watch their course, to see where we must steer so that she will pass the channel in safety. And so we hold, and let go, and pull, and lift, and ward—among rocks, around rocks, and over rocks.

And now we go on through this solemn, mysterious way. The river is very deep, the canyon very narrow, and still obstructed, so that there is no steady flow of the stream; but the waters reel and roll and boil, and we are scarcely able to determine where we can go. Now the boat is carried to the right, perhaps close to the wall; again, she is shot into the stream, and perhaps is dragged over to the other side, where, caught in a whirlpool, she spins about. We can neither land nor run as we please. The boats are entirely unmanageable; no order in their running can be preserved; now one, now another, is ahead, each crew laboring for its own preservation. In such a place we come to another rapid. Two of the boats run it perforce. One succeeds in landing, but there is no foothold by which to make a portage and she is pushed out again into the stream. The next minute a great reflex wave fills the open compart-ment; she is water-logged, and drifts unmanageable. Breaker after breaker rolls over her and one capsizes her. The men are thrown out; but they cling to the boat, and she drifts down some distance alongside of us and we are able to catch her. She is soon bailed out and the men are aboard once more; but the oars are lost, and so a pair from the "Emma Dean" is spared. Then for two miles we find smooth water.

Clouds are playing in the canyon to-day. Sometimes they roll down in great masses, filling the gorge with gloom; sometimes they hang aloft from wall to wall and cover the canyon with a roof of impending storm, and we can peer long distances up and down this canyon corridor, with its cloud-roof overhead, its walls of black granite, and its river bright with the sheen of broken waters. Then a gust of wind sweeps down a side gulch and, making a rift in the clouds, reveals the blue heavens, and a stream of sunlight pours in. Then the clouds drift away into the distance, and hang around crags and peaks and pinnacles and towers and walls, and cover them with a mantle that lifts from time to time and sets them all in sharp relief. Then baby clouds creep out of side canyons, glide around points, and creep back again into more distant gorges. Then clouds arrange in strata across the canyon, with intervening vista views to cliffs and rocks beyond. The clouds are children of the heavens, and when they play among the rocks they lift them to the region above.

It rains! Rapidly little rills are formed above, and these soon grow into brooks, and the brooks grow into creeks and tumble over the walls in innumerable cascades, adding their wild music to the roar of the river. When the rain ceases the rills, brooks, and creeks run dry. The waters that fall during a rain on these steep rocks are gathered at once into the river; they could scarcely be poured in more suddenly if some vast spout ran from the clouds to the stream itself. When a storm bursts over the canyon a side gulch is dangerous, for a sudden flood may come, and the inpouring waters will raise the river so as to hide the rocks.

Early in the afternoon we discover a stream entering from the north—a clear, beautiful creek, coming down through a gorgeous red canyon. We land and camp on a sand beach above its mouth, under a great, overspreading tree with willow-shaped leaves. * * *

SAMUEL CLEMENS
(MARK TWAIN)
1835–1910

The Mississippi flowed through Clemens's life, as through his art. When he met the pilot Horace Bixby in 1857 on a trip down to New Orleans, he decided to sign on under him as a "cub," learning the river. By 1859 he was fully licensed as a pilot himself; he served in this capacity until the outbreak of the Civil War put an end to river travel between North and South. He also recalled the signals with which the Mississippi's shifting depths were called out to the pilot when he chose Mark Twain as his pen name. Scenes along the Mississippi figure beautifully in Huckleberry Finn, *which was begun in 1876 and finally published in 1883. This same period saw Clemens's return to his piloting experiences in another form. In 1875 two articles entitled "Old Times on the Mississippi" appeared in the* Atlantic Monthly. *They became the seed for* Life on the Mississippi, *published in 1883.*

From LIFE ON THE MISSISSIPPI

[The Book of the River]

* * * The face of the water, in time, became a wonderful book—a book that was a dead language to the uneducated passenger, but which told its mind to me without reserve, delivering its most cherished secrets as clearly as if it uttered them with a voice. And it was not a book to be read once and thrown aside, for it had a new story to tell every day. Throughout the long twelve hundred miles there was never a page that was void of interest, never one that you could leave unread without loss, never one that you would want to skip, thinking you could find higher enjoyment in some other thing. There never was so wonderful a book written by man; never one whose interest was so absorbing, so unflagging, so sparklingly renewed with every re-perusal. The passenger who could not read it was charmed with a peculiar sort of faint dimple on its surface (on the rare occasion when he did not overlook it altogether); but to the pilot that was an *italicized* passage; indeed, it was more than that, it was a legend of the largest capitals, with a string of shouting exclamation points at the end of it; for it meant that a wreck or a rock was buried there that could tear the life out of the strongest vessel that ever floated. It is the faintest and simplest expression the water ever makes, and the most hideous to a pilot's eye. In truth, the passenger who could not read this book saw nothing but all manner of pretty pictures in it, painted by the sun and shaded by the clouds, whereas to the trained eye these were not pictures at all, but the grimmest and most dead-earnest of reading-matter.

Now when I had mastered the language of this water and had come to know every trifling feature that bordered the great river as familiarly as I knew the letters of the alphabet, I had made a valuable acquisition. But I had lost something, too. I had lost something which could never be restored to me while I lived. All the grace, the beauty, the poetry had gone out of the majestic river! I still keep in mind a certain wonderful sunset which I witnessed when steamboating was new to me. A broad expanse of the river was turned to blood; in the middle distance the red hue brightened into gold, through which a solitary log came floating, black and conspicuous; in one place a long, slanting mark lay sparkling upon the water; in another the surface was broken by boiling, tumbling

Life on the Mississippi (Boston: Osgood and Co., 1883).

rings, that were as many-tinted as an opal; where the ruddy flush was faintest, was a smooth spot that was covered with graceful circles and radiating lines, ever so delicately traced; the shore on our left was densely wooded, and the sombre shadow that fell from this forest was broken in one place by a long, ruffled trail that shone like silver; and high above the forest wall a clean-stemmed dead tree waved a single leafy bough that glowed like a flame in the unobstructed splendor that was flowing from the sun. There were graceful curves, reflected images, woody heights, soft distances; and over the whole scene, far and near, the dissolving lights drifted steadily, enriching it, every passing moment, with new marvels of coloring.

I stood like one bewitched. I drank it in, in a speechless rapture. The world was new to me, and I had never seen anything like this at home. But as I have said, a day came when I began to cease from noting the glories and the charms which the moon and the sun and the twilight wrought upon the river's face; another day came when I ceased altogether to note them. Then, if that sunset scene had been repeated, I should have looked upon it without rapture, and should have commented upon it, inwardly, after this fashion: This sun means that we are going to have wind to-morrow; that floating log means that the river is rising, small thanks to it; that slanting mark on the water refers to a bluff reef which is going to kill somebody's steamboat one of these nights, if it keeps on stretching out like that; those tumbling "boils" show a dissolving bar and a changing channel there; the lines and circles in the slick water over yonder are a warning that that troublesome place is shoaling up dangerously; that silver streak in the shadow of the forest is the "break" from a new snag, and he has located himself in the very best place he could have found to fish for steamboats; that tall dead tree, with a single living branch, is not going to last long, and then how is a body ever going to get through this blind place at night without the friendly old landmark?

No, the romance and the beauty were all gone from the river. All the value any feature of it had for me now was the amount of usefulness it could furnish toward compassing the safe piloting of a steamboat. Since those days, I have pitied doctors from my heart. What does the lovely flush in a beauty's cheek mean to a doctor but a "break" that ripples above some deadly disease? Are not all her visible charms sown thick with what are to him the signs and symbols of hidden decay? Does he ever see her beauty at all, or doesn't he simply view her professionally, and comment upon her unwholesome condition all to himself? And doesn't he sometimes wonder whether he has gained most or lost most by learning his trade?

CELIA THAXTER
1835–1894

Celia Thaxter's garden on the Isle of Shoals, offshore from Portsmouth, New Hampshire, was her most influential work of art during her lifetime. Painters like Childe Hassam were inspired by the lavishness of this fifty-by fifteen-foot plot which Thaxter cultivated each year after her April return to the island. In her lyrical but precise descriptions of the individual flowers, we can feel her affinity with her contemporaries like Robert Browning and John Ruskin; her powers of observing both plants and birds also recall Gilbert White, whom she admired and quoted. As the literature of gardens becomes an increasingly prominent part of American nature writing in our own day, Thaxter's intense engagement with her own island garden becomes an even more significant landmark of American nature writing.

From AN ISLAND GARDEN

The garden suffers from the long drought in this last week of July, though I water it faithfully. The sun burns so hot that the earth dries again in an hour, after the most thorough drenching I can give it. The patient flowers seem to be standing in hot ashes, with the air full of fire above them. The cool breeze from the sea flutters their drooping petals, but does not refresh them in the blazing noon. Outside the garden on the island slopes the baked turf cracks away from the heated ledges of rock, and all the pretty growths of Sorrel and Eyebright, Grasses and Crowfoot, Potentilla and Lion's-tongue, are crisp and dead. All things begin again to pine and suffer for the healing touch of the rain.

Toward noon on this last day of the month the air darkens, and around the circle of the horizon the latent thunder mutters low. Light puffs of wind eddy round the garden, and whirl aloft the weary Poppy

An Island Garden (Boston: Houghton Mifflin, 1894).

petals high in air, till they wheel like birds about the chimney-tops. Then all is quiet once more. In the rich, hot sky the clouds pile themselves slowly, superb white heights of thunder-heads warmed with a brassy glow that deepens to rose in their clefts toward the sun. These clouds grow and grow, showing like Alpine summits amid the shadowy heaps of looser vapor; all the great vault of heaven gathers darkness; soon the cloudy heights, melting, are suffused in each other, losing shape and form and color. Then over the coast-line the sky turns a hard gray-green, against which rises with solemn movement and awful deliberation an arch of leaden vapor spanning the heavens from southwest to northeast, livid, threatening, its outer edges shaped like the curved rim of a mushroom, gathering swiftness as it rises, while the water beneath is black as hate, and the thunder rolls peal upon peal, as faster and faster the wild arch moves upward into tremendous heights above our heads. The whole sky is dark with threatening purple. Death and destruction seem ready to emerge from beneath that flying arch of which the livid fringes stream like gray flame as the wind rends its fierce and awful edge. Under it afar on the black level water a single sail gleams chalk-white in the gloom, a sail that even as we look is furled away from our sight, that the frail craft which bears it may ride out the gale under bare poles, or drive before it to some haven of safety. Earth seems to hold her breath before the expected fury. Lightning scores the sky from zenith to horizon, and across from north to south "a fierce, vindictive scribble of fire" writes its blinding way, and the awesome silence is broken by the cracking thunder that follows every flash. A moment more, and a few drops like bullets strike us; then the torn arch flies over in tattered rags, a monstrous apparition lost in darkness; then the wind tears the black sea into white rage and roars and screams and shouts with triumph,—the floods and the hurricane have it all their own way. Continually the tempest is shot through with the leaping lightning and crashing thunder, like steady cannonading, echoing and reëchoing, roaring through the vast empty spaces of the heavens. In pauses of the tumult a strange light is fitful over sea and rocks, then the tempest begins afresh as if it had taken breath and gained new strength. One's whole heart rises responding to the glory and the beauty of the storm, and is grateful for the delicious refreshment of the rain. Every leaf rejoices in the life-giving drops. Through the dense sparkling rain-curtain the lightning blazes now in crimson and in purple sheets of flame. Oh, but the wind is wild! Spare my treasures, oh, do not slay utterly my beautiful, beloved flowers! The tall stalks bend and strain, the Larkspurs bow. I hold my breath while the danger lasts, thinking only of the wind's power to harm the garden; for the leaping lightning and the crashing thunder I love, but the gale fills me with dread for my flowers defenseless. Still down pour the refreshing floods; everything is drenched: where are the humming-birds? The boats toss madly on the moorings, the sea breaks wildly on the shore, the world is drowned and gone, there is nothing but tempest and tumult and rush and roar of wind and rain.

The long trailing sprays of the Echinocystus vine stretch and strain like pennons flying out in the blast, the Wistaria tosses its feathery plumes over the arch above the door. Alas, for my bank of tall Poppies and blue Cornflowers and yellow Chrysanthemums outside! The Poppies are laid low, never to rise again, but the others will gather themselves together by and by, and the many-colored fires of Nasturtiums will clothe the slope with new beauty presently. The storm is sweeping past, already the rain diminishes, the lightning pales, the thunder retreats till leagues and leagues away we hear it "moaning and calling out of other lands." The clouds break away and show in the west glimpses of pure, melting blue, the sun bursts forth, paints a rainbow in the east upon the flying fragments of the storm, and pours a flood of glory over the drowned earth; the pelted flowers take heart and breathe again, every leaf shines, dripping with moisture; the grassy slopes laugh in sweet color; the sea calms itself to vast tranquillity and answers back the touch of the sun with a million glittering smiles.

Though the outside bank of flowers is wrecked and the tall Poppies prone upon the ground, those inside the garden are safe because I took the precaution to run two rows of wire netting up and down through the beds for their support. So, when the winds are cruelly violent, the tall, brittle stalks lean against this light but strong bulwark and are unhurt.

After the storm, in the clear, beautiful morning, before sunrise I went as usual into the garden to gather my flowers. To and fro, up and down over the ruined bank I passed; the wind blew cool and keen from the west, though the sky was smiling. The storm had beaten the flowers flat all over the slope; in scarlet and white and blue and pink and purple and orange bloom they were prostrate everywhere, leaves, stalks, blossoms, and all tangled and matted in an inextricable confusion. Swiftly I made my way through it, finding a foothold here and there, and stooping for every freshly unfolded cup or star or bell whose bud the tempest had spared. As I neared the little western gate with my hands full of blossoms to enter the garden on my way to the house, I was stopped still as a statue before a most pathetic sight. There, straight across the way, a tall Poppy plant lay prone upon the ground, and clinging to the stem of one of its green seed-pods sat my precious pet humming-bird, the dearest of the flock that haunt the garden, the tamest of them all. His eyes were tightly closed, his tiny claws clasped the stem automatically, he had no feeling, he was rigid with cold. The chill dew loaded the gray-green Poppy leaves, the keen wind blew sharply over him,—he is dead, I thought with a pang, as I shifted my flowers in a glowing heap to my left arm, and clasped the frozen little body in the palm of my right hand. It was difficult to disengage his slender wiry claws from their close grip on the chilly stalk, but he never moved or showed a sign of life as I took him off. I held him most tenderly in my closed hand, very careful not to crush or even press his tiny perishing body, and breathed into the shut hollow of my palm upon him with a warm and loving breath. I was so very busy, there were so many things to be done that morning, I could not stop to sit down and nurse him back to life. But I held him safe, and as I went up and down the garden paths gathering the rest of my flowers, I breathed every moment into my hand upon him. Ten, fifteen, twenty minutes passed; he made no sign of life. Alas, I thought, he is truly dead; when all at once I felt the least little thrill pass through the still, cold form, an answering thrill of joy ran through me in response, and more softly, closely, tenderly yet I sent my warm breath to the tiny creature as I still went on with my work. In a few minutes more I began to feel the smallest fluttering pulse of life throbbing faintly within him; in yet a few moments more he stirred and stretched his wings, comforting himself in the genial heat. When at last I felt him all alive, I took a small shallow basket of yellow straw, very small and light, and in it put a tuft of soft cotton wool, filled a tiny glass cup with sugar and water, honey-thick, placed it in the basket by the cotton, then gently laid the wee bird on the warm fluff. His eyes were still closed, but he moved his head slowly from side to side. The sun had risen and was pouring floods of light and heat into the garden. I carried the basket out into the corner where the heavenly blue Larkspurs stood behind the snow-whiteness of the full blossoming Lilies, and among the azure spikes I hung the pretty cradle where the sunbeams lay hottest and brightest on the flowers. The wind, grown balmy and mild, rocked the tall flower-spikes gently, the basket swayed with them, and the heat was so reviving that the dear little creature presently opened his eyes and quietly looked about him. At that my heart rejoiced. It was delightful to watch his slow return to his old self as I still went on with my work, looking continually toward him to see how he was getting on. The ardent sunbeams sent fresh life through him; suddenly he rose, an emerald spark, into the air, and quivered among the blue flowers, diving deep into each winged blossom for his breakfast of honey.

All day and every day he haunts the garden, and when tired rests contentedly on the small twig of a dry pea-stick near the Larkspurs. The

rosy Peas blossom about him, the Hollyhock flowers unfold in glowing
pink with lace-like edges of white; the bees hum there all day in and
out of the many flowers; the butterflies hover and waver and wheel.
When one comes too near him, up starts my beauty and chases him
away on burnished wings, away beyond the garden's bounds, and
returns to occupy his perch in triumph,—the dry twig he has taken for
his home the whole sweet summer long. Other humming-birds haunt
the place, but he belongs there; they go and come, but he keeps to his
perch and his Larkspurs faithfully. He is so tame he never stirs from his
twig for anybody, no matter how near a person may come; he alights on
my arms and hands and hair unafraid; he rifles the flowers I hold, when
I am gathering them, and I sometimes think he is the very most charming thing in the garden. The jealous bees and the butterflies follow the
flowers I carry also, sometimes all the way into the house. The other
day, as I sat in the piazza which the vines shade with their broad green
leaves and sweet white flowers climbing up to the eaves and over the
roof, I saw the humming-birds hovering over the whole expanse of
green, to and fro, and discovered that they were picking off and devouring the large transparent aphides scattered, I am happy to say but spar-
ingly, over its surface, every little gnat and midge they snapped up with
avidity. I had fancied they lived on honey, but they appeared to like the
insects quite as well.

   In the sweet silence before sunrise, standing in the garden I watch
the large round shield of the full moon slowly fading in the west from
copper to brass and then to whitest silver, throwing across a sea of glass
its long, still reflection, while the deep, pure sky takes on a rosy warmth
of color from the approaching sun. Soon an insufferable glory burns on
the edge of the eastern horizon; up rolls the great round red orb and
sets the dew twinkling and sparkling in a thousand rainbows, sending
its first rejoicing rays over the wide face of the world. When in these
fresh mornings I go into my garden before any one is awake, I go for
the time being into perfect happiness.   *   *   *

JOHN BURROUGHS
1837–1921

John Muir and John Burroughs were undoubtedly the two most popular and successful American nature writers of the late nineteenth and early twentieth centuries, yet in their lives and writings they represented strikingly different approaches to nature. Unlike Muir, who left home early and spent much of his adult life wandering in the rugged mountain wildernesses of California and Alaska, Burroughs spent most of his eighty-four years living on a small New York farm overlooking the Hudson River, not far from the Catskill Mountains where he was born. Though not a conservation activist like Muir, his quiet, genial "nature rambles" did much to popularize amateur nature observation and increase the appreciation of local environments.

Burroughs worked in Washington, D.C., for ten years, where he became friends with Walt Whitman, who served as a nurse in army hospitals during the Civil War. Burroughs's first book, Notes on Walt Whitman, Poet and Person *(1867), was the first appreciative study of that major literary figure, whose cosmic optimism and celebration of the American landscape is reflected, in a quieter vein, in Burroughs's own writings: In 1873 Burroughs was able to move back to his beloved Catskill region, where he built Riverby, a stone house overlooking the Hudson River, and wrote more than two dozen books that sold over one and a half million copies. Later he built "Slabsides," a summer retreat which became in time a pilgrimage for his legions of admirers and which today is owned and maintained by the John Burroughs Society.*

Though Burroughs's geographical locus was small, his intellectual range was large. He saw Darwinian evolution as a liberating vision and was one of the first popular writers to embrace the new biological science. At the turn of the century he joined forces with Theodore Roosevelt and others in a war against the "nature fakers"—writers like Ernest Thompson Seton whose animal stories, Burroughs felt, were full of anthropomorphic fictions and unrealistic events. The dispute, which became quite public and often bitter, helped to rid much nature writing of the time of excessive sentimentality and to ground it more in scientific observation. Yet, as

Burroughs was fond of saying, "Knowledge is only half the task. The other half is love."

Burroughs's earlier books of genial essays on local birds and flowers tend to hold diminished interest for today's readers. Yet many of his descriptions of places remain fresh and provocative, and his later essays tend to confront larger, more philosophical issues raised by modern science, which remain central challenges for writers today.

IN MAMMOTH CAVE

Some idea of the impression which Mammoth Cave makes upon the senses, irrespective even of sight, may be had from the fact that blind people go there to see it, and are greatly struck with it. I was assured that this is a fact. The blind seem as much impressed by it as those who have their sight. When the guide pauses at the more interesting point, or lights the scene up with a great torch or with Bengal lights, and points out the more striking features, the blind exclaim, "How wonderful! how beautiful!" They can feel it, if they cannot see it. They get some idea of the spaciousness when words are uttered. The voice goes forth in these colossal chambers like a bird. When no word is spoken, the silence is of a kind never experienced on the surface of the earth, it is so profound and abysmal. This, and the absolute darkness, to a person with eyes makes him feel as if he were face to face with the primordial nothingness. The objective universe is gone; only the subjective remains; the sense of hearing is inverted, and reports only the murmurs from within. The blind miss much, but much remains to them. The great cave is not merely a spectacle to the eye; it is a wonder to the ear, a strangeness to the smell and to the touch. The body feels the presence of unusual conditions through every pore.

For my part, my thoughts took a decidedly sepulchral turn; I thought of my dead and of all the dead of the earth, and said to myself, the darkness and the silence of their last resting-place is like this; to this we must all come at last. No vicissitudes of earth, no changes of seasons, no sound of storm or thunder penetrate here; winter and summer, day and night, peace or war, it is all one; a world beyond the reach of change, because beyond the reach of life. What peace, what repose, what desolation! The marks and relics of the Indian, which disappear so quickly from the light of day above, are here beyond the reach of

Riverby (Boston: Houghton Mifflin, 1894).

natural change. The imprint of his moccasin in the dust might remain undisturbed for a thousand years. At one point the guide reaches his arm beneath the rocks that strew the floor and pulls out the burnt ends of canes, which were used, probably, when filled with oil or grease, by the natives to light their way into the cave doubtless centuries ago.

Here in the loose soil are ruts worn by cartwheels in 1812, when, during the war with Great Britain, the earth was searched to make saltpetre. The guide kicks corn-cobs out of the dust where the oxen were fed at noon, and they look nearly as fresh as ever they did. In those frail corn-cobs and in those wheel-tracks as if the carts had but just gone along, one seemed to come very near to the youth of the century, almost to overtake it.

At a point in one of the great avenues, if you stop and listen, you hear a slow, solemn ticking like a great clock in a deserted hall; you hear the slight echo as it fathoms and sets off the silence. It is called the clock, and is caused by a single large drop of water falling every second into a little pool. A ghostly kind of clock there in the darkness, that is never wound up and that never runs down. It seemed like a mockery where time is not, and change does not come, — the clock of the dead. This sombre and mortuary cast of one's thoughts seems so natural in the great cave, that I could well understand the emotions of a lady who visited the cave with a party a few days before I was there. She went forward very reluctantly from the first; the silence and the darkness of the huge mausoleum evidently impressed her imagination, so that when she got to the spot where the guide points out the "Giant's Coffin," a huge, fallen rock, which in the dim light takes exactly the form of an enormous coffin, her fear quite overcame her, and she begged piteously to be taken back. Timid, highly imaginative people, especially women, are quite sure to have a sense of fear in this strange underground world. The guide told me of a lady in one of the parties he was conducting through, who wanted to linger behind a little all alone; he suffered her to do so, but presently heard a piercing scream. Rushing back, he found her lying prone upon the ground in a dead faint. She had accidentally put out her lamp, and was so appalled by the darkness that instantly closed around her that she swooned at once.

Sometimes it seemed to me as if I were threading the streets of some buried city of the fore-world. With your little lantern in your hand, you follow your guide through those endless and silent avenues, catching glimpses on either hand of what appears to be some strange antique architecture, the hoary and crumbling walls rising high up into the darkness. Now we turn a sharp corner, or turn down a street which crosses our course at right angles; now we come out into a great circle, or spacious court, which the guide lights up with a quick-paper torch, or a colored chemical light. There are streets above you and streets below you. As this was a city where day never entered, no provision for light needed to be made, and it is built one layer above another to the number of four or five, or on the plan of an enormous ant-hill, the lowest avenues being several hundred feet beneath the uppermost. The main avenue leading in from the entrance is called the Broadway, and if Broadway, New York, were arched over and reduced to utter darkness and silence, and its roadway blocked with mounds of earth and fragments of rock, it would, perhaps, only lack that gray, cosmic, elemental look, to make it resemble this. A mile or so from the entrance we pass a couple of rude stone houses, built forty or more years ago by some consumptives, who hoped to prolong their lives by a residence in this pure, antiseptic air. Five months they lived here, poor creatures, a half dozen of them, without ever going forth into the world of light. But the long entombment did not arrest the disease; the mountain did not draw the virus out, but seemed to draw the strength and vitality out, so that when the victims did go forth into the light and air, bleached as white as chalk, they succumbed at once, and nearly all died before they could reach the hotel, a few hundred yards away.

Probably the prettiest thing they have to show you in Mammoth Cave is the Star Chamber. This seems to have made an impression upon Emerson when he visited the cave, for he mentions it in one of his essays, "Illusions." The guide takes your lantern from you and leaves you seated upon a bench by the wayside, in the profound cosmic darkness. He retreats along a side alley that seems to go down to a lower level, and at a certain point shades his lamp with his hat, so that the light falls upon the ceiling over your head. You look up, and the first thought is that there is an opening just there that permits you to look forth upon the midnight skies. You see the darker horizon line where the sky ends and the mountains begin. The sky is blue-black and is thickly studded with stars, rather small stars, but apparently genuine. At one point a long, luminous streak simulates exactly the form and effect of a comet. As you gaze, the guide slowly moves his hat, and a black cloud gradually creeps over the sky, and all is blackness again. Then you hear footsteps retreating and dying away in the distance. Presently all is still, save the ringing in your own ears. Then after a few moments, during which you have sat in silence like that of the interstellar spaces, you hear over your left shoulder a distant flapping of wings, followed by the crowing of a cock. You turn your head in that direction and behold a faint dawn breaking on the horizon. It slowly increases till you hear footsteps approaching, and your dusky companion, playing the part of Apollo, with lamp in hand ushers in the light of day. It is rather theatrical, but a very pleasant diversion nevertheless.

Another surprise was when we paused at a certain point, and the guide asked me to shout or call in a loud voice. I did so without any unusual effect following. Then he spoke in a very deep bass, and instantly the rocks all about and beneath us became like the strings of an Aeolian harp. They seemed transformed as if by enchantment. Then I tried, but did not strike the right key; the rocks were dumb; I tried again, but got no response; flat and dead the sounds came back as if in mockery; then I struck a deeper bass, the chord was hit, and the solid walls seemed to become as thin and frail as a drum-head or as the frame of a violin. They fairly seemed to dance about us, and to recede away from us. Such wild, sweet music I had never before heard rocks discourse. Ah, the magic of the right key! "Why leap ye, ye high hills?" why, but that they had been spoken to in the right voice? Is not the whole secret of life to pitch our voices in the right key? Responses come from the very rocks when we do so. I thought of the lines of our poet of Democracy: —

"Surely, whoever speaks to me in the right voice, him or her I shall follow,
As the water follows the moon, silently, with fluid steps, anywhere around
    the globe."

Where we were standing was upon an arch over an avenue which crossed our course beneath us. The reverberations on Echo River, a point I did not reach, can hardly be more surprising, though they are described as wonderful.

There are four or five levels in the cave, and a series of avenues upon each. The lowest is some two hundred and fifty feet below the entrance. Here the stream which has done all this carving and tunneling has got to the end of its tether. It is here on a level with Green River in the valley below, and flows directly into it. I say the end of its tether, though if Green River cuts its valley deeper, the stream will, of course, follow suit. The bed of the river has probably, at successive periods, been on a level with each series of avenues of the cave. The stream is now doubtless but a mere fraction of its former self. Indeed, every feature of the cave attests the greater volume and activity of the forces which carved it, in the earlier geologic ages. The waters have worn the rock as if it were but ice. The domes and pits are carved and fluted in precisely the way dripping water flutes snow or ice. The rainfall must have been enormous in those early days, and it must have had a much stronger and sharper tooth of carbonic acid gas than now. It has carved out enormous pits with perpendicular sides, two or three hundred feet deep. Goring Dome I remember particularly. You put your head through an irregularly shaped window in the wall at the side of one of the avenues, and there is this huge shaft or well, starting from some higher level and going down two hundred feet below you. There must have been such wells in the old glaciers, worn by a rill of water slowly eating its way down. It was probably ten feet across, still moist and dripping. The guide threw down a lighted torch, and it fell and fell, till I had to crane my neck far out to see it finally reach the bottom. Some of these pits are simply appalling, and where the way is narrow, have been covered over to prevent accidents.

No part of Mammoth Cave was to me more impressive than its entrance, probably because here its gigantic proportions are first revealed to you, and can be clearly seen. That strange colossal underworld here looks out into the light of day, and comes in contrast with familiar scenes and objects. When you are fairly in the cave, you cannot see it; that is, with your aboveground eyes; you walk along by the dim light of your lamp as in a huge wood at night; when the guide lights up the more interesting portions with his torches and colored lights, the effect is weird and spectral; it seems like a dream; it is an unfamiliar world; you hardly know whether this is the emotion of grandeur which you experience, or of mere strangeness. If you could have the light of day in there, you would come to your senses, and could test the reality of your impressions. At the entrance you have the light of day, and you look fairly in the face of this underground monster, yea, into his open mouth, which has a span of fifty feet or more, and down into his contracting throat, where a man can barely stand upright, and where the light fades and darkness begins. As you come down the hill through the woods from the hotel, you see no sign of the cave till you emerge into a small opening where the grass grows and the sunshine falls, when you turn slightly to the right, and there at your feet yawns this terrible pit; and you feel indeed as if the mountain had opened its mouth and was lying in wait to swallow you down, as a whale might swallow a shrimp. I never grew tired of sitting or standing here by this entrance and gazing into it. It had for me something of the same fascination that the display of the huge elemental forces of nature have, as seen in thunder-storms, or in a roaring ocean surf. Two phœbe-birds had their nests in little niches of the rocks, and delicate ferns and wild flowers fringed the edges.

Another very interesting feature to me was the behavior of the cool air which welled up out of the mouth of the cave. It simulated exactly a fountain of water. It rose up to a certain level, or until it filled the depression immediately about the mouth of the cave, and then flowing over at the lowest point, ran down the hill toward Green River, along a little water-course, exactly as if it had been a liquid. I amused myself by wading down into it as into a fountain. The air above was muggy and hot, the thermometer standing at about eighty-six degrees, and this cooler air of the cave, which was at a temperature of about fifty-two degrees, was separated in the little pool or lakelet which is formed from the hotter air above it by a perfectly horizontal line. As I stepped down into it I could feel it close over my feet, then it was at my knees, then I was immersed to my hips, then to my waist, then I stood neck-deep in it, my body almost chilled, while my face and head were bathed by a sultry, oppressive air. Where the two bodies of air came in contact, a slight film of vapor was formed by condensation; I waded in till I could look under this as under a ceiling. It was as level and as well defined as a sheet of ice on a pond. A few moments' immersion into this aerial fountain made one turn to the warmer air again. At the depression in the rim of the basin one had but to put his hand down to feel the cold air flowing over like water. Fifty yards below you could still wade into it as into a creek, and at a hundred yards it was still quickly perceptible, but broader and higher; it had begun to lose some of its coldness, and to mingle with the general air; all the plants growing on the margin of the water-course were in motion, as well as the leaves on the low branches of the trees near by. Gradually this cool current was dissipated and lost in the warmth of the day.

JOHN MUIR
1838–1914

Although John Muir has long been famous as an explorer and environmental activist, critical esteem for his writing has only recently caught up with admiration for his adventures. His style could be exuberant, even florid, and Muir once wryly complained about the necessity, in revising his work, of "slaughtering gloriouses." But his mountain narratives also convey the authority of an acute eye and a firm understanding of geological processes. The first man to ascribe a glacial origin to Yosemite Valley, he buttressed his view both with a careful record of the grooves on the rocks and with the knowledge of glaciology he learned at the University of Wisconsin, eventually prevailing over the opposing, cataclysmic theory of Josiah Whitney and the scientific establishment.

Born in Dunbar, Scotland, and raised as a youth on a Wisconsin homestead, Muir went on to explore wild places from Alaska to South

America. But Yosemite Valley, from the moment he strode into it, became his classic landscape. The effort to save Yosemite, and to preserve the nearby valley at Hetch Hetchy, was the focus of his articles for the Century magazine, and led to his founding of the Sierra Club in 1892. These articles also became the basis for The Mountains of California (1894). Muir's writing introduced the prophetic voice that resounds in contemporary authors like Edward Abbey—the fierce advocacy for wilderness that sometimes leads to searing critiques of society. Looking at the commercial interests establishing themselves in the Sierra Nevada at century's end, Muir could already proclaim, "The money-changers are in the temple."

A WIND-STORM IN THE FORESTS

The mountain winds, like the dew and rain, sunshine and snow, are measured and bestowed with love on the forests to develop their strength and beauty. However restricted the scope of other forest influences, that of the winds is universal. The snow bends and trims the upper forests every winter, the lightning strikes a single tree here and there, while avalanches mow down thousands at a swoop as a gardener trims out a bed of flowers. But the winds go to every tree, fingering every leaf and branch and furrowed bole; not one is forgotten; the Mountain Pine towering with outstretched arms on the rugged buttresses of the icy peaks, the lowliest and most retiring tenant of the dells; they seek and find them all, caressing them tenderly, bending them in lusty exercise, stimulating their growth, plucking off a leaf or limb as required, or removing an entire tree or grove, now whispering and cooing through the branches like a sleepy child, now roaring like the ocean; the winds blessing the forests, the forests the winds, with ineffable beauty and harmony as the sure result.

After one has seen pines six feet in diameter bending like grasses before a mountain gale, and ever and anon some giant falling with a crash that shakes the hills, it seems astonishing that any, save the lowest thickset trees, could ever have found a period sufficiently stormless to establish themselves; or, once established, that they should not, sooner or later, have been blown down. But when the storm is over, and we behold the same forests tranquil again, towering fresh and unscathed in erect majesty, and consider what centuries of storms have fallen upon them since they were first planted,—hail, to break the tender seedlings;

The Mountains of California (New York: Century, 1894).

lightning, to scorch and shatter; snow, winds, and avalanches, to crush and overwhelm,—while the manifest result of all this wild storm-culture is the glorious perfection we behold; then faith in Nature's forestry is established, and we cease to deplore the violence of her most destructive gales, or of any other storm-implement whatsoever.

There are two trees in the Sierra forests that are never blown down, so long as they continue in sound health. These are the Juniper and the Dwarf Pine of the summit peaks. Their stiff, crooked roots grip the storm-beaten ledges like eagles' claws, while their lithe, cord-like branches bend round compliantly, offering but slight holds for winds, however violent. The other alpine conifers—the Needle Pine, Mountain Pine, Two-leaved Pine, and Hemlock Spruce—are never thinned out by this agent to any destructive extent, on account of their admirable toughness and the closeness of their growth. In general the same is true of the giants of the lower zones. The kingly Sugar Pine, towering aloft to a height of more than 200 feet, offers a fine mark to storm-winds; but it is not densely foliaged, and its long, horizontal arms swing round compliantly in the blast, like tresses of green, fluent algæ in a brook; while the Silver Firs in most places keep their ranks well together in united strength. The Yellow or Silver Pine is more frequently overturned than any other tree on the Sierra, because its leaves and branches form a larger mass in proportion to its height, while in many places it is planted sparsely, leaving open lanes through which storms may enter with full force. Furthermore, because it is distributed along the lower portion of the range, which was the first to be left bare on the breaking up of the ice-sheet at the close of the glacial winter, the soil it is growing upon has been longer exposed to post-glacial weathering, and consequently is in a more crumbling, decayed condition than the fresher soils farther up the range, and therefore offers a less secure anchorage for the roots.

While exploring the forest zones of Mount Shasta, I discovered the path of a hurricane strewn with thousands of pines of this species. Great and small had been uprooted or wrenched off by sheer force, making a clean gap, like that made by a snow avalanche. But hurricanes capable of doing this class of work are rare in the Sierra, and when we have explored the forests from one extremity of the range to the other, we are compelled to believe that they are the most beautiful on the face of the earth, however we may regard the agents that have made them so.

There is always something deeply exciting, not only in the sounds of winds in the woods, which exert more or less influence over every mind, but in their varied waterlike flow as manifested by the movements of the trees, especially those of the conifers. By no other trees are they rendered so extensively and impressively visible, not even by the lordly tropic palms or tree-ferns responsive to the gentlest breeze. The waving of a forest of the giant Sequoias is indescribably impressive and sublime, but the pines seem to me the best interpreters of winds. They are mighty waving goldenrods, ever in tune, singing and writing wind-music all their long century lives. Little, however, of this noble tree-waving and tree-music will you see or hear in the strictly alpine portion of the forests. The burly Juniper, whose girth sometimes more than equals its height, is about as rigid as the rocks on which it grows. The slender lash-like sprays of the Dwarf Pine stream out in wavering ripples, but the tallest and slenderest are far too unyielding to wave even in the heaviest gales. They only shake in quick, short vibrations. The Hemlock Spruce, however, and the Mountain Pine, and some of the tallest thickets of the Two-leaved species bow in storms with considerable scope and gracefulness. But it is only in the lower and middle zones that the meeting of winds and woods is to be seen in all its grandeur.

One of the most beautiful and exhilarating storms I ever enjoyed in the Sierra occurred in December, 1874, when I happened to be exploring one of the tributary valleys of the Yuba River. The sky and the ground and the trees had been thoroughly rain-washed and were dry again. The day was intensely pure, one of those incomparable bits of California winter, warm and balmy and full of white sparkling sunshine, redolent of all the purest influences of the spring, and at the same time enlivened with one of the most bracing wind-storms conceivable. Instead of camping out, as I usually do, I then chanced to be stopping at the house of a friend. But when the storm began to sound, I lost no time in pushing out into the woods to enjoy it. For on such occasions Nature has always something rare to show us, and the danger to life and limb is hardly greater than one would experience crouching deprecatingly beneath a roof.

It was still early morning when I found myself fairly adrift. Delicious sunshine came pouring over the hills, lighting the tops of the pines, and setting free a stream of summery fragrance that contrasted strangely with the wild tones of the storm. The air was mottled with pine-tassels and bright green plumes, that went flashing past in the sunlight like birds pursued. But there was not the slightest dustiness, nothing less pure than leaves, and ripe pollen, and flecks of withered bracken and moss. I heard trees falling for hours at the rate of one every two or three minutes; some uprooted, partly on account of the loose, water-soaked condition of the ground; others broken straight across, where some weakness caused by fire had determined the spot. The gestures of the various trees made a delightful study. Young Sugar Pines, light and feathery as squirrel-tails, were bowing almost to the ground; while the grand old patriarchs, whose massive boles had been tried in a hundred storms, waved solemnly above them, their long, arching branches streaming fluently on the gale, and every needle thrilling and ringing and shedding off keen lances of light like a diamond. The Douglas Spruces, with long sprays drawn out in level tresses, and needles massed in a gray, shimmering glow, presented a most striking appearance as they stood in bold relief along the hilltops. The madroños in the dells, with their red bark and large glossy leaves tilted every way, reflected the sunshine in throbbing spangles like those one so often sees on the rippled surface of a glacier lake. But the Silver Pines were now the most impressively beautiful of all. Colossal spires 200 feet in height waved like supple goldenrods chanting and bowing low as if in worship, while the whole mass of their long, tremulous foliage was kindled into one continuous blaze of white sun-fire. The force of the gale was such that the most steadfast monarch of them all rocked down to its roots with a motion plainly perceptible when one leaned against it. Nature was holding high festival, and every fiber of the most rigid giants thrilled with glad excitement.

I drifted on through the midst of this passionate music and motion, across many a glen, from ridge to ridge; often halting in the lee of a rock for shelter, or to gaze and listen. Even when the grand anthem had swelled to its highest pitch, I could distinctly hear the varying tones of individual trees,—Spruce, and Fir, and Pine, and leafless Oak,—and even the infinitely gentle rustle of the withered grasses at my feet. Each was expressing itself in its own way,—singing its own song, and making its own peculiar gestures,—manifesting a richness of variety to be found in no other forest I have yet seen. The coniferous woods of Canada, and the Carolinas, and Florida, are made up of trees that resemble one another about as nearly as blades of grass, and grow close together in much the same way. Coniferous trees, in general, seldom possess individual character, such as is manifest among Oaks and Elms. But the California forests are made up of a greater number of distinct species than any other in the world. And in them we find, not only a marked differentiation into special groups, but also a marked individuality in almost every tree, giving rise to storm effects indescribably glorious.

Toward midday, after a long, tingling scramble through copses of hazel and ceanothus, I gained the summit of the highest ridge in the neighborhood; and then it occurred to me that it would be a fine thing to climb one of the trees to obtain a wider outlook and get my ear close to the Æolian music of its topmost needles. But under the circumstances the choice of a tree was a serious matter. One whose instep was not very strong seemed in danger of being blown down, or of being struck by others in case they should fall; another was branchless to a considerable height above the ground, and at the same time too large to be grasped with arms and legs in climbing; while others were not favorably situated for clear views. After cautiously casting about, I made choice of the tallest of a group of Douglas Spruces that were growing close together like a tuft of grass, no one of which seemed likely to fall unless all the rest fell with it. Though comparatively young, they were about 100 feet high, and their lithe, brushy tops were rocking and swirling in wild ecstasy. Being accustomed to climb trees in making botanical studies, I experienced no difficulty in reaching the top of this one, and never before did I enjoy so noble an exhilaration of motion. The slender tops fairly flapped and swished in the passionate torrent, bending and swirling backward and forward, round and round, tracing indescribable combinations of vertical and horizontal curves, while I clung with muscles firm braced, like a bobolink on a reed.

In its widest sweeps my tree-top described an arc of from twenty to thirty degrees, but I felt sure of its elastic temper, having seen others of the same species still more severely tried—bent almost to the ground indeed, in heavy snows—without breaking a fiber. I was therefore safe, and free to take the wind into my pulses and enjoy the excited forest from my superb outlook. The view from here must be extremely beautiful in any weather. Now my eye roved over the piny hills and dales as over fields of waving grain, and felt the light running in ripples and broad swelling undulations across the valleys from ridge to ridge, as the shining foliage was stirred by corresponding waves of air. Oftentimes these waves of reflected light would break up suddenly into a kind of beaten foam, and again, after chasing one another in regular order, they would seem to bend forward in concentric curves, and disappear on some hillside, like sea-waves on a shelving shore. The quantity of light reflected from the bent needles was so great as to make whole groves appear as if covered with snow, while the black shadows beneath the trees greatly enhanced the effect of the silvery splendor.

Excepting only the shadows there was nothing somber in all this wild sea of pines. On the contrary, notwithstanding this was the winter season, the colors were remarkably beautiful. The shafts of the pine and libocedrus were brown and purple, and most of the foliage was well tinged with yellow; the laurel groves, with the pale undersides of their leaves turned upward, made masses of gray; and then there was many a dash of chocolate color from clumps of manzanita, and jet of vivid crimson from the bark of the madroños, while the ground on the hillsides, appearing here and there through openings between the groves, displayed masses of pale purple and brown.

The sounds of the storm corresponded gloriously with this wild exuberance of light and motion. The profound bass of the naked branches and boles booming like waterfalls; the quick, tense vibrations of the pine-needles, now rising to a shrill, whistling hiss, now falling to a silky murmur; the rustling of laurel groves in the dells, and the keen metallic click of leaf on leaf—all this was heard in easy analysis when the attention was calmly bent.

The varied gestures of the multitude were seen to fine advantage, so that one could recognize the different species at a distance of several miles by this means alone, as well as by their forms and colors, and the way they reflected the light. All seemed strong and comfortable, as if really enjoying the storm, while responding to its most enthusiastic greetings. We hear much nowadays concerning the universal struggle for existence, but no struggle in the common meaning of the word was manifest here; no recognition of danger by any tree; no deprecation; but rather an invincible gladness as remote from exultation as from fear.

I kept my lofty perch for hours, frequently closing my eyes to enjoy the music by itself, or to feast quietly on the delicious fragrance that was streaming past. The fragrance of the woods was less marked than that produced during warm rain, when so many balsamic buds and leaves are steeped like tea; but, from the chafing of resiny branches against each other, and the incessant attrition of myriads of needles, the gale was spiced to a very tonic degree. And besides the fragrance from these local sources there were traces of scents brought from afar. For this wind came first from the sea, rubbing against its fresh, briny waves, then distilled through the redwoods, threading rich ferny gulches, and spreading itself in broad undulating currents over many a flower-enameled ridge of the coast mountains, then across the golden plains, up the purple foot-hills, and into these piny woods with the varied incense gathered by the way.

Winds are advertisements of all they touch, however much or little we may be able to read them; telling their wanderings even by their scents alone. Mariners detect the flowery perfume of land-winds far at sea, and sea-winds carry the fragrance of dulse and tangle far inland, where it is quickly recognized, though mingled with the scents of a thousand land-flowers. As an illustration of this, I may tell here that I breathed sea-air on the Firth of Forth, in Scotland, while a boy; then was taken to Wisconsin, where I remained nineteen years; then, without in all this time having breathed one breath of the sea, I walked quietly, alone, from the middle of the Mississippi Valley to the Gulf of Mexico, on a botanical excursion, and while in Florida, far from the coast, my attention wholly bent on the splendid tropical vegetation about me, I suddenly recognized a sea-breeze, as it came sifting through the palmettos and blooming vine-tangles, which at once awakened and set free a thousand dormant associations, and made me a boy again in Scotland, as if all the intervening years had been annihilated.

Most people like to look at mountain rivers, and bear them in mind; but few care to look at the winds, though far more beautiful and sublime, and though they become at times about as visible as flowing water. When the north winds in winter are making upward sweeps over the curving summits of the High Sierra, the fact is sometimes published with flying snow-banners a mile long. Those portions of the winds thus embodied can scarce be wholly invisible, even to the darkest imagination. And when we look around over an agitated forest, we may see something of the wind that stirs it, by its effects upon the trees. Yonder it descends in a rush of water-like ripples, and sweeps over the bending pines from hill to hill. Nearer, we see detached plumes and leaves, now speeding by on level currents, now whirling in eddies, or, escaping over the edges of the whirls, soaring aloft on grand, upswelling domes of air, or tossing on flame-like crests. Smooth, deep currents, cascades, falls, and swirling eddies, sing around every tree and leaf, and over all the varied topography of the region with telling changes of form, like mountain rivers conforming to the features of their channels.

After tracing the Sierra streams from their fountains to the plains, marking where they bloom white in falls, glide in crystal plumes, surge gray and foam-filled in boulder-choked gorges, and slip through the woods in long, tranquil reaches — after thus learning their language and forms in detail, we may at length hear them chanting all together in one grand anthem, and comprehend them all in clear inner vision, covering the range like lace. But even this spectacle is far less sublime and not a whit more substantial than what we may behold of these storm-streams of air in the mountain woods.

We all travel the milky way together, trees and men; but it never occurred to me until this stormday, while swinging in the wind, that trees are travelers, in the ordinary sense. They make many journeys, not extensive ones, it is true; but our own little journeys, away and back again, are only little more than tree-wavings — many of them not so much.

When the storm began to abate, I dismounted and sauntered down through the calming woods. The storm-tones died away, and, turning toward the east, I beheld the countless hosts of the forests hushed and tranquil, towering above one another on the slopes of the hills like a devout audience. The setting sun filled them with amber light, and seemed to say, while they listened, "My peace I give unto you."

As I gazed on the impressive scene, all the so-called ruin of the storm was forgotten, and never before did these noble woods appear so fresh, so joyous, so immortal.

THE WATER-OUZEL

The waterfalls of the Sierra are frequented by only one bird,—the Ouzel or Water Thrush (*Cinclus Mexicanus*, Sw.). He is a singularly joyous and lovable little fellow, about the size of a robin, clad in a plain waterproof suit of bluish gray, with a tinge of chocolate on the head and shoulders. In form he is about as smoothly plump and compact as a pebble that has been whirled in a pot-hole, the flowing contour of his body being interrupted only by his strong feet and bill, the crisp wing-tips, and the up-slanted wren-like tail.

Among all the countless waterfalls I have met in the course of ten years' exploration in the Sierra, whether among the icy peaks, or warm foot-hills, or in the profound yosemitic cañons of the middle region, not one was found without its Ouzel. No cañon is too cold for this little bird, none too lonely, provided it be rich in falling water. Find a fall, or cascade, or rushing rapid, anywhere upon a clear stream, and there you will surely find its complementary Ouzel, flitting about in the spray, diving in foaming eddies, whirling like a leaf among beaten foam-bells; ever vigorous and enthusiastic, yet self-contained, and neither seeking nor shunning your company.

If disturbed while dipping about in the margin shallows, he either sets off with a rapid whir to some other feeding-ground up or down the stream, or alights on some half-submerged rock or snag out in the current, and immediately begins to nod and courtesy like a wren, turning his head from side to side with many other odd dainty movements that never fail to fix the attention of the observer.

He is the mountain streams' own darling, the humming-bird of blooming waters, loving rocky ripple-slopes and sheets of foam as a bee loves flowers, as a lark loves sunshine and meadows. Among all the mountain birds, none has cheered me so much in my lonely wanderings,—none so unfailingly. For both in winter and summer he sings, sweetly, cheerily, independent alike of sunshine and of love, requiring no other inspiration than the stream on which he dwells. While water sings, so must he, in heat or cold, calm or storm, ever attuning his voice in sure accord; low in the drought of summer and the drought of winter, but never silent.

During the golden days of Indian summer, after most of the snow has been melted, and the mountain streams have become feeble,—a succession of silent pools, linked together by shallow, transparent currents and strips of silvery lacework,—then the song of the Ouzel is at its lowest ebb. But as soon as the winter clouds have bloomed, and the mountain treasuries are once more replenished with snow, the voices of the streams and ouzels increase in strength and richness until the flood season of early summer. Then the torrents chant their noblest anthems, and then is the flood-time of our songster's melody. As for weather, dark days and sun days are the same to him. The voices of most song-birds, however joyous, suffer a long winter eclipse; but the Ouzel sings on through all the seasons and every kind of storm. Indeed no storm can be more violent than those of the waterfalls in the midst of which he delights to dwell. However dark and boisterous the weather, snowing, blowing, or cloudy, all the same he sings, and with never a note of sadness. No need of spring sunshine to thaw *his* song, for it never freezes. Never shall you hear anything wintry from *his* warm breast; no pinched cheeping, no wavering notes between sorrow and joy; his mellow, fluty voice is ever tuned to downright gladness, as free from dejection as cock-crowing.

It is pitiful to see wee frost-pinched sparrows on cold mornings in the mountain groves shaking the snow from their feathers, and hopping about as if anxious to be cheery, then hastening back to their hidings out of the wind, puffing out their breast-feathers over their toes, and subsiding among the leaves, cold and breakfastless, while the snow continues to fall, and there is no sign of clearing. But the Ouzel never calls forth a single touch of pity; not because he is strong to endure, but rather because he seems to live a charmed life beyond the reach of every influence that makes endurance necessary.

One wild winter morning, when Yosemite Valley was swept its length from west to east by a cordial snow-storm, I sallied forth to see what I might learn and enjoy. A sort of gray, gloaming-like darkness filled the valley, the huge walls were out of sight, all ordinary sounds were smothered, and even the loudest booming of the falls was at times buried beneath the roar of the heavy-laden blast. The loose snow was already over five feet deep on the meadows, making extended walks impossible without the aid of snow-shoes. I found no great difficulty, however, in making my way to a certain ripple on the river where one of my ouzels lived. He was at home, busily gleaning his breakfast among the pebbles of a shallow portion of the margin, apparently unaware of anything extraordinary in the weather. Presently he flew out to a stone against which the icy current was beating, and turning his back to the wind, sang as delightfully as a lark in springtime.

After spending an hour or two with my favorite, I made my way across the valley, boring and wallowing through the drifts, to learn as definitely as possible how the other birds were spending their time. The Yosemite birds are easily found during the winter because all of them excepting the Ouzel are restricted to the sunny north side of the valley, the south side being constantly eclipsed by the great frosty shadow of the wall. And because the Indian Cañon groves, from their peculiar exposure, are the warmest, the birds congregate there, more especially in severe weather.

I found most of the robins cowering on the lee side of the larger branches where the snow could not fall upon them, while two or three of the more enterprising were making desperate efforts to reach the mistletoe berries by clinging nervously to the under side of the snow-crowned masses, back downward, like woodpeckers. Every now and then they would dislodge some of the loose fringes of the snow-crown, which would come sifting down on them and send them screaming back to camp, where they would subside among their companions with a shiver, muttering in low, querulous chatter like hungry children.

Some of the sparrows were busy at the feet of the larger trees gleaning seeds and benumbed insects, joined now and then by a robin weary of his unsuccessful attempts upon the snow-covered berries. The brave woodpeckers were clinging to the snowless sides of the larger boles and overarching branches of the camp trees, making short flights from side to side of the grove, pecking now and then at the acorns they had stored in the bark, and chattering aimlessly as if unable to keep still, yet evidently putting in the time in a very dull way, like storm-bound travelers at a country tavern. The hardy nut-hatches were threading the open furrows of the trunks in their usual industrious manner, and uttering their quaint notes, evidently less distressed than their neighbors. The Steller jays were of course making more noisy stir than all the other birds combined; ever coming and going with loud bluster, screaming as if each had a lump of melting sludge in his throat, and taking good care to improve the favorable opportunity afforded by the storm to steal from the acorn stores of the woodpeckers. I also noticed one solitary gray eagle braving the storm on the top of a tall pine-stump just outside the main grove. He was standing bolt upright with his back to the wind, a tuft of snow piled on his square shoulders, a monument of passive endurance. Thus every snow-bound bird seemed more or less uncomfortable if not in positive distress. The storm was reflected in every gesture, and not one cheerful note, not to say song, came from a single bill; their cowering, joyless endurance offering a striking contrast to the spontaneous, irrepressible gladness of the Ouzel, who could no more help exhaling sweet song than a rose sweet fragrance. He *must* sing though the heavens fall. I remember noticing the distress of a pair of robins during the violent earthquake of the year 1872, when the pines of the Valley, with strange movements, flapped and waved their branches, and beetling rock-brows came thundering down to the meadows in tremendous avalanches. It did not occur to me in the midst of the excitement of other observations to look for the ouzels, but I doubt not they were singing straight on through it all, regarding the terrible rock-thunder as fearlessly as they do the booming of the waterfalls.

What may be regarded as the separate songs of the Ouzel are exceedingly difficult of description, because they are so variable and at the same time so confluent. Though I have been acquainted with my favorite ten years, and during most of this time have heard him sing nearly every day, I still detect notes and strains that seem new to me. Nearly all of his music is sweet and tender, lapsing from his round breast like water over the smooth lip of a pool, then breaking farther on into a sparkling foam of melodious notes, which glow with subdued enthusiasm, yet without expressing much of the strong, gushing ecstasy of the bobolink or skylark.

The more striking strains are perfect arabesques of melody, composed of a few full, round, mellow notes, embroidered with delicate trills which fade and melt in long slender cadences. In a general way his music is that of the streams refined and spiritualized. The deep booming notes of the falls are in it, the trills of rapids, the gurgling of margin eddies, the low whispering of level reaches, and the sweet tinkle of separate drops oozing from the ends of mosses and falling into tranquil pools.

The Ouzel never sings in chorus with other birds, nor with his kind, but only with the streams. And like flowers that bloom beneath the surface of the ground, some of our favorite's best song-blossoms never rise above the surface of the heavier music of the water. I have often observed him singing in the midst of beaten spray, his music completely buried beneath the water's roar; yet I knew he was surely singing by his gestures and the movements of his bill.

His food, as far as I have noticed, consists of all kinds of water insects, which in summer are chiefly procured along shallow margins. Here he wades about ducking his head under water and deftly turning over pebbles and fallen leaves with his bill, seldom choosing to go into deep water where he has to use his wings in diving.

He seems to be especially fond of the larvæ of mosquitos, found in abundance attached to the bottom of smooth rock channels where the current is shallow. When feeding in such places he wades up-stream, and often while his head is under water the swift current is deflected upward along the glossy curves of his neck and shoulders, in the form of a clear, crystalline shell, which fairly incloses him like a bell-glass, the shell being broken and re-formed as he lifts and dips his head; while ever and anon he sidles out to where the too powerful current carries him off his feet; then he dexterously rises on the wing and goes gleaning again in shallower places.

But during the winter, when the stream-banks are embossed in snow, and the streams themselves are chilled nearly to the freezing-point, so that the snow falling into them in stormy weather is not wholly dissolved, but forms a thin, blue sludge, thus rendering the current opaque—then he seeks the deeper portions of the main rivers, where he may dive to clear water beneath the sludge. Or he repairs to some open lake or millpond, at the bottom of which he feeds in safety.

When thus compelled to betake himself to a lake, he does not plunge into it at once like a duck, but always alights in the first place upon some rock or fallen pine along the shore. Then flying out thirty or forty yards, more or less, according to the character of the bottom, he alights with a dainty glint on the surface, swims about, looks down, finally makes up his mind, and disappears with a sharp stroke of his wings. After feeding for two or three minutes he suddenly reappears, showers the water from his wings with one vigorous shake, and rises abruptly into the air as if pushed up from beneath, comes back to his perch, sings a few minutes, and goes out to dive again; thus coming and going, singing and diving at the same place for hours.

The Ouzel is usually found singly; rarely in pairs, excepting during the breeding season, and *very* rarely in threes or fours. I once observed three thus spending a winter morning in company, upon a small glacier lake, on the Upper Merced, about 7500 feet above the level of the sea. A storm had occurred during the night, but the morning sun shone unclouded, and the shadowy lake, gleaming darkly in its setting of fresh snow, lay smooth and motionless as a mirror. My camp chanced to be within a few feet of the water's edge, opposite a fallen pine, some of the branches of which leaned out over the lake. Here my three dearly welcome visitors took up their station, and at once began to embroider the frosty air with their delicious melody, doubly delightful to me that particular morning, as I had been somewhat apprehensive of danger in breaking my way down through the snow-choked cañons to the lowlands.

The portion of the lake bottom selected for a feeding-ground lies at a depth of fifteen or twenty feet below the surface, and is covered with a short growth of algæ and other aquatic plants,—facts I had previously determined while sailing over it on a raft. After alighting on the glassy surface, they occasionally indulged in a little play, chasing one another round about in small circles; then all three would suddenly dive together, and then come ashore and sing.

The Ouzel seldom swims more than a few yards on the surface, for, not being web-footed, he makes rather slow progress, but by means of his strong, crisp wings he swims, or rather flies, with celerity under the surface, often to considerable distances. But it is in withstanding the force of heavy rapids that his strength of wing in this respect is most strikingly manifested. The following may be regarded as a fair illustration of his power of sub-aquatic flight. One stormy morning in winter when the Merced River was blue and green with unmelted snow, I observed one of my ouzels perched on a snag out in the midst of a swift-rushing rapid, singing cheerily, as if everything was just to his mind; and while I stood on the bank admiring him, he suddenly plunged into the sludgy current, leaving his song abruptly broken off. After feeding a minute or two at the bottom, and when one would suppose that he must inevitably be swept far down-stream, he emerged just where he went down, alighted on the same snag, showered the water-beads from his feathers, and continued his unfinished song, seemingly in tranquil ease as if it had suffered no interruption.

The Ouzel alone of all birds dares to enter a white torrent. And though strictly terrestrial in structure, no other is so inseparably related to water, not even the duck, or the bold ocean albatross, or the stormy-petrel. For ducks go ashore as soon as they finish feeding in undisturbed places, and very often make long flights overland from lake to lake or field to field. The same is true of most other aquatic birds. But the Ouzel, born on the brink of a stream, or on a snag or boulder in the midst of it, seldom leaves it for a single moment. For, notwithstanding he is often on the wing, he never flies overland, but whirs with rapid, quail-like beat above the stream, tracing all its windings. Even when the stream is quite small, say from five to ten feet wide, he seldom shortens his flight by crossing a bend, however abrupt it may be; and even when disturbed by meeting some one on the bank, he prefers to fly over one's head, to dodging out over the ground. When, therefore, his flight along a crooked stream is viewed endwise, it appears most strikingly wavered—a description on the air of every curve with lightning-like rapidity.

The vertical curves and angles of the most precipitous torrents he traces with the same rigid fidelity, swooping down the inclines of cascades, dropping sheer over dizzy falls amid the spray, and ascending with the same fearlessness and ease, seldom seeking to lessen the steepness of the acclivity by beginning to ascend before reaching the base of the fall. No matter though it may be several hundred feet in height he holds straight on, as if about to dash headlong into the throng of booming rockets, then darts abruptly upward, and, after alighting at the top of the precipice to rest a moment, proceeds to feed and sing. His flight is solid and impetuous, without any intermission of wing-beats,—one homogeneous buzz like that of a laden bee on its way home. And while thus buzzing freely from fall to fall, he is frequently heard giving utterance to a long outdrawn train of unmodulated notes, in no way connected with his song, but corresponding closely with his flight in sustained vigor.

Were the flights of all the ouzels in the Sierra traced on a chart, they would indicate the direction of the flow of the entire system of ancient glaciers, from about the period of the breaking up of the ice-sheet until near the close of the glacial winter; because the streams which the ouzels so rigidly follow are, with the unimportant exceptions of a few side tributaries, all flowing in channels eroded for them out of the solid flank of the range by the vanished glaciers,—the streams tracing the ancient glaciers, the ouzels tracing the streams. Nor do we find so complete compliance to glacial conditions in the life of any other mountain bird, or animal of any kind. Bears frequently accept the pathways laid down by glaciers as the easiest to travel; but they often leave them and cross over from cañon to cañon. So also, most of the birds trace the moraines to some extent, because the forests are growing on them. But they wander far, crossing the cañons from grove to grove, and draw exceedingly angular and complicated courses.

The Ouzel's nest is one of the most extraordinary pieces of bird architecture I ever saw, odd and novel in design, perfectly fresh and beautiful, and in every way worthy of the genius of the little builder. It is about a foot in diameter, round and bossy in outline, with a neatly arched opening near the bottom, somewhat like an old-fashioned brick oven, or Hottentot's hut. It is built almost exclusively of green and yellow mosses, chiefly the beautiful fronded hypnum that covers the rocks and old drift-logs in the vicinity of waterfalls. These are deftly interwoven, and felted together into a charming little hut; and so situated that many of the outer mosses continue to flourish as if they had not been plucked. A few fine, silky-stemmed grasses are occasionally found interwoven with the mosses, but, with the exception of a thin layer lining the floor, their presence seems accidental, as they are of a species found growing with the mosses and are probably plucked with them. The site chosen for this curious mansion is usually some little rock-shelf within reach of the lighter particles of the spray of a waterfall, so that its walls are kept green and growing, at least during the time of high water.

No harsh lines are presented by any portion of the nest as seen in place, but when removed from its shelf, the back and bottom, and sometimes a portion of the top, is found quite sharply angular, because it is made to conform to the surface of the rock upon which and against which it is built, the little architect always taking advantage of slight crevices and protuberances that may chance to offer, to render his structure stable by means of a kind of gripping and dovetailing.

In choosing a building-spot, concealment does not seem to be taken into consideration; yet notwithstanding the nest is large and guilelessly exposed to view, it is far from being easily detected, chiefly because it swells forward like any other bulging moss-cushion growing naturally in such situations. This is more especially the case where the nest is kept fresh by being well sprinkled. Sometimes these romantic little huts have their beauty enhanced by rack-ferns and grasses that spring up around the mossy walls, or in front of the door-sill, dripping with crystal beads.

Furthermore, at certain hours of the day, when the sunshine is poured down at the required angle, the whole mass of the spray enveloping the fairy establishment is brilliantly irised; and it is through so glorious a rainbow atmosphere as this that some of our blessed ouzels obtain their first peep at the world.

Ouzels seem so completely part and parcel of the streams they inhabit, they scarce suggest any other origin than the streams themselves; and one might almost be pardoned in fancying they come direct from the living waters, like flowers from the ground. At least, from whatever cause, it never occurred to me to look for their nests until more than a year after I had made the acquaintance of the birds themselves, although I found one the very day on which I began the search. In making my way from Yosemite to the glaciers at the heads of the Merced and Tuolumne rivers, I camped in a particularly wild and romantic portion of the Nevada cañon where in previous excursions I had never failed to enjoy the company of my favorites, who were attracted here, no doubt, by the safe nesting-places in the shelving rocks, and by the abundance of food and falling water. The river, for miles above and below, consists of a succession of small falls from ten to sixty feet in height, connected by flat, plume-like cascades that go flashing from fall to fall, free and almost channelless, over waving folds of glacier-polished granite.

On the south side of one of the falls, that portion of the precipice which is bathed by the spray presents a series of little shelves and tablets caused by the development of planes of cleavage in the granite, and by the consequent fall of masses through the action of the water. "Now here," said I, "of all places, is the most charming spot for an Ouzel's nest." Then carefully scanning the fretted face of the precipice through the spray, I at length noticed a yellowish moss-cushion, growing on the edge of a level tablet within five or six feet of the outer folds of the fall. But apart from the fact of its being situated where one acquainted with the lives of ouzels would fancy an Ouzel's nest ought to be, there was nothing in its appearance visible at first sight, to distinguish it from other bosses of rock-moss similarly situated with reference to perennial spray; and it was not until I had scrutinized it again and again, and had removed my shoes and stockings and crept along the face of the rock within eight or ten feet of it, that I could decide certainly whether it was a nest or a natural growth.

In these moss huts three or four eggs are laid, white like foam-bubbles; and well may the little birds hatched from them sing water songs, for they hear them all their lives, and even before they are born.

I have often observed the young just out of the nest making their odd gestures, and seeming in every way as much at home as their experienced parents, like young bees on their first excursions to the flower fields. No amount of familiarity with people and their ways seems to change them in the least. To all appearance their behavior is just the same on seeing a man for the first time, as when they have seen him frequently.

On the lower reaches of the rivers where mills are built, they sing on through the din of the machinery, and all the noisy confusion of dogs, cattle, and workmen. On one occasion, while a wood-chopper was at work on the river-bank, I observed one cheerily singing within reach of the flying chips. Nor does any kind of unwonted disturbance put him in bad humor, or frighten him out of calm self-possession. In passing through a narrow gorge, I once drove one ahead of me from rapid to rapid, disturbing him four times in quick succession where he could not very well fly past me on account of the narrowness of the channel. Most birds under similar circumstances fancy themselves pursued, and become suspiciously uneasy; but, instead of growing nervous about it, he made his usual dippings, and sang one of his most tranquil strains. When observed within a few yards their eyes are seen to express remarkable gentleness and intelligence; but they seldom allow so near a view unless one wears clothing of about the same color as the rocks and trees, and knows how to sit still. On one occasion, while rambling along the shore of a mountain lake, where the birds, at least those born that season, had never seen a man, I sat down to rest on a large stone close to the water's edge, upon which it seemed the ouzels and sandpipers were in the habit of alighting when they came to feed on that part of the shore, and some of the other birds also, when they came down to wash or drink. In a few minutes, along came a whirring Ouzel and alighted on the stone beside me, within reach of my hand. Then suddenly observing me, he stooped nervously as if about to fly on the instant, but as I remained as motionless as the stone, he gained confidence, and looked me steadily in the face for about a minute, then flew quietly to the outlet and began to sing. Next came a sandpiper and gazed at me with much the same guileless expression of eye as the Ouzel. Lastly, down with a swoop came a Stellar's jay out of a fir-tree, probably with the intention of moistening his noisy throat. But instead of sitting confidingly as my other visitors had done, he rushed off at once, nearly tumbling heels over head into the lake in his suspicious confusion, and with loud screams roused the neighborhood.

Love for song-birds, with their sweet human voices, appears to be more common and unfailing than love for flowers. Every one loves flowers to some extent, at least in life's fresh morning, attracted by them as instinctively as humming-birds and bees. Even the young Digger Indians have sufficient love for the brightest of those found growing on the mountains to gather them and braid them as decorations for the hair. And I was glad to discover, through the few Indians that could be induced to talk on the subject, that they have names for the wild rose and the lily, and other conspicuous flowers, whether available as food or otherwise. Most men, however, whether savage or civilized, become apathetic toward all plants that have no other apparent use than the use of beauty. But fortunately one's first instinctive love of songbirds is never wholly obliterated, no matter what the influences upon our lives may be. I have often been delighted to see a pure, spiritual glow come into the countenances of hard business-men and old miners, when a song-bird chanced to alight near them. Nevertheless, the little mouthful of meat that swells out the breasts of some song-birds is too often the cause of their death. Larks and robins in particular are brought to market in hundreds. But fortunately the Ouzel has no enemy so eager to eat his little body as to follow him into the mountain solitudes. I never knew him to be chased even by hawks.

An acquaintance of mine, a sort of foot-hill mountaineer, had a pet cat, a great, dozy, overgrown creature, about as broad-shouldered as a lynx. During the winter, while the snow lay deep, the mountaineer sat in his lonely cabin among the pines smoking his pipe and wearing the dull time away. Tom was his sole companion, sharing his bed, and sitting beside him on a stool with much the same drowsy expression of eye as his master. The good-natured bachelor was content with his hard fare of soda-bread and bacon, but Tom, the only creature in the world acknowledging dependence on him, must needs be provided with fresh meat. Accordingly he bestirred himself to contrive squirrel-traps, and waded the snowy woods with his gun, making sad havoc among the few winter birds, sparing neither robin, sparrow, nor tiny nut-hatch, and the pleasure of seeing Tom eat and grow fat was his great reward.

One cold afternoon, while hunting along the river-bank, he noticed a plain-feathered little bird skipping about in the shallows, and immediately raised his gun. But just then the confiding songster began to sing, and after listening to his summery melody the charmed hunter turned away, saying, "Bless your little heart, I can't shoot you, not even for Tom."

Even so far north as icy Alaska, I have found my glad singer. When I was exploring the glaciers between Mount Fairweather and the Stikeen River, one cold day in November, after trying in vain to force a way through the innumerable icebergs of Sum Dum Bay to the great glaciers at the head of it, I was weary and baffled and sat resting in my canoe convinced at last that I would have to leave this part of my work for another year. Then I began to plan my escape to open water before the young ice which was beginning to form should shut me in. While I thus lingered drifting with the bergs, in the midst of these gloomy forebodings and all the terrible glacial desolation and grandeur, I suddenly heard the wellknown whir of an Ouzel's wings, and, looking up, saw my little comforter coming straight across the ice from the shore. In a second or two he was with me, flying three times round my head with a happy salute, as if saying, "Cheer up, old friend; you see I'm here, and all's well." Then he flew back to the shore, alighted on the topmost jag of a stranded iceberg, and began to nod and bow as though he were on one of his favorite boulders in the midst of a sunny Sierra cascade.

The species is distributed all along the mountain-ranges of the Pacific Coast from Alaska to Mexico, and east to the Rocky Mountains. Nevertheless, it is as yet comparatively little known. Audubon and Wilson did not meet it. Swainson was, I believe, the first naturalist to describe a specimen from Mexico. Specimens were shortly afterward procured by Drummond near the sources of the Athabasca River, between the fifty-fourth and fifty-sixth parallels; and it has been collected by nearly all of the numerous exploring expeditions undertaken of late through our Western States and Territories; for it never fails to engage the attention of naturalists in a very particular manner.

Such, then, is our little cinclus, beloved of every one who is so fortunate as to know him. Tracing on strong wind every curve of the most precipitous torrents from one extremity of the Sierra to the other; not fearing to follow them through their darkest gorges and coldest snow-tunnels; acquainted with every waterfall, echoing their divine music; and throughout the whole of their beautiful lives interpreting all that we in our unbelief call terrible in the utterances of torrents and storms, as only varied expressions of God's eternal love.

W(ILLIAM) H(ENRY) HUDSON
1841–1922

Born near Buenos Aires, Argentina, of American parents, W. H. Hudson grew up on his family farm and as a young man roamed widely through the South American countryside on horseback. When Hudson was fifteen, rheumatic fever curtailed his activities, and he remained in poor health most of his life. In 1874 he came to London, where he wrote and published over twenty books, many of them reflecting his intense, lifelong interest in birds. Though he achieved great popular success with the publication of Green Mansions *(1904)—a romantic fantasy whose heroine, Rima, lives in the trees of the Venezuelan jungles and communicates with birds—recognition as a serious writer came only in his last years. His reputation today rests on such vivid reminiscences of his Argentine boyhood as* Idle Days in Patagonia *(1893) and on his treatments of the English countryside and its people in such works as* Afoot in England *(1909) and* A Shepherd's Life *(1910). He is now regarded, with Richard Jefferies, as one of the most important English nature writers of the late Victorian period.*

Hudson viewed himself as "a naturalist in the old original sense of the word: one who is mainly concerned with the 'life and conversation of animals' and whose work is consequently more like play." If Thoreau could refer affectionately to his "fishy friend" in the Concord River, Hudson could befriend a pig and at the same time face simply and directly the persistent issue of "eating our fellow mortals." There is a Hardyesque, brooding quality to his writings, yet a largeness of outlook that enables him to utter stark pronouncements on the human condition in a strangely comforting voice. Like John Burroughs, he found in Darwinian theory an enriching vision: "For we are no longer isolated, standing like starry visitors on a mountain-top, surveying life from the outside; but are on a level with and part and parcel of it."

MY FRIEND THE PIG

Is there a man among us who on running through a list of his friends is unable to say that there is one among them who is a perfect pig? I think not; and if any reader says that he has no such an one for the simple reason that he would not and could not make a friend of a perfect pig, I shall maintain that he is mistaken, that if he goes over the list a second time and a little more carefully, he will find in it not only a pig, but a sheep, a cow, a fox, a cat, a stoat, and even a perfect toad.

But all this is a question I am not concerned with, seeing that the pig I wish to write about is a real one—a four-footed beast with parted hoofs. I have a friendly feeling towards pigs generally, and consider them the most intelligent of beasts, not excepting the elephant and the anthropoid ape—the dog is not to be mentioned in this connection. I also like his disposition and attitude towards all other creatures, especially man. He is not suspicious, or shrinkingly submissive, like horses, cattle, and sheep; nor an impudent devil-may-care like the goat; nor hostile like the goose; nor condescending like the cat; nor a flattering parasite like the dog. He views us from a totally different, a sort of democratic, standpoint as fellow-citizens and brothers, and takes it for granted, or grunted, that we understand his language, and without servility or insolence he has a natural, pleasant, camerados-all or hail-fellow-well-met air with us.

It may come as a shock to some of my readers when I add that I like him, too, in the form of rashers on the breakfast-table; and this I say with a purpose on account of much wild and idle talk one hears on this question even from one's dearest friends—the insincere horror expressed and denunciation of the revolting custom of eating our fellow-mortals. The other day a lady of my acquaintance told me that she went to call on some people who lived a good distance from her house, and was obliged to stay to luncheon. This consisted mainly of roast pork, and as if that was not enough, her host, when helping her, actually asked if she was fond of a dreadful thing called the crackling!

It is a common pose; but it is also something more, since we find it mostly in persons who are frequently in bad health and are restricted to a low diet; naturally at such times vegetarianism appeals to them. As their health improves they think less of their fellow-mortals. A little chicken broth is found uplifting; then follows the inevitable sole, then calves' brains, then a sweetbread, then a partridge, and so on,

The Book of a Naturalist (New York: George H. Doran, 1919).

progressively, until they are once more able to enjoy their salmon or turbot, veal and lamb cutlets, fat capons, turkeys and geese, sirloins of beef, and, finally, roast pig. That's the limit; we have outgrown cannibalism, and are not keen about haggis, though it is still eaten by the wild tribes inhabiting the northern portion of our island. All this should serve to teach vegetarians not to be in a hurry. Thoreau's "handful of rice" is not sufficient for us, and not good enough yet. It will take long years and centuries of years before the wolf with blood on his iron jaws can be changed into the white innocent lamb that nourishes itself on grass.

Let us now return to my friend the pig. He inhabited a stye at the far end of the back garden of a cottage or small farmhouse in a lonely little village in the Wiltshire downs where I was staying. Close to the stye was a gate opening into a long green field, shut in by high hedges, where two or three horses and four or five cows were usually grazing. These beasts, not knowing my sentiments, looked askance at me and moved away when I first began to visit them, but when they made the discovery that I generally had apples and lumps of sugar in my coat pockets they all at once became excessively friendly and followed me about, and would put their heads in my way to be scratched, and licked my hands with their rough tongues to show that they liked me. Every time I visited the cows and horses I had to pause beside the pig-pen to open the gate into the field; and invariably the pig would get up and coming towards me salute me with a friendly grunt. And I would pretend not to hear or see, for it made me sick to look at his pen in which he stood belly-deep in the fetid mire, and it made me ashamed to think that so intelligent and good-tempered an animal, so profitable to man, should be kept in such abominable conditions. Oh, poor beast, excuse me, but I'm in a hurry and have no time to return your greeting or even to look at you!

In this village, as in most of the villages in all this agricultural and pastoral county of Wiltshire, there is a pig-club, and many of the cottagers keep a pig; they think and talk a great deal about their pigs, and have a grand pig-day gathering and dinner, with singing and even dancing to follow, once a year. And no wonder that this is so, considering what they get out of the pig; yet in any village you will find it kept in this same unspeakable condition. It is not from indolence nor because they take pleasure in seeing their pig unhappy before killing him or sending him away to be killed, but because they cherish the belief that the filthier the state in which they keep their pig the better the pork will be! I have met even large prosperous farmers, many of them, who cling to this delusion. One can imagine a conversation between one of these Wiltshire pig-keepers and a Danish farmer. "Yes," the visitor would say, "we too had the same notion at one time, and thought it right to keep our pigs as you do; but that was a long time back, when English and Danes were practically one people, seeing that Canute was king of both countries. We have since then adopted a different system; we now believe, and the results prove that we are in the right way, that it is best to consider the animal's nature and habits and wants, and to make the artificial conditions imposed on him as little oppressive as may be. It is true that in a state of nature the hog loves to go into pools and wallow in the mire, just as stags, buffaloes, and many other beasts do, especially in the dog-days when the flies are most troublesome. But the swine, like the stag, is a forest animal, and does not love filth for its own sake, nor to be left in a miry pen, and though not as fastidious as a cat about his coat, he is naturally as clean as any other forest creature."

Here I may add that in scores of cases when I have asked a cottager why he didn't keep a pig, his answer has been that he would gladly do so, but for the sanitary inspectors, who would soon order him to get rid of it, or remove it to a distance on account of the offensive smell. It is probable that if it could be got out of the cottager's mind that there must need be an offensive smell, the number of pigs fattened in the villages would be trebled.

I hope now after all these digressions I shall be able to go on with the history of my friend the pig. One morning as I passed the pen he grunted—spoke, I may say—in such a pleasant friendly way that I had to stop and return his greeting; then, taking an apple from my pocket, I placed it in his trough. He turned it over with his snout, then looked up and said something like "Thank-you" in a series of gentle grunts. Then he bit off and ate a small piece, then another small bite, and eventually taking what was left in his mouth he finished eating it. After that he always expected me to stay a minute and speak to him when I went to the field; I knew it from his way of greeting me, and on such occasions I gave him an apple. But he never ate it greedily: he appeared more inclined to talk than to eat, until by degrees I came to understand what he was saying. What he said was that he appreciated my kind intentions in giving him apples. But, he went on, to tell the real truth, it is not a fruit I am particularly fond of. I am familiar with its taste as they sometimes give me apples, usually the small unripe or bad ones that fall from the trees. However, I don't actually dislike them. I get skim milk and am rather fond of it; then a bucket of mash, which is good enough for hunger; but what I enjoy most is a cabbage, only I don't get one very often now. I sometimes think that if they would let me out of this muddy pen to ramble like the sheep and other beasts in the field or on the downs I should be able to pick up a number of morsels which would taste better than anything they give me. Apart from the subject of food I hope you won't mind my telling you that I'm rather fond of being scratched on the back.

So I scratched him vigorously with my stick, and made him wriggle his body and wink and blink and smile delightedly all over his face. Then I said to myself: "Now what the juice can I do more to please him?" For though under sentence of death, he had done no wrong, but was a good, honest-hearted fellow-mortal, so that I felt bound to do something to make the miry remnant of his existence a little less miserable.

I think it was the word *juice* I had just used—for that was how I pronounced it to make it less like a swear-word—that gave me an inspiration. In the garden, a few yards back from the pen, there was a large clump of old elder-trees, now overloaded with ripening fruit—the biggest clusters I had ever seen. Going to the trees I selected and cut the finest bunch I could find, as big round as my cap, and weighing over a pound. This I deposited in his trough and invited him to try it. He sniffed at it a little doubtfully, and looked at me and made a remark or two, then nibbled at the edge of the cluster, taking a few berries into his mouth, and holding them some time before he ventured to crush them. At length he did venture, then looked at me again and made more remarks, "Queer fruit this! Never tasted anything quite like it before, but I really can't say yet whether I like it or not."

Then he took another bite, then more bites, looking up at me and saying something between the bites, till, little by little, he had consumed the whole bunch; then turning round, he went back to his bed with a little grunt to say that I was now at liberty to go on to the cows and horses.

However, on the following morning he hailed my approach in such a lively manner, with such a note of expectancy in his voice, that I concluded he had been thinking a great deal about elder-berries, and was anxious to have another go at them. Accordingly I cut him another bunch, which he quickly consumed, making little exclamations the while—"Thank you, thank you, very good—very good indeed!" It was a new sensation in his life, and made him very happy, and was almost as good as a day of liberty in the fields and meadows and on the open green downs.

From that time I visited him two or three times a day to give him huge clusters of elder-berries. There were plenty for the starlings as well; the clusters on those trees would have filled a cart.

Then one morning I heard an indignant scream from the garden, and peeping out saw my friend, the pig, bound hand and foot, being lifted by a dealer into his cart with the assistance of the farmer.

"Good-bye, old boy!" said I as the cart drove off; and I thought that by and by, in a month or two, if several persons discovered a peculiar and fascinating flavour in their morning rasher, it would be due to the elder-berries I had supplied to my friend the pig, which had gladdened his heart for a week or two before receiving his quietus.

From IDLE DAYS IN PATAGONIA

[Desert Solitude]

* * * If there be such a thing as historical memory in us, it is not strange that the sweetest moment in any life, pleasant or dreary, should be when Nature draws near to it, and, taking up her neglected instrument, plays a fragment of some ancient melody, long unheard on the earth.

It might be asked: If Nature has at times this peculiar effect on us, restoring instantaneously the old vanished harmony between organism and environment, why should it be experienced in a greater degree in the Patagonian desert than in other solitary places—a desert which is waterless, where animal voices are seldom heard, and vegetation is grey instead of green? I can only suggest a reason for the effect being so much greater in my own case. In sub-tropical woods and thickets, and in wild forests in temperate regions, the cheerful verdure and bright colours of flower and insects, if we have acquired a habit of looking closely at these things, and the melody and noises of bird-life, engages the senses; there is movement and brightness; new forms, animal and vegetable, are continually appearing, curiosity and expectation are excited, and the mind is so much occupied with novel objects that the effect of wild nature in its entirety is minimised. In Patagonia the monotony of the plains, or expanse of low hills, the universal unrelieved greyness of everything, and the absence of animal forms and objects new to the eye, leave the mind open and free to receive an impression of visible nature as a whole. One gazes on the prospect as on the sea, for it stretches away sea-like, without change, into infinitude; but without the sparkle of water, the changes of hue which shadow and sunlight and nearness and distance give, and motion of waves and white flash of

Idle Days in Patagonia (London: J.M. Dent, 1893).

foam. It has a look of antiquity, of desolation, of eternal peace, of a desert that has been a desert from of old and will continue a desert for ever; and we know that its only human inhabitants are a few wandering savages, who live by hunting as their progenitors have done for thousands of years. Again, in fertile savannahs and pampas there may appear no signs of human occupancy, but the traveller knows that eventually the advancing tide of humanity will come with its flocks and herds, and the ancient silence and desolation will be no more; and this thought is like human companionship, and mitigates the effect of nature's wildness on the spirit. In Patagonia no such thought or dream of the approaching changes to be wrought by human agency can affect the mind. There is no water there, the arid soil is sand and gravel—pebbles rounded by the action of ancient seas, before Europe was; and nothing grows except the barren things that nature loves—thorns, and a few woody herbs, and scattered tufts of wiry bitter grass.

Doubtless we are not all affected in solitude by wild nature in the same degree; even in the Patagonian wastes many would probably experience no such mental change as I have described. Others have their instincts nearer to the surface, and are moved deeply by nature in any solitary place; and I imagine that Thoreau was such a one. At all events, although he was without the Darwinian lights which we have, and these feelings were always to him "strange," "mysterious," "unaccountable," he does not conceal them. There is the "something uncanny in Thoreau" which seems inexplicable and startling to such as have never been startled by nature, nor deeply moved; but which, to others, imparts a peculiarly delightful aromatic flavour to his writings. It is his wish towards a more primitive mode of life, his strange abandonment when he scours the wood like a half-starved hound, and no morsel could be too savage for him; the desire to take a ranker hold on life and live more as the animals do: the sympathy with nature so keen that it takes his breath away; the feeling that all the elements were congenial to him, which made the wildest scenes unaccountably familiar, so that he came and went with a strange liberty in nature. Once only he had doubts, and thought that human companionship might be essential to happiness; but he was at the same time conscious of a slight insanity in the mood; and he soon again became sensible of the sweet beneficent society of nature, of an infinite and unaccountable friendliness all at once like an atmosphere sustaining him. ⁎   ⁎   ⁎

CLARENCE KING
1842–1901

Mountaineering in the Sierra Nevada (1872) was a pioneering work of literature about the Sierra. King's own background, temperament, and interests were such that his book represents a virtual anthology of early responses to the mountainous West. Like John Muir, he was a daring climber; the account of his ascent of Mount Tyndall (named by King) is one of the most thrilling tales in our literature of exploration. Like Mark Twain and Bret Harte, King fills his narrative with "local color" about the miners, settlers—both Anglo and Mexican—and native peoples of the West. Racism is an element in some of these anecdotes, as it often is in writing produced along the American frontier. Finally, as a Yale graduate, a scientist who would become the founding director of the U.S. Geological Survey, and the intimate friend of such men as John Hay and Henry Adams, King shows us the reaction of a cultivated easterner to the revelation of western mountains. He found the Sierra both a confirmation of Ruskin's aesthetics and an escape from books, both a chance to assert the dominance of his sex, race, and caste and an escape from the restrictions of such identity.

From MOUNTAINEERING IN THE SIERRA NEVADA

THE RANGE

The western margin of this continent is built of a succession of mountain chains folded in broad corrugations, like waves of stone upon whose seaward base beat the mild small breakers of the Pacific.

By far the grandest of all these ranges is the Sierra Nevada, a long and massive uplift lying between the arid deserts of the Great Basin and the Californian exuberance of grain-field and orchard; its eastern

Mountaineering in the Sierra Nevada (Boston: James R. Osgood, 1872).

slope, a defiant wall of rock plunging abruptly down to the plain; the western, a long, grand sweep, well watered and overgrown with cool, stately forests; its crest a line of sharp, snowy peaks springing into the sky and catching the *alpenglow* long after the sun has set for all the rest of America.

The Sierras have a structure and a physical character which are individual and unique. To Professor Whitney and his corps of the Geological Survey of California is due the honor of first gaining a scientific knowledge of the form, plan, and physical conditions of the Sierras. How many thousands of miles, how many toilsome climbs, we made, and what measure of patience came to be expended, cannot be told; but the general harvest is gathered in, and already a volume of great interest (the forerunner of others) has been published.

The ancient history of the Sierras goes back to a period when the Atlantic and Pacific were one ocean, in whose depths great accumulations of sand and powdered stone were gathering and being spread out in level strata.

It is not easy to assign the age in which these submarine strata were begun, nor exactly the boundaries of the embryo continents from whose shores the primeval breakers ground away sand and gravel enough to form such incredibly thick deposits.

It appears most likely that the Sierra region was submerged from the earliest Palæozoic, or perhaps even the Azoic, age. Slowly the deep ocean valley filled up, until, in the late Triassic period, the uppermost tables were in water shallow enough to drift the sands and clays into wave and ripple ridges. With what immeasurable patience, what infinite deliberation, has nature amassed the materials for these mountains! Age succeeded age; form after form of animal and plant life perished in the unfolding of the great plan of development, while the suspended sands of that primeval sea sunk slowly down and were stretched in level plains upon the floor of stone.

Early in the Jurassic period an impressive and far-reaching movement of the earth's crust took place, during which the bed of the ocean rose in crumpled waves towering high in the air and forming the mountain framework of the Western United States. This system of upheavals reached as far east as Middle Wyoming and stretched from Mexico probably into Alaska. Its numerous ridges and chains, having a general northeast trend, were crowded together in one broad zone whose western and most lofty member is the Sierra Nevada. During all of the Cretaceous period, and a part of the Tertiary, the Pacific beat upon its seaward foot-hills, tearing to pieces the rocks, crumbling and grinding the shores, and, drifting the powdered stone and pebbles beneath its waves, scattered them again in layers. This submarine table-land fringed the whole base of the range and extended westward an unknown distance under the sea. To this perpetual sea-wearing of the Sierra Nevada base was added the detritus made by the cutting out of cañons, which in great volumes continually poured into the Pacific, and was arranged upon its bottom by currents.

In the late Tertiary period a chapter of very remarkable events occurred. For a second time the evenly laid beds of the sea-bottom were crumpled by the shrinking of the earth. The ocean flowed back into deeper and narrower limits, and, fronting the Sierra Nevada, appeared the present system of Coast Ranges. The intermediate depression, or sea-trough as I like to call it, is the valley of California, and is therefore a more recent continental feature than the Sierra Nevada. At once then from the folded rocks of the Coast Ranges, from the Sierra summits and the inland plateaus, and from numberless vents caused by the fierce dynamical action, there poured out a general deluge of melted rock. From the bottom of the sea sprung up those fountains of lava whose cooled material forms many of the islands of the Pacific, and, all along the coast of America, like a system of answering beacons, blazed up volcanic chimneys. The rent mountains glowed with out-pourings of molten stone. Sheets of lava poured down the slopes of the Sierra, covering an immense proportion of its surface, only the high granite and metamorphic peaks reaching above the deluge. Rivers and lakes floated up in a cloud of steam and were gone forever. The misty sky of these volcanic days glowed with innumerable lurid reflections, and, at intervals along the crest of the range, great cones arose, blacken-ing the sky with their plumes of mineral smoke. At length, having exhausted themselves, the volcanoes burned lower and lower, and, at last, by far the greater number went out altogether. With a tendency to extremes which "development" geologists would hesitate to admit, nature passed under the dominion of ice and snow.

The vast amount of ocean water which had been vaporized floated over the land, condensed upon hill-tops, chilled the lavas, and finally buried beneath an icy covering all the higher parts of the mountain system. According to well-known laws, the overburdened summits unloaded themselves by a system of glaciers. The whole Sierra crest was one pile of snow, from whose base crawled out the ice-rivers, wear-ing their bodies into the rock, sculpturing as they went the forms of val-leys, and brightening the surface of their tracks by the friction of stones and sand which were bedded, armor-like, in their nether surface. Having made their way down the slope of the Sierra, they met a low-land temperature of sufficient warmth to arrest and waste them. At last, from causes which are too intricate to be discussed at present, they shrank slowly back into the higher summit fastnesses, and there gradually perished, leaving only a crest of snow. The ice melted, and upon the whole plateau, little by little, a thin layer of soil accumulated, and, replacing the snow; there sprang up a forest of pines, whose shadows fall pleasantly to-day over rocks which were once torrents of lava and across the burnished pathways of ice. Rivers, pure and sparkling, thread the bottom of these gigantic glacier valleys. The volcanoes are extinct, and the whole theatre of this impressive geological drama is now the most glorious and beautiful region of America.

As the characters of the *Zauberflöte* passed safely through the trial of fire and the desperate ordeal of water, so, through the terror of volcanic fires and the chilling empire of ice, had the great Sierra come into the present age of tranquil grandeur.

*   *   *

There are but few points in America where such extremes of physical condition meet. What contrasts, what opposed sentiments, the two views awakened! Spread out below us lay the desert, stark and glaring, its rigid hill-chains lying in disordered grouping, in attitudes of the dead. The bare hills are cut out with sharp gorges, and over their stone skeletons scanty earth clings in folds, like shrunken flesh; they are emaciated corses of once noble ranges now lifeless, outstretched as in a long sleep. Ghastly colors define them from the ashen plain in which their feet are buried. Far in the south were a procession of whirlwind columns slowly moving across the desert in spectral dimness. A white light beat down, dispelling the last trace of shadow, and above hung the burnished shield of hard, pitiless sky.

Sinking to the *west* from our feet the gentle golden-green *glacis* sloped away, flanked by rolling-hills covered with a fresh vernal carpet of grass, and relieved by scattered groves of dark oak-trees. Upon the distant valley were checkered fields of grass and grain just tinged with the first ripening yellow. The bounding Coast Ranges lay in the cool shadow of a bank of mist which drifted in from the Pacific, covering their heights. Flocks of bright clouds floated across the sky, whose blue was palpitating with light, and seemed to rise with infinite perspective. Tranquility, abundance, the slow, beautiful unfolding of plant life, dark shadowed spots to rest our tired eyes upon, the shade of giant oaks to lie down under, while listening to brooks, contralto larks, and the soft distant lowing of cattle.

I have given the outlines of aspect along our ride across the Chabazon, omitting many amusing incidents and some *genre* pictures of rare interest among the Kaweah Indiana, as I wished simply to illustrate the relations of the Sierra with the country bordering its east base, — the barrier looming above a desert.

In Nevada and California, farther north, this wall rises more grandly, but its face rests upon a modified form of desert plains of less extent than the Colorado, and usually covered with sage-plants and other brushy *compositæ* of equally pitiful appearance. Large lakes of complicated saline waters are dotted under the Sierra shadow, the ancient terraces built upon foot-hill and outlying volcanic ranges indicating their former expansion into inland seas; and farther north still, where plains extend east of Mount Shasta, level sheets of lava form the country, and open black, rocky channels, for the numerous branches of the Sacramento and Klamath.

Approaching the Sierras anywhere from the west, you will perceive a totally different topographical and climatic condition. From the Coast Range peaks especially one obtains an extended and impressive prospect. I had fallen behind the party one May evening of our march across Pacheco's Pass, partly because some wind-bent oaks trailing almost horizontally over the wild-oat surface of the hills, and marking, as a living record, the prevalent west wind, had arrested me and called out compass and note-book; and because there had fallen to my lot an incorrigibly deliberate mustang to whom I had abandoned myself to be carried along at his own pace, comforted withal that I should get in too late to have any hand in the cooking of supper. We reached the crest, the mustang coming to a conspicuous and unwarrantable halt; I yielded, however, and sat still in the saddle, looking out to the east.

Brown foot-hills, purple over their lower slopes with "fil-a-ree" blossoms, descended steeply to the plain of California, a great, inland, prairie sea, extending for five hundred miles, mountain-locked, between the Sierras and coast hills, and now a broad arabesque surface of colors. Miles of orange-colored flowers, cloudings of green and white, reaches of violet which looked like the shadow of a passing cloud, wandering in natural patterns over and through each other, sunny and intense along near our range, fading in the distance into pale bluish-pearl tones, and divided by long, dimly seen rivers, whose margins were edged by belts of bright emerald green. Beyond rose three hundred miles of Sierra half lost in light and cloud and mist, the summit in places sharply seen against a pale, beryl sky, and again buried in warm, rolling clouds. It was a mass of strong light, soft, fathomless shadows, and dark regions of forest. However, the three belts upon its front were tolerably clear. Dusky foot-hills rose over the plain with a coppery gold tone, suggesting the line of mining towns planted in its rusty ravines, — a suggestion I was glad to repel, and look higher into that cool, solemn realm where the pines stand, green-roofed, in infinite colonnade. Lifted above the bustling industry of the plains and the melodramatic mining theatre of the foot-hills, it has a grand, silent life of its own, refreshing to contemplate even from a hundred miles away.

While I looked the sun descended; shadows climbed the Sierras, casting a gloom over foot-hill and pine, until at last only the snow summits, reflecting the evening light, glowed like red lamps along the mountain wall for hundreds of miles. The rest of the Sierra became invisible. The snow burned for a moment in the violet sky, and at last went out.

GERARD MANLEY HOPKINS
1844–1889

Nature writing includes many outstanding contributions by writers primarily associated with other genres. With Hopkins, the pattern of gradual emergence as a nature writer is further complicated by his invisibility as a poet during his lifetime. When he entered the Jesuit novitiate in 1868, he burned all his early poetry. The groundbreaking work by which we know him today was not printed until Robert Bridges edited it for publication in 1918. Hopkins's artistic achievement has since drawn readers back to his notebooks and sketches, where his vividness and distinction as a nature writer are impressive. "Inscape," his concept for the integrity of every natural object or landscape, as of every authentic poem, is powerfully conveyed in those entries. Natural experience, for Hopkins, is always individual, a lyrical whole, stressed and startling.

From NOTEBOOKS AND PAPERS OF
GERARD MANLEY HOPKINS

EXTRACTS FROM EARLY DIARIES

[1866]

Drops of rain hanging on rails etc seen with only the lower rim
lighted like nails (of fingers). Screws of brooks and twines. Soft chalky
look with more shadowy middles of the globes of cloud on a night with
a moon faint or concealed. Mealy clouds with a not brilliant moon.
Blunt buds of the ash. Pencil buds of the beech. Lobes of the trees.
Cups of the eyes, Gathering back the lightly hinged eyelids. Bows of
the eyelids. Pencil of eyelashes. Juices of the eyeball. Eyelids like
leaves, petals, caps, tufted hats, handkerchiefs, sleeves, gloves. Also of
the bones sleeved in flesh. Juices of the sunrise. Joints and veins of the
same. Vermilion look of the hand held against a candle with the darker
parts as the middles of the fingers and especially the knuckles covered
with ash.

[1870]

I have no other word yet for that which takes the eye or mind in a
bold hand or effective sketching or in marked features or again in
graphic writing, which not being beauty nor true inscape yet gives
interest and makes ugliness even better than meaninglessness.—On
the Common the snow was channeled all in parallels by the sharp
driving wind and upon the tufts of grass (where by the dark colour
shewing through it looked greyish) it came to turret-like clusters or like
broken shafts of basalt.—In the Park in the afternoon the wind was
driving little clouds of snow-dust which caught the sun as they rose and
delightfully took the eyes: flying up the slopes they looked like breaks
of sunlight fallen through ravelled cloud upon the hills and again like
deep flossy velvet blown to the root by breath which passed all along.
Nearer at hand along the road it was gliding over the ground in white
wisps that between trailing and flying shifted and wimpled like so many
silvery worms to and from one another.

Notebooks and Papers of Gerard Manley Hopkins, ed. Humphrey House (London and
New York: Oxford University Press, 1937).

The squirrel was about in our trees all the winter. For instance about Jan. 2 I often saw it.

March 12—A fine sunset: the higher sky dead clear blue bridged by a broad slant causeway rising from right to left of wisped or grass cloud, the wisps lying across; the sundown yellow, moist with light but ending at the top in a foam of delicate white pearling and spotted with big tufts of cloud in colour russet between brown and purple but edged with brassy light. But what I note it all for is this: before I had always taken the sunset and the sun as quite out of gauge with each other, as indeed physically they are for the eye after looking at the sun is blunted to everything else and if you look at the rest of the sunset you must cover the sun, but today I inscaped them together and made the sun the true eye and ace of the whole, as it is. It was all active and tossing out light and started as strongly forward from the field as a long stone or a boss in the knop of the chalice-stem: it is indeed by stalling it so that it falls into scape with the sky.

Sept. 24—First saw the Northern Lights. My eye was caught by beams of light and dark very like the crown of horny rays the sun makes behind a cloud. At first I thought of silvery cloud until I saw that these were more luminous and did not dim the clearness of the stars in the Bear. They rose slightly radiating thrown out from the earthline. Then I saw soft pulses of light one after another rise and pass upwards arched in shape but waveringly and with the arch broken. They seemed to float, not following the warp of the sphere as falling stars look to do but free though concentrical with it. This busy working of nature wholly independent of the earth and seeming to go on in a strain of time not reckoned by our reckoning of days and years but simpler and as if correcting the preoccupation of the world by being preoccupied with and appealing to and dated to the day of judgment was like a new witness to God and filled me with delightful fear

Oct. 25—A little before 7 in the evening a wonderful Aurora, the same that was seen at Rome (shortly after its seizure by the Italian government) and taken as a sign of God's anger. It gathered a little below the zenith, to the S.E. I think—a knot or crown, not a true circle, of dull blood-coloured horns and dropped long red beams down the sky on every side, each impaling its lot of stars. An hour or so later its colour was gone but there was still a pale crown in the same place: the skies were then clear and ashy and fresh with stars and there were flashes of or like sheet-lightning. The day had been very bright and clear, dis-

tances smart, herds of towering pillow clouds, one great stack in particular over Pendle was knoppled all over in fine snowy tufts and pencilled
with bloom-shadow of the greatest delicacy. In the sunset all was big and
there was a world of swollen cloud holding the yellow-rose light like a
lamp while a few sad milky blue slips passed below it. At night violent
hailstorms and hail again next day, and a solar halo. Worth noticing too
perhaps the water-runs were then mulled and less beautiful than usual

[1871]

I have been watching clouds this spring and evaporation, for instance
over our Lenten chocolate. It seems as if the heat by *aestus*, throes/ one
after another threw films of vapour off as boiling water throws off steam
under films of water, that is bubbles. One query then is whether these
films contain gas or no. The film seems to be set with tiny bubbles which
gives it a grey and grained look. By throes perhaps which represent the
moments at which the evener stress of the heat has overcome the resistance of the surface or of the whole liquid. It would be reasonable then to
consider the films as the shell of gas-bubbles and the grain on them as a
network of bubbles condensed by the air as the gas rises.—Candle smoke
goes by just the same laws, the visible film being here of unconsumed
substance, not hollow bubbles. The throes can be perceived/ like the
thrills of a candle in the socket: this is precisely to *reech*, whence *reek*.
They may be a breath of air be laid again and then shew like grey wisps
on the surface—which shews their part-solidity. They seem to be drawn
off the chocolate as you might take up a napkin between your fingers
that covered something, not so much from here or there as from the
whole surface at one reach, so that the film is perceived at the edges and
makes in fact a collar or ring just within the walls all round the cup; it
then draws together in a cowl like a candleflame but not regularly or
without a break: the question is why. Perhaps in perfect stillness it would
not but the air breathing it aside entangles it with itself. The film seems
to rise not quite simultaneously but to peel off as if you were tearing
cloth; then giving an end forward like the corner of a handkerchief and
beginning to coil it makes a long wavy hose you may sometimes look
down, as a ribbon or a carpenter's shaving may be made to do. Higher
running into frets and silvering in the sun with the endless coiling, the
soft bound of the general motion and yet the side lurches sliding into
some particular pitch it makes a baffling and charming sight.—Clouds
however solid they may look far off are I think wholly made of film in the
sheet or in the tuft. The bright woolpacks that pelt before a gale in a clear
sky are in the tuft and you can see the wind unravelling and rending

them finer than any sponge till within one easy reach overhead they are morselled to nothing and consumed—it depends of course on their size. Possibly each tuft in forepitch or in origin is quained and a crystal. Rarer and wilder packs have sometimes film in the sheet, which may be caught as it turns on the edge of the cloud like an outlying eyebrow. The one in which I saw this was a north-east wind, solid but not crisp, white like the white of egg, and bloated-looking

What you look hard at seems to look hard at you, hence the true and the false in stress of nature. One day early in March when long streamers were rising from over Kemble End one large flake loop-shaped, not a streamer but belonging to the string, moving too slowly to be seen, seemed to cap and fill the zenith with a white shire of cloud. I looked long up at it till the tall height and the beauty of the scaping—regularly curled knots springing if I remember from fine stems, like foliation in wood or stone—had strongly grown on me. It changed beautiful changes, growing more into ribs and one stretch of running into branching like coral. Unless you refresh the mind from time to time you cannot always remember or believe how deep the inscape in things is

May 9—

* * *

Later—The Horned Violet is a pretty thing, gracefully lashed. Even in withering the flower ran through beautiful inscapes by the screwing up of the petals into straight little barrels or tubes. It is not that inscape does not govern the behavior of things in slack and decay as one can see even in the pining of the skin in the old and even in a skeleton but that horror prepossesses the mind, but in this case there was nothing in itself to show even whether the flower were shutting or opening

The 'pinion' of the blossom in the comfrey is remarkable for the beauty of the coil and its regular lessening to its centre. Perhaps the duller-coloured sorts shew it best

Oct. 5—A goldencrested wren had got into my room at night and circled round dazzled by the gaslight on the white cieling; when caught even and put out it would come in again. Ruffling the crest, which is mounted over the crown and eyes like beetle-brows; I smoothed and fingered the little orange and yellow feathers which are hidden in it. Next morning I found many of these about the room and enclosed them in a letter to Cyril on his wedding day.

Aug. 16—We rose at four, when it was stormy and I saw dun-coloured waves leaving trailing hoods of white breaking on the beach. Before going I took a last look at the breakers, wanting to make out how the comb is morselled so fine into string and tassel, as I have lately noticed it to be. I saw big smooth flinty waves, carved and scuppled in shallow grooves, much swelling when the wind freshened, burst on the rocky spurs of the cliff at the little cove and break into bushes of foam. In an enclosure of rocks the peaks of the water romped and wandered and a light crown of tufty scum standing high on the surface kept slowly turning round: chips of it blew off and gadded about without weight in the air. At eight we sailed for Liverpool in wind and rain. I think it is the salt that makes rain at sea sting so much. There was a good-looking young man on board that got drunk and sung "I want to go home to Mamma." I did not look much at the sea: the crests I saw ravelled up by the wind into the air in arching whips and straps of glassy spray and higher broken into clouds of white and blown away. Under the curl shone a bright juice of beautiful green. The foam exploding and smouldering under water makes a chrysoprase green. From Blackburn I walked: infinite stiles and sloppy fields, for there has been much rain. A few big shining drops hit us aslant as if they were blown off from eaves or leaves. Bright sunset: all the sky hung with tall tossed clouds, in the west with strong printing glass edges, westward lamping with tipsy buff-light, the colour of yellow roses. Parlick ridge like a pale goldish skin without body. The plain about Clitheroe was sponged out by a tall white storm of rain. The sun itself and the spot of "session" dappled with big laps and flowers-in-damask of cloud. But we hurried too fast and it knocked me up. We went to the College, the seminary being wanted for the secular priests' retreat: almost no gas, for the retorts are being mended; therefore candles in bottles, things not ready, darkness and despair. In fact being unwell I was quite downcast: nature in all her parcels and faculties gaped and fell apart, *fatiscebat*, like a clod cleaving and holding only by strings of root. But this must often be

Nov. 8—Walking with Wm. Splaine we saw a vast multitude of starlings making an unspeakable jangle. They would settle in a row of trees; then, one tree after another, rising at a signal they looked like a cloud of specks of black snuff or powder struck up from a brush or broom or shaken from a wig; then they would sweep round in whirlwinds—you could see the nearer and farther bow of the rings by the size and blackness; many would be in one phase at once, all narrow black flakes hurling round, then in another; then they would fall upon a field and so on. Splaine wanted a gun: then 'there it would rain meat' he said. I thought they must be full of enthusiasm and delight hearing their cries and stirring and cheering one another

RICHARD JEFFERIES
1848–1887

Jefferies was an unusually prolific writer, one who pursued journalism for a number of years and whose many books included both novels and political commentary. Today, though, he is remembered mainly for his accounts of rural life in an era when the social and economic bases of English agriculture life were changing rapidly. Books like The Gamekeeper at Home *(1878),* Wild Life in a Southern County *(1879), and* Field and Hedgerow *(1889) looked not only at the traditional farming landscapes of England but also at the human communities they supported and the kinds of individuals they produced. He both celebrated a passing way of life and conveyed to his urban readers the harsher elements of a traditional life "in nature."*

OUT OF DOORS IN FEBRUARY

The cawing of the rooks in February shows that the time is coming when their nests will be re-occupied. They resort to the trees, and perch above the old nests to indicate their rights; for in the rookery possession is the law, and not nine-tenths of it only. In the slow dull cold of winter even these noisy birds are quiet, and as the vast flocks pass over, night and morning, to and from the woods in which they roost, there is scarcely a sound. Through the mist their black wings advance in silence, the jackdaws with them are chilled into unwonted quiet, and unless you chance to look up the crowd may go over unnoticed. But so soon as the waters begin to make a sound in February, running in the ditches and splashing over stones, the rooks commence the speeches and conversations which will continue till late into the following autumn.

The general idea is that they pair in February, but there are some reasons for thinking that the rooks, in fact, choose their mates at the

The Open Air (London: Chatto and Windus, 1885).

end of the preceding summer. They are then in large flocks, and if only casually glanced at appear mixed together without any order or arrangement. They move on the ground and fly in the air so close, one beside the other, that at the first glance or so you cannot distinguish them apart. Yet if you should be lingering along the by-ways of the fields as the acorns fall, and the leaves come rustling down in the warm sunny autumn afternoons, and keep an observant eye upon the rooks in the trees, or on the fresh-turned furrows, they will be seen to act in couples. On the ground couples alight near each other, on the trees they perch near each other, and in the air fly side by side. Like soldiers each has his comrade. Wedged in the ranks every man looks like his fellow, and there seems no tie between them but a common discipline. Intimate acquaintance with barrack or camp life would show that every one had his friend. There is also the mess, or companionship of half a dozen, a dozen, or more, and something like this exists part of the year in the armies of the rooks. After the nest time is over they flock together, and each family of three or four flies in concert. Later on they apparently choose their own particular friends, that is the young birds do so. All through the winter after, say October, these pairs keep together, though lost in the general mass to the passing spectator. If you alarm them while feeding on the ground in winter, supposing you have not got a gun, they merely rise up to the nearest tree, and it may then be observed that they do this in pairs. One perches on a branch and a second comes to him. When February arrives, and they resort to the nests to look after or seize on the property there, they are in fact already paired, though the almanacs put down St. Valentine's day as the date of courtship.

There is very often a warm interval in February, sometimes a few days earlier and sometimes later, but as a rule it happens that a week or so of mild sunny weather occurs about this time. Released from the grip of the frost, the streams trickle forth from the fields and pour into the ditches, so that while walking along the footpath there is a murmur all around coming from the rush of water. The murmur of the poets is indeed louder in February than in the more pleasant days of summer, for then the growth of aquatic grasses checks the flow and stills it, whilst in February, every stone, or flint, or lump of chalk divides the current and causes a vibration. With this murmur of water, and mild time, the rooks caw incessantly, and the birds at large essay to utter their welcome of the sun. The wet furrows reflect the rays so that the dark earth gleams, and in the slight mist that stays farther away the light pauses and fills the vapour with radiance. Through this luminous mist the larks race after each other twittering, and as they turn aside, swerving in their swift flight, their white breasts appear for a moment. As while standing by a pool the fishes come into sight, emerging as they swim round from the shadow of the deeper water, so the larks dart over the low hedge, and through the mist, and pass before you, and are gone again. All at once one checks his pursuit, forgets the immediate object, and rises, singing as he soars. The notes fall from the air over the dark wet earth, over the dank grass, and broken withered fern of the hedges, and listening to them it seems for a moment spring. There is sunshine in the song: the lark and the light are one. He gives us a few minutes of summer in February days. In May he rises before as yet the dawn is come, and the sunrise flows down to us under through his notes. On his breast, high above the earth, the first rays fall as the rim of the sun edges up at the eastward hill. The lark and the light are as one, and wherever he glides over the wet furrows the glint of the sun goes with him. Anon alighting he runs between the lines of the green corn. In hot summer, when the open hillside is burned with bright light, the larks are then singing and soaring. Stepping up the hill laboriously, suddenly a lark starts into the light and pours forth a rain of unwearied notes overhead. With bright light, and sunshine, and sunrise, and blue skies the bird is so associated in the mind, that even to see him in the frosty days of winter, at least assures us that summer will certainly return.

Ought not winter, in allegorical designs, the rather to be represented with such things that might suggest hope than such as convey a cold and grim despair? The withered leaf, the snowflake, the hedging bill that cuts and destroys, why these? Why not rather the dear larks for one? They fly in flocks, and amid the white expanse of snow (in the south) their pleasant twitter or call is heard as they sweep along seeking some grassy spot cleared by the wind. The lark, the bird of the light, is there in the bitter short days. Put the lark then for winter, a sign of hope, a certainty of summer. Put, too, the sheathed bud, for if you search the hedge you will find the buds there, on tree and bush, carefully wrapped around with the case which protects them as a cloak. Put, too, the sharp needles of the green corn; let the wind clear it of snow a little way, and show that under cold clod and colder snow the green thing pushes up, knowing that summer must come. Nothing despairs but man. Set the sharp curve of the white new moon in the sky: she is white in true frost, and yellow a little if it is devising change. Set the new moon as something that symbols an increase. Set the shepherd's crook in a corner as a token that the flocks are already enlarged in number. The shepherd is the symbolic man of the hardest winter time. His work is never more important than then. Those that only roam the fields when they are pleasant in May, see the lambs at play in the meadow, and naturally think of lambs and May flowers. But the lamb was born in the adversity of snow. Or you might set the morning star, for it burns and burns and glitters in the winter dawn, and throws forth beams like those of metal consumed in oxygen. There is nought that I know by comparison with which I might indicate the glory of the morning star, while yet the dark night hides in the hollows. The lamb is born in the fold. The morning star glitters in the sky. The bud is alive in its sheath; the green corn under the snow; the lark twitters as he passes. Now these to me are the allegory of winter.

These mild hours in February check the hold which winter has been gaining, and as it were, tear his claws out of the earth, their prey. If it has not been so bitter previously, when this Gulf stream or current of warmer air enters the expanse it may bring forth a butterfly and tenderly woo the first violet into flower. But this depends on its having been only moderately cold before, and also upon the stratum, whether it is backward clay, or forward gravel and sand. Spring dates are quite different according to the locality, and when violets may be found in one district, in another there is hardly a woodbine-leaf out. The border line may be traced, and is occasionally so narrow, one may cross over it almost at a step. It would sometimes seem as if even the nut-tree bushes bore larger and finer nuts on the warmer soil, and that they ripened quicker. Any curious in the first of things, whether it be a leaf, or flower, or a bird, should bear this in mind, and not be discouraged because he hears some one else has already discovered or heard something.

A little note taken now at this bare time of the kind of earth may lead to an understanding of the district. It is plain where the plough has turned it, where the rabbits have burrowed and thrown it out, where a tree has been felled by the gales, by the brook where the bank is worn away, or by the sediment at the shallow places. Before the grass and weeds, and corn and flowers have hidden it, the character of the soil is evident at these natural sections without the aid of a spade. Going slowly along the footpath—indeed you cannot go fast in moist February—it is a good time to select the places and map them out where herbs and flowers will most likely come first. All the autumn lies prone on the ground. Dead dark leaves, some washed to their woody frames, short grey stalks, some few decayed hulls of hedge fruit, and among these the mars or stocks of the plants that do not die away, but lie as it were on the surface waiting. Here the strong teazle will presently stand high; here the ground-ivy will dot the mound with bluish-purple. But it will be necessary to walk slowly to find the ground-ivy flowers under the cover of the briers. These bushes will be a likely place for a blackbird's nest; this thick close hawthorn for a bullfinch; these bramble thickets with remnants of old nettle stalks will be frequented by the whitethroat after a while. The hedge is now but a lattice-work which will before long be hung with green. Now it can be seen through, and now is the time to arrange for future discovery. In May everything will be hidden, and unless the most promising places are selected beforehand, it will not be easy to search them out. The broad ditch will be arched over, the plants rising on the mound will meet the green boughs drooping, and all the vacancy will be filled. But having observed the spot in winter you can almost make certain of success in spring.

It is this previous knowledge which invests those who are always on the spot, those who work much in the fields or have the care of woods, with their apparent prescience. They lead the new comer to a hedge, or the corner of a copse, or a bend of the brook, announcing beforehand that they feel assured something will be found there; and so it is. This, too, is one reason why a fixed observer usually sees more than one who rambles a great deal and covers ten times the space. The fixed observer who hardly goes a mile from home is like the man who sits still by the edge of a crowd, and by-and-by his lost companion returns to him. To walk about in search of persons in a crowd is well known to be the worst way of recovering them. Sit still and they will often come by. In a far more certain manner this is the case with birds and animals. They all come back. During a twelvemonth probably every creature would pass over a given locality: every creature that is not confined to certain places. The whole army of the woods and hedges marches across a single farm in twelve months. A single tree—especially an old tree—is visited by four-fifths of the birds that ever perch in the course of that period. Every year, too, brings something fresh, and adds new visitors to the list. Even the wild sea birds are found inland, and some that scarce seem able to fly at all are cast far ashore by the gales. It is difficult to believe that one would not see more by extending the journey, but, in fact, experience proves that the longer a single locality is studied the more is found in it. But you should know the places in winter as well as in tempting summer, when song and shade and colour attract every one to the field. You should face the mire and slippery path. Nature yields nothing to the sybarite. The meadow glows with buttercups in spring, the hedges are green, the woods lovely; but these are not to be enjoyed in their full significance unless you have traversed the same places when bare, and have watched the slow fulfilment of the flowers.

The moist leaves that remain upon the mounds do not rustle, and the thrush moves among them unheard. The sunshine may bring out a rabbit, feeding along the slope of the mound, following the paths or runs. He picks his way, he does not like wet. Though out at night in the

dewy grass of summer, in the rain-soaked grass of winter, and living all
his life in the earth, often damp nearly to his burrows, no time, and no
succession of generations can make him like wet. He endures it, but he
picks his way round the dead fern and the decayed leaves. He sits in the
bunches of long grass, but he does not like the drops of rain or dew on
it to touch him. Water lays his fur close, and mats it, instead of running
off and leaving him sleek. As he hops a little way at a time on the
mound he chooses his route almost as we pick ours in the mud and
pools of February. By the shore of the ditch there still stand a few dry,
dead dock stems, with some dry reddish-brown seed adhering. Some
dry brown nettle stalks remain; some grey and broken thistles; some
teazles leaning on the bushes. The power of winter has reached its
upmost now, and can go no farther. These bines which still hang in the
bushes are those of the greater bindweed, and will be used in a month
or so by many birds as conveniently curved to fit about their nests. The
stem of wild clematis, grey and bowed, could scarcely look more dead.
Fibres are peeling from it, they come off at the touch of the fingers.
The few brown feathers that perhaps still adhere where the flowers
once were are stained and discoloured by the beating of the rain. It is
not dead: it will flourish again ere long. It is the sturdiest of creepers,
facing the ferocious winds of the hills, the tremendous rains that blow
up from the sea, and bitter frost, if only it can get its roots into soil that
suits it. In some places it takes the place of the hedge proper and
becomes itself the hedge. Many of the trunks of the elms are swathed
in minute green vegetation which has flourished in the winter, as the
clematis will in the summer. Of all, the brambles bear the wild works
of winter best. Given only a little shelter, in the corner of the hedges or
under trees and copses they retain green leaves till the buds burst
again. The frosts tint them in autumn with crimson, but not all turn
colour or fall. The brambles are the bowers of the birds; in these still
leafy bowers they do the courting of the spring, and under the brambles
the earliest arum, and cleaver, or avens, push up. Round about them
the first white nettle flowers, not long now; latest too, in the autumn.
The white nettle sometimes blooms so soon (always according to locality), and again so late, that there seems but a brief interval between, as
if it flowered nearly all the year round. So the berries on the holly if let
alone often stay till summer is in, and new berries begin to appear
shortly afterwards. The ivy, too, bears its berries far into the summer.
Perhaps if the country be taken at large there is never a time when
there is not a flower of some kind out, in this or that warm southern
nook. The sun never sets, nor do the flowers ever die. There is life
always, even in the dry fir-cone that looks so brown and sapless.
    The path crosses the uplands where the lapwings stand on the paral-

lel ridges of the ploughed field like a drilled company; if they rise they wheel as one, and in the twilight move across the fields in bands, invisible as they sweep near the ground, but seen against the sky in rising over the trees and the hedges. There is a plantation of fir and ash on the slope, and a narrow waggon-way enters it, and seems to lose itself in the wood. Always approach this spot quietly, for whatever is in the wood is sure at some time or other to come to the open space of the track. Wood-pigeons, pheasants, squirrels, magpies, hares, everything feathered or furred, down to the mole, is sure to seek the open way. Butterflies flutter through the copse by it in summer, just as you or I might use the passage between the trees. Towards the evening the partridges may run through to join their friends before roost-time on the ground. Or you may see a covey there now and then, creeping slowly with humped backs, and at a distance not unlike hedgehogs in their motions. The spot therefore should be approached with care; if it is only a thrush out it is a pleasure to see him at his ease and, as he deems, unobserved. If a bird or animal thinks itself noticed it seldom does much, some will cease singing immediately they are looked at. The day is perceptibly longer already. As the sun goes down, the western sky often takes a lovely green tint in this month, and one stays to look at it, forgetting the dark and miry way homewards. I think the moments when we forget the mire of the world are the most precious. After a while the green corn rises higher out of the rude earth.

Pure colour almost always gives the idea of fire, or rather it is perhaps as if a light shone through as well as colour itself. The fresh green blade of corn is like this, so pellucid, so clear and pure in its green as to seem to shine with colour. It is not brilliant—not a surface gleam or an enamel,—it is stained through. Beside the moist clods the slender flags arise filled with the sweetness of the earth. Out of the darkness under—that darkness which knows no day save when the ploughshare opens its chinks—they have come to the light. To the light they have brought a colour which will attract the sunbeams from now till harvest. They fall more pleasantly on the corn, toned, as if they mingled with it. Seldom do we realize that the world is practically no thicker to us than the print of our footsteps on the path. Upon that surface we walk and act our comedy of life, and what is beneath is nothing to us. But it is out from that under-world, from the dead and the unknown, from the cold moist ground, that these green blades have sprung. Yonder a steam-plough pants up the hill, groaning with its own strength, yet all that strength and might of wheels, and piston, and chains, cannot drag from the earth one single blade like these. Force cannot make it; it must grow—an easy word to speak or write, in fact full of potency. It is this mystery of growth and life, of beauty, and sweetness, and colour, starting forth from the clods that gives the corn its power over me. Somehow I identify myself with it; I live again as I see it. Year by year it is the same, and when I see it I feel that I have once more entered on a new life. And I think the spring, with its green corn, its violets, and hawthorn-leaves, and increasing song, grows yearly dearer and more dear to this our ancient earth. So many centuries have flown! Now it is the manner with all natural things to gather as it were by smallest particles. The merest grain of sand drifts unseen into a crevice, and by-and-by another; after a while there is a heap; a century and it is a mound, and then every one observes and comments on it. Time itself has gone on like this; the years have accumulated, first in drifts, then in heaps, and now a vast mound, to which the mountains are knolls, rises up and overshadows us. Time lies heavy on the world. The old, old earth is glad to turn from the cark and care of drifted centuries to the first sweet blades of green.

There is sunshine today after rain, and every lark is singing. Across the vale a broad cloud-shadow descends the hillside, is lost in the hollow, and presently, without warning, slips over the edge, coming swiftly along the green tips. The sunshine follows — the warmer for its momentary absence. Far, far down in a grassy coomb stands a solitary cornrick, conical roofed, casting a lonely shadow — marked because so solitary, and beyond it on the rising slope is a brown copse. The leafless branches take a brown tint in the sunlight; on the summit above there is furze; then more hill lines drawn against the sky. In the tops of the dark pines at the corner of the copse, could the glance sustain itself to see them, there are finches warming themselves in the sunbeams. The thick needles shelter them from the current of air, and the sky is bluer above the pines. Their hearts are full already of the happy days to come, when the moss yonder by the beech, and the lichen on the fir-trunk, and the loose fibres caught in the fork of an unbending bough, shall furnish forth a sufficient mansion for their young. Another broad cloud-shadow, and another warm embrace of sunlight. All the serried ranks of the green corn bow at the word of command as the wind rushes over them.

There is largeness and freedom here. Broad as the down and free as the wind, the thought can roam high over the narrow roofs in the vale. Nature has affixed no bounds to thought. All the palings, and walls, and crooked fences deep down yonder are artificial. The fetters and traditions, the routine, the dull roundabout which deadens the spirit like the cold moist earth, are the merest nothings. Here it is easy with the physical eye to look over the highest roof. The moment the eye of the mind is filled with the beauty of things natural an equal freedom and width of view come to it. Step aside from the trodden footpath of personal experience, throwing away the petty cynicism born of petty hopes disappointed. Step out upon the broad down beside the green corn, and let its freshness become part of life.

The wind passes, and it bends—let the wind, too, pass over the spirit. From the cloud-shadow it emerges to the sunshine—let the heart come out from the shadow of roofs to the open glow of the sky. High above, the songs of the larks fall as rain—receive it with open hands. Pure is the colour of the green flags, the slender-pointed blades—let the thought be pure as the light that shines through that colour. Broad are the downs and open the aspect—gather the breadth and largness of view. Never can that view be wide enough and large enough, there will always be room to aim higher. As the air of the hills enriches the blood, so let the presence of these beautiful things enrich the inner sense. One memory of the green corn, fresh beneath the sun and wind, will lift up the heart from the clods.

ABSENCE OF DESIGN IN NATURE—
THE PRODIGALITY OF NATURE
AND NIGGARDLINESS OF MAN

In the parlour to which I have retired from the heat there is a chair and a table, and a picture on the wall: the chair was made for an object and a purpose, to sit in; the table for a purpose, to write on; the picture was painted for a purpose, to please the eye. But outside, in the meadow, in the hedge, on the hill, in the water; or, looking still farther, to the sun, the moon, and stars, I see no such chair, or table, or picture.

Pondering deeply and for long upon the plants, the living things (myself, too, as a physical being): upon the elements, on the holy miracle, water; the holy miracle, sunlight; the earth, and the air, I come at last—and not without, for a while, sorrow—to the inevitable conclusion that there is no object, no end, no purpose, no design, and no plan; no anything, that is.

By a strong and continued effort, I compelled myself to see the world mentally: with my mind, as it were, abstracted; hold yourself, as it were, apart from it, and there is no object, and no plan; no law, and no rule.

From childhood we build up for ourselves an encyclopaedia of the world, answering all questions: we turn to Day, and the reply is Light;

The Old House at Coate, ed. Samuel J. Looker (London: Lutterworth Press, 1948).

to Night, and the reply is Darkness. It is difficult to burst through these fetters and to get beyond Day and Night: but, in truth, there is no Day and Night; the sun always shines. It is our minds which supply the purpose, the end, the plan, the law, and the rule. For the practical matters of life, these are sufficient—they are like conventional agreements. But if you wish to really know the truth, there is none. When you first realize this, the whole arch of thought falls in; the structure the brain has reared, or, rather, which so many minds have reared for it, becomes a crumbling ruin, and there seems nothing left. I felt crushed when I first saw that there was no chair, no table, no picture, in nature: I use 'nature' in the widest sense; in the cosmos then. Nothing especially made for man to sit on, to write on, to admire—not even the colour of the buttercups or the beautiful sun-gleam which had me spellbound glowing on the water in my hand in the rocky cell.

The rudest quern ever yet discovered in which the earliest man ground his wheat did not fall from the sky; even that poor instrument, the mere hollowed stone, was not thrown to him prepared for use; he had to make it himself. There neither is, nor has been, nor will be any chair, or table, or picture, or quern in the cosmos. Nor is there any plan even in the buttercups themselves, looked at for themselves: they are not geometrical, or mathematical; nor precisely circular, nor anything regular. A general pattern, as a common colour, may be claimed for them, a pattern, however, liable to modification under cultivation; but, fully admitting this, it is no more than saying that water is water: that one crystal is always an octahedron, another a dodecahedron; that one element is oxygen and another hydrogen; that the earth is the earth; and the sun, the sun. It is only stating in the simplest way the fact that a thing *is*: and, after the most rigid research, that is, in the end, all that can be stated.

To say that there is a general buttercup pattern is only saying that it is not a bluebell or violet. Perhaps the general form of the buttercup is not absolutely necessary to its existence; many birds can fly equally well if their tails be removed, or even a great part of their wings. There are some birds that do not fly at all. Some further illustrations presently will arise; indeed, nothing could be examined without affording some. I had forgotten that the parlour, beside the chair and table, had a carpet. The carpet has a pattern: it is woven; the threads can be discerned, and a little investigation shows beyond doubt that it was designed and made by a man. It is certainly pretty and ingenious. But the grass of my golden meadow has no design, and no purpose: it is beautiful, and more; it is divine.

When at last I had disabused my mind of the enormous imposture of a design, an object, and an end, a purpose or a system, I began to see dimly how much more grandeur, beauty and hope there is in a divine chaos—not chaos in the sense of disorder or confusion but simply the absence of order—than there is in a universe made by pattern. This draught-board universe my mind had laid out: this machine-made world and piece of mechanism; what a petty, despicable, micro-cosmus I had substituted for the reality.

Logically, that which has a design or a purpose has a limit. The very idea of a design or a purpose has since grown repulsive to me, on account of its littleness. I do not venture, for a moment, even to attempt to supply a reason to take the place of the exploded plan. I simply deliberately deny, or, rather, I have now advanced to that stage that to my own mind even the admission of the subject to discussion is impossible. I look at the sunshine and feel that there is no contracted order: there is divine chaos, and, in it, limitless hope and possibilities.

Without number, the buttercups crowd the mead: not one here and there, or sufficient only to tint the sward. There is not just enough for some purpose: there they are without number, in all the extravagance of uselessness and beauty. The apple-bloom—it is falling fast now as the days advance—who can count the myriad blossoms of the orchard? There are leaves upon the hedges which bound that single meadow on three sides (the fourth being enclosed by a brook) enough to occupy the whole summer to count; and before it was half done they would be falling. But that half would be enough for shadow—for use.

Half the rain that falls would be enough. Half the acorns on the oaks in autumn, more than enough. Wheat itself is often thrown into the sty. Famines and droughts occur, but whenever any comes it is in abundance—sow a grain of wheat, and the stalk, one stalk alone, of those that rise from it will yield forty times.

There is no *enough* in nature. It is one vast prodigality. It is a feast. There is no economy: it is all one immense extravagance. It is all giving, giving, giving: no saving, no penury; a golden shower of good things is for ever descending. I love beyond all things to contemplate this indescribable lavishness—I would it could be introduced into our human life. I know, none better, having gone through the personal experience myself, that it is at the present moment impossible to practise it: that each individual is compelled, in order to exist, to labour, to save, and to economize. I know, of course, as all do who have ever read a book, that attempts to distribute possessions, to live in community of goods, have each failed miserably. If I rightly judge, the human race would require a century of training before even an approximation to such a thing were possible. All this, and much more to the same effect,

I fully admit. But still the feeling remains and will not be denied. I dislike the word economy: I detest the word thrift; I hate the thought of saving. Maybe some scheme in the future may be devised whereby such efforts may be turned to a general end. This alone I am certain of: there is no economy, thrift, or saving, in nature; it is one splendid waste. It is that waste which makes it so beautiful, and so irresistible! Now nature was not made by man, and is a better exemplar than he can furnish: each thread in this carpet goes to form the pattern; but go out into my golden mead and gather ten thousand blades of grass, and it will not destroy it.

Perhaps there never were so many houses upon the face of the earth as at the present day: so luxuriously appointed, so comfortable, so handsomely furnished. Yet, with all this wealth and magnificence, these appointments and engineering: with all these many courses at dinner and array of wines, it has ever seemed to me a mean and penurious age. It is formal and in order; there is no heart in it. Food should be broadcast, open, free: wine should be in flagons, not in tiny glasses; in a word, there should be genial waste. Let the crumbs fall: there are birds enough to pick them up.

The greatest proof of the extreme meanness of the age is the long list of names appended to a subscription for a famine or a fashionable charity. Worthy as are these objects, the donors write down their own unutterable meanness. There are men in their warehouses, their offices, on their lands, who have served them honourably for years and have received for their wage just exactly as much as experience has proved can be made to support life. No cheque with a great flourishing signature has ever been presented to them.

I say that the entire labouring population—some skilled trades excepted as not really labouring—is miserably underpaid, not because there is a pressure or scarcity, a trouble, a famine, but from pure selfishness. This selfishness, moreover, is not intentional, but quite unconscious; and individuals are not individually guilty, because they are within their rights. A man has a hundred thousand pounds: he eats and drinks and pleases his little whims—likely enough quite innocent little whims—but he never gives to a friend, or a relation; never assists, does nothing with it. This is commercially right, but it is not the buttercups in the golden mead; it is not the grain of wheat that yielded forty times. It is not according to the exemplar of nature. Therefore I say that although I admit all attempts to adjust possessions have been and for the age at least must prove failures, yet my feeling remains the same. Thrift, economy, accumulation of wealth, are inventions; they are not nature. As there are more than enough buttercups in this single meadow for the pleasure of all the children in the hamlet, so too it is a fact, a very stubborn fact, that there is more than enough food in the world for all its human children. In the year 1880, it was found, on careful calculation made for strictly commercial purposes, that there was a surplus grain production of[1] bushels. That is to say, if every buttercup in this meadow represented a bushel of wheat, there would be all that over and above what was necessary. This is a very extraordinary fact. That the wheat has to be produced, to be distributed; that there are a thousand social complications to be considered, is, of course, incontrovertible. Still, there was the surplus; bushels of golden grain as numerous as the golden buttercups.

But that does not represent the capacity of the earth for production: it is not possible to guage that capacity—so practically inexhaustible is it.

Thrift and economy and accumulation, therefore, represent a state of things contrary to the exemplar of nature, and in individual life they destroy its beauty. There is no pleasure without waste: the banquet is a formality; the wine tasteless, unless the viands and the liquor are in prodigal quantities. Give me the lavish extravagance of the golden mead!

[1] There is a blank in the manuscript here.

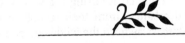

MABEL OSGOOD WRIGHT
1859–1934

The recent return of Mabel Osgood Wright's 1894 work The Friendship of the Seasons *into print clarifies certain continuities in our literary tradition. It demonstrates a deeply rooted American lineage for that celebration of gardens now associated with such contemporary authors as Michael Pollan and Jamaica Kincaid. It displays the continuing and decisive influence of the Concord sages, as transmitted through her Unitarian-minister father, a student of William Ellery Channing. And it presents the example of a woman whose love of birds and the flowering world made her a remarkably effective advocate—through the Connecticut Audubon Society and other vehicles—for the protection of habitat, the regulation of hunting, and the establishment of national parks.*

THE STORY OF A GARDEN

There is a garden that is not like the other gardens round about. In many of these gardens the flowers are only prisoners, forced to weave carpets on the changeless turf, and when the eye is sated and the impression palls, they become to their owners, who have no part in them, merely purchased episodes.

This garden that I know has a bit of green, a space of flowers, and a stretch of wildness, as Bacon says a garden should always have. At its birth the twelve months each gave to it a gift, that it might always yield an offering to the year, and presently it grew so lovable that there came to it a soul.

The song-sparrow knows that this is so; the mottled owl that lives in the hollow sassafras has told it to the night-hawk. Catbirds and robins, routed from other gardens by fusillades, still their quick-throbbing hearts, feeling its protection. The coward crow alone knows its exclusion, for he was unhoused from the tall pines and banished for fratricide. The purling bluebird, claiming the pole-top house as an ancestral bequest, repeats the story every springtime. The oriole and swallow whisper of it in their southward course, and, returning, bring with them willing colonists.

The rock polypody creeps along in confidence, with no ruthless hand to strip it off, and the first hepatica opens its eyes in safety, for tongues of flame or the grub-axe have not crippled it during the winter. Once the petted garden beauties looked askance, from their smooth beds in the tilled corner, and drew their skirts away from the wildwood company, but now, each receiving according to its need, they live in perfect concord.

The wild rose in the chinky wall peeps shyly at her glowing sisters, and the goldenrod bows over it to gossip with the pentstemon. And this is how it came to be, for the garden was no haphazard accident. Nature began it, and, following her master-touch, the hand and brain of a man, impelled by a reverent purpose, evolved its shaping.

This man, even when a little boy, had felt the potency of Nature's touch to soothe the heartache. One day, led by an older mate, he trudged a weary way to see a robber hanged. The child, not realizing the scene he was to witness, was shocked to nervous frenzy, and a pitying bystander, thinking to divert his mind, gave him a shilling. Spying a bird pedlar in the crowd, he bought a goldfinch and a pint of seeds, and the horror of the hanging was quite forgotten and effaced by the little bird, his first possession. To it he gave his confidence and told all his

The Friendship of Nature: A New England Chronicle of Birds and Flowers (New York: Macmillan Co., 1894).

small griefs and joys, and through the bird Nature laid her warm hand on his heart and gently drew it toward their mutual Master, and never after did he forget her consolation.

All this was more than seventy years ago. When the boy grew to manhood, following the student life, the spirit of the bird that had blotted out the scene of civil murder was still with him. Its song kept his thoughts single and led him toward green fields, that their breath might leaven lifeless things, strengthening the heart that felt a world-weariness, as all must feel at times when facing human limitations.

Love came, and home; then, following hand in hand, honour and disappointment; and again, with double purpose, he turned Natureward. Not to the goatish Pan, but to Nature's motherhood, to find a shrine upon her breast where he might keep his holiest thoughts, and watch them grow. A place apart, where the complete man might be at rest, and walking in the cool of day feel the peace of God.

At first the garden was a formless bit of waste, but Nature tangles things with a motive, and it was in the making that it came to win a soul, for the man's spirit grew so calm and strong that it gave its overplus to what it wrought.

The garden's growth was nowhere warped or stunted by tradition; there was no touch of custom's bondage to urge this or that. No rudeness had despoiled its primal wildness, and lovers, who had trodden paths under the trees, were its sole discoverers. It was rock-fenced and briar-guarded; the sharp shadows of the cedars dialed the hours, and the ground-pine felt its darkened way beneath them with groping fingers.

This happened before I was, but hearing of it often, sound has imparted its sense to sight, and it all seems visual. With my first consciousness, the days were filled with planting and with growth; the pines already hid the walls, and cattle tracks were widened into paths and wound among young maples, elms, and beeches. Then there grew in me a love that made the four garden walls seem like the boundaries of the world.

Nothing was troubled but to free it from the oppression of some other thing. The sparrow kept his bush, and between him and the hawkheadsman a hand was raised. The wood thrush, finding his haunts untouched, but that his enemies, the black snakes, might no longer boldly engulf his nestlings, raised his clear voice and sang "O Jubilate Deo!" The gardener who planted no longer watches the bird's flight, but the garden still tells its story. Will you come in? The gate is never closed except to violence.

Eight acres of rolling ground, and in the centre a plainly cheerful house decides the point of view. The location of a house much affects the inmates; here sunshine penetrates every room and a free current of air sweeps all about, and there is a well of sparkling water close at hand. This well is rock-drilled, deep and cold, and the patron divinity of all good wells, the north star, watches over it, and nightly Ursa Major's dipper circles above, as if offering a cooling draught to all the constellations.

For a space about the house the grass is cropped, and some plump beds of geraniums, Fuchsias, heliotropes, serve to grade the eye from indoor precision, to rest the vision before the trees and moving birds compel it to investigation. However much natural wildness may soothe and satisfy, the home is wholly a thing of man's making, and he may gather about it the growing things that need his constant ministry. The sight of such an open space gives the birds more confidence, and the worm enemies that always follow cultivation offer them a change of food.

The old queen-apple tree that casts its petals every May against the window-panes, like snow blushing at its own boldness, held many nests last spring. A bluebird spied a knot-hole where decay had left him an easy task; a pair of yellow warblers, with cinnamon-streaked breasts, fastened their tiny cup between a forked branch above the range of sight. For several days I watched these birds, fluttering about the window corners where cobwebs cling and spiders weave, and thought they searched for food, until, following the yellow flash they made among the leaves, I saw that they were building; and when I secured the empty nest in August, it proved to be a dainty thing woven of dry grass, the down of dandelions, cocoons, and cobwebs.

A robin raised two broods, building a new nest for the second, as the first one was too near the path to suit his partner's nerves. He spent his days in prying earth-worms from the lawn, singing at dawn and twilight so deliciously that he furnished one more proof that bird voices, even of the same species, have individual powers of expression, like those of men.

The fourth bird to build, a red-eyed vireo, was quite shy at first, yet hung the nest over the path, so that when I passed to and fro her ruby eyes were on a level with me. After the eggs were laid, she allowed me to bend down the branch, and a few days later, to smooth her head gently with my finger. A chipping sparrow added his wee nest to the collection, watching the horses as they passed, timidly craving a hair from each, and finally securing a tuft from an old mattress, with which he lined his home to his complete content.

If you would keep the wild birds in your garden, you must exclude from it four things: English sparrows, the usual gardeners, cats, and firearms. These sparrows, even if not belligerent, are antagonistic to song birds, and brawl too much; a cat of course, being a cat, carries its own condemnation; a gun aimed even at a target brings terror into bird-land; and a gardener, of the type that mostly bear the name, is a sort of bogyman, as much to Nature lovers as to the birds. The gardener wishes this, orders that, is rigid in point of rights and etiquette, and looks with scarcely veiled contempt at all wild things, flowers, birds, trees; would scrape away the soft pine needles from the footpaths and scatter stone dust in their place, or else rough, glaring pebbles. He would drive away the songsters with small shot, his one idea of a proper garden bird being a china peacock.

It is, of course, sadly true, that cherries, strawberries, grapes, and hungry birds cannot meet with safety to the fruit, but we should not therefore emulate the men of Killingworth. We may buy from a neighbouring farmer, for a little money, all the fruit we lack, but who for untold gold can fill the hedge with friendly birds, if once we grieve or frighten them away?

You may grow, however, tender peas in plenty, and all the vegetables that must go direct from earth to table to preserve their flavour; only remember when you plant the lettuce out, to dedicate every fourth head to the wild rabbits, who, even while you plant, are twitching their tawny ears under the bushes, and then you will suffer no disappointment. Once in a time a gardener-naturalist may drift to you, and your garden will then entertain a kindred spirit. Such a man came to this garden, a young Dane, full of northern legend and sentiment, recognizing through rough and varied work the motive of the place,—like drawing like; and with him, a blonde-haired, laughing wife, and a wee daughter called Zinnia, for the gay flowers, and he found time to steal among the trees in the June dawns to share in the bird's raptures, making his life in living.

It is a drowsy August afternoon; the birds are quiet, and the locusts express the heat by their intonation. The Japan lilies, in the border back of the house, are densely sweet, the geraniums mockingly red, and the lemon-verbena bushes are drooping. The smooth grass and trim edges stop before an arch that spans the path, and about it shrubs straggle, grouping around a tall ash. This ash, a veritable lodestone to the birds, is on the borderland of the wild and cultivated, and they regard it as the Mussulman does his minaret, repairing there to do homage. Before the leaves appear the wood thrush takes the topmost branch to sing his matins, as if, by doing so, he might, before his neighbours, give the sun greeting.

The robins light on it, *en route*, when they fear that their thefts in other gardens will find them out, and the polite cedar-birds, smoothing each other's feathers, sun themselves in it daily before the flocks break into pairs. Upon the other side, a hospitable dogwood spreads itself, a goodly thing from spring till frost, and from it spireas, Deutzias, weigelas, lilacs, the flowering quince, and strawberry shrub, follow the path that winds under the arch, past mats of ferns and laurel, to a tilled corner, a little inner garden, where plants are nursed and petted, and no shading tree or greedy root robs them of sun or nourishment.

Along the path between the pines, the black leaf mould of the woods has been strewn freely. The fern tribe is prolific in this neighbourhood, and a five-mile circuit encloses some twenty species, most of which may be transplanted, if you keep in mind their special needs. This spot is cool and shady, but the soil is dry from careful drainage. The aspidiums flourish well; A. *acrostichoides*, of two varieties, better known as the Christmas fern, with heavy varnished fronds, A. *marginàle*, with pinnate, dull-green fronds, A. *cristatum*, almost doubly pinnate and with them the fragrant *Dicksonia punctilobula*, whose straw-coloured lace carpets the autumn woods with sunlight, and the black-stemmed maidenhair grows larger every year, rearing its curving fronds two feet or more.

What endless possibilities creep into the garden with every barrow of wood earth! How many surprises cling about the roots of the plant you hope to transfer uninjured from its home! Bring a tuft of ferns, lo! there springs up a dozen unseen things—a pad of partridge vine, an umber of ginseng, a wind flower; in another year the round leaves of the pyrola may appear and promenade in pairs and trios quite at their ease, until the fern bed becomes a constant mystery. For many years some slow awaking seeds will germinate, the rarer violets, perhaps an orchis.

I brought a mat of club moss, with a good lump of earth, as was my habit, from the distant woods. Several years after, happening to stop to clear away some dead branches, I started in surprise, for enthroned in the centre of the moss, a very queen, was a dark pink cypripedium, the Indian moccasin. It is an orchid very shy of transposition, seldom living over the second season after its removal, seeming to grieve for its native home with the fatal Heimweh, so that the seed must have come with the moss and done its growing in the fern nook. ✻ ✻ ✻

ERNEST THOMPSON SETON
1860–1946

Born in England, Ernest Thompson Seton moved at the age of six with his family to Canada. There, as a hunter and tracker, he reveled in the great variety of wildlife on the prairies of Manitoba. While a young man he studied art in Paris and earned a living for a while as an animal illustrator, but with the publication of Wild Animals I Have Known *in 1898, he became the most popular author of animal stories in his day. Over the next fifty years he wrote over forty more books, including* Biography of a Grizzly *(1901) and* Lives of the Hunted *(1901).*

As these titles suggest, Seton identified strongly with the animals he described. He is generally credited with inventing the modern genre of "animal fiction writing," though he always insisted that "every incident in their biographies is true." Seton's writing is distinguished not only by carefully observed detail and narrative impetus but also by his attempts to enter into and portray the minds and personalities of his subjects. This made him one the prime targets of the more Darwinian naturalist John Burroughs, who attacked the "nature fakers" in a famous 1903 article in the Atlantic Monthly. *Many of Seton's observations, however, have been borne out by the research of contemporary animal behaviorists.*

The following excerpt is taken from one of his most famous stories and is based on Seton's own experience as a bounty hunter hired to trap a notorious and elusive wolf-pack leader in New Mexico in 1893. Its emotional power is enhanced by Seton's ambivalence about his own actions and his ability to shift the reader's sympathy.

From WILD ANIMALS I HAVE KNOWN

NOTE TO THE READER

These stories are true. Although I have left the strict line of historical truth in many places, the animals in this book were all real charac-

Wild Animals I Have Known (New York: Scribner's 1898).

ters. They lived the lives I have depicted, and showed the stamp of heroism and personality more strongly by far than it has been in the power of my pen to tell.

I believe that natural history has lost much by the vague general treatment that is so common. What satisfaction would be derived from a ten-page sketch of the habits and customs of Man? How much more profitable it would be to devote that space to the life of some one great man. This is the principle I have endeavored to apply to my animals. The real personality of the individual, and his view of life are my theme, rather than the ways of the race in general, as viewed by a casual and hostile human eye.

This may sound inconsistent in view of my having pieced together some of the characters, but that was made necessary by the fragmentary nature of the records. There is, however, almost no deviation from the truth in Lobo.   *   *   *

Lobo lived his wild romantic life from 1889 to 1894 in the Currumpaw region, as the ranchmen know too well, and died, precisely as related, on January 31, 1894.

*   *   *

The fact that these stories are true is the reason why all are tragic. The life of a wild animal *always has a tragic end.*

Such a collection of histories naturally suggests a common thought—a moral it would have been called in the last century. No doubt each different mind will find a moral to its taste, but I hope some will herein find emphasized a moral as old as Scripture—we and the beasts are kin. Man has nothing that the animals have not at least a vestige of, the animals have nothing that man does not in some degree share.

Since, then, the animals are creatures with wants and feelings differing in degree only from our own, they surely have their rights. This fact, now beginning to be recognized by the Caucasian world, was first proclaimed by Moses and was emphasized by the Buddhist over two thousand years ago.

LOBO

*   *   *   Some years before . . . I had been a wolf-hunter, but my occupations since then had been of another sort, chaining me to stool and desk. I was much in need of a change, and when a friend, who was also a ranch-owner on the Currumpaw, asked me to come to New Mexico and try if I could do anything with this predatory pack, I accepted the invitation and, eager to make the acquaintance of its king, was as soon as possible among the mesas of that region. I spent some

time riding about to learn the country, and at intervals, my guide would point to the skeleton of a cow to which the hide still adhered, and remark, "That's some of his work."

It became quite clear to me that, in this rough country, it was useless to think of pursuing Lobo with hounds and horses, so that poison or traps were the only available expedients. At present we had no traps large enough, so I set to work with poison.

I need not enter into the details of a hundred devices that I employed to circumvent this 'loup-garou'; there was no combination of strychnine, arsenic, cyanide, or prussic acid, that I did not essay; there was no manner of flesh that I did not try as bait; but morning after morning, as I rode forth to learn the result, I found that all my efforts had been useless. The old king was too cunning for me. * * *

At length the wolf traps arrived, and with two men I worked a whole week to get them properly set out. We spared no labor or pains, I adopted every device I could think of that might help to insure success. The second day after the traps arrived, I rode around to inspect, and soon came upon Lobo's trail running from trap to trap. In the dust I could read the whole story of his doings that night. He had trotted along in the darkness, and although the traps were so carefully concealed, he had instantly detected the first one. Stopping the onward march of the pack, he had cautiously scratched around it until he had disclosed the trap, the chain, and the log, then left them wholly exposed to view with the trap still unsprung, and passing on he treated over a dozen traps in the same fashion. Very soon I noticed that he stopped and turned aside as soon as he detected suspicious signs on the trail and a new plan to outwit him at once suggested itself. I set the traps in the form of an H; that is, with a row of traps on each side of the trail, and one on the trail for the cross-bar of the H. Before long, I had an opportunity to count another failure. Lobo came trotting along the trail, and was fairly between the parallel lines before he detected the single trap in the trail, but he stopped in time, and why or how he knew enough I cannot tell, the Angel of the wild things must have been with him, but without turning an inch to the right or left, he slowly and cautiously backed on his own tracks, putting each paw exactly in its old track until he was off the dangerous ground. Then returning at one side he scratched clods and stones with his hind feet till he had sprung every trap. This he did on many other occasions, and although I varied my methods and redoubled my precautions, he was never deceived, his sagacity seemed never at fault, and he might have been pursuing his career of rapine to-day, but for an unfortunate alliance that proved his ruin and added his name to the long list of heroes who, unassailable when alone, have fallen through the indiscretion of a trusted ally.

Once or twice, I had found indications that everything was not quite right in the Currumpaw pack. There were signs of irregularity, I thought; for instance there was clearly the trail of a smaller wolf running ahead of the leader, at times, and this I could not understand until a cowboy made a remark which explained the matter.

"I saw them to-day," he said, "and the wild one that breaks away is Blanca." Then the truth dawned upon me, and I added, "Now, I know that Blanca is a she-wolf, because were a he-wolf to act thus, Lobo would kill him at once."

This suggested a new plan. I killed a heifer, and set one or two rather obvious traps about the carcass. Then cutting off the head, which is considered useless offal, and quite beneath the notice of a wolf, I set it a little apart and around it placed two powerful steel traps properly deodorized and concealed with the utmost care. During my operations I kept my hands, boots, and implements smeared with fresh blood, and afterward sprinkled the ground with the same, as though it had flowed from the head; and when the traps were buried in the dust I brushed the place over with the skin of a coyote, and with a foot of the same animal made a number of tracks over the traps. The head was so placed that there was a narrow passage between it and some tussocks, and in this passage I buried two of my best traps, fastening them to the head itself.

Wolves have a habit of approaching every carcass they get the wind of, in order to examine it, even when they have no intention of eating of it, and I hoped that this habit would bring the Currumpaw pack within reach of my latest stratagem. I did not doubt that Lobo would detect my handiwork about the meat, and prevent the pack approaching it, but I did build some hopes on the head, for it looked as though it had been thrown aside as useless.

Next morning, I sallied forth to inspect the traps, and there, oh joy! were the tracks of the pack, and the place where the beef-head and its traps had been was empty. A hasty study of the trail showed that Lobo had kept the pack from approaching the meat, but one, a small wolf, had evidently gone on to examine the head as it lay apart and had walked right into one of the traps.

We set out on the trail, and within a mile discovered that the hapless wolf was Blanca. Away she went, however, at a gallop, and although encumbered by the beef-head, which weighted over fifty pounds, she speedily distanced my companion who was on foot. But we overtook her when she reached the rocks for the horns of the cow's head became caught and held her fast. She was the handsomest wolf I had ever seen. Her coat was in perfect condition and nearly white.

She turned to fight, and raising her voice in the rallying cry of her race, sent a long howl rolling over the cañon. From far away upon the mesa came a deep response, the cry of Old Lobo. That was her last call, for now we had closed in on her, and all her energy and breath were devoted to combat.

Then followed the inevitable tragedy, the idea of which I shrank from afterward more than at the time. We each threw a lasso over the neck of the doomed wolf, and strained our horses in opposite directions until the blood burst from her mouth, her eyes glazed, her limbs stiffened and then fell limp. Homeward then we rode, carrying the dead wolf, and exulting over this, the first death-blow we had been able to inflict on the Currumpaw pack.

At intervals during the tragedy, and afterward as we rode homeward, we heard the roar of Lobo as he wandered about on the distant mesas, where he seemed to be searching for Blanca. He had never really deserted her, but knowing that he could not save her, his deep-rooted dread of firearms had been too much for him when he saw us approaching. All that day we heard him wailing as he roamed in his quest, and I remarked at length to one of the boys, "Now, indeed, I truly know that Blanca was his mate."

As evening fell he seemed to be coming toward the home cañon, for his voice sounded continually nearer. There was an unmistakable note of sorrow in it now. It was no longer the loud, defiant howl, but a long, plaintive wail; "Blanca! Blanca!" he seemed to call. And as night came down, I noticed that he was not far from the place where we had overtaken her. At length he seemed to find the trail, and when he came to the spot where we had killed her, his heart-broken wailing was piteous to hear. It was sadder than I could possibly have believed. Even the stolid cowboys noticed it, and said they had "never heard a wolf carry on like that before." He seemed to know exactly what had taken place, for her blood had stained the place of her death.

Then he took up the trail of the horses and followed it to the ranch-house. Whether in hopes of finding her there, or in quest of revenge, I know not, but the latter was what he found, for he surprised our unfortunate watch-dog outside and tore him to little bits within fifty yards of the door. He evidently came alone this time, for I found but one trail next morning, and he had galloped about in a reckless manner that was very unusual with him. I had half expected this, and had set a number of additional traps about the pasture. Afterward I found that he had indeed fallen into one of these, but such was his strength, he had torn himself loose and cast it aside.

I believed that he would continue in the neighborhood until he found her body at least, so I concentrated all my energies on this one enterprise of catching him before he left the region, and while yet in this reckless mood. Then I realized what a mistake I had made in killing Blanca, for by using her as a decoy I might have secured him the next night.

I gathered in all the traps I could command, one hundred and thirty strong steel wolf-traps, and set them in fours in every trail that led into the cañon; each trap was separately fastened to a log, and each log was separately buried. In burying them, I carefully removed the sod and every particle of earth that was lifted we put in blankets, so that after the sod was replaced and all was finished the eye could detect no trace of human handiwork. When the traps were concealed I trailed the body of poor Blanca over each place, and made of it a drag that circled all about the ranch, and finally I took off one of her paws and made with a line of tracks over each trap. Every precaution and device known to me I used, and retired at a late hour to await the result.

Once during the night I thought I heard Old Lobo, but was not sure of it. Next day I rode around, but darkness came on before I completed the circuit of the north cañon, and I had nothing to report. At supper one of the cowboys said, "There was a great row among the cattle in the north cañon this morning, maybe there is something in the traps there." It was afternoon of the next day before I got to the place referred to, and as I drew near a great grizzly form arose from the ground, vainly endeavoring to escape, and there revealed before me stood Lobo, King of Currumpaw, firmly held in the traps. Poor old hero, he had never ceased to search for his darling, and when he found the trail her body had made he followed it recklessly, and so fell into the snare prepared for him. There he lay in the iron grasp of all four traps, perfectly help-less, and all around him were numerous tracks showing how the cattle had gathered about him to insult the fallen despot, without daring to approach within his reach. For two days and two nights he had lain there, and now was worn out with struggling. Yet, when I went near him, he rose up with bristling mane and raised his voice, and for the last time made the cañon reverberate with his deep bass roar, a call for help, the muster call of his band. But there was none to answer him, and, left alone in his extremity, he whirled about with all his strength and made a desperate effort to get at me. All in vain, each trap was a dead drag of over three hundred pounds, and in their relentless four-fold grasp, with great steel jaws on every foot, and the heavy logs and chains all entangled together, he was absolutely powerless. How his huge ivory tusks did grind on those cruel chains, and when I ventured to touch him with my rifle-barrel he left grooves on it which are there to this day. His eyes glared green with hate and fury, and his jaws snapped with a hollow 'chop,' as he vainly endeavored to reach me and my trembling horse. But he was worn out with hunger and struggling and loss of blood, and he soon sank exhausted to the ground.

Something like compunction came over me, as I prepared to deal out to him that which so many had suffered at his hands.

"Grand old outlaw, hero of a thousand lawless raids, in a few minutes you will be a great load of carrion. It cannot be otherwise." Then I swung my lasso and sent it whistling over his head. But not so fast; he was yet far from being subdued, and, before the supple coils had fallen on his neck he seized the noose and, with one fierce chop, cut through its hard thick strands, and dropped it in two pieces at his feet.

Of course I had my rifle as a last resource, but I did not wish to spoil his royal hide, so I galloped back to the camp and returned with a cowboy and a fresh lasso. We threw to our victim a stick of wood which he seized in his teeth, and before he could relinquish it our lassoes whistled through the air and tightened on his neck.

Yet before the light had died from his fierce eyes, I cried, "Stay, we will not kill him; let us take him alive to the camp." He was so completely powerless now that it was easy to put a stout stick through his mouth, behind his tusks, and then lash his jaws with a heavy cord which was also fastened to the stick. The stick kept the cord in, and the cord kept the stick in so he was harmless. As soon as he felt his jaws were tied he made no further resistance, and uttered no sound, but looked calmly at us and seemed to say, "Well, you have got me at last, do as you please with me." And from that time he took no more notice of us.

We tied his feet securely, but he never groaned, nor growled, nor turned his head. Then with our united strength were just able to put him on my horse. His breath came evenly as though sleeping, and his eyes were bright and clear again, but did not rest on us. Afar on the great rolling mesas they were fixed, his passing kingdom, where his famous band was now scattered. And he gazed till the pony descended the pathway into the cañon, and the rocks cut off the view.

By travelling slowly we reached the ranch in safety, and after securing him with a collar and a strong chain, we staked him out in the pasture and removed the cords. Then for the first time I could examine him closely, and proved how unreliable is vulgar report when a living hero or tyrant is concerned. He had *not* a collar of gold about his neck, nor was there on his shoulders an inverted cross to denote that he had leagued himself with Satan. But I did find on one haunch a great broad scar, that tradition says was the fang-mark of Juno, the leader of Tannerey's wolf-hounds—a mark which she gave him the moment before he stretched her lifeless on the sand of the cañon.

I set meat and water beside him, but he paid no heed. He lay calmly on his breast, and gazed with those steadfast yellow eyes away past me down through the gateway of the cañon, over the open plains—his plains—nor moved a muscle when I touched him. When the sun went down he was still gazing fixedly across the prairie. I expected he would call up his band when night came, and prepared for them, but he had called once in his extremity, and none had come; he would never call again.

A lion shorn of his strength, an eagle robbed of his freedom, or a dove bereft of his mate, all die, it is said, of a broken heart; and who will aver that this grim bandit could bear the three-fold brunt, heart-whole? This only I know, that when the morning dawned, he was lying there still in his position of calm repose, his body unwounded, but his spirit was gone—the old King-wolf was dead.

I took the chain from his neck, a cowboy helped me to carry him to the shed where lay the remains of Blanca, and as we laid him beside her, the cattle-man exclaimed: "There, you *would* come to her, now you are together again."

GENE STRATTON PORTER
1863–1924

Gene Stratton Porter became one of the most popular authors of her day with novels like Freckles *(1904),* A Girl of the Limberlost *(1909), and* The Harvester *(1911). While some critics found her plots to be simplistic or her characters sentimental, no one could deny the vividness of her natural settings, specifically of northeastern Indiana's Limberlost Swamp, to which she returned again and again in her writing. Stratton Porter seems to have possessed galvanic energy. In addition to producing a steady stream of best-selling novels, she wrote a series of nonfiction books, often describing her passionate interest in birds. She was also a self-taught photographer of distinction and began illustrating many of her nature studies in about 1910. Moths of the Limberlost (1912), from which the following selection comes, was one of these books. It is reported that one reason for the superiority of her photographs of moths was that, unlike those in most such studies, they were taken of living specimens and conveyed the insects' true bloom and grace. In her writing about moths, too, Stratton Porter evokes the keen observation of a naturalist willing to invest many hours and much hot walking in order to glimpse a representative of some rare and beautiful species.*

From MOTHS OF THE LIMBERLOST

THE ROBIN MOTH

So I lost my first Cecropia, and from that day until a woman grown and much of this material secured, in all my field work among the birds, flowers, and animals, I never had seen another. They had taunted me in museums, and been my envy in private collections, but find one, I could not. When in my field work among the birds, so many moths of other families almost had thrust themselves upon me that I began a collection of reproductions of them, I found little difficulty in securing almost anything else. I could picture Sphinx Moths in any position I chose, and Lunas seemed eager to pose for me. A friend carried to me a beautiful tan-coloured Polyphemus with transparent moons like isinglass set in its wings of softest velvet down, and as for butterflies, it was not necessary to go afield for them; they came to me. I could pick a Papilio Ajax, that some of my friends were years in securing, from the pinks in my garden. A pair of Antiopas spent a night, and waited to be pictured in the morning, among the leaves of my passion vine. Painted Beauties swayed along my flowered walks, and in September a Viceroy reigned in state on every chrysanthemum, and a Monarch was enthroned on every sunbeam. No luck was too good for me, no butterfly or moth too rare, except forever and always the coveted Cecropia, and by this time I had learned to my disgust that it was one of the commonest of all.

Then one summer, late in June, a small boy, having an earnest, eager little face, came to me tugging a large box. He said he had some-

Moths of the Limberlost (Garden City, N.Y.: Doubleday, Page & Company, 1912)

thing for me. He said "they called it a butterfly, but he was sure it never was." He was eminently correct. He had a splendid big Cecropia. I was delighted. Of course to have found one myself would have filled my cup to overflowing, but to secure a perfect, living specimen was good enough. For the first time my childish loss seemed in a measure compensated. Then, I only could study a moth to my satisfaction and set it free; now, I could make reproductions so perfect that every antler of its antennæ could be counted with the naked eye, and copy its colours accurately, before giving back its liberty.

I asked him whether he wanted money or a picture of it, and as I expected, he said "money," so he was paid. An hour later he came back and said he wanted the picture. On being questioned as to his change of heart, he said "mamma told him to say he wanted the picture, and she would give him the money." My sympathy was with her. I wanted the studies I intended to make of that Cecropia myself, and I wanted them very badly.

I opened the box to examine the moth, and found it so numb with the cold over night, and so worn and helpless, that it could not cling to a leaf or twig. I tried repeatedly, and fearing that it had been subjected to rough treatment, and soon would be lifeless, for these moths live only a short time, I hastily set up a camera focusing on a branch. Then I tried posing my specimen. Until the third time it fell, but the fourth it clung, and crept down a twig, settling at last in a position that far surpassed any posing that I could do. I was very pleased, and yet it made a complication. It had gone so far that it might be off the plate and from focus. It seemed so stupid and helpless that I decided to risk a peep at the glass, and hastily removing the plate and changing the shutter, a slight but most essential alteration was made, everything replaced, and the bulb caught up. There was only a breath of sound as I turned, and then I stood horrified, for my Cecropia was sailing over a large elm tree in a corner of the orchard, and for a block my gaze followed it skyward, flying like a bird before it vanished in the distance, so quickly had it recovered in fresh air and sunshine.

I have undertaken to describe some very difficult things, but I would not attempt to portray my feelings, and three days later there was no change. It was in the height of my season of field work, and I had several extremely interesting series of bird studies on hand, and many miscellaneous subjects. In those days some pictures were secured that I then thought, and yet feel, will live, but nothing mattered to me. There was a standing joke among my friends that I never would be satisfied with my field work until I had made a study of a "Ha-ha bird," but I doubt if even that specimen would have lifted the gloom of those days.

Everything was a drag, and frequently I would think over it all in detail, and roundly bless myself for taking a prize so rare, to me at least, into the open.

The third day stands lurid in my memory. It was the hottest, most difficult day of all my years of experience afield. The temperature ranged from 104 to 108 in the village, and in quarries open to the east, flat fields, and steaming swamps it certainly could have been no cooler. With set cameras I was working for a shot of a hawk that was feeding on all the young birds and rabbits in the vicinity of its nest. I also wanted a number of studies to fill a commission that was pressing me. Subjects for several pictures had been found, and exposures made on them when the weather was so hot that the rubber slide of a plate holder would curl like a horseshoe if not laid on a case, and held flat by a camera while I worked. Perspiration dried, and the landscape took on a sombre black velvet hue, with a liberal sprinkling of gold stars. I sank into a stupor going home, and an old farmer aroused me, and disentangled my horse from a thicket of wild briers into which it had strayed. He said most emphatically that if I did not know enough to remain indoors weather like that, my friends should appoint me a "guardeen."

I reached the village more worn in body and spirit than I ever had been. I felt that I could not endure another degree of heat on the back of my head, and I was much discouraged concerning my work. Why not drop it all, and go where there were cool forests and breezes sighing? Perhaps my studies were not half so good as I thought! Perhaps people would not care for them! For that matter, perhaps the editors and publishers never would give the public an opportunity to see my work at all!

I dragged a heavy load up the steps and swung it to the veranda, and there stood almost paralyzed. On the top step, where I could not reach the Cabin door without seeing it, newly emerged, and slowly exercising a pair of big wings, with every gaudy marking fresh with new life, was the finest Cecropia I ever had seen anywhere. Recovering myself with a start, I had it under my net—that had waited twenty years to cover it! Inside the door I dropped the net, and the moth crept on my fingers. What luck! What extra golden luck! I almost felt that God had been sorry for me, and sent it there to encourage me to keep on picturing the beauties and wonders of His creations for people who could not go afield to see for themselves, and to teach those who could to protect helpless, harmless things for their use and beauty.

I walked down the hall, and vaguely scanned the solid rows of books and specimens lining the library walls. I scarcely realized the thought that was in my mind, but what I was looking for was not there.

The dining-room then, with panelled walls and curtains of tapestry? It was not there! Straight to the white and gold music room I went. Then a realizing sense came to me. It was *Brussels lace* for which I was searching! On the most delicate, snowiest place possible, on the finest curtain there, I placed my Cecropia, and then stepped back and gazed at it with a sort of "Touch it over my dead body" sentiment in my heart. An effort was required to arouse myself, to realize that I was not dreaming. To search the fields and woods for twenty years, and then find the specimen I had sought awaiting me at my own door! Well might it have been a dream, but that the Cecropia, clinging to the meshes of the lace, slowly opening and closing its wings to strengthen them for flight, could be nothing but a delightful reality.

A few days later, in the valley of the Wood Robin, while searching for its nest I found a large cocoon. It was above my head, but afterward I secured it by means of a ladder, and carried it home. Shortly there emerged a yet larger Cecropia, and luck seemed with me. I could find them everywhere through June, the time of their emergence, later their eggs, and the tiny caterpillars that hatched from them. During the summer I found these caterpillars, in different stages of growth, until fall, when after their last moult and casting of skin, they reached the final period of feeding; some were over four inches in length, a beautiful shade of greenish blue, with red and yellow warty projections—tubercles, according to scientific works.

It is easy to find the cocoons these caterpillars spin, because they are the largest woven by any moth, and placed in such a variety of accessible spots. They can be found in orchards, high on branches, and on water sprouts at the base of trees. Frequently they are spun on swamp willows, box-elder, maple, or wild cherry. Mr. Black once found for me the largest cocoon I ever have seen; a pale tan colour with silvery lights, woven against the inside of a hollow log. Perhaps the most beautiful of all, a dull red, was found under the flooring of an old bridge crossing a stream in the heart of the swamp, by a girl not unknown to fiction, who brought it to me. In a deserted orchard close the Wabash, Raymond once found a pair of empty cocoons at the foot of a big apple tree, fastened to the same twigs, and within two inches of each other.

But the most wonderful thing of all occurred when Wallace Hardison, a faithful friend to my work, sawed a board from the roof of his chicken house and carried to me twin Cecropia cocoons, spun so closely together they were touching, and slightly interwoven. By the closest examination I could discover slight difference between them. The one on the right was a trifle fuller in the body, wider at the top, a shade lighter in colour and the inner case seemed heavier.

All winter those cocoons occupied the place of state in my collection. Every few days I tried them to see if they gave the solid thump indicating healthy pupæ, and listened to learn if they were moving. By May they were under constant surveillance. On the fourteenth I was called from home a few hours to attend the funeral of a friend. I think nothing short of a funeral would have taken me, for the moth from a single cocoon had emerged on the eleventh. I hurried home near noon, only to find that I was late, for one was out, and the top of the other cocoon heaving with the movements of the second.

The moth that had escaped was a male. It clung to the side of the board, wings limp, its abdomen damp. The opening from which it came was so covered with terra cotta coloured down that I thought at first it must have disfigured itself; but full development proved it could spare that much and yet appear all right.

In the fall I had driven a nail through one corner of the board, and tacked it against the south side of the Cabin, where I made reproductions of the cocoons. The nail had been left, and now it suggested the same place. A light stroke on the head of the nail, covered with cloth to prevent jarring, fastened the board on a log. Never in all my life did I hurry as on that day, and I called my entire family into service. The Deacon stood at one elbow, Molly-Cotton at the other, and the gardener in the rear. There was not a second to be lost, and no time for an unnecessary movement; for in the heat and bright sunshine those moths would emerge and develop with amazing rapidity.

Molly-Cotton held an umbrella over them to prevent this as much as possible; the Deacon handed plate holders, and Brenner ran errands. Working as fast as I could make my fingers fly in setting up the camera, and getting a focus, the second moth's head was out, its front feet struggling to pull up the body, and its antennæ beginning to lift, when I was ready for the first snap at half-past eleven.

By the time I inserted the slide, turned the plate holder and removed another slide, the first moth to appear had climbed up the board a few steps, and the second was halfway out. Its antennæ were nearly horizontal now, and from its position I decided that the wings as they lay in the pupa case were folded neither to the back nor to the front, but pressed against the body in a lengthwise crumpled mass, the heavy front rib, or costa, on top.

Again I changed plates with all speed. By the time I was ready for the third snap the male had reached the top of the board, its wings opened for the first time, and began a queer trembling motion. The second one had emerged and was running into the first, so I held my finger in the line of its advance, and when it climbed on I lowered it to the edge of the board beside the cocoons. It immediately clung to the wood. The big pursy abdomen and smaller antennæ, that now turned forward in position, proved this a female. The exposure was made not ten seconds after she cleared the case, and with her back to the lens, so the position and condition of the wings and antennæ on emergence can be seen clearly.

Quickly as possible I changed the plates again; the time that elapsed could not have been over half a minute. The male was trying to creep up the wall, and the increase in the length and expansion of the female's wings could be seen. The colours on both were exquisite, but they grew a trifle less brilliant as the moths became dry.

Again I turned to the business of plate changing. The heat was intense, and perspiration was streaming from my face. I called to Molly-Cotton to shield the moths while I made the change. "Drat the moths!" cried the Deacon. "Shade your mother!" Being an obedient girl, she shifted the umbrella, and by the time I was ready for business, the male was on the logs and travelling up the side of the Cabin. The female was climbing toward the logs also, so that a side view showed her wings already beginning to lift above her back.

I had only five snapshot plates in my holders, so I was compelled to stop. It was as well, for surely the record was complete, and I was almost prostrate with excitement and heat. Several days later I opened each of the cocoons and made interior studies. The one on the right was split down the left side and turned back to show the bed of spun silk of exquisite colour that covers the inner case. Some say this silk has no commercial value, as it is cut in lengths reaching from the top around the inner case and back to the top again; others think it can be used. The one on the left was opened down the front of the outer case, the silk parted and the heavy inner case cut from top to bottom to show the smooth interior wall, the thin pupa case burst by the exit of the moth, and the cast caterpillar skin crowded at the bottom.

The pair mated that same night, and the female began laying eggs by noon the following day. She dotted them in lines over the inside of her box, and on leaves placed in it, and at times piled them in a heap instead of placing them as do these moths in freedom. Having taken a picture of a full-grown caterpillar of this moth brought to me by Mr. Andrew Idlewine, I now had a complete Cecropia history; eggs, full-grown caterpillars, twin cocoons, and the story of the emergence of the moths that wintered in them. I do not suppose Mr. Hardison thought he was doing anything unusual when he brought me those cocoons, yet by bringing them, he made it possible for me to secure this series of twin Cecropia moths, male and female, a thing never before recorded by lepidopterist or photographer so far as I can learn.

The Cecropia is a moth whose acquaintance nature-loving city people can cultivate. In December of 1906, on a tree, maple I think, near No. 2230 North Delaware Street, Indianapolis, I found four cocoons of this moth, and on the next tree, save one, another. Then I began watching, and in the coming days I counted them by the hundred through the city. Several bushels of these cocoons could have been clipped in Indianapolis alone, and there is no reason why any other city that has maple, elm, catalpa, and other shade trees would not have as many; so that any one who would like can find them easily.

*   *   *

My heart goes out to Cecropia because it is such a noble, birdlike, big fellow, and since it has decided to be rare with me no longer, all that is necessary is to pick it up, either in caterpillar, cocoon, or moth, at any season of the year, in almost any location. The Cecropia moth resembles the robin among birds; not alone because he is gray with red markings, but also he haunts the same localities. The robin is the bird of the eaves, the back door, the yard and orchard. Cecropia is the moth. My doorstep is not the only one they grace; my friends have found them in like places. Cecropia cocoons are attached to fences, chicken-coops, barns, houses, and all through the orchards of old country places, so that their emergence at bloom time adds to May and June one more beauty, and frequently I speak of them as the Robin Moth.

In connection with Cecropia there came to me the most delightful experience of my life. One perfect night during the middle of May, all the world white with tree bloom, touched to radiance with brilliant moonlight, intoxicating with countless blending perfumes, I placed a female Cecropia on the screen of my sleeping-room door and retired. The lot on which the Cabin stands is sloping, so that, although the front foundations are low, my door is at least five feet above the ground, and opens on a circular porch, from which steps lead down between two apple trees, at that time sheeted in bloom. Past midnight I was awakened by soft touches on the screen, faint pullings at the wire. I went to the door and found the porch, orchard, and night-sky alive with Cecropias holding high carnival. I had not supposed there were so many in all this world. From every direction they came floating like birds down the moonbeams. I carefully removed the female from the door to a window close beside, and stepped on the porch. No doubt I was permeated with the odour of the moth. As I advanced to the top step, that lay even with the middle branches of the apple trees, the exquisite big creatures came swarming around me. I could feel them on my hair, my shoulders, and see them settling on my gown and out-

stretched hands. Far as I could penetrate the night-sky more were coming. They settled on the bloom-laden branches, on the porch pillars, on
me indiscriminately. I stepped inside the door with one on each hand
and five clinging to my gown. This experience, I am sure, suggested
Mrs. Comstock's moth hunting in the Limberlost. Then I went back to
the veranda and revelled with the moths until dawn drove them to shelter. One magnificent specimen, birdlike above all the others, I followed
across the orchard and yard to a grape arbour, where I picked him from
the under side of a leaf after he had settled for the coming day.
Repeatedly I counted close to a hundred, and then they would so con-
fuse me by flight I could not be sure I was not numbering the same one
twice. With eight males, some of them fine large moths, one superb,
from which to choose, my female mated with an insistent, frowsy little
scrub lacking two feet and having torn and ragged wings. I needed no
surer proof that she had very dim vision.

MARY AUSTIN
1868–1934

*Although an extremely prolific author and a celebrated figure in her own
day, Mary Austin's work lapsed into obscurity for almost half a century
after her death. Perhaps part of the problem was that she cultivated so
many fields that critics didn't know what to do with her. Austin was a
novelist, a poet, and an essayist; a student of Indian culture, an advocate
of native peoples, and a feminist; a pioneer author in the areas of science
fiction and the nature writing of the Southwest. Over the past decade,
however, increasing numbers of her works have been reissued, as readers
discover in the range of her interests an integrity responsive to the challenges of our own day. One of her finest works,* The Land of Little Rain
*(1903), evokes the high desert country of southern California, and has
reestablished itself as a classic of American nature writing.*

THE LAND OF LITTLE RAIN

East away from the Sierras, south from Panamint and Amargosa, east and south many an uncounted mile, is the Country of Lost Borders.

Ute, Paiute, Mojave, and Shoshone inhabit its frontiers, and as far into the heart of it as a man dare go. Not the law, but the land sets the limit. Desert is the name it wears upon the maps, but the Indian's is the better word. Desert is a loose term to indicate land that supports no man; whether the land can be bitted and broken to that purpose is not proven. Void of life it never is, however dry the air and villainous the soil.

This is the nature of that country. There are hills, rounded, blunt, burned, squeezed up out of chaos, chrome and vermilion painted, aspiring to the snow-line. Between the hills lie high level-looking plains full of intolerable sun glare, or narrow valleys drowned in a blue haze. The hill surface is streaked with ash drift and black, unweathered lava flows. After rains water accumulates in the hollows of small closed valleys, and, evaporating, leaves hard dry levels of pure desertness that get the local name of dry lakes. Where the mountains are steep and the rains heavy, the pool is never quite dry, but dark and bitter, rimmed about with the efflorescence of alkaline deposits. A thin crust of it lies along the marsh over the vegetating area, which has neither beauty nor freshness. In the broad wastes open to the wind the sand drifts in hummocks about the stubby shrubs, and between them the soil shows saline traces. The sculpture of the hills here is more wind than water work, though the quick storms do sometimes scar them past many a year's redeeming. In all the Western desert edges there are essays in miniature at the famed, terrible Grand Cañon, to which, if you keep on long enough in this country, you will come at last.

Since this is a hill country one expects to find springs, but not to depend upon them; for when found they are often brackish and unwholesome, or maddening, slow dribbles in a thirsty soil. Here you find the hot sink of Death Valley, or high rolling districts where the air has always a tang of frost. Here are the long heavy winds and breathless calms on the tilted mesas where dust devils dance, whirling up into a wide, pale sky. Here you have no rain when all the earth cries for it, or quick downpours called cloud-bursts for violence. A land of lost rivers, with little in it to love; yet a land that once visited must be come back to inevitably. If it were not so there would be little told of it.

This is the country of three seasons. From June on to November it lies hot, still, and unbearable, sick with violent unrelieving storms; then

The Land of Little Rain (Boston: Houghton Mifflin, 1903).

on until April, chill, quiescent, drinking its scant rain and scanter snows; from April to the hot season again, blossoming, radiant, and seductive. These months are only approximate; later or earlier the rain-laden wind may drift up the water gate of the Colorado from the Gulf, and the land sets its seasons by the rain.

The desert floras shame us with their cheerful adaptations to the seasonal limitations. Their whole duty is to flower and fruit, and they do it hardly, or with tropical luxuriance, as the rain admits. It is recorded in the report of the Death Valley expedition that after a year of abundant rains, on the Colorado desert was found a specimen of Amaranthus ten feet high. A year later the same species in the same place matured in the drought at four inches. One hopes the land may breed like qualities in her human offspring, not tritely to "try," but to do. Seldom does the desert herb attain the full stature of the type. Extreme aridity and extreme altitude have the same dwarfing effect, so that we find in the high Sierras and in Death Valley related species in miniature that reach a comely growth in mean temperatures. Very fertile are the desert plants in expedients to prevent evaporation, turning their foliage edgewise toward the sun, growing silky hairs, exuding viscid gum. The wind, which has a long sweep, harries and helps them. It rolls up dunes about the stocky stems, encompassing and protective, and above the dunes, which may be, as with the mesquite, three times as high as a man, the blossoming twigs flourish and bear fruit.

There are many areas in the desert where drinkable water lies within a few feet of the surface, indicated by the mesquite and the bunch grass (*Sporobolus airoides*). It is this nearness of unimagined help that makes the tragedy of desert deaths. It is related that the final breakdown of that hapless party that gave Death Valley its forbidding name occurred in a locality where shallow wells would have saved them. But how were they to know that? Properly equipped it is possible to go safely across that ghastly sink, yet every year it takes its toll of death, and yet men find there sun-dried mummies, of whom no trace or recollection is preserved. To underestimate one's thirst, to pass a given landmark to the right or left, to find a dry spring where one looked for running water—there is no help for any of these things.

Along springs and sunken watercourses one is surprised to find such water-loving plants as grow widely in moist ground, but the true desert breeds its own kind, each in its particular habitat. The angle of the slope, the frontage of a hill, the structure of the soil determines the plant. South-looking hills are nearly bare, and the lower tree-line higher here by a thousand feet. Cañons running east and west will have one wall naked and one clothed. Around dry lakes and marshes the herbage preserves a set and orderly arrangement. Most species have well-defined areas of growth, the best index the voiceless land can give the traveler of his whereabouts.

If you have any doubt about it, know that the desert begins with the creosote. This immortal shrub spreads down into Death Valley and up to the lower timberline, odorous and medicinal as you might guess from the name, wandlike, with shining fretted foliage. Its vivid green is grateful to the eye in a wilderness of gray and greenish white shrubs. In the spring it exudes a resinous gum which the Indians of those parts know how to use with pulverized rock for cementing arrow points to shafts. Trust Indians not to miss any virtues of the plant world!

Nothing the desert produces expresses it better than the unhappy growth of the tree yuccas. Tormented, thin forests of it stalk drearily in the high mesas, particularly in that triangular slip that fans out eastward from the meeting of the Sierras and coastwise hills where the first swings across the southern end of the San Joaquin Valley. The yucca bristles with bayonet-pointed leaves, dull green, growing shaggy with age, tipped with panicles of fetid, greenish bloom. After death, which is slow, the ghostly hollow network of its woody skeleton, with hardly power to rot, makes the moonlight fearful. Before the yucca has come to flower, while yet its bloom is a creamy cone-shaped bud of the size of a small cabbage, full of sugary sap, the Indians twist it deftly out of its fence of daggers and roast it for their own delectation. So it is that in those parts where man inhabits one sees young plants of *Yucca arborensis* infrequently. Other yuccas, cacti, low herbs, a thousand sorts, one finds journeying east from the coastwise hills. There is neither poverty of soil nor species to account for the sparseness of desert growth, but simply that each plant requires more room. So much earth must be preëmpted to extract so much moisture. The real struggle for existence, the real brain of the plant, is underground; above there is room for a rounded perfect growth. In Death Valley, reputed the very core of desolation, are nearly two hundred identified species.

Above the lower tree-line, which is also the snow-line, mapped out abruptly by the sun, one finds spreading growth of piñon, juniper, branched nearly to the ground, lilac and sage, and scattering white pines.

There is no special preponderance of self-fertilized or wind-fertilized plants, but everywhere the demand for and evidence of insect life. Now where there are seeds and insects there will be birds and small mammals and where these are, will come the slinking, sharp-toothed kind that prey on them. Go as far as you dare in the heart of a lonely land, you cannot go so far that life and death are not before you. Painted lizards slip in and out of rock crevices, and pant on the white hot sands. Birds, hummingbirds even, nest in the cactus scrub; woodpeckers befriend the demoniac yuccas; out of the stark, treeless waste rings the music of the night-singing mockingbird. If it be summer and the sun well down, there will be a burrowing owl to call. Strange, furry, tricksy things dart across the open places, or sit motionless in the conning towers of the creosote. The poet may have "named all the birds without a gun," but not the fairy-footed, ground-inhabiting, furtive, small folk of the rainless regions. They are too many and too swift; how many you would not believe without seeing the footprint tracings in the sand. They are nearly all night workers, finding the days too hot and white. In mid-desert where there are no cattle, there are no birds of carrion, but if you go far in that direction the chances are that you will find yourself shadowed by their tilted wings. Nothing so large as a man can move unspied upon in that country, and they know well how the land deals with strangers. There are hints to be had here of the way in which a land forces new habits on its dwellers. The quick increase of suns at the end of spring sometimes overtakes birds in their nesting and effects a reversal of the ordinary manner of incubation. It becomes necessary to keep eggs cool rather than warm. One hot, stifling spring in the Little Antelope I had occasion to pass and repass frequently the nest of a pair of meadowlarks, located unhappily in the shelter of a very slender weed. I never caught them sitting except near night, but at midday they stood, or drooped above it, half fainting with pitifully parted bills, between their treasure and the sun. Sometimes both of them together with wings spread and half lifted continued a spot of shade in a temperature that constrained me at last in a fellow feeling to spare them a bit of canvas for permanent shelter. There was a fence in that country shutting in a cattle range, and along its fifteen miles of posts one could be sure of finding a bird or two in every strip of shadow; sometimes the sparrow and the hawk, with wings trailed and beaks parted, drooping in the white truce of noon.

If one is inclined to wonder at first how so many dwellers came to be in the loneliest land that ever came out of God's hands, what they do there and why stay, one does not wonder so much after having lived there. None other than this long brown land lays such a hold on the affections. The rainbow hills, the tender bluish mists, the luminous radiance of the spring, have the lotus charm. They trick the sense of time, so that once inhabiting there you always mean to go away without quite realizing that you have not done it. Men who have lived there, miners and cattle-men, will tell you this, not so fluently, but emphatically, cursing the land and going back to it. For one thing there is the divinest, cleanest air to be breathed anywhere in God's world. Some day the world will understand that, and the little oases on the windy tops of hills will harbor for healing its ailing, house-weary broods. There is promise there of great wealth in ores and earths, which is no wealth by reason of being so far removed from water and workable conditions, but men are bewitched by it and tempted to try the impossible.

You should hear Salty Williams tell how he used to drive eighteen- and twenty-mule teams from the borax marsh to Mojave, ninety miles, with the trail wagon full of water barrels. Hot days the mules would go so mad for drink that the clank of the water bucket set them into an uproar of hideous, maimed noises, and a tangle of harness chains, while Salty would sit on the high seat with the sun glare heavy in his eyes, dealing out curses of pacification in a level, uninterested voice until the clamor fell off from sheer exhaustion. There was a line of shallow graves along that road; they used to count on dropping a man or two every new gang of coolies brought out in the hot season. But when he lost his swamper, smitten without warning at the noon halt, Salty quit his job; he said it was "too durn hot." The swamper he buried by the way with stones upon him to keep the coyotes from digging him up, and seven years later I read the penciled lines on the pine head-board, still bright and unweathered.

But before that, driving up on the Mojave stage, I met Salty again crossing Indian Wells, his face from the high seat, tanned and ruddy as a harvest moon, looming through the golden dust above his eighteen mules. The land had called him.

The palpable sense of mystery in the desert air breeds fables, chiefly of lost treasure. Somewhere within its stark borders, if one believes report, is a hill strewn with nuggets; one seamed with virgin silver; an old clayey water-bed where Indians scooped up earth to make cooking pots and shaped them reeking with grains of pure gold. Old miners drifting about the desert edges, weathered into the semblance of the tawny hills, will tell you tales like these convincingly. After a little sojourn in that land you will believe them on their own account. It is a question whether it is not better to be bitten by the little horned snake of the desert that goes sidewise and strikes without coiling, than by the tradition of a lost mine.

And yet—and yet—is it not perhaps to satisfy expectation that one falls into the tragic key in writing of desertness? The more you wish of it the more you get, and in the mean time lose much of pleasantness. In that country which begins at the foot of the east slope of the Sierras and spreads out by less and less lofty hill ranges toward the Great Basin, it is possible to live with great zest, to have red blood and delicate joys, to pass and repass about one's daily performance an area that would make an Atlantic seaboard State, and that with no peril, and, according to our way of thought, no particular difficulty. At any rate, it was not people who went into the desert merely to write it up who invented the fabled Hassaympa, of whose waters, if any drink, they can no more see fact as naked fact, but all radiant with the color of romance. I, who must have drunk of it in my twice seven years' wanderings, am assured that it is worth while.

For all the toll the desert takes of a man it gives compensations, deep breaths, deep sleep, and the communion of the stars. It comes upon one with new force in the pauses of the night that the Chaldeans were a desert-bred people. It is hard to escape the sense of mastery as the stars move in the wide clear heavens to risings and settings unobscured. They look large and near and palpitant; as if they moved on some stately service not needful to declare. Wheeling to their stations in the sky, they make the poor world-fret of no account. Of no account you who lie out there watching, nor the lean coyote that stands off in the scrub from you and howls and howls.

LUTHER STANDING BEAR
1868–1939

Luther Standing Bear was born in South Dakota, the son of an Oglala Sioux chief. Following his graduation from Pennsylvania's Carlisle Indian School (also attended by Jim Thorpe, among others), he traveled abroad with Buffalo Bill Cody's Wild West Show and later had a successful career in films as a "Hollywood Indian." Nonetheless, he returned to South Dakota where he was chosen chief of the Oglala and became an important spokesperson for the traditions and rights of his people. Not only was he an early interpreter of Sioux lifeways for the larger, English-reading audience, but he also sought to contrast those practices with what he found to be the less-grounded values and institutions of Euro-America. For Standing Bear, the land and culture of the Sioux were inseparable, and the religious beliefs that other Americans of his time saw as superstitious were in fact conducive to mindfulness in daily life and a spirit of generosity and gratitude toward the Earth and other human beings.

The two main books written by Luther Standing Bear were My People the Sioux *(1928) and* Land of the Spotted Eagle *(1933), from which the present selection comes. He also produced a book for children about his boyhood and brought together a collection of Sioux tales.*

NATURE

The Lakota was a true naturist—a lover of Nature. He loved the earth and all things of the earth, the attachment growing with age. The old people came literally to love the soil and they sat or reclined on the ground with a feeling of being close to a mothering power. It was good for the skin to touch the earth and the old people liked to remove their moccasins and walk with bare feet on the sacred earth. Their tipis were built upon the earth and their altars were made of earth. The birds that flew in the air came to rest upon the earth and it was the final abiding place of all things that lived and grew. The soil was soothing, strengthening, cleansing, and healing.

This is why the old Indian still sits upon the earth instead of propping himself up and away from its life-giving forces. For him, to sit or lie upon the ground is to be able to think more deeply and to feel more keenly; he can see more clearly into the mysteries of life and come closer in kinship to other lives about him.

The earth was full of sounds which the old-time Indian could hear, sometimes putting his ear to it so as to hear more clearly. The forefathers of the Lakotas had done this for long ages until there had come to them real understanding of earth ways. It was almost as if the man were still a part of the earth as he was in the beginning, according to the legend of the tribe. This beautiful story of the genesis of the Lakota people furnished the foundation for the love they bore for earth and all things of the earth. Wherever the Lakota went, he was with Mother Earth. No matter where he roamed by day or slept by night, he was safe with her. This thought comforted and sustained the Lakota and he was eternally filled with gratitude.

From Wakan Tanka there came a great unifying life force that

Land of the Spotted Eagle (Lincoln: University of Nebraska, 1933).

flowed in and through all things—the flowers of the plains, blowing winds, rocks, trees, birds, animals—and was the same force that had been breathed into the first man. Thus all things were kindred and brought together by the same Great Mystery.

Kinship with all creatures of the earth, sky, and water was a real and active principle. For the animal and bird world there existed a brotherly feeling that kept the Lakota safe among them. And so close did some of the Lakotas come to their feathered and furred friends that in true brotherhood they spoke a common tongue.

The animal had rights—the right of man's protection, the right to live, the right to multiply, the right to freedom, and the right to man's indebtedness—and in recognition of these rights the Lakota never enslaved the animal, and spared all life that was not needed for food and clothing.

This concept of life and its relations was humanizing and gave to the Lakota an abiding love. It filled his being with the joy and mystery of living; it gave him reverence for all life; it made a place for all things in the scheme of existence with equal importance to all. The Lakota could despise no creature, for all were of one blood, made by the same hand, and filled with the essence of the Great Mystery. In spirit the Lakota was humble and meek. 'Blessed are the meek: for they shall inherit the earth,' was true for the Lakota, and from the earth he inherited secrets long since forgotten. His religion was sane, normal, and human.

Reflection upon life and its meaning, consideration of its wonders, and observation of the world of creatures, began with childhood. The earth, which was called *Maka*, and the sun, called *Anpetuwi*, represented two functions somewhat analogous to those of male and female. The earth brought forth life, but the warming, enticing rays of the sun coaxed it into being. The earth yielded, the sun engendered.

In talking to children, the old Lakota would place a hand on the ground and explain: 'We sit in the lap of our Mother. From her we, and all other living things, come. We shall soon pass, but the place where we now rest will last forever.' So we, too, learned to sit or lie on the ground and become conscious of life about us in its multitude of forms. Sometimes we boys would sit motionless and watch the swallow, the tiny ants, or perhaps some small animal at its work and ponder on its industry and ingenuity; or we lay on our backs and looked long at the sky and when the stars came out made shapes from the various groups. The morning and evening star always attracted attention, and the Milky Way was a path which was traveled by the ghosts. The old people told us to heed *wa maka skan*, which were the 'moving things of earth.'

This meant, of course, the animals that lived and moved about, and the stories they told of *wa maka skan* increased our interest and delight. The wolf, duck, eagle, hawk, spider, bear, and other creatures, had marvelous powers, and each one was useful and helpful to us. Then there were the warriors who lived in the sky and dashed about on their spirited horses during a thunder storm, their lances clashing with the thunder and glittering with the lightning. There was *wiwila*, the living spirit of the spring, and the stones that flew like a bird and talked like a man. Everything was possessed of personality, only differing with us in form. Knowledge was inherent in all things. The world was a library and its books were the stones, leaves, grass, brooks, and the birds and animals that shared, alike with us, the storms and blessings of earth. We learned to do what only the student of nature ever learns, and that was to feel beauty. We never railed at the storms, the furious winds, and the biting frosts and snows. To do so intensified human futility, so whatever came we adjusted ourselves, by more effort and energy if necessary, but without complaint. Even the lightning did us no harm, for whenever it came too close, mothers and grandmothers in every tipi put cedar leaves on the coals and their magic kept danger away. Bright days and dark days were both expressions of the Great Mystery, and the Indian reveled in being close to the Big Holy. His worship was unalloyed, free from the fears of civilization.

I have come to know that the white mind does not feel toward nature as does the Indian mind, and it is because, I believe, of the difference in childhood instruction. I have often noticed white boys gathered in a city by-street or alley jostling and pushing one another in a foolish manner. They spend much time in this aimless fashion, their natural faculties neither seeing, hearing, nor feeling the varied life that surrounds them. There is about them no awareness, no acuteness, and it is this dullness that gives ugly mannerisms full play; it takes from them natural poise and stimulation. In contrast, Indian boys, who are naturally reared, are alert to their surroundings; their senses are not narrowed to observing only one another, and they cannot spend hours seeing nothing, hearing nothing, and thinking nothing in particular. Observation was certain in its rewards; interest, wonder, admiration grew, and the fact was appreciated that life was more than mere human manifestation; that it was expressed in a multitude of forms. This appreciation enriched Lakota existence. Life was vivid and pulsing; nothing was casual and commonplace. The Indian lived—lived in every sense of the word—from his first to his last breath.

The character of the Indian's emotion left little room in his heart for antagonism toward his fellow creatures, this attitude giving him what is sometimes referred to as "the Indian point of view." Every true student, every lover of nature has "the Indian point of view," but there are few such students, for few white men approach nature in the Indian manner. The Indian and the white man sense things differently because the white man has put distance between himself and nature; and assuming a lofty place in the scheme of order of things has lost for him both reverence and understanding. Consequently the white man finds Indian philosophy obscure—wrapped, as he says, in a maze of ideas and symbols which he does not understand. A writer friend, a white man whose knowledge of 'Injuns' is far more profound and sympathetic than the average, once said that he had been privileged, on two occasions, to see the contents of an Indian medicine-man's bag in which were bits of earth, feathers, stones, and various other articles of symbolic nature; that a "collector" showed him one and laughed, but a great and world-famous archeologist showed him the other with admiration and wonder. Many times the Indian is embarrassed and baffled by the white man's allusions to nature in such terms as crude, primitive, wild, rude, untamed, and savage. For the Lakota, mountains, lakes, rivers, springs, valleys, and woods were all finished beauty; winds, rain, snow, sunshine, day, night, and change of seasons brought interest; birds, insects, and animals filled the world with knowledge that defied the discernment of man.

But nothing the Great Mystery placed in the land of the Indian pleased the white man, and nothing escaped his transforming hand. Wherever forests have not been mowed down; wherever the animal is recessed in their quiet protection; wherever the earth is not bereft of four-footed life—that to him is an "unbroken wilderness." But since for the Lakota there was no wilderness; since nature was not dangerous but hospitable; not forbidding but friendly, Lakota philosophy was healthy—free from fear and dogmatism. And here I find the great distinction between the faith of the Indian and the white man. Indian faith sought the harmony of man with his surroundings; the other sought the dominance of surroundings. In sharing, in loving all and everything, one people naturally found a measure of the thing they sought; while, in fearing, the other found need of conquest. For one man the world was full of beauty; for the other it was a place of sin and ugliness to be endured until he went to another world, there to become a creature of wings, half-man and half-bird. Forever one man directed his Mystery to change the world He had made; forever this man pleaded with Him to chastise His wicked ones; and forever he implored his Wakan Tanka to send His light to earth. Small wonder this man could not understand the other.

But the old Lakota was wise. He knew that man's heart, away from nature, becomes hard; he knew that lack of respect for growing, living things soon led to lack of respect for humans too. So he kept his youth close to its softening influence.

EDWARD THOMAS
1878–1917

Born of Welsh parents in London, Edward Thomas had written numerous volumes of essays, criticism, and natural history before his pivotal meeting in 1914 with Robert Frost, who befriended Thomas and encouraged him to begin writing poetry. Three years later he died at the Battle of Arras in Northern France as the first edition of his Poems *was going to press.*

Thomas's rich descriptions of the Sussex and Hampshire countryside are in the mainstream of English nature writing as represented by Gilbert White, Richard Jefferies, and W. H. Hudson. Yet beneath Thomas's sonorous prose and often archaic diction there is a distinctively modern sense of the individual's isolation, the invasion of the rural landscape by the industrial revolution, the transience of the human species on Earth, and the ambivalent nature of our "incompatible desires." There is also a pervasive melancholic tone that makes much of his writing seem in large degree a preparation for death, yet carried out by one who relishes life to the full even as he prepares to leave it.

HAMPSHIRE

The beeches on the beech-covered hills roar and strain as if they would fly off with the hill, and anon they are as meek as a great horse leaning his head over a gate. If there is a misty day there is one willow in a coombe lifting up a thousand silver catkins like a thousand lamps,

The South Country (London: J. M. Dent, 1909).

when there is no light elsewhere. Another day, a wide and windy day, is the jackdaw's, and he goes straight and swift and high like a joyous rider crying aloud on an endless savannah, and, underneath, the rippled pond is as bright as a peacock, and millions of beech leaves drive across the open glades of the woods, rushing to their Acheron. The bush harrow stripes the moist and shining grass; the plough changes the pale stubble into a ridgy chocolate; they are peeling the young ash sticks for hop poles and dipping them in tar. At the dying of that windy day the wind is still; there is a bright pale half-moon tangled in the pink whirl of after-sunset cloud, a sound of blackbirds from pollard oaks against the silver sky, a sound of bells from hamlets hidden among beeches.

Towards the end of March there are six nights of frost giving birth to still mornings of weak sunlight, of an opaque yet not definitely misty air. The sky is of a milky, uncertain pale blue without one cloud. Eastward the hooded sun is warming the slope fields and melting the sparkling frost. In many trees the woodpeckers laugh so often that their cry is a song. A grassy ancient orchard has taken possession of the visible sunbeams, and the green and gold of the mistletoe glows on the silvered and mossy branches of apple trees. The pale stubble is yellow and tenderly lit, and gives the low hills a hollow light appearance as if they might presently dissolve. In a hundred tiers on the steep hill, the uncounted perpendicular straight stems of beech, and yet not all quite perpendicular or quite straight, are silver-grey in the midst of a haze, here brown, there rosy, of branches and swelling buds. Though but a quarter of a mile away in this faintly clouded air they are very small, aerial in substance, infinitely remote from the road on which I stand, and more like reflections in calm water than real things.

At the lower margin of the wood the overhanging branches form blue caves, and out of these emerge the songs of many hidden birds. I know that there are bland melodious blackbirds of easy musing voices, robins whose earnest song, though full of passion, is but a fragment that has burst through a more passionate silence, hedge-sparrows of liquid confiding monotone, brisk acid wrens, chaffinches and yellowhammers saying always the same thing (a dear but courtly praise of the coming season), larks building spires above spires into the sky, thrushes of infinite variety that talk and talk of a thousand things, never thinking, always talking of the moment, exclaiming, scolding, cheering, flattering, coaxing, challenging, with merry-hearted, bold voices that must have been the same in the morning of the world when the forest trees lay, or leaned, or hung, where they fell. Yet I can distinguish neither blackbird, nor robin, nor hedge-sparrow, nor any one voice. All are blent into one seething stream of song. It is one song, not many. It is one spirit that sings. Mixed with them is the myriad stir of unborn things, of leaf and blade and flower, many silences at heart and root of tree, voices of hope and growth, of love that will be satisfied though it leap upon the swords of life. Yet not during all the day does the earth truly awaken. Even in town and city the dream prevails, and only dimly lighted their chalky towers and spires rise out of the sweet mist and sing together beside the waters.

The earth lies blinking, turning over languidly and talking like a half-wakened child that now and then lies still and sleeps though with eyes wide open. The air is still full of the dreams of a night which this mild sun cannot dispel. The dreams are prophetic as well as reminiscent, and are visiting the woods, and that is why they will not cast aside the veil. Who would rise if he could continue to dream? It is not spring yet. Spring is being dreamed, and the dream is more wonderful and more blessed than ever was spring. What the hour of waking will bring forth is not known. Catch at the dreams as they hover in the warm thick air. Up against the grey tiers of beech stems and the mist of the buds and fallen leaves rise two columns of blue smoke from two white cottages among trees; they rise perfectly straight and then expand into a balanced cloud, and thus make and unmake continually two trees of smoke. No sound comes from the cottages. The dreams are over them, over the brows of the children and the babes, of the men and the women, bringing great gifts, suggestions, shadowy satisfactions, consolations, hopes. With inward voices of persuasion those dreams hover and say that all is to be made new, that all is yet before us, and the lots are not yet drawn out of the urn.

We shall presently set out and sail into the undiscovered seas and find new islands of the free, the beautiful, the young. As is the dimly glimmering changeless brook twittering over the pebbles, so is life. It is but just leaving the fount. All things are possible in the windings between fount and sea.

Never again shall we demand the cuckoo's song from the August silence. Never will July nip the spring and lengthen the lambs' faces and take away their piquancy, or June shut a gate between us and the nightingale, or May deny the promise of April. Hark! before the end of afternoon the owls hoot in their sleep in the ivied beeches. A dream has flitted past them, more silent of wing than themselves. Now it is between the wings of the first white butterfly, and it plants a smile in the face of the infant that cannot speak: and again it is with the brimstone butterfly, and the child who is gathering celandine and cuckoo flower and violet starts back almost in fear at the dream.

The grandmother sitting in her daughter's house, left all alone in silence, her hands clasped upon her knees, forgets the courage without hope that has carried her through eighty years, opens her eyes, unclasps her hands from the knot as of stiff rope, distends them and feels the air, and the dream is between her fingers and she too smiles, she knows not why. A girl of sixteen, ill-dressed, not pretty, has seen it also. She has tied up her black hair in a new crimson ribbon. She laughs aloud with a companion at something they know in common and in secret, and as she does so lifts her neck and is glad from the sole of her foot to the crown of her head. She is lost in her laughter and oblivious of its cause. She walks away, and her step is as firm as that of a ewe defending her lamb. She was a poor and misused child, and I can see her as a woman of fifty, sitting on a London bench grey-complexioned, in old black hat, black clothes, crouching over a paper bag of fragments, in the beautiful August rain after heat. But this is her hour. That future is not among the dreams in the air to-day. She is at one with the world, and a deep music grows between her and the stars. Her smile is one of those magical things, great and small and all divine, that have the power to wield universal harmonies. At sight or sound of them the infinite variety of appearances in the world is made fairer than before, because it is shown to be a many-coloured raiment of the one. The raiment trembles, and under leaf and cloud and air a window is thrown open upon the unfathomable deep, and at the window we are sitting, watching the flight of our souls away, away to where they must be gathered into the music that is being built.

Often upon the vast and silent twilight, as now, is the soul poured out as a rivulet into the sea and lost, not able even to stain the boundless crystal of the air; and the body stands empty, waiting for its return, and, poor thing, knows not what it receives back into itself when the night is dark and it moves away. For we stand ever at the edge of Eternity and fall in many times before we die. Yet even such thoughts live not long this day. All shall be healed, says the dream. All shall be made new. The day is a fairy birth, a foundling not fathered nor mothered by any grey yesterdays. It has inherited nothing. It makes of winter and of the old springs that wrought nothing fair a stale creed, a senseless tale: they are naught: I do not wonder any longer if the lark's song has grown old with the ears that hear it or if it still be unchanged.

What dreams are there for that aged child who goes tottering and reeling up the lane at mid-day? He carries a basket of watercress on his back. He has sold two-pennyworth, and he is tipsy, grinning through the bruises of a tipsy fall, and shifting his cold pipe from one side of his mouth to the other. Though hardly sixty he is very old, worn and thin and wrinkled, and bent sideways and forward at the waist and the shoulders. Yet he is very young. He is just what he was forty years ago when the thatcher found him lying on his back in the sun instead of combing out the straw and sprinkling it with water for his use. He laid no plans as a youth; he had only a few transparent tricks and easy lies. Never has he thought of the day after to-morrow. For a few years in his prime he worked almost regularly for one or two masters, leaving them only now and then upon long errands of his own and known only to himself. It was then perhaps that he earned or received as a gift, along with a broken nose, his one name, which is Jackalone. For years he was the irresponsible jester to a smug townlet which was privately amused and publicly scandalized, and rewarded him in a gaol, where, unlike Tasso, he never complained. Since then he has lived by the sale of a chance rabbit or two, of watercress, of greens gathered when the frost is on them and nobody looking, by gifts of broken victuals, by driving a few bullocks to a fair, by casual shelter in barns, in roofless cottages, or under hedges.

He has never had father or mother or brother or sister or wife or child. No dead leaf in autumn wind or branch in flooded brook seems more helpless. He can deceive nobody. He is in prison two or three times a year for little things: it seems a charity to put a roof over his head and clip his hair. He has no wisdom; by nothing has he soiled what gifts were given to him at his birth. The dreams will not pass him by. They come to give him that confidence by which he lives in spite of men's and children's contumely.

How little do we know of the business of the earth, not to speak of the universe; of time, not to speak of eternity. It was not by taking thought that man survived the mastodon. The acts and thoughts that will serve the race, that will profit this commonwealth of things that live in the sun, the air, the earth, the sea, now and through all time, are not known and never will be known. The rumour of much toil and scheming and triumph may never reach the stars, and what we value not at all, are not conscious of, may break the surface of eternity with endless ripples of good. We know not by what we survive. There is much philosophy in that Irish tale of the poor blind woman who recovered her sight at St. Brigit's well. "Did I say more prayers than the rest? Not a prayer. I was young in those days. I suppose she took a liking to me, maybe because of my name being Brigit the same as her own."[1] Others went unrelieved away that day. We are as ignorant still. Hence the batlike fears about immortality. We wish to prolong what we can see and touch and talk of,

---

[1] *A Book of Saints and Wonders*, by Lady Gregory. [Thomas's note]

and knowing that clothes and flesh and other perishing things may not pass over the borders of death with us, we give up all, as if forsooth the undertaker and the gravedigger had arch-angelic functions. Along with the undertaker and the gravedigger ranks the historian and others who seem to bestow immortality. Each is like a child planting flowers severed from their stalks and roots, expecting them to grow. I never heard that the butterfly loved the chrysalis; but I am sure that the caterpillar looks forward to an endless day of eating green leaves and of continually swelling until it would despise a consummation of the size of a railway train. We can do the work of the universe though we shed friends and country and house and clothes and flesh, and become invisible to mortal eyes and microscopes. We do it now invisibly, and it is not these things which are us at all. That maid walking so proudly is about the business of eternity.

And yet it would be vain to pretend not to care about the visible many-coloured raiment of which our houses, our ships, our gardens, our books are part, since they also have their immortal selves and their everlasting place, else should we not love them with more than sight and hearing and touch. For flesh loves flesh and soul loves soul. Yet on this March day the supreme felicity is born of the two loves, so closely interwoven that it is permitted to forget the boundaries of the two, and for soul to love flesh and flesh to love soul. And this ancient child is rid of his dishonours and flits through the land floating on a thin reed of the immortal laughter. This is "not altogether fool." He is perchance playing some large necessary part in the pattern woven by earth that draws the gods to lean forward out of the heavens to watch the play and say of him, as of other men, of birds, of flowers: "They also are of our company." . . .

In the warm rain of the next day the chiffchaff sings among the rosy blossoms of the leafless larches, a small voice that yet reaches from the valley to the high hill. It is a double, many times repeated note that foretells the cuckoo's. In the evening the songs are bold and full, but the stems of the beeches are faint as soft columns of smoke and the columns of smoke from the cottages are like them in the still air.

Yet another frost follows, and in the dim golden light just after sunrise the shadows of all the beeches lie on the slopes, dark and more tangible than the trees, as if they were the real and those standing upright were the returned spirits above the dead.

Now rain falls and relents and falls again all day, and the earth is hidden under it, and as from a land submerged the songs mount through the veil. The mists waver out of the beeches like puffs of smoke or hang upon them or in them like fleeces caught in thorns: in the just penetrating sunlight the long boles of the beeches shine, and the chaffinch, the yellowhammer and the cirl bunting sing songs of blissful drowsiness. The Downs, not yet green, rise far off and look, through the rain, like old thatched houses. * * *

THE END OF SUMMER

* * * All night—for a week—it rains, and at last there is a still morning of mist. A fire of weeds and hedge-clippings in a little flat field is smouldering. The ashes are crimson, and the bluish-white smoke flows in a divine cloudy garment round the boy who rakes over the ashes. The heat is great, and the boy, straight and well made, wearing close gaiters of leather that reach above the knees, is languid at his task, and often leans upon his rake to watch the smoke coiling away from him like a monster reluctantly fettered, and sometimes bursting into an anger of sprinkled sparks. He adds some wet hay, and the smoke pours out of it like milky fleeces when the shearer reveals the inmost wool with his shears. Above and beyond him the pale blue sky is dimly white-clouded over beech woods, whose many greens and yellows and yellow-greens are softly touched by the early light which cannot penetrate to the blue caverns of shade underneath. Athwart the woods rises a fount of cottage-smoke from among mellow and dim roofs. Under the smoke and partly scarfed at times by a drift from it is the yellow of sunflower and dahlia, the white of anemone, the tenderest green and palest purple of a thick cluster of autumn crocuses that have broken out of the dark earth and stand surprised, amidst their own weak light as of the underworld from which they have come. Robins sing among the fallen apples, and the cooing of wood-pigeons is attuned to the soft light and the colours of the bowers. The yellow apples gleam. It is the gleam of melting frost. Under all the dulcet warmth of the face of things lurks the bitter spirit of the cold. Stand still for more than a few moments and the cold creeps with a warning, and then a menace into the breast. That is the bitterness that makes this morning of all others in the year so mournful in its beauty. The colour and the grace invite to still contemplation and long draughts of dream; the frost compels to motion. The scent is that of wood-smoke, of fruit and of some fallen leaves. This is the beginning of the pageant of autumn, of that gradual pompous dying which has no parallel in human life, yet draws us to it with sure bonds. It is a dying of the flesh, and we see it pass through a

kind of beauty which we can only call spiritual, of so high and inaccessible a strangeness is it. The sight of such perfection as is many times achieved before the end awakens the never more than lightly sleeping human desire of permanence. Now, now is the hour; let things be thus; thus for ever; there is nothing further to be thought of; let these remain. And yet we have a premonition that remain they must not for more than a little while. The motion of the autumn is a fall, a surrender, requiring no effort, and therefore the mind cannot long be blind to the cycle of things as in the spring it can when the effort and delight of ascension veils the goal and the decline beyond. A few frosts now, a storm of wind and rain, a few brooding mists, and the woods that lately hung dark and massive and strong upon the steep hills are transfigured and have become cloudily light and full of change and ghostly fair; the crowing of a cock in the still misty morning echoes up in the many-coloured trees like a challenge to the spirits of them to come out and be seen, but in vain. For months the woods have been homely and kind, companions and backgrounds to our actions and thoughts, the wide walls of a mansion utterly our own. We could have gone on living with them for ever. We had given up the ardours, the extreme ecstasy of our first bridal affection, but we had not forgotten them. We could not become indifferent to the Spanish chestnut trees that grow at the top of the steep rocky banks on either side of the road and mingle their foliage overhead. Of all trees well-grown chestnuts are among the most pleasant to look up at. For the foliage is not dense and it is for the most part close to the large boughs, so that the light comes easily down through all the horizontal leaves, and the shape of each separate one is not lost in the multitude, while at the same time the bold twists of the branches are undraped or easily seen through such translucent green. The trunks are crooked, and the handsome deep furrowing of the bark is often spirally cut. The limbs are few and wide apart so as to frame huge delicately lighted and shadowed chambers of silence or of birds' song. The leaves turn all together to a leathern hue, and when they fall stiffen and display their shape on the ground and long refuse to be merged in the dismal trodden hosts. But when the first one floats past the eye and is blown like a canoe over the pond we recover once more our knowledge and fear of Time. All those ladders of goose-grass that scaled the hedges of spring are dead grey; they are still in their places, but they clamber no longer. The chief flower is the yellow bloom set in the dark ivy round the trunks of the ash trees; and where it climbs over the holly and makes a solid sunny wall, and in the hedges, a whole people of wasps and wasp-like flies are always at the bloom with crystal wings, except when a passing shadow disperses them for a moment with one buzz. But these cannot long detain the eye from the crumbling woods in the haze or under the large white clouds—from the amber and orange bracken about our knees and the blue recesses among the distant golden beeches when the sky is blue but beginning to be laden with loose rain-clouds, from the line of leaf-tipped poplars that bend against the twilight sky; and there is no scent of flowers to hide that of dead leaves and rotting fruit. We must watch it until the end, and gain slowly the philosophy or the memory or the forgetfulness that fits us for accepting winter's boon. Pauses there are, of course, or what seem pauses in the declining of this pomp; afternoons when the rooks waver and caw over their beechen town and the pigeons coo content; dawns when the white mist is packed like snow over the vale and the high woods take the level beams and a hundred globes of dew glitter on every thread of the spiders' hammocks or loose perpendicular nets among the thorns, and through the mist rings the anvil a mile away with a music as merry as that of the daws that soar and dive between the beeches and the spun white cloud; mornings full of the sweetness of mushrooms and blackberries from the short turf among the blue scabious bloom and the gorgeous brier; empurpled evenings before frost when the robin sings passionate and shrill and from the garden earth float the smells of a hundred roots with messages of the dark world; and hours full of the thrush's soft November music. The end should come in heavy and lasting rain. At all times I love rain, the early momentous thunderdrops, the perpendicular cataract shining, or at night the little showers, the spongy mists, the tempestuous mountain rain. I like to see it possessing the whole earth at evening, smothering civilization, taking away from me myself everything except the power to walk under the dark trees and to enjoy as humbly as the hissing grass, while some twinkling house-light or song sung by a lonely man gives a foil to the immense dark force. I like to see the rain making the streets, the railway station, a pure desert, whether bright with lamps or not. It foams off the roofs and trees and bubbles into the water-butts. It gives the grey rivers a demonic majesty. It scours the roads, sets the flints moving, and exposes the glossy chalk in the tracks through the woods. It does work that will last as long as the earth. It is about eternal business. In its noise and myriad aspect I feel the mortal beauty of immortal things. And then after many days the rain ceases at midnight with the wind, and in the silence of dawn and frost the last rose of the world is dropping her petals down to the glistering whiteness, and there they rest blood-red on the winter's desolate coast.

ROCKWELL KENT
1882–1971

Like George Catlin, Rockwell Kent was primarily a visual artist, best known for his woodcut illustrations of such classics as Shakespeare's plays and Moby-Dick. *But he also wrote several books, among them two brief but vivid accounts of wilderness adventures.* Wilderness *(1920) recorded a year of homesteading in an isolated Alaska valley with his young son. His* N by E *(1930), from which the following chapters are taken, is an account of a voyage taken by Kent and two friends in a small schooner from New York City to Greenland. The trip ended abruptly and tragically in a shipwreck along Greenland's west coast, killing one of the crew. Despite the hardships endured, Kent's descriptions of the sea, the landscape, and the people he encounters reflect his passionate energy and, as with his illustrations, a cosmic view of man's nature and destiny.*

From N BY E

    At four-thirty in the afternoon of June 17th we sailed. The exasperating delay that had put off our sailing until that date, and on that date until that hour, the misgivings I had felt about the mate, all were forgotten in that moment of leave taking. The bright sun shone upon us; the lake was blue under the westerly breeze, and luminous, how luminous! the whole far world of our imagination. How like a colored lens the colored present! through it we see the forward vista of our lives. Here, in the measure that the water widened in our wake and heart strings stretched to almost breaking, the golden future neared us and enfolded us, made us at last—how soon!—oblivious to all things but the glamour of adventure. And while one world diminished, narrowed and then disappeared, before us a new world unrolled and neared us to display itself. Who can deny the human soul its everlasting need to

N by E (New York: Random House, 1930).

make the unknown known; not for the sake of knowing, not to inform itself or be informed or wise, but for the need to exercise the need to know? What is that need but the imagination's hunger for the new and raw materials of its creative trade? Of things and facts assured to us and known we've got to make the best, and live with it. That humdrum is the price of living. We *live* for those fantastic and unreal moments of beauty which our thoughts may build upon the passing panorama of experience.

Soon all that we had ever seen before was left behind and a new land of fields and farms, pastures and meadows, woods and open lands and rolling hills was streaming by, all in the mellow splendor of late afternoon in June, all green and clean and beautiful. We stripped and plunged ahead into the blue water; and catching hold of a rope as it swept by, trailed in the wake. It was so warm—the water and the early summer air. So we shall live all summer naked, and get brown and magnificent!

I cooked supper: hot baking-powder biscuit and—I don't remember what. "You're a wonderful cook!" said everyone. So I washed the dishes and put the cabin in order.

"Oh," thought I, "people are nice! the world is grand! I'm happy! God is good!"

* * *

In the half light of the early morning of July the fifth all hands bestirred themselves, got up; we came on deck. It was cold. The silent town lay dark against the eastern sky; the land was black, and stranded bergs glowed pale against it. Clear heavens strewn with stars, and a fair wind S. by W.!

Noiselessly, as if stealing away, we hoisted sail, weighed anchor and bore out. And so, without tumult and the clamor of leave takings, quietly as the coming dawn, we entered the solitude of the ocean.

And if we were not annihilated by the contemplation of such vast adventure it was by grace of that wise providence of man's nature which, to preserve his reason, lets him be thoughtless before immensity.

* * *

But that was centuries ago. In Greenland the environment of nature dominates; and into the sparse settlements along its rim of shore, into men's thoughts and moods and lives has entered something of the eternal peacefulness of the wilderness. It had to be. Man is less entity than consequence and his being is but a derivation of a less subjective world, a synthesis of what he calls the elements. Man's very spirit is a sublimation of cosmic energy and worships it as God; and every faculty to feel, perceive and know serves only to relate him closer to what is. God is the Father, man his connatural progeny; and thus the elements at work become for man the pattern for his conduct, the look and feel and sound of them—sunshine and storm, peace and turmoil, lightning and thunder and the quiet interludes—the formulae for his poor imitative moods and their expression. But in the wilderness invariably peace predominates; and seeing the quiet uneventfulness of lives lived there, their ordered lawlessness, the loveliness and grace of bearing and of look and smile that it so often breeds and fosters we may indeed "lament what man has made of man" and hold those circumstances of congestion which are called civilization to be less friendly to beauty than opposed to it.

More than two hundred years ago there came a Christian militant, Hand Egede, to Greenland, and brought those heathen folk the Gospel law. And Greenland, its wilderness and wilderness's heathen folk, reached out and gently laid its peaceful spirit upon Egede and all who followed him and lived there. Thus came the natives of Greenland to learn somewhat of the virtues of cleanliness and industry and thrift, and the Christians to become more godly, decent, quiet, honorable, fair than any Christians aggregate I've met with elsewhere in the world. Christ, too, if I remember, sojourned in the wilderness.

                                      *   *   *

So, excursioning and voyaging about, landing to stroll the settlements or climb the neighboring hills, looking at everything and listening, I came at last to have been north to Seventy and south to Sixty-thirty. My lingual limitation served to sharpen the perceptive faculties and shield me from dependence on such facts as others might have told me. It was, in consequence, what *seemed* to be that made my Greenland world; and such conclusions as I ventured on were reached inversely, from effect to cause. All that *looked* beautiful to me, was good; and if I held the smile of the Eskimo to be evidence of his serenity of soul, the inter-racial courtesy of the Danes to show the virtue of their rulership, and both to prove that life in Greenland's solitudes was good for man my thought at least began where science stops.

VIRGINIA WOOLF
1882–1941

As one of the leading modernists—author of Mrs. Dalloway *(1925), To*
the Lighthouse *(1927), and* Between the Acts *(1941) among many other*
novels—Woolf brought a lyrical focus to the stream of consciousness. She
was also one of the most important essayists and critics of her generation,
as reflected in such collections as The Common Reader *(1925) and* The
Death of the Moth *(1942). But her special gifts as a nature writer*
became clearest with the publication of her five-volume Diary *between*
1977 and 1984. As many of her entries show, Woolf was as attentive to the
outer weathers, the outer tides and blossomings, as she was to the inner.
She was alert, too, to the terrifying and redemptive independence of
nature from human rationales and needs. As she wrote in "Time Passes,"
from To the Lighthouse, *"In spring, the garden urns, casually filled with*
wind-blown plants, were gay as ever."

THE DEATH OF THE MOTH

Moths that fly by day are not properly to be called moths; they do
not excite that pleasant sense of dark autumn nights and ivy-blossom
which the commonest yellow-underwing asleep in the shadow of the
curtain never fails to rouse in us. They are hybrid creatures, neither gay
like butterflies nor sombre like their own species. Nevertheless the present specimen, with his narrow hay-coloured wings, fringed with a tassel
of the same colour, seemed to be content with life. It was a pleasant
morning, mid-September, mild, benignant, yet with a keener breath
than that of the summer months. The plough was already scoring the
field opposite the window, and where the share had been, the earth was
pressed flat and gleamed with moisture. Such vigour came rolling in
from the fields and the down beyond that it was difficult to keep the

The Death of the Moth and Other Essays (New York: Harcourt Brace, 1942).

eyes strictly turned upon the book. The rooks too were keeping one of their annual festivities; soaring round the tree tops until it looked as if a vast net with thousands of black knots in it had been cast up into the air; which, after a few moments sank slowly down upon the trees until every twig seemed to have a knot at the end of it. Then, suddenly, the net would be thrown into the air again in a wider circle this time, with the utmost clamour and vociferation, as though to be thrown into the air and settle slowly down upon the tree tops were a tremendously exciting experience.

The same energy which inspired the rooks, the ploughmen, the horses, and even, it seemed, the lean bare-backed downs, sent the moth fluttering from side to side of his square of the window-pane. One could not help watching him. One was, indeed, conscious of a queer feeling of pity for him. The possibilities of pleasure seemed that morning so enormous and so various that to have only a moth's part in life, and a day moth's at that, appeared a hard fate, and his zest in enjoying his meagre opportunities to the full, pathetic. He flew vigorously to one corner of his compartment, and, after waiting there a second, flew across to the other. What remained for him but to fly to a third corner and then to a fourth? That was all he could do, in spite of the size of the downs, the width of the sky, the far-off smoke of houses, and the romantic voice, now and then, of a steamer out at sea. What he could do he did. Watching him, it seemed as if a fibre, very thin but pure, of the enormous energy of the world had been thrust into his frail and diminutive body. As often as he crossed the pane, I could fancy that a thread of vital light became visible. He was little or nothing but life.

Yet, because he was so small, and so simple a form of the energy that was rolling in at the open window and driving its way through so many narrow and intricate corridors in my own brain and in those of other human beings, there was something marvellous as well as pathetic about him. It was as if someone had taken a tiny bead of pure life and decking it as lightly as possible with down and feathers, had set it dancing and zig-zagging to show us the true nature of life. Thus displayed one could not get over the strangeness of it. One is apt to forget all about life, seeing it humped and bossed and garnished and cumbered so that it has to move with the greatest circumspection and dignity. Again, the thought of all that life might have been had he been born in any other shape caused one to view his simple activities with a kind of pity.

After a time, tired by his dancing apparently, he settled on the window ledge in the sun, and, the queer spectacle being at an end, I forgot about him. Then, looking up, my eye was caught by him. He was trying to resume his dancing, but seemed either so stiff or so awkward that he could only flutter to the bottom of the window-pane; and when he tried to fly across it he failed. Being intent on other matters I watched these futile attempts for a time without thinking, unconsciously waiting for him to resume his flight, as one waits for a machine, that has stopped momentarily, to start again without considering the reason of its failure. After perhaps a seventh attempt he slipped from the wooden ledge and fell, fluttering his wings, on to his back on the window sill. The helplessness of his attitude roused me. It flashed upon me that he was in difficulties; he could no longer raise himself; his legs struggled vainly. But, as I stretched out a pencil, meaning to help him to right himself, it came over me that the failure and awkwardness were the approach of death. I laid the pencil down again.

The legs agitated themselves once more. I looked as if for the enemy against which he struggled. I looked out of doors. What had happened there? Presumably it was midday, and work in the fields had stopped. Stillness and quiet had replaced the previous animation. The birds had taken themselves off to feed in the brooks. The horses stood still. Yet the power was there all the same, massed outside indifferent, impersonal, not attending to anything in particular. Somehow it was opposed to the little hay-coloured moth. It was useless to try to do anything. One could only watch the extraordinary efforts made by those tiny legs against an oncoming doom which could, had it chosen, have submerged an entire city, not merely a city, but masses of human beings; nothing, I knew, had any chance against death. Nevertheless after a pause of exhaustion the legs fluttered again. It was superb this last protest, and so frantic that he succeeded at last in righting himself. One's sympathies, of course, were all on the side of life. Also, when there was nobody to care or to know, this gigantic effort on the part of an insignificant little moth, against a power of such magnitude, to retain what no one else valued or desired to keep, moved one strangely. Again, somehow, one saw life, a pure bead. I lifted the pencil again, useless though I knew it to be. But even as I did so, the unmistakable tokens of death showed themselves. The body relaxed, and instantly grew stiff. The struggle was over. The insignificant little creature now knew death. As I looked at the dead moth, this minute wayside triumph of so great a force over so mean an antagonist filled me with wonder. Just as life had been strange a few minutes before, so death was now as strange. The moth having righted himself now lay most decently and uncomplainingly composed. O yes, he seemed to say, death is stronger than I am.

ISAK DINESEN
1885–1962

For many Europeans in the nineteenth and early twentieth centuries, the African continent was their New World, full of economic opportunity and imaginative possibility. During the years 1913 to 1930 the Baroness Karen Blixen managed a large coffee plantation in the Kenyan highlands, shot lions, and had a love affair with the English game hunter Denys Finch Hatton. When, after a series of economic reversals, she was forced to sell her farm and leave her beloved adopted homeland, she returned to her native Denmark. There, writing in English under the pen name Isak Dinesen, she published Out of Africa (1937), which has become the most-celebrated European account of colonial Africa. The book is a memoir of her years in Kenya, filtered through a strong narrative sensibility. Its lyrical, incantatory language gives her experiences a mythic stature and the place itself an Edenic quality, where animals "were being created before my eyes and sent out as they were finished."

From OUT OF AFRICA

THE NGONG FARM

I had a farm in Africa, at the foot of the Ngong Hills. The Equator runs across these highlands, a hundred miles to the North, and the farm lay at an altitude of over six thousand feet. In the day-time you felt that you had got high up, near to the sun, but the early mornings and evenings were limpid and restful, and the nights were cold.

The geographical position, and the height of the land combined to create a landscape that had not its like in all the world. There was no fat on it and no luxuriance anywhere; it was Africa distilled up through six thousand feet, like the strong and refined essence of a continent. The colours were dry and burnt, like the colours in pottery. The trees

Out of Africa (New York: Random House, 1938).

had a light delicate foliage, the structure of which was different from that of the trees in Europe; it did not grow in bows or cupolas, but in horizontal layers, and the formation gave to the tall solitary trees a likeness to the palms, or a heroic and romantic air like fullrigged ships with their sails clewed up, and to the edge of a wood a strange appearance as if the whole wood were faintly vibrating. Upon the grass of the great plains the crooked bare old thorn-trees were scattered, and the grass was spiced like thyme and bog-myrtle; in some places the scent was so strong, that it smarted in the nostrils. All the flowers that you found on the plains, or upon the creepers and liana in the native forest, were diminutive like flowers of the downs, — only just in the beginning of the long rains a number of big, massive heavy-scented lilies sprang out on the plains. The views were immensely wide. Everything that you saw made for greatness and freedom, and unequalled nobility.

The chief feature of the landscape, and of your life in it, was the air. Looking back on a sojourn in the African highlands, you are struck by your feeling of having lived for a time up in the air. The sky was rarely more than pale blue or violet, with a profusion of mighty, weightless, ever-changing clouds towering up and sailing on it, but it has a blue vigour in it, and at a short distance it painted the ranges of hills and the woods a fresh deep blue. In the middle of the day the air was alive over the land, like a flame burning; it scintillated, waved and shone like running water, mirrored and doubled all objects, and created great Fata Morgana. Up in this high air you breathed easily, drawing in a vital assurance and lightness of heart. In the highlands you woke up in the morning and thought: Here I am, where I ought to be.

*   *   *

Before I took over the management of the farm, I had been keen on shooting and had been out on many Safaris. But when I became a farmer I put away my rifles.

The Masai, the nomadic, cattle-owning nation, were neighbours of the farm and lived on the other side of the river; from time to time some of them would come to my house to complain about a lion that was taking their cows, and to ask me to go out and shoot it for them, and I did so if I could. Sometimes, on Saturday, I also walked out on the Orungi plains to shoot a Zebra or two as meat for my farm-labourers, with a long tail of optimistic young Kikuyu after me. I shot birds on the farm, spurfowl and guineafowl, that are very good to eat. But for many years I was not out on any shooting expedition.

Still, we often talked on the farm of the Safaris that we had been on. Camping-places fix themselves in your mind as if you had spent long periods of your life in them. You will remember a curve of your waggon track in the grass of the plain, like the features of a friend.

Out on the Safaris, I had seen a herd of Buffalo, one hundred and twenty-nine of them, come out of the morning mist under a copper sky, one by one, as if the dark and massive, iron-like animals with the mighty horizontally swung horns were not approaching, but were being created before my eyes and sent out as they were finished. I had seen a herd of Elephant travelling through dense Native forest, where the sunlight is strewn down between the thick creepers in small spots and patches, pacing along as if they had an appointment at the end of the world. It was, in giant size, the border of a very old, infinitely precious Persian carpet, in the dyes of green, yellow and black-brown. I had time after time watched the progression across the plain of the Giraffe, in their queer, inimitable, vegetative gracefulness, as if it were not a herd of animals but a family of rare, long-stemmed, speckled gigantic flowers slowly advancing. I had followed two Rhinos on their morning promenade, when they were sniffing and snorting in the air of the dawn, —which is so cold that it hurts in the nose, —and looked like two very big angular stones rollicking in the long valley and enjoying life together. I had seen the royal lion, before sunrise, below a waning moon, crossing the grey plain on his way home from the kill, drawing a dark wake in the silvery grass, his face still red up to the ears, or during the midday-siesta, when he reposed contentedly in the midst of his family on the short grass and in the delicate, spring-like shade of the broad Acacia trees of his park of Africa.

All these things were pleasant to think of when times were dull on the farm. And the big game was out there still, in their own country; I could go and look them up once more if I liked. Their nearness gave a shine and play to the atmosphere of the farm. *   *   *

D(AVID) H(ERBERT) LAWRENCE
1885–1930

In one sense D. H. Lawrence was always writing about nature, whether his setting was the hillsides of Tuscany, the coal mines of the English Midlands, or the drawing rooms of an aristocrat's London town house. In most of his fiction the contrast between the barren, self-conscious,

*self-willed, and mechanized life of modern civilization and the uncon-
scious and instinctive life of "blood knowledge" and "dark gods" is
expressed in images of natural rhythms, vivid landscapes, and untamed
animals. Many readers find Lawrence's nature and travel essays among
his most congenial work, for in them his vital response to life is generally
free of the didactic elements of his novels and more political essays. As a
boy Lawrence painted watercolors of his native Nottingham flora, and
few if any writers have created such passionately intense portraits of flowers as in the following selection from* Phoenix *(1936), a posthumous collection of his writings.*

FLOWERY TUSCANY

I

Each country has its own flowers, that shine out specially there. In
England it is daisies and buttercups, hawthorn and cowslips. In
America, it is goldenrod, stargrass, June daisies, Mayapple and asters,
that we call Michaelmas daisies. In India, hibiscus and dattura and
champa flowers, and in Australia mimosa, that they call wattle, and
sharp-tongued strange heath-flowers. In Mexico it is cactus flowers, that
they call roses of the desert, lovely and crystalline among many thorns;
and also the dangling yard-long clusters of the cream bells of the yucca,
like dropping froth.

But by the Mediterranean, now as in the days of the Argosy, and, we
hope, for ever, it is narcissus and anemone, asphodel and myrtle.
Narcissus and anemone, asphodel, crocus, myrtle, and parsley, they
leave their sheer significance only by the Mediterranean. There are
daisies in Italy too: at Pæstum there are white little carpets of daisies, in
March, and Tuscany is spangled with celandine. But for all that, the
daisy and the celandine are English flowers, their best significance is
for us and for the North.

The Mediterranean has narcissus and anemone, myrtle and asphodel and grape hyacinth. These are the flowers that speak and are understood in the sun round the Middle Sea.

Tuscany is especially flowery, being wetter than Sicily and more
homely than the Roman hills. Tuscany manages to remain so remote,
and secretly smiling to itself in its many sleeves. There are so many

Phoenix: The Posthumous Papers of D. H. Lawrence (New York: Viking, 1936).

hills popping up, and they take no notice of one another. There are so many little deep valleys with streams that seem to go their own little way entirely, regardless of river or sea. There are thousands, millions of utterly secluded little nooks, though the land has been under cultivation these thousands of years. But the intensive culture of vine and olive and wheat, by the ceaseless industry of naked human hands and winter-shod feet, and slow-stepping, soft-eyed oxen does not devastate a country, does not denude it, does not lay it bare, does not uncover its nakedness, does not drive away either Pan or his children. The streams run and rattle over wild rocks of secret places, and murmur through blackthorn thickets where the nightingales sing all together, unruffled and undaunted.

It is queer that a country so perfectly cultivated as Tuscany, where half the produce of five acres of land will have to support ten human mouths, still has so much room for the wild flowers and the nightingale. When little hills heave themselves suddenly up, and shake themselves free of neighbours, man has to build his garden and his vineyard, and sculp his landscape. Talk of hanging gardens of Babylon, all Italy, apart from the plains, is a hanging garden. For centuries upon centuries man has been patiently modelling the surface of the Mediterranean countries, gently rounding the hills, and graduating the big slopes and the little slopes into the almost invisible levels of terraces. Thousands of square miles of Italy have been lifted in human hands, piled and laid back in tiny little flats, held up by the drystone walls, whose stones came from the lifted earth. It is a work of many, many centuries. It is the gentle sensitive sculpture of all the landscape. And it is the achieving of the peculiar Italian beauty which is so exquisitely natural, because man, feeling his way sensitively to the fruitfulness of the earth, has moulded the earth to his necessity without violating it.

Which shows that it *can* be done. Man *can* live on the earth and by the earth without disfiguring the earth. It has been done here, on all these sculptured hills and softly, sensitively terraced slopes.

But, of course, you can't drive a steam plough on terraces four yards wide, terraces that dwindle and broaden and sink and rise a little, all according to the pitch and the breaking outline of the mother hill. Corn has got to grow on these little shelves of earth, where already the grey olive stands semi-invisible, and the grapevine twists upon its own scars. If oxen can step with that lovely pause at every little stride, they can plough the narrow field. But they will have to leave a tiny fringe, a grassy lip over the drystone wall below. And if the terraces are too narrow to plough, the peasant digging them will still leave the grassy lip, because it helps to hold the surface in the rains.

And here the flowers take refuge. Over and over and over and over has this soil been turned, twice a year, sometimes three times a year, for several thousands of years. Yet the flowers have never been driven out. There is a very rigorous digging and sifting, the little bulbs and tubers are flung away into perdition, not a weed shall remain.

Yet spring returns, and on the terrace lips, and in the stony nooks between terraces, up rise the aconites, the crocuses, the narcissus and the asphodel, the inextinguishable wild tulips. There they are, for ever hanging on the precarious brink of an existence, but for ever triumphant, never quite losing their footing. In England, in America, the flowers get rooted out, driven back. They become fugitive. But in the intensive cultivation of ancient Italian terraces, they dance round and hold their own.

Spring begins with the first narcissus, rather cold and shy and wintry. They are the little bunchy, creamy narcissus with the yellow cup like the yolk of the flower. The natives call these flowers *tazzette*, little cups. They grow on the grassy banks rather sparse, or push up among thorns.

To me they are winter flowers, and their scent is winter. Spring starts in February, with the winter aconite. Some icy day, when the wind is down from the snow of the mountains, early in February, you will notice on a bit of fallow land, under the olive trees, tight, pale-gold little balls, clenched tight as nuts, and resting on round ruffs of green near the ground. It is the winter aconite suddenly come.

The winter aconite is one of the most charming flowers. Like all the early blossoms, once her little flower emerges it is quite naked. No shutting a little green sheath over herself, like the daisy or the dandelion. Her bubble of frail, pale, pure gold rests on the round frill of her green collar, with the snowy wind trying to blow it away.

But without success. The *tramontana* ceases, comes a day of wild February sunshine. The clenched little nuggets of the aconite puff out, they become light bubbles, like small balloons, on a green base. The sun blazes on, with February splendour. And by noon, all under the olives are wide-open little suns, the aconites spreading all their rays; and there is an exquisitely sweet scent, honey-sweet, not narcissus-frosty; and there is a February humming of little brown bees.

Till afternoon, when the sun slopes, and the touch of snow comes back into the air.

But at evening, under the lamp on the table, the aconites are wide and excited, and there is a perfume of sweet spring that makes one almost start humming and trying to be a bee.

Aconites don't last very long. But they turn up in all odd places—on clods of dug earth, and in land where the broad-beans are thrusting up, and along the lips of terraces. But they like best land left fallow for one winter. There they throng, showing how quick they are to seize on an opportunity to live and shine forth.

In a fortnight, before February is over, the yellow bubbles of the aconite are crumpling to nothingness. But already in a cosy-nook the violets are dark purple, and there is a new little perfume in the air.

Like the debris of winter stand the hellebores, in all the wild places, and the butcher's broom is flaunting its last bright red berry. Hellebore is Christmas roses, but in Tuscany the flowers never come white. They emerge out of the grass towards the end of December, flowers wintry of winter, and they are delicately pale green, and of a lovely shape, with yellowish stamens. They have a peculiar wintry quality of invisibility, so lonely rising from the sere grass, and pallid green, held up like a little hand-mirror that reflects nothing. At first they are single upon a stem, short and lovely, and very wintry-beautiful, with a will not to be touched, not to be noticed. One instinctively leaves them alone. But as January draws towards February, these hellebores, these greenish Christmas roses become more assertive. Their pallid water-green becomes yellow, pale sulphur-yellow-green, and they rise up, they are in tufts, in throngs, in veritable bushes of greenish open flowers, assertive, bowing their faces with a hellebore assertiveness. In some places they throng among the bushes and above the water of the stream, giving the peculiar pale glimmer almost of primroses, as you walk among them. Almost of primroses, yet with a coarse hellebore leaf and an up-rearing hellebore assertiveness, like snakes in winter.

And as one walks among them, one brushes the last scarlet off the butcher's broom. This low little shrub is the Christmas holly of Tuscany, only a foot or so high, with a vivid red berry stuck on in the middle of its sharp hard leaf. In February the last red ball rolls off the prickly plume, and winter rolls with it. The violets already are emerging from the moisture.

But before the violets make any show, there are the crocuses. If you walk up through the pine-wood, that lifts its umbrellas of pine so high, up till you come to the brow of the hill at the top, you can look south, due south, and see snow on the Apennines, and on a blue afternoon, seven layers of blue-hilled distance.

Then you sit down on that southern slope, out of the wind, and there it is warm, whether it be January or February, *tramontana* or not. There the earth has been baked by innumerable suns, baked and baked again; moistened by many rains, but never wetted for long. Because it is rocky, and full to the south, and sheering steep in the slope.

And there, in February, in the sunny baked desert of that crumbly slope, you will find the first crocuses. On the sheer aridity of crumbled stone you see a queer, alert little star, very sharp and quite small. It has opened out rather flat, and looks like a tiny freesia flower, creamy, with a smear of yellow yolk. It has no stem, seems to have been just lightly dropped on the crumbled, baked rock. It is the first hill-crocus.

II

North of the Alps, the everlasting winter is interrupted by summers that struggle and soon yield; south of the Alps, the everlasting summer is interrupted by spasmodic and spiteful winters that never get a real hold, but that are mean and dogged. North of the Alps, you may have a pure winter's day in June. South of the Alps, you may have a midsummer day in December or January or even February. The in-between, in either case, is just as it may be. But the lands of the sun are south of the Alps, for ever.

Yet things, the flowers especially, that belong to both sides of the Alps, are not much earlier south than north of the mountains. Through all the winter there are roses in the garden, lovely creamy roses, more pure and mysterious than those of summer, leaning perfect from the stem. And the narcissus in the garden are out by the end of January, and the little simple hyacinths early in February.

But out in the fields, the flowers are hardly any sooner than English flowers. It is mid-February before the first violets, the first crocus, the first primrose. And in mid-February one may find a violet, a primrose, a crocus in England, in the hedgerows and the garden corner.

And still there is a difference. There are several kinds of wild crocus in this region of Tuscany: being little spiky mauve ones, and spiky little creamy ones, that grow among the pine-trees of the bare slopes. But the beautiful ones are those of a meadow in the corner of the woods, the low hollow meadow below the steep, shadowy pine-slopes, the secretive grassy dip where the water seeps through the turf all winter, where the stream runs between thick bushes, where the nightingale sings his mightiest in May, and where the wild thyme is rosy and full of bees, in summer.

Here the lavender crocuses are most at home—here sticking out of the deep grass, in a hollow like a cup, a bowl of grass, come the lilac-coloured crocuses, like an innumerable encampment. You may see them at twilight, with all the buds shut, in the mysterious stillness of the grassy underworld, palely glimmering like myriad folded tents. So the apaches still camp, and close their tepees, in the hollows of the great hills of the West, at night.

But in the morning it is quite different. Then the sun shines strong on the horizontal green cloud-puffs of the pines, the sky is clear and full of life, the water runs hastily, still browned by the last juice of crushed olives. And there the earth's bowl of crocuses is amazing. You cannot believe that the flowers are really still. They are open with such delight, and their pistil-thrust is so red-orange, and they are so many, all reaching out wide and marvellous, that it suggests a perfect ecstasy of radiant, thronging movement, lit-up violet and orange, and surging in some invisible rhythm of concerted, delightful movement. You cannot believe they do not move, and make some sort of crystalline sound of delight. If you sit still and watch, you begin to move with them, like moving with the stars, and you feel the sound of their radiance. All the little cells of the flowers must be leaping with flowery life and utterance.

And the small brown honey-bees hop from flower to flower, dive down, try, and off again. The flowers have been already rifled, most of them. Only sometimes a bee stands on his head, kicking slowly inside the flower, for some time. He has found something. And all the bees have little loaves of pollen, bee-bread, in their elbow-joints.

The crocuses last in their beauty for a week or so, and as they begin to lower their tents and abandon camp, the violets begin to thicken. It is already March. The violets have been showing like tiny dark hounds for some weeks. But now the whole pack comes forth, among the grass and the tangle of wild thyme, till the air all sways subtly scented with violets, and the banks above where the crocuses had their tents are now swarming brilliant purple with violets. They are the sweet violets of early spring, but numbers have made them bold, for they flaunt and ruffle till the slopes are a bright blue-purple blaze of them, full in the sun, with an odd late crocus still standing wondering and erect amongst them.

And now that it is March, there is a rush of flowers. Down by the other stream, which turns sideways to the sun, and has tangles of brier and bramble, down where the hellebore has stood so wan and dignified all winter, there are now white tufts of primroses, suddenly come. Among the tangle and near the water-lip, tufts and bunches of primroses, in abundance. Yet they look more wan, more pallid, more flimsy than English primroses. They lack some of the full wonder of the northern flowers. One tends to overlook them, to turn to the great, solemn-faced purple violets that rear up from the bank, and above all, to the wonderful little towers of the grape-hyacinth.

I know no flower that is more fascinating, when it first appears, than the blue grape-hyacinth. And yet, because it lasts so long, and keeps on coming so repeatedly, for at least two months, one tends later on to ignore it, even to despise it a little. Yet that is very unjust.

The first grape-hyacinths are flowers of blue, thick and rich and meaningful, above the unrenewed grass. The upper buds are pure blue, shut tight; round balls of pure, perfect warm blue, blue, blue; while the lower bells are darkish blue-purple, with the spark of white at the mouth. As yet, none of the lower bells has withered, to leave the greenish, separate sparseness of fruiting that spoils the grape-hyacinth later on, and makes it seem naked and functional. All hyacinths are like that in the seeding.

But, at first, you have only a compact tower of night-blue clearing to dawn, and extremely beautiful. If we were tiny as fairies, and lived only a summer, how lovely these great trees of bells would be to us, towers of night and dawn-blue globes. They would rise above us thick and succulent, and the purple globes would push the blue ones up, with white sparks of ripples, and we should see a god in them.

As a matter of fact, someone once told me they were the flowers of the many-breasted Artemis; and it is true, the Cybele of Ephesus, with her clustered breasts was like a grape-hyacinth at the bosom.

This is the time, in March, when the sloe is white and misty in the hedge-tangle by the stream, and on the slope of land the peach tree stands pink and alone. The almond blossom, silvery pink, is passing, but the peach, deep-toned, bluey, not at all ethereal, this reveals itself like flesh, and the trees are like isolated individuals, the peach and the apricot.

A man said this spring: "Oh, I *don't* care for peach blossom! It is such a vulgar pink!" One wonders what anybody means by a "vulgar" pink. I think pink flannelette is rather vulgar. But probably it's the flannelette's fault, not the pink. And peach blossom has a beautiful sensual pink, far from vulgar, most rare and private. And pink is so beautiful in a landscape, pink houses, pink almond, pink peach and purply apricot, pink asphodels.

It is so conspicuous and so individual, that pink among the coming green of spring, because the first flowers that emerge from winter seem always white or yellow or purple. Now the celandines are out, and along the edges of the *podere*, the big, sturdy, black-purple anemones, with black hearts.

They are curious, these great, dark-violet anemones. You may pass them on a grey day, or at evening or early morning, and never see them. But as you come along in the full sunshine, they seem to be baying at you with all their throats, baying deep purple into the air. It is because they are hot and wide open now, gulping the sun. Whereas when they are shut, they have a silkiness and a curved head, like the curve of an umbrella handle, and a peculiar outward colourlessness, that makes them quite invisible. They may be under your feet, and you will not see them.

Altogether anemones are odd flowers. On these last hills above the plain, we have only the big black-purple ones, in tufts here and there, not many. But two hills away, the young green corn is blue with the lilac-blue kind, still the broad-petalled sort with the darker heart. But these flowers are smaller than our dark-purple, and frailer, more silky. Ours are substantial, thickly vegetable flowers, and not abundant. The others are lovely and silky-delicate, and the whole corn is blue with them. And they have a sweet, sweet scent, when they are warm.

Then on the priest's *podere* there are the scarlet, Adonis-blood anemones: only in one place, in one long fringe under a terrace, and there by a path below. These flowers above all you will never find unless you look for them in the sun. Their silver silk outside makes them quite invisible, when they are shut up.

Yet, if you are passing in the sun, a sudden scarlet faces on to the air, one of the loveliest scarlet apparitions in the world. The inner surface of the Adonis-blood anemone is as fine as velvet, and yet there is no suggestion of pile, not as much as on a velvet rose. And from this inner smoothness issues the red colour, perfectly pure and unknown of earth, no earthiness, and yet solid, not transparent. How a colour manages to be perfectly strong and impervious, yet of a purity that suggests condensed light, yet not luminous, at least, not transparent, is a problem. The poppy in her radiance is translucent, and the tulip in her utter redness has a touch of opaque earth. But the Adonis-blood anemone is neither translucent nor opaque. It is just pure condensed red, of a velvetiness without velvet, and a scarlet without glow.

This red seems to me the perfect premonition of summer—like the red on the outside of apple blossom—and later, the red of the apple. It is the premonition in redness of summer and of autumn.

The red flowers are coming now. The wild tulips are in bud, hanging their grey leaves like flags. They come up in myriads, wherever they get a chance. But they are holding back their redness till the last days of March, the early days of April.

Still, the year is warming up. By the high ditch the common magenta anemone is hanging its silky tassels, or opening its great magenta daisy-shape to the hot sun. It is much nearer to red than the big-petalled anemones are; except the Adonis-blood. They say these anemones sprang from the tears of Venus, which fell as she went looking for Adonis. At that rate, how the poor lady must have wept, for the anemones by the Mediterranean are common as daisies in England.

The daisies are out here too, in sheets, and they too are red-mouthed.

The first ones are big and handsome. But as March goes on, they dwindle to bright little things, like tiny buttons, clouds of them together. That means summer is nearly here.

The red tulips open in the corn like poppies, only with a heavier red. And they pass quickly, without repeating themselves. There is little lingering in a tulip.

In some places there are odd yellow tulips, slender, spiky, and Chinese-looking. They are very lovely, pricking out their dulled yellow in slim spikes. But they too soon lean, expand beyond themselves, and are gone like an illusion.

And when the tulips are gone, there is a moment's pause, before summer. Summer is the next move.

III

In the pause towards the end of April, when the flowers seem to hesitate, the leaves make up their minds to come out. For some time, at the very ends of the bare boughs of fig trees, spurts of pure green have been burning like little cloven tongues of green fire vivid on the tips of the candelabrum. Now these spurts of green spread out, and begin to take the shape of hands, feeling for the air of summer. And tiny green figs are below them, like glands on the throat of a goat.

For some time, the long stiff whips of the vine have had knobby pink buds, like flower buds. Now these pink buds begin to unfold into greenish, half-shut fans of leaves with red in the veins, and tiny spikes of flower, like seed-pearls. Then, in all its down and pinky dawn, the vine-rosette has a frail, delicious scent of a new year.

Now the aspens on the hill are all remarkable with the translucent membranes of blood-veined leaves. They are gold-brown, but not like autumn, rather like the thin wings of bats when like birds—call them birds—they wheel in clouds against the setting sun, and the sun glows through the stretched membrane of their wings, as through thin, brown-red stained glass. This is the red sap of summer, not the red dust of autumn. And in the distance the aspens have the tender panting glow of living membrane just come awake. This is the beauty of the frailty of spring.

The cherry tree is something the same, but more sturdy. Now, in the last week of April, the cherry blossom is still white, but waning and passing away: it is late this year; and the leaves are clustering thick and softly copper in their dark, blood-filled glow. It is queer about fruit trees in this district. The pear and the peach were out together. But now the pear tree is a lovely thick softness of new and glossy green, vivid with a tender fullness of apple-green leaves, gleaming among all the other green of the landscape, the half-high wheat, emerald, and the grey olive, half-invisible, the browning green of the dark cypress, the black of the evergreen oak, the rolling, heavy green puffs of the stone-pines, the flimsy green of small peach and almond trees, the sturdy young green of horse-chestnut. So many greens, all in flakes and shelves and tilted tables and round shoulders and plumes and shaggles and uprisen bushes, of greens and greens, sometimes blindingly brilliant at evening, when the landscape looks as if it were on fire from inside, with green-ness and with gold.

The pear is perhaps the greenest thing in the landscape. The wheat may shine lit-up yellow, or glow bluish, but the pear tree is green in itself. The cherry has white, half-absorbed flowers, so has the apple. But the plum is rough with her new foliage, and inconspicuous, inconspicu-ous as the almond, the peach, the apricot, which one can no longer find in the landscape, though twenty days ago they were the distinguished pink individuals of the whole countryside. Now they are gone. It is the time of green, pre-eminent green, in ruffles and flakes and slabs.

In the wood, the scrub-oak is only just coming uncrumpled, and the pines keep their hold on winter. They are wintry things, stone-pines. At Christmas, their heavy green clouds are richly beautiful. When the cypresses raise their tall and naked bodies of dark green, and the osiers are vivid red-orange, on the still blue air, and the land is lavender, then, in mid-winter, the landscape is most beautiful in colour, surging with colour.

But now, when the nightingale is still drawing out his long, wistful, yearning, teasing plaint-note, and following it up with a rich and joyful burble, the pines and the cypresses seem hard and rusty, and the wood has lost its subtlety and its mysteriousness. It still seems wintry in spite of the yellowing young oaks, and the heath in flower. But hard, dull pines above, and hard, dull, tall heath below, all stiff and resistant, this is out of the mood of spring.

In spite of the fact that the stone-white heath is in full flower, and very lovely when you look at it, it does not, casually, give the impres-sion of blossom. More the impression of having its tips and crests all dipped in hoarfrost; or in a whitish dust. It has a peculiar ghostly colourlessness amid the darkish colourlessness of the wood altogether, which completely takes away the sense of spring.

Yet the tall white heath is very lovely, in its invisibility. It grows sometimes as tall as a man, lifting up its spires and its shadowy-white fingers with a ghostly fullness, amid the dark, rusty green of its lower bushiness; and it gives off a sweet honeyed scent in the sun, and a cloud of fine white stone-dust, if you touch it. Looked at closely, its little bells are most beautiful, delicate and white, with the brown-purple inner eye and the dainty pin-head of the pistil. And out in the sun at the edge of the wood, where the heath grows tall and thrusts up its spires of dim white next a brilliant, yellow-flowering vetch-bush, under a blue sky, the effect has a real magic.

And yet, in spite of all, the dim whiteness of all the flowering heath-fingers only adds to the hoariness and out-of-date quality of the pinewoods, now in the pause between spring and summer. It is the ghost of the interval.

Not that this week is flowerless. But the flowers are little lonely things, here and there: the early purple orchid, ruddy and very much alive, you come across occasionally, then the little groups of bee-orchid, with their ragged concerted indifference to their appearance. Also there are the huge bud-spikes of the stout, thick-flowering pink orchid, huge buds like fat ears of wheat, hard-purple and splendid. But already odd grains of the wheat-ear are open, and out of the purple hangs the delicate pink rag of a floweret. Also there are very lovely and choice cream-coloured orchids with brown spots on the long and delicate lip. These grow in the more moist places, and have exotic tender spikes, very rare-seeming. Another orchid is a little, pretty yellow one.

But orchids, somehow, do not make a summer. They are too aloof and individual. The little slate-blue scabious is out, but not enough to raise an appearance. Later on, under the real hot sun, he will bob into notice. And by the edges of the paths there are odd rosy cushions of wild thyme. Yet these, too, are rather samples than the genuine thing. Wait another month, for wild thyme.

The same with the irises. Here and there, in fringes along the upper edge of terraces, and in odd bunches among the stones, the dark-purple iris sticks up. It is beautiful, but it hardly counts. There is not enough of it, and it is torn and buffeted by too many winds. First the wind blows with all its might from the Mediterranean, not cold, but definitely wearying, with its rude and insistent pushing. Then, after a moment of calm, back comes a hard wind from the Adriatic, cold and disheartening. Between the two of them, the dark-purple iris flutters and tatters and curls as if it were burnt: while the little yellow rock-rose streams at the end of its thin stalk, and wishes it had not been in such a hurry to come out.

There is really no hurry. By May, the great winds will drop, and the great sun will shake off his harassments. Then the nightingale will sing an unbroken song, and the discreet, barely audible Tuscan cuckoo will be a little more audible. Then the lovely pale-lilac irises will come out in all their showering abundance of tender, proud, spiky bloom, till the air will gleam with mauve, and a new crystalline lightness will be everywhere.

The iris is half-wild, half-cultivated. The peasants sometimes dig up the roots, iris root, orris root (orris powder, the perfume that is still used). So, in May, you will find ledges and terraces, fields just lit up with the mauve light of irises, and so much scent in the air, you do not notice it, you do not even know it. It is all the flowers of iris, before the olive invisibly blooms.

There will be tufts of iris everywhere, rising up proud and tender. When the rose-coloured wild gladiolus is mingled in the corn, and the love-in-the-mist opens blue: in May and June, before the corn is cut.

But as yet it is neither May nor June, but end of April, the pause between spring and summer, the nightingale singing interruptedly, the bean-flowers dying in the bean-fields, the bean-perfume passing with spring, the little birds hatching in the nests, the olives pruned, and the vines, the last bit of late ploughing finished, and not much work to hand, now, not until the peas are ready to pick, in another two weeks or so. Then all the peasants will be crouching between the pea-rows, endlessly, endlessly gathering peas, in the long pea-harvest which lasts two months.

So the change, the endless and rapid change. In the sunny countries, the change seems more vivid, and more complete than in the grey countries. In the grey countries, there is a grey or dark permanency, over whose surface passes change ephemeral, leaving no real mark. In England, winters and summers shadowily give place to one another. But underneath lies the grey substratum, the permanency of cold, dark reality where bulbs live, and reality is bulbous, a thing of endurance and stored-up, starchy energy.

But in the sunny countries, change is the reality and permanence is artificial and a condition of imprisonment. In the North, man tends instinctively to imagine, to conceive that the sun is lighted like a candle, in an everlasting darkness, and that one day the candle will go out, the sun will be exhausted, and the everlasting dark will resume uninterrupted sway. Hence, to the northerner, the phenomenal world is essentially tragical, because it is temporal and must cease to exist. Its very existence implies ceasing to exist, and this is the root of the feeling of tragedy.

But to the southerner, the sun is so dominant that, if every phenomenal body disappeared out of the universe, nothing would remain but bright luminousness, sunniness. The absolute is sunniness; and shadow, or dark, is only merely relative: merely the result of something getting between one and the sun.

This is the instinctive feeling of the ordinary southerner. Of course, if you start to *reason*, you may argue that the sun is a phenomenal body. Therefore it came into existence, therefore it will pass out of existence, therefore the very sun is tragic in its nature.

But this is just argument. We think, because we have to light a candle in the dark, therefore some First Cause had to kindle the sun in the infinite darkness of the beginning.

The argument is entirely shortsighted and specious. We do not know in the least whether the sun ever came into existence, and we have not the slightest possible ground for conjecturing that the sun will ever pass out of existence. All that we do know, by actual experience, is that shadow comes into being when some material object intervenes between us and the sun, and that shadow ceases to exist when the intervening object is removed. So that, of all temporal or transitory or bound-to-cease things that haunt our existence, shadow or darkness, is the one which is purely and simply temporal. We can think of death, if we like, as of something permanently intervening between us and the sun: and this is at the root of the southern, under-world idea of death. But this doesn't alter the sun at all. As far as experience goes, in the human race, the one thing that is always there is the shining sun, and dark shadow is an accident of intervention.

Hence, strictly, there is no tragedy. The universe contains no tragedy, and man is only tragical because he is afraid of death. For my part, if the sun always shines, and always will shine, in spite of millions of clouds of words, then death, somehow, does not have many terrors. In the sunshine, even death is sunny. And there is no end to the sunshine.

That is why the rapid change of the Tuscan spring is utterly free, for me, of any sense of tragedy. "Where are the snows of yesteryear?" Why, precisely where they ought to be. Where are the little yellow aconites of eight weeks ago? I neither know nor care. They were sunny and the sun shines, and sunniness means change, and petals passing and coming. The winter aconites sunnily came, and sunnily went. What more? The sun always shines. It is our fault if we don't think so.

E. L. GRANT WATSON
1885–1970

In his long and distinguished career, Grant Watson served as a zoologist and anthropologist on an expedition to northwest Australia, pursued a career as a novelist (with encouragement from Joseph Conrad), and practiced Jungian psychoanalysis. All of these facets of his experience contributed to a distinctive approach to natural history. He described organisms and other phenomena with scientific authority and precision, yet he also shaped his observations into compelling narratives. Engaged as he was in the drama and variety of the physical creation, he invariably found in it as well a manifestation of the spirit. In the following essay, for instance, he suggests surprising parallels between the erotic encounters of slugs and the spiritual ideals of human love. Watson belongs in the company of such other scientist–nature writers as Rachel Carson, Loren Eiseley, and Lewis Thomas, for whom the ultimate products of hypotheses, experimentation, and data were a kindling of wonder and an enhanced sense of kinship with the larger community of life.

UNKNOWN EROS

The three of us sat on the step outside the kitchen door and watched the hermaphrodite copulation of slugs, while the summer night grew dark, and nightingales sang from the thorn thickets, and glow-worms hung curled in dewy grass stems. First with curiosity and surprise, then with wonder, and then in an awed and fascinated silence, we watched the slow, rhythmical movements of the two creatures. They were hanging on a thick, twisted thread of their own slime. One end was fastened to a ledge under the scullery window, and at the other, more than a foot distant, hung the yellow and black slugs, spinning slowly, first in one direction, then in the other. Their interlocked bodies formed a sin-

Descent of Spirit: Writings of E. L. Grant Watson (Sidney, Australia: Primavera Press, 1990).

gle sphere, in such intimacy of embrace as would not have been possible with any skeletoned animal. Their soft and yielding substances flowed and pressed and welded one against the other, every inner curve finding its counterpart, and as they pressed closer and closer, they hung so supported on their thread that every part that could be in contact, should be, and all their outer curves were touched only by air, with no earthly contact to offer any resistance to their movement.

After a period of twisting and turning, there came out from under the dorsal mantle of each slug a process which at first appeared single-ended, but which as it elongated grew to be the shape of a stag's antler. Of the shape, but of far different substance, for this process was soft and yielding, like a fungus, and the fronds of each separate point pressed and slid over the flesh of its companion. With tender and tentative yet rhythmic gropings did these organs progress, till each clinging antler-form, always moving, always penetrating, quivering and exploring, enfolded the body of its fellow, and pressed the foremost portions of that ramifying and pulsating system under the mantle folds and deep into the body of its mate.

By the light of a lantern we watched them for an hour and more, as they spun on their thread. Over their moist bodies there passed risings and swellings and sinkings, while the plastic penetration of the antler-fronds proceeded, binding them and enmeshing them one with the other. Not only the exposure of their inner organs but what actuated these organs from within seemed here outwardly revealed. The reproductive urge of life was displayed under our wondering eyes, and the soft embracings wrought of those gestures, made so strangely apparent, was for our astonished recognition. The sensual tenderness of those blendings was not to be mistaken. No crude, stiff or ill-adjusted mechanism was here present; all was fluid as the come and go of wavelets; yet the fire of love was in their moisture, and we watched spellbound in admiration, until at last the invading and caressing organs contracted and withdrew, while the mutual surgings of those mucous bodies moderated.

The slugs' embrace of love was ended. They parted, and each separately lowered itself on its own strand of slime to the earth.

Surely no physical union between higher animals or humans was ever consummated with such completeness. Hardened and formal limb-shapes are incapable of such blending; only with our invisible emotions and with the subtleties of our souls can we humans so harmoniously and intimately interpenetrate. For this reason, and perhaps for others, the physical-sensual cannot alone completely satisfy us, as it must satisfy such lowly-seeming creatures as slugs.

The marvel of material union, which we had watched that summer night, appeared to our speculative meditations as a physical presentation of love-emotions, which in human beings can find expression only in the realm of the invisible spiritual. Donne and Shakespeare have described these ecstasies; telling how, in similar manner, may human lovers be apart and suspended on some psychic cord of their own making, and there revolving to their own motions, touched only by the soft forces of air, and undisturbed by any contact with earth, they may for a time experience such interpenetration of their androgynous souls as their too-rigidified and formal bodies are incapable of achieving. Thus in the material, it is with slugs, creatures of the mire, that only rarely come within the scope of our recognition; yet that which works in them, in the deed of their reproduction, is more than that which is ordinarily expressed in their slow, earth-bound progress. May it not be that these humble manifestations, composed of parenchyma and slime, are the physical counterpart of our spiritual comprehensions?

And now that the act was over, what was to be done with the two hungry-looking slugs, each a good three inches long, which were making off towards my young lettuces? I could not be so ungrateful as to kill them after having watched their mating with so much appreciation, nor could I be so bad a gardener as to let them be free to spawn in my soil, and devour my seedlings. I put them in a box, and the next day took them into the country and let them go in a distant meadow.

WAVE AND CLIFF

That the sea-wave, as it surges with complex eddies, flowing and ebbing, should fit with such perfect adjustment to the rock-surface of the cliff is one of the most inevitable of natural things; so inevitable and so natural, that it would seem foolish to question why or how such close reciprocal adjustment is accomplished. Science will tell us that water flows to level, that waves are produced by the passing of air currents, and that rock can be moulded by constant minute frictions. Yet these facts, which science can measure and confirm, when we have valued them for what they are worth, still leave us, as we watch the waves lapping against the cliff-foot, with questions unanswered; they leave us with our essential questions almost unframed, and with a growing wonder, which speculates and half-accepts the message of the waves' movements. The recognition grows that this reciprocal relation, so exact, and, in the large, so unvarying, yet subject to so many varying

moods, is the norm of all interrelatedness, the condition of all existence. As the waves pass and change, and appear to come again and
again to change, they present conflict and adjustment, a duality forming a unity, and a unity, flowing and changing into a manifold diversity.

In Gruinard Bay there are peninsulas which jut out into clear water.
The rocks fall steeply to the sea, and near the base of these are places
where a man can sit, and from where, on a calm day, he may see the
wide sea-level extending far on either side, disturbed only by the undulations of a slow swell and the lesser ripples, slantingly furrowed by the
breeze. Beneath the pale, transparent green, the seaweeds are tossed and
swirled, first one way and then another, with what would seem a rhythmical movement, yet which, on exacter observation, records no definite
rhythm. Its comings and goings are unpredictable, and the haphazard
tossings of the seaweed declare what little one can know of the wind and
sea currents. Slight as that knowledge is, it is sufficient to start speculations concerning those forces of gravitation, of changeful wind, and of
varying watery momentums, of moon and sun. These are far too elaborate for analysis, yet though the mind remains baffled by that complex of
movements, it is possible to watch and to absorb into experience those
seemingly haphazard gestures. As the mood of attention deepens there
seems to arise a kind of kinship and recognition between the tossed
fronds of seaweed and the inner, secret happenings of life.

Those gesturing, brown tresses, which appear so much at the disposal
of the in-sucking and out-gushing water, are held firm at the root. They
swing in answer to the flows and ebbs, which, themselves, are so unpre-
dictably determined by the weight of the adjacent sea and atmosphere.
Thus they swing to and fro, and up and down, sometimes heaped
together in attitudes of lassitude and resignation, sometimes each spray
outflung from its centre, and sometimes swirled and twisted like serpents
in Medusa's hair. On rare occasions the fronds of any particular bunch
will dispose themselves in a gesture of brief ecstasy; a changeful happiness—yet the next moment the swirl and swing to and fro continues, up
and down, heaped and separated, again sucked under and cast forth.

The moments of ecstasy return in flashes and half flashes, and not
only from the tide-tossed weed but from ripples and shifting lights, and
they are heard in tones of the withdrawal and the pushing forward of
the innumerable, small, wet lips of the sea against the land, and in the
sounds of the breaking of waves on distant beaches. They evoke memories, diffusing and spreading faint recognitions, sinking and permeating
and, when almost lost amidst the less articulate melodies from which
they have been born, they rise again, quickening and baffling the heart
with fresh surprises.

A couple of oyster-catchers are beating their short, sharp-pointed wings in quick flight as they sweep past the promontory. They call out questioning cries at the stranger on the rock; they make wide circles and return, and as one of them passes, it also is touched by that same ecstatic accident, as the weed was touched and caught by; it checks in its flight, lifts its head and flutters its wings faster. A mood of out-gushing, irreflective delight possesses the bird. The next moment it is flying after its companion, uttering its piercing calls. Like the chance-tossed weed, like the ripples, on a wind-touched sea, it has answered to the joy concealed within the happenings which seem to men chance-born.

The mood of quickened attention, so happily caught, yet so short-lived, returns from the bird and from the weed, from the air-provoked ripples, and from the waves which come so gently, mildly from the ocean; it regards the cliff and the wave, and the fact, so obvious to science, yet so mysterious, that these, in their contact of antagonism and adjustment, *fit one another*. The cliff determines the shape that the wave must take at their meeting, and the waves, in their innumerable comings and goings, mould the cliff.

HENRY BESTON
1888–1968

In the fall of 1926 Henry Beston came to live in the Fo'castle, a dune cottage of his own design, on a Massachusetts barrier beach fronting the Atlantic Ocean. The Outermost House (1928), his account of "A Year of Life on the Great Beach of Cape Cod," has became a classic of the solitary sojourn form of nature writing, which includes Walden *and* Pilgrim at Tinker Creek. *Unlike Thoreau, however, Beston intended not so much to put into practice certain principles of living already held as to "know this coast and share its mysterious and elemental life." His style has a sensual and rhythmic richness unsurpassed in the genre, expressing his belief that "poetry is as necessary to comprehension as science" and conveying a vivid tactile sense of his surroundings. The book is infused with an extraordinary sense of human and natural drama, all encompassed by what*

Beston called "the burning ritual of the year." He had a gift for creating memorable utterances, and The Outermost House *contains some of the most-quoted passages of twentieth-century nature writing.*

From THE OUTERMOST HOUSE: A YEAR OF LIFE ON THE GREAT BEACH OF CAPE COD

AUTUMN, OCEAN, AND BIRDS

There is a new sound on the beach, and a greater sound. Slowly, and day by day, the surf grows heavier, and down the long miles of the beach, at the lonely stations, men hear the coming winter in the roar. Mornings and evenings grow cold, the northwest wind grows cold; the last crescent of the month's moon, discovered by chance in a pale morning sky, stands north of the sun. Autumn ripens faster on the beach than on the marshes and the dunes. Westward and landward there is colour; seaward, bright space and austerity. Lifted to the sky, the dying grasses on the dune tops' rim tremble and lean seaward in the wind, wraiths of sand course flat along the beach, the hiss of sand mingles its thin stridency with the new thunder of the sea.

I have been spending my afternoons gathering driftwood and observing birds. The skies being clear, noonday suns take something of the bite out of the wind, and now and then a warmish west-sou'westerly finds its way back into the world. Into the bright, vast days I go, shouldering home my sticks and broken boards and driving shore birds on ahead of me, putting up sanderlings and sandpipers, ringnecks and knots, plovers and killdeer, coveys of a dozen, little flocks, great flocks, compact assemblies with a regimented air. For a fortnight past, October 9th to October 23d, an enormous population of the migrants has been "stopping over" on my Eastham sands, gathering, resting, feeding, and commingling. They come, they go, they melt away, they gather again; for actual miles the intricate and inter-crisscross pattern of their feet runs unbroken along the tide rim of Cape Cod.

Yet it is no confused and careless horde through which I go, but an army. Some spirit of discipline and unity has passed over these countless little brains, waking in each flock a conscious sense of its collective self and giving each bird a sense of himself as a member of some

The Outermost House: A Year of Life on the Great Beach of Cape Cod (Garden City, N.Y.: Doubleday, 1928).

migrant company. Lone fliers are rare, and when seen have an air of being in pursuit of some flock which has overlooked them and gone on. Swift as the wind they fly, speeding along the breakers with the directness of a runner down a course, and I read fear in their speed. Sometimes I see them find their own and settle down beside them half a mile ahead, sometimes they melt away into a vista of surf and sky, still speeding on, still seeking.

The general multitude, it would seem, consists of birds who have spent the summer somewhere on the outer Cape and of autumn reinforcements from the north.

I see the flocks best when they are feeding on the edge of a tide which rises to its flood on the later afternoon. No summer blur of breaker mist or glassiness of heat now obscures these outer distances, and as on I stride, keeping to the lower beach when returning with a load, I can see birds and more birds and ever more birds ahead. Every last advance of a dissolved breaker, coursing on, flat and seething, has those who run away before it, turning its flank or fluttering up when too closely pursued; every retreating in-sucked slide has those who follow it back, eagerly dipping and gleaning. Having fed, the birds fly up to the upper beach and sit there for hours in the luke-cold wind, flock by flock, assembly by assembly. The ocean thunders, pale wisps and windy tatters of wintry cloud sail over the dunes, and the sandpipers stand on one leg and dream, their heads tousled deep into their feathers.

I wonder where these thousands spend the night. Waking the other morning just before sunrise, I hurried into my clothes and went down to the beach. North and then south I strolled, along an ebbing tide, and north and south the great beach was as empty of bird life as the sky. Far to the south, I remember now, a frightened pair of semipalmated sandpipers did rise from somewhere on the upper beach and fly toward me swift and voiceless, pass me on the flank, and settle by the water's edge a hundred yards or so behind. They instantly began to run about and feed, and as I watched them an orange sun floated up over the horizon with the speed and solemnity of an Olympian balloon.

The tide being high these days late in the afternoon, the birds begin to muster on the beach about ten o'clock in the morning. Some fly over from the salt meadows, some arrive flying along the beach, some drop from the sky. I startle up a first group on turning from the upper beach to the lower. I walk directly at the birds—a general apprehension, a rally, a scutter ahead, and the birds are gone. Standing on the beach, fresh claw marks at my feet, I watch the lovely sight of the group instantly turned into a constellation of birds, into a fugitive pleiades whose living stars keep their chance positions; I watch the spiralling flight, the momentary tilts of the white bellies, the alternate shows of the clustered, grayish backs. The group next ahead, though wary from the first, continues feeding. I draw nearer; a few run ahead as if to escape me afoot, others stop and prepare to fly; nearer still, the birds can stand no more; another rally, another scutter, and they are following their kin along the surges.

No aspect of nature on this beach is more mysterious to me than the flights of these shorebird constellations. The constellation forms, as I have hinted, in an instant of time, and in that same instant develops its own will. Birds which have been feeding yards away from each other, each one individually busy for his individual body's sake, suddenly fuse into this new volition and, flying, rise as one, coast as one, tilt their dozen bodies as one, and as one wheel off on the course which the new group will has determined. There is no such thing, I may add, as a lead bird or guide. Had I more space I should like nothing better than to discuss this new will and its instant or origin, but I do not want to crowd this part of my chapter, and must therefore leave the problem to all who study the psychic relations between the individual and a surrounding many. My special interest is rather the instant and synchronous obedience of each speeding body to the new volition. By what means, by what methods of communication does this will so suffuse the living constellation that its dozen or more tiny brains know it and obey it in such an instancy of time? Are we to believe that these birds, all of them, are *machina*, as Descartes long ago insisted, mere mechanisms of flesh and bone so exquisitely alike that each cogwheel brain, encountering the same environmental forces, synchronously lets slip the same mechanic ratchet? or is there some psychic relation between these creatures? Does some current flow through them and between them as they fly? Schools of fish, I am told, make similar mass changes of direction. I saw such a thing once, but of that more anon.

We need another and a wiser and perhaps a more mystical concept of animals. Remote from universal nature, and living by complicated artifice, man in civilization surveys the creature through the glass of his knowledge and sees thereby a feather magnified and the whole image in distortion. We patronize them for their incompleteness, for their tragic fate of having taken form so far below ourselves. And therein we err, and greatly err. For the animal shall not be measured by man. In a world older and more complete than ours they move finished and complete, gifted with extensions of the senses we have lost or never attained, living by voices we shall never hear. They are not brethren, they are not underlings; they are other nations, caught with ourselves in the net of life and time, fellow prisoners of the splendour and travail of the earth.

The afternoon sun sinks red as fire; the tide climbs the beach, its foam a strange crimson; miles out, a freighter goes north, emerging from the shoals. *   *   *

NIGHT ON THE GREAT BEACH

Our fantastic civilization has fallen out of touch with many aspects of nature, and with none more completely than with night. Primitive folk, gathered at a cave mouth round a fire, do not fear night; they fear, rather, the energies and creatures to whom night gives power; we of the age of the machines, having delivered ourselves of nocturnal enemies, now have a dislike of night itself. With lights and ever more lights, we drive the holiness and beauty of night back to the forests and the sea; the little villages, the crossroads even, will have none of it. Are modern folk, perhaps, afraid of night? Do they fear that vast serenity, the mystery of infinite space, the austerity of stars? Having made themselves at home in a civilization obsessed with power, which explains its whole world in terms of energy, do they fear at night for their dull acquiescence and the pattern of their beliefs? Be the answer what it will, to-day's civilization is full of people who have not the slightest notion of the character or the poetry of night, who have never even seen night. Yet to live thus, to know only artificial night, is as absurd and evil as to know only artificial day.

Night is very beautiful on this great beach. It is the true other half of the day's tremendous wheel; no lights without meaning stab or trouble it; it is beauty, it is fulfilment, it is rest. Thin clouds float in these heavens, islands of obscurity in a splendour of space and stars: the Milky Way bridges earth and ocean; the beach resolves itself into a unity of form, its summer lagoons, its slopes and uplands merging; against the western sky and the falling bow of sun rise the silent and superb undulations of the dunes.

My nights are at their darkest when a dense fog streams in from the sea under a black, unbroken floor of cloud. Such nights are rare, but are most to be expected when fog gathers off the coast in early summer; this last Wednesday night was the darkest I have known. Between ten o'clock and two in the morning three vessels stranded on the outer beach—a fisherman, a four-masted schooner, and a beam trawler. The fisherman and the schooner have been towed off, but the trawler, they say, is still ashore.

I went down to the beach that night just after ten o'clock. So utterly black, pitch dark it was, and so thick with moisture and trailing showers, that there was no sign whatever of the beam of Nauset; the sea was only a sound, and when I reached the edge of the surf the dunes themselves had disappeared behind. I stood as isolate in that immensity of rain and night as I might have stood in interplanetary space. The sea was troubled and noisy, and when I opened the darkness with an outlined cone of light from my electric torch I saw that the waves were washing up green coils of sea grass, all coldly wet and bright in the motionless and unnatural radiance. Far off a single ship was groaning its way along the shoals. The fog was compact of the finest moisture; passing by, it spun itself into my lens of light like a kind of strange, aërial, and liquid silk. Effin Chalke, the new coast guard, passed me going north, and told me that he had had news at the halfway house of the schooner at Cahoon's.

It was dark, pitch dark to my eye, yet complete darkness, I imagine, is exceedingly rare, perhaps unknown in outer nature. The nearest natural approximation to it is probably the gloom of forest country buried in night and cloud. Dark as the night was here, there was still light on the surface of the planet. Standing on the shelving beach, with the surf breaking at my feet, I could see the endless wild uprush, slide, and withdrawal of the sea's white rim of foam. The men at Nauset tell me that on such nights they follow along this vague crawl of whiteness, trusting to habit and a sixth sense to warn them of their approach to the halfway house.

Animals descend by starlight to the beach, North, beyond the dunes, muskrats forsake the cliff and nose about in the driftwood and weed, leaving intricate trails and figure eights to be obliterated by the day; the lesser folk—the mice, the occasional small sand-coloured toads, the burrowing moles—keep to the upper beach and leave their tiny footprints under the overhanging wall. In autumn skunks, beset by a shrinking larder, go beach combing early in the night. The animal is by preference a clean feeder and turns up his nose at rankness. I almost stepped on a big fellow one night as I was walking north to meet the first man south from Nauset. There was a scamper, and the creature ran up the beach from under my feet; alarmed he certainly was, yet was he contained and continent. Deer are frequently seen, especially north of the light. I find their tracks upon the summer dunes.

Years ago, while camping on this beach north of Nauset, I went for a stroll along the top of the cliff at break of dawn. Though the path followed close enough along the edge, the beach below was often hidden, and I looked directly from the height to the flush of sunrise at sea. Presently the path, turning, approached the brink of the earth precipice, and on the beach below, in the cool, wet rosiness of dawn, I saw three deer playing. They frolicked, rose on their hind legs, scampered off, and returned again, and were merry. Just before sunrise they trotted off north together down the beach toward a hollow in the cliff and the path that climbs it.

Occasionally a sea creature visits the shore at night. Lone coast guardsmen, trudging the sand at some deserted hour, have been startled by seals. One man fell flat on a creature's back, and it drew away from under him, flippering toward the sea, with a sound "halfway between a squeal and a bark." I myself once had rather a start. It was long after sundown, the light dying and uncertain, and I was walking home on the top level of the beach and close along the slope descending to the ebbing tide. A little more than halfway to the Fo'castle a huge unexpected something suddenly writhed horribly in the darkness under my bare foot. I had stepped on a skate left stranded by some recent crest of surf, and my weight had momentarily annoyed it back to life.

Facing north, the beam of Nauset becomes part of the dune night. As I walk toward it, I see the lantern, now as a star of light which waxes and wanes three mathematical times, now as a lovely pale flare of light behind the rounded summits of the dunes. The changes in the atmosphere change the colour of the beam; it is now whitish, now flame golden, now golden red; it changes its form as well, from a star to a blare of light, from a blare of light to a cone of radiance sweeping a circumference of fog. To the west of Nauset I often see the apocalyptic flash of the great light at the Highland reflected on the clouds or even on the moisture in the starlit air, and, seeing it, I often think of the pleasant hours I have spent there when George and Mary Smith were at the light and I had the good fortune to visit as their guest. Instead of going to sleep in the room under the eaves, I would lie awake, looking out of a window to the great spokes of light revolving as solemnly as a part of the universe.

All night long the lights of coastwise vessels pass at sea, green lights going south, red lights moving north. Fishing schooners and flounder draggers anchor two or three miles out, and keep a bright riding light burning on the mast. I see them come to anchor at sundown, but I rarely see them go, for they are off at dawn. When busy at night, these fishermen illumine their decks with a scatter of oil flares. From shore, the ships might be thought afire. I have watched the scene through a night glass. I could see no smoke, only the waving flares, the reddish radiance on sail and rigging, an edge of reflection overside, and the enormous night and sea beyond.

One July night, as I returned at three o'clock from an expedition north, the whole night, in one strange, burning instant, turned into a phantom day. I stopped and, questioning, stared about. An enormous meteor, the largest I have ever seen, was consuming itself in an effulgence of light west of the zenith. Beach and dune and ocean appeared out of nothing, shadowless and motionless, a landscape whose every tremor and vibration were stilled, a landscape in a dream.

The beach at night has a voice all its own, a sound in fullest harmony with its spirit and mood—with its little, dry noise of sand forever moving, with its solemn, overspilling, rhythmic seas, with its eternity of stars that sometimes seem to hang down like lamps from the high heavens—and that sound the piping of a bird. As I walk the beach in early summer my solitary coming disturbs it on its nest, and it flies away, troubled, invisible, piping its sweet, plaintive cry. The bird I write of is the piping plover, *Charadrius melodus*, sometimes called the beach plover or the mourning bird. Its note is a whistled syllable, the loveliest musical note, I think, sounded by any North Atlantic bird.

Now that summer is here I often cook myself a camp supper on the beach. Beyond the crackling, salt-yellow driftwood flame, over the pyramid of barrel staves, broken boards, and old sticks all atwist with climbing fire, the unseen ocean thunders and booms, the breaker sounding hollow as it falls. The wall of the sand cliff behind, with its rim of grass and withering roots, its sandy crumblings and erosions, stands gilded with flame; wind cries over it; a covey of sandpipers pass between the ocean and the fire. There are stars, and to the south Scorpio hangs curving down the sky with ringed Saturn shining in his claw.

Learn to reverence night and to put away the vulgar fear of it, for, with the banishment of night from the experience of man, there vanishes as well a religious emotion, a poetic mood, which gives depth to the adventure of humanity. By day, space is one with the earth and with man—it is his sun that is shining, his clouds that are floating past; at night, space is his no more. When the great earth, abandoning day, rolls up the deeps of the heavens and the universe, a new door opens for the human spirit, and there are few so clownish that some awareness of the mystery of being does not touch them as they gaze. For a moment of night we have a glimpse of ourselves and of our world islanded in its stream of stars—pilgrims of mortality, voyaging between horizons across eternal seas of space and time. Fugitive though the instant be, the spirit of man is, during it, ennobled by a genuine moment of emotional dignity, and poetry makes its own both the human spirit and experience. *   *   *

ORION RISES ON THE DUNES

So came August to its close, ending its last day with a night so luminous and still that a mood came over me to sleep out on the open beach under the stars. There are nights in summer when darkness and ebbing tide quiet the universal wind, and this August night was full of that quiet of absence, and the sky was clear. South of my house, between the bold fan of a dune and the wall of a plateau, a sheltered hollow opens seaward, and to this nook I went, shouldering my blankets sailorwise. In the star-shine the hollow was darker than the immense and solitary beach, and its floor was still pleasantly warm with the overflow of day.

I fell asleep uneasily, and woke again as one wakes out-of-doors. The vague walls about me breathed a pleasant smell of sand, there was no sound, and the broken circle of grass above was as motionless as something in a house. Waking again, hours afterward, I felt the air grown colder and heard a little advancing noise of waves. It was still night. Sleep gone and past recapture, I drew on my clothes and went to the beach. In the luminous east, two great stars aslant were rising clear of the exhalations of darkness gathered at the rim of night and ocean — Betelgeuse and Bellatrix, the shoulders of Orion. Autumn had come, and the Giant stood again at the horizon of day and the ebbing year, his belt still hidden in the bank of cloud, his feet in the deeps of space and the far surges of the sea.

My year upon the beach had come full circle; it was time to close my door. Seeing the great suns, I thought of the last time I marked them in the spring, in the April west above the moors, dying into the light and sinking. I saw them of old above the iron waves of black December, sparkling afar. Now, once again, the Hunter rose to drive summer south before him, once again autumn followed on his steps. I had seen the ritual of the sun; I had shared the elemental world. Wraiths of memories began to take shape. I saw the sleet of the great storm slanting down again into the grass under the thin seepage of moon, the blue-white spill of an immense billow on the outer bar, the swans in the high October sky, the sunset madness and splendour of the year's terns over the dunes, the clouds of beach birds arriving, the eagle solitary in the blue. And because I had known this outer and secret world, and been able to live as I had lived, reverence and gratitude greater and deeper than ever possessed me, sweeping every emotion else aside, and space and silence an instant closed together over life. Then time gathered again like a cloud, and presently the stars began to pale over an ocean still dark with remembered night.

During the months that have passed since that September morning some have asked me what understanding of Nature one shapes from so strange a year? I would answer that one's first appreciation is a sense that the creation is still going on, that the creative forces are as great and as active to-day as they have ever been, and that to-morrow's morning will be as heroic as any of the world. *Creation is here and now*. So near is man to the creative pageant, so much a part is he of the endless and incredible experiment, that any glimpse he may have will be but the revelation of a moment, a solitary note heard in a symphony thundering through debatable existences of time. Poetry is as necessary to comprehension as science. It is as impossible to live without reverence as it is without joy.

And what of Nature itself, you say—that callous and cruel engine, red in tooth and fang? Well, it is not so much of an engine as you think. As for "red in tooth and fang," whenever I hear the phrase or its intellectual echoes I know that some passer-by has been getting life from books. It is true that there are grim arrangements. Beware of judging them by whatever human values are in style. As well expect Nature to answer to your human values as to come into your house and sit in a chair. The economy of nature, its checks and balances, its measurements of competing life—all this is its great marvel and has an ethic of its own. Live in Nature, and you will soon see that for all its non-human rhythm, it is no cave of pain. As I write I think of my beloved birds of the great beach, and of their beauty and their zest of living. And if there are fears, know also that Nature has its unexpected and unappreciated mercies.

Whatever attitude to human existence you fashion for yourself, know that it is valid only if it be the shadow of an attitude to Nature. A human life, so often likened to a spectacle upon a stage, is more justly a ritual. The ancient values of dignity, beauty, and poetry which sustain it are of Nature's inspiration; they are born of the mystery and beauty of the world. Do no dishonour to the earth lest you dishonour the spirit of man. Hold your hands out over the earth as over a flame. To all who loved her, who open to her the doors of their veins, she gives of her strength, sustaining them with her own measureless tremor of dark life. Touch the earth, love the earth, honour the earth, her plains, her valleys, her hills, and her seas; rest your spirit in her solitary places. For the gifts of life are the earth's and they are given to all, and they are the songs of birds at daybreak, Orion and the Bear, and dawn seen over ocean from the beach.

ALDO LEOPOLD
1888–1948

Leopold was a professional conservationist—a forester who early under-
stood the concept and value of wilderness, a professor of wildlife manage-
ment at the University of Wisconsin who became a champion of the
predators' role within a healthy, stable ecosystem. But his chief impor-
tance within the environmental movement and the literature of nature
alike is as the author of A Sand County Almanac, *published in 1949,*
shortly after he had died fighting a fire.

As Leopold follows the year through its circle, at the rural Wisconsin
"shack" where his family spent weekends and vacations, he echoes
Thoreau's celebration of a quiet landscape. He gives to chickadees and
pine seedlings the same attentiveness other nature writers bring to a
sperm whale or a sequoia. At the same time, he is a magnificent teacher
about the way the natural environment has been impoverished and
about some of his own experiments at restoring fertility and diversity to
his "sand farm." The clarity of Aldo Leopold's observations and the
"Land Ethic" which emerges from them have made A Sand County
Almanac *a major influence on American attitudes toward our natural*
environment.

From A SAND COUNTY ALMANAC

MARSHLAND ELEGY

A dawn wind stirs on the great marsh. With almost imperceptible
slowness it rolls a bank of fog across the wide morass. Like the white
ghost of a glacier the mists advance, riding over phalanxes of tamarack,
sliding across bog-meadows heavy with dew. A single silence hangs
from horizon to horizon.

A Sand County Almanac (New York: Oxford University Press, 1949).

Out of some far recess of the sky a tinkling of little bells falls soft upon the listening land. Then again silence. Now comes a baying of some sweet-throated hound, soon the clamor of a responding pack. Then a far clear blast of hunting horns, out of the sky into the fog.

High horns, low horns, silence, and finally a pandemonium of trumpets, rattles, croaks, and cries that almost shakes the bog with its nearness, but without yet disclosing whence it comes. At last a glint of sun reveals the approach of a great echelon of birds. On motionless wing they emerge from the lifting mists, sweep a final arc of sky, and settle in clangorous descending spirals to their feeding grounds. A new day has begun on the crane marsh.

A sense of time lies thick and heavy on such a place. Yearly since the ice age it has awakened each spring to the clangor of cranes. The peat layers that comprise the bog are laid down in the basin of an ancient lake. The cranes stand, as it were, upon the sodden pages of their own history. These peats are the compressed remains of the mosses that clogged the pools, of the tamaracks that spread over the moss, of the cranes that bugled over the tamaracks since the retreat of the ice sheet. An endless caravan of generations has built of its own bones this bridge into the future, this habitat where the oncoming host again may live and breed and die.

To what end? Out on the bog a crane, gulping some luckless frog, springs his ungainly hulk into the air and flails the morning sun with mighty wings. The tamaracks re-echo with his bugled certitude. He seems to know.

Our ability to perceive quality in nature begins, as in art, with the pretty. It expands through successive stages of the beautiful to values as yet uncaptured by language. The quality of cranes lies, I think, in this higher gamut, as yet beyond the reach of words.

This much, though, can be said: our appreciation of the crane grows with the slow unraveling of earthly history. His tribe, we now know, stems out of the remote Eocene. The other members of the fauna in which he originated are long since entombed within the hills. When we hear his call we hear no mere bird. We hear the trumpet in the orchestra of evolution. He is the symbol of our untamable past, of that incredible sweep of millennia which underlies and conditions the daily affairs of birds and men.

And so they live and have their being—these cranes—not in the constricted present, but in the wider reaches of evolutionary time. Their annual return is the ticking of the geologic clock. Upon the place of their return they confer a peculiar distinction. Amid the endless mediocrity of the commonplace, a crane marsh holds a palentological patent of nobility, won in the march of aeons, and revocable only by shotgun. The sadness discernible in some marshes arises, perhaps, from their once having harbored cranes. Now they stand humbled, adrift in history.

Some sense of this quality in cranes seems to have been felt by sportsmen and ornithologists of all ages. Upon such quarry as this the Holy Roman Emperor Frederick loosed his gyrfalcons. Upon such quarry as this once swooped the hawks of Kublai Khan. Marco Polo tells us: "He derives the highest amusement from sporting with gyrfalcons and hawks. At Changanor the Khan has a great Palace surrounded by a fine plain where are found cranes in great numbers. He causes millet and other grains to be sown in order that the birds may not want."

The ornithologist Bengt Berg, seeing cranes as a boy upon the Swedish heaths, forthwith made them his life work. He followed them to Africa and discovered their winter retreat on the White Nile. He says of his first encounter: "It was a spectacle which eclipsed the flight of the roc in the Thousand and One Nights."

When the glacier came down out of the north, crunching hills and gouging valleys, some adventuring rampart of the ice climbed the Baraboo Hills and fell back into the outlet gorge of the Wisconsin River. The swollen waters backed up and formed a lake half as long as the state, bordered on the east by cliffs of ice, and fed by the torrents that fell from melting mountains. The shorelines of this old lake are still visible; its bottom is the bottom of the great marsh.

The lake rose through the centuries, finally spilling over east of the Baraboo range. There it cut a new channel for the river, and thus drained itself. To the residual lagoons came the cranes, bugling the defeat of the retreating winter, summoning the on-creeping host of living things to their collective task of marsh-building. Floating bogs of sphagnum moss clogged the lowered waters, filled them. Sedge and leatherleaf, tamarack and spruce successively advanced over the bog, anchoring it by their root fabric, sucking out its water, making peat. The lagoons disappeared, but not the cranes. To the moss-meadows that replaced the ancient waterways they returned each spring to dance and bugle and rear their gangling sorrel-colored young. These, albeit birds, are not properly called chicks, but *colts*. I cannot explain why. On some dewy June morning watch them gambol over their ancestral pastures at the heels of the roan mare, and you will see for yourself.

One year not long ago a French trapper in buckskins pushed his canoe up one of the moss-clogged creeks that thread the great marsh. At this attempt to invade their miry stronghold the cranes gave vent to loud and ribald laughter. A century or two later Englishmen came in covered wagons. They chopped clearings in the timbered moraines that border the marsh, and in them planted corn and buckwheat. They did not intend, like the Great Khan at Changanor, to feed the cranes. But the cranes do not question the intent of glaciers, emperors, or pioneers. They ate the grain, and when some irate farmer failed to concede their usufruct in his corn, they trumpeted a warning and sailed across the marsh to another farm.

There was no alfalfa in those days, and the hill-farms made poor hay land, especially in dry years. One dry year someone set a fire in the tamaracks. The burn grew up quickly to bluejoint grass, which, when cleared of dead trees, made a dependable hay meadow. After that, each August, men appeared to cut hay. In winter, after the cranes had gone South, they drove wagons over the frozen bogs and hauled the hay to their farms in the hills. Yearly they plied the marsh with fire and axe, and in two short decades hay meadows dotted the whole expanse.

Each August when the haymakers came to pitch their camps, singing and drinking and lashing their teams with whip and tongue, the cranes whinnied to their colts and retreated to the far fastnesses. "Red shitepokes" the haymakers called them, from the rusty hue which at that season often stains the battleship-gray of crane plumage. After the hay was stacked and the marsh again their own, the cranes returned, to call down out of October skies the migrant flocks from Canada. Together they wheeled over the new-cut stubbles and raided the corn until frosts gave the signal for the winter exodus.

These haymeadow days were the Arcadian age for marsh dwellers. Man and beast, plant and soil lived on and with each other in mutual toleration, to the mutual benefit of all. The marsh might have kept on producing hay and prairie chickens, deer and muskrat, crane-music and cranberries forever.

The new overlords did not understand this. They did not include soil, plants, or birds in their ideas of mutuality. The dividends of such a balanced economy were too modest. They envisaged farms not only around, but *in* the marsh. An epidemic of ditch-digging and land-booming set in. The marsh was gridironed with drainage canals, speckled with new fields and farmsteads.

But crops were poor and beset by frosts, to which the expensive ditches added an aftermath of debt. Farmers moved out. Peat beds dried, shrank, caught fire. Sun-energy out of the Pleistocene shrouded the countryside in acrid smoke. No man raised his voice against the waste, only his nose against the smell. After a dry summer not even the winter snows could extinguish the smoldering marsh. Great pockmarks were burned into field and meadow, the scars reaching down to the sands of the old lake, peat-covered these hundred centuries. Rank weeds sprang out of the ashes, to be followed after a year or two by aspen scrub. The cranes were hard put, their numbers shrinking with the remnants of unburned meadow. For them, the song of the power shovel came near being an elegy. The high priests of progress knew nothing of cranes, and cared less. What is a species more or less among engineers? What good is an undrained marsh anyhow?

For a decade or two crops grew poorer, fires deeper, wood-fields larger, and cranes scarcer, year by year. Only reflooding, it appeared, could keep the peat from burning. Meanwhile cranberry growers had, by plugging drainage ditches, reflooded a few spots and obtained good yields. Distant politicians bugled about marginal land, over-production, unemployment relief, conservation. Economists and planners came to look at the marsh. Surveyors, technicians, CCC's, buzzed about. A counter-epidemic of reflooding set in. Government bought land, reset-tled farmers, plugged ditches wholesale. Slowly the bogs are re-wetting. The fire-pocks become ponds. Grass fires still burn, but they can no longer burn the wetted soil.

All this, once the CCC camps were gone, was good for cranes, but not so the thickets of scrub popple that spread inexorably over the old burns, and still less the maze of new roads that inevitably follow gov-ernmental conservation. To build a road is so much simpler than to think of what the country really needs. A roadless marsh is seemingly as worthless to the alphabetical conservationist as an undrained one was to the empire-builders. Solitude, the one natural resource still undow-ered of alphabets, is so far recognized as valuable only by ornithologists and cranes.

Thus always does history, whether of marsh or market place, end in paradox. The ultimate value in these marshes is wildness, and the crane is wildness incarnate. But all conservation of wildness is self-defeating, for to cherish we must see and fondle, and when enough have seen and fondled, there is no wilderness left to cherish.

Some day, perhaps in the very process of our benefactions, perhaps in the fullness of geologic time, the last crane will trumpet his farewell and spiral skyward from the great marsh. High out of the clouds will fall the sound of hunting horns, the baying of the phantom pack, the tinkle of little bells, and then a silence never to be broken, unless per-chance in some far pasture of the Milky Way.

THINKING LIKE A MOUNTAIN

A deep chesty bawl echoes from rimrock to rimrock, rolls down the mountain, and fades into the far blackness of the night. It is an outburst of wild defiant sorrow, and of contempt for all the adversities of the world.

Every living thing (and perhaps many a dead one as well) pays heed to that call. To the deer it is a reminder of the way of all flesh, to the pine a forecast of midnight scuffles and of blood upon the snow, to the coyote a promise of gleanings to come, to the cowman a threat of red ink at the bank, to the hunter a challenge of fang against bullet. Yet behind these obvious and immediate hopes and fears there lies a deeper meaning, known only to the mountain itself. *Only the mountain has lived long enough to listen objectively to the howl of a wolf.*

Those unable to decipher the hidden meaning know nevertheless that it is there, for it is felt in all wolf country, and distinguishes that country from all other land. It tingles in the spine of all who hear wolves by night, or who scan their tracks by day. Even without sight or sound of wolf, it is implicit in a hundred small events: the midnight whinny of a pack horse, the rattle of rolling rocks, the bound of a fleeing deer, the way shadows lie under the spruces. Only the ineducable tyro can fail to sense the presence or absence of wolves, or the fact that mountains have a secret opinion about them.

My own conviction on this score dates from the day I saw a wolf die. We were eating lunch on a high rimrock, at the foot of which a turbulent river elbowed its way. We saw what we thought was a doe fording the torrent, her breast awash in white water. When she climbed the bank toward us and shook out her tail, we realized our error: it was a wolf. A half-dozen others, evidently grown pups, sprang from the willows and all joined in a welcoming mêlée of wagging tails and playful maulings. What was literally a pile of wolves writhed and tumbled in the center of an open flat at the foot of our rimrock.

In those days we had never heard of passing up a chance to kill a wolf. In a second we were pumping lead into the pack, but with more excitement than accuracy: how to aim a steep downhill shot is always confusing. When our rifles were empty, the old wolf was down, and a pup was dragging a leg into impassable slide-rocks.

We reached the old wolf in time to watch a fierce green fire dying in her eyes. I realized then, and have known ever since, that there was something new to me in those eyes—something known only to her and to the mountain. I was young then, and full of trigger-itch; I thought that because fewer wolves meant more deer, that no wolves would mean hunters' paradise. But after seeing the green fire die, I sensed that neither the wolf nor the mountain agreed with such a view.

Since then I have lived to see state after state extirpate its wolves. I have watched the face of many a newly wolfless mountain, and seen the south-facing slopes wrinkle with a maze of new deer trails. I have seen every edible bush and seedling browsed, first to anaemic desuetude, and then to death. I have seen every edible tree defoliated to the height of a saddlehorn. Such a mountain looks as if someone had given God a new pruning shears, and forbidden Him all other exercise. In the end the starved bones of the hoped-for deer herd, dead of its own too-much, bleach with the bones of the dead sage, or molder under the high-lined junipers.

I now suspect that just as a deer herd lives in mortal fear of its wolves, so does a mountain live in mortal fear of its deer. And perhaps with better cause, for while a buck pulled down by wolves can be replaced in two or three years, a range pulled down by too many deer may fail of replacement in as many decades.

So also with cows. The cowman who cleans his range of wolves does not realize that he is taking over the wolf's job of trimming the herd to fit the range. He has not learned to think like a mountain. Hence we have dustbowls, and rivers washing the future into the sea.

THE LAND ETHIC

When god-like Odysseus returned from the wars in Troy, he hanged all on one rope a dozen slave-girls of his household whom he suspected of misbehavior during his absence.

This hanging involved no question of propriety. The girls were property. The disposal of property was then, as now, a matter of expediency, not of right and wrong.

Concepts of right and wrong were not lacking from Odysseus' Greece: witness the fidelity of his wife through the long years before at last his black-prowed galleys clove the wine-dark seas for home. The ethical structure of that day covered wives, but had not yet been extended to human chattels. During the three thousand years which have since elapsed, ethical criteria have been extended to many fields of conduct, with corresponding shrinkages in those judged by expediency only.

THE ETHICAL SEQUENCE

This extension of ethics, so far studied only by philosophers, is actually a process in ecological evolution. Its sequences may be described in ecological as well as in philosophical terms. An ethic, ecologically, is a limitation on freedom of action in the struggle for existence. An ethic, philosophically, is a differentiation of social from anti-social conduct. These are two definitions of one thing. The thing has its origin in the tendency of interdependent individuals or groups to evolve modes of co-operation. The ecologist calls these symbioses. Politics and economics are advanced symbioses in which the original free-for-all competition has been replaced, in part, by co-operative mechanisms with an ethical content.

The complexity of co-operative mechanisms has increased with population density, and with the efficiency of tools. It was simpler, for example, to define the anti-social uses of sticks and stones in the days of the mastodons than of bullets and billboards in the age of motors.

The first ethics dealt with the relation between individuals; the Mosaic Decalogue is an example. Later accretions dealt with the relation between the individual and society. The Golden Rule tries to integrate the individual to society; democracy to integrate social organization to the individual.

There is as yet no ethic dealing with man's relation to land and to the animals and plants which grow upon it. Land, like Odysseus' slave-girls, is still property. The land-relation is still strictly economic, entailing privileges but not obligations.

The extension of ethics to this third element in human environment is, if I read the evidence correctly, an evolutionary possibility and an ecological necessity. It is the third step in a sequence. The first two have already been taken. Individual thinkers since the days of Ezekiel and Isaiah have asserted that the despoliation of land is not only inexpedient but wrong. Society, however, has not yet affirmed their belief. I regard the present conservation movement as the embryo of such an affirmation.

An ethic may be regarded as a mode of guidance for meeting ecological situations so new or intricate, or involving such deferred reactions, that the path of social expediency is not discernible to the average individual. Animal instincts are modes of guidance for the individual in meeting such situations. Ethics are possibly a kind of community instinct in-the-making.

THE COMMUNITY CONCEPT

All ethics so far evolved rest upon a single premise: that the individual is a member of a community of interdependent parts. His instincts prompt him to compete for his place in that community, but his ethics prompt him also to co-operate (perhaps in order that there may be a place to compete for).

The land ethic simply enlarges the boundaries of the community to include soils, waters, plants, and animals, or collectively: the land.

This sounds simple: do we not already sing our love for and obligation to the land of the free and the home of the brave? Yes, but just what and whom do we love? Certainly not the soil, which we are sending helter-skelter downriver. Certainly not the waters, which we assume have no function except to turn turbines, float barges, and carry off sewage. Certainly not the plants, of which we exterminate whole communities without batting an eye. Certainly not the animals, of which we have already extirpated many of the largest and most beautiful species. A land ethic of course cannot prevent the alteration, management, and use of these "resources," but it does affirm their right to continued existence, and, at least in spots, their continued existence in a natural state.

In short, a land ethic changes the role of *Homo sapiens* from conqueror of the land-community to plain member and citizen of it. It implies respect for his fellow-members, and also respect for the community as such.

In human history, we have learned (I hope) that the conqueror role is eventually self-defeating. Why? Because it is implicit in such a role that the conqueror knows, *ex cathedra*, just what makes the community clock tick, and just what and who is valuable, and what and who is worthless, in community life. It always turns out that he knows neither, and this is why his conquests eventually defeat themselves.

In the biotic community, a parallel situation exists. Abraham knew exactly what the land was for: it was to drip milk and honey into Abraham's mouth. At the present moment, the assurance with which we regard this assumption is inverse to the degree of our education.

The ordinary citizen today assumes that science knows what makes the community clock tick; the scientist is equally sure that he does not. He knows that the biotic mechanism is so complex that its workings may never be fully understood.

That man is, in fact, only a member of a biotic team is shown by an ecological interpretation of history. Many historical events, hitherto explained solely in terms of human enterprise, were actually biotic interactions between people and land. The characteristics of the land determined the facts quite as potently as the characteristics of the men who lived on it.

Consider, for example, the settlement of the Mississippi valley. In the years following the Revolution, three groups were contending for its control: the native Indian, the French and English traders, and the American settlers. Historians wonder what would have happened if the

English at Detroit had thrown a little more weight into the Indian side of those tipsy scales which decided the outcome of the colonial migration into the cane-lands of Kentucky. It is time now to ponder the fact that the cane-lands, when subjected to the particular mixture of forces represented by the cow, plow, fire, and axe of the pioneer, became bluegrass. What if the plant succession inherent in this dark and bloody ground had, under the impact of these forces, given us some worthless sedge, shrub, or weed? Would Boone and Kenton have held out? Would there have been any overflow into Ohio, Indiana, Illinois, and Missouri? Any Louisiana Purchase? Any transcontinental union of new states? Any Civil War?

Kentucky was one sentence in the drama of history. We are commonly told what the human actors in this drama tried to do, but we are seldom told that their success, or the lack of it, hung in large degree on the reaction of particular soils to the impact of the particular forces exerted by their occupancy. In the case of Kentucky, we do not even know where the bluegrass came from—whether it is a native species, or a stowaway from Europe.

Contrast the cane-lands with what hindsight tells us about the Southwest, where the pioneers were equally brave, resourceful, and persevering. The impact of occupancy here brought no bluegrass, or other plant fitted to withstand the bumps and buffetings of hard use. This region, when grazed by livestock, reverted through a series of more and more worthless grasses, shrubs, and weeds to a condition of unstable equilibrium. Each recession of plant types bred erosion; each increment to erosion bred a further recession of plants. The result today is a progressive and mutual deterioration, not only of plants and soils, but of the animal community subsisting thereon. The early settlers did not expect this: on the ciénegas of New Mexico some even cut ditches to hasten it. So subtle has been its progress that few residents of the region are aware of it. It is quite invisible to the tourist who finds this wrecked landscape colorful and charming (as indeed it is, but it bears scant resemblance to what it was in 1848).

This same landscape was "developed" once before, but with quite different results. The Pueblo Indians settled the Southwest in pre-Columbian times, but they happened *not* to be equipped with range livestock. Their civilization expired, but not because their land expired.

In India, regions devoid of any sod-forming grass have been settled, apparently without wrecking the land, by the simple expedient of carrying the grass to the cow, rather than vice versa. (Was this the result of some deep wisdom, or was it just good luck? I do not know.)

In short, the plant succession steered the course of history; the pioneer simply demonstrated, for good or ill, what successions inhered in the land. Is history taught in this spirit? It will be, once the concept of land as a community really penetrates our intellectual life.

THE ECOLOGICAL CONSCIENCE

Conservation is a state of harmony between men and land. Despite nearly a century of propaganda, conservation still proceeds at a snail's pace; progress still consists largely of letterhead pieties and convention oratory. On the back forty we still slip two steps backward for each forward stride.

The usual answer to this dilemma is "more conservation education." No one will debate this, but is it certain that only the *volume* of education needs stepping up? Is something lacking in the *content* as well?

It is difficult to give a fair summary of its content in brief form, but, as I understand it, the content is substantially this: obey the law, vote right, join some organizations, and practice what conservation is profitable on your own land; the government will do the rest.

Is not this formula too easy to accomplish anything worth-while? It defines no right or wrong, assigns no obligation, calls for no sacrifice, implies no change in the current philosophy of values. In respect of land-use, it urges only enlightened self-interest. Just how far will such education take us? An example will perhaps yield a partial answer.

By 1930 it had become clear to all except the ecologically blind that southwestern Wisconsin's topsoil was slipping seaward. In 1933 the farmers were told that if they would adopt certain remedial practices for five years, the public would donate CCC labor to install them, plus the necessary machinery and materials. The offer was widely accepted, but the practices were widely forgotten when the five-year contract period was up. The farmers continued only those practices that yielded an immediate and visible economic gain for themselves.

This led to the idea that maybe farmers would learn more quickly if they themselves wrote the rules. Accordingly the Wisconsin Legislature in 1937 passed the Soil Conservation District Law. This said to farmers, in effect: *We, the public, will furnish you free technical service and loan you specialized machinery, if you will write your own rules for land-use. Each county may write its own rules, and these will have the force of law.* Nearly all the counties promptly organized to accept the proffered help, but after a decade of operation, *no county has yet written a single rule.* There has been visible progress in such practices as strip-cropping, pasture renovation, and soil liming, but none in fencing woodlots against grazing, and none in excluding plow and cow from steep slopes.

The farmers, in short, have selected those remedial practices which were profitable anyhow, and ignored those which were profitable to the community, but not clearly profitable to themselves.

When one asks why no rules have been written, one is told that the community is not yet ready to support them; education must precede rules. But the education actually in progress makes no mention of obligations to land over and above those dictated by self-interest. The net result is that we have more education but less soil, fewer healthy woods, and as many floods as in 1937.

The puzzling aspect of such situations is that the existence of obligations over and above self-interest is taken for granted in such rural community enterprises as the betterment of roads, schools, churches, and baseball teams. Their existence is not taken for granted, nor as yet seriously discussed, in bettering the behavior of the water that falls on the land, or in the preserving of the beauty or diversity of the farm landscape. Land-use ethics are still governed wholly by economic self-interest, just as social ethics were a century ago.

To sum up: we asked the farmer to do what he conveniently could to save his soil, and he has done just that, and only that. The farmer who clears the woods off a 75 per cent slope, turns his cows into the clearing, and dumps its rainfall, rocks, and soil into the community creek, is still (if otherwise decent) a respected member of society. If he puts lime on his fields and plants his crops on contour, he is still entitled to all the privileges and emoluments of his Soil Conservation District. The District is a beautiful piece of social machinery, but it is coughing along on two cylinders because we have been too timid, and too anxious for quick success, to tell the farmer the true magnitude of his obligations. Obligations have no meaning without conscience, and the problem we face is the extension of the social conscience from people to land.

No important change in ethics was ever accomplished without an internal change in our intellectual emphasis, loyalties, affections, and convictions. The proof that conservation has not yet touched these foundations of conduct lies in the fact that philosophy and religion have not yet heard of it. In our attempt to make conservation easy, we have made it trivial.

SUBSTITUTES FOR A LAND ETHIC

When the logic of history hungers for bread and we hand out a stone, we are at pains to explain how much the stone resembles bread. I now describe some of the stones which serve in lieu of a land ethic.

One basic weakness in a conservation system based wholly on economic motives is that most members of the land community have no economic value. Wildflowers and songbirds are examples. Of the 22,000 higher plants and animals native to Wisconsin, it is doubtful whether more than 5 per cent can be sold, fed, eaten, or otherwise put to economic use. Yet these creatures are members of the biotic community, and if (as I believe) its stability depends on its integrity, they are entitled to continuance.

When one of these non-economic categories is threatened, and if we happen to love it, we invent subterfuges to give it economic importance. At the beginning of the century songbirds were supposed to be disappearing. Ornithologists jumped to the rescue with some distinctly shaky evidence to the effect that insects would eat us up if birds failed to control them. The evidence had to be economic in order to be valid.

It is painful to read these circumlocutions today. We have no land ethic yet, but we have at least drawn nearer the point of admitting that birds should continue as a matter of biotic right, regardless of the presence or absence of economic advantage to us.

A parallel situation exists in respect of predatory mammals, raptorial birds, and fish-eating birds. Time was when biologists somewhat overworked the evidence that these creatures preserve the health of game by killing weaklings, or that they control rodents for the farmer, or that they prey only on "worthless" species. Here again, the evidence had to be economic in order to be valid. It is only in recent years that we hear the more honest argument that predators are members of the community, and that no special interest has the right to exterminate them for the sake of a benefit, real or fancied, to itself. Unfortunately this enlightened view is still in the talk stage. In the field the extermination of predators goes merrily on: witness the impending erasure of the timber wolf by fiat of Congress, the Conservation Bureaus, and many state legislatures.

Some species of trees have been "read out of the party" by economics-minded foresters because they grow too slowly, or have too low a sale value to pay as timber crops: white cedar, tamarack, cypress, beech, and hemlock are examples. In Europe, where forestry is ecologically more advanced, the non-commercial tree species are recognized as members of the native forest community, to be preserved as such, within reason. Moreover some (like beech) have been found to have a valuable function in building up soil fertility. The interdependence of the forest and its constituent tree species, ground flora, and fauna is taken for granted.

Lack of economic value is sometimes a character not only of species or groups, but of entire biotic communities: marshes, bogs, dunes, and "deserts" are examples. Our formula in such cases is to relegate their conservation to government as refuges, monuments, or parks. The difficulty is that these communities are usually interspersed with more valuable private lands; the government cannot possibly own or control such scattered parcels. The net effect is that we have relegated some of them to ultimate extinction over large areas. If the private owner were ecologically minded, he would be proud to be the custodian of a reasonable proportion of such areas, which add diversity and beauty to his farm and to his community.

In some instances, the assumed lack of profit in these "waste" areas has proved to be wrong, but only after most of them had been done away with. The present scramble to reflood muskrat marshes is a case in point.

There is a clear tendency in American conservation to relegate to government all necessary jobs that private landowners fail to perform. Government ownership, operation, subsidy, or regulation is now widely prevalent in forestry, range management, soil and watershed management, park and wilderness conservation, fisheries management, and migratory bird management, with more to come. Most of this growth in governmental conservation is proper and logical, some of it is inevitable. That I imply no disapproval of it is implicit in the fact that I have spent most of my life working for it. Nevertheless the question arises: What is the ultimate magnitude of the enterprise? Will the tax base carry its eventual ramifications? At what point will governmental conservation, like the mastodon, become handicapped by its own dimensions? The answer, if there is any, seems to be in a land ethic, or some other force which assigns more obligation to the private landowner.

Industrial landowners and users, especially lumbermen and stockmen, are inclined to wail long and loudly about the extension of government ownership and regulation to land, but (with notable exceptions) they show little disposition to develop the only visible alternative: the voluntary practice of conservation on their own lands.

When the private landowner is asked to perform some unprofitable act for the good of the community, he today assents only with out-stretched palm. If the act costs him cash this is fair and proper, but when it costs only forethought, open-mindedness, or time, the issue is at least debatable. The overwhelming growth of land-use subsidies in recent years must be ascribed, in large part, to the government's own agencies for conservation education: the land bureaus, the agricultural colleges, and the extension services. As far as I can detect, no ethical obligation toward land is taught in these institutions.

To sum up: a system of conservation based solely on economic self-interest is hopelessly lopsided. It tends to ignore, and thus eventually to eliminate, many elements in the land community that lack commercial value, but that are (as far as we know) essential to its healthy functioning. It assumes, falsely, I think, that the economic parts of the biotic clock will function without the uneconomic parts. It tends to relegate to government many functions eventually too large, too complex, or too widely dispersed to be performed by government.

An ethical obligation on the part of the private owner is the only visible remedy for these situations.

THE LAND PYRAMID

An ethic to supplement and guide the economic relation to land presupposes the existence of some mental image of land as a biotic mechanism. We can be ethical only in relation to something we can see, feel, understand, love, or otherwise have faith in.

The image commonly employed in conservation education is "the balance of nature." For reasons too lengthy to detail here, this figure of speech fails to describe accurately what little we know about the land mechanism. A much truer image is the one employed in ecology: the biotic pyramid. I shall first sketch the pyramid as a symbol of land, and later develop some of its implications in terms of land-use.

Plants absorb energy from the sun. This energy flows through a circuit called the biota, which may be represented by a pyramid consisting of layers. The bottom layer is the soil. A plant layer rests on the soil, an insect layer on the plants, a bird and rodent layer on the insects, and so on up through various animal groups to the apex layer, which consists of the larger carnivores.

The species of a layer are alike not in where they came from, or in what they look like, but rather in what they eat. Each successive layer depends on those below it for food and often for other services, and each in turn furnishes food and services to those above. Proceeding upward, each successive layer decreases in numerical abundance. Thus, for every carnivore there are hundreds of his prey, thousands of their prey, millions of insects, uncountable plants. The pyramidal form of the system reflects this numerical progression from apex to base. Man shares an intermediate layer with the bears, raccoons, and squirrels which eat both meat and vegetables.

The lines of dependency for food and other services are called food chains. Thus soil-oak-deer-Indian is a chain that has now been largely converted to soil-corn-cow-farmer. Each species, including ourselves, is a link in many chains. The deer eats a hundred plants other than oak, and the cow a hundred plants other than corn. Both, then, are links in a hundred chains. The pyramid is a tangle of chains so complex as to seem disorderly, yet the stability of the system proves it to be a highly organized structure. Its functioning depends on the co-operation and competition of its diverse parts.

In the beginning, the pyramid of life was low and squat; the food chains short and simple. Evolution has added layer after layer, link after link. Man is one of thousands of accretions to the height and complexity of the pyramid. Science has given us many doubts, but it has given us at least one certainty: the trend of evolution is to elaborate and diversify the biota.

Land, then, is not merely soil; it is a fountain of energy flowing through a circuit of soils, plants, and animals. Food chains are the living channels which conduct energy upward; death and decay return it to the soil. The circuit is not closed; some energy is dissipated in decay, some is added by absorption from the air, some is stored in soils, peats, and long-lived forests; but it is a sustained circuit, like a slowly augmented revolving fund of life. There is always a net loss by downhill wash, but this is normally small and offset by the decay of rocks. It is deposited in the ocean and, in the course of geological time, raised to form new lands and new pyramids.

The velocity and character of the upward flow of energy depend on the complex structure of the plant and animal community, much as the upward flow of sap in a tree depends on its complex cellular organization. Without this complexity, normal circulation would presumably not occur. Structure means the characteristic numbers, as well as the characteristic kinds and functions, of the component species. This interdependence between the complex structure of the land and its smooth functioning as an energy unit is one of its basic attributes.

When a change occurs in one part of the circuit, many other parts must adjust themselves to it. Change does not necessarily obstruct or divert the flow of energy; evolution is a long series of self-induced changes, the net result of which has been to elaborate the flow mechanism and to lengthen the circuit. Evolutionary changes, however, are usually slow and local. Man's invention of tools has enabled him to make changes of unprecedented violence, rapidity, and scope.

One change is in the composition of floras and faunas. The larger predators are lopped off the apex of the pyramid; food chains, for the first time in history, become shorter rather than longer. Domesticated species from other lands are substituted for wild ones, and wild ones are moved to new habitats. In this world-wide pooling of faunas and floras, some species get out of bounds as pests and diseases, others are extinguished. Such effects are seldom intended or foreseen; they represent unpredicted and often untraceable readjustments in the structure. Agricultural science is largely a race between the emergence of new pests and the emergence of new techniques for their control.

Another change touches the flow of energy through plants and animals and its return to the soil. Fertility is the ability of soil to receive, store, and release energy. Agriculture, by overdrafts on the soil, or by too radical a substitution of domestic for native species in the super-structure, may derange the channels of flow or deplete storage. Soils depleted of their storage, or of the organic matter which anchors it, wash away faster than they form. This is erosion.

Waters, like soil, are part of the energy circuit. Industry, by polluting waters or obstructing them with dams, may exclude the plants and animals necessary to keep energy in circulation.

Transportation brings about another basic change: the plants or animals grown in one region are now consumed and returned to the soil in another. Transportation taps the energy stored in rocks, and in the air, and uses it elsewhere; thus we fertilize the garden with nitrogen gleaned by the guano birds from the fishes of seas on the other side of the Equator. Thus the formerly localized and self-contained circuits are pooled on a world-wide scale.

The process of altering the pyramid for human occupation releases stored energy, and this often gives rise, during the pioneering period, to a deceptive exuberance of plant and animal life, both wild and tame. These releases of biotic capital tend to becloud or postpone the penalties of violence.

This thumbnail sketch of land as an energy circuit conveys three basic ideas:

(1) That land is not merely soil.
(2) That the native plants and animals kept the energy circuit open; others may or may not.
(3) That man-made changes are of a different order than evolution-ary changes, and have effects more comprehensive than is intended or foreseen.

These ideas, collectively, raise two basic issues: Can the land adjust itself to the new order? Can the desired alterations be accomplished with less violence?

Biotas seem to differ in their capacity to sustain violent conversion. Western Europe, for example, carries a far different pyramid than Caesar found there. Some large animals are lost; swampy forests have become meadows or plow-land; many new plants and animals are introduced, some of which escape as pests; the remaining natives are greatly changed in distribution and abundance. Yet the soil is still there and, with the help of imported nutrients, still fertile; the waters flow normally; the new structure seems to function and to persist. There is no visible stoppage or derangement of the circuit.

Western Europe, then, has a resistant biota. Its inner processes are tough, elastic, resistant to strain. No matter how violent the alterations, the pyramid, so far, has developed some new *modus vivendi* which preserves its habitability for man, and for most of the other natives.

Japan seems to present another instance of radical conversion without disorganization.

Most other civilized regions, and some as yet barely touched by civilization, display various stages of disorganization, varying from initial symptoms to advanced wastage. In Asia Minor and North Africa diagnosis is confused by climatic changes, which may have been either the cause or the effect of advanced wastage. In the United States the degree of disorganization varies locally; it is worst in the Southwest, the Ozarks, and parts of the South, and least in New England and the Northwest. Better land-uses may still arrest it in the less advanced regions. In parts of Mexico, South America, South Africa, and Australia a violent and accelerating wastage is in progress, but I cannot assess the prospects.

This almost world-wide display of disorganization in the land seems to be similar to disease in an animal, except that it never culminates in complete disorganization or death. The land recovers, but at some reduced level of complexity, and with a reduced carrying capacity for people, plants, and animals. Many biotas currently regarded as "lands of opportunity" are in fact already subsisting on exploitative agriculture, i.e. they have already exceeded their sustained carrying capacity. Most of South America is overpopulated in this sense.

In arid regions we attempt to offset the process of wastage by reclamation, but it is only too evident that the prospective longevity of reclamation projects is often short. In our own West, the best of them may not last a century.

The combined evidence of history and ecology seems to support one general deduction: the less violent the man-made changes, the greater the probability of successful readjustment in the pyramid. Violence, in turn, varies with human population density; a dense population requires a more violent conversion. In this respect, North America has a better chance for permanence than Europe, if she can contrive to limit her density.

This deduction runs counter to our current philosophy, which assumes that because a small increase in density enriched human life, that an indefinite increase will enrich it indefinitely. Ecology knows of no density relationship that holds for indefinitely wide limits. All gains from density are subject to a law of diminishing returns.

Whatever may be the equation for men and land, it is improbable that we as yet know all its terms. Recent discoveries in mineral and vitamin nutrition reveal unsuspected dependencies in the up-circuit:

incredibly minute quantities of certain substances determine the value
of soils to plants, of plants to animals. What of the down-circuit? What
of the vanishing species, the preservation of which we now regard as an
esthetic luxury? They helped build the soil; in what unsuspected ways
may they be essential to its maintenance? Professor Weaver proposes
that we use prairie flowers to reflocculate the wasting soils of the dust
bowl; who knows for what purpose cranes and condors, otters and grizzlies may some day be used?

LAND HEALTH AND THE A-B CLEAVAGE

A land ethic, then, reflects the existence of an ecological conscience,
and this in turn reflects a conviction of individual responsibility for the
health of the land. Health is the capacity of the land for self-renewal.
Conservation is our effort to understand and preserve this capacity.

Conservationists are notorious for their dissensions. Superficially
these seem to add up to mere confusion, but a more careful scrutiny
reveals a single plane of cleavage common to many specialized fields.
In each field one group (A) regards the land as soil, and its function as
commodity-production; another group (B) regards the land as a biota,
and its function as something broader. How much broader is admittedly in a state of doubt and confusion.

In my own field, forestry, group A is quite content to grow trees like
cabbages, with cellulose as the basic forest commodity. It feels no inhibition against violence; its ideology is agronomic. Group B, on the
other hand, sees forestry as fundamentally different from agronomy
because it employs natural species, and manages a natural environment rather than creating an artificial one. Group B prefers natural
reproduction on principle. It worries on biotic as well as economic
grounds about the loss of species like chestnut, and the threatened loss
of the white pines. It worries about a whole series of secondary forest
functions: wildlife, recreation, watersheds, wilderness areas. To my
mind, Group B feels the stirrings of an ecological conscience.

In the wildlife field, a parallel cleavage exists. For Group A the basic
commodities are sport and meat; the yardsticks of production are
ciphers of take in pheasants and trout. Artificial propagation is acceptable as a permanent as well as a temporary recourse—if its unit costs
permit. Group B, on the other hand, worries about a whole series of
biotic side-issues. What is the cost in predators of producing a game
crop? Should we have further recourse to exotics? How can management restore the shrinking species, like prairie grouse, already hopeless
as shootable game? How can management restore the threatened rarities, like trumpeter swan and whooping crane? Can management prin-

ciples be extended to wildflowers? Here again it is clear to me that we have the same A-B cleavage as in forestry.

In the larger field of agriculture I am less competent to speak, but there seem to be somewhat parallel cleavages. Scientific agriculture was actively developing before ecology was born, hence a slower penetration of ecological concepts might be expected. Moreover the farmer, by the very nature of his techniques, must modify the biota more radically than the forester or the wildlife manager. Nevertheless, there are many discontents in agriculture which seem to add up to a new vision of "biotic farming."

Perhaps the most important of these is the new evidence that poundage or tonnage is no measure of the food-value of farm crops; the products of fertile soil may be qualitatively as well as quantitatively superior. We can bolster poundage from depleted soils by pouring on imported fertility, but we are not necessarily bolstering food-value. The possible ultimate ramifications of this idea are so immense that I must leave their exposition to abler pens.

The discontent that labels itself "organic farming," while bearing some of the earmarks of a cult, is nevertheless biotic in its direction, particularly in its insistence on the importance of soil flora and fauna.

The ecological fundamentals of agriculture are just as poorly known to the public as in other fields of land-use. For example, few educated people realize that the marvelous advances in technique made during recent decades are improvements in the pump, rather than the well. Acre for acre, they have barely sufficed to offset the sinking level of fertility.

In all of these cleavages, we see repeated the same basic paradoxes: man the conqueror *versus* man the biotic citizen; science the sharpener of his sword *versus* science the searchlight on his universe; land the slave and servant *versus* land the collective organism. Robinson's injunction to Tristram may well be applied, at this juncture, to *Homo sapiens* as a species in geological time:

> Whether you will or not
> You are a King, Tristram, for you are one
> Of the time-tested few that leave the world,
> When they are gone, not the same place it was.
> Mark what you leave.

THE OUTLOOK

It is inconceivable to me that an ethical relation to land can exist without love, respect, and admiration for land, and a high regard for its value. By value, I of course mean something far broader than mere economic value; I mean value in the philosophical sense.

Perhaps the most serious obstacle impeding the evolution of a land ethic is the fact that our educational and economic system is headed away from, rather than toward, an intense consciousness of land. Your true modern is separated from the land by many middlemen, and by innumerable physical gadgets. He has no vital relation to it; to him it is the space between cities on which crops grow. Turn him loose for a day on the land, and if the spot does not happen to be a golf links or a "scenic" area, he is bored stiff. If crops could be raised by hydroponics instead of farming, it would suit him very well. Synthetic substitutes for wood, leather, wool, and other natural land products suit him better than the originals. In short, land is something he has "outgrown."

Almost equally serious as an obstacle to a land ethic is the attitude of the farmer for whom the land is still an adversary, or a taskmaster that keeps him in slavery. Theoretically, the mechanization of farming ought to cut the farmer's chains, but whether it really does is debatable.

One of the requisites for an ecological comprehension of land is an understanding of ecology, and this is by no means co-extensive with "education"; in fact, much higher education seems deliberately to avoid ecological concepts. An understanding of ecology does not necessarily originate in courses bearing ecological labels; it is quite as likely to be labeled geography, botany, agronomy, history, or economics. This is as it should be, but whatever the label, ecological training is scarce.

The case for a land ethic would appear hopeless but for the minority which is in obvious revolt against these "modern" trends.

The "key-log" which must be moved to release the evolutionary process for an ethic is simply this: quit thinking about decent land-use as solely an economic problem. Examine each question in terms of what is ethically and esthetically right, as well as what is economically expedient. A thing is right when it tends to preserve the integrity, stability, and beauty of the biotic community. It is wrong when it tends otherwise.

It of course goes without saying that economic feasibility limits the tether of what can or cannot be done for land. It always has and it always will. The fallacy the economic determinists have tied around our collective neck, and which we now need to cast off, is the belief that economics determines *all* land-use. This is simply not true. An innumerable host of actions and attitudes, comprising perhaps the bulk of all land relations, is determined by the land-users' tastes and predilections, rather than by his purse. The bulk of all land relations hinges on investments of time, forethought, skill, and faith rather than on investments of cash. As a land-user thinketh, so is he.

I have purposely presented the land ethic as a product of social evolution because nothing so important as an ethic is ever "written." Only the most superficial student of history supposes that Moses "wrote" the Decalogue; it evolved in the minds of a thinking community, and Moses wrote a tentative summary of it for a "seminar." I say tentative because evolution never stops.

The evolution of a land ethic is an intellectual as well as emotional process. Conservation is paved with good intentions which prove to be futile, or even dangerous, because they are devoid of critical understanding either of the land, or of economic land-use. I think it is a truism that as the ethical frontier advances from the individual to the community, its intellectual content increases.

The mechanism of operation is the same for any ethic: social approbation for right actions: social disapproval for wrong actions.

By and large, our present problem is one of attitudes and implements. We are remodeling the Alhambra with a steam-shovel, and we are proud of our yardage. We shall hardly relinquish the shovel, which after all has many good points, but we are in need of gentler and more objective criteria for its successful use.

JOSEPH WOOD KRUTCH
1893–1970

This anthology contains several examples of biologists who, rather late in their career, became well known as writers of humanistic essays. A contrasting example is provided by the career of Joseph Wood Krutch. A New York drama critic and professor of literature in the 1920s and 1930s, Krutch's urbane and urban viewpoint of human nature was expressed in such books as The Modern Temper: A Study *and* A Confession *(1929)—a largely pessimistic view of modern civilization and its discontents. But in middle age he reread Thoreau and took his advice to heart, eventually moving to Arizona and concentrating his literary energies on the desert environment with which his name is now most strongly associated. His natural history essays, full of wit, wide-ranging allusions, and a compassion for all forms of life, made him one of the most popular and influential nature writers of his time. Yet he retained the intellectual's*

*restless curiosity about ultimate meanings, and his essays characteristi-
cally embrace an examination of such questions as the nature of life or
humankind's true place in the scheme of existence.*

LOVE IN THE DESERT

The ancients called love "the Mother of all things," but they didn't
know the half of it. They did not know, for instance, that plants as well
as animals have their love life and they supposed that even some of the
simpler animals were generated by sunlight on mud without the intervention of Venus.

Centuries later when Chaucer and the other medieval poets made
"the mystic rose" a euphemism for an anatomical structure not commonly mentioned in polite society, they too were choosing a figure of
speech more appropriate than they realized, because every flower really
is a group of sex organs which the plants have glorified while the animals—surprisingly enough, as many have observed—usually leave the
corresponding items of their own anatomy primitive, unadorned and
severely functional. The ape, whose behind blooms in purple and red,
represents the most any of the higher animals has achieved along this
line and even it is not, by human standards, any great aesthetic success.
At least no one would be likely to maintain that it rivals either the
poppy or the orchid.

In another respect also plants seem to have been more aesthetically
sensitive than animals. They have never tolerated that odd arrangement by which the same organs are used for reproduction and excre-
tion. Men, from St. Bernard to William Butler Yeats, have ridiculed or
scorned it and recoiled in distaste from the fact that, as Yeats put it,
"love has pitched his mansion in / The place of excrement." As a matter of fact, the reptiles are the only backboned animals who have a special organ used only in mating. Possibly—though improbably, I am
afraid—if this fact were better known it might be counted in favor of a
generally unpopular group.

All this we now know and, appropriately enough, much of it—especially concerning the sexuality of plants—was first discovered during
the eighteenth-century Age of Gallantry. No other age would have
been more disposed to hail the facts with delight and it was much
inclined to expound the new knowledge in extravagantly gallant terms.

The Voice of the Desert (New York: William Sloane, 1954).

One does not usually think of systematizers as given to rhapsody, but Linnaeus, who first popularized the fact that plants can make love, wrote rhapsodically of their nuptials:

> *The petals serve as bridal beds which the Great Creator has so gloriously arranged, adorned with such noble bed curtains and perfumed with so many sweet scents, that the bridegroom there may celebrate his nuptials with all the greater solemnity. When the bed is thus prepared, it is time for the bridegroom to embrace his beloved bride and surrender his gifts to her: I mean, one can see how* testiculi *open and emit* pulverem genitalem, *which falls upon* tubam *and fertilizes* ovarium.

In England, half a century later, Erasmus Darwin, distinguished grandfather of the great Charles, wrote even more exuberantly in his didactic poem, "The Loves of the Plants," where all sorts of gnomes, sylphs and other mythological creatures benevolently foster the vegetable *affaires de coeur*. It is said to have been one of the best-selling poems ever published, no doubt because it combined the newly fashionable interest in natural history with the long standing obsession with "the tender passion" as expressible in terms of cupids, darts, flames and all the other clichés which now survive only in St. Valentine's Day gifts.

Such romantic exuberance is not much favored today when the seamy side is likely to interest us more. We are less likely to abandon ourselves to a participation in the joys of spring than to be on our guard against "the pathetic fallacy" even though, as is usually the case, we don't know exactly what the phrase means or what is "pathetic" about the alleged fallacy. Nevertheless, those who consent, even for a moment, to glance at that agreeable surface of things with which the poets used to be chiefly concerned will find in the desert what they find in every other spring, and they may even be aware that the hare, which here also runs races with itself, is a good deal fleeter than any Wordsworth was privileged to observe in the Lake Country.

In this warm climate, moreover, love puts in his appearance even before "the young sonne hath in the Ram his halfe cours y-ronne" or, in scientific prose, ahead of the spring equinox. Many species of birds, which for months have done little more than chirp, begin to remember their songs. In the canyons where small pools are left from some winter rain, the subaqueous and most mysterious of all spring births begins and seems to recapitulate the first morning of creation. Though I have never noticed that either of the two kinds of doves which spend the whole year with us acquire that "livelier iris" which Tennyson celebrated, the lizard's belly turns turquoise blue, as though to remind his mate that even on their ancient level sex has its aesthetic as well as its biological aspect. Fierce sparrow hawks take to sitting side by side on telegraph wires, and the Arizona cardinal, who has remained all winter long more brilliantly red than his eastern cousin ever is, begins to think romantically of his neat but not gaudy wife. For months before, he had been behaving like an old married man who couldn't remember what he once saw in her. Though she had followed him about, he had sometimes driven her rudely away from the feeding station until he had had his fill. Now gallantry begins to revive and he may even graciously hand her a seed.

A little later the cactus wren and the curved-bill thrasher will build nests in the wicked heart of the cholla cactus and, blessed with some mysterious impunity, dive through its treacherous spines. Somewhere among the creosote bushes, by now yellow with blossoms, the jack rabbit—an unromantic looking creature if there ever was one—will be demonstrating that she is really a hare, not a rabbit at all, by giving birth to young furred babies almost ready to go it alone instead of being naked, helpless creatures like the infant cottontail. The latter will be born underground, in a cozily lined nest; the more rugged jack rabbit on the almost bare surface.

My special charge, the Sonoran spadefoot toad, will remain buried no one knows how many feet down for months still to come. He will not celebrate his spring until mid-July when a soaking rain penetrates deeply enough to assure him that on the surface a few puddles will form. Some of those puddles may just possibly last long enough to give his tadpoles the nine or ten days of submersion necessary, if they are to manage the metamorphosis which will change them into toadlets capable of repeating that conquest of the land which their ancestors accomplished so many millions of years ago. But while the buried spadefoots dally, the buried seeds dropped last year by the little six-week ephemerals of the desert will spring up and proceed with what looks like indecent haste to the business of reproduction, as though—as for them is almost the case—life were not long enough for anything except preparation for the next generation.

Human beings have been sometimes praised and sometimes scorned because they fall so readily into the habit of pinning upon their posterity all hope for a good life, of saying, "At least my children will have that better life which I somehow never managed to achieve." Even plants do that, as I know, because when I have raised some of the desert annuals under the unsuitable conditions of a winter living room, they have managed, stunted and sickly though they seemed, to seed. "At least," they seemed to say, "our species is assured another chance."

And if this tendency is already dominant in a morning glory, human beings will probably continue to accept it in themselves also, whether, by human standards, it is wise or not.

As I write this another spring has just come around. With a regularity in which there is something pleasantly comic, all the little romances, dramas and tragedies are acting themselves out once more, and I seize the opportunity to pry benevolently.

Yesterday I watched a pair of hooded orioles—he, brilliant in orange and black; she, modestly yellow green—busy about a newly constructed nest hanging from the swordlike leaves of a yucca, where one would have been less surprised to find the lemon yellow cousin of these birds which builds almost exclusively in the yucca. From this paradise I drove away the serpent—in this case a three-foot diamondback rattler who was getting uncomfortably close to the nesting site—and went on to flush out of the grass at least a dozen tiny Gambel's quail whose male parent, hovering close by, bobbed his head plume anxiously as he tried to rally them again. A quarter of a mile away a red and black Gila monster was sunning himself on the fallen trunk of another yucca, and, for all I know, he too may have been feeling some stirring of the spring, though I can hardly say that he showed it.

From birds as brightly colored as the orioles one expects only gay domesticity and lighthearted solicitude. For that reason I have been more interested to follow the home life of the road runner, that unbird-like bird whom we chose at the beginning as a desert dweller par excellence. One does not expect as much of him as one does of an oriole for two good reasons. In the first place, his normal manner is aggressive, ribald and devil-may-care. In the second place, he is a cuckoo, and the shirking of domestic responsibilities by some of the tribe has been notorious for so long that by some confused logic human husbands who are the victims of unfaithfulness not only wear the horns of the deer but are also said to be cuckolded. The fact remains, nevertheless, that though I have watched the developing domestic life of one road runner couple for weeks, I have observed nothing at which the most critical could cavil.

The nest—a rather coarse affair of largish sticks—was built in the crotch of a thorny cholla cactus some ten feet above the ground, which is rather higher than usual. When first found there were already in it two eggs, and both of the parent birds were already brooding them, turn and turn about. All this I had been led to expect because the road runner, unlike most birds, does not wait until all the eggs have been laid before beginning to incubate. Instead she normally lays them one by one a day or two apart and begins to set as soon as the first has arrived. In other words the wife follows the advice of the Planned Parenthood Association and "spaces" her babies—perhaps because lizards and snakes are harder to come by than insects, and it would be too much to try to feed a whole nest full of nearly grown infants at the same time. Moreover, in the case of my couple "self-restraint" or some other method of birth control had been rigorously practiced and two young ones were all there were.

Sixteen days after I first saw the eggs, both had hatched. Presently both parents were bringing in lizards according to a well-worked-out plan. While one sat on the nest to protect the young from the blazing sun, the other went hunting. When the latter returned with a catch, the brooding bird gave up its place, went foraging in its turn and presently came back to deliver a catch, after which it again took its place on the nest. One day, less than a month after the eggs were first discovered, one baby was standing on the edge of the nest itself, the other on a cactus stem a few feet away. By the next day both had disappeared.

Thus, despite the dubious reputation of the family to which he belongs, the road runner, like the other American cuckoos, seems to have conquered both the hereditary taint and whatever temptations his generally rascally disposition may have exposed him to. In this case at least, both husband and wife seemed quite beyond criticism, though they do say that other individuals sometimes reveal a not-too-serious sign of the hereditary weakness when a female will, on occasion, lay her eggs in the nest of another bird of her own species—which is certainly not so reprehensible as victimizing a totally different bird as the European cuckoo does.

Perhaps the superior moral atmosphere of America has reformed the cuckoo's habit and at least no American representative of the family regularly abandons its eggs to the care of a stranger. Nevertheless, those of us who are inclined to spiritual pride should remember that we do have a native immoralist, abundant in this same desert country and just as reprehensible as any to be found in decadent Europe—namely the cowbird, who is sexually promiscuous, never builds a home of his own and is inveterately given to depositing eggs in the nests of other birds. In his defense it is commonly alleged that his "antisocial conduct" should be excused for the same reason that such conduct is often excused in human beings—because, that is to say, it is actually the result not of original sin but of certain social determinants. It seems that long before he became a cowbird this fellow was a buffalo bird. And because he had to follow the wide ranging herds if he was to profit from the insects they started up from the grass, he could never settle down long enough to raise a family. Like Rousseau and like Walt Whitman, he had to leave his offspring (if any) behind.

However that may be, it still can hardly be denied that love in the desert has its still seamier side. Perhaps the moth, whom we have already seen playing pimp to a flower and profiting shamelessly from the affair, can also be excused on socio-economic grounds. But far more shocking things go on in dry climates as well as in wet, and to excuse them we shall have to dig deeper than the social system right down into the most ancient things-as-they-used-to-be. For an example which seems to come straight out of the most unpleasant fancies of the Marquis de Sade, we might contemplate the atrocious behavior of the so-called tarantula spider of the sandy wastes. Here, unfortunately, is a lover whom all the world will find it difficult to love.

This tarantula is a great hairy fellow much like the kind which sometimes comes north in a bunch of bananas and which most people have seen exhibited under a glass in some fruiterer's window. Most visitors to the desert hate and fear him at sight, even though he is disinclined to trouble human beings and is incapable even upon extreme provocation of giving more than a not-too-serious bite. Yet he does look more dangerous than the scorpion and he is, if possible, even less popular.

He has a leg spread of four or sometimes of as much as six inches, and it is said that he can leap for as much as two feet when pouncing upon his insect prey. Most of the time he spends in rather neat tunnels or burrows excavated in the sand, from which entomologists in search of specimens flush him out with water. And it is chiefly in the hottest months, especially after some rain, that one sees him prowling about, often crossing a road and sometimes waiting at a screen door to be let in. Except for man, his most serious enemy is the "tarantula hawk," a large black-bodied wasp with orange-red wings, who pounces upon his larger antagonist, paralyzes him with a sting and carries his now helpless body to feed the young wasps which will hatch in their own underground burrow.

Just to look at the tarantula's hairy legs and set of gleaming eyes is to suspect him of unconventionality or worse, and the suspicions are justified. He is one of those creatures in whom love seems to bring out the worst. Moreover, because at least one of the several species happens to have been the subject of careful study, the details are public. About the only thing he cannot be accused of is "infantile sexuality," and he can't be accused of that only because the male requires some eleven years to reach sexual maturity or even to develop the special organs necessary for his love making—if you can stretch this euphemistic term far enough to include his activities.

When at last he has come of age, he puts off the necessity of risking contact with a female as long as possible. First he spins a sort of web into which he deposits a few drops of sperm. Then he patiently taps the web for a period of about two hours in order to fill with the sperm the two special palps or mouth parts which he did not acquire until the molt which announced his maturity. Then, and only then, does he go off in search of his "mystic and somber Dolores" who will never exhibit toward him any tender emotions.

If, as is often the case, she shows at first no awareness of his presence, he will give her a few slaps until she rears angrily with her fangs spread for a kill. At this moment he then plays a trick which nature, knowing the disposition of his mate, has taught him and for which nature has provided a special apparatus. He slips two spurs conveniently placed on his forelegs over the fangs of the female, in such a way that the fangs are locked into immobility. Then he transfers the sperm which he has been carrying into an orifice in the female, unlocks her fangs and darts away. If he is successful in making his escape, he may repeat the process with as many as three other females. But by this time he is plainly senile and he slowly dies, presumably satisfied that his life work has been accomplished. Somewhat unfairly, the female may live for a dozen more years and use up several husbands. In general outline the procedure is the same for many spiders, but it seems worse in him, because he is big enough to be conspicuous.

It is said that when indiscreet birdbanders announced their discovery that demure little house wrens commonly swap mates between the first and second of their summer broods, these wrens lost favor with many old ladies who promptly took down their nesting boxes because they refused to countenance such loose behavior.

In the case of the tarantula we have been contemplating mores which are far worse. But there ought, it seems to me, to be some possible attitude less unreasonable than either that of the old ladies who draw away from nature when she seems not to come up to their very exacting standards of behavior, and the seemingly opposite attitude of inverted romantics who are prone either to find all beasts other than man completely beastly, or to argue that since man is biologically a beast, nothing should or can be expected of him that is not found in all his fellow creatures.

Such a more reasonable attitude will, it seems to me, have to be founded on the realization that sex has had a history almost as long as the history of life, that its manifestations are as multifarious as the forms assumed by living things and that their comeliness varies as much as do the organisms themselves. Man did not invent it and he was not the first to exploit either the techniques of love making or the emotional and aesthetic themes which have become associated with them. Everything either beautiful or ugly of which he has found himself capable is somewhere anticipated in the repertory of plant and animal behavior. In some creatures sex seems a bare and mechanical necessity; in others the opportunity for elaboration has been seized upon and developed in many different directions. Far below the human level, love can be a game on the one hand, or a self-destructive passion on the other. It can inspire tenderness or cruelty; it can achieve fulfillment through either violent domination or prolonged solicitation. One is almost tempted to say that to primitive creatures, as to man, it can be sacred or profane, love or lust.

The tarantula's copulation is always violent rape and usually ends in death for the aggressor. But over against that may be set not only the romance of many birds but also of other less engaging creatures in whom nevertheless a romantic courtship is succeeded by an epoch of domestic attachment and parental solicitude. There is no justification for assuming, as some romanticists do, that the one is actually more "natural" than the other. In one sense nature is neither for nor against what have come to be human ideals. She includes both what we call good and what we call evil. We are simply among her experiments, though we are, in some respects, the most successful.

Some desert creatures have come quite a long way from the tarantula—and in our direction, too. Even those who have come only a relatively short way are already no longer repulsive. Watching from a blind two parent deer guarding a fawn while he took the first drink at a water hole, it seemed that the deer at least had come a long way.

To be sure many animals are, if this is possible, more "sex obsessed" than we—intermittently at least. Mating is the supreme moment of their lives and for many, as for the male scorpion and the male tarantula, it is also the beginning of the end. Animals will take more trouble and run more risks than men usually will, and if the Strindbergs are right when they insist that the woman still wants to consume her mate, the biological origin of that grisly impulse is rooted in times which are probably more ancient than the conquest of dry land.

Our currently best-publicized student of human sexual conduct has argued that some of what are called "perversions" in the human being—homosexuality, for example—should be regarded as merely "normal variations" because something analogous is sometimes observed in the animal kingdom. But if that argument is valid then nothing in the textbooks of psychopathology is "abnormal." Once nature had established the fact of maleness and femaleness, she seems to have experimented with every possible variation on the theme. By comparison, Dr. Kinsey's most adventurous subjects were hopelessly handicapped by the anatomical and physiological limitations of the human being.

In the animal kingdom, monogamy, polygamy, polyandry and promiscuity are only trivial variations. Nature makes hermaphrodites, as well as Tiresiases who are alternately of one sex and then the other; also hordes of neuters among the bees and the ants. She causes some males to attach themselves permanently to their females and teaches others how to accomplish impregnation without ever touching them. Some embrace for hours; some, like Onan, scatter their seed. Many males in many different orders—like the seahorse and the ostrich, for example—brood the eggs, while others will eat them, if they get a chance, quite as blandly as many females will eat their mate, once his business is done. Various male spiders wave variously decorated legs before the eyes of a prospective spouse in the hope (often vain) that she will not mistake them for a meal just happening by. But husband-eating is no commoner than child-eating. Both should be classed as mere "normal variants" in human behavior if nothing except a parallel in the animal kingdom is necessary to establish that status. To her children nature seems to have said, "Copulate you must. But beyond that there is no rule. Do it in whatever way and with whatever emotional concomitants you choose. That you should do it somehow or other is all that I ask."

If one confines one's attention too closely to these seamy sides, one begins to understand why, according to Gibbon, some early Fathers of the Church held that sex was the curse pronounced upon Adam and that, had he not sinned, the human race would have been propagated "by some harmless process of vegetation." Or perhaps one begins to repeat with serious emphasis the famous question once asked by the Messrs. Thurber and White, "Is Sex Necessary?" And the answer is that, strictly speaking, it isn't. Presumably the very first organisms were sexless. They reproduced by a "process of vegetation" so harmless that not even vegetable sexuality was involved. What is even more impressive is the patent fact that it is not necessary today. Some of the most successful of all plants and animals—if by successful you mean abundantly surviving—have given it up either entirely or almost entirely. A virgin birth may require a miracle if the virgin is to belong to the human race, but there is nothing miraculous about it in the case of many of nature's successful children. Parthenogenesis, as the biologist calls it, is a perfectly normal event.

Ask the average man for a serious answer to the question what sex is "for" or why it is "necessary," and he will probably answer without thinking that it is "necessary for reproduction." But the biologist knows that it is not. Actually the function of sex is not to assure reproduction but to prevent it—if you take the word literally and hence to mean "exact duplication." Both animals and plants could "reproduce" or "duplicate" without sex. But without it there would be little or no variation, heredities could not be mixed, unexpected combinations could not arise, and evolution would either never have taken place at all or, at least, taken place so slowly that we might all still be arthropods or worse.

If in both the plant and animal kingdom many organisms are actually abandoning the whole of the sexual process, that is apparently because they have resigned their interest in change and its possibilities. Everyone knows how the ants and the bees have increased the single-minded efficiency of the worker majority by depriving them of a sexual function and then creating a special class of sexual individuals. But their solution is far less radical than that of many of the small creatures, including many insects, some of whom are making sexless rather than sexual reproduction the rule, and some of whom are apparently dispensing with the sexual entirely so that no male has ever been found.

In the plant world one of the most familiar and successful of all weeds produces its seeds without pollinization, despite the fact that it still retains the flower which was developed long ago as a mechanism for facilitating that very sexual process which it has now given up. That it is highly "successful" by purely biological standards no one who has ever tried to eliminate dandelions from a lawn is likely to doubt. As I have said before, they not only get along very well in the world, they have also been astonishing colonizers here, since the white man unintentionally brought them from Europe, probably in hay. Sexless though the dandelion is, it is inheriting the earth, and the only penalty which it has to pay for its sexlessness is the penalty of abandoning all hope of ever being anything except a dandelion, even of being a better dandelion than it is. It seems to have said at some point, "This is good enough for me. My tribe flourishes. We have found how to get along in the world. Why risk anything?"

But if, from the strict biological standpoint, sex is "nothing but" a mechanism for encouraging variation, that is a long way from saying that there are not other standpoints. It is perfectly legitimate to say that it is also "for" many other things. Few other mechanisms ever invented or stumbled upon opened so many possibilities, entailed so many unforeseen consequences. Even in the face of those who refuse to entertain the possibility that any kind of purpose or foreknowledge guided evolution, we can still find it permissible to maintain that every invention is "for" whatever uses or good results may come from it, that all things, far from being "nothing but" their origins, are whatever they have become. Grant that and one must grant also that the writing of sonnets is one of the things which sex is "for."

Certainly nature herself discovered a very long time ago that sex was—or at least could be used—"for" many things besides the production of offspring not too monotonously like the parent. Certainly also, these discoveries anticipated pretty nearly everything which man himself has ever found it possible to use sex "for." In fact it becomes somewhat humiliating to realize that we seem to have invented nothing absolutely new.

Marital attachment? Attachment to the home? Devotion to children? Long before us, members of the animal kingdom had associated them all with sex. Before us they also founded social groups on the family unit and in some few cases even established monogamy as the rule! Even more strikingly, perhaps, many of them abandoned *force majeure* as the decisive factor in the formation of a mating pair and substituted for it courtships, which became a game, a ritual and an aesthetic experience. Every device of courtship known to the human being was exploited by his predecessors: colorful costume display, song, dance and the wafted perfume. And like man himself, certain animals have come to find the preliminary ceremonies so engaging that they prolong them far beyond the point where they have any justification outside themselves. The grasshopper, for instance, continues to sing like a troubadour long after the lady is weary with waiting.

Even more humiliating, perhaps, than the fact that we have invented nothing is the further fact that the evolution has not been in a straight line from the lowest animal families up to us. The mammals, who are our immediate ancestors, lost as well as gained in the course of their development. No doubt because they lost the power to see colors (which was not recaptured until the primates emerged), the appeal of the eye plays little part in their courtship. In fact "love" in most of its manifestations tends to play a much lesser part in their lives than in that of many lower creatures—even in some who are distinctly less gifted than the outstandingly emotional and aesthetic birds. On the whole, mammalian sex tends to be direct, unadorned, often brutal, and not even the apes, despite their recovery from color blindness, seem to have got very far beyond the most uncomplicated erotic experiences and practices. Intellectually the mammals may be closer to us than any other order of animals, but emotionally and aesthetically they are more remote than some others—which perhaps explains the odd fact that most comparisons with any of them, and all comparisons with the primates, are derogatory. You may call a woman a "butterfly" or describe her as "birdlike." You may even call a man "leonine." But there is no likening with an ape which is not insulting.

How consciously, how poetically or how nobly each particular kind of creature may have learned to love, Venus only knows. But at this very moment of the desert spring many living creatures, plant as well as animal, are celebrating her rites in accordance with the tradition which happens to be theirs.

Fortunately, it is still too early for the tarantulas to have begun their amatory black mass, which, for all I know, may represent one of the oldest versions of the rituals still practiced in the worship of Mr. Swinburne's "mystic and somber Dolores." But this very evening as twilight falls, hundreds of moths will begin to stir themselves in the dusk and presently start their mysterious operations in the heart of those yucca blossoms which are just now beginning to open on the more precocious plants. Young jack rabbits not yet quite the size of an adult cottontail are proof that their parents went early about their business, and many of the brightly colored birds—orioles, cardinals and tanagers—are either constructing their nests or brooding their eggs. Some creatures seem to be worshiping only Venus Pandemos; some others have begun to have some inkling that the goddess manifests herself also as the atavist which the ancients called Venus Eurania. But it is patent to anyone who will take the trouble to look that they stand now upon different rungs of that Platonic ladder of love which man was certainly not the first to make some effort to climb.

Of this I am so sure that I feel it no betrayal of my humanity when I find myself entering with emotional sympathy into a spectacle which is more than a mere show, absorbing though it would be if it were no more than that. Modern knowledge gives me, I think, ample justification for the sense that I am not outside but a part of it, and if it did not give me that assurance, then I should probably agree that I would rather be "some pagan suckled in a creed outworn" than compelled to give it up.

Those very same biological sciences which have traced back to their lowly origins the emotional as well as the physiological characteristics of the sentient human being inevitably furnish grounds for the assumption that if we share much with the animals, they must at the same time share much with us. To maintain that all the conscious concomitants of our physical activities are without analogues in any creatures other than man is to fly in the face of the very evolutionary principles by which those "hardheaded" scientists set so much store. It is to assume that desire and joy have no origins in simpler forms of the same thing, that everything human has "evolved" except the consciousness which makes us aware of what we do. A Descartes, who held that man was an animal-machine differing from other animal-machines in that he alone possessed a gland into which God had inserted a soul, might consistently make between man and the other animals an absolute distinction. But the evolutionist is the last man who has a right to do anything of the sort.

He may, if he can consent to take the extreme position of the pure behaviorist, maintain that in man and the animals alike consciousness neither is nor can be anything but a phosphorescent illusion on the surface of physiological action and reaction, and without any substantial reality or any real significance whatsoever. But there is no choice between that extreme position and recognition of the fact that animals, even perhaps animals as far down in the scale as any still living or preserved in the ancient rocks, were capable of some awareness and of something which was, potentially at least, an emotion.

Either love as well as sex is something which we share with animals, or it is something which does not really exist in us. Either it is legitimate to feel some involvement in the universal Rites of Spring, or it is not legitimate to take our own emotions seriously. And even if the choice between the two possibilities were no more than an arbitrary one, I know which of the alternatives I should choose to believe in and to live by.

HENRY WILLIAMSON
1895–1977

Like Ernest Thompson Seton, Henry Williamson is considered one of the major developers of the modern "animal biography." Though his best-known books—Tarka the Otter (1927) and Salar the Salmon (1935)—are generally regarded as "children's classics" in this country, they were quickly praised in his native England by such literary luminaries as Thomas Hardy, Arnold Bennett, and Walter de la Mare.

After serving in France in World War I, Williamson moved into an ancient cottage in a rural village in North Devon and began to write prolifically about the surrounding countryside of moors, fens, rivers, and estu-

aries for which his imagination seems to have been made. Notable in these writings is the abundant use of indigenous regional terms to describe landscape and wildlife.

Though it was published as a novel, Williamson called Tarka the Otter *a work of "true imagination," in which he sought to be true to the facts of known and knowable animal behavior. Today the book is still regarded as a remarkably reliable chronicle of otter natural history. Williamson had the ability to create not only fully imagined, particular natural worlds but also, in the lives of his wild creatures, evocative physical counterparts to complex human emotions. In the following chapter, for instance, he describes Tarka's "slow slide" downstream after his defeat in love.*

From TARKA THE OTTER

He was alone, a young male of a ferocious and persecuted tribe whose only friends, except the Spirit that made it, were its enemies—the otter hunters. His cubhood was ended, and now indeed did his name fit his life, for he was a wanderer, and homeless, with nearly every man and dog against him.

Tarka fished the pools and guts of the Branton pill, eating what he caught among the feathery and aromatic leaves of the sea-wormwood plants which grew in the mudded cracks of the sloping stone wall with the sea-beet, the scentless sea-lavender, and the glasswort. One night a restlessness came over him, and he rode on the flood-tide to the head of the pill, which was not much wider than the gravel barges made fast to rusty anchors half-hidden in the grass, and to bollards of rotting wood. The only living thing that saw him arrive at the pill-head was a rat which was swarming down one of the mooring ropes, and when it smelled otter it let out a squeak and rapidly climbed over the sprig of furze tied to the rope to stop rats, and ran back into the ship. Tarka padded out of the mud, and along the footpath on the top of the sea-wall, often pausing with raised head and twitching nostrils, until he came to where the stream, passing through a culvert under the road, fell into a concrete basin and rushed thence down a stony slope into the pill. Entering the water above the fish-pass, he swam under the culvert, following the stream round bends and past a farmyard, through another culvert under a cart road, and on till he came to a stone bridge near a railway station. A horse and butt, or narrow farm cart, was cross-

Tarka the Otter: His Joyful Water-Life and Death in the Country of the Two Rivers (New York: Dutton, 1928).

ing the bridge, and he spread himself out beside a stone, so that three inches of water covered his head and back and rudder. When the butt had gone, he saw a hole, and crept up it. It was the mouth of an earthenware drain, broken at the joint. He found a dry place within. When it was quiet again, he went under the bridge and fished up the stream, returning at dawn to the drain.

He was awakened by the noise of pounding hooves; but the noises grew remote and he curled up again, using his thick rudder as a pillow on which to rest his throat. Throughout the day the noises of hooves recurred, for below the bridge was a ford where farm horses were taken to water. Twice he crept down the drain, but each time there was a bright light at the break in the pipe, and so he went back. At dusk he slipped out and went upstream again. Just above the bridge was a chestnut tree, and under it a shed, where ducks were softly quacking. He climbed on the bank, standing with his feet in sprays of ivy, his nose upheld, his head peering. The scents of the ducks were thick and luring as vivid colour is to a child. Juices flowed into his mouth, his heart beat fast. He moved forward, he thought of warm flesh, and his eyes glowed amber with the rays of a lamp in the farmhouse kitchen across the yard. The chestnut tree rustled its last few rusty leaves above him. Then across the vivid smear of duck scent strayed the taint of man; an ivy leaf trembled, a spider's web was broken, the river murmured, and the twin amber dots were gone.

Beside the stream was a public footpath and an illuminated building wherein wheels spun and polished connecting-rods moved to regular pulses which thudded in the air like the feet of men running on a bank. Tarka dived. He could not swim far, for by the electric power station the river slid over a fall. He swam to the right bank, but it was a steep wall of concrete. Again he dived, swimming upstream and crawling out on the bank. For many minutes he was afraid to cross the railway line, but at last he ran swiftly over the double track, and onwards until he reached the stream flowing deep under a foot-bridge.

He had been travelling for an hour, searching the uvvers of the banks for fish as he had learned in cubhood, when on a sandy scour he found the pleasing scent of otter. He whistled and hurried upstream, following the scent lying wherever the seals had been pressed. Soon he heard a whistle, and a feeling of joy warmed his being.

A small otter was waiting for him, sitting on a boulder, licking her coat with her tongue, the white tip of her rudder in the water. As Tarka approached, she looked at him, but she did not move from the boulder, nor did she cease to lick her neck when he placed his forepads on the stone and looked up into her face. He mewed to her and crawled out of the water to stand on hindlegs beside her and touch her nose. He licked her face, while his joy grew to a powerful feeling, so that when she continued to disregard him, he whimpered and struck her with one of his pads. White-tip yikkered and bit him in the neck. Then she slid into the water, and with a playful sweep of her rudder swam away from him.

He followed and caught her, and they rolled in play; and to Tarka returned a feeling he had not felt since the early days in the hollow tree, when he was hungry and cold and needing his mother. He mewed like a cub to White-tip, but she ran away. He followed her into a meadow. It was strange play, it was miserable play, it was not play at all, for Tarka was an animal dispirited. He pressed her, but she yikkered at him, and snapped at his neck whenever he tried to lick her face, until his mewing ceased altogether and he rolled her over, standing on her as though she were a salmon just lugged to land. With a yinny of anger she threw him off, and faced him with swishing rudder, tissing through her teeth.

Afterwards she ignored him, and returned to the river as though she were alone, to search under stones for mullyheads, or loach. He searched near her. He caught a black and yellow eel-like fish, whose round sucker-mouth was fastened to the side of a trout, but she would not take it. It was a lamprey. He dropped it before her again and again, pretending to have caught it anew each time. She swung away from his offering as though she had caught the lamprey and Tarka would seize it from her. The sickly trout, which had been dying for days with the lamprey fastened to it, floated down the stream; it had been a cannibal trout and had eaten more than fifty times its own weight of smaller trout. Tar from the road, after rain, had poisoned it. A rat ate the body the next day, and Old Nog speared and swallowed the rat three nights later. The rat had lived a jolly and murderous life, and died before it could fear.

The lamprey escaped alive, for Tarka dropped it and left White-tip in dejection. He had gone a few yards when he turned to see if she were following him. Her head was turned, she was watching. He was so thrilled that his whistle—a throat sound, like the curlew's—was low and flute-like. She answered. He was in love with White-tip, and as in all wild birds and animals, his emotions were as intense as they were quick. He felt neither hunger nor fatigue, and he would have fought for her until he was weak now that she had whistled to him. They galloped into the water-meadow, where in his growing desire he rushed at her, rolling her over and recoiling from her snapping of teeth. She sprang after him and they romped among the clumps of flowering rush, startling the rabbits at feed and sending up the woodcock which had just flown from the long low island seventeen miles off the estuary bar.

White-tip was younger than Tarka, and had been alone for three weeks before the old, grey-muzzled otter had met and taken care of her. Her mother had been killed by the otter-hounds, during the last meet of the otter-hunting season, at the end of September.

Tarka and White-tip returned to the stream, where among the dry stalks of angelica and hemlock they played hide-and-seek. But whenever his playfulness would change into a caress, she yinny-yikkered at him. She softened after a while, and allowed him to lick her head, once even licking his nose before running away. She was frightened of him, and yet was glad to be with him, for she had been lonely since she had lost Greymuzzle, when a marshman's dog had chased them out of a clump of rushes where they had been lying rough. Tarka caught her, and was prancing round her on a bank of gravel when down the stream came a dog-otter with three white ticks on his brow, a heavy, slow-moving, coarse-haired otter who had travelled down from the moor to find just such a mate as the one before him. Tarka cried *Ic-yang!* and ran at him, but the dog-otter, who weighed thirty pounds, bit him in the neck and shoulder. Tarka ran back, tissing, swinging and swaying his head before he ran forward and attacked. The older dog rolled him over, and bit him several times. Tarka was so mauled that he ran away. The dog followed him, but Tarka did not turn to fight. He was torn about the head and neck, and bitten thrice through the tongue and narrow lower jaw.

He stopped at the boulder where White-tip had been sitting when first she had seen him, and listened to the whistles of his enemy. The water sang its stone-song in the dark as it flowed its course to the sea. He waited, but White-tip never came, so he sank into the water and allowed himself to be carried down past bends and under stone arches of the little bridges which carried the lanes. He floated with hardly a paddle, listening to the song of the water and sometimes lapping to cool his tongue. The wheels and rods of the power station turned and gleamed behind glass windows like the wings of dragonflies; over the fall he slid, smooth as oil. Slowly and unseen he drifted, under the chestnut tree, under the bridge, past the quiet railway station, the orchards, the meadows, and so to the pill-head. The current dropped him into the basin of the fish-pass, and carried him down the slide to salt water. With the ebb he floated by ketches and gravel barges, while ring-plover and little stints running at the line of lapse cried their sweet cries of comradeship. The mooring kegs bobbed and turned in the ebb, the perches, tattered with sea-weed, leaned out of the trickling mud of the fairway, where curlew walked, sucking up worms in their long curved bills. Tarka rode on with the tide. It took him into the estuary, where the real sea was fretting the sandbanks. He heard a whistle, and answered it gladly. Greymuzzle was fishing in the estuary, and calling to White-tip.

The old otter, patient in life after many sorrows and fears, caressed his bitten face and neck and licked his hurts. They hunted together, and slept during the day in a drain in one of the dykes of the marsh, which was watered by a fresh stream from the hills lying northwards. Night after night they hunted in the sea, and often when the tide was low they played in the Pool opposite the fishing village that was built around the base of a hill. The north-east wind blew cold over the pans and sandy hillocks near the sea, but Greymuzzle knew a warm sleeping place in a clump of round-headed club-rush, near the day-hide of a bittern. She became dear to Tarka, and gave him fish as though he were her cub, and in the course of time she took him for her mate.

DONALD CULROSS PEATTIE
1898–1964

The moderns that Donald Culross Peattie wrote for in his An Almanac for Moderns *(1935) were a skeptical generation. They were the descendants of Darwin and Freud and the inheritors of World War I, who had seen "the trees blasted by the great guns and the birds feeding on men's eyes." A government botanist and freelance writer, Peattie went to live with his family in rural Illinois during the Depression. His deliberate choice of the archaic literary form of a daily almanac contrasted the stable natural order of the ancient philosophers and naturalists with the modern existential view of nature as soulless and purposeless. Its 365 short chapters not only pose many of the philosophical questions that have preoccupied contemporary nature writers but also contain an informal survey of natural science and evocative observations of seasonal life. In his combination of lyrical style and scientific observation, Peattie anticipates such writers as Lewis Thomas and Loren Eiseley.*

From AN ALMANAC
FOR MODERNS

MARCH TWENTY-FIRST

On this chill uncertain spring day, toward twilight, I have heard the first frog quaver from the marsh. That is a sound that Pharaoh listened to as it rose from the Nile, and it blended, I suppose, with his discontents and longings, as it does with ours. There is something lonely in that first shaken and uplifted trilling croak. And more than lonely, for I hear a warning in it, as Pharaoh heard the sound of plague. It speaks of the return of life, animal life, to the earth. It tells of all that is most unutterable in evolution—the terrible continuity and fluidity of protoplasm, the irrepressible forces of reproduction—not mystical human love, but the cold batrachian jelly by which we vertebrates are linked to the things that creep and writhe and are blind yet breed and have being. More than half it seems to threaten that when mankind has quite thoroughly shattered and eaten and debauched himself with his own follies, that voice may still be ringing out in the marshes of the Nile and the Thames and the Potomac, unconscious that Pharaoh wept for his son.

It always seems to me that no sooner do I hear the first frog trill than I find the first cloud of frog's eggs in a wayside pool, so swiftly does the emergent creature pour out the libation of its cool fertility. There is life where before there was none. It is as repulsive as it is beautiful, as silvery-black as it is slimy. Life, in short, raw and exciting, life almost in primordial form, irreducible element.

MARCH TWENTY-SECOND

For the ancients the world was a little place, bounded between Ind and Thule. The sky bent very low over Olympus, and astronomers had not yet taken the friendliness out of the stars. The shepherd kings of the desert called them by the names Job knew, Al-Debaran, Fomalhaut, Mizar, Al-Goth, Al-Tair, Deneb and Achernar. For the Greeks the glittering constellations made pictures of their heroes and heroines, and of beasts and birds. The heavenly truth of their Arcadian mythology blazed nightly in the skies for the simplest clod to read.

Through all this celestial splendor the sun plowed yearly in a broad

An Almanac for Moderns (New York: Putnam, 1935).

track that they called the zodiac. As it entered each constellation a new month with fresh significances and consequences was marked down by a symbol. Lo, in the months when the rains descended, when the Nile and the Tigris and Yangtse rose, the sun entered the constellations that were like Fishes, and like Water Carriers! In the hot dry months it was in the constellation that is unmistakably a scorpion, bane of the desert. Who could say that the stars in their orderly procession did not sway a man's destiny?

Best of all, the year began with spring, with the vernal equinox. It was a natural, a pastoral, a homely sort of year, which a man could take to his heart and remember; he could tell the date by the feeling in his bones. It is the year which green things, and the beasts and birds in their migrations, all obey, a year like man's life, from his birth cry to the snows upon the philosopher's head.

MARCH TWENTY-THIRD

The old almanacs have told off their years, and are dead with them. The weather-wisdom and the simple faith that cropped up through them as naturally as grass in an orchard, are withered now, and their flowers of homely philosophy and seasonable prediction and reflection are dry, and only faintly, quaintly fragrant. The significance of the Bull and the Crab and the Lion are not more dead, for the modern mind, than the Nature philosophy of a generation ago. This age has seen the trees blasted to skeletons by the great guns, and the birds feeding on men's eyes. Pippa has passed.

It is not that man alone is vile. Man is a part of Nature. So is the atomic disassociation called high explosive. So are violent death, rape, agony, and rotting. They were all here, and quite natural, before our day, in the sweet sky and the blowing fields.

There is no philosophy with a shadow of realism about it, save a philosophy based upon Nature. It turns a smiling face, a surface easily conquered by the gun, the bridge, the dynamite stick. Yet there is no obedience but to its laws. Hammurabi spoke and Rameses commanded, and the rat gnawed and the sun shone and the hive followed its multiplex and golden order. Flowers pushed up their child faces in the spring, and the bacteria slowly took apart the stuff of life. Today the Kremlin commands, the Vatican speaks. And tomorrow the rat will still be fattening, the sun be a little older, and the bacteria remain lords of creation, whose subtraction would topple the rest of life.

Now how can a man base his way of thought on Nature and wear so happy a face? How can he take comfort from withering grass where he lays his head, from a dying sun to which he turns his face, from a mortal woman's head pressed on his shoulder? To say how that might be, well might he talk the year around.

MARCH TWENTY-FOURTH

Perhaps in Tempe the wild lawns are thick with crocuses, and narcissus blows around Paestum, but here on this eastern shore of a western world, spring is a season of what the embittered call realism, by which they mean the spoiling of joy. Joy will come, as the joy of a child's birth comes—after the pains. So dry cold winds still walk abroad, under gray skies.

It is not that nothing blooms or flies; the honey bees were out for an hour, the one hour of sunlight, and above the pools where the salamander's eggs drift in inky swirls, the early midges danced. Down the runs and rills I can hear the calling of a red-bird entreating me to come and find him, come and find him! There is a black storm of grackles in the tree limbs where the naked maple flowers are bursting out in scarlet tips from their bud scales, and a song-sparrow sits on an alder that dangles out its little gold tails.

We are so used to flowers wrapped up in the pretty envelopes of their corollas and calyces, so softened in our taste for the lovely in Nature, that we scarcely rate an alder catkin as a flower at all. Yet it is nothing else—nothing but the male anthers, sowing the wind with their freight of fertilizing pollen. The small, compact female flowers, like tiny cones, wait in the chill wild air for the golden cargo.

So does our spring begin, in a slow flowering on the leafless wood of the bough of hazel and alder and poplar and willow, a hardy business, a spawning upon the air, like the spawning in the ponds, a flowering so primitive that it carries us back to ancient geologic times, when trees that are now fossils sowed the wind like these, their descendants—an epoch when the world, too, was in its naked springtime.

MARCH TWENTY-FIFTH

The beginnings of spring, the true beginnings, are quite unlike the springtides of which poets and musicians sing. The artists become conscious of spring in late April, or May, when it is not too much to say that the village idiot would observe that birds are singing and nesting, that fields bear up their freight of flowering and ants return to their proverbial industry.

But the first vernal days are younger. Spring steals in shyly, a tall, naked child in her pale gold hair, amidst us the un-innocent, skeptics in wool mufflers, prudes in gumshoes and Grundies with head-colds. Very secretly the old field cedars sow the wind with the freight of their ancient pollen. A grackle in the willow croaks and sings in the uncertain, ragged voice of a boy. The marshes brim, and walking is a muddy business. Oaks still are barren and secretive. On the lilac tree only the twin buds suggest her coming maturity and flowering. But there in the pond float the inky masses of those frog's eggs, visibly life in all its rawness, its elemental shape and purpose. Now is the moment when the secret of life could be discovered, yet no one finds it.

MARCH TWENTY-SIXTH

Out of the stoa, two thousand years ago, strode a giant to lay hold on life and explain it. He went down to the "primordial slime" of the seashore to look for its origin. There if anywhere he would find it, he thought, where the salt water and the earth were met, and the mud quivered like a living thing, and from it emerged strange shapeless primitive beings, themselves scarce more than animate bits of ooze. To Aristotle it seemed plain enough that out of the dead and the inanimate is made the living, and back to death are turned the bodies of all things that have lived, to be used over again. So nothing was wasted; all moved in a perpetual cycle. Out of vinegar, he felt certain, came vinegar eels, out of dung came blow-flies, out of decaying fruit bees were born, and out of the rain pool frogs spawned.

But the eye of even Aristotle was purblind in its nakedness. Of the spore and the sperm he never dreamed; he guessed nothing of bacteria. Now man can peer down through the microscope, up at the revealed stars. And behold, the lens has only multiplied the facts and deepened the mystery.

For now we know that spontaneous generation never takes place. Life comes only from life. Was not the ancient Hindu symbol for it a serpent with a tail in its mouth? Intuitive old fellows, those Aryan brothers of ours, wise in their superstitions, like old women. Life, we discover, is a closed, nay, a charmed circle. Wherever you pick it up, it has already begun; yet as soon as you try to follow it, it is already dying.

MARCH TWENTY-SEVENTH

First to grasp biology as a science, Aristotle thought that he had also captured the secret of life itself. From the vast and original body of his observation, he deduced a cosmology like a pure Greek temple, symmetrical and satisfying. For two millenniums it housed the serene intelligence of the race.

Here was an absolute philosophy; nothing need be added to it; detraction was heretical. It traced the ascent of life from the tidal ooze up to man, the plants placed below the animals, the animals ranged in order of increasing intelligence. Beyond man nothing could be imagined but God, the supreme intelligence. God was all spirit; the lifeless rock was all matter. Living beings on this earth were spirit infusing matter.

Still this conception provides the favorite text of poet or pastor, praising the earth and the fullness thereof. It fits so well with the grandeur of the heavens, the beauty of the flowers at our feet, the rapture of the birds! The Nature lover of today would ask nothing better than that it should be true.

Aristotle was sure of it. He points to marble in a quarry. It is only matter; then the sculptor attacks it with his chisel, with a shape in his mind. With form, soul enters into the marble. So all living things are filled with soul, some with more, some with less. But even a jellyfish is infused with that which the rock possesses not. Thus existence has its origin in supreme intelligence, and everything has an intelligent cause and serves its useful purpose. That purpose is the development of higher planes of existence. Science, thought its Adam, had but to put the pieces of the puzzle together, to expose for praise the cosmic design, all beautiful.

MARCH TWENTY-EIGHTH

The hook-nosed Averroës, the Spanish Arab born in Cordova in 1126, and one time cadi of Seville, shook a slow dissenting head. He did not like this simile of Aristotle's, of the marble brought to life and form by the sculptor. The simile, he keenly perceived, would be applicable at best if the outlines of the statue were already performed in the marble as it lay in the quarry. For that is precisely how we find life. The tree is preformed in the seed; the future animal already exists in the embryo. Wherever we look we find form, structure, adaptation, already present. Never has it been vouchsafed to us to see pure creation out of the lifeless.

And Galileo, also, ventured to shake the pillars of the Schoolmen's Aristotelian temple. Such a confirmed old scrutinizer was not to be drawn toward inscrutable will. The stars, nearest of all to Aristotle's God, should have moved with godlike precision, and Galileo, peering, found them erring strangely all across heaven. He shrugged, but was content. Nature itself was the miracle, Nature with all its imperfections. Futile for science to try to discover what the forces of Nature are; it can only discover how they operate.

MARCH TWENTY-NINTH

Comforting, sustaining, like the teat to the nursling, is Aristotle's beautiful idea that everything serves a useful purpose and is part of the great design. Ask, for instance, of what use is grass. Grass, the pietist assures us, was made in order to nourish cows. Cows are here on earth to nourish men. So all flesh is grass, and grass was put here for man.

But of what use, pray, is man? Would anybody, besides his dog, miss him if he were gone? Would the sun cease to shed its light because there were no human beings here to sing the praises of sunlight? Would there be sorrow among the little hiding creatures of the underwood, or loneliness in the hearts of the proud and noble beasts? Would the other simians feel that their king was gone? Would God, Jehovah, Zeus, Allah, miss the sound of hymns and psalms, the odor of frankincense and flattery?

There is no certainty vouchsafed us in the vast testimony of Nature that the universe was designed for man, nor yet for any purpose, even the bleak purpose of symmetry. The courageous thinker must look the inimical aspects of his environment in the face, and accept the stern fact that the universe is hostile and deathy to him save for a very narrow zone where it permits him, for a few eons, to exist.

MARCH THIRTIETH

Archaic and obsolete sounds the wisdom of the great old Greek. Life his pronouncement ran, is soul pervading matter. What, soul in a jellyfish, an oyster, a burdock? Then by soul he could not have meant that moral quality which Paul of Tarsus or Augustine of Hippo were to call soul. Aristotle is talking rather about that undefined but essential and precious something that just divides the lowliest microörganism from the dust; that makes the ugly thousand-legged creature flee from death; that makes the bird pour out its heart in morning rapture; that makes the love of man for woman a holy thing sacred to the carrying on of the race.

But what is this but life itself? In every instance Aristotle but affirms that living beings are matter pervaded by a noble, a palpitant and thrilling thing called life. This is the mystery, and his neat cosmology solves nothing of it. But it is not Aristotle's fault that he did not give us the true picture of things. It is Nature herself, as we grow in comprehension of her, who weans us from our early faith.

MARCH THIRTY-FIRST

Aristotle's rooms in the little temple of the Lyceum were the first laboratory, where dissection laid bare the sinews and bones of life. The Lyceum was a world closer to the marine biological station at Wood's Hole, Massachusetts, than it was to its neighbor the Parthenon. Its master did for marine biology what Euclid did for geometry; his work on the embryology of the chick still stands as a nearly perfect monograph of biological investigation. The originality, the scope of his works, the magnificence of his dream for biology as an independent science, have probably never been surpassed by any one who has lived since.

Unlike many of the more timid or less gifted investigators of today, Aristotle could not help coming to conclusions about it all. For his cosmology it should be said that it was the best, perhaps the only possible, philosophy of the origin and nature of life which the times permitted. We can all feel in our bones how agreeable it were to accept the notion of design, symmetry, purpose, an evolution toward a spiritual godhead such as Aristotle assures us exists.

But as it was Aristotle himself who taught us to observe, investigate, deduce what facts compel us to deduce, so we must concede that it is Nature herself, century after century—day after day, indeed, in the whirlwind progress of science—that propels us farther and farther away from Aristotelian beliefs. At every point she fails to confirm the grand old man's cherished picture of things. There are persons so endowed by temperament that they will assert that if Nature has no "soul," purpose, nor symmetry, we needs must put them in the picture, lest the resulting composition be scandalous, intolerable, and maddening. To such the scientist can only say, "Believe as you please."

APRIL FIRST

I say that it touches a man that his blood is sea water and his tears are salt, that the seed of his loins is scarcely different from the same cells in a seaweed, and that of stuff like his bones are coral made. I say that physical and biologic law lies down with him, and wakes when a child stirs in the womb, and that the sap in a tree, uprushing in the spring, and the smell of the loam, where the bacteria bestir themselves in darkness, and the path of the sun in the heaven, these are facts of first importance to his mental conclusions, and that a man who goes in no consciousness of them is a drifter and a dreamer, without a home or any contact with reality.

VLADIMIR NABOKOV
1899–1977

Nabokov's reputation as a subtle, cosmopolitan, and ironic writer rests upon his many novels, including Lolita (1955) and his autobiographical work, Speak Memory (revised 1966). But from his boyhood in an aristocratic Russian family through his long residence in the United States to his death in Switzerland, he was also passionately devoted to hunting, studying, and writing about butterflies. He produced dozens of technical papers in the field of lepidoptery, of which he was an acknowledged authority. In addition, butterflies figured in both his fiction and his memoirs as emblems of beauty and transience and of memory's delicate persistence.

BUTTERFLIES

1

On a summer morning, in the legendary Russia of my boyhood, my first glance upon awakening was for the chink between the white inner shutters. If it disclosed a watery pallor, one had better not open them at all, and so be spared the sight of a sullen day sitting for its picture in a puddle. How resentfully one would deduce, from a line of dull light, the leaden sky, the sodden sand, the gruel-like mess of broken brown blossoms under the lilacs—and that flat, fallow leaf (the first casualty of the season) pasted upon a wet garden bench!

But if the chink was a long glint of dewy brilliancy, then I made haste to have the window yield its treasure. With one blow, the room would be cleft into light and shade. The foliage of birches moving in the sun had the translucent green tone of grapes, and in contrast to this there was the dark velvet of fir trees against a blue of extraordinary

Speak, Memory: An Autobiography Revisited (New York: Putnam, 1966).

intensity, the like of which I rediscovered only many years later, in the montane zone of Colorado.

From the age of seven, everything I felt in connection with a rectangle of framed sunlight was dominated by a single passion. If my first glance of the morning was for the sun, my first thought was for the butterflies it would engender. The original event had been banal enough. On the honeysuckle, overhanging the carved back of a bench just opposite the main entrance, my guiding angel (whose wings, except for the absence of a Florentine limbus, resemble those of Fra Angelico's Gabriel) pointed out to me a rare visitor, a splendid, pale-yellow creature with black blotches, blue crenels, and a cinnabar eyespot above each chrome-rimmed black tail. As it probed the inclined flower from which it hung, its powdery body slightly bent, it kept restlessly jerking its great wings, and my desire for it was one of the most intense I have ever experienced. Agile Ustin, our town-house janitor, who for a comic reason (explained elsewhere) happened to be that summer in the country with us, somehow managed to catch it in my cap, after which it was transferred, cap and all, to a wardrobe, where domestic naphthalene was fondly expected by Mademoiselle to kill it overnight. On the following morning, however, when she unlocked the wardrobe to take something out, my Swallowtail, with a mighty rustle, flew into her face, then made for the open window, and presently was but a golden fleck dipping and dodging and soaring eastward, over timber and tundra, to Vologda, Viatka and Perm, and beyond the gaunt Ural range to Yakutsk and Verkhne Kolymsk, and from Verkhne Kolymsk, where it lost a tail, to the fair Island of St. Lawrence, and across Alaska to Dawson, and southward along the Rocky Mountains—to be finally overtaken and captured, after a forty-year race, on an immigrant dandelion under an endemic aspen near Boulder. In a letter from Mr. Brune to Mr. Rawlins, June 14, 1735, in the Bodleian collection, he states that one Mr. Vernon followed a butterfly nine miles before he could catch him (The Recreative Review or Eccentricities of Literature and Life. Vol. 1, p. 144, London, 1821).

Soon after the wardrobe affair I found a spectacular moth, marooned in a corner of a vestibule window, and my mother dispatched it with ether. In later years, I used many killing agents, but the least contact with the initial stuff would always cause the porch of the past to light up and attract that blundering beauty. Once, as a grown man, I was under ether during appendectomy, and with the vividness of a decalcomania picture I saw my own self in a sailor suit mounting a freshly emerged Emperor moth under the guidance of a Chinese lady who I knew was my mother. It was all there, brilliantly reproduced in my dream, while my own vitals were being exposed: the soaking, ice-cold absorbent cotton pressed to the insect's lemurian head; the subsiding spasms of its body; the satisfying crackle produced by the pin penetrating the hard crust of its thorax; the careful insertion of the point of the pin in the cork-bottomed groove of the spreading board; the symmetrical adjustment of the thick, strong-veined wings under neatly affixed strips of semitransparent paper.

2

* * * The mysteries of mimicry had a special attraction for me. Its phenomena showed an artistic perfection usually associated with man-wrought things. Consider the imitation of oozing poison by bubblelike macules on a wing (complete with pseudo-refraction) or by glossy yellow knobs on a chrysalis ("Don't eat me—I have already been squashed, sampled and rejected"). Consider the tricks of an acrobatic caterpillar (of the Lobster Moth) which in infancy looks like bird's dung, but after molting develops scrabbly hymenopteroid appendages and baroque characteristics, allowing the extraordinary fellow to play two parts at once (like the actor in Oriental shows who *becomes* a pair of intertwisted wrestlers): that of a writhing larva and that of a big ant seemingly harrowing it. When a certain moth resembles a certain wasp in shape and color, it also walks and moves its antennae in a waspish, unmothlike manner. When a butterfly has to look like a leaf, not only are all the details of a leaf beautifully rendered but markings mimicking grub-bored holes are generously thrown in. "Natural selection," in the Darwinian sense, could not explain the miraculous coincidence of imitative aspect and imitative behavior, nor could one appeal to the theory of "the struggle for life" when a protective device was carried to a point of mimetic subtlety, exuberance, and luxury far in excess of a predator's power of appreciation. I discovered in nature the nonutilitarian delights that I sought in art. Both were a form of magic, both were a game of intricate enchantment and deception.

3

I have hunted butterflies in various climes and disguises: as a pretty boy in knickerbockers and sailor cap; as a lanky cosmopolitan expatriate in flannel bags and beret; as a fat hatless old man in shorts. Most of my cabinets have shared the fate of our Vyra house. Those in our town house and the small addendum I left in the Yalta Museum have been destroyed, no doubt, by carpet beetles and other pests. A collection of South European stuff that I started in exile vanished in Paris during World War Two. All my American captures from 1940 to 1960 (several thousands of specimens including great rarities and types) are in the

Mus. of Comp. Zoology, the Am. Nat. Hist. Mus., and the Cornell Univ. Mus. of Entomology, where they are safer than they would be in Tomsk or Atomsk. Incredibly happy memories, quite comparable, in fact, to those of my Russian boyhood, are associated with my research work at the MCZ, Cambridge, Mass. (1941–1948). No less happy have been the many collecting trips taken almost every summer, during twenty years, through most of the states of my adopted country.

In Jackson Hole and in the Grand Canyon, on the mountain slopes above Telluride, Colo., and on a celebrated pine barren near Albany, N.Y., dwell, and will dwell, in generations more numerous than editions, the butterflies I have described as new. Several of my finds have been dealt with by other workers; some have been named after me. One of these, Nabokov's Pug (*Eupithecia nabokovi* McDunnough), which I boxed one night in 1943 on a picture window of James Laughlin's Alta Lodge in Utah, fits most philosophically into the thematic spiral that began in a wood on the Oredezh around 1910—or perhaps even earlier, on that Nova Zemblan river a century and a half ago.

Few things indeed have I known in the way of emotion or appetite, ambition or achievement, that could surpass in richness and strength the excitement of entomological exploration. From the very first it had a great many intertwinkling facets. One of them was the acute desire to be alone, since any companion, no matter how quiet, interfered with the concentrated enjoyment of my mania. Its gratification admitted of no compromise or exception. Already when I was ten, tutors and governesses knew that the morning was mine and cautiously kept away.

In this connection, I remember the visit of a schoolmate, a boy of whom I was very fond and with whom I had excellent fun. He arrived one summer night—in 1913, I think—from a town some twenty-five miles away. His father had recently perished in an accident, the family was ruined and the stouthearted lad, not being able to afford the price of a railway ticket, had bicycled all those miles to spend a few days with me.

On the morning following his arrival, I did everything I could to get out of the house for my morning hike without his knowing where I had gone. Breakfastless, with hysterical haste, I gathered my net, pill boxes, killing jar, and escaped through the window. Once in the forest, I was safe; but still I walked on, my calves quaking, my eyes full of scalding tears, the whole of me twitching with shame and self-disgust, as I visualized my poor friend, with his long pale face and black tie, moping in the hot garden—patting the panting dogs for want of something better to do, and trying hard to justify my absence to himself.

Let me look at my demon objectively. With the exception of my parents, no one really understood my obsession, and it was many years before I met a fellow sufferer. One of the first things I learned was not to depend on others for the growth of my collection. One summer afternoon, in 1911, Mademoiselle came into my room, book in hand, started to say she wanted to show me how wittily Rousseau denounced zoology (in favor of botany), and by then was too far gone in the gravitational process of lowering her bulk into an armchair to be stopped by my howl of anguish: on that seat I had happened to leave a glass-lidded cabinet tray with long, lovely series of the Large White. Her first reaction was one of stung vanity: her weight, surely, could not be accused of damaging what in fact it had demolished; her second was to console me: *Allons donc, ce ne sont que des papillons de potager!*—which only made matters worse. A Sicilian pair recently purchased from Staudinger had been crushed and bruised. A huge Biarritz example was utterly mangled. Smashed, too, were some of my choicest local captures. Of these, an aberration resembling the Canarian race of the species might have been mended with a few drops of glue; but a precious gynandromorph, left side male, right side female, whose abdomen could not be traced and whose wings had come off, was lost forever: one might re-attach the wings but one could not prove that all four belonged to that headless thorax on its bent pin. Next morning, with an air of great mystery, poor Mademoiselle set off for St. Petersburg and came back in the evening bringing me ("something better than your cabbage butterflies") a banal Urania moth mounted on plaster. "How you hugged me, how you danced with joy!" she exclaimed ten years later in the course of inventing a brand-new past.

Our country doctor, with whom I had left the pupae of a rare moth when I went on a journey abroad, wrote me that everything had hatched finely; but in reality a mouse had got at the precious pupae, and upon my return the deceitful old man produced some common Tortoiseshell butterflies, which, I presume, he had hurriedly caught in his garden and popped into the breeding cage as plausible substitutes (so *he* thought). Better than he, was an enthusiastic kitchen boy who would sometimes borrow my equipment and come back two hours later in triumph with a bagful of seething invertebrate life and several additional items. Loosening the mouth of the net which he had tied up with a string, he would pour out his cornucopian spoil—a mass of grasshoppers, some sand, the two parts of a mushroom he had thriftily plucked on the way home, more grasshoppers, more sand, and one battered Small White. * * *

4

When, having shaken off all pursuers, I took the rough, red road that ran from our Vyra house toward field and forest, the animation and luster of the day seemed like a tremor of sympathy around me.

Very fresh, very dark Arran Browns, which emerged only every second year (conveniently, retrospection has fallen here into line), flitted among the firs or revealed their red markings and checkered fringes as they sunned themselves on the roadside bracken. Hopping above the grass, a diminutive Ringlet called Hero dodged my net. Several moths, too, were flying—gaudy sun lovers that sail from flower to flower like painted flies, or male insomniacs in search of hidden females, such as that rust-colored Oak Eggar hurtling across the shrubbery. I noticed (one of the major mysteries of my childhood) a soft pale green wing caught in a spider's web (by then I knew what it was: part of a Large Emerald). The tremendous larva of the Goat Moth, ostentatiously segmented, flat-headed, flesh-colored and glossily flushed, a strange creature "as naked as a worm" to use a French comparison, crossed my path in frantic search for a place to pupate (the awful pressure of metamorphosis, the aura of a disgraceful fit in a public place). On the bark of that birch tree, the stout one near the park wicket, I had found last spring a dark aberration of Sievers' Carmelite (just another gray moth to the reader). In the ditch, under the bridgelet, a bright-yellow Silvius Skipper hobnobbed with a dragonfly (just a blue libellula to me). From a flower head two male Coppers rose to a tremendous height, fighting all the way up—and then, after a while, came the downward flash of one of them returning to his thistle. These were familiar insects, but at any moment something better might cause me to stop with a quick intake of breath. I remember one day when I warily brought my net closer and closer to an uncommon Hairstreak that had daintily settled on a sprig. I could clearly see the white W on its chocolate-brown underside. Its wings were closed and the inferior ones were rubbing against each other in a curious circular motion—possibly producing some small, blithe crepitation pitched too high for a human ear to catch. I had long wanted that particular species, and, when near enough, I struck. You have heard champion tennis players moan after muffing an easy shot. You may have seen the face of the world-famous grandmaster Wilhelm Edmundson when, during a simultaneous display in a Minsk café, he lost his rook, by an absurd oversight, to the local amateur and pediatrician, Dr. Schach, who eventually won. But that day nobody (except my older self) could see me shake out a piece of twig from an otherwise empty net and stare at a hole in the tarlatan.

5

Near the intersection of two carriage roads (one, well-kept, running north-south in between our "old" and "new" parks, and the other, muddy and rutty, leading, if you turned west, to Batovo) at a spot where aspens crowded on both sides of a dip, I would be sure to find in the third week of June great blue-black nymphalids striped with pure white, gliding and wheeling low above the rich clay which matched the tint of their undersides when they settled and closed their wings. Those were the dung-loving males of what the old Aurelians used to call the Poplar Admirable, or, more exactly, they belonged to its Bucovinan subspecies. As a boy of nine, not knowing that race, I noticed how much our North Russian specimens differed from the Central European form figured in Hofmann, and rashly wrote to Kuznetsov, one of the greatest Russian, or indeed world, lepidopterists of all time, naming my new subspecies "*Limenitis populi rossica.*" A long month later he returned my description and aquarelle of "*rossica* Nabokov" with only two words scribbled on the back of my letter: "*bucovinensis* Hormuzaki." How I hated Hormuzaki! And how hurt I was when in one of Kuznetsov's later papers I found a gruff reference to "schoolboys who keep naming minute varieties of the Poplar Nymph!" Undaunted, however, by the *populi* flop, I "discovered" the following year a "new" moth. That summer I had been collecting assiduously on moonless nights, in a glade of the park, by spreading a bedsheet over the grass and its annoyed glow-worms, and casting upon it the light of an acytelene lamp (which, six years later, was to shine on Tamara). Into that arena of radiance, moths would come drifting out of the solid blackness around me, and it was in that manner, upon that magic sheet, that I took a beautiful *Plusia* (now *Phytometra*) which, as I saw at once, differed from its closest ally by its mauve-and-maroon (instead of golden-brown) forewings, and narrower bractea mark and was not recognizably figured in any of my books. I sent its description and picture to Richard South, for publication in *The Entomologist*. He did not know it either, but with the utmost kindness checked it in the British Museum collection—and found it had been described long ago as *Plusia excelsa* by Kretschmar. I received the sad news, which was most sympathetically worded (". . . should be congratulated for obtaining . . . very rare Volgan thing . . . admirable figure . . .") with the utmost stoicism; but many years later, by a pretty fluke (I know I should not point out these plums to people), I got even with the first discoverer of *my* moth by giving his own name to a blind man in a novel. ✳ ✳ ✳

6

The "English" park that separated our house from the hayfields was an extensive and elaborate affair with labyrinthine paths, Turgenevian benches, and imported oaks among the endemic firs and birches. The struggle that had gone on since my grandfather's time to keep the park from reverting to the wild state always fell short of complete success. No gardener could cope with the hillocks of frizzly black earth that the pink hands of moles kept heaping on the tidy sand of the main walk. Weeds and fungi, and ridgelike tree roots crossed and recrossed the sun-flecked trails. Bears had been eliminated in the eighties, but an occasional moose still visited the grounds. On a picturesque boulder, a little mountain ash and a still smaller aspen had climbed, holding hands, like two clumsy, shy children. Other, more elusive trespassers—lost pic-nickers or merry villagers—would drive our hoary gamekeeper Ivan crazy by scrawling ribald words on the benches and gates. The disinte-grating process continues still, in a different sense, for when, nowadays, I attempt to follow in memory the winding paths from one given point to another, I notice with alarm that there are many gaps, due to obliv-ion or ignorance, akin to the terra-incognita blanks map makers of old used to call "sleeping beauties."

Beyond the park, there were fields, with a continuous shimmer of butterfly wings over a shimmer of flowers—daisies, bluebells, scabious, and others—which now rapidly pass by me in a kind of colored haze like those lovely, lush meadows, never to be explored, that one sees from the diner on a transcontinental journey. At the end of this grassy wonderland, the forest rose like a wall. There I roamed, scanning the tree trunks (the enchanted, the silent part of a tree) for certain tiny moths, called Pugs in England—delicate little creatures that cling in the daytime to speckled surfaces, with which their flat wings and turned-up abdomens blend. There, at the bottom of that sea of sunshot greenery, I slowly spun round the great boles. Nothing in the world would have seemed sweeter to me than to be able to add, by a stroke of luck, some remarkable new species to the long list of Pugs already named by others. And my pied imagination, ostensibly, and almost grotesquely, groveling to my desire (but all the time, in ghostly conspir-acies behind the scenes, coolly planning the most distant events of my destiny), kept providing me with hallucinatory samples of small print: ". . . the only specimen so far known . . ." ". . . the only specimen known of *Eupithecia petropolitanata* was taken by a Russian schoolboy . . ." ". . . by a young Russian collector . . ." ". . . by myself in the Government of St. Petersburg, Tsarskoe Selo District, in 1910 . . . 1911 . . . 1912 . . . 1913 . . ." And then, thirty years later, that blessed black night in the Wasatch Range.

At first—when I was, say, eight or nine—I seldom roamed farther than the fields and woods between Vyra and Batovo. Later, when aim-ing at a particular spot half-a-dozen miles or more distant, I would use a bicycle to get there with my net scrapped to the frame; but not many forest paths were passable on wheels; it was possible to ride there on horseback, of course, but, because of our ferocious Russian tabanids, one could not leave a horse haltered in a wood for any length of time: my spirited bay almost climbed up the tree it was tied to one day trying to elude them: big fellows with watered-silk eyes and tiger bodies, and gray little runts with an even more painful proboscis, but much more sluggish: to dispatch two or three of these dingy tipplers with one crush of the gloved hand as they glued themselves to the neck of my mount afforded me a wonderful empathic relief (which a dipterist might not appreciate). Anyway, on my butterfly hunts I always preferred hiking to any other form of locomotion (except, naturally, a flying seat gliding leisurely over the plant mats and rocks of an unexplored mountain, or hovering just above the flowery roof of a rain forest); for when you walk, especially in a region you have studied well, there is an exquisite pleasure in departing from one's itinerary to visit, here and there by the wayside, this glade, that glen, this or that combination of soil and flora—to drop in, as it were, on a familiar butterfly in his particular habitat, in order to see if he has emerged, and if so, how he is doing.

There came a July day—around 1910, I suppose—when I felt the urge to explore the vast marshland beyond the Oredezh. After skirting the river for three or four miles, I found a rickety footbridge. While crossing over, I could see the huts of a hamlet on my left, apple trees, rows of tawny pine logs lying on a green bank, and the bright patches made on the turf by the scattered clothes of peasant girls, who, stark naked in shallow water, romped and yelled, heeding me as little as if I were the discarnate carrier of my present reminiscences.

On the other side of the river, a dense crowd of small, bright blue male butterflies that had been tippling on the rich, trampled mud and cow dung through which I trudged rose all together into the spangled air and settled again as soon as I had passed.

After making my way through some pine groves and alder scrub I came to the bog. No sooner had my ear caught the hum of diptera around me, the guttural cry of a snipe overhead, the gulping sound of the morass under my foot, than I knew I would find here quite special arctic butterflies, whose pictures, or, still better, nonillustrated descriptions I had worshiped for several seasons. And the next moment I was among them. Over the small shrubs of bog bilberry with fruit of a dim, dreamy blue, over the brown eye of stagnant water, over moss and mire, over the flower spikes of the fragrant bog orchid (the *nochnaya fialka* of Russian poets), a dusky little Fritillary bearing the name of a Norse goddess passed in low, skimming flight. Pretty Cordigera, a gemlike moth, buzzed all over its uliginose food plant. I pursued rose-margined Sulphurs, gray-marbled Satyrs. Unmindful of the mosquitoes that furred my forearms, I stooped with a grunt of delight to snuff out the life of some silver-studded lepidopteron throbbing in the folds of my net. Through the smells of the bog, I caught the subtle perfume of butterfly wings on my fingers, a perfume which varies with the species—vanilla, or lemon, or musk, or a musty, sweetish odor difficult to define. Still unsated, I pressed forward. At last I saw I had come to the end of the marsh. The rising ground beyond was a paradise of lupines, columbines, and penstemons. Mariposa lilies bloomed under Ponderosa pines. In the distance, fleeting cloud shadows dappled the dull green of slopes above timber line, and the gray and white of Longs Peak.

I confess I do not believe in time. I like to fold my magic carpet, after use, in such a way as to superimpose one part of the pattern upon another. Let visitors trip. And the highest enjoyment of time-lessness—in a landscape selected at random—is when I stand among rare butterflies and their food plants. This is ecstasy, and behind the ecstasy is something else, which is hard to explain. It is like a momentary vacuum into which rushes all that I love. A sense of oneness with sun and stone. A thrill of gratitude to whom it may concern—to the contrapuntal genius of human fate or to tender ghosts humoring a lucky mortal.

SIGURD OLSON
1899–1982

Certain writers become the representatives for their chosen landscapes. Just as John Muir is identified with Yosemite and Mary Austin with the southwestern deserts, Sigurd Olson expresses, and assumes, the character of the northern Midwest. His home for much of his career as a teacher, guide, and writer was in Ely, Minnesota—near the Quetico–Superior wilderness which he helped to protect. In his essays and books Olson told the stories of that region of waters and forests. One of his special accomplishments was to integrate the lives of animals, plants, and changing sky with the tales of voyageurs and present-day homesteaders, to convey both the mystery and the human significance of his beloved North Woods. Olson's books include The Singing Wilderness *(1956),* Listening Point *(1958), and* The Lonely Land *(1961).*

NORTHERN LIGHTS

The lights of the aurora moved and shifted over the horizon. Sometimes there were shafts of yellow tinged with green, then masses of evanescence which moved from east to west and back again. Great streamers of bluish white zigzagged like a tremendous trembling curtain from one end of the sky to the other. Streaks of yellow and orange and red shimmered along the flowing borders. Never for a moment were they still, fading until they were almost completely gone, only to dance forth again in renewed splendor with infinite combinations and startling patterns of design.

The lake lay like a silver mirror before me, and from its frozen surface came subterranean rumblings, pressure groans, sharp reports from the newly forming ice. As far as I could see, the surface was clear and shining. That ice was something to remember here in the north, for most years the snows come quickly and cover the first smooth glaze of freezing almost as soon as it is formed, or else the winds ruffle the surface of the crystallizing water and fill it with ridges and unevenness. But this time there had been no wind or snow to interfere, and the ice everywhere was clear—seven miles of perfect skating, something to dream about in years to come.

Hurriedly I strapped on my skates, tightened the laces, and in a moment was soaring down the path of shifting light which stretched endlessly before me. Out in the open away from shore there were few cracks—stroke—stroke—stroke—long and free, and I knew the joy that skating and skiing can give, freedom of movement beyond myself. But to get the feel of soaring, there must be miles of distance and conditions must be right. As I sped down the lake, I was conscious of no effort, only of the dancing lights in the sky and a sense of lightness and exaltation.

Shafts of light shot up into the heavens above me and concentrated there in a final climactic effort in which the shifting colors seemed drained from the horizons to form one gigantic rosette of flame and

The Singing Wilderness *(New York: Knopf, 1956).*

yellow and greenish purple. Suddenly I grew conscious of the reflections from the ice itself and that I was skating through a sea of changing color caught between the streamers above and below. At that moment I was part of the aurora, part of its light and of the great curtain that trembled above me.

Those moments of experience are rare. Sometimes I have known them while swimming in the moonlight, again while paddling a canoe when there was no wind and the islands seemed inverted and floating on the surface. I caught it once when the surf was rolling on an ocean coast and I was carried on the crest of a wave that had begun a thousand miles away. Here it was once more—freedom of movement and detachment from the earth.

Down the lake I went straight in to the glistening path, speeding through a maze of changing color—stroke—stroke—stroke—the ringing of steel on ice, the sharp, reverberating rumbles of expansion below. Clear ice for the first time in years, and the aurora blazing away above it.

At the end of the lake I turned and saw the glittering lights of Winton far behind me. I lay down on the ice to rest. The sky was still bright and I watched the shifting lights come and go. I knew what the astronomers and the physicists said, that they were caused by sunspots and areas of gaseous disturbance on the face of the sun that bombarded the earth's stratosphere with hydrogen protons and electrons which in turn exploded atoms of oxygen, nitrogen, helium, and the other elements surrounding us. Here were produced in infinite combinations all the colors of the spectrum. It was all very plausible and scientific, but tonight that explanation left me cold. I was in no mood for practicality, for I had just come skating down the skyways themselves and had seen the aurora from the inside. What did the scientists know about what I had done? How could they explain what had happened to me and the strange sensations I had known?

Much better the poem of Robert Service telling of the great beds of radium emanating shafts of light into the northern darkness of the Yukon and how men went mad trying to find them. How infinitely more satisfying to understand and feel the great painting by Franz Johnson of a lone figure crossing a muskeg at night with the northern lights blazing above it. I stood before that painting in the Toronto Art Gallery one day and caught all the stark loneliness, all the beauty and the cold of that scene, and for a moment forgot the busy city outside.

I like to think of them as the ghost dance of the Chippewas. An Indian once told me that when a warrior died, he gathered with his fellows along the northern horizon and danced the war dances they had known on earth. The shifting streamers and the edgings of color came from the giant headdress they wore. I was very young when I first saw them that way, and there were times during those enchanted years when I thought I could distinguish the movements of individual bodies as they rushed from one part of the sky to another. I knew nothing then of protons or atoms and saw the northern lights as they should be seen. I knew, too, the wonderment that only a child can know and a beauty that is enhanced by mystery.

As I lay there on the ice and thought of these things I wondered if legendry could survive scientific truth, if the dance of the protons would replace the ghost dance of the Chippewas. I wondered as I began to skate toward home if anything—even knowing the physical truth—could ever change the beauty of what I had seen, the sense of unreality. Indian warriors, exploding atoms, beds of radium—what difference did it make? What counted was the sense of the north they gave me, the fact that they typified the loneliness, the stark beauty of frozen muskegs, lakes, and forests. Those northern lights were part of me and I of them.

On the way back I noticed that there was a half-moon over the cluster of lights in the west. I skirted the power dam at the mouth of the Kawishiwi River, avoiding the blaze of its light on the black water below the spillway. Then suddenly the aurora was gone and the moon as well.

Stroke—stroke—stroke—the shores were black now, pinnacled spruce and shadowed birch against the sky. At the landing I looked back. The ice was still grumbling and groaning, still shaping up to the mold of its winter bed.

EDWIN WAY TEALE
1899–1980

His many books on nature made Teale one of the most important teachers of his generation. Among the best known of them were The Lost Woods *(1945),* North with the Spring *(1951), and* A Walk Through the Year *(1978). He took excursions, in Thoreau's sense of the word, and, as a fine photographer, frequently focused his explorations with close-up*

*images. Teale was most closely associated with one magazine, Audubon,
of which he served as a contributing editor from 1942 to 1980. In addition
to being an important nature writer, Teale advanced and interpreted the
tradition through his popular editions of the works of Audubon, Thoreau,
Muir, and Henri Fabre.*

THE LOST WOODS

A back-country road was carrying us south, carrying us through a
snow-filled landscape and under the sullen gray of a December sky.
Minute by minute, the long chain of the Indiana dunes receded
behind us. Ahead, beyond the bobbing ears of the horses, I could catch
glimpses of the blue-white, far-away ridges of the Valparaiso moraine.

Our low bob-sled tilted and pitched over the frozen ruts. Beside me,
my bearded grandfather clung to the black strips of the taut reins and
braced himself with felt-booted feet widespread. At every lurch, my
own short, six-year-old legs, dangling below the seat, gyrated wildly like
the tail of an off-balance cat.

We had left Lone Oak, my grandfather's dune-country farm, that
winter morning, to drive to a distant woods. In the late weeks of
autumn, my grandfather had been busy there, felling trees and cutting
firewood. He was going after a load of this wood and he was taking me
along. At first, we drove through familiar country—past Gunder's big
red barn, the weed lot and the school house. Then we swung south and
crossed the right-of-way of the Pere Marquette railroad. Beyond, we
journeyed into a world that was, for me, new and unexplored. The road
ran on and on. We seemed traversing vast distances while the smell of
coming snow filled the air.

Eventually, I remember, we swung off the road into a kind of lane.
The fences soon disappeared and we rode out into open country,
onto a wide, undulating sea of whiteness with here and there the
island of a bush-clump. As we progressed, a ribbon of runner-tracks
and hoof-marks steadily unrolled, lengthened, and followed us across
the snow.

Winter trees, gray and silent, began to rise around us. They were old
trees, gnarled and twisted. We came to a frozen stream and turned to
follow its bank. The bob-sled, from time to time, would rear suddenly

The Lost Woods: Adventures of a Naturalist (New York: Dodd, Mead, 1945).

and then plunge downward as a front runner rode over a low stump or hidden log. Each time the sled seat soared and dropped away, I clung grimly to my place or clawed wildly at my grandfather's overcoat. He observed with a chuckle:

"'T takes a good driver t' hit *all* th' stumps."

Then, while the snow slipped backward beneath the runners and the great trees of that somber woods closed around us, we rode on in silence. As we advanced, the trees grew steadily thicker; the woods more dark and lonely. In a small clearing, my grandfather pulled up beside a series of low, snow-covered walls. Around us were great white mounds that looked like igloos. The walls were the corded stovewood; the igloos were the snow-clad piles of discarded branches.

Wisps of steam curled up from the sides of the heated horses and my grandfather threw blankets over their backs before he bent to the work of tossing stovewood into the lumber-wagon bed of the bob-sled. The hollow thump and crash of the frozen sticks reverberated through the still woods.

I soon tired of helping and wandered about, small as an atom, among the great trees—oak and beech, hickory and ash and sycamore. An air of strangeness and mystery enveloped the dark woods. I peered timidly down gloomy aisles between the trees. Branches rubbed together in the breeze with sudden shrieks or mournful wailings and the cawing of a distant crow echoed dismally. I was at once enchanted and fearful. Each time I followed one of the corridors away from the clearing, I hurried back to be reassured by the sight of my bundled-up grandfather stooping and rising as he picked up the cordwood and tossed it into the sled.

He stopped from time to time to point out special trees. In the hollow of one great beech, he had found two quarts of shelled nuts stored away by a squirrel. In another tree, with a gaping rectangular hole chopped in its upper trunk, the owner of the woods had obtained several milk-pails full of dark honey made by a wild swarm of bees. Still another hollow tree had a story to tell. It was an immense sycamore by the stream-bank. Its interior, smoke-blackened and cavernous, was filled with a damp and acrid odor. One autumn night, there, hunters had treed and smoked out a raccoon.

There were other exciting discoveries: the holes of owls and woodpeckers; the massed brown leaves of squirrel-nests high in the bare branches; the tracks of small wild animals that wound about among the trees, that crisscrossed on the ice, that linked together the great mounds of the discarded branches. In one place, the wing-feathers of an owl had left their imprint on the snow; and there, the trail of some small animal had ended and there, on the white surface, were tiny drops of red. From the dark mouth of a burrow, under the far bank of the frozen stream, tracks led away over the ice. I longed to follow them around a distant bend in the stream. But the reaches beyond, forbidding under the still tenseness of the ominous sky, slowed my steps to a standstill. However, my mind and imagination were racing.

Behind and beyond the silence and inactivity of the woods, there was a sense of action stilled by our presence; of standing in a charmed circle where all life paused, enchanted, until we passed on. I had the feeling that animals would appear, their interrupted revels and battles would recommence, with our departure. My imagination invested the woods with a fearful and delicious atmosphere of secrecy and wildness. It left me with an endless curiosity about this lonely tract and all of its inhabitants.

After nearly half an hour had gone by, my grandfather's long sled was full and he called me back to the seat. As we rode away, I looked back as long as I could see the trees, watching to the last this gloomy woods, under its gloomy sky, which had made such a profound impression on me. All the way home, I was silent, busy with my own speculations.

Thirty years later, I spent one whole summer's day driving my car over dirt roads of the region, searching for this old, remembered woods. But I never found it. Perhaps I took the wrong turns. Perhaps the woods had been felled and the land turned into cultivated fields. Perhaps I failed to recognize the wooded tract as seen through the eyes of a small boy. I know that, as I drove about, the great distances of childhood had greatly shrunk. How soon I came to the corners! How much smaller were the trees than I remembered them; how much lower the hills! Time seemed to have dwarfed the towering barns of boyhood and to have reduced the size of cornfields and pastures. At any rate, I never saw the ancient trees of that old woodland a second time. The Lost Woods of childhood remained lost forever.

In talking to others, I have come to believe that most of us have had some such experience—that some lonely spot, some private nook, some glen or streamside-scene impressed us so deeply that even today its memory recalls the mood of a lost enchantment. At the age of eighty, my grandmother used to recall with delight a lonely tract she called "The Beautiful Big South Woods." There, as a girl one spring day, she had seen the whole floor of the woods, acre on acre, carpeted with the blooms of bloodroot and spring beauties and blue and pink hepaticas. She had seen the woods only once but she never forgot it.

When Henry Thoreau was five, his parents, then living in the city of Boston, took him eighteen miles into the country to a woodland scene that he, too, never forgot. It was, he said, one of the earliest scenes stamped on the tablets of his memory. During succeeding years of childhood, that woodland formed the basis of his dreams. The spot to which he had been taken was Walden Pond, near Concord. Twenty-three years later, writing in his cabin on the shores of this same pond, Thoreau noted the unfading impression that "fabulous landscape" had made and how, even at that early age, he had given preference to this recess—"where almost sunshine and shadow were the only inhabitants that varied the scene"—over the tumultuous city in which he lived. .

John C. Merriam, at the time he was President of the Carnegie Institution of Washington, D.C., wrote of the profound effect a woodland hilltop, rising beyond the pastures of a valley at his childhood home, had had upon his early life. The margin of this forest seemed like an impenetrable wall beyond which lay a place of continuous night. Often at evening he could hear the howling of wolves among the trees. His imagination peopled the hilltop with strange creatures and, in his mind, the timbered tract became symbolical of all that is mysterious and awaiting solution. As the later years of his life passed, this eminent scientist wrote, the thing that led him on was the endless challenge of the unknown—a challenge that appeared to him first in the form of this dark and distant woods of his boyhood.

Such lasting impressions of life, such moments of far-reaching consequences, almost always arrive unbidden. Rarely can they be planned beforehand. A friend of mine, a writer who lived as a small boy in Mexico, once told me that the pivotal day in his life came when he was six. His mother imagined he was destined to be a great pianist. When she heard that Paderewski was to give a concert in Mexico City, she planned that that event would be the turning point in the child's life. The memory of the master's music was to spur him on to greatness. The evening came. The master played. The boy returned home in disgrace. At the end of the first number, he had fallen fast asleep. Thereafter, he was permitted to follow his own bent.

For me, the Lost Woods became a starting point and a symbol. It was a symbol of all the veiled and fascinating secrets of the out-of-doors. It was the starting point of my absorption in the world of Nature. The image of that somber woods returned a thousand times in memory. It aroused in my mind an interest in the ways and the mysteries of the wild world that a lifetime is not too long to satisfy.

E(LWYN) B(ROOKS) WHITE
1899–1985

White is a writer beloved for his voice—modest and humorous, ironic and forgiving. He honed this masterly, humane style during long years on the staff of The New Yorker. *Among the books for which White is especially remembered are his amusing and helpful* The Elements of Style *(1935, with William Strunk, Jr.) and his children's stories* Stuart Little *(1945),* Charlotte's Web *(1952), and* The Trumpet of the Swan *(1970). Collections of his essays, such as* Points of My Compass *(1962), confirm his pleasure in country things, especially gardening, farmyard animals, and sailing.*

A Slight Sound at Evening

Allen Cove, Summer, 1954

In his journal for July 10–12, 1841, Thoreau wrote: "A slight sound at evening lifts me up by the ears, and makes life seem inexpressibly serene and grand. It may be in Uranus, or it may be in the shutter." The book into which he later managed to pack both Uranus and the shutter was published in 1854, and now, a hundred years having gone by, *Walden*, its serenity and grandeur unimpaired, still lifts us up by the ears, still translates for us that language we are in danger of forgetting, "which all things and events speak without metaphor, which alone is copious and standard."

Walden is an oddity in American letters. It may very well be the oddest of our distinguished oddities. For many it is a great deal too odd, and for many it is a particular bore. I have not found it to be a well-liked book among my acquaintances, although usually spoken of

The Points of My Compass: Letters from the East, the West, the North, the South (New York: Harper and Row, 1962).

with respect, and one literary critic for whom I have the highest regard can find no reason for anyone's giving *Walden* a second thought. To admire the book is, in fact, something of an embarrassment, for the mass of men have an indistinct notion that its author was a sort of Nature Boy.

I think it is of some advantage to encounter the book at a period in one's life when the normal anxieties and enthusiasms and rebellions of youth closely resemble those of Thoreau in that spring of 1845 when he borrowed an ax, went out to the woods, and began to whack down some trees for timber. Received at such a juncture, the book is like an invitation to life's dance, assuring the troubled recipient that no matter what befalls him in the way of success or failure he will always be welcome at the party—that the music is played for him, too, if he will but listen and move his feet. In effect, that is what the book is—an invitation, unengraved; and it stirs one as a young girl is stirred by her first big party bid. Many think it a sermon; many set it down as an attempt to rearrange society; some think it an exercise in nature-loving; some find it a rather irritating collection of inspirational puffballs by an eccentric show-off. I think it none of these. It still seems to me the best youth's companion yet written by an American, for it carries a solemn warning against the loss of one's valuables, it advances a good argument for traveling light and trying new adventures, it rings with the power of positive adoration, it contains religious feeling without religious images, and it steadfastly refuses to record bad news. Even its pantheistic note is so pure as to be noncorrupting—pure as the flute-note blown across the pond on those faraway summer nights. If our colleges and universities were alert, they would present a cheap pocket edition of the book to every senior upon graduating, along with his sheepskin, or instead of it. Even if some senior were to take it literally and start felling trees, there could be worse mishaps: the ax is older than the Dictaphone and it is just as well for a young man to see what kind of chips he leaves before listening to the sound of his own voice. And even if some were to get no farther than the table of contents, they would learn how to name eighteen chapters by the use of only thirty-nine words and would see how sweet are the uses of brevity.

If Thoreau had merely left us an account of a man's life in the woods or if he had simply retreated to the woods and there recorded his complaints about society, or even if he had contrived to include both records in one essay, *Walden* would probably not have lived a hundred years. As things turned out, Thoreau, very likely without knowing quite what he was up to, took man's relation to Nature and man's dilemma in society and man's capacity for elevating his spirit and he beat all these matters together, in a wild free interval of self-justification and delight, and produced an original omelette from which people can draw nourishment in a hungry day. *Walden* is one of the first of the vitamin-enriched American dishes. If it were a little less good than it is, or even a little less queer, it would be an abominable book. Even as it is, it will continue to baffle and annoy the literal mind and all those who are unable to stomach its caprices and imbibe its theme. Certainly the plodding economist will continue to have rough going if he hopes to emerge from the book with a clear system of economic thought. Thoreau's assault on the Concord society of the mid-nineteenth century has the quality of a modern Western: he rides into the subject at top speed, shooting in all directions. Many of his shots ricochet and nick him on the rebound, and throughout the melee there is a horrendous cloud of inconsistencies and contradictions, and when the shooting dies down and the air clears, one is impressed chiefly by the courage of the rider and by how splendid it was that somebody should have ridden in there and raised all that ruckus.

When he went to the pond, Thoreau struck an attitude and did so deliberately, but his posturing was not to draw the attention of others to him but rather to draw his own attention more closely to himself. "I learned this at least by my experiment: that if one advances confidently in the direction of his dreams, and endeavors to live the life which he has imagined, he will meet with a success unexpected in common hours." The sentence has the power to resuscitate the youth drowning in his sea of doubt. I recall my exhilaration upon reading it, many years ago, in a time of hesitation and despair. It restored me to health. And now in 1954 when I salute Henry Thoreau on the hundredth birthday of his book, I am merely paying off an old score—or an installment on it.

In his journal for May 3–4, 1838—Boston to Portland—he wrote: "Midnight—head over the boat's side—between sleeping and waking—with glimpses of one or more lights in the vicinity of Cape Ann. Bright moonlight—the effect heightened by seasickness." The entry illuminates the man, as the moon the sea on that night in May. In Thoreau the natural scene was heightened, not depressed, by a disturbance of the stomach, and nausea met its match at last. There was a steadiness in at least one passenger if there was none in the boat. Such steadiness (which in some would be called intoxication) is at the heart of *Walden*—confidence, faith, the discipline of looking always at what is to be seen, undeviating gratitude for the life-everlasting that he found growing in his front yard. "There is nowhere recorded a simple and irrepressible satisfaction with the gift of life, any memorable praise of God." He worked to correct that deficiency. *Walden* is his acknowledgment of the gift of life. It is the testament of a man in a high state of indignation because (it seemed to him) so few ears heard the uninterrupted poem of creation, the morning wind that forever blows. If the man sometimes wrote as though all his readers were male, unmarried, and well-connected, it is because he gave his testimony during the callow years. For that matter, he never really grew up. To reject the book because of the immaturity of the author and the bugs in the logic is to throw away a bottle of good wine because it contains bits of the cork.

Thoreau said he required of every writer, first and last, a simple and sincere account of his own life. Having delivered himself of this chesty dictum, he proceeded to ignore it. In his books and even in his enormous journal, he withheld or disguised most of the facts from which an understanding of his life could be drawn. *Walden*, subtitled "Life in the Woods," is not a simple and sincere account of a man's life, either in or out of the woods; it is an account of a man's journey into the mind, a toot on the trumpet to alert the neighbors. Thoreau was well aware that no one can alert his neighbors who is not wide-awake himself, and he went to the woods (among other reasons) to make sure that he would stay awake during his broadcast. What actually took place during the years 1845–47 is largely unrecorded, and the reader is excluded from the private life of the author, who supplies almost no gossip about himself, a great deal about his neighbors and about the universe.

As for me, I cannot in this short ramble give a simple and sincere account of my own life, but I think Thoreau might find it instructive to know that this memorial essay is being written in a house that, through no intent on my part, is the same size and shape as his own domicile on the pond—about ten by fifteen, tight, plainly finished, and at a little distance from my Concord. The house in which I sit this morning was built to accommodate a boat, not a man, but by long experience I have learned that in most respects it shelters me better than the larger dwelling where my bed is, and which, by design, is a manhouse not a boathouse. Here in the boathouse I am a wilder and, it would appear, a healthier man, by a safe margin. I have a chair, a bench, a table, and I can walk into the water if I tire of the land. My house fronts a cove. Two fishermen have just arrived to spot fish from the air—an osprey and a man in a small yellow plane who works for the fish company. The man, I have noticed, is less well equipped than the hawk, who can dive directly on his fish and carry it away, without telephoning. A mouse and a squirrel share the house with me. The building is, in fact, a multiple dwelling, a semidetached affair. It is because I am semidetached while here that I find it possible to transact this private business with the fewest obstacles.

There is also a woodchuck here, living forty feet away under the wharf. When the wind is right, he can smell my house; and when the wind is contrary, I can smell his. We both use the wharf for sunning, taking turns, each adjusting his schedule to the other's convenience. Thoreau once ate a woodchuck. I think he felt he owed it to his readers, and that it was little enough, considering the indignities they were suffering at his hands and the dressing-down they were taking. (Parts of *Walden* are pure scold.) Or perhaps he ate the woodchuck because he believed every man should acquire strict business habits, and the woodchuck was destroying his market beans. I do not know. Thoreau had a strong experimental streak in him. It is probably no harder to eat a woodchuck than to construct a sentence that lasts a hundred years. At any rate, Thoreau is the only writer I know who prepared himself for his great ordeal by eating a woodchuck; also the only one who got a hangover from drinking too much water. (He was drunk the whole time, though he seldom touched wine or coffee or tea.)

Here in this compact house where I would spend one day as deliberately as Nature if I were not being pressed by the editor of a magazine, and with a woodchuck (as yet uneaten) for neighbor, I can feel the companionship of the occupant of the pond-side cabin in Walden woods, a mile from the village, near the Fitchburg right of way. Even my immediate business is no barrier between us: Thoreau occasionally batted out a magazine piece, but was always suspicious of any sort of purposeful work that cut into his time. A man, he said, should take care not to be thrown off the track by every nutshell and mosquito's wing that falls on the rails.

There has been much guessing as to why he went to the pond. To set it down to escapism is, of course, to misconstrue what happened. Henry went forth to battle when he took to the woods, and *Walden* is the report of a man torn by two powerful and opposing drives—the desire to enjoy the world (and not be derailed by a mosquito wing) and the urge to set the world straight. One cannot join these two successfully, but sometimes, in rare cases, something good or even great results from the attempt of the tormented spirit to reconcile them. Henry went forth to battle, and if he set the stage himself, if he fought on his own terms and with his own weapons, it was because it was his nature to do things differently from most men, and to act in a cocky fashion. If the pond and the woods seemed a more plausible site for a house than an intown location, it was because a cowbell made for him a sweeter sound than a churchbell. *Walden*, the book, makes the sound of a cowbell, more than a churchbell, and proves the point, although both sounds are in it, and both remarkably clear and sweet. He simply preferred his churchbell at a little distance.

I think one reason he went to the woods was a perfectly simple and commonplace one—and apparently he thought so, too. "At a certain season of our life," he wrote, "we are accustomed to consider every spot as the possible site of a house." There spoke the young man, a few years out of college, who had not yet broken away from home. He hadn't married, and he had found no job that measured up to his rigid standards of employment, and like any young man, or young animal, he felt uneasy and on the defensive until he had fixed himself a den. Most young men, of course, casting about for a site, are content merely to draw apart from their kinfolks. Thoreau, convinced that the greater part of what his neighbors called good was bad, withdrew from a great deal more than family: he pulled out of everything for a while, to serve everybody right for being so stuffy, and to try his own prejudices on the dog.

The house-hunting sentence above, which starts the chapter called "Where I Lived, and What I Lived For," is followed by another passage that is worth quoting here because it so beautifully illustrates the offbeat prose that Thoreau was master of, a prose at once strictly disciplined and wildly abandoned. "I have surveyed the country on every side within a dozen miles of where I live," continued this delirious young man. "In imagination I have bought all the farms in succession, for all were to be bought, and I knew their price. I walked over each farmer's premises, tasted his wild apples, discoursed on husbandry with him, took his farm at his price, at any price, mortgaging it to him in my mind; even put a higher price on it—took everything but a deed of it—took his word for his deed, for I dearly love to talk—cultivated it, and him too to some extent, I trust, and withdrew when I had enjoyed it long enough, leaving him to carry it on." A copy-desk man would get a double hernia trying to clean up that sentence for the management, but the sentence needs no fixing, for it perfectly captures the meaning of the writer and the quality of the ramble.

"Wherever I sat, there I might live, and the landscape radiated from me accordingly." Thoreau, the home-seeker, sitting on his hummock with the entire State of Massachusetts radiating from him, is to me the most humorous of the New England figures, and *Walden* the most humorous of the books, though its humor is almost continuously subsurface and there is nothing deliberately funny anywhere, except a few weak jokes and bad puns that rise to the surface like the perch in the pond that rose to the sound of the maestro's flute. Thoreau tended to write in sentences, a feat not every writer is capable of, and *Walden* is, rhetorically speaking, a collection of certified sentences, some of them, it would now appear, as indestructible as they are errant. The book is distilled from the vast journals, and this accounts for its intensity: he picked out bright particles that pleased his eye, whirled them in the kaleidoscope of his content, and produced the pattern that has endured—the color, the form, the light.

On this its hundredth birthday, Thoreau's *Walden* is pertinent and timely. In our uneasy season, when all men unconsciously seek a retreat from a world that has got almost completely out of hand, his house in the Concord woods is a haven. In our culture of gadgetry and the multiplicity of convenience, his cry "Simplicity, simplicity, simplicity!" has the insistence of a fire alarm. In the brooding atmosphere of war and the gathering radioactive storm, the innocence and serenity of his summer afternoons are enough to burst the remembering heart, and one gazes back upon that pleasing interlude—its confidence, its purity, its deliberateness—with awe and wonder, as one would look upon the face of a child asleep.

"This small lake was of most value as a neighbor in the intervals of a gentle rain-storm in August, when, both air and water being perfectly still, but the sky overcast, midafternoon had all the serenity of evening, and the wood-thrush sang around, and was heard from shore to shore." Now, in the perpetual overcast in which our days are spent, we hear with extra perception and deep gratitude that song, tying century to century.

I sometimes amuse myself by bringing Henry Thoreau back to life and showing him the sights. I escort him into a phone booth and let him dial Weather. "This is a delicious evening," the girl's voice says, "when the whole body is one sense, and imbibes delight through every pore." I show him the spot in the Pacific where an island used to be, before some magician made it vanish. "We know not where we are," I murmur. "The light which puts out our eyes is darkness to us. Only that day dawns to which we are awake." I thumb through the latest copy of *Vogue* with him. "Of two patterns which differ only by a few threads more or less of a particular color," I read, "the one will be sold readily, the other lie on the shelf, though it frequently happens that, after the lapse of a season, the latter becomes the most fashionable." Together we go outboarding on the Assabet, looking for what we've lost—a hound, a bay horse, a turtledove. I show him a distracted farmer who is trying to repair a hay baler before the thunder shower breaks. "This farmer," I remark, "is endeavoring to solve the problem of a livelihood by a formula more complicated than the problem itself. To get his shoestrings he speculates in herds of cattle."

I take the celebrated author to Twenty-One for lunch, so the waiters may study his shoes. The proprietor welcomes us. "The gross feeder," remarks the proprietor, sweeping the room with his arm, "is a man in the larva stage." After lunch we visit a classroom in one of those schools conducted by big corporations to teach their superannuated executives how to retire from business without serious injury to their health. (The shock to men's systems these days when relieved of the exacting routine of amassing wealth is very great and must be cushioned.) "It is not necessary," says the teacher to his pupils, "that a man should earn his living by the sweat of his brow, unless he sweats easier than I do. We are determined to be starved before we are hungry."

I turn on the radio and let Thoreau hear Winchell beat the red hand around the clock. "Time is but the stream I go a-fishing in," shouts Mr. Winchell, rattling his telegraph key. "Hardly a man takes a half hour's nap after dinner, but when he wakes he holds up his head and asks, 'What's the news?' If we read of one man robbed, or murdered, or killed by accident, or one house burned, or one vessel wrecked, or one steamboat blown up, or one cow run over on the Western Railroad, or one mad dog killed, or one lot of grasshoppers in the winter—we need never read of another. One is enough."

I doubt that Thoreau would be thrown off balance by the fantastic sights and sounds of the twentieth century. "The Concord nights," he once wrote, "are stranger than the Arabian nights." A four-engined airliner would merely serve to confirm his early views on travel. Everywhere he would observe, in new shapes and sizes, the old predicaments and follies of men—the desperation, the impedimenta, the meanness—along with the visible capacity for elevation of the mind and soul. "This curious world which we inhabit is more wonderful than it is convenient; more beautiful than it is useful; it is more to be admired and enjoyed than used." He would see that today ten thousand engineers are busy making sure that the world shall be convenient even if it is destroyed in the process, and others are determined to increase its usefulness even though its beauty is lost somewhere along the way.

At any rate, I'd like to stroll about the countryside in Thoreau's company for a day, observing the modern scene, inspecting today's snow-storm, pointing out the sights, and offering belated apologies for my sins. Thoreau is unique among writers in that those who admire him find him uncomfortable to live with—a regular hairshirt of a man. A little band of dedicated Thoreauvians would be a sorry sight indeed: fellows who hate compromise and have compromised, fellows who love wildness and have lived tamely, and at their side, censuring them and chiding them, the ghostly figure of this upright man, who long ago gave corroboration to impulses they perceived were right and issued warnings against the things they instinctively knew to be their enemies. I should hate to be called a Thoreauvian, yet I wince every time I walk into the barn I'm pushing before me, seventy-five feet by forty, and the author of *Walden* has served as my conscience through the long stretches of my trivial days.

Hairshirt or no, he is a better companion than most, and I would not swap him for a soberer or more reasonable friend even if I could. I can reread his famous invitation with undiminished excitement. The sad thing is that not more acceptances have been received, that so many decline for one reason or another, pleading some previous engagement or ill health. But the invitation stands. It will beckon as long as this remarkable book stays in print—which will be as long as there are August afternoons in the intervals of a gentle rainstorm, as long as there are ears to catch the faint sounds of the orchestra. I find it agreeable to sit here this morning, in a house of correct proportions, and hear across a century of time his flute, his frogs, and his seductive summons to the wildest revels of them all.

MERIDEL LeSUEUR
1900–1996

LeSueur has been described as a "radical" writer because of her identification with the labor movement and association with the Communist Party. But the word is also appropriate because of her attention to the agricultural roots of culture. Her early commitment to feminism, as well as her years of living in Iowa, Kansas, and Minnesota, led LeSueur to pay special attention to the myths of Demeter and Persephone. Several of the pieces included in Ripening: Selected Work, 1927–1980 (1982) *evoke the continuity of human life in the Americas by focusing on the motif of corn.*

THE ANCIENT PEOPLE AND THE NEWLY COME

Born out of the caul of winter in the north, in the swing and circle of the horizon, I am rocked in the ancient land. As a child I first read the scriptures written on the scroll of frozen moisture by wolf and rabbit, by the ancient people and the newly come. In the beginning of the century the Indian smoke still mingled with ours. The frontier of the whites was violent, already injured by vast seizures and massacres. The winter nightmares of fear poisoned the plains nights with psychic airs of theft and utopia. The stolen wheat in the cathedrallike granaries cried out for vengeance.

Most of all one was born into space, into the great resonance of space, a magnetic midwestern valley through which the winds clashed in lassoes of thunder and lightning at the apex of the sky, the very wrath of God.

The body repeats the landscape. They are the source of each other and create each other. We were marked by the seasonal body of earth, by the terrible migrations of people, by the swift turn of a century, verging on change never before experienced on this greening planet. I sensed the mound and swell above the mother breast, and from embryonic eye took sustenance and benediction, and went from mother enclosure to prairie spheres curving into eachother.

I was born in winter, the village snow darkened toward midnight, footsteps on boardwalks, the sound of horses pulling sleighs, and the ring of bells. The square wooden saltbox house held the tall shadows, thrown from kerosene lamps, of my grandmother and my aunt and uncle (missionaries home from India) inquiring at the door.

It was in the old old night of the North Country. The time of wood before metal. Contracted in cold, I lay in the prairie curves of my mother, in the planetary belly, and outside the vast horizon of the plains, swinging dark and thicketed, circle within circle. The round moon sinister reversed upside down in the sign of Neptune, and the twin fishes of Pisces swimming toward Aquarius in the dark.

But the house was New England square, four rooms upstairs and four rooms downstairs, exactly set upon a firm puritan foundation, surveyed on a level, set angles of the old geometry, and thrust up on the plains like an insult, a declamation of the conqueror, a fortress of our God, a shield against excess and sin.

Ripening: Selected Work, 1927–1980, ed. Elaine Hedges (Old Westbury, N.Y.: Feminist Press, 1982).

I had been conceived in the riotous summer and fattened on light and stars that fell on my underground roots, and every herb, corn plant, cricket, beaver, red fox leaped in me in the old Indian dark. I saw everything was moving and entering. The rocking of mother and prairie breast curved around me within the square. The field crows flew in my flesh and cawed in my dream.

Crouching together on Indian land in the long winters, we grew in sight and understanding, heard the rumbling of glacial moraines, clung to the edge of holocaust forest fires, below-zero weather, grasshopper plagues, sin, wars, crop failures, drouth, and the mortgage. The severity of the seasons and the strangeness of a new land, with those whose land had been seized looking in our windows, created a tension of guilt and a tightening of sin. We were often snowed in, the villages invisible and inaccessible in cliffs of snow. People froze following the rope to their barns to feed the cattle. But the cyclic renewal and strength of the old prairie earth, held sacred by thousands of years of Indian ritual, the guerrilla soil of the Americas, taught and nourished us.

We flowed through and into the land, often evicted, drouthed out, pushed west. Some were beckoned to regions of gold, space like a mirage throwing up pictures of utopias, wealth, and villages of brotherhood. Thousands passed through the villages, leaving their dead, deposits of sorrow and calcium, leaching the soil, creating and marking with their faces new wheat and corn, producing idiots, mystics, prophets, and inventors. Or, as an old farmer said, we couldn't move; nailed to the barn door by the wind, we have to make a windmill, figure out how to plow without a horse, and invent barbed wire. A Dakota priest said to me, "It will be from here that the prophets come."

Nowhere in the world can spring burst out of the iron bough as in the Northwest. When the plains, rising to the Rockies, swell with heat, and the delicate glow and silence of the melting moisture fills the pure space with delicate winds and the promise of flowers. We all came, like the crocus, out of the winter dark, out of the captive village where along the river one winter the whole population of children died of diphtheria. In the new sun we counted the dead, and at the spring dance the living danced up a storm and drank and ate heartily for the pain of it. They danced their alien feet into the American earth and rolled in the haymow to beget against the wilderness new pioneers.

All opened in the spring. The prairies, like a great fan, opened. The people warmed, came together in quilting bees, Ladies' Aid meetings, house raisings. The plowing and the planting began as soon as the thaw let the farmers into the fields. Neighbors helped each other. As soon as the seed was in, the churches had picnics and baptizings. The ladies donned their calico dresses and spread a great board of food, while the children ran potato races and one-legged races and the men played horseshoes and baseball. Children were born at home with the neighbor woman. Sometimes the doctor got there. When I was twelve, I helped the midwife deliver a baby. I held onto the screaming mother, her lips bitten nearly off, while she delivered in pieces a dead, strangled corpse. Some people who made it through the winter died in the spring, and we all gathered as survivors to sing "The Old Rugged Cross," "Shall We Gather at the River?" and "God Be with You Till We Meet Again."

The Poles and the Irish had the best parties, lasting for two or three days sometimes. But even the Baptist revival meetings were full of singing (dancing prohibited), and hundreds were forgiven, talking in tongues. Once I saw them break the ice to baptize a screaming woman into the water of life for her salvation.

On Saturday nights everybody would shoot the works, except the prohibitionists and the "good" people, mostly Protestant teetolers who would appear at church on Sunday morning. The frontier gamblers, rascals, and speculators filled the taverns—drink, women, and gambling consuming the wealth of the people and the land. There were gaming palaces for the rich, even horse racing in Stillwater. In St. Paul Nina Clifford, a powerful figure, had two whorehouses, one for gentlemen from "the Hill" and the other for lumberjacks coming in from the woods to spend their hard-earned bucks. It was said that three powers had divided St. Paul among them—Bishop Ireland took "the Hill," Jim Hill took the city for his trains, and Nina Clifford took all that was below "the Hill."

When the corn was "knee-high by the Fourth of July," and the rainfall was good and the sun just right, there was rejoicing in the great Fourth of July picnics that specialized in oratory. Without loudspeakers there were speeches that could be heard the length of the grove, delivered by orators who practiced their wind. When farm prices fell because of the speculation of the Grain Exchange in Minneapolis, the threatened farmers met on the prairie and in the park, the town plaza, and the courthouse to speak out against the power of monopoly. They came for miles, before and after the harvest, in farm wagons with the whole family. They passed out manifestos and spoke of organizing the people to protect themselves from the predators.

There is no place in the world with summer's end, fall harvest, and Indian summer as in Minnesota. They used to have husking bees. The wagons went down the corn rows, and the men with metal knives on their fingers cut the ears off the stalks and tossed them into the wagons. Then they husked the ears, dancing afterward, and if a man got a red ear he could kiss his girl. In August there were great fairs, and the farmers came in to show their crops and beasts, and the workers showed their new reapers and mowers.

There was the excitement of the fall, the terror of the winter coming on. In the winter we didn't have what we did not can, preserve, ferment, or bury in sand. We had to hurry to cut the wood and to get the tomatoes, beans, and piccalilli canned before frost in the garden. It was like preparing for a battle. My grandmother wrapped the apples in newspaper and put them cheek by jowl in the barrels. Cabbage was shredded and barreled for sauerkraut. Even the old hens were killed. I was always surprised to see my gentle grandmother put her foot on the neck of her favorite hen and behead her with a single stroke of a long-handled ax.

The days slowly getting shorter, the herbs hung drying as the woods turned golden. Everything changes on the prairies at the end of summer, all coming to ripeness, and the thunderheads charging in the magnetic moisture of the vast skies. The autumnal dances are the best medicine against the threat of winter, isolation again, dangers. The barns were turned into dance halls before the winter hay was cut. The women raised their long skirts and danced toward hell in schottisches, round dances, and square dances. The rafters rang with the music of the old fiddlers and the harmonica players.

When the golden leaves stacked Persian carpets on the ground and the cornfields were bare, we saw again the great hunched land naked, sometimes fall plowed or planted in winter wheat. Slowly the curve seemed to rise out of the glut of summer, and the earth document was visible script, readable in the human tenderness of risk and ruin.

The owl rides the meadow at his hunting hour. The fox clears out the pheasants and the partridges in the cornfield. Jupiter rests above Antares, and the fall moon hooks itself into the prairie sod. A dark wind flows down from Mandan as the Indians slowly move out of the summer campground to go back to the reservation. Aries, buck of the sky, leaps to the outer rim and mates with earth. Root and seed turn into flesh. We turn back to each other in the dark together, in the short days, in the dangerous cold, on the rim of a perpetual wilderness. *   *   *

RENÉ DUBOS
1901–1982

René Dubos was a microbiologist whose early work led to the commercial development of antibiotics. He was also an internationally known environmentalist and the author of over twenty books, including the Pulitzer Prize–winning So Human an Animal *(1969). Born in St. Brice, France, his concept of mankind's role in nature runs counter to much current American environmental thinking. Reflecting the Catholic agricultural heritage of his native countryside, Dubos challenges accepted beliefs such as that wilderness is the most desirable and inspiring of landscapes, or that human alterations of nature are necessarily destructive.*

A Family of Landscapes

Some of the landscapes that we most admire are the products of environmental degradation. The denuded islands of the Aegean Sea, the rocky shores of the Mediterranean basin, the semidesertic areas of the American Southwest are regions that appeal to countless people from all social and ethnic groups, as well as professional ecologists. Yet these landscapes derive much of their color and sculptural beauty from deforestation and erosion, the two cardinal sins of ecology. The immense majority of people, furthermore, elect to live in places from which the wilderness has been eradicated and which have been profoundly transformed by human habitation. Orthodox ecological criteria are therefore not adequate to evaluate the quality of a particular environment for human life.

Since the humanization of Earth inevitably results in destruction of the wilderness and of many living species that depend on it, there is a fundamental conflict between ecological doctrine and human cultures, a conflict whose manifestations are most glaring in Greece.

The Wooing of Earth (New York: Scribners, 1980).

On two occasions during the past few years, I visited the eleventh-century Byzantine monastery of Moni Kaisarianis, located some five miles southeast of Athens. The monastery is nestled on the slopes of Mount Hymettus at 1,100 feet elevation. A trail meanders from it toward the Hymettus mountain through an almost treeless landscape amidst thyme, lavender, sage, mint, and other aromatic plants. The rock formations of the area are denuded, but the luminous sky gives them an architectural quality particularly bewitching under the violet light of sunset.

A short distance from the monastery, the trail reaches an outcrop of rocks that affords a sudden view of the Acropolis, Mount Lycabettus, and the entire city of Athens. As is so often the case in Greece, the buildings—whether pagan or Christian—derive a dramatic quality independent of their architectural merit from their natural setting. But the landscape surrounding the monastery is not natural; it has been transformed by several thousand years of human occupation.

The grounds associated with the Moni Kaisarianis monastery are planted with almond and olive trees, two species that have long been part of the Greek flora but originated in south central or southeastern Asia. The road that leads from Athens to the monastery is shaded with eucalyptus trees introduced from Australia. Beyond the monastery, the Hymettus is stark and luminous but its rock formations were originally masked by earth and trees. Its bold architecture became clearly visible only during historical times as a result of deforestation and erosion.

Ecologists and historians agree that most of the Mediterranean world was wooded before human occupation. What we now regard as the typical Greek landscape, often stark and treeless, is the result of human activities. The rock structures were revealed only after the felling of the trees, which resulted in extensive erosion. The slopes have been kept denuded by rabbits, sheep, and goats that continuously destroy any new growth either of trees or grass. Erosion and overgrazing are the forces, inadvertently set in motion by human activities, that enable light to play its bewitching game on the white framework of Attica.

The humanization of the Greek wilderness has been achieved at great ecological loss. Writers of the classical, Hellenistic, and Roman periods were aware of the transformations brought about by deforestation in the Mediterranean world. In *Critias*, Plato compared the land of Attica to the "bones of a wasted body . . . the richer and softer parts of the soil having fallen away, and the mere skeleton being left." In ancient times, still according to Plato, the buildings had "roofs of timber cut from trees which were of a size sufficient to cover the largest houses." After deforestation, however, "the mountains only afforded sustenance to bees." The famous Hymettus honey is thus linked to deforestation, which permitted the growth of sun-loving aromatic plants.

As long as the mountain slopes were wooded, the land of Greece, as well as of other Mediterranean countries, was enriched by rainfall, but by Plato's time erosion caused the water to "flow off the bare earth into the sea. . . ." The sacred groves and other sanctuaries were originally established near springs and streams, but these progressively dried up as a consequence of deforestation.

The Ilissus River, which has its source on Mount Hymettus and runs through Athens, was still a lively stream in Plato's time. On a hot day in midsummer, Socrates and Phaedrus walked toward a tall plane tree on the banks of the Ilissus a short distance from the Agora. There, as reported in the famous dialogue, they discussed rhetoric, philosophy, and love while cooling their feet in the stream that they found "delightfully clear and bright." Today, the Ilissus is dry much of the year and, covered by a noisy roadway, serves as a sewer. There could not be a more dramatic symbol of the damage done by deforestation, erosion, and urban mismanagement.

The reforestation of Greece would certainly result in climatic and agricultural improvements. As Henry Miller writes in *The Colossus of Maroussi*, "The tree brings water, fodder, cattle, produce . . . shade, leisure, song. . . . Greece does not need archeologists—she needs arboriculturists." But a cover of trees would make the landscape very different from the image that we, and the Greeks themselves, have had of Greece since classical times. In his poem "The Satyr or the Naked Song," the Greek poet Kostes Palamas (1859–1943) sees in the stark eroded structures of the present landscape a symbol of the austerity and purity of the Greek genius; the landscape triumphantly proclaims the "divine nudity" of Greece. Henry Miller himself, a few pages before and after the passage quoted above in which he advocates reforestation, marvels at the quality given to the landscape by the rocks that "have been lying for centuries exposed to this divine illumination . . . nestling amid dancing colored shrubs in a blood-stained soil." In Miller's words, these rocks "are symbols of life eternal." He does not mention that they are visible only because of deforestation and erosion.

While visiting the Moni Kaisarianis monastery, I noticed a dark opaque zone on the slopes of Mount Hymettus; this area had been reforested with pines. To me, it looked like an inkblot on the luminous landscape, especially at sunset, when the subtle violet atmosphere suffuses the bare rocks throughout the mountain range. The "divine illumination" lost much of its magic where it was absorbed by the pine trees.

The mountains of Attica were probably difficult to penetrate and frightening when completely wooded, but they have now acquired some of the qualities of a park. The traveler can move on their open surfaces, and vision can extend into a distance of golden light. I have wondered whether the dark and ferocious divinities of the preclassical Greek period did not become more serene and more playful precisely because they had emerged from the dark forests into the open landscape. Would logic have flourished if Greece had remained covered with an opaque tangle of trees?

There is no doubt that people spoiled the water economy and impoverished the land when they destroyed the forests of the Mediterranean world. But it is true also that deforestation allowed the landscape to express certain of its potentialities that had remained hidden under the dense vegetation. Not only did removal of the trees permit the growth of sun-loving aromatic plants and favor the spread of honeybees, as Plato had recognized; more importantly, it revealed the underlying architecture of the area and perhaps helped the soaring of the human mind.

The full expression of the Mediterranean genius may require both the cool mysterious fountains in the sacred groves and the bright light shining on the sun-loving plants amid the denuded rocks. Ecology becomes a more complex but far more interesting science when human aspirations are regarded as an integral part of the landscape.

The wide range of natural and humanized environments of which we have knowledge calls to mind the collection of photographs that Edward Steichen published a few years ago under the title *The Family of Man*. In his book, Steichen created a panorama of the human species as it can be found today all over the Earth in its richly diversified types—the unloved and miserable as well as the loved and glorious, those conveying a sad resignation as well as those radiating a challenging beauty. Just as Steichen conceived his collection to be, in his own words, "a mirror of the universal elements and emotions in the everydayness of life," so I shall refer to many different aspects of the Earth, some that evolved under the influence of natural forces and others that have been completely transformed by human activities—whether these were ecologically constructive or destructive.

Many people reject Steichen's view that beauty can be found in all the visages of the family of man, and even more will be disturbed by my statement that deforestation and erosion have produced some of the most admired landscapes. It is obvious, of course, that the Earth and its atmosphere have been spoiled in many places by human carelessness and greed, but I feel nevertheless that almost any kind of scenery, even artificial and desolate, can be a source of interest and pleasure if one knows how to recognize in it patterns of visual organization, aspects of ecological curiosity, and matters of human concern. * * *

NORMAN MACLEAN
1902–1990

It has been remarked that nature writers start later and keep going longer than writers in most other fields. An English professor at the University of Chicago for more than four decades, Norman Maclean published A River Runs Through It and Other Stories in 1976, when he was 73 years old. This volume is a collection of two novellas and a short story closely based on Maclean's experiences growing up and working as a logger and on a U.S. Forest Service crew in the Idaho and Montana mountains during the 1920s. The title story, from which the following passage was excerpted, is a narrative memoir of his family, in which, as he put it, "there was no clear line between religion and fly fishing." The central relationship is between the narrator and his brother Paul, whose troubled life contrasts sharply with his consummate skill as a fly fisherman. The book contains stunning descriptions, not only of the canyon river country but of trout fishing itself, imbued with metaphoric and symbolic resonances reminiscent of Melville's descriptions of whales and whaling. Here, Maclean seems to suggest, in sympathetic contact with nature, we achieve a grace and unity in our lives sadly at odds with our dealings with one another.

From A River Runs Through It

* * * Paul and I fished a good many big rivers, but when one of us referred to "the big river" the other knew it was the Big Blackfoot. It isn't the biggest river we fished, but it is the most powerful, and per pound, so are its fish. It runs straight and hard—on a map or from an airplane it is almost a straight line running due west from its headwaters at Rogers Pass on the Continental Divide to Bonner, Montana, where it empties into the South Fork of the Clark Fork of the Columbia. It runs hard all the way.

Near its headwaters on the Continental Divide there is a mine with a thermometer that stopped at 69.7 degrees below zero, the lowest temperature ever officially recorded in the United States (Alaska omitted). From its headwaters to its mouth it was manufactured by glaciers. The first sixty-five miles of it are smashed against the southern wall of its valley by glaciers that moved in from the north, scarifying the earth; its lower twenty-five miles were made overnight when the great glacial lake covering northwestern Montana and northern Idaho broke its ice dam and spread the remains of Montana and Idaho mountains over hundreds of miles of the plains of eastern Washington. It was the biggest flood in the world for which there is geological evidence; it was so vast a geological event that the mind of man could only conceive of it but could not prove it until photographs could be taken from earth satellites.

The straight line on the map also suggests its glacial origins; it has no meandering valley, and its few farms are mostly on its southern tributaries which were not ripped up by glaciers; instead of opening into a wide flood plain near its mouth, the valley, which was cut overnight by a disappearing lake when the great ice dam melted, gets narrower and narrower until the only way a river, an old logging railroad, and an automobile road can fit into it is for two of them to take to the mountainsides.

It is a tough place for a trout to live—the river roars and the water is too fast to let algae grow on the rocks for feed, so there is no fat on the fish, which must hold most trout records for high jumping.

Besides, it is the river we knew best. My brother and I had fished the Big Blackfoot since nearly the beginning of the century—my father before then. We regarded it as a family river, as a part of us, and I surrender it now only with great reluctance to dude ranches, the unselected inhabitants of Great Falls, and the Moorish invaders from California.

A River Runs Through It and Other Stories (Chicago: University of Chicago Press, 1976).

Early next morning Paul picked me up in Wolf Creek, and we drove across Rogers Pass where the thermometer is that stuck at three-tenths of a degree short of seventy below. As usual, especially if it were early in the morning, we sat silently respectful until we passed the big Divide, but started talking the moment we thought we were draining into another ocean. Paul nearly always had a story to tell in which he was the leading character but not the hero.

He told his Continental Divide stories in a seemingly light-hearted, slightly poetical mood such as reporters often use in writing "human-interest" stories, but, if the mood were removed, his stories would appear as something about him that would not meet the approval of his family and that I would probably find out about in time anyway. He also must have felt honor-bound to tell me that he lived other lives, even if he presented them to me as puzzles in the form of funny stories. Often I did not know what I had been told about him as we crossed the divide between our two worlds.

"You know," he began, "it's been a couple of weeks since I fished the Blackfoot." At the beginning, his stories sounded like factual reporting. He had fished alone and the fishing had not been much good, so he had to fish until evening to get his limit. Since he was returning directly to Helene he was driving up Nevada Creek along an old dirt road that followed section lines and turned at right angles at section corners. It was moonlight, he was tired and feeling in need of a friend to keep him awake, when suddenly a jackrabbit jumped on to the road and started running with the headlights. "I didn't push him too hard," he said, "because I didn't want to lose a friend." He drove, he said, with his head outside the window so he could feel close to the rabbit. With his head in the moonlight, his account took on poetic touches. The vague world of moonlight was pierced by the intense white triangle from the headlights. In the center of the penetrating isosceles was the jackrabbit, which, except for the length of his jumps, had become a snowshoe rabbit. The phosphorescent jackrabbit was doing his best to keep in the center of the isosceles but was afraid he was losing ground and, when he looked back to check, his eyes shone with whites and blues gathered up from the universe. My brother said, "I don't know how to explain what happened next, but there was a right-angle turn in this section-line road, and the rabbit saw it, and I didn't."

Later, he happened to mention that it cost him $175.00 to have his car fixed, and in 1937 you could almost get a car rebuilt for $175.00. Of course, he never mentioned that, although he did not drink when he fished, he always started drinking when he finished.

I rode part of the way down the Blackfoot wondering whether I had been told a little human-interest story with hard luck turned into humor or whether I had been told he had taken too many drinks and smashed hell out of the front end of his car.

Since it was no great thing either way, I finally decided to forget it, and, as you see, I didn't. I did, though, start thinking about the canyon where we were going to fish.

The canyon above the old Clearwater bridge is where the Blackfoot roars loudest. The backbone of a mountain would not break, so the mountain compresses the already powerful river into sound and spray before letting it pass. Here, of course, the road leaves the river; there was no place in the canyon for an Indian trail; even in 1806 when Lewis left Clark to come up the Blackfoot, he skirted the canyon by a safe margin. It is no place for small fish or small fishermen. Even the roar adds power to the fish or at least intimidates the fisherman.

When we fished the canyon we fished on the same side of it for the simple reason that there is no place in the canyon to wade across. I could hear Paul start to pass me to get to the hole above, and, when I realized I didn't hear him anymore, I knew he had stopped to watch me. Although I have never pretended to be a great fisherman, it was always important to me that I was a fisherman and looked like one, especially when fishing with my brother. Even before the silence continued, I knew that I wasn't looking like much of anything.

Although I have a warm personal feeling for the canyon, it is not an ideal place for me to fish. It puts a premium upon being able to cast for distance, and yet most of the time there are cliffs or trees right behind the fisherman so he has to keep all his line in front of him. It's like a baseball pitcher being deprived of his windup, and it forces the fly fisherman into what is called a "roll cast," a hard cast that I have never mastered. The fisherman has to work enough line into his cast to get distance without throwing any line behind him, and then he has to develop enough power from a short arc to shoot it out across the water.

He starts accumulating the extra amount of line for the long cast by retrieving his last cast so slowly that an unusual amount of line stays in the water and what is out of it forms a slack semiloop. The loop is enlarged by raising the casting arm straight up and cocking the wrist until it points to 1:30. There, then, is a lot of line in front of the fisherman, but it takes about everything he has to get it high in the air and out over the water so that the fly and leader settle ahead of the line—the arm is a piston, the wrist is a revolver that uncocks, and even the body gets behind the punch. Important, too, is the fact that the extra amount of line remaining in the water until the last moment gives a semisolid bottom to the cast. It is a little like a rattlesnake striking, with a good piece of his tail on the ground as something to strike from. All this is easy for a rattlesnake, but has always been hard for me.

Paul knew how I felt about my fishing and was careful not to seem superior by offering advice, but he had watched so long that he couldn't leave now without saying something. Finally he said, "The fish are out farther." Probably fearing he had put a strain on family relations, he quickly added, "Just a little farther."

I reeled in my line slowly, not looking behind so as not to see him. Maybe he was sorry he had spoken, but, having said what he said, he had to say something more. "Instead of retrieving the line straight toward you, bring it in on a diagonal from the downstream side. The diagonal will give you a more resistant base to your loop so you can put more power into your forward cast and get a little more distance."

Then he acted as if he hadn't said anything and I acted as if I hadn't heard it, but as soon as he left, which was immediately, I started retrieving my line on a diagonal, and it helped. The moment I felt I was getting a little more distance I ran for a fresh hole to make a fresh start in life.

It was a beautiful stretch of water, either to a fisherman or a photographer, although each would have focused his equipment on a different point. It was a barely submerged waterfall. The reef of rock was about two feet under the water, so the whole river rose into one wave, shook itself into spray, then fell back on itself and turned blue. After it recovered from the shock, it came back to see how it had fallen.

No fish could live out there where the river exploded into the colors and curves that would attract photographers. The fish were in that slow backwash, right in the dirty foam, with the dirt being one of the chief attractions. Part of the speckles would be pollen from pine trees, but most of the dirt was edible insect life that had not survived the waterfall.

I studied the situation. Although maybe I had just added three feet to my roll cast, I still had to do a lot of thinking before casting to compensate for some of my other shortcomings. But I felt I had already made the right beginning—I had already figured out where the big fish would be and why.

Then an odd thing happened. I saw him. A black back rose and sank in the foam. In fact, I imagined I saw spines on his dorsal fin until I said to myself, "God, he couldn't be so big you could see his fins." I even added, "You wouldn't even have seen the fish in all that foam if you hadn't first thought he would be there." But I couldn't shake the conviction that I had seen the black back of a big fish, because, as someone often forced to think, I know that often I would not see a thing unless I thought of it first.

Seeing the fish that I first thought would be there led me to wondering which way he would be pointing in the river. "Remember, when you make the first cast," I thought, "that you saw him in the backwash where the water is circling upstream, so he will be looking downstream, not upstream, as he would be if he were in the main current."

I was led by association to the question of what fly I would cast, and to the conclusion that it had better be a large fly, a number four or six, if I was going after the big hump in the foam.

From the fly, I went to the other end of the cast, and asked myself where the hell I was going to cast from. There were only gigantic rocks at this waterfall, so I picked one of the biggest, saw how I could crawl up it, and knew from that added height I would get added distance, but then I had to ask myself, "How the hell am I going to land the fish if I hook him while I'm standing up there?" So I had to pick a smaller rock, which would shorten my distance but would let me slide down it with a rod in my hand and a big fish on.

I was gradually approaching the question all river fishermen should ask before they make the first cast, "If I hook a big one, where the hell can I land him?"

One great thing about fly fishing is that after a while nothing exists of the world but thoughts about fly fishing. It is also interesting that thoughts about fishing are often carried on in dialogue form where Hope and Fear—or, many times, two Fears—try to outweigh each other.

One fear looked down the shoreline and said to me (a third person distinct from the two fears), "There is nothing but rocks for thirty yards, but don't get scared and try to land him before you get all the way down to the first sandbar."

The Second Fear said, "It's forty, not thirty, yards to the first sandbar and the weather has been warm and the fish's mouth will be soft and he will work off the hook if you try to fight him forty yards downriver. It's not good but it will be best to try to land him on a rock that is closer."

The First Fear said, "There is a big rock in the river that you will have to take him past before you land him, but, if you hold the line tight enough on him to keep him this side of the rock, you will probably lose him."

The Second Fear said, "But if you let him get on the far side of the rock, the line will get caught under it, and you will be sure to lose him."

That's how you know when you have thought too much—when you become a dialogue between *You'll probably lose* and *You're sure to lose.* But I didn't entirely quit thinking, although I did switch subjects. It is not in the book, yet it is human enough to spend a moment before casting in trying to imagine what the fish is thinking, even if one of its eggs is as big as its brain and even if, when you swim underwater, it is hard to imagine that a fish has anything to think about. Still, I could never be talked into believing that all a fish knows is hunger and fear. I have tried to feel nothing but hunger and fear and don't see how a fish could ever grow to six inches if that were all he ever felt. In fact, I go so far sometimes as to imagine that a fish thinks pretty thoughts. Before I made the cast, I imagined the fish with the black back lying cool in the carbonated water full of bubbles from the waterfalls. He was looking downriver and watching the foam with food in it backing upstream like a floating cafeteria coming to wait on its customers. And he probably was imagining that the speckled foam was eggnog with nutmeg sprinkled on it, and, when the whites of eggs separated and he saw what was on shore, he probably said to himself, "What a lucky son of a bitch I am that this guy and not his brother is about to fish this hole."

I thought all these thoughts and some besides that proved of no value, and then I cast and I caught him.

I kept cool until I tried to take the hook out of his mouth. He was lying covered with sand on the little bar where I had landed him. His gills opened with his penultimate sighs. Then suddenly he stood up on his head in the sand and hit me with his tail and the sand flew. Slowly at first my hands began to shake, and, although I thought they made a miserable sight, I couldn't stop them. Finally, I managed to open the large blade to my knife which several times slid off his skull before it went through his brain.

Even when I bent him he was way too long for my basket, so his tail stuck out.

There were black spots on him that looked like crustaceans. He seemed oceanic, including barnacles. When I passed my brother at the next hole, I saw him study the tail and slowly remove his hat, and not out of respect to my prowess as a fisherman.

I had a fish, so I sat down to watch a fisherman.

He took his cigarettes and matches from his shirt pocket and put them in his hat and pulled his hat down tight so it wouldn't leak. Then he unstrapped his fish basket and hung it on the edge of his shoulder where he could get rid of it quick should the water get too big for him. If he studied the situation he didn't take any separate time to do it. He jumped off a rock into the swirl and swam for a chunk of cliff that had dropped into the river and parted it. He swam in his clothes with only his left arm—in his right hand, he held his rod high and sometimes all I could see was the basket and rod, and when the basket filled with water sometimes all I could see was the rod.

The current smashed him into the chunk of cliff and it must have hurt, but he had enough strength remaining in his left fingers to hang to a crevice or he would have been swept into the blue below. Then he still had to climb to the top of the rock with his left fingers and his right elbow which he used like a prospector's pick. When he finally stood on top, his clothes looked hydraulic, as if they were running off him.

Once he quit wobbling, he shook himself duck-dog fashion, with his feet spread apart, his body lowered and his head flopping. Then he steadied himself and began to cast and the whole world turned to water.

Below him was the multitudinous river, and, where the rock had parted it around him, big-grained vapor rose. The mini-molecules of water left in the wake of his line made momentary loops of gossamer, disappearing so rapidly in the rising big-grained vapor that they had to be retained in memory to be visualized as loops. The spray emanating from him was finer-grained still and enclosed him in a halo of himself. The halo of himself was always there and always disappearing, as if he were candlelight flickering about three inches from himself. The images of himself and his line kept disappearing into the rising vapors of the river, which continually circled to the tops of the cliffs where, after becoming a wreath in the wind, they became rays of the sun.

The river above and below his rock was all big Rainbow water, and he would cast hard and low upstream, skimming the water with his fly but never letting it touch. Then he would pivot, reverse his line in a great oval above his head, and drive his line low and hard downstream, again skimming the water with his fly. He would complete this grand circle four or five times, creating an immensity of motion which culminated in nothing if you did not know, even if you could not see, that now somewhere out there a small fly was washing itself on a wave. Shockingly, immensity would return as the Big Blackfoot and the air above it became iridescent with the arched sides of a great Rainbow.

He called this "shadow casting," and frankly I don't know whether to believe the theory behind it—that the fish are alerted by the shadows of flies passing over the water by the first casts, so hit the fly the moment it touches the water. It is more or less the "working up an appetite" theory, almost too fancy to be true, but then every fine fisherman has a few fancy stunts that work for him and for almost no one else. Shadow casting never worked for me, but maybe I never had the strength of arm and wrist to keep line circling over the water until fish imagined a hatch of flies was out.

My brother's wet clothes made it easy to see his strength. Most great casters I have known were big men over six feet, the added height certainly making it easier to get more line in the air in a bigger arc. My brother was only five feet ten, but he had fished so many years his body had become partly shaped by his casting. He was thirty-two now, at the height of his power, and he could put all his body and soul into a four-and-a-half-ounce magic totem pole. Long ago, he had gone far beyond my father's wrist casting, although his right wrist was always so important that it had become larger than his left. His right arm, which our father had kept tied to the side to emphasize the wrist, shot out of his shirt as if it were engineered, and it, too, was larger than his left arm. His wet shirt bulged and came unbuttoned with his pivoting shoulders and hips. It was also not hard to see why he was a street fighter, especially since he was committed to getting in the first punch with his right hand.

Rhythm was just as important as color and just as complicated. It was one rhythm superimposed upon another, our father's four-count rhythm of the line and wrist being still the base rhythm. But superimposed upon it was the piston two count of his arm and the long overriding four count of the completed figure eight of his reversed loop.

The canyon was glorified by rhythms and colors. * * *

JOHN STEINBECK
1902–1968

Steinbeck's novels are rooted in the landscape of California, particularly in the farming and fishing communities of the Salinas Valley and the Carmel Peninsula. One early novel, To a God Unknown *(1933) is essentially a modern retelling of a pagan fertility myth. In 1940 Steinbeck undertook a research voyage to the Gulf of Mexico with his close friend, the marine biologist Edward F. Ricketts. Their joint account of that voyage was published in* The Sea of Cortez *(1941). Steinbeck's portion was republished in 1951 as* The Log from the Sea of Cortez; *its introduction is one of the most concise statements ever made on the difference between the scientific and personal approaches to nature writing.*

From THE LOG FROM THE SEA OF CORTEZ

INTRODUCTION

The design of a book is the pattern of a reality controlled and shaped by the mind of the writer. This is completely understood about poetry or fiction, but it is too seldom realized about books of fact. And yet the impulse which drives a man to poetry will send another man into the tide pools and force him to try to report what he finds there. Why is an expedition to Tibet undertaken, or a sea bottom dredged? Why do men, sitting at the microscope, examine the calcareous plates of a sea-cucumber, and, finding a new arrangement and number, feel an exaltation and give the new species a name, and write about it possessively? It would be good to know the impulse truly, not to be confused by the "services to science" platitudes or the other little mazes into which we entice our minds so that they will not know what we are doing.

We have a book to write about the Gulf of California. We could do one of several things about its design. But we have decided to let it form itself: its boundaries a boat and a sea; its duration a six weeks' charter time; its subject everything we could see and think and even imagine; its limits — our own without reservation.

We made a trip into the Gulf; sometimes we dignified it by calling it an expedition. Once it was called the Sea of Cortez, and that is a better-sounding and a more exciting name. We stopped in many little harbors and near barren coasts to collect and preserve the marine invertebrates of the littoral. One of the reasons we gave ourselves for this trip — and when we used this reason, we called the trip an expedition — was to observe the distribution of invertebrates, to see and to record their kinds and numbers, how they lived together, what they ate, and how they reproduced. That plan was simple, straightforward, and only a part of the truth. But we did tell the truth to ourselves. We were curious. Our curiosity was not limited, but was as wide and horizonless as that of Darwin or Agassiz or Linnaeus or Pliny. We wanted to see everything our eyes would accommodate, to think what we could, and, out of our seeing and thinking, to build some kind of structure in modeled imitation of the observed reality. We knew that what we would see and record and construct would be warped, as all knowledge patterns are warped, first, by the collective pressure and stream of our time and race, second by the thrust of our individual personalities. But knowing

The Log from the Sea of Cortez (New York: Viking, 1951).

this, we might not fall into too many holes—we might maintain some balance between our warp and the separate things, the external reality. The oneness of these two might take its contribution from both. For example: the Mexican sierra has "XVII–15–IX" spines in the dorsal fin. These can easily be counted. But if the sierra strikes hard on the line so that our hands are burned, if the fish sounds and nearly escapes and finally comes in over the rail, his colors pulsing and his tail beating the air, a whole new relational externality has come into being—an entity which is more than the sum of the fish plus the fisherman. The only way to count the spines of the sierra unaffected by this second relational reality is to sit in a laboratory, open an evil-smelling jar, remove a stiff colorless fish from formalin solution, count the spines, and write the truth "D. XVII–15–IX." There you have recorded a reality which cannot be assailed—probably the least important reality concerning either the fish or yourself.

It is good to know what you are doing. The man with his pickled fish has set down one truth and has recorded in his experience many lies. The fish is not that color, that texture, that dead, nor does he smell that way.

Such things we had considered in the months of planning our expedition and we were determined not to let a passion for unassailable little truths draw in the horizons and crowd the sky down on us. We knew that what seemed to us true could be only relatively true anyway. There is no other kind of observation. The man with his pickled fish has sacrificed a great observation about himself, the fish, and the focal point, which is his thought on both the sierra and himself.

We suppose this was the mental provisioning of our expedition. We said, "Let's go wide open. Let's see what we see, record what we find, and not fool ourselves with conventional scientific strictures. We could not observe a completely objective Sea of Cortez anyway, for in that lonely and uninhabited Gulf our boat and ourselves would change it the moment we entered. By going there, we would bring a new factor to the Gulf. Let us consider that factor and not be betrayed by this myth of permanent objective reality. If it exists at all, it is only available in pickled tatters or in distorted flashes. Let us go," we said, "into the Sea of Cortez, realizing that we become forever a part of it; that our rubber boots slogging through a flat of eelgrass, that the rocks we turn over in a tide pool, make us truly and permanently a factor in the ecology of the region. We shall take something away from it, but we shall leave something too." And if we seem a small factor in a huge pattern, nevertheless it is of relative importance. We take a tiny colony of soft corals from a rock in a little water world. And that isn't terribly important to the tide pool. Fifty miles away the Japanese shrimp boats are dredging with overlapping scoops, bringing up tons of shrimps, rapidly destroying the species so that it may never come back, and with the species destroying the ecological balance of the whole region. That isn't very important in the world. And thousands of miles away the great bombs are falling and the stars are not moved thereby. None of it is important or all of it is.

We determined to go doubly open so that in the end we could, if we wished, describe the sierra thus: "D. XVII–15–IX; A. II–15–IX," but also we could see the fish alive and swimming, feel it plunge against the lines, drag it threshing over the rail, and even finally eat it. And there is no reason why either approach should be inaccurate. Spine-count description need not suffer because another approach is also used. Perhaps out of the two approaches, we thought, there might emerge a picture more complete and even more accurate than either alone could produce. And so we went.

LAURENS VAN DER POST
1906–1996

The range of Laurens Van Der Post's writing reflects his many careers as farmer, explorer, soldier, anthropologist, and naturalist. Following his childhood and youth in South Africa, he served the British government in various capacities, in such places as North Africa and Java. Van Der Post's voice is that of a seasoned, cosmopolitan writer, but is also alert to life's overtones of symbolism and mystery. The Lost World of the Kalahari (1958) and The Heart of the Hunter (1961) are works of sympathetic participation in the vision of the Kalahari Bushmen. Nature for Van Der Post, as for these endangered hunting cultures, is a holy world of quest, sacrifice, and rebirth.

From THE HEART OF THE HUNTER

A BRAVERY OF BIRDS

From Tsane we travelled fast eastwards towards the great pans of Kukong and Kakia along one of the oldest tracks across the desert. At one place where there was unexpectedly good shade in which we paused to rest at noon, Ben said, "You know, I came by here first as a boy of eight leading the oxen in my father's front wagon. We camped here, and almost at once some Bushmen came to see us. They had their shelters only a hundred yards from here, and came to offer us skins of game in exchange for tobacco. I passed this spot many times afterwards and always they reappeared for the same purpose, friendly, gay and excited. Ten years ago I came by again, but they had vanished. I was afraid they had died in the outbreak of pest which raged just then in the Kalahari, until I met an old Bushman I knew at the well at Kukong. I asked him, because you know it is amazing how they pass on news of one another from end to end of the desert. He seemed astonished that I had not heard of so shocking an event. Did I not know, he asked, that they had heard people passing along the track one day and rushed out as they always did to barter their skins? Well, the government was travelling by, and when the police with the government saw that among the skins there was one of a gemsbuck, they took the two best young hunters away with them. The remaining people immediately moved right away from the tracks. They were safe all right, but their hearts were troubled for the two young men who had not come back, as the hearts of all the people had been wide open to the young men. As you see, the Bushmen are not back on their ancient stand and will, I fear, never be again."

"But why did the police do that?" someone asked.

"Because the gemsbuck is royal game here and protected," Ben said, his voice edged with the irony of it, "and the Bushman is not."

"What happened to the two young hunters? Do you know?" I asked.

He shook his head sadly, saying he could only guess they had never returned because once such essentially innocent people had been punished and imprisoned they became so deeply confused that they seldom found the way back.

The same evening, camp made earlier than usual because of good progress that day, Ben and I took our guns to see if we could not shoot a

The Heart of the Hunter (New York: Morrow, 1961).

springbuck for our supper and breakfast. We found some in a pan nearby, but they were on the other side, too far for us to stalk effectively before dark. The moment we set foot on the floor of the pan they were aware of us, stopped grazing and stood, heads up, to watch us apprehensively. So we started to climb back on to the ridge overlooking the pan with the idea of sitting down there for a while to enjoy the stillness and beauty of the evening.

The end of the journey was very near now. It is remarkable how a sense of valediction heightens one's awareness of the beauty of the world. I think it is because beauty is a summons to journey, is both a hail and a farewell of the spirit, and since our deepest pattern is a round of departure and return, we never recognize it more clearly than at the beginning and end of our journeys. Indeed all the traffic and the travail in between may be directed just to that end. Besides, when one has lived as close to nature for as long as we had done, one is not tempted to commit the metropolitan error of assuming that the sun rises and sets, the day burns out and the night falls, in a world outside oneself. These are great and reciprocal events, which occur also in ourselves. In this moment of heightened sensibility, there on the lip of the pan, I was convinced that, just as the evening was happening in us, so were we in it, and the music of our participation in a single overwhelming event was flowing through us.

This sense of participation enclosed in one moment of time was increased by the presence of the pan itself. From the ridge where we seated ourselves we had an immense view of the desert. In that light it looked in terms of earth what the sea is in terms of water, without permanent form and without end. Then suddenly there was the pan at our feet, a shape which was definite and real, the ridge describing an almost perfect circle against the sky and presenting the waste around it with a flawless container. It was a geometrical paradigm of life's need for form, a demonstration of the proposition that unless life were contained it could not be. I remembered the excitement I felt on first seeing an Etruscan vase, perceiving why the summons for a renewal of the European spirit had to emerge in men's imagination as a vision of a greater container in the shape of the Holy Grail. But I got no further, for just then Ben interrupted. He asked, in a voice so much affected by the mood of the moment that it was barely more than a whisper, "Do you remember this place?"

"I do, Ben." I recognized it as country we had passed through several times before, but no more. Knowing from his tone he had something particular in mind, I said, "Why did you ask?"

"Because of the ostriches. We saw them first just by those bushes there. Surely you remember them?" He sounded somewhat disappointed at my vague response.

But I had it now! The pan was the scene of one of the loveliest deeds I have ever seen. Nearly ten years ago, Ben and I had come fast over the ridge one evening and without pause slipped down the side silently in our truck. As a result we surprised what looked like a lone couple of ostriches on the edge of the floor of the pan. The birds had panicked; they circled each other wildly until a clear design of action emerged. To our amazement the female broke away from her mate and came resolutely towards us. Ben, who was driving, halted the truck at once and said, "Please, don't move! Just watch this."

Knowing how ostriches hate and fear men, I do not think I have ever seen a braver deed. The bird was desperately afraid. Her heart beating visibly in her throat, she advanced towards us like a soldier against a machine-gun post. With the late afternoon sun making a halo round her feathers, which stood erect with the fearful tension in her, she came on pretending to be mortally hurt, limping badly and trailing one great wing as if it were broken. Then, trying to give the impression that she had only just seen us, she stopped, whisked about and skipped with broken steps sideways into the bush. Before she had gone far, however, she halted. The one great wing sagged more than ever, giving her a list like a ship about to founder. She looked fearfully over her shoulder to see if we were following her. When she saw we had not done so, she appeared baffled and dismayed, and once more came back towards us to repeat the performance, this time so close that the trailing wing nearly touched the bumper of the truck.

Meanwhile the male, in the shining black dress of a bountiful summer, hurried the other way in a zigzag fashion like a ship tacking into the wind. He would rush off in a few giant strides, stop, lower his head, flap his wings, look up to see how the female was getting on, then run off on the other tack again. When his rushes had presently taken him into a bare patch of sand higher up on the ridge, we saw the cause of it all: the male was trying to hustle out of danger nineteen little ostrich chickens, while the female distracted our attention by doing all she could to entice us into capturing her instead. The chickens were so new that the sheen of the yolk of the eggs from which they had been hatched was like silk upon them. All the time they were within sight their mother became increasingly reckless in her efforts to draw us away, and once she looked truly tragic with despair because we would not follow her. Not until her family was out of sight did she desist: even then she did not hasten to join them, but with her wing still trailing drew away from us in the opposite direction. How could I not remember?

I looked at the place where it had all happened, and in the light of the memory it looked like hallowed ground. I think Ben felt something similar, because he began speaking to me with a release of emotion he rarely allowed himself. And I report his words here in full because they helped me greatly later in understanding the imagery of birds in the first spirit of things.

People, he said, often asked him which of all the creatures encountered in his many years as a hunter and dweller, in far-away places of Africa, he found most impressive. Always he answered that it would have to be a bird of some kind. This never failed to surprise them, because people are apt to be dazzled by physical power, size, frightfulness, and they expected him to say an elephant, lion, buffalo or some other imposing animal. But he stuck to his answer; there was nothing more wonderful in Africa than its birds. I asked why precisely. He paused and drew a circle with his finger in the red sand in front of him before saying that it was for many reasons, but in the first place because birds flew. He said it in such a way that I felt I had never before experienced fully the wonder of birds flying.

I waited silently for him to find the next link in his chain of thought. In the second place, he remarked, because birds sang. He himself loved all natural sounds in the bush and the desert, but he had to admit none equalled the sounds of birds. It was as if the sky made music in their throats and one could hear the sun rise and set, the night fall and the first stars come out in their voices. Other animals were condemned to make only such noises as they must, but birds seemed free to utter the sounds they wanted to, to shape them at will and invent new ones to express all the emotions of living matter released on wings from its own dead weight. He knew of nothing so beautiful as the sight of a bird utterly abandoned to its song, every bit of its being surrendered to the music, the tip of the tiniest feather trembling like a tuning fork with sound. Sometimes too, birds danced to their own music. And they not only sang. They also conversed. There appeared to be little they could not convey to one another by sound. He himself had always listened with the greatest care to bird sound and never ceased to marvel at the variety of intelligence it conveyed to him.

Stranger still was their capacity of being aware of things before they happened. This was positively amazing. When the great earth tremor shook the northern Kalahari some years before, Ben was travelling with a herd of cattle along the fringes of the Okovango swamp. One day he was watching some old-fashioned storks, sacred ibis and giant herons along the edges of a stream. Suddenly the birds stopped feeding, looked uneasily about them, and then all at once took to their wings as if obedient to a single command. They rose quickly in the air and began wheeling over the river, making the strangest sounds. The sound had not fallen long on the still air before the ground under his feet started to shake, the cattle to bellow and run, and as far as his eyes could see the banks of the stream began to break away from the bush, as if sliced from it by a knife, and to collapse into the water. He had no doubt the birds knew what was coming, and he made a careful note of their behaviour and the sound they uttered.

Even more wonderful, however, was their beauty. Colour, for instance, lovely as it was in most animals, served the latter only for camouflage. But with birds it was much more. Of all the creatures, none dressed so well as the birds of Africa. They had summer and winter dresses, special silks for making love, coats and skirts for travel, and more practical clothes that did not show the dirt and wear and tear of domestic use. Even the soberest ones among them, which went about the country austere as elders of the Dutch Reformed Church collecting from parsimonious congregations on Sunday mornings—the old-fashioned storks in black and white, or the secretary birds with their stiff starched fronts and frock coats—their dress was always of an impeccable taste.

This beauty and good taste did not stop at dress. It showed in the building of their nests: no animals could rival the diversity and elegance of the home birds made for themselves. The worst builders were the carnivorous ones—the toughy-pants like the lamb-catchers, Batteleurs and white-breasted jackal birds. Just as the warrior races in Africa built the worst huts in the land, so did the fighting birds make the ugliest homes; but on the whole the nests of the birds were things of beauty and joy.

Most wonderful of all was the way this beauty appeared in their eggs—shaped and painted as if by an artist. Had I ever seen a bird's egg that was ugly? Compare these lovely speckled, dappled or sky-blue surfaces, slightly milky as if veiled by a remote cirrhus cloud, to the eggs laid by snakes, turtles and crocodiles. Even the drabbest of eggs laid by a bird was beautiful in comparison. Ben himself was always excited when a mere farmyard hen produced eggs with a gipsy tan and tiny sun-freckles on the shell. Again, birds collaborated, in forethought and purpose, with other living creatures. There was the bird that picked the crocodile's teeth clean for him; the bird that rode the rhinoceros, feeding on the parasites that troubled him and warning him of danger in return for his hospitality; the egret who did the same for cattle and buffalo; and perhaps the greatest of them all, the honey-diviner who, as I had seen for myself, even co-operated with the universally feared and mistrusted man.

Finally there was their quality of courage. I had witnessed an example of it that day years ago in a female ostrich; but all birds had it.

When one considered what tender, small, delicate and defenceless things most birds were, they were perhaps the bravest creatures in the world. He had seen far more moving instances of the courage of the birds of Africa than he could possibly relate, but he would mention only one of the most common—birds defending their nests against snakes. On those occasions they had a rallying cry, which was a mixture of faith and courage just keeping ahead of despair and fear. It would draw birds from all around to the point of danger, and the recklessness with which one little feathered body after another would hurl itself at the head of a snake, beating with its wings and shrieking its Valkyrian cry, had to be seen to be believed. Ben once saw a black mamba driven dazed out of a tree by only a score or so of resolute little birds. The mamba, which he killed, measured close on ten feet, and this snake is itself a creature of fiery courage and determination. No, all in all, he had no doubt that birds were the most wonderful of all living things.

Ben paused, and motioned to me to listen. The first of the night plover was calling from the far end of the pan, a long sort of wail like a ship's pipe mustering her crew to take her out to sea. It was nearly dark. I had not noticed the quick flight of time, so absorbed had I been in listening. When the plover's call died away, Ben jumped to his feet with a cat-like ease that never failed to astonish me in so big a man and asked with one of his rare smiles, had I not heard the referee's whistle? Light, he said, had stopped play. The game was over and it was time to get back to camp.

T(ERENCE) H(ANBURY) WHITE
1906–1964

White's novels based on the Arthurian legends were collected in 1958 into The Once and Future King. *During much of the research and writing for this major project he lived as a recluse, first in Ireland and then on the Channel Island of Alderney. While he sometimes preferred to isolate himself from human society, White responded intensely and lovingly to animals. The* Goshawk *(1951) conveyed his enduring interest in falcons and falconry. The selection here on snakes comes from* England Have My Bones *(1936).*

THE SNAKES ARE ABOUT

The snakes are about again. Last year I used to go out with Hughesdon to catch them, and then turn them loose in the sitting-room. At one time I had about a dozen. There are four in the room just now.

Grass snakes are fascinating pets. It is impossible to impose upon them, or to steal their affections, or to degrade either party in any way. They are always inevitably themselves, and with a separate silurian beauty. The plates of the jaw are fixed in an antediluvian irony. They move with silence, unless in crackling grass or with a scaly rustle over a wooden floor, pouring themselves over obstacles and round them. They are inquisitive. They live loose in the room, except that I lock them up at nights so that the maids can clean in the mornings without being frightened. The big open fireplace is full of moss and ferns, and there is an aquarium full of water in which they can soak themselves if they wish. But mostly they prefer to lie under the hot pipes of the radiator, or to burrow inside the sofa. We had to perform a Caesarian operation on the sofa last year, to get out a big male.

It is nice to come into the room and look quickly round it, to see what they are doing. Perhaps there is one behind Aldous Huxley on the bookshelves, and it is always worth moving the left-hand settle away from the wall. One of them has a passion for this place and generally falls out. Another meditates all day in the aquarium, and the fourth lives in the moss.

Or it is nice to be working in the arm-chair, and to look up suddenly at an imagined sound. A female is pouring from behind the sofa. As the floor is of polished wood she gets a poor grip on it (she prefers the sheepskin hearth-rug) and elects to decant herself along the angle between wall and floor. Here she can press sideways as well as downwards, and gets a better grip.

She saw our movement as we looked up, and now stops dead, her head raised in curiosity. Her perfect forked tongue flickers blackly out of its specially armoured hole (like the hole for the starting handle in a motor, but constructed so as to close itself when not in use) and waves itself like lightning in our direction. It is what she feels with in front of her, her testing antennae, and this is her mark of interrogation. An empathic movement: she can't reach us, but she is thinking Who or What? And so the tongue comes out. We sit quite still.

The tongue comes out two or three times (its touch on the hand is as delicate as the touch of a butterfly) and flickers in the air. It is a

England Have My Bones (New York: Macmillan Company, 1936).

beautiful movement, with more down in it than up. It can be faintly
reproduced by waggling the bent forefinger quickly in a vertical plane.
Then she goes on with her pour, satisfied, towards her objective in the
moss. We sit as still as a mouse.

I try to handle these creatures as little as possible. I do not want to
steal them from themselves by making them pets. The exchange of
hearts would degrade both of us. It is only that they are nice. Nice to
see the strange wild things loose, living their ancient unpredictable
lives with such grace. They are more ancient than the mammoth, and
infinitely more beautiful. They are dry, cool and strong. The fitting and
variation of the plates, the lovely colouring, the movement, their few
thoughts: one could meditate upon them like a jeweller for months.

It is exciting to catch them. You go to a good wood, and look for
snaky places in it. It is difficult to define these. There has to be undergrowth, but not overgrowth: a sunny patch, a glade or tiny clearing in
the trees: perhaps long grass and a bit of moss, but not too wet. You go
into it and there is a rustle. You can see nothing, but dive straight at the
sound. You see just a few inches of the back, deceptively fluid for
catching hold of, as it flashes from side to side. You must pounce on it
at once, for there is no time to think, holding it down or grabbing it by
head or tail or anywhere. There is no time to select. This is always
exciting to me, because I frighten myself by thinking that it might be
an adder. As a matter of fact, there are very few adders in the Shire, and
in any case they move differently. An adder would strike back at you, I
suppose, but a grass snake does not. It pretends to strike, with mouth
wide open and the most formidable-looking fangs; but it stops its head
within a millimetre of the threatened spot, a piece of bluff merely.

When you have grabbed your snake, you pick it up. Instantly it
curls round your hand and arm, hissing and lunging at you with the
almost obtuse angle of its jaw; exuding a white fluid from its vent,
which has a metallic stink like acetylene. Take no notice of it at all.
Like an efficient governess with a refractory child, you speak sharply to
the smelly creature and hold it firmly. You take hold of its tail, unwind
it, roll it in a ball (it is wriggling so much that it generally helps in
this), tie it up in your handkerchief, put it in your trouser pocket and
look for another.

When you loose it in your sitting-room it rushes off along the floor,
swishing frantically but making little progress on the polished wood,
and conceals itself in the darkest corner. At night, when you come to
lock it up, it makes a fuss. It produces the smell again, and the hiss. In
the morning it is the same. Next night perhaps the smell is omitted, or
fainter. In a few days there is only a dim hiss, a kind of grumble. This

goes as well, until there is only a gentle protesting undulation as it is lifted off the ground.

I remember particularly two of last year's snakes. One was a baby male (the yellow markings are brighter in the male) only about eight inches long. He was a confiding snake, and I once took him to church in my pocket, to make him a Christian and to comfort me during the sermon. I hope it was not an undue interference with his life: I never carried him about like that again, he seemed to like the warmth of my pocket, and I believe he did not change his creed.

Talking of Christians, I never christened the snakes. To have called them names would have been ridiculous, as it is with cars. A snake cannot have a name. If it had to be addressed I suppose it would be addressed by its generic title: Snake.

The other one, I regret to say, was nearly a pet. She was a well-grown female with a scar on her neck. I suppose this had been done to her by man. It was the scar that first attracted me to her, or rather made me take special notice of her, because she was easy to distinguish. I soon found that when the time came for putting her to bed she did not undulate. She never troubled to conceal herself at bedtime, nor to slide away from me when I approached. She would crawl right up to me, and pour over my feet while I was working. There was no horrible affection or prostration; only she was not afraid of me. She went over my feet because they were in a direct line with the place she was making for. She trusted, or at least was indifferent.

It was a temptation. One coldish afternoon she was sitting in my chair when I wanted to read. I picked her up and put her in my lap. She was not particularly comfortable, and began to go away. I held her gently by the tail. She decided that it was not worth a scene, and stayed. I put my free hand over her, and she curled up beneath it, the head sticking out between two fingers and the tongue flickering every now and then, when a thought of curiosity entered her slow, free mind.

After that I used sometimes to sit with my two hands cupped, and she would curl between them on cold days. My hands were warm, that was all.

It was not quite all. I am afraid a hideous tinge of possession is creeping into this account. When other people came into the room she used to hiss. I would be dozing with her tight, dry coils between my palms, and there would be a hiss. The door would have opened and somebody would have come in. Or again, if I showed her to people she would hiss at them. If they tried to catch her, she would pour away. But when I gave her to them she was quiet.

I think I succeeded in keeping my distance. At any rate, she had a love affair with one of the males. I remember finding them coiled together on the corner table: a double rope-coil of snake which looked like a single one, except that it had two heads. I did not realise that this was an affair of the heart, at the time.

Later on she began to look ill. She was lumpy and flaccid. I became worried about the commissariat. Snakes rarely eat—seldom more than once a fortnight—but when they do eat they are particular. The staple food is a live frog, swallowed alive and whole. Anybody who has ever kept snakes will know how difficult it is to find a frog. The whole of the Shire seems to be populated by toads: one can scarcely move without treading on a toad: but toads disagree with snakes. They exude something from the skin.

I had been short of frogs lately, and (as I merely kept them loose in the aquarium so that the snakes could help themselves when they wanted) did not know when she had last had a meal. I thought I was starving her and became agitated. I spent hours looking for frogs, and found one eventually, but she wouldn't touch it. I tried a gold-fish, but that was no good either. She got worse. I was afraid she was poisoned, or melancholic from her unnatural surroundings.

Then came the proud day. I got back at half-past twelve, and looked for her on the hearth-rug, but she was not there. She was in the aquarium, sunlit from the french windows. Not only she. I went closer and looked. There were twenty-eight eggs.

Poor old lady, she was in a dreadful state. Quite apathetic and powerless, she could scarcely lift her head. Her body had fallen in on itself, leaving two ridges, as if she were quite a slim snake dressed in clothes too big for her. When I picked her up she hung limp, as if she were actually dead; but her tongue flickered. I didn't know what to do.

I got a gold-fish bowl and half-filled it with fresh grass clippings. I put her in it, with the frog, and tied paper over the top as if it were a jam jar. I made holes in the paper and took it out on to the lawn, in the full glare of the summer sun. Snakes are woken up by heat, and the bowl would concentrate the sun's beams. It was all I could think of or do, before I went in to lunch.

I came back in half an hour. The bowl was warm with moisture, the grass clippings were browning, the frog was gone; and inside was Matilda (she positively deserved a name) as fit as a flea and twice as frisky.

The scarred snake may have been a good mistress, but she was a bad mother. If she had known anything about maternity, she would not have laid her eggs in the aquarium. It seems that water is one of the things that is fatal to the eggs of grass snakes. I picked them out, and put them in another gold-fish bowl, this time full of grass clippings that were already rotten. Then I left them in the sun. They only went mouldy.

She was completely tame, and the inevitable happened. The time came for me to go away for two months, so I gave her her liberty. I took her out into the fountain court (next time it shall be into the deepest and most unpopulated forest) and put her on the ground in the strong July sunlight. She was delighted by it, and pleased to go. I watched her to-froing away, till she slipped into the angle of a flowerbed, and then went resolutely indoors. There were plenty of other things in the future besides grass snakes.

That night I went down to the lake to bathe, and stepped over a dead snake in the moonlight. I guessed before I looked for the scar. I had kept my distance successfully, so that there were no regrets at parting, but I had destroyed a natural balance. She had lost her bitter fear of man: a thing which it is not wise to lose.

I feel some difficulty in putting this properly. Some bloody-minded human being had come across her on a path and gone for her with a stick. She was harmless, useless dead, very beautiful, easy prey. He slaughtered her with a stick, and grass snakes are not easy to kill. It is easy to maim them, to bash them on the head until the bones are pulp. The lower jaw no longer articulates with the upper one, but lies sideways under the crushed skull, shewing the beautiful colours of its unprotected inner side. The whole reserved face suddenly looks pitiful, because it has been spoilt and ravaged. The black tongue makes a feeble flicker still.

These things had been done, to a creature which was offering confidence, with wanton savagery. Why? Why the waste of beauty and the degradation to the murderer himself? He was not creating a beauty by destroying this one. He cannot even have considered himself clever.

RACHEL CARSON
1907–1964

Trained as a marine biologist, Rachel Carson pursued careers both as a specialist in commercial fisheries and as a writer. The Sea Around Us *(1951) portrayed the ocean as a single complex entity, changing the scale of her readers' perspectives in much the same way Lewis Thomas was to do with his view of earth as a living cell. The Edge of the Sea (1955) explored the special richness of "the marginal world." Her most influential book was* Silent Spring (1962), *in which Carson demonstrated the harmful effects of pesticides on the health of the environment. Her ability at once to specify the chemical and biological details of the problem and to evoke the devastation of wildlife in forest and stream gave her book an extraordinary impact on public opinion. Following its publication, President Kennedy called for a federal investigation, which resulted in much tighter controls on the use of DDT and other toxic products.*

THE MARGINAL WORLD

The edge of the sea is a strange and beautiful place. All through the long history of Earth it has been an area of unrest where waves have broken heavily against the land, where the tides have pressed forward over the continents, receded, and then returned. For no two successive days is the shore line precisely the same. Not only do the tides advance and retreat in their eternal rhythms, but the level of the sea itself is never at rest. It rises or falls as the glaciers melt or grow, as the floor of the deep ocean basins shifts under its increasing load of sediments, or as the earth's crust along the continental margins warps up or down in adjustment to strain and tension. Today a little more land may belong to the sea, tomorrow a little less. Always the edge of the sea remains an elusive and indefinable boundary.

The shore has a dual nature, changing with the swing of the tides, belonging now to the land, now to the sea. On the ebb tide it knows the harsh extremes of the land world, being exposed to heat and cold, to wind, to rain and drying sun. On the flood tide it is a water world, returning briefly to the relative stability of the open sea.

Only the most hardy and adaptable can survive in a region so mutable, yet the area between the tide lines is crowded with plants and animals. In this difficult world of the shore, life displays its enormous toughness and vitality by occupying almost every conceivable niche. Visibly, it carpets the intertidal rocks; or half hidden, it descends into fissures and crevices, or hides under boulders, or lurks in the wet gloom of sea caves. Invisibly, where the casual observer would say there is no life, it lies deep in the sand, in burrows and tubes and passage-

The Edge of the Sea (Boston: Houghton Mifflin, 1955).

ways. It tunnels into solid rock and bores into peat and clay. It encrusts weeds or drifting spars or the hard, chitinous shell of a lobster. It exists minutely, as the film of bacteria that spreads over a rock surface or a wharf piling; as spheres of protozoa, small as pinpricks, sparkling at the surface of the sea; and as Lilliputian beings swimming through dark pools that lie between the grains of sand.

The shore is an ancient world, for as long as there has been an earth and sea there has been this place of the meeting of land and water. Yet it is a world that keeps alive the sense of continuing creation and of the relentless drive of life. Each time that I enter it, I gain some new awareness of its beauty and its deeper meanings, sensing that intricate fabric of life by which one creature is linked with another, and each with its surroundings.

In my thoughts of the shore, one place stands apart for its revelation of exquisite beauty. It is a pool hidden within a cave that one can visit only rarely and briefly when the lowest of the year's low tides fall below it, and perhaps from that very fact it acquires some of its special beauty. Choosing such a tide, I hoped for a glimpse of the pool. The ebb was to fall early in the morning. I knew that if the wind held from the northwest and no interfering swell ran in from a distant storm the level of the sea should drop below the entrance to the pool. There had been sudden ominous showers in the night, with rain like handfuls of gravel flung on the roof. When I looked out into the early morning the sky was full of a gray dawn light but the sun had not yet risen. Water and air were pallid. Across the bay the moon was a luminous disc in the western sky, suspended above the dim line of distant shore—the full August moon, drawing the tide to the low, low levels of the threshold of the alien sea world. As I watched, a gull flew by, above the spruces. Its breast was rosy with the light of the unrisen sun. The day was, after all, to be fair.

Later, as I stood above the tide near the entrance to the pool, the promise of that rosy light was sustained. From the base of the steep wall of rock on which I stood, a moss-covered ledge jutted seaward into deep water. In the surge at the rim of the ledge the dark fronds of oarweeds swayed, smooth and gleaming as leather. The projecting ledge was the path to the small hidden cave and its pool. Occasionally a swell, stronger than the rest, rolled smoothly over the rim and broke in foam against the cliff. But the intervals between such swells were long enough to admit me to the ledge and long enough for a glimpse of that fairy pool, so seldom and so briefly exposed.

And so I knelt on the wet carpet of sea moss and looked back into the dark cavern that held the pool in a shallow basin. The floor of the

cave was only a few inches below the roof, and a mirror had been created in which all that grew on the ceiling was reflected in the still water
below.

Under water that was clear as glass the pool was carpeted with green
sponge. Gray patches of sea squirts glistened on the ceiling and colonies
of soft coral were a pale apricot color. In the moment when I looked
into the cave a little elfin starfish hung down, suspended by the merest
thread, perhaps by only a single tube foot. It reached down to touch its
own reflection, so perfectly delineated that there might have been, not
one starfish, but two. The beauty of the reflected images and of the
limpid pool itself was the poignant beauty of things that are ephemeral,
existing only until the sea should return to fill the little cave.

Whenever I go down into this magical zone of the low water of the
spring tides, I look for the most delicately beautiful of all the shore's
inhabitants—flowers that are not plant but animal, blooming on the
threshold of the deeper sea. In that fairy cave I was not disappointed.
Hanging from its roof were the pendent flowers of the hydroid
Tubularia, pale pink, fringed and delicate as the wind flower. Here were
creatures so exquisitely fashioned that they seemed unreal, their beauty
too fragile to exist in a world of crushing force. Yet every detail was functionally useful, every stalk and hydranth and petal-like tentacle fashioned for dealing with the realities of existence. I knew that they were
merely waiting, in that moment of the tide's ebbing, for the return of the
sea. Then in the rush of water, in the surge of surf and the pressure of
the incoming tide, the delicate flower heads would stir with life. They
would sway on their slender stalks, and their long tentacles would sweep
the returning water, finding in it all that they needed for life.

And so in that enchanted place on the threshold of the sea the realities that possessed my mind were far from those of the land world I had
left an hour before. In a different way the same sense of remoteness and
of a world apart came to me in a twilight hour on a great beach on the
coast of Georgia. I had come down after sunset and walked far out over
sands that lay wet and gleaming, to the very edge of the retreating sea.
Looking back across that immense flat, crossed by winding, water-filled
gullies and here and there holding shallow pools left by the tide, I was
filled with awareness that this intertidal area, although abandoned
briefly and rhythmically by the sea, is always reclaimed by the rising
tide. There at the edge of low water the beach with its reminders of the
land seemed far away. The only sounds were those of the wind and the
sea and the birds. There was one sound of wind moving over water,
and another of water sliding over the sand and tumbling down the faces
of its own wave forms. The flats were astir with birds, and the voice of

the willet rang insistently. One of them stood at the edge of the water and gave its loud, urgent cry; an answer came from far up the beach and the two birds flew to join each other.

The flats took on a mysterious quality as dusk approached and the last evening light was reflected from the scattered pools and creeks. Then birds became only dark shadows, with no color discernible. Sanderlings scurried across the beach like little ghosts, and here and there the darker forms of the willets stood out. Often I could come very close to them before they would start up in alarm—the sanderlings running, the willets flying up, crying. Black skimmers flew along the ocean's edge silhouetted against the dull, metallic gleam, or they went flitting above the sand like large, dimly seen moths. Sometimes they "skimmed" the winding creeks of tidal water, where little spreading surface ripples marked the presence of small fish.

The shore at night is a different world, in which the very darkness that hides the distractions of daylight brings into sharper focus the elemental realities. Once, exploring the night beach, I surprised a small ghost crab in the searching beam of my torch. He was lying in a pit he had dug just above the surf, as though watching the sea and waiting. The blackness of the night possessed water, air, and beach. It was the darkness of an older world, before Man. There was no sound but the all-enveloping, primeval sounds of wind blowing over water and sand, and of waves crashing on the beach. There was no other visible life—just one small crab near the sea. I have seen hundreds of ghost crabs in other settings, but suddenly I was filled with the odd sensation that for the first time I knew the creature in its own world—that I understood, as never before, the essence of its being. In that moment time was suspended; the world to which I belonged did not exist and I might have been an onlooker from outer space. The little crab alone with the sea became a symbol that stood for life itself—for the delicate, destructible, yet incredibly vital force that somehow holds its place amid the harsh realities of the inorganic world.

The sense of creation comes with memories of a southern coast, where the sea and the mangroves, working together, are building a wilderness of thousands of small islands off the southwestern coast of Florida, separated from each other by a tortuous pattern of bays, lagoons, and narrow waterways. I remember a winter day when the sky was blue and drenched with sunlight; though there was no wind one was conscious of flowing air like cold clear crystal. I had landed on the surf-washed tip of one of those islands, and then worked my way around to the sheltered bay side. There I found the tide far out, exposing the broad mud flat of a cove bordered by the mangroves with their twisted branches, their glossy leaves, and their long prop roots reaching down, grasping and holding the mud, building the land out a little more, then again a little more.

The mud flats were strewn with the shells of that small, exquisitely colored mollusk, the rose tellin, looking like scattered petals of pink roses. There must have been a colony nearby, living buried just under the surface of the mud. At first the only creature visible was a small heron in gray and rusty plumage—a reddish egret that waded across the flat with the stealthy, hesitant movements of its kind. But other land creatures had been there, for a line of fresh tracks wound in and out among the mangrove roots, marking the path of a raccoon feeding on the oysters that gripped the supporting roots with projections from their shells. Soon I found the tracks of a shore bird, probably a sanderling, and followed them a little; then they turned toward the water and were lost, for the tide had erased them and made them as though they had never been.

Looking out over the cove I felt a strong sense of the interchangeability of land and sea in this marginal world of the shore, and of the links between the life of the two. There was also an awareness of the past and of the continuing flow of time, obliterating much that had gone before, as the sea had that morning washed away the tracks of the bird.

The sequence and meaning of the drift of time were quietly summarized in the existence of hundreds of small snails—the mangrove periwinkles—browsing on the branches and roots of the trees. Once their ancestors had been sea dwellers, bound to the salt waters by every tie of their life processes. Little by little over the thousands and millions of years the ties had been broken, the snails had adjusted themselves to life out of water, and now today they were living many feet above the tide to which they only occasionally returned. And perhaps, who could say how many ages hence, there would be in their descendants not even this gesture of remembrance for the sea.

The spiral shells of other snails—these quite minute—left winding tracks on the mud as they moved about in search of food. They were horn shells, and when I saw them I had a nostalgic moment when I wished I might see what Audubon saw, a century and more ago. For such little horn shells were the food of the flamingo, once so numerous on this coast, and when I half closed my eyes I could almost imagine a flock of these magnificent flame birds feeding in that cove, filling it with their color. It was a mere yesterday in the life of the earth that they were there; in nature, time and space are relative matters, perhaps most truly perceived subjectively in occasional flashes of insight, sparked by such a magical hour and place.

There is a common thread that links these scenes and memories—the spectacle of life in all its varied manifestations as it has appeared, evolved, and sometimes died out. Underlying the beauty of the spectacle there is meaning and significance. It is the elusiveness of that meaning that haunts us, that sends us again and again into the natural world where the key to the riddle is hidden. It sends us back to the edge of the sea, where the drama of life played its first scene on earth and perhaps even its prelude; where the forces of evolution are at work today, as they have been since the appearance of what we know as life; and where the spectacle of living creatures faced by the cosmic realities of their world is crystal clear.

LOREN EISELEY
1907–1977

Few writers so effectively fuse personal and professional perspectives in their work as does Eiseley, and his nature essays introduced a strong autobiographical element into contemporary nature writing. Born in the bleak Nebraska plains country and raised as a solitary child by a deaf and mentally unstable mother, he began writing poetry and stories at an early age. After a circuitous and interrupted academic career, including several years as a drifter through the West during the Depression, he eventually became a professor of anthropology and curator of early man at the University of Pennsylvania. His first book, The Immense Journey *(1957), was an unexpected best-seller and has proved to be one of the most influential collections of American essays published since the mid–twentieth century. Its evolutionary perspective, emphasizing the long hidden past of human nature, is colored by a dark and brooding temperament, infusing ordinary natural events with symbolic resonance. Though Eiseley's outlook is often characterized by loneliness and pessimism, he finds comfort in our shared condition with other animals and in man's capacity for "rare and hidden communion with nature."*

THE JUDGMENT OF THE BIRDS

It is a commonplace of all religious thought, even the most primitive, that the man seeking visions and insight must go apart from his fellows and live for a time in the wilderness. If he is of the proper sort, he will return with a message. It may not be a message from the god he set out to seek, but even if he has failed in that particular, he will have had a vision or seen a marvel, and these are always worth listening to and thinking about.

The world, I have come to believe, is a very queer place, but we have been part of this queerness for so long that we tend to take it for granted. We rush to and fro like Mad Hatters upon our peculiar errands, all the time imagining our surroundings to be dull and ourselves quite ordinary creatures. Actually, there is nothing in the world to encourage this idea, but such is the mind of man, and this is why he finds it necessary from time to time to send emissaries into the wilderness in the hope of learning of great events, or plans in store for him, that will resuscitate his waning taste for life. His great news services, his world-wide radio network, he knows with a last remnant of healthy distrust will be of no use to him in this matter. No miracle can withstand a radio broadcast, and it is certain that it would be no miracle if it could. One must seek, then, what only the solitary approach can give—a natural revelation.

Let it be understood that I am not the sort of man to whom is entrusted direct knowledge of great events or prophecies. A naturalist, however, spends much of his life alone, and my life is no exception. Even in New York City there are patches of wilderness, and a man by himself is bound to undergo certain experiences falling into the class of which I speak. I set mine down, therefore: a matter of pigeons, a flight of chemicals, and a judgment of birds, in the hope that they will come to the eye of those who have retained a true taste for the marvelous, and who are capable of discerning in the flow of ordinary events the point at which the mundane world gives way to quite another dimension.

New York is not, on the whole, the best place to enjoy the downright miraculous nature of the planet. There are, I do not doubt, many remarkable stories to be heard there and many strange sights to be seen, but to grasp a marvel fully it must be savored from all aspects. This cannot be done while one is being jostled and hustled along a crowded street. Nevertheless, in any city there are true wildernesses where a man can be alone. It can happen in a hotel room, or on the high roofs at dawn.

The Immense Journey (New York: Random House, 1957).

One night on the twentieth floor of a midtown hotel I awoke in the dark and grew restless. On an impulse I climbed upon the broad old-fashioned window sill, opened the curtains and peered out. It was the hour just before dawn, the hour when men sigh in their sleep, or, if awake, strive to focus their wavering eyesight upon a world emerging from the shadows. I leaned out sleepily through the open window. I had expected depths, but not the sight I saw.

I found I was looking down from that great height into a series of curious cupolas or lofts that I could just barely make out in the darkness. As I looked, the outlines of these lofts became more distinct because the light was being reflected from the wings of pigeons who, in utter silence, were beginning to float outward upon the city. In and out through the open slits in the cupolas passed the white-winged birds on their mysterious errands. At this hour the city was theirs, and quietly, without the brush of a single wing tip against stone in that high, eerie place, they were taking over the spires of Manhattan. They were pouring upward in a light that was not yet perceptible to human eyes, while far down in the black darkness of the alleys it was still midnight.

As I crouched half asleep across the sill, I had a moment's illusion that the world had changed in the night, as in some immense snowfall, and that if I were to leave, it would have to be as these other inhabitants were doing, by the window. I should have to launch out into that great bottomless void with the simple confidence of young birds reared high up there among the familiar chimney pots and interposed horrors of the abyss.

I leaned farther out. To and fro went the white wings, to and fro. There were no sounds from any of them. They knew man was asleep and this light for a little while was theirs. Or perhaps I had only dreamed about man in this city of wings—which he could surely never have built. Perhaps I, myself, was one of these birds dreaming unpleasantly a moment of old dangers far below as I teetered on a window ledge.

Around and around went the wings. It needed only a little courage, only a little shove from the window ledge to enter that city of light. The muscles of my hands were already making little premonitory lunges. I wanted to enter that city and go away over the roofs in the first dawn. I wanted to enter it so badly that I drew back carefully into the room and opened the hall door. I found my coat on the chair, and it slowly became clear to me that there was a way down through the floors, that I was, after all, only a man.

I dressed then and went back to my own kind, and I have been rather more than usually careful ever since not to look into the city of light. I had seen, just once, man's greatest creation from a strange inverted angle, and it was not really his at all. I will never forget how those wings went round and round, and how, by the merest pressure of the fingers and a feeling for air, one might go away over the roofs. It is a knowledge, however, that is better kept to oneself. I think of it sometimes in such a way that the wings, beginning far down in the black depths of the mind, begin to rise and whirl till all the mind is lit by their spinning, and there is a sense of things passing away, but lightly, as a wing might veer over an obstacle.

To see from an inverted angle, however, is not a gift allotted merely to the human imagination. I have come to suspect that within their degree it is sensed by animals, though perhaps as rarely as among men. The time has to be right; one has to be, by chance or intention, upon the border of two worlds. And sometimes these two borders may shift or interpenetrate and one sees the miraculous.

I once saw this happen to a crow.

This crow lives near my house, and though I have never injured him, he takes good care to stay up in the very highest trees and, in general, to avoid humanity. His world begins at about the limit of my eyesight.

On the particular morning when this episode occurred, the whole countryside was buried in one of the thickest fogs in years. The ceiling was absolutely zero. All planes were grounded, and even a pedestrian could hardly see his outstretched hand before him.

I was groping across a field in the general direction of the railroad station, following a dimly outlined path. Suddenly out of the fog, at about the level of my eyes, and so closely that I flinched, there flashed a pair of immense black wings and a huge beak. The whole bird rushed over my head with a frantic cawing outcry of such hideous terror as I have never heard in a crow's voice before, and never expect to hear again.

He was lost and startled, I thought, as I recovered my poise. He ought not to have flown out in this fog. He'd knock his silly brains out.

All afternoon that great awkward cry rang in my head. Merely being lost in a fog seemed scarcely to account for it—especially in a tough, intelligent old bandit such as I knew that particular crow to be. I even looked once in the mirror to see what it might be about me that had so revolted him that he had cried out in protest to the very stones.

Finally, as I worked my way homeward along the path, the solution came to me. It should have been clear before. The borders of our worlds had shifted. It was the fog that had done it. That crow, and I knew him well, never under normal circumstances flew low near men. He had been lost all right, but it was more than that. He had thought he was high up, and when he encountered me looming gigantically through the fog, he had perceived a ghastly and, to the crow mind, unnatural sight. He had seen a man walking on air, desecrating the very heart of the crow kingdom, a harbinger of the most profound evil a crow mind could conceive of—air-walking men. The encounter, he must have thought, had taken place a hundred feet over the roofs.

He caws now when he sees me leaving for the station in the morning, and I fancy that in that note I catch the uncertainty of a mind that has come to know things are not always what they seem. He has seen a marvel in his heights of air and is no longer as other crows. He has experienced the human world from an unlikely perspective. He and I share a viewpoint in common: our worlds have interpenetrated, and we both have faith in the miraculous.

It is a faith that in my own case has been augmented by two remarkable sights. As I have hinted previously, I once saw some very odd chemicals fly across a waste so dead it might have been upon the moon, and once, by an even more fantastic piece of luck, I was present when a group of birds passed a judgment upon life.

On the maps of the old voyageurs it is called *Mauvaises Terres*, the evil lands, and, slurred a little with the passage through many minds, it has come down to us anglicized as the Badlands. The soft shuffle of moccasins has passed through its canyons on the grim business of war and flight, but the last of those slight disturbances of immemorial silences died out almost a century ago. The land, if one can call it a land, is a waste as lifeless as that valley in which lie the kings of Egypt. Like the Valley of the Kings, it is a mausoleum, a place of dry bones in what once was a place of life. Now it has silences as deep as those in the moon's airless chasms.

Nothing grows among its pinnacles; there is no shade except under great toadstools of sandstone whose bases have been eaten to the shape of wine glasses by the wind. Everything is flaking, cracking, disintegrating, wearing away in the long, imperceptible weather of time. The ash of ancient volcanic outbursts still sterilizes its soil, and its colors in that waste are the colors that flame in the lonely sunsets on dead planets. Men come there but rarely, and for one purpose only, the collection of bones.

It was a late hour on a cold, wind-bitten autumn day when I climbed a great hill spined like a dinosaur's back and tried to take my bearings. The tumbled waste fell away in waves in all directions. Blue air was darkening into purple along the bases of the hills. I shifted my knapsack, heavy with the petrified bones of long-vanished creatures, and studied my compass. I wanted to be out of there by nightfall, and already the sun was going sullenly down in the west.

It was then that I saw the flight coming on. It was moving like a little close-knit body of black specks that danced and darted and closed again. It was pouring from the north and heading toward me with the undeviating relentlessness of a compass needle. It streamed through the shadows rising out of monstrous gorges. It rushed over towering pinnacles in the red light of the sun, or momentarily sank from sight within their shade. Across that desert of eroding clay and wind-worn stone they came with a faint wild twittering that filled all the air about me as those tiny living bullets hurtled past into the night.

It may not strike you as a marvel. It would not, perhaps, unless you stood in the middle of a dead world at sunset, but that was where I stood. Fifty million years lay under my feet, fifty million years of bellowing monsters moving in a green world now gone so utterly that its very light was travelling on the farther edge of space. The chemicals of all that vanished age lay about me in the ground. Around me still lay the shearing molars of dead titanotheres, the delicate sabers of soft-stepping cats, the hollow sockets that had held the eyes of many a strange, outmoded beast. Those eyes had looked out upon a world as real as ours; dark, savage brains had roamed and roared their challenges into the steaming night.

Now they were still here, or, put it as you will, the chemicals that made them were here about me in the ground. The carbon that had driven them ran blackly in the eroding stone. The stain of iron was in the clays. The iron did not remember the blood it had once moved within, the phosphorus had forgot the savage brain. The little individual moment had ebbed from all those strange combinations of chemicals as it would ebb from our living bodies into the sinks and runnels of oncoming time.

I had lifted up a fistful of that ground. I held it while that wild flight of south-bound warblers hurtled over me into the oncoming dark. There went phosphorus, there went iron, there went carbon, there beat the calcium in those hurrying wings. Alone on a dead planet I watched that incredible miracle speeding past. It ran by some true compass over field and waste land. It cried its individual ecstasies into the air until the gullies rang. It swerved like a single body, it knew itself and, lonely, it bunched close in the racing darkness, its individual entities feeling about them the rising night. And so, crying to each other their identity, they passed away out of my view.

I dropped my fistful of earth. I heard it roll inanimate back into the gully at the base of the hill: iron, carbon, the chemicals of life.

Like men from those wild tribes who had haunted these hills before me seeking visions, I made my sign to the great darkness. It was not a mocking sign, and I was not mocked. As I walked into my camp late that night, one man, rousing from his blankets beside the fire, asked sleepily, "What did you see?"

"I think, a miracle," I said softly, but I said it to myself. Behind me that vast waste began to glow under the rising moon.

I have said that I saw a judgment upon life, and that it was not passed by men. Those who stare at birds in cages or who test minds by their closeness to our own may not care for it. It comes from far away out of my past, in a place of pouring waters and green leaves. I shall never see an episode like it again if I live to be a hundred, nor do I think that one man in a million has ever seen it, because man is an intruder into such silences. The light must be right, and the observer must remain unseen. No man sets up such an experiment. What he sees, he sees by chance.

You may put it that I had come over a mountain, that I had slogged through fern and pine needles for half a long day, and that on the edge of a little glade with one long, crooked branch extending across it, I had sat down to rest with my back against a stump. Through accident I was concealed from the glade, although I could see into it perfectly.

The sun was warm there, and the murmurs of forest life blurred softly away into my sleep. When I awoke, dimly aware of some commotion and outcry in the clearing, the light was slanting down through the pines in such a way that the glade was lit like some vast cathedral. I could see the dust motes of wood pollen in the long shaft of light, and there on the extended branch sat an enormous raven with a red and squirming nestling in his beak.

The sound that awoke me was the outraged cries of the nestling's parents, who flew helplessly in circles about the clearing. The sleek black monster was indifferent to them. He gulped, whetted his beak on the dead branch a moment and sat still. Up to that point the little tragedy had followed the usual pattern. But suddenly, out of all that area of woodland, a soft sound of complaint began to rise. Into the glade fluttered small birds of half a dozen varieties drawn by the anguished outcries of the tiny parents.

No one dared to attack the raven. But they cried there in some instinctive common misery, the bereaved and the unbereaved. The glade filled with their soft rustling and their cries. They fluttered as though to point their wings at the murderer. There was a dim intangible ethic he had violated, that they knew. He was a bird of death.

And he, the murderer, the black bird at the heart of life, sat on there, glistening in the common light, formidable, unmoving, unperturbed, untouchable.

The sighing died. It was then I saw the judgment. It was the judgment of life against death. I will never see it again so forcefully presented. I will never hear it again in notes so tragically prolonged. For in the midst of protest, they forgot the violence. There, in that clearing, the crystal note of a song sparrow lifted hesitantly in the hush. And finally, after painful fluttering, another took the song, and then another, the song passing from one bird to another, doubtfully at first, as though some evil thing were being slowly forgotten. Till suddenly they took heart and sang from many throats joyously together as birds are known to sing. They sang because life is sweet and sunlight beautiful. They sang under the brooding shadow of the raven. In simple truth they had forgotten the raven, for they were the singers of life, and not of death.

I was not of that airy company. My limbs were the heavy limbs of an earthbound creature who could climb mountains, even the mountains of the mind, only by a great effort of will. I knew I had seen a marvel and observed a judgment, but the mind which was my human endowment was sure to question it and to be at me day by day with its heresies until I grew to doubt the meaning of what I had seen. Eventually darkness and subtleties would ring me round once more.

And so it proved until, on the top of a stepladder, I made one more observation upon life. It was cold that autumn evening, and, standing under a suburban street light in a spate of leaves and beginning snow, I was suddenly conscious of some huge and hairy shadows dancing over the pavement. They seemed attached to an odd, globular shape that was magnified above me. There was no mistaking it. I was standing under the shadow of an orb-weaving spider. Gigantically projected against the street, she was about her spinning when everything was going underground. Even her cables were magnified upon the sidewalk and already I was half-entangled in their shadows.

"Good Lord," I thought, "she has found herself a kind of minor sun and is going to upset the course of nature."

I procured a ladder from my yard and climbed up to inspect the situation. There she was, the universe running down around her, warmly arranged among her guy ropes attached to the lamp supports—a great black and yellow embodiment of the life force, not giving up to either frost or stepladders. She ignored me and went on tightening and improving her web.

I stood over her on the ladder, a faint snow touching my cheeks, and surveyed her universe. There were a couple of iridescent green beetle cases turning slowly on a loose strand of web, a fragment of luminescent eye from a moth's wing and a large indeterminable object, perhaps a cicada, that had struggled and been wrapped in silk. There were also little bits and slivers, little red and blue flashes from the scales of anonymous wings that had crushed there.

Some days, I thought, they will be dull and gray and the shine will be out of them; then the dew will polish them again and drops hang on the silk until everything is gleaming and turning in the light. It is like a mind, really, where everything changes but remains, and in the end you have these eaten-out bits of experience like beetle wings.

I stood over her a moment longer, comprehending somewhat reluctantly that her adventure against the great blind forces of winter, her seizure of this warming globe of light, would come to nothing and was hopeless. Nevertheless it brought the birds back into my mind, and that faraway song which had traveled with growing strength around a forest clearing years ago—a kind of heroism, a world where even a spider refuses to lie down and die if a rope can still be spun on to a star. Maybe man himself will fight like this in the end, I thought, slowly realizing that the web and its threatening yellow occupant had been added to some luminous store of experience, shining for a moment in the fogbound reaches of my brain.

The mind, it came to me as I slowly descended the ladder, is a very remarkable thing; it has gotten itself a kind of courage by looking at a spider in a street lamp. Here was something that ought to be passed on to those who will fight our final freezing battle with the void. I thought of setting it down carefully as a message to the future: *In the days of the frost seek a minor sun.*

But as I hesitated, it became plain that something was wrong. The marvel was escaping—a sense of bigness beyond man's power to grasp, the essence of life in its great dealings with the universe. It was better, I decided, for the emissaries returning from the wilderness, even if they were merely descending from a stepladder, to record their marvel, not to define its meaning. In that way it would go echoing on through the minds of men, each grasping at that beyond out of which the miracles emerge, and which, once defined, ceases to satisfy the human need for symbols.

In the end I merely made a mental note: One specimen of Epeira observed building a web in a street light. Late autumn and cold for spiders. Cold for men, too. I shivered and left the lamp glowing there in my mind. The last I saw of Epeira she was hauling steadily on a cable. I stepped carefully over her shadow as I walked away.

RICHARD WRIGHT
1908–1960

Richard Wright's enduring place in American letters rests primarily on two works: Native Son *(1940), a pioneering and powerful novel about young African American in Chicago convicted of murder, and* Black Boy *(1945), a moving and evocative "autobiographical novel" of his own Mississippi childhood, which was marked by poverty and parental neglect. The following passages from* Black Boy *are significant for their remarkable evocation of the sensuous delight and wonder with which the natural world can enter and shape the mind of a child, even in the midst of an impoverished, oppressive, and often violent human environment.*

From BLACK BOY

*   *   *   Each event spoke with a cryptic tongue. And the moments of living slowly revealed their coded meanings. There was the wonder I felt when I first saw a brace of mountainlike, spotted, black-and-white horses clopping down a dusty road through clouds of powdered clay.

There was the delight I caught in seeing long straight rows of red and green vegetables stretching away in the sun to the bright horizon.

There was the faint, cool kiss of sensuality when dew came on to my cheeks and shins as I ran down the wet green garden paths in the early morning.

There was the vague sense of the infinite as I looked down upon the yellow, dreaming waters of the Mississippi River from the verdant bluffs of Natchez.

There were the echoes of nostalgia I heard in the crying strings of wild geese winging south against a bleak, autumn sky.

There was the tantalizing melancholy in the tingling scent of burning hickory wood.

Black Boy: A Record of Childhood and Youth (New York: Harper, 1945).

There was the teasing and impossible desire to imitate the petty pride of sparrows wallowing and flouncing in the red dust of country roads.

There was the yearning for identification loosed in me by the sight of a solitary ant carrying a burden upon a mysterious journey.

There was the disdain that filled me as I tortured a delicate, blue-pink crawfish that huddled fearfully in the mudsill of a rusty tin can.

There was the aching glory in masses of clouds burning gold and purple from an invisible sun.

There was the liquid alarm I saw in the blood-red glare of the sun's afterglow mirrored in the squared panes of whitewashed frame houses.

There was the languor I felt when I heard green leaves rustling with a rainlike sound.

There was the incomprehensible secret embodied in a whitish toad-stool hiding in the dark shade of a rotting log.

There was the experience of feeling death without dying that came from watching a chicken leap about blindly after its neck had been snapped by a quick twist of my father's wrist.

There was the great joke that I felt God had played on cats and dogs by making them lap their milk and water with their tongues.

There was the thirst I had when I watched clear, sweet juice trickle from sugar cane being crushed.

There was the hot panic that welled up in my throat and swept through my blood when I first saw the lazy, limp coils of a blue-skinned snake sleeping in the sun.

There was the speechless astonishment of seeing a hog stabbed through the heart, dipped into boiling water, scraped, split open, gutted, and strung up gaping and bloody.

There was the love I had for the mute regality of tall, moss-clad oaks.

There was the hint of cosmic cruelty that I felt when I saw the curved timbers of a wooden shack that had been warped in the summer sun.

There was the saliva that formed in my mouth whenever I smelt clay dust potted with fresh rain.

There was the cloudy notion of hunger when I breathed the odor of new-cut, bleeding grass.

And there was the quiet terror that suffused my senses when vast hazes of gold washed earthward from star-heavy skies on silent nights . . .

* * *

The days and hours began to speak now with a clearer tongue. Each experience had a sharp meaning of its own.

There was the breathlessly anxious fun of chasing and catching flitting fireflies on drowsy summer nights.

There was the drenching hospitality in the pervading smell of sweet magnolias.

There was the aura of limitless freedom distilled from the rolling sweep of tall green grass swaying and glinting in the wind and sun.

There was the feeling of impersonal plenty when I saw a boll of cotton whose cup had spilt over and straggled its white fleece toward the earth.

There was the pitying chuckle that bubbled in my throat when I watched a fat duck waddle across the back yard.

There was the suspense I felt when I heard the taut, sharp song of a yellow-black bee hovering nervously but patiently above a white rose.

There was the drugged, sleepy feeling that came from sipping glasses of milk, drinking them slowly so that they would last a long time, and drinking enough for the first time in my life.

There was the bitter amusement of going into town with Granny and watching the baffled stares of white folks who saw an old white woman leading two undeniably Negro boys in and out of stores on Capitol Street.

There was the slow, fresh, saliva-stimulating smell of cooking cotton seeds.

There was the excitement of fishing in muddy country creeks with my grandpa on cloudy days.

There was the fear and awe I felt when Grandpa took me to a sawmill to watch the giant whirring steel blades whine and scream as they bit into wet green logs.

There was the puckery taste that almost made me cry when I ate my first half-ripe persimmon.

There was the greedy joy in the tangy taste of wild hickory nuts.

There was the dry hot summer morning when I scratched my bare arms on briers while picking blackberries and came home with my fingers and lips stained black with sweet berry juice.

There was the relish of eating my first fried fish sandwich, nibbling at it slowly and hoping that I would never eat it up.

There was the all-night ache in my stomach after I had climbed a neighbor's tree and eaten stolen, unripe peaches.

There was the morning when I thought I would fall dead from fear after I had stepped with my bare feet upon a bright little green garden snake.

And there were the long, slow, drowsy days and nights of drizzling rain . . .

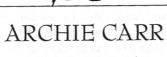

ARCHIE CARR
1909–1987

Born in Mobile, Alabama, Archie Carr taught biology at the University of Florida for many years. *He conducted pioneering research on the remarkable ocean migrations of sea turtles and became an internationally known leader of conservation efforts to protect these endangered marine reptiles.* *His book* So Excellent A Fishe: A Natural History of Sea Turtles *(1967) is considered a classic in the field. Carr's style is witty and full of sharp observations of local cultures. In this chapter from* The Windward Road: Adventures of a Naturalist on Remote Caribbean Shores *(1957), an account of his research on ridley turtles, the author demonstrates two critical requirements of a good nature writer: the ability to turn frustrations into opportunities, and the willingness to be diverted from one's intended pursuits by new and unexpected distractions, in this case, the love life of sloths.*

THE LIVELY PETES OF PARQUE VARGAS

It was midday outside, but in Parque Vargas it was twilight. On the hot streets around the plaza the people of Puerto Limón had given up for the day. There were a few bicycles still and an occasional taxi clattered by, and the little shoeshine boys still shrieked at intervals. Inside the shady park most of the sloths had given up hours before and hung quietly now, back down, from the high limbs of the Indian laurel trees. A casual observer would have said that *all* the sloths were sleeping; but I knew better, or hoped I did.

There was one sloth up there that I felt sure was, relatively speaking, seething with emotion. It was not the sort of emotion that stirs the sur-

The Windward Road: Adventures of a Naturalist on Remote Carribean Shores (New York: Knopf, 1967).

face appreciably, but I was certain it was there all the same. It was this sloth and its supposed emotion, and my suspense over the outcome of it, that held me in the dim interior of the plaza on that fifth day of my stay in Puerto Limón.

To walk into Parque Vargas from the searing sunshine on the streets around it was like diving into a deep spring. The light was the same and the feel of the air on your skin was too. The limbs of the tall trees interlocked so closely that only a stray splash of sunlight reached the ground, and so high that eddies of breeze, geared by the bay wind, swirled and drifted among the trunks. It is hard to imagine the appeal of such a place unless you have waited for time to pass in a Caribbean town. For five days I had been doing a great deal of waiting here in midtown Limón, and I should have been on a low limb indeed if it had not been for the sloths and the shade and the wind-drift in Parque Vargas.

The first day I entered the tiny office of Aerovías Costarricenses, I was struck by the unpretentious scale of the enterprise. There was a counter across the single room and a platform scale in front of it. Behind the counter was a table, a typewriter, and a girl, and the girl was pretty like any Tica; but she seemed depressed when I asked if I could charter a plane for the next day.

"Our plane is discomposed," she said in Spanish.

"Your plane?" I said. "Do you have only one plane? What is the matter with it?"

"*¿Quién sabe?*" the girl said. "Paco is trying to find out. He flew to San José Wednesday and the trip is high, over the volcano; too high. He flies over 12,500 feet, and this is bad for a plane like ours; and when he got back it was discomposed. All the passengers and freight for Sixaola and la Barra are waiting. *Lástima* — a pity."

"Then you have no idea when I might be able to hire the plane?"

"Where do you want to go?" she said.

"To Tortuguero. Maybe to la Barra."

"But we go to those places. Why hire the plane? Why do you not just go as a passenger? It's all the same, only cheaper."

"But I want to fly low and look for turtles."

"Paco likes to fly low and look for turtles."

"Well — would he circle when I asked him to?"

"That would cost more, I suppose. A little. But if you will fly with the guaro you can go at almost the passenger rate."

"The guaro?" I said. "What guaro?" *Guaro* is a corruption of the Castilian word *aguardiente* and is used in most parts of Central America for locally made sugarcane rum. The girl held a hand out three feet above the floor. She held it knuckles-down, not palm-down as a gringo would.

"It is a big can of guaro for the company laborers," she said. "The Atlantic Trading Company at Tortuguero. The company employs a great many Mosquitos, and they need guaro. If Saturday night comes and the guaro lacks, they are sad; and they sometimes go away."

"All right," I said. "I don't mind riding with the guaro if we can fly low. When do you think the plane might possibly be ready?"

"This is all I can tell you: the first flight will have to be for Sixaola and the second for Barra del Colorado. You should be ready to leave the day after the plane is composed—or if it is finished late, the second day after. Where can I locate you, noontimes?"

I told her. Then I went away and waited five days. Mornings I hired a bicycle or a horse, and once, at prohibitive cost, a taxi, and went up the coast or back into the fecund hinterland catching frogs and lizards and snakes; or swam in the rock pools behind the reef at the point; or visited the turtle crawls and fish-landings across the river.

Every day at noon I came back to the plaza and waited and felt the coolness and watched the sloths in the trees. If you should ever have to do any waiting where there are sloths, I can recommend watching them as a way to pass the time. It is as good as reading *War and Peace*—it never gives out on you. You can while away a whole half-hour finding out whether a given sloth intends to reach for another twigful of leaves, or to scratch himself again. No matter which course he chooses, you can count on spending another half-hour watching him carry it out.

In Parque Vargas there are nine sloths. One man I talked to insisted that there are twenty-five, but I counted them every day for five days and always came out with nine. The consensus among the people I questioned was that this number was probably right and that it had remained about the same ever since they could remember. The nine sloths live in twenty-eight *laurel de la India* trees. The trees, which are a species of *Ficus* (fig), are wound about with philodendrons and their limbs sweep up and interlace to form a closed canopy that is a fine continuous roadway and pasture for the sloths.

The sloths in Parque Vargas are Gray's three-toed sloth, a race confined to Panama and Costa Rica and belonging to a genus that is found in lowland woods from Honduras to Bolivia. Sloths are among the strangest of all mammals. If you work at it you can make out a case for placing them in the order Edentata, along with the anteaters and armadillos. But no anteater or armadillo can begin to match the sloth for eccentricity.

In the first place, sloths *look* strange before they make a move of any kind. The mature animal weighs about twenty-five pounds and has a round, earless head on a long neck, bulging eyes, and long legs that seem to have no joints. Each hand and foot has three permanently flexed claws with which the creature hangs upside down beneath the limbs he lives on. The fur is coarse and bristly, and during the wet season green algae grow among the hairs and give it a greenish color thought to be useful as camouflage.

But nothing in the appearance of this animal is anywhere near as curious as its incredible sloth. I use the term in its original sense of "slowness"—with none of its acquired connotation of *reprehensible* slowness. The sloth of these animals is one of the marvels of nature. It is a mockery of motion, an eerily mechanical, nerve-racking slowness that contractile protoplasm was never meant to support. The cytoplasm of an amœba streams faster than a sloth flees from a hungry boa constrictor. And besides being thus pointlessly, unbearably slow in everything it undertakes, the sloth is hesitant and vacillating in undertaking anything.

For example, a sloth may initiate some simple, straightforward move—like reaching for another handhold, say—and you may find that you must wait many minutes before it is clear whether he is carrying out the act or has stopped to reconsider the whole plan.

In spite of all this it seems to me unfair to brand the sloth as stupid and of "primitive mentality," as writers on the subject are inclined to do. It is quite possible that the animal is not stupid at all, and that its physical slowness is just that and nothing more, or even possibly a useful adaptation that we have not the wit to understand. Apparently no animal psychologist has turned his attention to sloths. They have never been put through a maze, for example, to get really reliable data on their learning ability. Putting a sloth through a maze would be quite a technical feat, I imagine, and very time consuming; but it must be done sooner or later, in fairness to the sloth. Meantime I shall go on wondering if the sloth may not be every bit as bright as its way of life calls for.

The slowness of sloths must be, in some way that I have never heard explained, part of a pattern of adaptation for life in the treetops. Some arboreal vertebrates are not slow, to be sure. Squirrels, for instance, are paragons of agility. But on the other hand there are three very different groups of backboned animals—sloths, chameleons, and the slow loris—that do live in treetops and that share the same odd retardation of motion and locomotion. Since these animals are not closely related, the loris being a kind of lemur—a primate like ourselves—the chameleon a lizard, and the sloth a climbing relative of the anteaters, and since their terrestrial relatives are not in any case notably slothful, the only sensible conclusion is that slowing down your muscles may be one way of fitting yourself for life in trees. The Pleistocene relatives of the modern tree sloths were ground-dwelling creatures, some of which were as big as oxen. We call them ground "sloths," but this does not mean that they were any more deliberate in their movements than, say, a bear. They are sloths only by a kind of reverse inheritance from their arboreal descendants.

Simply talking to people in the park, I was unable to find out either when the trees there were planted, or when the sloths came to live in them. Some said both had been there always. This is clearly a mistake, since the tree, as its Spanish name indicates, is native to Asia. Others said that a man who liked sloths put a pair there during the 1890's, and that the trees were already well matured then and formed the continuous canopy requisite to flourishing sloth life. An old woman sitting on a bench said that she remembered the time when a mayor of the town decided the sloths were a public nuisance and a menace to the future of the trees and should be shot. A policeman was detailed to kill them, with a military rifle that made a great *bulla* here in the center of the city. I told the woman that I had counted nine sloths only that morning.

"Exactly," she said with emphasis. "Now they are many. *Abundan*. It is my belief that the soldier missed two of them, or perhaps missed a pregnant female."

There may be records somewhere that would help establish the date of founding of the sloth-*laurel* community in Parque Vargas, but this is not essential. The important thing is that a native animal and an exotic plant have come together there and have worked out a natural equilibrium, with the sloth population completely dependent on the trees for food and yet not damaging them beyond the point of tolerance. It would be interesting to know exactly how the balance is maintained from one generation to the next. The sloths eat the leaves and fruit of the trees, and have no other food, and it seems safe to assume that the supply of food is the dominant factor controlling the size of the colony. Such uncluttered examples of population balance are easy to arrange among weevils in a flour bin or among protozoa in a jar of broth; but they are hard to find in nature, and especially among vertebrates. No zoologist of my acquaintance would have ventured to predict that two sloths placed in a grove of twenty-eight trees, of a sort untasted by sloths before and surrounded by city streets, would give rise to a population stabilized at nine individuals after thirty or forty years, or whatever the period has been.

It is curious how little this colony of extraordinary creatures intrudes upon the life of the city. Once in a great while a baby sloth, which normally clings spraddle-legged to its mother's upturned belly, falls out of a tree. Very rarely, for no discernible reason, a grown one descends and shuffles painfully across the bare earth among the plants and buttressed tree trunks. At such times a crowd of urchins and idle people gathers about the pitiful form. The boys shout and push one another toward the sloth, and when they try to stir up trouble between it and a jaded bitch more interested in sleep, the policeman strolls over from the sidewalk and tells them to stop molesting animals.

The only real excitement the sloths occasion comes when one of them swings from one handhold to another across the busy street on one of the electric wires that are strung through the trees. These are the times when everybody downtown stops to watch and talk about the sloths. "The lively Pete will be electrocuted," the Latins tell one another. The Castilian word for the sloth is *perezoso*, meaning about the same as ours; but in Central America it is everywhere known as *perico ligero*, "lively Pete," which is a fine example of vernacular irony. A windowful of Latin and creole girls across the street shrieks at the policeman at the corner of the park to stop the sloth before he gets to the insulator nailed to the angle of their building. A growing gang of boys leaning on bicycles howls and whistles with joy and mock lust.

"Yes, mon," one of them says. "I glod-ly walk this wy-ah upside down to get in that room. I do it fas-teah than the *perico!*"

After a while someone phones the power company, a truck comes with a ladder, and the sloth is plucked down and returned to the park. The people wander away or hang around in knots talking about the elections that are coming. The sloths are forgotten by everybody but me.

I was watching them still, as I said, on my fifth day in Limón. It was noon, and I was lying on my back and looking straight up to where two sloths were hanging in the deep shadows. There were several people on neighboring benches, but they had long ago got used to me and were paying me no attention. One of the sloths was scratching, dragging a foot across its shaggy side with the speed and regularity of the pendulum of a grandfather's clock. There was no likelihood that it would do anything else for hours. I had already lost interest in that sloth. But the other one, as I have said, excited me.

For fifteen minutes this second sloth had seemed to me to be moving nearer the first. If this was true, and the approach not just fortuitous, then perhaps the first sloth was a female and the second a male. It could be that I was at last to be allowed to witness what I had hoped for five days to see — the love-making of the lively Pete. Slow as the animal is, and upside down . . . I was intensely, perhaps even morbidly curious, and this approach was the first hint of sex that had crept into the activities of the sloths.

I waited longer. The advancing sloth — the supposed rutting

male—had reached a point no more than two feet from what I hoped was his goal and he was still moving. I had every reason to feel that there was purpose in his advance, and to hope that within a few minutes I would learn whether his motive was really sex or just pugnacity or sociability instead. When the gap had narrowed to ten or twelve inches, the sloth stopped dead still, then unhooked one fore foot, moved it slowly to one side, and held it poised there. For a full minute he hung motionless. Then he began turning his head with barely perceptible motion—a slow swoop and swing to the left, then back one hundred and eighty degrees to the right, then forward again, where it stopped. He seemed to be staring fixedly at the sloth in front. I was not able to make out his expression, but I can tell you I was on tenterhooks. Something was bound to happen now.

It did. A little boy touched my shoulder. "*Señor,*" he piped, "*está compuesto el avión*—the airplane is fixed!" His voice trembled with excitement at the news he bore.

"Wait a minute," I said. "Look up there. Look at those lively Petes." I pointed with tense concern, and the boy looked up into the dome of the grove and saw two sloths on a limb, one of them scratching itself rhythmically, the other hanging immobile from three legs.

"Yes, sir," he said. "You can see them any time. They live there."

"Sure, I know that," I said, still excited. "But what are they going to do?"

"Do?" the boy said.

"Yes. What are they going to do now?"

He hovered uncertainly, trying to pin down the joke that must be eluding him. "What are they going to do?" he said timidly.

"*Sí, hombre!*" I said. "What's the matter with you?"

He shrugged and spread his hands and rolled his eyes up toward the sloths and then back at me with mild reproach.

"They will hang there till it gets too hot. After that they will go to sleep." Then he tried once more to bring order into his world: "But, señor, the lady at the Aerovías says you must come now. Paco is making the Sixaola flight this afternoon and tomorrow you can have the airplane, but he has to see you now before he goes. *Ahorita!*"

With a sigh I rolled off the bench and onto my feet and found a coin for the boy.

"*Bueno. Muy bien,*" I said.

But the boy was tough, really. He was going to try once more to get at the bottom of this thing.

"You are not content?" he said, eyeing me narrowly. "Though the airplane is composed?"

I was not content. I was cheated. Five days' waiting and hardly a foot between two sloths and their sex rites; and suddenly the wretched plane was composed.

"Well, yes and no." I said.

The little boy looked at me with widening eyes. He did not know whether it was being grown up or being a gringo that had warped my soul. But he was sure life was becoming a complex thing here in this city where he was born.

WALLACE STEGNER
1909–1993

As a novelist and short-story writer, Wallace Stegner explored the experience of pioneers, mountain men, and settlers in the intermountain West. His essays, too, collected in The Sound of Mountain Water *(1969), investigated the issue of Western identity. Like Mary Austin, he asked his questions while looking at the landscape and felt the winds that blow away our easy, familiar answers. During his years as professor of english and director of the writing program at Stanford, Stegner advocated the work of western writers like John Wesley Powell and A. B. Guthrie, Jr., while encouraging a new generation of authors working in that always-new terrain.*

GLEN CANYON SUBMERSUS

Glen Canyon, once the most serenely beautiful of all the canyons of the Colorado River, is now Lake Powell, impounded by the Glen Canyon Dam. It is called a great recreational resource. The Bureau of Reclamation promotes its beauty in an attempt to counter continuing criticisms of the dam itself, and the National Park Service, which manages the Recreation Area, is installing or planning facilities for all the

The Sound of Mountain Water (Garden City, N.Y.: Doubleday, 1969).

boating, water skiing, fishing, camping, swimming, and plain sightseeing that should now ensue.

But I come back to Lake Powell reluctantly and skeptically, for I remember Glen Canyon as it used to be.

Once the river ran through Glen's two hundred miles in a twisting, many-branched stone trough eight hundred to twelve hundred feet deep, just deep enough to be impressive without being overwhelming. Awe was never Glen Canyon's province. That is for the Grand Canyon. Glen Canyon was for delight. The river that used to run here cooperated with the scenery by flowing swift and smooth, without a major rapid. Any ordinary boatman could take anyone through it. Boy Scouts made annual pilgrimages on rubber rafts. In 1947 we went through with a party that contained an old lady of seventy and a girl of ten. There was superlative camping anywhere, on sandbars furred with tamarisk and willow, under cliffs that whispered with the sound of flowing water.

Through many of those two hundred idyllic miles the view was shut in by red walls, but down straight reaches or up side canyons there would be glimpses of noble towers and buttes lifting high beyond the canyon rims, and somewhat more than halfway down there was a major confrontation where the Kaiparowits Plateau, seventy-five hundred feet high, thrust its knife-blade cliff above the north rim to face the dome of Navajo Mountain, more than ten thousand feet high, on the south side. Those two uplifts, as strikingly different as if designed to dominate some gigantic world's fair, added magnificence to the intimate colored trough of the river.

Seen from the air, the Glen Canyon country reveals itself as a barestone, salmon-pink tableland whose surface is a chaos of domes, knobs, beehives, baldheads, hollows, and potholes, dissected by the deep cork-screw channels of streams. Out of the platform north of the main river rise the gray-green peaks of the Henry Mountains, the last-discovered mountains in the contiguous United States. West of them is the bloody welt of the Waterpocket Fold, whose westward creeks flow into the Escalante, the last-discovered river. Northward rise the cliffs of Utah's high plateaus. South of Glen Canyon, like a great period at the foot of the fifty-mile exclamation point of the Kaiparowits, is Navajo Mountain, whose slopes apron off on every side into the stone and sand of the reservation.

When cut by streams, the Navajo sandstone which is the country rock forms monolithic cliffs with rounded rims. In straight stretches the cliffs tend to be sheer, on the curves undercut, especially in the narrow side canyons. I have measured a six-hundred-foot wall that was undercut a good five hundred feet—not a cliff at all but a musical shell for the multiplication of echoes. Into these deep scoured amphitheaters on the outside of bends, the promontories on the inside fit like thighbones into a hip socket. Often, straightening bends, creeks have cut through promontories to form bridges, as at Rainbow Bridge National Monument, Gregory Bridge in Fiftymile Canyon, and dozens of other places. And systematically, when a river cleft has exposed the rock to the lateral thrust of its own weight, fracturing begins to peel great slabs from the cliff faces. The slabs are thinner at top than at bottom, and curve together so that great alcoves form in the walls. If they are near the rim, they may break through to let a window-wink of sky down on a canyon traveler, and always they make panels of fresh pink in weathered and stained and darkened red walls.

Floating down the river one passed, every mile or two on right or left, the mouth of some side canyon, narrow, shadowed, releasing a secret stream into the taffy-colored, whirlpooled Colorado. Between the mouth of the Dirty Devil and the dam, which is a few miles above the actual foot of the Glen Canyon, there are at least three dozen such gulches on the north side, including the major canyon of the Escalante; and on the south nearly that many more, including the major canyon of the San Juan. Every such gulch used to be a little wonder, each with its multiplying branches, each as deep at the mouth as its parent canyon. Hundreds of feet deep, sometimes only a few yards wide, they wove into the rock so sinuously that all sky was shut off. The floors were smooth sand or rounded stone pavement or stone pools linked by stone gutters, and nearly every gulch ran, except in flood season, a thin clear stream. Silt pockets out of reach of flood were gardens of fern and redbud; every talus and rockslide gave footing to cottonwood and willow and single-leafed ash; ponded places were solid with watercress; maidenhair hung from seepage cracks in the cliffs.

Often these canyons, pursued upward, ended in falls, and sometimes the falls came down through a slot or a skylight in the roof of a domed chamber, to trickle down the wall into a plunge pool that made a lyrical dunk bath on a hot day. In such chambers the light was dim, reflected, richly colored. The red rock was stained with the dark manganese exudations called desert varnish, striped black to green to yellow to white along horizontal lines of seepage, patched with the chemical, sunless green of moss. One such grotto was named Music Temple by Major John Wesley Powell on his first exploration, in 1869; another is the so-called Cathedral in the Desert, at the head of Clear Water Canyon off the Escalante.

That was what Glen Canyon was like before the closing of the dam in 1963. What was flooded here was potentially a superb national park. It had its history, too, sparse but significant. Exploring the gulches, one came upon ancient chiseled footholds leading up the slickrock to mortared dwellings or storage cysts of the Basket Makers and Pueblos who once inhabited these canyons. At the mouth of Padre Creek a line of chiseled steps marked where Fathers Escalante and Dominguez, groping back toward Santa Fe in 1776, got their animals down to the fjord that was afterward known as the Crossing of the Fathers. In Music Temple men from Powell's two river expeditions had scratched their names. Here and there on the walls near the river were names and initials of men from Robert Brewster Stanton's party that surveyed a water-level railroad down the canyon in 1889–90, and miners from the abortive goldrush of the 1890's. There were Mormon echoes at Lee's Ferry, below the dam, and at the slot canyon called Hole-in-the-Rock, where a Mormon colonizing party got their wagons down the cliffs on their way to the San Juan in 1880.

Some of this is now under Lake Powell. I am interested to know how much is gone, how much left. Because I don't much like the thought of power boats and water skiers in these canyons, I come in March, before the season has properly begun, and at a time when the lake (stabilized they say because of water shortages far downriver at Lake Mead) is as high as it has ever been, but is still more than two hundred feet below its capacity level of thirty-seven hundred feet. Not everything that may eventually be drowned will be drowned yet, and there will be none of the stained walls and exposed mudflats that make a drawdown reservoir ugly at low water.

Our boat is the Park Service patrol boat, a thirty-four-foot diesel work-horse. It has a voice like a bulldozer's. As we back away from the dock and head out deserted Wahweap Bay, conversing at the tops of our lungs with our noses a foot apart, we acknowledge that we needn't have worried about motor noises among the cliffs. We couldn't have heard a Chriscraft if it had passed us with its throttle wide open.

One thing is comfortingly clear from the moment we back away from the dock at Wahweap and start out between the low walls of what used to be Wahweap Creek toward the main channel. Though they have diminished it, they haven't utterly ruined it. Though these walls are lower and tamer than they used to be, and though the whole sensation is a little like looking at a picture of Miss America that doesn't show her legs, Lake Powell *is* beautiful. It isn't Glen Canyon, but it is something in itself. The contact of deep blue water and uncompromising stone is bizarre and somehow exciting. Enough of the canyon feeling is left so that traveling up-lake one watches with a sense of discovery as every bend rotates into view new colors, new forms, new vistas: a great glowing wall with the sun on it, a slot side canyon buried to the eyes in water and inviting exploration, a half-drowned cave on whose roof dance the little flames of reflected ripples.

Moreover, since we float three hundred feet or more above the old river, the views out are much wider, and where the lake broadens, as at Padre Creek, they are superb. From the river, Navajo Mountain used to be seen only in brief, distant glimpses. From the lake it is often visible for minutes, an hour, at a time—gray-green, snow-streaked, a high mysterious bubble rising above the red world, incontrovertibly the holy mountain. And the broken country around the Crossing of the Fathers was always wild and strange as a moon landscape, but you had to climb out to see it. Now, from the bay that covers the crossing and spreads into the mouths of tributary creeks, we see Gunsight Butte, Tower Butte, and the other fantastic pinnacles of the Entrada formation surging up a sheer thousand feet above the rounding platform of the Navajo. The horizon reels with surrealist forms, dark red at the base, gray from there to rimrock, the profiles rigid and angular and carved, as different as possible from the Navajo's filigreed, ripple-marked sandstone.

We find the larger side canyons, as well as the deeper reaches of the main canyon, almost as impressive as they used to be, especially after we get far enough up-lake so that the water is shallower and the cliffs less reduced in height. Navajo Canyon is splendid despite the flooding of its green bottom that used to provide pasture for the stolen horses of raiders. Forbidden Canyon that leads to Rainbow Bridge is lessened, but still marvelous: it is like going by boat to Petra. Rainbow Bridge itself is still the place of magic that it used to be when we walked the six miles up from the river, and turned a corner to see the great arch framing the dome of Navajo Mountain. The canyon of the Escalante, with all its tortuous side canyons, is one of the stunning scenic experiences of a lifetime, and far easier to reach by lake than it used to be by foot or horseback. And all up and down these canyons, big or little, is the constantly changing, nobly repetitive spectacle of the cliffs with their contrasts of rounding and sheer, their great blackboard faces and their amphitheaters. Streaked with desert varnish, weathered and lichened and shadowed, patched with clean pink fresh-broken stone, they are as magically colored as shot silk.

And there is God's plenty of it. This lake is already a hundred and fifty miles long, with scores of tributaries. If it ever fills—which its critics guess it will not—it will have eighteen hundred miles of shoreline. Its fishing is good and apparently getting better, not only catfish and perch but rainbow trout and largemouth black bass that are periodically sown broadcast from planes. At present its supply and access points are few and far apart—at Wahweap, Hall's Crossing, and Hite—but when floating facilities are anchored in the narrows below Rainbow Bridge and when boat ramps and supply stations are developed at Warm Creek, Hole-in-the-Rock, and Bullfrog Basin, this will draw people. The prediction of a million visitors in 1965 is probably enthusiastic, but there is no question that as developed facilities extend the range of boats and multiply places of access, this will become one of the great water playgrounds.

And yet, vast and beautiful as it is, open now to anyone with a boat or the money to rent one, available soon (one supposes) to the quickie tour by float-plane and hydrofoil, democratically accessible and with its most secret beauties captured on color transparencies at infallible exposures, it strikes me, even in my exhilaration, with the consciousness of loss. In gaining the lovely and the usable, we have given up the incomparable.

The river's altitude at the dam was about 3150 feet. At 3490 we ride on 340 feet of water, and that means that much of the archaeology and most of the history, both of which were concentrated at the river's edge or near it, are as drowned as Lyonesse. We chug two hundred feet over the top of the square masonry tower that used to guard the mouth of Forbidden Canyon. The one small ruin that we see up Navajo Canyon must once have been nearly inaccessible, high in the cliff. Somehow (though we do not see them) we think it ought to have a line of footholds leading down into and under the water toward the bottom where the squash and corn gardens used to grow.

The wildlife that used to live comfortably in Glen Canyon is not there on the main lake. Except at the extreme reach of the water up side canyons, and at infrequent places where the platform of the Navajo sandstone dips down so that the lake spreads in among its hollows and baldheads, this reservoir laps vertical cliffs, and leaves no home for beaver or waterbird. The beaver have been driven up the side canyons, and have toppled whole groves of cottonwoods ahead of the rising water. While the water remains stable, they have a home; if it rises, they move upward; if it falls, the Lord knows what they do, for falling water will leave long mud flats between their water and their food. In the side canyons we see a few mergansers and redheads, and up the Escalante Arm the blue herons are now nesting on the cliffs, but as a future habitat Lake Powell is as unpromising for any of them as for the beaver.

And what has made things difficult for the wildlife makes them difficult for tourists as well. The tamarisk and willow bars are gone, and finding a campsite, or even a safe place to land a boat, is not easy. When the stiff afternoon winds sweep up the lake, small boats stay in shelter, for a swamping could leave a man clawing at a vertical cliff, a mile from any crawling-out place.

Worst of all are the places I remember that are now irretrievably gone. Surging up-lake on the second day I look over my shoulder and recognize the swamped and truncated entrance to Hidden Passage Canyon, on whose bar we camped eighteen years ago when we first came down this canyon on one of Norman Nevills' river trips. The old masked entrance is swallowed up, the water rises almost over the shoulder of the inner cliffs. Once that canyon was a pure delight to walk in; now it is only another slot with water in it, a thing to poke a motorboat into for five minutes and then roar out again. And if that is Hidden Passage, and we are this far out in the channel, then Music Temple is straight down.

The magnificent confrontation of the Kaiparowits and Navajo Mountain is still there, possibly even more magnificent because the lake has lifted us into a wider view. The splendid sweep of stained wall just below the mouth of the San Juan is there, only a little diminished. And Hole-in-the-Rock still notches the north rim, though the cove at the bottom where the Mormons camped before rafting the river and starting across the bare rock-chaos of Wilson Mesa is now a bay, with sunfish swimming among the tops of drowned trees. The last time I was here, three years ago, the river ran in a gorge three hundred feet below where our boat ties up for the night, and the descent from rim to water was a longer, harder way. The lake makes the feat of those Mormons look easier than it was, but even now, no one climbing the thousand feet of cliff to the slot will ever understand how they got their wagons down there.

A mixture of losses, diminishments, occasional gains, precariously maintained by the temporary stabilization of the lake. There are plenty of people willing to bet that there will never be enough water in the Colorado to fill both Lake Mead, now drawn far down, and Lake Powell, still 210 feet below its planned top level, much less the two additional dams proposed for Marble and Bridge canyons,[1] between these two. If there ever is — even if there is enough to raise Lake Powell fifty or a hundred feet — there will be immediate drastic losses of beauty. Walls now low, but high enough to maintain the canyon feeling, will go under, walls now high will be reduced. The wider the lake spreads, the less character it will have. Another fifty feet of water would sub-

---

[1] Since given up, at least temporarily [Stegner's note].

merge the Gregory Natural Bridge and flood the floor of the Cathedral in the Desert; a hundred feet would put both where Music Temple is; two hundred feet would bring water and silt to the very foot of Rainbow Bridge. The promontories that are now the most feasible camping places would go, as the taluses and sandbars have already gone. Then indeed the lake would be a vertical-walled fjord widening in places to a vertical-walled lake, neither as beautiful nor as usable as it still is. And the moment there is even twenty or thirty feet of drawdown, every side canyon is a slimy stinking mudflat and every cliff is defaced at the foot by a band of mud and minerals.

By all odds the best thing that could happen, so far as the recreational charm of Lake Powell is concerned, would be a permanently stabilized lake, but nobody really expects that. People who want to see it in its diminished but still remarkable beauty should go soon. And people who, as we do, remember this country before the canyons were flooded, are driven to dream of ways by which some parts of it may still be saved, or half-saved.

The dream comes on us one evening when we are camped up the Escalante. For three days we have been deafened by the noise of our diesel engines, and even when that has been cut off there has been the steady puttering of the generator that supplies our boat with heat, light, and running water. Though we weakly submit to the comforts, we dislike the smell and noise: we hate to import into this rock-and-water wilderness the very things we have been most eager to escape from. A wilderness that must be approached by power boat is no wilderness any more, it has lost its magic. Now, with the engines cut and the generator broken down, we sit around a campfire's more primitive light and heat and reflect that the best moments of this trip have been those in which the lake and its powerboat necessities were least dominant—eating a quiet lunch on a rock in Navajo Canyon, walking the 1.7 miles of sandy trail to the Rainbow Bridge or the half mile of creek bottom to the Cathedral in the Desert, climbing up the cliff to Hole-in-the-Rock. Sitting on our promontory in the Escalante canyon without sign or sound of the mechanical gadgetry of our civilization, we feel descending on us, as gentle as evening on a blazing day, the remembered canyon silence. It is a stillness like no other I have experienced, for at the very instant of bouncing and echoing every slight noise off cliffs and around bends, the canyons swallow them. It is as if they accentuated them, briefly and with a smile, as if they said, "Wait!" and suddenly all sound has vanished, there is only a hollow ringing in the ears.

We find that whatever others may want, we would hate to come here in the full summer season and be affronted with the constant roar and wake of power boats. We are not, it seems, water-based in our pleasures; we can't get a thrill out of doing in these marvelous canyons what one can do on any resort lake. What we have most liked on this trip has been those times when ears and muscles were involved, when the foot felt sand or stone, when we could talk in low voices, or sit so still that a brilliant collared lizard would come out of a crack to look us over. For us, it is clear, Lake Powell is not a recreational resource, but only a means of access; it is the canyons themselves, or what is left of them, that we respond to.

Six or seven hundred feet above us, spreading grandly from the rim, is the Escalante Desert, a basin of unmitigated stone furrowed by branching canyons as a carving platter is furrowed by gravy channels. It is, as a subsidiary drainage basin, very like the greater basin in whose trough once lay Glen Canyon, now the lake. On the north this desert drains from the Circle Cliffs and the Aquarius Plateau, on the east from the Waterpocket Fold, on the west from the Kaiparowits. In all that waste of stone fifty miles long and twenty to thirty wide there is not a resident human being, not a building except a couple of cowboy shelter shacks, not a road except the washed-out trail that the Mormons of 1880 established from the town of Escalante to Hole-in-the-Rock. The cattle and sheep that used to run on this desert range have ruined it and gone. That ringing stillness around us is a total absence of industrial or civilized decibels.

Why not, we say, sitting in chilly fire-flushed darkness under mica stars, why not throw a boom across the mouth of the Escalante Canyon and hold this one precious arm of Lake Powell for the experiencing of silence? Why not, giving the rest of that enormous water to the motorboats and the waterskiers, keep one limited tributary as a canoe or rowboat wilderness? There is nothing in the way of law or regulation to prevent the National Park Service from managing the Recreation Area in any way it thinks best, nothing that forbids a wilderness or primitive or limited-access area within the larger recreational unit. The Escalante Desert is already federal land, virtually unused. It and its canyons are accessible by packtrain from the town of Escalante, and will be accessible by boat from the facility to be developed at Hole-in-the-Rock. All down the foot of the Kaiparowits, locally called Fiftymile Mountain or Wild Horse Mesa, the old Mormon road offers stupendous views to those who from choice or necessity want only to drive to the edge of the silence and look in.

I have been in most of the side gulches off the Escalante—Coyote Gulch, Hurricane Wash, Davis Canyon, and the rest. All of them have bridges, windows, amphitheaters, grottoes, sudden pockets of green. And some of them, including the superlative Coyote Gulch down which even now it is possible to take a packtrain to the river, will never be drowned even if Lake Powell rises to its planned thirty-seven-hundred-foot level. What might have been done for Glen Canyon as a whole may still be done for the higher tributaries of the Escalante. Why not? In the name of scenery, silence, sanity, why not?

For awe pervades that desert of slashed and channeled stone overlooked by the cliffs of the Kaiparowits and the Aquarius and the distant peaks of the Henrys; and history, effaced through many of the canyons, still shows us its dim marks here: a crude *mano* discarded by an ancient campsite, a mortared wall in a cave, petroglyphs picked into a cliff face, a broken flint point glittering on its tee of sand on some blown mesa, the great rock where the Mormons danced on their way to people Desolation. This is country that does not challenge our identity as creatures, but it lets us shed most of our industrial gadgetry, and it shows us our true size.

Exploring the Escalante basin on a trip in 1961, we probed for the river through a half dozen quicksand gulches and never reached it, and never much cared because the side gulches and the rims gave us all we could hold. We saw not a soul outside our own party, encountered not a vehicle, saw no animals except a handful of cows and one mule that we scared up out of Davis Gulch when we rolled a rock over the rim. From every evening camp, when the sun was gone behind the Kaiparowits rim and the wind hung in suspension like a held breath and the Henrys northeastward and Navajo Mountain southward floated light as bubbles on the distance, we watched the eastern sky flush a pure, cloudless rose, darker at the horizon, paler above; and minute by minute the horizon's darkness defined itself as the blue-domed shadow of the earth cast on the sky, thinning at its upward arc to violet, lavender, pale lilac, but clearly defined, steadily darkening upward until it swallowed all the sky's light and the stars pierced through it. Every night we watched the earth-shadow climb the hollow sky, and every dawn we watched the same blue shadow sink down toward the Kaiparowits, to disappear at the instant when the sun splintered sparks off the rim.

In that country you cannot raise your eyes—unless you're in a canyon—without looking a hundred miles. You can hear coyotes who have somehow escaped the air-dropped poison baits designed to exterminate them. You can see in every sandy pocket the pug tracks of wildcats, and every waterpocket in the rock will give you a look backward into geologic time, for every such hole swarms with triangular crablike creatures locally called tadpoles but actually first cousins to the trilobites who left their fossil skeletons in the Paleozoic.

In the canyons you do not have the sweep of sky, the long views, the freedom of movement on foot, but you do have the protection of cliffs, the secret places, cool water, arches and bridges and caves, and the sunken canyon stillness into which, musical as water falling into a plunge pool, the canyon wrens pour their showers of notes in the mornings.

Set the Escalante Arm aside for the silence, and the boatmen and the water skiers can have the rest of that lake, which on the serene, warm, sun-smitten trip back seems more beautiful than it seemed coming up. Save this tributary and the desert back from it as wilderness, and there will be something at Lake Powell for everybody. Then it may still be possible to make expeditions as rewarding as the old, motorless river trips through Glen Canyon, and a man can make his choice between forking a horse and riding down Coyote Gulch or renting a houseboat and chugging it up somewhere near the mouth of the Escalante to be anchored and used as a base for excursions into beauty, wonder, and the sort of silence in which you can hear the swish of falling stars.

Coda: Wilderness Letter

Los Altos, Calif. Dec. 3, 1960

David E. Pesonen
Wildland Research Center
Agricultural Experiment Station
243 Mulford Hall
University of California
Berkeley 4, Calif.

Dear Mr. Pesonen:

I believe that you are working on the wilderness portion of the Outdoor Recreation Resources Review Commission's report. If I may, I should like to urge some arguments for wilderness preservation that involve recreation, as it is ordinarily conceived, hardly at all. Hunting, fishing, hiking, mountain-climbing, camping, photography, and the enjoyment of natural scenery will all, surely, figure in your report. So will the wilderness as a genetic reserve, a scientific yardstick by which we may measure the world in its natural balance against the world in its man-made imbalance. What I want to speak for is not so much the wilderness uses, valuable as those are, but the wilderness *idea*, which is a resource in itself. Being an intangible and spiritual resource, it will seem mystical to the practical-minded—but then anything that cannot be moved by a bulldozer is likely to seem mystical to them.

I want to speak for the wilderness idea as something that has helped form our character and that has certainly shaped our history as a people. It has no more to do with recreation than churches have to do with recreation, or than the strenuousness and optimism and expansiveness of what historians call the "American Dream" have to do with recreation. Nevertheless, since it is only in this recreation survey that the values of wilderness are being compiled, I hope you will permit me to insert this idea between the leaves, as it were, of the recreation report.

Something will have gone out of us as a people if we ever let the remaining wilderness be destroyed; if we permit the last virgin forests to be turned into comic books and plastic cigarette cases; if we drive the few remaining members of the wild species into zoos or to extinction; if we pollute the last clear air and dirty the last clean streams and push our paved roads through the last of the silence, so that never again will Americans be free in their own country from the noise, the exhausts, the stinks of human and automotive waste. And so that never again can we have the chance to see ourselves single, separate, vertical and individual in the world, part of the environment of trees and rocks and soil, brother to the other animals, part of the natural world and competent to belong in it. Without any remaining wilderness we are committed wholly, without chance for even momentary reflection and rest, to a headlong drive into our technological termite-life, the Brave New World of a completely man-controlled environment. We need wilderness preserved—as much of it as is still left, and as many kinds—because it was the challenge against which our character as a people was formed. The reminder and the reassurance that it is still there is good for our spiritual health even if we never once in ten years set foot in it. It is good for us when we are young, because of the incomparable sanity it can bring briefly, as vacation and rest, into our insane lives. It is important to us when we are old simply because it is there—important, that is, simply as idea.

We are a wild species, as Darwin pointed out. Nobody ever tamed or domesticated or scientifically bred us. But for at least three millennia we have been engaged in a cumulative and ambitious race to modify and gain control of our environment, and in the process we have come close to domesticating ourselves. Not many people are likely, any more, to look upon what we call "progress" as an unmixed blessing. Just as surely as it has brought us increased comfort and more material goods, it has brought us spiritual losses, and it threatens now to become the Frankenstein that will destroy us. One means of sanity is to retain a hold on the natural world, to remain, insofar as we can, good animals. Americans still have that chance, more than many peoples; for while we were demonstrating ourselves the most efficient and ruthless environment-busters in history, and slashing and burning and cutting our way through a wilderness continent, the wilderness was working on us. It remains in us as surely as Indian names remain on the land. If the abstract dream of human liberty and human dignity became, in America, something more than an abstract dream, mark it down at least partially to the fact that we were in subtle ways subdued by what we conquered.

The Connecticut Yankee, sending likely candidates from King Arthur's unjust kingdom to his Man Factory for rehabilitation, was overoptimistic, as he later admitted. These things cannot be forced, they have to grow. To make such a man, such a democrat, such a believer in human individual dignity, as Mark Twain himself, the frontier was necessary, Hannibal and the Mississippi and Virginia City, and reaching out from those the wilderness; the wilderness as opportunity and as idea, the thing that has helped to make an American different from and, until we forget it in the roar of our industrial cities, more fortunate than other men. For an American, insofar as he is new and different at all, is a civilized man who has renewed himself in the wild. The American experience has been the confrontation by old peoples and cultures of a world as new as if it had just risen from the sea. That gave us our hope and our excitement, and the hope and excitement can be passed on to newer Americans, Americans who never saw any phase of the frontier. But only so long as we keep the remainder of our wild as a reserve and a promise—a sort of wilderness bank.

As a novelist, I may perhaps be forgiven for taking literature as a reflection, indirect but profoundly true, of our national consciousness. And our literature, as perhaps you are aware, is sick, embittered, losing its mind, losing its faith. Our novelists are the declared enemies of their society. There has hardly been a serious or important novel in this century that did not repudiate in part or in whole American technological culture for its commercialism, its vulgarity, and the way in which it has dirtied a clean continent and a clean dream. I do not expect that the preservation of our remaining wilderness is going to cure this condition. But the mere example that we can as a nation apply some other criteria than commercial and exploitative considerations would be heartening to many Americans, novelists or otherwise. We need to demonstrate our acceptance of the natural world, including ourselves;

we need the spiritual refreshment that being natural can produce. And one of the best places for us to get that is in the wilderness where the fun houses, the bulldozers, and the pavements of our civilization are shut out.

Sherwood Anderson, in a letter to Waldo Frank in the 1920's, said it better than I can. "Is it not likely that when the country was new and men were often alone in the fields and the forest they got a sense of bigness outside themselves that has now in some way been lost. . . . Mystery whispered in the grass, played in the branches of trees overhead, was caught up and blown across the American line in clouds of dust at evening on the prairies. . . . I am old enough to remember tales that strengthen my belief in a deep semi-religious influence that was formerly at work among our people. The flavor of it hangs over the best work of Mark Twain. . . . I can remember old fellows in my home town speaking feelingly of an evening spent on the big empty plains. It had taken the shrillness out of them. They had learned the trick of quiet. . . ."

We could learn it too, even yet; even our children and grandchildren could learn it. But only if we save, for just such absolutely non-recreational, impractical, and mystical uses as this, all the wild that still remains to us.

It seems to me significant that the distinct downturn in our literature from hope to bitterness took place almost at the precise time when the frontier officially came to an end, in 1890, and when the American way of life had begun to turn strongly urban and industrial. The more urban it has become, and the more frantic with technological change, the sicker and more embittered our literature, and I believe our people, have become. For myself, I grew up on the empty plains of Saskatchewan and Montana and in the mountains of Utah, and I put a very high valuation on what those places gave me. And if I had not been able periodically to renew myself in the mountains and deserts of western America I would be very nearly bughouse. Even when I can't get to the back country, the thought of the colored deserts of southern Utah, or the reassurance that there are still stretches of prairie where the world can be instantaneously perceived as disk and bowl, and where the little but intensely important human being is exposed to the five directions and the thirty-six winds, is a positive consolation. The idea alone can sustain me. But as the wilderness areas are progressively exploited or "improved," as the jeeps and bulldozers of uranium prospectors scar up the deserts and the roads are cut into the alpine timberlands, and as the remnants of the unspoiled and natural world are progressively eroded, every such loss is a little death in me. In us.

I am not moved by the argument that those wilderness areas which have already been exposed to grazing or mining are already deflowered, and so might as well be "harvested." For mining I cannot say much good except that its operations are generally short-lived. The extractable wealth is taken and the shafts, the tailings, and the ruins left, and in a dry country such as the American West the wounds men make in the earth do not quickly heal. Still, they are only wounds; they aren't absolutely mortal. Better a wounded wilderness than none at all. And as for grazing, if it is strictly controlled so that it does not destroy the ground cover, damage the ecology, or compete with the wildlife it is in itself nothing that need conflict with the wilderness feeling or the validity of the wilderness experience. I have known enough range cattle to recognize them as wild animals; and the people who herd them have, in the wilderness context, the dignity of rareness; they belong on the frontier, moreover, and have a look of rightness. The invasion they make on the virgin country is a sort of invasion that is as old as Neolithic man, and they can, in moderation, even emphasize a man's feeling of belonging to the natural world. Under surveillance, they can belong; under control, they need not deface or mar. I do not believe that in wilderness areas where grazing has never been permitted, it should be permitted; but I do not believe either that an otherwise untouched wilderness should be eliminated from the preservation plan because of limited existing uses such as grazing which are in consonance with the frontier condition and image.

Let me say something on the subject of the kinds of wilderness worth preserving. Most of those areas contemplated are in the national forests and in high mountain country. For all the usual recreational purposes, the alpine and forest wilderness are obviously the most important, both as genetic banks and as beauty spots. But for the spiritual renewal, the recognition of identity, the birth of awe, other kinds will serve every bit as well. Perhaps, because they are less friendly to life, more abstractly non-human, they will serve even better. On our Saskatchewan prairie, the nearest neighbor was four miles away, and at night we saw only two lights on all the dark rounding earth. The earth was full of animals—field mice, ground squirrels, weasels, ferrets, badgers, coyotes, burrowing owls, snakes. I knew them as my little brothers, as fellow creatures, and I have never been able to look upon animals in any other way since. The sky in that country came clear down to the ground on every side, and it was full of great weathers, and clouds, and winds, and hawks. I hope I learned something from knowing intimately the creatures of the earth; I hope I learned something from looking a long way, from looking up, from being much alone. A prairie like that, one big enough to carry the eye clear to the sinking, rounding horizon, can be as lonely and grand and simple in its forms as the sea. It is as good a place as any for the wilderness experience to happen; the vanishing prairie is as worth preserving for the wilderness idea as the alpine forests.

So are great reaches of our western deserts, scarred somewhat by prospectors but otherwise open, beautiful, waiting, close to whatever God you want to see in them. Just as a sample, let me suggest the Robbers' Roost country in Wayne County, Utah, near the Capitol Reef National Monument. In that desert climate the dozer and jeep tracks will not soon melt back into the earth, but the country has a way of making the scars insignificant. It is a lovely and terrible wilderness, such a wilderness as Christ and the prophets went out into; harshly and beautifully colored, broken and worn until its bones are exposed, its great sky without a smudge or taint from Technocracy, and in hidden corners and pockets under its cliffs the sudden poetry of springs. Save a piece of country like that intact, and it does not matter in the slightest that only a few people every year will go into it. That is precisely its value. Roads would be a desecration, crowds would ruin it. But those who haven't the strength or youth to go into it and live can simply sit and look. They can look two hundred miles, clear into Colorado; and looking down over the cliffs and canyons of the San Rafael Swell and the Robbers' Roost they can also look as deeply into themselves as anywhere I know. And if they can't even get to the places on the Aquarius Plateau where the present roads will carry them, they can simply contemplate the *idea,* take pleasure in the fact that such a timeless and uncontrolled part of earth is still there.

These are some of the things wilderness can do for us. That is the reason we need to put into effect, for its preservation, some other principle than the principles of exploitation or "usefulness" or even recreation. We simply need that wild country available to us, even if we never do more than drive to its edge and look in. For it can be a means of reassuring ourselves of our sanity as creatures, a part of the geography of hope.

Very sincerely yours,

Wallace Stegner

JACQUETTA HAWKES
1910–1996

*Daughter of a Cambridge University biochemist and cousin to poet
Gerard Manley Hopkins, Jacquetta Hawkes always considered science
and art to be complementary approaches to understanding nature and
human culture. A professional archaeologist, she conducted excavations in England, Europe, and the Middle East in the 1930s, served in
the War Cabinet Offices during World War II, founded the British
Commission for UNESCO, and was governor of the British Film
Institute. Her many publications include archeological works, poetry,
biographies, and novels, as well as plays written in collaboration with
her husband, J. B. Priestley.*

*A Land, published in 1952 with illustrations by the sculptor Henry
Moore, is on one level a fascinating geological and archeological history
of the British Isles. But on another level it is an extended poetic and
philosophical meditation on time, the growth of human consciousness,
and especially the interplay between landscapes and the cultures that
form and are formed by them. No less than geologists and archaeologists,
Hawkes asserts, poets, painters, and musicians have helped shape our
vision of landscape. Hawkes's gift for story-telling, her intelligent, intimate style and broad knowledge of her subject, combine with the singular insularity and continuity of England's long history to produce a
compelling narrative with mythic dimensions.*

From A LAND

TWO THEMES

When I have been working late on a summer night, I like to go out
and lie on the patch of grass in our back garden. This garden is a

A Land (New York: Random House, 1952).

square of about twenty feet, so that to lie in it is like exposing oneself in an open box or tray. Not far below the topsoil is the London Clay which, as Primrose Hill, humps up conspicuously at the end of the road. The humus, formed by the accumulations first of forest and then of meadow land, must once have been fertile enough, but nearly a century in a back garden has exhausted it. After their first season, plants flower no more, and are hard put to it each year even to make a decent show of leaves. The only exceptions are the lilies of the valley, possessors of some virtue that enables them to draw their tremendous scent from the meanest soils. The sunless side of the garden has been abandoned to them, and now even in winter it is impossible to fork the earth there, so densely is it matted with the roots and pale nodes from which their flowers will rise.

Another result of the impoverishment of the soil is that the turf on which I lie is meagre and worn, quite without buoyancy. I would not have it otherwise, for this hard ground presses my flesh against my bones and makes me agreeably conscious of my body. In bed I can sleep, here I can rest awake. My eyes stray among the stars, or are netted by the fine silhouettes of the leaves immediately overhead and from them passed on to the black lines of neighbouring chimney pots, misshapen and stolid, yet always inexplicably poignant. Cats rustle in the creeper on the end wall. Sometimes they jump down so softly that I do not hear them alight and yet am aware of their presence in the garden with me. Making their silken journeys through the dark, the cats seem as untamed, as remote, as the creatures that moved here before there were any houses in the Thames valley.

By night I have something of the same feeling about cats that I have always, and far more strongly, about birds: that perfectly formed while men were still brutal, they now represent the continued presence of the past. Once birds sang and flirted among the leaves while men, more helpless and less accomplished, skulked between the trunks below them. Now they linger in the few trees that men have left standing, or fit themselves into the chinks of the human world, into its church towers, lamp-posts and gutters. It is quite illogical that this emotion should be concentrated on birds; insects, for example, look, and are, more ancient. Perhaps it is evoked by the singing, whistling and calling that fell into millions of ancestral ears and there left images that we all inherit. The verses of medieval poets are full of birds as though in them these stored memories had risen to the surface. Once in the spring I stood at the edge of some Norfolk ploughland listening to the mating calls of the plover that were tumbling ecstatically above the fields. The delicious effusions of turtle doves bubbled from a coppice at my back.

It seemed to me that I had my ear to a great spiral shell and that these sounds rose from it. The shell was the vortex of time, and as the birds themselves took shape, species after species, so their distinctive songs were formed within them and had been spiralling up ever since. Now, at the very lip of the shell, they reached my present ear.

As I lie looking at the stars with that blend of wonder and familiarity they alone can suggest, a barge turning the bend in Regent's Park Canal hoots, a soft wedge of sound in the darkness that is cut across by the long rumble of a train drawing out from Euston Station. Touched by these sounds, like a snail I retract my thoughts from the stars and banish the picture of the earth and myself hanging among them. Instead I become conscious of the huge city spreading for miles on all sides, of the innumerable fellow creatures stretched horizontally a few feet above the ground in their upstairs bedrooms, and of the railways, roads and canals rayed out towards all the extremities of Britain. The people sitting in those lighted carriages, even the bargee leaning sleepily on his long tiller, are not individuals going to board meetings in Manchester or bringing in coal for London furnaces. For the instant they are figures moved about the map by unknown forces, as helpless as the shapes of history that can be seen behind them, all irresistibly impelled to the achievement of this moment.

The Thames flows widening towards the city it has created; the coastline of Britain encloses me within a shape as familiar as the constellations of the stars, and as consciously felt as the enclosing walls of this garden. The coast with its free, sweeping lines among the young formations of the east and south, and its intricate, embattled line of headland and bay among the ancient rocks of the west and north. The shape seems constant in its familiarity yet in fact is continuously changing. Even the stern white front that Albion turns to the Continent is withdrawing at the rate of fifteen inches a year. I remember as a small child being terrified by a big fall of cliff at Hunstanton, and I am certain that my terror was not so much due to the thought of being crushed—the fall had happened some days before, as by some inkling of impermanence. It was the same knowledge, though in a sadder and less brutal form, that came stealing in from the submerged forest, also to be seen at Hunstanton, a dreary expanse of blackened tree stumps exposed at low tide.

Always change, and yet at this moment, at every given moment, the outline of Britain, like all outlines, has reality and significance. It is the endless problem of the philosophers; either they give process, energy, its due and neglect its formal limitations, or they look only at forms and forget the irresistible power of change. The answers to all the great secrets are hidden somewhere in this thicket, those of ethics and aesthetics as well as of metaphysics.

I know of no philosophy that can disprove that this land, having achieved this moment, was not always bound to achieve it, or that I, because I exist, was not always inevitably coming into existence. It is therefore as an integral part of the process that I claim to tell the story of the creation of what is at present known as Britain, a land which has its own unmistakable shape at this moment of time.

There are many ways in which this story can be told, just as a day in the life of this house behind me could be described in terms of its intake of food and fuel, and its corresponding output through drains, dustbins and chimneys, or in terms of the movement in space and time of its occupants, or of their emotional relationships. All these forms, even the most material, would be in some sense creations of the story-teller's mind, and for this reason the counterpoint to the theme of the creation of a land shall be the growth of consciousness, its gradual concentration and intensification within the human skull.

That consciousness has now reached a stage in its growth at which it is impelled to turn back to recollect happenings in its own past which it has, as it were, forgotten. In the history of thought, this is the age of history. Some forms of these lost memories lie in the unconscious strata of mind itself, these dark, rarely disturbed layers that have accumulated, as mould accumulates in a forest, through the shedding of innumerable lives since the beginning of life. In its search for these forms consciousness is working, not always I think very sensitively, through its psychologists. I am certainly involved in their findings, but as narrator am not concerned with them. Instead I am concerned with other forms of memory, those recollections of the world and of man that are pursued on behalf of consciousness by geologists and archaeologists.

Unfortunately they have not yet gone far enough to recall the formation of the planet Earth. In my own childhood I drew a crude picture in my mind of a fragment flying off from the side of the sun, much as a piece of clay, carelessly handled, flies from a pot revolving on the potter's wheel. Then there were other, conflicting, pictures of the formation of planets by awe-inspiring cosmic road accidents, immense collisions. It seems that both were fanciful. Yet as we have not yet remembered what did happen, I must begin with a white-hot young earth dropping into its place like a fly into an unseen four-dimensional cobweb, caught up in a delicate tissue of forces where it assumed its own inevitable place, following the only path, the only orbit that was open to it.

At first the new planet was hot enough to shine with its own light, but so small a particle, lacking the nuclear energy that allows the sun to shine gloriously for billions of years at the expense only of some slight change in girth, could not keep its heat for very long. Its rays turned from white to red, then faded till Earth was lit only from without, from the sun round which it swung on an invisible thread. From that time night and day were established, the shadow of the Earth pointing into space like a huge black tent.

Writing in 1949 I say that night and day were established. It is, I know, foolish to use these words for a time before consciousness had grown in men and had formed the image of night and day as the spinning globe sent them from sunlight under the cone of shadow and out again at dawn. I should wait to use these words until this procession of light and darkness had formed one of the most deep-set images in the mind of man. But the concept is now so familiar that I cannot express myself otherwise.

I lie here and feel Earth rustling through space, its rotundity between me and the sun, the shadow above me acting as a searchlight to reveal the stars whose light left them long before there were eyes on this planet to receive it. Now the two little globes of my eyes, unlit in the darkness, look up at their shining globes, and who shall say that we do not gaze at one another, affect one another?

The first pallor of the rising moon dimming the stars over the chimneys reminds me of our modest satellite. I have known her for so long that she is an accepted part of the night, yet were I lying on Jupiter the sky would be radiant with ten moons, while on Saturn the rings would glisten day and night in a glorious bow. Now she has risen into sight, our one familiar moon. A beautiful world to our eyes, but cold and lifeless; without water or atmosphere she is a presage of what Earth might become. I should like to know whether in those icy rocks there are the fossils of former life, organisms that had gone some way in the process in which we are involved before they were cut short by an eternal drought. Do they lie in the rocks beneath the rays of a sun that once gave them life but now beats meaninglessly on a frigid landscape?

I feel them at their employment, the sun, moon, Earth and all the rest, even while more intimately I am aware of Britain moving through the night which, like a candle extinguisher, has put out her ordinary life. But if, which heaven forbid, I were at this moment to leap into a jet aeroplane we could catch up with day in a few hours, or could plunge into winter in a few days. It is difficult to remember for how great a part of history these thoughts and images would have appeared as the wildest delusions of a madman. We felt more secure when we believed ourselves to be standing on a plate under the protective dome of heaven with day and night given for work and sleep. If we were less confident in Athens it was only by intuition and native courage. Now knowledge of material facts imposes humility upon us, willy nilly. Not that I would allow myself to repent the divine curiosity that has led to this knowledge. Like everyone else within the walls of these islands I am a European, and as a European committed utterly to *la volonté de la conscience et la volonté de la découverte*. To enjoy, to create (which is to love) and to try to understand is all that at the moment I can see of duty. As for apparent material facts, I hope that in time we shall have come to know so many, and to have seen through so many, that they will no longer appear as important as they now do.

At present, certainly, they are powerful; we have allowed them to become our masters. Yet, strangely, as I lie here in my ignorance under the stars, I am aware of awe but not of terror, of humility but not of insignificance.

Meanwhile the moon has drawn clear of the chimneys. How ungrateful we have been to call her inconstant when she is the only body in the heavens to have remained faithful to us in spite of our intelligence, the only body that still revolves about us. She is riding high and I must go to bed before first the Isle of Thanet noses out, and then London itself emerges on the other side of night.

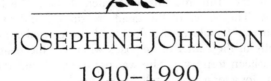

JOSEPHINE JOHNSON
1910–1990

Josephine Johnson won the Pulitzer Prize for her novel Now in November, *published in 1934. Over the course of a long career, she continued to produce novels, stories, and poems. As a nature writer, she is best known for two books:* The Inland Island *(1969) and* Seven Houses: A Memoir of Time and Places *(1973). The former volume, from which the present selection is taken, follows the circle of the seasons through the woods and fields around her Ohio farm. Within this traditional structure, there is an edginess previously uncommon in the genre. Beyond the precision of her observations and the sober authority of her voice, this is a book marked by the pain of the Vietnam War. Johnson's interests in world peace and conservation collided as she took excursions around her home-ground. So far from Vietnam, she nonetheless found her composure shaken by distant rumbles of battle. Her knowledge of atrocities that were being committed, as she felt, in her name caused her to question the validity of her own solitude and peace. At the same time, they brought a tragic magnitude to her experience of the woods and may have made her especially alert to the incessant competition of animals and plants with each other for a moment in the light.*

From THE INLAND ISLAND

MARCH

The hear-hear of the cardinals in the cold is an icy sound. Knuckles on glass. It will sound different in April. The great wicker cage of the for-sythia bush, which covers thirty feet now, is turning yellow. It has moved outward from a moderate bush to this wild tangle—as the banyan tree expands—up and then down, to root in an ever-widening circle of curved earth-seeking arches and wild, unrooted shoots. A beautiful mess, a witch's hair. An open cage, through which the birds pour all day long, or sit in sullen lumps to feel the morning sun. Rabbits and opossum move along its floor; it is the quails' way station in their endless circling of the land—a line of bright, speckled globes, brown heads, they run into this wild witch shelter. The hawk beats silently away. The voices of the quail are strange and sweet. A cat pounced on one female in the flock and she carried a grey ragged patch directly in the center of her back where all the neat, exquisite feathers had been torn away. She was not crippled, but the purplish skin was bare for a long time, circled with grey down.

The hill edge is rimmed with daffodil shoots, falsely called forth, betrayed by snow and ice. Yellow-sheathed buds surviving grimly. This is a new month because a little bunch of squares on the calendar says so. Once it was the year's beginning. The Saxons called it *Hlydmonath*, loud and stormy month. Around and around the seasons go, and every year the gaping mouth of March, the windy month that breaks the season open. Mars, god of war, lengthening month, and out of this month flow length-ening days, wind, leaves, and war, torn crows and promises. O promises.

There are brick-red elm buds against the sky. The far hills appear

The Inland Island (New York: Simon and Schuster, 1969).

lavender instead of grey. Or do I imagine this because I turned the calendar page over? Month of the bloodstone, that strange stone of polished plasma, green with red spots, those drops that represent the blood of Christ in our mythology of gems.

How long can I go on pretending the world is not changed beyond recognition? The blood from that one nation which we ravage flows all over the world. It flows into my mind night and day. Blood, despair and rage. It's not frozen into a stone. It's living blood and it's turning black, not red.

The traditional March winds have been around for several months, but now they come again, leaf lifting, tree shaking. It is terrifying to see a huge tree move and shake clear to its roots. Or does it only *seem* to move? The branches whip and bend, they strain, and the roar of the wind strikes terror to our hearts and we think the tree has moved from bud to root. The wind is glorious and pagan. It blows the dust of our lives a thousand miles. We read the dust. If there's a deadly message in it, that's for us to know. The wind comes bearing things with it, lifting up, sweeping on. The great invisible river that has no need or knowledge of us.

No, I don't like to walk in the wind and the rain. The rain gets on my glasses and down my neck. It chills me and there's no fire inside. The wind's an enemy, too. Except the joy of hiding from it, finding a hollow haven in the leeward side of a ravine. It seldom turns corners. Like the ancient Chinese ghosts, it must flow straight ahead. Weak, furry creatures watch from holes. Birds flatten against the bark. The crow locks his knees. One doesn't find the small perching birds in heaps like tumbleweeds, so they survive somehow. Shoulder up in the heart of the prickly cedar. Cover their eyes with their wings or hole down in their feathers, flatten their crests. The leaves, not mashed by ice and heavy snow, break up, turn on end. Dry leaves start awkwardly on brief journeys, wild risings, brown and ghostly dispersions to a traveling nowhere and a huddling under the thorns.

Rain turning to snow.

Terrible headache probably caught from sick buzzard at the zoo.

The quail come through the snow, which is steady and slanted, not aimless; thoughtful, as in slower snows. The quail rouse up the leaves, make disorderly beds. The snow covers them. The grass sticks up green hairs, The daffodils stick up, their pale heads watery and discouraged. The delicate cold petals of forsythia open in the house. A triumph. A success. It makes one feel among the great gardeners at last. Didn't rot; didn't slump; didn't wait until its brethren outside were also in full bloom. These frosty fingers I regard with awe again.

The near trees are stark black. The far trees a net of white webs, the

far hills grey. Now the snow shifts its pattern. The steady slant turns to a
spiral. The flakes seem larger, rounder. Some trees are turning green
with lichen. Yesterday the fields were alive with flocking robins. When
the robin flies down he does not flutter, he drops. When he eats, he pulls
at his food, as though pulling a worm out of the ground. He lifts the
leaves in his beak and tosses them aside. He jerks up the crusts of bread.

The forsythia cage has turned pure white. An arched net of snow. The
delicate grey juncos line its inner branches. Its buds are white. It has
caught and caged every flake of snow that's fallen. And under the snow
run the thin gold lines of sap. The forsythia came from China and is
related to the olive. *Forsythia suspensa* has hollow upright branches,
Forsythia intermedia has arched branches, and most forsythia are hybrids
of *F. intermedia*, which was the offshoot of *F. viridissima* and *F. suspensa*,
the wild olive ancestors from Asia. *Intermedia* is a dull name for these
great arching bridges and branches, half in heaven, half in earth. It's
viridissima—virere, to be green, *vire*, the arrow-feathered, to achieve a
rotary motion . . . *vir*, to turn . . . virent, freshness . . . to shoot up . . . up
and turn . . . burrow down and rise again . . . *virbiasses* . . . being twice a
man (and *sississimus* is Swahili for red ants—are you with us still?).

This is *suspensa-viridissima* that we have here, not just a squatty
intermedia, studded with barn sparrows, huge brown buds on every
hoop.

The snow is slowing down. A trilling comes. Some early amorous
bird. The cardinals feed each other. A courting gesture. He offers her a
sunflower seed. Her beak is full already. He jabs away. Eventually his
loving peck is received. Usually it's better handled, better beaked, a
sweet reciprocity.

He stretches lean in his courting song. His crest stretched up, his
whole body thinned out as he sings, watching the female, tung tung
tung . . . the strained persistence of a single note. She flies away. He
flies down to eat, Light, but not heat.

The ground is whiter, whiter, the bark greener. The mournful
sound of the Norfolk and Western hoots up the valley. It rolls up two
miles through the snow.

The quail are eating and down comes a squirrel, snow-frosted
pink-rimmed eyes, icy little claws. What is the quail attitude toward the
squirrel? They stop eating, they huddle together, and each and every
quail turns its back toward the squirrel and stares off into the distance.
He rummages about, digs up snow-covered bread, runs away. They
turn around and resume their eating. A curious sight. How does he
feel? He doesn't give a damn. He is long gone with his greasy bread.

It appears to be snowing birds—sparrows and juncos, tomtits and chickadees, woodpeckers and cardinals, and suddenly a robin is snowed down.

May God make the snow fall on this terrible war. Bury the armor. Cool the burned.

The year moves on toward spring. The wicker cage of the forsythia grows more yellow every day. They used to call these changeable days "the dangerous weather." Every day is dangerous now, climatewise and otherwise. For three months there have been these shiftings, these extremes.

Woke one morning to an extraordinary snow. Deep soft blanket world. Great white elephants looming everywhere. The honeysuckle and the fence a hairy mammoth. The wind released the pine branches slowly—a long exploding tip of pine moves in a circle, like the sensitive tip of an elephant's trunk.

On the green-grey ice of the pond a caravan of camels is kneeling. Strange shapes everywhere. The garbage cans like giant popovers. Two men with guns. "We didn't see no signs." One was kneeling in the road, his gun pointed at the field's "No Hunting" sign.

A world too beautiful, too strange for words. After awhile the sun and wind turned all the snow into something mad and messy. Suds, batter, dashed against the trees. The snow on the ground was pitted from odd falling clumps off the branches. Once a load descended like a falling bird with outstretched wings that shattered into light.

Showers of snow fell in a veil before the sun. All through the woods white birds flying and falling. And yet it was silent. Lovely and silent and blazing white.

Then, in a few days, the warming began. The invisible melting.

Spongy ground. The dry hummocks of grass squashing under foot into the snow ooze. Saw tadpoles in the brown water of the pond and, dimly, mossy encrusted crawfish crawling near the shore. The algae rose in green clumps with curious rootlike hangings, green Medusas. The ground is pitted with mouse and mole and shrew holes. It looks like a big brown cheese around the bird feeder. Bees are out; mosquitoes. Big black ants appear on the sink.

That kingfisher sound was a flicker. I should have remembered that peculiar call. It paused briefly, high in black cherry. (Was it a flicker the opossum gnawed, at the Beckers, sitting preposterously calm on the flood-lighted lawn, his little nasty hands shuffling yellow feathers, spreading the wings like a fan before his face, and gnawing the bloody bones?)

Two doves flew down. Indescribable softness of their color, pink over grey; the shape of doves is maddeningly beautiful. Their feet were pink in the moss. They have tiny little heads. Their descent, awkward, nervous, twittering, is a disaster more than a descent. Their rising is labored and noisy. They are so slow only their extreme nervousness saves them. But beautiful—oh, beautiful—dove grey, dove rose, dove blue, and the exquisite lines of their stupid little heads. They drive the bluejays away.

Very warm in the woods. The big creek filthy—a greasy grey like an enormous flowing of dirty dishwater with tiny particles as from thousands of human teeth pickings. Hardly wanted to get my boots wet in it.

But the second creek is pure and flowing over the stones. Sweet wet earth smell. Sat by the large granite boulder, almost green now, orange where the water edged around it. Found two silvery nests of vireos still knitted well with the moss between. No birds. No tracks. Only the leaves on the move. Leaping up, tugging at the thin frozen blanket of themselves, moving about like squirrels. Got scared by a rabbit bolting out from his shelter. Sounded louder than a gunshot.

Stroked a fallen log covered with moss as though it were a green, living thing. A cloud of yellow pollen blew from my hand. Other logs had every crack filled with the oyster-shaped fungus of *Stereum versicolor*, some small and greyish, upright as oysters, shoved in the cracks, and the larger spread as the tails of drumming pheasants, beautiful with copper bands, brown, grey and mossy-green bands, damp and velvety smooth to touch. Another fungus grew with them, less distinctly banded, whose underside was ridged and folded, *Daedalea quercina*, which means worked cunningly, and labyrinthine—after Daedalus who made the Minotuar's deep winding labyrinth. A heavy name for this crimped and greyish bit of stuff! I would like to find a *Polyporus squamosus*, which they say grows seven feet in circumference, forty-two pounds in weight, and all this colorful expansion in four weeks alone. Found, instead, a root so resembling the head of a goat that horns, ears and beard were clearly there, a rude yet delicate resemblance that startled.

It grows almost hot in the ravine. The air is soft and damp, a soft light breeze at times. The creek is running well. The cold water flowed through hollow cow bones on the pebbles, the air full of little white things flying. And on a rock a lively coal-black springtail brisked about. The bright-green varnished shoots of star-of-Bethlehem came up in clumps, and the pure-white bulbs were exposed along the creek where the earth had washed away. Moss everywhere—green moss on stones, on logs, bright emerald green. The tree across the creek that slants up skyward is covered with moss now, snow all gone. The only whiteness, odd, bleached rocks stranded here and there, returning to their old shell color. And there are flat square stones, grey, ominous, chiseled with laws, like some mosaic tablet (or one could think of them as stone sandwiches instead, very old and stale).

Someone has upturned a big stone under a tree, its earth side raw and exposed. And has ripped a great vine from the tree and knocked it aside.

The stream is crossed and recrossed by fallen logs. And on one a mourning-cloak butterfly sits in the sun, slowly folding and opening his dark wings edged with cream. A red so dark it is black. A phoebe has come, but it is silent. The only sounds are the crows and the creek.

Almost invisible under the leaves, the salt-and-pepper flowers are opening with a faint honey scent. And from a fresh hole comes a puff of skunk. The scarlet cup fungus are open. Bright cardinal mouths among the leaves.

And then comes a more believable March day. Harsh, cold. The week of warmth and sun has urged everything forward in preparation for this grey descent, this icy air, this icy mist. Fold on fold, over every bud, every gold daffodil. Every premature arrival. Get them up. Freeze them out. Chill. Chill. It's some ghastly gothic joke that's played each year.

The forsythia fountain looks alive though, runs yellow in all the hoops, and all the hoops are hung with yellow drops, the unopened flowers arrested by the cold. Above them the elms are a mass of brown buds, brown lace. The redbirds are shrill, deceptively cheerful.

It's a good cold day for burning. Three thousand scraps of paper bearing things of importunity and word of man's sinning, cries for help, and stuff for sale. The very things you save accumulate until they press you down. Too much of everything. I am obsessed with the thought of starting over. Starting again with everything gone. A March thought. Fifty-seven years of accumulation. But man is not earth. He does not change to marble under pressure. Just assumes a curious shape. Living conglomerate.

> "Conglomerates represent consolidated gravel, and always indicate an aqueous origin, often the delta of an ancient stream or the invasion of sea over land. Important to geologists in interpreting past events."

Well, here we stand. A mass of rounded pebbles of varying sizes cemented together. A few of boulder size, but all rounded in water and transported a great distance. Waiting to be interpreted. Blue clay, yellow clay, and some geodes, hollow and lined, surprisingly, with beautiful crystals. These lumps gathered around some hard and undissolving memory. Love, failure, hate—or that lost ice-cream soda on the streetcar ride. The refused, the ungiven, the undone. These lie inside like fossil insects, but in some no nucleus can be found. No leaf. No shell. No bone.

The brown buds are cold and so is the mist. The spongy earth is cold. The green moss is icy. The frogs have sunk back in the mud again. (Did you ever feel that the cold feet of frogs were walking over you? No. Only hot insects biting at random all night long.) Birds should be migrating this way soon. Something to add to our unkept life-list.

Now is the time for taking stock. What do we have to deal with here? Shut the door. Or put your shoulders against the wind. Stomp on stuff coming up. Hold it down. We've got to have a quiet moment here. Tides of life and tides of decay, and all more silent than the sea.

What can we learn from the past? What teach? Our experience is vinegar. Mother-of-vinegar. *Mother*, this strange transparent stuff. Membrane of yeast cells and bacteria, added to wine and cider to start vinegar (and also—for the record—mother is *modder*, is mud). How can I live in two worlds? As you have been living. Torn, divided, growing by shreds and wounds, growing by ridicule. Until one resembles the astonishing corn plant emerging from rolled and shredded newspaper. Behold the poet, the ragged and upright soul of man.

The poets were honored men in the old days in Ireland. They had a place then. Voices of wisdom, bearers of tidings, good, bad, true, invented, newsmen and singers. We're a small bunch today. Fellow tapirs. We cannot understand each other. Gather seven hundred together in a kingdom? One hundred in a castle? I wouldn't spend the night there. But things were different then.

I had a strange thought this month in the wind and snow. A vision of the Coming—not the Second Coming, that dream of Christians, but a Coming of a man from the Arctic, marching slowly down. Brown, broad, majestic. Great, flat, moon-shaped face. Black eyes. A pagan man to deliver us from our bloody and binding dreams. To deliver us from our stinking wars of religion, wars of patriotism. Our wars in which we use the bodies of burned children to ward off our childish nightmares of a Communist world. Our war for democracy in which we blind, burn, starve, and cripple children so that they may vote at twenty-one.

The bud ends of box-elder twigs look like hard green roses. Green stem, green rose, tough buds. Smooth and wiry branches. Very tough. Formed like the horsetails, the ancient plants. Scarred already, and only a year alive. There is a chill wind, and the clouds are edgeless and thin.

LEWIS THOMAS
1913–1993

Born in Flushing, New York, the son of a general practitioner, Dr. Thomas had a long and distinguished medical career, including the presidency of the Memorial Sloane-Kettering Cancer Center from 1973 to 1980. His essays first appeared in the New England Journal of Medicine *and were collected in* The Lives of a Cell: Notes of a Biology Watcher *(1974), which won the National Book Award. Two other collections followed:* The Medusa and the Snail: More Notes of a Biology Watcher *(1979) and* Late Night Thoughts on Listening to Mahler's Ninth Symphony *(1983).*

The compassion and concern for the earth and its creatures that pervade Lewis Thomas's writing would be enough to earn him inclusion in the company of nature writers. But it was his novel application of the concepts and language of modern cellular biology to traditional environmental subjects that significantly enlarged and enriched the genre and earned him a place as one of the most innovative and important writers in the field today. Characterized by wit, tough-minded optimism, and an extraordinary ability to make specialized concepts accessible to the general reader, his books take challenging ecological positions. He claims, for instance, that the community of life is not fragile but "the toughest membrane imaginable in the universe, opaque to probability, impermeable to death"—an assertion that demonstrates the importance of metaphor to scientific as well as to artistic thought.

DEATH IN THE OPEN

Most of the dead animals you see on highways near the cities are dogs, a few cats. Out in the countryside, the forms and coloring of the dead are strange; these are the wild creatures. Seen from a car window

Lives of a Cell: Notes of a Biology Watcher (New York: Viking, 1974).

they appear as fragments, evoking memories of woodchucks, badgers, skunks, voles, snakes, sometimes the mysterious wreckage of a deer.

It is always a queer shock, part a sudden upwelling of grief, part unaccountable amazement. It is simply astounding to see an animal dead on a highway. The outrage is more than just the location; it is the impropriety of such visible death, anywhere. You do not expect to see dead animals in the open. It is the nature of animals to die alone, off somewhere, hidden. It is wrong to see them lying out on the highway; it is wrong to see them anywhere.

Everything in the world dies, but we only know about it as a kind of abstraction. If you stand in a meadow, at the edge of a hillside, and look around carefully, almost everything you can catch sight of is in the process of dying, and most things will be dead long before you are. If it were not for the constant renewal and replacement going on before your eyes, the whole place would turn to stone and sand under your feet.

There are some creatures that do not seem to die at all; they simply vanish totally into their own progeny. Single cells do this. The cell becomes two, then four, and so on, and after a while the last trace is gone. It cannot be seen as death; barring mutation, the descendants are simply the first cell, living all over again. The cycles of the slime mold have episodes that seem as conclusive as death, but the withered slug, with its stalk and fruiting body, is plainly the transient tissue of a developing animal; the free-swimming amebocytes use this organ collectively in order to produce more of themselves.

There are said to be a billion billion insects on the earth at any moment, most of them with very short life expectancies by our standards. Someone has estimated that there are 25 million assorted insects hanging in the air over every temperate square mile, in a column extending upward for thousands of feet, drifting through the layers of the atmosphere like plankton. They are dying steadily, some by being eaten, some just dropping in their tracks, tons of them around the earth, disintegrating as they die, invisibly.

Who ever sees dead birds, in anything like the huge numbers stipulated by the certainty of the death of all birds? A dead bird is an incongruity, more startling than an unexpected live bird, sure evidence to the human mind that something has gone wrong. Birds do their dying off somewhere, behind things, under things, never on the wing.

Animals seem to have an instinct for performing death alone, hidden. Even the largest, most conspicuous ones find ways to conceal themselves in time. If an elephant missteps and dies in an open place, the herd will not leave him there; the others will pick him up and carry the body from place to place, finally putting it down in some inexplicably suitable location. When elephants encounter the skeleton of an elephant out in the open, they methodically take up each of the bones and distribute them, in a ponderous ceremony, over neighboring acres.

It is a natural marvel. All of the life of the earth dies, all of the time, in the same volume as the new life that dazzles us each morning, each spring. All we see of this is the odd stump, the fly struggling on the porch floor of the summer house in October, the fragment on the highway. I have lived all my life with an embarrassment of squirrels in my backyard, they are all over the place, all year long, and I have never seen, anywhere, a dead squirrel.

I suppose it is just as well. If the earth were otherwise, and all the dying were done in the open, with the dead there to be looked at, we would never have it out of our minds. We can forget about it much of the time, or think of it as an accident to be avoided, somehow. But it does make the process of dying seem more exceptional than it really is, and harder to engage in at the times when we must ourselves engage.

In our way, we conform as best we can to the rest of nature. The obituary pages tell us of the news that we are dying away, while the birth announcements in finer print, off at the side of the page, inform us of our replacements, but we get no grasp from this of the enormity of scale. There are 3 billion of us on the earth, and all 3 billion must be dead, on a schedule, within this lifetime. The vast mortality, involving something over 50 million of us each year, takes place in relative secrecy. We can only really know of the deaths in our households, or among our friends. These, detached in our minds from all the rest, we take to be unnatural events, anomalies, outrages. We speak of our own dead in low voices; struck down, we say, as though visible death can only occur for cause, by disease or violence, avoidably. We send off for flowers, grieve, make ceremonies, scatter bones, unaware of the rest of the 3 billion on the same schedule. All of that immense mass of flesh and bone and consciousness will disappear by absorption into the earth, without recognition by the transient survivors.

Less than a half century from now, our replacements will have more than doubled the numbers. It is hard to see how we can continue to keep the secret, with such multitudes doing the dying. We will have to give up the notion that death is catastrophe, or detestable, or avoidable, or even strange. We will need to learn more about the cycling of life in the rest of the system, and about our connection to the process. Everything that comes alive seems to be in trade for something that dies, cell for cell. There might be some comfort in the recognition of synchrony, in the formation that we all go down together, in the best of company.

THE WORLD'S BIGGEST MEMBRANE

Viewed from the distance of the moon, the astonishing thing about the earth, catching the breath, is that it is alive. The photographs show the dry, pounded surface of the moon in the foreground, dead as an old bone. Aloft, floating free beneath the moist, gleaming membrane of bright blue sky, is the rising earth, the only exuberant thing in this part of the cosmos. If you could look long enough, you would see the swirling of the great drifts of white cloud, covering and uncovering the half-hidden masses of land. If you had been looking for a very long, geologic time, you could have seen the continents themselves in motion, drifting apart on their crustal plates, held afloat by the fire beneath. It has the organized, self-contained look of a live creature, full of information, marvelously skilled in handling the sun.

It takes a membrane to make sense out of disorder in biology. You have to be able to catch energy and hold it, storing precisely the needed amount and releasing it in measured shares. A cell does this, and so do the organelles inside. Each assemblage is poised in the flow of solar energy, tapping off energy from metabolic surrogates of the sun. To stay alive, you have to be able to hold out against equilibrium, maintain imbalance, bank against entropy, and you can only transact this business with membranes in our kind of world.

When the earth came alive it began constructing its own membrane, for the general purpose of editing the sun. Originally, in the time of pre-biotic elaboration of peptides and nucleotides from inorganic ingredients in the water on the earth, there was nothing to shield out ultraviolet radiation except the water itself. The first thin atmosphere came entirely from the degassing of the earth as it cooled, and there was only a vanishingly small trace of oxygen in it. Theoretically, there could have been some production of oxygen by photo-dissociation of water vapor in ultraviolet light, but not much. This process would have been self-limiting, as Urey showed, since the wave lengths needed for photolysis are the very ones screened out selectively by oxygen; the production of oxygen would have been cut off almost as soon as it occurred.

The formation of oxygen had to await the emergence of photosynthetic cells, and these were required to live in an environment with sufficient visible light for photosynthesis but shielded at the same time against lethal ultraviolet. Berkner and Marshall calculate that the green cells must therefore have been about ten meters below the surface of water, probably in pools and ponds shallow enough to lack strong convection currents (the ocean could not have been the starting place).

You could say that the breathing of oxygen into the atmosphere was the result of evolution, or you could turn it around and say that evolution was the result of oxygen. You can have it either way. Once the photosynthetic cells had appeared, very probably counterparts of today's blue-green algae, the future respiratory mechanism of the earth was set in place. Early on, when the level of oxygen had built up to around 1 per cent of today's atmospheric concentration, the anaerobic life of the earth was placed in jeopardy, and the inevitable next stage was the emergence of mutants with oxidative systems and ATP. With this, we were off to an explosive developmental stage in which great varieties of respiring life, including the multicellular forms, became feasible.

Berkner has suggested that there were two such explosions of new life, like vast embryological transformations, both dependent on the threshold levels of oxygen. The first, at 1 per cent of the present level, shielded out enough ultraviolet radiation to permit cells to move into the surface layers of lakes, rivers, and oceans. This happened around 600 million years ago, at the beginning of the Paleozoic era, and accounts for the sudden abundance of marine fossils of all kinds in the record of this period. The second burst occurred when oxygen rose to 10 per cent of the present level. At this time, around 400 million years ago, there was a sufficient canopy to allow life out of the water and onto the land. From here on it was clear going, with nothing to restrain the variety of life except the limits of biologic inventiveness.

It is another illustration of our fantastic luck that oxygen filters out the very bands of ultraviolet light that are most devastating for nucleic acids and proteins, while allowing full penetration of the visible light needed for photosynthesis. If it had not been for this semipermeability, we could never have come along.

The earth breathes, in a certain sense. Berkner suggests that there may have been cycles of oxygen production and carbon dioxide consumption, depending on relative abundances of plant and animal life, with the ice ages representing periods of apnea. An overwhelming richness of vegetation may have caused the level of oxygen to rise above today's concentration, with a corresponding depletion of carbon dioxide. Such a drop in carbon dioxide may have impaired the "greenhouse" property of the atmosphere, which holds in the solar heat otherwise lost by radiation from the earth's surface. The fall in temperature would in turn have shut off much of living, and, in a long sigh, the level of oxygen may have dropped by 90 per cent. Berkner speculates that this is what happened to the great reptiles; their size may have been all right for a richly oxygenated atmosphere, but they had the bad luck to run out of air.

Now we are protected against lethal ultraviolet rays by a narrow rim of ozone, thirty miles out. We are safe, well ventilated, and incubated, provided we can avoid technologies that might fiddle with that ozone, or shift the levels of carbon dioxide. Oxygen is not a major worry for us, unless we let fly with enough nuclear explosives to kill off the green cells in the sea; if we do that, of course, we are in for strangling.

It is hard to feel affection for something as totally impersonal as the atmosphere, and yet there it is, as much a part and product of life as wine or bread. Taken all in all, the sky is a miraculous achievement. It works, and for what it is designed to accomplish it is as infallible as anything in nature. I doubt whether any of us could think of a way to improve on it, beyond maybe shifting a local cloud from here to there on occasion. The word "chance" does not serve to account well for structures of such magnificence. There may have been elements of luck in the emergence of chloroplasts, but once these things were on the scene, the evolution of the sky became absolutely ordained. Chance suggests alternatives, other possibilities, different solutions. This may be true for gills and swim-bladders and forebrains, matters of detail, but not for the sky. There was simply no other way to go.

We should credit it for what it is: for sheer size and perfection of function, it is far and away the grandest product of collaboration in all of nature.

It breathes for us, and it does another thing for our pleasure. Each day, millions of meteorites fall against the outer limits of the membrane and are burned to nothing by the friction. Without this shelter, our surface would long since have become the pounded powder of the moon. Even though our receptors are not sensitive enough to hear it, there is comfort in knowing that the sound is there overhead, like the random noise of rain on the roof at night.

JOHN HAY
b. 1915

In his foreword to The Run (1959), John Hay states, "This book mirrors an attempt to go farther afield, from one man's center." Hay's center, for most of his writing career, has been the sandy peninsula of Cape Cod, a fertile ground for many nature writers. Grandson and namesake of Abraham Lincoln's personal secretary and Theodore Roosevelt's secretary of state, Hay was born in Ipswich, Massachusetts. He first came to Cape Cod in 1942 to study with the poet Conrad Aiken. After service in World War II he built a home in the town of Brewster, where he has lived ever since. The Run, a personal account of the life cycle of alewives, or migratory herring, was the first of a distinguished series of nature books which reflect Hay's dual emphasis on locality—identification of and with a known home ground—and a recognition of the global connections in local phenomena: "All landscape contains the potential world." His work has traced a widening gyre of settings over the years, including Maine, New Hampshire, Greenland, and Costa Rica. Hay's quietly suggestive and elegant style masks one of the most innovative and daring of contemporary writers in the genre. His books explore not only various outer environments he has encountered but also the inner, hidden connections between human and other lives, expressing his belief that the deepest sources of our being lie in a nature we have done our best to ignore. Among John Hay's recent books are The Bird of Light (1991) and A Beginner's Faith in Things Unseen (1995).

THE COMMON NIGHT

Did the alewives choose the night or late evening hours to come in by? So I had been told. By daylight evidence, the fish population increased at the Herring Run on the mornings following a nightly high

The Run (New York: Doubleday, 1959).

tide. I had also heard that there were more alewives running during the tides of the full moon, in the farthest monthly reaches of ebb and flood; but this was a correlation that would be hard for me to make without more years to judge by. In the middle of May on the days just after the first quarter of the moon, which came on the sixteenth, the fish seemed to be running just about as hard as they did during the days preceding the full moon in April, which had appeared on the twenty-fourth. Judging accordingly, it seemed as though their migration had its own ebb and flood during those months. All this was not much better than impression plus hearsay, but there seemed to be some justice to the night tide theory, so, to begin with, I went down to the shore late one evening during an incoming tide to see if there might be any sign of the alewives.

About eight o'clock, an hour before high tide, the tide was running strongly in at Paine's Creek. The channel in the marshes flooded over its banks and marsh grasses were floating and stirring as the swaying waters rose around them. It was near dark. I could see some seaweed flinging by against the sandy bottom at the mouth of the creek, and a big, ghostly green eel slithered up at the edge of the bank the waves were licking, seemed to look up at me, looped back into the water, and disappeared; but it was too dark to see much more than those black clumps of seaweed racing by. I saw a group of gulls standing in shoal waters beyond the beach, where waves were rolling in hard under a steady northwest wind. The sun's cauldron had dropped down, a raw, glistening orange-red, into the sea and back of the curved horizon, leaving its horizontal flush behind.

I walked back under the lee of the sand banks bordering the curving creek. The tide was pulsing and roaring, its waters loping in to the creek which began to turn a harder, darker blue under the sky. Then I began to hear the innumerable soft slaps of fish breaking the surface. The alewives were making their entry from the sea.

And the gulls proclaimed their coming. Out in the Bay, they began to gather by the hundreds, clambering up with a scrambled yelping and hollering. The last smoky, red line of sunset was disappearing and they hovered over it in a maddened, high, wide swarm like huge bees. It grew darker, and a black-crowned night heron, or "quawk," sometimes "quok," a name true to the sound it makes, flew by with rounded wings against a star. The gulls began to disappear, streaming faintly like ashes against the last fires on the sea, but still crying vastly and collectively toward a world of distances. And in terrible simplicity, the alewives were swimming toward the inland gauntlet they would have to run, having a title, by their common, wild, and ancient advent, to all great kindled things. Who will see more than that in his short life, with its many meetings and separations?

I by an old and natural right felt a fierce water-deep wonder of the spirit. The beyondness in me went back to its beginnings. I thought of the nights on which children I have known were born, and of the voyages of war, leave-takings at railroad stations and at ports of embarkation, and of dreams in which I struggled toward new meetings and other lives. The wind blew through the arches of the stars, and the surfaces of the dipping earth, water, and sky in their lasting communion made me dizzy. I felt a cold inevitable grandeur, below consciousness, a swim and go in an uttermost wild world, past home or my life's memory.

So by this evidence the alewives came in at night, and, as a further discovery not to be denied, so had I. Perhaps it was the closest I would ever get to the non-human fish in a darkness where all the components of existence ran the same race. That real depth, fish-oriented, nakedly omnipotent, fills men when they recognize it with more awe than their limited worlds can encompass.

As I started back, about a quarter of an hour before the full tide, headlights swept over where the road ended at the shore, and in a minute or two a couple of fishermen lurched down the sand with high rubber waders on, carrying their casting rods. They stood on the beach in the dark, one of them coaching the other in baiting his hook. I came up and spoke to them, hardly able to make out their faces. The older one, he who did the coaching, told me that they had just got a pail of herrin' from the Brewster run to use as bait. They had hopes that there would be some bass here, the famous "stripers" chasing the alewives in. They brought their long rods sideways and back to sling the bait out into the black and silver waves. The older man spoke low words against the wind, and I strained to hear him. Suddenly he thought he felt what must be alewives nosing his line and bumping against it on their way by into the mouth of the inlet. Last year, he told me, he had seen hundreds of them dead on the flats, and the gulls, he said, had slit their sides open as if with knives to get the roe. The waves had begun to slacken off when I left, and the fishermen were still casting, but without much hope of a strike.

THE DOVEKIE AND THE OCEAN SUNFISH

It was in December and a slight snow had started. Inland of Cape Cod waters there was very little wind, and the flakes were spaced far apart. They seemed to hover. They would lift and dip, then slant down

In Defense of Nature (Boston: Little, Brown, 1969).

and touch slowly to the ground. The day was gray and quiet, tempered by the even recital of falling flakes, with a dull sky and a strip of gray sea in the distance. I watched the snow flakes acting in the space around me, over and inside and by each other in slow periods, and then I drove down to the Outer Beach.

The surf there was roaring, and hard wet flakes of snow came in heavily, driven by a seaward wind, and made my eyes smart. The curled-over, breaking waves were glassy green in the gray day and pushed in soapy sheets of foam along the sands. The ocean itself seemed full of kingly mountains meeting or withholding, conflicting, pushing each other aside, part of a collective immensity that could express itself in the last little bursting spit of a salt bubble flung out of the foam that seethed into the beach, a kind of statement of articulated aim out of a great language still untamed.

Out on those cold looming and receding waters seabirds rode. There were flocks of dark brown and black-and-white eiders a few hundred yards beyond the shoreline. Gulls beat steadily along the troughs of the waves. Then I caught sight of thirty or forty dovekies, the chunky black and white "little auks" that come down from the far north during the winter to feed off the coast. Their color showed clearly against the green water, white under black heads, black backs and wings, with little upturned tail feathers. They dipped continually into the water, so quickly as to escape my notice much of the time. They made fish flips into those tumultuous sea surfaces as easily as minnows in a gentle pond.

The dovekie (called little auk in Europe) breeds by the millions in the high Arctic, principally in North Greenland, Spitzbergen, northern Novaya Zemlya, and Franz Joseph Land, with scattered colonies in subarctic areas. Drift ice is their principal habitat. They like areas where the ice is not too densely packed, and they feed in openings or leads between the ice floes, on small crustaceans in the plankton. They seem to avoid the warm waters of the Gulf Stream, following the Arctic currents instead. Especially during periods of great population growth, emigration flights of the dovekies take them on long journeys which scatter them to many regions, and occasionally take them inland in dovekie "wrecks" where they may be killed or injured, and become vulnerable to many kinds of predators. Their short narrow wings make it difficult, though not impossible, for them to take off from inland areas. One winter I saw many dead ones scattered down the Cape Cod highway, little "penguin-like" birds unknown to most car drivers.

These dovekies off Cape Cod waters were flying and diving, casting their bodies back and forth between air and water like so many balls. Their speed is not great, but when they fly off over the water their wing-beat is very quick, and their landing and diving is done with a dash and play that belies their stumpy appearance. In fact a dovekie in flight, though it lacks the agility and swinging maneuverability of a swallow, is at the same time somewhat reminiscent of those birds, with its short wings alternately beating fast and gliding, like a swift. When they swam under water these chunky little "pineknots" did so with a quick, supple, almost fishlike beat, making me think that fins after all were not so far removed from wings.

Not visible, though they brought it with them in some measure these vast distances down to New England, was the grandeur of the Arctic, whose auroras, growling ice and sunsets I could only imagine. Dovekies breed in bare heaps of rock that cover great screes, hillsides of stone, sloping down to the sea. They arrive in there in the spring when green begins to show through frozen ground and rivulets and waterfalls begin to sound while the dovekies themselves utter a watery, twittering, trilling call. There they dot the slopes in great numbers "like pepper and salt," it says in *The Birds of Greenland,* and fly about "in huge flocks which resemble at a distance swarms of mosquitoes or drifting smoke."

To the Eskimos these little birds are of primary importance. They net them with long-handled nets as they fly over their breeding cliffs and store them in the frozen ground for winter food. Eskimo women gather their eggs, where their nests are found between the rocks and stones, beginning in June. They also use their skins for a birdskin coat called a *tingmiaq.* For one such coat about fifty skins are needed. Admiral Donald B. MacMillan, the Arctic explorer, has described—in a passage quoted by Forbush—how much the spring arrival of the dovekies has meant to the Eskimos: "But what is that great, pulsating, musical note which seems to fill all space? Now loud and clear, now diminishing to a low hum, the sound proclaims the arrival of the true representative of bird life in the Arctic, the dovekie, or little auk *(Plautus alle).* The long dark winter has at last passed away. The larder open to all is empty. The sun is mounting higher into the heavens day by day. Now and then a seal is seen sunning himself at his hole. The Eskimos are living from hand to mouth. And then that glad cry, relieving all anxiety for the future, bringing joy to every heart, *'Ark-pood-e-ark-suit! Ark-pood-e-ark-suit!'* (Little auks! Little auks!)."

Arctic foxes also depend on the dovekies for food, as well as their immaculately bluish white eggs. Dovekies are preyed upon by ravens and gyrfalcons too, but the big, persistent enemy of which they are in mortal fear is the glaucous or burgomaster gull.

The nitrogen-rich guano of the dovekies, filtering down through the massive hillsides of loose bare stone, fertilizes a deep growth of moss in brilliant green bands. This and other plant life supports hare and ptarmigan and is probably good pasturage for what was once an abundant population of caribou. And in the sea this little bird that makes for provision in depth, both good and cruel, is often eaten by white whales, large fish, and seals.

What a major burden for one so small! Millions of dovekies feed their young during the nesting season and a whole Arctic world depends that they be successful, on and on into the future. The little ones are constantly hungry and chirp in a shrill impatient way until: "the old bird feeds them by disgorging into their bills the content of its well-filled pouch. The consoling, soothing murmur of the old bird to the young, and the satisfied chirping of the young shows how solicitous the one and how grateful the other." Sweet domesticity still keeps the terrible world in order. The little auk is also a tough, all-weather bird, capable of survival in extreme circumstances, and so far it has the room it needs and the isolation and a major range.

How can we predators dare count the bodies, human and other than human, of those on whom we prey! How dare we wipe out whole populations in the first place, either by pulling a trigger or spreading our wastes and poisons by the ton. Still, "dare" may not be the right word for a brutal carelessness so widespread that many men do not dare do anything but trade on it. How many tons of waste go into the air every minute? How much sewage goes into the earth's once pure and flowing waters, or oil's dirty devastation is spread on the seas? How much over-killing will it take for us to come face to face with what is left of the vital innocence that has upheld the world without us up to now, like the dovekie and its Arctic pyramid of needs?

As I watched the dovekies, I saw one come in out of the water, letting itself be washed ashore with the broad sheet of foam sent in ahead of the breakers. It staggered and flopped ahead up the slope of sand beyond the water's reach and stood there. I walked up to the little bird and it made only a slight effort to get away. Its thick breast feathers were coated with oil. They were also stained with the red of its blood. Since oil fouling causes birds to lose the natural insulation of their feathers, and poisons them when they preen, I supposed that this one had been weakened and perhaps flung against a rock by the waves. I picked it up, and what a trembling there was in it, what a whirring heart! I carried it back to my car, thinking to clean the oil off later and see if I could bring it back to health, always a precarious job. But after a few minutes in the heated car the bird's blood began to flow from its breast, to my great dismay, and there seemed to be little I could do about it. I began to feel like a terrible meddler and a coward. With that I took it back to calmer waters, on the bay side of the Cape, and let it go without illusions about the cruel mother that might take care of it better than I, but I felt that my fruitless attempts to save it would be worse. I remember the little beak with its pink lining, open and threatening, as the dovekie protested my picking it up, the black eyes blinking and glistening, the feel of its heart. It was more than a small bird in mortal trouble. It had in it the greatness of its northern range. I had, I suppose, with no excuses for what I failed to do, given it back to its own latitude, to the spirit of choking, roaring waters, of skies with smoky clouds where the little auks themselves looked like smoke in the distance, and of swinging winds and draws between the ice floes, and the rivulets of spring. That bird, which had bloodied me and been so close and warm in my hand, left me on the beach to shake with the weight of human ignorance. *   *   *

THOMAS MERTON
1915–1968

From 1941 to the end of his life, Thomas Merton lived as a Catholic monk at the Abbey of Our Lady of Gethsemani near Bardstown, Kentucky. A member of the Trappist order, which strictly observes vows of silence, he nonetheless became a leading writer of his generation, publishing dozens of books of poetry, essays, religious meditations, plays, and novels, as well as his well-known autobiography, The Seven Storey Mountain *(1948). Because of its exploratory rather than doctrinaire approach toward religious questions, Merton's work has appealed to a wide audience. Though he led the life of a solitary contemplative, his writing reflects a deep awareness of current political and cultural issues. In the following essay, from* Raids on the Unspeakable *(1964), his description of a rainshower at night in the woods—"this wonderful, unintelligible, perfectly innocent speech"—gradually expands into a meditation on solitude, personal freedom, and the nature of joy. In 1968, while traveling in Thailand, Merton was accidentally electrocuted.*

RAIN AND THE RHINOCEROS

Let me say this before rain becomes a utility that they can plan and distribute for money. By "they" I mean the people who cannot understand that rain is a festival, who do not appreciate its gratuity, who think that what has no price has no value, that what cannot be sold is not real, so that the only way to make something *actual* is to place it on the market. The time will come when they will sell you even your rain. At the moment it is still free, and I am in it. I celebrate its gratuity and its meaninglessness.

The rain I am in is not like the rain of cities. It fills the woods with an immense and confused sound. It covers the flat roof of the cabin and its porch with insistent and controlled rhythms. And I listen, because it reminds me again and again that the whole world runs by rhythms I have not yet learned to recognize, rhythms that are not those of the engineer.

I came up here from the monastery last night, sloshing through the cornfield, said Vespers, and put some oatmeal on the Coleman stove for supper. It boiled over while I was listening to the rain and toasting a piece of bread at the log fire. The night became very dark. The rain surrounded the whole cabin with its enormous virginal myth, a whole world of meaning, of secrecy, of silence, of rumor. Think of it: all that speech pouring down, selling nothing, judging nobody, drenching the thick mulch of dead leaves, soaking the trees, filling the gullies and crannies of the wood with water, washing out the places where men have stripped the hillside! What a thing it is to sit absolutely alone, in the forest, at night, cherished by this wonderful, unintelligible, perfectly innocent speech, the most comforting speech in the world, the talk that rain makes by itself all over the ridges, and the talk of the watercourses everywhere in the hollows!

Nobody started it, nobody is going to stop it. It will talk as long as it wants, this rain. As long as it talks I am going to listen.

But I am also going to sleep, because here in this wilderness I have learned how to sleep again. Here I am not alien. The trees I know, the night I know, the rain I know. I close my eyes and instantly sink into the whole rainy world of which I am a part, and the world goes on with me in it, for I am not alien to it. I am alien to the noises of cities, of people, to the greed of machinery that does not sleep, the hum of power that eats up the night. Where rain, sunlight and darkness are contemned, I cannot sleep. I do not trust anything that has been fabri-

Raids on the Unspeakable (New York: New Directions, 1964).

cated to replace the climate of woods or prairies. I can have no confidence in places where the air is first fouled and then cleansed, where the water is first made deadly and then made safe with other poisons. There is nothing in the world of buildings that is not fabricated, and if a tree gets in among the apartment houses by mistake it is taught to grow chemically. It is given a precise reason for existing. They put a sign on it saying it is for health, beauty, perspective; that it is for peace, for prosperity; that it was planted by the mayor's daughter. All of this is mystification. The city itself lives on its own myth. Instead of waking up and silently existing, the city people prefer a stubborn and fabricated dream; they do not care to be a part of the night, or to be merely of the world. They have constructed a world outside the world, against the world, a world of mechanical fictions which contemn nature and seek only to use it up, thus preventing it from renewing itself and man.

Of course the festival of rain cannot be stopped, even in the city. The woman from the delicatessen scampers along the sidewalk with a newspaper over her head. The streets, suddenly washed, became transparent and alive, and the noise of traffic becomes a plashing of fountains. One would think that urban man in a rainstorm would *have* to take account of nature in its wetness and freshness, its baptism and its renewal. But the rain brings no renewal to the city, only to tomorrow's weather, and the glint of windows in tall buildings will then have nothing to do with the new sky. All "reality" will remain somewhere inside those walls, counting itself and selling itself with fantastically complex determination. Meanwhile the obsessed citizens plunge through the rain bearing the load of their obsessions, slightly more vulnerable than before, but still only barely aware of external realities. They do not see that the streets shine beautifully, that they themselves are walking on stars and water, that they are running in skies to catch a bus or a taxi, to shelter somewhere in the press of irritated humans, the faces of advertisements and the dim, cretinous sound of unidentified music. But they must know that there is wetness abroad. Perhaps they even *feel* it. I cannot say. Their complaints are mechanical and without spirit.

Naturally no one can believe the things they say about the rain. It all implies one basic lie: *only the city is real.* That weather, not being planned, not being fabricated, is an impertinence, a wen on the visage of progress. (Just a simple little operation, and the whole mess may become relatively tolerable. Let business *make* the rain. This will give it meaning.)

Thoreau sat in *his* cabin and criticized the railways. I sit in mine and wonder about a world that has, well, progressed. I must read *Walden* again, and see if Thoreau already guessed that he was part of what he thought he could escape. But it is not a matter of "escaping." It is not even a matter of protesting very audibly. Technology is here, even in the cabin. True, the utility line is not here yet, and so G.E. is not here yet either. When the utilities and G.E. enter my cabin arm in arm it will be nobody's fault but my own. I admit it. I am not kidding anybody, even myself. I will suffer their bluff and patronizing complacencies in silence. I will let them think they know what I am doing here.

They are convinced that I *am having fun.*

This has already been brought home to me with a wallop by my Coleman lantern. Beautiful lamp: It burns white gas and sings viciously but gives out a splendid green light in which I read Philoxenos, a sixth-century Syrian hermit. Philoxenos fits in with the rain and the festival of night. Of this, more later. Meanwhile: what does my Coleman lantern tell me? (Coleman's philosophy is printed on the cardboard box which I have (guiltily) not shellacked as I was supposed to, and which I have tossed in the woodshed behind the hickory chunks.) Coleman says that the light is good, and has a reason: it *"Stretches days to give more hours of fun."*

Can't I just be in the woods without any special reason? Just being in the woods, at night, in the cabin, is something too excellent to be justified or explained! It just *is.* There are always a few people who are in the woods at night, in the rain (because if there were not the world would have ended), and I am one of them. We are not having fun, we are not "having" anything, we are not *"stretching our days,"* and if we had fun it would not be measured by hours. Though as a matter of fact that is what fun seems to be: a state of diffuse excitation that can be measured by the clock and "stretched" by an appliance.

There is no clock that can measure the speech of this rain that falls all night on the drowned and lonely forest.

Of course at three-thirty A.M. the SAC plane goes over, red light winking low under the clouds, skimming the wooded summits on the south side of the valley, loaded with strong medicine. Very strong. Strong enough to burn up all these woods and stretch our hours of fun into eternities.

And that brings me to Philoxenos, a Syrian who had fun in the sixth century, without benefit of appliances, still less of nuclear deterrents.

Philoxenos in his ninth *memra* (on poverty) to dwellers in solitude, says that there is no explanation and no justification for the solitary life, since it is without a law. To be a contemplative is therefore to be an outlaw. As was Christ. As was Paul.

One who is not "alone," says Philoxenos, has not discovered his identity. He seems to be alone, perhaps, for he experiences himself as "individual." But because he is willingly enclosed and limited by the laws and illusions of collective existence, he has no more identity than an unborn child in the womb. He is not yet conscious. He is alien to his own truth. He has senses, but he cannot use them. He has life, but no identity. To have an identity, he has to be awake, and aware. But to be awake, he has to accept vulnerability and death. Not for their own sake: not out of stoicism or despair—only for the sake of the invulnerable inner reality which we cannot recognize (which we can only *be*) but to which we awaken only when we see the unreality of our vulnerable shell. The discovery of this inner self is an act and affirmation of solitude.

Now if we take our vulnerable shell to be our true identity, if we think our mask is our true face, we will protect it with fabrications even at the cost of violating our own truth. This seems to be the collective endeavor of society: the more busily men dedicate themselves to it, the more certainly it becomes a collective illusion, until in the end we have the enormous, obsessive, uncontrollable dynamic of fabrications designed to protect mere fictitious identities—"selves," that is to say, regarded as objects. Selves that can stand back and see themselves having fun (an illusion which reassures them that they are real).

Such is the ignorance which is taken to be the axiomatic foundation of all knowledge in the human collectivity: in order to experience yourself as real, you have to suppress the awareness of your contingency, your unreality, your state of radical need. This you do by creating an awareness of yourself as *one who has no needs that he cannot immediately fulfill.* Basically, this is an illusion of omnipotence: an illusion which the collectivity arrogates to itself, and consents to share with its individual members in proportion as they submit to its more central and more rigid fabrications.

You have needs; but if you behave and conform you can participate in the collective power. You can then satisfy all your needs. Meanwhile, in order to increase its power over you, the collectivity increases your needs. It also tightens its demand for conformity. Thus you can become all the more committed to the collective illusion in proportion to becoming more hopelessly mortgaged to collective power.

How does this work? The collectivity informs and shapes your will to happiness ("have fun") by presenting you with irresistible images of yourself as you would like to be: having *fun that is so perfectly credible that it allows no interference of conscious doubt.* In theory such a good time can be so convincing that you are no longer aware of even a remote possibility that it might change into something less satisfying. In practice, expensive fun always admits of a doubt, which blossoms out into another full-blown need, which then calls for a still more credible and more costly refinement of satisfaction, which again fails you. The end of the cycle is despair.

Because we live in a womb of collective illusion, our freedom remains abortive. Our capacities for joy, peace, and truth are never liberated. They can never be used. We are prisoners of a process, a dialectic of false promises and real deceptions ending in futility.

"The unborn child," says Philoxenos, "is already perfect and fully constituted in his nature, with all his senses, and limbs, but he cannot make use of them in their natural functions, because, in the womb, he cannot strengthen or develop them for such use."

Now, since all things have their season, there is a time to be unborn. We must begin, indeed, in the social womb. There is a time for warmth in the collective myth. But there is also a time to be born. He who is spiritually "born" as a mature identity is liberated from the enclosing womb of myth and prejudice. He learns to think for himself, guided no longer by the dictates of need and by the systems and processes designed to create artificial needs and then "satisfy" them.

This emancipation can take two forms: first that of the active life, which liberates itself from enslavement to necessity by considering and serving the needs of others, without thought of personal interest or return. And second, the contemplative life, which must not be construed as an escape from time and matter, from social responsibility and from the life of sense, but rather, as an advance into solitude and the desert, a confrontation with poverty and the void, a renunciation of the empirical self, in the presence of death, and nothingness, in order to overcome the ignorance and error that spring from the fear of "being nothing." The man who dares to be alone can come to see that the "emptiness" and "usefulness" which the collective mind fears and condemns are necessary conditions for the encounter with truth.

It is in the desert of loneliness and emptiness that the fear of death and the need for self-affirmation are seen to be illusory. When this is faced, then anguish is not necessarily overcome, but it can be accepted and understood. Thus, in the heart of anguish are found the gifts of peace and understanding: not simply in personal illumination and liberation, but by commitment and empathy, for the contemplative must assume the universal anguish and the inescapable condition of mortal man. The solitary, far from enclosing himself in himself, becomes every man. He dwells in the solitude, the poverty, the indigence of every man.

It is in this sense that the hermit, according to Philoxenos, imitates

Christ. For in Christ, God takes to Himself the solitude and dereliction of man: every man. From the moment Christ went out into the desert to be tempted, the loneliness, the temptation and the hunger of every man became the loneliness, temptation and hunger of Christ. But in return, the gift of truth with which Christ dispelled the three kinds of illusion offered him in his temptation (security, reputation and power) can become also our own truth, if we can only accept it. It is offered to us also in temptation. "You too go out into the desert," said Philoxenos, "having with you nothing of the world, and the Holy Spirit will go with you. See the freedom with which Jesus has gone forth, and go forth like Him—see where he has left the rule of men; leave the rule of the world where he has left the law, and go out with him to fight the power of error."

And where is the power of error? We find it was after all not in the city, but in *ourselves*.

Today the insights of a Philoxenos are to be sought less in the tracts of theologians than in the meditations of the existentialists and in the Theater of the Absurd. The problem of Berenger, in Ionesco's *Rhinoceros*, is the problem of the human person stranded and alone in what threatens to become a society of monsters. In the sixth century Berenger might perhaps have walked off into the desert of Scete, without too much concern over the fact that all his fellow citizens, all his friends, and even his girl Daisy, had turned into rhinoceroses.

The problem today is that there are no deserts, only dude ranches.

The desert islands are places where the wicked little characters in the *Lord of the Flies* come face to face with the Lord of the Flies, form a small, tight, ferocious collectivity of painted faces, and arm themselves with spears to hunt down the last member of their group who still remembers with nostalgia the possibilities of rational discourse.

When Berenger finds himself suddenly the last human in a rhinoceros herd he looks into the mirror and says, humbly enough, "After all, man is not as bad as all that, is he?" But his world now shakes mightily with the stampede of his metamorphosed fellow citizens, and he soon becomes aware that the very stampede itself is the most telling and tragic of all arguments. For when he considers going out into the street "to try to convince them," he realizes that he "would have to learn their language." He looks in the mirror and sees that *he no longer resembles anyone*. He searches madly for a photograph of people as they were before the big change. But now humanity itself has become incredible, as well as hideous. To be the last man in the rhinoceros herd is, in fact, to be a monster.

Such is the problem which Ionesco sets us in his tragic irony: solitude and dissent become more and more impossible, more and more absurd. That Berenger finally accepts his absurdity and rushes out to challenge the whole herd only points up the futility of a commitment to rebellion. At the same time in *The New Tenant (Le Nouveau Locataire)* Ionesco portrays the absurdity of a logically consistent individualism which, in fact, is a self-isolation by the pseudo-logic of proliferating needs and possessions.

Ionesco protested that the New York production of *Rhinoceros* as a farce was a complete misunderstanding of his intention. It is a play not merely against *conformism* but about *totalitarianism*. The rhinoceros is not an amiable beast, and with him around the fun ceases and things begin to get serious. Everything has to make sense and be totally useful to the totally obsessive operation. At the same time Ionesco was criticized for not giving the audience "something positive" to take away with them, instead of just "refusing the human adventure." (Presumably "rhinoceritis" is the latest in human adventure!) He replied: "They [the spectators] leave in a void—and that was my intention. It is the business of a free man to pull himself out of this void by his own power and not by the power of other people!" In this Ionesco comes very close to Zen and to Christian eremitism.

"In all the cities of the world, it is the same," says Ionesco. "The universal and modern man is the man in a rush (i.e. a rhinoceros), a man who has no time, who is a prisoner of necessity, who cannot understand that *a thing might perhaps be without usefulness*; nor does he understand that, at bottom, it is the useful that may be a useless and back-breaking burden. If one does not understand the usefulness of the useless and the uselessness of the useful, one cannot understand art. And a country where art is not understood is a country of slaves and robots. . . ." (*Notes et Contre Notes*, p. 129) Rhinoceritis, he adds, is the sickness that lies in wait "for those who *have lost the sense and the taste for solitude.*"

The love of solitude is sometimes condemned as "hatred of our fellow men." But is this true? If we push our analysis of collective thinking a little further we will find that the dialectic of power and need, of submission and satisfaction, ends by being a dialectic of hate. Collectivity needs not only to absorb everyone it can, but also implicitly to hate and destroy whoever cannot be absorbed. Paradoxically, one of the needs of collectivity is to reject certain classes, or races, or groups, in order to strengthen its own self-awareness by hating them instead of absorbing them.

Thus the solitary cannot survive unless he is capable of loving everyone, without concern for the fact that he is likely to be regarded by all of them as a traitor. Only the man who has fully attained his own spiritual identity can live without the need to kill, and without the need of a doctrine that permits him to do so with a good conscience. There will always be a place, says Ionesco, *"for those isolated consciences who have stood up for the universal conscience"* as against the mass mind. But their place is solitude. They have no other. Hence it is the solitary person (whether in the city or in the desert) who does mankind the inestimable favor of reminding it of its true capacity for maturity, liberty and peace.

It sounds very much like Philoxenos to me.

And it sounds like what the rain says. We still carry this burden of illusion because we do not dare to lay it down. We suffer all the needs that society demands we suffer, because if we do not have these needs we lose our "usefulness" in society—the usefulness of suckers. We fear to be alone, and to be ourselves, and so to remind others of the truth that is in them.

"I will not make you such rich men as have need of many things," said Philoxenos (putting the words on the lips of Christ), "but I will make you true rich men who have need of nothing. Since it is not he who has many possessions that is rich, but he who has no needs." Obviously, we shall always have *some* needs. But only he who has the simplest and most natural needs can be considered to be without needs, since the only needs he has are real ones, and the real ones are not hard to fulfill if one is a free man!

The rain has stopped. The afternoon sun slants through the pine trees: and how those useless needles smell in the clear air!

A dandelion, long out of season, has pushed itself into bloom between the smashed leaves of last summer's day lilies. The valley resounds with the totally uninformative talk of creeks and wild water.

Then the quails begin their sweet whistling in the wet bushes. Their noise is absolutely useless, and so is the delight I take in it. There is nothing I would rather hear, not because it is a better noise than other noises, but because it is the voice of the present moment, the present festival.

Yet even here the earth shakes. Over at Fort Knox the Rhinoceros is having fun.

FAITH McNULTY
b. 1918

Like E. B. White, Faith McNulty is a nature writer nurtured by associa-
tion with that most urbane of magazines, The New Yorker, *where she first*
became a staff writer in 1953. Many of her articles relate to the dangers of
extinction, as do two of her books: The Whooping Crane: The Bird That
Defies Extinction *(1966) and* Must They Die?: The Strange Case of the
Prairie Dog and the Black-Footed Ferret *(1971). Regardless of the specific*
environmental dangers she has sometimes reported on, animals have
always been her first love. She is the author of many books for children that
focus on the life stories of individual creatures. She has often also produced
portraits of particular animals, collected as The Wildlife Stories of Faith
McNulty *(1980)—the volume from which the present, charming selection*
comes. It is interesting to note that McNulty's best-known book represents a
departure from the main themes of her writing. The Burning Bed *(1980),*
which tells the story of a woman who murdered her abusive ex-husband as
he slept, was made into a television movie in 1984.

MOUSE

On a sunny morning in early September my husband called me out
to the barn of our Rhode Island farm. I found him holding a tin can
and peering into it with an expression of foolish pleasure. He handed
me the can as though it contained something he had just picked up at
Tiffany's. Crouched at the bottom was a young mouse, not much bigger than a bumblebee. It stared up with eyes like polished seeds. Its
long whiskers vibrated like a hummingbird's wings. It was a beautiful
little creature and clearly still too small to cope with a wide and dangerous world.

The Wildlife Stories of Faith McNulty (New York: Doubleday, 1980).

I don't know how old Mouse was when Richard found her, but I doubt it was a fortnight. She was not only tiny, but weak. He told me he had found her on the doorstep and that when he picked her up he thought she was done for. By chance he had a gumdrop in his pocket. He placed it on his palm beside the limp mouse. The smell acted as a quick stimulant. She struggled to her feet and flung herself upon the gumdrop, ate voraciously, and was almost instantly restored to health.

Richard made a wire cage for Mouse (we never found a name more fitting), and we made a place for it on a table in the kitchen. Here I could watch her while I was peeling vegetables but I found that I often simply watched while uncounted minutes went by. I had had no idea that there were so many things to notice about a mouse.

For the first few days Mouse had the gawkiness of a puppy. Her head and feet looked too big. Her hind legs had a tendency to spraddle. But she had fine sharp teeth and a striking air of manful competence. She cleaned herself, all over, with serious pride. Her method was oddly cat-like. Sitting on her small behind (she could have sat on a postage stamp without spilling over), she licked her flanks, then moistened her paws to go over her ears, neck, and face. She would grasp a hind leg with suddenly simian hands while she licked the extended toes. For the finale she would pick up her tail, and as though eating corn on the cob, wash its inch and a half of threadlike length with her tongue.

Mouse's baby coat was dull gunmetal gray. It soon changed to a bright reddish brown. Her belly remained white. She had dark gray anklets and white feet. I had thought of mouse tails as hairless and limp. Not so. Mouse's tail was furred, and rather than trailing it behind her like a piece of string, she held it quite stiffly. Sometimes it rose over her back like a quivering question mark.

When Mouse's coat turned red I was able to identify her from a book—*The Mammals of Rhode Island*—which said that although *leucopus*, or white-footed mice, are easily confused with *maniculatus*, or deer mice, there are only *leucopus* in Rhode Island. The book also said that white-footed mice are found everywhere, from hollow trees to bureau drawers; that they are nocturnal and a favorite food of owls. Judging by Mouse's enthusiasm for chicken, owl, if she could get it, would be one of *her* favorite foods. Her range of taste was wide. Though grains were a staple, she liked meat, fruit, and vegetables. I usually offered her a tiny bit of whatever was on the chopping board. She tasted and considered each item, rejecting some and seizing others with delight. A melon seed was a great prize. To this day, when I throw melon seeds in the garbage I feel sad to waste them and wish I had a mouse to give them to.

Mouse became tame within a few days of her capture. She nibbled my fingers and batted them with her paws like a playful puppy. She liked to be stroked. If I held her in my hand and rubbed gently with a forefinger, she would raise her chin, the way a cat will, to be stroked along the jawbone, then raise a foreleg and wind up lying flat on her back in the palm of my hand, eyes closed, paws hanging limp, and nose pointed upward in apparent bliss.

Mouse could distinguish people. If, when she was asleep, I poked my finger into her nest, she licked and nibbled it as though grooming it. If my husband offered his finger, she would sniff it and then give it a firm little bite accompanied by an indignant chirp. When, for a time, she was in my sister's care, she accepted my sister but bit anyone else. In one respect, however, she never trusted even me. She suspected me of intending to steal her food. If I approached while she was eating she assumed a protective crouch. I think she was uttering tiny ultrasonic growls.

When I looked up scientific studies of mice I was disappointed to find that most investigators had been interested, not in the mice, but in using them as a tool to study something related to human physiology. In one paper, however, I read that "in mice the rate of defecation and urination is an index of emotionality." Dedicated scientists had spent days harassing mice and counting the resultant hail of tiny turds. I had assumed that mice have no control over these functions; an inference based on the careless behavior of certain anonymous mice that sometimes visited my kitchen shelves. When I handled Mouse, however, nothing of the sort ever happened. It could not have been sheer luck. She must have exercised some restraint.

I was surprised by still other aspects of Mouse's behavior. She was a heavy sleeper. She slept in a plastic cup from a thermos, covering herself with a bedding of rags that she shredded into fluff as soft as a down quilt. If I pushed aside the covers I would find her curled up on her side like a doughnut, dead to the world. As I touched her, her eyes would open. Then she would raise her chin, stretch herself, and yawn enormously, showing four wicked front teeth and a red tongue that curled like a wolf's. She would rise slowly, carefully stretching her hind legs and long toes, then suddenly pull herself together, fan out her whiskers, and be ready for anything. Her athletic ability was astounding. As she climbed around her cage she became incredibly flexible, stretching this way and that like a rubber band. She could easily stand on her hind legs to reach something dangled above her. Her jumping power was tremendous. Once I put her in an empty garbage pail while I cleaned her cage. She made a straight-upward leap of fifteen inches and neatly cleared the rim.

Mouse's cage was equipped with an exercise wheel, on which she traveled many a league to nowhere. Richard and I racked our brains for a way to utilize "one mouse-power." Her cage was also furnished with twigs that served as a perch. After a while I replaced her plastic cup with half a coconut shell, inverted and with a door cut in the lower edge. It made a most attractive mouse house; quite tropical in feeling. She stuffed her house from floor to ceiling with fluff. She kept some food in the house, but her major storehouse was a small aluminum can screwed to the wall of the cage. We called it (and beg the generous reader to forgive the cuteness), the First Mouse National Bank. If I sprinkled birdseed on the floor of her cage, Mouse would work diligently to transport it, stuffed in her cheeks, for deposit in her bank.

Mouse was full of curiosity and eager to explore. When I opened the door of her cage she ran about the tabletop in short bursts of motion, looking, somehow, as though she were on roller skates. I feared she might skate right over the edge of the table, but she always managed to stop in time. All objects she met—books, pencils, ashtrays, rubber bands, and such odds and ends—were subjected to a taste test. If a thing was portable, a pencil for instance, she might haul it a short way. Her attention was brief; a quick nibble and on to the next. One day she encountered a chicken bone. She grabbed it in her teeth and began to tug. As she danced around, pulling and hauling, she looked like a terrier struggling to retrieve the thighbone of an elephant. Alas, the task was too great. She had to settle for a fragment of meat and leave the bone behind.

What fascinated me most was Mouse's manual dexterity. Her front paws had four long fingers and a rudimentary thumb. She used them to hold, to manipulate, and to stuff things into her mouth for carrying. Her paws were equally equipped for climbing. They had small projections, like the calluses inside a man's hand, that helped her to cling, fly fashion, to vertical surfaces.

Though Mouse kept busy, I feared her life might be warped by loneliness and asked a biologist I knew for help. He not only determined Mouse's sex (this is not easy; to the layman the rear end of a mouse is quite enigmatic) but provided a laboratory mouse as a companion.

I found the new mouse unattractive. He had a mousy smell, whereas Mouse was odorless; I named him Stinky. His coat was like dusty black felt. He was careless about grooming. He had small, squinty eyes, a Roman nose, a fat-hipped, lumpy shape, and a ratty, hairless tail. Nature would not have been likely to create such a mouse without the help of man. With some misgivings I put him in Mouse's cage. She mounted to the top of her perch and sat shivering and staring, ears cocked so that her face looked like that of a little red fox.

Stinky lumbered about the cage, squinting at nothing in particular. Stumbling over some of Mouse's seeds, he made an enthusiastic buck-toothed attack on these goodies. This stirred Mouse to action. She flashed down the branch and, cautiously approaching from the rear, nibbled Stinky's tail. He paid no heed, but continued to gobble up whatever he found. Mouse nibbled him more boldly, working up from his tail to the fur of his back. I began to fear she would depilate him before he realized it. Finally she climbed on his back and nibbled his ears. He showed a certain baffled resistance, but made no other response. Disgusted, Mouse ate a few seeds and went to bed.

From this unpromising start a warm attachment bloomed. The two mice slept curled up together. Mouse spent a great deal of time licking Stinky, holding him down and kneading him with her paws. He returned her caresses, but with less ardor, reserving his more passionate interest for food. Food was a source of strife. In a contest Stinky was domineering but dumb. Mouse was quick and clever.

Stinky's greed prompted Richard to fashion Mouse's bank, tailoring the opening to fit her slim figure and exclude Stinky's chubby one—or at least most of it. He could get his head and shoulders inside, but not his fat belly. Richard fastened the bank near the top of the cage. When Stinky got his head in, his hind part was left dangling helplessly, and he soon gave up attempts at robbery.

One day, as an experiment, I put the bank down on the floor of the cage. Stinky sniffed at the opening. Mouse watched, whiskers quivering, and I had the distinct impression of consternation on her face. With a quickness of decision that amazed me, she seized a wad of bedding, dragged it across the floor of the cage, and stuffed it into the door of the bank, effectively corking up her treasure. It was a brilliant move. Baffled, Stinky lumbered away.

In spite of their ungenerous behavior toward each other, I felt that Stinky made a real contribution to Mouse's happiness. Once or twice I separated them for a day or so. Their reunions were joyous, with Mouse scrambling all over Stinky and just about licking him to pieces, and even stolid Stinky showing excitement. They lived together for about a year. I wasn't aware that Stinky was ill, but one day I saw Mouse sitting trembling on her branch when she should have been asleep. I looked in the cup and found Stinky stone-cold dead. "Mouse will miss you," I thought as I heaved his crummy little body into the weeds. I got another laboratory mouse called Pinky to take Stinky's place, but he bullied Mouse so relentlessly that I sent him back. Mouse lived alone for the rest of her days.

Before closing Mouse's story I would like to tell about an episode that took place in Mouse's first weeks with me. After I had had her about ten days I found a lump on her belly near the hind leg. I first noticed it as she lay on her back in my palm while I stroked her. The lump grew larger each day and I feared she had some fatal disease—a tumor of some sort. During the ensuing search for help for Mouse I discovered a peculiar fact about human nature; for some reason people laugh at a mouse.

I live near the University of Rhode Island. I phoned and said I wanted to talk to an expert on mice. The response was laughter. I said my mouse had a mysterious ailment. More laughter, but I was given the name of a woman, Dr. C., I'll call her. Carrying Mouse in her cage, I found the professor in her office. I explained my trouble. When Dr. C. stopped laughing she said she couldn't touch my mouse lest it have germs that would contaminate her laboratory mice. She suggested I ask Dr. H. to examine Mouse. We went to his office. Dr. H. chuckled patronizingly and agreed to look at Mouse's lump. I took her out of her cage and held her belly up for inspection. Both professors gasped at the sight of the lump. Both were baffled.

Dr. H. offered a shot of penicillin, but admitted he had no idea how to measure a dose for a patient weighing half an ounce. Dr. C. forthrightly suggested autopsy. I thanked them and left. Crossing the campus, I passed the library. On an impulse I borrowed a manual of veterinary medicine and took it home.

That evening I skimmed through descriptions of disease after disease looking for symptoms that might fit. Nothing sounded similar to Mouse's trouble until I came to "Cuterebra Infestation" on page 929 and knew I had found the answer. The larva of the botfly, the manual said, lives in a pocket that it forms under the skin of its host, which may be any mammal, most often young. When the parasite reaches full size it emerges through the skin. The manual said that the cure was a simple matter of opening the lump and removing the larva. I hurried to the telephone and dialed a local veterinarian. His wife answered and insisted that I give her the message. Foreseeing difficulty, I replied evasively, "I have an animal that needs a slight operation."

"What kind of animal?" persisted Mrs. Vet.

"A mouse."

There was a long, cold silence. Then the woman asked, in icy tones, "Is it a *white* mouse?"

"No," I admitted. "It is brown."

Mrs. Vet said that her husband did not include mice in his practice and hung up.

The next day my young son hit his toe a glancing blow with an ax while chopping wood, inflicting a wound that needed stitches. As it happened, I had been just about to take Mouse to our local animal shelter for further consultation. She and her cage were in the car as I drove my son to the emergency room at the hospital. While he was being stitched a plan formed in my mind.

The moment John limped out of the operating room I buttonholed the young doctor, a nice soap opera type. I told him I had a mouse that needed surgery. He laughed. With his eye on some pretty nurses standing nearby, he made jokes about calling in the anesthetist, scrubbing up, and so on. The nurses giggled and my cause was won. "Don't go away," I cried, and ran out to get Mouse.

When I got back the doctor had the sheepish look of a man who has been trapped by his own jest, but when he saw Mouse's problem his eyes widened with pure scientific amazement. He studied the lump. We gravely discussed the operating procedure. I held Mouse tightly, on her back, in the palm of my hand. A nurse applied a dab of antiseptic. The doctor made a small incision. Mouse squeaked, but there was no blood. The doctor called for forceps and pulled forth a big, horrible, wiggling grub. There was a babble of astonishment, congratulations, and, inevitably, laughter from the crowd that had gathered around us. The doctor looked pleased and put the grub in a bottle of alcohol as a medical curiosity.

I put Mouse in her cage. She ran about lightly, showing no ill effects. I took her and my limping son home. Both patients healed quickly. I paid a large bill for John's toe, but there was no charge for the mouse.

Mouse lived with me for over three years, which is, I believe, a good deal beyond the span usually allotted to mice. She showed no sign of growing old or feeble, but one day I found her dead. As I took her almost weightless body in my hand and carried it out to the meadow, I felt a genuine sadness. In a serious sense she had given me so much. She had stirred my imagination and opened a window on a Lilliputian world no less real than my world for all its miniature dimensions. By watching her I had learned and changed. Beyond that there had been moments, elusive of description, when I had felt a contact between her tiny being and my own. Sometimes when I touched her lovingly and she nibbled my fingers in return, I felt as though an affectionate message were passing between us. The enormous distance between us seemed to be bridged momentarily by faint but perceptible signals.

I put Mouse's body down in the grass and walked back to the house. I knew that there may be as many mice as there are visible stars. They are given life and extinguished as prodigally as leaves unfolding and falling from trees. I was not sad for Mouse because of her death, but sad for me because I knew I would miss her.

FARLEY MOWAT
b. 1921

One of Canada's most popular writers, Farley Mowat is a gifted story-teller whose material comes in large part from his own experiences as a naturalist and anthropologist. His work has included such juvenile books as The Dog That Wouldn't Be (1957) and Owls in the Family (1961), *popular histories such as* Westviking: The Ancient Norse in Greenland and North America (1965), *and short stories (The Snow Walker, 1975).*

In his work Mowat has focused primarily on the Arctic, specifically on its wildlife and native peoples and their mistreatment at the hands of government and industry. In his later books — A Whale for the Killing (1972), The Great Betrayal: Arctic Canada Now (1976), And No Birds Sang (1979) — *his sympathy for threatened species and cultures has been expressed in an increasingly angry and cynical voice.* The Farley Mowat Reader *was published in 1997.*

His best-known work, Never Cry Wolf (1963), *is an account of his misadventures as an Arctic field biologist for the Dominion Wildlife Service and has become a classic in the field. Often unashamedly anthropomorphic, it sharply satirizes governmental and scientific bureaucracy and is one of the first books to paint a sympathetic, even affectionate portrait of the wolf. The chapter excerpted here dramatizes Mowat's process of self-education among these animals, which eventually results in his determination to "know the wolves, not for what they were supposed to be but for what they actually were."*

From NEVER CRY WOLF

The lack of sustained interest which the big male wolf had displayed toward me was encouraging enough to tempt me to visit the den again the next morning; but this time, instead of the shotgun and the hatchet

Never Cry Wolf (Boston: Little, Brown, 1963).

(I still retained the rifle, pistol and hunting knife) I carried a high-powered periscopic telescope and a tripod on which to mount it.

It was a fine sunny morning with enough breeze to keep the mosquito vanguard down. When I reached the bay where the esker was, I chose a prominent knoll of rock some four hundred yards from the den, behind which I could set up my telescope so that its objective lenses peered over the crest, but left me in hiding. Using consummate fieldcraft, I approached the chosen observation point in such a manner that the wolves could not possibly have seen me and, since the wind was from them to me, I was assured that they would have had no suspicion of my arrival.

When all was in order, I focused the telescope; but to my chagrin I could see no wolves. The magnification of the instrument was such that I could almost distinguish the individual grains of sand in the esker; yet, though I searched every inch of it for a distance of a mile on each side of the den, I could find no indication that wolves were about, or had ever been about. By noon, I had a bad case of eyestrain and a worse one of cramps, and I had almost concluded that my hypothesis of the previous day was grievously at fault and that the "den" was just a fortuitous hole in the sand.

This was discouraging, for it had begun to dawn on me that all of the intricate study plans and schedules which I had drawn up were not going to be of much use without a great deal of co-operation on the part of the wolves. In country as open and as vast as this one was, the prospects of getting within visual range of a wolf except by the luckiest of accidents (and I had already had more than my ration of these) were negligible. I realized that if this was not a wolves' den which I had found, I had about as much chance of locating the actual den in this faceless wilderness as I had of finding a diamond mine.

Glumly I went back to my unproductive survey through the telescope. The esker remained deserted. The hot sand began sending up heat waves which increased my eyestrain. By 2:00 P.M. I had given up hope. There seemed no further point in concealment, so I got stiffly to my feet and prepared to relieve myself.

Now it is a remarkable fact that a man, even though he may be alone in a small boat in mid-ocean, or isolated in the midst of the trackless forest, finds that the very process of unbuttoning causes him to become peculiarly sensitive to the possibility that he may be under observation. At this critical juncture none but the most self-assured of men, no matter how certain he may be of his privacy, can refrain from casting a surreptitious glance around to reassure himself that he really is alone.

To say I was chagrined to discover I was *not* alone would be an understatement; for sitting directly behind me, and not twenty yards away, were the missing wolves.

They appeared to be quite relaxed and comfortable, as if they had been sitting there behind my back for hours. The big male seemed a trifle bored; but the female's gaze was fixed on me with what I took to be an expression of unabashed and even prurient curiosity.

The human psyche is truly an amazing thing. Under almost any other circumstances I would probably have been panic-stricken, and I think few would have blamed me for it. But these were not ordinary circumstances and my reaction was one of violent indignation. Outraged, I turned my back on the watching wolves and with fingers which were shaking with vexation, hurriedly did up my buttons. When decency, if not my dignity, had been restored, I rounded on those wolves with a virulence which surprised even me.

"Shoo!" I screamed at them. "What the hell do you think you're at, you . . . you . . . peeping Toms! Go away, for heaven's sake!"

The wolves were startled. They sprang to their feet, glanced at each other with a wild surmise, and then trotted off, passed down a draw, and disappeared in the direction of the esker. They did not once look back.

With their departure I experienced a reaction of another kind. The realization that they had been sitting almost within jumping distance of my unprotected back for God knows how long set up such a turmoil of the spirit that I had to give up all thought of carrying on where my discovery of the wolves had forced me to leave off. Suffering from both mental and physical strain, therefore, I hurriedly packed my gear and set out for the cabin.

My thoughts that evening were confused. True, my prayer had been answered, and the wolves had certainly co-operated by reappearing; but on the other hand I was becoming prey to a small but nagging doubt as to just *who* was watching *whom*. I felt that I, because of my specific superiority as a member of *Homo sapiens*, together with my intensive technical training, was entitled to pride of place. The sneaking suspicion that this pride had been denied and that, in point of fact, *I* was the one who was under observation, had an unsettling effect upon my ego.

In order to establish my ascendancy once and for all, I determined to visit the wolf esker itself the following morning and make a detailed examination of the presumed den. I decided to go by canoe, since the rivers were now clear and the rafting lake ice was being driven offshore by a stiff northerly breeze.

It was a fine, leisurely trip to Wolf House Bay, as I had now named it. The annual spring caribou migration north from the forested areas of Manitoba toward the distant tundra plains near Dubawnt Lake was under way, and from my canoe I could see countless skeins of caribou crisscrossing the muskegs and the rolling hills in all directions. No wolves were in evidence as I neared the esker, and I assumed they were away hunting a caribou for lunch.

I ran the canoe ashore and, fearfully laden with cameras, guns, binoculars and other gear, laboriously climbed the shifting sands of the esker to the shadowy place where the female wolf had disappeared. En route I found unmistakable proof that this esker was, if not the home, at least one of the favorite promenades of the wolves. It was liberally strewn with scats and covered with wolf tracks which in many places formed well-defined paths.

The den was located in a small wadi in the esker, and was so well concealed that I was on the point of walking past without seeing it, when a series of small squeaks attracted my attention. I stopped and turned to look, and there, not fifteen feet below me, were four small, gray beasties engaged in a free-for-all wrestling match.

At first I did not recognize them for what they were. The fat, fox faces with pinprick ears; the butterball bodies, as round as pumpkins; the short, bowed legs and the tiny upthrust sprigs of tails were so far from my conception of a wolf that my brain refused to make the logical connection.

Suddenly one of the pups caught my scent. He stopped in the midst of attempting to bite off a brother's tail and turned smoky blue eyes up toward me. What he saw evidently intrigued him. Lurching free of the scrimmage, he padded toward me with a rolling, wobbly gait; but a flea bit him unexpectedly before he had gone far, and he had to sit down to scratch it.

At this instant an adult wolf let loose a full-throated howl vibrant with alarm and warning, not more than fifty yards from me.

The idyllic scene exploded into frenzied action.

The pups became gray streaks which vanished into the gaping darkness of the den mouth. I spun around to face the adult wolf, lost my footing, and started to skid down the loose slope toward the den. In trying to regain my balance I thrust the muzzle of the rifle deep into the sand, where it stuck fast until the carrying-strap dragged it free as I slid rapidly away from it. I fumbled wildly at my revolver, but so cluttered was I with cameras and equipment straps that I did not succeed in getting the weapon clear as, accompanied by a growing avalanche of sand, I shot past the den mouth, over the lip of the main ridge and down the full length of the esker slope. Miraculously, I kept my feet; but only by dint of superhuman contortions during which I was alternately bent forward like a skier going over a jump, or leaning backward at such an acute angle I thought my backbone was going to snap.

It must have been quite a show. When I got myself straightened out and glanced back up the esker, it was to see *three* adult wolves ranged side by side like spectators in the Royal Box, all peering down at me with expressions of incredulous delight.

I lost my temper. This is something a scientist seldom does, but I lost mine. My dignity had been too heavily eroded during the past several days and my scientific detachment was no longer equal to the strain. With a snarl of exasperation I raised the rifle but, fortunately, the thing was so clogged with sand that when I pressed the trigger nothing happened.

The wolves did not appear alarmed until they saw me begin to dance up and down in helpless fury, waving the useless rifle and hurling imprecations at their cocked ears; whereupon they exchanged quizzical looks and silently withdrew out of my sight.

I too withdrew, for I was in no fit mental state to carry on with my exacting scientific duties. To tell the truth, I was in no fit mental state to do anything except hurry home to Mike's and seek solace for my tattered nerves and frayed vanity in the bottom of a jar of wolf-juice.

I had a long and salutary session with the stuff that night, and as my spiritual bruises became less painful under its healing influence, I reviewed the incidents of the past few days. Inescapably, the realization was being borne in upon my preconditioned mind that the centuries-old and universally accepted human concept of wolf character was a palpable lie. On three separate occasions in less than a week I had been completely at the mercy of these "savage killers"; but far from attempting to tear me limb from limb, they had displayed a restraint verging on contempt, even when I invaded their home and appeared to be posing a direct threat to the young pups.

This much was obvious, yet I was still strangely reluctant to let the myth go down the drain. Part of this reluctance was no doubt due to the thought that, by discarding the accepted concepts of wolf nature, I would be committing scientific treason; part of it to the knowledge that recognition of the truth would deprive my mission of its fine aura of danger and high adventure; and not the least part of that reluctance was probably due to my unwillingness to accept the fact that I had been made to look like a blithering idiot—not by my fellow man, but by mere brute beasts.

Nevertheless I persevered.

When I emerged from my session with the wolf-juice the following morning I was somewhat the worse for wear in a physical sense; but I was cleansed and purified spiritually. I had wrestled with my devils and I had won. I had made my decision that, from this hour onward, I would go open-minded into the lupine world and learn to see and know the wolves, not for what they were supposed to be, but for what they actually were.

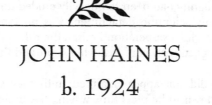

JOHN HAINES
b. 1924

From his early work as a sculptor, to his experience as a homesteader in Alaska between 1954 and 1969, to his poetry, there is a stark integrity to John Haines's work. His desire has been to strip life down to its essential forms and colors. Reviewers of his books of poems—which include Winter News *(1966),* The Stone Harp *(1971), and* News from the Glacier *(1982)—have noticed his ability to evoke a cold, white world that seems in some way absolute. In addition to his poetry, Haines has produced several volumes of essays treating his experiences of homesteading and trapping in Alaska and meditating on the meaning of place.* Living Off the Country: Essays on Poetry and Place *appeared in 1981;* The Stars, The Snow, The Fire: Twenty-Five Years in the Northern Wilderness *in 1989; and* Fables and Distances *in 1996.*

MOMENTS AND JOURNEYS

The movement of things on this earth has always impressed me. There is a reassuring vitality in the annual rise of a river, in the return of the Arctic sun, in the poleward flight of spring migrations, in the seasonal trek of nomadic peoples. A passage from Edwin Muir's autobiography speaks to me of its significance.

> I remember . . . while we were walking one day on the Mönchsberg—a smaller hill on the opposite side of the river—looking down on a green

Living Off the Country (Ann Arbor: University of Michigan, 1981).

plain that stretched away to the foothills, and watching in the distance people moving along the tiny roads. Why do such things seem enormously important to us? Why, seen from a distance, do the casual journeys of men and women, perhaps going on some trivial errand, take on the appearance of a pilgrimage? I can only explain it by some deep archetypal image in our minds of which we become conscious only at the rare moments when we realize that our own life is a journey. [Edwin Muir, *An Autobiography* (Sommers, Conn.: Seabury, 1968), p. 217].

This seems to me like a good place to begin, not only for its essential truth, but because it awakens in me a whole train of images—images of the journey as I have come to understand it, moments and stages in existence. Many of these go back to the years I lived on my homestead in Alaska. That life itself, part of the soil and weather of the place, seemed to have about it much of the time an aura of deep and lasting significance. I wasn't always aware of this, of course. There were many things to be struggled with from day to day, chores of one sort or another—cabins to be built, crops to be looked after, meat to kill, and wood to cut—all of which took a kind of passionate attention. But often when I was able to pause and look up from what I was doing, I caught brief glimpses of a life much older than mine.

Some of these images stand out with great force from the continual coming and going of which they were part—Fred Campbell, the old hunter and miner I had come to know, that lean, brown man of patches and strange fits. He and I and my first wife, Peg, with seven dogs—five of them carrying packs—all went over Buckeye Dome one day in the late summer of 1954. It was a clear, hot day in mid-August, the whole troop of us strung out on the trail. Campbell and his best dog, a yellow bitch named Granny, were in the lead. We were in a hurry, or seemed to be, the dogs pulling us on, straining at their leashes for the first two or three miles, and then, turned loose, just panting along, anxious not to be left behind. We stopped only briefly that morning, to adjust a dog pack and to catch our wind. Out of the close timber with its hot shadows and swarms of mosquitoes, we came into the open sunlight of the dome. The grass and low shrubs on the treeless slopes moved gently in the warm air that came from somewhere south, out of the Gulf of Alaska.

At midday we halted near the top of the dome to look for water among the rocks and to pick blueberries. The dogs, with their packs removed, lay down in the heat, snapping at flies. Buckeye Dome was the high place nearest to home, though it was nearly seven miles by trail from Richardson. It wasn't very high, either—only 3,000 feet—but it rose clear of the surrounding hills. From its summit you could see in any direction, as far west as Fairbanks when the air was clear enough. We saw other high places, landmarks in the distance, pointed out to us and named by Campbell: Banner Dome, Cockscomb, Bull Dome, and others I've forgotten. In the southeast, a towering dust cloud rose from the Delta River. Campbell talked to us of his trails and camps, of years made of such journeys as ours, an entire history told around the figure of one man. We were new to the North and eager to learn all we could. We listened, sucking blueberries from a tin cup.

And then we were on the move again. I can see Campbell in faded jeans and red felt hat, bending over one of the dogs as he tightened a strap, swearing and saying something about the weather, the distance, and himself getting too old to make such a trip. We went off down the steep north slope of the dome in a great rush, through miles of windfalls, following that twisting, root-grown trail of his. Late in the evening, wading the shallows of a small creek, we came tired and bitten to his small cabin on the shore of a lake he had named for himself.

That range of images is linked with another at a later time. By then I had my own team, and with our four dogs we were bound uphill one afternoon in the cool September sunlight to pick cranberries on the long ridge overlooking Redmond Creek. The tall, yellow grass on the partly cleared ridge bent over in the wind that came easily from the west. I walked behind, and I could see, partly hidden by the grass, the figures of the others as they rounded the shoulder of a little hill and stopped to look back toward me. The single human figure there in the sunlight under moving clouds, the dogs with their fur slightly ruffled, seemed the embodiment of an old story.

And somewhere in the great expanse of time that made life in the wilderness so open and unending, other seasons were stations on the journey. Coming across the Tanana River on the midwinter ice, we had three dogs in harness and one young female running loose beside us. We had been three days visiting a neighbor, a trapper living on the far side of the river, and were returning home. Halfway across the river we stopped to rest; the sled was heavy, the dogs were tired and lay down on the ice.

Standing there, leaning on the back of the sled, I knew a vague sense of remoteness and peril. The river ice always seemed a little dangerous, even when it was thick and solid. There were open stretches of clear, blue water, and sometimes large, deep cracks in the ice where the river could be heard running deep and steady. We were heading downriver into a cloudy December evening. Wind came across the ice, pushing a little dry snow, and no other sound—only the vast presence of snow and ice, scattered islands, and the dark crest of Richardson Hill in the distance.

To live by a large river is to be kept in the heart of things. We become involved in its life, the heavy sound of it in the summer as it wears away silt and gravel from its cutbanks, pushing them into sandbars that will be islands in another far off year. Trees are forever tilting over the water, to fall and be washed away, to lodge in a drift pile somewhere downstream. The heavy gray water drags at the roots of willows, spruce, and cotton-woods; sometimes it brings up the trunk of a tree buried in sand a thousand years before, or farther back than that, in the age of ice. The log comes loose from the fine sand, heavy and dripping, still bearing the tunnel marks made by the long dead insects. Salmon come in midsummer, then whitefish, and salmon again in the fall; they are caught in our nets and carried away to be smoked and eaten, to be dried for winter feed. Summer wears away into fall; the sound of the river changes. The water clears and slowly drops; pan ice forms in the eddies. One morning in early winter we wake to a great and sudden silence: the river is frozen.

We stood alone there on the ice that day, two people, four dogs, and a loaded sled, and nothing before us but land and water into Asia. It was time to move on again. I spoke to the dogs and gave the sled a push.

Other days. On a hard-packed trail home from Cabin Creek, I halted the dogs part way up a long hill in scattered spruce. It was a clear evening, not far below zero. Ahead of us, over an open ridge, a full moon stood clear of the land, enormous and yellow in the deep blue of the Arctic evening. I recalled how Billy Melvin, an old miner from the early days at Richardson, had once described to me a moonrise he had seen, a full moon coming up ahead of him on the trail, "big as a rain barrel." And it was very much like that—an enormous and rusty rain barrel into which I looked, and the far end of the barrel was open. I stood there, thinking it might be possible to go on forever into that snow and yellow light, with no sound but my own breathing, the padding of the dog's feet, and the occasional squeak of the sled runners. The moon whitened and grew smaller; twilight deepened, and we went on to the top of the hill.

What does it take to make a journey? A place to start from, something to leave behind. A road, a trail, or a river. Companions, and something like a destination: a camp, an inn, or another shore. We might imagine a journey with no destination, nothing but the act of going, and with never an arrival. But I think we would always hope to find *something* or *someone*, however unexpected and unprepared for.

Seen from a distance or taken part in, all journeys may be the same, and we arrive exactly where we are.

One late summer afternoon, near the road to Denali Park, I watched the figures of three people slowly climb the slope of a mountain in the northeast. The upper part of the mountain was bare of trees, and the small alpine plants there were already red and gold from the early frost. Sunlight came through broken rain clouds and lit up the slope and its three moving figures. They were so far away that I could not tell if they were men or women, but the red jacket worn by one of them stood out brightly in the sun. They climbed higher and higher, bound for a ridge where some large rocks broke through the thin soil. A shadow kept pace with them, slowly darkening the slope below them, as the sun sank behind another mountain in the southwest. I wondered where they were going—perhaps to hunt mountain sheep—or they were climbing to a berry patch they knew. It was late in the day; they would not get back by dark. I watched them as if they were figures in a dream, who bore with them the destiny of the race. They stopped to rest for a while near the skyline, but were soon out of sight beyond the ridge. Sunlight stayed briefly on the high rock summit, and then a rain cloud moved in and hid the mountaintop.

When life is simplified, its essence becomes clearer, and we know our lives as part of some ancient human activity in a time measured not by clocks and calendars but by the turning of a great wheel, the positions of which are not wage-hours, nor days and weeks, but immense stations called Spring, Summer, Autumn, and Winter. I suppose it will seem too obvious to say that this sense of things will be far less apparent to people closed off in the routine of a modern city. I think many people must now and then be aware of such moments as I have described, but do not remember them, or attach no special significance to them. They are images that pass quickly from view because there is no place for them in our lives. We are swept along by events we cannot link together in a significant pattern, like a flood of refugees pushed on by the news of a remote disaster. The rush of conflicting impressions keeps away stillness, and it is in stillness that the images arise, as they will, fluently and naturally, when there is nothing to prevent them.

There is the dream journey and the actual life. The two seem to touch now and then, and perhaps when men lived less complicated and distracted lives the two were not separate at all, but continually one thing. I have read somewhere that this was once true for the Yuma Indians who lived along the Colorado River. They dreamed at will, and moved without effort from waking into dreaming life; life and dream were bound together. And in this must be a kind of radiance, a very old and deep assurance that life has continuity and meaning, that things are somehow in place. It is the journey resolved into one endless present.

And the material is all around us. I retain strong images from treks with my stepchildren: of a night seven years ago when we camped on a mountaintop, a night lighted by snow patches and sparks from a windy fire going out. Sleeping on the frozen ground, we heard the sound of an owl from the cold, bare oak trees above us. And there was a summer evening I spent with a small class of schoolchildren near Painted Rock in central California. We had come to learn about Indians. The voices of the children carried over the burned fields under the red glare of that sky, and the rock gave back heat in the dusk like an immense oven. There are ships and trains that pull away, planes that fly into the night; or the single figure of a man crossing an otherwise empty lot. If such moments are not as easily come by, as clear and as resonant as they once were in the wilderness, it may be because they are not so clearly linked to the life that surrounds them and of which they are part. They are present nonetheless, available to imagination, and of the same character.

One December day a few years ago, while on vacation in California, I went with my daughter and a friend to a place called Pool Rock. We drove for a long time over a mountain road, through meadows touched by the first green of the winter rains, and saw few fences or other signs of people. Leaving our car in a small campground at the end of the road, we hiked four miles up a series of canyons and narrow gorges. We lost our way several times but always found it again. A large covey of quail flew up from the chaparral on a slope above us; the tracks of deer and bobcat showed now and then in the sand under our feet. An extraordinary number of coyote droppings scattered along the trail attracted our attention. I poked one of them with a stick, saw that it contained much rabbit fur and bits of bone. There were patches of ice in the streambed, and a few leaves still yellow on the sycamores.

We came to the rock in mid-afternoon, a great sandstone pile rising out of the foothills like a sanctuary or a shrine to which one comes yearly on a pilgrimage. There are places that take on symbolic value to an individual or a tribe, "soul-resting places," a friend of mine has called them. Pool Rock has become that to me, symbolic of that hidden, original life we have done so much to destroy.

We spent an hour or two exploring the rock, a wind and rain-scoured honeycomb stained yellow and rose by a mineral in the sand. Here groups of the Chumash Indians used to come, in that time of year when water could be found in the canyons. They may have come to gather certain foods in season, or to take part in magic rites whose origin and significance are no longer understood. In a small cave at the base of the rock, the stylized figures of headless reptiles, insects, and strange birdmen are painted on the smoke-blackened walls and ceiling. These and some bear paw impressions gouged in the rock, and a few rock mortars used for grinding seeds, are all that is left of a once-flourishing people.

We climbed to the summit of the rock, using the worn footholds made long ago by the Chumash. We drank water from the pool that gave the rock its name, and ate our lunch, sitting quietly in the cool sunlight. And then the wind came up, whipping our lunchbag over the edge of the rock; a storm was moving in from the coast. We left the rock by the way we had come, and hiked down the gorge in the windy, leaf-blown twilight. In the dark, just before the rain, we came to the campground, laughing, speaking of the things we had seen, and strangely happy.

MAXINE KUMIN
b. 1925

Maxine Kumin won the Pulitzer Prize in 1973 for her collection of poetry entitled Up Country. *She has published eleven other volumes of poetry, as well as fiction, children's literature, and essays on the natural and cultural landscape surrounding her New Hampshire farm. Kumin is a lover of the deep rhythms of life in northern New England and a sympathetic observer of her human and animal neighbors in that region. Among the works containing her reflections on home, family, and place are* To Make a Prairie: Essays on Poets, Poetry, and Country Living *(1980);* In Deep: Country Essays *(1987); and* Women, Animals, and Vegetables: Essays and Stories *(1994). In the following essay, which explores the strong bond between horses and people, especially women, Kumin, like Vicki Hearne, considers our relationship with domesticated animals and the possibilities of interspecies communication.*

SILVER SNAFFLES

The best present of my childhood, a British storybook called *Silver Snaffles*, by Primrose Cumming, arrived on my tenth birthday. In this fantasy tale, a little girl walks through the dark corner of a draft pony's stall into a sunlit world where articulate ponies with good English country-squire manners and highly individual personalities give lessons in equitation and stable management to some eager, horseless young-sters. After several instructive episodes the story culminates in a joyous hunt staged by the foxes themselves, who are also great conversational-ists. Jenny, launched as a rider, learns that she is to be given a pony of her own. Now she must relinquish her right to the magic password and to the dark corner of Tattles's stall, through which she has melted every evening into a better world. I thought it the saddest ending in the whole history of literature.

By this age I had already begun to ride—an hour a week, one dollar an hour—on one of those patient livery horses who put up with the clumsiness of the novice. A year or two later, when I could wield a manure fork and manage a full-size wheelbarrow, I began to learn something about stable management. It didn't especially matter to me whether I rode or not; I was happy just to be in the presence of horses. I wanted to inhale them, and I wanted them to take me in. Days that I wasn't allowed to walk two miles to the rental stable, I skulked above stairs and reread *Silver Snaffles*.

Eventually, I wore this book out with ritual rereadings; somehow it disappeared from my life. When, forty-odd years later, I assumed my duties at the Library of Congress, it occurred to me that *Silver Snaffles* was undoubtedly housed in that enormous repository of knowledge. The day I found it in the cavernous, dusty stacks, I sat down on the marble floor and turned the familiar pages in disbelief. The book was real, after all! Then I carried it up to the Poetry Room that overlooks the Capitol and the Mall and read the story all over again, savoring the parts I had remembered verbatim. While serious affairs of state were being conducted only a few hundred yards away, I patted the sturdy, faded blue cover and wept.

I think I was crying for my lost childhood, but I'm still not sure. The profound, cathartic effect of the book was not assuaged for me by Jenny's acquiring a pony of her own. Nothing would compensate for that loss of innocence, the unworldliness that had enabled her to speak

In Deep: Country Essays (New York: Viking Penguin, 1987)

the password and walk into a kingdom of virtue and honor. There, all the ponies were Platonic philosopher-kings, and the children, their willing pupils, rose with pure hearts through the ranks of Becoming to the heightened state of Being represented by well-mucked stalls, well-ridden and chatty equines, and steaming bran mashes.

At this remove, I can now understand my fascination with the text. The deprivation that I felt when the heroine was denied continued access to Paradise was the loss of something I can only call interspecies communication. *Pace,* Anna Freud; *pace,* D. H. Lawrence; this bonding, I believe, lies at the heart of horse fever for any age or gender.

Interaction with the horse takes place on several levels. Physical communication is foremost. You learn its body language and it learns to respond to a body language you use to ask for changes in gait, direction, and body frame. You may spend weeks and weeks going around the dressage ring or beaten track in your pasture, riding a stiff, unbalanced, green horse, thinking *round, round,* and asking with your body aids for flexion, for bending, for the horse to use its hind legs, to track up under its body, to come onto the bit instead of lagging behind it. And then one day there is a little breakthrough. For two or three strides you and your horse move in absolute harmony. Later, another breakthrough and another. You are ecstatic. The horse is pretty happy, too. He/she has understood you, made the effort, developed the supporting musculature, and goes forward confidently, in balance. From this two-way exchange evolves a shared precision, a shared joy in fluidity of motion, rhythm and cadence not unlike the delight an athlete takes in executing, for example, figures on skates, or completing a slalom course on skis or performing intricate high dives. We have, blessedly, an atavistic need to let our bodies speak. To achieve a bonding with a horse you have raised and schooled from its gawky weanling stage, to arrive at control without exploitation, to work toward new goals that mean stretching limits for both horse and rider is to employ that body language, and to do so with equestrian tact. Like poetic tact, also based on a sense of touch, it is arrived at only gradually and with self-discipline.

Horses, like people, enjoy enlarging their scope. Horses asked only to perform the same routine day after day grow dull, bored, apathetic. Faced with new fields, a variety of trails, different cross-country courses and dressage rings, the supple, athletic, healthy horse takes pleasure in meeting the challenge.

Growth and change without brutality, achieving goals by skilled effort on both parts, depend on interspecies communication. At the risk of sounding sexist—some wonderful trainers and handlers are male—I profoundly believe that women work best with problem horses. (This is not the same thing as riding them best over seven-foot *puissance* walls in Madison Square Garden.) Women's empathy, subtlety, ability to read the nuance of difference that leads to change; women's gift of timing; women's instinct for nurturing, all contribute to their considerable successes, for example, in the large Thoroughbred and Standardbred breeding-farm operations; in the show ring; as exercise riders, grooms, and jockeys in the racing world; and as instructors at levels ranging from the local day-camp riding program to training establishments like Morven Park or the Potomac Horse Center.

Clinics I have attended, featuring a holistic approach to behavioral problems in a variety of performance horses, from Olympic contenders to backyard ponies, attract audiences that are 90 percent female. Somehow the notion of improving performance by developing a trusting partnership between horse and human elicits a readier response from women than from men. Old attitudes die hard; the macho concept of muscling a horse into submission still has its adherents.

Granted, many elements enter into the human passion for horses, and developing sexuality may be one of these. But the old stereotype about girls and horses has always seemed to me too facile to be trusted. For one thing, it does not take into account that the girl-horse affection is much more an East Coast phenomenon than a Western one. As you move into Colorado, Montana, Utah, and so on, you encounter a greater ratio of boys struck with the same fever. In a culture where the cow pony represents not only the freedom to explore space, but also the means to develop such working skills as cutting and roping and such recreational skills as rodeo riding or barrel-racing, males are in the majority. Eastern pony clubs and 4-H horse groups have been for the most part the purview of mothers, much as Little League has belonged for the most part to competitive fathers.

I further mistrust the Freudian concept because it does not adequately explain the substantial number of adult females who, despite their comfortable adaptation to sex roles involving marriage and child-rearing, continue to lease, own, care for, ride and/or raise horses. Some of us even do so in collusion with riding husbands, who are also unabashed nurturers and can be found pitching hay, sweeping up, and otherwise taking an active part in horsekeeping and riding.

I do not blink the fact that the horse often comes into a young girl's life at a time when she feels a need to take control in some measure of what is essentially an uncontrollable environment. The key factor is that an animal's responses can be counted on. When all else shifts, changes and disappoints, the horse can remain her one constant.

I asked Julia, my favorite twelve-year-old working visitor, why she loves riding horseback. These are her unrehearsed responses. "It gives you a sense of freedom. You're sort of out of touch because you're higher up than everybody else." And then she added, "Taking care of horses makes you feel good because you're making somebody else [*sic*] feel good. It's sort of a comforting feeling when you get done [with the evening barn chores] because you've put them away and you know you've made them feel cozy and secure."

The bond with horses, no matter its origin, becomes at some stage functionally autonomous. However it began, it is enjoyed ultimately for its own worth. Why else do I trudge through four feet of snow and ice at below-zero temperatures, winter after winter, to tend a barn full of mares and foals? Intrinsic worth is what keeps me creosoting fences at the peak of blackfly season, when merely to speak out of doors is to inhale swarms of the nasty, biting gnats. In summer heat waves when the horses doze in cool stalls all day and are turned out to graze all night, the barn must be cleaned at nine p.m., dinner guests or no. In the searing electrical storms that pounce in July and August, when fear of lightning striking in the open pastures hurries me out to round up the scattered group, intrinsic worth keeps me going. Under my breath I may mutter, "Virtue is its own reward, and that's all it is," a dictum offered by the poet Howard Nemerov, but clearly I would not have it otherwise. And in the fall, when mosquitoes and deer-flies wane and woodland trails are carpeted with alternating patches of pine needles and still-bright leaves, intrinsic worth mingles with all the sweetness of delayed gratification.

"The Horse," Gervase Markham wrote in 1614, "will take such delight in his keeper's company, that he shall never approach him but the horse will with a kind of cheerful or inward neighing show the joy he takes to behold him, and where this mutual love is knit and combined, there the beast must needs prosper and the rider reap reputation and profit."

Cheerful or inward neighing is what I would cross-stitch on a sampler, if I were assigned to work one, as my mother was. Hers read, in the astonishingly even stitches of a twelve-year-old, *He maketh me to lie down in green pastures.* An appropriate piety, under the circumstances.

ANN HAYMOND ZWINGER
b. 1925

Born in Muncie, Indiana, Ann Zwinger studied art history at Wellesley College and Indiana University. Most of her books are illustrated with her own graceful and meticulous drawings of plants, which she considers integral complements to the writing itself. Her first book, Beyond the Aspen Grove *(1970), is an account of learning how to live on a wild tract of land high in the Colorado Rockies. She has also written books about Wyoming, Utah, the Baja Peninsula, the Grand Canyon, and the deserts of the Southwest, and coauthored* A Conscious Stillness: Two Naturalists on Thoreau's Rivers *(1982) with Edwin Way Teale. The following selection, set in the Sonoran Desert, is from* The Mysterious Lands *(1989), a personal account of the four great desert regions of the American West. Rarely do Zwinger's accounts contain dramatic events or sweeping philosophical statements. But her careful attention to the texture and color of the natural world, combined with an intelligent and amiable voice, have cumulative strength. Her books succeed in what Joseph Conrad described as the most important and most difficult task of the artist: "to make you see."*

OF RED-TAILED HAWKS AND
BLACK-TAILED GNATCATCHERS

Three concerns haunted me before I came on this bighorn sheep count: that I would be uneasy alone, that time would hang heavy, that I could not endure the heat. Instead I have felt at home, there have not been enough hours in the day, and the heat has become a bearable if not always welcome companion. The words of Joseph Wood Krutch, also writing about the Sonoran Desert, come to mind: "Not to have

The Mysterious Lands (New York: Dutton, 1989).

known—as most men have not—either the mountain or the desert is not to have known one's self. Not to have known one's self is to have known no one."

At noontime I am concentrating so hard on taking notes that when a cicada lets off a five-second burst like a bandsaw going through metal I jump. When there has been no activity at the tank for over an hour I opt for a can of tuna sprinkled with the juice of half a lemon.

No sooner do I open the can and get my fork out than all the birds explode from the rocks on which they've congregated. A red-tailed hawk bullets straight toward me, talons extended, tail spread. No sound, no screaming. It breaks off, rises with no rodent in its talons, wheels, spirals upward, and swoops again. Again no luck. It makes no third try. Its disappearance is followed by a great shocked silence. Half an hour passes before the doves venture, one by one, back to their sentinel rock.

I see only this one hawk stoop. The only time redtails are quiet is on the attack. Otherwise I hear their eerie *KEEEeeeeer KEEeeer* that ricochets off the sky itself long before they come in to water. One afternoon a redtail sits on the steep cliff to the right; a couple of feet below it perch some house finches; a black-throated sparrow searches the bush beside it; a pair of ash-throated flycatchers rest above it; and across the tank, a batch of doves roost peacefully on their rock. All birds must be vulnerable to attack at the water hole and many avoid too much exposure by being able to drink very quickly, or coming in early and late when raptors are not hunting. Yet there are also these moments when the lion and the lamb lie down together.

I put aside my field glasses and lift my fork. This time a turkey vulture alights on the big rock. The rock is nearly vertical on the side overlooking the tank and this is where it chooses to descend to water. It gets about a quarter of the way down, contorted in an awkward position, big feet splayed out on the rock, tail pushed up behind at a painful angle, as it looks intently down at the water, a hilarious study in reluctance. Gingerly it inches down (vultures have no claw-grasping capability as birds of prey like eagles and hawks do) until it can hold no longer and crash-lands so ridiculously onto the apron to drink that I laugh out loud. The closeness of the rocks around the pool impedes the bird's maneuvering space because of its broad wingspan, and so it edges as close as it can to ensure dropping upon the only place where it can stand and drink.

Four species of this misanthropic-looking bird existed in the Pleistocene, and this creature looks like one of the originals. On the ground, vultures are hunched and awkward bundles of feathers, but in the air, where I watch them during much of the day, they are magnificent, graceful soarers. They ascend with the updrafts coming off the hot desert floor, floating and lifting to cooler air, where visibility is superb; at five hundred feet the visible horizon is twenty-seven miles away, and at two thousand feet, fifty-five miles. Such big raptors generally get as much water as they need from the carrion they consume.

One more try on the tuna fish. A Harris' antelope squirrel scuttles down the wash to nibble on saguaro seeds. Tail held high over its back for shade, a single white stripe on each flank, it has quick, jerky movements typical of the ever-watchful. It stands slightly hunkered up on its hind legs to eat, but its rear end never touches the hot sand. It spends little time feeding and soon tucks back into the brush near the blind, where it undoubtedly has a burrow. There it spread-eagles on the cool soil and unloads its body heat, before taking on the desert again. Still, it withstands unusually high heat loads because of a lower basal metabolism.

The next day, emboldened, it hops up on the iceless ice chest in the blind and puts its head in my empty plastic cup, which tips over with a clatter, sending the squirrel flying.

I eat lunch at a fashionable three o'clock. By four o'clock, shadows cover the tank and the blind is in sun. A cloud cover that has kept the temperature relatively low all day has disappeared and the sun has an unobstructed shot at my back. Until the sun drops behind the ridge, it is the most miserable time of the day.

The resident robber fly alights in front of me, makes a slapdash attempt at a wandering fly, misses, and returns to watching. Flies are widely dispersed in the desert, more prevalent than any other insect order, and of these, bee flies and robber flies are the most numerous. The huge eyes of robber flies give them peripheral vision; with streamlined stilettolike bodies that allow swift flight, and needlelike beaks, they are efficient predators on the desert wafters and drifters. I've watched this one impale a smaller fly in flight, almost too quickly for the eye to follow, then alight to suck out the juices. Today it seems scarcely to care.

In the evening a caterwauling of Gambel's quail issues from the mesquite trees, where they perch in the branches. Their fussing is interspersed with a short soapsudsy cluck, embellished by a silvery *tink* at the end. The first night I was here they scolded and fumed about the stranger in their midst. The second night they gossiped and fussed, and the third night I awoke to find them within a few feet of the cot.

They visit the big mesquites only in the evening, always after sunset, and they are always noisy. I would think a predator could hear them a mile off. Their drinking patterns have evolved to avoid predators: they commonly come in to water twice a day, one period beginning at dawn, the other ending at dusk. Birds of prey tend to arrive around noon, so the quail's watering time does not overlap.

Quail have been termed "annual" birds because of their variable yearly populations. In the Sonoran Desert, the number of young quail per adult found in the fall correlates to rainfall during the previous December to April; in the Mojave, the same high correlation exists between young and October-to-March precipitation (a relationship that exists in other desert animals, among them bighorn sheep). Their reproductive activity begins before green vegetation becomes a part of their diet, the amount of which would give them clues to the amount of nourishment available, and hence the clutch size that could survive. Although quail do not breed at all in exceptionally dry years, they are one of the most prolific of birds.

They remind me of charming windup toys, painted wooden birds bustling about with staccato movements, officiously giving each other directions as they bustle among the creosote bushes. As I watch them, I remember the Kawaiisu Indian story about the tear marks on Quail's face because her young died, one after the other, when she made her cradles out of sandbar willow—a wood that the Indians therefore do not use for cradles.

The day dims and I stretch out to count the stars framed in a triangle of mesquite branches. Content, I realize I have reached, as Sigurd Olsen wrote, "the point where days are governed by daylight and dark, rather than by schedules, where one eats if hungry and sleeps when tired, and becomes completely immersed in the ancient rhythms, then one begins to live."

Yes.

Early in the morning, and again in the evening, bees create an unholy cantillation around the blind, some working the few mesquite and creosote flowers that remain, but most of them just circling in holding patterns of their own devising.

The number of creosote bushes in the western deserts make it a prime source for pollen and nectar. Several bee species are closely associated pollinators of the creosote bush although it does not depend on any single species for successful pollination. The most numerous pollinator in the summertime is *Perdita larrea*, a tiny bee just an eighth of an inch long. Because of its small size *Perdita* can harvest pollen larger bees cannot, but smallness may also curtail its value as a pollinator. Smallness prevents it from making reliable contact with a stigma, and its foraging range is modest and limited.

While the bees are outside the blind I remain quietly inside, hoping that the shade will discourage them from visiting and, most of all, from stinging at some imagined insult. With great relief each day I watch the bees move toward the water tank as the day heats. A thin black cloud of them is visible through the spotting scope. More than once I see a bighorn sheep make a hurried withdrawal from the tank, shaking its head vigorously from what I assume to be a bee's umbrage.

This morning bighorns remain at the tank, either drinking or standing around, for more than an hour before they leave. I follow them with the field glasses as long as I can. When I begin recording my stopwatch notations on the work sheet, I hear a strong bleating at erratic intervals. Search as I may, I can neither see around the saguaro to identify the activity nor locate the source by sound.

By default, I see a great deal of the old saguaro for the next quarter hour. The cactus is badly riddled about eight feet off the ground, so that daylight shows through the ribs, undoubtedly the work of the resident white-throated wood rat. Wood rats are the only animals that consistently eat cactus; other species may feed on it occasionally but cannot make it a steady diet because of the high content of malic and oxalic acids, created by the cactus's CAM photosynthesis. Oxalic acid is in the form of insoluble calcium oxalate crystals that, in humans, cause severe renal problems. For six months cactus is the wood rat's major source of food, reaching a peak in late May when it may comprise more than 90 percent of its diet.

The saguaro is closely pleated, waiting to expand when the rains come. The bellowslike action, which permits expansion of the stem without tearing the inelastic skin, allows precise adjustment to water storage, an action that starts with even very light rainfall. Water loss during the dry season reduces the volume of storage tissue, shrinking the stem; as the diameter becomes smaller, the ribs draw closer together, hence the pleated look. After a rain the process reverses rapidly, with more intensity on the south side of the stem, probably because water-conducting tissue is more prevalent there; the north side does not begin to swell until a few days later. About three feet off the ground on the north side facing me, an extra rib has been added to the basic number of twelve ribs. This onset of radial growth, or bifurcation, doesn't occur until the trunk grows to at least twelve inches in diameter.

A cactus of this immense size can absorb 95 percent of its total weight in water, sometimes up to a ton, expanding for up to three weeks after rains. Confined to the upper three inches of soil, tough ropelike roots may extend outward fifteen or twenty feet, placed to suck up as much water as possible before it evaporates. Lacking a stabilizing taproot, a saguaro topples if these lateral roots are severed.

The south and north sides of the cactus are measurably different. Ribs on the south side are deeper. Receiving the greater amount of direct sunlight, the deeper furrows provide a modicum of shade, reducing the time that direct sunlight heats the surface. Spine lengths on the south side are longer, and may provide an insulating layer of air. Fruits ripen first on the south side. Branch ends are usually colder than the main stalk during the night, and since single flowers form at the tips only, more rapid development occurs on the warmer side. Dryness promotes the formation of flower buds; plants growing with a favorable water supply make handsome vegetative growth but tend not to bloom.

Flowering is, after all, not an aesthetic contribution, but a survival mechanism.

The variety of birds in the Cabeza Prieta has been a surprise, especially at this hot, dry time of year. A pair of ash-throated flycatchers, tails a bright reddish brown, forage in the mesquite. Only gullies and washes with larger shrubs and trees support enough insects to attract the flycatchers. Three verdins squabble in a saltbush, their yellow caps bright but their bodies blending and disappearing in the fretted background. Feeding on insects (as most migratory birds do) they can exist without free water as long as insects are available. Subject to more water loss because of a less advantageous ratio of surface area to volume than that enjoyed by larger birds, they spend the day closeted in shade and shadow. Females weave ball-like nests in the thorniest of bushes, catclaw and sometimes cholla, and line them with down and feathers for the young, while the male builds a simpler nest for protection against the chill of desert nights. A wreath of spiny twigs around each entrance protects the relatively low-placed and vulnerable nests.

Wings whir close by as a Costa's hummingbird checks out my hanging red nylon stuff sack, then perches on a mesquite twig. This three-gram female loses nearly half of her body weight on hot days, making dependence on surface water and succulent food greater than that of large birds, which lose much less. When nectar from ocotillo and Indian paintbrush, as well as agaves, is available, hummingbirds obtain moisture in their food, but at this dry time of the year, when no flowers bloom, they must have access to free water instead.

At dusk a desert cottontail rollicks down the wash, which is now in partial shade. This cottontail is lighter than the mountain cottontails with which I am familiar, illustrating the tendency of desert mammals to be paler in pelage, somewhat smaller, and with longer ears and legs than their cooler-climate counterparts. It does not hop; its front and back feet move together in a rocking motion like a hobbyhorse canter.

It nibbles some fallen saguaro fruit, then beds down on the other side of the sandy strip in a thicket of saltbush. Another one appears. The first returns. They face each other three feet apart, feinting. The first one dashes for the other, who levitates straight up while the first dashes beneath, and then both bucket off. To me it looks like great good fun and I enjoy their silly antics.

My band of bighorn sheep prefer morning watering. They come with some precision to the tank. This morning there is a considerable amount of amatory exercise going on between the young rams and the ewes. Breeding season is not far off and mating behavior patterns briefly interrupt the more unstructured ambles to the tank.

A ram follows a young ewe downslope, nose so close to her tail that he trips for not watching where his feet are going; the scent of a ewe's urine communicates whether or not she is in estrus. The ewe appears to ignore him, stopping to browse along the way or simply look about. I watch three different pairs, and each female, at different times, stops and urinates, a behavior that occurs only when a ram closely follows.

A group of six sheep come in early today, with some I've not seen before. They remain two hours, most of it standing around and mountain watching, before they move back up the hill. Busy recording, I hear the cracking of branches close by, and suddenly there are five right in front of the blind, noisily cracking mesquite pods. Although they prefer grass, there is precious little of that here and so they have become more opportunistic in their feeding. The group includes both a young ram and a big mature ram, two lambs, and a female. They are obviously aware I am here, look at me with less than the curiosity I feel I deserve, but otherwise pay no more attention to me than to the saguaro. I hear them crunching mesquite pods as they move downslope.

Sweeping the slope with binoculars, I spot more sheep on the hillside. As I switch to the spotting scope to watch them more closely, I hear a jet coming in low. A sonic boom rips the air and reverberates so close the whole blind vibrates, and even though I know it's coming, I still jump when the sound hits.

The bighorns on the hillside never turn a hair. The ten white-winged doves on the sentinel rock never move. The mourning doves never stop calling. I later find this to be generally true, that the frequency of a sonic boom resembles that of a thunderstorm, and that wildlife here generally is not disturbed by either. This observer is.

At midday, when tank activity has closed down and the temperature reads 106 degrees F., the metal of the spotting scope burns my fingers.

The water I take from the jerry can to drink is hot enough to brew tea
with. I can't complain. Early travelers had it much worse; John
Durivage, crossing the same desert in 1849, found

> the water was detestable and at any other time would have proved a pow-
> erful emetic, but now it was *agua dulce*. A tincture of bluelick, iodides of
> sulfur, Epsom salts, and a strong decoction of decomposed mule flesh
> were the component parts of this delectable compound.

A western whiptail lizard patters out from behind a water can and
prowls the edge of the blind, snuffling the dirt floor as it goes. I puzzle
over the dark color, an allover deep smoky brown with a checkered back
pattern, a good-size lizard with a long tail, the last three inches of which
are almost black—until I realize that the soil under the mesquite tree
where it forages is brown from the humus of leaves and pods and twigs
built up over the years. Western whiptails are very variable in color,
tending to match the ground upon which they travel. When the lizard
reaches my bare right foot, to my delight, it tickles right over my toes.

I pour four inches of water into the bucket, put my shirt in, wet it
thoroughly, and put it back on wet. Even though the water is hot it
chills immediately and sits on my skin like salvation. It refreshes and
energizes me enough to take a walk. When I return, hundreds of bees
have found the water I unadvisedly left in the bucket. My blue sleeping
bag is the dearest object of their desire. They cling to my blue shirt
with ecstasy or joyously explore every object in the blind.

I sulk outside in the flimsy shade of a creosote bush for forty-five
minutes while they enjoy my grudging hospitality.

Behind me, the next morning, are the subtle sounds in the brush that
indicate small-animal movement, but peer as I may, I see nothing.
Finally I pick out a dull-brown robin-size bird, which scratches the dirt,
pauses, disappears, pops up on a branch and down again, difficult to
see and to follow. I finally get a good glimpse of a fierce yellow eye and
an almost raptorlike gaze: a curve-billed thrasher. With something in
its beak it flies to a nearby cholla and solves a mystery: at last I know
who built the large messy nest in the cholla in which I found three
fully-feathered, grayish-brown young, eyes tightly closed, beaks like
hand-drawn brackets too big for their heads, so nondescript as to be
unidentifiable. A high percentage of desert birds nest in inaccessible
cavities, spiny trees and shrubs and cacti, symptomatic of the strong
predatory pressures on nesting birds here, but only cactus wrens and
curve-billed thrashers regularly brave the cholla.

By noon the temperature is 102 degrees F. and climbing. I was going to walk, for there are things I want to check out. I was going to draw, but even with a paper towel beneath my arm I still perspire enough to buckle the paper, and the drawings are out of proportion and awkward. I was going to do a lot of things, but the heat, combined with my usual after-lunch low metabolism, saps my ambition. A white-winged dove calls, repetitive, insistent, annoying. A thankless breeze comes through the blind. The little ground squirrel arrives, unnecessarily spry and lively and perky. I feel listless to the point of stupor.

Actually I may have reached the point where I can live with this heat, this everywhere, this without-respite hot. My skin feels cool so evaporative cooling is working. It would be heaven to pour a pail of cool water over my overheated head, but all the water is hot. Not warm. Hot. And I yearn for every ice cube I ever heedlessly rinsed down a drain.

The silence of noon is palpable, more an onerous, enveloping physical presence than a lack of sound. A female Gila woodpecker lands on a mesquite branch, mouth agape. When a mourning dove coos, it has a lunatic overtone. A fly's drone sounds like a freight train.

The white-winged doves sit on the sentinel boulder above the tank, in full sun. The heat radiating off the rocks must make the cul-de-sac in which the tank sits unbearable, yet the doves appear unperturbed. The thrasher, beak agape, sits on her nest, lit by full sun. In this high-heat time of day she must be there to shield the young, enjoying none of the comforts of a shaded nest but also not vulnerable to attack, a brutal trade-off.

Rather than evaporating water to lower body temperature, most desert birds allow a passive rise in body temperature. Birds ordinarily have a higher body temperature than mammals, well above 100 degrees F., and this higher temperature allows them to dissipate heat by radiation. Birds, with the exception of burrowing owls, cannot utilize burrows for cooling, nor can they sweat. They can sit in the shade and extend their wings to expose bare patches of skin; they can compress their plumage to reduce its insulating value; or they can gape, which is a means of evaporative cooling.

When ambient temperature rises above body temperature, there is one more option: gular flutter—fluttering the thin floor of the mouth and the soft skin under the throat. Gular flutter increases evaporation and is highly developed in several desert bird species, among them doves, quails, nighthawks, and roadrunners.

But sometimes even that doesn't suffice. On the last afternoon I check the thermometer at three o'clock, already 108 degrees on its way up to 112 degrees F. Heat rolls down off the surrounding ridges like a *nuée ardente*, consuming everything in its path.

Hopping up on a bough of the mesquite against which the blind is built is a tiny black-tailed gnatcatcher, not much bigger than a hummingbird. Illustrations in bird-identification books show such plump, neatly feathered creatures. This little soul is waiflike and thin, feathers disheveled and tufting out, the Edith Piaf of the bird world.

She walks up the branch in front of me, less than an arm's length away. She stands with her body high off the branch, tail quivering. As I watch her, her head nods forward in that familiar "I can't hold my head up another minute" droop. As soon as her head drops she jerks it up and opens her eyes in a gesture so reminiscent of a child fighting sleep I have to smile. She moves to stand beside my duffel bag, which is wedged up against the tree trunk and, as I watch, slowly lists until she leans completely against it as if exhausted by the heat. Her head drops forward and she starts awake several times more. Finally her head remains down. She sleeps.

In time my head falls forward and I too jerk awake, and finally, very quietly, scoot down in the chair so that my neck can rest on the back, and then I doze as well. I awake with a stiff neck just as the little gnatcatcher stirs. She pulls herself upright, shakes her feathers flat, looks about perkily for a moment, and in leisurely fashion, hops into the brush.

My last evening here. I walk up to the divide that separates my valley from the desert flats to the west, winding to the top, flushing a lizard, climbing over granite boulders, avoiding the barbed and the spurred. When I reach the top I climb up on a boulder that is set out of the wind that hollows through the defile. Below, saguaros stalk down a dry wash and then disappear as the land levels out into a wild, open emptiness.

Looking out over the pure sweep of seamless desert, I am surprised to realize that the easy landscapes stifle me—closed walls of forests, ceilings of boughs, neat-trimmed lawns, and ruffled curtains of trees hide the soft horizons. I prefer the absences and the big empties, where the wind ricochets from sand grain to mountain. I prefer the crystalline dryness and an unadulterated sky strewn from horizon to horizon with stars. I prefer the raw edges and the unfinished hems of the desert landscape.

Desert is where I want to be when there are no more questions to ask.

J(OHN) A(LEC) BAKER
b. 1926

In 1967 an unknown 41-year-old writer published an extraordinary book chronicling an intense, ten-year exploration of the lives and nature of peregrine falcons along the tidal coastlines of eastern England. The Peregrine, *J. A. Baker's account of these supreme predators, condensed into the diary of a single winter, is a combination of accurate detailed observation and a willed identification with his subject. His style is full of hard, Saxon, richly imagistic language reminiscent of Gerard Manley Hopkins. Like many contemporary nature writers he believes that "honest observation" is not enough, that the "emotions and behavior of the watcher are also facts, and they must be truthfully recorded." In* The Peregrine, *from which the following introductory chapter is taken, and in his subsequent book,* The Hill of Summer *(1969), he evokes his native Essex landscape with the same parochial devotion as Gilbert White and Richard Jefferies. Yet Baker's view of nature is more disturbing than these earlier writers', and his countryside is almost devoid of human figures.*

BEGINNINGS

East of my home, the long ridge lies across the skyline like the low hull of a submarine. Above it, the eastern sky is bright with reflections of distant water, and there is a feeling of sails beyond land. Hill trees mass together in a dark-spired forest, but when I move towards them they slowly fan apart, the sky descends between, and they are solitary oaks and elms, each with its own wide territory of winter shadow. The calmness, the solitude of horizons lures me towards them, through them, and on to others. They layer the memory like strata.

From the town, the river flows north-east, bends east round the

The Peregrine (New York: Harper and Row, 1967).

north side of the ridge, turns south to the estuary. The upper valley is a flat open plain, lower down it is narrow and steep-sided, near the estuary it is again flat and open. The plain is like an estuary of land, scattered with island farms. The river flows slowly, meanders; it is too small for the long, wide estuary, which was once the mouth of a much larger river that drained most of middle England.

Detailed descriptions of landscape are tedious. One part of England is superficially so much like another. The differences are subtle, coloured by love. The soil here is clay: boulder clay to the north of the river, London clay to the south. There is gravel on the river terraces, and on the higher ground of the ridge. Once forest, then pasture, the land is now mainly arable. Woods are small, with few large trees; chiefly oak standards with hornbeam or hazel coppice. Many hedges have been cut down. Those that still stand are of hawthorn, blackthorn, and elm. Elms grow tall in the clay; their varying shapes contour the winter sky. Cricket-bat willows mark the river's course, alders line the brook. Hawthorn grows well. It is a country of elm and oak and thorn. People native to the clay are surly and slow to burn, morose and smouldering as alder wood, laconic, heavy as the land itself.

There are four hundred miles of tidal coast, if all the creeks and islands are included; it is the longest and most irregular county coastline. It is the driest county, yet watery-edged, flaking down to marsh and salting and mud-flat. The drying sandy mud of the ebb-tide makes the sky clear above; clouds reflect water and shine it back inland.

Farms are well ordered, prosperous, but a fragrance of neglect still lingers, like a ghost of fallen grass. There is always a sense of loss, a feeling of being forgotten. There is nothing else here; no castles, no ancient monuments, no hills like green clouds. It is just a curve of the earth, a rawness of winter fields. Dim, flat, desolate lands that cauterise all sorrow.

I have always longed to be a part of the outward life, to be out there at the edge of things, to let the human taint wash away in emptiness and silence as the fox sloughs his smell into the cold unworldliness of water; to return to the town as a stranger. Wandering flushes a glory that fades with arrival.

I came late to the love of birds. For years I saw them only as a tremor at the edge of vision. They know suffering and joy in simple states not possible for us. Their lives quicken and warm to a pulse our hearts can never reach. They race to oblivion. They are old before we have finished growing.

The first bird I searched for was the nightjar, which used to nest in the valley. Its song is like the sound of a stream of wine spilling from a height into a deep and booming cask. It is an odorous sound, with a bouquet that rises to the quiet sky. In the glare of day it would seem thinner and drier, but dusk mellows it and gives it vintage. If a song could smell, this song would smell of crushed grapes and almonds and dark wood. The sound spills out, and none of it is lost. The whole wood brims with it. Then it stops. Suddenly, unexpectedly. But the ear hears it still, a prolonged and fading echo, draining and winding out among the surrounding trees. Into the deep stillness, between the early stars and the long afterglow, the nightjar leaps up joyfully. It glides and flutters, dances and bounces, lightly, silently away. In pictures it seems to have a frog-like despondency, a mournful aura, as though it were sepulchred in twilight, ghostly and disturbing. It is never like that in life. Through the dusk, one sees only its shape and its flight, intangibly light and gay, graceful and nimble as a swallow.

Sparrowhawks were always near me in the dusk, like something I meant to say but could never quite remember. Their narrow heads glared blindly through my sleep. I pursued them for many summers, but they were hard to find and harder to see, being so few and so wary. They lived a fugitive, guerrilla life. In all the overgrown neglected places the frail bones of generations of sparrowhawks are sifting down now into the deep humus of the woods. They were a banished race of beautiful barbarians, and when they died they could not be replaced.

I have turned away from the musky opulence of the summer woods, where so many birds are dying. Autumn begins my season of hawk-hunting, spring ends it, winter glitters between like the arch of Orion.

I saw my first peregrine on a December day at the estuary ten years ago. The sun reddened out of the white river mist, fields glittered with rime, boats were encrusted with it; only the gently lapping water moved freely and shone. I went along the high river-wall towards the sea. The stiff crackling white grass became limp and wet as the sun rose through a clear sky into dazzling mist. Frost stayed all day in shaded places, the sun was warm, there was no wind.

I rested at the foot of the wall and watched dunlin feeding at the tide-line. Suddenly they flew upstream, and hundreds of finches fluttered overhead, whirling away with a 'hurr' of desperate wings. Too slowly it came to me that something was happening which I ought not to miss. I scrambled up, and saw that the stunted hawthorns on the inland slope of the wall were full of fieldfares. Their sharp bills pointed to the northeast, and they clacked and spluttered in alarm. I followed their point, and saw a falcon flying towards me. It veered to the right, and passed inland. It was like a kestrel, but bigger and yellower, with a more bullet-shaped head, longer wings, and greater zest and buoyancy of flight. It did not glide till it saw starlings feeding in stubble, then it swept down and was hidden among them as they rose. A minute later it rushed overhead and was gone in a breath into the sunlit mist. It was flying much higher than before, flinging and darting forwards, with its sharp wings angled back and flicking like a snipe's.

This was my first peregrine. I have seen many since then, but none has excelled it for speed and fire of spirit. For ten years I spent all my winters searching for that restless brilliance, for the sudden passion and violence that peregrines flush from the sky. For ten years I have been looking upward for that cloud-biting anchor shape, that crossbow flinging through the air. The eye becomes insatiable for hawks. It clicks towards them with ecstatic fury, just as the hawk's eye swings and dilates to the luring food-shapes of gulls and pigeons.

To be recognised and accepted by a peregrine you must wear the same clothes, travel by the same way, perform actions in the same order. Like all birds, it fears the unpredictable. Enter and leave the same fields at the same time each day, soothe the hawk from its wildness by a ritual of behavior as invariable as its own. Hood the glare of the eyes, hide the white tremor of the hands, shade the stark reflecting face, assume the stillness of a tree. A peregrine fears nothing he can see clearly and far off. Approach him across open ground with a steady unfaltering movement. Let your shape grow in size but do not alter its outline. Never hide yourself unless concealment is complete. Be alone. Shun the furtive oddity of man, cringe from the hostile eyes of farms. Learn to fear. To share fear is the greatest bond of all. The hunter must become the thing he hunts. What is, is now, must have the quivering intensity of an arrow thudding into a tree. Yesterday is dim and monochrome. A week ago you were not born. Persist, endure, follow, watch.

Hawk-hunting sharpens vision. Pouring away behind the moving bird, the land flows out from the eye in deltas of piercing colour. The angled eye strikes through the surface dross as the obliqued axe cuts to the heart of a tree. A vivid sense of place grows like another limb. Direction has colour and meaning. South is a bright, blocked place, opaque and stifling; West is a thickening of the earth into trees, a drawing together, the great beef side of England, the heavenly haunch; North is open, bleak, a way to nothing; East is a quickening in the sky, a beckoning of light, a storming suddenness of sea. Time is measured by a clock of blood. When one is active, close to the hawk, pursuing, the pulse races, time goes faster; when one is still, waiting, the pulse quietens, time is slow. Always, as one hunts for the hawk, one has an oppressive sense of time contracting inwards like a tightening spring.

One hates the movement of the sun, the steady alteration of the light, the increase of hunger, the maddening metronome of the heart-beat. When one says 'ten o'clock' or 'three o'clock,' this is not the grey and shrunken time of towns; it is the memory of a certain fulmination or declension of light that was unique to that time and that place on that day, a memory as vivid to the hunter as burning magnesium. As soon as the hawk-hunter steps from his door he knows the way of the wind, he feels the weight of the air. Far within himself he seems to see the hawk's day growing steadily towards the light of their first encounter. Time and the weather hold both hawk and watcher between their turning poles. When the hawk is found, the hunter can look lovingly back at all the tedium and misery of searching and waiting that went before. All is transfigured, as though the broken columns of a ruined temple had suddenly resumed their ancient splendour.

I shall try to make plain the bloodiness of killing. Too often this has been slurred over by those who defend hawks. Flesh-eating man is in no way superior. It is so easy to love the dead. The word 'predator' is baggy with misuse. All birds eat living flesh at some time in their lives. Consider the cold-eyed thrush, that springy carnivore of lawns, worm stabber, basher to death of snails. We should not sentimentalise his song, and forget the killing that sustains it.

In my diary of a single winter I have tried to preserve a unity, binding together the bird, the watcher, and the place that holds them both. Everything I describe took place while I was watching it, but I do not believe that honest observation is enough. The emotions and behavior of the watcher are also facts, and they must be truthfully recorded.

For ten years I followed the peregrine. I was possessed by it. It was a grail to me. Now it has gone. The long pursuit is over. Few peregrines are left, there will be fewer, they may not survive. Many die on their backs, clutching insanely at the sky in their last convulsions, withered and burnt away by the filthy, insidious pollen of farm chemicals. Before it is too late, I have tried to recapture the extraordinary beauty of this bird and to convey the wonder of the land he lived in, a land to me as profuse and glorious as Africa. It is a dying world, like Mars, but glowing still.

JOHN FOWLES
b. 1926

Best known as the author of such novels as The Collector *(1963),* The Magus *(1966), and* The French Lieutenant's Woman *(1969), Fowles has also collaborated with landscape photographers in such works as* Islands *(1979) and* The Tree *(1979). Myth and local natural history play important roles in much of his fiction.* The *protagonist of* The French Lieutenant's Woman, *for instance, is a collector of local fossils, and Fowles himself was a curator of a local natural history museum in Dorset. The following essay, first included as the central section of* The Tree, *examines the pervasive cultural myth of "the green man" in order to confront the question of what ultimately constitutes our need for contact with nature's presence, or "green chaos"—a need that for Fowles seems to reside in the basic structure of the human psyche.*

From THE TREE

A few years ago I stood in a historic place. It was not a great battle-field, a house, a square, the site of one famous event; but the site only of countless very small ones—a neat little eighteenth-century garden, formally divided by gravel walks into parterres, with a small wooden house in one corner where the garden's owner had once lived. There is only one other garden to compare with it in human history, and that is the one in the Book of Genesis, which never existed outside words. The one in which I stood is very real, and it lies in the old Swedish university town of Uppsala. Its owner was the great warehouse clerk and indexer of nature, Carl Linnaeus, who between 1730 and 1760 docketed, or attempted to docket, most of animate being. Perhaps nothing is more moving at Uppsala than the actual smallness and ordered simplicity of that garden (my father would have loved it) and

The Tree (Boston: Little, Brown, 1979).

the immense consequences that sprung from it in terms of the way we see and think about the external world. It is something more than one more famous shrine for lovers of nature, like Selborne or Coate Farm or Walden Pond. In fact, for all its air of gentle peace, it is closer to a nuclear explosion, whose radiations and mutations inside the human brain were incalculable and continue to be so: the place where an intellectual seed landed, and is now grown to a tree that shadows the entire globe.

I am a heretic about Linnaeus, and find nothing less strange, or more poetically just, than that he should have gone mad at the end of his life. I do not dispute the value of the tool he gave to natural science—which was in itself no more than a shrewd extension of the Aristotelian system and which someone else would soon have elaborated, if he had not; but I have doubts about the lasting change it has effected in ordinary human consciousness.

I sense in my collaborator on this book one likeness with my father. It is not that I don't share some of Frank Horvat's fertile attachment—and my father's—to the single tree, the tree in itself, and the art of cultivating it, literally or artistically. But I must confess my own love is far more of trees, more exactly of the complex internal landscapes they form when left to themselves. In the colonial organism, the green coral, of the wood or forest, experience, adventure, aesthetic pleasure, I think I could even say truth, all lie for me beyond the canopy and exterior wall of leaves, and beyond the individual.

Evolution has turned man into a sharply isolating creature, seeing the world not only anthropocentrically but singly, mirroring the way we like to think of our private selves. Almost all our art before the Impressionist—or their St. John the Baptist, William Turner—betrays our love of clearly defined boundaries, unique identities, of the individual thing released from the confusion of background. This power of detaching an object from its surroundings and making us concentrate on it is an implicit criterion in all our judgements on the more realistic side of visual art; and very similar, if not identical, to what we require of optical instruments like microscopes and telescopes—which is to magnify, to focus sharper, to distinguish better, to single from the ruck. A great deal of science is devoted to this same end: to providing specific labels, explaining specific mechanisms and ecologies, in short for sorting and tidying what seems in the mass indistinguishable one from the other. Even the simplest knowledge of the names and habits of flowers or trees starts this distinguishing or individuating process, and removes us a step from total reality towards anthropocentrism; that is, it acts mentally as an equivalent of the camera view-finder. Already it destroys or curtails certain possibilities of seeing, apprehending and experiencing. And that is the bitter fruit from the tree of Uppsalan knowledge.

It also begs very considerable questions as to the realities of the boundaries we impose on what we see. In a wood the actual visual "frontier" of any one tree is usually impossible to distinguish, at least in summer. We feel, or think we feel, nearest to a tree's "essence" (or that of its species) when it chances to stand like us, in isolation; but evolution did not intend trees to grow singly. Far more than ourselves they are social creatures, and no more natural as isolated specimens than man is as a marooned sailor or a hermit. Their society in turn creates or supports other societies of plants, insects, birds, mammals, micro-organisms; all of which we may choose to isolate and section off, but which remain no less the ideal entity, or whole experience, of the wood—and indeed are still so seen by most of primitive mankind.

Scientists restrict the word symbiotic to those relationships between species that bring some detectable mutual benefit; but the true wood, the true place of any kind, is the sum of all its phenomena. They are all in some sense symbiotic, being together in a togetherness of beings. It is only because such a vast sum of interactions and coincidences in time and place is beyond science's calculation (a scientist might say, beyond useful function, even if calculable) that we so habitually ignore it, and treat the flight of the bird and the branch it flies from, the leaf in the wind and its shadow on the ground, as separate events, or riddles—what bird? which branch? what leaf? which shadow? These question-boundaries (where do I file that?) are ours, not of reality. We are led to them, caged by them not only culturally and intellectually, but quite physically, by the restlessness of our eyes and their limited field and acuity of vision. Long before the glass lens and the movie-camera were invented, they existed in our eyes and minds, both in our mode of perception and in our mode of analysing the perceived: endless short sequence and jump-cut, endless need to edit and range this raw material.

I spent all my younger life as a more or less orthodox amateur naturalist; as a pseudo-scientist, treating nature as some sort of intellectual puzzle, or game, in which being able to name names and explain behaviourisms—to identify and to understand machinery—constituted all the pleasures and the prizes. I became slowly aware of the inadequacy of this approach: that it insidiously cast nature as a kind of opponent, an opposite team to be outwitted and beaten; that in a number of very important ways it distracted from the total experience and the total meaning of nature—and not only of what I personally needed from nature, not only as I had long, if largely unconsciously, begun to feel it (which was neither scientifically nor sentimentally, but in a way for which I had, and still have, no word). I came to believe that this approach represented a major human alienation, affecting all of us, both personally and socially; moreover, that such alienation had much more ancient roots behind the historical accident of its present scientific, or pseudo-scientific, form.

Naming things is always implicitly categorizing and therefore collecting them, attempting to own them; and because man is a highly acquisitive creature, brainwashed by most modern societies into believing that the act of acquisition is more enjoyable than the fact of having acquired, that getting beats having got, mere names and the objects they are tied to soon become stale. There is a constant need, or compulsion, to seek new objects and names—in the context of nature, new species and experiences. Everyday ones grow mute with familiarity, so known they become unknown. And not only in non-human nature: only fools think our attitude to our fellow-men is a thing distinct from our attitude to "lesser" life on this planet.

All this is an unhappy legacy from Victorian science, which was so characteristically obsessed with both the machine and exact taxonomy. I came only the other day on a letter in a forgotten drawer of the little museum of which I am curator. It was from a well-known Victorian fern expert, concerning some twenty or so specimens he had been sent from Dorset—all reducible, to a modern botanist, to three species. But this worthy gentleman felt obliged, in a welter of Latin polysyllables, to grant each specimen some new sub-specific or varietal rank, as if they were unbaptized children and might all go to hell if they were not given individual names. It would be absurd to deny the Victorians their enormous achievements in saner scientific fields, and I am not engaging in some sort of Luddite fantasy, wishing the machine they invented had been different, or even not at all. But we are far better at seeing the immediate advantages of such gains in knowledge of the exterior world than at assessing the costs of them. The particular cost of understanding the mechanism of nature, of having so successfully itemized and pigeon-holed it, lies most of all in the ordinary person's perception of it, in his or her ability to live with and care for it—and not to see it as challenge, defiance, enemy. Selection from total reality is no less necessary in science than it is in art; but outside those domains (in both of which the final test of selection is utility, or yield, to our own species) it seriously distorts and limits any worthwhile relationship.

I caused my hosts at Uppsala, where I went to lecture on the novel, some puzzlement by demanding (the literary business once over) to see Linnaeus's garden rather than the treasures of one of the most famous libraries in Europe. The feeling that I was not behaving as a decent writer should was familiar. Again and again in recent years I have told visiting literary academics that the key to my fiction, for what it is worth, lies in my relationship with nature—I might almost have said, for reasons I will explain, in trees. Again and again I have seen, under varying degrees of politeness, this assertion treated as some sort of irrelevant quirk, eccentricity, devious evasion of what must be the real truth: literary influences and theories of fiction, all the rest of that purely intellectual midden which faculty hens and cocks so like scratching over. Of course such matters are a part of the truth; but they are no more the whole truth than that the tree we see above ground is the whole tree. Even if we do discuss nature, I soon sense that we are talking about two different things: on their side some abstract intellectual concept, and on mine an experience whose deepest value lies in the fact that it cannot be directly described by any art . . . including that of words.

One interrogator even accused me of bad faith: that if I sincerely felt so deeply on the matter, I should write more about it. But what I gain most from nature is beyond words. To try to capture it verbally immediately places me in the same boat as the namers and would-be owners of nature: that is, it exiles me from what I most need to learn. It is a little as it is in atomic physics, where the very act of observation changes what is observed; though here the catch lies in trying to describe the observation. To enter upon such a description is like trying to capture the uncapturable. Its only purpose can be to flatter the vanity of the describer—a function painfully obvious in many of the more sentimental natural history writers.

But I think the most harmful change brought about by Victorian science in our attitude to nature lies in the demand that our relation with it must be purposive, industrious, always seeking greater knowledge. This dreadfully serious and puritanical approach (nowhere better exhibited in the nineteenth century than in the countless penny magazines aimed at young people) has had two very harmful effects. One is that it turned the vast majority of contemporary Western mankind away from what had become altogether too much like a duty, or a school lesson; the second is that the far saner eighteenth-century attitude, which viewed nature as a mirror for philosophers, as an evoker of emotion, as a pleasure, a poem, was forgotten. There are intellectual reasons as well for this. Darwin made sentimental innocence, nature as mainly personal or aesthetic experience, vaguely wicked. Not only did he propose a mechanism seemingly as iron as the steam-engine, but his very method of discovery, and its success in solving a great conundrum, offered an equally iron or one-sided model for the amateur naturalist himself, and made the older and more humanist approach seem childish. A "good" amateur naturalist today merely means one whose work is valued by the professional scientists in his field.

An additional element of alienation has come with the cinema and television, which are selective in another way. They present natural reality not only through other eyes, but a version of it in which the novelty or rarity of the subject plays a preponderant part in choice and treatment. Of course the nature film or programme has an entertainment value; of course there are some social goods in the now ubiquitous availability of copies of other people's images and opinions of actual things and events; but as with the Linnaean system, there is a cost. Being taken by camera into the deepest African jungle, across the Arctic wastes, thirty fathoms deep in the sea, may seem a "miracle of modern technology", but it will no more bring the viewer nearer the reality of nature, or a proper human relationship with the actual nature around him, than merely reading novels is likely to teach the writing of them. The most one can say is that it may help; a much more common result is to be persuaded of the futility of even trying.

Increasingly we live (and not only in terms of nature and novels) by the old tag, *Aut Caesar, aut nullus*. If I can't be Caesar, I'll be no one. If I can't have the knowledge of a scientist, I'll know nothing. If I can't have superb close-ups and rare creatures in the nature around me, to hell with it. Perhaps any representation of nature is better, to those remote from it in their daily lives, than none. Yet a great deal of such representation seems to me to descend straight from the concept of the menagerie, another sadly alienating selection, or reduction, from reality. Poking umbrellas through iron bars did not cease with the transition from the zoo to the screen.

Much of seventeenth- and eighteenth-century science and erudition is obsolete nonsense in modern scientific terms: in its personal interpolations, its diffuse reasoning, its misinterpreted evidence, its frequent blend of the humanities with science proper—its quotations from Horace and Virgil in the middle of a treatise on forestry. But one general, if unconscious, assumption lying behind almost all pre-Victorian science—that it is being presented by an entire human being, with all his complexities, to an audience of other entire human beings—has been much too soon dismissed as a mere historical phenomenon, at best exhibiting an engaging amateurishness, at worst sheer stupidity, from neither of which we have anything to learn. It is not of course the fault of modern scientists that most of their formal discourse is now of so abstruse a nature that only their fellow specialists can hope to understand it; that the discourse itself is increasingly mechanical, with words reduced to cogs and treated as poor substitutes for some more purely scientific formulation; nor is it directly their fault that their vision of empirical knowledge, the all-important value they put upon proven or demonstrable fact, has seeped down to dominate the popular view of nature—and our education about it. Our fallacy lies in supposing that the limiting nature of scientific method corresponds to the nature of ordinary experience.

Ordinary experience, from waking second to second, is in fact highly synthetic (in the sense of combinative or constructive), and made of a complexity of strands, past memories and present perceptions, times and places, private and public history, hopelessly beyond science's powers to analyse. It is quintessentially "wild," in the sense my father disliked so much: unphilosophical, irrational, uncontrollable, incalculable. In fact it corresponds very closely—despite our endless efforts to "garden," to invent disciplining social and intellectual systems—with wild nature. Almost all the richness of our personal existence derives from this synthetic and eternally present "confused" consciousness of both internal and external reality, and not least because we know it is beyond the analytical, or destructive, capacity of science.

Half by its principles, half by its inventions, science now largely dictates and forms our common, or public, perception of and attitudes to external reality. One can say of an attitude that it is generally held by society; but society itself is an abstraction, a Linnaeus-like label we apply to a group of individuals seen in a certain context and for a certain purpose; and before the attitude can be generally held, it must pass through the filter of the individual consciousness, where this irreducible "wild" component lies—the one that may agree with science and society, but can never be wholly plumbed, predicted or commanded by them.

One of the oldest and most diffused bodies of myth and folklore has accreted round the idea of the man in the trees. In all his manifestations, as dryad, as stag-headed Herne, as outlaw, he possesses the characteristic of elusiveness, a power of "melting" into the trees, and I am certain the attraction of the myth is so profound and universal because it is constantly "played" inside every individual consciousness.

This notion of the green man—or green woman, as W. H. Hudson made her—seen as emblem of the close connection between the actuality of present consciousness (not least in its habitual flight into a mental greenwood) and what seems to me lost by science in man's attitude to nature—that is, the "wild" side of his own, his inner feeling as opposed to the outer, fact-bound, conforming face imposed by fashion—helped me question my old pseudo-scientist self. But it also misled me for a time. In the 1950s I grew interested in the Zen theories of "seeing" and of aesthetics: of learning to look beyond names at things-in-themselves. I stopped bothering to identify species new to me, I concentrated more and more on the familiar, daily nature around me, where I then lived. But living without names is impossible, if not downright idiocy, in a writer; and living without explanation or speculation as to causality, little better—for Western man, at least. I discovered, too, that there was less conflict than I had imagined between nature as external assembly of names and facts and nature as internal feeling; that the two modes of seeing or knowing could in fact marry and take place almost simultaneously, and enrich each other.

Achieving a relationship with nature is both a science and an art, beyond mere knowledge or mere feeling alone; and I now think beyond oriental mysticism, transcendentalism, "meditation techniques" and the rest—or at least as we in the West have converted them to our use, which seems increasingly in a narcissistic way: to make ourselves feel more positive, more meaningful, more dynamic. I do not believe nature is to be reached that way either, by turning it into a therapy, a free clinic for admirers of their own sensitivity. The subtlest of our alienations from it, the most difficult to comprehend, is our eternal need to use it in some way, to derive some personal yield. We shall never fully understand nature (or ourselves), and certainly never respect it, until we dissociate the wild from the notion of usability—however innocent and harmless the use. For it is the general uselessness of so much of nature that lies at the root of our ancient hostility and indifference to it.

There is a kind of coldness, I would rather say a stillness, an empty space, at the heart of our forced co-existence with all the other species of the planet. Richard Jefferies coined a word for it: the ultra-humanity of all that is not man . . . not with us or against us, but outside and beyond us, truly alien. It may sound paradoxical, but we shall not cease to be alienated—by our knowledge, by our greed, by our vanity—from nature until we grant it its unconscious alienation from us.

I am not one of those supreme optimists who think all the world's ills, and especially this growing divide between man and nature, can be cured by a return to a quasi-agricultural, ecologically "caring" society. It is not that I doubt it might theoretically be so cured; but the possibility of the return defeats my powers of imagination. The majority of Western man is now urban, and the whole world will soon follow suit. A very significant tilt of balance in human history is expected by the end of the coming decade: over half of all mankind will by then have moved inside

towns and cities. Any hope of reversing that trend, short of some universal catastrophe, is as tiny and precarious as the Monarch butterflies I
watched, an autumn or two ago, migrating between the Fifth Avenue
skyscrapers in central Manhattan. All chance of a close acquaintance
with nature, be it through intellect and education, be it in the simplest
way of all, by having it near at hand, recedes from the many who already
effectively live in a support system in outer space, a creation of science,
and without means to escape it, culturally or economically.

But the problem is not, or only minimally, that nature itself is in
imminent danger or that we shall lose touch with it simply because we
have less access to it. A number of species, environments, unusual
ecologies are in danger, there are major pollution problems; but even
in our most densely populated countries the ordinary wild remains far
from the brink of extinction. We may not exaggerate the future threats
and dangers, but we do exaggerate the present and actual state of this
global nation—underestimate the degree to which it is still surviving
and accessible to those who want to experience it. It is far less nature
itself that is yet in true danger than our attitude to it. Already we behave
as if we live in a world that holds only a remnant of what there actually
is; in a world that may come, but remains a black hypothesis, not a
present reality.

I believe the major cause of this more mental than physical rift lies
less in the folly or onesidedness of our societies and educational systems, or in the historical evolution of man into a predominantly urban
and industrial creature, a thinking termite, than in the way we have,
during these last hundred and fifty years, devalued the kind of experience or knowledge we loosely define as art; and especially in the way
we have failed to grasp its deepest difference from science. No art is
truly teachable in its essence. All the knowledge in the world of its
techniques can provide in itself no more than imitations or replicas of
previous art. What is irreplaceable in any object of art is never, in the
final analysis, its technique or craft, but the personality of the artist, the
expression of his or her unique and individual feeling. All major
advances in technique have come about to serve this need. Techniques
in themselves are always reducible to sciences, that is, to learnability.
Once Joyce has written, Picasso painted, Webern composed, it requires
only a minimal gift, besides patience and practice, to copy their techniques exactly; yet we all know why this kind of technique-copy, even
when it is so painstakingly done—for instance, in painting—that it
deceives museum and auction-house experts, is counted worthless
beside the work of the original artist. It is not *of* him or her; it is not art,
but imitation.

As it is with the true "making" arts, so it is with the other aspects of human life of which we say that full knowledge or experience also requires an art—some inwardly creative or purely personal factor beyond the power of external teaching to instil or science to predict. Attempts to impart recipes or set formulae as to practice and enjoyment are always two-edged, since the question is not so much whether they may or may not enrich the normal experience of that abstract thing, the normal man or woman, but the certainty that they must in some way damage that other essential component of the process, the contribution of the artist in this sense—the individual experiencer, the "green man" hidden in the leaves of his or her unique and once-only being.

Telling people why, how and when they ought to feel this or that—whether it be with regard to the enjoyment of nature, of food, of sex, or anything else—may, undoubtedly sometimes does, have a useful function in dispelling various kinds of socially harmful ignorance. But what this instruction cannot give is the deepest benefit of any art, be it of making, or of knowing, or of experiencing: which is self-expression and self-discovery. The last thing a sex-manual can be is an *ars amoris*—a science of coupling, perhaps, but never an art of love. Exactly the same is true of so many nature-manuals. They may teach you how and what to look for, what to question in external nature; but never in your own nature.

In science greater knowledge is always and indisputably good; it is by no means so throughout all human existence. We know it from art proper, where achievement and great factual knowledge, or taste, or intelligence, are in no way essential companions; if they were, our best artists would also be our most learned academics. We can know it by reducing the matter to the absurd, and imagining that God, or some Protean visitor from outer space, were at one fell sweep to grant us all knowledge. Such omniscience would be worse than the worst natural catastrophe, for our species as a whole; would extinguish its soul, lose it all pleasure and reason for living.

This is not the only area in which, like the rogue computer beloved of science fiction fans, some socially or culturally consecrated proposition—which may be true or good in its social or cultural context—extends itself to the individual; but it is one of the most devitalizing. Most mature artists know that great general knowledge is more a hindrance than a help. It is only innately mechanical, salami-factory novelists who set such great store by research; in nine cases out of ten what natural knowledge and imagination cannot supply is in any case precisely what needs to be left out. The green man in all of us is well aware of this. In practice we spend far more time rejecting knowledge than trying to gain it, and wisely. But it is in the nature of all society, let alone one deeply imbued with a scientific and technological ethos, to bombard us with ever more knowledge—and to make any questioning or rejection of it unpatriotic and immoral.

Art and nature are siblings, branches of the one tree; and nowhere more than in the continuing inexplicability of many of their processes, and above all those of creation and of effect on their respective audiences. Our approach to art, as to nature, has become increasingly scientized (and dreadfully serious) during this last century. It sometimes seems now as if it is principally there not for itself but to provide material for labelling, classifying, analysing—specimens for "setting," as I used to set moths and butterflies. This is of course especially true of—and pernicious in—our schools and universities. I think the first sign that I might one day become a novelist (though I did not then realize it) was the passionate detestation I developed at my own school for all those editions of examination books that began with a long introduction: an anatomy lesson that always reduced the original text to a corpse by the time one got to it, a lifeless demonstration of a pre-established proposition. It took me years to realize that even geniuses, the Shakespeares, the Racines, the Austens, have human faults.

Obscurity, the opportunity a work of art gives for professional explainers to show their skills, has become almost an aesthetic virtue; at another extreme the notion of art as vocation (that is, something to which one is genetically suited) is dismissed as non-scientific and inegalitarian. It is not a gift beyond personal choice, but one that can be acquired, like knowledge of science, by rote, recipe and hard work. Elsewhere we become so patterned and persuaded by the tone of the more serious reviewing of art in our magazines and newspapers that we no longer notice their overwhelmingly scientific tone, or the paradox of this knowing-naming technique being applied to a non-scientific object—one whose production the artist himself cannot fully explain, and one whose effect the vast majority of the non-reviewing audience do not attempt to explain.

The professional critic or academic would no doubt say this is mere ignorance, that both artists and audiences have to be taught to understand themselves and the object that links them, to make the relationship articulate and fully conscious; defoliate the wicked green man, hunt him out of his trees. Of course there is a place for the scientific, or quasi-scientific, analysis of art, as there is (and far greater) for that of nature. But the danger, in both art and nature, is that all emphasis is placed on the created, not the creation.

All artefacts, all bits of scientific knowledge, share one thing in common: that is, they come to us from the past, they are relics of something already observed, deduced, formulated, created, and as such qualify to go through the Linnaean and every other scientific mill. Yet we cannot say that the "green" or creating process does not happen or has no importance just because it is largely private and beyond lucid description and rational analysis. We might as well argue that the young wheat-plant is irrelevant because it can yield nothing to the miller and his stones. We know that in any sane reality the green blade is as much the ripe grain as the child is father to the man. Nor of course does the simile apply to art alone, since we are all in a way creating our future out of our present, our "published" outward behavior out of our inner green being. One main reason we may seldom feel this happening is that society does not want us to. Such random personal creativity is offensive to all machines.

I began this wander through the trees—we shall come to them literally, by the end—in search of that much looser use of the word "art" to describe a way of knowing and experiencing and enjoying outside the major modes of science and art proper . . . a way not concerned with scientific discovery and artefacts, a way that is internally rather than externally creative, that leaves very little public trace; and yet which for those very reasons is almost wholly concentrated in its own creative process. It is really only the qualified scientist or artist who can escape from the interiority and constant nowness, the green chaos of this experience, by making some aspect of it exterior and so fixing it in past time, or known knowledge. Thereby they create new, essentially parasitical orders and categories of phenomena that in turn require both a science and an art of experiencing.

But nature is unlike art in terms of its product—what we in general know it by. The difference is that it is not only created, an external object with a history, and so belonging to a past; but also creating in the present, as we experience it. As we watch, it is so to speak rewriting, reformulating, repainting, rephotographing itself. It refuses to stay fixed and fossilized in the past, as both the scientist and the artist feel it somehow ought to; and both will generally try to impose this fossilization on it.

Verbal tenses can be very misleading here: we stick adamantly in speech to the strict protocol of actual time. Of and in the present we speak in the present, of the past in the past. But our psychological tenses can be very different. Perhaps because I am a writer (and nothing is more fictitious than the past in which the first, intensely alive and present, draft of a novel goes down on the page), I long ago noticed this in my naturalist self: that is, a disproportionately backward element in any present experience of nature, a retreat or running-back to past knowledge and experience, whether it was the definite past of personal memory or the indefinite, the imperfect, of stored "ological" knowledge and proper scientific behaviour. This seemed to me often to cast a mysterious veil of deadness, of having already happened, over the actual and present event or phenomenon.

I had a vivid example of it only a few years ago in France, long after I thought I had grown wise to this self-imposed brainwashing. I came on my first Military Orchid, a species I had long wanted to encounter, but hitherto never seen outside a book. I fell on my knees before it in a way that all botanists will know. I identified, to be quite certain, with Professors Clapham, Tutin and Warburg in hand (the standard British *Flora*), I measured, I photographed, I worked out where I was on the map, for future reference. I was excited, very happy, one always remembers one's "first" of the rarer species. Yet five minutes after my wife had finally (other women are not the only form of adultery) torn me away, I suffered a strange feeling. I realized I had not actually *seen* the three plants in the little colony we had found. Despite all the identifying, measuring, photographing, I had managed to set the experience in a kind of present past, a having-looked, even as I was temporally and physically still looking. If I had had the courage, and my wife the patience, I would have asked her to turn and drive back, because I knew I had just fallen, in the stupidest possible way, into an ancient trap. It is not necessarily too little knowledge that causes ignorance; possessing too much, or wanting to gain too much, can produce the same result.

There is something in the nature of nature, in its presentness, its seeming transcience, its creative ferment and hidden potential, that corresponds very closely with the wild, or green man, in our psyches; and it is a something that disappears as soon as it is relegated to an automatic pastness, a status of merely classifiable *thing*, image taken *then*. "Thing" and "then" attract each other. If it is thing, it was then; if it was then, it is thing. We lack trust in the present, this moment, this actual seeing, because our culture tells us to trust only the reported back, the publicly framed, the edited, the thing set in the clearly artistic or the clearly scientific angle of perspective. One of the deepest lessons we have to learn is that nature, of its nature, resists this. It waits to be seen otherwise, in its individual presentness and from our individual presentness.

I come now near the heart of what seems to me to be the single greatest danger in the rich legacy left us by Linnaeus and the other founding fathers of all our sciences and scientific mores and methods—or more fairly, left us by our leaping evolutionary ingenuity in the invention of tools. All tools, from the simplest word to the most advanced space probe, are disturbers and rearrangers of primordial nature and reality—

are, in the dictionary definition, "mechanical implements for working upon something." What they have done, and I suspect in direct proportion to our ever-increasing dependence on them, is to addict us to purpose: both to looking for purpose in everything external to us and to looking internally for purpose in everything we do—to seek explanation of the outside world by purpose, to justify our seeking by purpose. This addiction to finding a reason, a function, a quantifiable yield, has now infiltrated all aspects of our lives—and become effectively synonymous with pleasure. The modern version of hell is purposelessness.

Nature suffers particularly in this, and our indifference and hostility to it is closely connected with the fact that its only purpose appears to be being and surviving. We may think that this comprehends all animate existence, including our own; and so it must, ultimately; but we have long ceased to be content with so abstract a motive. A scientist would rightly say that all form and behaviour in nature is highly purposive, or strictly designed for the end of survival—specific or genetic, according to theory. But most of this functional purpose is hidden to the non-scientist, indecipherable; and the immense variety of nature appears to hide nothing, nothing but a green chaos at the core—which we brilliantly purposive apes can use and exploit as we please, with a free conscience.

A green chaos. Or a wood.

FRANKLIN RUSSELL
b. 1926

Franklin Russell is a native of New Zealand, a Canadian citizen, and a long-time resident of New York City. In the early 1960s Russell found himself "stifled by the synthetic nature of life" in his Manhattan apartment and set off to discover the world of islands—specifically, the seabird colonies of Maine and the Canadian maritimes. The result was The Secret Islands (1965), a series of adventures and encounters not only with birds but with colorful human figures from remote communities whose way of life had changed little over two centuries. The author's approach

*is personal and confessional, recording his failures as well as his achieve-
ments. Russell is not the first writer to find nature's impersonal fecundity
overwhelming and even appalling; but few have described it in such vivid
and dramatic detail as in the following account of the birds of Funk
Island, an isolated rock fifty miles off the coast of Newfoundland. It is a
place where ordinary categories of meaning seem to break down, where
"Chaos is order. Order is a mystery. Time is meaningless"—an experience at once disconcerting and strangely vitalizing.*

THE ISLAND OF AUKS

Arthur Sturge was caught in an immortal moment, straining back
on his oar as he moved the heavy dory toward Funk Island. This was
the last lunge of the journey; the longliner heaved behind us; Uncle
Jacob had bellowed his final exhortation of good luck.

I saw the island close up as I glanced over Sturge's shoulder, and I
knew I was duplicating the experience of a thousand men before me.
From the boat it seemed incredible that such a stream of humanity—
explorers, Indians, sealers, whalers, codfishermen—had ever reached
this lonely place. Yet the island, and its auks, had drawn them as it was
drawing me.

The island was a blank wall of rock, thirty feet high, suave and
bland, and topped by a thin, fast-moving frieze of murres who, presumably, were anxiously watching our boat. I have read about a moment of
truth, even written about it, but not until I was in this dory, in this
place, did I really understand what it meant. It was the final throw of
the dice. Would we be able to land?

"Hard to say. Them waves is risin' high. . . ."

I looked toward the island and saw the water pitching, silent and
ominous, up the blank rock. I had come this far, but now I could think
only of being capsized—dashed against the rocks or maimed under the
boat's keel. I had already talked to a dozen men who had traveled thousands of miles only to be turned back at this point.

I sat in the back of the dory, carried forward by the momentum of a
determination long sustained. In a moment, the boat was rising and
falling against the rock face on six-foot waves.

"Get up front," Arthur Sturge said.

The Secret Islands (New York: Norton, 1965).

He eased the boat toward the rock. Willie hunched in the bows. At the peak of a wave, he jumped, grabbed at the rock face, and clung. I could see that the rock was gouged with handholds into which my fingers must fit as I jumped.

"Arl roight!" Arthur shouted.

Willie now had his back to the rock; he was facing me so that he might try to seize me if I fell. The boat rose and wobbled at its peak; I jumped, hit the rocks, and felt my fingers slip into the grooves. As I clung there, I realized the grooves were man-made. Of course. Other men had met the same problem. Of course they had done something about it. These grooves into which my fingers fitted so neatly might have been cut in Drake's time, or before, when the Beothuk Indians came to plunder the island; or had they been cut by Eskimos, a thousand years before Christ?

Muscles knotted, and I strained upward. The problems of landing on Funk Island have remained unchanged. Rockets to the moon and the splitting of the atom mean nothing when it comes to landing on Funk Island. The equation for success is constant: an open boat with a skillful oarsman and a man willing to jump.

Willie had disappeared over the top of the rock while I was still absorbed in finding hand- and footholds in rock slippery with algae and bird excrement. I mounted the crest of the rock and Funk Island spread out, an explosion of sight, sound, and smell. I saw, but I did not see; I saw dark masses of murres in the distance; I saw curtains of buzzing kittiwakes interposing themselves like thousands of pretty white butterflies; I saw rolling hummocks of bare rock. But it was the sound that came to me first. We walked forward over intransigent, bare granite, and the sound swelled like thunder. A literate biologist has described it as "a rushing of waters," but that description does not satisfy poet or artist. It is orchestral, if a million players can be imagined: rich, sensuous, hypnotic.

When we came to the edge of the first great concourse of birds, perhaps two hundred thousand of them staining the rock densely black and white, the orchestral analogy became even more vivid because I could hear, among those thousands of voices, rippling spasms of pathos and melancholy—Brahmsian. The adult birds cried *ehr-ehr-ehr*, crescendo, diminuendo, gushes of emotion. The cries of the flying birds—and there were thousands in the air—swelled and faded in haunting harmony as they passed low overhead. Buried in this amalgam of voices were the piping screams of the young murres, sounds so piercing they hurt the ear.

We moved around the periphery of the murres. With every step, I was conscious of new expansions in the scope of sound. A sibilant undertone to the massive main theme was faintly discernible, the sound of innumerable wings beating: *flacka-flacka-flacka-flacka.* Wings struck each other—*clack, clack, clack*—as birds, flying in thick layers, collided in mid-air. Then another buried sound, a submelody, a counterpoint: *gaggla-gaggla-gaggla.* The gannets were hidden somewhere among the murre hordes. Other sounds were reduced to minutiae in the uproar: the thin cries of herring gulls, the rasping moans of kittiwakes hovering high overhead.

I had been on the island an hour and only now was I really registering the sound of it.

Next, overwhelmingly, came real vision. The murres were massed so thickly they obscured the ground. The birds stood shoulder to shoulder, eyeball to eyeball. In places, they were so densely packed that if one bird stretched or flapped her wings, she sent a sympathetic spasm rippling away from her on all sides.

All life was in constant, riotous motion. Murre heads wavered and darted; wings beat; birds landed clumsily among the upraised heads of their comrades; birds took off and thrashed passageways through the birds ahead of them, knocking them down. Chicks ran from adult to adult; eggs rolled across bare rock, displaced by kicking feet.

The murres heeded me, yet they did not. I approached them and a rising roar of protest sounded, a concentration of the general uproar, which seemed not directed at me at all but at the outrage of intrusion. I walked away from them and the roar died instantly.

The sun was well up, a brilliant star in an azure sky, and I walked to the quiet shore, away from the main masses of murres. Willie had disappeared into a gully. Perhaps I needed time to assimilate. But there was no time. The multiple dimensions of the sight came pouring in. The air streamed with birds coming at me. I threw up my hands, shouted at them, but the shout was lost, ineffectual, not causing a single bird to swerve or otherwise acknowledge me.

At least a hundred thousand birds were aloft at once. They circled the island endlessly, like fighter bombers making strafing runs on a target, flying the full length of the island, then turning out to sea and sweeping back offshore to begin another run. They came on relentlessly and the sky danced with them.

This was not, I realized, the hostile reaction of individual birds who saw their nests threatened. Instead, the murres were a tribe of animals resisting a threat to their island. Individually, they intimidated nothing. Collectively, they emanated power and strength. I looked into a thousand cold eyes and felt chill, impersonal hostility in the air.

I climbed to the top of a ridge and looked down the length of the island, looked into the masses of birds hurtling toward me, looked down to the grounded hordes, a living, writhing backbone of murres, murres, murres. Then, after the visual shock came the olfactory impact. The smell of Funk Island is the smell of death. It is probably the source of the island's name, which in various languages means "to steam," "to create a great stench," "to smoke"; it may also mean "fear." The island certainly smells ghastly. No battlefield could ever concentrate such a coalition of dead and dying.

A change of wind brought the smell to us, choking, sickening. As I walked down a slope and out of the force of the wind, the air clotted with the smell. In a hollow at the bottom of the slope, it had collected in such concentration that I gagged and my throat constricted. The fishermen *knew* the smell was poisonous. Uncle Jacob Sturge had told me how one fisherman who tried to run through the concentration of birds was nearly gassed unconscious.

The smell of Funk Island comes from a combination of corruption. There are no scavengers, except bacteria, so dead bodies lie where they have fallen. The debris of a million creatures has nowhere to go. Eggs by the scores of thousands lie everywhere, so that I could not see which were being brooded, which were rotten. In one small gully, unwanted or untended eggs had been kicked together in a one hundred foot driftline by the constantly moving feet of the birds.

The smell of the island came in diminishing waves as the sea breeze died and the heat rose from the rocks. A ripple of explosions fled away among a nearby concentration of birds. I listened; the sound was manlike. It reminded me of a popgun I had used when I was a boy. From a nearby hill packed with murres, another flurry of explosions, then single shots haphazardly firing all around me. If the smell of the island needed an exemplifying sound, this was it. The explosions were the sound of rotten eggs bursting in the growing heat.

I walked, while the smell gathered in my nostrils and took on various identities. It was the thin, sour smell of bird excrement: acidic, astringent, more than a hundred tons of it splashed on the island every day. Underlying that smell was the stench of the rotting fish which lay everywhere after being vomited up by the parent murres but not eaten by the nestlings. The smell was of putrescence, of oil, of fish, and of an indescribable other thing: the stench of a million creatures packed together in a small place.

I walked halfway down the length of the island, a distance of perhaps five hundred feet, but my progress was slow because of the difficulty I had in assimilating everything I saw.

In my mind were scraps of history. I was thinking, for instance, of how Newfoundland's ancient Indians, the Beothuks, camped in a gulch when they were on bird- and egg-hunting expeditions; of how, until recently, the gulch was an archaeological repository of old knives, spoons, belaying pins, and broken pots, testifying to more than two hundred years of exploitation of the island by hunters of meat, oil, eggs, and feathers.

Willie appeared on a far ridge. He was standing at the edge of Indian Gulch. I walked toward him along the rim of a concourse of murres. A feeble spring flowed into the gulch and created a small pond, which was also fed by the sea during heavy swells. Sea water belched up into it through a narrow crack in the rock. Into this pond poured a ceaseless flow of excrement, coughed-up fish, bodies, rotten eggs, and live nestlings. By midsummer, the water was mucid, pea-green, fermenting, almost bubbling with corruption.

The murres were not distressed by this putrid mess; as Willie walked along the top of the gulch, hundreds of them dropped down to the water and floated. Suddenly, the pond was roiled into green foam as a group of birds took off. Their departure triggered another flight, which because murres fly poorly, was a failure. The birds crashed on top of each other or piled into heaps along the steep banks. This drew a sympathetic flight from murres perched precariously on the cliffs and a cloud of birds took off. Their departure sent eggs and nestlings spilling off the cliffs into the water.

But on Funk Island, nothing matters. Death is nothing. Life is nothing. Chaos is order. Order is a mystery. Time is meaningless. The deep-throated roar of the colony cries out to a heedless sky. The human observer, cowed by its primitive energy, by its suggestion of the unnameable, stumbles on blindly.

As I walked, I examined my growing sense of reality and sought a guidepost to what it all meant. I had thought (an hour before? two hours? it was nearer to four hours) that a sweep of Brahmsian rhetoric could describe the island. Already, the image was obsolete. Now, I felt a mechanistic sense, Prokofievian, an imperative monotone, the sound of Mars. The struggling, homuncular forms piled together in such utter, inhuman chaos denied any ordered view of the universe.

I had to wonder whether a poet had preceded me to the island; or did the island have a counterpart elsewhere? De la Mare's disgust at the massacre?

> And silence fell: the rushing sun
> Stood still in paths of heat,
> Gazing in waves of horror on
> The dead about my feet.

It was nearly noon, and sun flames reached for the island and scorched it. I was enervated, but I was also recovering normal sensibility, which brought me *details* of the life of the murre colony. Everywhere I looked now, young murres looked back. In places, they were packed thirty and forty together among the adults. There is only one word to describe them and it is not in any dictionary. They are murrelings: tiny, rotund, dusky balls of fluff with the most piercing voices ever given a young bird. Their piping screams must be essential for them to assert themselves above the roar of the adults. How else could they identify themselves to their parents? Yet, bafflement grew as I watched them. One murreling in that featureless mass of birds was infinitely smaller than one needle in a stack of hay. How contact is kept with the parent birds remains a mystery of biology.

Warned by the scope of destruction in the colony, I was not surprised to find that the murrelings were expendable. Life moves to and from the colony at high speed. A murreling fell from a rock, bounced into an evil puddle, and was trampled by a throng of adults. Hearing screams from a rock I disengaged a murreling jammed in a crevice, looked down, saw a mass of fluffy bodies wedged deeper in the crevice. Murrelings fell from cliffs into the sea, rose and floated in foam, screaming. Murrelings lay dead among pustular eggs; they lay in heaps and windrows in olivaceous puddles.

I knew from murre literature that the murrelings often became shocked by prolonged rain and died by the thousands. That, in the context of this island, was not surprising. An *individual* death was shocking. Willie had walked back up the other side of the island and we met at the edge of a group of murres. Willie groaned.

"I don't feel well," he said, rolling his eyes in mock nausea.

As he spoke, I looked over his shoulder in horror. In the middle of the murre mass, standing on slightly higher ground, was a group of gannets. These birds, though inferior in number, occupied the best territory. Though dominant, they seemed to have an amicable relationship with the murres. In places, murres and gannets were mixed together; murrelings gathered around gannets as though they were murres.

A gannet on a nest had reached down casually for one of these nearby murrelings, and as I watched, upended it and swallowed the struggling youngster. It was not the sight of such casual destruction that was shocking; it was the sound of the murreling dying.

It screamed when it was seized by the gannet's beak, which was bigger than the murreling's entire body. It screamed as it was hoisted into the air. Horrifyingly, it screamed loudest as it was being swallowed. The gannet, though a big, powerful bird, had to swallow hard to get the murreling down. Its neck writhed and its beak gaped and all the time the awful screams of the murreling came up out of the gannet's throat. The cries became fainter and fainter.

"Horrible," Willie said. "Oi never gets used to it."

The roar of the birds became a lamentation, a collusion of agony and sorrow. The flying creatures seemed to be in streaming retreat. Why *that* murreling and not any of the others still around at the feet of the gannet? If gannets really relished murre flesh, surely they would quickly wipe out all the murrelings near them. But they do not.

It was now afternoon and the sun plashed white and pitiless light on rock. For some time, I had been aware of a growing disorientation. I took a picture to the east, seventy thousand birds; to the west, one hundred thousand; in the air, twenty thousand. The noise, the smell, the screams, the corpses, the green puddles, pushed bonily into my chest. I fumbled with film but could not decide how to reload the camera or, indeed, remember what setting to use, or how to release the shutter.

"Oi t'ink oi'll go and sit behind dat rock," Willie said. "Oi goes funny in de head after a while here."

Uncle Jacob had mentioned that the island could drive a man mad. I was being sickened by the pressure of it. Once, in Australia, I watched men systematically kill several hundred thousand rabbits they had penned against a fence. The steady thocking of cudgels hammering rabbit skulls continued hour after hour, eventually dulling the eye and diminishing the hearing. On Funk Island, my observing sense was losing its ability to see and to record.

I sought release in reverie and walked, half-conscious of what I was doing, toward an incongruous green field that lay alone in the middle of the island. Its bareness suggested another place of personal memory, and an association of ideas. My ancestors were Scottish and fought the English at Culloden. When I went to Culloden, two hundred years after the battle, I was overwhelmed by those long, sinister mounds of mass burial of the clans.

This Funk Island field was also a midden of slain creatures. It was a great natural-history site, as significant to an ornithologist as Ashurbanipal's palace would be to an archaeologist. Here, generations of flightless great auks had flocked to breed after eight months of oceanic wandering. Their occupancy built up soil. Here, also, they were slain throughout the eighteenth century until they became extinct, probably early in the nineteenth.

A puffin bolted out of the ground ahead and flipped a bone from her burrow entrance as she left. I knelt and clawed a handful of bones out of the burrow. I saw other burrows, bones spilling out of them as though they were entrances to a disorderly catacomb.

This was not fantasy. These were great-auk bones, still oozing out of the earth nearly a hundred and fifty years after the last bird had gone.

The bones permeated the ground under my feet; puffins dug among them and kicked them aside to find graveyard sanctuary. Life in the midst of death.

All at once, walking across bare rock, the murres well distant, I felt a release. Willie was not in sight; the longliner was off fishing somewhere. The granite underfoot changed texture, became a desert I had walked, then a heath, a moor I had tramped, and eventually, all the bare and empty places of earth I had ever known. I felt the presence of friends and heard their voices. But something was wrong. Some were still friends but others had closed, deceitful faces. Inhibition and self-deception fell away; flushes of hate and love passed as the faces moved back and forth. Forgotten incidents came to mind. What was happening?

Uncle Jacob's voice: "A man could go mad on the Funks."

This was enough. I turned toward the shore, to the *Doris and Lydia*, which had appeared from nowhere. Willie leaped eagerly from his place of refuge behind a rock. The roar of the murres receded. I imagined the island empty during much of the long year, naked as a statue against the silent hiss of mist coming out of the Labrador Current, or the thunder of an Atlantic gale piling thirty-foot waves up the sides of the island.

"She be a sight to see in the winter," Uncle Benny had said.

The seasons of millennia switched back and forth. The island suffocated in the original gases of earth: argon, radon, krypton, xenon, neon. The island disappeared in yellow fog, and water slid down its sides. The island was a corpse, dead a million years, its surface liquefied, with rot running into its granite intestines. The island festered, and rivulets of pus coursed down its sides. The island was death. The island was life.

Only in retrospect could the island become real. Later, I was to return to the island and actually live on it in order to turn my disbelief into lasting memory. Willie moved parallel to me, jumping from rock to rock and displacing a fluttering canopy of kittiwakes. He was a different man now as he met me at the landing site, beaming and lighthearted.

"So dat's de Funks, eh?" he said, and he was proud that I had seen it.

In the boat below us, Arthur and Cyril smiled. Arthur was relaxed now, in contrast to his silent, absorbed intensity when he was trying to get me to the island, and on it. Both men, and Willie, poised at the top of the cliff, were caught for a moment by the camera, like toreros who have survived a bloody afternoon and will hear the bugles again tomorrow.

With a final look over my shoulder at the silently fleeing birds, I slid down the cliff to the boat and Funk Island became a part of the history of my life.

EDWARD ABBEY
1927–1989

Over fifteen years as a fire lookout and park ranger in the Southwest, Abbey developed the passions that pervade and enliven much of his writing—deep love of the desert and bitterness about its desecration by the miners, dammers, developers, and tourists. One of his marks as a writer is the fierceness with which he pursues his polemic on behalf of the desert, declaring war against the atrocities he has witnessed. Abbey's novel The Monkey Wrench Gang *(1975) portrays a band of environmental guerillas so vividly that it encouraged the founding of the radical environmental movement* Earth First! *Desert Solitaire (1968), his best-known work, describes Abbey's experiences as a ranger at Utah's Arches National Monument. With sardonic honesty Abbey shows, reflects on, and rails against his own implication in the culture of "industrial tourism." But the central source of his book's power is its evocation of the desert landscape itself, in its overwhelming, threatening presence and its elusive beauty.*

THE SERPENTS OF PARADISE

The April mornings are bright, clear and calm. Not until the afternoon does the wind begin to blow, raising dust and sand in funnelshaped twisters that spin across the desert briefly, like dancers, and then collapse—whirlwinds from which issue no voice or word except the forlorn moan of the elements under stress. After the reconnoitering dust-devils comes the real, the serious wind, the voice of the desert rising to a demented howl and blotting out sky and sun behind yellow clouds of dust, sand, confusion, embattled birds, last year's scrub-oak leaves, pollen, the husks of locusts, bark of juniper. . . .

Time of the red eye, the sore and bloody nostril, the sand-pitted

Desert Solitaire (New York: McGraw-Hill, 1968).

windshield, if one is foolish enough to drive his car into such a storm. Time to sit indoors and continue that letter which is never finished—while the fine dust forms neat little windows under the edge of the door and on the windowsills. Yet the springtime winds are as much a part of the canyon country as the silence and the glamorous distances; you learn, after a number of years, to love them also.

The mornings therefore, as I started to say and meant to say, are all the sweeter in the knowledge of what the afternoon is likely to bring. Before beginning the morning chores I like to sit on the sill of my doorway, bare feet planted on the bare ground and a mug of hot coffee in hand, facing the sunrise. The air is gelid, not far above freezing, but the butane heater inside the trailer keeps my back warm, the rising sun warms the front, and the coffee warms the interior.

Perhaps this is the loveliest hour of the day, though it's hard to choose. Much depends on the season. In midsummer the sweetest hour begins at sundown, after the awful heat of the afternoon. But now, in April, we'll take the opposite, that hour beginning with the sunrise. The birds, returning from wherever they go in winter, seem inclined to agree. The pinyon jays are whirling in garrulous, gregarious flocks from one stunted tree to the next and back again, erratic exuberant games without any apparent practical function. A few big ravens hang around and croak harsh clanking statements of smug satisfaction from the rimrock, lifting their greasy wings now and then to probe for lice. I can hear but seldom see the canyon wrens singing their distinctive song from somewhere up on the cliffs: a flutelike descent—never ascent—of the whole-tone scale. Staking out new nesting claims, I understand. Also invisible but invariably present at some indefinable distance are the mourning doves whose plaintive call suggests irresistibly a kind of seeking-out, the attempt by separated souls to restore a lost communion:

Hello . . . they seem to cry, *who . . . are . . . you?*

And the reply from a different quarter. *Hello* . . . (pause) *where . . . are . . . you?*

No doubt this line of analogy must be rejected. It's foolish and unfair to impute to the doves, with serious concerns of their own, an interest in questions more appropriate to their human kin. Yet their song, if not a mating call or a warning, must be what it sounds like, a brooding meditation on space, on solitude. The game.

Other birds, silent, which I have not yet learned to identify, are also lurking in the vicinity, watching me. What the ornithologist terms l.g.b.'s—little gray birds—they flit about from point to point on noiseless wings, their origins obscure.

As mentioned before, I share the housetrailer with a number of mice. I don't know how many but apparently only a few, perhaps a single family. They don't disturb me and are welcome to my crumbs and leavings. Where they came from, how they got into the trailer, how they survived before my arrival (for the trailer had been locked up for six months), these are puzzling matters I am not prepared to resolve. My only reservation concerning the mice is that they do attract rattlesnakes.

I'm sitting on my doorstep early one morning, facing the sun as usual, drinking coffee, when I happen to look down and see almost between my bare feet, only a couple of inches to the rear of my heels, the very thing I had in mind. No mistaking that wedgelike head, that tip of horny segmented tail peeping out of the coils. He's under the doorstep and in the shade where the ground and air remain very cold. In his sluggish condition he's not likely to strike unless I rouse him by some careless move of my own.

There's a revolver inside the trailer, a huge British Webley .45, loaded, but it's out of reach. Even if I had it in my hands I'd hesitate to blast a fellow creature at such close range, shooting between my own legs at a living target flat on solid rock thirty inches away. It would be like murder; and where would I set my coffee? My cherrywood walking stick leans against the trailerhouse wall only a few feet away but I'm afraid that in leaning over for it I might stir up the rattler or spill some hot coffee on his scales.

Other considerations come to mind. Arches National Monument is meant to be among other things a sanctuary for wildlife—for all forms of wildlife. It is my duty as a park ranger to protect, preserve and defend all living things within the park boundaries, making no exceptions. Even if this were not the case I have personal convictions to uphold. Ideals, you might say. I prefer not to kill animals. I'm a humanist; I'd rather kill a *man* than a snake.

What to do. I drink some more coffee and study the dormant reptile at my heels. It is not after all the mighty diamondback, *Crotalus atrox*, I'm confronted with but a smaller species known locally as the horny rattler or more precisely as the Faded Midget. An insulting name for a rattlesnake, which may explain the Faded Midget's alleged bad temper. But the name is apt: he is small and dustylooking, with a little knob above each eye—the horns. His bite though temporarily disabling would not likely kill a full-grown man in normal health. Even so I don't really want him around. Am I to be compelled to put on boots or shoes every time I wish to step outside? The scorpions, tarantulas, centipedes, and black widows are nuisance enough.

I finish my coffee, lean back and swing my feet up and inside the doorway of the trailer. At once there is a buzzing sound from below and the rattler lifts his head from his coils, eyes brightening, and extends his narrow black tongue to test the air.

After thawing out my boots over the gas flame I pull them on and come back to the doorway. My visitor is still waiting beneath the doorstep, basking in the sun, fully alert. The trailerhouse has two doors. I leave by the other and get a long-handled spade out of the bed of the government pickup. With this tool I scoop the snake into the open. He strikes; I can hear the click of the fangs against steel, see the strain of venom. He wants to stand and fight, but I am patient; I insist on herding him well away from the trailer. On guard, head aloft—that evil slit-eyed weaving head shaped like the ace of spades—tail whirring, the rattler slithers sideways, retreating slowly before me until he reaches the shelter of a sandstone slab. He backs under it.

You better stay there, cousin, I warn him; if I catch you around the trailer again I'll chop your head off.

A week later he comes back. If not him, his twin brother. I spot him one morning under the trailer near the kitchen drain, waiting for a mouse. I have to keep my promise.

This won't do. If there are midget rattlers in the area there may be diamondbacks too—five, six or seven feet long, thick as a man's wrist, dangerous. I don't want *them* camping under my home. It looks as though I'll have to trap the mice.

However, before being forced to take that step I am lucky enough to capture a gopher snake. Burning garbage one morning at the park dump, I see a long slender yellow-brown snake emerge from a mound of old tin cans and plastic picnic plates and take off down the sandy bed of a gulch. There is a burlap sack in the cab of the truck which I carry when plucking Kleenex flowers from the brush and cactus along the road; I grab that and my stick, run after the snake and corner it beneath the exposed roots of a bush. Making sure it's a gopher snake and not something less useful, I open the neck of the sack and with a great deal of coaxing and prodding get the snake into it. The gopher snake, *Drymarchon corais couperi*, or bull snake, has a reputation as the enemy of rattlesnakes, destroying or driving them away whenever encountered.

Hoping to domesticate this sleek, handsome and docile reptile, I release him inside the trailerhouse and keep him there for several days. Should I attempt to feed him? I decide against it—let him eat mice. What little water he may need can also be extracted from the flesh of his prey.

The gopher snake and I get along nicely. During the day he curls up like a cat in the warm corner behind the heater and at night he goes about his business. The mice, singularly quiet for a change, make themselves scarce. The snake is passive, apparently contented, and makes no resistance when I pick him up with my hands and drape him over an arm or around my neck. When I take him outside into the wind and sunshine his favorite place seems to be inside my shirt, where he wraps himself around my waist and rests on my belt. In this position he sometimes sticks his head out between shirt buttons for a survey of the weather, astonishing and delighting any tourists who may happen to be with me at the time. The scales of a snake are dry and smooth, quite pleasant to the touch. Being a cold-blooded creature, of course, he takes his temperature from that of the immediate environment—in this case my body.

We are compatible. From my point of view, friends. After a week of close association I turn him loose on the warm sandstone at my doorstep and leave for patrol of the park. At noon when I return he is gone. I search everywhere beneath, nearby and inside the trailerhouse, but my companion has disappeared. Has he left the area entirely or is he hiding somewhere close by? At any rate I am troubled no more by rattlesnakes under the door.

The snake story is not yet ended.

In the middle of May, about a month after the gopher snake's disappearance, in the evening of a very hot day, with all the rosy desert cooling like a griddle with the fire turned off, he reappears. This time with a mate.

I'm in the stifling heat of the trailer opening a can of beer, barefooted, about to go outside and relax after a hard day watching cloud formations. I happen to glance out the little window near the refrigerator and see two gopher snakes on my verandah engaged in what seems to be a kind of ritual dance. Like a living caduceus they wind and unwind about each other in undulant, graceful, perpetual motion, moving slowly across a dome of sandstone. Invisible but tangible as music is the passion which joins them—sexual? combative? both? A shameless *voyeur*, I stare at the lovers, and then to get a closer view run outside and around the trailer to the back. There I get down on hands and knees and creep toward the dancing snakes, not wanting to frighten or disturb them. I crawl to within six feet of them and stop, flat on my belly, watching from the snake's-eye level. Obsessed with their ballet, the serpents seem unaware of my presence.

The two gopher snakes are nearly identical in length and coloring; I cannot be certain that either is actually my former household pet. I cannot even be sure that they are male and female, though their performance resembles so strongly a *pas de deux* by formal lovers. They intertwine and separate, glide side by side in perfect congruence, turn like mirror images of each other and glide back again, wind and unwind again. This is the basic pattern but there is a variation: at regular intervals the snakes elevate their heads, facing one another, as high as they can go, as if each is trying to outreach or overawe the other. Their heads and bodies rise, higher and higher, then topple together and the rite goes on.

I crawl after them, determined to see the whole thing. Suddenly and simultaneously they discover me, prone on my belly a few feet away. The dance stops. After a moment's pause the two snakes come straight toward me, still in flawless unison, straight toward my face, the forked tongues flickering, their intense wild yellow eyes staring directly into my eyes. For an instant I am paralyzed by wonder; then, stung by a fear too ancient and powerful to overcome I scramble back, rising to my knees. The snakes veer and turn and race away from me in parallel motion, their lean elegant bodies making a soft hissing noise as they slide over the sand and stone. I follow them for a short distance, still plagued by curiosity, before remembering my place and the requirements of common courtesy. For godsake let them go in peace, I tell myself. Wish them luck and (if lovers) innumerable offspring, a life of happily ever after. Not for their sake alone but for your own.

In the long hot days and cool evenings to come I will not see the gopher snakes again. Nevertheless I will feel their presence watching over me like totemic deities, keeping the rattlesnakes far back in the brush where I like them best, cropping off the surplus mouse population, maintaining useful connections with the primeval. Sympathy, mutual aid, symbiosis, continuity.

How can I descend to such anthropomorphism? Easily—but is it, in this case entirely false? Perhaps not. I am not attributing human motives to my snake and bird acquaintances. I recognize that when and where they serve purposes of mine they do so for beautifully selfish reasons of their own. Which is exactly the way it should be. I suggest, however, that it's a foolish, simple-minded rationalism which denies any form of emotion to all animals but man and his dog. This is no more justified than the Moslems are in denying souls to women. It seems to me possible, even probable, that many of the nonhuman undomesticated animals experience emotions unknown to us. What do the coyotes mean when they yodel at the moon? What are the dolphins trying so patiently to tell us? Precisely what did those two enraptured gopher snakes have in mind when they came gliding toward my eyes over the naked sandstone? If I had been as capable of trust as I am sus-

ceptible to fear I might have learned something new or some truth so
very old we have all forgotten it.

> They do not sweat and whine about their condition,
> They do not lie awake in the dark and weep for their sins. . . .

All men are brothers, we like to say, half-wishing sometimes in secret
it were not true. But perhaps it is true. And is the evolutionary line from
protozoan to Spinoza any less certain? That also may be true. We are
obliged, therefore, to spread the news, painful and bitter though it may
be for some to hear, that all living things on earth are kindred.

THE GREAT AMERICAN DESERT

In my case it was love at first sight. This desert, all deserts, any
desert. No matter where my head and feet may go, my heart and my
entrails stay behind, here on the clean, true, comfortable rock, under
the black sun of God's forsaken country. When I take on my next incarnation, my bones will remain bleaching nicely in a stone gulch under
the rim of some faraway plateau, way out there in the back of beyond.
An unrequited and excessive love, inhuman no doubt but painful anyhow, especially when I see my desert under attack. "The one death I
cannot bear," said the Sonoran-Arizonan poet Richard Shelton. The
kind of love that makes a man selfish, possessive, irritable. If you're
thinking of a visit, my natural reaction is like a rattlesnake's—to warn
you off. What I want to say goes something like this.

Survival Hint #1: Stay out of there. Don't go. Stay home and read a
good book, this one for example. The Great American Desert is an
awful place. People get hurt, get sick, get lost out there. Even if you survive, which is not certain, you will have a miserable time. The desert is
for movies and God-intoxicated mystics, not for family recreation.

Let me enumerate the hazards. First the Walapai tiger, also known
as conenose kissing bug. *Triatoma protracta* is a true bug, black as sin,
and it flies through the night quiet as an assassin. It does not attack
directly like a mosquito or deerfly, but alights at a discreet distance,
undetected, and creeps upon you, its hairy little feet making not the
slightest noise. The kissing bug is fond of warmth and like Dracula

The Journey Home: Some Words in Defense of the American West (New York: Dutton,
1977).

requires mammalian blood for sustenance. When it reaches you the bug crawls onto your skin so gently, so softly that unless your senses are hyperacute you feel nothing. Selecting a tender point, the bug slips its conical proboscis into your flesh, injecting a poisonous anesthetic. If you are asleep you will feel nothing. If you happen to be awake you may notice the faintest of pinpricks, hardly more than a brief ticklish sensation, which you will probably disregard. But the bug is already at work. Having numbed the nerves near the point of entry the bug proceeds (with a sigh of satisfaction, no doubt) to withdraw blood. When its belly is filled, it pulls out, backs off, and waddles away, so drunk and gorged it cannot fly.

At about this time the victim awakes, scratching at a furious itch. If you recognize the symptoms at once, you can sometimes find the bug in your vicinity and destroy it. But revenge will be your only satisfaction. Your night is ruined. If you are of average sensitivity to a kissing bug's poison, your entire body breaks out in hives, skin aflame from head to toe. Some people become seriously ill, in many cases requiring hospitalization. Others recover fully after five or six hours except for a hard and itchy swelling, which may endure for a week.

After the kissing bug, you should beware of rattlesnakes; we have half a dozen species, all offensive and dangerous, plus centipedes, millipedes, tarantulas, black widows, brown recluses, Gila monsters, the deadly poisonous coral snakes, and giant hairy desert scorpions. Plus an immense variety and near-infinite number of ants, midges, gnats, bloodsucking flies, and blood-guzzling mosquitoes. (You might think the desert would be spared at least mosquitoes? Not so. Peer in any water hole by day: swarming with mosquito larvae. Venture out on a summer's eve: The air vibrates with their mournful keening.) Finally, where the desert meets the sea, as on the coasts of Sonora and Baja California, we have the usual assortment of obnoxious marine life: sandflies, ghost crabs, stingrays, electric jellyfish, spiny sea urchins, maneating sharks, and other creatures so distasteful one prefers not even to name them.

It has been said, and truly, that everything in the desert either stings, stabs, stinks, or sticks. You will find the flora here as venomous, hooked, barbed, thorny, prickly, needled, saw-toothed, hairy, stickered, mean, bitter, sharp, wiry, and fierce as the animals. Something about the desert inclines all living things to harshness and acerbity. The soft evolve out. Except for sleek and oily growths like the poison ivy—oh yes, indeed—that flourish in sinister profusion on the dank walls above the quicksand down in those corridors of gloom and labyrinthine monotony that men call canyons.

We come now to the third major hazard, which is sunshine. Too much of a good thing can be fatal. Sunstroke, heatstroke, and dehydration are common misfortunes in the bright American Southwest. If you can avoid the insects, reptiles, and arachnids, the cactus and the ivy, the smog of the southwestern cities, and the lung fungus of the desert valleys (carried by dust in the air), you cannot escape the desert sun. Too much exposure to it eventually causes, quite literally, not merely sunburn but skin cancer.

Much sun, little rain also means an arid climate. Compared with the high humidity of more hospitable regions, the dry heat of the desert seems at first not terribly uncomfortable—sometimes even pleasant. But that sensation of comfort is false, a deception, and therefore all the more dangerous, for it induces overexertion and an insufficient consumption of water, even when water is available. This leads to various internal complications, some immediate—sunstroke, for example—and some not apparent until much later. Mild but prolonged dehydration, continued over a span of months or years, leads to the crystallization of mineral solutions in the urinary tract, that is, to what urologists call urinary calculi or kidney stones. A disability common in all the world's arid regions. Kidney stones, in case you haven't met one, come in many shapes and sizes, from pellets smooth as BB shot to highly irregular calcifications resembling asteroids. Vietcong shrapnel, and crown-of-thorns starfish. Some of these objects may be "passed" naturally; others can be removed only by means of the Davis stone basket or by surgery. Me—I was lucky; I passed mine with only a groan, my forehead pressed against the wall of a pissoir in the rear of a Tucson bar that I cannot recommend.

You may be getting the impression by now that the desert is not the most suitable of environments for human habitation. Correct. Of all the Earth's climatic zones, excepting only the Antarctic, the deserts are the least inhabited, the least "developed," for reasons that should now be clear.

You may wish to ask, Yes, okay, but among North American deserts which is the *worst?* A good question—and I am happy to attempt to answer.

Geographers generally divide the North American desert—what was once termed "the Great American Desert"—into four distinct regions or subdeserts. These are the Sonoran Desert, which comprises southern Arizona, Baja California, and the state of Sonora in Mexico; the Chihuahuan Desert, which includes west Texas, southern New Mexico, and the states of Chihuahua and Coahuila in Mexico; the Mojave Desert, which includes southeastern California and small portions of Nevada, Utah, and Arizona; and the Great Basin Desert, which includes most of Utah and Nevada, northern Arizona, northwestern New Mexico, and much of Idaho and eastern Oregon.

Privately, I prefer my own categories. Up north in Utah somewhere is the canyon country—places like Zeke's Hole, Death Hollow, Pucker Pass, Buckskin Gulch, Nausea Crick, Wolf Hole, Mollie's Nipple, Dirty Devil River, Horse Canyon, Horseshoe Canyon, Lost Horse Canyon, Horsethief Canyon, and Horseshit Canyon, to name only the more classic places. Down in Arizona and Sonora there's the cactus country; if you have nothing better to do, you might take a look at High Tanks, Salome Creek, Tortilla Flat, Esperero ("Hoper") Canyon, Holy Joe Peak, Depression Canyon, Painted Cave, Hell Hole Canyon, Hell's Half Acre, Iceberg Canyon, Tiburon (Shark) Island, Pinacate Peak, Infernal Valley, Sykes Crater, Montezuma's Head, Gu Oidak, Kuakatch, Pisinimo, and Baboquivari Mountain, for example.

Then there's The Canyon. *The Canyon*. The Grand. That's one world. And North Rim—that's another. And Death Valley, still another, where I lived one winter near Furnace Creek and climbed the Funeral Mountains, tasted Badwater, looked into the Devil's Hole, hollered up Echo Canyon, searched for and never did find Seldom Seen Slim. Looked for *satori* near Vane, Nevada, and found a ghost town named Bonnie Claire. Never made it to Winnemucca. Drove through the Smoke Creek Desert and down through Big Pine and Lone Pine and home across the Panamints to Death Valley again—home sweet home that winter.

And which of these deserts is the worst? I find it hard to judge. They're all bad—not half bad but all bad. In the Sonoran Desert, Phoenix will get you if the sun, snakes, bugs, and arthropods don't. In the Mojave Desert, it's Las Vegas, more sickening by far than the Glauber's salt in the Death Valley sinkholes. Go to Chihuahua and you're liable to get busted in El Paso and sandbagged in Ciudad Juárez—where all old whores go to die. Up north in the Great Basin Desert, on the Plateau Province, in the canyon country, your heart will break, seeing the strip mines open up and the power plants rise where only cowboys and Indians and J. Wesley Powell ever roamed before.

Nevertheless, all is not lost; much remains, and I welcome the prospect of an army of lug-soled hiker's boots on the desert trails. To save what wilderness is left in the American Southwest—and in the American Southwest only the wilderness is worth saving—we are going to need all the recruits we can get. All the hands, heads, bodies, time, money, effort we can find. Presumably—and the Sierra Club, the Wilderness Society, the Friends of the Earth, the Audubon Society, the

Defenders of Wildlife operate on this theory—those who learn to love what is spare, rough, wild, undeveloped, and unbroken will be willing to fight for it, will help resist the strip miners, highway builders, land developers, weapons testers, power producers, tree chainers, clear cutters, oil drillers, dam beavers, subdividers—the list goes on and on—before that zinc-hearted, termite-brained, squint-eyed, nearsighted, greedy crew succeeds in completely californicating what still survives of the Great American Desert.

So much for the Good Cause. Now what about desert hiking itself, you may ask. I'm glad you asked that question. I firmly believe that one should never—I repeat *never*—go out into that formidable wasteland of cactus, heat, serpents, rock, scrub, and thorn without careful planning, thorough and cautious preparation, and complete—never mind the expense!—*complete* equipment. My motto is: Be Prepared.

That is my belief and that is my motto. My practice, however, is a little different. I tend to go off in a more or less random direction myself, half-baked, half-assed, half-cocked, and half-ripped. Why? Well, because I have an indolent and melancholy nature and don't care to be bothered getting all those *things* together—all that bloody *gear*—maps, compass, binoculars, poncho, pup tent, shoes, first-aid kit, rope, flashlight, inspirational poetry, water, food—and because anyhow I approach nature with a certain surly ill-will, daring Her to make trouble. Later when I'm deep into Natural Bridges National Moneymint or Zion National Parkinglot or say General Shithead National Forest Land of Many Abuses why then, of course, when it's a bit late, then I may wish I had packed that something extra: matches perhaps, to mention one useful item, or maybe a spoon to eat my gruel with.

If I hike with another person it's usually the same; most of my friends have indolent and melancholy natures too. A cursed lot, all of them. I think of my comrade John De Puy, for example, sloping along for mile after mile like a goddamned camel—indefatigable—with those J. C. Penny hightops on his feet and that plastic pack on his back he got with five books of Green Stamps and nothing inside it but a sketchbook, some homemade jerky and a few cans of green chiles. Or Douglas Peacock, ex-Green Beret, just the opposite. Built like a buffalo, he loads a ninety-pound canvas pannier on his back at trailhead, loaded with guns, ammunition, bayonet, pitons and carabiners, cameras, field books, a 150-foot rope, geologist's sledge, rock samples, assay kit, field glasses, two gallons of water in steel canteens, jungle boots, a case of C-rations, rope hammock, pharmaceuticals in a pig-iron box, raincoat, overcoat, two-man mountain tent, Dutch oven, hibachi, shovel, ax, inflatable boat, and near the top of the load and distributed through side and back pockets, easily accessible, a case of beer. Not because he enjoys or needs all that weight—he may never get to the bottom of that cargo on a ten-day outing—but simply because Douglas uses his packbag for general storage both at home and on the trail and perfers not to have to rearrange everything from time to time merely for the purposes of a hike. Thus my friends De Puy and Peacock; you may wish to avoid such extremes.

A few tips on desert etiquette:

1. Carry a cooking stove, if you must cook. Do not burn desert wood, which is rare and beautiful and required ages for its creation (an iron-wood tree lives for over 1,000 years and juniper almost as long).
2. If you must, out of need, build a fire, then for God's sake allow it to burn itself out before you leave—do not bury it, as Boy Scouts and Campfire Girls do, under a heap of mud or sand. Scatter the ashes; replace any rocks you may have used in constructing a fireplace; do all you can to obliterate the evidence that you camped here. (The Search & Rescue Team may be looking for you.)
3. Do not bury garbage—the wildlife will only dig it up again. Burn what will burn and pack out the rest. The same goes for toilet paper: Don't bury it, *burn it.*
4. Do not bathe in desert pools, natural tanks, *tinajas,* potholes. Drink what water you need, take what you need, and leave the rest for the next hiker and more important for the bees, birds, and animals—bighorn sheep, coyotes, lions, foxes, badgers, deer, wild pigs, wild horses—whose *lives* depend on that water.
5. Always remove and destroy survey stakes, flagging, advertising signboards, mining claim markers, animal traps, poisoned bait, seismic exploration geophones, and other such artifacts of industrialism. The men who put those things there are up to no good and it is our duty to confound them. Keep America Beautiful. Grow a Beard. Take a Bath. Burn a Billboard.

Anyway—why go into the desert? Really, why do it? That sun, roaring at you all day long. The fetid, tepid, vapid little water holes slowly evaporating under a scum of grease, full of cannibal beetles, spotted toads, horsehair worms, liver flukes, and down at the bottom, inevitably, the pale cadaver of a ten-inch centipede. Those pink rattlesnakes down in The Canyon, those diamondback monsters thick as a truck driver's wrist that lurk in shady places along the trail, those unpleasant solpugids and unnecessary Jerusalem crickets that scurry on dirty claws across your face at night. Why? The rain that comes down like lead shot and wrecks the trail, those sudden rockfalls of obscure origin that crash like thunder ten feet behind you in the heart of a dead-still afternoon. The ubiquitous buzzard, so patient—but only so patient. The sullen and hostile Indians, all on welfare. The ragweed, the tumbleweed, the Jimson weed, the snakeweed. The scorpion in your shoe at dawn. The dreary wind that blows all spring, the psyche-delic Joshua trees waving their arms at you on moonlight nights. Sand in the soup de jour. Halazone tablets in your canteen. The barren hills that always go up, which is bad, or down, which is worse. Those canyons like catacombs with quicksand lapping at your crotch. Hollow, mummified horses with forelegs casually crossed, dead for ten years, leaning against the corner of a barbed-wire fence. Packhorses at night, iron-shod, clattering over the slickrock through your camp. The last tin of tuna, two flat tires, not enough water and a forty-mile trek to Tule Well. An osprey on a cardón cactus, snatching the head off a living fish—always the best part first. The hawk sailing by at 200 feet, a squirming snake in its talons. Salt in the drinking water. Salt, selenium, arsenic, radon and radium in the water, in the gravel, in your bones. Water so hard it bends light, drills holes in rock and chokes up your radiator. Why go there? Those places with the hardcase names: Starvation Creek, Poverty Knoll, Hungry Valley, Bitter Springs, Last Chance Canyon, Dungeon Canyon, Whipsaw Flat, Dead Horse Point, Scorpion Flat, Dead Man Draw, Stinking Spring, Camino del Diablo, Jornado del Muerto . . . Death Valley.

Well then, why indeed go walking into the desert, that grim ground, that bleak and lonesome land where, as Genghis Khan said of India, "the heat is bad and the water makes men sick"?

Why the desert, when you could be strolling along the golden beaches of California? Camping by a stream of pure Rocky Mountain spring water in colorful Colorado? Loafing through a laurel slick in the misty hills of North Carolina? Or getting your head mashed in the greasy alley behind the Elysium Bar and Grill in Hoboken, New Jersey? Why the desert, given a world of such splendor and variety?

A friend and I took a walk around the base of a mountain up beyond Coconino County, Arizona. This was a mountain we'd been planning to circumambulate for years. Finally we put on our walking shoes and did it. About halfway around this mountain, on the third or fourth day, we paused for a while—two days—by the side of a stream, which the Navajos call Nasja because of the amber color of the water. (Caused perhaps by juniper roots—the water seems safe enough to drink.) On our second day there I walked down the stream, alone, to look at the canyon beyond. I entered the canyon and followed it for half the after-noon, for three or four miles, maybe, until it became a gorge so deep, narrow and dark, full of water and the inevitable quagmires of quicksand, that I turned around and looked for a way out. A route other than the way I'd come, which was crooked and uncomfortable and buried—I wanted to see what was up on top of this world. I found a sort of chimney flue on the east wall, which looked plausible, and sweated and cursed my way up through that until I reached a point where I could walk upright, like a human being. Another 300 feet of scrambling brought me to the rim of the canyon. No one, I felt certain, had ever before departed Nasja Canyon by that route.

But someone had. Near the summit I found an arrow sign, three feet long, formed of stones and pointing off into the north toward those same old purple vistas, so grand, immense, and mysterious, of more canyons, more mesas and plateaus, more mountains, more cloud-dappled sun-spangled leagues of desert sand and desert rock, under the same old wide and aching sky.

The arrow pointed into the north. But what was it pointing *at*? I looked at the sign closely and saw that those dark, desert-varnished stones had been in place for a long, long, time; they rested in compacted dust. They must have been there for a century at least. I followed the direction indicated and came promptly to the rim of another canyon and a drop-off straight down of a good 500 feet. Not that way, surely. Across this canyon was nothing of any unusual interest that I could see—only the familiar sun-blasted sandstone, a few scrubby clumps of blackbrush and prickly pear, a few acres of nothing where only a lizard could graze, surrounded by a few square miles of more nothingness interesting chiefly to horned toads. I returned to the arrow and checked again, this time with field glasses, looking away for as far as my aided eyes could see toward the north, for ten, twenty, forty miles into the distance. I studied the scene with care, looking for an ancient Indian ruin, a significant cairn, perhaps an abandoned mine, a hidden treasure of some inconceivable wealth, the mother of all mother lodes. . . .

But there was nothing out there. Nothing at all. Nothing but the desert. Nothing but the silent world.

That's why.

PETER MATTHIESSEN
b. 1927

Peter Matthiessen has written novels, including At Play in the Fields of
the Lord *(1965) and* Far Tortuga *(1975), and works of nonfiction, including* The Tree Where Man Was Born *(1972),* The Wind Birds *(1973),*
The Snow Leopard *(1978), and* African Silences *(1991). Many of his
works reflect his own expeditions to wild places around the world, from
Kenya to New Guinea, Nepal to the Northwest Territories. Matthiessen
bears witness to the damage inflicted on wilderness and on isolated cultures alike by rampant exploitation. But he also celebrates the excellence,
and elusiveness, of wildness.*

From THE TREE WHERE MAN WAS BORN

RITES OF PASSAGES

*   *   * One morning the dog pile broke apart before daylight and
headed off toward the herds under Naabi Hill. Unlike lions, which
often go hungry, the wild dogs rarely fail to make a kill, and this time
they were followed from the start by three hyenas that had waited near
the den. The three humped along behind the pack, and one of the
dogs paused to sniff noses with a hyena by way of greeting. In the distance, zebras yelped like dogs, and the dogs chittered quietly like birds
as they loped along. As the sun rose out of the Gol Mountains, they
faked an attack on a string of wildebeest and moved on.

A mile and a half east of the den, the pack cut off a herd of zebra
and ran it in tight circles. There were foals in this herd, but the dogs
had singled out a pregnant mare. When the herd scattered, they closed
in, streaming along in the early light, and almost immediately she fell
behind and then gave up, standing motionless as one dog seized her

The Tree Where Man Was Born (New York: Dutton, 1972).

nose and others ripped at her pregnant belly and others piled up under her tail to get at her entrails at the anus, surging at her with such force that the flesh of her uplifted quarters quaked in the striped skin. Perhaps in shock, their quarry shares the detachment of the dogs, which attack it peaceably, ears forward, with no slightest sign of snapping or snarling. The mare seemed entirely docile, unafraid, as if she had run as she had been hunted, out of instinct, and without emotion: only rarely will a herd animal attempt to defend itself with the hooves and teeth used so effectively in battles with its own kind, though such resistance might well spare its life. The zebra still stood a full half-minute after her guts had been snatched out, then sagged down dead. Her unborn colt was dragged into the clear and snapped apart off to one side.

The morning was silent but for the wet sound of eating; a Caspian plover and a band of sand grouse picked at the mute prairie. The three hyenas stood in wait, and two others appeared after the kill. One snatched a scrap and ran with it; the meat, black with blood and mud, dragged on the ground. Chased by the rest, the hyena made a shrill sound like a pig squeal. When their spirit is up, hyenas will take on a lion, and if they chose, could bite a wild dog in half, but in daylight, they seem ill at ease; they were scattered by one tawny eagle, which took over the first piece of meat abandoned by the dogs. The last dog to leave, having finished with the fetus, drove the hyenas off the carcass of the mare on its way past, then frisked on home. In a day and a night, when lions and hyenas, vultures and marabous, jackals, eagles, ants, and beetles have all finished, there will be no sign but the stained pressed grass that a death ever took place.

All winter in the Serengeti damp scrawny calves and afterbirths are everywhere, and old or diseased animals fall in the night. Fat hyenas, having slaked their thirst, squat in the rain puddles, and gaping lions lie belly to the sun. On Naabi Hill the requiem birds, digesting carrion, hunch on the canopies of low acacia. Down to the west, a young zebra wanders listlessly by itself. Unlike topi and kongoni, which are often seen alone, the zebra and wildebeest prefer the herd; an animal by itself may be sick or wounded, and draws predators from all over the plain. This mare had a deep gash down her right flank, and a slash of claws across the striping of her quarters; red meat gleamed on right foreleg and left fetlock. It seemed strange that an attacking lion close enough to maul so could have botched the job, but the zebra pattern makes it difficult to see at night, when it is most vulnerable to attack by lions, and zebra are strong animals; a thin lioness that I saw once at

Ngorongoro had a broken incisor hanging from her jaw that must have been the work of a flying hoof.

Starvation is the greatest threat to lions, which are inefficient hunters and often fail to make a kill. Unlike wild dog packs, which sometimes overlap in their wide hunting range, lions will attack and even eat another lion that has entered their territory, snapping and snarling in the same antagonistic way with which they join their pride mates on a kill, whereas when hunting, they are silent and impassive. In winter when calves of gazelle and gnu litter the plain, the lions are well fed, but at other seasons they may be so hungry that their own cubs are driven from the kills: ordinarily, however, the lion will permit cubs to feed even when the lioness that made the kill is not permitted to approach. Until it is two, the cub is a dependent, and less than half of those born in the Serengeti survive the first year of life. In hard times, cubs may be eaten by hyenas, or by the leopard, which has a taste for other carnivores, including domestic cats and dogs.

A former warden of the Serengeti who feels that plains game should be killed to feed these starving cubs is opposed by George Schaller on the grounds that such artificial feeding would interfere with the balance of lion numbers as well as with the natural selection that maintains the vitality of the species. Dr. Schaller is correct, I think, and yet my sympathies are with the predator, not with the hunted, perhaps because a lion is perceived as an individual, whereas one member of a herd of thousands seems but a part of a compound organism, with little more identity than one termite in a swarm. Separated from the herd, it gains identity, like the zebra killed by the wild dogs, but even so I felt more pity for an injured lion that I saw near the Seronera River in the hungry months of summer, a walking husk of mane and bone, so weak that the dry weather wind threatened to knock it over.

The death of any predator is disturbing. I was startled one day to see a hawk in the talons of Verreaux's eagle-owl; perhaps it had been killed in the act of killing. Another day, by a korongo, I helped Schaller collect a dying lioness. She had emaciated hindquarters and the staggers, and at our approach, she reeled to her feet, then fell. In the interests of science as well as mercy, for he wished an autopsy, George shot her with an overdose of tranquilizer. Although she twitched when the needle struck, and did not rise, she got up after a few minutes and weaved a few feet more and fell again as if defeated by the obstacle of the korongo, where frogs trilled in oblivion of unfrogly things. I had the strong feeling that the lioness, sensing death, had risen to escape it, like the vultures I had heard of somewhere that flew up from the poisoned meat set out for lions, circling higher and higher into the sky, only to fall like stones as life forsook them. A moment later, her head rose up, then flopped for the last time, but she would not die. Sprinkled with hopping lion flies and the fat ticks that in lions are a sign of poor condition, she lay there in a light rain, her gaunt flanks twitching.

The episode taught me something about George Schaller, who is single-minded, not easy to know. George is a stern pragmatist, unable to muster up much grace in the face of unscientific attitudes; he takes a hard-eyed look at almost everything. Yet at this moment his boyish face was openly upset, more upset than I had ever thought to see him. The death of the lioness was painless, far better than being found by the hyenas, but it was going on too long; twice he returned to the Land Rover for additional dosage. We stood there in a kind of vigil, feeling more and more depressed, and the end, when it came at last, was shocking. The poor beast, her life going, began to twitch and tremble. With a little grunt, she turned onto her back and lifted her hind legs into the air. Still grunting, she licked passionately at the grass, and her haunches shuddered in long spasms, and this last abandon shattered the detachment I had felt until that moment. I was swept by a wave of feeling, then a pang so sharp that, for a moment, I felt sick, as if all the waste and loss in life, the harm one brings to oneself and others, had been drawn to a point in this lonely passage between light and darkness.

Mid-March when the long rains were due was a time of wind and dry days in the Serengeti, with black trees in iron silhouette on the hard sunsets and great birds turning forever on a silver sky. A full moon rose in a night rainbow, but the next day the sun was clear again, flat as a disc in the pale universe.

Two rhino and a herd of buffalo had brought up the rear of the eastward migration. Unlike the antelope, which blow with the wind and grasses, the dark animals stood earthbound on the plain. The antelope, all but a few, had drifted east under the Crater Highlands, whereas the zebra, in expectation of the rains, were turning west again toward the woods. Great herds had gathered at the Seronera River, where the local prides of lion were well fed. Twenty lions together, dozing in the golden grass, could sometimes be located by the wave of a black tail tuft or the black ear tips of a lifted head that gazed through the sun shimmer of the seed heads. Others gorged in uproar near the river crossings, tearing the fat striped flanks on fresh green beds—now daytime kills were common. Yet for all their prosperity, there was an air of doom about the lions. The males, especially, seemed too big, and they walked too slowly between feast and famine, as if in some dim intuition that the time of the great predators was running out.

Pairs of male lions, unattached to any pride, may hunt and live together in great harmony, with something like demonstrative affection. But when two strangers meet, there seems to be a waiting period, while fear settles. One sinks into the grass at a little distance, and for a long time they watch each other, and their sad eyes, unblinking, never move. The gaze is the warning, and it is the same gaze, wary but unwavering, with which lions confront man. The gold cat eyes shimmer with hidden lights, eyes that see everything and betray nothing. When the lion is satisfied that the threat is past, the head is turned, as if ignoring it might speed the departure of an unwelcome and evil-smelling presence. In its torpor and detachment, the lion sometimes seems the dullest beast in Africa, but one has only to watch a file of lions setting off on the evening hunt to be awed anew by the power of this animal.

One late afternoon of March, beyond Maasai Kopjes, eleven lionesses lay on a kill, and the upraised heads, in a setting sun, were red. With their grim visages and flat glazed eyes, these twilight beasts were ominous. Then the gory heads all turned as one, ear tips alert. No animal was in sight, and their bellies were full, yet they glared steadfastly away into the emptiness of plain, as if something that no man could sense was imminent.

Not far off there was a leopard; possibly they scented it. The leopard lay on an open rise, in the shadow of a wind-worn bush, and unlike the lions, it lay gracefully. Even stretched on a tree limb, all four feet hanging, as it is seen sometimes in the fever trees, the leopard has the grace of complete awareness, with all its tensions in its pointed eyes. The lion's gaze is merely baleful; that of the leopard is malevolent, a distillation of the trapped fear that is true savagery.

Under a whistling thorn the leopard lay, gold coat on fire in the sinking sun, as if imagining that so long as it lay still it was unseen. Behind it was a solitary thorn tree, black and bony in the sunset, and from a crotch in a high branch, turning gently, torn hide matted with caked blood, the hollow form of a gazelle hung by the neck. At the insistence of the wind, the delicate black shells of the turning hoofs, on tiptoe, made a dry clicking in the silence of the plain.

ELEPHANT KINGDOMS

One morning a great company of elephants came from the woodlands, moving eastward toward the Togoro Plain. "It's like the old Africa, this," Myles Turner said, coming to fetch me. "It's one of the greatest sights a man can see."

We flew northward over the Orangi River. In the wake of the elephant herds, stinkbark acacia were scattered like sticks, the haze of yellow blossoms bright in the killed trees. Through the center of the destruction, west to east, ran a great muddied thoroughfare of the sort described by Selous in the nineteenth century. Here the center of the herd had passed. The plane turned eastward, coming up on the elephant armies from behind. More than four hundred animals were pushed together in one phalanx; a smaller group of one hundred and another of sixty were nearby. The four hundred moved in one slow-stepping swaying mass, with the largest cows along the outer ranks and big bulls scattered on both sides. "Seventy and eighty pounds, some of those bulls," Myles said. (Trophy elephants are described according to the weight of a single tusk; an eighty-pound elephant would carry about twice that weight in ivory. "Saw an eighty today." "*Did* you!")

Myles said that elephants herded up after heavy rains, but that this was an enormous congregation for the Serengeti. In 1913, when the first safari came here, the abounding lions and wild dogs were shot as vermin, but no elephants were seen at all. Even after 1925, when the plains were hunted regularly by such men as Philip Percival and the American, Martin Johnson, few elephants were reported. Not until after 1937, it is said, when the Serengeti was set aside as a game reserve (it was not made a national park until 1951), did harried elephants from the developing agricultural country of west Kenya move south into this region, but it seems more likely that they were always present in small numbers, and merely increased as a result of human pressures in suitable habitats outside the park.

Elephants, with their path-making and tree-splitting propensities, will alter the character of the densest bush in very short order; probably they rank with man and fire as the greatest force for habitat change in Africa. In the Serengeti, the herds are destroying many of the taller trees which are thought to have risen at the beginning of the century, in a long period without grass fires that followed plague, famine, and an absence of the Maasai. Dry season fires, often set purposely by poachers and pastoral peoples, encourage grassland by suppressing new woody growth; when accompanied by drought, and fed by a woodland tinder of elephant-killed trees, they do lasting damage to the soil and the whole environment. Fires waste the dry grass that is used by certain animals, and the regrowth exhausts the energy in the grass roots that is needed for good growth in the rainy season. In the Serengeti in recent years, fire and elephants together have converted miles and miles of acacia wood to grassland, and damaged the stands of yellowbark acacia or fever tree along the water courses. The range of the plains game has increased, but the much less numerous woodland species such as the roan antelope and oribi become ever more difficult to see.

Beneath the plane, the elephant mass moved like gray lava, leaving behind a ruined bog of mud and twisted trees. An elephant can eat as much as six hundred pounds of grass and browse each day, and it is a destructive feeder, breaking down many trees and shrubs along the way. The Serengeti is immense, and can absorb this damage, but one sees quickly how an elephant invasion might affect more vulnerable areas. Ordinarily the elephant herds are scattered and nomadic, but pressure from settlements, game control, and poachers sometimes confines huge herds to restricted habitats which they may destroy. Already three of Tanzania's new national parks—Serengeti, Manyara, and Ruaha—have more elephants than is good for them. The elephant problem, where and when and how to manage them, is a great controversy in East Africa, and its solution must affect the balance of animals and man throughout the continent.

Anxious to see the great herd from the ground, I picked up George Schaller at Seronera and drove northwest to Banagi, then westward on the Ikoma-Musoma track to the old northwest boundary of the park, where I headed across country. I had taken good bearings from the air, but elephants on the move can go a long way in an hour, and even for a vehicle with four-wheel drive, this rough bush of high grass, potholes, rocks, steep brushy streams, and swampy mud is very different from the hardpan of the plain. The low hot woods lacked rises or landmarks, and for a while it seemed that I had actually misplaced four hundred elephants.

Then six bulls loomed through the trees, lashing the air with their trunks, ears blowing, in a stiff-legged swinging stride; they forded a steep gully as the main herd, ahead of them, appeared on a wooded rise. Ranging up and down the gully, we found a place to lurch across, then took off eastward, hoping to find a point downwind of the herd where the elephants would pass. But their pace had slowed as the sun rose; we worked back to them, upwind. The elephants were destroying a low wood—this is not an exaggeration—with a terrible cracking of trees, but after a while they moved out onto open savanna. In a swampy stream they sprayed one another and rolled in the water and coated their hides with mud, filling the air with a thick sloughing sound like the wet meat sound made by predators on a kill. Even at rest the herd flowed in perpetual motion, the ears like delicate great petals, the ripple of the mud-caked flanks, the coiling trunks—a dream rhythm, a rhythm of wind and trees. "It's a nice life," Schaller said. "Long, and without fear." A young one could be killed by a lion, but only a desperate lion would venture near a herd of elephants, which are among the few creatures that reach old age in the wild.

There has been much testimony to the silence of the elephant, and all of it is true. At one point there came a cracking sound so small that had I not been alert for the stray elephants all around, I might never have seen the mighty bull that bore down on us from behind. A hundred yards away, it came through the scrub and deadwood like a cloud shadow, dwarfing the small trees of the open woodland. I raised binoculars to watch him turn when he got our scent, but the light wind had shifted and instead the bull was coming fast, looming higher and higher, filling the field of the binoculars, forehead, ears, and back agleam with wet mud dredged up from the donga. There was no time to reach the car, nothing to do but stand transfixed. A froggish voice said, "What do you think, George?" and got no answer.

Then the bull scented us—the hot wind was shifting every moment—and the dark wings flared, filling the sky, and the air was split wide by that ultimate scream that the elephant gives in alarm or agitation, that primordial warped horn note out of oldest Africa. It altered course without missing a stride, not in flight but wary, wideeared, passing man by. Where first aware of us, the bull had been less than one hundred feet away—I walked it off—and he was somewhat nearer where he passed. "He was pretty close," I said finally to Schaller. George cleared his throat. "You don't want them any closer than that," he said. "Not when you're on foot." Schaller, who has no taste for exaggeration, had a very respectful look upon his face.

Stalking the elephants, we were soon a half-mile from my Land Rover. What little wind there was continued shifting, and one old cow, getting our scent, flared her ears and lifted her trunk, holding it upraised for a long time like a question mark. There were new calves with the herd, and we went no closer. Then the cow lost the scent, and the sloughing sound resumed, a sound that this same animal has made for four hundred thousand years. Occasionally there came a brief scream of agitation, or the crack of a killed tree back in the wood, and always the *thuck* of mud and water, and a rumbling of elephantine guts, the deepest sound made by any animal on earth except the whale.

Africa. Noon. The hot still waiting air. A hornbill, gnats, the green hills in the distance, wearing away west toward Lake Victoria.

* * *

Of all African animals, the elephant is the most difficult for man to live with, yet its passing—if this must come—seems the most tragic of all. I can watch elephants (and elephants alone) for hours at a time, for sooner or later the elephant will do something very strange such as mow grass with its toenails or draw the tusks from the rotted carcass of another elephant and carry them off into the bush. There is mystery behind that masked gray visage, an ancient life force, delicate and mighty, awesome and enchanted, commanding the silence ordinarily reserved for mountain peaks, great fires, and the sea. I remember a remark made by a girl about her father, a businessman of narrow sensibilities who, casting about for a means of self-gratification, traveled to Africa and slew an elephant. Standing there in his new hunting togs in a vast and hostile silence, staring at the huge dead bleeding thing that moments before had borne such life, he was struck for the first time in his headlong passage through his days by his own irrelevance. "Even *he*," his daughter said, "knew he'd done something stupid."

The elephant problem, still unresolved, will eventually affect conservation policies throughout East Africa, where even very honest governments may not be able to withstand political pressure to provide meat for the people. Already there is talk of systematic game-cropping in the parks on a sustained yield basis, especially since park revenues from meat and hides and tusks could be considerable, and this temptation may prove impossible to resist for the new governments. Or an outbreak of political instability might wreck the tourist industry that justifies the existence of the parks, thus removing the last barrier between the animals and a hungry populace. African schoolchildren are now taught to appreciate their wild animals and the land, but public attitudes may not change in time to spare the wildlife in the next decades, when the world must deal with the worst consequences of overpopulation and pollution. And a stubborn fight for animal preservation in disregard of people and their famine-haunted future would only be the culminating failure of the western civilization that, through its blind administration of vaccines and quinine, has upset the ecologies of a whole continent. Thus wildlife must be treated in terms of resource management in this new Africa which includes, besides gazelles, a growing horde of tattered humans who squat for days and weeks and months and years on end, in a seeming trance, awaiting hope. In the grotesque costumes of Africa roadsides—rag-wrapped heads and the wool greatcoats and steel helmets of old white man's wars are worn here in hundred-degree heat—the figures look like survivors of a cataclysm. Once, in Nanyuki, I saw a legless man, lacking all means of locomotion, who had been installed in an old auto tire in a ditch at the end of town. Fiercely, eyes bulging, oblivious of the rush of exhaust fumes spinning up the dust around his ears, he glared at an ancient newspaper, as if deciphering the news of doomsday. *   *   *

RED GOD

* * * Sun, heat, stillness were all one. The dying sun in the Ngurumans gave color to the cooking fire, and after dark came a hot wind that fanned night fires all around the horizon, and drove one tongue of flame onto the ridge above the lifeless lake. Though ready to break camp at a moment's notice, I slept poorly—the moon and wind and fire made me restless. But in a red dawn, the wind died again, and the fire sank into the grass, waiting for night.

South of Magadi the road scatters, and wandering tracks cross the white lake bed. There is water where the wading birds are mirrored, and in the liquid shimmer of the heat, a still wildebeest wavers in its own reflection. An hour later, from the west, the ghostly beast was still in sight; it had not moved.

The track winds southwest toward Shombole. Huge termitaria slouch here and there in the dry scrub, and over toward the Nguruman Escarpment, a whirlwind spins a plume of desert dust up the Rift's dark face into the smoky sky of East African summer. Eventually the track descends again, between the dead volcano and the marsh of Uaso Ngiro. In a water gleam that parts the fierce bright reeds, a woman and a man are bathing. The woman squats, her small shoulders demure, but the man stands straight as a gazelle and gazes, body shining, the archetypal man of Africa that I first saw in the Sudan.

The Shombole track comes to an end at three shacks under the volcano, where a duka serves the outlying Maasai with beads and wire for ornament, red cloth, sweet drinks, and cocoa. I gave a ride to a young morani who guided me with brusque motions through the bush to a stony cattle trail that winds between hill and marsh, around Shombole. Farther on, we picked up two Maasai women, and all four of us were squashed into the front when, in the full heat of the desert afternoon, on hot rocky ground at the mud edge of a rotting swamp in this lowest and hottest pit of the Rift Valley floor, my faithful Land Rover, thirty-five miles from Magadi and ninety-five beyond Nairobi, gave a hellish clang and, dragging its guts over the stones, lurched to a halt.

In a bad silence, the Maasai women thanked me and departed. The boy stood by, less out of expectation of reward or even curiosity, I decided, than some sense of duty toward a stranger in Maasai Land. Squatting on my heels and swatting flies, I peered dizzily at the heavy iron shaft, the sand and stone and thorn stuck to raw grease where the shaft had sheared at the universal coupling, cutting off the transmission of power to the rear wheels. In front-wheel drive, the car would move forward weakly, but my limited tools were not able to detach the revolving shaft from the transmission: dragging and clanging in an awful din of steel and rocks, it threatened to shake the car to pieces.

To cool my nerves, I drank a quart of Tusker beer. The Land Rover had picked a poor place to collapse, but at least it had got me to my destination, and the sun if not the heat would soon be gone. Any time now, the airplane of Douglas-Hamilton, coming to meet me, would be landing on the bare mud flats at the north end of Natron. Tomorrow we were to climb Shombole, and after that, if no repairs seemed possible, Iain could fly out to Magadi, and leave word of my straits and whereabouts. But as it happened, Iain and Oria were never to appear: they had sent word to Nairobi that has not reached me to this day. Next morning I rigged a whole series of rope slings, held in place by stay lines from the side, that carried the rotating shaft just off the ground, although they burned through regularly from friction. Setting off at sunrise at three miles an hour, with the frequent stops to repair or replace the sling giving the straining car an opportunity to cool off, I arrived in two hours at the duka. A length of soft iron wire presented me by the proprietor was better than the rope, but not much better, and the last of it wore through as I reached Magadi in mid-afternoon, having made not less than fifteen trips beneath the car, in terrific heat, measuring my length in the fine volcanic ash that a hellish wind impacted in hair, lungs, and fingernails. The kind Asian manager of the Magadi Store and his driver-mechanic replaced the sheared bevel pinion with an ingenious makeshift rig that would see the car safely to Nairobi, but all of this still lay ahead as I stood there looking as stupid as I felt under the gaze of that young herdsman by the shores of Natron.

* * *

Since the disabled car was inland from the lake, it seemed best to walk the last mile to the flats, to greet Iain and Oria and to make certain that I was not overlooked; already I was listening for the droning of the motor that would draw to a point the misty distances down toward Lengai. Accompanied by the morani, I followed a cattle trail between the marsh and a thorny rock strangely swollen by thick pink blossoms of the desert rose. Near the mouth of the Uaso Ngiro, green reeds give way to open flats where the Natron leaves a crust crisscrossed by ostrich tracks. Here the young warrior, mounting the rock, made a grand sweeping gesture of his cape toward the horizons of Maasai Land, and sighed with all his being. The red and blue beads swinging from his ears stood for sun and water, but now the sun was out of balance with the rain, and the grass was thin. The Maasai speak of the benevolent Black God who brings rain, and the malevolent Red God who begrudges it, the Black God living in dark thunderheads and the Red in the merciless dry-season sun; Black God and Red are different tempers of Ngai, for God is embodied in the rain and the fierce heat, besides ruling the great pastures of the sky. Looming thunder is feared: the Red God seeks to pierce the Black God's kingdoms, in hope of bringing harm to man. But in distant thunder the Maasai hear the Black God saying, "Let man be. . . ."

From where we stood, awed by the view, white flats extended a half-mile to the water's edge, where the heat waves rose in a pink fire of thousands upon thousands of flamingos. All around the north end of the lake the color shimmered, and for some distance down both shores; on the west shore, under the dark Sonjo escarpments, an upside-down forest was reflected. Southeast, the outline of Gelai was a phantom mountain in an amorphous sky, and in the south, the lake vanished in brown vapors that shrouded Ol Doinyo Lengai.

In this somber kingdom of day shadows and dead smokes, the fresh pinks of flamingos and the desert rose appeared unnatural. What belonged here were those tracks of giant birds, like black crosses in the crystalline white soda, and this petrified white bone dung of hyena, and the hieroglyph of a gazelle in quest of salt that had followed some dim impulse far out onto the flats. I remembered the Grant's gazelles on the Chalbi Desert, and the rhino that had climbed Lengai, and the wildebeest at a dead halt for want of impulse, in the shimmer of the soda lake, at noon. What drives such animals away from life-giving conditions into the wasteland—what happens in those rigid clear-eyed heads? How did the hippopotamus find its way up into the Crater Highlands, to blunder into the waters of Ngorongoro? Today one sees them there with wonder, encircled by steep walls, and the mystery deepens when a fish eagle plummets to the springs east of the lake and rises once more against the sky, in its talons a gleam of unknown life from the volcano.

We walked out into the silence of the flats. Somewhere on the mud, our footprints crossed the border of Tanzania, for Natron lies entirely in that country. I listened for the airplane but there was nothing, only the buzzing of these birds that fed with their queer heads upside down, straining diatoms and algae from the stinking waters even as they squirted it with the guano that kept the algae reproducing—surely one of the shortest and most efficient life chains in all nature, at once exhilarating and oppressive in the mindlessness of such blind triumphal life in a place so poisonous and dead. A string of flamingos rose from the pink gases, restoring sharpness to the sky, then sank again into the oblivion of their millions.

Twilight was coming. The boy pointed to a far en-gang under Shombole. *"Aia,"* he said, by way of parting—So be it—and stalked away in fear of the African night, his red cape darkening against the white. *"Aia,"* I said, watching him go. Soon he vanished under the volcano. This age-set of moran may be the last, for the Maasai of Kenya, upset at being left behind by tribes they once considered worthless, voted this year to discontinue the moran system and send young Maasai to school. But in Maasai Land all change comes slowly, whether in Kenya or Tanzania. The month before, in the region of Ol Alilal, in the Crater Highlands, there was a new age-set of circumcised boys dressed in the traditional black garments bound with broad bead belts and wearing the spectral white paint around the eyes that signifies death and rebirth as a man, and on their shaved heads, arranged on a wood frame that looked from afar like an informal halo, black ostrich plumes danced in the mountain wind. When their hair grew out again, the boys would be young warriors, perhaps the last age-set of moran.

One of the Ol Alilal moran was very sick, and we took him in to the government dispensary at Nainokanoka. This tall boy of seventeen or eighteen could no longer walk; I carried his light body in my arms to the dark shack where to judge from his face, he thought that he would die. Yet here at least he had a chance that he might not have had at Ol Alilal. Though the Maasai have little faith in witchcraft, they recognize ill provenance and evil spirits, and a person dying is removed outside the fences so that death will not bring the village harm. Eventually the body is taken to the westward toward the setting sun, and laid on its left side with knees drawn up, head to the north and face to the east, right arm crossing the breast and left cushioning the head. There it is left to be dealt with by hyenas. Should someone die inside a hut, then the whole village must be moved, and it is said that the people listen for the howl of the hyena, and establish the new village in that direction. The Maasai are afraid of death, though not afraid to die.

For a long time I stood motionless on the white desert, numbed by these lowering horizons so oblivious of man, understanding at last the stillness of the lone animals that stand transfixed in the distances of Africa. Perhaps because I was alone, and therefore more conscious of my own insignificance under the sky, and aware, too, that the day was dying, and that the airplane would not appear, I felt overwhelmed by the age and might of this old continent, and drained of strength: all seemed pointless in such emptiness, there was nowhere to go. I wanted to lie flat out on my back on this almighty mud, but instead I returned slowly into Kenya, pursued by the mutter of primordial birds. The flamingo sound, rising and falling with the darkening pinks of the gathering birds, was swelling again like an oncoming rush of motley wings—birds, bats, ancient flying things, thick insects.

The galumphing splosh of a pelican, gathering tilapia from the freshwater mouth of the Uaso Ngiro, was the first sound to rise above the wind of the flamingos. Next came a shrill whooping of the herdsmen, hurrying the last cattle across delta creeks to the bomas in the foothills of Shombole. A Maasai came running from the hills to meet me, bearing tidings of two dangerous lions—"*Simba! Simba mbili!*"—that haunted this vicinity. He asked nothing of me except caution, and as soon as his warning was delivered, ran back a mile or more in the near-darkness to the shelter of his en-gang. Perhaps the earliest pioneers were greeted this way almost everywhere by the wild peoples—the thought was saddening, but his act had made me happy.

I built a fire and broiled the fresh beef I had brought for three, to keep it from going bad, and baked a potato in the coals, and fried tomatoes, and drank another beer, all the while keeping an eye out for bad lions. I also made tea and boiled two eggs for breakfast, to dispense with fire-making in the dawn. As yet I had no energy to think about tomorrow, much less attempt makeshift repairs; the cool of first light would be time enough for that. Moving slowly so as not to stir the heat, I brushed my teeth and rigged my bed roll and climbed out on the car roof, staring away over Lake Natron. I was careful to be quiet: the night has ears, as the Maasai say.

From the Crater Highlands rose the Southern Cross; the Pleiades, which the Maasai associate with rains, had waned in early June. July is the time of wind and quarrels, and now, in August, the grass was dry and dead. In August, September, and October, called the Months of Hunger, the people pin grass to their clothes in hope of rain, for grass is sign of prosperity and peace, but not until the Pleiades returned, and the southeast monsoon, would the white clouds come that bring the precious water. (The Mbugwe of the southern flats of Lake Manyara resort to rainmakers, and formerly, in time of drought, so it is said, would sacrifice an unblemished black bull, then an unblemished black man, and finally the rainmaker himself.)

The light in my small camp under Shombole was the one light left in all the world. Staring up at the black cone that filled the night sky to the east, I knew I would never climb it. There was a long hard day ahead with nothing certain at the end of it, and I had no heart for the climb alone, especially here in this sullen realm that had held me at such a distance. The ascent of Lengai and the descent into Embagai Crater had both been failures, and the great volcanoes of the Crater Highlands had remained lost in the clouds. At Natron, my friends had failed to come and my transport had broken down, and tomorrow I would make a slow retreat. And perhaps this came from the pursuit of some fleeting sense of Africa, seeking to fix in time the timeless, to memorize the immemorial, instead of moving gently, in awareness, letting the sign, like the crimson bird, become manifest where it would.

From where I watched, a sentinel in the still summer, there rose and fell the night highlands of two countries, from the Loita down the length of the Ngurumans to the Sonjo scarps that overlook Lake Natron. In the Loita, so the Maasai say, lives Enenauner, a hairy giant, one side flesh, the other stone, who devours mortal men lost in the forests; Enenauner carries a great club, and is heard tokking on trees as it moves along. A far hyena summoned the night feeders, and flamingos in crescent moved north across a crescent moon toward Naivasha and Nakuru. Down out of the heavens came their calls, a remote electric sound, as if in this place, in such immensities of silence, one had heard heat lightning.

Toward midnight, in the Sonjo Hills, there leapt up two sudden fires. Perhaps this was sign of the harvest festival, Mbarimbari, for these were not the grass fires that leap along the night horizons in the dry season; the twin flames shone like leopard's eyes from the black hills. At this time of year God comes to the Sonjo from Ol Doinyo Lengai, and a few of their ancient enemies, the Maasai, bring goats to be slaughtered at Mbarimbari, where they howl to Ngai for rain and children.

The Sonjo, isolated from the world, know that it is coming to an end. Quarrels and warfare will increase, and eventually the sky will be obscured by a horde of birds, then insect clouds, and finally a shroud of dust. Two suns will rise from the horizons, one in the east, one in the west, as a signal to man that the end of the world is near. At the ultimate noon, when the two suns meet at the top of the sky, the earth will shrivel like a leaf, and all will die.

From THE WIND BIRDS

* * * The restlessness of shorebirds, their kinship with the distance and swift seasons, the wistful signal of their voices down the long coastlines of the world make them, for me, the most affecting of wild creatures. I think of them as birds of wind, as "wind birds." To the traveler confounded by exotic birds, not to speak of exotic specimens of his own

The Wind Birds (New York: Viking, 1973).

kind, the voice of the wind birds may be the lone familiar note in a strange land, and I have many times been glad to find them; meeting a whimbrel one fine summer day of February in Tierra del Fuego, I wondered if I had not seen this very bird a half-year earlier, at home. The spotted and white-rumped sandpipers, the black-bellied and golden plovers are birds of Sagaponack, but the spotted sandpiper has cheered me with its jaunty teeter on the Amazon and high up in the Andes (and so has its Eurasian counterpart, the common sandpiper, on the White Nile and in Galway and in the far-off mountains of New Guinea); one bright noon at the Straits of Magellan, the white-rump passed along the shore in flocks. I have seen golden plover on Alaskan tundra and in the cane-fields of Hawaii, and heard the black-belly's wild call on wind-bright seacoast afternoons from Yucatán to the Great Barrier Reef.

The voice of the black-bellied plover carries far, a fluting, melancholy *toor-a-lee* or *pee-ur-ee* like a sea bluebird's, often heard before the bird is seen. In time of storm, it sometimes seems to be the only bird aloft, for, with its wing span of two feet or more, the black-bellied plover is a strong flier; circumpolar and almost cosmopolitan, it migrates down across the world from breeding grounds within the Arctic Circle. Yet as a wanderer it is rivaled by several shorebirds, not least of all the sanderling of the Sagaponack beach, which ranks with the great skua and the Arctic tern as one of the most far-flung birds on earth.

The sanderling is the white sandpiper or "peep" of summer beaches, the tireless toy bird that runs before the surf. Because of the bold role it plays in its immense surroundings, it is the one sandpiper that most people have noticed. Yet how few notice it at all, and few of the fewer still who recognize it will ever ask themselves why it is there or where it might be going. We stand there heedless of an extraordinary accomplishment: the diminutive creature making way for us along the beaches of July may be returning from an annual spring voyage which took it from central Chile to nesting grounds in northeast Greenland, a distance of eight thousand miles. One has only to consider the life force packed tight into that puff of feathers to lay the mind wide open to the mysteries—the order of things, the why and the beginning. As we contemplate that sanderling, there by the shining sea, one question leads inevitably to another, and all questions come full circle to the questioner, paused momentarily in his own journey under the sun and sky. ＊ ＊ ＊

NOEL PERRIN
b. 1927

Like fellow New Yorkers E. B. White and Edward Hoagland, Noel Perrin formed his bond with the New England landscape as an adult, when he came to Dartmouth College to teach in 1959. Shortly after moving to New Hampshire, he bought a hundred-acre farm in nearby Thetford Center, Vermont, where he has pursued a second career as a "sometime farmer." His four collections of essays—First Person Rural (1978), Second Person Rural (1980), Third Person Rural (1983), and Last Person Rural (1991)—exhibit a characteristic urbane wit and an unsentimental but warmly appreciative view of the pleasures of rural life: raising crops and animals, sugaring, the use of tools, small-town neighborliness, and the recurring seasons. In his later essays, there is an increased consciousness of and protest against the growing threats to his beloved countryside from urban sprawl, fossil fuel pollution, nuclear arms, and agribusiness.

PIG TALES

I. THE PIG AS ARTIST

Pigs get a bad press. Pigs are regarded as selfish and greedy—as living garbage pails. Pigs are the villains in George Orwell's *Animal Farm*. Pigs have little mean eyes.

There is truth in this account—not that it's entirely the fault of the pigs. For perhaps five thousand generations pigs have been deliberately bred to be gluttonous. Out of each litter men have picked the piglets that were least solicitous of their brothers and sisters when nursing; have set aside the shoats that hogged the trough most and were most thoroughly selfish; have butchered the adult pigs that showed any signs of moderation, and kept the ones that had the most ravenous appetites

Second Person Rural: More Essays of a Sometime Farmer (Boston: Godine, 1980).

and gained weight the fastest. This was their breeding stock. Do the same thing with human beings for five thousand generations, and it would be interesting to see what kind of people resulted.

The marvel is that pigs have been able to keep so many interesting character traits despite fifty centuries of such single-minded breeding. For example, I once had two pigs who were artists. At least, I think they were. At a minimum, they were ingenious carpenters.

These two I had bought as piglets, and they were my first livestock, my first year living on a farm. I treated them handsomely. They had a fenced pen forty feet square (it was going to be our garden the next year), and they had a little house. The house was an old wooden shipping crate I had found in the barn, to which I had added a peaked, wood-shingled roof. There was no need to give their house a roof of shining cedar shingles—I had just happened to find about a third of a square of unused shingles when I was cleaning the barn, and it amused me to make an elaborate house for the piglets.

While they were little, they paid the roof no attention at all—didn't seem to notice it. They slept in the house, close together in the hay, and otherwise they were out digging in their yard. But when they got to be about three months old, and were already good-sized shoats, they began to help each other climb up on the roof of the house. As far as I could tell, they did this for the same reason that Sir Edmund Hillary climbed Everest. The challenge interested them, and they liked the view. I minded slightly, because they tracked mud all over the clean shingles, but mostly I enjoyed seeing them up there, staring calmly into the distance.

Then one early-fall evening, when the pigs were about five months old, I came home and found the roof gone. Well, not the roof, but every single shingle, and most of the nails. The pigs had not only pulled them off, they had split them into pieces averaging four to five inches wide, and they had made a cedar shelf all along one side of their pen. A cedar shelf forty feet long. It was made of dirt raised about four inches above the ground level, and topped with a continuous row of split shingles. It must have taken them most of the day, especially since they had neatly removed the nails.

Clearly a cedar shelf had nothing to do with food, nor could it have been part of a plan to escape. By raising the ground level four inches on that side, they were in fact making escape harder, since pigs escape by burrowing.

I see only three possibilities. One is that the little sow (I had a barrow and a sow) wanted to improve the housekeeping arrangements, and got her brother to help her put up the shelf for storage. But as they never kept anything on it, I consider that unlikely. The second is that it had religious significance. It *was* the eastern side of the pen they chose, and the autumn sun rose right over that shelf. But as they never performed any ritual whatsoever, I discount that possibility, too. I think they were making a work of art.

The shelf lasted five days, and then one afternoon I came home to find it completely gone—the dirt as well as the shingles. The shingles were in a pile, ready for some further use later; the dirt had been merged back into the general garden. But, then, much twentieth-century art *is* ephemeral, by design. I think Bingo and Bacon weren't interested in a work for the ages, but in a Happening.

This was their masterpiece, but not their only work of art. They never did anything more with the shingles. (I eventually used them for kindling.) But in the two months more that they lived, they three times filled their troughs with mud. Each time, they got the mud exactly even with the top of the trough, and smoothed it until there was almost a patina. I can't say much for the color values, but the texture was stunning. Judging by the last trough especially, earth art is really more a pig form than a human one.

II. FREUDIAN PIGS

My next pair of pigs were also a brother and sister. They had no artistic leanings, but one of them at least had considerable psychological complexity. That year I was getting most of my pig food from a big commercial bakery. The bakery had a thrift shop, where you could buy store returns for half price. What didn't sell there was available to pig owners for two cents a pound. Before you could buy, you had to sign a pledge that you wouldn't feed any of it to human beings. (I broke the pledge almost immediately. My second or third trip, instead of the usual pallid bread, I got about fifty pounds of English muffins, and a hundred pounds of little pies and cakelets. I tried everything. Stale but edible.)

These two pigs had been named by my six-year-old daughter, and they were called simply Sow and Boar (pronounced boor). Boor was greedy even for a pig, and he normally allowed his sister no more than a third of the food. When I came back from the bakery with pies or muffins, he did his level best to keep her from getting anything at all.

What I should have done, of course, was to make another trough and put it on the far side of the pen. Boor would have tried hard to keep her away from both troughs, but since they would have been forty feet apart, I don't think he could have succeeded.

What I actually did was to play policeman. On pie-and-cake days, I would fill their trough and then stand there with a good-sized stick as the two pigs came hurtling in. Both would get their heads deep in the pastry. As soon as he was swallowing regularly, Boor would turn and give Sow a battering-ram blow with his head, and she would squeal and back up about three feet. Then keeping one small, mean eye on her, to make sure she didn't try to sneak back, he would set about eating the whole vast troughful himself.

At this point I would interfere. Since I am one of those who think that policemen should give warnings before they start making arrests, I would give Boor a light tap. That meant: Move over, boy, and let your sister have some. He would respond by speeding up his already rapid feeding pace.

After the warning, the summons. My next step was to give him a good whack in the ribs. He would then back up just long enough for Sow to race in for a bite—and then, looking like a pink animated tank, he would come thundering back, wham Sow out of the way with one mighty blow, and plunge his snout in. At this point he got two whacks. That caused him to retreat about ten feet, where he would stand with blazing eyes, cursing shrilly in his throat. As far as I could tell, the curses were aimed equally at Sow and me, as joint conspirators in the plot to do him out of his dinner.

After a minute he would control himself, though, and it was at this point that he became a Freudian pig. He had the sense to sublimate. First he would trot across the pen and take a good drink of water. Then he would lean against one of the fence posts, close his eyes, and begin swaying back and forth in such a way as to give himself a lovely scratch. His eyes weren't *really* closed, of course, because if I now turned to tiptoe away, he was back across the pen at rocket speed, and once again battering Sow with his head.

But if I stayed, he would finish his rub, have another leisurely drink of water, and then, setting that powerful head of his like a dozer blade, he would move deliberately into the center of the pen and start excavating. He put the full force of a strong personality into these excavations. I have known him to get so involved that I could actually leave, and be as much as thirty yards away, before he noticed and sped back to the trough. His craters and trenches were not art, they were only displacement, but they worked. Boar could deal creatively with the frustration of his deepest interests, which is something I can seldom manage myself.

III. THE DARING YOUNG PIG

One thinks of pigs, or at least *I* think of pigs, as essentially earth-bound creatures. They have massive bodies on small legs. They spend a lot of time lying down. Their escape technique is invariably the tunnel.

Once even half grown, they are terrified of heights. I have caught an escaped shoat of under two hundred pounds, because he had reached a retaining wall less than thirty inches high and stood trembling at the edge, afraid that if he went over he might break a leg.

It was for this reason that I was so long in figuring out how one of my next pair of pigs kept getting out.

At that time I was experimenting with a portable pigpen. The theory was that I would keep a pair of young pigs in it all summer, moving the pen about to wherever I wanted brier bushes or burdocks or milkweed dug up. Then in the fall, when we had harvested the big fenced garden, we would pop the pigs in there. They would plow and fertilize it for next year.

The theory was not a sound one. For one thing, there is not enough level land in Vermont for a portable pigpen to be tight at the bottom. For another, a pig-plowed garden looks more like a battlefield where both sides had heavy artillery than it does like a smooth planting surface. The one really successful period out of that whole summer and fall was the first three days after the pigs had entered the garden. Then, appearing and reappearing among the cornstalks and the tall weeds, they looked like a painting by Henri Rousseau.

My story, though, takes place the second week in May. I had just built the portable pigpen, and just put the piglets in it. They were exactly a month old, pink with black spots, slightly larger than footballs. The pen was one day old, made of rough-cut hemlock boards spaced two inches apart. I made it eight by five, and two and a half feet high. As good as a football field to pigs that size.

A couple of hours after I had put the pigs in, one of them was out and trotting on her little hooves over to my wife's flower beds. She wasn't too hard to catch. A little milk in a dish, and she came right over, docile as a kitten. I took a careful look around the bottom of the pen, saw there was a tiny dip in the land, decided she must have squeezed out there. I moved the pen to one of the few absolutely level spots we have. Twenty minutes later the same little pig was out again. Absolutely no sign of digging. There was no way she could slip through two-inch gaps between boards. Logic said she must therefore have gone over the top. Could she be climbing, using the gaps as footholds? It didn't seem likely. But she clearly *was* getting out, so I raised the sides one more board, this time not leaving a space. The pen was now three feet high—the same height as a fence for adult cattle. The top foot was solid smooth board.

It amazes me now that I could have thought I had the problem solved, but I did think it. I calmly went in to lunch, sure that the piglets were now secure.

Before lunch was on the table, the little sow was back in the flowers.

Being quite full of milk by now, she was harder to catch. In fact, it took two people. This time, though quite hungry myself, I decided to linger near the pen and see what her escape route was. I got behind a tree and waited.

In about two minutes she came sailing over the top of the pen, made a four-point landing, and trotted briskly off toward the peonies. She had jumped three times her own height.

I had to put a chicken-wire ceiling on the pen before she would stay in, and she had to make twenty or thirty jumps before she would believe the wire would hold her.

What if we had spent five thousand generations breeding pigs to jump, instead of to stuff themselves and lie down? By now the most sluggish hog among them would be able to enter barns by the second-story windows, and farming would be an even more picturesque business than it already is.

IV. PROTECTOR OF PIGS

Mickey Jamieson and I were on a pig-buying trip. We had a pile of burlap sacks in the back of the truck, some baler twine to tie them with, and a little money. Our destination was Mr. Harrington's barn in Sharon, Vermont, reputed to contain more pigs than any other structure within fifty miles. Neither of us had ever been in it, though we had both driven by in the spring, and been awed by the sight of thirty or forty brood sows out in the fields grazing.

Crossing Sharon Ridge on the way, we had stopped to visit with a farmer Mickey knew. Before we left, Mickey had parted with two ten-dollar bills and acquired two very small piglets. This was why one of the sacks in back was tightly tied, and had some tendency to move under its own power.

We pulled up in front of Harrington's barn, and turned off the engine. We did not get out. The reason was that an absolutely giant Saint Bernard had appeared from inside the barn. He came bounding up to the truck, put his front paws on the rear bumper (the truck sagged perceptibly), and shoved his head in the back. He seemed to be looking for something. When we tried opening a door, he raised his head and growled deep in his throat.

After a minute Mr. Harrington emerged from the barn. He too, came over and looked in the back of the truck. Then he smiled. 'He's worried you're hurting them piglets,' he said. 'He can smell 'em, and he can hear 'em squeaking. It upsets him. If he was sure it was something you're doing, you'd *never* get out of that truck.'

It turned out that the dog was one Mr. Harrington looked after for

some summer people. 'But he don't know that. He thinks he belongs
here. And he thinks God put him on earth to look after pigs.'

Inside, the barn was a pig buyer's dream. There was a seven-hundred-pound boar taking a nap on a pile of hay. Ranks of sows in stalls.
In a big space at one end, something like fifty piglets flashing around
like a school of minnows. Many of them the black-spotted kind that
Mickey and I both favored.

The dog paced with us as we walked up and down the barn with Mr.
Harrington. As we walked, Mr. Harrington told us stories about different
ways a Saint Bernard eases life for pigs. Some of the stories could only be
repeated in a bar toward the end of the evening (or in a pig barn at any
time). These were the ones about the big dog's failure to understand mating behavior in the pig world. A young sow of three hundred pounds is
sometimes coy and willing simultaneously. When that happened at
Harrington's, the Saint Bernard took her squeals for help as completely
serious. He would interpose his massive body between the loving couple.
He was even prepared to bite where it hurt most. More than once a fully
aroused seven-hundred-pound boar had found himself trotting meekly
back to his end of the barn, pretending that all he wanted was a quiet rest.

Some of the stories were about tourists. Harrington got a lot. Not
regular tourists, but country people from twenty or thirty miles away
who just wanted to see that many pigs in one place. He ran a kind of
all-pig zoo. In winter, when farmers have the most free time, he averaged a set of visitors a day.

One set consisted of a young couple and their four-year-old son. The
mother and the boy were nice enough, Harrington said, but the father
was what he called ugly. The boy kept asking questions, and the father
kept telling him to shut up. Soon the father announced that one more
question and the kid would be damn sorry he'd asked it. There was
silence. But after a few minutes the little boy turned and asked Mr.
Harrington if he could get in and play with the piglets.

The father must have been waiting for this, because his hand moved
back instantly, ready to cuff. The blow never struck, however. The dog
had been pacing behind, and when the hand swung back, he caught it
firmly and apparently rather painfully in his teeth. The whole party
stopped. The father looked back cautiously over his shoulder. 'What do
I do now?' he asked.

'Mister, I don't know,' Harrington answered. 'But I know there's one
thing you ain't going to do.'

Pigs may have small mean eyes. But I rather think I prefer life in a pig
barn to life out here in the big wide world.

URSULA K. LE GUIN
b. 1929

Ursula Le Guin is the author of such highly acclaimed works of science fiction and fantasy as The Left Hand of Darkness *(1969),* The Lathe of Heaven *(1971), and* The Earth-sea Trilogy *(1977); more recent works include* Buffalo Gals and Other Animal Presences *(1987) and* Dancing at the Edge of the World *(1989). Her nature writing, like her fiction, transcends the traditional genre boundaries to examine such subjects as the cultural roots of human nature, sexual politics, and the power of myth and dream. In the following essay, which describes the 1980 eruption of Mount Saint Helens near her home in Portland, Oregon, she explores how, in an age without communal myth, we try to reduce natural disasters to human scale by creating personal metaphors out of them.*

A VERY WARM MOUNTAIN

> An enormous region extending from north-central
> Washington to northeastern California and including
> most of Oregon east of the Cascades is covered by basalt
> lava flows. . . . The unending cliffs of basalt along the
> Columbia River . . . 74 volcanoes in the Portland
> area . . . A blanket of pumice that averages about 50 feet
> thick . . .
> —Roadside Geology of Oregon
> Alt and Hyndman, 1978.

Everybody takes it personally. Some get mad. Damn stupid mountain went and dumped all that dirty gritty glassy gray ash that flies like flour and lies like cement all over their roofs, roads, and rhododendrons. Now they have to clean it up. And the scientists are a real big help, all

Parabola, 5, no. 4, Fall 1980.

they'll say is we don't know, we can't tell, she might dump another load of ash on you just when you've got it all cleaned up. It's an outrage.

Some take it ethically. She lay and watched her forests being cut and her elk being hunted and her lakes being fished and fouled and her ecology being tampered with and the smoky, snarling suburbs creeping closer to her skirts, until she saw it was time to teach the White Man's Children a lesson. And she did. In the process of the lesson, she blew her forests to matchsticks, fried her elk, boiled her fish, wrecked her ecosystem, and did very little damage to the cities: so that the lesson taught to the White Man's Children would seem, at best, equivocal.

But everybody takes it personally. We try to reduce it to human scale. To make a molehill out of the mountain.

Some got very anxious, especially during the dreary white weather that hung around the area after May 18 (the first great eruption, when she blew 1300 feet of her summit all over Washington, Idaho, and points east) and May 25 (the first considerable ashfall in the thickly populated Portland area west of the mountain). Farmers in Washington State who had the real fallout, six inches of ash smothering their crops, answered the reporters' questions with polite stoicism; but in town a lot of people were cross and dull and jumpy. Some erratic behavior, some really weird driving. "Everybody on my bus coming to work these days talks to everybody else, they never used to." "Everybody on my bus coming to work sits there like a stone instead of talking to each other like they used to." Some welcomed the mild sense of urgency and emergency as bringing people together in mutual support. Some—the old, the ill—were terrified beyond reassurance. Psychologists reported that psychotics had promptly incorporated the volcano into their private systems; some thought they were controlling her, and some thought she was controlling them. Businessmen, whom we know from the Dow Jones Reports to be an almost ethereally timid and emotional breed, read the scare stories in Eastern newspapers and cancelled all their conventions here; Portland hotels are having a long cool summer. A Chinese Cultural Attaché, evidently preferring earthquakes, wouldn't come farther north than San Francisco. But many natives were irrationally exhilarated, secretly, heartlessly welcoming every steam-blast and earth-tremor: Go it, mountain!

Everybody read in the newspapers everywhere that the May 18 eruption was "five hundred times greater than the bomb dropped on Hiroshima." Some reflected that we have bombs much more than five hundred times more powerful than the 1945 bombs. But these are never mentioned in the comparisons. Perhaps it would upset people in

Moscow, Idaho or Missoula, Montana, who got a lot of volcanic ash dumped on them, and don't want to have to think, what if that stuff had been radioactive? It really isn't nice to talk about, is it. I mean, what if something went off in New Jersey, say, and *was* radioactive—Oh, stop it. That volcano's way out west there somewhere anyhow.

Everybody takes it personally.

I had to go into hospital for some surgery in April, while the mountain was in her early phase—she jumped and rumbled, like the Uncles in *A Child's Christmas in Wales,* but she hadn't done anything spectacular. I was hoping she wouldn't perform while I couldn't watch. She obliged and held off for a month. On May 18 I was home, lying around with the cats, with a ringside view: bedroom and study look straight north about forty-five miles to the mountain.

I kept the radio tuned to a good country western station and listened to the reports as they came in, and wrote down some of the things they said. For the first couple of hours there was a lot of confusion and contradiction, but no panic, then or later. Late in the morning a man who had been about twenty miles from the blast described it: "Pumice-balls and mud-balls began falling for about a quarter of an hour, then the stuff got smaller, and by nine it was completely and totally black dark. You couldn't see ten feet in front of you!" He spoke with energy and admiration. Falling mud-balls, what next? The main West Coast artery, I-5, was soon closed because of the mud and wreckage rushing down the Toutle River towards the highway bridges. Walla Walla, 160 miles east, reported in to say their street lights had come on automatically at about ten in the morning. The Spokane-Seattle highway, far to the north, was closed, said an official expressionless voice, "on account of darkness."

At one-thirty that afternoon, I wrote:

It has been warm with a white high haze all morning, since six A.M., when I saw the top of the mountain floating dark against yellow-rose sunrise sky above the haze.

That was, of course, the last time I saw or will ever see that peak.

Now we can see the mountain from the base to near the summit. The mountain itself is whitish in the haze. All morning there has been this long, cobalt-bluish drift to the east from where the summit would be. And about ten o'clock there began to be visible clots, like cottage cheese curds, above the summit. Now the eruption cloud is visible from the summit of the mountain till obscured by a cloud layer at about twice the height of the mountain, i.e., 25–30,000 feet. The eruption cloud is very solid-looking, like sculptured marble, a beautiful blue in the deep relief of baroque curls, sworls, curled-cloud-shapes—darkening towards the top—a wonderful color. One is aware of motion, but (being shaky, and looking through shaky binoculars) I don't actually see the carven-blue-sworl-shapes move. Like the shadow on a sundial. It is enormous. Forty-five miles away. It is so much bigger than the mountain itself. It is silent, from this distance. Enormous, silent. It looks not like anything earthy, from the earth, but it does not look like anything atmospheric, a natural cloud, either. The blue of it is storm-cloud blue but the shapes are far more delicate, complex, and immense than stormcloud shapes, and it has this solid look; a weightiness, like the capital of some unimaginable column—which in a way indeed it is, the pillar of fire being underground.

At four in the afternoon a reporter said cautiously, "Earthquakes are being felt in the metropolitan area," to which I added, with feeling, "I'll say they are!" I had decided not to panic unless the cats did. Animals are supposed to know about earthquakes, aren't they? I don't know what our cats know; they lay asleep in various restful and decorative poses on the swaying floor and the jiggling bed, and paid no attention to anything except dinner time. I was not allowed to panic.

At four-thirty a meteorologist, explaining the height of that massive, storm-blue pillar of cloud, said charmingly, "You must understand that the mountain is very warm. Warm enough to lift the air over it to 75,000 feet."

And a reporter: "Heavy mud flow on Shoestring Glacier, with continuous lightning." I tried to imagine that scene. I went to the television, and there it was. The radio and television coverage, right through, was splendid. One forgets the joyful courage of reporters and cameramen when there is something worth reporting, a real Watergate, a real volcano.

On the 19th, I wrote down from the radio, "A helicopter picked the logger up while he was sitting on a log surrounded by a mud flow." This rescue was filmed and shown on television: the tiny figure crouching hopeless in the huge abomination of ash and mud. I don't know if this man was one of the loggers who later died in the Emanuel Hospital burn center, or if he survived. They were already beginning to talk about the "killer eruption," as if the mountain had murdered with intent. Taking it personally . . . Of course she killed. Or did they kill themselves? Old Harry who wouldn't leave his lodge and his whiskey and his eighteen cats at Spirit Lake, and quite right too, at eighty-three; and the young cameraman and the young geologist, both up there on the north side on the job of their lives; and the loggers who went back to work because logging was their living; and the tourists who thought a volcano is like Channel Six, if you don't like the show you turn it off, and took their RVs and their kids up past the roadblocks and the reasonable warnings and the weary county sheriffs sick of arguing: they were all there to keep the appointment. Who made the appointment?

A firefighter pilot that day said to the radio interviewer, "We do what the mountain says. It's not ready for us to go in."

On the 21st I wrote:

> Last night a long, strange, glowing twilight; but no ash has yet fallen west of the mountain. Today, fine, gray, mild, dense Oregon rain. Yesterday afternoon we could see her vaguely through the glasses. Looking appallingly lessened—short, flat—That is painful. She was so beautiful. She hurled her beauty in dust clear to the Atlantic shore, she made sunsets and sunrises of it, she gave it to the western wind. I hope she erupts magma and begins to build herself again. But I guess she is still unbuilding. The Pres. of the U.S. came today to see her. I wonder if he thinks he is on her level. Of course he could destroy much more than she has destroyed if he took a mind to.

On June 4 I wrote:

> Could see her through the glasses for the first time in two weeks or so. It's been dreary white weather with a couple of hours sun in the afternoons.— Not the new summit, yet; that's always in the roil of cloud/plume. But both her long lovely flanks. A good deal of new snow has fallen on her (while we had rain), and her SW face is white, black, and gray, much seamed, in unfamiliar patterns.
> "As changeless as the hills—"
> Part of the glory of it is being included in an event on the geologic scale. Being enlarged. "I shall lift up mine eyes unto the hills," yes: "whence cometh my help."

In all the Indian legends dug out by newspaper writers for the occasion, the mountain is female. Told in the Dick-and-Jane style considered appropriate for popular reportage of Indian myth, with all the syllables hyphenated, the stories seem even more naive and trivial than myths out of context generally do. But the theme of the mountain as woman—first ugly, then beautiful, but always a woman—is consistent. The mapmaking whites of course named the peak after a man, an Englishman who took his title, Baron St. Helens, from a town in the North Country: but the name is obstinately feminine. The Baron is forgotten, Helen remains. The whites who lived on and near the mountain called it The Lady. Called her The Lady. It seems impossible not to take her personally. In twenty years of living through a window from her I guess I have never really thought of her as "it."

She made weather, like all single peaks. She put on hats of cloud, and took them off again, and tried a different shape, and sent them all skimming off across the sky. She wore veils: around the neck, across the breast: white, silver, silver-gray, gray-blue. Her taste was impeccable. She knew the weathers that became her, and how to wear the snow.

Dr. William Hamilton of Portland State University wrote a lovely piece for the college paper about "volcano anxiety," suggesting that the silver cone of St. Helens had been in human eyes a breast, and saying:

> St. Helens' real damage to us is not . . . that we have witnessed a denial of the trustworthiness of God (such denials are our familiar friends). It is the perfection of the mother that has been spoiled, for part of her breast has been removed. Our metaphor has had a mastectomy.
>
> At some deep level, the eruption of Mt. St. Helens has become a new metaphor for the very opposite of stability—for that greatest of twentieth-century fears—cancer. Our uneasiness may well rest on more elusive levels than dirty windshields.

This comes far closer to home than anything else I've read about the "meaning" of the eruption, and yet for me it doesn't work. Maybe it would work better for men. The trouble is, I never saw St. Helens as a breast. Some mountains, yes: Twin Peaks in San Francisco, of course, and other round, sweet California hills—breasts, bellies, eggs, anything maternal, bounteous, yielding. But St. Helens in my eyes was never part of a woman; she is a woman. And not a mother but a sister.

These emotional perceptions and responses sound quite foolish when written out in rational prose, but the fact is that, to me, the eruption was all mixed up with the women's movement. It may be silly but there it is; along the same lines, do you know any woman who wasn't rooting for Genuine Risk to take the Triple Crown? Part of my satisfaction and exultation at each eruption was unmistakably feminist solidarity. You men think you're the only ones can make a really nasty mess? You think you got all the firepower, and God's on your side? You think you run things? Watch this, gents. Watch the Lady act like a woman.

For that's what she did. The well-behaved, quiet, pretty, serene, domestic creature peaceably yielding herself to the uses of man all of a sudden said NO. And she spat dirt and smoke and steam. She blackened half her face, in those first March days, like an angry brat. She fouled herself like a mad old harridan. She swore and belched and farted, threatened and shook and swelled, and then she spoke. They heard her voice two hundred miles away. Here I go, she said. I'm doing my thing now. Old Nobodaddy you better JUMP!

Her thing turns out to be more like childbirth than anything else, to my way of thinking. But not on our scale, not in our terms. Why should she speak in our terms or stoop to our scale? Why should she bear any birth that we can recognize? To us it is cataclysm and destruction and deformity. To her—well, for the language for it one must go to the scientists or to the poets. To the geologists. St. Helens is doing exactly what she "ought" to do—playing her part in the great pattern of events perceived by that noble discipline. Geology provides the only time-scale large enough to include the behavior of a volcano without deforming it. Geology, or poetry, which can see a mountain and a cloud as, after all, very similar phenomena. Shelley's cloud can speak for St. Helens:

> I silently laugh
> At my own cenotaph . . .
> And arise, and unbuild it again.

So many mornings waking I have seen her from the window before any other thing: dark against red daybreak, silvery in summer light, faint above river-valley fog. So many times I have watched her at evening, the faintest outline in mist, immense, remote, serene: the center, the central stone. A self across the air, a sister self, a stone. "The stone is at the center," I wrote in a poem about her years ago. But the poem is impertinent. All I can say is impertinent.

When I was writing the first draft of this essay in California, on July 23, she erupted again, sending her plume to 60,000 feet. Yesterday, August 7, as I was typing the words "the 'meaning' of the eruption," I checked out the study window and there it was, the towering blue cloud against the quiet northern sky—the fifth major eruption. How long may her labor be? A year, ten years, ten thousand? We cannot predict what she may or might or will do, now, or next, or for the rest of our lives, or ever. A threat: a terror: a fulfillment. This is what serenity is built on. This unmakes the metaphors. This is beyond us, and we must take it personally. This is the ground we walk on.

EDWARD O. WILSON
b. 1929

One of the foremost authorities on social insects, E. O. Wilson gained notoriety in 1975 with the publication of Sociobiology: The New Synthesis, *an attempt to give a biological and evolutionary basis to social and individual behavior, including that of human beings. The controversy set off by the book and fueled by Wilson's subsequent publications (including* On Human Nature, *which won the Pulitzer Prize in 1978) remains unabated. His critics accuse him of a reductionist and mechanistic view of culture and morality, with dangerous eugenic and racist overtones, while his admirers claim that his work is Darwinian in scope and import and that, like his nineteenth-century predecessor, he is the target of philistines who object to his conclusions on ethical and emotional, rather than scientific, grounds. By contrast,* Biophilia: The Human Bond to Other Species *(1984) is a collection of personal humanistic essays written in elegant, often rapt prose.* Biophilia *is a term coined by Wilson which he defines as "the innate tendency to focus on life and lifelike processes." A central concept in the book, as in his* Consilience *(1998), is that humanistic values are strengthened, not weakened, by understanding their evolutionary origins. Wilson believes that our humanity is rooted in our kinship with other animals and that "the key to saving life is learning to love it more." The following selection from* Biophilia *argues, as does much of nature writing, new and old, that an ongoing synthesis of scientific and imaginative approaches to life is integral to full human understanding.*

THE BIRD OF PARADISE

Come with me now to another part of the living world. The role of science, like that of art, is to blend exact imagery with more distant

Biophilia: The Human Bond to Other Species (Cambridge, Mass.: Harvard University Press, 1984).

meaning, the parts we already understand with those given as new into larger patterns that are coherent enough to be acceptable as truth. The biologist knows this relation by intuition during the course of field work, as he struggles to make order out of the infinitely varying patterns of nature.

Picture the Huon Peninsula of New Guinea, about the size and shape of Rhode Island, a weathered horn projecting from the northeastern coast of the main island. When I was twenty-five, with a fresh Ph.D. from Harvard and dreams of physical adventure in far-off places with unpronounceable names, I gathered all the courage I had and made a difficult and uncertain trek directly across the peninsular base. My aim was to collect a sample of ants and a few other kinds of small animals up from the lowlands to the highest part of the mountains. To the best of my knowledge I was the first biologist to take this particular route. I knew that almost everything I found would be worth recording, and all the specimens collected would be welcomed into museums.

Three days' walk from a mission station near the southern Lae coast brought me to the spine of the Sarawaget range, 12,000 feet above sea level. I was above treeline, in a grassland sprinkled with cycads, squat gymnospermous plants that resemble stunted palm trees and date from the Mesozoic Era, so that closely similar ancestral forms might have been browsed by dinosaurs 80 million years ago. On a chill morning when the clouds lifted and the sun shone brightly, my Papuan guides stopped hunting alpine wallabies with dogs and arrows, I stopped putting beetles and frogs in bottles of alcohol, and together we scanned the rare panoramic view. To the north we could make out the Bismarck Sea, to the south the Markham Valley and the more distant Herzog Mountains. The primary forest covering most of this mountainous country was broken into bands of different vegetation according to elevation. The zone just below us was the cloud forest, a labyrinth of interlocking trunks and branches blanketed by a thick layer of moss, orchids, and other epiphytes that ran unbroken off the tree trunks and across the ground. To follow game trails across this high country was like crawling through a dimly illuminated cave lined with a spongy green carpet.

A thousand feet below, the vegetation opened up a bit and assumed the appearance of typical lowland rain forest, except that the trees were denser and smaller and only a few flared out into a circle of blade-thin buttresses at the base. This is the zone botanists call the mid-mountain forest. It is an enchanted world of thousands of species of birds, frogs, insects, flowering plants, and other organisms, many found nowhere else. Together they form one of the richest and most nearly pure segments of the Papuan flora and fauna. To visit the mid-mountain forest is to see life as it existed before the coming of man thousands of years ago.

The jewel of the setting is the male Emperor of Germany bird of paradise *(Paradisaea guilielmi)*, arguably the most beautiful bird in the world, certainly one of the twenty or so most striking in appearance. By moving quietly along secondary trails you might glimpse one on a lichen-encrusted branch near the tree tops. Its head is shaped like that of a crow—no surprise because the birds of paradise and crows have a close common lineage—but there the outward resemblance to any ordinary bird ends. The crown and upper breast of the bird are metallic oil-green and shine in the sunlight. The back is glossy yellow, the wings and tail deep reddish maroon. Tufts of ivory-white plumes sprout from the flanks and sides of the breast, turning lacy in texture toward the tips. The plume rectrices continue on as wirelike appendages past the breast and tail for a distance equal to the full length of the bird. The bill is blue-gray, the eyes clear amber, the claws brown and black.

In the mating season the male joins others in leks, common courtship arenas in the upper tree branches, where they display their dazzling ornaments to the more somberly caparisoned females. The male spreads his wings and vibrates them while lifting the gossamer flank plumes. He calls loudly with bubbling and flutelike notes and turns upside down on the perch, spreading the wings and tail and pointing his rectrices skyward. The dance then reaches a climax as he fluffs up the green breast feathers and opens out the flank plumes until they form a brilliant white circle around his body, with only the head, tail, and wings projecting beyond. The male sways gently from side to side, causing the plumes to wave gracefully as if caught in an errant breeze. Seen from a distance his body now resembles a spinning and slightly out-of-focus white disk.

This improbable spectacle in the Huon forest has been fashioned by millions of generations of natural selection in which males competed and females made choices, and the accouterments of display were driven to a visual extreme. But this is only one trait, seen in physiological time and thought about at a single level of causation. Beneath its plumed surface, the Emperor of Germany bird of paradise possesses an architecture culminating an ancient history, with details exceeding those that can be imagined from the naturalist's simple daylight record of color and dance.

Consider one such bird for a moment in the analytic manner, as an object of biological research. Encoded within its chromosomes is the developmental program that led with finality to a male *Paradisaea guilielmi*. The completed nervous system is a structure of fiber tracts more complicated than any existing computer, and as challenging as all the rain forests of New Guinea surveyed on foot. A microscopic study will someday permit us to trace the events that culminate in the electric commands carried by the efferent neurons to the skeletal-muscular system and reproduce, in part, the dance of the courting male. This machinery can be dissected and understood by proceeding to the level of the cell, to enzymatic catalysis, microfilament configuration, and active sodium transport during electric discharge. Because biology sweeps the full range of space and time, there will be more discoveries renewing the sense of wonder at each step of research. By altering the scale of perception to the micrometer and millisecond, the laboratory scientist parallels the trek of the naturalist across the land. He looks out from his own version of the mountain crest. His spirit of adventure, as well as personal history of hardship, misdirection, and triumph, are fundamentally the same.

Described this way, the bird of paradise may seem to have been turned into a metaphor of what humanists dislike most about science: that it reduces nature and is insensitive to art, that scientists are conquistadors who melt down the Inca gold. But bear with me a minute. Science is not just analytic; it is also synthetic. It uses artlike intuition and imagery. In the early stages, individual behavior can be analyzed to the level of genes and neurosensory cells, whereupon the phenomena have indeed been mechanically reduced. In the synthetic phase, though, even the most elementary activity of these biological units creates rich and subtle patterns at the levels of organism and society. The outer qualities of *Paradisaea guilielmi*, its plumes, dance, and daily life, are functional traits open to a deeper understanding through the exact description of their constituent parts. They can be redefined through the exact description of their constituent parts. They can be redefined as holistic properties that alter our perception and emotion in surprising and pleasant ways.

There will come a time when the bird of paradise is reconstituted by the synthesis of all the hard-won analytic information. The mind, bearing a newfound power, will journey back to the familiar world of seconds and centimeters. Once again the glittering plumage takes form and is viewed at a distance through a network of leaves and mist. Then we see the bright eye open, the head swivel, the wings extend. But the familiar motions are viewed across a far greater range of cause and effect. The species is understood more completely; misleading illusions have given way to light and wisdom of a greater degree. One turn of the cycle of intellect is then complete. The excitement of the scientist's search for the true material nature of the species recedes, to be replaced in part by the more enduring responses of the hunter and poet.

What are these ancient responses? The full answer can only be given through a combined idiom of science and the humanities, whereby the investigation turns back into itself. The human being, like the bird of paradise, awaits our examination in the analytic-synthetic manner. As always by honored tradition, feeling and myth can be viewed at a distance through physiological time, idiosyncratically, in the manner of traditional art. But they can also be penetrated more deeply than ever was possible in the prescientific age, to their physical basis in the processes of mental development, the brain structure, and indeed the genes themselves. It may even be possible to trace them back through time past cultural history to the evolutionary origins of human nature. With each new phase of synthesis to emerge from biological inquiry, the humanities will expand their reach and capability. In symmetric fashion, with each redirection of the humanities, science will add dimensions to human biology.

GARY SNYDER
b. 1930

Gary Snyder is one of the most versatile and vital presences in American letters. His association with Alan Ginsburg, Jack Kerouac, and Philip Whalen in the Beat Movement, his early significance as a conduit for the Zen tradition into twentieth-century poetry in English, and his respectful, informed interest in shamanism are just three of the areas in which he has both helped to renew our literary culture and influenced the larger orientation of our society. Turtle Island, which won the Pulitzer Prize in 1974, is one of his collections of poetry deserving special note — as the volume where his vision of authentic life on this continent found notably full expression. Rivers and Mountains Without End (1996) is a volume of poetry on which Snyder worked for almost four decades, and which also synthesizes many of his themes in a particularly forceful way. Because of Snyder's influence over so many years on the wilderness movement and on American thinking about the spiritual meaning of nature, his 1990 collection of essays The Practice of the Wild was immediately greeted with special excitement. The essays in this volume are dense and challenging, yet are also offered in the voice of an entrancing storyteller. In them, Snyder has made a major contribution to the tradition of American nature writing.

ANCIENT FORESTS OF THE FAR WEST

But ye shall destroy their altars, break their images, and cut down their groves.

EXODUS 34:13

AFTER THE CLEARCUT

We had a tiny dairy farm between Puget Sound and the north end of Lake Washington, out in the cut-over countryside. The bioregionalists call that part of northwestern Washington state "Ish" after the suffix that means "river" in Salish. Rivers flowing into Puget Sound are the Snohomish Skykomish, Samamish, Duwamish, Stillaguamish.

I remember my father dynamiting stumps and pulling the share out with a team. He cleared two acres and fenced it for three Guernseys. He built a two-story barn with stalls and storage for the cows below and chickens above. He and my mother planted fruit trees, kept geese, sold milk. Behind the back fence were the woods: a second-growth jungle of alder and cascara trees with native blackberry vines sprawling over the stumps. Some of the stumps were ten feet high and eight or ten feet in diameter at the ground. High up the sides were the notches the fallers had chopped in to support the steel-tipped planks, the springboards, they stood on while felling. This got them above the huge swell of girth at the bottom. Two or three of the old trees had survived—small ones by comparison—and I climbed those, especially one western red cedar (*xelpai'its* in Snohomish) that I fancied became my advisor. Over the years I roamed the second-growth Douglas fir, western hemlock, and cedar forest beyond the cow pasture, across the swamp, up a long slope, and into a droughty stand of pines. The woods were more of a home than home. I had a permanent campsite where I would sometimes cook and spend the night.

When I was older I hiked into the old-growth stands of the foothill valleys of the Cascades and the Olympics where the shade-tolerant skunk cabbage and devil's club underbrush is higher than your head and the moss carpets are a foot thick. Here there is always a deep aroma

The Practice of the Wild (San Francisco: North Point, 1990).

of crumbled wet organisms—fungus—and red rotten logs and a few bushes of tart red thimbleberries. At the forest edges are the thickets of salal with their bland seedy berries, the yellow salmonberries, and the tangles of vine-maples. Standing in the shade you look out into the burns and the logged-off land and see the fireweed in bloom.

A bit older, I made it into the high mountains. The snowpeaks were visible from near our place: in particular Mt. Baker and Glacier Peak to the north and Mt. Rainier to the south. To the west, across Puget Sound, the Olympics. Those unearthly glowing floating snowy summits are a promise to the spirit. I first experienced one of those distant peaks up close at fifteen, when I climbed Mt. Saint Helens. Rising at 3 A.M. at timberline and breaking camp so as to be on glacier ice by six; standing in the rosy sunrise at nine thousand feet on a frozen slope to the crisp tinkle of crampon points on ice—these are some of the esoteric delights of mountaineering. To be immersed in ice and rock and cold and upper space is to undergo an eery, rigorous initiation and transformation. Being above all the clouds with only a few other high mountains also in the sunshine, the human world still asleep under its gray dawn cloud blanket, is one of the first small steps toward Aldo Leopold's "think like a mountain." I made my way to most of the summits of the Northwest—Mt. Hood, Mt. Baker, Mt. Rainier, Mt. Adams, Mt. Stuart, and more—in subsequent years.

At the same time, I became more aware of the lowlands. Trucks ceaselessly rolled down the river valleys out of the Cascades loaded with great logs. Walking the low hills around our place near Lake City I realized that I had grown up in the aftermath of a clearcut, and that it had been only thirty-five or forty years since all those hills had been logged. I know now that the area had been home to some of the largest and finest trees the world has ever seen, an ancient forest of hemlock and Douglas fir, a temperate-zone rainforest since before the glaciers. And I suspect that I was to some extent instructed by the ghosts of those ancient trees as they hovered near their stumps. I joined the Wilderness Society at seventeen, subscribed to *Living Wilderness*, and wrote letters to Congress about forestry issues in the Olympics.

But I was also instructed by the kind of work done by my uncles, our neighbors, the workers of the whole Pacific Northwest. My father put me on one end of a two-man crosscut saw when I was ten and gave me the classic instruction of "don't ride the saw"—don't push, only pull—and I loved the clean swish and ring of the blade, the rhythm, the comradeship, the white curl of the wood that came out with the rakers, the ritual of setting the handles, and the sprinkle of kerosene (to dissolve pitch) on the blade and into the kerf. We cut rounds out of down logs to split for firewood. (Unemployed men during the Depression felled the tall cedar stumps left from the first round of logging to buck them into blanks and split them with froes for the hand-split cedar shake trade.) We felled trees to clear pasture. We burned huge brush-piles.

People love to do hard work together and to feel that the work is real; that is to say primary, productive, needed. Knowing and enjoying the skills of our hands and our well-made tools is fundamental. It is a tragic dilemma that much of the best work men do together is no longer quite right. The fine information on the techniques of hand-whaling and all the steps of the flensing and rendering described in *Moby Dick* must now, we know, be measured against the terrible specter of the extinction of whales. Even the farmer or the carpenter is uneasy: pesticides, herbicides, creepy subsidies, welfare water, cheap materials, ugly subdivisions, walls that won't last. Who can be proud? And our conservationist-environmentalist-moral outrage is often (in its frustration) aimed at the logger or the rancher, when the real power is in the hands of people who make unimaginably larger sums of money, people impeccably groomed, excellently educated at the best universities—male and female alike—eating fine food and reading classy literature, while orchestrating the investment and legislation that ruin the world. As I grew into young manhood in the Pacific Northwest, advised by a cedar tree, learning the history of my region, practicing mountaineering, studying the native cultures, and inventing the little rituals that kept my spirit sane, I was often supporting myself by the woodcutting skills I learned on the Depression stump-farm.

AT WORK IN THE WOODS

In 1952 and '53 I worked for the Forest Service as a lookout in the northern Cascades. The following summer, wanting to see new mountains, I applied to a national forest in the Mt. Rainier area. I had already made my way to the Packwood Ranger Station and purchased my summer's supply of lookout groceries when the word came to the district (from Washington, D.C.) that I should be fired. That was the McCarthy era and the Velde Committee hearings were taking place in Portland. Many of my acquaintances were being named on TV. It was the end of my career as a seasonal forestry worker for the government.

I was totally broke, so I decided to go back to the logging industry. I hitched east of the Oregon Cascades to the Warm Springs Indian Reservation and checked in with the Warm Springs Lumber Company. I had scaled timber here the summer of '51, and now they hired me on as a chokersetter. This is the lava plateau country south of the Columbia River and in the drainage of the Deschutes, up to the head-waters of the Warm Springs River. We were cutting old-growth Ponderosa Pine on the middle slopes of the east side, a fragrant open forest of massive straight-trunked trees growing on volcanic soils. The upper edge verged into the alpine life-zone, and the lower edge — far-ther and farther out into the desert — became sagebrush by degrees. The logging was under contract with the tribal council. The proceeds were to benefit the people as a whole.

11 August '54
Chokersetting today. Madras in the evening for beer. Under the shadow of Mt. Jefferson. Long cinnamon-colored logs. This is "pine" and it belongs to "Indians" — what a curious knotting-up. That these Indians & these trees, that coexisted for centuries, should suddenly be possessor and possessed. Our concepts to be sure.

I had no great problem with that job. Unlike the thick-growing Douglas fir rainforests west of the Cascades, where there are arguments for clearcutting, the drier pine forests are perfect for selective cutting. Here the slopes were gentle and they were taking no more than 40 per-cent of the canopy. A number of healthy mid-sized seed trees were left standing. The D8 Cats could weave their way through without barking the standing trees.

Chokersetting is part of the skidding operation. First into the woods are the timber cruisers who estimate the standing board feet and mark the trees. Then come the road-building Cats and graders. Right on their heels are the gypo fallers — gypos get paid for quantity produced rather than a set wage — and then comes the skidding crew. West-of-the-mountains skidding is typically a high-lead or skyline cable operation where the logs are yarded in via a cable system strung out of a tall spar tree. In the eastside pine forest the skidding is done with top-size Caterpillar tractors. The Cat pulls a crawler-tread "arch" trailer behind it with a cable running from the Cat's aft winch up and over the pulley-wheel at the top of the arch, and then down where the cable divides into three massive chains that end in heavy steel hooks, the butt-hooks. I was on a team of two that worked behind one Cat. It was a two-Cat show.

Each Cat drags the felled and bucked logs to the landing — where they are loaded on trucks — from its own set of skid trails. While it is dragging a load of logs in, the chokersetters (who stay behind up the skid trails) are studying the next haul. You pick out the logs you'll give the Cat next trip, and determine the sequence in which you'll hook them so they will not cross each other, flip, twist over, snap live trees down, hang up on stumps, or make other dangerous and complicating moves. Chokersetters should be light and wiry. I wore White's caulked logger boots with steel points like tiny weasel-fangs set in the sole. I was thus enabled to run out and along a huge log or up its slope with perfect footing, while looking at the lay and guessing the physics of its mass in motion. The Cat would be coming back up the skid trail dragging the empty choker cables and would swing in where I signaled. I'd pluck two or three chokers off the butt-hooks and drag the sixteen-foot cables behind me into the logs and brush. The Cat would go on out to the other chokersetter who would take off his cables and do the same.

As the Cat swung out and was making its turnaround, the chokersetters would be down in dirt and duff, ramming the knobbed end of the choker under the log, bringing it up and around, and hooking it into the sliding steel catch called a "bell" that would noose up on the log when the choker pulled taut. The Cat would back its arch into where I stood, holding up chokers. I'd hook the first "D"—the ring on the free end of the choker—over the butt-hook and send the Cat to the next log. It could swing ahead and pull alongside while I leaped atop another load and hung the next choker onto the butt-hook. Then the winch on the rear of the Cat would wind in, and the butts of the logs would be lifted clear of the ground, hanging them up in the arch between the two crawler-tread wheels.

> Stood straight
> > holding the choker high
> As the Cat swung back the arch
> > piss-firs falling,
> Limbs snapping on the tin hat
> > bright D caught on
> Swinging butt-hooks
> > ringing against cold steel.
>
> from *Myths and Texts*

The next question was, how would they fan out? My Cat-skinner was Little Joe, nineteen and just recently married, chewing plug and always joking. I'd give him the highball sign and at the same time run back out the logs, even as he started pulling, to leap off the back end. Never stand between a fan of lying logs, they say. When the tractor hauls out they might swing in and snap together—"Chokersetters lose their legs that way." And don't stand anywhere near a snag when the load goes out. If the load even lightly brushes it, the top of the snag, or the whole thing, might come down. I saw a dead schoolmarm (a tree with a crotch in its top third) snap off and fall like that, grazing the tin hat of a chokersetter called Stubby. He was lucky.

> The D8 tears through piss-fir
> Scrapes the seed-pine
>   chipmunks flee,
> A black ant carries an egg
> Aimlessly from the battered ground.
> Yellowjackets swarm and circle
> Above the crushed dead log, their home.
> Pitch oozes from barked
>   trees still standing,
> Mashed bushes make strange smells.
> Lodgepole pines are brittle.
> Camprobbers flutter to watch.

I learned tricks, placements, pulls from the experienced choker-setters—ways to make a choker cable swing a log over, even to make it jump out from under. Ways and sequences of hooking on chokers that when first in place looked like a messy spiderweb, but when the Cat pulled out, the tangle of logs would right itself and the cables mysteriously fan out into a perfect pull with nothing crossed. We were getting an occasional eight-foot-diameter tree and many five and six footers: these were some of the most perfect ponderosa pine I have ever seen. We also had white fir, Douglas fir, and some larch.

I was soon used to the grinding squeaking roar and rattle of the Cat, the dust, and the rich smells that rose from the bruised and stirred-up soil and plant life. At lunchtime, when the machinery was silent, we'd see deer picking their way through the torn-up woods. A black bear kept breaking into the crummy truck to get at the lunches until someone shot him and the whole camp ate him for dinner. There was no rancor about the bear, and no sense of conquest about the logging work. The men were stoic, skillful, a bit overworked, and full of terrible (but funny!) jokes and expressions. Many of them were living on the Rez, which was shared by Wasco, Wishram, and Shoshone people. The lumber company gave priority to the Native American locals in hiring.

> Ray Wells, a big Nisqually, and I
>   each set a choker
> On the butt-logs of two big Larch
> In a thornapple thicket and a swamp.
>   waiting for the Cat to come back,
> "Yesterday we gelded some ponies

"My father-in-law cut the skin on the balls
"He's a Wasco and don't speak English
"He grabs a handful of tubes and somehow
                    cuts the right ones.
"The ball jumps out, the horse screams
"But he's all tied up.
The Caterpillar clanked back down.
In the shadow of that racket
                    diesel and iron tread
I thought of Ray Wells' tipi out on the sage flat
The gelded ponies
Healing and grazing in the dead white heat.

There were also old white guys who had worked in the lumber industry all their lives: one had been active in the Industrial Workers of the
World, the "Wobblies," and had no use for the later unions. I told him
about my grandfather, who had soapboxed for the Wobblies in Seattle's
Yesler Square, and my Uncle Roy, whose wife Anna was also the chief
cook at a huge logging camp at Gray's Harbor around World War I. I told
him of the revived interest in anarchosyndicalism among some circles in
Portland. He said he hadn't had anyone talk Wobbly talk with him in
twenty years, and he relished it. His job, knotbumper, kept him at the
landing where the skidding Cats dropped the logs off. Although the buckers cut the limbs off, sometimes they left stubs which would make the logs
hard to load and stack. He chopped off stubs with a double-bitted axe. Ed
had a circular wear-mark impressed in the rear pocket of his stagged jeans:
it was from his round axe-sharpening stone. Between loads he constantly
sharpened his axe, and he could shave a paper-thin slice off a Day's Work
plug, his chew, with the blade.

Ed McCullough, a logger for thirty-five years
Reduced by the advent of chainsaws
To chopping off knots at the landing:
"I don't have to take this kind of shit,
Another twenty years
                    and I'll tell 'em to shove it"
                    (he was sixty-five then)
In 1934 they lived in shanties
At Hooverville, Sullivan's Gulch.
When the Portland-bound train came through
The trainmen tossed off coal.

"Thousands of boys shot and beat up
For wanting a good bed, good pay,
                    decent food, in the woods—"

No one knew what it meant:
"Soldiers of Discontent."

On one occasion a Cat went to the landing pulling only one log, and not the usual 32-foot length but a 16. Even though it was only half-length the Cat could barely drag it. We had to rig two chokers to get around it, and there was not much pigtail left. I know now that the tree had been close to being of record size. The largest Ponderosa Pine in the world, near Mt. Adams, which I went out some miles of dust dirt roads to see, isn't much larger around than was that tree.

How could one not regret seeing such a massive tree go out for lumber? It was an elder, a being of great presence, a witness to the centuries. I saved a few of the tan free-form scales from the bark of that log and placed them on the tiny altar I kept on a box by my bunk at the logging camp. It and the other offerings (a flicker feather, a bit of broken bird's-egg, some obsidian, and a postcard picture of the Bodhisattva of Transcendent Intelligence, Manjusri) were not "my" offerings to the forest, but the forest's offerings to all of us. I guess I was just keeping some small note of it.

All of the trees in the Warm Springs forest were old growth. They were perfect for timber, too, most of them rot-free. I don't doubt that the many seed-trees and smaller trees left standing have flourished, and that the forest came back in good shape. A forester working for the Bureau of Indian Affairs and the tribal council had planned that cut.

Or did it come back in good shape? I don't know if the Warm Springs timber stands have already been logged again. They should not have been, but—

There was a comforting conservationist rhetoric in the world of forestry and lumber from the mid-thirties to the late fifties. The heavy clearcutting that has now devastated the whole Pacific slope from the Kern River to Sitka, Alaska, had not yet begun. In those days forestry professionals still believed in selective logging and actually practiced sustained yield. Those were, in hindsight, the last years of righteous forest management in the United States.

EVERGREEN

The raw dry country of the American West had an odd effect on American politics. It transformed and even radicalized some people. Once the West was closed to homesteading and the unclaimed lands became public domain, a few individuals realized that the future of these lands was open to public discussion. Some went from exploration and appreciation of wilderness to political activism.

Daoist philosophers tell us that surprise and subtle instruction might come forth from the Useless. So it was with the wastelands of the American West—inaccessible, inhospitable, arid, and forbidding to the eyes of most early Euro-Americans. The Useless Lands became the dreaming place of a few nineteenth- and early-twentieth-century men and women (John Wesley Powell on matters of water and public lands, Mary Austin on Native Americans, deserts, women) who went out into the space and loneliness and returned from their quests not only to criticize the policies and assumptions of the expanding United States but, in the name of wilderness and the commons, to hoist the sails that are filling with wind today. Some of the newly established public lands did have potential uses for lumber, grazing, and mining. But in the case of timber and grass, the best lands were already in private hands. What went into the public domain (or occasionally into Indian reservation status) was—by the standards of those days—marginal land. The off-limits bombing ranges and nuclear test sites of the Great Basin are public domain lands, too, borrowed by the military from the BLM.

So the forests that were set aside for the initial Forest Reserves were not at that time considered prime timber land. Early-day lumber interests in the Pacific Northwest went for the dense, low-elevation conifer forests like those around the house I grew up in or those forests right on saltwater or near rivers. This accessible land, once clearcut, became real estate, but the farther reaches were kept by the big companies as commercial forest. Much of the Olympic Peninsula forest land is privately held. Only by luck and chance did an occasional low-elevation stand such as the Hoh River forest in Olympic National Park, or Jedediah Smith redwoods in California, end up in public domain. It is by virtue of these islands of forest survivors that we can still see what the primeval forest of the West Coast—in its densest and most concentrated incarnation—was like. "Virgin forest" it was once called, a telling term. Then it was called "old growth" or in certain cases "climax." Now we begin to call it "ancient forest."

On the rainy Pacific slope there were million-acre stands that had been coevolving for millennia, possibly for over a million years. Such forests are the fullest examples of ecological process, containing as they do huge quantities of dead and decaying matter as well as the new green and preserving the energy pathways of both detritus and growth. An ancient forest will have many truly large old trees—some having craggy, broken-topped, mossy "dirty" crowns with much organic accumulation, most with holes and rot in them. There will be standing snags and tons of dead down logs. These characteristics, although not delightful to lumbermen ("overripe"), are what make an ancient forest more than a stand of timber: it is a palace of organisms, a heaven for many beings, a temple where life deeply investigates the puzzle of itself. Living activity goes right down to and under the "ground"—the litter, the duff. There are termites, larvae, millipedes, mites, earthworms, springtails, pillbugs, and the fine threads of fungus woven through. "There are as many as 5,500 individuals (not counting the earthworms and nematodes) per square foot of soil to a depth of 13 inches. As many as 70 different species have been collected from less than a square foot of rich forest soil. The total animal population of the soil and litter together probably approaches 10,000 animals per square foot" (Robinson, 1988, 87).

The dominant conifers in this forest, Douglas fir, western red cedar, western hemlock, noble fir, Sitka spruce, and coastal redwood, are all long-lived and grow to great size. They are often the longest-lived of their genera. The old forests of the western slopes support some of the highest per-acre biomass—total living matter—the world has seen, approached only by some of the Australian eucalyptus forests. An old-growth temperate hardwood forest, and also the tropical forests, average around 153 tons per acre. The west slope forests of the Oregon Cascades averaged 433 tons per acre. At the very top of the scale, the coastal redwood forests have been as high as 1,831 tons per acre (Waring and Franklin, 1979).

Forest ecologists and paleoecologists speculate on how such a massive forest came into existence. It seems the western forest of twenty or so million years ago was largely deciduous hardwoods—ash, maple, beech, oak, chestnut, elm, gingko—with conifers only at the highest elevations. Twelve to eighteen million years ago, the conifers began to occupy larger areas and then made continuous connection with each other along the uplands. By a million and a half years ago, in the early Pleistocene, the conifers had completely taken over and the forest was essentially as it is now. Forests of the type that had prevailed earlier, the hardwoods, survive today in the eastern United States and were also the original vegetation (before agriculture and early logging) of China and Japan. Visiting Great Smoky Mountains National Park today might give you an idea of what the mountain forests outside the old Chinese capital of Xian, known earlier as Ch'ang-an, looked like in the ninth century.

In the other temperate-zone forests of the world, conifers are a secondary and occasional presence. The success of the West Coast conifers can be attributed, it seems, to a combination of conditions: relatively cool and quite dry summers (which do not serve deciduous trees so well) combined with mild wet winters (during which the conifers continue to photosynthesize) and an almost total absence of typhoons. The enormous size of the trunks helps to store moisture and nutrients against drought years. The forests are steady-growing and productive (from a timber standpoint) while young, and these particular species keep growing and accumulating biomass long after most other temperate-zone trees have reached equilibrium.

Here we find the northern flying squirrel (which lives on truffles) and its sacred enemy the spotted owl. The Douglas squirrel (or chickaree) lives here, as does its sacred enemy the treetop-dashing pine marten that can run a squirrel down. Black bear seeks the grubs in long-dead logs in her steady ambling search. These and hosts of others occupy the deep shady stable halls—less wind, less swing of temperature, steady moisture—of the huge tree groves. There are treetop-dwelling red-backed voles who have been two hundred feet high in the canopy for hundreds of generations, some of whom have never descended to the ground (Maser, 1989). In a way the web that holds it all together is the mycelia, the fungus-threads that mediate between root-tips of plants and chemistry of soils, bringing nutrients in. This association is as old as plants with roots. The whole of the forest is supported by this buried network.

The forests of the maritime Pacific Northwest are the last remaining forests of any size left in the temperate zone. Plato's *Critias* passage (¶III) says: "In the primitive state of the country [Attica] its mountains were high hills covered with soil . . . and there was abundance of wood in the mountains. Of this last the traces still remain, for although some of the mountains now only afford sustenance to bees, not so very long ago there were still to be seen roofs of timber cut from trees growing there . . . and there were many other high trees. . . . Moreover the land reaped the benefit of the annual rainfall, not as now losing the water which flows off the bare earth into the sea." The cautionary history of the Mediterranean forests is well known. Much of this destruction has taken place in recent centuries, but it was already well under way, especially in the lowlands, during the classical period. In neolithic times the whole basin had perhaps 500 million acres of forest. The higher-elevation forests are all that survive, and even they occupy only 30 percent of the mountain zone—about 45 million acres. Some 100 million acres of land once densely covered with pine, oak, ash, laurel, and myrtle now have only traces of vegetation. There is a more sophisticated vocabulary in the Mediterranean for postforest or nonforest plant communities than we have in California (where everything scrubby is called chaparral). *Maquis* is the term for oak, olive, myrtle, and juniper scrub. An assembly of low waxy drought-resistant shrubs is called *garrigue*. *Batha* is open bare rock and eroding ground with scattered low shrubs and annuals.

People who live there today do not even know that their gray rocky hills were once rich in groves and wildlife. The intensified destruction was a function of the *type* of agriculture. The small self-sufficient peasant farms and their commons began to be replaced by the huge slave-run *latifundia* estates owned in absentia and planned according to central markets. What wildlife was left in the commons might then be hunted out by the new owners, the forest sold for cash, and field crops extended for what they were worth. "The cities of the Mediterranean littoral became deeply involved in an intensive region-wide trade, with cheap manufactured products, intensified markers and factory-like industrial production. . . . These developments in planned colonization, economic planning, world currencies and media for exchange had drastic consequences for the natural vegetation from Spain through to India" (Thirgood, 1981, 29).

China's lowland hardwood forests gradually disappeared as agriculture spread and were mostly gone by about thirty-five hundred years ago. (The Chinese philosopher Meng-zi commented on the risks of clearcutting in the fourth century B.C.) The composition of the Japanese forest has been altered by centuries of continuous logging. The Japanese sawmills are now geared down to about eight-inch logs. The original deciduous hardwoods are found only in the most remote mountains. The prized aromatic hinoki (the Japanese chamaecypress), which is essential to shrine and temple buildings, is now so rare that logs large enough for renovating traditional structures must be imported from the West Coast. Here it is known as Port Orford cedar, found only in southern Oregon and in the Siskiyou Mountains of northern California. It was used for years to make arrow shafts. Now Americans cannot afford it. No other softwood on earth commands such prices as the Japanese buyers are willing to pay for this species.

Commercial West Coast logging started around the 1870s. For decades it was all below the four-thousand-foot level. That was the era of the two-man saw, the double-bitted axe-cut undercuts, springboards, the kerosene bottle with a hook wired onto it stuck in the bark. Gypo hand-loggers felled into the saltwater bays of Puget Sound and rafted their logs to the mills. Then came steam donkey-engine yarders and ox teams, dragging the huge logs down corduroy skidroads or using immense wooden logging wheels that held the butt end aloft as the tail of the log dragged. The ox teams were replaced by narrow-gauge trains, and the steam donkeys by diesel. The lower elevations of the West Coast were effectively totally clearcut.

Chris Maser (1989, xviii) says: "Every increase in the technology of logging and the utilization of wood fiber has expedited the exploitation of forests; thus from 1935 through 1980 the annual volume of timber cut has increased geometrically by 4.7% per year. . . . By the 1970s, 65% of the timber cut occurred above 4,000 feet in elevation, and because the average tree harvested has become progressively younger and smaller, the increase in annual acreage cut has been five times greater than the increase in volume cut during the last 40 years."

During these years the trains were replaced by trucks, and the high-lead yarders in many cases were replaced by the more mobile crawler-tread tractors we call Cats. From the late forties on, the graceful, musical Royal Chinook two-man falling saws were hung up on the walls of the barns, and the gasoline chainsaw became the faller's tool of choice. By the end of World War II the big logging companies had (with a few notable exceptions) managed to overexploit and mismanage their own timberlands and so they now turned to the federal lands, the people's forests, hoping for a bailout. So much for the virtues of private forest landowners—their history is abysmal—but there are still ill-informed privatization romantics who argue that the public lands should be sold to the highest bidders.

> San Francisco 2 x 4s
> were the woods around Seattle:
> Someone killed and someone built, a house,
> a forest, wrecked or raised
> All America hung on a hook
> & burned by men in their own praise.

Before World War II the U.S. Forest Service played the role of a true conservation agency and spoke against the earlier era of clearcutting. It usually required its contractors to do selective logging to high standards. The allowable cut was much smaller. It went from 3.5 billion board feet in 1950 to 13.5 billion feet in 1970. After 1961 the new Forest Service leadership cosied up to the industry, and the older conservation-oriented personnel were washed out in waves through the sixties and seventies. The USFS now hires mostly road-building engineers. Their silviculturists think of themselves as fiber-growing engineers, and some profess to see no difference between a monoculture plantation of even-age seedlings and a wild forest (or so said Tahoe National Forest silviculturist Phil Aune at a public hearing on the management plan in 1986). The public relations people still cycle the conservation rhetoric of the thirties, as though the Forest Service had never permitted a questionable clearcut or sold old-growth timber at a financial loss.

The legislative mandate of the Forest Service leaves no doubt about its responsibility to manage the forest lands *as forests*, which means that lumber is only one of the values to be considered. It is clear that the forests must be managed in a way that makes them permanently sustainable. But Congress, the Department of Agriculture, and business combine to find ways around these restraints. *Renewable* is confused with *sustainable* (just because certain organisms keep renewing themselves does not mean they will do so—especially if abused—forever), and *forever*—the length of time a forest should continue to flourish—is changed to mean "about a hundred and fifty years." Despite the overwhelming evidence of mismanagement that environmental groups have brought against the Forest Service bureaucracy, it arrogantly and stubbornly resists what has become a clear public call for change. So much for the icon of "management" with its uncritical acceptance of the economic speed-trip of modern times (generating faster and faster logging rotations in the woods) as against: slow cycles.

We ask for slower rotations, genuine streamside protection, fewer roads, no cuts on steep slopes, only occasional shelterwood cuts, and only the most prudent application of the appropriate smaller clearcut. We call for a return to selective logging, and to all-age trees, and to serious heart and mind for the protection of endangered species. (The spotted owl, the fisher, and the pine marten are only part of the picture.) There should be *absolutely no more logging* in the remaining ancient forests. In addition we need the establishment of habitat corridors to keep the old-growth stands from becoming impoverished biological islands.

Many of the people in the U.S. Forest Service would agree that such practices are essential to genuine sustainability. They are constrained by the tight net of exploitative policies forced on them by Congress and industry. With good practices North America could maintain a lumber industry and protect a halfway decent amount of wild forest for ten thousand years. That is about the same number of years as the age of the continuously settled village culture of the Wei River valley in China, a span of time that is not excessive for humans to consider and plan by. As it is, the United States is suffering a net loss of 900,000 acres of forest per year (*Newsweek*, 2 October 1989). Of that loss, an estimated 60,000 acres is ancient forest (Wilson, 1989, 112).

The deep woods turn, turn, and turn again. The ancient forests of the West are still around us. All the houses of San Francisco, Eureka, Corvallis, Portland, Seattle, Longview, are built with those old bodies: the 2 x 4s and siding are from the logging of the 1910s and 1920s. Strip the paint in an old San Francisco apartment and you find prime-quality coastal redwood panels. We live out our daily lives in the shelter of ancient trees. Our great-grandchildren will more likely have to live in the shelter of riverbed-aggregate. Then the forests of the past will be truly entirely gone.

Out in the forest it takes about the same number of years as the tree lived for a fallen tree to totally return to the soil. If societies could learn to live by such a pace there would be no shortages, no extinctions. There would be clear streams, and the salmon would always return to spawn.

> A virgin
> Forest
> Is ancient; many-
> Breasted,
> Stable; at
> Climax.

EXCURSUS: SAILOR MEADOW, SIERRA NEVADA

We were walking in mid-October down to Sailor Meadow (about 5,800 feet), to see an old stand on a broad bench above the north fork of the American River in the northern Sierra Nevada. At first we descended a ridge-crest through chinquapin and manzanita, looking north to the wide dome of Snow Mountain and the cliffs above Royal Gorge. The faint trail leveled out and we left it to go to the stony hills at the north edge of the hanging basin. Sitting beneath a cedar growing at the top of the rocks we ate lunch.

Then we headed southwest over rolls of forested stony formations and eventually more gentle slopes into a world of greater and greater trees. For hours we were in the company of elders.

Sugar pines predominate. There are properly mature symmetrical trees a hundred and fifty feet high that hold themselves upright and keep their branches neatly arranged. But then *beyond* them, *above* them, loom the *ancient trees:* huge, loopy, trashy, and irregular. Their bark is redder and the plates more spread, they have fewer branches, and those surviving branches are great in girth and curve wildly. Each one is unique and goofy. Mature incense cedar. Some large red fir. An odd Douglas fir. A few great Jeffrey pine. (Some of the cedars have cat-face burn marks from some far back fire at their bases—all on the northwest side. None of the other trees show these burn marks.)

And many snags, in all conditions: some just recently expired with red or brown dead needles still clinging, some deader yet with plates of bark hanging from the trunk (where bats nest), some pure white smooth dead ones with hardly any limbs left, but with an occasional neat woodpecker hole; and finally the ancient dead: all soft and rotten while yet standing.

Many have fallen. There are freshly fallen snags (which often take a few trees with them) and the older fallen snags. Firm down logs you must climb over, or sometimes you can walk their length, and logs that crumble as you climb them. Logs of still another age have gotten soft and begun to fade, leaving just the pitchy heartwood core and some pitchy rot-proof limbs as signs. And then there are some long subtle hummocks that are the last trace of an old gone log. The straight line of mushrooms sprouting along a smooth ground surface is the final sign, the last ghost, of a tree that "died" centuries ago.

A carpet of young trees coming in—from six inches tall to twenty feet, all sizes—waiting down here on the forest floor for the big snags standing up there dead to keel over and make more canopy space. Sunny, breezy, warm, open, light—but the great trees are all around us. Their trunks fill the sky and reflect a warm golden light. The whole canopy has that sinewy look of ancient trees. Their needles are distinctive tiny patterns against the sky—the red fir most strict and fine.

The forests of the Sierra Nevada, like those farther up the West Coast, date from that time when the earlier deciduous hardwood forests were beginning to fade away before the spreading success of the conifers. It is a million years of "family" here, too, the particular composition of local forest falling and rising in elevation with the ice age temperature fluctuations, advancing or retreating from north and south slope positions, but keeping the several plant communities together even as the boundaries of their zones flowed uphill or down through the centuries. Absorbing fire, adapting to the summer drought, flowing through the beetle-kill years; always a web reweaving. Acorns feeding deer, manzanita feeding robins and raccoons, madrone feeding band-tailed pigeon, porcupine gnawing young cedar bark, bucks thrashing their antlers in the willows.

The middle-elevation Sierra forest is composed of sugar pine, ponderosa pine, incense cedar, Douglas fir, and at slightly higher elevations Jeffrey pine, white fir, and red fir. All of these trees are long-lived. The sugar pine and ponderosa are the largest of all pines. Black oak, live oak, tanbark oak, and madrone are the common hardwoods.

The Sierra forest is sunny-shady and dry for fully half the year. The loose litter, the crackliness, the dustiness of the duff, the curl of crisp madrone leaves on the ground, the little coins of fallen manzanita leaves. The pine-needle floor is crunchy, the air is slightly resinous and aromatic, there is a delicate brushing of spiderwebs everywhere. Summer forest: intense play of sun and the vegetation in still steady presence—not giving up water, not wilting, not stressing, just quietly holding. Shrubs with small, aromatic, waxy, tough leaves. The shrub color is often blue-gray.

The forest was fire-adapted over the millennia and is extremely resistant to wildfire once the larger underbrush has burnt or died away. The early emigrants described driving their wagons through parklike forest of great trees as they descended the west slope of the range. The early logging was followed by devastating fires. Then came the suppression of fires by the forest agencies, and that led to the brushy understory that is so common to the Sierra now. The Sailor Meadow forest is a spacious, open, fireproof forest from the past.

At the south end of the small meadow the area is named for, beyond a thicket of aspen, standing within a grove of flourishing fir, is a remarkably advanced snag. It once was a pine over two hundred feet tall. Now around the base all the sapwood has peeled away, and what's holding the bulky trunk up is a thin column of heartwood, which is itself all punky, shedding, and frazzled. The great rotten thing has a lean as well! Any moment it might go.

How curious it would be to die and then remain standing for another century or two. To enjoy "dead verticality." If humans could do it we would hear news like, "Henry David Thoreau finally toppled over." The human community, when healthy, is like an ancient forest. The little ones are in the shade and shelter of the big ones, even rooted in their lost old bodies. All ages, and all together growing and dying. What some silviculturists call for—"even-age management," plantations of trees the same size growing up together—seems like rationalistic utopian totalitarianism. We wouldn't think of letting our children live in regimented institutions with no parental visits and all their thinking shaped by a corps or professionals who just follow official manuals (written by people who never raised kids). Why should we do it to our forests?

"All-age unmanaged"—that's a natural community, human or other. The industry prizes the younger and middle-aged trees that keep their symmetry, keep their branches even of length and angle. But let there also be really old trees who can give up all sense of propriety and begin throwing their limbs out in extravagant gestures, dancelike poses, displaying their insouciance in the face of mortality, holding themselves available to whatever the world and the weather might propose. I look up to them: they are like the Chinese Immortals, they are Han-shan and Shi-de sorts of characters—to have lived that long is to have permission to be eccentric, to be the poets and painters among trees, laughing, ragged, and fearless. They make me almost look forward to old age.

In the fir grove we can smell mushrooms, and then we spot them along the base of rotten logs. A cluster of elegant polypores, a cortinarius, and in the open, pushing up dry needles from below, lots of russula and boletus. Some scooped-out hollows where the deer have dug them out. Deer love mushrooms.

We tried to go straight across the southern end of the meadow but it was squishy wet beneath the dry-looking collapsed dead plants and grasses, so we went all the way around the south end through more aspen and found (and saved) more mushrooms. Clouds started blowing in from the south and the breeze filled the sky with dry pine needles raining down. It was late afternoon, so we angled up steep slopes cross-country following deer-paths for an hour and found the overgrown trail to an abandoned mine, and it led us back to the truck.

US YOKELS

This little account of the great forests of the West Coast can be taken as a model of what has been happening elsewhere on the planet. All the natural communities of the world have been, in their own way, "ancient" and every natural community, like a family, includes the infants, the adolescents, the mature adults, the elders. From the corner of the forest that has had a recent burn, with its fireweed and blackberries, to the elder moist dark groves—this is the range of the integrity of the whole. The old stands of hoary trees (or half-rotten saguaro in the Sonoran Desert or thick-boled well-established old manzanita in the Sierra foothills) are the grandparents and information-holders of their communities. A community needs its elders to continue. Just as you could not grow culture out of a population of kindergarten children, a forest cannot realize its own natural potential without the seed-reservoirs, root-fungus threads, birdcalls, and magical deposits of tiny feces that are the gift from the old to the young. Chris Maser says, "We need ancient forests for the survival of ancient forests."

When the moldboard plows of the early midwestern farmers "cut the grass roots—a sound that reminded one of a zipper being opened or closed—a new way of life opened, which simultaneously closed, probably forever, a long line of ecosystems stretching back thirty million years" (Jackson, 1987, 78). But the oldest continuous ecosystems on earth are the moist tropical forests, which in Southeast Asia are estimated to date back one hundred million years.

> Thin arching buttressing boles of the white-barked tall
> straight trees, Staghorn ferns leaning out from the limbs
> and the crotches up high. Trees they call brushbox,

coachwood, crabapple, Australian red cedar (names
brought from Europe)—and Red carrabeen, Yellow
carrabeen, Stinging-trees, Deep blue openings leaning
onward.

Light of green arch of leaves far above
Drinking the water that flows through the roots
Of the forest, Terania creek, flowing out of Pangaia,
Down from Gondwanaland,
Stony soil, sky bottom shade

Long ago stone deep
Roots from the sky
Clear water down through the roots
Of the trees that reach high in the shade
Birdcalls bring us awake
Whiplash birdcalls laugh us awake—

Booyong, Carrabeen, Brushbox, Black butt, Wait-a-while
(Eucalypts dry land thin soil succeeders
Searching scrabbly ground for seventy million years—)

            But these older tribes of trees
            Travel always as a group.
            Looking out from the cliffs
            On the ridge above treetops,
            Sitting up in the dust ledge shelter
            Where we lived all those lives.
                              Queensland, 1981

A multitude of corporations are involved in the deforestation of the tropics. Some got their start logging in Michigan or the Pacific Northwest—
Georgia Pacific and Scott Paper are now in the Philippines, Southeast
Asia, or Latin America with the same bright-colored crawler tractors
and the buzzing yellow chainsaws. In the summer of 1987 in Brazil's
western territory of Rondonia—as part of the chaotic "conversion" of
Amazonia to other uses—an area of forest the size of Oregon was in
flames. One sometimes hears the innocent opinion that everyone is a
city-dweller now. That time may be coming, but at the moment the
largest single population in the world is people of several shades of
color farming in the warmer zones. Until recently a large part of that
realm was in trees, and the deep-forest dwelling cultures had diverse
and successful ways to live there. In those times of smaller population,
the long-rotation slash-and-burn style of farming mixed with foraging

posed no ecological threat. Today a combination of large-scale logging, agribusiness development, and massive dam projects threatens every corner of the backcountry.

In Brazil there is a complex set of adversaries. On one side the national government with its plans for development is allied with multinationals, wealthy cattle interests, and impoverished mainstream peasants. On the other side, resisting deforestation, are the public and private foresters and scientists making cause with the small local lumber firms, the established jungle-edge peasants, environmental organizations, and the forest-dwelling tribes. The Third World governments usually deny "native title" and the validity of communal forest ownership histories, such as the *adat* system of the Penan of Sarawak, a sophisticated multidimensional type of commons. The Penan people must put their bodies in the road to protest logging trucks *in their own homeland* and then go to jail as criminals.

Third World policies in regard to wilderness all too often run a direction set by India in 1938 when it opened the tribal forest lands of Assam to outside settlement saying "indigenous people alone would be unable, without the aid of immigrant settlers, to develop the province's enormous wasteland resources within a reasonable period" (Richards and Tucker, 1988, 107). All too many people in power in the governments and universities of the world seem to carry a prejudice against the natural world—and also against the past, against history. It seems Americans would live by a Chamber-of-Commerce Creationism that declares itself satisfied with a divinely presented Shopping Mall. The integrity and character of our own ancestors is dismissed with "I couldn't live like that" by people who barely know how to live *at all*. An ancient forest is seen as a kind of overripe garbage, not unlike the embarrassing elderly.

> Forestry. "How
> Many people
> Were harvested
> In Viet-nam?"
>
> Clear-cut. "Some
> Were children,
> Some were over-ripe."

The societies that live by the old ways (Snyder, 1977) had some remarkable skills. For those who live by foraging—the original forest botanists and zoologists—the jungle is a rich supply of fibers, poisons, medicines, intoxicants, detoxicants, containers, water-proofing, food, dyes, glues, incense, amusement, companionship, inspiration, and also stings, blows, and bites. These primary societies are like the ancient forests of our human history, with similar depths and diversities (and simultaneously "ancient" and "virgin"). The *lore* of wild nature is being lost along with the inhabitory human cultures. Each has its own humus of custom, myth, and lore that is now being swiftly lost—a tragedy for us all.

Brazil provides incentives for this kind of destructive development. Even as some mitigations are promised, there are policies in place that actively favor large corporations, displace natives, and at the same time do nothing for the mainstream poor. America disempowers Third World farmers by subsidizing overproduction at home. Capitalism plus big government often looks like welfare for the rich, providing breaks to companies that clearcut timber at a financial loss to the public. The largest single importer of tropical hardwoods is Japan (Mazda, Mitsubishi) and the second largest is the USA.

We must hammer on the capitalist economies to be at least capitalist enough to see to it that the corporations that buy timber off our public lands pay a fair market price for it. We must make the hard-boiled point that the world's trees are virtually worth more standing than they would be as lumber, because of such diverse results of deforestation as life-destroying flooding in Bangladesh and Thailand, the extinction of millions of species of animals and plants, and global warming. And, finally, we are not speaking only of forest-dwelling cultures or endangered species like voles or lemurs when we talk of ecological integrity and sustainability. We are looking at the future of our contemporary urban-industrial society as well. Not so long ago the forests were our depth, a sun-dappled underworld, an inexhaustible timeless source. Now they are vanishing. We are all endangered yokels. (*Yokel:* some English dialect, originally meaning "a green woodpecker or yellowhammer.")

JOHN McPHEE
b. 1931

A pioneer of the "new journalism," John McPhee first published most of his numerous books as essays in The New Yorker. *Though celebrated for the great variety of his subjects—ranging from the history of oranges to profiles of athletes to the development of experimental aircraft—a preponderance of his books focuses on natural or environmental topics.* The Pine Barrens *(1968) describes a large, little-known wild area not far from his home in Princeton, New Jersey.* Encounters with the Archdruid *(1971) is an in-depth study of radical conservationist David Brower.* Coming Into the Country *(1977) is a study of the Alaskan wilderness and the political issues surrounding it. Starting with* Basin and Range *(1981), he has written an acclaimed series of books based on his explorations with North American geologists. McPhee has been widely praised for his vivid descriptions, evocations of character, and uncanny ability to take difficult or arcane subjects and render them of compelling interest to the general reader. Like a good novelist, McPhee creates a strong point of view not by overtly expressed opinions but by the careful choice of fact and detail and the sheer possessive energy of his writing.*

UNDER THE SNOW

When my third daughter was an infant, I could place her against my shoulder and she would stick there like velvet. Only her eyes jumped from place to place. In a breeze, her bright-red hair might stir, but she would not. Even then, there was profundity in her repose.

When my fourth daughter was an infant, I wondered if her veins were full of ants. Placing her against a shoulder was a risk both to her and to the shoulder. Impulsively, constantly, everything about her moved. Her head seemed about to revolve as it followed the bestirring world.

Table of Contents (New York: Farrar, Straus and Giroux, 1985).

These memories became very much alive some months ago when—one after another—I had bear cubs under my vest. Weighing three, four, 5.6 pounds, they were wild bears, and for an hour or so had been taken from their dens in Pennsylvania. They were about two months old, with fine short brown hair. When they were made to stand alone, to be photographed in the mouth of a den, they shivered. Instinctively, a person would be moved to hold them. Picked up by the scruff of the neck, they splayed their paws like kittens and screamed like baby bears. The cry of a baby bear is muted, like a human infant's heard from her crib down the hall. The first cub I placed on my shoulder stayed there like a piece of velvet. The shivering stopped. Her bright-blue eyes looked about, not seeing much of anything. My hand, cupped against her back, all but encompassed her rib cage, which was warm and calm. I covered her to the shoulders with a flap of down vest and zipped up my parka to hold her in place.

I was there by invitation, an indirect result of work I had been doing nearby. Would I be busy on March 14th? If there had been a conflict—if, say, I had been invited to lunch on that day with the Queen of Scotland and the King of Spain—I would have gone to the cubs. The first den was a rock cavity in a lichen-covered sandstone outcrop near the top of a slope, a couple of hundred yards from a road in Hawley. It was on posted property of the Scrub Oak Hunting Club—dry hardwood forest underlain by laurel and patches of snow—in the northern Pocono woods. Up in the sky was Buck Alt. Not long ago, he was a dairy farmer, and now he was working for the Keystone State, with directional antennae on his wing struts angled in the direction of bears. Many bears in Pennsylvania have radios around their necks as a result of the summer trapping work of Alt's son Gary, who is a wildlife biologist. In winter, Buck Alt flies the country listening to the radio, crissing and crossing until the bears come on. They come on stronger the closer to them he flies. The transmitters are not omnidirectional. Suddenly, the sound cuts out. Buck looks down, chooses a landmark, approaches it again, on another vector. Gradually, he works his way in, until he is flying in ever tighter circles above the bear. He marks a map. He is accurate within two acres. The plane he flies is a Super Cub.

The den could have served as a set for a Passion play. It was a small chamber, open on one side, with a rock across its entrance. Between the freestanding rock and the back of the cave was room for one large bear, and she was curled in a corner on a bed of leaves, her broad head plainly visible from the outside, her cubs invisible between the rock and a soft place, chuckling, suckling, in the wintertime tropics of their own mammalian heaven. Invisible they were, yes, but by no means inaudible. What biologists call chuckling sounded like starlings in a tree.

People walking in woods sometimes come close enough to a den to cause the mother to get up and run off, unmindful of her reputation as a fearless defender of cubs. The cubs stop chuckling and begin to cry: possibly three, four cubs—a ward of mewling bears. The people hear the crying. They find the den and see the cubs. Sometimes they pick them up and carry them away, reporting to the state that they have saved the lives of bear cubs abandoned by their mother. Wherever and whenever this occurs, Gary Alt collects the cubs. After ten years of bear trapping and biological study, Alt has equipped so many sows with radios that he has been able to conduct a foster-mother program with an amazingly high rate of success. A mother in hibernation will readily accept a foster cub. If the need to place an orphan arises somewhat later, when mothers and their cubs are out and around, a sow will kill an alien cub as soon as she smells it. Alt has overcome this problem by stuffing sows' noses with Vicks VapoRub. One way or another, he has found new families for forty-seven orphaned cubs. Forty-six have survived. The other, which had become accustomed over three weeks to feedings and caresses by human hands, was not content in a foster den, crawled outside, and died in the snow.

With a hypodermic jab stick, Alt now drugged the mother, putting her to sleep for the duration of the visit. From deeps of shining fur, he fished out cubs. One. Two. A third. A fourth. Five! The fifth was a foster daughter brought earlier in the winter from two hundred miles away. Three of the four others were male—a ratio consistent with the heavy preponderance of males that Alt's studies have shown through the years. To various onlookers he handed the cubs for safekeeping while he and several assistants carried the mother into the open and weighed her with block and tackle. To protect her eyes, Alt had blindfolded her with a red bandanna. They carried her upside down, being extremely careful lest they scrape and damage her nipples. She weighed two hundred and nineteen pounds. Alt had caught her and weighed her some months before. In the den, she had lost ninety pounds. When she was four years old, she had had four cubs; two years later, four more cubs; and now, after two more years, four cubs. He knew all that about her, he had caught her so many times. He referred to her as Daisy. Daisy was as nothing compared with Vanessa, who was sleeping off the winter somewhere else. In ten seasons, Vanessa had given birth to twenty-three cubs, and had lost none. The growth and reproductive rates of black bears are greater in Pennsylvania than anywhere else. Black bears in Pennsylvania grow more rapidly than griz-

zlies in Montana. Eastern black bears are generally much larger than
Western ones. A seven-hundred-pound bear is unusual but not rare in
Pennsylvania. Alt once caught a big boar like that who had a
thirty-seven-inch neck and was a hair under seven feet long.

This bear, nose to tail, measured five feet five. Alt said, "That's a
nice long sow." For weighing the cubs, he had a small nylon stuff sack.
He stuffed it with bear and hung it on a scale. Two months before,
when the cubs were born, each would have weighed approximately
half a pound—less than a newborn porcupine. Now the cubs weighed
3.4, 4.1, 4.4, 4.6, 5.6—cute little numbers with soft tan noses and erec-
tile pyramid ears. Bears have sex in June and July, but the mother's system holds the fertilized egg away from the uterus until November,
when implantation occurs. Fetal development lasts scarcely six weeks.
Therefore, the creatures who live upon the hibernating mother are so
small that everyone survives.

The orphan, less winsome than the others, looked like a
chocolate-covered possum. I kept her under my vest. She seemed content there and scarcely moved. In time, I exchanged her for 5.6—the
big boy in the litter. Lifted by the scruff and held in the air, he bawled,
flashed his claws, and curled his lips like a woofing boar. I stuffed him
under the vest, where he shut up and nuzzled. His claws were already
more than half an inch long. Alt said that the family would come out
of the den in a few weeks but that much of the spring would go by
before the cubs gained weight. The difference would be that they were
no longer malleable and ductile. They would become pugnacious and
scratchy, not to say vicious, and would chew up the hand that caressed
them. He said, "If you have an enemy, give him a bear cub."

Six men carried the mother back to the den, the red bandanna still
tied around her eyes. Alt repacked her into the rock. "We like to return
her to the den as close as possible to the way we found her," he said.
Someone remarked that one biologist can work a coon, while an army
is needed to deal with a bear. An army seemed to be present. Twelve
people had followed Alt to the den. Some days, the group around him
is four times as large. Alt, who is in his thirties, was wearing a visored
khaki cap with a blue-and-gold keystone on the forehead, and a khaki
cardigan under a khaki jump suit. A lithe and light-bodied man with
tinted glasses and a blond mustache, he looked like a lieutenant in the
Ardennes Forest. Included in the retinue were two reporters and a news
photographer. Alt encourages media attention, the better to soften the
image of the bears. He says, "People fear bears more than they need to,
and respect them not enough." Over the next twenty days, he had
scheduled four hundred visitors—state senators, representatives, com-

missioners, television reporters, word processors, biologists, friends—to go along on his rounds of dens. Days before, he and the denned bears had been hosts to the BBC. The Brits wanted snow. God was having none of it. The BBC brought in the snow.

In the course of the day, we made a brief tour of dens that for the time being stood vacant. Most were rock cavities. They had been used before, and in all likelihood would be used again. Bears in winter in the Pocono Plateau are like chocolate chips in a cookie. The bears seldom go back to the same den two years running, and they often change dens in the course of a winter. In a forty-five-hundred-acre housing development called Hemlock Farms are twenty-three dens known to be in current use and countless others awaiting new tenants. Alt showed one that was within fifteen feet of the intersection of East Spur Court and Pommel Drive. He said that when a sow with two cubs was in there he had seen deer browsing by the outcrop and ignorant dogs stopping off to lift a leg. Hemlock Farms is expensive, and full of cantilevered cypress and unencumbered glass. Houses perch on high flat rock. Now and again, there are bears in the rock—in, say, a floor-through cavity just under the porch. The owners are from New York. Alt does not always tell them that their property is zoned for bears. Once, when he did so, a "FOR SALE" sign went up within two weeks.

Not far away is Interstate 84. Flying over it one day, Buck Alt heard an oddly intermittent signal. Instead of breaking off once and cleanly, it broke off many times. Crossing back over, he heard it again. Soon he was in a tight turn, now hearing something, now nothing, in a pattern that did not suggest anything he had heard before. It did, however, suggest the interstate. Where a big green sign says, "MILFORD 11, PORT JERVIS 20," Gary hunted around and found the bear. He took us now to see the den. We went down a steep slope at the side of the highway and, crouching, peered into a culvert. It was about fifty yards long. There was a disc of daylight at the opposite end. Thirty inches in diameter, it was a perfect place to stash a body, and that is what the bear thought, too. On Gary's first visit, the disc of daylight had not been visible. The bear had denned under the eastbound lanes. She had given birth to three cubs. Soon after he found her, heavy rains were predicted. He hauled the family out and off to a vacant den. The cubs weighed less than a pound. Two days later, water a foot deep was racing through the culvert.

Under High Knob, in remote undeveloped forest about six hundred metres above sea level, a slope falling away in an easterly direction contained a classic excavated den: a small entrance leading into an intimate ovate cavern, with a depression in the center for a bed—in all, about twenty-four cubic feet, the size of a refrigerator-freezer. The den had not been occupied in several seasons, but Rob Buss, a district game protector who works regularly with Gary Alt, had been around to check it three days before and had shined his flashlight into a darkness stuffed with fur. Meanwhile, six inches of fresh snow had fallen on High Knob, and now Alt and his team, making preparations a short distance from the den, scooped up snow in their arms and filled a big sack. They had nets of nylon mesh. There was a fifty-fifty likelihood of yearling bears in the den. Mothers keep cubs until their second spring. When a biologist comes along and provokes the occupants to emerge, there is no way to predict how many will appear. Sometimes they keep coming and coming, like clowns from a compact car. As a bear emerges, it walks into the nylon mesh. A drawstring closes. At the same time, the den entrance is stuffed with a bag of snow. That stops the others. After the first bear has been dealt with, Alt removes the sack of snow. Out comes another bear. A yearling weighs about eighty pounds, and may move so fast that it runs over someone on the biological team and stands on top of him sniffing at his ears. Or her ears. Janice Gruttadauria, a research assistant, is a part of the team. Bear after bear, the procedure is repeated until the bag of snow is pulled away and nothing comes out. That is when Alt asks Rob Buss to go inside and see if anything is there.

Now, moving close to the entrance, Alt spread a tarp on the snow, lay down on it, turned on a five-cell flashlight, and put his head inside the den. The beam played over thick black fur and came to rest on a tiny foot. The sack of snow would not be needed. After drugging the mother with a jab stick, he joined her in the den. The entrance was so narrow he had to shrug his shoulders to get in. He shoved the sleeping mother, head first, out of the darkness and into the light.

While she was away, I shrugged my own shoulders and had a look inside. The den smelled of earth but not of bear. The walls were dripping with roots. The water and protein metabolism of hibernating black bears has been explored by the Mayo Clinic as a research model for, among other things, human endurance on long flights through space and medical situations closer to home, such as the maintenance of anephric human beings who are awaiting kidney transplants.

Outside, each in turn, the cubs were put in the stuff sack—a male and a female. The female weighed four pounds. Greedily, I reached for her when Alt took her out of the bag. I planted her on my shoulder while I wrote down facts about her mother: weight, a hundred and ninety-two pounds; length, fifty-eight inches; some toes missing; severe frostbite from a bygone winter evidenced along the edges of the ears.

Eventually, with all weighing and tagging complete, it was time to go. Alt went into the den. Soon he called out that he was ready for the mother. It would be a tight fit. Feet first, she was shoved in, like a safe-deposit box. Inside, Alt tugged at her in close embrace, and the two of them gradually revolved until she was at the back and their positions had reversed. He shaped her like a doughnut—her accustomed den position. The cubs go in the center. The male was handed in to him. Now he was asking for the female. For a moment, I glanced around as if looking to see who had her. The thought crossed my mind that if I bolted and ran far enough and fast enough I could flag a passing car and keep her. Then I pulled her from under the flap of my vest and handed her away.

Alt and others covered the entrance with laurel boughs, and covered the boughs with snow. They camouflaged the den, but that was not the purpose. Practicing wildlife management to a fare-thee-well, Alt wanted the den to be even darker than it had been before; this would cause the family to stay longer inside and improve the cubs' chances when at last they faced the world.

In the evening, I drove down off the Pocono Plateau and over the folded mountains and across the Great Valley and up the New Jersey Highlands and down into the basin and home. No amount of intervening terrain, though—and no amount of distance—could remove from my mind the picture of the covered entrance in the Pennsylvania hillside, or the thought of what was up there under the snow.

EDWARD HOAGLAND
b. 1932

For many years Edward Hoagland spent his winters in New York City, where he was born, and his summers in the small town of Barton in northern Vermont. This city-country rhythm is strongly present in his work and seems deliberately to dramatize the divided attitude of modern humans toward nature. Hoagland is a tough-minded optimist, not about specific environmental issues but in the belief that life in general "can and ought to be good, and is even meant *to be good.*" *Unlike most writers of his generation who have deplored and fought against the destruction of the natural world, Hoagland often assumes, and even accepts, the impending loss of much that he loves, both in civilization and in the wild. By doing so, he is able to rejoice in and capture what is vanishing with unfettered relish, keen perception, and a kind of fatalistic good humor. Such collections as* The Courage of Turtles (1971), Walking the Dead Diamond River *(1973), and* Red Wolves and Black Bears *(1976) celebrate with equal vitality and admiration the life of circus clowns, turtles, city streets, wilderness canoeing, tugboats, wolves, and other writers. His autobiography,* Compass Points, *appeared in 2001.*

HAILING THE ELUSORY MOUNTAIN LION

The swan song sounded by the wilderness grows fainter, ever more constricted, until only sharp ears can catch it at all. It fades to a nearly inaudible level, and yet there never is going to be any one time when we can say right *now* it is gone. Wolves meet their maker in wholesale lots, but coyotes infiltrate eastward, northward, southeastward. Woodland caribou and bighorn sheep are vanishing fast, but moose have expanded their range in some areas.

Mountain lions used to have practically the run of the Western Hemisphere, and they still do occur from Cape Horn to the Big Muddy River at the boundary of the Yukon and on the coasts of both oceans, so that they are the most versatile land mammal in the New World, probably taking in more latitudes than any other four-footed wild creature anywhere. There are perhaps only four to six thousand left in the United States, though there is no place that they didn't once go, eating deer, elk, pikas, porcupines, grasshoppers, and dead fish on the beach. They were called mountain lions in the Rockies, pumas (originally an Incan word) in the Southwestern states, cougars (a naturalist's corruption of an Amazonian Indian word) in the Northwest, panthers in the traditionalist East—"painters" in dialect-proud New England—or catamounts. The Dutchmen of New Netherland called them tigers, red tigers, deer tigers, and the Spaniards *leones* or *leopardos*. They liked to eat horses—wolves preferred beef and black bears favored pork—but as adversaries of mankind they were overshadowed at first because bears appeared more formidable and wolves in their howling packs were more flamboyant

Walking the Dead Diamond River (New York: Random House, 1973).

and more damaging financially. Yet this panoply of names is itself quite
a tribute, and somehow the legends about "panthers" have lingered
longer than bear or wolf tales, helped by the animal's own limber,
far-traveling stealth and as a carry-over from the immense mythic force
of the great cats of the Old World. Though only Florida among the
Eastern states is known for certain to have any left, no wild knot of
mountains or swamp is without rumors of panthers; nowadays people
delight in these, keeping their eyes peeled. It's wishful, and the wandering, secretive nature of the beast ensures that even Eastern panthers will
not soon be certifiably extinct. An informal census among experts in
1963 indicated that an island of twenty-five or more may have survived
in the New Brunswick-Maine-Quebec region, and Louisiana may still
have a handful, and perhaps eight live isolated in the Black Hills of
South Dakota, and the Oklahoma panhandle may have a small
colony—all outside the established range in Florida, Texas, and the Far
West. As with the blue whale, who will be able to say when they have
been eliminated?

"Mexican lion" is another name for mountain lions in the border
states—a name that might imply a meager second-best rating there yet
ties to the majestic African beasts. Lions are at least twice as big as
mountain lions, measuring by weight, though they are nearly the same
in length because of the mountain lion's superb long tail. Both animals
sometimes pair up affectionately with mates and hunt in tandem, but
mountain lions go winding through life in ones or twos, whereas the
lion is a harem-keeper, harem-dweller, the males eventually becoming
stay-at-homes, heavy figureheads. Lions enjoy the grassy flatlands,
forested along the streams, and they stay put, engrossed in communal
events—roaring, grunting, growling with a racket like the noise of gears
being stripped—unless the game moves on. They sun themselves, pre-
side over the numerous kibbutz young, sneeze from the dust, and bask
in dreams, occasionally waking up to issue reverberating, guttural pronouncements which serve notice that they are now awake.

Mountain lions spirit themselves away in saw-toothed canyons and
on escarpments instead, and when conversing with their mates they
coo like pigeons, sob like women, emit a flat slight shriek, a popping
bubbling growl, or mew, or yowl. They growl and suddenly caterwaul
into falsetto—the famous scarifying, metallic scream functioning as a
kind of hunting cry close up, to terrorize and start the game. They ramble as much as twenty-five miles in a night, maintaining a large loop of
territory which they cover every week or two. It's a solitary, busy life,
involving a survey of several valleys, many deer herds. Like tigers and
leopards, mountain lions are not sociably inclined and don't converse

at length with the whole waiting world, but they are even less noisy; they seem to speak most eloquently with their feet. Where a tiger would roar, a mountain lion screams like a castrato. Where a mountain lion hisses, a leopard would snarl like a truck stuck in snow.

Leopards are the best counterpart to mountain lions in physique and in the tenor of their lives. Supple, fierce creatures, skilled at concealment but with great self-assurance and drive, leopards are bolder when facing human beings than the American cats. Basically they are hot-land beasts and not such remarkable travelers individually, though as a race they once inhabited the broad Eurasian land mass all the way from Great Britain to Malaysia, as well as Africa. As late as the 1960s, a few were said to be still holding out on the shore of the Mediterranean at Mount Mycale, Turkey. (During a forest fire twenty years ago a yearling swam the narrow straits to the Greek island Samos and holed up in a cave, where he was duly killed—perhaps the last leopard ever to set foot in Europe on his own.) Leopards are thicker and shorter than adult mountain lions and seem to lead an athlete's indolent, incurious life much of the time, testing their perfected bodies by clawing tree trunks, chewing on old skulls, executing acrobatic leaps, and then rousing themselves to the semiweekly antelope kill. Built with supreme hardness and economy, they make little allowance for man—they don't see him as different. They relish the flesh of his dogs, and they run up a tree when hunted and then sometimes spring down, as heavy as a chunk of iron wrapped in a flag. With stunning, gorgeous coats, their tight, dervish faces carved in a snarl, they head for the hereafter as if it were just one more extra-emphatic leap—as impersonal in death as the crack of the rifle was.

The American leopard, the jaguar, is a powerfully built, serious fellow, who, before white men arrived, wandered as far north as the Carolinas, but his best home is the humid basin of the Amazon. Mountain lions penetrate these ultimate jungles too, but rather thinly, thriving better in the cooler, drier climate of the untenanted pampas and on the mountain slopes. They are blessed with a pleasant but undazzling coat, tan except for a white belly, mouth and throat, and some black behind the ears, on the tip of the tail and at the sides of the nose, and so they are hunted as symbols, not for their fur. The cubs are spotted, leopardlike, much as lion cubs are. If all of the big cats developed from a common ancestry, the mountain lions' specialization has been unpresumptuous—away from bulk and savagery to traveling light. Toward deer, their prey, they may be as ferocious as leopards, but not toward chance acquaintances such as man. They sometimes break their necks, their jaws, their teeth, springing against the necks of quarry

they have crept close to—a fate in part resulting from the circumstance
that they can't ferret out the weaker individuals in a herd by the device
of a long chase, the way wolves do; they have to take the luck of the
draw. None of the cats possess enough lung capacity for gruelling runs.
They depend upon shock tactics, bursts of speed, sledge-hammer leaps,
strong collarbones for hitting power, and shearing dentition, whereas
wolves employ all the advantages of time in killing their quarry, as well
as the numbers and gaiety of the pack, biting the beast's nose and
rump—the technique of a thousand cuts—lapping the bloody snow.
Wolves sometimes even have a cheering section of flapping ravens
accompanying them, eager to scavenge after the brawl.

It's a risky business for the mountain lion, staking the strength and
impact of his neck against the strength of the prey animal's neck.
Necessarily, he is concentrated and fierce; yet legends exist that mountain lions have irritably defended men and women lost in the wilderness against marauding jaguars, who are no friends of theirs, and (with
a good deal more supporting evidence) that they are susceptible to an
odd kind of fascination with human beings. Sometimes they will tentatively seek an association, hanging about a campground or following a
hiker out of curiosity, perhaps, circling around and bounding up on a
ledge above to watch him pass. This mild modesty has helped preserve
them from extinction. If they have been unable to make any adjustments to the advent of man, they haven't suicidally opposed him
either, as the buffalo, wolves, and grizzlies did. In fact, at close quarters
they seem bewildered. When treed, they don't breathe a hundred-proof
ferocity but puzzle over what to do. They're too light-bodied to bear
down on the hunter and kill him easily, even if they should attack—a
course they seem to have no inclination for. In this century in the
United States only one person, a child of thirteen, has been killed by a
mountain lion; that was in 1924. And they're informal animals. Lolling
in an informal sprawl on a high limb, they can't seem to summon any
Enobarbus-like front of resistance for long. Daring men occasionally
climb up and toss lassos about a cat and haul him down, strangling him
by pulling from two directions, while the lion, mortified, appalled,
never does muster his fighting aplomb. Although he could fight off a
pack of wolves, he hasn't worked out a posture to assume toward man
and his dogs. Impotently, he stiffens, as the dinosaurs must have when
the atmosphere grew cold.

Someday hunting big game may come to be regarded as a form of
vandalism, and the remaining big creatures of the wilderness will skulk
through restricted reserves wearing radio transmitters and numbered
collars, or bearing stripes of dye, as many elephants already do, to aid

the busy biologists who track them from the air. Like a vanishing race of trolls, more report and memory than a reality, they will inhabit children's books and nostalgic articles, a special glamour attaching to those, like mountain lions, that are geographically incalculable and may still be sighted away from the preserves. Already we've become enthusiasts. We want game about us—at least at a summer house; it's part of privileged living. There is a precious privacy about seeing wildlife, too. Like meeting a fantastically dressed mute on the road, the fact that no words are exchanged and that *he's* not going to give an account makes the experience light-hearted; it's wholly ours. Besides, if anything out of the ordinary happened, we know we can't expect to be believed, and since it's rather fun to be disbelieved—fishermen know this—the privacy is even more complete. Deer, otter, foxes are messengers from another condition of life, another mentality, and bring us tidings of places where we don't go.

Ten years ago at Vavenby, a sawmill town on the North Thompson River in British Columbia, a frolicsome mountain lion used to appear at dusk every ten days or so in a bluegrass field alongside the river. Deer congregated there, the river was silky and swift, cooling the summer air, and it was a festive spot for a lion to be. She was thought to be a female, and reputedly left tracks around an enormous territory to the north and west—Raft Mountain, Battle Mountain, the Trophy Range, the Murtle River, and Mahood Lake—territory on an upended, pelagic scale, much of it scarcely accessible to a man by trail, where the tiger lilies grew four feet tall. She would materialize in this field among the deer five minutes before dark, as if checking in again, a habit that may have resulted in her death eventually, though for the present the farmer who observed her visits was keeping his mouth shut about it. This was pioneer country; there were people alive who could remember the time when poisoning the carcass of a cow would net a man a pile of dead predators—a family of mountain lions to bounty, maybe half a dozen wolves, and both black bears and grizzlies. The Indians considered lion meat a delicacy, but they had clans which drew their origins at the Creation from ancestral mountain lions, or wolves or bears, so these massacres amazed them. They thought the outright bounty hunters were crazy men.

Even before Columbus, mountain lions were probably not distributed in saturation numbers anywhere, as wolves may have been. Except for the family unit—a female with her half-grown cubs—each lion seems to occupy its own spread of territory, not as a result of fights with intruders but because the young transient share the same instinct for solitude and soon sheer off to find vacant mountains and valleys. A mature lion kills only one deer every week or two, according to a study by Maurice Hornocker in Idaho, and therefore is not really a notable factor in controlling the local deer population. Rather, it keeps watch contentedly as that population grows, sometimes benefitting the herds by scaring them onto new wintering grounds that are not overbrowsed, and by its very presence warding off other lions.

This thin distribution, coupled with the mountain lion's taciturn habits, make sighting one a matter of luck, even for game officials located in likely country. One warden in Colorado I talked to had indeed seen a pair of them fraternizing during the breeding season. He was driving a jeep over an abandoned mining road, and he passed two brown animals sitting peaceably in the grass, their heads close together. For a moment he thought they were coyotes and kept driving, when all of a sudden the picture registered that they were *cougars!* He braked and backed up, but of course they were gone. He was an old-timer, a man who had crawled inside bear dens to pull out the cubs, and knew where to find clusters of buffalo skulls in the recesses of the Rockies where the last bands had hidden; yet this cryptic instant when he was turning his jeep round a curve was the only glimpse—unprovable—that he ever got of a mountain lion.

Such glimpses usually are cryptic. During a summer I spent in Wyoming in my boyhood, I managed to see two coyotes, but both occasions were so fleeting that it required an act of faith on my part afterward to feel sure I had seen them. One of the animals vanished between rolls of ground; the other, in rougher, stonier, wooded country, cast his startled gray face in my direction and simply was gone. Hunching, he swerved for cover, and the brush closed over him. I used to climb to a vantage point above a high basin at twilight and watch the mule deer steal into the meadows to feed. The grass grew higher than their stomachs, the steep forest was close at hand, and they were as small and fragile-looking as filaments at that distance, quite human in coloring, gait and form. It was possible to visualize them as a naked Indian hunting party a hundred years before—or not to believe in their existence at all, either as Indians or deer. Minute, aphid-sized, they stepped so carefully in emerging, hundreds of feet below, that, straining my eyes, I needed to tell myself constantly that they were deer; my imagination, left to its own devices with the dusk settling down, would have made of them a dozen other creatures.

Recently, walking at night on the woods road that passes my house in Vermont, I heard footsteps in the leaves and windfalls. I waited, listening—they sounded too heavy to be anything less than a man, a large deer or a bear. A man wouldn't have been in the woods so late, my dog stood respectfully silent and still, and they did seem to shuffle porten-tously. Sure enough, after pausing at the edge of the road, a fully grown bear appeared, visible only in dimmest outline, staring in my direction for four or five seconds. The darkness lent a faintly red tinge to his coat; he was well built. Then, turning, he ambled off, almost immediately lost to view, though I heard the noise of his passage, interrupted by sev-eral pauses. It was all as concise as a vision, and since I had wanted to see a bear close to my own house, being a person who likes to live in a melting pot, whether in the city or country, and since it was too dark to pick out his tracks, I was grateful when the dog inquisitively urinated along the bear's path, thereby confirming that at least I had witnessed *something*. The dog seemed unsurprised, however, as if the scent were not all that remarkable, and, sure enough, the next week in the car I encountered a yearling bear in daylight two miles downhill, and a cub a month later. My farmer neighbors were politely skeptical of my accounts, having themselves caught sight of only perhaps a couple of bears in all their lives.

So it's with sympathy as well as an awareness of the tricks that enthu-siasm and nightfall may play that I have been going to nearby towns seeking out people who have claimed at one time or another to have seen a mountain lion. The experts of the state—game wardens, taxider-mists, the most accomplished hunters—emphatically discount the claims, but the believers are unshaken. They include some summer people who were enjoying a drink on the back terrace when the appari-tion of a great-tailed cat moved out along the fringe of the woods on a deer path; a boy who was hunting with his 22 years ago near the village dump and saw the animal across a gully and fired blindly, then ran away and brought back a search party, which found a tuft of toast-colored fur; and a state forestry employee, a sober woodsman, who caught the cat in his headlights while driving through Victory Bog in the wildest corner of the Northeast Kingdom. Gordon Hickok, who works for a furniture factory and has shot one or two mountain lions on hunting trips in the West, saw one cross U.S. 5 at a place called Auger Hole near Mount Hor. He tracked it with dogs a short distance, finding a fawn with its head gnawed off. A high-school English teacher reported seeing a mountain lion cross another road, near Runaway Pond, but the hunters who quickly went out decided that the prints were those of a big bobcat, splayed impressively in the mud and snow. Fifteen years ago a watch-man in the fire tower on top of Bald Mountain had left grain scattered in the grooves of a flat rock under the tower to feed several deer. One night, looking down just as the dusk turned murky, he saw two slim long-tailed lions creep out of the scrubby border of spruce and inspect the rock, sniffing deer droppings and dried deer saliva. The next night, when he was in his cabin, the dog barked and, looking out the window, again he saw the vague shape of a lion just vanishing.

A dozen loggers and woodsmen told me such stories. In the Adirondacks I've also heard some persuasive avowals—one by an old dog-sled driver and trapper, a French Canadian; another by the owner of a tourist zoo, who was exhibiting a Western cougar. In Vermont perhaps the most eager rumor buffs are some of the farmers. After all, now that packaged semen has replaced the awesome farm bull and so many procedures have been mechanized, who wants to lose *all* the adventure of farming? Until recently the last mountain lion known to have been killed in the Northeast was recorded in 1881 in Barnard, Vermont. However, it has been learned that probably another one was shot from a tree in 1931 in Mundleville, New Brunswick, and still another trapped seven years later in Somerset County in Maine. Bruce S. Wright, director of the Northeastern Wildlife Station (which is operated at the University of New Brunswick with international funding), is convinced that though they are exceedingly rare, mountain lions are still part of the fauna of the region; in fact, he has plaster casts of tracks to prove it, as well as a compilation of hundreds of reported sightings. Some people may have mistaken a golden retriever for a lion, or may have intended to foment a hoax, but all in all the evidence does seem promising. Indeed, after almost twenty years of search and study, Wright himself finally saw one.

The way these sightings crop up in groups has often been pooh-poohed as greenhorn fare or as a sympathetic hysteria among neighbors, but it is just as easily explained by the habit mountain lions have of establishing a territory that they scout through at intervals, visiting an auspicious deer-ridden swamp or remote ledgy mountain. Even at such a site a successful hunt could not be mounted without trained dogs, and if the population of the big cats was extremely sparse, requiring of them long journeys during the mating season, and yet with plenty of deer all over, they might not stay for long. One or two hundred miles is no obstacle to a Western cougar. The cat might inhabit a mountain ridge one year, and then never again.

Fifteen years ago, Francis Perry, who is an ebullient muffin of a man, a farmer all his life in Brownington, Vermont, saw a mountain lion "larger and taller than a collie, and grayish yellow" (he had seen them in circuses). Having set a trap for a woodchuck, he was on his way to visit the spot when he came over a rise and, at a distance of fifty yards, saw the beast engaged in eating the dead woodchuck. It bounded off, but Perry set four light fox traps for it around the woodchuck.

Apparently, a night or two later the cat returned and got caught in three of these, but they couldn't hold it; it pulled free, leaving the marks of a struggle. Noel Perry, his brother, remembers how scared Francis looked when he came home from the first episode. Noel himself saw the cat (which may have meant that Brownington Swamp was one of its haunts that summer), once when it crossed a cow pasture on another farm the brothers owned, and once when it fled past his rabbit dogs through underbrush while he was training them—he thought for a second that its big streaking form was one of the dogs. A neighbor, Robert Chase, also saw the animal that year. Then again last summer, for the first time in fifteen years, Noel Perry saw a track as big as a bear's but round like a mountain lion's, and Robert's brother, Larry Chase, saw the actual cat several times one summer evening, playing a chummy hide-and-seek with him in the fields.

Elmer and Elizabeth Ambler are in their forties, populists politically, and have bought a farm in Glover to live the good life, though he is a truck driver in Massachusetts on weekdays and must drive hard in order to be home when he can. He's bald, with large eyebrows, handsome teeth and a low forehead, but altogether a strong-looking, clear, humane face. He is an informational kind of man who will give you the history of various breeds of cattle or a talk about taxation in a slow and musical voice, and both he and his wife, a purposeful, self-sufficient redhead, are fascinated by the possibility that they live in the wilderness. Beavers inhabit the river that flows past their house. The Amblers say that on Black Mountain nearby hunters "disappear" from time to time, and bears frequent the berry patches in their back field—they see them, their visitors see them, people on the road see them, their German shepherds meet them and run back drooling with fright. They've stocked their farm with horned Herefords instead of the polled variety so that the creatures can "defend themselves." Ambler is intrigued by the thought that apart from the danger of bears, someday "a cat" might prey on one of his cows. Last year, looking out the back window, his wife saw through binoculars an animal with a flowing tail and "a cat's gallop" following a line of trees where the deer go, several hundred yards uphill behind the house. Later, Ambler went up on snowshoes and found tracks as big as their shepherds'; the dogs obligingly ran alongside. He saw walking tracks, leaping tracks and deer tracks marked with blood going toward higher ground. He wonders whether the cat will ever attack him. There are plenty of bobcats around, but they both say they know the difference. The splendid, nervous *tail* is what people must have identified in order to claim they have seen a mountain lion.

I, too, cherish the notion that I may have seen a lion. Mine was crouched on an overlook above a grass-grown, steeply pitched wash in the Alberta Rockies—a much more likely setting than anywhere in New England. It was late afternoon on my last day at Maligne Lake, where I had been staying with my father at a national-park chalet. I was twenty; I could walk forever or could climb endlessly in a sanguine scramble, going out every day as far as my legs carried me, swinging around for home before the sun went down. Earlier, in the valley of the Athabasca, I had found several winter-starved or wolf-killed deer, well picked and scattered, and an area with many elk antlers strewn on the ground where the herds had wintered safely, dropping their antlers but not their bones. Here, much higher up, in the bright plenitude of the summer, I had watched two wolves and a stately bull moose in one mountain basin, and had been up on the caribou barrens on the ridge west of the lake and brought back the talons of a hawk I'd found dead on the ground. Whenever I was watching game, a sort of stopwatch in me started running. These were moments of intense importance and intimacy, of new intimations and aptitudes. Time had a jam-packed character, as it does during a mile run.

I was good at moving quietly through the woods and at spotting game, and was appropriately exuberant. The finest, longest day of my stay was the last. Going east, climbing through a luxuriant terrain of up-and-down boulders, brief brilliant glades, sudden potholes fifty feet deep—a forest of moss-hung lodgepole pines and firs and spare, gaunt spruce with the black lower branches broken off—I came upon the remains of a young bear, which had been torn up and shredded. Perhaps wolves had cornered it during some imprudent excursion in the early spring. (Bears often wake up while the snow is still deep, dig themselves out and rummage around in the neighborhood sleepily for a day or two before bedding down again under a fallen tree.) I took the skull along so that I could extract the teeth when I got hold of some tools. Discoveries like this represent a superfluity of wildlife and show how many beasts there are scouting about.

I went higher. The marmots whistled familially; the tall trees wilted to stubs of themselves. A pretty stream led down a defile from a series of openings in front of the ultimate barrier of a vast mountain wall which I had been looking at from a distance each day on my outings. It wasn't too steep to be climbed, but it was a barrier because my energies were not sufficient to scale it and bring me back the same night. Besides, it stretched so majestically, surflike above the lesser ridges, that I liked to think of it as the Continental Divide.

On my left as I went up this wash was an abrupt, grassy slope that enjoyed a southern exposure and was sunny and windblown all winter, which kept it fairly free of snow. The ranger at the lake had told me it served as a wintering ground for a few bighorn sheep and for a band of mountain goats, three of which were in sight. As I approached laboriously, these white, pointy-horned fellows drifted up over a rise, managing to combine their retreat with some nippy good grazing as they went, not to give any pursuer the impression that they had been pushed into flight. I took my time too, climbing to locate the spring in a precipitous cleft of rock where the band did most of its drinking, and finding the shallow, high-ceilinged cave where the goats had sheltered from storms, presumably for generations. The floor was layered with rubbery droppings, tramped down and sprinkled with tufts of shed fur, and the back wall was checkered with footholds where the goats liked to clamber and perch. Here and there was a horn lying loose—a memento for me to add to my collection from an old individual that had died a natural death, secure in the band's winter stronghold. A bold, thriving family of pack rats emerged to observe me. They lived mainly on the nutritives in the droppings, and were used to the goats' tolerance; they seemed astonished when I tossed a stone.

I kept scrabbling along the side of the slope to a section of outcroppings where the going was harder. After perhaps half an hour, crawling around a corner, I found myself faced with a bighorn ram who was taking his ease on several square yards of bare earth between large rocks, a little above the level of my head. Just as surprised as I, he stood up. He must have construed the sounds of my advance to be those of another sheep or goat. His horns had made a complete curl and then some; they were thick, massive and bunched together like a high Roman helmet, and he himself was muscly and military, with a grave-looking nose. A squared-off, middle-aged, trophy-type ram, full of imposing professionalism, he was at the stage of life when rams sometimes stop herding and live as rogues.

He turned and tried a couple of possible exits from the pocket where I had found him, but the ground was badly pitched and would require a reeling gait and loss of dignity. Since we were within a national park and obviously I was unarmed, he simply was not inclined to put himself to so much trouble. He stood fifteen or twenty feet above me, pushing his tongue out through his teeth, shaking his head slightly and dipping it into charging position as I moved closer by a step or two, raising my hand slowly toward him in what I proposed as a friendly greeting. The day had been a banner one since the beginning, so while I recognized immediately that this meeting would be a valued memory, I felt as natural in his company as if he were a friend of mine reincarnated in a shag suit. I saw also that he was going to knock me for a loop, head over heels down the steep slope, if I sidled nearer, because he did not by any means feel as expansive and exuberant at our encounter as I did. That was the chief difference between us. I was talking to him with easy gladness, and beaming; he was not. He was unsettled and on his mettle, waiting for me to move along, the way a bighorn sheep waits for a predator to move on in wildlife movies when each would be evenly matched in a contest of strength and position. Although his warlike nose and high bone helmet, blocky and beautiful as weaponry, kept me from giving in to my sense that we were brothers, I knew I could stand there for a long while. His coat was a down-to-earth brown, edgy with muscle, his head was that of an unsmiling veteran standing to arms, and despite my reluctance to treat him as some sort of boxed-in prize, I might have stayed on for half the afternoon if I hadn't realized that I had other sights to see. It was not a day to dawdle.

I trudged up the wash and continued until, past tree line, the terrain widened and flattened in front of a preliminary ridge that formed an obstacle before the great roaring, silent, surflike mountain wall that I liked to think of as the Continental Divide, although it wasn't. A cirque separated the preliminary ridge from the ultimate divide, which I still hoped to climb to and look over. The opening into this was roomy enough, except for being littered with enormous boulders, and I began trying to make my way across them. Each was boat-sized and rested upon underboulders; it was like running in place. After tussling with this landscape for an hour or two, I was limp and sweating, pinching my cramped legs. The sun had gone so low that I knew I would be finding my way home by moonlight in any case, and I could see into the cirque, which was big and symmetrical and presented a view of sheer barbarism; everywhere were these cruel boat-sized boulders.

Giving up and descending to the goats' draw again, I had a drink from the stream and bathed before climbing farther downward. The grass was green, sweet-smelling, and I felt safely close to life after that sea of dead boulders. I knew I would never be physically younger or in finer country; even then the wilderness was singing its swan song. I had no other challenges in mind, and though very tired, I liked looking up at the routes where I'd climbed. The trio of goats had not returned, but I could see their wintering cave and the cleft in the rocks where the spring was. Curiously, the bighorn ram had not left; he had only withdrawn upward, shifting away from the outcroppings to an open sweep of space where every avenue of escape was available. He was lying on a carpet of grass and, lonely pirate that he was, had his head turned in my direction.

It was from this same wash that looking up, I spotted the animal I took to be a mountain lion. He was skulking among some outcroppings at a point lower on the mountainside than the ledges where the ram originally had been. A pair of hawks or eagles were swooping at him by turns, as if he were close to a nest. The slant between us was steep, but the light of evening was still more than adequate. I did not really see the wonderful tail—that special medallion—nor was he particularly big for a lion. He was gloriously catlike and slinky, however, and so indifferent to the swooping birds as to seem oblivious of them. There are plenty of creatures he wasn't: he wasn't a marmot, a goat or other grass-eater, a badger, a wolf or coyote or fisher. He *may* have been a big bobcat or a wolverine, although he looked ideally lion-colored. He had a cat's strong collarbone structure for hitting, powerful haunches for vaulting, and the almost mystically small head mountain lions possess, with the gooseberry eyes. Anyway, I believed him to be a mountain lion, and standing quietly I watched him as he inspected in leisurely fashion the ledge that he was on and the one under him savory with every trace of goat—frosty-colored with the white hairs they'd shed. The sight was so dramatic that it seemed to be happening close to me, though in fact he and the hawks or eagles, whatever they were, were miniaturized by distance.

If I'd kept motionless, eventually I could have seen whether he had the proper tail, but such scientific questions had no weight next to my need to essay some kind of communication with him. It had been exactly the same when I'd watched the two wolves playing together a couple of days before. They were above me, absorbed in their game of noses-and-paws. I had recognized that I might never witness such a scene again, yet I couldn't hold myself in. Instead of talking and raising my arm to them, as I had with the ram, I'd shuffled forward impetuously as if to say *Here I am!* Now, with the lion, I tried hard to dampen my impulse and restrained myself as long as I could. Then I stepped toward him, just barely squelching a cry in my throat but lifting my hand—as clumsy as anyone is who is trying to attract attention.

At that, of course, he swerved aside instantly and was gone. Even the two birds vanished. Foolish, triumphant and disappointed, I hiked on down into the lower forests, gargantuanly tangled, another life zone—not one which would exclude a lion but one where he would not be seen. I'd got my second wind and walked lightly and softly, letting the silvery darkness settle around me. The blowdowns were as black as whales; my feet sank in the moss. Clearly this was as crowded a day as I would ever have, and I knew my real problem would not be to make myself believed but rather to make myself understood at all, sim-

ply in reporting the story, and that I must at least keep the memory
straight for myself. I was so happy that I was unerring in distinguishing
the deer trails going my way. The forest's night beauty was supreme in
its promise, and I didn't hurry.

THOUGHTS ON RETURNING TO
THE CITY AFTER FIVE MONTHS ON
A MOUNTAIN WHERE THE WOLVES HOWLED

City people are more supple than country people, and the sanest
city people, being more tested and more broadly based in the world of
men, are the sanest people on earth. As to honesty, though, or good
sense, no clear-cut distinction exists either way.

I like gourmets, even winetasters. In the city they correspond to the
old-timers who knew all the berries and herbs, made money collecting
the roots of the ginseng plant, and knew the taste of each hill by its
springs. Alertness and adaptability in the city are transferable to the
country if you feel at home there, and alertness there can quickly be
transmuted into alertness here. It is not necessary to choose between
being a country man and a city man, as it is to decide, for instance,
some time along in one's thirties, whether one is an Easterner or a
Westerner. (Middle Westerners, too, make the choice: people in
Cleveland consider themselves Easterners, people in Kansas City know
they are Western.) But one can be both a country man and a city man.
Once a big frog in a local pond, now suddenly I'm tiny again, and
delighted to be so, kicking my way down through the water, swimming
along my anchor chains and finding them fast in the bottom.

Nor must one make a great sacrifice in informational matters. I
know more about bears and wolves than anybody in my town or the
neighboring towns up there and can lead lifelong residents in the
woods, yet the fierce, partisan block associations in my neighborhood in
New York apparently know less than I do about the closer drug-peddling
operations or they surely would have shut them down. This is not to say
that such information is of paramount importance, however. While,
lately, I was tasting the October fruit of the jack-in-the-pulpit and
watching the club-moss smoke with flying spores as I walked in the
woods, my small daughter, who had not seen me for several weeks,

Red Wolves and Black Bears (New York: Random House, 1976).

missed me so much that when I did return, she threw up her arms in helpless and choked excitement to shield her eyes, as if I were the rising sun. The last thing I wish to be, of course, is the sun—being only a guilty father.

But what a kick it is to be back, seeing newspaperman friends; newspapermen are the best of the city. There are new restaurants down the block, and today I rescued an actual woodcock—New York is nothing if not cosmopolitan. Lost, it had dived for the one patch of green in the street, a basket of avocados in the doorway of Shanvilla's Grocery, and knocked itself out. I'd needed to drag myself back from that mountain where the wolves howl, and yet love is what I feel now; the days are long and my eyes and emotions are fresh.

The city is dying irreversibly as a metropolis. We who love it must recognize this if we wish to live in it intelligently. All programs, all palliatives and revenue-sharing, can only avail to ease what we love into oblivion a little more tenderly (if a tender death is ever possible for a city). But to claim that the city is dying, never to "turn the corner," is not to announce that we should jump for the lifeboats. There are still no better people than New Yorkers. No matter where I have been, I rediscover this every fall. And my mountain is dying too. The real estate ads up in that country put it very succinctly. "Wealth you can walk on," they say. As far as that goes, one cannot live intelligently without realizing that we and our friends and loved ones are all dying. But one's ideals, no: no matter what currently unfashionable ideals a person may harbor in secret, from self-sacrifice and wanting to fall in love to wanting to fight in a war, there will continue to be opportunities to carry them out.

My country neighbor is dying right now, wonderfully fiercely—nothing but stinging gall from his lips. The wolves' mountain bears his name, and at eighty-six he is dying almost on the spot where he was born, in the oneroom schoolhouse in which he attended first grade, to which he moved when his father's house burned. This would not be possible in the city. In the city we live by being supple, bending with the wind. He lived by bending with the wind too, but his were the north and west winds.

You New Yorkers will excuse me for missing my barred owls, ruffed grouse and snowshoe rabbits, my grosbeaks and deer. I love what you love too. In the city and in the country there is a simple, underlying basis to life which we forget almost daily: that life is good. We forget because losing it or wife, children, health, friends is so awfully painful, and because life is hard, but we know from our own experience as well as our expectations that it can and ought to be good, and is even *meant* to be good. Any careful study of living things, whether wolves, bears or man, reminds one of the same direct truth; also of the clarity of the fact that evolution itself is obviously not some process of drowning beings clutching at straws and climbing from suffering and travail and virtual expiration to tenuous, momentary survival. Rather, evolution has been a matter of days well-lived, chameleon strength, energy, zappy sex, sunshine stored up, inventiveness, competitiveness, and the whole fun of busy brain cells. Watch how a rabbit loves to run; watch him set scenting puzzles for the terrier behind him. Or a wolf's amusement at the anatomy of a deer. Tug, tug, he pulls out the long intestines: ah, Yorick, how *long* you are!   •

An acre of forest will absorb six tons of carbon dioxide in a year.

Wordsworth walked an estimated 186,000 miles in his lifetime.

Robert Rogers' Twenty-first Rule of Ranger warfare was: "If the enemy pursue your rear, take a circle till you come to your own tracks, and there form an ambush to receive them, and give them the first fire."

Rain-in-the-face, a Hunkpapa Sioux, before attacking Fort Totten in the Dakota Territory in 1866: "I prepared for death. I painted as usual like the eclipse of the sun, half black and half red."

WILLIAM KITTREDGE
b. 1932

William Kittredge has had a significant impact on both the literature and the environmental discourse of the West. With his background in ranching and his distinguished career as a writer and a professor of English at the University of Montana, he brings a broad perspective to the real prospects, human and natural, of the astonishing landscape where he makes his home. He is not a dogmatist but rather a poser of hard questions. These are often rooted in the tensions within his own experience yet also reflect the larger conflicted releationships between Americans and their land. His principal works of nonfiction include Owning It All *(1987),* Hole in the Sky: A Memoir *(1992), and* Who Owns the West *(1996). He is also the coeditor, with Annick Smith, of* The Last Best

Place (1988), an anthology of literature by Montana writers about their home state. This collection has become the inspiration and prototype for numerous other books attempting to collect the writing, across diverse genres, that will evoke the stories, character, and possibilities of particular states or bioregions.

OWNING IT ALL

Imagine the slow history of our country in the far reaches of southeastern Oregon, a backlands enclave even in the American West, the first settlers not arriving until a decade after the end of the Civil War. I've learned to think of myself as having had the luck to grow up at the tail end of a way of existing in which people lived in everyday proximity to animals on territory they knew more precisely than the patterns in the palms of their hands.

In Warner Valley we understood our property as others know their cities, a landscape of neighborhoods, some sacred, some demonic, some habitable, some not, which is as the sea, they tell me, is understood by fishermen. It was only later, in college, that I learned it was possible to understand Warner as a fertile oasis in a vast featureless sagebrush desert.

Over in that other world on the edge of rain-forests which is the Willamette Valley of Oregon, I'd gone to school in General Agriculture, absorbed in a double-bind sort of learning, studying to center myself in the County Agent/Corps of Engineers mentality they taught and at the same time taking classes from Bernard Malamud and wondering with great romantic fervor if it was in me to write the true history of the place where I had always lived.

Straight from college I went to Photo Intelligence work in the Air Force. The last couple of those years were spent deep in jungle on the island of Guam, where we lived in a little compound of cleared land, in a quonset hut.

The years on Guam were basically happy and bookish: we were newly married, with children. A hundred or so yards north of our quonset hut, along a trail through the luxuriant undergrowth between coconut palms and banana trees, a ragged cliff of red porous volcanic rock fell directly to the ocean. When the Pacific typhoons came roaring in, our hut was washed with blowing spray from the great breakers. On

Owning It All (St. Paul, Minn.: Graywolf, 1987)

calm days we would stand on the cliff at that absolute edge of our jungle and island, and gaze out across to the island of Rota, and to the endlessness of ocean beyond, and I would marvel at my life, so far from southeastern Oregon.

And then in the late fall of 1958, after I had been gone from Warner Valley for eight years, I came back to participate in our agriculture. The road in had been paved, we had Bonneville Power on lines from the Columbia River, and high atop the western rim of the valley there was a TV translator, which beamed fluttering pictures from New York and Los Angeles direct to us.

And I had changed, or thought I had, for a while. No more daydreams about writing the true history. Try to understand my excitement as I climbed to the rim behind our house and stood there by our community TV translator. The valley where I had always seen myself living was open before me like another map and playground, and this time I was an adult, and high up in the War Department. Looking down maybe 3,000 feet into Warner, and across to the high basin and range desert where we summered our cattle, I saw the beginnings of my real life as an agricultural manager. The flow of watercourses in the valley was spread before me like a map, and I saw it as a surgeon might see the flow of blood across a chart of anatomy, and saw myself helping to turn the fertile homeplace of my childhood into a machine for agriculture whose features could be delineated with the same surgeon's precision in my mind.

It was work which can be thought of as craftsmanlike, both artistic and mechanical, creating order according to an ideal of beauty based on efficiency, manipulating the forces of water and soil, season and seed, manpower and equipment, laying out functional patterns for irrigation and cultivation on the surface of our valley. We drained and leveled, ditched and pumped, and for a long while our crops were all any of us could have asked. There were over 5,000 water control devices. We constructed a perfect agricultural place, and it was sacred, so it seemed.

Agriculture is often envisioned as an art, and it can be. Of course there is always survival, and bank notes, and all that. But your basic bottom line on the farm is again and again some notion of how life should be lived. The majority of agricultural people, if you press them hard enough, even though most of them despise sentimental abstractions, will admit they are trying to create a good place, and to live as part of that goodness, in the kind of connection which with fine reason we call *rootedness*. It's just that there is good art and bad art.

These are thoughts which come back when I visit eastern Oregon. I park and stand looking down into the lava-rock and juniper-tree canyon where Deep Creek cuts its way out of the Warner Mountains, and the great turkey buzzard soars high in the yellow-orange light above the evening. The fishing water is low, as it always is in late August, unfurling itself around dark and broken boulders. The trout, I know, are hanging where the currents swirl across themselves, waiting for the one entirely precise and lucky cast, the Renegade fly bobbing toward them.

Even now I can see it, each turn of water along miles of that creek. Walk some stretch enough times with a fly rod and its configurations will imprint themselves on your being with Newtonian exactitude. Which is beyond doubt one of the attractions of such fishing—the hours of learning, and then the intimacy with a living system that carries you beyond the sadness of mere gaming for sport.

What I liked to do, back in the old days, was pack in some spuds and an onion and corn flour and spices mixed up in a plastic bag, a small cast-iron frying pan in my wicker creel and, in the late twilight on a gravel bar by the water, cook up a couple of rainbows over a fire of snapping dead willow and sage, eating alone while the birds flitted through the last hatch, wiping my greasy fingers on my pants while the heavy trout began rolling at the lower ends of the pools.

The canyon would be shadowed under the moon when I walked out to show up home empty-handed, to sit with my wife over a drink of whiskey at the kitchen table. Those nights I would go to bed and sleep without dreams, a grown-up man secure in the house and the western valley where he had been a child, enclosed in a topography of spirit he assumed he knew more closely than his own features in the shaving mirror.

So, I ask myself, if it was such a pretty life, why didn't I stay? The peat soil in Warner Valley was deep and rich, we ran good cattle, and my most sacred memories are centered there. What could run me off?

Well, for openers, it got harder and harder to get out of bed in the mornings and face the days, for reasons I didn't understand. More and more I sought the comfort of fishing that knowable creek. Or in winter the blindness of television.

My father grew up on a homestead place on the sagebrush flats outside Silver Lake, Oregon. He tells of hiding under the bed with his sisters when strangers came to the gate. He grew up, as we all did in that country and era, believing that the one sure defense against the world was property. I was born in 1932, and recall a life before the end of World War II in which it was possible for a child to imagine that his family owned the world.

Warner Valley was largely swampland when my grandfather bought the MC Ranch with no downpayment in 1936, right at the heart of the Great Depression. The outside work was done mostly by men and horses and mules, and our ranch valley was filled with life. In 1937 my father bought his first track-layer, a secondhand RD6 Caterpillar he used to build a 17-mile diversion canal to carry the spring floodwater around the east side of the valley, and we were on our way to draining all swamps. The next year he bought an RD7 and a John Deere 36 combine which cut an 18-foot swath, and we were deeper into the dream of power over nature and men, which I had begun to inhabit while playing those long-ago games of war.

The peat ground left by the decaying remnants of ancient tule beds was diked into huge undulating grainfields—Houston Swamp with 750 irrigated acres, Dodson Lake with 800—a final total of almost 8,000 acres under cultivation, and for reasons of what seemed like common sense and efficiency, the work became industrialized. Our artistry worked toward a model whose central image was the machine.

The natural patterns of drainage were squared into drag-line ditches, the tules and the aftermath of the oat and barley crops were burned—along with a little more of the combustible peat soil every year. We flood-irrigated when the water came in spring, drained in late March, and planted in a 24-hour-a-day frenzy which began around April 25 and ended—with luck—by the 10th of May, just as leaves on the Lombardy poplar were breaking from their buds. We summered our cattle on more than a million acres of Taylor Grazing Land across the high lava rock and sagebrush desert out east of the valley, miles of territory where we owned most of what water there was, and it was ours. We owned it all, or so we felt. The government was as distant as news on the radio.

The most intricate part of my job was called "balancing water," a night and day process of opening and closing pipes and redwood headgates and running the 18-inch drainage pumps. That system was the finest plaything I ever had.

And despite the mud and endless hours, the work remained play for a long time, the making of a thing both functional and elegant. We were doing God's labor and creating a good place on earth, living the pastoral yeoman dream—that's how our mythology defined it, although nobody would ever have thought to talk about work in that way.

And then it all went dead, over years, but swiftly.

You can imagine our surprise and despair, our sense of having been profoundly cheated. It took us a long while to realize some unnamable thing was wrong, and then we blamed it on ourselves, our inability to manage enough. But the fault wasn't ours, beyond the fact that we had all been educated to believe in a grand bad factory-land notion as our prime model of excellence.

We felt enormously betrayed. For so many years, through endless efforts, we had proceeded in good faith, and it turned out we had wrecked all we had not left untouched. The beloved migratory rafts of waterbirds, the green-headed mallards and the redheads and canvasbacks, the cinnamon teal and the great Canadian honkers, were mostly gone along with their swampland habitat. The hunting, in so many ways, was no longer what it had been.

We wanted to build a reservoir, and litigation started. Our laws were being used against us, by people who wanted a share of what we thought of as our water. We could not endure the boredom of our mechanical work, and couldn't hire anyone who cared enough to do it right. We baited the coyotes with 1080, and rodents destroyed our alfalfa; we sprayed weeds and insects with 2-4-D Ethyl and Malathion, and Parathion for clover mite, and we shortened our own lives.

In quite an actual way we had come to victory in the artistry of our playground warfare against all that was naturally alive in our native home. We had reinvented our valley according to the most persuasive ideal given us by our culture, and we ended with a landscape organized like a machine for growing crops and fattening cattle, a machine that creaked a little louder each year, a dreamland gone wrong.

One of my strongest memories comes from a morning when I was maybe 10 years old, out on the lawn before our country home in spring, beneath a bluebird sky. I was watching the waterbirds coming off the valley swamps and grainfields where they had been feeding overnight. They were going north to nesting grounds on the Canadian tundra, and that piece of morning, inhabited by the sounds of their wings and their calling in the clean air, was wonder-filled and magical. I was enclosed in a living place.

No doubt that memory has persisted because it was a sight of possibility which I will always cherish—an image of the great good place rubbed smooth over the years like a river stone, which I touch again as I consider why life in Warner Valley went so seriously haywire. But never again in my lifetime will it be possible for a child to stand out on a bright spring morning in Warner Valley and watch the waterbirds come through in enormous, rafting vee-shaped flocks of thousands—and I grieve.

My father is a very old man. A while back we were driving up the Bitterroot Valley of Montana, and he was gazing away to the mountains. "They'll never see it the way we did," he said, and I wonder what he saw.

We shaped our piece of the West according to the model provided by our mythology, and instead of a great good place such order had given us enormous power over nature, and a blank perfection of fields.

A Mythology can be understood as a story that contains a set of implicit instructions from a society to its members, telling them what is valuable and how to conduct themselves if they are to preserve the things they value.

The teaching mythology we grew up with in the American West is a pastoral story of agricultural ownership. The story begins with a vast innocent continent, natural and almost magically alive, capable of inspiring us to reverence and awe, and yet savage, a wilderness. A good rural people come from the East, and they take the land from its native inhabitants, and tame it for agricultural purposes, bringing civilization: a notion of how to live embodied in law. The story is as old as invading armies, and at heart it is a racist, sexist, imperialist mythology of conquest; a rationale for violence—against other people and against nature.

At the same time, that mythology is a lens through which we continue to see ourselves. Many of us like to imagine ourselves as honest yeomen who sweat and work in the woods or the mines or the fields for a living. And many of us are. We live in a real family, a work-centered society, and we like to see ourselves as people with the good luck and sense to live in a place where some vestige of the natural world still exists in working order. Many of us hold that natural world as sacred to some degree, just as it is in our myth. Lately, more and more of us are coming to understand our society in the American West as an exploited colony, threatened by greedy outsiders who want to take our sacred place away from us, or at least to strip and degrade it.

In short, we see ourselves as a society of mostly decent people who live with some connection to a holy wilderness, threatened by those who lust for power and property. We look for Shane to come riding out of the Tetons, and instead we see Exxon and the Sierra Club. One looks virtually as alien as the other.

And our mythology tells us we own the West, absolutely and morally—we own it because of our history. Our people brought law to this difficult place, they suffered and they shed blood and they survived, and they earned this land for us. Our efforts have surely earned us the right to absolute control over the thing we created. The myth tells us this place is ours, and will always be ours, to do with as we see fit.

That's a most troubling and enduring message, because we want to believe it, and we do believe it, so many of us, despite its implicit ironies and wrongheadedness, despite the fact that we took the land from someone else. We try to ignore a genocidal history of violence against the Native Americans.

In the American West we are struggling to revise our dominant mythology, and to find a new story to inhabit. Laws control our lives, and they are designed to preserve a model of society based on values learned from mythology. Only after re-imagining our myths can we coherently remodel our laws, and hope to keep our society in a realistic relationship to what is actual.

In Warner Valley we thought we were living the right lives, creating a great precise perfection of fields, and we found the mythology had been telling us an enormous lie. The world had proven too complex, or the myth too simpleminded. And we were mortally angered.

The truth is, we never owned all the land and water. We don't even own very much of them, privately. And we don't own anything absolutely or forever. As our society grows more and more complex and interwoven, our entitlement becomes less and less absolute, more and more likely to be legally diminished. Our rights to property will never take precedence over the needs of society. Nor should they, we all must agree in our grudging hearts. Ownership of property has always been a privilege granted by society, and revokable.

Down by the slaughterhouse my grandfather used to keep a chicken-wire cage for trapping magpies. The cage was as high as a man's head, and mounted on a sled so it could be towed off and cleaned. It worked on the same principle as a lobster trap. Those iridescent black-and-white birds could get in to feed on the intestines of butchered cows—we never butchered a fat heifer or steer for our own consumption, only aged dry cows culled from the breeding herd—but they couldn't get out.

Trapped under the noontime sun, the magpies would flutter around in futile exploration for a while, and then would give in to a great sullen presentiment of their fate, just hopping around picking at left-overs and waiting.

My grandfather was Scots-English, and a very old man by then, but his blue eyes never turned watery and lost. He was one of those cow-men we don't see so often anymore, heedless of most everything out-side his playground, which was livestock and seasons and property, and, as the seasons turned, more livestock and more property, a game which could be called accumulation.

All the notes were paid off, and you would have thought my grandfa-ther would have been secure, and released to ease back in wisdom.

But no such luck. It seemed he had to keep proving his ownership.

This took various forms, like endless litigation, which I have heard described as the sport of kings, but the manifestation I recall most vividly was that of killing magpies.

In the summer the ranch hands would butcher in the after-supper cool of an evening a couple of times a week. About once a week, when a number of magpies had gathered in the trap, maybe 10 or 15, my grandfather would get out his lifetime 12-gauge shotgun and have someone drive him down to the slaughterhouse in his dusty, ancient gray Cadillac, so he could look over his catch and get down to the business at hand. Once there, the ritual was slow and dignified, and always inevitable as one shoe after another.

The old man would sit there a while in his Cadillac and gaze at the magpies with his merciless blue eyes, and the birds would stare back with their hard black eyes. The summer dust would settle around the Cadillac, and the silent confrontation would continue. It would last several minutes.

Then my grandfather would sigh, and swing open the door on his side of the Cadillac, and climb out, dragging his shotgun behind him, the pockets of his gray gabardine suit-coat like a frayed uniform bulging with shells. The stock of the shotgun had been broken sometime deep in the past, and it was wrapped with fine brass wire, which shone golden in the sunlight while the old man thumbed shells into the magazine. All this without saying a word.

In the ear of my mind I try to imagine the radio playing softly in the Cadillac, something like "Room Full of Roses" or "Candy Kisses," but there was no radio. There was just the ongoing hum of insects and the clacking of the mechanism as the old man pumped a shell into the firing chamber.

He would lift the shotgun, and from no more than 12 feet, sighting down that barrel where the bluing was mostly worn off, through the chicken wire into the eyes of those trapped magpies, he would kill them one by one, taking his time, maybe so as to prove that this was no accident.

He would fire and there would be a minor explosion of blood and feathers, the huge booming of the shotgun echoing through the flattening light of early afternoon, off the sage-covered hills and down across the hay meadows and the sloughs lined with dagger-leafed willow, frightening great flights of blackbirds from the fence lines nearby, to rise in flocks and wheel and be gone.

"Bastards," my grandfather would mutter, and then he would take his time about killing another, and finally he would be finished and turn without looking back, and climb into his side of the Cadillac, where the door still stood open. Whoever it was whose turn it was that day would drive him back up the willow-lined lane through the meadows to the ranch house beneath the Lombardy poplar, to the cool shaded living room with its faded linoleum where the old man would finish out his day playing pinochle with my grandmother and anyone else he could gather, sometimes taking a break to retune a favorite program on the Zenith Trans-Oceanic radio.

No one in our family, so far as I ever heard, knew any reason why the old man had come to hate magpies with such specific intensity in his old age. The blackbirds were endlessly worse, the way they would mass together in flocks of literally thousands, to strip and thrash in his oat and barley fields, and then feed all fall in the bins of grain stockpiled to fatten his cattle.

"Where is the difference?" I asked him once, about the magpies.

"Because they're mine," he said. I never did know exactly what he was talking about, the remnants of entrails left over from the butchering of culled stocker cows, or the magpies. But it became clear he was asserting his absolute lordship over both, and over me, too, so long as I was living on his property. For all his life and most of mine the notion of property as absolute seemed like law, even when it never was.

Most of us who grew up owning land in the West believed that any impairment of our right to absolute control of that property was a taking, forbidden by the so-called "taking clause" of the Constitution. We believed regulation of our property rights could never legally reduce the value of our property. After all, what was the point of ownership if it was not profitable? Any infringement on the control of private property was a communist perversion.

But all over the West, as in all of America, the old folkway of property as an absolute right is dying. Our mythology doesn't work anymore.

We find ourselves weathering a rough winter of discontent, snared in the uncertainties of a transitional time and urgently yearning to inhabit a story that might bring sensible order to our lives—even as we know such a story can only evolve through an almost literally infinite series of recognitions of what, individually, we hold sacred. The liberties our people came seeking are more and more constrained, and here in the West, as everywhere, we hate it.

Simple as that. And we have to live with it. There is no more running away to territory. This is it, for most of us. We have no choice but to live in community. If we're lucky we may discover a story that teaches us to abhor our old romance with conquest and possession.

My grandfather died in 1958, toppling out of his chair at the pinochle table, soon after I came back to Warner, but his vision dominated our lives until we sold the ranch in 1967. An ideal of absolute ownership that defines family as property is the perfect device for driving people away from one another. There was a rule in our family. "What's good for the property is good for you."

"Every time there was more money we bought land," my grandmother proclaimed after learning my grandfather had been elected to the Cowboy Hall of Fame. I don't know if she spoke with pride or bitterness, but I do know that, having learned to understand love as property, we were all absolutely divided at the end; relieved to escape amid a litany of divorce and settlements, our family broken in the getaway.

I cannot grieve for my grandfather. It is hard to imagine, these days, that any man could ever again think he owns the birds.

Thank the lord there were other old men involved in my upbringing. My grandfather on my mother's side ran away from a Germanic farmstead in Wisconsin the year he was fourteen, around 1900, and made his way to Butte. "I was lucky," he would say. "I was too young to go down in the mines, so they put me to sharpening steel."

Seems to me such a boy must have been lucky to find work at all, wandering the teeming difficult streets of the most urban city in the American West. "Well, no," he said. "They put you to work. It wasn't like that. They were good to me in Butte. They taught me a trade. That's all I did was work. But it didn't hurt me any."

After most of ten years on the hill—broke and on strike, still a very young man—he rode the rails south to the silver mines in what he called "Old Mexico," and then worked his way back north through the mining country of Nevada in time for the glory days in Goldfield and Rhyolite and Tonopah. At least those are the stories he would tell. "This Las Vegas," he would say. "When I was there you could have bought it all for a hundred and fifty dollars. Cost you ten cents for a drink of water."

To my everlasting sadness, I never really quizzed him on the facts. Now I look at old photographs of those mining camps, and wonder. It's difficult for me to imagine the good gentle man I knew walking those tough dusty streets. He belonged, at least in those Butte days, to the International Brotherhood of Blacksmiths and Helpers. I still have his first dues card. He was initiated July 11, 1904, and most of the months of 1904 and 1905 are stamped, DUES PAID.

Al died in an old folks' home in Eugene, Oregon. During the days of his last summer, when he knew the jig was up, a fact he seemed to regard with infallible good humor, we would sit in his room and listen

to the aged bemused woman across the hall chant her litany of childhood, telling herself that she was somebody and still real.

It was always precisely the same story, word by particular word. I
wondered then how much of it was actual, lifting from some deep
archive in her memory, and now I wonder how much of it was pure
sweet invention, occasioned by the act of storytelling and by the gener-
ative, associative power of language. I cannot help but think of ancient
fires, light flickering on the faces of children and storytellers detailing
the history of their place in the scheme of earth.

The story itself started with a screen-door slamming and her mother
yelling at her when she was a child coming out from the back porch of
a white house, and rotting apples on the ground under the trees in the
orchard, and a dog which snapped at the flies. "Mother," she would
exclaim in exasperation, "I'm fine."

The telling took about three minutes, and she told it like a story for
grandchildren. "That's nice," she would say to her dog. "That's nice."

Then she would lapse into quiet, rewinding herself, seeing an old
time when the world contained solace enough to seem complete, and
she would start over again, going on until she had lulled herself back
into sleep. I would wonder if she was dreaming about that dog amid the
fallen apples, snapping at flies and yellowjackets.

At the end she would call the name of that dog over and over in a
quavering, beseeching voice—and my grandfather would look to me
from his bed and his eyes would be gleaming with laughter, such an
old man laughing painfully, his shoulders shaking, and wheezing.

"Son of a bitch," he would whisper, when she was done calling the
dog again, and he would wipe the tears from his face with the sleeve of
his hospital gown. *Son of a bitch.* He would look to me again, and
other than aimless grinning acknowledgment that some mysterious
thing was truly funny, I wouldn't know what to do, and then he would
look away to the open window, beyond which a far-off lawn mower
droned, like this time he was the one who was embarrassed. Not long
after that he was dead, and so was the old woman across the hall.

"Son of a bitch," I thought, when we were burying Al one bright
afternoon in Eugene, and I found myself suppressing laughter. Maybe
it was just a way of ditching my grief for myself, who did not know him
well enough to really understand what he thought was funny. I have
Al's picture framed on my wall, and I can still look to him and find
relief from the old insistent force of my desire to own things. His laughter is like a gift.

WENDELL BERRY
b. 1934

Our approach to agriculture lays the foundation for our culture as a whole; our attentiveness to our immediate physical environments reflects the clarity of our vision and self-expression. These convictions, developed most systematically in The Unsettling of America *(1977), appear throughout Wendell Berry's life and writing. They are expressed through his decision to reclaim a worn-out hill farm in his native Kentucky; through his novels, such as* The Memory of Old Jack *(1974), which celebrate the virtues and struggles of his ancestors in that land; through his volumes of poetry including* Farming: A Handbook *(1970) and* Clearing *(1977); and through his essays on agriculture, wilderness, and the need for different attitudes toward the land. Like Aldo Leopold, Berry understands that respectful, joyful work is a valid and constructive form of relation to nature. The farm, as well as the wilderness, is precious. Among his recent works of nonfiction developing such values are* Another Turn of the Crank *(1998) and* Life Is a Miracle *(2000).*

AN ENTRANCE TO THE WOODS

On a fine sunny afternoon at the end of September I leave my work in Lexington and drive east on I-64 and the Mountain Parkway. When I leave the Parkway at the little town of Pine Ridge I am in the watershed of the Red River in the Daniel Boone National Forest. From Pine Ridge I take Highway 715 out along the narrow ridgetops, a winding tunnel through the trees. And then I turn off on a Forest Service Road and follow it to the head of a foot trail that goes down the steep valley wall of one of the tributary creeks. I pull my car off the road and lock it, and lift on my pack.

It is nearly five o'clock when I start walking. The afternoon is bril-

Recollected Essays 1965–1980 (San Francisco: North Point, 1981).

liant and warm, absolutely still, not enough air stirring to move a leaf. There is only the steady somnolent trilling of insects, and now and again in the woods below me the cry of a pileated woodpecker. Those, and my footsteps on the path, are the only sounds.

From the dry oak woods of the ridge I pass down into the rock. The foot trails of the Red River Gorge all seek these stony notches that little streams have cut back through the cliffs. I pass a ledge overhanging a sheer drop of the rock, where in a wetter time there would be a waterfall. The ledge is dry and mute now, but on the face of the rock below are the characteristic mosses, ferns, liverwort, meadow rue. And here where the ravine suddenly steepens and narrows, where the shadows are long-lived and the dampness stays, the trees are different. Here are beech and hemlock and poplar, straight and tall, reaching way up into the light. Under them are evergreen thickets of rhododendron. And wherever the dampness is there are mosses and ferns. The faces of the rock are intricately scalloped with veins of ironstone, scooped and carved by the wind.

Finally from the crease of the ravine I am following there begins to come the trickling and splashing of water. There is a great restfulness in the sounds these small streams make; they are going down as fast as they can, but their sounds seem leisurely and idle, as if produced like gemstones with the greatest patience and care.

A little later, stopping, I hear not far away the more voluble flowing of the creek. I go on down to where the trail crosses and begin to look for a camping place. The little bottoms along the creek here are thickety and weedy, probably having been kept clear and cropped or pastured not so long ago. In the more open places are little lavender asters, and the even smaller-flowered white ones that some people call beeweed or farewell-summer. And in low wet places are the richly flowered spikes of great lobelia, the blooms an intense startling blue, exquisitely shaped. I choose a place in an open thicket near the stream, and make camp.

It is a simple matter to make camp. I string up a shelter and put my air mattress and sleeping bag in it, and I am ready for the night. And supper is even simpler, for I have brought sandwiches for this first meal. In less than an hour all my chores are done. It will still be light for a good while, and I go over and sit down on a rock at the edge of the stream.

And then a heavy feeling of melancholy and lonesomeness comes over me. This does not surprise me, for I have felt it before when I have been alone at evening in wilderness places that I am not familiar with. But here it has a quality that I recognize as peculiar to the narrow hollows of the Red River Gorge. These are deeply shaded by the trees and by the valley walls, the sun rising on them late and setting early; they are more dark than light. And there will often be little rapids in the stream that will sound, at a certain distance, exactly like people talking. As I sit on my rock by the stream now, I could swear that there is a party of campers coming up the trail toward me, and for several minutes I stay alert, listening for them, their voices seeming to rise and fall, fade out and lift again, in happy conversation. When I finally realize that it is only a sound the creek is making, though I have not come here for company and do not want any, I am inexplicably sad.

These are haunted places, or at least it is easy to feel haunted in them, alone at nightfall. As the air darkens and the cool of the night rises, one feels the immanence of the wraiths of the ancient tribesmen who used to inhabit the rock houses of the cliffs; of the white hunters from east of the mountains; of the farmers who accepted the isolation of these nearly inaccessible valleys to crop the narrow bottoms and ridges and pasture their cattle and hogs in the woods; of the seekers of quick wealth in timber and ore. For though this is a wilderness place, it bears its part of the burden of human history. If one spends much time here and feels much liking for the place, it is hard to escape the sense of one's predecessors. If one has read of the prehistoric Indians whose flint arrowpoints and pottery and hominy holes and petroglyphs have been found here, then every rock shelter and clifty spring will suggest the presence of those dim people who have disappeared into the earth. Walking along the ridges and the stream bottoms, one will come upon the heaped stones of a chimney, or the slowly filling depression of an old cellar, or will find in the spring a japonica bush or periwinkles or a few jonquils blooming in a thicket that used to be a dooryard. Wherever the land is level enough there are abandoned fields and pastures. And nearly always there is the evidence that one follows in the steps of the loggers.

That sense of the past is probably one reason for the melancholy that I feel. But I know that there are other reasons.

One is that, though I am here in body, my mind and my nerves too are not yet altogether here. We seem to grant to our high-speed roads and our airlines the rather thoughtless assumption that people can change places as rapidly as their bodies can be transported. That, as my own experience keeps proving to me, is not true. In the middle of the afternoon I left off being busy at work, and drove through traffic to the freeway, and then for a solid hour or more I drove sixty or seventy miles an hour, hardly aware of the country I was passing through, because on the freeway one does not have to be. The landscape has been subdued so that one may drive over it at seventy miles per hour without any concession whatsoever to one's whereabouts. One might as well be flying. Though one is in Kentucky one is not experiencing Kentucky; one is experiencing the highway, which might be in nearly any hill country east of the Mississippi.

Once off the freeway, my pace gradually slowed, as the roads became progressively more primitive, from seventy miles an hour to a walk. And now, here at my camping place, I have stopped altogether. But my mind is still keyed to seventy miles an hour. And having come here so fast, it is still busy with the work I am usually doing. Having come here by the freeway, my mind is not so fully here as it would have been if I had come by the crookeder, slower state roads; it is incalculably farther away than it would have been if I had come all the way on foot, as my earliest predecessors came. When the Indians and the first white hunters entered this country they were altogether here as soon as they arrived, for they had seen and experienced fully everything between here and their starting place, and so the transition was gradual and articulate in their consciousness. Our senses, after all, were developed to function at foot speeds; and the transition from foot travel to motor travel, in terms of evolutionary time, has been abrupt. The faster one goes, the more strain there is on the senses, the more they fail to take in, the more confusion they must tolerate or gloss over—and the longer it takes to bring the mind to a stop in the presence of anything. Though the freeway passes through the very heart of this forest, the motorist remains several hours' journey by foot from what is living at the edge of the right-of-way.

But I have not only come to this strangely haunted place in a short time and too fast. I have in that move made an enormous change: I have departed from my life as I am used to living it, and have come into the wilderness. It is not fear that I feel; I have learned to fear the everyday events of human history much more than I fear the everyday occurrences of the woods; in general, I would rather trust myself to the woods than to any government that I know of. I feel, instead, an uneasy awareness of severed connections, of being cut off from all familiar places and of being a stranger where I am. What is happening at home? I wonder, and I know I can't find out very easily or very soon.

Even more discomforting is a pervasive sense of unfamiliarity. In the places I am most familiar with—my house, or my garden, or even the woods near home that I have walked in for years—I am surrounded by associations; everywhere I look I am reminded of my history and my hopes; even unconsciously I am comforted by any number of proofs that my life on the earth is an established and a going thing. But I am in this hollow for the first time in my life. I see nothing that I recognize. Everything looks as it did before I came, as it will when I am gone. When I look over at my little camp I see how tentative and insignificant it is. Lying there in my bed in the dark tonight, I will be absorbed in the being of this place, invisible as a squirrel in his nest.

Uneasy as this feeling is, I know it will pass. Its passing will produce a deep pleasure in being there. And I have felt it often enough before that I have begun to understand something of what it means:

Nobody knows where I am. I don't know what is happening to anybody else in the world. While I am here I will not speak, and will have no reason or need for speech. It is only beyond this lonesomeness for the places I have come from that I can reach the vital reality of a place such as this. Turning toward this place, I confront a presence that none of my schooling and none of my usual assumptions have prepared me for: the wilderness, mostly unknowable and mostly alien, that is the universe. Perhaps the most difficult labor for my species is to accept its limits, its weakness and ignorance. But here I am. This wild place where I have camped lies within an enormous cone widening from the center of the earth out across the universe, nearly all of it a mysterious wilderness in which the power and the knowledge of men count for nothing. As long as its instruments are correct and its engines run, the airplane now flying through this great cone is safely within the human freehold; its behavior is as familiar and predictable to those concerned as the inside of a man's living room. But let its instruments or its engines fail, and at once it enters the wilderness where nothing is foreseeable. And these steep narrow hollows, these cliffs and forested ridges that lie below, are the antithesis of flight.

Wilderness is the element in which we live encased in civilization, as a mollusk lives in his shell in the sea. It is a wilderness that is beautiful, dangerous, abundant, oblivious of us, mysterious, never to be conquered or controlled or second-guessed, or known more than a little. It is a wilderness that for most of us most of the time is kept out of sight, camouflaged, by the edifices and the busyness and the bothers of human society.

And so, coming here, what I have done is strip away the human facade that usually stands between me and the universe, and I see more clearly where I am. What I am able to ignore much of the time, but find undeniable here, is that all wildernesses are one: there is a profound joining between this wild stream deep in one of the folds of my native country and the tropical jungles, the tundras of the north, the oceans and the deserts. Alone here, among the rocks and the trees, I see that I am alone also among the stars. A stranger here, unfamiliar with my surroundings, I am aware also that I know only in the most relative terms my whereabouts within the black reaches of the universe. And because the natural processes are here so little qualified by anything human, this fragment of the wilderness is also joined to other times; there flows over it a nonhuman time to be told by the growth and death of the forest and the wearing of the stream. I feel drawing out beyond my comprehension perspectives from which the growth and the death of a large poplar would seem as continuous and sudden as the raising and the lowering of a man's hand, from which men's history in the world, their brief clearing of the ground, will seem no more than the opening and shutting of an eye.

And so I have come here to enact—not because I want to but because, once here, I cannot help it—the loneliness and the humbleness of my kind. I must see in my flimsy shelter, pitched here for two nights, the transience of capitols and cathedrals. In growing used to being in this place, I will have to accept a humbler and a truer view of myself than I usually have.

A man enters and leaves the world naked. And it is only naked—or nearly so—that he can enter and leave the wilderness. If he walks, that is; and if he doesn't walk it can hardly be said that he has entered. He can bring only what he can carry—the little that it takes to replace for a few hours or a few days an animal's fur and teeth and claws and functioning instincts. In comparison to the usual traveler with his dependence on machines and highways and restaurants and motels—on the economy and the government, in short—the man who walks into the wilderness is naked indeed. He leaves behind his work, his household, his duties, his comforts—even, if he comes alone, his words. He immerses himself in what he is not. It is a kind of death.

The dawn comes slow and cold. Only occasionally, somewhere along the creek or on the slopes above, a bird sings. I have not slept well, and I waken without much interest in the day. I set the camp to rights, and fix breakfast, and eat. The day is clear, and high up on the points and ridges to the west of my camp I can see the sun shining on the woods. And suddenly I am full of an ambition: I want to get up where the sun is; I want to sit still in the sun up there among the high rocks until I can feel its warmth in my bones.

I put some lunch into a little canvas bag, and start out, leaving my jacket so as not to have to carry it after the day gets warm. Without my jacket, even climbing, it is cold in the shadow of the hollow, and I have a long way to go to get to the sun. I climb the steep path up the valley wall, walking rapidly, thinking only of the sunlight above me. It is as though I have entered into a deep sympathy with those tulip poplars that grow so straight and tall out of the shady ravines, not growing a branch worth the name until their heads are in the sun. I am so concentrated on the sun that when some grouse flush from the undergrowth ahead of me, I am thunderstruck; they are already planing down into the underbrush again before I can get my wits together and realize what they are.

The path zigzags up the last steepness of the bluff and then slowly levels out. For some distance it follows the backbone of a ridge, and then where the ridge is narrowest there is a great slab of bare rock lying full in the sun. This is what I have been looking for. I walk out into the center of the rock and sit, the clear warm light falling unobstructed all around. As the sun warms me I begin to grow comfortable not only in my clothes, but in the place and the day. And like those light-seeking poplars of the ravines, my mind begins to branch out.

Southward, I can hear the traffic on the Mountain Parkway, a steady continuous roar—the corporate voice of twentieth-century humanity, sustained above the transient voices of its members. Last night, except for an occasional airplane passing over, I camped out of reach of the sounds of engines. For long stretches of time I heard no sounds but the sounds of the woods.

Near where I am sitting there is an inscription cut into the rock:

A · J · SARGENT
fEB · 24 · 1903

Those letters were carved there more than sixty-six years ago. As I look around me I realize that I can see no evidence of the lapse of so much time. In every direction I can see only narrow ridges and narrow deep hollows, all covered with trees. For all that can be told from this height by looking, it might still be 1903—or, for that matter, 1803 or 1703, or 1003. Indians no doubt sat here and looked over the country as I am doing now; the visual impression is so pure and strong that I can almost imagine myself one of them. But the insistent, the overwhelming, evidence of the time of my own arrival is in what I can hear—that roar of the highway off there in the distance. In 1903 the continent was still covered by a great ocean of silence, in which the sounds of machinery were scattered at wide intervals of time and space. Here, in 1903, there were only the natural sounds of the place. On a day like this, at the end of September, there would have been only the sounds of a few faint crickets, a woodpecker now and then, now and then the wind. But today, two-thirds of a century later, the continent is covered by an ocean of engine noise, in which silences occur only sporadically and at wide intervals.

From where I am sitting in the midst of this island of wilderness, it is as though I am listening to the machine of human history—a huge fly-wheel building speed until finally the force of its whirling will break it in pieces, and the world with it. That is not an attractive thought, and yet I find it impossible to escape, for it has seemed to me for years now that the doings of men no longer occur within nature, but that the natural places which the human economy has so far spared now survive almost accidentally within the doings of men. This wilderness of the Red River now carries on its ancient processes *within* the human climate of war and waste and confusion. And I know that the distant roar of engines, though it may *seem* only to be passing through this wilderness, is really bearing down upon it. The machine is running now with a speed that produces blindness—as to the driver of a speeding automobile the only thing stable, the only thing not a mere blur on the edge of the retina, is the automobile itself—and the blindness of a thing with power promises the destruction of what cannot be seen. That roar of the highway is the voice of the American economy; it is sounding also wherever strip mines are being cut in the steep slopes of Appalachia, and wherever cropland is being destroyed to make roads and suburbs, and wherever rivers and marshes and bays and forests are being destroyed for the sake of industry or commerce.

No. Even here where the economy of life is really an economy— where the creation is yet fully alive and continuous and self-enriching, where whatever dies enters directly into the life of the living—even here one cannot fully escape the sense of an impending human catastrophe. One cannot come here without the awareness that this is an island surrounded by the machinery and the workings of an insane greed, hungering for the world's end—that ours is a "civilization" of which the work of no builder or artist is symbol, nor the life of any good man, but rather the bulldozer, the poison spray, the hugging fire of napalm, the cloud of Hiroshima.

Though from the high vantage point of this stony ridge I see little hope that I will ever live a day as an optimist, still I am not desperate. In fact, with the sun warming me now, and with the whole day before me to wander in this beautiful country, I am happy. A man cannot despair if he can imagine a better life, and if he can enact something of its possibility. It is only when I am ensnarled in the meaningless ordeals and the ordeals of meaninglessness, of which our public and political life is now so productive, that I lose the awareness of something better, and feel the despair of having come to the dead end of possibility.

Today, as always when I am afoot in the woods, I feel the possibility, the reasonableness, the practicability of living in the world in a way that would enlarge rather than diminish the hope of life. I feel the possibility of a frugal and protective love for the creation that would be unimaginably more meaningful and joyful than our present destructive and wasteful economy. The absence of human society, that made me so uneasy last night, now begins to be a comfort to me. I am afoot in the woods. I am alive in the world, this moment, without the help or the interference of any machine. I can move without reference to anything except the lay of the land and the capabilities of my own body. The necessities of foot travel in this steep country have stripped away all superfluities. I simply could not enter into this place and assume its quiet with all the belongings of a family man, property holder, etc. For the time, I am reduced to my irreducible self. I feel the lightness of body that a man must feel who has just lost fifty pounds of fat. As I leave the bare expanse of the rock and go in under the trees again, I am aware that I move in the landscape as one of its details.

Walking through the woods, you can never see far, either ahead or behind, so you move without much of a sense of getting anywhere or of moving at any certain speed. You burrow through the foliage in the air much as a mole burrows through the roots in the ground. The views that open out occasionally from the ridges afford a relief, a recovery of orientation, that they could never give as mere "scenery," looked at from a turnout at the edge of a highway.

The trail leaves the ridge and goes down a ravine into the valley of a creek where the night chill has stayed. I pause only long enough to drink the cold clean water. The trail climbs up onto the next ridge.

It is the ebb of the year. Though the slopes have not yet taken on the bright colors of the autumn maples and oaks, some of the duller trees are already shedding. The foliage has begun to flow down the cliff faces and the slopes like a tide pulling back. The woods is mostly quiet, subdued, as if the pressure of survival has grown heavy upon it, as if above the growing warmth of the day the cold of winter can be felt waiting to descend.

At my approach a big hawk flies off the low branch of an oak and out over the treetops. Now and again a nuthatch hoots, off somewhere in the woods. Twice I stop and watch an ovenbird. A few feet ahead of me there is a sudden movement in the leaves, and then quiet. When I slip up and examine the spot there is nothing to be found. Whatever passed there has disappeared, quicker than the hand that is quicker than the eye, a shadow fallen into a shadow.

In the afternoon I leave the trail. My walk so far has come perhaps three-quarters of the way around a long zig-zagging loop that will eventually bring me back to my starting place. I turn down a small unnamed branch of the creek where I am camped, and I begin the loveliest part of the day. There is nothing here resembling a trail. The best way is nearly always to follow the edge of the stream, stepping from one stone to another. Crossing back and forth over the water, stepping on or over rocks and logs, the way ahead is never clear for more than a few feet. The stream accompanies me down, threading its way under boulders and logs and over little falls and rapids. The rhododendron overhangs it so closely in places that I can go only by stopping. Over the rhododendron are the great dark heads of the hemlocks. The streambanks are ferny and mossy. And through this green tunnel the voice of the stream changes from rock to rock; subdued like all the other autumn voices of the woods, it seems sunk in a deep contented meditation on the sounds of *l*.

The water in the pools is absolutely clear. If it weren't for the shadows and ripples you would hardly notice that it is water; the fish would seem to swim in the air. As it is, where there is no leaf floating, it is impossible to tell exactly where the plane of the surface lies. As I walk up on a pool the little fish dart every which way out of sight. And then after I sit still a while, watching, they come out again. Their shadows flow over the rocks and leaves on the bottom. Now I have come into the heart of the woods. I am far from the highway and can hear no sound of it. All around there is a grand deep autumn quiet, in which a few insects dream their summer songs. Suddenly a wren sings way off in the underbrush. A red-breasted nuthatch walks, hooting, headfirst down the trunk of a walnut. An ovenbird walks out along the limb of a hemlock and looks at me, curious. The little fish soar in the pool, turning their clean quick angles, their shadows seeming barely to keep up. As I lean and dip my cup in the water, they scatter. I drink, and go on.

When I get back to camp it is only the middle of the afternoon or a little after. Since I left in the morning I have walked something like eight miles. I haven't hurried—have mostly poked along, stopping often and looking around. But I am tired, and coming down the creek I have got both feet wet. I find a sunny place, and take off my shoes and socks and set them to dry. For a long time then, lying propped against the trunk of a tree, I read and rest and watch the evening come.

All day I have moved through the woods, making as little noise as possible. Slowly my mind and my nerves have slowed to a walk. The quiet of the woods has ceased to be something that I observe; now it is something that I am a part of. I have joined it with my own quiet. As the twilight draws on I no longer feel the strangeness and uneasiness of the evening before. The sounds of the creek move through my mind as they move through the valley, unimpeded and clear.

When the time comes I prepare supper and eat, and then wash kettle and cup and spoon and put them away. As far as possible I get things ready for an early start in the morning. Soon after dark I go to bed, and I sleep well.

I wake long before dawn. The air is warm and I feel rested and wide awake. By the light of a small candle lantern I break camp and pack. And then I begin the steep climb back to the car.

The moon is bright and high. The woods stands in deep shadow, the light falling soft through the openings of the foliage. The trees appear immensely tall, and black, gravely looming over the path. It is windless and still; the moonlight pouring over the country seems more potent than the air. All around me there is still that constant low singing of the insects. For days now it has continued without letup or inflection, like ripples on water under a steady breeze. While I slept it went on through the night, a shimmer on my mind. My shoulder brushes a low tree overhanging the path and a bird that was asleep on one of the branches startles awake and flies off into the shadows, and I go on with the sense that I am passing near to the sleep of things.

In a way this is the best part of the trip. Stopping now and again to rest, I linger over it, sorry to be going. It seems to me that if I were to stay on, today would be better than yesterday, and I realize it was to renew the life of that possibility that I came here. What I am leaving is something to look forward to.

THE MAKING OF A MARGINAL FARM

One day in the summer of 1956, leaving home for school, I stopped on the side of the road directly above the house where I now live. From there you could see a mile or so across the Kentucky River Valley, and perhaps six miles along the length of it. The valley was a green trough full of sunlight, blue in its distances. I often stopped here in my comings and goings, just to look, for it was all familiar to me from before the time my memory began: woodlands and pastures on the hillsides; fields and croplands, wooded slew-edges and hollows in the bottoms; and through the midst of it the tree-lined river passing down from its headwaters near the Virginia line toward its mouth at Carrollton on the Ohio.

Standing there, I was looking at land where one of my great-great-great-grandfathers settled in 1803, and at the scene of some of the happiest times of my own life, where in my growing-up years I camped, hunted, fished, boated, swam, and wandered—where, in short, I did whatever escaping I felt called upon to do. It was a place where I had happily been, and where I always wanted to be. And I remember gesturing toward the valley that day and saying to the friend who was with me: "That's all I need."

I meant it. It was an honest enough response to my recognition of its beauty, the abundance of its lives and possibilities, and of my own love for it and interest in it. And in the sense that I continue to recognize all that, and feel that what I most need is here, I can still say the same thing.

And yet I am aware that I must necessarily mean differently—or at least a great deal more—when I say it now. Then I was speaking mostly from affection, and did not know, by half, what I was talking about. I was speaking of a place that in some ways I knew and in some ways cared for, but did not live in. The differences between knowing a place and living in it, between cherishing a place and living responsibly in it, had not begun to occur to me. But they are critical differences, and understanding them has been perhaps the chief necessity of my experience since then.

I married in the following summer, and in the next seven years lived in a number of distant places. But, largely because I continued to feel that what I needed was here, I could never bring myself to want to live in any other place. And so we returned to live in Kentucky in the summer of 1964, and that autumn bought the house whose roof my friend and I had looked down on eight years before, and with it "twelve acres more or less." Thus I began a profound change in my life. Before, I had lived according to expectation rooted in ambition. Now I began to live according to a kind of destiny rooted in my origins and in my life. One should not speak too confidently of one's "destiny;" I use the word to refer to causes that lie deeper in history and character than mere intention or desire. In buying the little place known as Lanes Landing, it seems to me, I began to obey the deeper causes.

We had returned so that I could take a job at the University of Kentucky in Lexington. And we expected to live pretty much the usual academic life: I would teach and write; my "subject matter" would be, as it had been, the few square miles in Henry County where I grew up. We bought the tiny farm at Lanes Landing, thinking that we would use it as a "summer place," and on that understanding I began, with the help of two carpenter friends, to make some necessary repairs on the house. I no longer remember exactly how it was decided, but that work had hardly begun when it became a full-scale overhaul.

By so little our minds had been changed: this was not going to be a house to visit, but a house to live in. It was as though, having put our hand to the plow, we not only did not look back, but could not. We renewed the old house, equipped it with plumbing, bathroom, and oil furnace, and moved in on July 4, 1965.

Once the house was whole again, we came under the influence of the "twelve acres more or less." This acreage included a steep hillside pasture, two small pastures by the river, and a "garden spot" of less than half an acre. We had, besides the house, a small barn in bad shape, a good large building that once had been a general store, and a small garage also in usable condition. This was hardly a farm by modern standards, but it was land that could be used, and it was unthinkable that we would not use it. The land was not good enough to afford the possibility of a cash income, but it would allow us to grow our food—or most of it. And that is what we set out to do.

In the early spring of 1965 I had planted a small orchard; the next spring we planted our first garden. Within the following six or seven years we reclaimed the pastures, converted the garage into a henhouse, rebuilt the barn, greatly improved the garden soil, planted berry bushes, acquired a milk cow—and were producing, except for hay and grain for our animals, nearly everything that we ate: fruit, vegetables, eggs, meat, milk, cream, and butter. We built an outbuilding with a meat room and a food-storage cellar. Because we did not want to pollute our land and water with sewage, and in the process waste nutrients that should be returned to the soil, we built a composting privy. And so we began to attempt a life that, in addition to whatever else it was, would be responsibly agricultural. We used no chemical fertilizers. Except for a little rotenone, we used no insecticides. As our land and our food became healthier, so did we. And our food was of better quality than any that we could have bought.

We were not, of course, living an idyll. What we had done could not have been accomplished without difficulty and a great deal of work. And we had made some mistakes and false starts. But there was great satisfaction, too, in restoring the neglected land, and in feeding ourselves from it.

Meanwhile, the forty-acre place adjoining ours on the downriver side had been sold to a "developer," who planned to divide it into lots for "second homes." This project was probably doomed by the steep-

ness of the ground and the difficulty of access, but a lot of bulldoz-
ing—and a lot of damage—was done before it was given up. In the fall
of 1972, the place was offered for sale and we were able to buy it.

We now began to deal with larger agricultural problems. Some of this
new land was usable; some would have to be left in trees. There were
perhaps fifteen acres of hillside that could be reclaimed for pasture, and
about two and a half acres of excellent bottomland on which we would
grow alfalfa for hay. But it was a mess, all of it badly neglected, and a considerable portion of it badly abused by the developer's bulldozers. The
hillsides were covered with thicket growth; the bottom was shoulder high
in weeds; the diversion ditches had to be restored; a bulldozed gash
meant for "building sites" had to be mended; the barn needed a new
foundation, and the cistern a new top; there were no fences. What we
had bought was less a farm than a reclamation project—which has now,
with a later purchase, grown to seventy-five acres.

While we had only the small place, I had got along very well with a
Gravely "walking tractor" that I owned, and an old Farmall A that I
occasionally borrowed from my Uncle Jimmy. But now that we had
increased our acreage, it was clear that I could not continue to depend
on a borrowed tractor. For a while I assumed that I would buy a tractor
of my own. But because our land was steep, and there was already talk
of a fuel shortage—and because I liked the idea—I finally decided to
buy a team of horses instead. By the spring of 1973, after a lot of inquiring and looking, I had found and bought a team of five-year-old sorrel
mares. And—again by the generosity of my Uncle Jimmy, who has
never thrown any good thing away—I had enough equipment to make
a start.

Though I had worked horses and mules during the time I was growing up, I had never worked over ground so steep and problematical as
this, and it had been twenty years since I had worked a team over
ground of any kind. Getting started again, I anticipated every new task
with uneasiness, and sometimes with dread. But to my relief and
delight, the team and I did all that needed to be done that year, getting
better as we went along. And over the years since then, with that team
and others, my son and I have carried on our farming the way it was
carried on in my boyhood, doing everything with our horses except bal-
ing the hay. And we have done work in places and in weather in which
a tractor would have been useless. Experience has shown us—or
re-shown us—that horses are not only a satisfactory and economical
means of power, especially on such small places as ours, but are probably *necessary* to the most conservative use of steep land. Our farm, in

fact, is surrounded by potentially excellent hillsides that were maintained in pasture until tractors replaced the teams.

Another change in our economy (and our lives) was accomplished
in the fall of 1973 with the purchase of our first wood-burning stove.
Again the petroleum shortage was on our minds, but we also knew
that from the pasture-clearing we had ahead of us we would have an
abundance of wood that otherwise would go to waste — and when that
was gone we would still have our permanent wood lots. We thus
expanded our subsistence income to include heating fuel, and since
then have used our furnace only as a "backup system" in the coldest
weather and in our absences from home. The horses also contribute
significantly to the work of fuel-gathering; they will go easily into difficult places and over soft ground or snow where a truck or a tractor
could not move.

As we have continued to live on and from our place, we have slowly
begun its restoration and healing. Most of the scars have now been
mended and grassed over, most of the washes stopped, most of the
buildings made sound; many loads of rocks have been hauled out of the
fields and used to pave entrances or fill hollows; we have done perhaps
half of the necessary fencing. A great deal of work is still left to do, and
some of it — the rebuilding of fertility in the depleted hillsides — will take
longer than we will live. But in doing these things we have begun a
restoration and a healing in ourselves.

I should say plainly that this has not been a "paying proposition." As
a reclamation project, it has been costly both in money and in effort. It
seems at least possible that, in any other place, I might have had little
interest in doing any such thing. The reason I have been interested in
doing it here, I think, is that I have felt implicated in the history, the
uses, and the attitudes that have depleted such places as ours and made
them "marginal."

I had not worked long on our "twelve acres more or less" before I
saw that such places were explained almost as much by their human
history as by their nature. I saw that they were not "marginal" because
they ever were unfit for human use, but because in both culture and
character *we* had been unfit to use them. Originally, even such steep
slopes as these along the lower Kentucky River Valley were deep-soiled
and abundantly fertile; "jumper" plows and generations of carelessness
impoverished them. Where yellow clay is at the surface now, five feet
of good soil may be gone. I once wrote that on some of the nearby
uplands one walks as if "knee-deep" in the absence of the original soil.
On these steeper slopes, I now know, that absence is shoulder-deep.

That is a loss that is horrifying as soon as it is imagined. It happened easily, by ignorance, indifference, "a little folding of the hands to sleep." It cannot be remedied in human time; to build five feet of soil takes perhaps fifty or sixty thousand years. This loss, once imagined, is potent with despair. If a people in adding a hundred and fifty years to itself subtracts fifty thousand from its land, what is there to hope?

And so our reclamation project has been, for me, less a matter of idealism or morality than a kind of self-preservation. A destructive history, once it is understood as such, is a nearly insupportable burden. Understanding it is a disease of understanding, depleting the sense of efficacy and paralyzing effort, unless it finds healing work. For me that work has been partly of the mind, in what I have written, but that seems to have depended inescapably on work of the body and of the ground. In order to affirm the values most native and necessary to me—indeed, to affirm my own life as a thing decent in possibility—I needed to know in my own experience that this place did not have to be abused in the past, and that it can be kindly and conservingly used now.

With certain reservations that must be strictly borne in mind, our work here has begun to offer some of the needed proofs.

Bountiful as the vanished original soil of the hillsides may have been, what remains is good. It responds well—sometimes astonishingly well—to good treatment. It never should have been plowed (some of it never should have been cleared), and it never should be plowed again. But it can be put in pasture without plowing, and it will support an excellent grass sod that will in turn protect it from erosion, if properly managed and not overgrazed.

Land so steep as this cannot be preserved in row crop cultivation. To subject it to such an expectation is simply to ruin it, as its history shows. Our rule, generally, has been to plow no steep ground, to maintain in pasture only such slopes as can be safely mowed with a horse-drawn mower, and to leave the rest in trees. We have increased the numbers of livestock on our pastures gradually, and have carefully rotated the animals from field to field, in order to avoid overgrazing. Under this use and care, our hillsides have mended and they produce more and better pasturage every year.

As a child I always intended to be a farmer. As a young man, I gave up that intention, assuming that I could not farm and do the other things I wanted to do. And then I became a farmer almost unintentionally and by a kind of necessity. That wayward and necessary becoming—along with my marriage, which has been intimately a part of it—is the major event of my life. It has changed me profoundly from the man and the writer I would otherwise have been.

There was a time, after I had left home and before I came back, when this place was my "subject matter." I meant that too, I think, on the day in 1956 when I told my friend, "That's all I need." I was regarding it, in a way too easy for a writer, as a mirror in which I saw myself. There was obviously a sort of narcissism in that—and an inevitable superficiality, for only the surface can reflect.

In coming home and settling on this place, I began to *live* in my subject, and to learn that living in one's subject is not at all the same as "having" a subject. To live in the place that is one's subject is to pass through the surface. The simplifications of distance and mere observation are thus destroyed. The obsessively regarded reflection is broken and dissolved. One sees that the mirror was a blinder; one can now begin to see where one is. One's relation to one's subject ceases to be merely emotional or esthetical, or even merely critical, and becomes problematical, practical, and responsible as well. Because it must. It is like marrying your sweetheart.

Though our farm has not been an economic success, as such success is usually reckoned, it is nevertheless beginning to make a kind of economic sense that is consoling and hopeful. Now that the largest expenses of purchase and repair are behind us, our income from the place is beginning to run ahead of expenses. As income I am counting the value of shelter, subsistence, heating fuel, and money earned by the sale of livestock. As expenses I am counting maintenance, newly purchased equipment, extra livestock feed, newly purchased animals, reclamation work, fencing materials, taxes, and insurance.

If our land had been in better shape when we bought it, our expenses would obviously be much smaller. As it is, once we have completed its restoration, our farm will provide us a home, produce our subsistence, keep us warm in winter, and earn a modest cash income. The significance of this becomes apparent when one considers that most of this land is "unfarmable" by the standards of conventional agriculture, and that most of it was producing nothing at the time we bought it.

And so, contrary to some people's opinion, it *is* possible for a family to live on such "marginal" land, to take a bountiful subsistence and some cash income from it, and, in doing so, to improve both the land and themselves. (I believe, however, that, at least in the present economy, this should not be attempted without a source of income other than the farm. It is now extremely difficult to pay for the best of farmland by farming it, and even "marginal" land has become unreasonably expensive. To attempt to make a living from such land is to impose a severe strain on land and people alike.)

I said earlier that the success of our work here is subject to reservations. There are only two of these, but both are serious.

The first is that land like ours—and there are many acres of such land in this country—can be conserved in use only by competent knowledge, by a great deal more work than is required by leveler land, by a devotion more particular and disciplined than patriotism, and by ceaseless watchfulness and care. All these are cultural values and resources, never sufficiently abundant in this country, and now almost obliterated by the contrary values of the so-called "affluent society."

One of my own mistakes will suggest the difficulty. In 1974 I dug a small pond on a wooded hillside that I wanted to pasture occasionally. The excavation for that pond—as I should have anticipated, for I had better reason than I used—caused the hillside to slump both above and below. After six years the slope has not stabilized, and more expense and trouble will be required to stabilize it. A small hillside farm will not survive many mistakes of that order. Nor will a modest income.

The true remedy for mistakes is to keep from making them. It is not in the piecemeal technological solutions that our society now offers, but in a change of cultural (and economic) values that will encourage in the whole population the necessary respect, restraint, and care. Even more important, it is in the possibility of settled families and local communities, in which the knowledge of proper means and methods, proper moderations and restraints, can be handed down, and so accumulate in place and stay alive; the experience of one generation is not adequate to inform and control its actions. Such possibilities are not now in sight in this country.

The second reservation is that we live at the lower end of the Kentucky River watershed, which has long been intensively used, and is increasingly abused. Strip mining, logging, extractive farming, and the digging, draining, roofing, and paving that go with industrial and urban "development," all have seriously depleted the capacity of the watershed to retain water. This means not only that floods are higher and more frequent than they would be if the watershed were healthy, but that the floods subside too quickly, the watershed being far less a sponge, now, than it is a roof. The floodwater drops suddenly out of the river, leaving the steep banks soggy, heavy, and soft. As a result, great strips and blocks of land crack loose and slump, or they give way entirely and disappear into the river in what people here call "slips."

The flood of December 1978, which was unusually high, also went down extremely fast, falling from banktop almost to pool stage within a couple of days. In the aftermath of this rapid "drawdown," we lost a block of bottom-land an acre square. This slip, which is still crumbling, severely damaged our place, and may eventually undermine two buildings. The same flood started a slip in another place, which threatens a third building. We have yet another building situated on a huge (but, so far, very gradual) slide that starts at the river and, aggravated by two state highway cuts, goes almost to the hilltop. And we have serious river bank erosion the whole length of our place.

What this means is that, no matter how successfully we may control erosion on our hillsides, our land remains susceptible to a more serious cause of erosion that we cannot control. Our river bank stands literally at the cutting edge of our nation's consumptive economy. This, I think, is true of many "marginal" places—it is true, in fact, of many places that are not marginal. In its consciousness, ours is an upland society; the ruin of watersheds, and what that involves and means, is little considered. And so the land is heavily taxed to subsidize an "affluence" that consists, in reality, of health and goods stolen from the unborn.

Living at the lower end of the Kentucky River watershed is what is now known as "an educational experience"—and not an easy one. A lot of information comes with it that is severely damaging to the reputation of our people and our time. From where I live and work, I never have to look far to see that the earth does indeed pass away. But however that is taught, and however bitterly learned, it is something that should be known, and there is a certain good strength in knowing it. To spend one's life farming a piece of the earth so passing is, as many would say, a hard lot. But it is, in an ancient sense, the human lot. What saves it is to love the farming.

N. SCOTT MOMADAY
b. 1934

Momaday manages in his writing both to celebrate and extend his Kiowa heritage and to enrich the traditions of fiction, poetry, and nature writing in English. Among his books are The Way to Rainy Mountain *(1969), retelling Kiowa folktales; a novel,* House Made of Dawn *(1968); and* The Gourd Dancer *(1976), a volume of poems. One of the central challenges he sets for himself as an artist is to convey Native American insights and symbols in new language and forms. The key is to understand the power, and dangers, of naming, rightly understood. Momaday has said in an interview that "I believe that the Indian has an understanding of the physical world and of the earth as a spiritual entity that is his, very much his own. The non-Indian can benefit a good deal by having that perception revealed to him."*

THE WAY TO RAINY MOUNTAIN

A single knoll rises out of the plain in Oklahoma, north and west of the Wichita Range. For my people, the Kiowas, it is an old landmark, and they gave it the name Rainy Mountain. The hardest weather in the world is there. Winter brings blizzards, hot tornadic winds arise in the spring, and in summer the prairie is an anvil's edge. The grass turns brittle and brown, and it cracks beneath your feet. There are green belts along the rivers and creeks, linear groves of hickory and pecan, willow and witch hazel. At a distance in July or August the steaming foliage seems almost to writhe in fire. Great green and yellow grasshoppers are everywhere in the tall grass, popping up like corn to sting the flesh, and tortoises crawl about on the red earth, going nowhere in the plenty of time. Loneliness is an aspect of the land. All things in the plain are isolate; there is no confusion of objects in the eye, but *one* hill or *one* tree

The Way to Rainy Mountain (Albuquerque: University of New Mexico Press, 1969).

or *one* man. To look upon that landscape in the early morning, with the sun at your back, is to lose the sense of proportion. Your imagination comes to life, and this, you think, is where Creation was begun.

I returned to Rainy Mountain in July. My grandmother had died in the spring, and I wanted to be at her grave. She had lived to be very old and at last infirm. Her only living daughter was with her when she died, and I was told that in death her face was that of a child.

I like to think of her as a child. When she was born, the Kiowas were living the last great moment of their history. For more than a hundred years they had controlled the open range from the Smoky Hill River to the Red, from the headwaters of the Canadian to the fork of the Arkansas and Cimarron. In alliance with the Comanches, they had ruled the whole of the southern Plains. War was their sacred business, and they were among the finest horsemen the world has ever known. But warfare for the Kiowas was preeminently a matter of disposition rather than of survival, and they never understood the grim, unrelenting advance of the U.S. Cavalry. When at last, divided and ill-provisioned, they were driven onto the Staked Plains in the cold rains of autumn, they fell into panic. In Palo Duro Canyon they abandoned their crucial stores to pillage and had nothing then but their lives. In order to save themselves, they surrendered to the soldiers at Fort Sill and were imprisoned in the old stone corral that now stands as a military museum. My grandmother was spared the humiliation of those high gray walls by eight or ten years, but she must have known from birth the affliction of defeat, the dark brooding of old warriors.

Her name was Aho, and she belonged to the last culture to evolve in North America. Her forebears came down from the high country in western Montana nearly three centuries ago. They were a mountain people, a mysterious tribe of hunters whose language has never been positively classified in any major group. In the late seventeenth century they began a long migration to the south and east. It was a journey toward the dawn, and it led to a golden age. Along the way the Kiowas were befriended by the Crows, who gave them the culture and religion of the Plains. They acquired horses, and their ancient nomadic spirit was suddenly free of the ground. They acquired Tai-me, the sacred Sun Dance doll, from that moment the object and symbol of their worship, and so shared in the divinity of the sun. Not least, they acquired the sense of destiny, therefore courage and pride. When they entered upon the southern Plains they had been transformed. No longer were they slaves to the simple necessity of survival; they were a lordly and dangerous society of fighters and thieves, hunters and priests of the sun. According to their origin myth, they entered the world through a hollow log. From one point of view, their migration was the fruit of an old prophecy, for indeed they emerged from a sunless world.

Although my grandmother lived out her long life in the shadow of Rainy Mountain, the immense landscape of the continental interior lay like memory in her blood. She could tell of the Crows, whom she had never seen, and of the Black Hills, where she had never been. I wanted to see in reality what she had seen more perfectly in the mind's eye, and traveled fifteen hundred miles to begin my pilgrimage.

Yellowstone, it seemed to me, was the top of the world, a region of deep lakes and dark timber, canyons and waterfalls. But, beautiful as it is, one might have the sense of confinement there. The skyline in all directions is close at hand, the high wall of the woods and deep cleavages of shade. There is a perfect freedom in the mountains, but it belongs to the eagle and the elk, the badger and the bear. The Kiowas reckoned their stature by the distance they could see, and they were bent and blind in the wilderness.

Descending eastward, the highland meadows are a stairway to the plain. In July the inland slope of the Rockies is luxuriant with flax and buckwheat, stonecrop and larkspur. The earth unfolds and the limit of the land recedes. Clusters of trees, and animals grazing far in the distance, cause the vision to reach away and wonder to build upon the mind. The sun follows a longer course in the day, and the sky is immense beyond all comparison. The great billowing clouds that sail upon it are the shadows that move upon the grain like water, dividing light. Farther down, in the land of the Crows and Blackfeet, the plain is yellow. Sweet clover takes hold of the hills and bends upon itself to cover and seal the soil. There the Kiowas paused on their way; they had come to the place where they must change their lives. The sun is at home on the plains. Precisely there does it have the certain character of a god. When the Kiowas came to the land of the Crows, they could see the dark lees of the hills at dawn across the Bighorn River, the profusion of light on the grain shelves, the oldest deity ranging after the solstices. Not yet would they veer southward to the caldron of the land that lay below; they must wean their blood from the northern winter and hold the mountains a while longer in their view. They bore Tai-me in procession to the east.

A dark mist lay over the Black Hills, and the land was like iron. At the top of a ridge I caught sight of Devil's Tower upthrust against the gray sky as if in the birth of time the core of the earth had broken through its crust and the motion of the world was begun. There are things in nature that engender an awful quiet in the heart of man; Devil's Tower is one of them. Two centuries ago, because they could not do otherwise, the Kiowas made a legend at the base of the rock. My grandmother said:

> *Eight children were there at play, seven sisters and their brother. Suddenly the boy was struck dumb; he trembled and began to run upon his hands and feet. His fingers became claws, and his body was covered with fur. Directly there was a bear where the boy had been. The sisters were terrified; they ran, and the bear after them. They came to the stump of a great tree, and the tree spoke to them. It bade them climb upon it, and as they did so it began to rise into the air. The bear came to kill them, but they were just beyond its reach. It reared against the tree and scored the bark all around with its claws. The seven sisters were borne into the sky, and they became the stars of the Big Dipper.*

From that moment, and so long as the legend lives, the Kiowas have kinsmen in the night sky. Whatever they were in the mountains, they could be no more. However tenuous their well-being, however much they had suffered and would suffer again, they had found a way out of the wilderness.

My grandmother had a reverence for the sun, a holy regard that now is all but gone out of mankind. There was a wariness in her, and an ancient awe. She was a Christian in her later years, but she had come a long way about, and she never forgot her birthright. As a child she had been to the Sun Dances; she had taken part in those annual rites, and by them she had learned the restoration of her people in the presence of Tai-me. She was about seven when the last Kiowa Sun Dance was held in 1887 on the Washita River above Rainy Mountain Creek. The buffalo were gone. In order to consummate the ancient sacrifice—to impale the head of a buffalo bull upon the medicine tree—a delegation of old men journeyed into Texas, there to beg and barter for an animal from the Goodnight herd. She was ten when the Kiowas came together for the last time as a living Sun Dance culture. They could find no buffalo; they had to hang an old hide from the sacred tree. Before the dance could begin, a company of soldiers rode out from Fort Sill under orders to disperse the tribe. Forbidden without cause the essential act of their faith, having seen the wild herds slaughtered and left to rot upon the ground, the Kiowas backed away forever from the medicine tree. That was July 20, 1890, at the great bend of the Washita. My grandmother was there. Without bitterness, and for as long as she lived, she bore a vision of deicide.

Now that I can have her only in memory, I see my grandmother in the several postures that were peculiar to her: standing at the wood stove on a winter morning and turning meat in a great iron skillet; sitting at the south window, bent above her beadwork, and afterwards, when her vision failed, looking down for a long time into the fold of her hands; going out upon a cane, very slowly as she did when the weight of age came upon her; praying. I remember her most often at prayer. She made long, rambling prayers out of suffering and hope, having seen many things. I was never sure that I had the right to hear, so exclusive where they of all mere custom and company. The last time I saw her she prayed standing by the side of her bed at night, naked to the waist, the light of a kerosene lamp moving upon her dark skin. Her long, black hair, always drawn and braided in the day, lay upon her shoulders and against her breasts like a shawl. I do not speak Kiowa, and I never understood her prayers, but there was something inherently sad in the sound, some merest hesitation upon the syllables of sorrow. She began in a high and descending pitch, exhausting her breath to silence; then again and again—and always the same intensity of effort, of something that is, and is not, like urgency in the human voice. Transported so in the dancing light among the shadows of her room, she seemed beyond the reach of time. But that was illusion; I think I knew then that I should not see her again.

Houses are like sentinels in the plain, old keepers of the weather watch. There, in a very little while, wood takes on the appearance of great age. All colors wear soon away in the wind and rain, and then the wood is burned gray and the grain appears and the nails turn red with rust. The windowpanes are black and opaque; you imagine there is nothing within, and indeed there are many ghosts, bones given up to the land. They stand here and there against the sky, and you approach them for a longer time than you expect. They belong in the distance; it is their domain.

Once there was a lot of sound in my grandmother's house, a lot of coming and going, feasting and talk. The summers there were full of excitement and reunion. The Kiowas are a summer people; they abide the cold and keep to themselves, but when the season turns and the land becomes warm and vital they cannot hold still; an old love of going returns upon them. The aged visitors who came to my grandmother's house when I was a child were made of lean and leather, and they bore themselves upright. They wore great black hats and bright ample shirts that shook in the wind. They rubbed fat upon their hair and wound their braids with strips of colored cloth. Some of them painted their faces and carried the scars of old and cherished enmities. They were an old council of warlords, come to remind and be reminded of who they were. Their wives and daughters served them well. The women might indulge themselves; gossip was at once the mark and compensation of their servitude. They made loud and elaborate talk among themselves, full of jest and gesture, fright and false alarm. They went abroad in fringed and flowered shawls, bright beadwork and German silver. They were at home in the kitchen, and they prepared meals that were banquets.

There were frequent prayer meetings, and great nocturnal feasts. When I was a child I played with my cousins outside, where the lamp-light fell upon the ground and the singing of the old people rose up around us and carried away into the darkness. There were a lot of good things to eat, a lot of laughter and surprise. And afterwards, when the quiet returned, I lay down with my grandmother and could hear the frogs away by the river and feel the motion of the air.

Now there is a funeral silence in the rooms, the endless wake of some final word. The walls have closed in upon my grandmother's house. When I returned to it in mourning, I saw for the first time in my life how small it was. It was late at night, and there was a white moon, nearly full. I sat for a long time on the stone steps by the kitchen door. From there I could see out across the land; I could see the long row of trees by the creek, the low light upon the rolling plains, and the stars of the Big Dipper. Once I looked at the moon and caught sight of a strange thing. A cricket had perched upon the handrail, only a few inches away from me. My line of vision was such that the creature filled the moon like a fossil. It had gone there, I thought, to live and die, for there, of all places, was its small definition made whole and eternal. A warm wind rose up and purled like the longing within me.

The next morning I awoke at dawn and went out on the dirt road to Rainy Mountain. It was already hot, and the grasshoppers began to fill the air. Still, it was early in the morning, and the birds sang out of the shadows. The long yellow grass on the mountain shone in the bright light, and a scissortail hied above the land. There, where it ought to be, at the end of a long and legendary way, was my grandmother's grave. Here and there on the dark stones were ancestral names. Looking back once, I saw the mountain and came away.

SUE HUBBELL
b. 1935

A former librarian, Sue Hubbell earned her living for many years as a beekeeper in the Ozark Mountains, where she moved in 1973. A Country Year (1987), her first book, is as much an account of a middle-aged woman coming into her own identity as it is a careful record of natural and human life in rural Missouri. Beginning with the pain of a divorce, the book describes her choice to forge an independent life in the country and her discovery of new enthusiasms: "Wild things and wild places pull me more strongly than they did a few years ago, and domesticity, dusting and cookery not at all." Her viewpoint is both clear-eyed and compassionate, but she deliberately shies away from grand statements and "nobler quests—white whales and Holy Grails—" making a case for small things well done and clearly seen. Most recently she has turned her attention to life along the Maine shore in Waiting for Aphrodite: Journeys into the Time Before Bones *(1999).*

From A COUNTRY YEAR

SPRING

* * * I met Paul, the boy who was to become my husband, when he was sixteen and I was fifteen. We were married some years later, and the legal arrangement that is called marriage worked well enough while we were children and while we had a child. But we grew older, and the son went off to school, and marriage did not serve as a structure for our lives as well as it once had. Still, he was the man in my life for all those years. There was no other. So when the legal arrangement was ended, I had a difficult time sifting through the emotional debris that was left after the framework of an intimate, thirty-year association had broken.

A *Country Year* (New York: Harper and Row, 1987).

I went through all the usual things: I couldn't sleep or eat, talked feverishly to friends, plunged recklessly into a destructive affair with a man who had more problems than I did but who was convenient, made a series of stupid decisions about my honey business and pretty generally botched up my life for several years running. And for a long, long time, my mind didn't work. I could not listen to the news on the radio with understanding. My attention came unglued when I tried to read anything but the lightest froth. My brain spun in endless, painful loops, and I could neither concentrate nor think with any semblance of order. I had always rather enjoyed having a mind, and I missed mine extravagantly. I was out to lunch for three years.

I mused about structure, framework, schemata, system, classification and order. I discovered a classification Jorge Luis Borges devised, claiming that

> A *certain Chinese encyclopedia divides animals into:*
> a. Belonging to the Emperor
> b. Embalmed
> c. Tame
> d. Sucking pigs
> e. Sirens
> f. Fabulous
> g. Stray dogs
> h. Included in the present classification
> i. Frenzied
> j. Innumerable
> k. Drawn with a very fine camel-hair brush
> l. Et cetera
> m. Having just broken the water pitcher
> n. That from a long way off look like flies.

Friends and I laughed over the list, and we decided that the fact that we did so tells more about us and our European, Western way of thinking than it does about a supposed Oriental world view. We believe we have a more proper concept of how the natural world should be classified, and when Borges rumples that concept it amuses us. That I could join in the laughter made me realize I must have retained some sense of that order, no matter how disorderly my mind seemed to have become.

My father was a botanist. When I was a child he reserved Saturday afternoons for me, and we spent many of them walking in woods and rough places. He would name the plants we came upon by their Latin binomials and tell me how they grew. The names were too hard for me, but I did understand that plants had names that described their relationships one to another and found this elegant and interesting even when I was six years old.

So after reading the Borges list, I turned to Linnaeus. Whatever faults the man may have had as a scientist, he gave us a beautiful tool for thinking about diversity in the world. The first word in his scheme of Latin binomials tells the genus, grouping diverse plants which nevertheless share a commonality; the second word names the species, plants alike enough to regularly interbreed and produce offspring like themselves. It is a framework for understanding, a way to show how pieces of the world fit together.

I have no Latin, but as I began to botanize, to learn to call the plants around me up here on my hill by their Latin names, I was diverted from my lack of wits by the wit of the system.

Commelina virginica, the common dayflower, is a rangy weed bearing blue flowers with unequal sepals, two of them showy and rounded, the third hardly noticeable. After I identified it as that particular *Commelina*, named from a sample taken in Virginia, I read in one of my handbooks, written before it was considered necessary to be dull to be taken seriously:

> *Delightful Linnaeus, who dearly loved his little joke, himself confesses to have named the day-flowers after three brothers Commelyn, Dutch botanists, because two of them—commemorated in the showy blue petals of the blossom—published their works; the third, lacking application and ambition, amounted to nothing, like the third inconspicuous whitish third petal.*

There is a tree growing in the woodland with shiny, oval leaves that turn brilliant red early in the fall, sometimes even at summer's end. It has small clusters of white flowers in June that bees like, and later blue fruits that are eaten by bluebirds and robins. It is one of the tupelos, and people in this part of the country call it black-gum or sour-gum. When I was growing up in Michigan I knew it as pepperidge. Its botanic name is *Nyssa sylvatica*. *Nyssa* groups the tupelos, and is derived from the Nyseides—the Greek nymphs of Mount Nysa who cared for the infant Dionysus. *Sylvatica* means "of the woodlands." *Nyssa sylvatica*, a wild, untamed name. The trees, which are often hollow when old, served as beehives for the first American settlers, who cut sections of them, capped them and dumped in the swarms that they found. To this day some people still call beehives "gums," unknowingly acknowledging the common name of the tree. The hollow logs were also used for making pipes that carried salt water to the salt works in

Syracuse in colonial days. The ends of the wooden pipes could be fitted together without using iron bands, which would rust.

This gives me a lot to think about when I come across *Nyssa sylvatica* in the woods.

I botanized obsessively during that difficult time. Every day I learned new plants by their Latin names. I wandered about the woods that winter, good for little else, examining the bark of leafless trees. As wildflowers began to bloom in the spring, I carried my guidebooks with me, and filled a fat notebook as I identified the plants, their habitats, habits and dates of blooming. I had to write them down, for my brain, unaccustomed to exercise, was now on overload.

One spring afternoon, I was walking back down my lane after getting the mail. I had two fine new flowers to look up when I got back to the cabin. Warblers were migrating, and I had been watching them with binoculars; I had identified one I had never before seen. The sun was slanting through new leaves, and the air was fragrant with wild cherry (*Prunus serotina: Prunus*—plum, *serotina*—late blooming) blossoms, which my bees were working eagerly. I stopped to watch them, standing in the sunbeam. The world appeared to have been running along quite nicely without my even noticing it. Quietly, gratefully, I discovered that a part of me that had been off somewhere nursing grief and pain had returned. I had come back from lunch.

Once back, I set about doing all the things that one does when one returns from lunch. I cleared the desk and tended to the messages that others had left. I had been gone for a long time, so there was quite a pile to clear away before I could settle down to the work of the afternoon of my life, the work of building a new kind of order, a structure on which a fifty-year-old woman can live her life alone, at peace with herself and the world around her. * * *

WINTER

* * * A group of people concerned about a proposal to dam the river came over to my place last evening to talk. The first to arrive was my nearest neighbor. He burst excitedly into the cabin, asking me to bring a flashlight and come back to his pickup; he had something to show me. I followed him to his truck, where he took the flashlight and switched it on to reveal a newly killed bobcat stretched out in the bed of his truck. The bobcat was a small one, probably a female. Her broad face was set off by longer hair behind her jaws, and her pointed ears ended in short tufts of fur. Her tawny winter coat, heavy and full, was spotted with black, and her short stubby tail had black bars. Her body

was beginning to stiffen in death, and I noticed a small trickle of blood from her nostrils.

"They pay thirty-five dollars a pelt now over at the country seat," my neighbor explained. "That's groceries for next week," he said proudly. None of us back here on the river has much money, and an opportunity to make next week's grocery money was fortunate for him, I knew. "And I guess you'll thank me because that's surely the varmint that's been getting your chickens," he added, for I had said nothing yet.

But I wasn't grateful. I was shocked and sad in a way that my neighbor would not have understood.

I had not heard a shot and didn't see the gun that he usually carries in the rack in his pickup, so I asked him how he had killed her.

"It was just standing there in the headlights when I turned the corner before your place," he said, "so I rammed it with the pickup bumper and knocked it out, and then I got out and finished it off with the tire iron."

His method of killing sounds more savage than it probably was. Animals in slaughterhouses are stunned before they are killed. Once stunned, the important thing was to kill the bobcat quickly, and I am sure my neighbor did so, for he is a practiced hunter.

Others began to arrive at the meeting and took note of the kill. One of them, a trapper, said that the going price of $35 a pelt was a good one. Not many years ago, the pelt price was under $2. Demand for the fur, formerly scorned for its poor quality, was created by a ban on imported cat fur and a continuing market for fur coats and trim.

My neighbor and the trapper are both third-generation Ozarkers. They could have gone away from here after high school, as did many of their classmates, and made easy money in the cities, but they stayed because they love the land. This brings us together in our opposition to damming the river to create a recreational lake, but our sensibilities are different, the product of different personalities and backgrounds. They come from families who have lived off the land from necessity; they have a deep practical knowledge of it and better skills than I have for living here with very little money. The land, the woods and the rivers, and all that are in and on them are resources to be used for those who have the knowledge and skills. They can cut and sell timber, clear the land for pasture, sell the gravel from the river. Ozarkers pick up wild black walnuts and sell them to the food-processing companies that bring hulling machines to town in October. There are fur buyers, too, so they trap animals and sell the pelts. These Ozarkers do not question the happy fact that they are at the top of the food chain, but kill to eat what swims in the river and walks in the woods, and accept as a matter of course that it takes life to maintain life. In this they are more responsible than I am; I buy my meat in neat sanitized packages from the grocery store.

Troubled by this a few years back, I raised a dozen chickens as meat birds, then killed and dressed the lot, but found that killing chicken Number Twelve was no easier than killing chicken Number One. I didn't like taking responsibility for killing my own meat, and went back to buying it at the grocery store. I concluded sourly that righteousness and consistency are not my strong points, since it bothered me not at all to pull a carrot from the garden, an act quite as life-ending as shooting a deer.

I love this land, too, and I was grateful that we could all come together to stop it from being destroyed by an artificial lake. But my aesthetic is a different one, and comes from having lived in places where beauty, plants and animals are gone, so I place a different value on what remains than do my Ozark friends and neighbors. Others at the meeting last night had lived at one time in cities, and shared my prejudices. In our arrogance, we sometimes tell one another that we are taking a longer view. But in the very long run I'm not so sure, and as in most lofty matters, like my failed meat project, I suspect that all our opinions are simply an expression of a personal sense of what is fitting and proper.

Certainly my reaction to seeing the dead bobcat was personal. I knew that bobcat, and she probably knew me somewhat better, for she would have been a more careful observer than I.

Four or five years ago, a man from town told me he had seen a mountain lion on Pigeon Hawk Bluff, the cliffs above the river just to the west of my place. There is a rocky outcropping there, and he had left his car on the road and walked out to it to look at the river two hundred and fifty feet below. He could see a dead turkey lying on a rock shelf, and climbed down to take a closer look. As he reached out to pick up the bird, he was attacked by a mountain lion who came out of a small cave he had not been able to see from above. He showed me the marks along his forearm—scars, he claimed, where the mountain lion had raked him before he could scramble away. There were marks on his arm, to be sure, but I don't know that a mountain lion or any other animal put them there. I suspect that the story was an Ozark stretcher, for the teller, who logs in many hours with the good old boys at the café in town, is a heavy and slow-moving man; it is hard to imagine him climbing nimbly up or down a steep rock face. Nor would I trust his identification of a mountain lion, an animal more talked of at the café than ever seen in this country.

Mountain lions are large, slender, brownish cats with long tails and small rounded ears. This area used to be part of their range, but as men moved in to cut timber and hunt deer, the cats' chief prey, their habitat was destroyed and they retreated to the west and south. Today they are seen regularly in Arkansas, but now and again there are reports of mountain lions in this part of the Ozarks. With the deer population growing, as it has in recent years under the Department of Conservation's supervision, wildlife biologists say that mountain lions will return to rocky and remote places to feed on them.

After the man told me his story, I watched around Pigeon Hawk Bluff on the outside chance that he might really have seen a mountain lion but in the years since I have never spotted one. I did, however, see a bobcat one evening, near the rock outcropping. This part of the Ozarks is still considered a normal part of bobcat range, but they are threatened by the same destruction of habitat that pushed the mountain lion back to wilder places, and they are uncommon.

Bobcats also kill and feed on deer, but for the most part they eat smaller animals: mice, squirrels, opossums, turkey, quail and perhaps some of my chickens. They are night hunters, and seek out caves or other suitable shelters during the day. In breeding season, the females often chose a rocky cliff cave as a den. I never saw the bobcat's den, but it may have been the cave below the lookout point on the road, although that seems a trifle public for a bobcat's taste. The cliff is studded with other caves of many sizes, and most are inaccessible to all but the most surefooted. I saw the bobcat several times after that, walking silently along the cliff's edge at dusk. Sometimes in the evening I heard the piercing scream of a bobcat from that direction, and once, coming home late at night, I caught her in the road in the pickup's headlight beam. She stood there, blinded, until I switched off the headlights. Then she padded away into the shadows.

That stretch of land along the river, with its thickets, rocky cliffs and no human houses, would make as good a home ground as any for a bobcat. Females are more particular about their five miles or so of territory than are males, who sometimes intrude upon one another's bigger personal ranges, but bobcats all mark their territories and have little contact with other adults during their ten years or so of life.

I don't know for sure that the bobcat I have seen and heard over the past several years was always the same one, but it probably was, and last night probably I saw her dead in the back of my neighbor's pickup truck. *   *   *

TIM ROBINSON
b. 1935

Tim Robinson is an Englishman who has devoted himself to the Aran Islands and adjacent islands of western Ireland, recording and mapping every aspect of their geology, topography, history, and culture. Setting Foot on the Shores of Connemara *(1984) relates his infatuation and growing familiarity with this long-settled and dramatic world of rock and sea. The two volumes of his* Stones of Aran — Pilgrimage *(1986) and* Labyrinth *(1995) — explore the terrain of that island with a cartographic precision and intricacy that take a reader with him, literally step by step.*

TIMESCAPE WITH SIGNPOST

*   *   * As with thousands of others, it was a mild curiosity engendered by Flaherty's film[1] that first brought us (my wife and myself) to Aran, in the summer of 1972. On the day of our arrival we met an old man who explained the basic geography: 'The ocean', he told us, 'goes all around the island.' We let the remark direct our rambles on that brief holiday, and found indeed that the ocean encircles Aran like the rim of a magnifying glass, focusing attention to the point of obsession. A few months later we determined to leave London and the career in the visual arts I was pursuing there, and act on my belief in the virtue of an occasional brusque and even arbitrary change in mode of life. (I mention these personal details only as being the minimum necessary for the definition of the moment on which this narrative will converge, the point in physical and cultural space from which this timescape is observed and on which this book stands.) On that previous summer holiday Aran had presented itself, not at all as Flaherty's

---

Stones of Aran: Pilgrimage (Dublin: Lilliput, 1986).

[1] *Man of Aran* (1932) by American director Robert Flaherty. [Editors' note]

pedestal of rock on which to strike a heroic stance, but rather as a bed of flower-scented sunlight and breezes on which one might flirt delectably with alternative futures. But on our definitive arrival in November we found that bed canopied with hailstorms and full of all the damps of the Atlantic. The closing-in of that winter, until the days seemed like brief and gloomy dreams interrupting ever intenser nights, was accompanied by an unprecedented sequence of deaths, mainly by drowning or by falls and exposure on the crags, that perturbed and depressed the island, quite extinguished the glow of Christmas, and ceased only with the turn of the year, the prayers of the priest and the sinister total of seven. It was a severe induction but it left us with a knowledge of the dark side of this moon that has controlled the tides of our life ever since.

For my part (M's being her own story), what captivated me in that long winter were the immensities in which this little place is wrapped: the processions of grey squalls that stride in from the Atlantic horizon, briefly lash us with hail and go sailing off towards the mainland trailing rainbows; the breakers that continue to arch up, foam and fall across the shoals for days after a storm has abated; the long, wind-rattled nights, untamed then by electricity below, wildly starry above. Then I was dazzled by the minutiae of spring, the appearance each in its season of the flowers, starting with the tiny, white whitlow-grass blossoms hardly distinguishable from the last of the hailstones in the scant February pastures, and culminating by late May in paradisal tapestry-work across every meadow and around every rock. The summer had me exploring the honeysuckled boreens and the breezy clifftops; autumn proposed the Irish language, the blacksmith's quarter-comprehended tales, the intriguing gossip of the shops, and the discovery that there existed yet another literature it would take four or five years to begin to make one's own. This cycle could have spun on, the writings I had come here to do having narrowed themselves into a diary of intoxication with Aran, but that some way of contributing to this society and of surviving financially had to be found.

A suggestion from the post mistress in the western village of Cill Mhuirbhigh gave me the form of this contribution: since I seemed to have a hand for the drawing, an ear for the placenames and legs for the boreens, why should I not make a map of the islands, for which endless summersful of visitors would thank and pay me? The idea appealed to me so deeply that I began work that same day. My conceptions of what could be expressed through a map were at that time sweeping but indefinite; maps of a very generalized and metaphorical sort had been latent in the abstract paintings and environmental constructions I had shown in London, in that previous existence that already seemed so long ago, but I had not engaged myself to such a detailed relationship with an actual place before. The outcome, published in 1975, was a better image of my ignorance than of my knowledge of Aran, but it was generously received by the islanders, prospered moderately with the tourists, and brought me into contact with the specialists in various fields who visited Aran. During the subsequent years of accumulation towards the second version of the map, published in 1980, I have walked the islands in companionship with such visiting experts as well as with the custodians of local lore whom I sought out in every village, and have tried to see Aran through variously informed eyes — and then, alone again, I have gone hunting for those rare places and times, the nodes at which the layers of experience touch and may be fused together. But I find that in a map such points and the energy that accomplishes such fusions (which is that of poetry, not some vague 'interdisciplinary' fervour) can, at the most, be invisible guides, benevolent ghosts, through the tangles of the explicit; they cannot themselves be shown or named. So, chastened in my expectations of them, I now regard the Aran maps as preliminary storings and sortings of material for another art, the world-hungry art of words.

However, although the maps underlie this book, the conception of the latter dates from a moment in the preparation for the former. I was on a summer's beach one blinding day watching the waves unmaking each other, when I became aware of a wave, or a recurrent sequence of waves, with a denser identity and more purposeful momentum than the rest. This appearance, which passed by from east to west and then from west to east and so on, resolved itself under my stare into the fins and backs of two dolphins (or were there three?), the follower with its head close by the flank of the leader. I waded out until they were passing and repassing within a few yards of me; it was still difficult to see the smoothly arching succession of dark presences as a definite number of individuals. Yet their unity with their background was no jellyfish-like dalliance with dissolution; their mode of being was an intensification of their medium into alert, reactive self-awareness; they were wave made flesh, with minds solely to ensure the moment-by-moment reintegration of body and world.

This instance of a wholeness beyond happiness made me a little despondent, standing there thigh deep in Panthalassa (for if Pangaea is shattered and will not be mended by our presence on it, the old ocean holds together throughout all its twisting history): a dolphin may be its own poem, but we have to find our rhymes elsewhere, between words in literature, between things in science, and our way back to the world involves us in an endless proliferation of detours. Let the problem be symbolized by that of taking a single step as adequate to the ground it clears as is the dolphin's arc to its wave. Is it possible to think towards a *human* conception of this 'good step'? (For the dolphin's ravenous cybernetics and lean hydrodynamics induce in me no nostalgia for imaginary states of past instinctive or future theological grace. Nor is the ecological imperative, that we learn to tread more lightly on the earth, what I have in mind—though that commandment, which is always subject to challenge on pragmatic grounds if presented as a mere facilitation of survival, might indeed acquire some authority from the attitude to the earth I would like to hint at with my step.) But our world has nurtured in us such a multiplicity of modes of awareness that it must be impossible to bring them to a common focus even for the notional duration of a step. The dolphin's world, for all that its inhabitants can sense Gulf Streams of diffuse beneficences, freshening influences of rivers and perhaps a hundred other transparent gradations, is endlessly more continuous and therefore productive of unity than ours, our craggy, boggy, overgrown and overbuilt terrain, on which every step carries us across geologies, biologies, myths, histories, politics, etcetera, and trips us with the trailing *Rosa spinosissima* of personal associations. To forget these dimensions of the step is to forgo our honour as human beings, but an awareness of them equal to the involuted complexities under foot at any given moment would be a crushing back-load to have to carry. Can such contradictions be forged into a state of consciousness even fleetingly worthy of its ground? At least one can speculate that the structure of condensation and ordering necessary to pass from such various types of knowledge to such an instant of insight would have the characteristics of a work of art, partaking of the individuality of the mind that bears it, yet with a density of content and richness of connectivity surpassing any state of that mind. So the step lies beyond a certain work of art; it would be like a reading of that work. And the writing of such a work? Impossible, for many reasons, of which the brevity of life is one.

However, it will already be clear that Aran, of the world's countless facets one of the most finely carved by nature, closely structured by labour and minutely commented by tradition, is *the* exemplary terrain upon which to dream of that work, the guide-book to the adequate step. *Stones of Aran* is all made up of steps, which lead in many directions but perpetually return to, loiter near, take short-cuts by, stumble over or impatiently kick aside that ideal. (Otherwise, it explores and takes its form from a single island, Árainn itself; the present work makes a circuit of the coast, whose features present themselves as stations of a

Pilgrimage, while the sequel will work its way through the interior, tracing out the *Labyrinth.*) And although I am aware that that moment on the beach, like all moments one remembers as creative, owes as much to the cone of futurity opening out from it as to the focusing of the past it accomplished, I will take it as the site of my book, so that when at last it is done I will have told the heedless dolphins how it is, to walk this paradigm of broken, blessed, Pangaea.

CHET RAYMO
b. 1936

Chet Raymo continues a distinguished line of nature writers who have also been teachers of science. He particularly resembles one of his predecessors, Loren Eiseley, in his impulse to convey the grandeur of creation, to stimulate awe. In addition to teaching physics and astronomy at Stonehill College, in Easton, Massachusetts, he has written columns on science for The Boston Globe *for over fifteen years. In those columns, as in the following essay from* The Soul of the Night *(1985), his characteristics as a writer are descriptive precision, narrative energy, and alertness to the spiritual meanings of the physical universe. His other books include* Honey from Stone: A Naturalist's Search for God *(1987) and a novel,* The Dork of Cork *(1993).*

THE SILENCE

Yesterday on Boston Common I saw a young man on a skateboard collide with a child. The skateboarder was racing down the promenade and smashed into the child with full force. I saw this happen from a considerable distance. It happened without a sound. It happened in dead silence. The cry of the terrified child as she darted to avoid the skate-

The Soul of the Night (Englewood Cliffs, N.J.: Prentice-Hall, 1985).

board and the scream of the child's mother at the moment of impact were absorbed by the gray wool of the November day. The child's body simply lifted up into the air and, in slow motion, as if in a dream, floated above the promenade, bounced twice like a rubber ball, and lay still.

All of this happened in perfect silence. It was as if I were watching the tragedy through a telescope. It was as if the tragedy were happening on another planet. I have seen stars exploding in space, colossal, planet-shattering, distanced by light-years, framed in the cold glass of a telescope, utterly silent. It was like that.

During the time the child was in the air, the spinning Earth carried her half a mile to the east. The motion of the Earth about the sun carried her back again forty miles westward. The drift of the solar system among the stars of the Milky Way bore her silently twenty miles toward the star Vega. The turning pinwheel of the Milky Way Galaxy carried her 300 miles in a great circle about the galactic center. After that huge flight through space she hit the ground and bounced like a rubber ball. She lifted up into the air and flew across the Galaxy and bounced on the pavement.

It is a thin membrane that separates us from chaos. The child sent flying by the skateboarder bounced in slow motion and lay still. There was a long pause. Pigeons froze against the gray sky. Promenaders turned to stone. Traffic stopped on Beacon Street. The child's body lay inert on the asphalt like a piece of crumpled newspaper. The mother's cry was lost in the space between the stars.

How are we to understand the silence of the universe? They say that certain meteorites, upon entering the Earth's atmosphere, disintegrate with noticeable sound, but beyond the Earth's skin of air the sky is silent. There are no voices in the burning bush of the Galaxy. The Milky Way flows across the dark shoals of the summer sky without an audible ripple. Stars blow themselves to smithereens; we hear nothing. Millions of solar systems are sucked into black holes at the centers of the galaxies; they fall like feathers. The universe fattens and swells in a Big Bang, a fireball of Creation exploding from a pinprick of infinite energy, the ultimate firecracker; there is no soundtrack. The membrane is ruptured, a child flies through the air, and the universe is silent.

In Catholic churches between Good Friday and Easter Eve the bells are stilled. Following a twelfth-century European custom, the place of the bells is taken by *instruments des ténèbres* (instruments of darkness), wooden clackers and other noisemakers that remind the faithful of the terrifying sounds that were presumed to have accompanied the death of Christ. It was unthinkable that a god should die and the heavens remain silent. Lightning crashed about the darkened hill of Calvary.

The veil of the temple was loudly rent. The Earth quaked and rocks split. Stars boomed in their courses. This din and thunder, according to medieval custom, are evoked by the wooden instruments.

Yesterday on Boston Common a child flew through the air, and there was no protest from the sky. I listened. I turned the volume of my indignation all the way up, and I heard nothing.

There is a scene in Michelangelo Antonioni's film *Red Desert* in which a woman approaches a construction site where men are building a large linear-array radio telescope. "What is it for?" she asks. One of the workmen replies, "It is for listening to the stars." "Oh," she exclaims with innocent enthusiasm. "Can I listen?"

Let us listen. Let us connect the multimillion-dollar telescopes to our kitchen radios and convert the radiant energy of the stars into sound. What would we hear? The random crackle of the elements. The static of electrons fidgeting between energy levels in the atoms of stellar atmospheres. The buzz of hydrogen. The hiss and sputter of matter intent upon obeying the stochastic laws of quantum physics. Random, statistical, indifferent noise. It would be like the hum of a beehive or the clatter of shingle slapped by a wave.

In high school we did an experiment with an electric bell in a glass jar. The bell was suspended inside the jar, and the wires carrying electricity were led in through holes in the rubber stopper that closed the jar's mouth. The bell was set clanging. Then the air was pumped from the jar. Slowly, the sound of the bell was snuffed out. The clapper beat a silent tattoo. We watched the clapper thrashing silently in the vacuum, like a moth flailing its soft wings against the outside of a window pane.

Even by the standard of the vacuum in the bell jar, the space between the stars is empty. The emptiness between the stars is unimaginably vast. If the sun were a golf ball in Boston, the Earth would be a pinpoint twelve feet away, and the nearest star, Alpha Centauri, would be another golf ball (two golf balls, really, two golf balls and a pea; it is a triple star) in Cincinnati. The distances between the stars are huge compared with the sizes of the stars: a golf ball in Boston, two golf balls and a pea in Cincinnati, a marble in Miami, a basketball in San Francisco. The trackless trillions of miles between the stars are a vacuum more perfect than any vacuum that has yet been created on Earth. In our part of the Milky Way Galaxy, interstellar space contains about one atom of matter per cubic centimeter, one atom in every volume of space equal to the size of a sugar cube. The silent vacuum of the bell jar was a million times inferior to the vacuum of space. In the almost perfect vacuum of interstellar space, stars detonate, meteors blast craters on moons, and planets split at their seams with no more sound than the pulsing clapper of the bell in the evacuated jar.

Once I saw the Crab Nebula through a powerful telescope. The nebula is the expanding debris of an exploded star, a wreath of shredded star-stuff eight light-years wide and 5000 light-years away. What I saw in the telescope was hardly more than a blur of light, more like a smudge of dust on the mirror of the scope than the shards of a dying star. But seeing through a telescope is 50 percent vision and 50 percent imagination. In the blur of light I could easily imagine the outrushing shock wave, the expanding envelope of high-energy radiation, the torn filaments of gas, the crushed and pulsing remnant of the skeletal star. I stood for a quarter of an hour with my eye glued to the eyepiece of the scope. I felt a powerful sensation of energy unleashed, of an old building collapsing onto its foundations in a roar of dust at the precise direction of a demolition expert. As I watched the Crab Nebula, I felt as if I should be wearing earplugs, like an artillery man or the fellow who operates a jackhammer. But there was no sound.

The Chinese saw the Crab when it blew up. In A.D. 1054 a new star appeared in Taurus. For weeks it burned more brightly than Venus, bright enough to be seen in broad daylight. Then the star gradually faded from sight. The Chinese recorded the "guest star" in their annals. Nine hundred years later the explosion continues. We point our telescopes to the spot in Taurus where the "guest star" appeared in 1054, and we see the bubble of furious gases still rushing outward.

Doris Lessing began her fictional chronicle of space with this dedication: *For my father, who used to sit, hour after hour, night after night, outside our home in Africa, watching the stars. "Well," he would say, "if we blow ourselves up, there's plenty more where we came from."* Yes, there's plenty more, all right, even if one or two blow themselves up now and then. A billion billion stars scattered in the vacuum of space. A star blew up for the Chinese in 1054. A star blew up for Tycho Brahe in 1572, and another for Kepler in 1604. They go in awesome silence.

The physical silence of the universe is matched by its moral silence. A child flies through the air toward injury, and the galaxies continue to whirl on well-oiled axis. But why should I expect anything else? There are no Elysian Fields up there beyond the seventh sphere where gods pause in their revels to glance down aghast at our petty tragedies. What's up there is just one galaxy after another, magnificent in their silent turning, sublime in their huge indifference. The number of galaxies may be infinite. Our indignation is finite. Divide any finite number by infinity and you get zero.

Only a few hundred yards from the busy main street of my New England village, the Queset Brook meanders through a marsh as apparently remote as any I might wish for. To drift down that stream in November is to enter primeval silence. The stream is dark and sluggish. It pushes past the willow roots and the thick green leaves of the arrowhead like syrup. The wind hangs dead in the air. The birds have fled south. Trail bikes are stacked away for the winter, and snowmobiles are still buried at the backs of garages. For a few weeks in November the marsh near Queset Brook is as silent as the space between the stars.

How fragile as our hold on silence. The creak of the wagon on a distant highway was sometimes noise enough to interrupt Thoreau's reverie. Thoreau was perceptive enough to know that the whistle of the Fitchburg Railroad (whose track lay close by Walden Pond) heralded something more than the arrival of the train, but he could hardly have imagined the efficiency with which technology has intruded upon our world of natural silence. Thoreau rejoiced in owls; their hoot, he said, was a sound well suited to swamps and twilight woods. The interval between the hoots was a deepened silence suggesting, said Thoreau, "a vast and undeveloped nature which men have not recognized." Thoreau rejoiced in that silent interval, as I rejoice in the silence of the November marsh.

As a student, I came across a book by Max Picard called *The World of Silence*. The book offered an insight that seems more valuable to me now than it did then. Silence, said Picard, is the source from which language springs, and to silence language must constantly return to be recreated. Only in relation to silence does sound have significance. It is for this silence, so treasured by Picard, that I turn to the marsh near Queset Brook in November. It is for this silence that I turn to the stars, to the ponderous inaudible turning of galaxies, to the clanging of God's great bell in the vacuum. The silence of the stars is the silence of creation and re-creation. It is the silence of that which cannot be named. It is a silence to be explored alone. Along the shore of Walden Pond the owl hooted a question whose answer lay hidden in the interval. The interval was narrow but infinitely deep, and in that deep hid the soul of the night.

I drift in my canoe down the Queset Brook and I listen, ears alert, like an animal that sniffs a meal or a threat on the wind. I'm not sure what it is that I want to hear out of all this silence, out of this palpable absence of sound. A scrawny cry, perhaps, to use a phrase of the poet Wallace Stevens: "a scrawny cry from outside . . . a chorister whose c preceded the choir . . . still far away." Is that too much to hope for? I don't ask for the full ringing of the bell. I don't ask for a clap of thunder that would rend the veil in the temple. A scrawny cry will do, from far off there among the willows and the cattails, from far off there among the galaxies.

The child sent flying by the young man on the skateboard bounced on the pavement and lay still. The pigeons rose against the gray sky. Promenaders turned to stone. How long was it that the child's body lay there like a piece of crumpled newspaper? How long did my heart thrash silently in my chest like the clapper of a bell in a vacuum? Perhaps it was a minute, perhaps only a fraction of a second. Then the world's old rhythms began again. A crowd gathered. Someone lifted the injured child into his arms and rushed with the mother toward help. Gawkers milled about distractedly and dispersed. The clamor of the city engulfed the common. Traffic moved again on Beacon Street.

JIM HARRISON
b. 1937

Jim Harrison is best known for his fiction—including A Good Day to Die *(1973),* Legends of the Fall *(1979), and* The Road Home *(1998). But he has also published ten volumes of poetry, written screenplays for Warner Brothers, and produced a number of memorable essays on the land and culture of Michigan's Upper Peninsula. In this region, remote from cities but richly storied, he has found a corollary and inspiration for his own rugged and dramatic imagination. The piece included here, describing a camping trip taken in New Mexico with fellow writer Doug Peacock, is told with an engaging irreverence reminiscent of Edward Abbey.*

THE BEGINNER'S MIND

At the onset it has occurred to me as a novelist and poet that I could not write a legitimate natural history essay at gunpoint. As indicated earlier in my life by my grades in high school or college in the life sciences and geology, my mind was either elsewhere or nowhere in particular. I was apt at metaphor but a zygote resembled a question mark without a question. After setting a new record for a low grade in the one-hundred-rock identification test in the Natural Sciences Department at Michigan State University, the professor gazed at me with the intense curiosity owned by the man who discovered the duckbill platypus. At the time, nineteen, my mind had been diverted by Rimbaud and Dostoyevsky, Mozart and Stravinsky, and if Rilke had said in his "Letters to a Young Poet" to study invertebrate zoology I would have done so, only he never broached the subject.

Curiously, I'm still trying. There's an old wood Burgundy carton in my four-wheel drive that contains a dozen or so natural history guide-books that I use frequently. Once in the sandhills of Nebraska I sat on a knoll on a June afternoon and identified all the weeds and grasses around me using Van Bruggen's *Wildflowers, Grasses and Other Plants of the Northern Plains and Black Hills*. I also fell asleep and saw Crazy Horse, who helped me dream up the heroine for my then unwritten novel, *Dalva*. On waking, it dawned on me as it had dozens of times before that everything goes together or we're in real trouble. Mozart and the loon belong to the same nature, as does the mind of Lorca and the gray hawk I'm lucky to have nesting near our adobe in Patagonia, Arizona. The coyote's voice and the petroglyph of the lizard king near Baboquivari marry in a purer voice than any of our current machineries of joy. The elf owls that flocked into the black oak above our campfire on the Gray Ranch made me feel more at home on earth than my farmhouse of twenty-five years. That many elf owls in one tree lifts your skull so you may see them with another eye that more closely resembles their own. William Blake's line is appropriate, "How do we know but that every bird that cuts the airy way is an immense world of delight closed to our senses five?" This is the opposite of the anthropomorphism so properly scorned in literary types of scientists. I simply agree with the visionary notion of Neil Claremon that reality is the aggregate of the perceptions of all creatures.

But back to the not-so-ordinary earth and the Gray Ranch. On my

Originally published in *Heart of the Land: Essays on the Last Great Places* (New York: Pantheon, 1995).

first trip there a few years ago I realized it takes a golden eagle or a bush pilot to make a quick read of five hundred square miles. I was thrown directly back into the dozens of Zane Grey novels I had read in my youth, which was not a bad place to be considering the direction of current events toward chaos and fungoid tribalism. There was an urge to yodel "purple mountain's majesty," or reconcoct that Rousseauian fantasy that far up some distant arroyo in the Animas, now shrouded in January shadows, all the local creatures were drinking milk from the same golden bowl. It was, and is, that kind of place. Of course I wondered why Dad didn't own this rather than the three-acre Michigan swamp which, nonetheless, was good birding. My science aversion did not include the birds that were introduced in the third grade by Audubon cards which had a specific leg up on baseball cards.

The fact is the Gray Ranch is breathtaking, that is, you forget to breathe, the vision before when you come over the back road from Douglas is vertiginous, surreal, the vast expanse of valley before you not quite convincingly real. Frankly, the only thing that could improve on it would be an Apache village, but that one has been kissed permanently good-bye. I have never quite understood why much of our Bureau of Land Management and Forest Service land could not be returned to its original owners. There is firm evidence that they would do a better job of managing it.

From painful experience I mentally rattled off a number of cautionary notes. There is a wonderful quote in Huanchu Daoren's reflections on the Tao, *Back to Beginnings*. "Mountain forests are beautiful places, but once you become attached to them, they become cities." What is meant by *attached* here is a desperate clinging, an obsessiveness that finally blinds you to the wilderness before you, at which point you may as well be in Times Square or touring the Pentagon. More importantly, in this state of mind you cannot competently defend the wilderness you presume to love.

On the first trip my camping partner, Doug Peacock of grizzly bear renown, was intent on sleeping out in the really high lonesome despite the warnings at ranch headquarters that it was going to be "mighty cool" in the high country. This turned out to be a cowboy euphemism for a temperature of 15 degrees. The tip-off about the cold front had actually come the day before when we were looking for water birds out on the Wilcox playa in a snowstorm. The bedrooms at ranch headquarters looked rather attractive and so did the idea of central heating, but then I had just come down from northern Michigan where it is truly frigid in January and it was unseemly for me to hedge. That night when

the temperature plummeted it occurred to me I hadn't slept outside in
Michigan during the winter since I won my Polar Bear merit badge in
the boy scouts, after which I was booted out as a malcontent.

Peacock, however, is the ultimate camper, in some years spending
over half his nights under the stars. We simply used two sleeping bags
apiece, one stuffed in another, and wore stocking caps. I had been having the most intense of Hollywood screenplay problems but they
drifted away in the face of stars that glittered barely above the treetops
and sycamores so burnished by the moonglow they kept rearranging
themselves as if their roots were underground legs. Our only real problem was the olive oil congealed around the edges of the frying pan and
the Bordeaux was overchilled in our gloved hands. There was a mighty
chorus from a nearby bobcat who was treating the new odors of garlic
and Italian sausage with noisy surprise. The most recurrent thought
during those two days was wishing for a seven-year vacation so I could
adequately walk out the ranch, slowly identifying everything that wasn't
underground. I might even memorize the clouds.

Two years later on the eve of our return there is a specific freight of
confusion about the Gray Ranch. In the interim I had been assured in
both Montana and Michigan that the ranch had been sold to Ted
Turner. Since I'm quite a fibber myself, what with being a novelist, I
tend to believe other fibbers whole cloth. I fully understood that The
Nature Conservancy might not wish to keep that much capital in one
basket, the mildest of understatements, and though Turner is indeed an
environmentalist, I feared his interest in buffalo that do not belong in
the area.

Another, rather astounding, rumor arose that Drum Hadley was buying the ranch through his Animas Foundation, with the Conservancy
retaining large easements in the higher altitudes. As an option to sitting
around in a dither, I checked the rumor out and found it was true.
"Astounding" is not too strong a word, as I knew of Hadley only
through his poetry which had been recommended to me by Charles
Olson one sunny spring afternoon in Gloucester, Massachusetts, long
ago, and later by Gary Snyder. In the religious world this would be similar to being lauded by Pope John XXIII and Gandhi. I had always
thought of Hadley as a Black Mountain populist who had holed up on
a ranch in a canyon near Douglas, Arizona, and certainly hadn't
guessed that he could muster the wherewithal to buy a ranch of this
awesome proportion.

I recalled a quote from Hadley's mentor, Olson, in his book on
Melville, *Call Me Ishmael,* "I take space to be the central fact to
man born in America, from Folsom cave to now." This, whether illu-

sion or reality, is a whopper of a statement, but it was more true when Olson wrote the book forty years ago, and certainly purer truth in Melville's time.

Why, then, should I be such an ardent claustrophobe, despite the fact that I spend nine months of the year in Michigan's Upper Peninsula and Patagonia, Arizona, the coenvirons of bear, mountain lion, all sorts of creatures, and in each place, the stray wolf still passes by? It wasn't just the hearing of the dark wings of the madness of overpopulation in the future. More real is the prospect that developers buy wild regions and dice them into parcels for us who love the outdoors and have the cash to buy them. The Forbes ranch in northern New Mexico was a dire portent, and one could, properly informed, add a thousand other places this was happening. There is a nearly spiritual truth in Edward Abbey's comment, "It's not the beer cans I mind, it's the roads." With Hadley's purchase and the Conservancy's easements, this immense ranch would remain intact, and I could stop mentally turning it into a city.

Late on an April afternoon we set up camp with Peacock in a hurry to take a walk for another look at some petroglyphs he had noted two years before. What he thinks of as a stroll is an aerobic nightmare for the less hardy. Ten miles is not improbable for this geezer in the Michigan woods, but in the rumpled West I go my own slower way when camping with my partner who is thought by many to be the world's largest billy goat. I also fall with some frequency, my feet refusing to acknowledge a terrain where you have to watch where your feet are stepping. The tendinitis in my bursae was throbbing, the result of doing the splits in a fifty-yard skid down an arroyo near Patagonia, so I made my way slowly up a creek bed that owned an aura of mystery. The notion arose that I was a flatlander down to my very zygotes, my feet requiring moss, ferns, deadfalls, tamarack bogs, and osier-choked gullies.

Not so long ago, only a few minutes in geologic time, we attacked the wilds with implements of greed and domination. Now, or so it appears, we are having run at it with sporting equipment, none of it as friendly to the earth as the human foot or the hooves of horses. Walking makes the world its own size and a scant hour in a forty-acre woodlot is liable to dissipate the worst case of claustrophobia. The same hour in the high country of the Gray Ranch and you're ready to levitate. I remind myself again not to burden the air with requests from the wild but to see what I can see with the attentiveness of the creature world. I scout the creek canyon just far enough to see an enormous opening which I'll save for the morning.

We had our customary first-night camping dinner of thick rare Delmonicos wrapped in tortillas, accompanied by Bordeaux, which increases goodwill as proven by the French, those kindly souls. It was that first night that the curious elf owls gathered in the black oak branches above us. Doug had only seen them grouped this way once before down in the remotest Pinacates. Such splendor is humbling and properly so. It was the equivalent of wandering the Upper Peninsula for twenty years hoping to see a wolf and then seeing one a scant hundred yards from my cabin. When the owls left there were the nightjars, a song closer to the loon for the resonance of the memories it evokes. We were camped in the same place that we had been the cold night two years before, but now the dark was soft and dulcet and I watched the entire arc of the moon until it burst against Animas Peak, the last golden light shedding down the talus.

At dawn, for eccentric reasons, I scoured my guidebooks for something odd to look for on my walk, deciding on the rare night-blooming cereus. My hip pain was a torment, so every hundred yards or so I'd go blank and lay down like a tired deer. There was a wan hope to see the enormous male mountain lion that was said to live in the area. He kept himself as hidden as the night-blooming cereus though at one point I had the feeling I was being observed. Since I'm a somewhat goofy poet, I do not feel obligated like the scientist to regard these intuitions as nonsense. It is easy to forget that we are, above all else, mammals. An anthropology text has curled the hair of many an aesthete.

I mostly crawled up a steep hill that would have been regarded as Michigan's only mountain. It was rocky but in the crevices wild flowers bloomed and far above was a bona fide golden eagle. Two years before we had seen several at once in the area called the "flats," which is a single seventy-square-mile pasture, sort of an Ur-pasture still in the condition that pioneers had found it, along with the Apache. I'm not cattle shy as most amateur environmentalists, but my father was an agronomist and soil conservationist and I know overgrazing when I see it. You don't look sideways at grass, you look down. Cattle exposure that precipitates erosion is a good start.

Any sort of contentiousness was far from my mind, though, when I reached the mountaintop. One boulder was smooth and I imagined it was a habitual sitting spot for those of the Casas Grandes culture who had preceded me there by more than a thousand years. The area is visited by violent thunderstorms and I could see lightning had struck the place numerous times, shattering boulders into small chunks of crystal. The place would be a New Ager fantasy but then I was not in the mood to dislike anyone. Back home the anishinabe (Chippewa Indians) favor lightning trees and this place had endured godly punishment way beyond trees which burn and half-explode. It would be a good place, finally, to die, and we don't find many such locations in a lifetime. This is an utterly normal thought rather than a sad one. I'm unaware of anyone who has gotten off this beauteous earth alive save the Lord, and that is disputed by many.

The natural world had so grasped me that morning that I forgot lunch, but far up the draw I could smell it on my way back with an ursine wag to my head and a crinkled nose. I share with Peacock a love for all the simple pleasures, not just a few of them, and that dawn we had put together lamb shanks, a few heads of garlic, cascabels, and a pound of white tepary beans got from Gary Nabhan's Native Seed Search. Trail mix and freeze-dried offal doesn't soothe the imagination. We had heaped coals around the Dutch oven, and I judged by the odor a quarter-mile distant that it was ready. This kind of lunch is necessary if you are to take a nap, and if you don't take a nap you are not fresh for the day's second half. You become a conniving eco-ward heeler with fatigued ideas how you would run the West if you were king of the cordillera. By taking a nap I stay put as plain old Jim, who occasionally has something fit for the collective suggestion box. A nap can give you an hour's break from needing to be right all the time, an affliction leading to blindness to the natural world, not to speak of your wife and children.

Late that afternoon, after studying petroglyphs and flycatchers, we had a long jouncing drive to ranch headquarters to meet Mr. Hadley for dinner. I had prepared a list of questions about everything from the BLM, the Savory grazing methods, Wes Jackson, Bruce Babbitt (hooray, at last), the Gray Ranch's carrying capacity, methane, the flavor of local beef (excellent), none of which I asked because we started talking about twentieth-century poetry. All tolled, your putative reporter did not put forward a single germane question about the ranch, somewhat in the manner of my beloved Omaha Indians to whom it is impolite to ask questions of anyone about anything, so they don't. There is also a specific ranch etiquette I learned in the sandhills, certainly the best managed grazing area in the United States, where information is volunteered rather than extracted.

After dinner we took a walk down a moonlit road and Hadley quoted the third of Rilke's *Duino Elegies* in its entirety in German, the sort of act that raised his credibility in my belief system up there with Thomas Jefferson, whether he likes the comparison or not. Though I was modestly groggy at the time, it seemed reasonable that a poet could run a huge ranch better than anyone else, especially as in Hadley's case, he had thirty years of experience.

On the slow ride back, which was much shortened by Peacock's braying of every blues tune in his head, the moon lit up Animas Peak so it looked a short trot away, and as we gained altitude the wind stiffened. The sand and grit in the air yellowed the moon and the landscape. I guessed the wind by my Great Lakes standards to be about forty knots and we secured our campsite with difficulty. I turned my sleeping bag so that it would stop billowing like a wind sock, and looked out from our grassy bench at the landscape, which now was shimmering and haunted. Spirits were afoot. First came the Natives, then the turn-of-the-century cowboys, the night and day laborers of the cattle empires. A hundred years ago, or thereabouts, 400,000 cattle had perished to starvation in this two-hundred-mile-wide neighborhood between Cloverdale (population none) and Nogales on the Mexican border of Arizona. Despite the legion of naysayers, we're doing much better now. In fact, I was sleeping on a heretofore improbable experiment whether the natural and the man-organized communities could not only coexist but thrive to the mutual benefit of both. This was the teeter-totter that needed to be balanced between radical environmentalists and the stock associations, neither of which was going to go away. I was pretty much in the camp of the former and retained the right to shoot off my mouth about public grazing, but it was a splendid tonic that night to see what the private initiative that surrounded me with sure and certain hope had accomplished. The Gray Ranch was still here, big as all outdoors.

FREEMAN HOUSE
b. 1937

Although Totem Salmon *(1999) is Freeman House's first book, it is also a culmination of his decades fighting to save one of California's last species of native salmon. He writes from his background both as a commercial salmon fisherman and as cofounder of the Mattole Watershed Salmon Support Group and the Mattole Restoration Council. The story he tells includes natural history, ethnography, and personal narrative. It also eloquently addresses two issues of increasing importance to American environmentalism. One of these is a recognition that we need* not only to preserve but also in many cases to restore wild nature. The other is an awareness that true conservation inevitably depends on developing healthy and inclusive human communities, joyfully grounded in their own bioregions.

IN SALMON'S WATER

* * * The coevolution of humans and salmon on the North Pacific Rim fades into antiquity so completely that it is difficult to imagine a first encounter between the two species. Salmon probably arrived first. Their presence can be understood as one of the necessary preconditions for human settlement. Pacific salmon species became differentiated from their Atlantic ancestors no more than half a million years ago. Such adaptations were a response to their separation from their Atlantic salmon parent stock by land bridges such as the one that has periodically spanned the Bering Strait. By the time the Bering Sea land bridge last emerged, twelve thousand to twenty-five thousand years ago, in the Pleistocene epoch, the six species of Pacific salmon had arrived at their present characteristics and had attained their distribution over the vast areas of the North Pacific. As the ice pack retreated, the species continued to adapt ever more exactly as stocks or races—each finely attuned to one of the new rivers and to recently arrived human predators. If indeed humans first arrived in North America after crossing that land bridge from Asia, the sight of salmon pushing up the rivers of this eastern shore would have served as proof that this place too was livable.

On this mindblown midnight in the Mattole I could be any human at any time during the last few millennia, stunned by the lavish design of nature. The knowledge of the continuous presence of salmon in this river allows me to know myself for a moment as an expression of the continuity of human residence in this valley. Gone for a moment is my uncomfortable identity as part of a recently arrived race of invaders with doubtful title to the land; this encounter is one between species, human and salmonid. Such encounters have been happening as long as anyone can remember: the fish arrive to feed us and they do so at the same time every year and they do so with an obvious sense of intention. They come at intervals to feed us. They are very beautiful. What if they stopped coming?—which they must if we fail to relearn how to celebrate the true nature of the relationship.

Totem Salmon: Life Lessons from Another Species (Boston: Beacon, 1999).

For most of us, the understanding of how it might have been to live in a lavish system of natural provision is dim and may be obscured further by the scholarship that informs us. Our understanding of biology has been formulated during a time of less diversity and abundance in nature; our sense of relationship is replaced by fear of scarcity. By the time the anthropologists Alfred Kroeber and Erna Gunther were collecting their impressions of the life of the Native Americans of the Pacific Northwest, early in the twentieth century, the great salmon runs that had been an integral part of that life had already been systematically reduced. It may be this factor that makes the rituals described in their published papers seem transcendent and remote: ceremonial behavior that had evolved during a long period of dynamic balance has become difficult to understand in the period of swift decline that has followed.

It seems that in this part of the world, salmon have always been experienced by humans very directly as food, and food as relationship: the Yurok word for salmon, *nepu*, means "that which is eaten"; for the Ainu, the indigenous people of Hokkaido Island, the word is *shipe*, meaning "the real thing we eat." Given the abundance and regularity of the provision, one can imagine a relationship perceived as being between the feeder and those fed rather than between hunted and hunter. Villages in earlier times were located on the banks of streams, at the confluence of tributaries, because that is where the food delivers itself. The food swims up the stream each year at much the same time and gives itself, alive and generous.

It is not difficult to capture a salmon for food. My own first memory of salmon is of my father dressed for work as a radio dispatcher, standing on the low check dam across the Sacramento River at Redding and catching a king salmon in his arms, almost accidentally. The great Shasta Dam, which when completed would deny salmon access to the headwaters of the river, was still under construction. Twenty years later, as an urbanized young man, I found myself standing with a pitchfork, barefooted, in an inland tributary of the Klamath River, California's second largest river system. The salmon were beating their way upstream in the shallow water between my legs. Almost blindly, my comrades and I speared four or five of them. When the salmon come up the river, they come as food and they come as gift.

Salmon were also experienced as *connection*. At the time of year when the salmon come back, drawn up the rivers by spring freshets or fall rains, everyone in the old villages must have gained a renewal of their immediate personal knowledge of why the village was located where it was, of how tightly the lives of the people were tied to the lives of the salmon. The nets and drying racks were mended and ready.

Everyone had a role to play in the great flood of natural provision that followed. The salmon runs were the largest annual events for the village community. The overarching abundance of salmon—their sheer numbers—is difficult to imagine from our vantage point in the late twentieth century. Nineteenth-century firsthand accounts consistently describe rivers filled from bank to bank with ascending salmon: "You could walk across the rivers on their backs!" In the memory of my neighbor Russell Chambers, an octogenarian, there are stories of horses refusing to cross the Mattole in the fall because the river had for a time become a torrent of squirming, flashing, silvery salmon light.

It is equally difficult to imagine a collective life informed and infused by the exuberant seasonal pulses of surrounding nature over a lifetime, over the lifetime of generations. But for most of the years in tribal memory of this region's original inhabitants, the arrival of salmon punctuated, at least once annually, a flow of provision that included acorn and abalone in the south, clams and berries and smelt in the north, venison and mussels and tender greens everywhere. Humans lived on the northwest coasts of North America for thousands of years in a state of lavish natural provision inseparable from any concept of individual or community life and survival. Human consciousness organized the collective experience as an unbroken field of being: there is no separation between people and the multitudinous expressions of place manifested as food.

But each annual cycle is punctuated also by winter and the hungry time of early spring, and in the memory of each generation there are larger discontinuities of famine and upheaval. Within the memory of anyone's grandmother's grandfather, there is a catastrophe that has broken the cycle of abundance and brought hard times. California has periodic droughts that have lasted as long as a human generation. And there are cycles that have longer swings than can be encompassed by individual human lifetimes. Within any hundred-year period, floods alter the very structure of rivers. Along the Cascadian subduction zone, which stretches from Vancouver Island to Cape Mendocino in California, earthquakes and tidal waves three to five hundred years apart change the very nature of the landscape along its entire length. Whole new terraces rise up out of the sea in one place; the land drops away thirty feet in another. Rivers find new channels, and the salmon become lost for a time.

Even larger cycles include those long fluctuations of temperature in the air and water which every ten or twenty thousand years capture the water of the world in glaciers and the ice caps. Continents are scoured, mountain valleys deepened, coastlines reconfigured, human histories interrupted. These events become myths of a landscape in a state of perpetual creation; they are a part of every winter's storytelling. The stories cast a shadow on the psyche and they carry advice which cannot be ignored. Be attentive. Watch your step. Everything's alive and moving.

On a scale equivalent to that of the changes caused by ice ages and continental drift are the forces set loose by recent European invasions and conquests of North America, the exponential explosion of human population that drives this history, and the aberrant denial of the processes of interdependence which has come to define human behavior during this period.

Somewhere between these conflicting states of wonder—between natural provision erotic in its profligacy and cruel in its sometimes sudden and total withdrawal—lies the origins of the old ways. Somewhere beyond our modern notions of religion and regulation but partaking of both, human engagement with salmon—and the rest of the natural world—has been marked by behavior that is respectful, participatory, and ceremonial. And it is in this way that most of the human species has behaved most of the time it has been on the planet.

King salmon and I are together in the water. The basic bone-felt nature of this encounter never changes, even though I have spent parts of a lifetime seeking the meeting and puzzling over its meaning, trying to find for myself the right place in it. It is a *large* experience, and it has never failed to contain these elements, at once separate and combined: empty-minded awe; an uneasiness about my own active role both as a person and as a creature of my species; and a looming existential dread that sometimes attains the physicality of a lump in the throat, a knot in the abdomen, a constriction around the temples. They seem important, these various elements of response, like basic conditions of existence. I am smack in the middle of the beautiful off-handed description of our field of being that once flew up from my friend David Abram's mouth: that we are many sets of eyes staring out at each other from the same living body. For the instant, there is a part of that living body which is a cold wet darkness containing a pure burst of salmon muscle and intelligence, and containing also a clumsy human pursuing the ghost of a relationship.

I have left the big dip net leaning against the trailer up above the river. I forget that the captured fish is probably confused and will not quickly find its way out of the river pen. I race up the steep bank of the gorge as if everything depends upon my speed. My wader boots, half a size too large, catch on a tree root and I am thrown on my face in the mud. The bank is steep and I hit the ground before my body expects to, and with less force. I am so happy to be unhurt that I giggle absurdly. Why, tonight, am I acting like a hunter? All my training, social and intellectual, as well as my genetic predisposition, moves me to act like a predator rather than a grateful, careful guest at Gaia's table. Why am I acting as if this is an encounter that has a winner and a loser, even though I am perfectly aware that the goal of the encounter is to keep the fish alive?

I retrieve the dip net and return more slowly down the dark bank to the river. Flashing the beam of my headlamp on the water in the enclosure, I can see a shape darker than the dark water. The shape rolls as it turns to flash the pale belly. The fish is large—three or maybe four years old. It seems as long as my leg.

Several lengths of large PVC pipe are strewn along the edge of the river, half in the water and half out. These sections of heavy white or aquamarine tubing, eight, ten, and twelve inches in diameter, have been cut to length to provide temporary holding for a salmon of any of the various sizes that might arrive: the more closely contained the captured creature, the less it will thrash about and do injury to itself. I remove from the largest tube the perforated Plexiglas endplate held in place by large cotter pins.

I wade into the watery pen. Nowhere is the water deeper than my knees; the trap site has been selected for the rare regularity of its bottom and for its gentle gradient. The pen is small enough so that anywhere I stand I dominate half its area. Here, within miles of its headwaters, the river is no more than thirty feet across. The pen encloses half its width. I wade slowly back and forth to get a sense of the fish's speed and strength. This one seems to be a female, recently arrived. When she swims between my feet I can see the gentle swollen curve from gill to tail where her three to five thousand eggs are carried. She explores this new barrier to her upstream migration powerfully and methodically, surging from one side of the enclosure to another. Using the handle of the net to balance myself against the current, I find the edge of the pen farthest from the shore, turn off the headlamp, and stand quietly, listening again.

The rain has stopped. Occasionally I can hear her dorsal fin tear the surface of the water. After a few minutes I point my headlamp downward and flick the switch. Again the surface of the water seems to leap toward me. The fish is irritated or frightened by the light, and each of her exploratory surges moves her farther away from me, closer to the shore.

The great strength of her thrusts pushes her into water that is shallower than the depth of her body and she flounders. Her tail seeks purchase where there is none and beats the shallow water like a fibrillating heart. The whole weight of the river seems to tear against my legs as I take the few steps toward her. I reach over her with the net so that she lies between me and the mesh hoop. I hold the net stationary and kick at the water near her tail; she twists away from me and into the net. Now I can twist the mouth of the net up toward the air and she is completely encircled by the two-inch mesh. I move her toward deeper water and rest.

There are sparks of light rotating behind my eyes. The struggle in the net translates up my arms like low-voltage electricity. The weight of the fish amplified by the length of the net's handle is too much. I use two hands to grasp the aluminum rim at either side of the mouth of the net, and I rest and breathe. After a bit, I can release one side of the frame and hold the whole net jammed against my leg with one hand. I reach for the PVC tube and position its open mouth where I want it, half submerged and with the opening pointing toward us. I move the net and the fish around to my left side and grasp through the net the narrow part of her body just forward of her tail—the peduncle—where she is still twice the thickness of my wrist.

I only have enough strength to turn the fish in one direction or another; were I to try and lift her out of the water against her powerful lateral thrashing, I would surely drop her. The fish is all one long muscle from head to tail, and that muscle is longer, and stronger, than any muscle I can bring to bear. I direct her head toward the tube, and enclose tube and fish within the net. I drop the handle of the net, and move the fish forward, toward the tube.

There is a moment while I am holding the salmon and mesh entwined in elbow-deep water when everything goes still. Her eyes are utterly devoid of expression. Her gills pump and relax, pump and relax, measured and calmly regular. There is in that reflex an essence of aquatic creaturehood, a reality to itself entire. And there is a sense of great peacefulness, as when watching the rise and fall of a sleeping lover's chest. When I loosen my grasp, she swims out of the net and into the small enclosure.

Quickly, trembling, I lift the tail end of the tube so that her head is facing down into the river. I slide the Plexiglas endplate into place and fasten it, and she lies quietly, the tube just submerged and tethered to a stout willow. I sit down beside the dark and noisy river, beside the captured female salmon. I am sweating inside my rubber gear. The rain has begun again. I think about the new year and the promise of the eggs inside her. I am surrounded by ghosts that rise off the river like scant fog.

WILLIAM LEAST HEAT-MOON (WILLIAM TROGDON) b. 1939

William Least Heat-Moon writes books that resemble what Leslie Marmon Silko has described, in her own Pueblo tradition, as "story-maps": narratives with a topographic specificity that gives them "survival value." His two most-celebrated volumes seem initially to pursue radically different approaches. Blue Highways: A Journey into America *(1982) records a thirteen-thousand-mile drive around the United States on secondary roads that took him to the small towns and regions where rooted, traditional cultures still survived. In 1991, Heat-Moon published* PrairyErth *(a deep map), which by contrast focused exclusively on Chase County, at the heart of Kansas and America. In both of these works, however, his concerns have been to describe as precisely as possible his own location on the surface of the Earth; to integrate the geological, ecological, and human histories of each place into his own awareness in the present; and to make visible and memorable for his readers the enduring legacies of those deeper, intertwined stories.*

UNDER OLD NELL'S SKIRT

I know a man, a Maya in the Yucatán, who can call up wind: he whistles a clear, haunting, thirteen-note melody set in the Native American pentatonic scale. He whistles, the wind moves, and for some moments the heat of the tropical forest eases. It's a talent there to appreciate. But does he summon the wind, or does he know just the right time to whistle before the wind moves? He says, in effect, that he is on speaking terms with the wind, and by that he means it is a phenomenon, yes, but also a presence, and it has a name, Ik, and it is Ik that brings the sea-

PrairyErth (a deep map) (Boston: Houghton Mifflin, 1991).

sonal rain to Yucatán. You may call such a notion pantheism or primitivism or mere personification: he wouldn't care, because for him, for the Maya, for all of tribal America, the wind, the life bringer, is something to heed, to esteem: Ik.

In Kansas I've not heard any names for the nearly constant winds, the oldest of things here. When the Kansa Indians were pushed out of the state, they carried with them the last perception of wind as anything other than a faceless force, usually for destruction, the power behind terrible prairie wildfires, the clout in blizzards and droughts, and, most of all, in tornadoes that will take up everything, even fenceposts. But people here know wind well, they often speak of it, yet, despite the several names in other places for local American winds, in this state, whose very name may mean "wind-people," it has no identity but a direction, no epithet but a curse. A local preacher told me: *Giving names to nature is unchristian.* I said that it might help people connect with things and who knows where that might lead, and he said, *To idolatry.* Yet the fact remains: these countians are more activated by weather than religion.

Almost everything I see in this place sooner or later brings me back to the grasses; after all, this is the prairie, a topography that so surprised Anglo culture when it began arriving that it found for this grand-beyond no suitable word in its immense vocabulary, and it resorted to the French of illiterate trappers: prairie. Except in accounts of novice travelers, these grasslands have never been meadows, heaths, moors, downs, wolds. A woman in Boston once said to me, *Prairie is such a lovely word—and for so grim a place.*

More than all other things here, the grasses are the offspring of the wind, the power that helps evaporation equal precipitation to the detriment of trees, the power that breaks off leaves and branches, shakes crowns and rigid trunks to tear roots and disrupt transpiration, respiration, nutrient assimilation. But grasses before the wind bend and straighten and bend and keep their vital parts underground, and, come into season, they release their germ, spikelets, and seeds to the wind, the invisible sea that in this place must carry the code, the directions from the unfaced god, carry the imprint of rootlet and rhizome, blade and sheath, culm and rachis: the wind, the penisless god going and coming everywhere, the intercourse of the grasses, the sprayer of seed across the opened sex risen and waiting for the pattern set loose on the winds today of no name; and so the grasses pull the energy from the wind, the offspring of sunlight, to transmute soil into more grasses that ungulates eat into flesh that men turn into pot roasts and woolen socks.

Now: I am walking a ridge in the southern end of Saffordville quadrangle, and below me in the creek bottom are oaks of several kinds, cottonwood, hackberry, walnut, hickory, sycamore. Slippery elms, once providing a throat emulcent, try to climb the hills by finding rock crevices to shield their seed, and, if one sprouts, it will grow straight for a time, only to lose its inborn shape to the prevailing southerlies so that the windward sides of elms seem eaten off but the lee sides spread north like tresses unloosed in March. If a seedling succeeds on a ridge top, it will spread low as if to squat under the shears of windrush, and everywhere the elm trunks lean to the polestar and make the county appear as if its southern end had been lifted and tilted before the land could dry and set. A windmill must stand straight and turn into the wind to harvest water; but the slippery elm turns away to keep the wind from its wet pulp.

And there is another face to this thing from which life proceeds. Yesterday I walked down a ridge to get out of the November wind while I ate a sandwich, and I came upon a house foundation on a slope bereft of anything but grasses and knee-high plants. It was absolutely exposed, an oddity here, since most of the homes sit in the shelter of wooded vales. This one faced east—or it would have, had it still been there—and the only relief from the prevailing winds that the builder had sought was to set the back of the house to them. There was the foundation, some broken boards, a few rusting things, and, thirty feet away, a storm cellar, its door torn off, and that was all except for a rock road of two ruts. The cave, as people here call tornado cellars, was of rough-cut native stone with an arched roof, wooden shelves, and a packed-earth floor with Mason jar fragments glinting blue in the sunlight; one had been so broken that twin pieces at my feet said:

The shards seemed to be lost voices locked in silica and calling still.

These cellars once kept cool home-canned food (and rat snakes), and, when a tornado struck like a fang from some cloud-beast, they kept families that mocked their own timorousness by calling them *'fraidy holes*, and it did take nerve to go into the dim recesses with their spidered corners and dark, reptilian coils. I stepped down inside and sat on a stone fallen from the wall and ate safely in the doorway, but, even with the sun shafts, there was something dismal and haunted in the shadowed dust of dry rot here and dank of wet rot there. Things lay silent inside, the air quite stilled, and I felt something, I don't know what: something waiting.

Was there a connection between this cave and that house absent but for its foundation? The site, sloping southwest, seemed placed to catch a cyclone in a county in the heart of the notorious Tornado Alley of the Middle West, a belt that can average 250 tornadoes a year, more than anywhere else in the world. A hundred and sixty miles from here, Codell, Kansas, got thumped by a tornado every twentieth of May for three successive years, and five months ago a twister "touched down," mashed down really, a mile north of Saffordville at the small conglomeration of houses and trailers called Toledo, and the newspaper caption for a photograph of that crook'd finger of a funnel cloud was HOLY TOLEDO! Years earlier a cyclone wrecked a Friends meetinghouse there, but this time it skipped over the Methodists' church and went for their houses. In Chase County I've found a nonchalance about natural forces born of fatalism: *If it's gonna get me, it'll get me.* In Cottonwood Falls, on a block where a house once sat, the old cave remains, collapsing, yet around it are six house trailers. Riding out a tornado in a mobile home is like stepping into combine blades: trailers can become airborne chambers full of flying knives of aluminum and glass. No: if there is a dread in the county, it is not of dark skies but of the opposite, of clear skies, days and days of clear skies, of a drought nobody escapes, not even the shopkeepers. That any one person will suffer losses from a tornado, however deadly, goes much against the odds, and many residents reach high school before they first see a twister; yet, nobody who lives his full span in the county dies without a tornado story.

Tornado: a Spanish past participle meaning turned, from a verb meaning to turn, alter, transform, repeat, *and* to restore. Meteorologists speak of the reasons why the Midlands of the United States suffer so many tornadoes: a range of high mountains west of a great expanse of sun-heated plains at a much lower altitude, where dry and cold northern air can meet warm and moist southern air from a large body of water to combine with a circulation pattern mixing things up: that is to say, the jet stream from Arctic Canada crosses the Rockies to meet a front from the Gulf of Mexico over the Great Plains in the center of which sits Kansas, where, since 1950, people have sighted seventeen hundred tornadoes. It is a place of such potential celestial violence that the meteorologists at the National Severe Storms Forecast Center in Kansas City, Missouri, are sometimes called the Keepers of the Gates of Hell. Countians who have smelled the fulminous, cyclonic sky up close, who have felt the ground shake and heard the earth itself roar and have taken to a storm cellar that soon filled with a loathsome greenish air, find the image apt. The Keepers of the Gates of Hell have, in recent years, become adept at forecasting tornadoes, and they might even be able to suggest cures for them if only they could study them up close. Years ago a fellow proposed sending scientists into the eye of a tornado in an army tank until he considered the problem of transporting the machine to a funnel that usually lasts only minutes, and someone else suggested flying into a cyclone, whereupon a weather-research pilot said, yes, it was feasible if the aviator would first practice by flying into mountains.

Climatologists speak of thunderstorms pregnant with tornadoes, storm-breeding clouds more than twice the height of Mount Everest; they speak of funicular envelopes and anvil clouds with pendant mammati and of thermal instability of winds in cyclonic vorticity, of rotatory columns of air torquing at velocities up to three hundred miles an hour (although no anemometer in the direct path of a storm has survived), funnels that can move over the ground at the speed of a strolling man or at the rate of a barrel-assing semi on the turnpike; they say the width of the destruction can be the distance between home plate and deep center field and its length the hundred miles between New York City and Philadelphia. A tornado, although more violent than a much longer lasting hurricane, has a life measured in minutes, and weather-casters watch it snuff out as it was born: unnamed.

I know here a grandfather, a man as bald as if a cyclonic wind had taken his scalp—something witnesses claim has happened elsewhere—who calls twisters Old Nell, and he threatens to set crying children outside the back door for her to carry off. People who have seen Old Nell close, up under her skirt, talk about her colors: pastel-pink, black, blue, gray, and a survivor said this: *All at once a big hole opened in the sky with a mass of cherry-red, a yellow tinge in the center,* and another said: *a funnel with beautiful electric-blue light,* and a third person: *It was glowing like it was illuminated from the inside.* The witnesses speak of shapes: a formless black mass, a cone, cylinder, tube, ribbon, pendant, thrashing hose, dangling lariat, writhing snake, elephant trunk. They tell of ponds being vacuumed dry, eyes of geese sucked out, chickens clean-plucked from beak to bum, water pulled straight up out of toilet bowls, a woman's clothes torn off her, a wife killed after being jerked through a car window, a child carried two miles and set down with only scratches, a Cottonwood Falls mother (fearful of wind) cured of chronic headaches when a twister passed harmlessly within a few feet of her house, and, just south of Chase, a woman blown out of her living room window and dropped unhurt sixty feet away and falling unbroken beside her a phonograph record of "Stormy Weather."

London Harness, an eighty-five-year-old man who lives just six miles north of the county line, told me: *I knew a family years ago that was* crossing open country here in a horse and wagon. A bad storm come on fast, and the man run to a dug well and said, "I'm going down in here—you do the best you can!" The wife hollered and screamed and run to a ditch and laid down with their two little kids. That funnel dropped right in on them. After the storm passed over, she and the kids went to the well to say, "Come on up, Pappy," but there weren't no water down there, and he weren't down there. If you're in that path, no need of running.

Yesterday: in the sun the broken words on the Mason jar glinted and, against the foundation, the wind whacked dry grasses and seed pods, tap-tap-tap, rasp-rasp, and a yellow light lay over the November slope, and Ma and son: did they one afternoon come out of the cave to see what I see, an unhoused foundation, some twisted fence wire, and a sky turning golden in all innocence?

ATOP THE MOUND

What I cherish I've come to slowly, usually blindly, not seeing it for some time, and that's just how I discovered Jacobs' Mound, a truncated cone sitting close to the center of the Gladstone quadrangle. This most obvious old travelers' marker shows up clearly from two of the three highways, yet I was here several days before I noticed it, this isolated frustum so distinct. I must have been looking too closely and narrowly, but once I saw its volcano-cone symmetry (at night in the fire season, its top can flame and smolder) I was drawn to it as western travelers have always been to lone protuberances—Independence Rock, Pompey's Pillar, Chimney Rock—and within a day I headed down the Bloody Creek Road until the lane played out in a grassed vale. Some two aerial miles west of the mound, I climbed a ridge and sat down and watched it as if it might disappear like a flock of rare birds. That morning four people told me four things, one of them, the last, accurate: the regular sides and flattened top of the knob prove Indians built it for a burial mound; Colorado prospectors hid gold in it; an oil dome lay beneath it; and, none of those notions was true.

I walked down the hawk-harried ridge and struck out toward the mound, seemingly near enough to reach before sunset. Its sea-level elevation is fifteen hundred feet, but it rises only about a hundred from its base and three hundred above the surrounding humped terrain. In places the October grasses, russet-colored like low flames as if revealing their union with fire, reached to my belt and stunted my strides, and there were also aromatic asters and false indigo, both now dried to scratching stiffness. From the tall heads of Indian grass and the brown stalks of gayfeather, gossamer strung out in the slow wind like pennants ten and twelve feet long and silver in the sun, and these web lines snagged my trousers and chest and head until, after a mile, I was bestrung and on my way to becoming cocooned. Gray flittings rose from the ground like winged stones and threw themselves immediately into invisibility—I think they were vesper sparrows. Twice, prairie chickens broke noisily and did their sweet, dihedral-winged glides to new cover (Audubon said their bent-down wings enable the birds to turn their heads to see behind as they fly). I stopped to watch small events but never for long because the mound was drawing me as if it were a stone vortex in a petrified sea.

There are several ways not to walk in the prairie, and one of them is with your eye on a far goal, because you then begin to believe you're not closing the distance any more than you would with a mirage. My woodland sense of scale and time didn't fit this country, and I started wondering whether I could reach the summit before dark. On the prairie, distance and the miles of air turn movement to stasis and openness to a wall, a thing as difficult to penetrate as dense forest. I was hiking in a chamber of absences where the near was the same as the far, and it seemed every time I raised a step the earth rotated under me so that my foot fell just where it had lifted from. Limits and markers make travel possible for people: circumscribe our lines of sight and we can really get somewhere. Before me lay the Kansas of popular conception from Coronado on—that place you have to get through, that purgatory of mileage.

But I kept walking, and, when I dropped into hollows and the mound disappeared, I focused on a rock or a tuft of grass to keep from convoluting my track. Hiking in woods allows a traveler to imagine comforting enclosures, one leading to the next, and the walker can possess those little encompassed spaces, but the prairie and plains permit no such possession. Whatever else prairie is—grass, sky, wind—it is most of all a paradigm of infinity, a clearing full of many things except boundaries, and its power comes from its apparent limitlessness; there is no such thing as a small prairie any more than there is a little ocean, and the consequence of both is this challenge: try to take yourself seriously out here, you bipedal plodder, you complacent cartoon.

I came up out of a hollow, Jacobs' Mound big now on the horizon, and I could feel its swell in my legs, and then I was in the steep climb up its slope, and: I was on top. From the highway I'd guessed the summit to be the size of a city block, but it was less than a baseball infield, its elliptical perimeter just a hundred strides. So, its power lay not in size but rather in shape and dominion and its thrust into the imagination.

I sat and looked. The thousands of acres that lay encircled around the knob I really didn't see, not at first. I saw air, and I said, good god, look at all this air, and I recalled a woman saying, *Seems the air here hasn't ever been used before.* From a plane you look down, and from a mountain you look down, but from Jacobs' Mound you look out, out into. You're not up in the sky and you're not on the ground: you're nicely in between, at the altitude of those who fly in their dreams and skim roofs and treetops. Jacobs' Mound is thrush-flight high.

And then I understood: I like this prairie county because of its illusion of being away, out of, and I like how its unpopulousness seems to isolate it. Seventy percent of Americans live on two percent of the land, but in front of me, no percentage of them lived. Yet, in the far southeast, I could see trucks inching out the turnpike miles, the turbulence of their passage silenced by distance. And I could see fence lines, transmission towers, and dug ponds, things the pioneers would have viewed as marks of a progressive civilization but which to me, a grousing neo-primitivist, were signs of the continuing onslaught. The view I had homesteaders would have loved, and the one they had of unbroken vegetation and its diversities I would cherish. On top of the mound, insects whirred steadily, and the wind blew in easy continuousness, a drone like that in a seashell at the ear. In the nineteenth century, the Kansas clergyman and author William Quayle (who once wrote, *In a purely metaphorical sense I am a turnip*) traded his autograph for an acre of prairie, and, yesterday, I thought him a thief, but now, seeing the paltriness of an acre, I figured he was the one swindled.

On his great western expedition of 1806, Zebulon Pike crossed the Flint Hills just south of this big knob, and he surely couldn't have resisted climbing this rise for a good look around. In later years, perpendicular to his course ran an old freighter road and stage line that cut between here and Phenis Mound across the county line and five miles east. Near its base, a century ago, farmer John Buckingham plowed up a small redwood chest, took it home, pried it open, and found some old parchments, one marked in crude characters of eccentric orthography advising that nearby a buried sword pointed to the spot on Phenis Mound where lay a cache of golden nuggets. Buckingham thought it a prank until he remembered plowing up a rusted saber the year before; but his and others' diggings yielded only what the inland sea put down a quarter billion year ago.

People connect themselves to the land as their imaginations allow. The links of Chase countians to Jacobs' Mound, at least in an earlier time, were more calligraphic than auricular, and at my feet lay proof: a piece of limestone, palm-sized and flattened like a slate and cut into it a reversed J surmounted by an upside-down V: perhaps a cattle brand. In the days of first white settlement, people rode out here in buggies and hayracks, filled their jugs at one of the springs below the mound, picnicked on the summit, and scratched their names into the broken stones. I looked for more: nothing. Then I turned over a small rock and there, in faint relief under the low sunlight, JOHNY, and on another, MAE, and then I began turning stones, their hardness against one another striking out a strange and musical ringing, and I found more intaglios weathered to near invisibility but the letters uncommonly adroit. The mound was so covered with bits of alphabet it was as if Moses had here thrown down his tablets.

And then from the dark, granular soil I turned one that froze me: WAKONDA. In several variants, Wakonda is a Plains Indian name for the Great Mysterious, the Four-Winds-Source-of-All. Then my sense returned: an ancient Indian writing in Roman characters? I looked closely and could barely make out W KENDA running to the fractured right edge. Perhaps once: W. KENDALL. I put it again face-down so that it might continue its transfer into the mound.

Across America, lone risings have been sacred places to tribal Americans, places to reach out for the infinite. Where whites saw this knob and dreamed gold, aboriginal peoples (it's my guess) found it and dreamed God, and it must have belonged to their legends and gramarye, and they surely came to this erosional ellipse as leaves to the eddy.

BRUCE CHATWIN
1940–1989

After a distinguished early career as director of modern art at Sotheby and Company, Bruce Chatwin suddenly lost his sight. The doctor whom he consulted (according to the profile of Chatwin in Contemporary Authors) *"diagnosed his illness as a psychomatic conditon caused by scrutinizing pictures too closely and concluded that he would be cured by searching out long horizons." When his vision returned, Chatwin began the series of journeys—to the Sahara, Patagonia, and Australia, among other landscapes—from which his most memorable books emerged. Like John Muir, and before him the Apostle Paul, the writer arose from temporary blindness determined not just to see more vividly but also to keep moving. For the rest of his life, he was particularly fascinated with nomadic cultures. Chatwin's most-celebrated work, The Songlines (1987), combines elements of autobiography, travel writing, fiction, and the speculative essay in pondering the deep affinities between the lifeways and ceremonial practices of Australia's Aboriginal people with their home landscape. It has become an important reference for readers who, in many other parts of the Earth, are trying to understand more deeply their own creative role in the living cycles of the land. Chatwin died of AIDS in 1989.*

From THE SONGLINES

Arkady ordered a couple of cappuccinos in the coffee-shop. We took them to a table by the window and he began to talk.

I was dazzled by the speed of his mind, although at times I felt he sounded like a man on a public platform, and that much of what he said had been said before.

The Aboriginals had an earthbound philosophy. The earth gave life to a man; gave him his food, language and intelligence; and the earth took him back when he died. A man's 'own country', even an empty stretch of spinifex, was itself a sacred ikon that must remain unscarred.

'Unscarred, you mean, by roads or mines or railways?'

'To wound the earth', he answered earnestly, 'is to wound yourself, and if others wound the earth, they are wounding you. The land should be left untouched: as it was in the Dreamtime when the Ancestors sang the world into existence.'

'Rilke', I said, 'had a similar intuition. He also said song was existence.'

'I know,' said Arkady, resting his chin on his hands. '"Third Sonnet to Orpheus."'

The Aboriginals, he went on, were a people who trod lightly over the earth; and the less they took from the earth, the less they had to give in return. They had never understood why the missionaries forbade their

The Songlines (New York: Viking Penguin, 1987).

innocent sacrifices. They slaughtered no victims, animal or human. Instead, when they wished to thank the earth for its gifts, they would simply slit a vein in their forearms and let their own blood spatter the ground.

'Not a heavy price to pay,' he said. 'The wars of the twentieth century are the price for having taken too much.'

'I see,' I nodded doubtfully, 'but could we get back to the Songlines?'

'We could.'

My reason for coming to Australia was to try to learn for myself, and not from other men's books, what a Songline was—and how it worked. Obviously, I was not going to get to the heart of the matter, nor would I want to. I had asked a friend in Adelaide if she knew of an expert. She gave me Arkady's phone number.

'Do you mind if I use my notebook?' I asked.

'Go ahead.'

I pulled from my pocket a black, oilcloth-covered notebook, its pages held in place with an elastic band.

'Nice notebook,' he said.

'I used to get them in Paris,' I said. 'But now they don't make them any more.'

'Paris?' he repeated, raising an eyebrow as if he'd never heard anything so pretentious.

Then he winked and went on talking.

To get to grips with the concept of the Dreamtime, he said, you had to understand it as an Aboriginal equivalent of the first two chapters of Genesis—with one significant difference.

In Genesis, God first created the 'living things' and then fashioned Father Adam from clay. Here in Australia, the Ancestors created themselves from clay, hundreds and thousands of them, one for each totemic species.

'So when an Aboriginal tells you, "I have a Wallaby Dreaming," he means, "My totem is Wallaby. I am a member of the Wallaby Clan."'

'So a Dreaming is a clan emblem? A badge to distinguish "us" from "them"? "Our country" from "their country"?'

'Much more than that,' he said.

Every Wallaby Man believed he was descended from a universal Wallaby Father, who was the ancestor of all other Wallaby Men and of all living wallabies. Wallabies, therefore, were his brothers. To kill one for food was both fratricide and cannibalism.

'Yet,' I persisted, 'the man was no more wallaby than the British are lions, the Russians bears, or the Americans bald eagles?'

'Any species', he said 'can be a Dreaming. A virus can be a Dreaming. You can have a chickenpox Dreaming, a rain Dreaming, a desert-orange Dreaming, a lice Dreaming. In the Kimberleys they've now got a money Dreaming.'

'And the Welsh have leeks, the Scots thistles and Daphne was changed into a laurel.'

'Same old story,' he said.

He went on to explain how each totemic ancestor, while travelling through the country, was thought to have scattered a trail of words and musical notes along the line of his footprints, and how these Dreaming-tracks lay over the land as 'ways' of communication between the most far-flung tribes.

'A song', he said, 'was both map and direction-finder. Providing you knew the song, you could always find your way across country.'

'And would a man on "Walkabout" always be travelling down one of the Songlines?'

'In the old days, yes,' he agreed. 'Nowadays, they go by train or car.'

'Suppose the man strayed from his Songline?'

'He was trespassing. He might get speared for it.'

'But as long as he stuck to the track, he'd always find people who shared his Dreaming? Who were, in fact, his brothers?'

'Yes.'

'From whom he could expect hospitality?'

'And vice versa.'

'So song is a kind of passport and meal-ticket?'

'Again, it's more complicated.'

In theory, at least, the whole of Australia could be read as a musical score. There was hardly a rock or creek in the country that could not or had not been sung. One should perhaps visualise the Songlines as a spaghetti of Iliads and Odysseys, writhing this way and that, in which every 'episode' was readable in terms of geology.

'By episode', I asked, 'you mean "sacred site"?'

'I do.'

'The kind of site you're surveying for the railway?'

'Put it this way,' he said. 'Anywhere in the bush you can point to some feature of the landscape and ask the Aboriginal with you, "What's the story there?" or "Who's that?" The chances are he'll answer "Kangaroo" or "Budgerigar" or "Jew Lizard", depending on which Ancestor walked that way.'

'And the distance between two such sites can be measured as a stretch of song?'

'That', said Arkady, 'is the cause of all my troubles with the railway people.'

It was one thing to persuade a surveyor that a heap of boulders were the eggs of the Rainbow Snake, or a lump of reddish sandstone was the liver of a speared kangaroo. It was something else to convince him that a feature-less stretch of gravel was the musical equivalent of Beethoven's Opus 111.

By singing the world into existence, he said, the Ancestors had been poets in the original sense of *poesis*, meaning 'creation'. No Aboriginal could conceive that the created world was in any way imperfect. His religious life had a single aim: to keep the land the way it was and should be. The man who went 'Walkabout' was making a ritual journey. He trod in the footprints of his Ancestor. He sang the Ancestor's stanzas without changing a word or note—and so recreated the Creation.

'Sometimes,' said Arkady, 'I'll be driving my "old men" through the desert, and we'll come to a ridge of sandhills, and suddenly they'll all start singing. "What are you mob singing?" I'll ask, and they'll say, "Singing up the country, boss. Makes the country come up quicker."'

Aboriginals could not believe the country existed until they could see and sing it—just as, in the Dreamtime, the country had not existed until the Ancestors sang it.

'So the land', I said, 'must first exist as a concept in the mind? Then it must be sung? Only then can it be said to exist?'

'True.'

'In other words, "to exist" is "to be perceived"?'

'Yes.'

'Sounds suspiciously like Bishop Berkeley's Refutation of Matter.'

'Or Pure Mind Buddhism,' said Arkady, 'which also sees the world as an illusion.'

'Then I suppose these three hundred miles of steel, slicing through innumerable songs, are bound to upset your "old men's" mental balance?'

'Yes and no,' he said. 'They're very tough, emotionally, and very pragmatic. Besides, they've seen far worse than a railway.'

Aboriginals believed that all the 'living things' had been made in secret beneath the earth's crust, as well as all the white man's gear—his aeroplanes, his guns, his Toyota Land Cruisers—and every invention that will ever be invented; slumbering below the surface, waiting their turn to be called.

'Perhaps,' I suggested, 'they could sing the railway back into the created world of God?'

'You bet,' said Arkady.

* * *

The song still remains which names the land over which it sings.
Martin Heidegger, What Are Poets For?

Before coming to Australia I'd often talk about the Songlines, and people would inevitably be reminded of something else.

'Like the "ley-lines"?' they'd say: referring to ancient stone circles, menhirs and graveyards, which are laid out in lines across Britain. They are of great antiquity but are visible only to those with eyes to see.

Sinologists were reminded of the 'dragon-lines' of feng-shui, or traditional Chinese geomancy: and when I spoke to a Finnish journalist, he said the Lapps had 'singing stones', which were also arranged in lines.

To some, the Songlines were like the Art of Memory in reverse. In Frances Yates's wonderful book, one learned how classical orators, from Cicero and earlier, would construct memory palaces; fastening sections of their speech on to imaginary architectural features and then, after working their way round every architrave and pillar, could memorise colossal lengths of speech. The features were known as loci or 'places'. But in Australia the loci were not a mental construction, but had existed for ever, as events of the Dreamtime.

Other friends were reminded of the Nazca 'lines', which are etched into the meringue-like surface of the central Peruvian Desert and are, indeed, some kind of totemic map.

We once spent a hilarious week with their self-appointed guardian, Maria Reich. One morning, I went with her to see the most spectacular of all the lines, which was only visible at sunrise. I carried her photographic equipment up a steep hill of dust and stones while Maria, in her seventies, strode ahead. I was horrified to watch her roll straight past me to the bottom.

I expected broken bones, but she laughed, 'My father used to say that once you start to roll, you must keep on rolling.'

No. These were not the comparisons I was looking for. Not at this stage. I was beyond that.

Trade means friendship and co-operation; and for the Aboriginal the principal object of trade was song. Song, therefore, brought peace. Yet I felt the Songlines were not necessarily an Australian phenomenon, but universal: that they were the means by which man marked out his territory, and so organised his social life. All other successive systems were variants—or perversions—of this original model.

The main Songlines in Australia appear to enter the country from the north or the north-west—from across the Timor Sea or the Torres Strait—and from there weave their way southwards across the continent. One has the impression that they represent the routes of the first Australians—and that they have come from somewhere else.

How long ago? Fifty thousand years? Eighty or a hundred thousand years? The dates are insignificant compared to those from African prehistory.

And here I must take a leap into faith: into regions I would not expect anyone to follow.

I have a -vision of the Songlines stretching across the continents and ages; that wherever men have trodden they have left a trail of song (of which we may, now and then, catch an echo); and that these trails must reach back, in time and space, to an isolated pocket in the African savannah, where the First Man opening his mouth in defiance of the terrors that surrounded him, shouted the opening stanza of the World Song, 'I AM!'

MAXINE HONG KINGSTON
b. 1940

Woman Warrior: Memoirs of a Girlhood among Ghosts *(1976) established Maxine Hong Kingston as a powerful and original new voice in American literature. As she seeks to draw connections between the Stockton, California, of her own girlhood and the China of her parents and ancestors, Kingston comes to possess her own compound identity and also to achieve an encompassing sense of place.* China Men, *the 1980 successor to this book, continues to unfold the familial saga and to explore the role of gender in the complex traditions into which the author was born.*

Kingston has also lived and taught in Hawaii for a number of years. The selection included here comes from Hawai'i One Summer *(1987). The voice of this essay is a lighter and more playful one than is often the case in her two family chronicles. But it still conveys the wry sense of multiple perspectives engrained in her by her cross-cultural upbringing and offers revelations of the grotesque or delightful in scenes that others might find merely ordinary.*

A City Person Encountering Nature

A city person encountering nature hardly recognizes it, has no patience for its cycles, and disregards animals and plants unless they roar and exfoliate in spectacular aberrations. Preferring the city myself,

Hawai'i One Summer (San Francisco: Meadow, 1987).

I can better discern natural phenomena when books point them out; I also need to verify what I think I've seen, even though charts of phyla and species are orderly whereas nature is wild, unruly.

Last summer, my friend and I spent three days together at a beach cottage. She got up early every morning to see what "critters" the ocean washed up. The only remarkable things I'd seen at that beach in years were Portuguese man-o-war and a flightless bird, big like a pelican; the closer I waded toward it, the farther out to sea the bird bobbed.

We found flecks of whitish gelatin, each about a quarter of an inch in diameter. The wet sand was otherwise clean and flat. The crabs had not yet dug their holes. We picked up the blobs on our fingertips and put them in a saucer of sea water along with seaweeds and some branches of coral.

One of the things quivered, then it bulged, unfolded, and flipped over or inside out. It stretched and turned over like a human being getting out of bed. It opened and opened to twice its original size. Two arms and two legs flexed, and feathery wings flared, webbing the arms and legs to the body, which tapered to a graceful tail. Its ankles had tiny wings on them—like Mercury. Its back muscles were articulated like a comic book superhero's—blue and silver metallic leotards outlined with black racing stripes. It's a spaceman, I thought. A tiny spaceman in a spacesuit.

I felt my mind go wild. A little spaceship had dropped a spaceman on to our planet. The other blob went through its gyrations and also metamorphosed into a spaceman. I felt as if I were having the flying dream where I watch two perfect beings wheel in the sky.

The two critters glided about, touched the saucer's edges. Suddenly, the first one contorted itself, turned over, made a bulge like an octopus head, then flipped back, streamlined again. A hole in its side—a porthole, a vent—opened and shut. The motions happened so fast, we were not certain we had seen them until both creatures had repeated them many times.

I had seen similar quickenings: dry strawberry vines and dead trout revive in water. Leaves and fins unfurl; colors return.

We went outside to catch more, and, our eyes accustomed, found a baby critter. So there were more than a pair of these in the universe. So they grew. The baby had apparently been in the sun too long, though, and did not revive.

The next morning, bored that the critters were not performing more tricks, we blew on them to get them moving. By accident, their eyes or mouths faced, and sucked together. There was a churning. They wrapped their arms, legs, wings around one another.

Not knowing whether they were killing each other or mating, we tried unsuccessfully to part them. Guts, like two worms, came out of the portholes. Intestines, I thought; they're going to die. But the two excrescences braided together like DNA strands, then whipped apart, turned pale, and smokily receded into the holes. The critters parted, flipped, and floated away from each other.

After a long time, both of them fitted their armpits between the coral branches; we assumed that they were depositing eggs.

When we checked the clock, four hours had gone by. We'd both thought it had only been about twenty minutes.

That afternoon, the creatures seemed less distinct, their sharp lines blurring. I rubbed my eyes; the feathers were indeed melting. The beings were disintegrating in the water. I threw the coral as far out as I could into the ocean.

Later, back in town, we showed our biologist friend our sketches, I burbling about visitors from outer space. He said they were nudibranchs. This was our friend who as a kid had vowed that he would study Nature, but in college, he specialized in marine biology, and in graduate school, he studied shrimps. He was now doing research on one species of shrimp that he had discovered on one reef off O'ahu.

A new climate helps me to see nature. Here are some sights upon moving to Hawai'i:

Seven black ants, led by an orange one, dismembered a fly.

I peeled sunburn off my nose, and later recognized it as the flake of something an ant was marching away with.

A mushroom grew in a damp corner of the living room.

Giant philodendrons tear apart the cars abandoned in the jungle. Tendrils crawl out of the hoods; they climb the shafts of the steam shovels that had dug the highway. Roofs and trunks break open, turn red, orange, brown, and sag into the dirt.

Needing to read explanations of such strangeness, we bought an English magazine, *The Countryman*, which reports "The Wild Life and Tame" news.

"STAMPED TO DEATH—A hitherto peaceful herd of about fifty cows, being fetched in from pasture, suddenly began to rush around, and bellow in a most alarming manner. The source of their interest was a crippled gull, which did its best to escape; but the cows, snorting and bellowing, trampled it to death. They then quieted down and left the field normally.—Charles Brudett, Hants."

Also: "BIG EYE, Spring, 1967—When I was living in the Karoo, a man brought me a five-foot cape cobra which he had just killed. It had been unusually sluggish and the tail of another snake protruded from its mouth. This proved to be a boomslang, also poisonous but back-fanged; it was 1½ inches longer than the cobra and its head-end had been partly digested. — J. S. Taylor, Fife."

I took some students to the zoo after reading Blake's "Tyger, Tyger burning bright," Stevens's "Thirteen Ways of Looking at a Blackbird," and Lorenz's *King Solomon's Ring*. They saw the monkeys catch a pigeon and tear it apart. I kept reminding them that that was extraordinary. "Watch an animal going about its regular habits," I said, but then they saw an alligator shut its jaws on a low-flying pigeon. I remembered that I don't see ordinary stuff either.

I've watched ants make off with a used Band-Aid. I've watched a single termite bore through a book, a circle clean through. I saw a pigeon vomit milk, and didn't know whether it was sick, or whether its babies had died and the milk sacs in its throat were engorged. I have a friend who was pregnant at the same time as her mare, and, just exactly like the Chinese superstition that only one of the babies would live, the horse gave birth to a foal in two pieces.

When he was about four, we took our son crabbing for the "crabs with no eyes," as he called them. They did have eyes, but they were on stalks. The crabs fingered the bait as if with hands; very delicately they touched it, turned it, swung it. One grabbed hold of the line, and we pulled it up. But our son, a Cancer, said, "Let's name him Linda." We put Linda back in the river and went home.

JOHN HANSON MITCHELL
b. 1940

Four major threads run through John Mitchell's works of literary nonfiction: the ecological and human history of "Scratch Flats," a one square mile section of his hometown of Littleton, Massachusetts; his own personal and family history; non-Western concepts of time and space; and the abiding presence of his predecessor in that corner of New England, Henry David Thoreau. In Ceremonial Time: Fifteen Thousand Years on One Square Mile (1984), *he explores both the deep history of his home ground and the traditional Native American concept of time in which "past, present, and future can all be perceived in a single moment, generally during some dance or sacred ritual." In* Living at the End of Time *(1990), he chronicles a year he spent living in a small cabin on a remnant of rural landscape not far from a huge new computer plant. Mitchell's work is difficult to categorize for, just as he distrusts the conventional concepts of time and space, his narratives move effortlessly between natural history, human stories, and personal history.*

From LIVING AT THE END OF TIME

Once, years ago, I spent some time hiking in the Cévennes Mountains in southern France. One day I came to a ravine with a stream running through it. I had been walking all day, and the brook and the ferny cliff face beside it were wonderfully inviting, so I took off my shoes, cooled my feet in the waters, and then lay back on the mossy bank to rest. It was a dry, hot Provencal summer day, the type of day that brings out the resinous odors of the local herbs and sets the cicadas calling. I could hear their brittle whispers in the groves and dry hillsides beyond the ravine. By the stream the air was cooler and smelled of moisture and rank plants, and I lay there, listening to the ripple of the water over the riffles and the incessant dry rattle of the cicadas. I think I fell asleep for a minute or two, and in this half-state I sensed that something had come into the ravine. I opened my eyes and saw, or dreamed I saw, a hideous bearded face with loose lips, yellowed teeth, and a curving sweep of ringed horns curling beside its head. I leaped up. But there was only the tripping of the stream, the overpowering, shushing call of the insects. Then I heard a scrambling and the clatter of dry stones on the hill behind me. I was alone again.

The area I was hiking through was rich in ruins. Everywhere I went I would come across the remnants of various cultures that had existed in that part of the world for more than two thousand years—medieval Christian churches, Roman walls, towers, and aqueducts. In the dry valleys and hills there was an aura of the classic pastoral of goatherds and shepherds. In fact what I had probably seen—if indeed I was awake—was a goat. But I couldn't shake the strong impression that Pan or some other deity, some resident spirit of the ravine, had been watching me. I went back into the brush on the hill-

Living at the End of Time (Boston: Houghton Mifflin, 1990).

side and looked for goat signs—droppings or tracks or nipped-off branches. There were none.

I thought about that event for years. It seemed to say something about the nature of a landscape, the nature or sense of a place. I used to walk around the woods in North America with a feeling that something was missing, that somehow the land was incomplete or lonely. The event in the Cévennes identified the thing that was missing from the American landscape—namely the human element, the feeling of a land that had been lived in and worshipped by a people. During the time that I lived on the ridge, I came to realize my mistake. I had been expecting the wrong thing.

If indeed the American land has a spirit, if there is, as cultures throughout history have believed, some god overseeing the forests and hills, then here in North America it must be one of the fifteen-thousand-year-old deities of the people who inhabited the continent for the better part of its human history. It must be the T'chi Manitou, the creative force of the Algonquian people. It is Hobomacho, Menobohzo, and all the other spirits, giants, wood dwarfs, and monsters of the American Indian pantheon. Time and technology have not yet managed to obliterate them.

I had further evidence of this that spring. With the onset of warmer weather, almost every day I would walk back to the hemlock grove, sometimes just to visit the place, sometimes to use it as a starting point for a longer walk on the ridge. For me the grove was like one of those little processional chapels where penitents would stop during festivals and ceremonial parades. Nothing ever happened there. The trees were generally empty of bird life, and there were no flowers or shrubs growing on the darkened forest floor; but the place drew me in nonetheless.

One morning toward the end of May, while I was sitting in the grove, I heard a mysterious bird singing in the high branches, a bright yellow form whose markings I could not see clearly. I do not like hearing songs I cannot recognize, and I spent a long time outside the grove with binoculars trying to get a better view. Finally the bird left the cover of the branches and flew down the west slope of the ridge, through the oak trees, still singing. I followed after it, lost it, followed it some more, and then lost it completely; but all across the hillside, from a thousand trees and shrubs, I could hear the great rolling dawn chorus of other birds.

This particular bird had led me to a small terrace of pines, an area I did not recognize, but since it was still early and the day was fine, I continued down the slope with no particular destination in mind. I crossed a stone wall and walked downhill, through a tangle of brush, thinking that I should come in due course to the old wagon road that led to the lake where Megan Lewis and Emil and Minna lived. The road wasn't there. Neither was the oak tree that stood near it, towering above the other, lesser trees of the ridge. I was lost again.

I could not say how much of my habit of getting lost in the woods on the ridge was intentional. Elsewhere in the world I seem to have an unerring sense of direction, but here the landmarks moved. The sun changed course; the walls and old roads disappeared. On that day I simply wandered deeper into the maze, forced my way through a tangle of brush, and then broke out into a flat ground of white pine where the floor of the woods was clear of undergrowth and spread over with yellowish fallen needles. I lay down on the soft ground and stared up at the network of branches, listening to the birds sing. There were robins and buntings and thrushes, and I could hear close by the eerie descending song of the veery. Warblers were singing from the oaks behind me; a blue jay screamed. Somewhere to the south a flock of crows began to call, and then, suddenly, I heard the bird I had followed earlier. I did not move this time. I won't say I had lost interest. But the effort of getting up from the soft bed of pine to follow its song through the snagging tangle of vegetation beyond the pines seemed too much work for so fine a spring morning. The bird passed overhead, calling incessantly, a repeated series of chips. It stopped singing for a minute. I closed my eyes, opened them, and there it was in front of me. It was a Canada warbler, a bird I had not seen or heard since I lived in southern Connecticut some twenty years earlier. The sight of it brought on a surge of memories of walks in spring woodlands. I got up to follow.

The warbler moved out of the pines into some undergrowth, and then, still singing, flew into the trees again. I tried to catch up but got tangled in brush and decided to quit. It was getting hot now, and the air was very still. Just beyond the undergrowth, somewhere high among the leaves, the Canada warbler continued to sing, as if daring me to find it, but I had a sense suddenly that I should not move. The warbler sounded out; a blue jay called, and far off the barking of crows continued. I waited. A vast stillness descended over the ridge like a blanket. I felt like a hunter on the verge of a kill, spear drawn back, muscles tightening for the cast. Someone or something was nearby.

In the midst of the silence the Canada warbler called out again, and then in front of me, not ten yards away, I saw a beautiful red fox looking at me, its tail curled around its forelegs. As soon as our eyes met, it disappeared without a sound. I hardly had time to realize it had been there; it simply spun and fled up the ridge. I stepped out of the tangle and unexpectedly found myself on the old wagon road, not far from the Pawtucket burial ground. The fox was standing in the road, looking back over its shoulder. But as soon as it saw me, it streaked up the hill toward the boulders where the Indians were supposedly buried. I could see the rocks in front of me, gray-green, rounded shapes among the trees, standing like the broken columns of a ruined temple, and there among them I counted five white-tailed deer, brown fur against the green moss, their ears turned toward me, their eyes large and curious and serene.

I waited. They waited. I stepped forward. They twitched their ears. I took another step, and they spun on their hooves and dashed full tilt up the hill, tails flashing white, hoofbeats thumping the ground. In mid-flight one of them stopped and turned to face me. It stood alone, head held high, ears pointed forward. Then slowly, as I watched unmoving, it lifted its right leg elegantly, and with all the grace of some proud fla-menco dancer, stamped its hoof hard against the ground. It walked forward a few paces, head still high, and repeated the stamp, slower this time, and with more grace. We faced each other in this manner for a full two minutes, and in the space of that time a single name came into my head—T'chi Manitou. This was more than a thought; it seemed that the words actually rang out among the trees, and in fact had the deer not barked sharply, stamped again, and then followed the others up the hill, I would have said that it spoke the name. But then perhaps I had been thinking too much about Doctor John's story of the tiger in India. I went in among the boulders and sat down for a while, listening to the stillness and the periodic cries of the blue jays. There was a rustle of wind, and then a deeper quiet descended on the ridge. The sense of a haunted land was everywhere.

The fireflies hatched early in the meadow that June. I saw the first one just after dusk on the eighteenth, a bright spark against the black wall of the trees. It was at first a mere glow, a reflection of dew I thought, until I saw it rise above the grass, flashing, and spirit off to the east. Several others appeared that night, and after two or three days they were every-where, filling the meadow with light. Henry saw fireflies on the sixteenth of June in 1852 on an evening walk. There was heat lightning that evening on the horizon, and somewhere in the village someone was playing a flute. Beyond on the river, a mile distant, he could hear the rolling chorus of the bullfrogs. The night was hot, the air close, and in the meadows around him the bright flashes of the fireflies were sparkling. They were like fallen stars, and in their light he envisioned a union of sky and earth, each showing its light "for love," as he wrote.

In 1852 he had been gone from his cabin at Walden Pond for five years. He left, he said, for the same reason that he went there, because he had other lives to lead. Perhaps he realized that staying on would have dulled the experience. He continued to explore the natural world of Concord after he moved out, but he did not have that many years left.

In spite of his outdoor, elemental life, Henry was not entirely healthy. His lungs had probably been weakened by his work in his father's dusty pencil factory. Late in 1860 Henry was visited by his friend Bronson Alcott, who was suffering at the time from a bad cold. Henry caught the cold and it worsened into bronchitis, as his colds often did. He spent the winter housebound and weakened. He was still weak in the spring, and his doctor suggested that he go to a better climate, so Henry went off to—of all places—Minnesota. He spent two months there, studying the natural history of the region and, among other things, attending an Indian dance and ceremony. He was back in Concord by July, uncured, and that fall he was still in poor health. The problem by this time was not bronchitis but tuberculosis.

Henry had developed a more scientific interest in the natural world in his later years; he was studying and classifying trees and their history, and that winter, sick as he was, his work continued unabated. But it was not a good winter. The journal, the real work of his short life, ended in November with relatively light entries about the behavior of kittens and a storm.

January was cold that year, with high winds. An old man who was a friend of Henry's and a source of much of the folklore of the town, died. Henry developed pleurisy and was confined to his bed. By the end of winter it was clear that he was dying.

His was not a quick death. He was so weak he could speak only in whispers; the ominous flush of tuberculosis tinted his cheeks. He was living at home at this time with his mother and his last living sister, Sophia, and he insisted on having his bed moved downstairs to the living room so he could enjoy the life of the household till the end. In spite of his weakness, he continued to work, and, as everyone who visited him at this period remarked, his spirits were good—high almost, some said. His former jailer, Sam Staples, commented that he never saw a man die in such pleasure and peace. Henry was on the brink of the next world, and true to form, he seemed curious to enter it; he was ever the explorer. He continued to be attached to the world of the living, however. A friend visiting him, a man who was not afraid to speak directly, asked whether he might perhaps be able to see the "other side of the dark river." "One world at a time," Henry answered. Another visitor asked him if he had finally made his peace with God. "I was not aware," said Henry, "that we have ever quarreled." This was not a man trembling in the face of dissolution.

Early on the morning of May 6, 1862, he and his sister were reading sections of the manuscript of *A Week on the Concord and Merrimack Rivers.* It was a fitting project for a man whose life had, in some ways, been inspired and shaped by his relationship with his brother and their early years together. He was obviously preoccupied by events of that period of his life. Even on his deathbed, when Ellen Sewall's name came up Henry said, "I have always loved her."

Henry was too weak by this time to hold the manuscript, to read or make corrections, so Sophia would read aloud to him. She had come to the last chapter, the point at which Henry and his brother are approaching the mouth of the Nashua River, where it flows into the Merrimack. The narrative digresses at this point.

> There is a pleasant tract on the bank of the Concord, called Conantum, which I have in my mind,—the old deserted farmhouse. The desolate pasture with its bleak cliff, the open wood, the river-reach, the green meadow in the midst, and the moss-grown wild-apple orchard,—places where one may have many thoughts and not decide anything. It is a scene which I can not only remember, as I might a vision, but when I will can bodily revisit.

She read on. Henry was still, his breathing ever more shallow.

"There is something even in the lapse of time by which time recovers itself," she read.

It was a cool and breezy day on the Merrimack. The two brothers sat muffled in their cloaks while the river and the wind carried them along. They passed farms and homesteads where women and children stood on the bank, staring at them until they had swept out of sight.

The river rippled and bounded.

"We glided past the mouth of the Nashua, and not long after, of Salmon Brook, without more pause than the wind."

Sophia stopped reading. The two brothers were carried ever downstream, and downwind to Concord, a wild, easy ride. Henry knew the narrative by heart. "Now comes good sailing," he said.

After that he said something else. Sophia could not catch the full sentence, but Henry was somewhere back in Maine, somewhere in wilderness.

"Moose," he whispered. "Indian."

It was still early morning. Outside, spring was reawakening. The black and white warblers had returned, the flickers were whinnying from the distant wood lots, the smell of fresh soil filled the air, and the new leaves of the oaks and the maples made a lace work of the morning sky. Inside, the chronicler of such events was dead. *   *   *

RICHARD K. NELSON
b. 1941

Anthropology, when carried out with the sympathetic and imaginative intensity of a writer like Richard Nelson, expands the English tradition of nature writing. It opens up an Anglo-American worldview to a new possibility of identification with animals and broadens our definition of culture within the circling of the seasons. Nelson has often lived and studied with northern peoples over the past twenty years and has, in books such as Make Prayers to the Raven *(1983), connected descriptions of the cycle of water and light in that challenging terrain with the myths and techniques through which the native peoples have both endured and deepened their wisdom in the land. Increasingly, the key for him has been the hunting through which peoples like the Koyukon have confirmed their kinship with all of life. Through hunting Nelson too has learned a sacramental approach to the taking of life. It is a gift from the animal, the transmission of life and beauty in the future of our planet. A revised version of the following selection appears in Nelson's 1989 book* The Island Within. Heart and Blood: Living with Deer in America *appeared in 1997.*

THE GIFTS

Cold, clear, and calm in the pale blue morning. Snow on the high peaks brightening to amber. The bay a sheet of gray glass beneath a faint haze of steam. A November sun rises with the same fierce, chill stare of an owl's eye.

I stand at the window watching the slow dawn, and my mind fixes on the island. Nita comes softly down the stairs as I pack gear and complain of having slept too late for these short days. A few minutes later, Ethan trudges out on to the cold kitchen floor, barefoot and half asleep. We do not speak directly about hunting, to avoid acting proud

On Nature (San Francisco: North Point Press, 1987).

or giving offense to the animals. I say only that I will go to the island and look around; Ethan says only that he would rather stay at home with Nita. I wish he would come along so I could teach him things, but I know it will be quieter in the woods with just the dog.

They both wave from the window as I ease the skiff away from shore, crunching through cakes of freshwater ice the tide has carried in from Salmon River. It is a quick run through Windy Channel and out onto the freedom of the Sound, where the slopes of Mt. Sarichef bite cleanly into the frozen sky. The air stings against my face, but the rest of me is warm inside thick layers of clothes. Shungnak whines, paces, and looks over the gunwale toward the still-distant island.

Broad swells looming off the Pacific alternately lift the boat and drop it between smooth-walled canyons of water. Midway across the Sound a dark line of wind descends swiftly from the north, and within minutes we are surrounded by whitecaps. There are two choices: either beat straight up into them or cut an easier angle across the waves and take the spray. I vacillate for a while, then choose the icy spray over the intense pounding. Although I know it is wrong to curse the wind, I do it anyway.

A kittiwake sweeps over the water in great, vaulting arcs, its wings flexed against the touch and billow of the air. As it tilts its head passing over the boat, I think how clumsy and foolish we must look. The island's shore lifts slowly in dark walls of rock and timber that loom above the apron of snow-covered beach. As I approach the shelter of Low Point, the chop fades and the swell is smaller. I turn up along the lee, running between the kelp beds and the surf, straining my eyes for deer that may be feeding at the tide's edge.

Near the end of the point is a narrow gut that opens to a small. shallow anchorage. I ease the boat between the rocks, with lines of surf breaking close on either side. The waves rise and darken, their sharp edges sparkle in the sun, then long manes of spray whirl back as they turn inside out and pitch onto the shallow reef. The anchor slips down through ten feet of crystal water to settle among the kelp fronds and urchin-covered rocks. On a strong ebb the boat would go dry here, but today's tide change is only six feet. Before launching the punt I meticulously glass the broad, rocky shore and the sprawls of brown grass along the timber's edge. A tight bunch of rock sandpipers flashes up from the shingle and an otter loops along the windows of drift logs, but there is no sign of deer. I can't help feeling a little anxious, because the season is drawing short and our year's supply of meat is not yet in. Throughout the fall, deer have been unusually wary, haunting the dense underbrush and slipping away at the least disturbance. I've come near a few, but these were young ones that I stalked only for the luxury of seeing them from close range.

Watching deer is the same pleasure now that it was when I was younger, when I loved animals only with my eyes and judged hunting to be outside the bounds of morality. Later, I tried expressing this love through studies of zoology, but this only seemed to put another kind of barrier between humanity and nature—the detachment of science and abstraction. Then, through anthropology, I encountered the entirely different views of nature found in other cultures. The hunting peoples were most fascinating because they had achieved deepest intimacy with their wild surroundings and had made natural history the focus of their lives. At the age of twenty-two, I went to live with Eskimos on the arctic coast of Alaska. It was my first year away from home, I had scarcely held a rifle my hands, and the Eskimos—who call themselves the Real People—taught me their hunter's way.

The experience of living with Eskimos made very clear the direct, physical connectedness between all humans and the environments they draw existence from. Some years later, living with Koyukon Indians in Alaska's interior, I encountered a rich new dimension of that connectedness, and it profoundly changed my view of the world. Traditional Koyukon people follow a code of moral and ethical behavior that keeps a hunter in right relationship to the animals. They teach that all of nature is spiritual and aware, that it must be treated with respect, and that humans should approach the living world with restraint and humility. Now I struggle to learn if these same principles can apply in my own life and culture. Can we borrow from an ancient wisdom to structure a new relationship between ourselves and the environment? Or is Western society irreversibly committed to the illusion that humanity is separate from and dominant over the natural world?

A young bald eagle watches nervously from the peak of a tall hemlock as we bob ashore in the punt. Finally the bird lurches out, scoops its wings full of dense, cold air, and soars away beyond the line of trees. While I trudge up the long tide flat with the punt, Shungnak prances excitedly back and forth hunting for smells. The upper reaches are layered in slabbed with ice; slick cobbles shine like steel in the sun; frozen grass crackles underfoot. I lean the punt on a snow-covered log, pick up my rifle and small pack, and slip through the leafless alders into the forest.

My eyes take a moment adjusting to the sudden darkness, the deep green of boughs, and the somber, shadowy trunks. I feel safe and hidden here. The entire forest floor is covered with deep moss that should sponge gently beneath my feet. But today the softness is gone: frozen moss crunches with each step and brittle twigs snap, ringing out in the crisp air like strangers' voices. It takes a while to get used to this harshness in a forest that is usually so velvety and wet and silent. I listen to the clicking of gusts in the high branches and think that winter has come upon us like a fist.

At the base of a large nearby tree is a familiar patch of white—a scatter of deer bones—ribs, legs, vertebrae, two pelvis bones, and two skulls with half-bleached antlers. I put them here last winter, saying they were for the other animals, to make clear that they were not being thoughtlessly wasted. The scavengers soon picked them clean, the deer mice have gnawed them, and eventually they will be absorbed into the forest again. Koyukon elders say it shows respect, putting animal bones back in a clean, wild place instead of throwing them away with trash or scattering them in a garbage dump. The same obligations of etiquette that bind us to our human community also bind us to the natural community we live within.

Shungnak follows closely as we work our way back through a maze of windfalls, across clear disks of frozen ponds, and around patches of snow beneath openings in the forest canopy. I step and wait, trying to make no sound, knowing we could see deer at any moment. Deep snow has driven them down off the slopes and they are sure to be distracted with the business of the mating season.

We pick our way up the face of a high, steep scarp, then clamber atop a fallen log for a better view ahead. I peer into the semi-open understory of twiggy bushes, probing each space with my eyes. A downy woodpecker's call sparks from a nearby tree. Several minutes pass. Then a huckleberry branch moves, barely twitches, without the slightest noise . . . not far ahead.

Amid the scramble of brush where my eyes saw nothing a few minutes ago, a dim shape materializes, as if its own motion had created it. A doe steps into an open space, deep brown in her winter coat, soft and striking and lovely, dwarfed among the great trees, lifting her nose, looking right toward me. For perhaps a minute we are motionless in each other's gaze; then her head jerks to the left, her ears twitch back and forth, her tail flicks up, and she turns away in the stylized gait deer always use when alarmed.

Quick as a breath, quiet as a whisper, the doe glides off into the forest. Sometimes when I see a deer this way I know it is real at the moment, but afterward it seems like a daydream.

As we work our way back into the woods, I keep hoping for another look at her and thinking that a buck might have been following nearby. Any deer is legal game and I could almost certainly have taken her, but

I would rather wait for larger buck and let the doe bring on next year's young. Shungnak savors the ghost of her scent that hangs in the still air, but she has vanished.

Farther on, the snow deepens to a continuous cover beneath smaller trees, and we cross several sets of deer tracks, including some big prints with long toe drags. The snow helps to muffle our steps, but it is hard to see very far because the bushes are heavily loaded with powder. The thicket becomes a latticed maze of white on black, every branch hung and spangled in the thick fur of jeweled snow. We move through it like eagles cleaving between tumbled columns of cloud. New siftings occasionally drift down when the treetops are touched by the breeze.

Slots between the trunks up ahead shiver with blue where a muskeg opens. I angle toward it, feeling no need to hurry, picking every footstep carefully, stopping often to stare into the dizzying crannies, listening for any splinter of sound, keeping my senses tight and concentrated. A raven calls from high above the forest, and as I catch a glimpse of it an old question runs through my mind: Is this only the bird we see, or does it have the power and awareness Koyukon elders speak of? It lifts and plays on the wind far aloft, then folds up and rolls halfway over, a strong sign of luck in hunting. Never mind the issue of knowing; we should assume that power is here and let ourselves be moved by it.

I turn to look at Shungnak, taking advantage of her sharper hearing and magical sense of smell. She lifts her nose to the fresh but nebulous scent of several deer that have moved through here this morning. I watch her little radar ears, waiting for her to focus in one direction and hold it, hoping to see her body tense as it does when something moves nearby. But so far she only hears the twitching of red squirrels on dry bark. Shungnak and I have very different opinions of the squirrels. They excite her more than any other animal because she believes she will catch one someday. But for the hunter they are deceptive spurts of movement and sound, and their sputtering alarm calls alert the deer.

We approach a low, abrupt rise, covered with the obscuring brush and curtained with snow. A lift of wind hisses in the high trees, then drops away and leaves us in near-complete silence. I pause to choose a path through a scramble of blueberry bushes and little windfalls ahead, then glance back at Shungnak. She has her eyes and ears fixed off toward our left, almost directly across the current of breeze. She stands very stiff, quivering slightly, leaning forward as if she has already started to run but cannot release her muscles. I shake my finger at her as a warning to stay.

I listen as closely as possible, but hear nothing. I work my eyes into every dark crevice and slot among the snowy branches, but see nothing. I stand perfectly still and wait, then look again at Shungnak. Her head turns so slowly that I can barely detect the movement, until finally she is looking straight ahead. Perhaps it is just another squirrel. . . . I consider taking a few steps for a better view.

Then I see it.

A long, dark body appears among the bushes, moving deliberately upwind, so close I can scarcely believe I didn't see it earlier. Without looking away, I carefully slide the breech closed and lift the rifle to my shoulder, almost certain that a deer this size will be a buck. Shungnak, now forgotten behind me, must be contorted with the suppressed urge to give chase.

The deer walks easily, silently, along the little rise, never looking our way. Then he makes a sharp turn straight toward us. Thick tines of his antlers curve over the place where I have the rifle aimed. Koyukon elders teach that animals will come to those who've shown them respect, and will allow themselves to be taken in what is only a temporary death. At a moment like this, it is easy to sense that despite my abiding doubt there is a shared world beyond the one we know directly, a world the Koyukon people empower with spirits, a world that demands recognition and exacts a price from those who ignore it.

This is a very large buck. It comes so quickly that I have no chance to shoot, and then it is so close that I haven't the heart to do it. Fifty feet away, the deer lowers his head almost to the ground and lifts a slender branch that blocks his path. Snow shakes down onto his neck and clings to the fur of his shoulders as he slips underneath. Then he half-lifts his head and keeps coming. I ease the rifle down to watch, wondering how much closer he will get. Just now he makes a long, soft rutting call, like the bleating of sheep except lower and more hollow. His hooves tick against dry twigs hidden by the snow.

In the middle of a step he raises his head all the way up, and he sees me standing there—a stain against the pure white of the forest. A sudden spasm runs through his entire body, his front legs jerk apart, and he freezes all akimbo, head high, nostrils flared, coiled and hard. I can only look at him and wait, my mind snarled with irreconcilable emotions. Here is a perfect buck deer. In the Koyukon way, he has come to me; but in my own he has come too close. I am as congealed and transfixed as he is, as devoid of conscious thought. It is as if my mind has ceased to function and I only have eyes.

But the buck has no choice. He suddenly unwinds in a burst of ignited energy, springs straight up from the snow, turns in mid-flight, stabs the frozen earth again, and makes four great bounds off to the left.

His thick body seems to float, relieved of its own weight, as if a deer has the power to unbind itself from gravity.

The same deeper impulse that governs the flight of a deer governs the predator's impulse to pursue it. I watch the first leaps without moving a muscle. Then, not pausing for an instant of deliberation, I raise the rifle back to my shoulder, follow the movement of the deer's fleeing form, and wait until it stops to stare back. Almost at that instant, still moving without conscious thought, freed of the ambiguities that held me before, now no less animal than the animal I watch, my hands warm and steady and certain, acting from a more elemental sense than the ones that brought me to this meeting, I carefully align the sights and let go the sudden power.

The gift of the deer falls like a feather in the snow. And the rifle's sound has rolled off through the timber before I hear it.

I walk to the deer, now shaking a bit with swelling emotion. Shungnak is beside it already, whining and smelling, racing from one side to the other, stuffing her nose down in snow full of scent. She looks off into the brush, searching back and forth, as if the deer that ran is somewhere else, still running. She tries to lick at the blood that trickles down, but I stop her out of respect for the animal. Then, I suppose to consummate her own frustrated predatory energy, she takes a hard nip at its shoulder, shuns quickly away, and looks back as if she expects it to leap to its feet again.

As always, I whisper thanks to the animal for giving itself to me. The words are my own, not something I've learned from the Koyukon. Their elders might say that the words we use in prayer to the spirits of the natural world do not matter. Nor, perhaps, does it matter what form the spirits take in our own thoughts. What truly matters is only that prayer be made, to affirm our humility in the presence of nurturing power. Most of humanity throughout history has said prayers to the powers of surrounding nature, which they have recognized as their source of life. Surely it is not too late to recover this ancestral wisdom.

It takes a few minutes before I settle down inside and can begin the other work. Then I hang the deer with rope strung over a low branch and back twice through pulley-loops. I cut away the dark, pungent scent glands on its legs, and next make a careful incision along its belly, just large enough to reach the warm insides. The stomach and intestines come easily and cleanly; I cut through the diaphragm, and there is a hollow sound as the lungs pull free. Placing them on the soft snow, I whisper that these parts are left here for the other animals. Shungnak wants to take some for herself but I tell her to keep away. It is said that the life and awareness leaves an animal's remains slowly, and there are rules about what should be eaten by a dog. She will have her share of the scraps later on, when more of the life is gone.

After the blood has drained out, I sew the opening shut with a piece of line to keep the insides clean, and then toggle the deer's forelegs through a slit in a hind leg joint, so it can be carried like a pack. I am barely strong enough to get it up on my back, but there is plenty of time to work slowly toward the beach, stopping often to rest and cool down. During one of these stops I hear two ravens in an agitated exchange of croaks and gurgles, and I wonder if those black eyes have already spotted the remnants. No pure philanthropist, the raven gives a hunter luck only as a way of creating luck for himself.

Finally, I push through the low boughs of the beachside trees and ease my burden down. Afternoon sun throbs off the water, but a chill north wind takes all the warmth from it. Little gusts splay in dark patterns across the anchorage; the boat paces on its mooring lines; the Sound is racing with whitecaps. I take a good rest, watching the fox sparrow flit among the drift logs and a bunch of crows hassling over some bit of food at the water's edge.

Though I feel utterly satisfied, grateful, and contented, there is much to do and the day will slope away quickly. We are allowed more than one deer, so I will stay on the island for another look around tomorrow. It takes two trips to get everything out to the skiff, then we head up the shore toward the little cabin and secure anchorage at Bear Creek. By the time the boat is unloaded and tied off, the wind has faded and a late afternoon chill sinks down in the pitched, hard shadow of Sarichef.

Half-dry wood hisses and sputters, giving way reluctantly to flames in the rusted stove. It is nearly dusk when I bring the deer inside and set to work on it. Better to do this now than to wait, in case tomorrow is another day of luck. The animal hangs from a low beam, dim-lit by the kerosene lamp. I feel strange in its presence, as if it still watches, still glows with something of its life, still demands that nothing be done or spoken carelessly. A hunter should never let himself be deluded by pride or a false sense of dominance. It is not through our own power that we take life in nature; it is through the power of nature that life is given to us.

The soft hide peels away slowly from shining muscles, and the inner perfection of the deer's body is revealed. Koyukon and Eskimo hunters teach a refined art of taking an animal into its component parts, easing blade through crisp cartilage where bone joins bone, following the body's own design until it is disarticulated. There is no ugliness in it, only hands moving in concert with the beauty of an animal's making. Perhaps we have been too removed from this to understand, and we have lost touch with the process of one life being passed on to another. As my hands worked inside the deer, it is as if something has already begun to flow into me.

When the work is finished, I take two large slices from the hind quarter and put them in a pan atop the now-crackling stove. In a separate pot, I boil scraps of meat and fat for Shungnak, who has waited with as much patience as possible for a husky raised in a hunter's team up north. When the meat is finished cooking I sit on a sawed log and eat straight from the pan.

A meal could not be simpler, more satisfying, or more directly a part of the living process. I wish Ethan was here to share it, and I would explain to him again that when we eat the deer its flesh is then our flesh. The deer changes form and becomes us, and we in turn become creatures made of deer. Each time we eat the deer we should remember it and feel gratitude for what it has given us. And each time, we should carry thought like a prayer inside: "Thanks to the animal and to all that made it—the island and the forest, the air, and the rain . . ." We should remember that in the course of things, we are all generations of deer and of the earth-life that feeds us.

Warm inside my sleeping bag, I let the fire ebb away to coals. The lamp is out. The cabin roof creaks in the growing cold. I drift toward sleep, feeling pleased that there is no moon, so the deer will wait until dawn to feed. On the floor beside me, Shungnak jerks and whimpers in her dog's dreams.

Next morning we are in the woods with the early light. We follow yesterday's tracks, and just beyond the place of the buck, a pair of does drifts at the edge of sight and disappears. For an hour we angle north, then come slowly back somewhat deeper in the woods, moving crosswise to a growing easterly breeze. In two separate places, deer snort and pound away, invisible beyond a shroud of brush. Otherwise there is nothing.

Sometime after noon we come to a narrow muskeg with scattered lodgepole pines and a ragged edge of bushy, low-growing cedar. I squint against the sharp glare of snow. It has that peculiar look of old powder, a bit settled and touched by wind, very lovely but without the airy magic of a fresh fall. I gaze up the muskeg's easy slope, and above the encroaching wall of timber, seamed against the deep blue sky, is the brilliant peak of Sarichef with a great plume of snow streaming off in what must be a shuddering gale. It has a contradictory look of absoluteness and unreality about it, like a Himalayan summit suspended in mid-air over the saddle of a low ridge.

I move very slowly up the muskeg's east side, away from the breeze and in the sun's full warmth. Deer tracks crisscross the opening, but none of the animals stopped here to feed. Next to the bordering trees, the tracks follow a single, hard-packed trail, showing the deer's preference for cover. Shungnak keeps her nose to the thickly scented snow. We come across a pine sapling that a buck has torn with his antlers, scattering twigs and flakes of bark all around. But his tracks are hardened, frosted, and lack sharpness, so they are at least a day old.

We slip through a narrow point of trees, then follow the open edge again, pausing long moments between each footstep. A mixed tinkle of crossbills and siskins moves through the high timber, and a squirrel rattles from deep in the woods, too far off to be scolding us. Shungnak begins to pick up a strong ribbon of scent, but she hears nothing. I stopp for several minutes to study the muskeg's long, raveled fringe, the tangle of shade and thicket, the glaze of mantled boughs.

Then my eye barely catches a fleck of movement up ahead, near the ground and almost hidden behind the trunk of a leaning pine, perhaps a squirrel's tail or a bird. I lift my hand slowly to shade the sun, stand dead still, and wait to see if something is there. Finally it moves again.

At the very edge of the trees, almost out of sight in a little swale, small and furry and bright-tinged, turning one direction and then another, is the funnel of a single ear. Having seen this, I soon make out the other ear and the slope of a doe's forehead. Her neck is behind the leaning pine, but on the other side I can barely see the soft, dark curve of her back above the snow. She is comfortably bedded, gazing placidly into the distance, chewing her cud.

Shungnak has stopped twenty yards behind me in the point of trees and has no idea about the deer. I shake my finger at her until she lays her ears back and sits. Then I watch the doe again. She is fifty yards ahead of me, ten yards beyond the leaning tree, and still looking off at an angle. Her left eye is clearly visible and she refuses to turn her head away, so it might be impossible to get any closer. Perhaps I should just wait here, in case a buck is attending her nearby. But however improbable it might be under these circumstances, a thought is lodged in my mind: I can get near her.

My first step sinks down softly, but the second makes a loud budging sound. She snaps my way, stops chewing, and stares for several minutes. It seems hopeless, especially out here in an open field of crisp snow with only the narrow treetrunk for a screen. But she slowly turns away and starts to chew again. I move just enough so the tree blocks her eye and the rest of her head, but I can still see her ears. Every time she chews they shake just a bit, so I can watch them and step when her hearing is obscured by the sound of her own jaws.

Either this works or the deer has decided to ignore me, because after a short while I am near enough so the noise of my feet has to reach her easily. She should have jumped up and run long ago, but instead she lays there in serene repose. I deliberate on every step, try for the softest snow, wait long minutes before the next move, stalking like a cat toward ambush. I watch beyond her, into the surrounding shadows and across to the muskeg's farther edge, for the shape of a buck deer; but there is nothing. I feel ponderous, clumsy-footed, out-of-place, inimical. I should turn and run away, take fear on the deer's behalf, flee the mirrored image in my mind. But I clutch the cold rifle at my side and creep closer.

The wind refuses to blow and my footsteps seem like thunder in the still sunshine. But the doe only turns once to look my way, without even pointing her ears toward me, then stares off and begins to chew again.

I am ten feet from the leaning tree. My heart pounds so hard, I think those enchanted ears should hear the rush of blood in my temples. Yet a strange certainty has come into me, a quite unmystical confidence. Perhaps she has decided I am another deer, a buck attracted by her musk or a doe feeding gradually toward her. My slow pace and lapses of stillness would not seem human. For myself, I have lost awareness of elapsed time; I have no feeling of patience or impatience. It is as if the deer has moved slowly toward me on a cloud of snow, and I am adrift in the pure motion of experience.

I take the last step to the trunk of the leaning pine. It is bare of branches, scarcely wider than my hand, but perfectly placed to break my odd profile. There is no hope of getting any closer, so I slowly poke my head out to watch. She has an ideal spot: screened from the wind, warmed by the sun, and with a clear view of the muskeg. I can see muscles working beneath the close fur of her jaw, the rise and fall of her side each time she breathes, the shining edge of her ebony eye.

I hold absolutely still, but her body begins to stiffen, she lifts her head higher, and her ears twitch anxiously. Then instead of looking at me she turns her face to the woods, shifting her ears toward a sound I cannot hear. A few seconds later, the unmistakable voice of a buck drifts up, strangely disembodied, as if it comes from an animal somewhere underneath the snow. I huddle as close to the tree as I can, press against the hard, dry bark, and peek out around its edge.

There is a gentle rise behind the doe, scattered with sapling pines and clusters of juniper bushes. A rhythmic crunching of snow comes invisibly from the slope, then a bough shakes . . . and a buck walks easily into the open sunshine.

Focusing his attention completely on the doe, he comes straight toward her and never sees my intrusive shape just beyond. He slips through a patch of small trees, stops a few feet from where she lies, lowers his head and stretches it toward her, then holds this odd pose for a long moment. She reaches her muzzle out to one side, trying to find his scent. When he starts to move up behind her she stands quickly, bends her body into a strange sideways arc, and stares back at him. A moment later she walks off a bit, lifts her tail, and puts droppings in her tracks. The buck moves to the warm ground of her bed and lowers his nose to the place where her female scent is strongest.

Inching like a reptile on a cold rock, I have stepped out from the tree and let my whole menacing profile become visible. The deer are thirty feet away and stand well apart, so they can both see me easily. I am a hunter hovering near his prey and a watcher craving inhuman love, torn between the deepest impulses, hot and shallow-breathed and seething with unreconciled intent, hidden from opened eyes that look into the nimbus of sun and see nothing but the shadow they have chosen for themselves. In this shadow now, the hunter has vanished and only the watcher remains.

Drawn by the honey of the doe's scent, the buck steps quickly toward her. And now the most extraordinary thing happens. The doe turns away from him and walks straight for me. There is no hesitation, only a wild deer coming along the trail of hardened snow where the other deer have passed, the trail in which I stand at this moment. She raises her head, looks at me, and steps without hesitation.

My existence is reduced to a pair of eyes; a rush of unbearable heat flushes through my cheeks; and a sense of absolute certainty fuses in my mind.

The snow blazes so brightly that my head aches. The deer is a dark form growing larger. I look up at the buck, half embarrassed, as if to apologize that she has chosen me over him. He stares at her for a moment, turns to follow, then stops and watches anxiously. I am struck by how gently her narrow hooves touch the trail, how little sound they make as she steps, how thick the fur is on her flank and shoulder, how unfathomable her eyes look. I am consumed with a sense of her perfect elegance in the brilliant light. And then I am lost again in the whirling intensity of experience.

The doe is now ten feet from me. She never pauses or looks away. Her feet punch down mechanically into the snow, coming closer and closer, until they are less than a yard from my own. Then she stops, stretches her neck calmly toward me, and lifts her nose.

There is not the slightest question in my mind, as if this was certain to happen and I have known all along exactly what to do. I slowly raise my hand and reach out . . .

And my fingers touch the soft, dry, gently needling fur on top of the deer's head, and press down to the living warmth of flesh underneath.

She makes no move and shows no fear, but I can feel the flaming strength and tension that flow in her wild body as in no other animal I have ever touched. Time expands and I am suspended in the clear reality of that moment.

Then, by the flawed conditioning of a lifetime among fearless domesticated things, I instinctively drop my hand and let the deer smell it. Her dark nose, wet and shining, touches gently against my skin at the exact instant I realize the absoluteness of my error. And a jolt runs through her entire body as she realizes hers. Her muscles seize and harden; she seems to wrench her eyes away from me but her body remains, rigid and paralyzed. Having been deceived by her other senses, she keeps her nose tight against my hand for one more moment.

Then all the energy inside her triggers in a series of exquisite bounds. She flings out over the hummocks of snow-covered moss, suspended in effortless flight like fog blown over the muskeg in a gale. Her body leaps with such power that the muscles should twang aloud like a bowstring; the earth should shudder and drum; but I hear no sound. In the center of the muskeg she stops to look back, as if to confirm what must seem impossible. The buck follows in more earthbound undulations; they dance away together, and I am left in the meeting-place alone.

There is a blur of rushing feet behind me. No longer able to restrain herself, Shungnak dashes past, buries her nose in the soft tracks, and then looks back to ask if we can run after them. I had completely forgotten her, sitting near enough to watch the whole encounter, somehow resisting what must have been a prodigious urge to explode in chase. When I reach out to hug her, she smells the hand that touched the deer. And it seems as if it happened long ago.

For the past year I have kept a secret dream, that I would someday come close enough to touch a deer on this island. But since the idea came it seemed harder than ever to get near them. Now, totally unexpected and in a strange way, it has happened. Was the deer caught by some reckless twinge of curiosity? Had she never encountered a human on this wild island? Did she yield to some odd amorous confusion? I really do not care. I would rather accept this as pure experience and not give in to the notion that everything must be explained.

Nor do I care to think that I was chosen to see some manifestation of power, because I have little tolerance for such dreams of self-importance. I have never asked that nature open any doors to reveal the truth of spirit or mystery; I aspire to no shaman's path; I expect no visions, no miracles except the ones that fill every instant of ordinary life.

But there are vital lessons in the experience of moments such as these, if we live them in the light of wisdom taken from the earth and shaped by generations of elders. Two deer came and gave the choices to me. One deer I took and we will now share a single body. The other deer I touched and we will now share that moment. These events could be seen as opposites, but they are in fact identical. Both are founded in the same principles, the same relationship, the same reciprocity.

Move slowly, stay quiet, watch carefully . . . and be ever humble. Never show the slightest arrogance or disrespect. Koyukon elders would explain, in words quite different from my own, that I moved into two moments of grace, or what they would call luck. This is the source of success for a hunter or a watcher, not skill, not cleverness, not guile. Something is only given in nature, never taken.

I have heard the elders say that everything in nature has its own spirit and possesses a power beyond ours. There is no way to prove them right or wrong, though the beauty and interrelatedness of things should be evidence enough. We need not ask for shining visions as proof, or for a message from a golden deer glowing in the sky of our dreams. Above all else, we should assume that power moves in the world around us and act accordingly. If it is a myth, then spirit is within the myth and we should live by it. And if there is a commandment to follow, it is to approach all of earth-life, of which we are a part, with humility and respect.

Well soaked and shivering from a rough trip across the Sound, we pull into the dark waters of the bay. Sunset burns on Twin Peaks and the spindled ridge of Antler Mountain. The little house is warm with lights that shimmer on the calm near shore. I see Nita looking from the window and Ethan dashes out to wait by the tide, pitching rocks at the mooring buoy. He strains to see inside the boat, knowing that a hunter who tells his news aloud may offend the animals by sounding boastful. But when he sees the deer his excited voice seems to roll up and down the mountainside.

He runs for the house with Shungnak, carrying a load of gear, and I know he will burst inside with the news. Ethan, joyous and alive, boy made of deer.

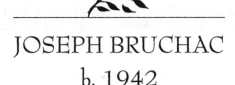

JOSEPH BRUCHAC
b. 1942

Joseph Bruchac's varied career as an author, editor, publisher, speaker, and storyteller has been notable for its steady focus on the role of literature in expressing and perpetuating Native American traditions. He lives with his family in Saratoga, New York, where he runs the Greenfield Review Press and produces his own novels, essays, children's books, and collections of stories. Both the Iroquois traditions of upper New York and his own Abenaki heritage are central to Bruchac's work. But in such coauthored books as Keepers of the Earth *(1988) and* Thirteen Moons on Turtle's Back *(1992) he has also drawn from a wider range of Native American cultures and has exercised considerable influence on environmental education in this country.*

In "The Circle Is the Way to See," Bruchac offers a commentary on our society's environmental attitudes and practices that is at once deeply rooted in his own Abenaki tradition and immediately pertinent to our contemporary situation.

THE CIRCLE IS THE WAY TO SEE

Waudjoset nudatlokugan bizwakamigwi alnabe. My story was out walking around, a wilderness lodge man. *Wawigit nudatlokugan.* Here lives my story. *Nudatlokugan Gluskabe.* It is a story of Gluskabe.

One day, Gluskabe went out to hunt. He tried hunting in the woods, but the game animals were not to be seen. Hunting is slow, he thought, and he returned to the wigwam where he lived with his grandmother, Woodchuck. He lay down on his bed and began to sing:

> *I wish for a game bag*
> *I wish for a game bag*

Story Earth: Native Voices of the Environment (San Francisco: Mercury House, 1993).

I wish for a game bag
To make it easy to hunt

He sang and sang until his grandmother could stand it no longer.
She made him a game bag of deer hair and tossed it to him. But he did
not stop singing:

I wish for a game bag
I wish for a game bag
I wish for a game bag
To make it easy to hunt

So she made him a game bag of caribou hair. She tossed it to him,
but still he continued to sing:

I wish for a game bag
I wish for a game bag
I wish for a game bag
To make it easy to hunt

She tried making a game bag of moose hair, but Gluskabe ignored
that as well. He sang:

I wish for a game bag
I wish for a game bag
I wish for a game bag
Of Woodchuck hair

Then Grandmother Woodchuck plucked the hair from her belly
and made a game bag. Gluskabe sat up and stopped singing.
"*Oleohneh, nohkemes,*" he said. "Thank you, Grandmother."

He went into the forest and called the animals. "Come," he said.
"The world is going to end and all of you will die. Get into my game
bag and you will not see the end of the world."

Then all of the animals came out of the forest and into his game
bag. He carried it back to the wigwam of his grandmother and said,
"Grandmother, I have brought game animals. Now we will not have a
hard time hunting."

Grandmother Woodchuck saw all the animals in the game bag.
"You have not done well, Grandson," she said. "In the future, our small
ones, our children's children, will die of hunger. You must not do this.
You must do what will help our children's children."

So Gluskabe went back into the forest with his game bag. He opened it. "Go, the danger is past," he said. Then the animals came out of the game bag and scattered throughout the forest. *Nedali medabegazu.*

There my story ends.

The story of Gluskabe's game bag has been told many times. A version much like this one was given to the anthropologist Frank Speck in 1918 by an elderly Penobscot man named Newell Lion. This and other Gluskabe stories that illustrate the relationship of human beings to the natural order are told to this day among the Penobscot and Sokokl, the Passamaquoddy and the Mississquoi, the Micmac and the other Wabanaki peoples whose place on this continent is called Ndakinna in the Abenaki language. Ndakinna—Our Land. A land that owns us and a land we must respect.

Gluskabe's game bag is a story that is central for an understanding of the native view of the place of human beings in the natural order, and it is a story with many, many meanings. Gluskabe, the Trickster, is the ultimate human being and also an old one who was here before human beings came. He contains both the Good Mind, which can benefit the people and help the Earth, and that other Twisted Mind, a mind governed by selfish thoughts that can destroy the natural balance and bring disaster.

He is greater than we are, but his problems and his powers are those of human beings. Because of our cunning and our power—a magical power—to make things, we can affect the lives of all else that lives around us. Yet when we overuse that power, we do not do well.

We must listen to the older and wiser voices of the earth—like the voice of Grandmother Woodchuck—or our descendants will, quite literally, starve. It is not so much a mystical as a practical relationship. Common sense.

Though my own native ancestry is Abenaki, and I regard the teachings and traditions of my Abenaki friends and elders, like the tales of Gluskabe, as a central part of my existence, I have also spent much of the last thirty-two years of my life learning from the elders of the Haudenosaunee nations, the People of the Longhouse—those nations of the Mohawk, Oneida, Onondaga, Cayuga, Seneca, and Tuscarora—commonly referred to today as the Iroquois.

We share this endangered corner of our continent, the area referred to on European-made maps as New York and New England. In fact, I live within a few hours' drive of the place where a man regarded as a messenger from the Creator and known as the Peacemaker joined with

Hiawatha—perhaps a thousand years ago—to bring together five war-
ring tribal nations into a League of Peace and plant a great pine tree as
the living symbol of that green and growing union of nations.

That Great League is now recognized by many historians as a direct
influence on the formation of modern ideas of democracy and on the
Constitution of the United States.

I think it right to recall here some of the environmental prophecies
of the Haudenosaunee people, not as an official representative of any
native nation, but simply as a humble storyteller. I repeat them not as a
chief nor as an elder, but as one who has listened and who hopes to
convey the messages he has heard with accuracy and honesty.

According to Iroquois traditions, some of which were voiced by the
prophet Ganio-dai-yo in the early 1800s, a time would come when the
elm trees would die. And then the maple, the leader of all the trees,
would also begin to die, from the top down.

In my own early years, I saw the elms begin to die. I worked as a tree
surgeon in my early twenties, cutting those great trees in the Finger
Lakes area of New York State, the traditional lands of the Cayuga
Nation of the Iroquois.

As I cut them, I remembered how their bark had once been used to
cover the old longhouses and how the elm was a central tree for the
old-time survival of the Iroquois. But an insect, introduced inadvertently, like the flus and measles and smallpox and the other diseases of
humans that killed more than 90 percent of the natives of North
America in the sixteenth and seventeenth centuries, brought with it
Dutch elm disease and spelt the end of the great trees.

Those trees were so beautiful, their limbs so graceful, their small
leaves a green fountain in the springtime, a message that it was time to
plant the corn as soon as they were the size of a squirrel's ear. And now
they are all gone because of the coming of the Europeans. Now, in the
last few years, the maple trees of New York and New England have
begun to die, from the top down—weakened, some say, by the acid rain
that falls, acid blown into the clouds by the smokestacks of the industries of the Ohio Valley, smoke carried across the land to fall as poison.

Is the Earth sick? From a purely human perspective, the answer
must certainly be yes. Things that humans count on for survival—basic
things such as clean water and clean air—have been affected.

The Iroquois prophecies also said a time would come when the air
would be harmful to breathe and the water harmful to drink. That time is
now. The waters of the St. Lawrence River are so full of chemicals from
industries, like Kaiser and Alcoa, on its shores that the turtles are covered
with cancers. (In the story of Creation as told by the Haudenosaunee, it

was the Great Turtle that floated up from the depths and offered its back as a place to support the Earth.)

Tom Porter, a Bear Clan chief of the Mohawks, used to catch fish from that same river to feed his family. The water that flowed around their island, part of the small piece of land still legally in the hands of the Mohawk people and called the St. Regis Reservation, that water brought them life. But a few years ago, he saw that the fish were no longer safe to eat. They would poison his children. He left his nets by the banks of the river. They are still there, rotting.

If we see "the Earth" as the web of life that sustains us, then there is no question that the web is weakened, that the Earth is sick. But if we look at it from another side, from the view of the living Earth itself, then the sickness is not that of the planet, the sickness is embodied in human beings, and, if carried to its illogical conclusion, the sickness will not kill the Earth, it will kill us.

Human self-importance is a big part of the problem. It is because we human beings have one power that no other creatures have—the power to upset the natural balance—that we are so dangerous to ourselves. Because we have that great power, we have been given ceremonies and lesson stories (which in many ways are ceremonies in and of themselves) to remind us of our proper place.

We are not the strongest of all the beings in Creation. In many ways, we are the weakest. We were given original instructions by the Creator. Those instructions, to put them as simply as possible, were to be kind to each other and to respect the Earth. It is because we human beings tend to forget those instructions that the Creator gave us stories like the tales of Gluskabe and sends teachers like the Peacemaker and Handsome Lake every now and then to help us remember and return us to the path of the Good Mind.

I am speaking now not of Europeans but of native people themselves. There are many stories in the native traditions of North America—like the Hopi tales of previous worlds being destroyed when human beings forgot those instructions—that explain what can happen when we lose sight of our proper place. Such stories and those teachers exist to keep human beings in balance, to keep our eyes focused, to help us recognize our place as part of the circle of Creation, not above it. When we follow our original instructions, we are equal to the smallest insects and the greatest whales, and if we take the lives of any other being in this circle of Creation it must be for the right reason—to help the survival of our own people, not to threaten the survival of the insect people or the whale people.

If we gather medicinal herbs, we must never take all that we find, only a few. We should give thanks and offer something in exchange, perhaps a bit of tobacco, and we should always loosen the earth and plant seeds so that more will grow.

But we, as humans, are weak and can forget. So the stories and the teachers who have been given the message from Creation come to us and we listen and we find the right path again.

That had been the way on this continent for tens of thousands of years before the coming of the Europeans. Ten thousand years passed after the deaths of the great beasts on this continent—those huge beings like the cave bear and the mammoth and the giant sloth, animals that my Abenaki people remember in some of our stories as monsters that threatened the lives of the people—before another living being on this continent was brought to extinction.

If it was native people who killed off those great animals ten thousand years ago, then it seems they learned something from that experience. The rattlesnake is deadly and dangerous, the grizzly and the polar bear have been known to hunt and kill human beings, but in native traditions those creatures are honored even as they are feared; the great bear is seen as closely related to human beings, and the rattlesnake is sometimes called Grandfather.

Then, with the coming of the Europeans, that changed. In the five hundred years since the arrival of Columbus on the shores of Hispaniola, hundreds of species have been exterminated. It has been done largely for profit, not for survival. And as the count goes higher, not only the survival of other species is in question but also the survival of the human species.

Part of my own blood is European because, like many native Americans today, many of my ancestors liked the new white people and the new black people (some of whom escaped from slavery and formed alliances and even, for a time, African/Indian maroon nations on the soils of the two American continents—such as the republic of Palmares in northeastern Brazil, which lasted most of the seventeenth century). I am not ashamed of any part of my racial ancestry. I was taught that it is not what is in the blood but what is carried in the culture that makes human beings lose their balance and forget their rightful place.

The culture of those human beings from Europe, however, had been at war with nature for a long time. They cut down most of their forests and killed most of the wild animals. For them, wildness was something to be tamed. To the native peoples of North America, wilderness was home, and it was not "wild" until the Europeans made it so. Still, I take heart at the thought that many of those who came to this hemisphere from Europe quickly learned to see with a native eye. So much so that the leaders of the new colonies (which were the first multinational corporations and had the express purpose of making money for the mother country—not of seeking true religious freedom, for they forbade any religions but their own) just as quickly passed laws to keep their white colonists from "going native."

If you do not trust my memory, then take a look at the words written by those colonizing Europeans themselves. You will find laws still on the books in Massachusetts that make it illegal for a man to have long hair. Why? Because it was a sign of sympathy with the Indians who wore their hair long. You will find direct references to colonists "consorting with the devil" by living like the "savages."

The native way of life, the native way of looking at the world and the way we humans live in that world, was attractive and meaningful. It was also more enjoyable. It is simple fact that the native people of New England, for example, were better fed, better clothed, and healthier than the European colonists. They also had more fun. European chroniclers of the time often wrote of the way in which the Indians made even work seem like play. They turned their work, such as planting a field or harvesting, into a communal activity with laughter and song.

Also, the lot of native women was drastically different from that of the colonial women. Native women had control over their own lives. They could decide who they would or would not marry, they owned their own land, they had true reproductive freedom (including herbal methods of birth control), and they had political power. In New England, women chiefs were not uncommon, and throughout the Northeast there were various arrangements giving women direct control in choosing chiefs. (To this day, among the Haudenosaunee, it is the women of each clan who choose the chiefs to represent them in the Grand Council of the League.)

In virtually every aspect of native life in North America—and I realize this is a huge generalization, for there were more than four hundred different cultures in North America alone in the fifteenth century and great differences between them—the idea of the circle, in one form or another, was a guiding principle. There was no clock time, but cyclical time. The seasons completed a circle, and so too did our human lives.

If we gather berries or hunt game in one place this year, then we may return to that place the following year to do the same. We must take care of that place properly—burning off the dry brush and dead berry bushes so that the ashes will fertilize the ground and new canes will grow, while at the same time ensuring that there will still be a clearing there in the forest with new green growth for the deer to eat.

The whole idea of wildlife conservation and ecology, in fact, was common practice among the native peoples of this continent. (There is also very sound documented evidence of the direct influence of native people and native ideas of a "land ethic" on people such as Henry David Thoreau, George Bird Grinnell, Ernest Thompson Seton, and others who were the founders of organizations like the Audubon Society, the Boy Scouts of America, and the whole modern conservation movement itself.) There was not, therefore, the European idea of devastating your own backyard and then moving on to fresh ground—to a new frontier (the backyard of your weaker neighbor).

If you see things in terms of circles and cycles, and if you care about the survival of your children, then you begin to engage in commonsense practices. By trial and error, over thousands of years, perhaps, you learn how to do things right. You learn to live in a way that keeps in mind, as native elders put it, seven generations. You ask yourself—as an individual and as a nation—how will the actions I take affect the seven generations to come? You do not think in terms of a four-year presidency or a yearly national budget, artificial creations that mean nothing positive in terms of the health of the Earth and the people. You say to yourself, what will happen if I cut these trees and the birds can no longer nest there? What will happen if I kill the female deer who has a fawn so that no animals survive to bring a new generation into the world? What will happen if I divert the course of this river or build a dam so that the fish and animals and plants downstream are deprived of water? What will happen if I put all the animals in my game bag?

And then, as the cycles of the seasons pass, you explain in the form of lesson stories what will happen when the wrong actions are taken. Then you will remember and your children's children will remember. There are thousands of such lesson stories still being kept by the native people of North America, and it is time for the world as a whole to listen.

The circle is the way to see. The circle is the way to live, always keeping in mind the seven generations to come, always asking: how will my deeds affect the lives of my children's children's children?

This is the message I have heard again and again. I give that message to you. My own "ethnic heritage" is a mixture of European and native, but the messages I have heard best and learned the most from spring from this native soil.

If someone as small and pitiful as I am can learn from those ancient messages and speak well enough to touch the lives of others, then it seems to me that any human being—native or nonnative—has the ability to listen and to learn. It is because of that belief that I share these words, for all the people of the Earth.

FRANKLIN BURROUGHS
b. 1942

Franklin Burroughs has taught literature and writing at Bowdoin College for over three decades. During that period he has been active in conservation efforts related to the Maine coast. But in Billy Watson's Croker Sack (1991), while evoking his present home, he also returned to the landscape of his boyhood in South Carolina. In Horry and the Waccamaw and River Home (both published in 1992), he wrote about that same Carolina country and about the canoeing and hunting that have been such a vivid part of his experiences there, as well as in Maine.

OF MOOSE AND A MOOSE HUNTER

When I first moved to Maine, I think I must have assumed that moose were pretty well extinct here, like the wolf or the caribou or the Abenaki Indian. But we had scarcely been in our house a week when a neighbor called us over to see one. She had a milk cow, and a yearling moose had developed a sort of fixation on it. The moose would come to the feedlot every afternoon at dusk and lean against the fence, moving along it when the cow did, staying as close to her as possible. Spectators made it skittish, and it would roll its eyes at us nervously and edge away from the lot, but never very far. It was gangly and ungainly; it held its head high, and had a loose, disjointed, herky-jerky trot that made it look like a puppet on a string.

The young moose hung around for a couple of weeks, and it became a small ritual to walk over in the summer evenings and watch it. My neighbor, Virginia Foster, had reported it to the warden, and the warden told her not to worry: the yearling had probably been driven off by its mother when the time had come for her to calve again, and it was just looking for a surrogate. It would soon give up and wander away, he

Billy Watson's Croker Sack (New York: Norton, 1991).

said, and he was right. But until that happened, I felt that Susan and I, at the beginning of our own quasi-rural existence, were seeing something from the absolute beginnings of all rural existence—a wild creature, baffled and intrigued by the dazzling peculiarities of humankind, was tentatively coming forward as a candidate for domestication. Mrs. Foster said that if the moose planned to hang around and mooch hay all winter, he'd damn well better expect to find himself in the traces and pulling a plough come spring.

First encounters mean a lot, and in the years that followed, moose never became for me what they are for many people in Maine: the incarnation and outward projection of that sense of wilderness and wildness that is inside you, like an emotion. As soon as I began going up into the northern part of the state whenever I could, for canoeing and trout fishing, the sight of them came to be familiar and ordinary, hardly worth mentioning. You would see one browsing along the shoulder of a busy highway or standing unconcerned in a roadside bog, while cars stopped and people got out and pointed and shutters clicked. Driving out on a rough logging road at dusk, after a day of trout fishing, you would get behind one, and it would lunge down the road ahead of you. Not wanting to panic it or cause it to hurt itself—a running moose looks out of kilter and all akimbo, like a small boy trying to ride a large bicycle—you'd stop, to allow the moose to get around the next curve, compose itself, and step out of the road. Then you'd go forward, around the curve, and there would be the moose, standing and waiting for the car to catch up to it and scare it out of its wits again. Sometimes you could follow one for half a mile like that, the moose never losing its capacity for undiluted primal horror and amazement each time the car came into sight. Finally it would turn out of the road, stand at the fringe of the woods, and, looking stricken and crestfallen as a lost dog, watch you go past.

Of course you also see them in postcard situations: belly deep in a placid pond, against a backdrop of mountains and sunset, or wading across the upper Kennebec, effortlessly keeping their feet in tumbling water that would knock a man down. Once two of them, a bull and a cow, materialized in a duck marsh as dawn came, and I watched them change from dim, looming silhouettes that looked prehistoric, like something drawn by the flickering illuminations of firelight on the walls of a cave, into things of bulk and substance, the bull wonderfully dark coated and, with his wide sweep of antlers and powerfully humped shoulders, momentarily regal.

But even when enhanced by the vast and powerful landscape they inhabit, moose remained for me animals whose ultimate context was somehow pastoral. An eighteenth- or nineteenth-century English or American landscape painting, showing cattle drinking at dusk from a gleaming river, or standing patiently in the shade of an oak, conveys a serenity that is profound and profoundly fragile. The cattle look sacred, and we know that they are not. To the extent that they epitomize mildness, peace, and contentment, they, and the paintings in which they occur, tacitly remind us that our allegiance to such virtues is qualified and unenduring, existing in the context of our historical violence, our love of excitement, motion, risk, and change. When I would be hunting or fishing, and a moose would present itself, it would not seem to come out of the world of predator and prey, where grim Darwinian rules determine every action. That world and those rules allow the opposite ends of our experience to meet, connecting our conception of the city to our conception of the wilderness. The moose would seem to come from some place altogether different, and that place most resembled the elegiac world of the pastoral painting, an Arcadian daydream of man and nature harmoniously oblivious to the facts of man and nature.

I suppose it would be more accurate to say that the moose came from wherever it came from, but that it seemed to enter the Arcadian region of the imagination. I found it a difficult animal to respond to. It was obviously wild, but it utterly lacked the poised alertness and magical evanescence that wild animals have. If by good fortune you manage to see a deer or fox or coyote before it sees you, and can watch it as it goes about its business unawares, you hold your breath and count the seconds. There is the sensation of penetrating a deep privacy, and there is something of Actaeon and Artemis in it—an illicit and dangerous joy in this spying. The animal's momentary vulnerability, despite all its watchfulness and wariness, brings your own life very close to the surface. But when you see a moose, it is always unawares. It merely looks peculiar, like something from very far away, a mild, displaced creature that you might more reasonably expect to encounter in a zoo.

In 1980, for the first time in forty-five years, Maine declared an open season on moose. Given the nature of the animal, this was bound to be a controversial decision. People organized, circulated petitions, collected signatures, and forced a special referendum. There were televised debates, bumper stickers, advertising campaigns, and letters to editors. The major newspapers took sides; judicious politicians commissioned polls. One side proclaimed the moose to be the state's sacred and official animal. The other side proclaimed moose hunting to be an ancient and endangered heritage, threatened by officious interlopers who had no understanding of the state's traditional way of life. Each side accused the other of being lavishly subsidized by alien organizations with sinister agendas: the Sierra Club, the National Rifle Association. The argument assumed ideological overtones: doves *vs.* hawks; newcomers *vs.* natives; urban Maine *vs.* rural Maine; liberals *vs.* conservatives.

At first this seemed to be just the usual rhetoric and rigmarole of public controversy. But as the debate continued, the moose seemed to become a test case for something never wholly articulated. It was as though we had to choose between simplified definitions of ourselves as a species. Moose hunters spoke in terms of our biology and our deep past. They maintained that we are predators, carnivores, of the earth earthly; that the killing and the eating of the moose expressed us as we always had been. The other side saw us as creatures compelled by civilization to evolve: to choose enlightenment over atavism, progress over regression, the hope of a gentler world to come over the legacy of instinctual violence. Both sides claimed the sanction of Nature—the moose hunters by embodying it, their opponents by protecting it. Each side dismissed the other's claim as sentimental nonsense.

I knew all along that when it came to moose hunting I was a prohibitionist, an abolitionist, a protectionist, but not a terribly zealous one. When the votes were counted and the attempt to repeal the moose season had been defeated, I doubted that much had been lost, in any practical way. The hunt was to last only a week, and only a thousand hunters, their names selected by lottery, would receive permits each year. It had been alleged that once moose were hunted, they would become as wild and wary as deer, but they have proved to be entirely ineducable. Hunter success ran close to 90 percent in that first year, and has been just as high in the years that followed; and the moose I continue to see each summer are no smarter or shyer than the one that had mooned around Mrs. Foster's feedlot, yearning to be adopted by her cow.

Late one afternoon, toward the end of September, the telephone rang, and there was a small voice, recognizably Terri Delisle's. "Liz there?" So I went and got Liz. She's old enough to have overcome all but the very last, genetically encoded traces of telephobia—just a momentary look of worry when she hears that it's for her, and a tentativeness in her "Hullo?" as though she were speaking not into the receiver but into a dark and possibly empty room.

Terri is her friend, her crony. The two of them get together—both polite, reticent, and normally quiet little girls—and spontaneously constitute between themselves a manic, exuberant subculture. It possesses them. They are no longer Terri and Liz but something collective: a swarm, a gang, a pack, or a carnival, having its own unruly gusts of volition. They glitter with mischief, laugh at everything, giggle, romp, and frolic; and I believe that, with each other's help, they actually lose for a moment all consciousness of the adult world that watches from within, waiting for children to draw toward it. They aren't destructive or insubordinate—that, after all, would be a backhanded acknowledgment of civilization, maturity, and responsibility. They are simply beyond the reach of reproof, like colts or puppies.

But on the telephone, with distance between them, self-conscious circumspection took over. I heard Liz's guarded and rigorously monosyllabic responses: "Yep." "He did?" "Sure—I'll have to ask Dad." "OK. Bye." And so she told me that Terri's father Henry had killed a moose. Would we like to go over and see it? "Sure," I say, all adult irony, "I'll have to ask Mom."

I knew Henry Delisle in a small and pleasant way. There were a lot of Delisles in town, and Henry, like his brother and most of his male cousins, worked over in Bath, at the shipyard—a welder, I think. But like many other natives of Bowdoinham, he had farming in the blood. The old Delisle farm, up on the Carding Machine Road, had long since been subdivided and sold, and Henry's neat, suburban-looking house sat on a wooded lot of only two or three acres. Even so, he had built himself a barn and a stock pen, and he kept a few pigs, a milk cow, and an old draft horse named Homer. There couldn't have been much economic sense to it, just a feeling that a house wasn't a home without livestock squealing or lowing or whickering out back. He plainly liked the whole life that livestock impose upon their owners—harnessing Homer up for a day of cutting and hauling firewood; making arrangements with local restaurants and grocery stores to get their spoiled and leftover food for pig fodder; getting the cow serviced every so often, and fattening the calf for the freezer. He had an antiquated Allis-Chalmers tractor, with a sickle bar and a tedder and a bailer. There are a lot of untended fields in Bowdoinham, and plenty of people were glad to let Henry have the hay if he would keep them mown.

That was how I had met him for the first time. He had come rattling up to the house in his big dilapidated flatbed truck to ask me if anybody planned to cut my fields that summer. In fact somebody did, and I told him so, but Henry had too much small-town civility, which coexists comfortably with small-town curiosity, simply to turn around and drive off. I appreciated that, and so we chatted for a while—Henry sitting up in his truck, talking with an abrupt and fidgety energy, and I standing down beside it.

He remembered my house from his boyhood: "Used to be a reg'lar old wreck of a place. They didn't have no electricity down here or nothing. Winters, they'd cut ice from the pond. Had a icehouse dug into the bank there; kept ice all through summer. Hard living." He told me a story I'd heard even before we bought the house, how one winter the eldest son had gone out to the barn to milk, as he did every morning, and had found his younger brother there, hanging from a ceiling joist. "Never a word or a note. That was a terrible thing to happen. Unfriendly people, but they didn't deserve that."

He laughed. "But they was *some* unfriendly, I want to tell you. I slipped down to the pond and set muskrat traps one fall. But they musta seen me. They pulled 'em every one out and kept 'em. I was afraid to ask—just a kid, you know. Probably still lying around in your barn somewhere." He looked at me and sized me up: "But I ain't afraid to ask now, and don't you be afraid to turn me down—would you mind me setting a few traps in that pond this fall? It used to be about lousy with muskrats." I hesitated at this—the pond was still full of muskrats, and I enjoyed seeing them sculling across it, pushing little bundles of cut grass ahead of them, or sitting out on a log, grooming themselves with a quick, professional adroitness. But I liked him for the way he had asked it, and there was something else. His country-bred practicality and local knowledge gave him an obscure claim—he was less indigenous than the muskrats, but far more so than I was. "Sure," I told him, "go ahead."

All this had taken place on a bright, airy morning in late July or early August, with the kind of high sky that would make anybody think of haying. Henry said he was glad he'd stopped by, and that I'd see him again once the trapping season opened. I reached up; we shook hands, and he backed the truck down the driveway. His windshield caught the sun for a moment and blinded me and then, as the truck swung up into the yard to turn around, I could see through the glass what I had not been able to see before. He had a passenger—a little girl sitting in the middle of the seat, right at his elbow. She did not look in my direction at all, but stared at the dashboard with that look of vacancy and suspended animation that you see on the faces of children watching Saturday morning cartoons. Henry grinned at me, waved goodbye, and the big truck went lumbering off.

That first meeting with Henry had been the summer before Elizabeth and Terri started school. Later, when they had become classmates and best friends, I learned that the girl I had seen in the truck was Stephanie, whom everybody called Tadpole. She was three years older than Terri, but that was a technicality.

Bowdoinham is a small, spread-out town. It tries to hold onto the idealized ethos of the New England village, but is in fact well on its way to becoming a bedroom community, a pucker-brush suburb. Like the state as a whole, it is full of outsiders moving in, old-timers dying out, and the uneasy sense of a lost distinctiveness.

The elementary school is the nearest thing to an agora that such a town has. Parents are separated by their backgrounds and expectations, and by the devious anxieties of people who feel that, in appearing to belong to the little unglamorous place they inhabit, they may misrepresent or compromise themselves. But children go to school, and it stands for the world. They make friends and enemies, and suddenly populate your household with unfamiliar names. It is as though you had sent them off as members and worshipers of a stable, self-sufficient Trinity consisting of Mama, Daddy, and themselves; and then had them return as rampant polytheists, blissfully rejoicing or wailing despairingly about the favors and sulks of capricious gods and goddesses named Tommy Blanchard, Vera Sedgely, Joanie Dinsmore, Nikki Toothacre, and Willie Billings. At school functions you would meet the parents of these entities, or, prodded by your child, would nervously call up Joan's or Nikki's mom, and arrange for that child to come over and play. And slowly, with no direct intention of doing so, you would find out about other families in the town—who they were and how they lived, how they regarded themselves and how they were regarded.

So we learned that Tadpole suffered from Down's syndrome. She was the first child of Henry and Debbie Delisle, born to them within a year of their marriage, when they themselves were just out of high school. Perhaps if they had had more education and experience they would have accepted the irremediable fact of their daughter's condition. As it was, they were mistrustful of the state and the school system and all the experts who came to help them and warn them and in some way to deprive them of the right to raise their daughter as they saw fit. Against every recommendation, they were determined to try to give Tadpole all the circumstances of an ordinary childhood.

When time came for Tadpole to go to school, Henry wrangled with the school board and the superintendent and the Department of Mental Health and Retardation. And finally everybody agreed that, for as long as it didn't create any disturbance, Tadpole could go to school with Terri. Word of that sort of thing gets around, and some parents didn't like it, fearing that what Henry had gained for his daughter would diminish the education and attention that their own children would receive. But I believe that most of us admired Henry and wished him well. He was his own man; in his battered old truck, with a tottering load of hay on it, or with Homer tethered to the headboard, he implied an old-fashioned resourcefulness and independence, which we could praise even if we couldn't emulate. It was heartening to see a man like that acting out of the best and simplest human impulse, and sticking to his guns, even if, in the long run, the case were hopeless.

And of course the case was hopeless, although at first it didn't appear to be. Tadpole was docile and affectionate, and in her first year and a half of school, she enjoyed an almost privileged status among her classmates. It was as though she were their mascot, like the wheezy old bulldog or jessed eagle you might find on the sidelines at a college football game. You would see a crowd of children fussing over her in the schoolyard, alternately courting her as though she were a potentate to be appeased, or babying her with bossy solicitude. Liz would report on all that Tadpole had done or said, as though she were a celebrity, in whom we should take a communal pride. And we did take a kind of pride in her. Her being at the school with the other children seemed proof that humane flexibility, sympathy, and tolerance were still operative in this overgrown country. There was something quaint about it, something from the putative innocence of the past.

But by the end of the second grade, Liz was bringing home bad news. Tadpole had begun to balk at going to school, and would misbehave when she was there. She was bigger than her classmates, and her truculence threatened them. They retaliated as children would, by teasing and persecution. She regressed, growing more withdrawn and morose, and would go through days of not speaking, or of only muttering to herself. Public opinion hardened. I don't think there were any petitions or formal proceedings to have Tadpole removed; it was just one of those sad things that had become plain and obvious. Henry and Debbie had no choice; they had to give in to the fact that confronted them every day. The next year, Tadpole and Terri were separated, and Tadpole was sent to school in Topsham, where there was a class for what the state calls Special Children.

When Terri would come over to play, she seemed untroubled by the change. She was as quick and inventive as ever. I did not know Henry well enough or see him often enough to speak to him about the matter, and hardly knew what I would or could have said. He got himself transferred to the night shift at the shipyard that fall, and he must have kept Tadpole out of the special class a good deal. I would regularly see the two of them together in the truck—usually first thing in the morning, when he'd just gotten off work. But he told me one morning, when he'd come to check the muskrat traps, that he had changed shifts purely to give himself more time for the woodcutting, haying, trapping, ice-fishing, and hunting that seemed to be his natural vocations.

So on the September afternoon in question, Liz and I got into the car—none of the rest of the household had any interest in a dead moose—and drove over. It was nearly dark when we turned up into Henry's driveway. His garage lights were on. He had set up a worktable of planks and sawhorses along the rear wall; the moose was hanging by the neck squarely in the center of the garage. From the driveway, it looked like a shrine or a crèche—the brightly lit space, clean and spare as an Edward Hopper interior; Henry and four other men standing chatting; and, just behind them, the lynched moose. Terri came running out, excited as on Christmas morning, and took us in to see.

From the outside, the moose's head appeared to go right up through the low ceiling of the garage, but once inside I could see that, when he had built the garage, Henry had left out one four-by-eight ceiling panel, to give him access to the attic. He had put an eye bolt in a collar tie, centered above the opening, so that he could rig a hoist from it. It was a smart arrangement, enabling him to convert an ordinary two-car garage into an abattoir whenever he had a cow or pig or deer to slaughter. The moose he had shot was a cow, and she was a big animal, hanging with her head in the attic, her rump scarcely a foot above the concrete floor. A big animal but not, Henry said, a big moose: "She'll dress out about five-fifty. Just a heifer. She'd have calved next spring."

Henry introduced me to the other men—neighbors who had wandered over out of curiosity, and his cousin Paul, who had been his partner in the hunt.

We were somehow an uncomfortably self-conscious group; it was as though we were all trying to ignore something. Perhaps it was that Paul and Henry were still dressed in their stained and ragged hunting gear, and were grubby and unshaven. The rest of us were in our ordinary street clothes, and only a few minutes ago were watching television or pottering around the house or having a drink and getting ready for supper. We had been in our familiar cocoons of routine and obligation, where the only world that matters is the human one. And now we were talking to men who were in another role, and we were abruptly confronting a large, dead animal, a thing from far beyond our lives.

I think it was more this awkwardness than aggression that made the man next to me, a bank manager new to town, speak the way he did: "Well, Henry. That's a weird damned animal. You sure it's not a camel?" Everybody laughed, but uneasily.

"Tell us about it," the man said. "How'd you bag the wily moose?"

Henry said there wasn't a whole lot to tell. The man asked him if he'd hired a guide. Henry said he hadn't.

"Well maybe you should have," the bank manager said. "If you had, you might have gotten yourself a bull. Then you'd have something to hang in your den."

Henry didn't answer. He got busy with a knife, whetting it against a butcher's steel. The man walked around the moose, looking at her appraisingly, as though she were an item in a yard sale. Then he said he had to get on back home, and left, and there was a general relaxing. Henry looked up.

Now he was going to tell us how you kill a moose, or how he had killed this one. None of us knew anything about moose hunting. The tradition of it had died out, and hunters—even very experienced ones like Henry and Paul—don't know moose in the way that they know deer. The hunt was limited to the upper third of the state, and a lot of people up there had set themselves up as moose guides, offering what was supposedly their deep-woods wisdom to anybody lucky enough to have a permit.

Henry snorted: "Hire a guide. You know what a moose guide is? He's a guy with a skidder, that's all. You go to his house and he'll take you out and leave you somewheres where he thinks there might be a moose, and charge you so much for that. Then you kill a moose and he'll charge you a arm and a leg to hook it up to the skidder and drag it out to your truck. So I go to this guy that's listed as a guide, and he explains it to me. And I say to him, 'Look, Don't tell me a word about where to find a moose Now if I get one, what'll you charge to drag him out?' 'Hundred dollars for the first hour; fifty dollars per hour after that,' he says. See, they got you. Law don't let you kill a moose less than fifty yards from the road. So I says to him, 'You prorate that first hour?' 'Fifty dollar minimum,' he says to me: 'Take it or leave it.' Musta thought I was from Massachusetts. 'See you later,' I says. And that fifty dollar, hundred dollar shit ain't from the time he drives his skidder off his trailer into the woods. It's from the time he gets in his truck right there in his front yard."

Paul quietly removed himself from Henry's audience and went into the kitchen. It wasn't his story, and there was a lot of work still to do.

"We had topo maps, and I seen some good bogs. Day before the season opened we drove and scouted all day. I don't know much about moose, but I know a moose'll walk on a log road or a skidder track if he can, instead of bustin' through the bushes. About suppertime we see a cow cross the road ahead of us, and go down a skidder trail. We followed her down on foot. There was a bog in there at the end of the trail, about a quarter mile in off the road, and there she was, feeding. Her and another cow too. That skidder trail was rough, but I figured we might be able to get the truck down it.

"Opening day it was raining. We parked a ways off and walked up to the skidder track and down to the bog. Got there before day. When it come day, one cow was there. I looked at her. She looked good, but not extra good. Animal like a moose got a lot of waste on 'em. Big bones, big body cavity—not as much meat as you'd think. That's what they tell me. And they told me when you see a cow, wait. It's rut, and a big bull might come along any time.'"

Paul came out from the house with his arms full—wrapping paper, freezer tape, a roll of builder's plastic. He spread the plastic over the table, and he didn't make any effort to be unobtrusive about it. But Henry was occupied by his story. It was like something he wanted to get off his chest or conscience. Maybe he just couldn't get over the strangeness of the moose.

"It ain't like a deer. A cow moose calls a bull. That's what they say and it's the truth. We watched her all day, and ever so often she'd set right down on her butt and beller, like a cow that ain't been milked. So we set there too, and waited, but no bull showed. By dark she'd worked over to the other side of the bog. Shoot her there and you'd have to cut her up and pack her out."

Henry was standing in front of the moose. Her chin was elevated and her long-lashed eyes were closed. All of the things that had so splendidly adapted her to her world of boreal forest, bog, and swamp made her look grotesque here: the great hollow snout, the splayed feet and overlong, knob-kneed legs. In whatever consciousness she had had, it was still the Ice Age—she was incapable of grasping human purposes or adjusting to human proximity. Her death was almost the first ritual of civilization, yet she was in our present, suspended in the naked light of a suburban garage, and we could only stand, hands in pockets, as though it were something we did every day.

"So we come back the next day, a little earlier even, and I sent Paul around to the far side of the bog. This time I hear her walking in on that skidder track just before day, and she got out in the bog and bellered some more. We was going to give her 'til noon. I figured if a bull showed, he'd come up the track too, and I could get him before he hit the bog.

"By noon she was all the way out in the middle of the bog again, but Paul stepped out of the bushes, easy, so's not to scare her too much. Took her the longest time even to notice him. Then she started trotting toward me, but she'd keep stopping to beller some more. It was almost like she was mad."

One of the men chuckled "More like she was desperate, if you ask me. If she didn't call herself up a boyfriend pretty quick, she was a dead duck."

"Well. Anyway, Paul had to slog out after her, keep shooing her along. I wanted her to get all the way out on the trail, but she musta smelt me. Stopped about ten foot from the woods and started throwing her head around and acting jumpy, like she might bolt. So I shot her there.

"We had a little work with the chain saw to clear the skidder trail out wide enough for the truck. Then we backed in and put a rope around her and dragged her out to dry ground. Used a come-along to hoist her up on a tree limb and dressed her out right there. Then cranked her up some more, backed the truck under, and lowered her in. On the way out, we stopped by that guy's house. I went in and told him we wouldn't be needing his damn skidder this year."

The whole time Henry talked, Paul kept coming and going, bringing out knives, a cleaver, a meat saw, and a plastic tarp. Elizabeth and Terri had examined the moose and then gone inside. I had been worried about Elizabeth. She was at least as sentimental as the average ten-year-old about animals; at country fairs she would lean against the stalls and gaze with pure yearning at Suffolk sheep or Highland cattle and especially at horses of any description. But she and Terri had looked the moose over as though she were a display in a museum of natural history, something interesting but remote. They had walked around her, rubbed the coarse, stiff hair, and inspected the big cloven feet, and then gone about their business.

Now, as Henry finished his story, they returned, giggling. Terri was carrying a child's chair, and Liz looked from her to me, trying not to laugh. Terri ran up to the moose and slipped the chair under her rump, and then the two of them stood back and waited on our reaction.

It was comic relief or comic desecration. Because the moose's hindquarters were so near the floor, her hind legs were spread stiffly out in front of her. With the addition of the chair, you suddenly saw her in a human posture, or really in two human postures. From the waist down, she looked like a big man sprawled back on a low seat. Above the waist, she had the posture of a well-bred lady of the old school, her back very straight, her head aloof, and her whole figure suggesting a strenuous and anxious rectitude.

In the ready, makeshift way of country people, Henry had taken one of Debbie's old worn-out lace curtains with him, and when he had killed and cleaned the moose, he had pinned the curtain across the body cavity, to keep out debris and insects and to allow the air to circulate and cool the animal while he and Paul drove back home. The curtain was longer than it needed to be, and now Terri picked up one end of it, brought it like a diaper up between the moose's legs, wrapped it around the hips, and tucked it in, so that it would stay up. The effect was funny in a way I don't like to admit I saw—the garment looked like something between a pinafore and a tutu. It was as though the moose had decided, in some moment of splendid delusion, to take up tap dancing or ballet, and was now waiting uncomfortably to go on stage for her first recital.

Terri and Liz admired the moose. "She needs a hat," Terri pronounced, and they ducked into the house. What they came out with was better than a hat—a coronet of plastic flowers, left over from some beauty pageant or costume.

"Daddy, could you put this on her? She's too high for us."

She was too high for Henry too, but he pulled the little chair from beneath the moose, then picked Terri up and set her on his shoulders. He stood on the chair and Terri, leaning out daringly, like a painter on a stepladder, managed to loop the coronet over one of the long ears, so that it hung lopsided. She slid down Henry to the ground, stepped back and dusted her hands together:

"There. That'll just have to do. I think Momma needs to see this. Maybe she'll lend us some mittens and a scarf. Let's go get her and Tadpole to come see."

"Terri, Paul and me got to get to work on that moose right now," Henry called after her, but she was already gone. The other two men who had come over to see the moose said they had to go, and left, one on foot and one in his car. Terri and Liz came back out with Debbie and Tadpole. Debbie looked at the moose and laughed. Terri was pleased.

"Don't you think she looks like a beauty queen, Mom? We could enter her in the Miss Bowdoinham contest."

"Well I guess so." Debbie turned to Tadpole: "Look at Daddy's moose that he brought us, honey." Tadpole looked at it and walked over as though she wanted to touch it, but didn't. Her face had that puffy, numbed look of someone just wakened from a deep sleep, and her movements were slow and labored.

Debbie called over to Terri. "Now your Daddy and Paul have to start, and I've got to run buy some more freezer paper. You and Stephanie come with me, and we can let Liz get home for her supper."

Terri gave the moose a comradely whack on the rump: "Goodbye, moose. You're going in the freezer." Liz patted the moose too, but more tentatively. Then they all trooped out.

I stood talking to Henry for a few minutes longer. He looked at the moose with her cockeyed halo and tried to make a joke of it. "If she'd been dressed that way this morning, maybe I'd have got a bull." But his laughter was awkward, apologetic. His remark about how little useable meat there really was on a moose, for all its great size, had not been lost on me, and yet I felt that it would be right to ask him for something, as a way of restoring to him a vestige of the old role of hunter as public benefactor, bringer home of the bacon. So I asked him if I could have some of the long hair from the nape of her neck, for trout flies.

"Sure thing," he said, all business "Tell you what: I won't cut it off now—don't want no more loose hair than I can help when we go to skin her. But when she's done, I'll clip some off and drop it by, next time I'm down your way. You can count on it."

I thanked him and left. Liz was subdued as we drove back toward home. You might have asked an older child what she was thinking, but not Liz, not for a few years yet. Besides, I wasn't so certain what *I* was thinking just then: two scenes alternated in my mind. One was a recollection, back from the previous November, a morning when heavy frost had sparkled white as mica on the dead grass, and I had been driving to work. I saw Henry walking across a stubble field, a big fox slung over his shoulder. He held the fox by its hind legs; its tail, curved over and lying along its back, was luxuriant and soft as an ostrich plume, and it stirred lightly in the breeze. I felt some sadness for the dead beauty of the fox, but it was Henry I remembered. He ought to have looked like a mighty hunter before the Lord, holding the bounty of his skill and cunning and knowledge of the ways of wild animals in his hand. But he was walking with a shambling hesitation, to keep pace with the daughter clinging to his other hand and trudging glumly at his side, beyond any idea of triumph or happiness.

The other image was of something that had not happened yet. June would come again, and I would be up north fishing again—this time with a fly that would have, wrapped in tight spirals around the shank of the hook to imitiate the segmented body of a nymph or mayfly, one or two strands of mane from Henry's moose. And I would look up from the water, almost dizzy with staring for so long at nothing but the tiny fly drifting in the current, and there they would be—maybe a cow and a calf—standing on the other bank, watching me watch them, trying to fathom it.

DOUG PEACOCK
b. 1942

With Grizzly Years: In Search of the American Wilderness *(1990),*
Doug Peacock made a major addition to the literature of the western
landscape, just as earlier (in the person of Hayduke in Abbey's The
Monkey Wrench Gang*), he had inspired one of the main characters in*
another of the region's classic books. Grizzly Years *records the author's*

flight to the mountains and his encounters with its most formidable denizen in the aftermath of his experiences in the Vietnam War. Not only did he find healing of his wartime wounds, but he also gained insights into the lives of the grizzlies that have made him one of their most tireless advocates.

THE BIG SNOW

NOVEMBER (1980s)

It was mid-November and a winter storm was coming to the mountains of northwest Wyoming. The wind was gentle, chinooklike, swaying the bare branches of an aspen grove against a gray sky. The trees' leaves, already drawn to the forest floor by the October frosts of the brief Rocky Mountain autumn, lay silent under crusted drifts of November snow. I struggled carrying a bulky rucksack up the open slope toward groves of mixed fir, spruce, and pine. I had a 9,000-foot ridge to climb, another valley to drop into and cross, then a steep north-facing mountainside to crawl up until I reached the 9,200-foot contour.

At that elevation, under the roots of a large, lightning-struck whitebark pine, was a five-foot-long tunnel dug into the side of the forty-degree mountain slope. The hole had been dug by a young grizzly bear. I knew because I had watched him dig it. He had started on the twentieth of September, removing hundreds of pounds of gravelly loam. He planned on sleeping out the winter in there. When he wasn't working on his den, he fed on the whitebark pine nut caches of red squirrels. In October he left the area. If he had not yet returned, this storm would, I thought, bring him back.

The upland country of the Yellowstone Plateau was open, pleasant country. I stepped across a tiny creek and saw, hidden behind a boulder, a layer of golden aspen leaves placered against the bottom of a dark pool. The tops of yellow grasses were still exposed in the meadows; under the shadows of the conifers, low banks of windblown snow awaited winter. The breeze drifted north of me on the lee side of the ridge, while a high wind ran over the rocky spine above.

Through my binoculars, far below me, a bull moose stood motionless in the bushy willow bottomland. The elk and a small herd of mountain sheep on the distant slope had bedded. My blood felt sluggish too. The barometric lows preceding major storms heralded lethargic times: the ungulates brushed up, the fish did not bite, and the

Grizzly Years: In Search of the American Wilderness (New York: Holt, 1990).

grizzlies hightailed it to denning sites, where they moped around and waited for the big snow. Grizzlies could sense winter storms days in advance. My three-year-old bear was probably putting the finishing touches on his den just then, raking it out one last time before adding bedding—grass, moss, or boughs of young firs. He would then withdraw to his porch, a dish-shaped depression in the loam just in front of the three-foot-wide opening to the tunnel, and lie like a sleepy puppy watching the darkening skies for the whiteness that would seal him within his mountain.

I walked uphill under the canopy of mature whitebark pines. Under a large tree, next to the trunk where the snow had melted, was a small pile of pinecones. A bear, probably a grizzly, had dug up a number of squirrel caches and raked out the seeds with his claws. Red squirrels were the middlemen; bears did not harvest the cones directly but were dependent on the arboreal rodents. Even in years when there were lots of pine nuts, if the squirrel population was down, the bears would not get many nuts. These pine nuts were a major source of food for Yellowstone's grizzlies. The three-year-old grizzly had been feeding on mast when I had stumbled across him digging his den six weeks earlier.

I climbed over a string of mossy ledges and topped out the ridge. Before me, to the south, lay the gentle, rolling up-country of the Snowy Range: willow bottoms, sagebrush meadows, and undulating grassy hillsides patched with groves of aspen and stands of pine and fir. The den site of the three-year-old bear lay four miles away, up on the steep side of the next low mountain. I could have gotten there by dark if I had pushed. But I had not come out here to hurry. Instead, I would hole up for the night and wait for the snow to begin falling.

Grizzlies at their denning sites are extremely shy; if they are disturbed they may abandon the area altogether and be forced to dig another den somewhere else. Once the first major storm of the fall begins, undisturbed bears become sluggish and are much less likely to be bothered by my presence. But I didn't plan on letting this little grizzly know I was around.

The high wind had died. The air felt heavy, still warm under its blanket of monotonous blue-gray sky. The snowy front pushed before it a low-pressure trough of languid creatures, a chorus of yawns. I stumbled down the mountain slope, through the dead grasses and winter trees, toward a narrow creek that meandered through willow thickets. By the time I reached the valley, just at dark, the wind had stopped altogether. The air was still and snow had begun to fall. Nothing moved except the large white flakes and the small creek, whose dark currents gathered the silent snow.

I followed the tiny creek up into the trees where the waters pooled behind the roots of a giant spruce. An eerie calm settled in over the mountains as I located a spot to sit against the huge spruce. I gathered wood from a nearby whitebark and kindled it with dry twigs off the lee side of a smaller spruce. I dug a down parka and wool stocking cap out of my backpack and prepared for a long night of staring into the fire. The temperature was dropping. The snow would be dry, and the spruce boughs would keep most of it off me. I carried no tent or sleeping bag this trip. I planned on sitting up all night tending the fire.

I spread out a small groundcloth—raingear, actually—dug into the bottom of the pack, and pulled out a foot-long oblong bundle wrapped in a spare wool sweater. I unwrapped a skull and placed it next to me facing the fire, balancing the upper jaw carefully upon the mandible. It was the skull of an adult grizzly, a female. I had come by it in the White Swan Saloon in the town north of here. A local horn hunter, a friend of mine, had bought it for me from a rancher who had poached the bear three months earlier on a grazing allotment in the national forest a few miles from the border of Yellowstone National Park. The same sheepherder had also shot at, but missed, another grizzly who was hanging around the female. That much was common knowledge. What was not known was that the two bears were related, and the previous winter they had denned together up the hill a mile south of my fire.

I didn't know what to do with the skull at first. I just did not want the sheepherder to have it. He had already made enough money selling the bear's paws and gallbladder. The sow[1] grizzly had never killed any sheep that I knew about, though that didn't mean she would not have started at any time. At the time she was killed, the female grizzly was almost eight, fairly old by Yellowstone standards. She had successfully mated once, probably when she was four and a half. The following winter she had given birth to a single cub—at least there had been only one cub with her the next spring, when I had gotten to know her. She had emerged from the den in late April, dropped down to the valley I had crossed an hour earlier, and fed on an elk carcass. I had backtracked her to her winter denning site. She and her cub had made a highly distinctive family: the sow's fur had a slight golden hue with a darker stripe running down her back. The cub had been nearly black with a silver collar extending into a light-colored chest yoke, which had faded sometime during his second year. They had been back on these slopes feeding on pine nuts the same fall, and I found their den the fol-

---

[1] I have retained the normal usage of "boar" and "sow" for male and female adult bears. A few experts object to this usage because of the association with domestic pigs. I agree with the spirit of this objection, but the term "boar" has been traced to the same origin as "bear" and its Icelandic cognate means "man." [Peacock's note]

lowing spring. Altogether I had found five dens on this same hillside within a few hundred yards of one another. All but the first could have been dug by the same female.

I consider the mountainside a special place, a place with power, as I do certain other valleys and basins there and up in northern Montana where grizzlies still roam. I return to these places year after year, to keep track of the bears and to log my life. The bears provided a calendar for me when I got back from Vietnam, when one year would fade into the next and I would lose great hunks of time to memory with no events or people to recall their passing. I had trouble with a world whose idea of vitality was anything other than the naked authenticity of living or dying. The world paled, as did all that my life had been before, and I found myself estranged from my own time. Wild places and grizzly bears solved this problem.

When I ran into the goldish mother bear and her dark cub on this mountainside I was more than a decade away from the war zone, and my seasonal migration to grizzly country had become a pattern. I had come to this place in the spring to greet the grizzlies as they emerged from their winter sleep and again late in the fall to see them into their dens. Since the female always denned in the same small area, it was easy. What I hadn't known was whether the young grizzly would return to den there after his mother was killed, or whether he knew how or where to dig a winter den. On September twentieth I had found my answer. Besides what he learned from his mother, this young grizzly had his own instincts.

The cub was now back on the family estate. I wondered what he would have done if the sow had still been alive—move off to another mountain? I was curious about these things, although I had come here this time for other reasons. I poked another stick into the flickering fire.

"Payback was a motherfucker," the grunts in Vietnam used to say. Meaning something about the difficulty of getting what you deserve—a sort of Stone Age notion of justice. Over there, believing nonsense in defiance of the blatant absence of any just distribution of earthly rewards and punishments helped you get through the night. After Vietnam, I caught myself saluting birds and tipping my watch cap to sunsets. I talked a lot when no one was around, especially to bears.

I tied a woolen scarf around my neck and held the skull up to the flames, staring beyond it where huge snowflakes glistened in the reflected light of the fire. Strings of connective tissue clung to the poorly scraped bone. The sheepherder had done a lousy job. I heard he had buried the grizzly's hide. He only dug up the skull 'cause someone offered a bundle of money. I should have gone back and blown a dozen of his stinking, bleating sheep into woolly heaven.

I felt the corrugated bark of the spruce tree pressing against my shoulders and looked back at the bear skull. "I wonder what you know, bear," I said to no one. Where had she spent her summers? Had she ever consorted with the great Bitter Creek Griz or fished the cutthroat spawning streams? I had never seen her play with her cub, although she had been a very protective mother. She had probably been pregnant when she was killed, having mated just subsequent to weaning her cub. Even in death, she was better off here on the mountain than mounted on some asshole's wall.

I set the skull down and threw a large deadfall on the coals. The log popped and sputtered, showering sparks that rose into the lower boughs of the spruce. I pulled my coat tighter around my shoulders, glad for the dead calm, which felt almost warm even though the night temperature had plunged far below freezing. I had a sense of urgency, even danger—the need to finish my business as soon as possible and get out. This was the storm that would begin winter. November blizzards had been known to dump over a foot of snow a day for several days. By the next evening, walking out would be difficult. All the roads on the plateau would be closed. In three days it would be all I could do to slide my pickup across the passes behind the snowplow. An accident or miscalculation could mean freezing, or wintering up here. But the predicament was familiar. Danger was part of what attracted me to grizzlies in the first place—danger married to great beauty.

My seasonal calendar was often tied to these blizzards: they told me when to leave the mountains. Big snows make winter. They send the grizzlies into their dens. At least, they do on this mountainside. Grizzlies do not all den at the same time; it all depends on sex, altitude, and how far south they live. For instance, the last Mexican grizzly in the Sierra Madre may not den at all if the winter is mild. South of Canada, females who are pregnant or living with young at higher elevations den first.

I nodded off briefly, leaning my head against the gnarled trunk of the spruce, thinking I could feel the weight of the snow piling up on it. I wrapped up the skull and packed it away, watching the big flakes filter through the branches like feathers of snow geese. Sitting in a major mountain storm in search of what some people regard as the fiercest animal on the continent instills a certain humility, an attitude that pries open in me a surprising receptiveness. My friend Gage, who was here with me when I stumbled on the first den on the mountainside, could find humility before nature in his backyard. I cannot: I need to confront several large, fierce animals who sometimes make meat of man to help recall the total concentration of the hunter. Then the old rusty senses, dulled by urban excesses, spring back to life, probing the shadows for shapes, sounds, and smells. Sometimes I am graced by a new insight into myself, a new combination of thoughts, a metaphor, that knocks on the door of mystery.

The fire's glow cast a halo of light in the falling snow, and I conjured an aura of reverence surrounding my mission.

VIETNAM NOTEBOOK

We walked point for the 101st Airborne during the summer of 1967. The operation centered on the country just north of the village of Ba An on the Song Tra Na in Quang Ngai Province. I was the only American green beret with our point platoon of mixed Vietnamese and Montagnard troops out of our A-camp at Bato. Behind us were three platoons of U.S. paratroopers.

The operation was not going well. Each unit had taken casualties. We had lost our platoon leader the night before. He caught a carbine round through his head just under and behind his eyes, which paralyzed his respiratory system. While I was keeping him alive by giving him mouth-to-mouth, the Americans mistakenly called in gunships on our position. I was the only one in the point platoon who spoke English, and by the time I could call off the air strike we had two more wounded. The platoon leader died while I was screaming fucking stop on the radio.

The next morning we led out into the rice paddies, walking along the low dikes. There were half a dozen water buffalo near the far side but no people except one nine- or ten-year-old boy in shorts. The boy might have been tending the buffalo. We walked across the paddy without incident. The airborne troops followed close behind.

The boy stood in the rice paddy thirty meters away watching me and the twenty irregulars as we walked by. When the boy saw the Americans he ran. Why he decided to run I would never know, but when he did the Americans opened up on him, first one or two, then an entire platoon, tearing chunks off his small body with M-16 rounds. My people watched, silent and grim-faced.

The truth was that any last vestige of religion had been choked out of me during the last two months in Vietnam by scenes of dead children. To this day, I cannot bear the image of a single dead child. In the years that followed, I had found it easier to talk to bears than priests. I had no talent for reentering society. Others of my generation marched and expanded their consciousness; I retreated to the woods and pushed my mind toward sleep with cheap wine.

By daylight I was cold and cramped, anxious to start moving up the hill. Five inches of fresh snow covered everything except the ground under the thickest trees. The air was still, no wind yet. Once it begins to blow, beware. This easy late-season stroll through the woods could quickly grow dangerous and I would have to get out fast.

The dry snow fell softly though harder now from the gray sky. I shuffled up through the pines in the morning gloom. Visibility was a couple hundred feet and decreasing. I figured the den lay half a mile directly upslope, and, although I thought I knew exactly how to approach it, it was possible to get turned around in this increasingly white landscape where every view looked the same. I pulled the thick wool cap down over my forehead to shield my eyes from the snowy glare, which even in low light could cause snow blindness.

At the base of a whitebark I saw the remains of a squirrel's nut cache scattered onto the fresh snow. I stepped over to the debris: ear-shaped flakes of whitebark cones, pieces of cones, and whole pinecones scattered over the snowdrifts. A frozen bear scat lay near the pinecone midden in the snowless crescent on the lee side of a tree. I poked at it with a stick, finding a couple of undigested red berries from a mountain ash tree. This sort of scat is common just before grizzlies den up, when they empty out their digestive tracts for the long sleep. The ash berries may act as a purgative, although I wondered where they came from, since I had not seen any *Sorbus* bushes for days.

The grizzly's carnivore gut, although elongated for a carnivore, is not made for digesting cellulose or the kinds of vegetable foods available in winter. Neither can he count on his skills as an opportunistic predator to keep him in food. So he must den up and hibernate. Some early springs I find the first scat a grizzly has dropped after emerging from the den: a cemented plug of hair. Bears don't eat, defecate, or urinate during their winter sleep. They slowly metabolize their own body fat. Their bodily functions slow, although bears may wake if disturbed or warmed by an unusually mild winter day. The sleep of bears is not the true hibernation of rodents, but it neatly solves the problem of winter survival.

The vertical fall of snow from invisible skies formed a gentle arc with the rising wind that blew in my face. About a foot of fluff had accumulated in the open and it continued to fall. The wind picked up and gusts blew bursts of powder snow off the tops of the pines.

I recognized a stark, dead fir with a double-pronged branching top. The den should be just uphill, across a rocky gully, maybe two hundred feet away. I stopped short to make sure my scent was not drifting in the direction of the den site. No problem, the wind was still in my face. I moved silently in the muffling snowfall directly downwind from the place where I thought the den was. I stood motionless for several minutes, sniffing the air. I could make out the faint but distinctive odor of the young grizzly. Until that moment I had not been certain I could relocate the spot or that the bear had not moved to another den.

Clark's nutcrackers cackled harshly just ahead, the first birdcalls all day. They were scolding, probably at the young bear, who may have lifted his head; the grizzly must have been moving. I waited for the jabbering to stop, then quietly crept uphill. I reached the trunk of a large whitebark pine and glimpsed a patch of bare dirt and stomped snow. I froze and slowly pulled my field glasses from under my outer sweater. A hundred feet up, I could see two brown ears protruding over a bench of loamy talus: the grizzly was asleep on the porch just outside his den. The bear raised his head and peered into the falling snow. His eyes closed; he yawned and dropped his muzzle again.

The last time I had seen this grizzly so lethargic was in the summer of his first year, when he and his mother had escaped the clouds of summer insects by bedding out the afternoon on a wedge of high snowfield. The cub had tired himself out by prancing up and down the angle of snow, suddenly folding his legs beneath him and rolling down to the icy edge of the snow patch, where he tried to bite off chunks without success. Finally he had turned inward in his frustration and bitten his rear paw. He had done this for a full third of an hour, once biting himself so hard he had bellowed in pain. His mother had watched him sympathetically and leaned back into the snow on her hind feet, offering her paps. The strange, rhythmic puttering sounds of nursing filled the air.

I thought about the days spent in the company of these two grizzlies, the one stretched out on the porch of his winter home, the other encapsulated in memory, her skull wrapped up in my pack. I needed to put this small part of the universe back in order.

The young grizzly stirred. He rose, shook off the snow, and turned, disappearing into his den. He may have known I was about but was too sluggish to do anything other than retreat within the mountain now. The season was very late. I crawled on up the slope using the trees as cover until I reached a tree across the gully almost opposite the den. I could clearly see the pile of excavated gravel and loam. In the crotch of the tree, at eye level, a crude scaffold of willow was tied to the branches. The platform faced eastward across the open ground in front of the den.

It was a child's idea. My little daughter had explained that bringing the skull back here would make a new bear.

The snow was blowing so hard I could barely make out the bear's porch forty feet away. I dug into my pack; my anger evaporated and all my attention focused on the present. Moving quickly, I set the skull on the framework of woven willow facing the den. I slipped a small bear paw of silver and turquoise off my neck and draped it over the skull; your fur against the cold, bear. When my skull lies with yours will you sing for me? The long sleep heals. We will find new life in the spring.

Only the black eye of the den entrance peered into the face of the blizzard. I shook the snow off my wool cap and shouldered my pack. I turned and half ran back down the slope, the soft snow tugging at the tops of my gaiters. In fifteen minutes I hit the valley bottom. The meadow was a whiteout. I turned east and walked with the storm as the wind slapped at my back. I checked my compass: a piece of cake. The big snow would lead me out.

ROBERT FINCH
b. 1943

As a Cape Cod nature writer, Robert Finch lives and works at the confluence of powerful currents. Henry Thoreau, Henry Beston, and John Hay have all testified to the revelations of sky and dunes in that sea-surrounded land. Finch's own books, including Common Ground *(1981),* The Primal Place *(1983),* Outlands *(1986), and* Death of a Hornet *(2000), perpetuate this rich tradition. But when he writes about natural processes that have been memorably evoked by his predecessors, Finch imparts his own contemporary slant. Though he has been active as an environmentalist, his books do not draw bitter lines between the older tides of nature and the current boom of tourism and construction. Rather, Finch is a writer who discovers and conveys saving continuities. Pausing amid the bustling tourist commerce of Chatham, just after he has returned from a solitary week on a barrier beach, he can describe himself as "content to be where I am, standing in that place of migrations and appetites."*

Death of a Hornet

For the past half hour I have been watching a remarkable encounter between a spider and a yellow hornet, for which I was the unwitting catalyst. I have found several of these hornets in my study recently, buzzing and beating themselves against the glass doors and windows, having crawled out, I presume, from the cracks between the still-unplastered sheets of rock lath on the ceiling. Usually I have managed to coax them out the door with a piece of paper or a book, but this morning my mind was abstracted with innumerable small tasks, so that when another of these large insects appeared buzzing violently, like a yellow-and-black column of electricity slowly sizzling up the window pane above my desk, I rather absentmindedly whacked it with a rolled-up bus schedule until it fell, maimed but still alive, onto the window sill.

My sill is cluttered with natural objects and apparatus used for studying and keeping insects and other forms of local wildlife—various small jars, a microscope box, a dissecting kit, an ancient phoebe's nest that was once built on our front door light, an aquarium pump, pieces of coral and seaweed, etc.—none of which has been used for several months. They now serve largely as an eclectic substrate for several large messy spider webs.

In one corner is a rather large, irregular, three-dimensional web occupying a good quarter-cubic-foot of space. It was into this web that the stricken hornet fell, catching about halfway down into the loose mesh and drawing out from her reclusiveness in the corner a nondescript brownish house spider with a body about three-eights of an inch long. The hornet hung tail-down, twirling tenuously from a single web-thread, while its barred yellow abdomen throbbed and jabbed repeatedly in instinctive attack. The motion could not really be called defensive, as the hornet was surely too far gone to recover, but it was as if it were determined to inflict whatever injury it could on whatever might approach it in its dying. Defense in insects, as with us, seems to be founded not on the ability to survive but on the resolution to keep from forgiving as long as possible.

The spider rushed out along her strands to investigate the commotion and stopped about an inch short of this enormous creature, three or four times her own size, with what seemed an "Oh, Lord, why me?" attitude, the stance of a fisherman who suddenly realizes he has hooked a wounded shark on his flounder line.

Death of a Hornet and Other Cape Cod Essays (Washington, D.C.: Counterpoint, 2000).

Whether or not her momentary hesitation reflected any such human emotion, the spider immediately set out to secure her oversized prey. After making a few tentative jabs toward the hornet, and apparently seeing that it could do no more than ineffectually thrust its stinger back and forth, she approached more deliberately, made a complete circuit around the hanging beast, and suddenly latched onto it at its "neck."

At this point I went and got the magnifying glass from my compact edition of the *Oxford English Dictionary* and stationed myself near the window corner to observe more closely. The spider did indeed seem to be fastening repeatedly onto the thin connection between the hornet's head and thorax—a spot, I theorized, that might be more easily injected with the spider's paralyzing venom.

While she remained attached to the hornet's neck, all motion in the spider's legs and body ceased, adding to the impression that some intense, concealed activity was taking place at the juncture. If so, it proved effective, for within a very few minutes almost all throbbing in the hornet's abdomen had stopped, and only the flickering of its rear legs indicated that any life remained.

During this process, the spider's movements were still cautious, but also somehow gentle, never violent or awkward as my whacking had been. She was almost solicitous, as if ministering to the stricken hornet, as carefully and as kindly as possible ending its struggles and its agony. Her graceful arched legs looked, through the glass, like miniature, transparent, bent soda straws, with dark spots of pigment at the joints, like bits of sediment clogging the leg segments.

Now the spider seemed to have made the hornet *hers*—her object, her possession—and her movements became more confident, proprietary, almost perfunctory in contrast. She no longer seemed aware of the hornet as something apart from her, foreign to the web, but rather as a part of her now, ready to be assimilated. She appeared to begin dancing around the paralyzed insect, her rear legs moving rapidly and rhythmically in a throwing motion toward the object in the center. I did not actually see any silk coming out of her abdomen, nor did her legs actually appear to touch the spinnerets there, but gradually a light film of webbing, a thin, foggy sheen, became visible around the hornet's midsection.

She would spin for several seconds, then climb an inch or two and attach a strand to a piece of webbing overhead. I thought at first that she was merely securing the hornet from its rather unstable attachment, but after she had done this a few times, I saw that, with each climb upwards, the hornet itself also moved a small fraction of an inch up and to the side. It was soon clear that the spider was *maneuvering* this enormous insect, in a very definite and deliberate manner, using her spun cables like a system of block and tackle, hoisting and moving her prey among and through the seemingly random network of spun silk.

In between these bouts of spinning and hoisting, the spider occasionally stopped and approached the hornet, now totally motionless and with one of its darkly veined wings bound to its barred side. She would place herself head down (the usual position for a spider in a web when not spinning) just above the hornet's head and, again becoming totally motionless as if in some paralysis of ecstasy, seemed to attach her mouth parts to those of her prey, as though engaged in some long, drawn-out death kiss. The two, insect and arachnid, remained attached so for ten to fifteen seconds at a time, after which the spider again resumed her hoisting and fastening. Was this some further injection of venom taking place, or was she beginning to suck the juices from the wasp's still-living body even as she was moving it somewhere? I was struck, mesmerized by this alternation of intimate, motionless contact of prey and predator, and the businesslike, bustling manipulation of an inert object by its possessor.

All in all the spider has moved the hornet about two inches to the side and one inch upward from the point where it landed, out of the center portion of the web and nearer the window frame, where now she crouches motionless behind it, perhaps using it to conceal herself while waiting for another prey. I pull myself away from the corner and put down the magnifying glass, feeling strangely drained from having been drawn in so strongly to watch such concentrated activity and dispassionate energy. There is something about spiders that no insect possesses, that makes it seem right that they are not true insects, but belong to a more ancient order of being. I like them in my home, but they will not bear too close watching.

I look back at the window corner and see that the characters of the drama are still there, once more in miniature tableau. All is quiet again; the spider remains crouched motionless behind its mummified prey, in that waiting game that spiders have perfected, where memory and hope play no part. There is only the stillness of an eternal present and the silent architecture of perfectly strung possibilities.

LINDA HASSELSTROM
b. 1943

One of the notable trends in contemporary nature writing is the increasing attention to working landscapes. Ranchers, like the South Dakota writer Linda Hasselstrom, often complement the revelations and reveries of more transcendentalist writers with the wind, grit, and labor of their own experience. They offer a regional perspective that has less to do with exuberantly falling in love with a new landscape than with dogged loyalty to a land and an inherited way of life that grow harder and harder to hold on to. In addition to her volumes of poetry and local history, Hasselstrom has published several books focusing not only on the rancher's life but, more specifically, on what it means to pursue such a livelihood as a woman. The selection here comes from Feels Like Far *(2000), a book which builds on and extends Hasselstrom's work in* Windbreak: A Woman Rancher on the Northern Plains *(1987),* Going over East: Reflections of a Woman Rancher *(1987), and* Land Circle: Writings Collected from the Land *(1991).*

NIGHTHAWKS FLY IN THUNDERSTORMS

* * * My first evening on the ranch, when I was nine years old, nighthawks plummeted like cannonballs into my life. My mother had just married John Hasselstrom, and I was helping her carry our suitcases and boxes to a tall house in the middle of a hay field. Its white paint weathered to gray, the old house had been lifted from the foundation where our new house would stand and deposited on railroad ties. We lived there all that summer of 1952, while the carpenters finished the new house near the garage and barn. When the wind blew hard behind a hailstorm, the house swayed like the cottonwoods beside it.

Knowing his instant family was used to lawns instead of hay, my

Feels Like Far: A Rancher's Life on the Great Plains (New York: The Lyons Press, 1999).

father had hitched up the team of horses to mow the lanky alfalfa around the gray house the day before we arrived, raking it into fat windrows. The green hay was so heavy I tripped crossing it every time I carried a load of belongings into the house. Outside the trimmed rectangle, plants stood high as my armpits, shimmering in the breeze.

That first night, while Mother stayed in the house unpacking and cooking supper, I sat on the railroad-tie steps and watched my new father drive through the dense green in his pickup. I went down the steps to meet him and we stood together enjoying the sun set, a ritual he has practiced all his life. Hundreds of birds dipped and swooped over my head, dark crescents against the sunset's gold. When I asked about the explosive sounds from the birds, he told me they were nighthawks.

At that moment, nighthawks became my favorite prairie bird. Even then, I knew the Lakota called them thunderbirds, but how I knew remains a mystery to me. The name's metaphor explained itself during the first thunderstorm I watched with my father, a tradition we began a few weeks later that summer.

Once we settled into the routine of ranch life, haying took up most of every day in August. The fields were often baked by the sun and bludgeoned by savage thunderstorms, hail and lightning in late afternoon. My father told stories of neighborhood men killed by lightning on tractors or horses, but he seemed to stay out in the storms longer than necessary for a man so sure of their power to kill.

As an adult, I wondered if his challenge to lightning was the only way he allowed himself to shake his fist at heaven, a fatalistic dare. Above all other lessons, ranchers learn patience with the unalterable: with blizzards during calving season, with lightning that kills one expensive bull in a pasture full of elderly cows, with rains that tear out fences. Ranch life is so crowded with chances for disaster that counting the awful possibilities could paralyze us if we allowed ourselves fear.

One of my father's mottoes was "What can happen, will." He never screamed curses when machinery broke down or calves brought less money in fall than we spent raising them. He faced whatever came, allowing himself only one terror, one chance to act out the anxiety he must have felt. A single wasp buzzing around his head on a summer day could move him out of his regular pace and make him dance. Ducking and weaving, he'd wave his handkerchief at the insect, hopping back and forth, his loose-jointed height almost graceful. He admitted his fear of wasps without embarrassment, my first lesson that even the brave are allowed fright.

But he declined to fear thunderstorms. At fifteen, I once refused to ride my horse after a cow because the black clouds were low and webbed with lightning, and I'd already heard a lot of stories about lightning striking the highest moving object, usually a horse and rider. My father's lips tightened, and we got in the pickup to find the cow. He ordered me to bring her in on foot and drove away as I ducked lightning flashes. When hail started a few minutes later, the cow tried to turn her back to it. I slapped her in the face, yelled, ducked in front of her. When I finally got the cow to the corral, both of us were covered with mud.

If my father was haying when a thunderstorm advanced, he kept wrestling the tractor with its creaking stacker around the field. On the mower, I'd forget to watch the falling alfalfa and fix my eyes on the bubbling underside of the dark clouds overhead. Each time his tractor neared the truck, I'd reach for my ignition key only to see the stacker teeth dropping again, scooping up another windrow of hay. As the machine lumbered back toward the stack, I'd shove the throttle ahead, driving faster on the next round, shoulders hunched expecting a blast between my shoulder blades.

I always noticed that more birds swarmed around my head just before a storm. When my mower drove insects out of the protective vegetation into the winds before the storm, they dived into the falling alfalfa to snatch them. I felt like a whirlwind, beating up insects to be devoured.

My father would finally pull up beside the stack only after a lightning bolt had plummeted into the hill less than a mile from us. I'd lift the sickle bar, shift into fifth gear, and race toward him. Shouting to be heard over the thunder, we'd cover the tractors' exhaust stacks with tin cans and dash to the truck. Heading home through the field, he would drive around stands of uncut alfalfa, unwilling even in his hurry to damage a single stalk. Beside the truck in the garage, we'd stand panting a minute, hoping the rain might let up before we bolted for the house. By the time we stumbled inside, electricity would be bursting into fireworks over the barn roof and my mother would be huddled in the dark bedroom, shrieking every time the thunder rolled.

Once inside, Father would walk from window to window, watching lightning buffet the hillsides, afraid of a prairie fire. I'd follow him. In a year when ample winter snowfall and spring rain meant a heavy harvest, he'd tell yarns about storms he'd seen as a boy, how the horses ran away with one of his brothers and tipped the mower over. If the clouds dropped down gray and boiling with water, bringing a flood that scoured the earth instead of soaking into the dry soil, he'd say, "Well, anyway, it filled the big dam. Tore out the fence, though."

Watching a windy storm in a dry year, he'd set his mouth grim and stare out the windows as a bolt struck. "That's a fire starter," he'd say. I'd strain my eyes watching for smoke. Either lightning would start a fire, or it wouldn't. If it did, either the rain would put it out or it wouldn't. If the rain didn't put it out, either the volunteer fire fighters would or they wouldn't. In several ways, we might lose the summer's hay already in the stack or the standing hay we hadn't yet cut. Neither anger nor fear would change the odds or the outcome.

Any summer storm could bring hail. During the hours my father paced the rooms with me trotting at his side, we'd listen rigidly for the first thump on the roof. A truly destructive hail always seemed to begin with one loud whack, followed by a long pause. We'd look at one another. One of us would guess at the size of that single hailstone. My father always raised his eyebrows and tilted his head as if the next stone's impact might be inaudible.

The second blow always shocked me. I'd jump as my mother screamed. Others followed quickly, like a drummer testing his skins before the concert, warming up for the *1812 Overture*. We'd stand silent at the big picture window as jagged chunks of ice battered leaves and branches from the trees, bounced on the driveway, chopped at the base of the tall yard grass we hadn't had time to mow. We never stated the obvious: that any crops not yet gathered were gone.

At such times, the prairie outside the window looked like an ocean with wet gusts rattling the windows and gray sheets of water sweeping across the landscape. We couldn't see more than a few feet beyond the glass. Occasionally, the murk would lift so we could glimpse one of the huge old cottonwoods east of the house, branches flailing against the wind.

Above the clothesline during that first storm, I glimpsed movement, the outline of a bird, wings beating furiously. I cried out, sure it would be battered to death by the hail.

"Oh, that's a nighthawk," my father said. A moment later I saw the distinctive white spots on its wings. Instead of flying into shelter or letting the wind take it south, the bird continued to flap in place. Watching, I thought of standing in a hot shower, how I raised both arms, bent my neck, turned in every direction so blessed heat could pound muscle aches. The thunderbird seemed to be doing the same, except that it required tremendous energy to remain stable against a fifty-mile-an-hour wind. For a quarter hour it raced and capered furiously in one position.

After that, I always saw the nighthawks cavorting in gales, noticed how they appeared before the rain to hunt above the wind-beaten grass.

As the nighthawks' flight became familiar, I was able to spot them any-where on the plains with one sweeping glance at the broad sky. Down low, they fly like other birds, with even, rhythmic wing strokes against the air. But hunting nighthawks alter the tempo. Using several brisk strokes alternating with slower wing beats to lift their bodies, they mount the sky with a jerky motion, uttering a single high note, until they are barely visible. Then from high overhead the nighthawk drops silently. At the end of the dive it swoops up and a peculiar boom reverberates.

How, I wondered, do they make that eerie tone? Does it come from their throats or is it produced by the intricate wing motions demanded to curb that furious dive and soar again? One source says the wings produce "a peculiar musical hum" in the courtship dive, but I also heard the sound on summer nights as dozens of the birds, both adults and adolescents, hunted around my eaves. Several books I consulted fail to mention the sound at all, yet it's the surest way to identify the flying acrobat of the prairie, even in darkness. No other prairie creature makes a similar noise.

Birds of prey—true hawks—drop from the sky hard and fast to drive spiked talons into the spines of rabbits, the shock killing animals larger than themselves. Nighthawks dive to catch insects on the wing, so the practice is called hawking.

Riding after cattle, I sometimes saw a nighthawk's refuge—no one could sensibly call it a nest. The bird simply lays two eggs blotched with shades of green on an outcropping of limestone. The rough, lichen-covered stone camouflages the eggs and perhaps holds warmth while the parents hunt.

When George and I walked in our windbreak, a single nighthawk might flutter up from the ground just ahead, and we'd glimpse one mottled egg beside a downy chick, both yellow splotched with green. Invariably, the adult drifting just above the tallest grass tricked our eyes away. When we looked again, the chick was invisible, yet two more steps might crush it. Once we knew the locale of a nest, we avoided it for a month. One evening in late summer we'd notice nighthawks that seemed clumsy, then see a hovering adult and know another generation had survived.

During long twilights without George, I sat by the stone cairn on the hillside, watching the sun suck daylight past the western hills. Hidden in the tall grass, I became invisible—or insignificant—to nighthawks swooping after millers, mosquitoes, and grasshoppers. As darkness deepened, it seemed to breathe nighthawk sounds. ⁂ ⁂ ⁂

TRUDY DITTMAR
b. 1944

Though Trudy Dittmar's first book will be published later this year by the University of Iowa Press, it has had, as Emerson observed about Whitman's Leaves of Grass, "a long foreground." Dittmar grew up on a farm in the rural community of Colt's Neck, New Jersey, where she spent her free time observing and exploring the swamps, meadows, and woodlots that surrounded it. After receiving an M.F.A. from Columbia University, she lived in Venice, Italy, for a time, then returned to New Jersey to teach. In 1991 she settled in northwestern Wyoming, began to take science courses, and developed her own distinctive form of the nature essay. It is a form which combines personal experience, scientific description, biological and physical theory, and aesthetic interpretation in nonlinear narratives that create unexpected connections. The following essay will be included in her forthcoming collection, Fauna and Flora, Earth and Sky: Brushes with Nature's Wisdom, *which she describes as having "the overarching theme of . . . the resonances between the human condition and nature-at-large . . . centered on how nature has taught me the meaning of my own life."*

MOOSE

It was early May and the woman was out saying a final goodbye to winter. She'd gone up into the high country where the snow was still firm, snowshoeing one last time. She crossed a little meadow and was tromping along not too far from the edge of the forest. It was very bright and hot in this meadow high up in the sun, and she'd just stopped to take off her jacket and tie it around her waist, when suddenly she noticed a cow moose about fifty feet ahead, at the edge of the trees.

The woman saw no sign of a calf, but it's likely that somewhere nearby there was one, because the cow was coming at her. Though she

North American Review, September-October 1996.

saw this, such details as the flailing forelegs, the boot-sized hoofs flashing down through the air at her, did not really register—it all happened too fast—and when the moose suddenly jerked to a halt and swerved away, she had no idea why. All the woman knew was that at that moment her feet were swept out from under her. She was plowing through open snow for some seconds and then she was plowing along through the trees. Fortunately the blanket of snow was still fairly thick, because by the time the moose got its leg free of the webbing of her snowshoe, the woman had been dragged a good distance. She was scraped and lacerated on all her exposed parts, and she had two broken bones.

You hear stories like these in every barroom in moose country. You hear them told in the morning gatherings of cronies in the local cafés. There are stories of bull moose in rutting season charging head-on into cars, trains, and bulldozers. There are stories of moose in no particular season at all turning on a dime and driving someone up a tree. Sometimes there's a warning: the ears flatten, the mane goes up, the moose does a thing that looks like he's sticking his tongue out at you. Other times there's nothing; the moose just comes. Anyone who spends much time in moose country hears lots of these stories. Though there's often a comical edge to them, they always involve a good bit of damage, and anyone who hears enough of them learns to be circumspect.

One crisp morning in rutting season I see a bull moose in the willows with a rack as wide as a redtail's wingspread. In an instant he's pricked his ears up, and with that giant rack bouncing in the air above him he's taking one of those big one-two-three kind of trots toward me. One-two-three, he stops, looks, and I'm off into the trees, scanning hard for one I can climb. Late one afternoon in the forest in summer, I'm lost and retracing my steps to a fork in the trail where I suspect I went wrong, when I see a big cow and her calf up ahead just where the crucial fork ought to be, and she's watching me. When the calf tries to nurse, she jerks from him, head tossing, and, backstepping respectfully, I show her emphatically that I'm off. The light's lowering, I'm not sure where I am, and to figure that out I need to check the very spot she stands on, but instead I migrate right off the trail and go floundering off through the trees.

Another time, emerging spider fashion from a steep climb through deadfall, I come face to face with a moose on a mountain top, his rump two feet from my destination, the door of my jeep. And still another time, a moose blocks my way on a granite trail, wide enough at his end for him to turn comfortably, but at my end a tad too narrow for me to turn with my pack on—a cliff wall jutting up inches away on one side of me, on the other a precipitous drop. These moose don't threaten, but neither do they make haste to yield. In both cases all I can do is wait for them to get bored and vacate the only path I can take, but I'm spared the boredom that usually accompanies idle waiting by one hearty spritz after another of anxiety.

Moose can be difficult. You try to give them a wide berth. But at the same time they're unpredictable—there's no standard m.o. with a moose. Despite all the bar and café stories, and despite those few times when I felt I was about to be grist for one of those stories myself, in the gamut of moose ways the moments of bluster are far less rule than exception, and almost all my encounters with moose have been very different from what the stories depict. They'll surprise you by what they will do, but what they *won't* do can surprise you more. A moose is enigmatic. A moose is, at times, a bottomless thing.

Of course on many counts a moose is a perfectly fathomable thing, no enigma, but quite explicable in terms of its adaptations to its environment. Many of these adaptations are extraordinarily apt and resourceful, prime emblems of how cunning nature's workings can be. And they're all the more prime perhaps, and all the more cunning, because they're frequently effected through features that seem so awkward and nonsensical, even comical sometimes.

No need to go further than the moose's physique, for an example. People who live in moose country have a particularly keen sense of the animal's drollness, and to borrow one local depiction, a moose looks like it got caught in a crusher and smooshed end to end. Its head and neck escaped the crusher, but the big compression pads of the crusher caught it right at the chest and right smack on the rear and squeezed hard, and when the moose came out he had a short little body humped up at the shoulders, his head much too big for it, his legs way too long. In contrast to his black chocolate trunk, these legs are grizzled like an old dog's muzzle, and his neck has a thing called a "bell" hanging from it, a clapper-like furry flap dangling down. The nose end of his face looks too big for the rest of it—his face is nose-heavy, wide and huge-nostrilled, finished off below with a pendulous upper lip—and against the bigness of the nose end of the face, the smallness of his eyes way up back off the muzzle is unsettling. He looks disproportioned and ungainly, a ragtag mix of a lot of things, none of them fully realized—the head an early attempt at something equine; the slope of the back from butt up to shoulder hump suggesting a start on a giraffe, abandoned early, before the designer had the courage to take the design all the way.

But a moose body is far from the work of a crusher, and there's method in the madness of these oddities. The modified giraffe aspect angles the neck well for eating from trees. The rangy legs help here too, for rising on its hind legs a moose can reach branches twelve feet up, and it can straddle saplings, riding them down between its long forelegs to get at the top shoots and leaves. And these four-and-a-half foot high moose legs maneuver easily in all sorts of elements and conditions. Moose can run soundlessly over a littered forest floor at 35 miles an hour; they can move with little effort in snow nearly two feet deep; they can plod over quaking muskegs where other large animals would flounder, swim 16-mile stretches of Great Lakes at a clip, and they've been seen crossing mud holes, buried to the withers in soupy mire. As for the big head, long faces are a standard adaptation of large herbivores, providing space for large grinding teeth to deal with the silica and cellulose of plants, and the moose's extra-long jaws enable him to include in his diet not just the tender herbaceous forbs and grasses and aquatics, but the tough woody plants as well. Then too, the pendulous muzzle, being nearly all muscle, enables him casually to strip shoots of their leaves, and the long snout contains phenomenal numbers of nerve endings, enabling him to detect the faintest smells. And though a moose isn't averse to sticking his head under water (they've reportedly uprooted aquatic plants from 18 feet deep) the long face lets him browse much underwater vegetation while keeping his eyes on what's going on around him, if he likes. Although his vision isn't reputed to be the best at midday, it's good in dim light, and he can rotate his eyes independently to front and rear, so he can even watch behind him without turning his head. Even the strange "bell" or dewlap has a beneficial function, for in addition to being a visual status symbol for males (the bigger the better), it's also thought to be a retainer for a special saliva containing sex pheromones.

It's always a little startling, the moose's oddness, the silliness of his disproportions, as if nature had played a joke on him, but all these idiosyncracies give the moose flexibility, they expand his ecological niche, so as curious as they are, you can get to the bottom of them. But though these oddities, once deciphered as deft adaptations, explain the moose rather nicely, they don't cover all the ground for me. You can see a moose every day and not fail to be struck by these oddnesses, these sillinesses, even once they've been explained; but eternally striking or not, they're not enough to account for the discomposure I feel whenever I notice a moose watching me.

Whenever, during my rambles, a dark place in the landscape materializes into a moose, what really startles me is the way he just stands there, looking. There's a thicket of willows up ahead, or a little pool choked with water plants, or a shadowy jumble of evergreen trunks, and suddenly the image of a moose coalesces within it, and I'm unnerved by the exotic stillness of him. Some would attribute his composure at these moments to bad eyesight. He's not just standing looking at you, they'd say; he doesn't see you at all. But in all these cases, whether he's seen me or not, he's heard me, he's smelled me—whatever. For who knows how long, he's known I was there. In all these cases, a deer would have frozen, then bolted. Ears sharp, an elk would have focused intently for an instant, and then he'd have been only a rusty flash among the trees. Other members of the family Cervidae are always nervously vigilant, but the moose just stands, looking. He's focused enough, sometimes quite alert, even stirred up. But most often he is merely casual, not frozen as if hoping for camouflage, but fidgeting nonchalantly, nipping buds or munching aquatics as he watches, chewing cud, flicking his ears at flies. His stillness is not lack of motion so much as cool silence and calm—and it unsettles me.

It's as if moose have a big still place in them, like a deep lake whose surface is never ruffled, its waters dark, cold, shining, and smooth. They are like the rocks in a landscape, the cliffs, the outcroppings. It sets a jolt in my blood, this quality, so alien in a wild animal. It triggers an eerie thrill.

Of course, you might chalk this quality up to a position of sureness. Only wolves and bears are their enemies—bears only when they're wounded, and in this country there are no more wolves. But in the fall I see the trucks go through town with moose dead in the back, long grizzled legs jutting up out from under the tarps, big palmate antlers jutting out just beside them, their heads now where their butts should be. And so the position-of-sureness theory doesn't quite work for me.

A moose can be dangerous, I know that, but when all is said and done he rarely threatens, rarely uses his formidable store of fierceness on anyone; and as if this were some kind of Eden—as if god were in his heaven and all was right with the world—when faced with a gun a moose many times just stands there, looking on. It is as if moose were informed with a simplicity so pristine that they don't grasp life's dark complexities. It is as if they are victims of a kind of innocence, cruelly exploited. How do you *really* explain such behavior?

On the coldest of days in a very cold time, a moose came to my doorstep. He came in a dream. It was not a sleeping dream, and not a daydream either, but a waking dream that was focused, although I, the dreamer, had no idea on what. But there was a focusing, a searching on the part of the dreamer, a stretching of all the invisible insides of the dreamer toward something. It was a focusing not *on* a thing, but *of* a thing: the dreamer's heart.

I was in my cabin, a dark speck on a bare knoll over 8000 feet up the leeward slope of the Wind River Mountains, just a few miles from the pass that traverses the western end of the range. Through the fine snow sifting down outside the windows you could see no farther than a few feet. Without sensory clue, I knew there was a visitor. I opened the door. He stood close by the doorstep, a huge dark form suspended in an ether of snow.

He didn't speak, but I heard him. He said to follow him and I did. Down the knoll and over the bench to the riverbottom. Across the river and up through the trees. Over the ribbon of snow that in summer is the road to the pass, and into the trees beyond it, climbing through them up sharp-planed slopes and down their nether sides, then up and down again, over wave upon wave of trees. Climbing and impossibly climbing, effortlessly, the moose form just ahead of me slow and indifferent and steady, like the constant procession of the planets traveling their ellipses, the perpetual spinning of the atoms in their orbitals.

Then we were at the top of Union Peak, on the crest of the highest slab of all the raw jagged slabs of granite that protrude from that mountain's top like massive knife blades standing on edge. And suddenly, as if I were two people, observer and observed, I saw myself from a long distance, as if through a telescope back in the cabin I'd left behind. I saw myself there on the slab at the moose's side, the two of us distant and tiny yet oversized in proportion to our surroundings, and despite all the snow sifting down through the miles between observed and observer, silhouetted sharply against the sky.

Standing beside him, I touched the moose's neck. I didn't stroke it, but simply laid my hand upon the side of it, flat. He stood motionless, made no reaction, but his lack of reaction itself was a kind of response. He was a moose and he let me touch him, and I felt the life of him.

He never betrayed the faintest awareness of me. You would have thought he was alone. It didn't matter; alone or not, he guided me. He stood at the top of an undistinguished peak in the storm of winter and looked out from it, and I followed his gaze and saw what he saw.

For a few moments the land was rushing before us, like a fast-forward of terrain photographed from an airplane, and us flying in it, our eyes the camera's eye. We stood stationary on our mountaintop and the land rushed south to north at us as we scanned down the southwestern slope of the Wind River Mountains, into the plains below, on down over their vast sweeps to Rock Springs and past, toward the southern border of the state. We scanned south or it rushed north, one or the other, all of it snow, all of it empty landscape, bleak and cold and lifeless with snow.

And then the rushing had stopped and there was the huge white curve of the top of the world, not south any longer, but the cold arc of ice which held the North Pole. Then that curve of ice lengthened, we saw a vast arc of the globe, many degrees, and then we had a still more distant view. We saw down past the vast cap of ice to where the land-forms started, the Scandinavian peninsula, its great lobes flat and edged with intricate indentations, sharp-edged as a sawblade and totally white. Beyond, a vast expanse of Eurasia, cold and still and white. A third of the globe white and inert and silent encompassed in our view, even as we stood on the rock blades of the mountain looking out over ridges of forest behind vast veils of falling snow.

White and white, nearly the whole eastern part of the Northern Hemisphere, every peninsula of it, its coastlines hard-edged and sharp white against winter gray sea. I kept looking, expecting something, a bit of motion, a spot of color, something to hook onto, but all the way down, as the curve lengthened, as the expanse grew, revealing more and more of the world, there was nothing but snow-covered continent and grey water all the way down.

It lulled me, that dream. There was a peace in it. All still, all white. Beautiful. But cold. So cold and still it frightened me that this should be my vision, that when I called out for a truth, this was what the moose showed me.

There's a theory that evolution is based on a struggle among genes, not bodies. It represents a twist on the traditional Darwinian view. In the competition for fitness (i.e., to leave more offspring), natural selection rewards individual bodies with variations best adapted to the environment, with the result that as their descendants increase in numbers, their species evolves into one which embodies universally these particular adaptations. Individuals are the basic unit of selection, say traditional Darwinians, while genes, which furnish the ingredients of variation, are just outfitters, as it were. But biologist Richard Dawkins' theory of genes as the agent of Darwinian processes would have us recognize that the gene is the true replicator on this planet, not the moose, not the fruit fly, not us. After all, in sexual reproduction an organism doesn't make a precise copy of itself, a gene does. And whereas the particular organism is relatively short-lived, the tiny bit of hereditary material Dawkins defines as a gene is long-lived, as it (or the information encoded within it) passes intact not only through generations of a species, but often beyond species, sometimes even throughout all the five kingdoms of life, informing everything from humans to willows to morels to kelp to streptococci. The bodies genes pass through are just vehicles, says this theory, each a unique and temporary combination of genes; but the genes (unless zapped by mutation) are constant, reproduced unaltered again and again. And since the true replicator is the gene, not the organism, says the theory, it's not the organism, but the gene that's the contender in the struggle for fitness, while bodies like us, moose and humans, are secondary.

Ultimately, the gene works not for individuals or species or any other taxonomic division, but for itself. "Natural selection favors those genes that manipulate their own propagation," says Dawkins, and so the gene programs us organisms, its vehicles, to do whatever is necessary to increase its numbers in the gene pool. And if we humans have deceived ourselves in this matter, thinking that we bodies are the ones with the stake in the struggle, it's because in organisms as complex as we are, the most effective programming plan hit upon by genes so far in the course of their evolution is to help us cope with limitless unpredictable environments by setting us up with an enormous plasticity of mind. Subject as our intelligence makes us to vast combinations of circumstances, it behooves our gene passengers, in short, to provide us with a consciousness so sophisticated that as decision-makers—at many levels of action, at least—we're emancipated from our master genes. That we're deluded in this matter is to be expected. It's hard to detect that we're mastered, when we are also master. Hard to detect that we're vehicles when we also have a big share of free will.

I hadn't yet heard this theory the day I had my moose dream, but many of its implications go hand in hand with intimations that raked my soul that day, and it may well have been the tip of some such iceberg that I was intuiting when the moose showed me the picture of a winterbound globe. The theory has its critics, and being no more than a casual student of such things, I can make no claims for its validity other than to note that some fine scientific minds (among them Francis Crick, of the double helix) have received it enthusiastically enough to build upon it. All I can say is that here and now, many moons past the day of the moose dream, I find the paradoxical logic of it all quite compelling, even rather lovely. But if it's true, the world is a colder place.

And we are cold in it, if it's true; and not just in what openly passes as our darker side—we've been inexorably forced to acknowledge the predatory protoreptilian parts of ourselves—their work is everywhere and hard to escape—but even by virtue of what we call our goodness, we are cold. For "natural selection favors those genes that manipulate the world to ensure their own propagation" and if a "superficially selfless" gene (as another writer has called it) will do better at getting its information passed on, then even altruism rears a selfish head.

There are all sorts of animal behaviors that involve altruism. Mice, for example, engage in grooming each other, and it seems a benevolent thing for them to do. But mice separated from other mice develop nasty sores on parts of their heads they can't reach, and so when one grooms another it's likely not out of good will alone. Support is offered in exchange for support. Altruism is reciprocal. Myriad examples of such behavior have been the focus of extensive, distinguished research. A lot of this behavior occurs among kin (ground squirrels screaming to warn relatives of danger); but a lot of it extends beyond kin, occurring among kind, as in the case of the mice mentioned above; and some extends beyond kind, across species (one ant species protects aphids in exchange for sugar they harvest), and even across kingdoms sometimes (another ant species living in the bull's horn acacia attacks all that tree's enemies in exchange for nutrients it produces, and even clips surrounding vegetation competing with the acacia for growing space and light). Throughout the web of living creatures, one after the other is doing something for another in expectation that the favor will be paid back.

I think of an acquaintance who was always flattering others—very effectively generally, but if you took close note, suspiciously much. "She's a compliment junky," a friend said, explaining her. "She puts out as much of it as she can in an effort to get as much as possible of the same thing back." Richard Dawkins wouldn't extrapolate from the evidence of reciprocal altruism in animals to human beings, but some of the biologists whose work he draws on would, and so would a number of evolutionary psychologists. According to them we've evolved a stake in good reputations on this sort of basis: if you do good to others, they'll do you good back. At some point in the labyrinthine course of human evolution, they say, via some subtle turn it became important not just to be good, but to *appear* to be good; just leaving the impression of goodness could gain the desired "reciprocation" for us. And the subtle adaptive adjustments didn't stop there. For as humans developed in astuteness, it came about that if we tried knowingly to deceive others we showed our hand in small ways—the others just might see through us—and so it became adaptive behavior to mime our concept of goodness to our own selves, deceiving ourselves about our goodness in order to do a better job of deceiving others about it. Ground-breaking researcher R. L. Trivers has related his genetic model of *reciprocal altruism* (the term is his coinage) to many of our overtly fine moral sentiments, suggesting that sympathy, gratitude, generosity, guilt, righteousness, and others are not as purely virtuous as we've thought all these centuries, but instead have been targeted by natural selection for improving our ability to deceive, to discern deceivers, and to escape having our own deceptions discerned. And all of this is related to getting our genes to the next generation—or, as Dawkins would have it, to the genes getting themselves there.

In spite of myself (and the anguish that invoked the dream moose), when I read of this research and the hypotheses it generates, my intellectual excitement is exquisite. My mind takes in these theories with bated breath. It's all so amazing, this business of the gene that shapes us to shape ourselves for the good of its survival—if valid, it's another gigantic strike for the elegance of the workings of nature—but it's a cold world. With the immense flexibility of our natures we're given the choice of goodness, unselfishness. And yet even our unselfishness is selfish. And even our choice isn't a choice.

You can flip that back on itself, of course. If the choice isn't our choice, you can say, the selfishness isn't our selfishness. It's the gene that's selfish, not the carrier of the gene. If I'm deviously supportive, only superficially altruistic—if I'm too hotly engaged in this struggle for fitness to achieve the disinterest necessary to qualify even my highest feelings as true love, it's not me but the genes in me that are behind it, and I'm off the hook.

The arctic tern, though indigenous to the coldest latitudes, flies 10,000 miles to avoid the hibernal extremes of its native habitat. Some bats drop their body temperatures to as low as 29 degrees Fahrenheit, and hang themselves up for the winter just about dead. Other mammals increase their body heat. Some tiny ones raise it considerably, insulating themselves underground or undersnow right at the onset of cold weather, but the arctic fox needn't do so until the outside temperature drops below minus 40° Fahrenheit, and even then, with only a minimal increase in body temperature he can sleep safely on open snow at minus 80° Fahrenheit for up to an hour, so well insulated is he, fur covering even the pads of his feet. Some amphibians burrow down into the mud all winter so as not to freeze, managing minimal respiration through their cloacas, even through their skin, while the wood frog buries itself in shallow soil of the forest and, by a miracle of biochemistry, freezes till spring. Although as the water in the spaces between his cells freezes he gets hard as a board, as long as the glucose his cells are packed with keeps the living matter of those cells free of ice, when he thaws out in spring he will start jumping again. As the days grow shorter many plants begin to dehydrate their cells. Consequently, although water by necessity freezes inside the plant, as in the wood frog it does so only in the spaces between the living cells, and so the plant, like the frog, goes on living in the frozen state. In some cases even the nature of the freezing within plants is different from usual; in a process called vitrification, the ice forms without crystallizing, so there are no sharp edges to puncture and destroy the cells. And some plants, like spring beauties and snowbank buttercups, even manage to develop while deep within snowdrifts, utilizing the meager light that penetrates snowpack to do some minimal photosynthesis.

Dealing with cold isn't easy. It requires ingenious biological plans. But even the cleverest accommodations to winter fail to exempt the plant or animal from the rigors of cold. There's always a sacrifice; winter exacts a harsh toll. It inflicts brutal physical hardship. Sometimes it dictates total suspension of activity, even of consciousness; the very life of some organisms is in abeyance for months at a time. Even then, winter kill is a fact of the season; whatever plan a species follows, there is always a percentage of the population that doesn't make it through. But tough on the world as it is, winter is not evil. It has nothing to do with morality. It's just a neutral coldness, part of the cycle of things. While one pole of the earth has its turn tilting toward the sun, the other tilts away from it, that's all.

The moose pays his winter dues like everyone. He makes a modest migration, not far south generally, mostly just down from the mountaintops, though because of his long flexibly jointed legs he needn't even make the trip too early, deferring departure from subalpine bogs and creek bottoms, if he wishes, till the snowpack is close to two feet. He makes a dietary adjustment, switching from nutritious pond weeds, sedges, and tender leafy shrubs to woody shrubs and trees, and this requires increased fermentation time in his rumen to deal with the heavy cellulose load. As his dry weight consumption reduces by half, he metabolizes body fat stored during the summer, losing weight slowly but steadily till spring. To compensate for lowered nutrition, he sheds the weighty antlers that would drain his energy (70 to 85 pounds of calcium grown in one four-month season, exceeding all other antler growth), and without going into torpor, he lowers his body temperature, reducing his basal metabolism and thus the energy demand on his food. He starves in bad seasons, if snow is too high; if he's weakened by too poor nutrition, he's susceptible to pneumonia, parasites, other disease, wolves.

Most of his adaptations aren't spectacular, but more often than not they cut the mustard. He has ten-inch guard hairs and a dense woolly underfur—in the wintertime an inch or more thickness of it on the inch-thick hide of his back—insulation to suit him for life in some of the coldest places, the high places in altitude and latitude, boreal forests of circumpolar lands. When the grasses and ferns and low shrubs of summer are two feet under in winter, he rips the bark off aspens with his lower incisors (like all cervids, he has no front teeth in his upper jaw), and straddling aspens weighed down by ice, he bends them down farther, to get at the finer twigs at the top of the tree. With his long, loose-hinged legs he moves fast even through deep snow, and when the snow surpasses the comfortable limit, moose, like deer, do a thing called *yarding*, several of them staying in a small area, or yard, and continually packing the snow down in a number of criss-crossing trails. Snow, in fact, is his winter comforter; when things are at their roughest he burrows down and insulates himself with a covering of it.

It's said that in the Middle Ages European moose were sometimes used as draft animals in Scandinavia, drawing sleds long distances through deep snow much faster than relays of horses could. It's even been reported that American moose were occasionally broken to harness. I don't have too much trouble believing it. Until not long ago some miles west of me there was a couple who hosted gatherings of moose every winter for years. All they did was put out plenty of hay and the moose came back and stayed every winter, munching around in their yard from November to April, clomping around on the hay bestrewn wooden deck that belted their house, and if the couple had wished it, I bet at least some of their moose guests would have stood for being harnessed up.

The moose is as winter a creature as almost any. His adaptations aren't as dramatic as those of some organisms, but in their understated way they're about as effective—in fact, in allowing him to go about the business of living almost as freely and fully in winter as in summer, they're more effective than most. When winter comes, the moose doesn't go around it or away from it, he doesn't switch to a whole new game plan, he doesn't shut down; he just lives in it, taking it as it comes as straightforwardly as he takes summer. He does more than survive the cold world; he makes it his home. Like my dream moose, he moves through it with something like the sang-froid and disinterest that characterize nature itself.

Last spring I stood on a rise at the border of forest. The pink petals of least lewisias hid beneath the new blades of grass at my feet. At the bottom of the long, steep slope below me there was a trickle of a stream and a deep streambank, eroded—an expanse of watery mud, black and textured with pocks. I saw something moving, and then he materialized, as usual. It was a bull moose in velvet, in mud above his knees. He was stuck, and even those long, loose-hinged legs wouldn't get him out of it.

He plunged and plunged, and it was to no avail. His feet were tangled in submerged roots perhaps, or perhaps due to illness or age he was simply too weak to defeat the suction's drag. But the moose showed no signs of panic. He worked for a while and then rested with complete unconcern. I'd once read an account of something similar, but hadn't quite believed it. The writer was a high level national parks official who'd watched a moose plunging in a quicksand of volcanic ash, and that's how he'd put it: between times he "rested with complete unconcern."

I stayed for a long time and watched him. But I was far from camp, and finally I had to leave. Still, I stayed for close to two hours and watched him, as he plunged and rested and plunged and rested, and when I left he didn't look any closer to getting out. Perhaps when the temperature dropped that night the mud would firm up and, the suction reduced, he'd get a foothold. If a grizzly didn't happen upon him first. I left reassuring myself with that rather strained notion. Perhaps the coldness he knew so well how to live with would save him here.

I philosophize on the neutrality of winter. On the beauty of the cold world the moose showed me. But when a bear surges over the top of the hill before me and, in his rolling gait, pours down its side toward the trees where I stand, it's all I can do to keep my wits about me. My heart can soar at the notion of the vast indifferent plan of nature, I can theorize what I theorize, I can know what I know, but when even just the metaphor for death comes, my heart freezes in me. I'm as far from the peace of that dispassionate power as I can be in those moments. Is the rest all delusion, hypocrisy?

The image of the moose in the mud says it isn't. Even the bones on the porch of my cabin this bitter fall morning say no. I went up to look before the snow should be final and found a profusion of them helter skelter in the willows around that stream, some flecked with matter, dried gut like scraps of rawhide. A long jawbone, a large femur, but I couldn't find a skull. I'd left the jeep on the log road and hiked in on impulse, no pack on, no water. While I was searching, I got more and more nervous as a strange leaden sky filled the east. By the time I got back down to the jeep my whole body was shaking. A fall blizzard on the mountain is beautiful as long as you're not caught in it.

I'd gathered the large ones, the whitest. Back home the books seemed to confirm that some were moose bones. I've spread them in a line under the porch rail. That way, most of the day they catch the sun.

ALICE WALKER
b. 1944

Alice Walker's fame rests primarily on her novels, one of which, The
Color Purple, *won the Pulitzer Prize in 1983. She has also created
important bodies of work in both poetry and nonfiction, however. In the
latter, Walker's lifetime of political activism often shines through. In
addition to her significant involvement in the civil rights movement and
her continuing advocacy of women's issues, she is a strong proponent of
animal rights. Sympathy is the core of all these convictions—an identification with others and a determination to make their lives better in any
way she can. Such identification is visceral and immediate, rather than
an abstract value. When Walker's gaze meets the eyes of a horse, as in the
selection that follows, she feels the same jolt of connection that Aldo
Leopold found in the green fire of the dying wolf's eyes.*

AM I BLUE?

> "Ain't these tears in these
> eyes tellin' you"?

For about three years my companion and I rented a small house in
the country that stood on the edge of a large meadow that appeared to
run from the end of our deck straight into the mountains. The mountains, however, were quite far away, and between us and them there
was, in fact, a town. It was one of the many pleasant aspects of the
house that you never really were aware of this.

It was a house of many windows, low, wide, nearly floor to ceiling in
the living room, which faced the meadow, and it was from one of these
that I first saw our closest neighbor, a large white horse, cropping grass,

Being in the World: An Environmental Reader for Writers, eds. Scott H. Slovic and
Terrell F. Dixon (New York: MacMillan, 1993).

flipping its mane, and ambling about—not over the entire meadow, which stretched well out of sight of the house, but over the five or so fenced-in acres that were next to the twenty-odd that we had rented. I soon learned that the horse, whose name was Blue, belonged to a man who lived in another town, but was boarded by our neighbors next door. Occasionally, one of the children, usually a stocky teen-ager, but sometimes a much younger girl or boy, could be seen riding Blue. They would appear in the meadow, climb up on his back, ride furiously for ten or fifteen minutes, then get off, slap Blue on the flanks, and not be seen again for a month or more.

There were many apple trees in our yard, and one by the fence that Blue could almost reach. We were soon in the habit of feeding him apples, which he relished, especially because by the middle of summer the meadow grasses—so green and succulent since January—had dried out from lack of rain, and Blue stumbled about munching the dried stalks half-heartedly. Sometimes he would stand very still just by the apple tree, and when one of us came out he would whinny, snort loudly, or stamp the ground. This meant, of course: I want an apple.

It was quite wonderful to pick a few apples, or collect those that had fallen to the ground overnight, and patiently hold them, one by one, up to his large, toothy mouth. I remained as thrilled as a child by his flexible dark lips, huge, cubelike teeth that crunched the apples, core and all, with such finality, and his high, broad-breasted *enormity*; beside which, I felt small indeed. When I was a child, I used to ride horses, and was especially friendly with one named Nan until the day I was riding and my brother deliberately spooked her and I was thrown, head first, against the trunk of a tree. When I came to, I was in bed and my mother was bending worriedly over me; we silently agreed that perhaps horseback riding was not the safest sport for me. Since then I have walked, and prefer walking to horseback riding—but I had forgotten the depth of feeling one could see in horses' eyes.

I was therefore unprepared for the expression in Blue's. Blue was lonely. Blue was horribly lonely and bored. I was not shocked that this should be the case; five acres to tramp by yourself, endlessly, even in the most beautiful of meadows—and his was—cannot provide many interesting events, and once rainy season turned to dry that was about it. No, I was shocked that I had forgotten that human animals and non-human animals can communicate quite well; if we are brought up around animals as children we take this for granted. By the time we are adults we no longer remember. However, the animals have not changed. They are in fact *completed* creations (at least they seem to be, so much more than we) who are not likely *to* change; it is their nature to express themselves. What else are they going to express? And they do. And, generally speaking, they are ignored.

After giving Blue the apples, I would wander back to the house, aware that he was observing me. Were more apples not forthcoming then? Was that to be his sole entertainment for the day? My partner's small son had decided he wanted to learn how to piece a quilt; we worked in silence on our respective squares as I thought. . . .

Well, about slavery: about white children, who were raised by black people, who knew their first all-accepting love from black women, and then, when they were twelve or so, were told they must "forget" the deep levels of communication between themselves and "mammy" that they knew. Later they would be able to relate quite calmly, "My old mammy was sold to another good family." "My old mammy was _____ _____." Fill in the blank. Many more years later a white woman would say: "I can't understand these Negroes, these blacks. What do they want? They're so different from us."

And about the Indians, considered to be "like animals" by the "settlers" (a very benign euphemism for what they actually were), who did not understand their description as a compliment.

And about the thousands of American men who marry Japanese, Korean, Filipina, and other non-English-speaking women and of how happy they report they are, *"blissfully,"* until their brides learn to speak English, at which point the marriages tend to fall apart. What then did the men see, when they looked into the eyes of the women they married, before they could speak English? Apparently only their own reflections.

I thought of society's impatience with the young. "Why are they playing the music so loud?" Perhaps the children have listened to much of the music of oppressed people their parents danced to before they were born, with its passionate but soft cries for acceptance and love, and they have wondered why their parents failed to hear.

I do not know how long Blue had inhabited his five beautiful, boring acres before we moved into our house; a year after we had arrived—and had also traveled to other valleys, other cities, other worlds—he was still there.

But then, in our second year at the house, something happened in Blue's life. One morning, looking out the window at the fog that lay like a ribbon over the meadow, I saw another horse, a brown one, at the other end of Blue's field. Blue appeared to be afraid of it, and for several days made no attempt to go near. We went away for a week. When we returned, Blue had decided to make friends and the two horses ambled or galloped along together, and Blue did not come nearly as often to the fence underneath the apple tree.

When he did, bringing his new friend with him, there was a different look in his eyes. A look of independence, of self-possession, of inalienable *horseness*. His friend eventually became pregnant. For months and months there was, it seemed to me, a mutual feeling between me and the horses of justice, of peace. I fed apples to them both. The look in Blue's eyes was one of unabashed "this is *itness*."

It did not, however, last forever. One day, after a visit to the city, I went out to give Blue some apples. He stood waiting, or so I thought, though not beneath the tree. When I shook the tree and jumped back from the shower of apples, he made no move. I carried some over to him. He managed to half-crunch one. The rest he let fall to the ground. I dreaded looking into his eyes—because I had of course noticed that Brown, his partner, had gone—but I did look. If I had been born into slavery, and my partner had been sold or killed, my eyes would have looked like that. The children next door explained that Blue's partner had been "put with him" (the same expression that old people used, I had noticed, when speaking of an ancestor during slavery who had been impregnated by her owner) so that they could mate and she conceive. Since that was accomplished, she had been taken back by her owner, who lived somewhere else.

Will she be back? I asked.

They didn't know.

Blue was like a crazed person. Blue *was*, to me, a crazed person. He galloped furiously, as if he were being ridden, around and around his five beautiful acres. He whinnied until he couldn't. He tore at the ground with his hooves. He butted himself against his single shade tree. He looked always and always toward the road down which his partner had gone. And then, occasionally, when he came up for apples, or I took apples to him, he looked at me. It was a look so piercing, so full of grief, a look so *human*, I almost laughed (I felt too sad to cry) to think there are people who do not know that animals suffer. People like me who have forgotten, and daily forget, all that animals try to tell us. "Everything you do to us will happen to you; we are your teachers, as you are ours. We are one lesson" is essentially it, I think. There are those who never once have even considered animals' rights: those who have been taught that animals actually want to be used and abused by us, as small children "love" to be frightened, or women "love" to be mutilated and raped. . . . They are the great-grandchildren of those who honestly thought, because someone taught them this: "Women can't think," and "niggers can't faint." But most disturbing of all, in Blue's large brown eyes was a new look, more painful than the look of despair: the look of disgust with human beings, with life; the look of hatred. And it was odd what the look of hatred did. It gave him, for the first time, the look of a beast. And what that meant was that he had put up a barrier within to protect himself from further violence; all the apples in the world wouldn't change that fact.

And so Blue remained, a beautiful part of our landscape, very peaceful to look at from the window, white against the grass. Once a friend came to visit and said, looking out on the soothing view: "And it *would* have to be a *white* horse; the very image of freedom." And I thought, yes, the animals are forced to become for us merely "images" of what they once so beautifully expressed. And we are used to drinking milk from containers showing "contented" cows, whose real lives we want to hear nothing about, eating eggs and drumsticks from "happy" hens, and munching hamburgers advertised by bulls of integrity who seem to command their fate.

As we talked of freedom and justice one day for all, we sat down to steaks. I am eating misery, I thought, as I took the first bite. And spit it out.

ANNIE DILLARD
b. 1945

Annie Dillard possesses one of the most recognizable voices in contemporary American prose. Her energetic and eclectic style, ranging from that of the religious mystic to that of the stand-up comedian, reflects her perception that existence embraces both the sublime and the absurd. Born in Pittsburgh and educated at Hollins College in Virginia, Dillard describes herself as "a poet and a walker with a background in theology and a penchant for quirky facts." Her books, she says, are not consciously about nature so much as "about what it feels like to be alive." Yet Pilgrim at Tinker Creek, *which won the Pulitzer Prize for Non-Fiction in 1974, has become one of the most influential and widely imitated works of contemporary nature writing. Its account of a year in the Roanoke Valley of the Blue Ridge Mountains explores the nature of human consciousness as much as local natural history, and gravitates toward mystery rather than conclusions. In contrast to other contemporary writers like*

Edward Abbey and Wendell Berry, Dillard's intent is not to discover or champion an environmental ethic but to bear witness equally to the beauty and terror of existence. Her other works include Tickets for a Prayer Wheel *(1974),* Teaching a Stone to Talk *(1982),* An American Childhood *(1987),* The Writing Life *(1989), and* For the Time Being *(1999).*

HEAVEN AND EARTH IN JEST

I used to have a cat, an old fighting tom, who would jump through the open window by my bed in the middle of the night and land on my chest. I'd half-awaken. He'd stick his skull under my nose and purr, stinking of urine and blood. Some nights he kneaded my bare chest with his front paws, powerfully, arching his back, as if sharpening his claws, or pummeling a mother for milk. And some mornings I'd wake in daylight to find my body covered with paw prints in blood; I looked as though I'd been painted with roses.

It was hot, so hot the mirror felt warm. I washed before the mirror in a daze, my twisted summer sleep still hung about me like sea kelp. What blood was this, and what roses? It could have been the rose of union, the blood of murder, or the rose of beauty bare and the blood of some unspeakable sacrifice or birth. The sign on my body could have been an emblem or a stain, the keys to the kingdom or the mark of Cain. I never knew. I never knew as I washed, and the blood streaked, faded, and finally disappeared, whether I'd purified myself or ruined the blood sign of the passover. We wake, if we ever wake at all, to mystery, rumors of death, beauty, violence. . . . "Seem like we're just set down here," a woman said to me recently, "and don't nobody know why."

These are morning matters, pictures you dream as the final wave heaves you up on the sand to the bright light and drying air. You remember pressure, and a curved sleep you rested against, soft, like a scallop in its shell. But the air hardens your skin; you stand; you leave the lighted shore to explore some dim headland, and soon you're lost in the leafy interior, intent, remembering nothing.

I still think of that old tomcat, mornings, when I wake. Things are tamer now; I sleep with the window shut. The cat and our rites are

Pilgrim at Tinker Creek (New York: Harper and Row, 1974).

gone and my life is changed, but the memory remains of something powerful playing over me. I wake expectant, hoping to see a new thing. If I'm lucky I might be jogged awake by a strange birdcall. I dress in a hurry, imagining the yard flapping with auks, or flamingos. This morning it was a wood duck, down at the creek. It flew away.

I live by a creek, Tinker Creek, in a valley in Virginia's Blue Ridge. An anchorite's hermitage is called an anchor-hold; some anchor-holds were simple sheds clamped to the side of a church like a barnacle to a rock. I think of this house clamped to the side of Tinker Creek as an anchor-hold. It holds me at anchor to the rock bottom of the creek itself and it keeps me steadied in the current, as a sea anchor does, facing the stream of light pouring down. It's a good place to live; there's a lot to think about. The creeks—Tinker and Carvin's—are an active mystery, fresh every minute. Theirs is the mystery of the continuous creation and all that providence implies: the uncertainty of vision, the horror of the fixed, the dissolution of the present, the intricacy of beauty, the pressure of fecundity, the elusiveness of the free, and the flawed nature of perfection. The mountains—Tinker and Brushy, McAfee's Knob and Dead Man—are a passive mystery, the oldest of all. Theirs is the one simple mystery of creation from nothing, of matter itself, anything at all, the given. Mountains are giant, restful, absorbent. You can heave your spirit into a mountain and the mountain will keep it, folded, and not throw it back as some creeks will. The creeks are the world with all its stimulus and beauty; I live there. But the mountains are home.

The wood duck flew away. I caught only a glimpse of something like a bright torpedo that blasted the leaves where it flew. Back at the house I ate a bowl of oatmeal; much later in the day came the long slant of light that means good walking.

If the day is fine, any walk will do; it all looks good. Water in particular looks its best, reflecting blue sky in the flat, and chopping it into graveled shallows and white chute and foam in the riffles. On a dark day, or a hazy one, everything's washed-out and lackluster but the water. It carries its own lights. I set out for the railroad tracks, for the hill the flocks fly over, for the woods where the white mare lives. But I go to the water.

Today is one of those excellent January partly cloudies in which light chooses an unexpected part of the landscape to trick out in gilt, and then shadow sweeps it away. You know you're alive. You take huge steps, trying to feel the planet's roundness arc between your feet. Kazantzakis says that when he was young he had a canary and a globe. When he freed the canary, it would perch on the globe and sing. All his life, wandering the earth, he felt as though he had a canary on top of his mind, singing.

West of the house, Tinker Creek makes a sharp loop, so that the creek is both in back of the house, south of me, and also on the other side of the road, north of me. I like to go north. There the afternoon sun hits the creek just right, deepening the reflected blue and lighting the sides of trees on the banks. Steers from the pasture across the creek come down to drink; I always flush a rabbit or two there; I sit on a fallen trunk in the shade and watch the squirrels in the sun. There are two separated wooden fences suspended from cables that cross the creek just upstream from my tree-trunk bench. They keep the steers from escaping up or down the creek when they come to drink. Squirrels, the neighborhood children, and I use the downstream fence as a swaying bridge across the creek. But the steers are there today.

I sit on the downed tree and watch the black steers slip on the creek bottom. They are all bred beef: beef heart, beef hide, beef hocks. They're a human product like rayon. They're like a field of shoes. They have cast-iron shanks and tongues like foam insoles. You can't see through to their brains as you can with other animals; they have beef fat behind their eyes, beef stew.

I cross the fence six feet above the water, walking my hands down the rusty cable and tightroping my feet along the narrow edge of the planks. When I hit the other bank and terra firma, some steers are bunched in a knot between me and the barbed-wire fence I want to cross. So I suddenly rush at them in an enthusiastic sprint, flailing my arms and hollering, "Lightning! Copperhead! Swedish meatballs!" They flee, still in a knot, stumbling across the flat pasture. I stand with the wind on my face.

When I slide under a barbed-wire fence, cross a field, and run over a sycamore trunk felled across the water, I'm on a little island shaped like a tear in the middle of Tinker Creek. On one side of the creek is a steep forested bank; the water is swift and deep on that side of the island. On the other side is the level field I walked through next to the steers' pasture; the water between the field and the island is shallow and sluggish. In summer's low water, flags and bulrushes grow along a series of shallow pools cooled by the lazy current. Water striders patrol the surface film, crayfish hump along the silt bottom eating filth, frogs shout and glare, and shiners and small bream hide among roots from the sulky green heron's eye. I come to this island every month of the year. I walk around it, stopping and staring, or I straddle the sycamore log over the creek, curling my legs out of the water in winter, trying to read. Today I sit on dry grass at the end of the island by the slower side of the creek.

I'm drawn to this spot. I come to it as to an oracle; I return to it as a man years later will seek out the battlefield where he lost a leg or an arm.

A couple of summers ago I was walking along the edge of the island to see what I could see in the water, and mainly to scare frogs. Frogs have an inelegant way of taking off from invisible positions on the bank just ahead of your feet, in dire panic, emitting a froggy "Yike!" and splashing into the water. Incredibly, this amused me, and, incredibly, it amuses me still. As I walked along the grassy edge of the island, I got better and better at seeing frogs both in and out of the water. I learned to recognize, slowing down, the difference in texture of the light reflected from mudbank, water, grass, or frog. Frogs were flying all around me. At the end of the island I noticed a small green frog. He was exactly half in and half out of the water, looking like a schematic diagram of an amphibian, and he didn't jump.

He didn't jump; I crept closer. At last I knelt on the island's winter-killed grass, lost, dumbstruck, staring at the frog in the creek just four feet away. He was a very small frog with wide, dull eyes. And just as I looked at him, he slowly crumpled and began to sag. The spirit vanished from his eyes as if snuffed. His skin emptied and drooped; his very skull seemed to collapse and settle like a kicked tent. He was shrinking before my eyes like a deflating football. I watched the taut, glistening skin on his shoulders ruck, and rumple, and fall. Soon, part of his skin, formless as a pricked balloon, lay in floating folds like bright scum on top of the water: it was a monstrous and terrifying thing. I gaped bewildered, appalled. An oval shadow hung in the water behind the drained frog; then the shadow glided away. The frog skin bag started to sink.

I had read about the giant water bug, but never seen one. "Giant water bug" is really the name of the creature, which is an enormous, heavy-bodied brown beetle. It eats insects, tadpoles, fish, and frogs. Its grasping forelegs are mighty and hooked inward. It seizes a victim with these legs, hugs it tight, and paralyzes it with enzymes injected during a vicious bite. That one bite is the only bite it ever takes. Through the puncture shoots the poisons that dissolve the victim's muscles and bones and organs—all but the skin—and through it the giant water bug sucks out the victim's body, reduced to a juice. This event is quite common in warm fresh water. The frog I saw was being sucked by a giant water bug. I had been kneeling on the island grass; when the unrecognizable flap of frog skin settled on the creek bottom, swaying, I stood up and brushed the knees of my pants. I couldn't catch my breath.

Of course, many carnivorous animals devour their prey alive. The usual method seems to be to subdue the victim by downing or grasping it so it can't flee, then eating it whole or in a series of bloody bites. Frogs eat everything whole, stuffing prey into their mouths with their thumbs. People have seen frogs with their wide jaws so full of live dragonflies they couldn't close them. Ants don't even have to catch their prey: in the spring they swarm over newly hatched, featherless birds in the nest and eat them tiny bite by bite.

That it's rough out there and chancy is no surprise. Every live thing is a survivor on a kind of extended emergency bivouac. But at the same time we are also created. In the Koran, Allah asks, "The heaven and the earth and all in between, thinkest thou I made them *in jest?*" It's a good question. What do we think of the created universe, spanning an unthinkable void with an unthinkable profusion of forms? Or what do we think of nothingness, those sickening reaches of time in either direction? If the giant water bug was not made in jest, was it then made in earnest? Pascal uses a nice term to describe the notion of the creator's, once having called forth the universe, turning his back to it: *Deus Absconditus.* Is this what we think happened? Was the sense of it there, and God absconded with it, ate it, like a wolf who disappears round the edge of the house with the Thanksgiving turkey? "God is subtle," Einstein said, "but not malicious." Again, Einstein said that "nature conceals her mystery by means of her essential grandeur, not by her cunning." It could be that God has not absconded but spread, as our vision and understanding of the universe have spread, to a fabric of spirit and sense so grand and subtle, so powerful in a new way, that we can only feel blindly of its hem. In making the thick darkness a swaddling band for the sea, God "set bars and doors" and said, "Hitherto shalt thou come, but no further." But have we come even that far? Have we rowed out to the thick darkness, or are we all playing pinochle in the bottom of the boat?

Cruelty is a mystery, and the waste of pain. But if we describe a world to compass these things, a world that is a long, brute game, then we bump against another mystery: the inrush of power and light, the canary that sings on the skull. Unless all ages and races of men have been deluded by the same mass hypnotist (who?), there seems to be such a thing as beauty, a grace wholly gratuitous. About five years ago I saw a mockingbird make a straight vertical descent from the roof gutter of a four-story building. It was an act as careless and spontaneous as the curl of a stem or the kindling of a star.

The mockingbird took a single step into the air and dropped. His wings were still folded against his sides as though he were singing from a limb and not falling, accelerating thirty-two feet per second per second, through empty air. Just a breath before he would have been dashed to the ground, he unfurled his wings with exact, deliberate care, revealing the broad bars of white, spread his elegant, white-banded tail, and so floated onto the grass. I had just rounded a corner when his insouciant step caught my eye; there was no one else in sight. The fact of his free fall was like the old philosophical conundrum about the tree that falls in the forest. The answer must be, I think, that beauty and grace are performed whether or not we will or sense them. The least we can do is try to be there.

Another time I saw another wonder: sharks off the Atlantic coast of Florida. There is a way a wave rises above the ocean horizon, a triangular wedge against the sky. If you stand where the ocean breaks on a shallow beach, you see the raised water in a wave is translucent, shot with lights. One late afternoon at low tide a hundred big sharks passed the beach near the mouth of a tidal river in a feeding frenzy. As each green wave rose from the churning water, it illuminated within itself the six- or eight-foot-long bodies of twisting sharks. The sharks disappeared as each wave rolled toward me; then a new wave would swell above the horizon, containing in it, like scorpions in amber, sharks that roiled and heaved. The sight held awesome wonders: power and beauty, grace tangled in a rapture with violence.

We don't know what's going on here. If these tremendous events are random combinations of matter run amok, the yield of millions of monkeys at millions of typewriters, then what is it in us, hammered out of those same typewriters, that they ignite? We don't know. Our life is a faint tracing on the surface of mystery, like the idle, curved tunnels of leaf miners on the face of a leaf. We must somehow take a wider view, look at the whole landscape, really see it, and describe what's going on here. Then we can at least wail the right question into the swaddling band of darkness, or, if it comes to that, choir the proper praise.

At the time of Lewis and Clark, setting the prairies on fire was a well-known signal that meant, "Come down to the water." It was an extravagant gesture, but we can't do less. If the landscape reveals one certainty, it is that the extravagant gesture is the very stuff of creation. After the one extravagant gesture of creation in the first place, the universe has continued to deal exclusively in extravagances, flinging intricacies and colossi down aeons of emptiness, heaping profusions on profligacies with ever-fresh vigor. The whole show has been on fire from the word go. I come down to the water to cool my eyes. But everywhere I look I see fire; that which isn't flint is tinder, and the whole world sparks and flames.

I have come to the grassy island late in the day. The creek is up; icy water sweeps under the sycamore log bridge. The frog skin, of course, is utterly gone. I have stared at that one spot on the creek bottom for so

long, focusing past the rush of water, that when I stand, the opposite
bank seems to stretch before my eyes and flow grassily upstream. When
the bank settles down I cross the sycamore log and enter again the big
plowed field next to the steers' pasture.

The wind is terrific out of the west; the sun comes and goes. I can
see the shadow on the field before me deepen uniformly and spread
like a plague. Everything seems so dull I am amazed I can even distinguish objects. And suddenly the light runs across the land like a
comber, and up the trees, and goes again in a wink: I think I've gone
blind or died. When it comes again, the light, you hold your breath,
and if it stays you forget about it until it goes again.

It's the most beautiful day of the year. At four o'clock the eastern sky
is a dead stratus black flecked with low white clouds. The sun in the
west illuminates the ground, the mountains, and especially the bare
branches of trees, so that everywhere silver trees cut into the black sky
like a photographer's negative of a landscape. The air and the ground
are dry; the mountains are going on and off like neon signs. Clouds
slide east as if pulled from the horizon, like a tablecloth whipped off a
table. The hemlocks by the barbed-wire fence are flinging themselves
east as though their backs would break. Purple shadows are racing east;
the wind makes me face east, and again I feel the dizzying, drawn sensation I felt when the creek bank reeled.

At four-thirty the sky in the east is clear; how could that big blackness
be blown? Fifteen minutes later another darkness is coming overhead
from the northwest; and it's here. Everything is drained of its light as if
sucked. Only at the horizon do inky black mountains give way to distant,
lighted mountains—lighted not by direct illumination but rather paled
by glowing sheets of mist hung before them. Now the blackness is in the
east; everything is half in shadow, half in sun, every clod, tree, mountain,
and hedge. I can't see Tinker Mountain through the line of hemlock, till
it comes on like a streetlight, ping, *ex nihilo*. Its sandstone cliffs pink and
swell. Suddenly the light goes; the cliffs recede as if pushed. The sun hits
a clump of sycamores between me and the mountains; the sycamore
arms light up, and *I can't see the cliffs*. They're gone. The pale network of
sycamore arms, which a second ago was transparent as a screen, is suddenly opaque, glowing with light. Now the sycamore arms snuff out, the
mountains come on, and there are the cliffs again.

I walk home. By five-thirty the show has pulled out. Nothing is left
but an unreal blue and a few banked clouds low in the north. Some
sort of carnival magician has been here, some fast-talking worker of
wonders who has the act backwards. "Something in this hand," he
says, "something in this hand, something up my sleeve, something

behind my back . . ." and abracadabra, he snaps his fingers, and it's all gone. Only the bland, blank-faced magician remains, in his unruffled coat, bare-handed, acknowledging a smattering of baffled applause. When you look again the whole show has pulled up stakes and moved on down the road. It never stops. New shows roll in from over the mountains and the magician reappears unannounced from a fold in the curtain you never dreamed was an opening. Scarves of clouds, rabbits in plain view, disappear into the black hat forever. Presto chango. The audience, if there is an audience at all, is dizzy from head-turning, dazed.

Like the bear who went over the mountain, I went out to see what I could see. And, I might as well warn you, like the bear, all that I could see was the other side of the mountain: more of same. On a good day I might catch a glimpse of another wooded ridge rolling under the sun like water, another bivouac. I propose to keep here what Thoreau called "a meteorological journal of the mind," telling some tales and describing some of the sights of this rather tamed valley, and exploring, in fear and trembling, some of the unmapped dim reaches and unholy fastnesses to which those tales and sights so dizzyingly lead.

I am no scientist. I explore the neighborhood. An infant who has just learned to hold his head up has a frank and forthright way of gazing about him in bewilderment. He hasn't the faintest clue where he is, and he aims to learn. In a couple of years, what he will have learned instead is how to fake it: he'll have the cocksure air of a squatter who has come to feel he owns the place. Some unwonted, taught pride diverts us from our original intent, which is to explore the neighborhood, view the landscape, to discover at least *where* it is that we have been so startlingly set down, if we can't learn why.

So I think about the valley. It is my leisure as well as my work, a game. It is a fierce game I have joined because it is being played anyway, a game of both skill and chance, played against an unseen adversary—the conditions of time—in which the payoffs, which may suddenly arrive in a blast of light at any moment, might as well come to me as anyone else. I stake the time I'm grateful to have, the energies I'm glad to direct. I risk getting stuck on the board, so to speak, unable to move in any direction, which happens enough, God knows; and I risk the searing, exhausting nightmares that plunder rest and force me face down all night long in some muddy ditch seething with hatching insects and crustaceans.

But if I can bear the nights, the days are a pleasure. I walk out; I see something, some event that would otherwise have been utterly missed and lost; or something sees me, some enormous power brushes me with its clean wing, and I resound like a beaten bell.

I am an explorer, then, and I am also a stalker, or the instrument of the hunt itself. Certain Indians used to carve long grooves along the wooden shafts of their arrows. They called the grooves "lightning marks," because they resembled the curved fissure lightning slices down the trunks of trees. The function of lightning marks is this: if the arrow fails to kill the game, blood from a deep wound will channel along the lightning mark, streak down the arrow shaft, and spatter to the ground, laying a trail dripped on broad-leaves, on stones, that the barefoot and trembling archer can follow into whatever deep or rare wilderness it leads. I am the arrow shaft, carved along my length by unexpected lights and gashes from the very sky, and this book is the straying trail of blood.

Something pummels us, something barely sheathed. Power broods and lights. We're played on like a pipe; our breath is not our own. James Houston describes two young Eskimo girls sitting cross-legged on the ground, mouth on mouth, blowing by turns each other's throat cords, making a low, unearthly music. When I cross again the bridge that is really the steers' fence, the wind has thinned to the delicate air of twilight; it crumples the water's skin. I watch the running sheets of light raised on the creek's surface. The sight has the appeal of the purely passive, like the racing of light under clouds on a field, the beautiful dream at the moment of being dreamed. The breeze is the merest puff, but you yourself sail headlong and breathless under the gale force of the spirit. . . .

LIVING LIKE WEASELS

A weasel is wild. Who knows what he thinks? He sleeps in his underground den, his tail draped over his nose. Sometimes he lives in his den for two days without leaving. Outside, he stalks rabbits, mice, muskrats, and birds, killing more bodies than he can eat warm, and often dragging the carcasses home. Obedient to instinct, he bites his prey at the neck, either splitting the jugular vein at the throat or crunching the brain at the base of the skull, and he does not let go. One naturalist refused to kill a weasel who was socketed into his hand deeply as a rattlesnake. The man could in no way pry the tiny weasel

Teaching a Stone to Talk: Expeditions and Encounters (New York: Harper and Row, 1982).

off, and he had to walk half a mile to water, the weasel dangling from his palm, and soak him off like a stubborn label.

And once, says Ernest Thompson Seton—once, a man shot an eagle out of the sky. He examined the eagle and found the dry skull of a weasel fixed by the jaws to his throat. The supposition is that the eagle had pounced on the weasel and the weasel swiveled and bit as instinct taught him, tooth to neck, and nearly won. I would like to have seen that eagle from the air a few weeks or months before he was shot: was the whole weasel still attached to his feathered throat, a fur pendant? Or did the eagle eat what he could reach, gutting the living weasel with his talons before his breast, bending his beak, cleaning the beautiful airborne bones?

I have been reading about weasels because I saw one last week. I startled a weasel who startled me, and we exchanged a long glance.

Twenty minutes from my house, through the woods by the quarry and across the highway, is Hollins Pond, a remarkable piece of shallowness, where I like to go at sunset and sit on a tree trunk. Hollins Pond is also called Murray's Pond; it covers two acres of bottomland near Tinker Creek with six inches of water and six thousand lily pads. In winter, brown-and-white steers stand in the middle of it, merely dampening their hooves; from the distant shore they look like miracle itself, complete with miracle's nonchalance. Now, in summer, the steers are gone. The water lilies have blossomed and spread to a green horizontal plane that is terra firma to plodding blackbirds, and tremulous ceiling to black leeches, crayfish, and carp.

This is, mind you, suburbia. It is a five-minute walk in three directions to rows of houses, though none is visible here. There's a 55 mph highway at one end of the pond, and a nesting pair of wood ducks at the other. Under every bush is a muskrat hole or a beer can. The far end is an alternating series of fields and woods, fields and woods, threaded everywhere with motorcycle tracks—in whose bare clay wild turtles lay eggs.

So. I had crossed the highway, stepped over two low barbed-wire fences, and traced the motorcycle path in all gratitude through the wild rose and poison ivy of the pond's shoreline up into high grassy fields. Then I cut down through the woods to the mossy fallen tree where I sit. This tree is excellent. It makes a dry, upholstered bench at the upper, marshy end of the pond, a plush jetty raised from the thorny shore between a shallow blue body of water and a deep blue body of sky.

The sun had just set. I was relaxed on the tree trunk, ensconced in the lap of lichen, watching the lily pads at my feet tremble and part dreamily over the thrusting path of a carp. A yellow bird appeared to my right and flew behind me. It caught my eye; I swiveled around—and the next instant, inexplicably, I was looking down at a weasel, who was looking up at me.

Weasel! I'd never seen one wild before. He was ten inches long, thin as a curve, a muscled ribbon, brown as fruitwood, soft-furred, alert. His face was fierce, small and pointed as a lizard's; he would have made a good arrowhead. There was just a dot of chin, maybe two brown hairs' worth, and then the pure white fur began that spread down his underside. He had two black eyes I didn't see, any more than you see a window.

The weasel was stunned into stillness as he was emerging from beneath an enormous shaggy wild rose bush four feet away. I was stunned into stillness twisted backward on the tree trunk. Our eyes locked, and someone threw away the key.

Our look was as if two lovers, or deadly enemies, met unexpectedly on an overgrown path when each had been thinking of something else: a clearing blow to the gut. It was also a bright blow to the brain, or a sudden beating of brains, with all the charge and intimate grate of rubbed balloons. It emptied our lungs. It felled the forest, moved the fields, and drained the pond; the world dismantled and tumbled into that black hole of eyes. If you and I looked at each other that way, our skulls would split and drop to our shoulders. But we don't. We keep our skulls. So.

He disappeared. This was only last week, and already I don't remember what shattered the enchantment. I think I blinked, I think I retrieved my brain from the weasel's brain, and tried to memorize what I was seeing, and the weasel felt the yank of separation, the careening splashdown into real life and the urgent current of instinct. He vanished under the wild rose. I waited motionless, my mind suddenly full of data and my spirit with pleadings, but he didn't return.

Please do not tell me about "approach-avoidance conflicts." I tell you I've been in that weasel's brain for sixty seconds, and he was in mine. Brains are private places, muttering through unique and secret tapes—but the weasel and I both plugged into another tape simultaneously, for a sweet and shocking time. Can I help it if it was a blank?

What goes on in his brain the rest of the time? What does a weasel think about? He won't say. His journal is tracks in clay, a spray of feathers, mouse blood and bone: uncollected, unconnected, loose-leaf, and blown.

I would like to learn, or remember, how to live. I come to Hollins Pond not so much to learn how to live as, frankly, to forget about it. That is, I don't think I can learn from a wild animal how to live in particular—shall I suck warm blood, hold my tail high, walk with my footprints precisely over the prints of my hands?—but I might learn something of mindlessness, something of the purity of living in the physical senses and the dignity of living without bias or motive. The weasel lives in necessity and we live in choice, hating necessity and dying at the last ignobly in its talons. I would like to live as I should, as the weasel lives as he should. And I suspect that for me the way is like the weasel's: open to time and death painlessly, noticing everything, remembering nothing, choosing the given with a fierce and pointed will.

I missed my chance. I should have gone for the throat. I should have lunged for that streak of white under the weasel's chin and held on, held on through mud and into the wild rose, held on for a dearer life. We could live under the wild rose wild as weasels, mute and uncomprehending. I could very calmly go wild. I could live two days in the den, curled, leaning on mouse fur, sniffing bird bones, blinking, licking, breathing musk, my hair tangled in the roots of grasses. Down is a good place to go, where the mind is single. Down is out, out of your ever-loving mind and back to your careless senses. I remember muteness as a prolonged and giddy fast, where every moment is a feast of utterance received. Time and events are merely poured, unremarked, and ingested directly, like blood pulsed into my gut through a jugular vein. Could two live that way? Could two live under the wild rose, and explore by the pond, so that the smooth mind of each is as everywhere present to the other, and as received and as unchallenged, as falling snow?

We could, you know. We can live any way we want. People take vows of poverty, chastity, and obedience—even of silence—by choice. The thing is to stalk your calling in a certain skilled and supple way, to locate the most tender and live spot and plug into that pulse. This is yielding, not fighting. A weasel doesn't "attack" anything; a weasel lives as he's meant to, yielding at every moment to the perfect freedom of single necessity.

I think it would be well, and proper, and obedient, and pure, to grasp your one necessity and not let it go, to dangle from it limp wherever it takes you. Then even death, where you're going no matter how you live, cannot you part. Seize it and let it seize you up aloft even, till your eyes burn out and drop; let your musky flesh fall off in shreds, and let your very bones unhinge and scatter, loosened over fields, over fields and woods, lightly, thoughtless, from any height at all, from as high as eagles.

TOTAL ECLIPSE

I

It had been like dying, that sliding down the mountain pass. It had been like the death of someone, irrational, that sliding down the mountain pass and into the region of dread. It was like slipping into fever, or falling down that hole in sleep from which you wake yourself whimpering. We had crossed the mountains that day, and now we were in a strange place—a hotel in central Washington, in a town near Yakima. The eclipse we had traveled here to see would occur early the next morning.

I lay in bed. My husband, Gary, was reading beside me. I lay in bed and looked at the painting on the hotel room wall. It was a print of a detailed and lifelike painting of a smiling clown's head, made out of vegetables. It was a painting of the sort which you do not intend to look at, and which, alas, you never forget. Some tasteless fate presses it upon you; it becomes part of the complex interior junk you carry with you wherever you go. Two years have passed since the total eclipse of which I write. During those years I have forgotten, I assume, a great many things I wanted to remember—but I have not forgotten that clown painting or its lunatic setting in the old hotel.

The clown was bald. Actually, he wore a clown's tight rubber wig, painted white; this stretched over the top of his skull, which was a cabbage. His hair was bunches of baby carrots. Inset in his white clown makeup, and in his cabbage skull, were his small and laughing human eyes. The clown's glance was like the glance of Rembrandt in some of the self-portraits: lively, knowing, deep, and loving. The crinkled shadows around his eyes were string beans. His eyebrows were parsley. Each of his ears was a broad bean. His thin, joyful lips were red chili peppers; between his lips were wet rows of human teeth and a suggestion of a real tongue. The clown print was framed in gilt and glassed.

To put ourselves in the path of the total eclipse, that day we had driven five hours inland from the Washington coast, where we lived. When we tried to cross the Cascades range, an avalanche had blocked the pass.

A slope's worth of snow blocked the road; traffic backed up. Had the avalanche buried any cars that morning? We could not learn. This highway was the only winter road over the mountains. We waited as highway crews bulldozed a passage through the avalanche. With two-by-fours and walls of plyboard, they erected a one-way, roofed tunnel through the avalanche. We drove through the avalanche tunnel, crossed the pass, and descended several thousand feet into central Washington and the broad Yakima valley, about which we knew only that it was orchard country. As we lost altitude, the snows disappeared; our ears popped; the trees changed, and in the trees were strange birds. I watched the landscape innocently, like a fool, like a diver in the rapture of the deep who plays on the bottom while his air runs out.

The hotel lobby was a dark, derelict room, narrow as a corridor, and seemingly without air. We waited on a couch while the manager vanished upstairs to do something unknown to our room. Beside us on an overstuffed chair, absolutely motionless, was a platinum-blond woman in her forties wearing a black silk dress and a strand of pearls. Her long legs were crossed; she supported her head on her fist. At the dim far end of the room, their backs toward us, sat six bald old men in their shirtsleeves, around a loud television. Two of them seemed asleep. They were drunks. "Number six!" cried the man on television, "Number six!"

On the broad lobby desk, lighted and bubbling, was a ten-gallon aquarium containing one large fish; the fish tilted up and down in its water. Against the long opposite wall sang a live canary in its cage. Beneath the cage, among spilled millet seeds on the carpet, were a decorated child's sand bucket and matching sand shovel.

Now the alarm was set for six. I lay awake remembering an article I had read downstairs in the lobby, in an engineering magazine. The article was about gold mining.

In South Africa, in India, and in South Dakota, the gold mines extend so deeply into the earth's crust that they are hot. The rock walls burn the miners' hands. The companies have to air-condition the mines; if the air conditioners break, the miners die. The elevators in the mine shafts run very slowly, down, and up, so the miners' ears will not pop in their skulls. When the miners return to the surface, their faces are deathly pale.

Early the next morning we checked out. It was February 26, 1979, a Monday morning. We would drive out of town, find a hilltop, watch the eclipse, and then drive back over the mountains and home to the coast. How familiar things are here; how adept we are; how smoothly and professionally we check out! I had forgotten the clown's smiling head and the hotel lobby as if they had never existed. Gary put the car in gear and off we went, as off we have gone to a hundred other adventures.

It was before dawn when we found a highway out of town and drove into the unfamiliar countryside. By the growing light we could see a band of cirrostratus clouds in the sky. Later the rising sun would clear these clouds before the eclipse began. We drove at random until we came to a range of unfenced hills. We pulled off the highway, bundled up, and climbed one of these hills.

II

The hill was five hundred feet high. Long winter-killed grass covered it, as high as our knees. We climbed and rested, sweating in the cold; we passed clumps of bundled people on the hillside who were setting up telescopes and fiddling with cameras. The top of the hill stuck up in the middle of the sky. We tightened our scarves and looked around.

East of us rose another hill like ours. Between the hills, far below, was the highway which threaded south into the valley. This was the Yakima valley; I had never seen it before. It is justly famous for its beauty, like every planted valley. It extended south into the horizon, a distant dream of a valley, a Shangri-la. All its hundreds of low, golden slopes bore orchards. Among the orchards were towns, and roads, and plowed and fallow fields. Through the valley wandered a thin, shining river; from the river extended fine, frozen irrigation ditches. Distance blurred and blued the sight, so that the whole valley looked like a thickness or sediment at the bottom of the sky. Directly behind us was more sky, and empty lowlands blued by distance, and Mount Adams. Mount Adams was an enormous, snow-covered volcanic cone rising flat, like so much scenery.

Now the sun was up. We could not see it; but the sky behind the band of clouds was yellow, and, far down the valley, some hillside orchards had lighted up. More people were parking near the highway and climbing the hills. It was the West. All of us rugged individualists were wearing knit caps and blue nylon parkas. People were climbing the nearby hills and setting up shop in clumps among the dead grasses. It looked as though we had all gathered on hilltops to pray for the world on its last day. It looked as though we had all crawled out of spaceships and were preparing to assault the valley below. It looked as though we were scattered on hilltops at dawn to sacrifice virgins, make rain, set stone stelae in a ring. There was no place out of the wind. The straw grasses banged our legs.

Up in the sky where we stood the air was lusterless yellow. To the west the sky was blue. Now the sun cleared the clouds. We cast rough shadows on the blowing grass; freezing, we waved our arms. Near the sun, the sky was bright and colorless. There was nothing to see.

It began with no ado. It was odd that such a well-advertised public event should have no starting gun, no overture, no introductory speaker. I should have known right then that I was out of my depth. Without pause or preamble, silent as orbits, a piece of the sun went away. We looked at it through welders' goggles. A piece of the sun was missing; in its place we saw empty sky.

I had seen a partial eclipse in 1970. A partial eclipse is very interesting. It bears almost no relation to a total eclipse. Seeing a partial eclipse bears the same relation to seeing a total eclipse as kissing a man does to marrying him, or as flying in an airplane does to falling out of an airplane. Although the one experience precedes the other, it in no way prepares you for it. During a partial eclipse the sky does not darken—not even when 94 percent of the sun is hidden. Nor does the sun, seen colorless through protective devices, seem terribly strange. We have all seen a sliver of light in the sky; we have all seen the crescent moon by day. However, during a partial eclipse the air does indeed get cold, precisely as if someone were standing between you and the fire. And blackbirds do fly back to their roosts. I had seen a partial eclipse before, and here was another.

What you see in an eclipse is entirely different from what you know. It is especially different for those of us whose grasp of astronomy is so frail that, given a flashlight, a grapefruit, two oranges, and fifteen years, we still could not figure out which way to set the clocks for Daylight Saving Time. Usually it is a bit of a trick to keep your knowledge from blinding you. But during an eclipse it is easy. What you see is much more convincing than any wild-eyed theory you may know.

You may read that the moon has something to do with eclipses. I have never seen the moon yet. You do not see the moon. So near the sun, it is as completely invisible as the stars are by day. What you see before your eyes is the sun going through phases. It gets narrower and narrower, as the waning moon does, and, like the ordinary moon, it travels alone in the simple sky. The sky is of course background. It does not appear to eat the sun; it is far behind the sun. The sun simply shaves away; gradually, you see less sun and more sky.

The sky's blue was deepening, but there was no darkness. The sun was a wide crescent, like a segment of tangerine. The wind freshened and blew steadily over the hill. The eastern hill across the highway grew dusky and sharp. The towns and orchards in the valley to the south were dissolving into the blue light. Only the thin river held a trickle of sun.

Now the sky to the west deepened to indigo, a color never seen. A dark sky usually loses color. This was a saturated, deep indigo, up in the air. Stuck up into that unwordly sky was the cone of Mount Adams, and the alpenglow was upon it. The alpenglow is that red light of sunset which holds out on snowy mountaintops long after the valleys and tablelands are dimmed. "Look at Mount Adams," I said, and that was the last sane moment I remember.

I turned back to the sun. It was going. The sun was going, and the world was wrong. The grasses were wrong; they were platinum. Their every detail of stem, head, and blade shone lightless and artificially distinct as an art photographer's platinum print. This color has never been seen on earth. The hues were metallic; their finish was matte. The hillside was a nineteenth-century tinted photograph from which the tints had faded. All the people you see in the photograph, distinct and detailed as their faces look, are now dead. The sky was navy blue. My hands were silver. All the distant hills' grasses were finespun metal which the wind laid down. I was watching a faded color print of a movie filmed in the Middle Ages; I was standing in it, by some mistake. I was standing in a movie of hillside grasses filmed in the Middle Ages. I missed my own century, the people I knew, and the real light of day.

I looked at Gary. He was in the film. Everything was lost. He was a platinum print, a dead artist's version of life. I saw on his skull the darkness of night mixed with the colors of day. My mind was going out; my eyes were receding the way galaxies recede to the rim of space. Gary was light-years away, gesturing inside a circle of darkness, down the wrong end of a telescope. He smiled as if he saw me; the stringy crinkles around his eyes moved. The sight of him, familiar and wrong, was something I was remembering from centuries hence, from the other side of death: yes, *that* is the way he used to look, when we were living. When it was our generation's turn to be alive. I could not hear him; the wind was too loud. Behind him the sun was going. We had all started down a chute of time. At first it was pleasant; now there was no stopping it. Gary was chuting away across space, moving and talking and catching my eye, chuting down the long corridor of separation. The skin on his face moved like thin bronze plating that would peel.

The grass at our feet was wild barley. It was the wild einkorn wheat which grew on the hilly flanks of the Zagros Mountains, above the

Euphrates valley, above the valley of the river we called *River.* We harvested the grass with stone sickles, I remember. We found the grasses on the hillsides; we built our shelter beside them and cut them down. That is how he used to look then, that one, moving and living and catching my eye, with the sky so dark behind him, and the wind blowing. God save our life.

From all the hills came screams. A piece of sky beside the crescent sun was detaching. It was a loosened circle of evening sky, suddenly lighted from the back. It was an abrupt black body out of nowhere; it was a flat disk; it was almost over the sun. That is when there were screams. At once this disk of sky slid over the sun like a lid. The sky snapped over the sun like a lens cover. The hatch in the brain slammed. Abruptly it was dark night, on the land and in the sky. In the night sky was a tiny ring of light. The hole where the sun belongs is very small. A thin ring of light marked its place. There was no sound. The eyes dried, the arteries drained, the lungs hushed. There was no world. We were the world's dead people rotating and orbiting around and around, embedded in the planet's crust, while the earth rolled down. Our minds were light-years distant, forgetful of almost everything. Only an extraordinary act of will could recall to us our former, living selves and our contexts in matter and time. We had, it seems, loved the planet and loved our lives, but could no longer remember the way of them. We got the light wrong. In the sky was something that should not be there. In the black sky was a ring of light. It was a thin ring, an old, thin silver wedding band, an old, worn ring. It was an old wedding band in the sky, or a morsel of bone. There were stars. It was all over.

III

It is now that the temptation is strongest to leave these regions. We have seen enough; let's go. Why burn our hands any more than we have to? But two years have passed; the price of gold has risen. I return to the same buried alluvial beds and pick through the strata again.

I saw, early in the morning, the sun diminish against a backdrop of sky. I saw a circular piece of that sky appear, suddenly detached, blackened, and backlighted; from nowhere it came and overlapped the sun. It did not look like the moon. It was enormous and black. If I had not read that it was the moon, I could have seen the sight a hundred times and never thought of the moon once. (If, however, I had not read that it was the moon—if, like most of the world's people throughout time, I had simply glanced up and seen this thing—then I doubtless would not have speculated much, but would have, like Emperor Louis of Bavaria in 840, simply died of fright on the spot.) It did not look like a dragon, although it looked more like a dragon than the moon. It looked like a lens cover, or the lid of a pot. It materialized out of thin air—black, and flat, and sliding, outlined in flame.

Seeing this black body was like seeing a mushroom cloud. The heart screeched. The meaning of the sight overwhelmed its fascination. It obliterated meaning itself. If you were to glance out one day and see a row of mushroom clouds rising on the horizon, you would know at once that what you were seeing, remarkable as it was, was intrinsically not worth remarking. No use running to tell anyone. Significant as it was, it did not matter a whit. For what is significance? It is significance for people. No people, no significance. This is all I have to tell you.

In the deeps are the violence and terror of which psychology has warned us. But if you ride these monsters deeper down, if you drop with them farther over the world's rim, you find what our sciences cannot locate or name, the substrate, the ocean or matrix or ether which buoys the rest, which gives goodness its power for good, and evil its power for evil, the unified field: our complex and inexplicable caring for each other, and for our life together here. This is given. It is not learned.

The world which lay under darkness and stillness following the closing of the lid was not the world we know. The event was over. Its devastation lay round about us. The clamoring mind and heart stilled, almost indifferent, certainly disembodied, frail, and exhausted. The hills were hushed, obliterated. Up in the sky, like a crater from some distant cataclysm, was a hollow ring.

You have seen photographs of the sun taken during a total eclipse. The corona fills the print. All of those photographs were taken through telescopes. The lenses of telescopes and cameras can no more cover the breadth and scale of the visual array than language can cover the breadth and simultaneity of internal experience. Lenses enlarge the sight, omit its context, and make of it a pretty and sensible picture, like something on a Christmas card. I assure you, if you send any shepherds a Christmas card on which is printed a three-by-three photograph of the angel of the Lord, the glory of the Lord, and a multitude of the heavenly host, they will not be sore afraid. More fearsome things can come in envelopes. More moving photographs than those of the sun's corona can appear in magazines. But I pray you will never see anything more awful in the sky.

You see the wide world swaddled in darkness; you see a vast breadth of hilly land, and an enormous, distant, blackened valley; you see towns' lights, a river's path, and blurred portions of your hat and scarf;

you see your husband's face looking like an early black-and-white film; and you see a sprawl of black sky and blue sky together, with unfamiliar stars in it, some barely visible bands of cloud, and over there, a small white ring. The ring is as small as one goose in a flock of migrating geese—if you happen to notice a flock of migrating geese. It is one 360th part of the visible sky. The sun we see is less than half the diameter of a dime held at arm's length.

The Crab Nebula, in the constellation Taurus, looks, through binoculars, like a smoke ring. It is a star in the process of exploding. Light from its explosion first reached the earth in 1054; it was a supernova then, and so bright it shone in the daytime. Now it is not so bright, but it is still exploding. It expands at the rate of seventy million miles a day. It is interesting to look through binoculars at something expanding seventy million miles a day. It does not budge. Its apparent size does not increase. Photographs of the Crab Nebula taken fifteen years ago seem identical to photographs of it taken yesterday. Some lichens are similar. Botanists have measured some ordinary lichens twice, at fifty-year intervals, without detecting any growth at all. And yet their cells divide; they live.

The small ring of light was like these things—like a ridiculous lichen up in the sky, like a perfectly still explosion 4,200 light-years away: it was interesting, and lovely, and in witless motion, and it had nothing to do with anything.

It had nothing to do with anything. The sun was too small, and too cold, and too far away, to keep the world alive. The white ring was not enough. It was feeble and worthless. It was as useless as a memory; it was as off kilter and hollow and wretched as a memory.

When you try your hardest to recall someone's face, or the look of a place, you see in your mind's eye some vague and terrible sight such as this. It is dark; it is insubstantial; it is all wrong.

The white ring and the saturated darkness made the earth and the sky look as they must look in the memories of the careless dead. What I saw, what I seemed to be standing in, was all the wrecked light that the memories of the dead could shed upon the living world. We had all died in our boots on the hilltops of Yakima, and were alone in eternity. Empty space stoppered our eyes and mouths; we cared for nothing. We remembered our living days wrong. With great effort we had remembered some sort of circular light in the sky—but only the outline. Oh, and then the orchard trees withered, the ground froze, the glaciers slid down the valleys and overlapped the towns. If there had ever been people on earth, nobody knew it. The dead had forgotten those they had loved. The dead were parted one from the other and could no longer remember the faces and lands they had loved in the light. They seemed to stand on darkened hilltops, looking down.

IV

We teach our children one thing only, as we were taught: to wake up. We teach our children to look alive there, to join by words and activities the life of human culture on the planet's crust. As adults we are almost all adept at waking up. We have so mastered the transition we have forgotten we ever learned it. Yet it is a transition we make a hundred times a day, as, like so many will-less dolphins, we plunge and surface, lapse and emerge. We live half our waking lives and all of our sleeping lives in some private, useless, and insensible waters we never mention or recall. Useless, I say. Valueless, I might add—until some-one hauls their wealth up to the surface and into the wide-awake city, in a form that people can use.

I do not know how we got to the restaurant. Like Roethke, "I take my waking slow." Gradually I seemed more or less alive, and already forgetful. It was now almost nine in the morning. It was the day of a solar eclipse in central Washington, and a fine adventure for everyone. The sky was clear; there was a fresh breeze out of the north.

The restaurant was a roadside place with tables and booths. The other eclipse-watchers were there. From our booth we could see their cars' California license plates, their University of Washington parking stickers. Inside the restaurant we were all eating eggs or waffles; people were fairly shouting and exchanging enthusiasms, like fans after a World Series game. Did you see . . . ? Did you see . . . ? Then some-body said something which knocked me for a loop.

A college student, a boy in a blue parka who carried a Hasselblad, said to us, "Did you see that little white ring? It looked like a Life Saver. It looked like a Life Saver up in the sky."

And so it did. The boy spoke well. He was a walking alarm clock. I myself had at that time no access to such a word. He could write a sen-tence, and I could not. I grabbed that Life Saver and rode it to the sur-face. And I had to laugh. I had been dumbstruck on the Euphrates River, I had been dead and gone and grieving, all over the sight of something which, if you could claw your way up to that level, you would grant looked very much like a Life Saver. It was good to be back among people so clever; it was good to have all the world's words at the mind's disposal, so the mind could begin its task. All those things for which we have no words are lost. The mind—the culture—has two little tools, grammar and lexicon: a decorated sand bucket and a matching shovel. With these we bluster about the continents and do all the world's work. With these we try to save our very lives.

There are a few more things to tell from this level, the level of the restaurant. One is the old joke about breakfast. "It can never be satisfied, the mind, never." Wallace Stevens wrote that, and in the long run he was right. The mind wants to live forever, or to learn a very good reason why not. The mind wants the world to return its love, or its awareness; the mind wants to know all the world, and all eternity, and God. The mind's sidekick, however, will settle for two eggs over easy.

The dear, stupid body is as easily satisfied as a spaniel. And, incredibly, the simple spaniel can lure the brawling mind to its dish. It is everlastingly funny that the proud, metaphysically ambitious, clamoring mind will hush if you give it an egg.

Further: while the mind reels in deep space, while the mind grieves or fears or exults, the workaday senses, in ignorance or idiocy, like so many computer terminals printing out market prices while the world blows up, still transcribe their little data and transmit them to the warehouse in the skull. Later, under the tranquilizing influence of fried eggs, the mind can sort through this data. The restaurant was a halfway house, a decompression chamber. There I remembered a few things more.

The deepest, and most terrifying, was this: I have said that I heard screams. (I have since read that screaming, with hysteria, is a common reaction even to expected total eclipses.) People on all the hillsides, including, I think, myself, screamed when the black body of the moon detached from the sky and rolled over the sun. But something else was happening at that same instant, and it was this, I believe, which made us scream.

The second before the sun went out we saw a wall of dark shadow come speeding at us. We no sooner saw it than it was upon us, like thunder. It roared up the valley. It slammed our hill and knocked us out. It was the monstrous swift shadow cone of the moon. I have since read that this wave of shadow moves 1,800 miles an hour. Language can give no sense of this sort of speed—1,800 miles an hour. It was 195 miles wide. No end was in sight—you saw only the edge. It rolled at you across the land at 1,800 miles an hour, hauling darkness like plague behind it. Seeing it, and knowing it was coming straight for you, was like feeling a slug of anesthetic shoot up your arm. If you think very fast, you may have time to think, "Soon it will hit my brain." You can feel the deadness race up your arm; you can feel the appalling, inhuman speed of your own blood. We saw the wall of shadow coming, and screamed before it hit.

This was the universe about which we have read so much and never before felt: the universe as a clockwork of loose spheres flung at stupefying, unauthorized speeds. How could anything moving so fast not crash, not veer from its orbit amok like a car out of control on a turn?

Less than two minutes later, when the sun emerged, the trailing edge of the shadow cone sped away. It coursed down our hill and raced eastward over the plain, faster than the eye could believe; it swept over the plain and dropped over the planet's rim in a twinkling. It had clobbered us, and now it roared away. We blinked in the light. It was as though an enormous, loping god in the sky had reached down and slapped the earth's face.

Something else, something more ordinary, came back to me along about the third cup of coffee. During the moments of totality, it was so dark that drivers on the highway below turned on their cars' headlights. We could see the highway's route as a strand of lights. It was bumper-to-bumper down there. It was eight-fifteen in the morning, Monday morning, and people were driving into Yakima to work. That it was as dark as night, and eerie as hell, an hour after dawn, apparently meant that in order to *see* to drive to work, people had to use their headlights. Four or five cars pulled off the road. The rest, in a line at least five miles long, drove to town. The highway ran between hills; the people could not have seen any of the eclipsed sun at all. Yakima will have another total eclipse in 2086. Perhaps, in 2086, businesses will give their employees an hour off.

From the restaurant we drove back to the coast. The highway crossing the Cascades range was open. We drove over the mountain like old pros. We joined our places on the planet's thin crust; it held. For the time being, we were home free.

Early that morning at six, when we had checked out, the six bald men were sitting on folding chairs in the dim hotel lobby. The television was on. Most of them were awake. You might drown in your own spittle, God knows, at any time; you might wake up dead in a small hotel, a cabbage head watching TV while snows pile up in the passes, watching TV while the chili peppers smile and the moon passes over the sun and nothing changes and nothing is learned because you have lost your bucket and shovel and no longer care. What if you regain the surface and open your sack and find, instead of treasure, a beast which jumps at you? Or you may not come back at all. The winches may jam, the scaffolding buckle, the air conditioning collapse. You may glance up one day and see by your headlamp the canary keeled over in its cage. You may reach into a cranny for pearls and touch a moray eel. You yank on your rope; it is too late.

Apparently people share a sense of these hazards, for when the total eclipse ended, an odd thing happened.

When the sun appeared as a blinding bead on the ring's side, the eclipse was over. The black lens cover appeared again, backlighted, and slid away. At once the yellow light made the sky blue again; the black lid dissolved and vanished. The real world began there. I remember now: we all hurried away. We were born and bored at a stroke. We rushed down the hill. We found our car; we saw the other people streaming down the hillsides; we joined the highway traffic and drove away.

We never looked back. It was a general vamoose, and an odd one, for when we left the hill, the sun was still partially eclipsed—a sight rare enough, and one which, in itself, we would probably have driven five hours to see. But enough is enough. One turns at last even from glory itself with a sigh of relief. From the depths of mystery, and even from the heights of splendor, we bounce back and hurry for the latitudes of home.

JAN ZITA GROVER
b. 1945

Jan Zita Grover spent eight years as an AIDS worker in San Francisco during the height of the epidemic there. Emotionally exhausted, she moved to a remote cabin in northwestern Wisconsin. Here she witnessed the ravaged "sand counties" earlier described in Aldo Leopold's Sand County Almanac *and began to perceive correspondences between damaged landscapes and damaged bodies. The result was* North Enough: AIDS and Other Clear-Cuts *(1996), a hard-edged and unsentimental*

*look at environmental and social crises. In the following essay, taken
from that book, Grover examines the validity of natural metaphors and
the meaning of such human concepts as health and beauty.*

CUTOVER

It's the sort of logged out, burned over district that makes westward
migration seem like a good idea. A land so used, so brutalized, that to
stay with it, to endure it, must often have seemed like a penance. Miles
of black oak and jack pine, much of it dead and down. Sand roads
lined with scrub. Beaten-down trailers and perpetually unfinished
houses, their composition walls dulling to grey as the seasons pass. Hills
and grades scraped nude, gullies branded into the thin sand soils by
ATVs. A place visited mostly when the bogs freeze over and hunters
from cities to the south—the Twins, Milwaukee, Chicago—spread
across the scratchy hills in search of bear and white-tails. The white
and red pine here is anything but natural, the result of Civilian
Conservation Corps reforestation in the thirties, when much of the
Cutover was replanted to pine as the only practicable solution to the
region's depopulation and failure as farmland.

The Wisconsin Cutover is a profoundly altered land, a profoundly
damaged culture. Logged over two to three times between 1860–1920,
the northern tier of Wisconsin's counties became all but depopulated
by humans and forests alike. Sold in the 1910s and '20s to naive
would-be farmers by the railroad and timber companies that had felled
the forests, the Cutover's soils were too thin, its growing season too
short for farming. By 1921, taxes on a million acres in the Cutover were
delinquent and over 40 percent of the tax deeds remained unsold. By
1927, over 2,500,000 acres were in tax delinquency and 80 percent of tax
deeds were unsold. In 1927, the University of Wisconsin Experiment
Station reported that the total acreage under cultivation in the "reset-
tled" Cutover was only 6 percent.

Up here in the extreme northwestern corner of Wisconsin, on the
pine barrens of Douglas County, there's seemingly a bar for every resident—bars hidden back on sand roads, bars tucked back in the trees.
My neighbor says the impressive ratio of bars to people gives fresh
meaning to the term "Build it, and they will come."

They're out there for a reason.

North Enough: AIDS and Other Clear-Cuts (St. Paul, Minn.: Graywolf, 1997).

The Cutover isn't pristine wilderness. It's the topography of more than a century of relentless abuse and adaptation to that abuse. The long glacial hills sliding away from the road are densely covered in knee-high popple. Beyond lie moraines bereft even of seedlings, a denuded pine barrens of sandy orange soil, piled slash, and the crisscross indicia of earthmovers' treads. Stumps like broken yellow beaver teeth. Only under snow cover is it conventionally beautiful. I could call it damaged, but that would be to emphasize only its scars; what surprises and moves me is the nimbleness and unexpectedness of its recovery.

The sand road to the last cabin the agent shows me winds past pulp tree plantations of jack pine. Turning onto County Road 50, we dip past thickets of oak and popple toward a low, boggy appendix of Crystal Lake. This is no managed landscape. Instead, it bears all the signs of a neglect neither benign nor malign, merely indifferent. Downed pine and oak everywhere. The few small birch choked by anonymous, weedy shrubs. With the exception of several near-dead red pines, not a tree is over 15 feet tall, so the landscape looks dwarfed, ignoble. A dead porcupine lies across the road, its viscera turned out on the tar surface like items at a yard sale. The hole in its abdomen is as smoothly incised as an eye tuck.

We flash past him, my doubts increasing. Up ahead someone has planted idiot strips of young red and white pine. Their soughing beauty makes the jack pine and black oak behind them look even more scrofulous and less North Woods-idyllic.

The northern black oak (*Quercus ellipsoidalis*) of the Great Lakes states isn't meant to stand alone; it's a forest player, inconspicuous and comfortable *en masse*. It doesn't spread like the bur oak of the prairies, luxuriating in wild space. Instead, it reaches, in an apologetic, arthritic way, just high enough to wave its tips at the sun and provide room beneath for browse. Lichens crowd its trunk and branches; its brittle twigs snap easily in wind, returning accommodatingly to the ground. Tree fanciers have nothing good to say about it. Donald Culross Peattie, author of the magisterial *A Natural History of Western and Eastern Trees*, is typical in his dismissal: he calls the black oak "peculiarly unkempt and formless in its winter nakedness," a graceless tree "you will never see it in cultivation . . . for it has no charms to recommend it."

The jack pine, the black oak's companion in northern Wisconsin, grows with equal humility: here "a mere runt as to height and grace, a weed in the opinion of the lumberman, fit for nothing but pulpwood," sniffs Peattie. Farther north, it's a straight, stately tree, but here on

Superior's south shore, jack pine looks self-effacing. It lacks the breath-taking height of reds and whites; as it ages, its branches rise popplelike toward the sky, diminishing its profile. At sixty, a jack pine is ancient, ready to fall; at sixty, a red or white pine is just attaining adulthood. Unlike the reds and whites, with their soft fans of needles in threes and fives, jacks produce blunt, short-bristled clusters and cones that recoil on themselves in tight gnarls. The French-Canadian *voyageurs* regarded the jack as unnatural, as bad luck—a conifer whose cones were mysteriously sealed shut. But jacks are the first conifers to reestablish themselves after a fire. Intense heat melts the resinous glue of their crescent-shaped cones, which open then like blowsy flowers, scattering their seed to the hot winds. The result, write Clifford and Isabel Ahlgren, "is extremely heavy jack pine reproduction, 'thick as hair on a dog's back.'" Jacks are the toughest and most adaptable of northern cone bearers, boreal trees that can thrive on the thinnest sand soils left behind by glaciers. Rangers celebrate jacks' tenacity and homeliness in doggerel—"There, there, lit-tle jack pine, don't you sigh. You'll be a white pine by and by"—but the Cutover's first loggers despised them. Their wood is soft and light, unsuit-able for timber, unworthy of the loggers' art, useful only for pulping.

Above all else, jacks are survivors.

Seeing the cabin for the first time, I know none of this. I am innocent of the temptation to metaphorize every tree, shrub, lichen. I know only that the landscape seems vaguely distressing and ugly, the forest mournful and neglected—not at all what I have imagined and hoped for. Against the hard white April snow, the forest lacks any beauty I can understand. I see trees with scoriatic bark, rheumatoid branches, the torn flags of last season's leaves. I sense that such trees are the result of damage done here, but I am not sympathetic to their homeliness. Like most city-dwellers dreaming of a forest retreat, I am seeking an unblem-ished North Woods of tall, stately trees. I want no part of these scarred veterans or of the opened earth, the trailer camps back in the trees, the sandbanks riven by ATVs. I am eager to move on without wasting any more time.

The cabin lies at the end of a sand road by a bay still opaque with ice. Far out on its horizon, something black and liquid dips and loops, pouring itself into the ice, reappearing in a skivvying line. With binoc-ulars, the dark coil resolves into an otter. The agent brightens at my new show of interest and quickly piles on other points in the property's favor: bald eagles, bobcat, bear.

I hear her as if from a great distance. Already something else com-pletely unexpected is working in me: a slight, not yet traceable intima-tion that this ruined land can be my teacher if only I will agree to become its pupil. What surrounds me I can see now only through the eye of convention, but I sense that through the eye of love and knowledge, I may one day find this place beautiful. I am eager to be schooled, to nurse what twitches of hope, of feeling, I can.

There are dangers in reading landscapes and other cultural artifacts as texts. The meaning of any text greatly exceeds the words used to constitute it: this is what intertextuality is about—the excess of cultural baggage we bring to reading something seemingly circumscribed and specific. The references we bring tend to be from other textual systems—films, music, literature—which for all their differences are still a particular kind of human artifact: symbolic representations of real acts.

Treating landscape as text is a dangerous project because land is not merely a representation. It is also a physical palimpsest of complex human, animal, and geologic acts, most of which are not primarily symbolic but written in flesh and soil and rock. While most landscapes are unquestionably cultural, it doesn't follow that theories devised for analyzing cultural representations are particularly applicable to reading them. The Cutover is a deep cultural landscape: even if I look back no further than the arrival of the first documented Europeans, the *voyageurs*, that still leaves almost four centuries of European, Ojibway, and Dakota actions on the land to account for and interpret. These woodlands have been fretted by the pathways of peoples west and south, then north and east, then south and west again. If the European settlers' arrival and displacement of earlier inhabitants seems to us now somehow more decisive, more tragic, than the Ojibway's displacement of the Dakota, it is partly because it is more recent and better documented *as text*. Anguish is kept alive in writing as well as in landscape by both displacer and displaced, heightened by the cultural differences between victor and vanquished.

Suppose I choose to look deeply at the area surrounding the cabin: what do I call such a search? Is it a textual reading? Because it involves the ways cultures shape land, I might instead call it landscape study or a species of cultural studies, with all that the latter term implies about eclectic methods and intentions. Does it matter at all what I call this project? Well, yes: depending on how I conceive it, certain data, certain methods, suggest themselves. My observations might be turned toward the jack pine's life cycle in one case and the history of European-American logging in another.

I ask myself why I find a landscape this damaged so beautiful, or at any rate so touching. Answering this question brings me to the lip of a personal abyss—the eight years I spent under the brand, the whip of AIDS.

I no longer believe there will be time, and time enough, for every-thing I want to do. That I can control many events. That my culture's standards of beauty and virtue are attainable or even desirable. I know how easy it is to stand outside my own body and watch it strain toward feeling, any feeling, at whatever cost. I've learned to find beauty in places where I never would have searched for or found it before—in an edema-tous face, a lesioned and smelly body, a mind rubbed numb by pain. Pain. A burned-over district. Mortal lessons: the beauty of a ravished landscape. Now middle-aged, I find mortality doubly my possession, keeper and kept.

The diminishment of this landscape mortifies and therefore disci-plines me. Its scars will outlast me, bearing witness for decades beyond my death of the damage done here. Fat-tired ATVs and their helmeted riders lay the land bare, pock and deface it until it runs red and open, as disease has defaced the bodies of my friends. I am learn-ing to love what has been defaced, learning to cherish it for reasons other than easy beauty. I walk after the ATVs, collecting beer cans and plastic leech tubs from the banks of the bass hole, tutoring myself in the difficult art of loving what is superficially ugly. Beauty flashes out unexpectedly. I try not to anticipate its location, try only to trust its imminence.

There are exceptions: curving in a hook southwest of the cabin is a bog that ends in a point forested with a stand of ancient red and white pine, immense and still, grave with age and uninterruption. These pines shelter an eagle couple who wheel over the bog every afternoon. The bog is thick with the improbable feeders who thrive on a peat-acid tea: tamarack, sundews, leatherleaf, pitcher plants, bog rosemary, Labrador tea. The pines beyond the bog on the point survived the felling of millions of their fellows because they were too difficult to log out. Crystal Lake is spring-fed; no rivers to merrily lead away fallen giants. Thus they stand on the point still, one of the few remaining stands of ancient whites in Douglas County.

There's no more lesson in this than in why some people with HIV survive ten, twelve years while others die after three. These things happen. I have learned to be deeply suspicious of metaphor, resist-ant to the pretty conceits I once used to explain pain and disaster. When I gaze south toward the black rampart of the surviving pines, I try to resist reading a moral into it—try to abjure the lessons that spin so readily to mind, like files summoned from a whirling disk. The pines have no more intrinsic meaning than the eagles or me. If I choose to read particular lessons in any of us, I must remember that they are *my* meanings, just as the comforts I drag from friends' deaths — heavy, cold, resistant as wet laundry — are not for or by them but for and by myself.

It's common to associate damaged landscape with open dumps, with suburbs shorn of the forests that preceded them, with prairies plowed under for four-bedroom, three-bath strip malls. There's little evidence of such damage here. Other than hillsides altered by the all too aptly named all-terrain vehicles, pines continue to double over in the northwest winds and blueberries to fruit underfoot. Deer flash like glimpsed dreams across the bog, and in late fall, old beater pickups prowl the sand roads, jammed with galvanized kennels and bawling coon hounds, sound as ancient as the cry of cranes. The woods are reputedly full of bear, the sky is thick with waterfowl, the lakes, so clear and deep, filled with muskie and pike. Oak leaves, ox-blood in fall, flutter against the navy sky, the bay water is black with cold — natural, natural, all so natural. So *what's the problem?* Why my heart-stopping conviction of measureless damage?

If explaining this is hard, it's because landscape presents itself as an epistemological puzzle. Can we understand a landscape by recurring to what it once was? The sentimental response to this would be yes: merely invoke the "preinvasion" or pre-European forest as a measure of what's been lost, and the job's apparently done. But it's not: *which* pre-European forest do we mourn the loss of? Forests in this sand-skinned country succeed each other with a slowness beyond human scale. Only pollen core samples taken from peat bogs provide a scale of change over forests' time sufficient for tracking how this land has responded to human and other alterations. If the Cutover's most visible recent damage was caused by European-American logging, it's also true that lightning-caused fires have altered these northern forests as dramatically, as conclusively, as loggers have. Or wind: in 1977, a 200-mile-an-hour straightline wind flattened miles of forest just south of my cabin as low and ugly as any logging operation ever did. The hills down there are covered now with six-foot popple indistinguishable from what succeeds a clear-cut. In the spin of centuries, forest succession barely registers the damage done by European settlers and loggers. So which forest am I mourning? Whose deaths?

Today on my dawn walk, I see for the first time a small meadow obscured all summer by the deciduous undergrowth along the road. Now I crash through the leafless shrubs to look at it more closely. It's perfectly round, knee deep in frost-stiffened grass. A former beaver pond, silted up, wind- and animal-seeded, moving through the lists of succession on its way to becoming a forest clearing, then a patch of forest. But what kind of forest? Jack pine and black oak? The trees around its edge are birch, suggesting that soil here is deeper, more moist, than in the surrounding pine barrens. So a hardwood thicket, perhaps: a small puzzle for people a hundred years from now, who will wonder at this unexplained ring of deep-soil hardwoods surrounded by dryland jack and oak.

Like that small meadow, the bay outside my door is slowly transmuting, silting up to become a bog. Already it is lapidary with peat eruptions marooning unwary canoeists when the water draws down in midsummer. If the next century is unusually dry and warm, the breakdown of water plants in the bay will accelerate and the bog along the western shore will expand; eventually the bog will dry to meadow. At that point, trees will begin to move in from the edges and form a swamp forest.

Should I call this process damage, should I call it succession? What model—too inappropriate, too human—do I use when I embrace this landscape as altered, imperfect? According to what and to whose sense of time?

I am watching the resident vulture soar on wings like ironing boards, rocking faintly on a thermal. I think about Perry's leg.

Shortly before he died, I got my first look at Perry's dying leg. I'd been uncomfortably aware of it for several weeks—a faintly sweet, overripe smell in the house, an undertone of rot.

He was reluctant to let me change the bandage. "Are you sure you want to do this? It can wait until morning. Are you sure? Are you *very* sure?"

I wanted to do it: It comforted me to think of him going to bed dry and clean when there was so little else I could do for him. I knelt in front of him like a subject before a king and slowly peeled his pant leg away from the soaked bandage. Yards of puke-green gauze, which I unwound and threw into a reeking pile.

How much of the world can I find in something so altered—in a leg no longer smooth, intact, encased in a tan skin but instead burst open, eruptive, returning to orderless matter?

I am very tempted to touch it, to find out what something so formless-looking can possibly feel like. Are there still nerve endings in this mass of dead and sloughing cells? Does it feel, this leg?

Is there a sense in which this leg can be viewed as a creation instead of only an annihilation? Its world is an entropic one, moist, swirling with energy turned on itself, no longer producing orderly structures.

Dermis, epidermis, capillary, vein, artery, ganglion. Instead, hyperbolic replication that guts out needed systems, floods cells, drowning them. The surface looks like deep night sky, dark, light-absorbing, starred with drops of serum winking back my reflection, the room, me kneeling there.

Light gathers up chaos, shapes it. Perry's leg shapes death: here is where it most visibly enters my friend, through this swollen leg. He hauls his death around with him; it comes this way. The leg, or what used to be the leg, midwifes urgent talk. *Talk death*, it urges. *It's present; you can smell it, you can see it.* It creates a faint sweet stench, deep as formalin, as ineradicable, as deeply remembered.

I debride Perry's leg with hydrogen peroxide, much as I would pour soda over a ham or meatloaf. The bubbles wink back at me, catch light; they might be stars wheeling in an unfamiliar galaxy.

Perry talks disparagingly about his leg.

I ask him, "How does it feel, seeing your leg like that?"

"Sometimes I simply can't bear it." As if he can't believe it's his.

"What do you do then?" I ask.

And the curious thing is that as soon as he answers, I forget what he has said. I have tucked away his reply, an unopened valentine, a lost letter. I have tried to remember, have fallen to sleep hoping I'll catch his answer when it bobs unguarded from sleep's deep hole, bursts through the skin of resisting consciousness. But I can't. His reply lies in some shaded place, guarded against memory.

Perhaps he doesn't answer me at all. Soon afterward he says he has developed a high tolerance for pain. And perhaps that is his answer: he has learned to dissociate himself from the slow dying of his body. But that can't be right. Perry hasn't apportioned himself into the comfortingly disengaged blocs of Body and Soul. He knows that the KS festering in his left leg has also laid siege to his lungs, his liver, his esophagus, his soul. None of him is unaffected by what now macerates his leg's flesh. He is turning into something else, rich and strange—a dead organism, human peat. Dear bog.

I rewind his leg's burial sheet.

BARRY LOPEZ
b. 1945

In his books Barry Lopez has looked long and hard at the uses, and mis-uses, to which the human mind has put its natural environment. His early collections of essays—Desert Notes (1976), Giving Birth to Thunder (1978), and River Notes (1979)—contain surrealistic fictional narratives and retellings of western Indian myths, presenting nonwestern, unconventional visions of landscape and animals. Of Wolves and Men (1978), winner of the John Burroughs Medal, is a definitive and fascinating exploration of how these social predators have been portrayed and treated by Native American cultures, European myth, American folklore, and contemporary biologists.

With the publication of Arctic Dreams: Imagination and Desire in A Northern Landscape (1986), Lopez became a leading contemporary spokesman for an ethical revaluation of our ecological behavior. Based on several years of personal research, the book surveys Arctic landscape, wildlife, culture, exploration, and exploitation to demonstrate how we project our dreams and desires on Earth's natural areas, more often than not to our own as well as the environment's detriment. One pervasive theme of the book is how native folklore and myth generally reflect the reality of a landscape more accurately than the history and science of a culture bent on conquest. Yet Lopez does not believe we must abandon our own cultural heritage—a fact which separates him from more absolute critics of Western civilization. In fact the wide-ranging and pen-etrating intellect which informs his writing stems from a loyalty to the best traditions of Western ethics and philosophy. Rather he asserts that we have an urgent need to understand our past behavior toward animals and to "renegotiate the contracts" with them. Above all, his books are elo-quent and informed appeals for tolerance and dignity in our dealings with all of Earth's inhabitants. Among his more recent works is About This Life (1998) from which the final selection here is taken.

From ARCTIC DREAMS

LANCASTER SOUND

* * * The first narwhals I ever saw lived far from here, in Bering Strait. The day I saw them I knew that no element of the earth's natural history had ever before brought me so far, so suddenly. It was as though something from a bestiary had taken shape, a creature strange as a giraffe. It was as if the testimony of someone I had no reason to doubt, yet could not quite believe, a story too farfetched, had been verified at a glance.

I was with a bowhead whale biologist named Don Ljungblad, flying search transects over Bering Sea. It was May, and the first bowheads of spring were slowly working their way north through Bering Strait toward their summer feeding grounds in the Chukchi and Beaufort seas. Each day as we flew these transects we would pass over belukha whale and walrus, ringed, spotted, and ribbon seals, bearded seals, and flocks of birds migrating to Siberia. I know of no other region in North America where animals can be met with in such numbers. Bering Sea itself is probably the richest of all the northern seas, as rich as Chesapeake Bay or the Grand Banks at the time of their discovery. Its bounty of crabs, pollock, cod, sole, herring, clams, and salmon is set down in wild numbers, the rambling digits of guesswork. The numbers of birds and marine mammals feeding here, to a person familiar with anything but the Serengeti or life at the Antarctic convergence, are magical. At the height of migration in the spring, the testament of life in Bering Sea is absolutely stilling in its dimensions.

The two weeks I spent flying with Ljungblad, with so many thousands of creatures moving through the water and the air, were a heady experience. Herds of belukha whale glided in silent shoals beneath transparent sheets of young ice. Squadrons of fast-flying sea ducks flashed beneath us as they banked away. We passed ice floes stained red in a hundred places with the afterbirths of walrus. Staring all day into the bright light reflected from the ice and water, however, and the compression in time of these extraordinary events, left me dazed some evenings.

Aspects of the arctic landscape that had become salient for me — its real and temporal borders; a rare, rich oasis of life surrounded by vast stretches of deserted land; the upending of conventional kinds of time;

Arctic Dreams: Imagination and Desire in a Northern Landscape (New York: Scribners, 1986).

biological vulnerability made poignant by the forgiving light of summer—all of this was evoked over Bering Sea.

The day we saw the narwhals we were flying south, low over Bering
Strait. The ice in Chukchi Sea behind us was so close it did not seem
possible that bowheads could have penetrated this far; but it is good to
check, because they can make headway in ice as heavy as this and they
are able to come a long way north undetected in lighter ice on the
Russian side. I was daydreaming about two bowheads we had seen that
morning. They had been floating side by side in a broad lane of unusually clear water between a shelf of shorefast ice and the pack ice—the
flaw lead. As we passed over, they made a single movement together, a
slow, rolling turn and graceful glide, like figure skaters pushing off,
these 50-ton leviathans. Ljungblad shouted in my earphones:
"Waiting." They were waiting for the ice in the strait to open up.
Ljungblad saw nearly 300 bowheads waiting calmly like this one year,
some on their backs, some with their chins resting on the ice.

The narwhals appeared in the middle of this reverie. Two males, with
ivory tusks spiraling out of their foreheads, the image of the unicorn with
which history has confused them. They were close to the same size and
light-colored, and were lying parallel and motionless in a long, straight
lead in the ice. My eye was drawn to them before my conscious mind, let
alone my voice, could catch up. I stared dumbfounded while someone
else shouted. Not just to see the narwhals, but here, a few miles northwest
of King Island in Bering Sea. In all the years scientists have kept records
for these waters, no one had ever seen a narwhal alive in Bering Sea.
Judging from the heaviness of the ice around them, they must have spent
the winter here.[1] They were either residents, a wondrous thought, or they
had come from the nearest population centers the previous fall, from
waters north of Siberia or from northeastern Canada.

The appearance of these animals was highly provocative. We made
circle after circle above them, until they swam away under the ice and
were gone. Then we looked at each other. Who could say what this
was, really?

Because you have seen something doesn't mean you can explain it.
Differing interpretations will always abound, even when good minds
come to bear. The kernel of indisputable information is a dot in space;
interpretations grow out of the desire to make this point a line, to give it a
direction. The directions in which it can be sent, the uses to which it can

---

[1] The narwhal is not nearly as forceful in the ice as the bowhead. It can break through
only about 6 inches of ice with its head. A bowhead, using its brow or on occasion its more
formidable chin, can break through as much as 18 inches of sea ice. [Lopez's note]

be put by a culturally, professionally, and geographically diverse society, are almost without limit. The possibilities make good scientists chary. In a region like the Arctic, tense with a hunger for wealth, with fears of plunder, interpretation can quickly get beyond a scientist's control. When asked to assess the meaning of a biological event—What were those animals doing out there? Where do they belong?—they hedge. They are sometimes reluctant to elaborate on what they saw, because they cannot say what it means, and they are suspicious of those who say they know. Some even distrust the motives behind the questions.

I think along these lines in this instance because of the animal. No large mammal in the Northern Hemisphere comes as close as the narwhal to having its very existence doubted. For some, the possibility that this creature might actually live in the threatened waters of Bering Sea is portentous, a significant apparition on the eve of an era of disruptive oil exploration there. For others, those with the leases to search for oil and gas in Navarin and Norton basins, the possibility that narwhals may live there is a complicating environmental nuisance. Hardly anyone marvels solely at the fact that on the afternoon of April 16, 1982, five people saw two narwhals in a place so unexpected that they were flabbergasted. They remained speechless, circling over the animals in a state of wonder. In those moments the animals did not have to mean anything at all. *   *   *

MIGRATION

*   *   * I visited Anaktuvuk Pass in 1978 with a friend, a wolf biologist who had made a temporary home there and who was warmly regarded for his tact, his penchant for listening, and his help during an epidemic of flu in the village. We spent several days watching wolves and caribou in nearby valleys and visiting at several homes. The men talked a lot about hunting. The evenings were full of stories. There were moments of silence when someone said something very true, peals of laughter when a man told a story expertly at his own expense. One afternoon we left and traveled far to the west to the headwaters of the Utukok River.

The Alaska Department of Fish and Game had a small field camp on the Utukok, at the edge of a gravel-bar landing strip. Among the biologists there were men and women studying caribou, moose, tundra grizzly, wolverine, and, now that my companion had arrived, wolves. The country around the Utukok and the headwaters of the Kokolik River is a wild and serene landscape in summer. Parts of the Western Arctic caribou herd are drifting over the hills, returning from the calving grounds. The sun is always shining, somewhere in the sky. For a week or more we had very fine, clear weather. Golden eagles circled high over the tundra, hunting. Snowy owls regarded us from a distance from their tussock perches. Short-eared owls, a gyrfalcon. Familiar faces.

A few days after we arrived, my companion and I went south six or seven miles and established a camp from which we could watch a distant wolf den. In that open, rolling country without trees, I had the feeling, sometimes, that nothing was hidden. It was during those days that I went for walks along Ilingnorak Ridge and started visiting ground-nesting birds, and developed the habit of bowing to them out of regard for what was wonderful and mysterious in their lives.

The individual animals we watched tested their surroundings, tried things they had not done before, or that possibly no animal like them had ever done before—revealing their capacity for the new. The preservation of this capacity to adapt is one of the central mysteries of evolution.

We watched wolves hunting caribou, and owls hunting lemmings. Arctic ground squirrel eating *irok*, the mountain sorrel. I thought a great deal about hunting. In 1949, Robert Flaherty told an amazing story, which Edmund Carpenter was later successful in getting published. It was about a man named Comock. In 1902, when he and his family were facing starvation, Comock decided to travel over the sea ice to an island he knew about, where he expected they would be able to find food (a small island off Cape Wolstenholme, at the northern tip of Quebec's Ungava Peninsula). On the journey across, they lost nearly all their belongings—all of Comock's knives, spears, and harpoons, all their skins, their stone lamps, and most of their dogs—when the sea ice suddenly opened one night underneath their camp. They were without hunting implements, without a stone lamp to melt water to drink, without food or extra clothing. Comock had left only one sled, several dogs, his snow knife, with which he could cut snow blocks to build a snow house, and stones to make sparks for a fire.

They ate their dogs. The dogs they kept ate the other dogs, which were killed for them. Comock got his family to the island. He fashioned, from inappropriate materials, new hunting weapons. He created shelter and warmth. He hunted successfully. He reconstructed his entire material culture, almost from scratch, by improvising and, where necessary, inventing. He survived. His family survived. His dogs survived and multiplied.

Over the years they carefully collected rare bits of driftwood and bone until Comock had enough to build the frame for an umiak. They saved bearded-seal skins, from which Comock's wife made a waterproof hull. And one summer day they sailed away, back toward Ungava Peninsula. Robert Flaherty, exploring along the coast, spotted Comock and his family and dogs approaching across the water. When they came close, Flaherty, recognizing the form of an umiak and the cut of Eskimo clothing but, seeing that the materials were strange and improvised, asked the Eskimo who he was. He said his name was Comock. "Where in the world have you come from?" asked Flaherty. "From far away, from big island, from far over there," answered Comock, pointing. Then he smiled and made a joke about how poor the umiak must appear, and his family burst into laughter.

I think of this story because at its heart is the industry and competence, the determination and inventiveness of a human family. And because it is about people who lived resolutely in the heart of every moment they found themselves in, disastrous and sublime.

During those days I spent on Ilingnorak Ridge, I did not know what I know now about hunting; but I had begun to sense the outline of what I would learn in the years ahead with Eskimos and from being introduced, by various people, to situations I could not have easily found my way to alone. The insights I felt during those days had to do with the nature of hunting, with the movement of human beings over the land, and with fear. The thoughts grew out of watching the animals.

The evidence is good that among all northern aboriginal hunting peoples, the hunter saw himself bound up in a sacred relationship with the larger animals he hunted. The relationship was full of responsibilities—to the animals, to himself, and to his family. Among the great and, at this point, perhaps tragic lapses in the study of aboriginal hunting peoples is a lack of comprehension about the role women played in hunting. We can presume, I think, that in the same way the hunter felt bound to the animals he hunted, he felt the contract incomplete and somehow even inappropriate if his wife was not part of it. In no hunting society could a man hunt successfully alone. He depended upon his wife for obvious reasons—for the preparation of food and clothing, companionship, humor, subtle encouragement—and for things we can only speculate about, things of a religious nature, bearing on the mutual obligations and courtesies with which he approached the animals he hunted.

Hunting in my experience—and by hunting I simply mean being out on the land—is a state of mind. All of one's faculties are brought to bear in an effort to become fully incorporated into the landscape. It is more than listening for animals or watching for hoofprints or a shift in the weather. It is more than an analysis of what one *senses*. To hunt means to have the land around you like clothing. To engage in a wordless dialogue with it, one so absorbing that you cease to talk with your human companions. It means to release yourself from rational images of what something "means" and to be concerned only that it "is." And then to recognize that things exist only insofar as they can be related to other things. These relationships—fresh drops of moisture on top of rocks at a river crossing and a raven's distant voice—become patterns. The patterns are always in motion. Suddenly the pattern—which includes physical hunger, a memory of your family, and memories of the valley you are walking through, these particular plants and smells—takes in the caribou. There is a caribou standing in front of you. The release of the arrow or bullet is like a word spoken out loud. It occurs at the periphery of your concentration.

The mind we know in dreaming, a nonrational, nonlinear comprehension of events in which slips in time and space are normal, is, I believe, the conscious working mind of an aboriginal hunter. It is a frame of mind that redefines patience, endurance, and expectation.

The focus of a hunter in a hunting society was not killing animals but attending to the myriad relationships he understood bound him into the world he occupied with them. He tended to those duties carefully because he perceived in them everything he understood about survival. This does not mean, certainly, that every man did this, or that good men did not starve. Or that shamans whose duty it was to intercede with the forces that empowered these relationships weren't occasionally thinking of personal gain or subterfuge. It only means that most men understood how to behave.

A fundamental difference between our culture and Eskimo culture, which can be felt even today in certain situations, is that we have irrevocably separated ourselves from the world that animals occupy. We have turned all animals and elements of the natural world into objects. We manipulate them to serve the complicated ends of our destiny. Eskimos do not grasp this separation easily, and have difficulty imagining themselves entirely removed from the world of animals. For many of them, to make this separation is analogous to cutting oneself off from light or water. It is hard to imagine how to do it.

A second difference is that, because we have objectified animals, we are able to treat them impersonally. This means not only the animals that live around us but animals that live in distant lands. For Eskimos, most relationships with animals are local and personal. The animals one encounters are part of one's community, and one has obligations to them. A most confusing aspect of Western culture for Eskimos to grasp is our depersonalization of relationships with the human and animal members of our communities. And it is compounded, rather than simplified, by their attempting to learn how to objectify animals.

Eskimos do not maintain this intimacy with nature without paying a certain price. When I have thought about the ways in which they differ from people in my own culture, I have realized that they are more afraid than we are. On a day-to-day basis, they have more fear. Not of being dumped into cold water from an umiak, not a debilitating fear. They are afraid because they accept fully what is violent and tragic in nature. It is a fear tied to their knowledge that sudden, cataclysmic events are as much a part of life, of really living, as are the moments when one pauses to look at something beautiful. A Central Eskimo shaman named Aua, queried by Knud Rasmussen about Eskimo beliefs, answered, "We do not believe. We fear."

To extend these thoughts, it is wrong to think of hunting cultures like the Eskimo's as living in perfect harmony or balance with nature. Their regard for animals and their attentiveness to nuance in the landscape were not rigorous or complete enough to approach an idealized harmony. No one knew that much. No one would say they knew that much. They faced nature with fear, with *ilira* (nervous awe) and *kappia* (apprehension). And with enthusiasm. They accepted hunting as a way of life—its violence, too, though they did not seek that out. They were unsentimental, so much so that most outsiders thought them cruel, especially in their treatment of dogs. Nor were they innocent. There is murder and warfare and tribal vendetta in their history; and today, in the same villages I walked out of to hunt, are families shattered by alcohol, drugs, and ambition. While one cannot dismiss culpability in these things, any more than one can hold to romantic notions about hunting, it is good to recall what a *struggle* it is to live with dignity and understanding, with perspicacity or grace, in circumstances far better than these. And it is helpful to imagine how the forces of life must be construed by people who live in a world where swift and fatal violence, like *ivu*, the suddenly leaping shore ice, is inherent in the land. The land, in a certain, very real way, compels the minds of the people.

A good reason to travel with Eskimo hunters in modern times is that, beyond nettlesome details—foods that are not to one's liking, a loss of intellectual conversation, a consistent lack of formal planning—in spite of these things, one feels the constant presence of people who know something about surviving. At their best they are resilient, practical, and enthusiastic. They pay close attention in realms where they feel a capacity for understanding. They have a quality of *nuannaarpoq*, of taking extravagant pleasure in being alive; and they delight in finding it in other people. Facing as we do our various Armageddons, they are a good people to know.

In the time I was in the field with Eskimos I wondered at the basis for my admiration. I admired an awareness in the men of providing for others, and the soft tone of voice they used around bloodshed. I never thought I could understand, from their point of view, that moment of preternaturally heightened awareness, and the peril inherent in taking a life; but I accepted it out of respect for their seriousness toward it. In moments when I felt perplexed, that I was dealing with an order outside my own, I discovered and put to use a part of my own culture's wisdom, the formal divisions of Western philosophy—metaphysics, epistemology, ethics, aesthetics, and logic—which pose, in order, the following questions. What is real? What can we understand? How should we behave? What is beautiful? What are the patterns we can rely upon?

As I traveled, I would say to myself, What do my companions see where I see death? Is the sunlight beautiful to them, the way it sparkles on the water? Which for the Eskimo hunter are the patterns to be trusted? The patterns, I know, could be different from ones I imagined were before us. There could be other, remarkably different insights.

Those days on Ilingnorak Ridge, when I saw tundra grizzly tearing up the earth looking for ground squirrels, and watched wolves hunting, and horned lark sitting so resolutely on her nest, and caribou crossing the river and shaking off the spray like diamonds before the evening sun, I was satisfied only to watch. This was the great drift and pause of life. These were the arrangements that made the land ring with integrity. Somewhere downriver, I remembered, a scientist named Edward Sable had paused on a trek in 1947 to stare at a Folsom spear point, a perfectly fluted object of black chert resting on a sandstone ledge. People, moving over the land. *   *   *

ICE AND LIGHT

*   *   * During the sea-lift passage of the *Soodoc* north through Davis Strait, en route to Little Cornwallis Island, I got in the habit of spending afternoons in the cab of a large front-loader that was chained down on the deck alongside other pieces of heavy machinery. I could sit there out of the wind and occasional rain, looking out through its spacious windows at the sea and ice. Sometimes I would read in the *Pilot of Arctic Canada*. Or I would read arctic history with a map spread out in my lap.

The days among the icebergs passed slowly. I sat in my makeshift catbird seat on the deck, or stood watching in the bows, or up on the bridge with my binoculars and sketchbook.

The icebergs were like pieces of Montana floating past. A different geography, I thought, from the one I grew up knowing.

Icebergs create an unfamiliar sense of space because the horizon retreats from them and the sky rises without any lines of compression behind them. It is this perspective that frightened pioneer families on the treeless North American prairies. Too much space, anchored only now and then by a stretch of bur oak savanna. Landscape painting of the T'ang and Sung dynasties (seventh to twelfth centuries) used this arrangement of space to create the sense of a large presence beyond. Indeed, the subject of such paintings was often their apparent emptiness.

American landscape painting in the nineteenth century, to return to an earlier thought, reveals a struggle with light and space that eventually set it apart from a contemporary European tradition of pastoral landscapes framed by trees, the world viewed from a carriage window. American painters meant to locate an actual spiritual presence in the North American landscape. Their paintings, according to art historians of the period, were the inspirations of men and women who "saw the face of God" in the prairies and mountains and along the river bottoms. One of the clearest expressions of this recasting of an understanding of what a landscape is were the almost austere compositions of the luminists. The atmosphere of these paintings is silent and contemplative. They suggest a private rather than a public encounter with the land. Several critics, among them Barbara Novak in her study of this period in American art, *Nature and Culture,* have described as well a peculiar "loss of ego" in the paintings. The artist disappears. The authority of the work lies, instead, with the land. And the light in them is like a creature, a living, integral part of the scene. The landscape is numinous, imposing, real. It ceases to be, as it was in Europe, merely symbolic.

At the height of his critical and popular acclaim in 1859, Frederic Edwin Church, one of the most prominent of the luminists, set sail for waters off the Newfoundland coast. He wanted to sketch the icebergs there. They seemed to him the very embodiment of light in nature. Following a three-week cruise, he returned to his studio in New York to execute a large painting.

The small field sketches he made—some are no larger than the palm of your hand—have a wonderful, working intimacy about them. He captures both the monolithic inscrutability of icebergs and the weathered, beaten look they have by the time they arrive that far south in the Labrador Sea. Looking closely at one drawing, made on July 1, I noticed that Church had penciled underneath it the words "strange supernatural."

The oil painting he produced from these sketches came to be called *The Icebergs.* It is so imposing—6 feet by 10 feet wide—a viewer feels he can almost step into it, which was Church's intent. In the foreground is a shelf of ice, part of an iceberg that fills most of the painting and which rises abruptly in the left foreground. On the right, the flooded ice shelf becomes part of a wave-carved grotto. In the central middle ground is a becalmed embayment, opening onto darker ocean waters to the left, which continue to a stormy horizon and other, distant icebergs. Dominating the background on the far side of the embayment is a high wall of ice and snow that carries all the way to the right of the painting. In the ocean air above is a rolling mist. The shading and forms of the icebergs are expertly limned—Church was an avid naturalist, and conscientious about such accuracy—and the colors, though slightly embellished, are true.

There are two oddities about this now very famous American landscape painting. When it was undraped at Gaupil's Gallery in New York on April 24, 1861, the reaction was more reserved than the lionized Church had anticipated. But *The Icebergs* differed from the rest of Church's work in one, crucial aspect: there was no trace of man in it. Convinced that he had perhaps made a mistake, Church took the work back to his studio and inserted in the foreground a bit of flotsam from a shipwreck, a portion of the main-topmast with the crow's nest. The painting was then exhibited in Boston, where it was no better received than it had been in New York. Only when it arrived in London did critics and audiences marvel. "A most weird and beautiful picture," wrote a reviewer in the *Manchester Guardian*. England, with its longer history of arctic exploration and whaling and but a few years removed from the tragedy of Sir John Franklin, was certainly more appreciative, at least, of its subject matter.

The second oddity is that Church's painting "disappeared" for 116 years. It was purchased in 1863 by a Sir Edward Watkin, after the London showing, to hang at his estate outside Manchester, called Rose Hill. It then passed by inheritance through Watkins' son to a purchaser of the estate; and then, by donation, to Saint Wilfred's Church nearby (which returned it to Rose Hill with regrets about its size). By 1979 Rose Hill had become the Rose Hill Remand Home for Boys, and *The Icebergs*, hanging without a frame in a stairwell, had been signed by one of the boys. Unaware of its value and seeking funds for the reform school's operation, the owners offered it for sale. The painting was brought back to New York and sold at auction on October 25, 1979, for $2.5 million, the highest price paid to that time for a painting in America. It now hangs in the Dallas Museum of Fine Arts, in Texas.

Church's decision to add the broken mast to *The Icebergs* speaks, certainly, to his commercial instincts, but the addition, I think, is more complex than this; and such a judgment is both too cynical and too simple.

Try as we might, we ultimately can make very little sense at all of nature without resorting to such devices. Whether they are such bald assertions of human presence as Church's cruciform mast or the intangible, metaphorical tools of the mind—contrast, remembrance, analogy—we bring our own worlds to bear in foreign landscapes in order to clarify them for ourselves. It is hard to imagine that we could do otherwise. The risk we take is of finding our final authority in the metaphors rather than in the land. To inquire into the intricacies of a distant landscape, then, is to provoke thoughts about one's own interior landscape, and the familiar landscapes of memory. The land urges us to come around to an understanding of ourselves.

A comparison with cathedrals has come to many Western minds in searching for a metaphor for icebergs, and I think the reasons for it are deeper than the obvious appropriateness of line and scale. It has to do with our passion for light.

Cathedral architecture signaled a quantum leap forward in European civilization. The gothic cathedral churches, with their broad bays of sunshine, flying buttresses that let windows rise where once there had been stone in the walls, and harmonious interiors—this "architecture of light" was a monument to a newly created theology. "God is light," writes a French cultural historian of the era, Georges Duby, and "every creature stems from that initial, uncreated, creative light." Robert Grosseteste, the twelfth-century founder of Oxford University, wrote that "physical light is the best, the most delectable, the most beautiful of all the bodies that exist."

Intellectually, the eleventh and twelfth centuries were an age of careful dialectics, a working out of relationships that eventually became so refined they could be expressed in the mathematics of cathedrals. Not only was God light but the *relationship* between God and man was light. The cathedrals, by the very way they snared the sun's energy, were an expression of God and of the human connection with God as well. The aesthetics of this age, writes Duby, was "based on light, logic, lucidity, and yearning for a God in a human form." Both the scholastic monks in their exegetical disquisitions and the illiterate people who built these churches, who sent these structures soaring into the sky—157 feet at Beauvais before it fell over on them—both, writes Duby, were "people trying to rise above their poverty through dreams of light."

It was an age of mystics. When Heinrich Suso, a Dominican monk, prayed at night in church, "it often seemed as if he were floating on air or sailing between time and eternity, on the deep tide of the unsoundable marvels of God." And it was an age of visionaries who spoke of the New Jerusalem of the Apocalypse, where there would be no darkness.

The erection of these monuments to spiritual awareness signaled a revival of cities, without which these edifices could not have survived. (The money to build them came largely from an emerging class of merchants and tradesmen, not royalty.) In time, however, the cathedrals became more and more esoteric, so heavily intellectualized an enterprise that, today, the raw, spiritual desire that was their original impetus seems lost. To the modern visitor, familiar with an architecture more facile and clever with light, the cathedrals now seem dark. Their stone has been eaten away by the acids and corrosives of industrial air. The age of mystics that bore them gave way rather too quickly to an age of rational intellects, of vast, baroque theological abstraction.

A final, ironic point: the mathematics that made the building of the cathedrals possible was carefully preserved by Arabs and Moors, by so-called infidels.

By the thirteenth century, Europe was starting to feel the vastness of Asia, the authority of other cultures. "The dissemination of knowledge," writes Duby, "and the strides made in the cultural sphere had opened [European] eyes and forced them to face facts: the world was infinitely larger, more various, and less docile than it had seemed to their forefathers; it was full of men who had not received the word of God, who refused to hear it, and who would not be easily conquered by arms. In Europe the days of holy war were over. The days of the explorers, traders, and missionaries had begun. After all, why persist in struggling against all those infidels, those expert warriors, when it was more advantageous to negotiate and attempt to insinuate oneself in those invincible kingdoms by business transactions and peaceful preaching?"

This was the philosophy that carried the Portuguese to India, the Spaniards to Peru, and the French and British into the hinterlands of northern North America. Hundreds of years later, a refinement on this philosophy of acquisition propelled Americans, Canadians, and Russians into the Arctic.

The conventional wisdom of our time is that European man has advanced by enormous strides since the age of cathedrals. He has landed on the moon. He has cured smallpox. He has harnessed the power in the atom. Another argument, however, might be made in the opposite direction, that all European man has accomplished in 900 years is a more complicated manipulation of materials, a more astounding display of his grasp of the physical principles of matter. That we are dazzled by mere styles of expression. That ours is not an age of mystics but of singular adepts, of performers. That the erection of the cathedrals was the last wild stride European man made before falling back into the confines of his intellect.

Of the sciences today, quantum physics alone seems to have found its way back to an equitable relationship with metaphors, those fundamental tools of the imagination. The other sciences are occasionally so bound by rational analysis, or so wary of metaphor, that they recognize and denounce anthropomorphism as a kind of intellectual cancer, instead of employing it as a tool of comparative inquiry, which is perhaps the only way the mind works, that parallelism we finally call narrative.

There is a word from the time of the cathedrals: agape, an expression of intense spiritual affinity with the mystery that is "to be sharing life with other life." Agape is love, and it can mean "the love of another for the sake of God." More broadly and essentially it is a humble, impassioned embrace of something outside the self, in the name of that which we refer to as God, but which also includes the self and *is* God. We are clearly indebted as a species to the play of our intelligence; we trust our future to it; but we do not know whether intelligence is reason or whether intelligence is this desire to embrace and be embraced in the pattern that both theologians and physicists call God. Whether intelligence, in other words, is love.

One day, sitting in my accustomed spot on the cargo deck of the *Soodoc*, I turned to see the second engineer, who had brought two cups of coffee. He was from Guyana. We talked about Guyana, and about the icebergs, some forty or fifty of which were then around us. He raised his chin to indicate and said, "How would you like to live up there? A fellow could camp up there, sail all the way to Newfoundland. Get off at Saint John's. How about it?" He laughed.

We laughed together. We searched the horizon for mirages with the binoculars, but we were not successful. When his break was over, the engineer went back below decks. I hung over the bow, staring into the bow wave at the extraordinary fluidity of that geometry on the calm waters of Melville Bay. I looked up at the icebergs. They so embodied the land. Austere. Implacable. Harsh but not antagonistic. Creatures of pale light. Once, camped in the Anaktiktoak Valley of the central Brooks Range in Alaska, a friend had said, gazing off across that broad glacial valley of soft greens and straw browns, with sunlight lambent on Tulugak Lake and the Anaktuvuk River in the distance, that it was so beautiful it made you cry.

I looked out at the icebergs. They were so beautiful they also made you afraid.

THE AMERICAN GEOGRAPHIES

It has become commonplace to observe that Americans know little of the geography of their country, that they are innocent of it as a landscape of rivers, mountains, and towns. They do not know, supposedly, the location of the Delaware Water Gap, the Olympic Mountains, or the Piedmont Plateau; and, the indictment continues, they have little conception of the way the individual components of this landscape are imperiled, from a human perspective, by modern farming practices or industrial pollution.

I do not know how true this is, but it is easy to believe that it is truer than most of us would wish. A recent Gallup Organization and National Geographic Society survey found Americans woefully ignorant of world geography. Three out of four couldn't locate the Persian Gulf. The implication was that we knew no more about our own homeland, and that this ignorance undermined the integrity of our political processes and the efficiency of our business enterprises.

As Americans, we profess a sincere and fierce love for the American landscape, for our rolling prairies, free-flowing rivers, and "purple mountains' majesty"; but it is hard to imagine, actually, where this particular landscape is. It is not just that a nostalgic landscape has passed away—Mark Twain's Mississippi is now dammed from Illinois to Louisiana and the prairies have all been sold and fenced. It is that it's always been a romantic's landscape. In the attenuated form in which it is presented on television today, in magazine articles and in calendar photographs, the essential wildness of the American landscape is reduced to attractive scenery. We look out on a familiar, memorized landscape that portends adventure and promises enrichment. There are no distracting people in it and few artifacts of human life. The animals are all beautiful, diligent, one might even say well behaved. Nature's unruliness, the power of rivers and skies to intimidate, and any evidence of disastrous human land management practices are all but invisible. It is, in short, a magnificent garden, a colonial vision of paradise imposed on a real place that is, at best, only selectively known.

The real American landscape is a face of almost incomprehensible depth and complexity. If one were to sit for a few days, for example, among the ponderosa pine forests and black lava fields of the Cascade Mountains in western Oregon, inhaling the pines' sweet balm on an

About This Life (New York: Knopf, 1998).

evening breeze from some point on the barren rock, and then were to step off to the Olympic Peninsula in Washington, to those rain forests with sphagnum moss floors soft as fleece underfoot and Douglas firs too big around for five people to hug, and then head south to walk the ephemeral creeks and sun-blistered playas of the Mojave Desert in southern California, one would be reeling under the sensations. The contrast is not only one of plants and soils, a different array, say, of brilliantly colored beetles. The shock to the senses comes from a different shape to the silence, a difference in the very quality of light, in the weight of the air. And this relatively short journey down the West Coast would still leave the traveler with all that lay to the east to explore—the anomalous sand hills of Nebraska, the heat and frog voices of Okefenokee Swamp, the fetch of Chesapeake Bay, the hardwood copses and black bears of the Ozark Mountains.

No one of these places, of course, can be entirely fathomed, biologically or aesthetically. They are mysteries upon which we impose names. Enchantments. We tick the names off glibly but lovingly. We mean no disrespect. Our genuine desire, though we may be skeptical about the time it would take and uncertain of its practical value to us, is to actually know these places. As deeply ingrained in the American psyche as the desire to conquer and control the land is the desire to sojourn in it, to sail up and down Pamlico Sound, to paddle a canoe through Minnesota's boundary waters, to walk on the desert of the Great Salt Lake, to camp in the stony hardwood valleys of Vermont.

To do this well, to really come to an understanding of a specific American geography, requires not only time but a kind of local expertise, an intimacy with place few of us ever develop. There is no way around the former requirement: if you want to know you must take the time. It is not in books. A specific geographical understanding, however, can be sought out and borrowed. It resides with men and women more or less sworn to a place, who abide there, who have a feel for the soil and history, for the turn of leaves and night sounds. Often they are glad to take the outlander in tow.

These local geniuses of American landscape, in my experience, are people in whom geography thrives. They are the antithesis of geographical ignorance. Rarely known outside their own communities, they often seem, at the first encounter, unremarkable and anonymous. They may not be able to recall the name of a particular wildflower—or they may have given it a name known only to them. They might have forgotten the precise circumstances of a local historical event. Or they can't say for certain when the last of the Canada geese passed through in the fall, or can't differentiate between two kinds of trout in the same creek. Like all of us, they have fallen prey to the fallacies of memory and are burdened with ignorance; but they are nearly flawless in the respect they bear these places they love. Their knowledge is intimate rather than encyclopedic, human but not necessarily scholarly. It rings with the concrete details of experience.

America, I believe, teems with such people. The paradox here, between a faulty grasp of geographical knowledge for which Americans are indicted and the intimate, apparently contradictory familiarity of a group of largely anonymous people, is not solely a matter of confused scale. (The local landscape is easier to know than a national landscape—and many local geographers, of course, are relatively ignorant of a national geography.) And it is not simply ironic. The paradox is dark. To be succinct: the politics and advertising that seek a national audience must project a national geography; to be broadly useful that geography must, inevitably, be generalized and it is often romantic. It is therefore frequently misleading and imprecise. The same holds true with the entertainment industry, but here the problem might be clearer. The same films, magazines, and television features that honor an imaginary American landscape also tout the worth of the anonymous men and women who interpret it. Their affinity for the land is lauded, their local allegiance admired. But the rigor of their local geographies, taken as a whole, contradicts a patriotic, national vision of unspoiled, untroubled land. These men and women are ultimately forgotten, along with the details of the landscapes they speak for, in the face of more pressing national matters. It is the chilling nature of modern society to find an ignorance of geography, local or national, as excusable as an ignorance of hand tools; and to find the commitment of people to their home places only momentarily entertaining. And finally naive.

If one were to pass time among Basawara people in the Kalahari Desert, or with Kreen-Akrora in the Amazon Basin, or with Pitjantjatjara Aborigines in Australia, the most salient impression they might leave is of an absolutely stunning knowledge of their local geography—geology, hydrology, biology, and weather. In short, the extensive particulars of their intercourse with it.

In forty thousand years of human history, it has only been in the last few hundred years or so that a people could afford to ignore their local geographies as completely as we do and still survive. Technological innovations from refrigerated trucks to artificial fertilizers, from sophisticated cost accounting to mass air transportation, have utterly changed concepts of season, distance, soil productivity, and the real cost of drawing sustenance from the land. It is now possible for a resident of Boston to bite into a fresh strawberry in the dead of winter; for someone in San Francisco to travel to Atlanta in a few hours with no worry of how formidable might be crossings of the Great Basin Desert or the Mississippi River; for an absentee farmer to gain a tax advantage from a farm that leaches poisons into its water table and on which crops are left to rot. The Pitjantjatjara might shake their heads in bewilderment and bemusement, not because they are primitive or ignorant people, not because they have no sense of irony or are incapable of marveling, but because they have not (many would say not yet) realized a world in which such manipulation of the land—surmounting the imperatives of distance it imposes, for example, or turning the large-scale destruction of forests and arable land into wealth—is desirable or plausible.

In the years I have traveled through America, in cars and on horseback, on foot and by raft, I have repeatedly been brought to a sudden state of awe by some gracile or savage movement of animal, some odd wrapping of a tree's foliage by the wind, an unimpeded run of dew-laden prairie stretching to a horizon flat as a coin where a pin-dot sun pales the dawn sky pink. I know these things are beyond intellection, that they are the vivid edges of a world that includes but also transcends the human world. In memory, when I dwell on these things, I know that in a truly national literature there should be odes to the Triassic reds of the Colorado Plateau, to the sharp and ghostly light of the Florida Keys, to the aeolian soils of southern Minnesota and the Palouse in Washington, though the modern mind abjures the literary potential of such subjects. (If the sand and floodwater farmers of Arizona and New Mexico were to take the black loams of Louisiana in their hands they would be flabbergasted, and that is the beginning of literature.) I know there should be eloquent evocations of the cobbled beaches of Maine, the plutonic walls of the Sierra Nevada, the orange canyons of the Kaibab Plateau. I have no doubt, in fact, that there are. They are as numerous and diverse as the eyes and fingers that ponder the country—it is that only a handful of them are known. The great majority are to be found in drawers and boxes, in the letters and private journals of millions of workaday people who have regarded their encounters with the land as an engagement bordering on the spiritual, as being fundamentally linked to their state of health.

One cannot acknowledge the extent and the history of this kind of testimony without being forced to the realization that something strange, if not dangerous, is afoot. Year by year, the number of people with firsthand experience in the land dwindles. Rural populations continue to shift to the cities. The family farm is in a state of demise, and government and industry continue to apply pressure on the native peoples of North America to sever their ties with the land. In the wake of this loss of personal and local knowledge, the knowledge from which a real geography is derived, the knowledge on which a country must ultimately stand, has come something hard to define but I think sinister and unsettling—the packaging and marketing of land as a form of entertainment. An incipient industry, capitalizing on the nostalgia Americans feel for the imagined virgin landscapes of their fathers, and on a desire for adventure, now offers people a convenient though sometimes incomplete or even spurious geography as an inducement to purchase a unique experience. But the line between authentic experience and a superficial exposure to the elements of experience is blurred. And the real landscape, in all its complexity, is distorted even further in the public imagination. No longer innately mysterious and dignified, a ground from which experience grows, it becomes a curiously generic backdrop on which experience is imposed.

In theme parks the profound, subtle, and protracted experience of running a river is reduced to a loud, quick, safe equivalence, a pleasant distraction. People only able to venture into the countryside on annual vacations are, increasingly, schooled in the belief that wild land will, and should, provide thrills and exceptional scenery on a timely basis. If it does not, something is wrong, either with the land itself or possibly with the company outfitting the trip.

People in America, then, face a convoluted situation. The land itself, vast and differentiated, defies the notion of a national geography. If applied at all it must be applied lightly, and it must grow out of the concrete detail of local geographies. Yet Americans are daily presented with, and have become accustomed to talking about, a homogenized national geography, one that seems to operate independently of the land, a collection of objects rather than a continuous bolt of fabric. It appears in advertisements, as a background in movies, and in patriotic calendars. The suggestion is that there *can* be a national geography because the constituent parts are interchangeable and can be treated as commodities. In day-to-day affairs, in other words, one place serves as well as another to convey one's point. On reflection, this is an appalling condescension and a terrible imprecision, the very antithesis of knowledge. The idea that either the Green River in Utah or the Salmon River in Idaho will do, or that the valleys of Kentucky and West Virginia are virtually interchangeable, is not just misleading. For people still dependent on the soil for their sustenance, or for people whose memories tie them to those places, it betrays a numbing casualness, a utilitarian, expedient, and commercial frame of mind. It heralds a society in which it is no longer necessary for human beings to know where they live, except as those places are described and fixed by numbers. The truly difficult and lifelong task of discovering where one lives is finally disdained.

If a society forgets or no longer cares where it lives, then anyone with the political power and the will to do so can manipulate the landscape to conform to certain social ideals or nostalgic visions. People may hardly notice that anything has happened, or assume that whatever happens—a mountain stripped of timber and eroding into its creeks—is for the common good. The more superficial a society's knowledge of the real dimensions of the land it occupies becomes, the more vulnerable the land is to exploitation, to manipulation for short-term gain. The land, virtually powerless before political and commercial entities, finds itself finally with no defenders. It finds itself bereft of intimates with indispensable, concrete knowledge. (Oddly, or perhaps not oddly, while American society continues to value local knowledge as a quaint part of its heritage, it continues to cut such people off from any real political power. This is as true for small farmers and illiterate cowboys as it is for American Indians, native Hawaiians, and Eskimos.)

The intense pressure of imagery in America, and the manipulation of images necessary to a society with specific goals, means the land will inevitably be treated like a commodity; and voices that tend to contradict the proffered image will, one way or another, be silenced or discredited by those in power. This is not new to America; the promulgation in America of a false or imposed geography has been the case from the beginning. All local geographies, as they were defined by hundreds of separate, independent native traditions, were denied in the beginning in favor of an imported and unifying vision of America's natural history. The country, the landscape itself, was eventually defined according to dictates of Progress like Manifest Destiny, and laws like the Homestead Act which reflected a poor understanding of the physical lay of the land.

When I was growing up in southern California, I formed the rudiments of a local geography—eucalyptus trees, February rains, Santa Ana winds. I lost much of it when my family moved to New York City, a move typical of the modern, peripatetic style of American life, responding to the exigencies of divorce and employment. As a boy I felt a hunger to know the American landscape that was extreme; when I was finally able to travel on my own, I did so. Eventually I visited most of the United States, living for brief periods of time in Arizona, Indiana,

Alabama, Georgia, Wyoming, New Jersey, and Montana before settling twenty years ago in western Oregon.

The astonishing level of my ignorance confronted me everywhere I went. I knew early on that the country could not be held together in a few phrases, that its geography was magnificent and incomprehensible, that a man or woman could devote a lifetime to its elucidation and still feel in the end that he had but sailed many thousands of miles over the surface of the ocean. So I came into the habit of traversing landscapes I wanted to know with local tutors and reading what had previously been written about, and in, those places. I came to value exceedingly novels and essays and works of nonfiction that connected human enterprise to real and specific places, and I grew to be mildly distrustful of work that occurred in no particular place, work so cerebral and detached as to be refutable only in an argument of ideas.

These sojourns in various corners of the country infused me, somewhat to my surprise on thinking about it, with a great sense of hope. Whatever despair I had come to feel at a waning sense of the real land and the emergence of false geographies—elements of the land being manipulated, for example, to create erroneous but useful patterns in advertising—was dispelled by the depth of a single person's local knowledge, by the serenity that seemed to come with that intelligence. Any harm that might be done by people who cared nothing for the land, to whom it was not innately worthy but only something ultimately for sale, I thought, would one day have to meet this kind of integrity, people with the same dignity and transcendence as the land they occupied. So when I traveled, when I rolled my sleeping bag out on the shores of the Beaufort Sea or in the high pastures of the Absaroka Range in Wyoming, or at the bottom of the Grand Canyon, I absorbed those particular testaments to life, the indigenous color and songbird song, the smell of sun-bleached rock, damp earth, and wild honey, with some crude appreciation of the singular magnificence of each of those places. And the reassurance I felt expanded in the knowledge that there were, and would likely always be, people speaking out whenever they felt the dignity of the earth imperiled in these places.

The promulgation of false geographies, which threaten the fundamental notion of what it means to live somewhere, is a current with a stable and perhaps growing countercurrent. People living in New York City are familiar with the stone basements, the cratonic geology, of that island and have a feeling for birds migrating through in the fall, their sequence and number. They do not find the city alien but human, its attenuated natural history merely different from that of rural Georgia or Kansas. I find the countermeasure, too, among Eskimos who cannot read but who might engage you for days on the subtleties of sea-ice topography. And among men and women who, though they have followed in the footsteps of their parents, have come to the conclusion that they cannot farm or fish or log in the way their ancestors did; the finite boundaries to this sort of wealth have appeared in their lifetime. Or among young men and women who have taken several decades of book-learned agronomy, zoology, silviculture, and horticulture, ecology, ethnobotany, and fluvial geomorphology and turned it into a new kind of local knowledge, who have taken up residence in a place and sought, both because of and in spite of their education, to develop a deep intimacy with it. Or they have gone to work, idealistically, for the National Park Service or the fish and wildlife services or for a private institution like The Nature Conservancy. They are people to whom the land is more than politics or economics. These are people for whom the land is alive. It feeds them, directly, and that is how and why they learn its geography.

In the end, then, if one begins among the blue crabs of Chesapeake Bay and wanders for several years, down through the Smoky Mountains and back to the bluegrass hills, along the drainages of the Ohio and into the hill country of Missouri, where in summer a chorus of cicadas might drown out human conversation, then up the Missouri itself, reading on the way the entries of Meriwether Lewis and William Clark and musing on the demise of the plains grizzly and the sturgeon, crosses west into the drainage of the Platte and spends the evenings with Gene Weltfish's *The Lost Universe,* her book about the Pawnee who once thrived there, then drops south to Palo Duro Canyon and the irrigated farms of the Llano Estacado in Texas, turns west across the Sangre de Cristo, southern-most of the Rocky Mountain ranges, and moves north and west up onto the slickrock mesas of Utah, those browns and oranges, the ocherous hues reverberating in the deep canyons, then goes north, swinging west to the insular ranges that sit like battleships in the pelagic space of Nevada, camps at the steaming edge of sulphur springs in the Black Rock Desert, where alkaline pans are glazed with a ferocious light, a heat to melt iron, then crosses the northern Sierra Nevada, waist-deep in summer snow in the passes, to descend to the valley of the Sacramento, and rises through groves of elephantine redwoods in the Coast Range, to arrive at Cape Mendocino, before Balboa's Pacific, cormorants and gulls, gray whales headed north for Unimak Pass in the Aleutians, the winds crashing down on you, facing the ocean over the blue ocean that gives the scene its true vastness, making this crossing, having been so often astonished at the line and the color of the land, the ingenious lives of its plants and animals, the varieties of its darknesses, the intensity of the stars overhead, you would be ashamed to discover, then, in yourself, any capacity to focus on ravages in the land that left you unsettled. You would have seen so much, breathtaking, startling, and outsize, that you might not be able for a long time to break the spell, the sense, especially finishing your journey in the West, that the land had not been as rearranged or quite as compromised as you had first imagined.

After you had slept some nights on the beach, however, with that finite line of the ocean before you and the land stretching out behind you, the wind first battering then cradling you, you would be compelled by memory, obligated by your own involvement, to speak of what left you troubled. To find the rivers dammed and shrunken, the soil washed away, the land fenced, a tracery of pipes and wires and roads laid down everywhere, blocking and channeling the movement of water and animals, cutting the eye off repeatedly and confining it—you had expected this. It troubles you no more than your despair over the ruthlessness, the insensitivity, the impetuousness of modern life. What underlies this obvious change, however, is a less noticeable pattern of disruption: acidic lakes, skies empty of birds, fouled beaches, the poisonous slags of industry, the sun burning like a molten coin in ruined air.

It is a tenet of certain ideologies that man is responsible for all that is ugly, that everything nature creates is beautiful. Nature's darkness goes partly unreported, of course, and human brilliance is often perversely ignored. What is true is that man has a power, literally beyond his comprehension, to destroy. The lethality of some of what he manufactures, the incompetence with which he stores it or seeks to dispose of it, the cavalier way in which he employs in his daily living substances that threaten his health, the leniency of the courts in these matters (as though products as well as people enjoyed the protection of the Fifth Amendment), and the treatment of open land, rivers, and the atmosphere as if, in some medieval way, they could still be regarded as disposal sinks of infinite capacity, would make you wonder, standing face to in the wind at Cape Mendocino, if we weren't bent on an errand of madness.

The geographies of North America, the myriad small landscapes that make up the national fabric, are threatened—by ignorance of what makes them unique, by utilitarian attitudes, by failure to include them in the moral universe, and by brutal disregard. A testament of minor voices can clear away an ignorance of any place, can inform us of its special qualities; but no voice, by merely telling a story, can cause the poisonous wastes that saturate some parts of the land to decompose, to evaporate. This responsibility falls ultimately to the national community, a vague and fragile entity to be sure, but one that, in America, can be ferocious in exerting its will.

Geography, the formal way in which we grapple with this areal mystery, is finally knowledge that calls up something in the land we recognize and respond to. It gives us a sense of place and a sense of community. Both are indispensable to a state of well-being, an individual's and a country's.

One afternoon on the Siuslaw River in the Coast Range of Oregon, in January, I hooked a steelhead, a sea-run trout, that told me, through the muscles of my hands and arms and shoulders, something of the nature of the thing I was calling "the Siuslaw River." Years ago I had stood under a pecan tree in Upson County, Georgia, idly eating the nuts, when slowly it occurred to me that these nuts would taste different from pecans growing somewhere up in South Carolina. I didn't need a sharp sense of taste to know this, only to pay attention at a level no one had ever told me was necessary. One November dawn, long before the sun rose, I began a vigil at the Dumont Dunes in the Mojave Desert in California, which I kept until a few minutes after the sun broke the horizon. During that time I named to myself the colors by which the sky changed and by which the sand itself flowed like a rising tide through grays and silvers and blues into yellows, pinks, washed duns, and fallow beiges.

It is through the power of observation, the gifts of eye and ear, of tongue and nose and finger, that a place first rises up in our mind; afterwards it is memory that carries the place, that allows it to grow in depth and complexity. For as long as our records go back, we have held these two things dear, landscape and memory. Each infuses us with a different kind of life. The one feeds us, figuratively and literally. The other protects us from lies and tyranny. To keep landscapes intact and the memory of them, our history in them, alive, seems as imperative a task in modern time as finding the extent to which individual expression can be accommodated, before it threatens to destroy the fabric of society.

If I were to now visit another country, I would ask my local companion, before I saw any museum or library, any factory or fabled town, to walk me in the country of his or her youth, to tell me the names of things and how, traditionally, they have been fitted together in a community. I would ask for the stories, the voice of memory over the land. I would ask to taste the wild nuts and fruits, to see their fishing lures, their bouquets, their fences. I would ask about the history of storms there, the age of the trees, the winter color of the hills. Only then would I ask to see the museums. I would want first the sense of a real place, to know that I was not inhabiting an idea. I would want to know the lay of the land first, the real geography, and take some measure of the love of it in my companion before I stood before the paintings or read works of scholarship. I would want to have something real and remembered against which I might hope to measure their truth.

SCOTT RUSSELL SANDERS
b. 1945

Scott Russell Sanders had already established himself as a fine writer of fiction and of books for children when A Paradise of Bombs *appeared in 1987. That was the collection of nonfiction that introduced a broad audience to the qualities of authenticity, elegance, personal modesty, and moral seriousness that have since made Sanders one of our most admired and valued essayists. Six other volumes of essays have followed between 1991 and 2001—* Secrets of the Universe, Staying Put: Making a Home in a Restless World, Writing from the Center, Hunting for Hope, The Country of Language, *and* The Force of Spirit. *There is remarkable continuity from one of these books to the next. Scott Russell Sanders celebrates the grounded pleasures of home and family, as well as the potential alliance between reading and our experiences of nature. He remains alert for the unanticipated, luminous moments that are our day's truest content, even as he ponders the grave challenges to ecological and social sustainability in our time. His readers revel in the crispness of his perceptions and feel grateful for the maturity and faithfulness of his example.*

BUCKEYE

Years after my father's heart quit, I keep in a wooden box on my desk the two buckeyes that were in his pocket when he died. Once the size of plums, the brown seeds are shriveled now, hollow, hard as pebbles, yet they still gleam from the polish of his hands. He used to reach for them in his overalls or suit pants and click them together, or he would draw them out, cupped in his palm, and twirl them with his blunt carpenter's fingers, all the while humming snatches of old tunes.

"Do you really believe buckeyes keep off arthritis?" I asked him more than once.

Writing from the Center (Bloomington: Indiana University Press, 1995).

He would flex his hands and say, "I do so far."

My father never paid much heed to pain. Near the end, when his worn knee often slipped out of joint, he would pound it back in place with a rubber mallet. If a splinter worked into his flesh beyond the reach of tweezers, he would heat the blade of his knife over a cigarette lighter and slice through the skin. He sought to ward off arthritis not because he feared pain but because he lived through his hands, and he dreaded the swelling of knuckles, the stiffening of fingers. What use would he be if he could no longer hold a hammer or guide a plow? When he was a boy he had known farmers not yet forty years old whose hands had curled into claws, men so crippled up they could not tie their own shoes, could not sign their names.

"I mean to tickle my grandchildren when they come along," he told me, "and I mean to build doll houses and turn spindles for tiny chairs on my lathe."

So he fondled those buckeyes as if they were charms, carrying them with him when our family moved from Ohio at the end of my childhood, bearing them to new homes in Louisiana, then Oklahoma, Ontario, and Mississippi, carrying them still on his final day when pain a thousand times fiercer than arthritis gripped his heart.

The box where I keep the buckeyes also comes from Ohio, made by my father from a walnut plank he bought at a farm auction. I remember the auction, remember the sagging face of the widow whose home was being sold, remember my father telling her he would prize that walnut as if he had watched the tree grow from a sapling on his own land. He did not care for pewter or silver or gold, but he cherished wood. On the rare occasions when my mother coaxed him into a museum, he ignored the paintings or porcelain and studied the exhibit cases, the banisters, the moldings, the parquet floors.

I remember him planing that walnut board, sawing it, sanding it, joining piece to piece to make foot stools, picture frames, jewelry boxes. My own box, a bit larger than a soap dish, lined with red corduroy, was meant to hold earrings and pins, not buckeyes. The top is inlaid with pieces fitted so as to bring out the grain, four diagonal joints converging from the corners toward the center. If I stare long enough at those converging lines, they float free of the box and point to a center deeper than wood.

I learned to recognize buckeyes and beeches, sugar maples and shagbark hickories, wild cherries, walnuts, and dozens of other trees while tramping through the Ohio woods with my father. To his eyes, their shapes, their leaves, their bark, their winter buds were as distinctive as the set of a friend's shoulders. As with friends, he was partial to some, craving their company, so he would go out of his way to visit particular trees, walking in a circle around the splayed roots of a sycamore, laying his hand against the trunk of a white oak, ruffling the feathery green boughs of a cedar.

"Trees breathe," he told me. "Listen."

I listened, and heard the stir of breath.

He was no botanist; the names and uses he taught me were those he had learned from country folks, not from books. Latin never crossed his lips. Only much later would I discover that the tree he called ironwood, its branches like muscular arms, good for axe handles, is known in the books as hophornbeam; what he called tuliptree or canoewood, ideal for log cabins, is officially the yellow poplar; what he called hoop ash, good for barrels and fence posts, appears in books as hackberry.

When he introduced me to the buckeye, he broke off a chunk of the gray bark and held it to my nose. I gagged.

"That's why the old-timers called it stinking buckeye," he told me. "They used it for cradles and feed troughs and peg legs."

"Why for peg legs?" I asked.

"Because it's light and hard to split, so it won't shatter when you're clumping around."

He showed me this tree in late summer, when the fruits had fallen and the ground was littered with prickly brown pods. He picked up one, as fat as a lemon, and peeled away the husk to reveal the shiny seed. He laid it in my palm and closed my fist around it so the seed peeped out from the circle formed by my index finger and thumb. "You see where it got the name?" he asked.

I saw: what gleamed in my hand was the eye of a deer, bright with life. "It's beautiful," I said.

"It's beautiful," my father agreed, "but also poisonous. Nobody eats buckeyes, except maybe a fool squirrel."

I knew the gaze of deer from living in the Ravenna Arsenal, in Portage County, up in the northeastern corner of Ohio. After supper we often drove the Arsenal's gravel roads, past the munitions bunkers, past acres of rusting tanks and wrecked bombers, into the far fields where we counted deer. One June evening, while mist rose from the ponds, we counted three hundred and eleven, our family record. We found the deer in herds, in bunches, in amorous pairs. We came upon lone bucks, their antlers lifted against the sky like the bare branches of dogwood. If you were quiet, if your hands were empty, if you moved slowly, you could leave the car and steal to within a few paces of a grazing deer, close enough to see the delicate lips, the twitching nostrils, the glossy, fathomless eyes.

The wooden box on my desk holds these grazing deer, as it holds the buckeyes and the walnut plank and the farm auction and the munitions bunkers and the breathing forests and my father's hands. I could lose the box, I could lose the polished seeds, but if I were to lose the memories I would become a bush without roots, and every new breeze would toss me about. All those memories lead back to the northeastern corner of Ohio, the place where I came to consciousness, where I learned to connect feelings with words, where I fell in love with the earth.

It was a troubled love, for much of the land I knew as a child had been ravaged. The ponds in the Arsenal teemed with bluegill and beaver, but they were also laced with TNT from the making of bombs. Because the wolves and coyotes had long since been killed, some of the deer, so plump in the June grass, collapsed on the January snow, whittled by hunger to racks of bones. Outside the Arsenal's high barbed fences, many of the farms had failed, their barns caving in, their topsoil gone. Ravines were choked with swollen couches and junked washing machines and cars. Crossing fields, you had to be careful not to slice your feet on tin cans or shards of glass. Most of the rivers had been dammed, turning fertile valleys into scummy playgrounds for boats.

One free-flowing river, the Mahoning, ran past the small farm near the Arsenal where our family lived during my later years in Ohio. We owned just enough land to pasture three ponies and to grow vegetables for our table, but those few acres opened onto miles of woods and creeks and secret meadows. I walked that land in every season, every weather, following animal trails. But then the Mahoning, too, was doomed by a government decision; we were forced to sell our land, and a dam began to rise across the river.

If enough people had spoken for the river, we might have saved it. If enough people had believed that our scarred country was worth defending, we might have dug in our heels and fought. Our attachments to the land were all private. We had no shared lore, no literature, no art to root us there, to give us courage, to help us stand our ground. The only maps we had were those issued by the state, showing a maze of numbered lines stretched over emptiness. The Ohio landscape never showed up on postcards or posters, never unfurled like tapestry in films, rarely filled even a paragraph in books. There were no mountains in that place, no waterfalls, no rocky gorges, no vistas. It was a country of low hills, cut over woods, scoured fields, villages that had lost their purpose, roads that had lost their way.

"Let us love the country of here below," Simone Weil urged. "It is real; it offers resistance to love. It is this country that God has given us to love. He has willed that it should be difficult yet possible to love it." Which is the deeper truth about buckeyes, their poison or their beauty? I hold with the beauty; or rather, I am held by the beauty, without forgetting the poison. In my corner of Ohio the gullies were choked with trash, yet cedars flickered up like green flames from cracks in stone; in the evening bombs exploded at the ammunition dump, yet from the darkness came the mating cries of owls. I was saved from despair by knowing a few men and women who cared enough about the land to clean up trash, who planted walnuts and oaks that would long outlive them, who imagined a world that would have no call for bombs.

How could our hearts be large enough for heaven if they are not large enough for earth? The only country I am certain of is the one here below. The only paradise I know is the one lit by our everyday sun, this land of difficult love, shot through with shadow. The place where we learn this love, if we learn it at all, shimmers behind every new place we inhabit.

A family move carried me away from Ohio thirty years ago; my schooling and marriage and job have kept me away ever since, except for visits in memory and in flesh. I returned to the site of our farm one cold November day, when the trees were skeletons and the ground shone with the yellow of fallen leaves. From a previous trip I knew that our house had been bulldozed, our yard and pasture had grown up in thickets, and the reservoir had flooded the woods. On my earlier visit I had merely gazed from the car, too numb with loss to climb out. But on this November day, I parked the car, drew on my hat and gloves, opened the door, and walked.

I was looking for some sign that we had lived there, some token of our affection for the place. All that I recognized, aside from the contours of the land, were two weeping willows that my father and I had planted near the road. They had been slips the length of my forearm when we set them out, and now their crowns rose higher than the telephone poles. When I touched them last, their trunks had been smooth and supple, as thin as my wrist, and now they were furrowed and stout. I took off my gloves and laid my hands against the rough bark. Immediately I felt the wince of tears. Without knowing why, I said hello to my father, quietly at first, then louder and louder, as if only shouts could reach him through the bark and miles and years.

Surprised by sobs, I turned from the willows and stumbled away toward the drowned woods, calling to my father. I sensed that he was nearby. Even as I called, I was wary of grief's deceptions. I had never seen his body after he died. By the time I reached the place of his death, a furnace had reduced him to ashes. The need to see him, to let go of him, to let go of this land and time, was powerful enough to summon mirages; I knew that. But I also knew, stumbling toward the woods, that my father was here.

At the bottom of a slope where the creek used to run, I came to an expanse of gray stumps and withered grass. It was a bay of the reservoir from which the water had retreated, the level drawn down by engineers or drought. I stood at the edge of this desolate ground, willing it back to life, trying to recall the woods where my father had taught me the names of trees. No green shoots rose. I walked out among the stumps. The grass crackled under my boots, breath rasped in my throat, but otherwise the world was silent.

Then a cry broke overhead and I looked up to see a red-tailed hawk launching out from the top of an oak. I recognized the bird from its band of dark feathers across the creamy breast and the tail splayed like rosy fingers against the sun. It was a red-tailed hawk for sure; and it was also my father. Not a symbol of my father, not a reminder, not a ghost, but the man himself, right there, circling in the air above me. I knew this as clearly as I knew the sun burned in the sky. A calm poured through me. My chest quit heaving. My eyes dried.

Hawk and father wheeled above me, circle upon circle, wings barely moving, head still. My own head was still, looking up, knowing and being known. Time scattered like fog. At length, father and hawk stroked the air with those powerful wings, three beats, then vanished over a ridge.

The voice of my education told me then and tells me now that I did not meet my father, that I merely projected my longing onto a bird. My education may well be right; yet nothing I heard in school, nothing I've read, no lesson reached by logic has ever convinced me as utterly or stirred me as deeply as did that red-tailed hawk. Nothing in my education prepared me to love a piece of the earth, least of all a humble, battered country like northeastern Ohio; I learned from the land itself.

Before leaving the drowned woods, I looked around at the ashen stumps, the wilted grass, and for the first time since moving from this place I was able to let it go. This ground was lost; the flood would reclaim it. But other ground could be saved, must be saved, in every watershed, every neighborhood. For each home ground we need new maps, living maps, stories and poems, photographs and paintings, essays and songs. We need to know where we are, so that we may dwell in our place with a full heart.

DAVID RAINS WALLACE
b. 1945

David Rains Wallace calls evolution "the great myth of modern times." Like all myths, it is double edged: a potential source of self-delusion but also a means of liberating us from obsolete and destructive notions about ourselves and our place in the world. In his John Burroughs Medal–winning book, The Klamath Knot (1984), Wallace explores an unusual tract of wilderness on the California-Oregon border in order to consider not only the complexities of current evolutionary theory but also the way in which evolutionary ideas, by recognizing the uniformity of all life, have the capacity to change human behavior for the better. In this he follows writers like John Burroughs, Aldo Leopold, and Lewis Thomas who see in evolution's very lack of a clear ethical code a useful model, for "In a world where such ambiguities reign, the idea that there can be no predetermined future may be a salutary one." By looking at evolution in the light of older myths—such as medieval alchemy and the legends of Bigfoot—he articulates one of the major aims of contemporary nature writers: to refashion the parables of our race in the light of new scientific understanding.

THE HUMAN ELEMENT

> There were giants in the earth in those days;
> and also after that.
>
> *Genesis 6:4*

Humanity has always been hard to define, and evolution hasn't made it easier. Older myths generally placed humans on a scale midway between animals and gods. This position had a comfortable stability. It gave people something to look down on and something to look up to. Some evolutionary myths have repeated this formula, with the idea that

The Klamath Knot (San Francisco: Sierra Club, 1983).

humans, having evolved from animals, will presently evolve into super-intelligent beings somewhat like the gods of earlier myth. This is understandably the most popular kind of evolutionary myth. It takes dozens of forms, from Teilhard de Chardin's noosphere to Nazi super-man eugenics.

A godlike future for the human race may be possible, indeed desir-able, assuming our future godlike omnipotence and immortality are accompanied by better behavior than that of, say, the Olympian gods. Evolution's four billion years on this planet do not foreshadow such a future, however. The symmetry of transformation from animal to god is not reflected in evolutionary evidence. Humans have not evolved from animals; we *are* animals, no less dependent on plant photosynthesis and bacterial decomposition for our survival than the lowliest flatworm. The ancient thinkers who developed the animal-human-god hierarchy were not aware of what we *have* evolved from. Like all animals, we have evolved from an intricate, fortuitous symbiosis of single-celled organisms. If there is symmetry to evolution, the future will not see us dominating all other life as gods. It will see us become part of a greater organism which we cannot imagine.

Evolutionary humanity is a truer microcosm of nature than medieval philosophers dreamed. The human body does not merely resemble nature in its parts, it recapitulates the history of life, as much a living reenactment of evolutionary dramas as the Klamath Mountains. Corpuscles float in a primal nutrient bath of blood; intestines crawl about absorbing food in the manner of primitive worms; lungs absorb and excrete gases as do gills and leaves. No human organ would look out of place if planted in some Paleozoic sponge bed or coral reef. Even our brain is an evolutionary onion, the core we share with fish and rep-tiles, the secondary layer we share with other mammals, and the outer layer we share with other primates.

Humanity can't be defined apart from the intricacies of natural selec-tion, mutation, symbiosis, preadaptation, and neoteny that formed it. It can't be defined apart from the millions of other species on the earth. Our evolutionary myths have been greatly oversimplified in our attempts to follow the thread of humanity into the past (and the future) while we ignore its entanglements with the threads of grasses, trees, snakes, and other beings. Such myths make a falsely passive background of an actively evolving world. To say that humanity descended from the trees, adopted a grassland hunting life, then invented agriculture and civilization, is racial solipsism. It would be quite as accurate to say that the forest abandoned the hominids, that the grasslands adopted them, that the first domestic plants and animals chose to live with our

Neolithic ancestors. I could write an evolutionary history of the human species in which its main significance is not as an inventor of language or builder of cities but as an ally of grasslands in their thirty-million-year struggle with forests. An extraterrestrial observer of the human colonization of North America would have seen more of spreading grasslands than of spreading cities, grasslands spread first by Indians with fire, then by whites with axes and plows.

It is possible to look upon humans and their civilization as a biological and geological force not qualitatively different from the volcanic eruptions, glaciations, and other catastrophes that have disturbed organic evolution. Nuclear war and wholesale industrial pollution may do life on earth more damage than a billion years of exploding volcanoes, but anthropoid greed and convection currents in the earth's mantle seem about equally random and senseless. Molecules simmering in the skull of a primate or sixty miles underground—what's the difference? Both explode when pressures get critical.

Such a view falls into the error of seeing evolution as a predetermined phenomenon, though. A humanity destined for demonic holocaust by its manipulative cleverness is a mirror image of the more popular evolutionary myth of a humanity destined for godlike triumph. If the four billion years of evolution demonstrate one thing, it is that humanity is not *destined* for anything. Evolution has always been open to new possibilities, which is why it has been so chaotic and devious. Every organism continually confronts a galaxy of evolutionary choices.

The difference between humans and other organisms is that humans, having discerned something of how evolution works, are now able to confront their choices consciously. This is not the same as saying that we now can *control* evolution. I don't know how much of a difference it is in effect: we may be able to perceive our choices and still be unable to choose and act. By overpopulating the planet as we are now doing, for example, we are making an evolutionary choice just as unplanned as that of our hominid ancestors when they began cracking antelope and other hominids over the head with sticks. Nevertheless, we do differ from the first hominids in our having some notion of the implications of our behavior. In Biblical terms we have heeded the serpent, eaten of the tree of knowledge, and lost our innocence. We now must face the possibility of choosing between good and evil, or, in evolutionary terms, between survival and extinction.

In other words humans have some degree of free will. As two millennia of theologians have been telling us, this is a perilous position. Pride is the great danger to the soul consciously seeking salvation. I think it is the great danger to the species consciously seeking survival too. In both cases pride can transform the best of virtues into the worst of vices. It can transform an individual's high intelligence into arrogance, and it can transform a species' considerable understanding of nature into stupid plundering.

The King of Phrygia tied the Gordian Knot in the temple of Apollo and prophesied that whoever untied it would become Lord of Asia. Alexander, proud young conqueror, cleverly cut the knot with his sword, became Lord of Asia, and died at age thirty-three of alcoholism, disease, or poison. Apollo was a god of the serpent as well as of the lyre. Conquering civilization could cut the Klamath knot, and that of every other wilderness: dam every river, log every forest, plow every meadow, until the last gasp of splendor subsides from the earth. "What now?" the serpent might whisper, as it perhaps whispered to Alexander on the banks of the Ganges.

The mythic resonance that evolution has given the natural world expresses a multitude of choices that humans have not consciously faced before. In old myths wherein nature remained the same from the world's creation until its end, our relationship with nature was much less laden with choices. Men couldn't change what the gods had made: "saving the planet" would have seemed an impertinence. But evolution, wherein a bear is not simply a black, shaggy animal but a wave of animals surging up through abysses of time from the original one-celled beings, raises troubling questions. Should we follow its competitive trend, manifested in natural selection, and try to survive by destroying everything that seems to get in our way? Should we follow its cooperative trend, manifested in symbiosis, and try to coexist with our parasites and our hosts (whoever *they* may be) in hope of some new synthesis? Should we try to do both? After all, that is what evolution does.

Much of the disquiet that has beset our thinking in the past two centuries seems related to evolution's burden of new choices. If life has taken such different shapes in the past, who can feel any assurance about the future? Entire realms of confident human activity begin to seem absurd. This is not necessarily a bad thing, of course. Some of the worst atrocities have been committed in pursuit of assurance, in flight from anxiety. If the future is essentially unknowable, at least good ends can no longer justify evil means. If it dispels our dreams of heaven, a world without destiny also wakens us from nightmares of hell.

The giants who left their tracks near Bluff Creek are eloquent mythic expressions of evolutionary uncertainty. Are they competitive lords of the snow forest? Cooperative children of the ancestral forest? Are they human? Are they alive? In a sense the giants are the missing link that Victorian society demanded Darwin and Huxley produce before it would bow to the new version of genesis. (Newspaper articles of 1884 tell of a young giant captured in British Columbia and shipped to London alive. It never arrived, victim, perhaps, of some conspiracy of Anglican divines?) Giants express our familial relationship to the rest of life. If we found them, could we rightfully continue to clear cut the giants' forests, dam their rivers, and trample their meadows? Even giant-hunters who consider them "just animals" advocate creating large preserves for giants.

With their elusiveness to civilized knowledge, giants express a gap that has arisen between our thinking habits, which are expressed in everyday speech, and our very recent awareness of evolutionary evidence. We condemn "brutality" and scorn murderers as "animals." We fear the sight of a shaggy beast shaped like a human. Yet we know that no wild animal is remotely capable of the deliberate torture and mass extermination that have become common in this most civilized of centuries. Knowing how unprecedented these horrors are, we no longer can blame them on our "lower" animal instincts, or hope to escape them by "rising above" our animal nature. We will not rise above our animal nature until we begin to live without food, water, and air. We are more protected by the timidity of the wild animal that remains in us than we are threatened by its aggressiveness.

We are fortunate to have the self-consciousness that allows us the possibility of free will. But we no longer can assume that our consciousness imbues us with a predetermined destiny separate from the rest of life. The only way we can separate ourselves from our animal, plant, and fungus relatives is to stop living, a viable and popular evolutionary option (considering the millions of extinct species) but one we're self-consciously averse to. Yet as we develop from species-exterminating hunters to land-eroding farmers to biosphere-polluting industrialists, we increasingly separate ourselves.

Few organisms survive in rapidly changing environments, and the world is changing faster than ever before. The fact that we've set these changes in motion doesn't mean we can control them. We must change to survive. No biological change will be fast enough now, though; we can't evolve as fast as the insects or rodents or microorganisms we've "conquered" because we reproduce so much more slowly. We must depend on cultural evolution. If our behavior is to change, our myths will have to change.

Myths began as imaginative projections of human consciousness onto nature. Trees had language, birds had thoughts, spiders had technology. When science found that nature does not, in fact, have a human consciousness, some thinkers concluded that myth was dead, that there was no further need for imaginative views of a world which, they thought, had no consciousness at all. But they misunderstood science. That nonhuman life has no human consciousness doesn't mean it has no consciousness. Science has opened a potential for imaginative interpretation of nature that is enormously greater than the simple projection of human thoughts and feelings onto the nonhuman. It has allowed us to begin to imagine states of consciousness quite different from our own. We can begin to see trees, birds, and spiders not as masks concealing humanlike spirits but as beings in their own right, beings that are infinitely more mysterious and wonderful than the nymphs and sprites of the old myths.

Science has raised the possibility that there are as many different consciousnesses in the world as there are organisms capable of perception. It also has raised the possibility that consciousness may arise in ways that seem very alien to us. The symbiotic superconsciousness I vaguely sense in forests is not outside scientific possibility.

The age of myth is not dead; it is just beginning, if humans can survive to inhabit it. Only, instead of myths peopled with talking trees, we must begin to create the opposite. (The fact that such myths—inhabited by "treeing talks"—aren't fully expressible with our present syntax and vocabulary is one measure of the magnitude of the enterprise.) Instead of inflating our human consciousness to fill trees, we must let the trees into our minds. It is not a sentimental undertaking. When science found that we don't have thoughts and feelings in common with the nonhuman, it also found we do have something equally important in common—origins. We are very different from trees, but we also are like them. As we learn how they live, we learn a great deal of how we live.

Learning does not occur only in the mind. High towers of intellectual learning require deep foundations of emotional knowledge, or they lack stability. The more we know about trees, the more we need to feel about them. The human element has grown too large and powerful for petty or trivial feelings about the nonhuman. What we feel about pettily, we begin to destroy, as we are destroying forests to produce junk mail and other trivialities.

Future myths will be different from past myths, but their function will be the same—to sustain life. When the human element was small, when there were billions of trees and only thousands of people, it was sustaining to imagine that trees contained spirits humans could talk to, propitiate, befriend. It gave proportion to the world. Now, when there are billions of people, and not so many trees, it is sustaining to imagine what it might be like to open one's flowers on a spring afternoon, or to stand silently, making food out of sunlight, for a thousand years. It gives proportion to the world.

Of course, imagination can only go so far. The incompleteness of scientific knowledge also limits emotional knowledge. We can't fully imagine a tree's existence because we don't know how, or if, a tree experiences its life. So something of the old mythological imagination probably will linger for a long time. We will continue to project our human feelings onto other organisms, as we try to imagine their non-human experience.

As with organisms, new myths don't appear fully formed, but evolve imprecisely out of old myths. Giants may be an example of such evolution. Giants seem to have originated as a way of giving human form to all that is titanic and inchoate in nature. In human form the awesomeness of rocks, waters, and tangled vegetation could be wrestled into submission, even befriended, by heroes and gods. Today's Klamath giants have something of this. In their dominance of the awesome snow forest, the giants affirm a desire for human power over wilderness, for a linkage with nature that is advantageous, albeit peaceable. If we found the Klamath giants, we would grasp some essence of the titanic knot of rocks, waters, and trees, as Beowulf and Gilgamesh grasped their ancient lands by defeating Grendel and Enkidu.

But the Klamath giants also have become more than shaggy, beetle-browed projections of human desire. We begin to see in them the possibility of a consciousness quite different from our own, of a being that may be very close to us in hominid origins, but that may have evolved in mysterious ways. We imagine an animal that somehow has understood the world more deeply than we have, and that thus inhabits it more comfortably and freely, while eluding our self-involved attempts to capture it.

Giants might be seen as a kind of preadapted myth that can help us to survive the world we've created. Giants have hovered for thousands of years in the backgrounds of our dreams of immortality and omniscience, large shadows humans cast behind them as they moved toward brilliant visions of limitless power. But now the visions are fading into a natural world that has proved much deeper than we ever had imagined. Giants can have a new function in an evolutionary myth. They link us to lakes, rivers, forests, and meadows that are our home as well as theirs. They lure us into the wilderness, as they lured me, not to devour us but to remind us where we are, on a living planet. If giants do not exist, to paraphase Voltaire, it is necessary to invent them.

ALISON HAWTHORNE DEMING
b. 1946

Over the past decade Alison Deming has emerged as an important figure in environmental literature. After fifteen years in the area of women's and public health, she dedicated herself wholeheartedly to writing. From 1983 on, she has won a stream of awards for her work, including the Pablo Neruda Prize, a Stegner Fellowship at Stanford, a Pushcart Prize for Nonfiction, and the Walt Whitman Award from the Academy of American Poets. Since 1990, she has served as Director of the Poetry Center at the University of Arizona. Volumes of poetry and essays both appeared in 1994: Science and Other Poems *and* Temporary Homelands. *The following excerpt is the final section of a long three-part essay from that latter collection. Her robust, engaging voice carries implications that may hit long after a first reading. Such caginess is perhaps not so surprising in a descendant of Nathaniel Hawthorne.*

WOLF, EAGLE, BEAR: AN ALASKA NOTEBOOK

* * * A beautiful spring day, warm and partly sunny. At sixty-five degrees Sitkans break out their tank tops and shorts. After a week of this atypical weather, everyone in town is high on the freedom of the outdoors. Joanne, Carolyn, and I decide on a hike up the Indian River Trail. We meet at the bookstore and, while we browse and chat with the owner, overhear the announcement on the radio that a brown bear has been sighted near the trailhead just where we had planned all week to hike. Two visitors and one local—a sizing-up. We defer to Carolyn, the local, who makes no bones about it—she's not comfortable knowing there's a bear in the area. What if we didn't know? If we hadn't come into the store and heard the broadcast? But we did and we have. And it is May—the month when the bears wake from hibernation. The

Temporary Homelands (San Francisco: Mercury House, 1994).

thought of crossing paths with a thousand-pound grizzly who hasn't eaten or taken a shit for six months, and may have hungry cubs bothering at her side, is enough to convince me that we should change our plans. But I'm reticent to say so, not wanting to appear lacking in courage. Wildlife researchers estimate that there are 150,000 brown, black, and polar bears in Alaska. That makes the distribution a little more than one bear for every three human inhabitants. Here in the southeast as many as 8,000 brown (or grizzly) bears populate the islands strewn between Ketchikan and Yakutat, in some areas living as densely as one bear per square mile. They thrive in the region, fattening on spawning salmon in the fall and finding the winters so moderate that some years they don't bother to hibernate at all. Knowing that the browns live here brings out an alertness and caution in some people, an aroused state of curiosity in others, and a challenge to the predatory instinct in yet others. No one can ignore the bear. One either hungers to encounter one or not to. And when I ask, "Should I be worried about the bears?" everyone has a bear story to tell.

The man who runs the sporting goods store tells me I'd be lucky ever to see a brown bear. He assures me I'm more likely to be killed by a mosquito or by tripping on a root than to be harmed by a bear. His wife tells me that she fears them—that a friend of theirs lost some leg muscle to a bear. "Of course," she qualifies the danger, "the man had been hunting and he had half a deer slung over his back." The woodsmen tell me that a brown bear will never attack unprovoked. But what is one to think about the respect for life of a species that, the biologists claim, will routinely hunt down and kill its own cubs? Someone else tells of the man hunting alone at the southern end of Baranof Island—"All the search party found were his boots . . . and his feet were still in them." "Hang bells on your pack," says one eavesdropper. "Don't carry food," says another. "Don't go in the woods when you're menstruating," the women say. The men tend to laugh when they tell bear stories, the women to look earnest. Others take the bear's side—telling with contempt of rich hunters from the lower forty-eight who charter a private plane and backcountry guide, down a brown bear and take only the claws as evidence of the kill. All bear partisans prefer the seriousness and respect with which Native people traditionally end an animal's life—eating the meat and the brain, making rope of the gut, a shirt of the mesenteries, a blanket of the hide, a necklace of the teeth and the claws.

We decide to drive by the house of a friend who is both hunter and naturalist, one who has spent more time in the wildlands than in town, and whose judgment we trust. He is sober about the danger of bears and provides us with an aerosol can of capsicum—the chemical that makes chili peppers hot and that has been marketed recently as a bear deterrent. It works only if sprayed from a distance of ten feet or less. In some cases, it merely serves to enrage an already testy animal. And he tells us his scariest bear story. While deer hunting in a remote forest of a remote island, he noticed fresh bear tracks underfoot. He quieted his dog, readied his gun, and walked on, deciding to follow the track. When he found himself circling back on his own tracks, he realized that the bear was stalking him. He stood still. There was quiet. Then his dog let loose the yelps only a bear could arouse; the bordering woods erupted with rampage, an ungodly roar ripping the air as the animal tore through the scrub. In a rage of ferocious proportion, the bear stormed back and forth at the edge of the muskeg. The hunter, locked in place, knew that even his rifle would be worthless if the bear decided to charge. But strangely it did not. It drew the territorial line with its anger, then pounded off into deep woods.

After this tale, the aerosol can strapped to Carolyn's knapsack provides me with little reassurance. Having known the terror of tooth and claw, I'm not titillated by the threat of enduring such an experience again—or a worse one. It's all well and good for Robinson Jeffers to rhapsodize about being eaten by a red-tailed hawk—"what an enskyment; what a life after death." I'll admit there is romance to the notion of my genetic material living in the ecosystem more freely and wildly than I, as a human being, would dare to do. But I prefer sticking around with my domestic companions to experiencing the enwildment of being eaten by a bear. My bear story, I begin to realize, soon will intersect with everyone else's like ripples on a pond.

"The Koyukon people," Richard Nelson writes in *Make Prayers to the Raven*, "have great respect for the unpredictability, aggressiveness, tenacity and physical power of brown bears. Unnecessary encounters are strictly avoided, and any approach to them is made with utmost caution, very different from the confidence hunters show in pursuing the milder, more predictable black bear. When a man comes across brown bear tracks on the fall snow, he is likely to leave the area immediately. People emphasize that this is a very difficult animal to kill, that a man alone could shoot it many times and still be attacked."

Special prohibitions govern the Koyukon woman's relationship to the brown bear. She must not eat the animal's meat and she must take care never to breathe the steam from the cooking bear meat. Violating the taboo would make a woman mean, insane, or ill, so great are the spiritual powers of the animal. She could not hunt them, should look away if one came into her sight, and she should not speak the animal's name.

We decide to steer clear of the bear sighting and take the shorter hike to Beaver Lake. We drive out past the pulp mill to a scrappy backwoods parking lot. The trail starts out as a steep course of switchbacks, reinforced on the most severe inclines with rough-cut timber stairs. Huge cedars, the tallest I have ever seen, and hemlocks climb the mountainsides below and above us. Beyond—steeper faces of sheer mountain, fractured peaks, silver blades of snowmelt slicing down through rock, snowpack blotching the high rocky crests. Topping the ridge, the trail levels off into muskeg and our breathing begins to relax.

A narrow boardwalk leads out from under the dank cedars across a grassy meadow. A scattering of blossoms brightens the field—clusters of yellow skunk cabbage, white bunchberries, and a single fuchsia orchid no bigger than a deerfly (a Jeffrey shooting star?). In the mud patches along the trailside appear clean tracks of the day's prior traffic—pointed deer hooves, waffled hiking boots, and dog paws. The farther into the woods we hike, the quieter we get. Quiet draws the forest closer, as if by stilling the mind's busy language center some vestigial territorial sense comes back to us. Why should I feel such pleasure to be in the wilds? Some inarticulate kind of memory calls me—not a personal memory, but something deeper—perhaps a species memory, a cellular sense of human origin. Emerson said it: "The mind loves its old home."

About a half mile out on the muskeg, we start single file into another shadowed cedar grove, the heady scent inviting. Suddenly, *bear* enters my mind—a ripping and real presence that sees me as either danger or food. The others move on ahead. I stop and stare at the dark wet mud to the left of the boardwalk. The boggy ground is a perfect medium for making an accurate textbook animal track—except this print is three-dimensional. The massive pressure of the animal's weight makes a bowl of the pad. Five oval toe marks press above it into the mud, above the toes five deeply gouged claw scars line up in a row. Brown bear. Black bear track would show claw marks closer to the toes and spread in an arc. The track is fresh. Water has not seeped into the print nor have the edges softened. The bear has left at the sound or the scent of our approach.

Never have I wished more to be without fear, to be the old Tlingit woman I read about who flipped the bears off with a switch of alder. Clearly the bear's track was headed away from us. But I couldn't stop thinking that fear might not be the bear's only response to our presence. What if her cubs gamboled obliviously into our midst? What if, in her winter-long hunger, one of us three women looked like the most delicious thing, and the slowest on our feet, for miles?

Among my tribe—the children of science—fear is a thing to resist and control. As far back into our tradition as I've looked, from the Book of Proverbs to Montaigne to Thoreau, I've found an echo of Franklin Roosevelt's well-heeled one-liner—"The only thing we have to fear is fear itself." As if all human fear were an emotional error, rather than a survival-enhancing product of evolution. Women in our culture understand and respond to fear differently than men do. For one thing, we face it more often, since we inherit a historical legacy of men's social and physical power over us. Fear is our radar. Walking alone down a city street at night or answering our own door to greet a stranger, we make a quick assessment and chart the appropriate course. It is not a question of overcoming fear so much as attuning the instrument of perception so that we can detect a real danger in time to avoid it. Street-smart fear is a woman's friend and companion.

A woman carries that emotional resonance with her when she goes to the woods. A man, we are taught to believe, goes to the woods to test his courage—scaling the heights, running the rapids, hooking and shooting the prey, downing or at least facing down the brown bear. A woman goes to the forest not to face her fear, but to escape it. She picks berries, mushrooms, mosses, and vines. If she imagines building a cabin there, she pictures its garden, its window boxes, and its raggedly domestic yard—not the moose rack mounted over the front door, not the revolver she will carry when berrying in bear country. But no matter how sweet and pastoral her longing for the wild, once there, she finds her old companion fear has come along. One woman tunes her radar to the psychopathic mountain man, another to the rampaging bear.

"Hey, you guys," I whisper ahead to the others. "Look at this." We all lean over the bear's imprint, quietly taking in the detail and remapping our sense of the territory.

On our way back down the trail, I take the lead, embarrassed by my eagerness to depart—but not so embarrassed as to change my mind. I wonder whether the others would have gone on if my concern had not been so apparent. Neither Carolyn nor Joanne had tried to cajole me onward, and I feel grateful that there is room in our friendship for an honest fear. We brush past the overarching devil's club, the alders fully fledged with new leaves, the salmonberry canes dropping their pink flowers to form green nuggets to proto-fruit.

I've read that early humans shared caves with hibernating bears and stacked cave-bear skulls in high-mountain sanctuaries. But since then the relationship between *ursus arctos* and *homo sapiens* has taken a dive. Our interactions now seem limited to two—avoidance or slaughter. Friendships that cross the borders between species are not impossible. The covenant between people and dogs is so strong, archaic, and cellular (lodged in our respective chromosomal codes) that it can transcend, as in my own case, traumatic experience that suggests canines are not to be trusted. Far more children die or are maimed each year by dog bites than from bear attack. Yet the attraction between human and canine species remains. So important was the dog to the ancient Egyptians that Anubis—the god of death, master of mummification, companion to the soul in death—was portrayed with the head of a jackal. The dog appears with Diana or Artemis as companion on the hunt, as mediator between the domestic and the wild, the earthly and the divine.

The friendship between people and dogs continues to be, even in our dense urban constellations, the most affectionate connection between our species and another. What other animal is so welcome in our homes, so responsive a companion, so interested in understanding our language and moods, so capable of bringing out playful, patient, and nurturing qualities in us? Konrad Lorenz speculated about how this friendship developed during the lives of our earliest ancestors. Jackals may have followed in the tracks of hunting parties, scavenging for spoils once the people had eaten their fill and made camp for the night. "The stone-age hunters," writes Lorenz, "must have found it quite agreeable to know that their camp was watched by a broad circle of jackals which, at the approach of a sabre-toothed tiger or a marauding cave-bear, gave tongue to the wildest tones." Perhaps on one restless night, a Neanderthal, unable to sleep because of his fears, got up and lay a trail of scraps leading toward his camp to draw the jackals near, knowing that if the pack slept soundly, the people could do the same. In the geologically slow motion of evolution (combined with the fast-forward of human will), that action led to poodles, Dobermans, and golden retrievers. It is our heritage to feel safe (already safe in our armed and alarmed urban homes) when we see a resting animal.

These days when we think and talk about our relationships with other animals, we never get much beyond questions of survival and turf. Which creature deserves its livelihood more, the spotted owl or the Oregon logger? It seems a given that we cannot stop the accelerating loss of species and, as those that remain are crowded onto shrinking islands of wildness, our questions become more godlike and less considerate of the godly complexity of nature. Should we shoot the coyote to save the whooping crane, kill the arctic fox to save the Canadian goose?

We save the freeze-dried tissue samples of the extinct ones, so far only imagining the biotechnology that might bring them back to life.

> Recipe For a Dinosaur:
> A. Find a bead of amber that contains a blood-sucking insect from the age of dinosaurs.
> B. Extract genetic material from the blood cells, preserved in the bead, of a bug-bitten dinosaur, and amplify the DNA using the polymerase chain reaction technique.
> C. Process and inject into the embryo of an alligator.
> D. Wait until it hatches.
> — *New York Times*, June 25, 1991

I suppose that to revive the dinosaur would be a kind of friendship with that species. After all the damage people have caused on the planet, any act of kindness will improve the value of our biological stock in the world. But I wonder what kind of planet we might live on now if the human project (not the rhetoric of our project) had been friendship, rather than greed.

We drive to another lake, this one accessible by car over a primitive dirt road. Blue Lake is not as wild a picnic spot as we had hoped for. A man follows us on foot, down the final rutted incline to the water's edge, carrying an outboard motor over his shoulder. Beer and McDonald's trash, five or six fishermen dot the bouldered shore. The water is clear blue-green, fed by waterfalling mountain streams, riled by a stout wind, slapping the rocks, harboring thousands of unseen life forms, too cold to enter, too beautiful not to wish to. We settle for the battered wildness, a little sorry for ourselves, unpack our lunches and eat.

I am glad to have left the high woods to the bear. I imagine the music that such a forest might have sung to me and I am sorry never to have heard it. But it's beautiful to think that a moody creature walks there, one so powerful its presence made us hasten away. I am glad to have left the mark of my scent in her brain, as she has left the mark of her paw in mine. Focusing a zoom lens would not have made me feel closer to her wildness. Whatever constitutes my scent in the mind of a bear is lumbering along the muskeg up by Beaver Lake. She and I exist as interruptions in each other's afternoon, and that is close enough to knowing her for me.

GRETEL EHRLICH
b. 1946

When The Solace of Open Spaces *was published in 1985, Gretel Ehrlich immediately established herself as one of the outstanding nature writers of our day. It tells the story of how, as she says in the Preface, she "was able to take up residence on earth with no alibis, no self-promoting schemes." Her road back to the earth led her to the arduous work and invigorating life of a Wyoming rancher. As she narrates her passage into that role, she also paints portraits of a land and of people who came into it along paths different from her own. Ehrlich brings out forcefully the fact that, where extremes of weather are so sudden and so great, no one, ranchers or animals, can survive many mistakes. Such an awareness of Wyoming's basic demands for alertness lends an ironic element to her writing that complements her lyrical quality. Reading about Ehrlich's Wyoming, like reading about Norman Maclean's Montana, is exciting for readers who have not lived in those sparsely settled, dramatic lands. Avidly, we turn the pages of lives we will not live, of landscapes we will not forget. Among Ehrlich's recent books are* Match to the Heart *(1994) and* Any Clear Thing That Blinds Us with Surprise *(1995).*

FRIENDS, FOES, AND WORKING ANIMALS

I used to walk in my sleep. On clear nights when the seals barked and played in phosphorescent waves, I climbed out the window and slept in a horse stall. Those "wild-child" stories never seemed odd to me; I had the idea that I was one of them, refusing to talk, sleeping only on the floor. Having become a city dweller, the back-to-the-land fad left me cold and I had never thought of moving to Wyoming. But here I am, and unexpectedly, my noctambulist's world has returned. Not in the sense that I still walk in my sleep—such restlessness has left me—but rather, the intimacy with what is animal in me has returned.

The Solace of Open Spaces (New York: Viking, 1985).

To live and work on a ranch implicates me in new ways: I have blood on my hands and noises in my throat that aren't human.

Animals give us their constant, unjaded faces and we burden them with our bodies and civilized ordeals. We're both humbled by and imperious with them. We're comrades who save each other's lives. The horse we pulled from a boghole this morning bucked someone off later in the day; one stock dog refuses to work sheep, while another brings back a calf we had overlooked while trailing cattle to another pasture; the heifer we doctored for pneumonia backed up to a wash and dropped her newborn calf over the edge; the horse that brings us home safely in the dark kicks us the next day. On and on it goes. What's stubborn, secretive, dumb, and keen in us bumps up against those same qualities in them. Their births and deaths are as jolting and random as ours, and because ranchers are food producers, we give ourselves as wholly to the sacrament of nurturing as to the communion of eating their flesh. What develops in this odd partnership is a stripped-down compassion, one that is made of frankness and respect and rigorously excludes sentimentality.

What makes westerners leery of "outsiders"—townspeople and city-slickers—is their patronizing attitude toward animals. "I don't know what in the hell makes those guys think they're smarter than my horse. Nothing I see them do would make me believe it," a cowboy told me. "They may like their steaks, but they sure don't want to help out when it comes to butchering. And their damned back-yard horses are spoiled. They make it hard for a horse to do something right and easy for him to do everything wrong. They're scared to get hot and tired and dirty out here like us; then they don't understand why a horse won't work for them."

On a ranch, a mother cow must produce calves, a bull has to perform, a stock dog and working horse should display ambition, savvy, and heart. If they don't, they're sold or shot. But these relationships of mutual dependency can't be dismissed so briskly. An animal's wordlessness takes on the cleansing qualities of space: we freefall through the beguiling operations of our own minds with which we calculate our miseries to responses that are immediate. Animals hold us to what is present: to who we are at the time, not who we've been or how our bank accounts describe us. What is obvious to an animal is not the embellishment that fattens our emotional résumés but what's bedrock and current in us: aggression, fear, insecurity, happiness, or equanimity. Because they have the ability to read our involuntary tics and scents, we're transparent to them and thus exposed—we're finally ourselves.

Living with animals makes us redefine our ideas about intelligence.

Horses are as mischievous as they are dependable. Stupid enough to let us use them, they are cunning enough to catch us off guard. We pay for their loyalty: they can be willful, hard to catch, dangerous to shoe, and buck on frosty mornings. In turn, they'll work themselves into a lather cutting cows, not for the praise they'll get but for the simple glory of out-dodging a calf or catching up with an errant steer. The outlaws in a horse herd earn their ominous names—the red roan called Bonecrusher, the sorrel gelding referred to as Widowmaker. Others are talented but insist on having things their own way. One horse used only for roping doesn't like to be tied up by the reins. As soon as you jump off he'll rub the head-stall over his ears and let the bit drop from his mouth, then just stand there as if he were tied to the post. The horses that sheepherders use become chummy. They'll stick their heads into a wagon when you get the cookies out, and eat the dogfood. One sheepherder I knew, decked out in bedroom slippers and baggy pants, rode his gelding all summer with nothing but bailing string tied around the horse's neck. They pic-nicked together every day on the lunch the herder had fixed: two sand-wiches and a can of beer for each of them.

A dog's reception of the jolts and currents of life comes in more clearly than a horse's. Ranchers use special breeds of dogs to work live-stock—blue and red heelers, border collies, Australian shepherds, and kelpies. Heelers, favored by cattlemen, are small, muscular dogs with wide heads and short, blue-gray hair. Their wide and deep chests enable them—like the quarter horse—to run fast for a short distance and endow them with extra lung capacity to work at high altitudes. By instinct they move cows, not by barking at them but by nipping their heels. What's uncanny about all these breeds is their responsiveness to human beings: we don't shout commands, we whisper directions, and because of their unshakable desire to please us, they can be called back from chasing a cow instantaneously. Language is not an obstacle to these dogs; they learn words very quickly. I know several dogs who are bilingual: they under-stand Spanish and English. Others are whizzes with names. On a pack trip my dog learned the names of ten horses and remembered the horse and the sound of his name for years. One friend taught his cowdog to jump onto the saddle so he could see the herd ahead, wait for a com-mand with his front feet riding the neck of the horse, then leap to the ground and bring a calf back or turn the whole herd.

My dog was born under a sheep wagon. He's a blue heeler–kelpie cross with a natural bobbed tail. Kelpies, developed in Australia in the nineteenth century, are also called dingoes, though they're part Scottish sheepdog too. While the instinct to work livestock is apparent from the time they are puppies, they benefit from further instruction, the way anyone with natural talent does. They're not sent to obedience school; these dogs learn from each other. A pup, like mine was, lives at sheep camp and is sent out with an older dog to learn his way around a band of sheep. They learn to turn the herd, to bring back strays, and to stay behind the horse when they're not needed.

Dogs who work sheep have to be gentler than cowdogs. Sheep are skittish and have a natural fear of dogs, whereas a mother cow will turn and fight a dog who gets near her calf. If kelpies, border collies, and Australian shepherds cower, they do so from timidity and because they've learned to stay low and out of sight of the sheep. With their pointed ears and handsome, wolfish faces, their resemblance to coyotes is eerie. But their instinct to work sheep is only a refinement of the desire to kill; they lick their chops as they approach the herd.

After a two-year apprenticeship at sheep camp, Rusty came home with me. He was carsick all the way, never having ridden in a vehicle, and, once home, there were more firsts: when I flushed the toilet, he ran out the door; he tried to lick the image on the screen of the television; when the phone rang he jumped on my lap, shoving his head under my arm. In April the ewes and lambs were trailed to spring range and Rusty rejoined them. By his second birthday he had walked two hundred miles behind a horse, returning to the mountain top where he had been born.

Dogs read minds and also maps. Henry III's greyhound tracked the king's coach from Switzerland to Paris, while another dog found his owner in the trenches during World War I. They anticipate comings and goings and seem to possess a prescient knowledge of danger. The night before a sheep foreman died, his usually well-behaved blue heeler acted strangely. All afternoon he scratched at the windows in an agony of panic, yet refused to go outside. The next day Keith was found dead on the kitchen floor, the dog standing over the man's chest as if shielding the defective heart that had killed his master.

While we cherish these personable working animals, we unfairly malign those that live in herds. Konrad Lorenz thinks of the anonymous flock as the first society, not unlike early medieval cities: the flock works as a wall of defense protecting the individual against aggressors. Herds are democratic, nonhierarchical. Wyoming's landscapes are so wide they can accommodate the generality of a herd. A band of fifteen hundred sheep moves across the range like a single body of water. To work them in a corral means opposing them: if you walk back through the middle of the herd, they will flow forward around you as if you were a rock in a stream. Sheep graze up a slope, not down the way cows do, as if they were curds of cream rising.

Cows are less herd-smart, less adhesive, less self-governing. On long treks, they travel single file, or in small, ambiguous crowds from which individuals veer off in a variety of directions. That's why cowboying is more arduous than herding sheep. On a long circle, cowboys are assigned positions and work like traffic cops directing the cattle. Those that "ride point" are the front men. They take charge of the herd's course, turning the lead down a draw, up a ridge line, down a creek, galloping ahead to chase off steers or bulls from someone else's herd, then quickly returning to check the speed of the long column. The cowboys at the back "ride drag." They push the cows along and pick up stragglers and defectors, inhaling the sweet and pungent perfume of the animals — a mixture of sage, sweet grass, milk, and hide, along with gulps of dust. What we may miss in human interaction here we make up for by rubbing elbows with wild animals. Their florid, temperamental lives parallel ours, as do their imperfect societies. They fight and bicker, show off, and make love. I watched a Big Horn ram in rut chase a ewe around a tree for an hour. When he caught and mounted her, his horns hit a low branch and he fell off. She ran away with a younger ram in pursuit. The last I saw of them, she was headed for a dense thicket of willows and the old ram was peering through the maze looking for her.

When winter comes there is a sudden population drop. Frogs, prairie dogs, rattlesnakes, and rabbits go underground, while the mallards and cinnamon teal, as well as scores of songbirds, fly south because they are smarter than we are. One winter day I saw a coyote take a fawn down on our frozen lake where in summer I row through fragrant flowers. He jumped her, grabbed her hind leg, and hung on as she ran. Halfway across the lake the fawn fell and the coyote went for her jugular. In a minute she was dead. Delighted with his catch, he dragged her here and there on the ice, then lay down next to her in a loving way and rubbed his silvery ruff in her hair before he ate her.

In late spring, which here, at six thousand feet, is June, the cow elk become proud mothers. They bring their day-old calves to a hill just above the ranch so we can see them. They're spotted like fawns but larger, and because they are so young, they wobble and fall when they try to play.

Hot summer weather brings the snakes and bugs. It's said that 80 percent of all animal species are insects, including six thousand kinds of ants and ten thousand bugs that sing. Like the wild ducks that use our lake as a flyaway, insects come and go seasonally. Mosquitoes come early and stay late, followed by black flies, gnats, Stendhalian red-and-black ants, then yellow jackets and wasps.

I know it does no good to ask historical questions — why so many insects exist—so I content myself with the cold ingenuity of their lives. In winter ants excavate below their hills and live snugly in subterranean chambers. Their heating system is unique. Worker ants go above ground and act as solar collectors, descending frequently to radiate heat below. They know when spring has come because the workers signal the change of seasons with the sudden increase of body heat: it's time to reinhabit the hill.

In a drought year rattlesnakes are epidemic. I sharpen my shovel before I irrigate the alfalfa fields and harvest vegetables carrying a shotgun. Rattlesnakes have heat sensors and move toward warm things. I tried nude sunbathing once: I fell asleep and woke just in time to see the grim, flat head of a snake angling toward me. Our new stock dog wasn't as lucky. A pup, he was bitten three times in one summer. After the first bite he staggered across the hayfield toward me, then keeled over, his eyes rolling back and his body shaking. The cure for snakebite is the same for animals as it is for humans: a costly antiserum must be injected as quickly as possible. I had to carry the dog half a mile to my pickup. By the time I had driven the thirty miles to town, his head and neck had swollen to a ghoulish size, but two days later he was heeling cows again.

Fall brings the wildlife down from the mountains. Elk and deer migrate through our front yard while in the steep draws above us, mountain lions and black bears settle in for the winter. Last night, while I was sleeping on the veranda, the sound of clattering dishes turned out to be two buck deer sparring in front of my bed. Later, a porcupine and her baby waddled past: "Meeee . . . meeee . . . meeee," the mother squeaked to keep the young one trundling along. From midnight until dawn I heard the bull elk bugle—a whistling, looping squeal that sounds porpoiselike at first, and then like a charging elephant. The screaming catlike sound that wakes us every few nights is a bobcat crouched in the apple tree.

Bobcats are small, weighing only twenty pounds or so, with short tails and long, rabbity back feet. They can nurse two small litters of kittens a year. "She's meaner than a cotton sack full of wildcats," I heard a cowboy say about a woman he'd met in the bar the night before. A famous riverman's boast from the paddlewheel days on the Mississippi goes this way: "I'm all man, save what's wildcat and extra lightning." *Les chats sauvages*, the French call them, but their savagery impresses me much less than their acrobatic skills. Bobcats will kill a doe by falling on her from a tree and riding her shoulders as she runs, reaching around and scratching her face until she falls. But just as I was falling asleep again, I thought I heard the bobcat purring.

ELLEN MELOY
b. 1946

Ellen Meloy, who lives on the San Juan River in southern Utah, is best known as a celebrant of the slickrock desert of the West. Her books The Last Cheater's Waltz: Beauty and Violence in the Desert Southwest *and* Raven's Exile: A Season on the Green River *convey her refreshing and energetic vision of this austere landscape. Meloy was also a contributor to* Testimony; *this collection, assembled to promote a higher level of protection for southwestern wilderness, has become recognized as a model for the way in which literary art may directly influence public policy in the area of conservation.*

THE FLORA AND FAUNA OF LAS VEGAS

> Human domination over nature is quite simply an illusion, a passing dream by a naïve species. It is an illusion that has cost us much, ensnared us in our own designs, given us a few boasts to make about our courage and genius, but all the same it is an illusion. Do what we will, the Colorado will one day find an unimpeded way to the sea.
>
> Donald Worster, *Under Western Skies*

Ascent. Summit. Descent. The interstate highway, the asphalt river, slips off the Colorado Plateau, rises and falls over the Great Basin's rhythmic contours of basin and range, and flows southwest toward the Mojave Desert. In basin, the highway crosses the Sevier River, which the 1776 Domínguez-Escalante expedition, ever hoping to find a Pacific passage, erroneously linked with the Green River in the Uinta Basin. Through range, the meticulously graveled and graded highway slopes bury Fremont village sites, their remains relocated to museums to make way for the four-lane. A few petroglyph panels are visible from

Raven's Exile: A Season on the Green River (New York: Henry Holt, 1994).

the road. We cannot study them. We cannot get off the highway. No exit. The panels pass in a blur, ancient peeps drowned by billboard shouts: IT'S THE REAL THING.

The flanks of the Tushar and Pavant ranges tip us into the Parowan Valley, where we nose the truck south into the current of traffic through Mormon farm towns, each with identical, master design brick churches surrounded by weekday-empty aprons of tarmac. Only Kmart has more parking lot. Solid and impervious, the churches may be rocket ships in disguise. When the Rapture comes, the Saints will simply hop in and blast off, smothering the apron in the dense vapor of afterburners without singeing a leaf on God's flora. Near Cedar City I glimpse a road kill that may or may not be a poodle flung from a recreational vehicle. At a rest stop a teenager lifts his muscle shirt and stares at his navel. We're closer, I think. We have entered the gravitational field.

Most of the billboards in St. George advertise Nevada casinos, luring Utahans over the nearby state line to Mesquite or Las Vegas, the pull on their retirement dollar stronger than the pull of their faith. Flanked by the Beaver Dam Mountains and Hurricane Cliffs, St. George hemorrhages subdivisions and factory outlet malls and a lunatic compulsion to have the most golf courses in the universe, irrigated by the Virgin River, soon to be dammed, IMAXed, and deflowered of rare desert tortoises. Perhaps St. Georgians deserve all the golf they can muster. Many are Downwinders, human receptacles of nuclear fallout that scars their lives with seemingly endless tragedy.

During the atmospheric nuclear testing in the Nevada desert west of St. George from 1951 to 1962, it was the Atomic Energy Commission's practice to wait until the wind blew toward Utah before detonating its "shots" in order to avoid contaminating populous Las Vegas or Los Angeles. An AEC memo declassified two decades after the test era described the people living in the fallout's path as "a low-use segment of the population." Loyal to a government they believed to be divinely inspired, taught by their church never to challenge authority, assured by that authority that the radiation was harmless, Utah's patriotic Mormons endured the toxic showers with little objection.

When Utahans and Nevadans reported their symptoms and fears, public health officials told them that only their "neurosis" about the bombs would make them ill. When women reported burns, peeling skin, nausea, and diarrhea—all symptoms of radiation sickness—when they said their hair, fingernails, and toenails fell out after a cloud of fallout passed over them, their doctors wrote "change of life" or "housewife's syndrome" or "recent hysterectomy" on their charts. The dangers of radiation were known but suppressed, a "noble lie" deemed a necessary cost of national security and the fight against communism. Bomb after bomb exploded, some of them, like Shot Harry in 1953, extremely "dirty" and lethal, showering fallout throughout the West. Each nuclear test released radiation in amounts comparable to the radiation released at Chernobyl in 1986. In at least two ways the Nevada tests were nothing like Chernobyl: There were 126 detonations. None was an accident.

At the Nevada state line we cast aside Utah's wholesome aura for its nemesis. Behind: Leave It to Beaver. Ahead: Sodom and Gomorrah. In dusk that sizzles at 103 degrees, the land sprawls in bowls of creosote bush cupped by serrated ribs of rock. Over a long rise, past a convoy of trucks afloat in mirages of diesel and heat, we top the crest of the final ridge and behold the valley below, an island of neon capped in sludgy brown smog, ringed by a rabid housing boom. Las Vegas. The meadows.

We grind down the freeway past warehouses and a cinder block wall over which a life-size white plaster elephant, rogue prop from a theme park, curls its trunk, flares its ears, and rests ivory tusks on the barrier that separates its lunging charge from the highway's shoulder. Oleander bushes, carbon monoxide-tolerant but poisonous in their own right—they once offed a few Boy Scouts who peeled their thin branches, impaled hot dogs on their tips, and roasted a lethal meal—line the freeway then surrender to a chute of concrete, where we fly without air-conditioning in the gridlock of an exit bottleneck, surrounded by chilled limousines and Porsches. No one leaps out to save our lives. The ambient light is pale yellow, like the inside of a banana peel.

Why this pilgrimage from Desolation Canyon, our home on Utah's Green River, to Glitter Gulch, from cougar-blessed red-rock wilderness to the apex of engineered fantasy, from mesmerization to masochism? Why have we ventured so far from the river? Because our river is here beneath our smoldering, heat-frayed, about-to-explode radials. Only in Egypt are more people dependent on the flow of one river than the people of Clark County, Nevada. By controlling the Colorado River through the state's southern tip, Nevadans freed themselves from the constraints posed by puny, ill-timed rainfall that otherwise barely sustained darkling beetles, chockwallas, and creosote bushes. In this century no place has been too remote or too parched to reach with a lifeline, and the Colorado River, by this point carrying water from the Green, San Juan, Virgin, and other tributaries, is Las Vegas's intravenous feeding, its umbilical to prosperity, the force that pulsates the neon through the tubes. Here the River immolates its wild treasures on the altar of entrepreneurial spirit. We have chosen to devote much of the

West's greatest waterway to this city. Las Vegas is the twentieth century's ultimate perversion of the River and the site of a twenty-first-century water war.

For every river rat this visit is mandatory. We cannot know the River until we know this place. Our pilgrimage also carries corollary missions. I hope to learn what Las Vegans know about their water. There is field research to be done. And I want everyone in the Excalibur Hotel and Casino, a massive, pseudomedieval, castellated grotesquerie with jousting matches, banquets, and 4,032 hotel rooms—4,032 *toilets*—to flush their toilets at precisely the same moment.

I wait in the truck while my husband, Mark, registers at the hotel, the only vehicle-enclosed human in Nevada without a veneer of tinted safety glass between her and the rude assault of Real Air. I cannot go into the hotel because Real Air has fused my skin to the Naugahyde panel inside the truck door. My earrings, a Hopi man-in-the-maze design inlaid in silver, conduct so much heat, they sear man-in-the-maze-shaped burns on my neck.

The second thing Mark says to the waitress as we pump freon through our organs inside an air-conditioned restaurant: "Are you real?" She has a practiced tolerance for stupid questions and a tattoo on her left breast. The menu offers an entree called Heavy Trim Beef Primals. "I'd like a cheeseburger, please, hold the onions," Mark says, Green River sand spilling from his cuffs as he passes her the menu. "Are you real?"

The restaurant seethes with slick-baited bloodsuckers in shark-skin suits on cappuccino breaks from their drug harems and sieges of women wearing very short skirts who should not, Vegas being the one place where they can get away with this. The bun-grazing skirt on the cigarette girl remains immobile as she vigorously diversifies her cigarette-shy market by peddling illuminated Yo-Yos. The diners' sunburns, freshly acquired while powerboating on nearby Lake Mead, radiate sufficient heat to melt the ice in our water glasses. While we played Lost Tribe of the Oligocene on the river, male strippers became passé and musical revues with full-figured dancers became the rage: SENSATIONAL. TALENTED. PUDGY, proclaims one flashing Strip marquee.

"What are you in the mood for?" Mark asks about the evening's casino crawl. "Knights? Rome? The circus? The tropics? Urban South American festivities?" We settle on the Tropicana, an island-theme concoction whose grand entry sprouts the huge plaster heads of tiki gods from tidy plots of stale-smelling hothouse petunias, ferns, fountains, and sprinkler heads pumping liquid no faster than the desert air can evaporate it. The fountains, a bartender informs us, use wastewater recycled from guests' rooms. Despite his admonitions and fervent offers of bottled designer

water, we down tap water by the gallons, never slaking our thirst. The bartender knows where his water comes from: Lake Mead, he says. We slug it
down. Chlorine Lite with a bouquet of Evinrude.

In Las Vegas, the best survival strategy is a wholesale reduction of
Self to imbecilic dipstick, easily managed in these clockless, window-
less mazes of flashing lights and blaring gaming devices with nary a
molecule of The Environment allowed across the transom. The idea is
complete disconnection from Earth, a realignment of the senses
through a techno-collage of myths and fantasies conjured by corporate
hacks. At the Tropicana, I inspect each potted palm for signs of life.
Then we transfer to the Río, Where It Is Always Carnival and not much
different from the other casinos save for the Brazilian motif and the
tiny televisions mounted above each video poker machine. I peer into
the foliage of potted banana trees, expecting at least a cricket. No palm,
no leaf, no pot is real, only the cigarette butts.

Mark disappears, mumbling about the anthropology of dentalfloss
bikinis and a stripper named Bunny Fajitas. Before I'm trampled to
death by a shriek of Rotarians from Pocatello, I duck away to rest on an
outskirt, unused stair step. From there I watch a terrified woman in
bright native African dress clutch the rail of a descending escalator in a
death grip. At the escalator's foot, her family nurses her down in their
melodic native tongue—from Senegal, perhaps, evidently an escalator-
less nation. She survives. Everyone hugs. Hoover Dam's turbines juice
the guitars and keyboards of a live band in the lobby. Smurf Intellect,
Los Deli Meats, Heavy Trim Beef Primals, I didn't catch the name but
the lyrics concern whips. A man in a crisp white shirt and dark slacks
(waiter? missionary?) tells me I cannot sit on this step. I cannot sit anywhere, he asserts officiously, except on the stools at bars, poker and slot
machines, and blackjack tables. He stares down his nose at me as if I
had dripped cobra spit on his shoes and barks, "You must leave."
Where's the river? Take me to the river. Take me to Senegal. At the
Excalibur no one can be persuaded to induce hydro-gridlock by a
simultaneous political flush of their toilets. Water simply seems too
bountiful; it fills hoses, sprinklers, fountains, waterfalls, water slides,
swimming pools, wishing wells, moats, fish tanks, and artificial lakes; it
greens an epidemic of golf courses and chills a million cocktails.

A grown man in scarlet doublet and mustard yellow panty hose
plops a tinsel wreath on my head and recites a sonnet in bad
high-school Chaucer, prologue to a halfhearted sell on tickets to a
jousting tournament. Somehow he knows I'm not the jousting type, but
he lets me keep the wreath. A woman standing next to a video poker
machine catches my eye: Liv Ullmann face, shorts, running shoes, a

thick blond braid down her back, a dippy smile across a tanned face. She is singing from *The Sound of Music.* In strikingly muscular arms she clutches a grocery bag filled with folded newspapers. She rivets her gaze on the video machine as if it were Christopher Plummer or an Alp and hefts out, "The hills are alive . . ."

Daft with the sheer profusion of man-made matter, Mark and I return to our hotel room and fling ourselves onto the bed, hot, weighty sheets draped over our fantasy-stuffed bodies, our feet protruding like Jesus' under the shroud in Mantegna's painting *The Lamentation Over the Dead Christ.* Sometime in the fitful night, a voice crackles over the intercom box above the bathroom doorway. "Please do not panic," the voice urges us. "The fire alarms mean nothing. Please stay in your rooms."

The river of traffic streaming down the Strip will kill me if I back up three feet off the boulevard curb, where I'm in the bushes risking my life to study nature in Vegas's endangered vacant lots, its postage-stamp plots of unpaved Mojave. The inventory so far: crickets, ants, pigeons, wind-strewn "escort girl" flyers as numerous as scutes on a pit viper, and a playing card (the king of spades). Cowbirds (those toxic parents!) chase kazooing cicadas through muffler-sizzled oleander bushes too spindly in foliage to hide the random upturned shopping cart or shade me from sunlight intensified by its infinite reflection off chrome and windshields. I observe one stunted specimen of Aleppo pine, *Pinus halepensis,* a drought-tolerant Mediterranean import largely relegated to freeways and residential areas. I find few bugs in the bush and plenty in the yellow pages under "Pests": termites, earwigs, roaches, pill bugs, silverfish, scorpions, plus rodents and a category called "olive control." Physiographically the Mojave Desert is a transitional province between the Great Basin to the north and the Sonoran Desert to the south. Biological boundaries of all three deserts mix here, so one would expect creosote bush, catclaw, mesquite, yucca, geckos, horned lizards, and the like. But hardly a particle of native flora or fauna lives in Strip habitat. I crawl out of the bushes and hike to safety. Off to find the meadows, *las vegas.*

Negligible rainfall, barely four inches annually, comes to the austere bowl of desert in which Las Vegas spreads. Over a century and a half ago, a carpet of spring-fed grasslands grew in this basin, an oasis in a sea of thorns, alkali, and dust. Except for an occasional flash flood through the washes, the nearby mountains flushed little moisture from their peaks. The basin's water came from an underground aquifer created during the Pleistocene, when rainfall was abundant. Big Springs surfaced in a mad gurgle to form the headwaters of Las Vegas Creek, which flowed easterly along the valley floor, then disappeared into the sand. An exploration party in 1844 recorded the creek's temperature at 115 degrees. Eleven years later a Mormon mission watered travelers between Salt Lake City and California settlements. The missionaries also mined lead from an ore vein along the nearby Colorado River and shipped it north to be made into bullets by the church's public works unit. The missionaries took it upon themselves—these were busy people—to teach the Indians, mostly Paiute, "farming and hygiene," although no one bothered to ask the Indians if they cared to farm or needed help in attending to their bodies. Nineteenth-century zealotry seemed obsessed with putting natives behind plows, in pants. "Discontent with the teepee and the Indian camp," claimed Merrill Gates of the U.S. Board of Indian Commissioners in the 1880s, "is needed to get the Indian out of the blanket and into trousers—and trousers with a pocket in them, and with a pocket that aches to be filled with dollars!"

By 1907 wells tapped much of the groundwater. Their strength— good water at constant pressure—and cheap land lured more settlers, who drained the meadows for crops and pasture. For nearly fifty years water flowed into farm, pipe, and oblivion; no one capped the wells until 1955. Las Vegas Creek had dried up five years before. Big Springs, now under pavement and the lock and key of the municipal water district, surfaced no more, and parts of the Las Vegas Valley had subsided as much as five feet, so much water had been mined. The meadows disappeared but for a trace, I was told, at Lions Club and Fantasy Park near downtown Las Vegas.

I drive to Fantasy Park on a boulevard that parallels a brief stretch of creek straitjacketed by concrete riprap. The creek begins and ends in enormous culverts; it merely belches aboveground for a few blocks so people can throw their litter into it. Fantasy Park grows limp-leafed trees in even rows, and despite a posting that the park is for children twelve and under, a few prostrate bodies of napping transients drop bombs of drool into a rather seedy lawn. Casino blitz envelops the park, buffered by mortuaries. Downtown Las Vegas, once heartland of the economy of sin, is now an outlier to the upscale Strip. Unless razed, it has no space for the entertainment mall, the computer-programmed volcano, artificial rain forest, concourse of Roman statuary, circus, castle, or thirty-story pyramid.

However outstripped by the illusion vendors of the nineties, surely downtown Las Vegas scores highest for the Stupidity of Man exhibit's best archival photograph. The 1951 photograph shows Vegas Vic, a landmark, sixty-foot-high neon cowboy on the cornice of the Pioneer Club, beckoning the pilgrims to girls, gambling, and glitz. His thumb is up, his cigarette dangles from his lips. Behind Vegas Vic and the cityscape rises a white-hot cloud on a slender stem, one of the atom wranglers' earliest nuclear bombs, popped off on ground zero less than a hundred miles away.

In Fantasy Park the homeless nappers awaken and roll off what would be the meadows' last stand had a lawn not replaced them. One of the men zombie-walks across the turf to the Binary Plasma Center. Two others approach me for spare change, grass clippings stuck to their sweaty T-shirts. I donate my Fun Book, a collection of courtesy coupons for drinks, playing chips, and discounts at beauty parlors. Casually I ask them where Las Vegas water comes from. The answer is unanimous: the faucet.

Las Vegas's faucets feed one of the highest per-capita water consumption rates in the nation, serving over 800,000 residents, twenty million visitors a year, and a monthly influx of several thousand new residents, most of them quality-of-life refugees from California. To feed the housing boom and the gaming industry's insatiable quest for the next great attraction, Las Vegas will likely be using every last drop of its legal share of Colorado River by the year 2002. It has considered buying water from a desalination plant in Santa Barbara, California, to trade with Los Angeles for rights to more Colorado River water. Las Vegas secured the last of the unappropriated groundwater in its own valley and seeks unclaimed water from the nearby Virgin River. It has also applied to import water from aquifers beneath the "empty" basins in Nevada's outback—fossil water, the ancient rain stored since the Pleistocene and rationed to the surface in spring creeks and seeps that give life to bighorn sheep, fish, lizards, plants, birds, and ranchers. The controversy pits rural Nevada against Las Vegas, sparking memories of a water grab by another lifestyle-obsessed megalopolis: the plumbing of eastern Sierra Nevada runoff by the city of Los Angeles during the early century, an exportation that drained the Owens Valley nearly dry. Sierra water, stored in snowpack, renews itself. Nevada's aquifers would be mined.

While everyone tries to predict the nature of a twenty-first century water war, thousands more newcomers unpack and scream for faucets. Unless a tarantula leaps up and bites off their lips, few seem to notice they live in a desert. At the Las Vegas Natural History Museum, my next research stop, the feature exhibit is a three-hundred-gallon tank swarming with those fascinating Mojave Desert endemics: live sharks.

        *   *   *

Las Vegas makes no bones about its premier commodity—honest fraud—but I don't care much for the place. The exceptions, however, are the pink tongues on the pudgy white tigers in their all-white new-Babylonian habitat box on the entry concourse of the Mirage Hotel and Casino. Each time I visit the tigers, they sleep behind their plate-glass shield, their languid, potbellied bodies sprawled across elevated benches, the sweet tongues drowsily lolling below exquisitely whiskered cheeks. The Mirage sucks a river of people off the Strip onto its moving sidewalks, channels them past the narcoleptic cats and a wall-sized aquarium of parrot fish, wrasses, angelfish, sharks, and other tropical prisoners, and spills them into the tributaries that flow to gaming rooms, bars, shops, and restaurants. Earlier I had seen the Sound of Music woman sleeping on a patch of Strip lawn, a bag lady with one grocery bag and the body of a marathon runner. Now she is here, singing to the poker machines, and I would gleefully join her had I not the singing voice of gargled bats. Like mobile tide pools, a shoal of Frenchmen in bright aloha shirts riffles noisily forward with the stream. Perched on bar stools like herons on a riverbank are Vegas's sunset women, hard-fleshed, sinewy women in crayon makeup, pink stilettos, and gazes to convince the most egocentric lout that they know far more than he does. These women should be allowed to run Las Vegas. They probably do.

In the bar beneath the Mirage's artificial rain forest, Mark sips a herbivore's daiquiri afloat with Chinese parasols, fruit, carrots, celery, and other verdure. He scouts for naysaying casino personnel while I dive under the table and crawl around the rain forest in search of wildlife. The thicket grows bromeliads, ferns, philodendrons, cricket noises, and roof-raking palm trees that thrust fat boles up through the epoxied floor. The philodendrons are real. I emerge, harvest the crop from my daiquiri, and study the couple across from us, whose furtive looks reveal that some outlaw love may soon be consummated.

Our cocktail server, who thinks her water comes from California but is not sure, enlightens us about the construction crews that were furiously ingesting the Strip's remnant open spaces. We had seen the activity earlier in the day, and we wondered about the new building in the parking lot behind the Circus Circus Casino.

"What are they building at Circus Circus?" Mark asks.

"That's the Grand Slam Canyon," she tells us, clearing the table of peach pits, orange rinds, celery leaves, kelp.

Grand Slam Canyon promises the Grand Canyon without the Grand Canyon's pesky discomforts—its infernal heat, wind, roadlessness, and size that defies the three-day vacation, its cacti, lizards, snakes, biting insects, burro poop, boulders, rapids, the possibility of death. Amidst hundred-foot peaks, swimming pools, water slides, pueblos, and a replica of the Grand Canyon's Havasu Falls, inside a climate-controlled, vented, pink womb of a dome, Grand Slam Canyon visitors will fly through rapids and waterfalls in a roller coaster. The River made better than itself.

By midnight my tongue is furry and dry, as if I had swallowed a mouthful of casino carpet. We walk outside the Mirage, where a hundred or more spectators watch a volcano erupt in the palm garden, upstaging a rising moon, spewing fire from propane burners and sloshing wastewater down its tiered slopes. Out from nowhere a single, frantic female mallard duck, her underside lit to molten gold by the tongues of flame, tries desperately to land in the volcano's moat. Mark and I stare incredulously at the duck, two faces pointed skyward among hundreds pointed volcano-ward. Unable to land in this perilous jungle of people, lights, and fire, the duck veers down the block toward Caesars Palace. With a sudden *ffzzt* and a shower of sparks barely distinguishable from the ambient neon, the duck incinerates in the web of transmission lines slicing through a seventy-foot gap in the Strip high-rises, a skein of wire and cable that surges with the power of the River.

EMILY HIESTAND
b. 1947

Trained as a visual artist at the Philadelphia College of Art, Emily Hiestand began her literary career as a poet (Green the Witch-Hazel Wood, *1989*). In The Very Rich Hours: Travels in Orkney, Belize, the Everglades, and Greece *(1992) she uses the form of the extended travel essay as a way of "testing premises" about human nature and how culture and environment interrelate. In her second book of essays,* Angela the Upside-Down Girl and Other Domestic Travels *(1998), her travels are closer to her home in Cambridge, Massachusetts, where she celebrates the natural and human spirit in an urban setting. But whatever and wherever her subject, Hiestand's writing is characterized by a passionate appetite and response, a remarkably astute and cultivated intelligence, and a style at once sensuous, elegant, mischievous, wry, generous, and visionary.*

ZIP-A-DEE-DO-DAH

For more than fifteen years, my husband, Peter, and I have lived on the top floor of a triple-decker in a small urban village in metropolitan Boston. As you may know, a triple-decker is a house style indigenous to New England, a big, boxy design that stacks three identical apartments, then adds enough Doric column, balcony, and parapet so the imposing front façade resembles a domestic temple for the American worker. Three-deckers emerged in our city in the late 19th century, largely in response to gangs of do-right Boston women protesting tenement slums so miserable they provoked the word "Calcutta." Simple, but airy and livable, the triple-decker was an enormous step forward for workers, a house of bread and roses. Not long after our triple-decker was built, someone had the good idea to surround it with trees. These many generations later, Peter and I look out from our third floor flat directly into an arabesque of limbs and leaves, into the upper canopies of an oak, an aging apple, and a sturdy Norway maple. And, closest to our rooms, so close its limbs brush the windowpanes and blur the boundary between tree and house—a tall, willowy wild black cherry tree.

We have always loved this tree especially, tending and observing it in all weathers, and seasons, and moods. Each spring the black cherry has also appealed to a pair of nesting blue jays; as I am writing it is May, and this year's jays have already arrived and are commenced on the project of building their nest, a task for which ornithologists have the naturally comic word *nidification*. As I watch the blue jays making their nest, moving in slo-mo so as to not startle them, I am fluctuating between a quiet panic—at having a life so marginal I can spend most of a morning watching blue jays nest—and the feeling that this event is happening at the hub of the universe. That stabilizing feeling comes to me courtesy of Black Elk, the Oglala Sioux medicine man who said of this world, "The circumference is nowhere; the center is everywhere." The Oglala idea is stunning; it conveys dignity and mystery on every place and life, and counters the swirly, centrifugal energies of a culture in which the center seems always *out there somewhere*, perhaps wherever Jennifer Lopez is.

Until we moved into this house, I had never seen a wild black cherry. It is a beautiful tree that grows in the eastern half of the country as far south as northern Florida. The youngest limbs are wrapped in a smooth, lustrous bark flecked with ruddy gold nicks. Over time, the swelling of the cambium layer causes the youthful bark to burst and split into sepa-

Angela the Upside-Down Girl and Other Domestic Travels (Boston: Beacon, 1998).

rate sections. As they age, these sections thicken, maturing finally into the rough, curling plates in which the older limbs and all of the central trunk are clad. The leaves are slender, the shape of canoes: a sea of bright green boats in spring, in fall, a fleet of yellow. When fully leafed, the black cherry canopy filters the afternoon sun so that warm western light shimmers over the plaster walls of our kitchen, making of a solid something more like water. The cherries themselves emerge in late spring as hard green dots; by early August, the fruits are plump, and as shiny and purple-black as the poison berries of fairy tales.

No sooner are they ripe than the black cherries are discovered and set upon by a horde of immensely happy grackles, two hundred birds, maybe more, swirling suddenly out of the hot August sky, descending in droves on the tree by our window. The grackles rustle and dart through the leaves, flashing their dark iridescence, which looks so much like motor oil on water. Wild with excitement, the birds unleash a high, twittering ruckus that can fill an entire urban block. They will haunt the tree for a fortnight, until they have plucked it clean, and all that time the grackles make as much joyful noise as the juiced-up cars that sweeten our streets on summer nights, pumping out rap rhymes and dance jams on Bazooka brand speakers.

In May, however, when the grackles are not yet on the scene, the black cherry tree belongs to a pair of jays. This year, as always, the birds have chosen to nest in a junction where three limbs meet and make a shallow pocket. Everything about this little wooden platform—its shape, and size, and location—must speak to the blue jays, must say in their pattern language, "This is perfect." For three days, the birds labor over their nest, pausing only now and then to cock their heads, to review construction, and to emit their suite of calls: a musical *queedle queedle*, a slurring *jeeah*, the sharp, namesake *jay*. In that decadent moment before you remember your scientific manners, you might think the two birds resemble a married couple building a barbecue pit over a long weekend. When the pair has gathered up enough material, the female plops herself in the middle of the mound, and begins squirming and shifting around, molding the interior to the shape of her breast. I believe the female is the only one to fit the nest to her body. I am not sure about that. I am not a student of birds, although I have on occasion traveled with serious birders to blinds and sanctuaries, and have watched them—the birders—for many hours and been very moved by their behavior.

Still, I am going to go out on a limb here, and guess that the kind of nest these blue jays are making has never provoked anyone to an encomium to nature's symmetry and perfection. The thing taking shape outside our window is no chambered nautilus shell, with its faultless, secreted spiral of form, often invoked when someone wants to take seriously the notion of a great designer, the This Is All Just Too Exquisite To Be Random argument. Here are some other things the blue jays' nest does not resemble: the fine teacup made by the ruby-throated hummingbird; the public works facility engineered by the rufous-breasted castle builder; the evening bag of the Baltimore oriole, a nest that any Upper East Side woman would carry to the opera proudly; the Greek vases turned out by cliff swallows; the flying saucer of the hammerkop, eight thousand twigs formed into a dome strong enough for an adult to stand on. None of these possibilities for ordering and smoothing out chaos has much impressed the blue jay, and at the end of all their labors, what the blue jays' nest most closely resembles is—a heap of trash.

And that is what I like about it. The blue jay's nest is a fantasia of refuse, a temporary, provisional architecture made of discarded materials plucked from the sidewalks, gutters, and yards of our neighborhood, a landscape that teems with the detritus so attractive to a blue jay eye: foil-coated lottery scratch tickets ("Set For Life," "Pharaoh's Gold"); Popsicle sticks still sticky with grape goo; twine from the morning papers; the silvery inner liner of the Kit-Kat candy bar. In his *Guide to Eastern Birds*, the great man of aves, Roger Tory Peterson, places the blue jay on the same page as the black-billed magpie, the creature of this and that. In another kind of guide, I would place the blue jay with the great collage artists, with Kurt Schwitters, Joseph Cornell, Jean Arp, and Louise Nevelson, all those bricoleurs who take the occasion of a fragmenting world to practice a recombinatory art, linking decorum and glitz, high and lo, the funny and elegiac, making a moody frisson of the commonplace.

And however motley its heap, the blue jay is surely guided, no less than the meticulous nautilus, by some inscribed-in-nucleic-acid knowledge. This is the blue jay way, the way the blue jay continues its kind. So these bricoleurs of the upper canopy know a good heap when they see one; and they know when that heap is fully realized. When it is, the female takes up residence, and at some point—I've never seen this—she lays her eggs. Over the next few days, for the rare, fleeting moments when she hops off her nest, anyone close by our window, waiting or just lucky, will see four tiny oval eggs. There is only a glimpse: the eggs are smooth, with a faint gloss, some years an olive color, others, the gray-blue-green of the Atlantic on a cloudy day.

And does the blue jays' affection for the motley and discarded give them an edge in the urban world, an advantage over fussier birds like the chickadee and lazuli bunting? I must ask a serious birder, a man named

Emerson Blake, known as Chip. The answer from Chip is yes—an elaborate yes that builds into a riff on why birds are particular in the first place; why some are disadvantaged by a metropolitan scene; why others are having a hard time making any nests at all these days. One of the hard cases is the situation of the spotted owl, a bird that wants peace and quiet, and wants it over an immense territory, over a great hushed swath of forest. As that kind of forest disappears, the spotted owl's endocrine system begins to shut down; its hormones cease to deliver the old imperative. Other birds are very particular about building materials; if the right twig or grass is not present, they simply do not nest. It's not whim, of course. The absence of proper materials is a leading indicator, a potent sign that larger conditions for life are not right, that the energy and effort to create young would likely fail. Birds that require certain materials or foods to breed are known as specialists. The advantage of being a specialist can be great: it often allows a bird to thrive in some uncontested niche. Think of it—a strategy that allows a creature to smoke, not only its competitors, but also *competition itself*. Ingenious. Thus Bachman's warbler prevailed in southern canebrakes. Thus the ivory-billed woodpecker lived undisturbed in virgin pine forests. Thus snail kites in Florida eat only the apple snail, for which purpose they have grown a special beak. But the risk of being a specialist is the highest risk imaginable, a double or nothing shot, for if canebrake, or virgin forest, or apple snails disappear, the specialist is, as Chip puts it, "out of business."

That was the fate of the dusky seaside sparrow, a small bird perfectly adapted to the tidal salt marshes along the Florida coast, and not counting on Cape Canaveral, nor the draining of the marshes for mosquito control. It is likely that a little planning by our species would have spared the niche of this sparrow—a bird whose name alone, the *dusky seaside sparrow*, so beautiful to say—made it worth sparing. A specialist like the California condor is another story. The condor's idea of what Southern California should be is now so profoundly at odds with what Southern California has become that the condor can probably no longer survive without perpetual human assistance.

Our birds, the blue jays outside our living room window, are neither specialists nor maladapted to this century. In fact, blue jays are the most general of generalists. More intelligent than many birds, they are able to withstand competition on several fronts, and if out-maneuvered, think nothing of taking up life in another site. "The blue jay," Chip muses, "is almost too adaptable." By which he means, he explains, that the success of generalists can mask the demise of specialists, who have the more sensitive bonds to place. Birders like Chip know what once existed, and they miss it. I believe that the adaptable blue jay may make them feel the way I do when I find that a Taco Bell hut has replaced a Southern diner that served grits and red-eye gravy. (It was an aluminum diner, with pale green and black booths, and a palm tree painted on the side.)

Adaptability, however, is what allows blue jays to nest on the margins of an old metropolis, and I am glad for that. Naturally I gasp at the colorful gems that live in the dappled suburbs ringing our city—the golden birds, the wax-winged, and cerulean birds. But on this street, where the airborne population is pigeons, grackles, and the occasional promotional blimp, the bright blue jay passes for beauty. And although I (who require cups of Earl Grey tea every morning, who will go miles for grits and red-eye gravy) am much in sympathy with the specialists, I deeply admire the blue jay's resiliency. Indeed, I study the blue jay's ways. This is the bird of buoyant making-do, the bird of inventive recycling, of flexibility and found art.

And, of course, this is the bird that comes to our window. It has come to us, like the puppy that toddles across the room from the cardboard birthing box, who puts its head in your lap, and chooses you. When life comes to you like that, you refuse it at your own peril. So I am a partisan of the jays, I take their part and root for their eggs. And however successful blue jays are in the larger picture, on this street their eggs are greatly endangered. The danger comes in the form of squirrels, those Visigoths of the urban forest who travel by telephone cable, who can raid a nest and turn four blue eggs into a glob of yellow slime in under a minute. In less time than it takes to watch Samuel Beckett's shivery play *Breath*, in which the complete stage action is: "Curtain rise; Cry and one breath; Curtain fall."

It can be just that swift and iconic with these nests. Sometimes the wild black cherry is a nursery, sometimes a bare ruined choir. In ruin, however, the blue jays are exemplary. They do not stand around in stunned silence and in mute disbelief, as we did recently absorbing some mournful news. They do not reel between the great cosmic equanimity where all must be balance, and the immediate realm where they have been roughed up in the worst possible way, according to Darwin. They seem not to mull the more-than-human scheme of justice, variously felt as a benevolence whose eye is on the sparrow, as a magisterial indifference, as a mocking voice in a whirlwind. The blue jays whose eggs have been eaten, whose nest has been upturned, merely fly away, on those coveted wings.

Eggs, they know very well, are fair game in the gulping world. The eager mouth of the ocean swallows most of its own children, and as it turns out, the blue jays' own favorite food is—other bird's eggs. Our very own jays, with whose loss I empathize, may have been out destroying other nests. "Trash birds," another of my bird world informants says of the blue jay, sniffing. "They're like the roughnecks of Dickensian London," he adds dismissively, "doing whatever they can to get by, and not thinking too much about the ethics of it." Listening to this friend get pretty steamed about the riff-raff of the bird republic, I start to wonder if we Americans have decided to deny that class exists in our society mostly so that we can have the fun of unconsciously projecting it onto flora and fauna.

Well, obviously I do not defend the blue jays' eating other bird's eggs. They should stop that, and also they should start eating more bright-orange and leafy green vegetables, more soy, and less fat. I honestly cannot vouch for our birds. I don't know what trashy things they might be doing when they are not close by our window, working, brooding, conveying warmth, being brave and patient, being all that parents can be. Let's just say that when a creature lays four speckled eggs close by your house, you like for those eggs to hatch.

So I go on rooting for the blue jays' eggs, and because the odds for the eggs are long, I root with a certain kind of hope. It is not the usual kind, defined in my OED as "desire combined with expectation." And it is certainly not the hope described as "expectation *without* desire," which becomes the near-certainty of *prospect*. The kind of hope I must hold for these eggs is hope without any expectation at all. This is the kind of hope that will be recognized at once by seasoned Red Sox fans, and fans of the woebegone Chicago Cubs. It is a brand of hope far from pie-eyed optimism, and very close to the appealing, grown-up mood that the French name *une douce resignation*.

In our new world, the adjective that most often appears before "resignation" is "bitter." But *douce resignation* is not the bitter, defeated mood so repugnant to the American spirit. Although it is, of course, *resigned* to the fact that the world is "a funny old world" (as Margaret Thatcher put it the week she was—hooray—booted out), this mood is sweet. A sweet resignation arises from the formidable effects of time and kindness, chaos and tragedy. It grants the unpredictable play of these great forces. The Belgian born poet, my friend Laure-Anne Bosselaar, tells me she finds Americans too doggedly, even eerily optimistic for *douce resignation* (she doesn't know about baseball), while her people, the Flemish, are too taciturn for the mood. The spirit of sweet resignation flourishes most in Mediterranean regions, in southern France and Italy, in Corsica and Spain. And a Buddhist pal tells me that *douce resignation* seems a close cousin to his practice of *detachment*—a way that fuses passionate caring with letting-be, a way of moving within the world's own startling motley of loveliness, nests, violence, sadness to the bone, and summer games.

This spring, in the penultimate year of the millennium, the nest outside our window has grown into the most extravagant heap yet. This year's birds have made the nest equivalent of Sabato Rodia's Watts Towers, those *sui generis* assemblies of steel, seashells, and broken glass. The masters Cornell and Schwitters would smile on this year's nest. Joseph Cornell might even want it for one of his elegant boxes. *Regardez*. Interwoven with the staples of twigs and dried grasses are these nesting materials: green and red telephone wire; a gum wrapper; part of a pre-tied drugstore bow; most of a label from a pretty good Beaujolais, George de Boeuf's Brouilly; and as the *pièce de résistance*, an object the birds have *queedled* and *queedled* over for hours—a large, white plastic picnic fork. The birds wrestled long with this prize, patiently nudging it into place, dropping it to the ground, flying it up again, finally getting the handle end embedded into the nest, with the tines pointing outwards so the whole utensil bristles from the nest at a jaunty angle, like a miniature pitchfork.

Is it something to give the destroyers pause? Will a plastic fork help this pair and their eggs make it over the Darwinian divide? We shall see. I can say with confidence that the fork came from one of two nearby spots: from Marcella's on the avenue, which makes the Classico sandwich of prosciutto sliced to translucency; or from the House O' Pizza, where, if you ask, Sal will make a sub that does not appear on the menu, a sub that is meatless and cheeseless but with all the condiments, including hot peppers—a superior sandwich that Sal and I have settled on calling a Nothing With Everything.

LINDA HOGAN
b. 1947

Like many outstanding Native American writers, Linda Hogan combines a highly localized sense of tradition with a global perspective. Her Chickasaw background is a subject explored in Hogan's poetry and fiction. But the question to which she often returns is, How are all of us, both human beings and the other species who are our kindred, to live? *Over the past decade, she also has produced a number of powerful essays that speak to these central themes. The essay reprinted here, "The Bats,"* comes from a collection of her essays entitled Dwellings: A Spiritual History of the Living World *(1995).*

THE BATS

The first time I was fortunate enough to catch a glimpse of mating bats was in the darkest corner of a zoo. I was held spellbound, seeing the fluid movement of the bats as they climbed each other softly and closed their wings together. They were an ink black world hanging from a rafter. The graceful angles of their dark wings opened and jutted out like an elbow or knee poking through a thin, dark sheet. A moment later it was a black, silky shawl pulled tight around them. Their turning was beautiful, a soundless motion of wind blowing great dark dunes into new configurations.

A few years later, in May, I was walking in a Minneapolis city park. The weather had been warm and humid. For days it had smelled of spring, but the morning grew into a sudden cold snap, the way Minnesota springs are struck to misery by a line of cold that travels in across the long, gray plains. The grass was crisp. It cracked beneath my feet. Chilled to the bone, and starting home, I noticed what looked like a brown piece of fur lying among the frosted blades of new grass. I walked toward it and saw the twiglike legs of a bat, wings folded like a black umbrella whose inner wires had been broken by a windstorm.

The bat was small and brown. It had the soft, furred body of a mouse with two lines of tiny black nipples exposed on the stomach. At first I thought it was dead, but as I reached toward it, it turned its dark, furrowed face to me and bared its sharp teeth. A fierce little mammal, it looked surprisingly like an angry human being. I jumped back. I would have pulled back even without the lightning fast memory of tales about rabid bats that tangle in a woman's hair.

In this park, I'd seen young boys shoot birds and turtles. Despite the bat's menacing face, my first thought was to protect it. Its fangs were still bared, warning me off. When I touched it lightly with a stick, it clamped down like it would never let go. I changed my mind; I decided it was the children who needed protection. Still, I didn't want to leave it lying alone and vulnerable in the wide spiny forest of grass blades that had turned sharp and brittle in the cold.

Rummaging through the trash can I found a lidded box and headed back toward the bat when I came across another bat. This bat, too, was lying brown and inert on the grass. That's when it occurred to me that the recent warm spell had been broken open by the cold and the bats,

Dwellings: A Spiritual History of the Living World (New York: Norton, 1995).

shocked back into hibernation, had stopped dead in flight, rendered inactive by the quick drop in temperature.

I placed both bats inside the box and carried them home. Now and then the weight would shift and there was the sound of scratching and clawing. I wondered if the warmth from my hands was enough heat to touch them back to life.

At home, I opened the box. The two bats were mating. They were joined together, their broken umbrella wings partly open, then moving, slumping, and opening again. These are the most beautiful turnings, the way these bodies curve and glide together, fold and open. It's elegant beyond compare, more beautiful than eels circling each other in the dark waters.

I put them in a warm corner outside, nestled safe in dry leaves and straw. I looked at them several times a day. Their fur, in the springtime, was misted with dewy rain. They mated for three days in the moldering leaves and fertile earth, moving together in that liquid way, then apart, like reflections on a mirror, a four-chambered black heart beating inside the closed tissue of wings. Between their long, starry finger bones were dark webbings of flesh, wings for sailing jagged across the evening sky. The black wing membranes were etched like the open palm of a human hand, stretched open, offering up a fortune for the reading. As I watched, the male stretched out, opened his small handlike claws to scratch his stomach, closed them again, and hid the future from my eyes.

By the fourth day, the male had become thin and exhausted. On that day he died and the female flew away with the new life inside her body.

For months after that, the local boys who terrorized the backyards of neighbors would not come near where I lived. I'd shown one the skeleton of the male and told them there were others. I could hear them talking in the alley late at night, saying, "Don't go in there. She has bats in her yard." So they'd smoke their cigarettes in a neighbor's yard while the neighbor watched them like a hawk from her kitchen window. My house escaped being vandalized.

My family lived in Germany once when I was a child. One day, exploring a forest with a friend, we came across a cave that went back into the earth. The dark air coming from inside the cave was cool, musty, and smelled damp as spring, but the entryway itself was dark and forboding, the entrance to a world we didn't know. Gathering our courage, we returned the next day with flashlights and stolen matches. It was late afternoon, almost dusk, when we arrived back at the cave. We had no more than just sneaked inside and held up the light when suddenly there was a roaring tumult of sound. Bats began to fly. We ran outside to twilight just as the sky above us turned gray with a fast-moving cloud of their ragged wings, flying up, down, whipping air, the whole sky seething. Afraid, we ran off toward the safety of our homes, half-screaming, half-laughing through the path in the forest. Not even our skirts catching on the brambles slowed us down.

Later, when we mentioned the cave of bats, we were told that the cave itself had been an ammunition depot during World War II, and that bat guano was once used in place of gunpowder. During the war, in fact, the American military had experimented with bats carrying bombs. It was thought that they could be used to fly over enemy lines carrying explosives that would destroy the enemy. Unfortunately, one of them flew into the attic of the general's house and the bomb exploded. Another blew up a colonel's car. They realized that they could not control the direction a bat would fly and gave up on their strategy of using life to destroy life.

Recently I visited a cave outside of San Antonio with writer and friend Naomi Nye. It was only a small mouth of earth, but once inside it, the sanctuaries stretched out for long distances, a labyrinth of passageways. No bats have inhabited that cave since people began to intrude, but it was still full of guano. Some of it had been taken out in the 1940s to be used as gunpowder. Aside from that, all this time later, the perfect climate inside the cave preserved the guano in its original form, with thick gray layers, as fresh and moist as when the bats had lived there.

Bats hear their way through the world. They hear the sounds that exist at the edges of our lives. Leaping through blue twilight they cry out a thin language, then listen for its echo to return. It is a dusky world of songs a pitch above our own. For them, the world throws back a language, the empty space rising between hills speaks an open secret then lets the bats pass through, here or there, in the dark air. Everything answers, the corner of a house, the shaking leaves on a wind-blown tree, the solid voice of bricks. A fence post talks back. An insect is located. A wall sings out its presence. There are currents of air loud as ocean waves, a music of trees, stones, charred stovepipes. Even our noisy silences speak out in a dark dimension of sound that is undetected by our limited hearing in the loud, vibrant land in which we live.

Once, Tennessee writer Jo Carson stuck her hearing aid in my ear and said, "Listen to this." I could hear her speak, listening through the device. I could hear the sound of air, even the noise of cloth moving against our skin, and a place in the sky. All of it drowned out the voices of conversation. It was how a bat must hear the world, I thought, a world alive in its whispering songs, the currents of air loud as waves of an ocean, a place rich with the music of trees and stones.

It is no wonder that bats have been a key element in the medicine bundles of some southern tribes. Bats are people from the land of souls, land where moon dwells. They are listeners to our woes, hearers of changes in earth, predictors of earthquake and storm. They live with the goddess of night in the lusty mouth of earth.

Some of the older bundles, mistakenly opened by non-Indians, were found to contain the bones of a bat, wrapped carefully in brain-tanned rawhide. The skeletons were intact and had been found naturally rather than killed or trapped by people, which would have rendered them neutral; they would have withdrawn their assistance from people. Many Indian people must have waited years, searching caves and floors and the ground beneath trees where insects cluster, in order to find a small bony skull, spine, and the long finger bones of the folded wings. If a bat skeleton were found with meat still on it, it was placed beside an anthill, and the ants would pick the bones clean.

I believe it is the world-place bats occupy that allows them to be of help to people, not just because they live inside the passageways between earth and sunlight, but because they live in double worlds of many kinds. They are two animals merged into one, a milk-producing rodent that bears live young, and a flying bird. They are creatures of the dusk, which is the time between times, people of the threshold, dwelling at the open mouth of inner earth like guardians at the womb of creation.

The bat people are said to live in the first circle of holiness. Thus, they are intermediaries between our world and the next. Hearing the chants of life all around them, they are listeners who pass on the language and songs of many things to human beings who need wisdom, healing, and guidance through our lives, we who forget where we stand in the world. Bats know the world is constantly singing, know the world inside the turning and twisting of caves, places behind and beneath our own. As they scuttle across cave ceilings, they leave behind their scratch marks on the ceiling like an ancient alphabet written by diviners who traveled through and then were gone from the thirteen-month world of light and dark.

And what curing dwells at the center of this world of sounds above our own? Maybe it's as if earth's pole to the sky lives in a weightless cave, poking through a skin of dark and night and sleep.

At night, I see them out of the corner of my eye, like motes of dust, as secret as the way a neighbor hits a wife, a ghost cat slinks into a basement, or the world is eaten through by rust down to the very heart of nothing. What an enormous world. No wonder it holds our fears and desires. It is all so much larger than we are.

I see them through human eyes that turn around a vision, eyes that see the world upside down before memory rights it. I don't hear the high-pitched language of their living, don't know if they have sorrow or if they tell stories longer than a rainstorm's journey, but I see them. How can we get there from here, I wonder, to the center of the world, to the place where the universe carries down the song of night to our human lives. How can we listen or see to find our way by feel to the heart of every yes or no? How do we learn to trust ourselves enough to hear the chanting of earth? To know what's alive or absent around us, and penetrate the void behind our eyes, the old, slow pulse of things, until a wild flying wakes up in us, a new mercy climbs out and takes wing in the sky?

ROBERT MICHAEL PYLE
b. 1947

Robert Pyle first became known as the author of six books on butterflies, including The Audubon Field Guide to North American Butterflies. *A Coloradan by birth, he now lives in southwestern Washington State.* Wintergreen: Listening to the Land's Heart *(1986), winner of the John Burroughs Medal, is an ecological and philosophical survey of the Willapa Hills region near his home, an area that has been heavily logged-over in recent decades. The book confronts a central question in contemporary nature writing, namely, how does one construct a viable environmental ethic in the face of apparently incorrigible human behavior and a nature which doesn't care? For Pyle, as for many other current writers, the answer lies in the adoption of what he calls "cosmic optimism," an evolutionary "frame of reference that does not encompass human fortunes alone." While not rejecting the idea of responsible stewardship toward the earth, Pyle takes heart in the view that, in the long run, the actions of human beings, for good or ill, matter little. Yet the author's own good humor and zest in natural experience make this attitude seem anything but cynical and nihilistic. Among Pyle's recent books are* Where Bigfoot Walks *(1995) and* Chasing Monarchs *(1999).*

AND THE COYOTES WILL LIFT A LEG

I wish I had said it first. Someone else did—who knows who?—and won't get credit for it. Spinners of clever quotations share the relative immortality of their words sometimes; platitude-makers, almost never. Even so, I still wish I'd said it first; "nature bats last."

On the other hand, maybe this isn't a platitude. The thing about a "good" platitude is that it should be self-evident. I'm not at all sure that "nature bats last" is self-evident to very many people at all. Perhaps it's just a platitude for pantheists, a byword for Earth Firsters and others who sometimes seem as willing to exclude humans from their concept of nature as most people are to neglect the other species. In which case, it misses the point altogether.

The point of a platitude, as I see it, is to preach a point to people who already know it but act as if they don't. I doubt that those who need to know that nature bats last have any clue at all as to what it means, even lack a cosmology in which it could make any sense. That renders it an esoteric idea, an impossibility for a platitude. I guess it isn't one after all.

An aphorism, then; a verbal balm, a tonic thought. It was meant, of course, as a warning, a shaken finger; but falling on mostly deaf ears as such, it recycles pretty well as a curse of revenge.

What does it mean? "Nature bats last." It means, we may be in the lead now, the natural world may seem the underdog and down in points as well. But when we've finished our act, hit the grand slam, or struck out (which may be the same thing), nature has an extra inning coming—all to herself, unopposed, unending. No one will be keeping score anymore, and guess who wins?

Let me be clear from the start. This is not a threat. I am not writing another admonishment to repent before the day of ecological reckoning. It is late in the day for us to clean up our environmental act, and I am assuming we will not. On the local scale (as in Willapa) and increasingly on the larger, the major decisions have already been made, and we can only live with them. But I'm not preaching doom—what a waste of time to preach the inevitable! We all die; all species die. The only question is, when will we pull ourselves off the respirator?

To me, nature's batting last is neither a warning nor a threat. It is a cheerfully flip recognition of a certainty. And a comforting certainty it is: imagine, the glory of the universe going on and on, free at last of the bad bet that was man on earth! When John Lennon wrote "Imagine,"

Wintergreen: Listening to the Land's Heart (Boston: Houghton Mifflin, 1986).

he could have added a verse: "Imagine there's no people." My humanism ends where we become so fond of ourselves that we cannot imagine the mortality of mankind.

But supposing, against all odds, we began to run the world right (a phrase that contains in its emphasis the seeds of its own defeat)? Couldn't we then change the batting order? Wouldn't I at least want to hope for the endlessness of the human race?

Sure. But that's vain. The best we could do would be to postpone our departure. Any time on for good behavior would just amount to a stay of execution. To think we could indefinitely put off the end of the age of man by acting right toward the earth for a change is like taking up running in dissipated middle age in the hope of cheating death: it might work for a while. You can't prolong life forever, not for an organism, not for a species. But you can sure as hell hasten its demise.

This is harsh stuff, and there has been some harshness in some of the previous essays. I have criticized and taken account of what I feel to have been mistakes. An ungenerous reader could mutter "Cynic!" and close the book, so near the end. But I am not cynical about humans and the rest of nature. When I insist upon the mortality of all species, including our own, it is not an unhappy thought. And when I invoke that aphorism of uncertain category and origin, "nature bats last," it is in good cheer that I do so. My outlook, ultimately, is not a pessimistic one. But then my frame of reference does not encompass human fortunes alone.

Let's look at outlook, for nature, humans, and otherwise. I have always been a short-term optimist, by nature. Whether that has a genetic element or comes from example, I cannot know. But I have always believed that more good things were likely to occur than bad. (This may have to do with my rather catholic tastes as to what constitutes "good.") It is a matter of being open to possibility and aware of serendipity's whisper. Everyone is invited to serendipity's picnic, but only a few bother to attend. Positive thinking? Are we headed toward platitudes? It's more than that. It's being willing to conspire with the physics of fate (chance, really) to harvest luck from happenstance.

Jung called coincidence "synchronicity" and it happens to us all if we are only aware. Co-incidence—happening with. You must be ready to see it and do more than say "wow" when you do. To pluck a plum when you pass beneath the bough, you've got to be looking up. To catch the glisten of the green snail beneath the plum tree, you must regard the ground. To capture more good than bad, you scan the whole and, mantislike, snatch the happy moment before it springs away, out of reach.

I am not a fatalist, and when some great coincidence brings me joy I try not to say it was "meant to happen." Strings of bad "luck" do sometimes befall people, even those who watch for the good. My brother has had a lifelong run of bad breaks, more than his share, while I feel I've had more than my fair share of good ones. Stochastically (a word I learned to toss around in graduate school that means "chances are") one is as likely to be felled by lightning as lifted by the lottery. Life *is* a lottery. But somehow, seekers after something often seem to get better breaks than others who fail to look around. Or do they simply find more compensations?

Of course, another reason for short-term optimism lies in our ability to apply will and thought and action to effect change in our time. We can create a nature reserve and enjoy it for the rest of our lives. We can vote the bums out. We can live selectively, choosing that which we wish to experience. And there are, after all, far too many pleasures available to be able to sample them all: too many wild and intriguing places to ever visit, people to meet, birds to watch, symphonies to hear, and so on. The riches embarrass our poor ability to enjoy them. Pessimism in the short-term is its own punishment, since it vitiates the will and makes one a pawn of circumstance.

Looking out toward the midterm, however, my attitude rotates. Beyond the here and now, a cautiously pessimistic outlook seems only reasonable and realistic. I suppose this means shifting out of my own life and into the many other lives on earth. Speaking of the world, there is no gravity; the earth sucks (whoever said this first probably wouldn't own to it). My, how it sucks these days. Admittedly the *Wahkiakum County Eagle* gives one a less jaundiced view than *The New York Times* might, but I also see the *Longview Daily News* occasionally, listen to *All Things Considered*, and watch what the cat brings in. How anyone can be honestly optimistic over the next century, regarding mankind, I cannot divine. I won't repeat the litany; it's there for all to see, who read any papers at all, or the walls, between the lines, tea leaves, sweaty palms, tarot cards, or the weather. Even the Bible seems to have it about right, somewhere toward the back (if not in the "to have dominion" part in the beginning).

Come to think of it, the Bible does get it right at both ends. Humans took dominion over the earth, now they face Armageddon. The story is rather circuitous from A to B and the cause-and-effect gets a bit mixed up, but it's all there. The sad part is that it gives people the idea that someone else is going to clean up after them. If they're not responsible for the outcome, if they're not culpable for their mess, how can people be expected to function with the future in mind?

The present and near future could get downright depressing if it weren't for nature. As John Hay more elegantly put it in his small classic, *In Defense of Nature*, "What is there to be optimistic about, especially in the face of enduring human perversity? Not a great deal that is predictable; but if enough of us are willing to walk out and meet nature instead of bypassing it, then we will at last belong. And when all is said and done, real stature comes from an attachment to the unknown."

Yes. And this brings me to the long-term, where for me optimism swings round like a major moon to again eclipse the darker view. Unreservedly, I am optimistic in the long run. Not necessarily for *Homo sapiens*, whose puny fate fails to concern the cosmos. But for nature, which is everything, the whole to which our greater allegiance belongs. And for the earth, which is all most of us shall ever directly know of the universe, finally to be freed from human bondage.

Here is where I differ from many deists. They see salvation from earthly dross in an afterlife for the soul. I see afterlife as salvation of earthly dross that is the soul. The perpetuation of my matter in crocus, coal, or comet is all I need know about the next act—that atoms continue in nature. We both see something coming that ratifies what has gone before and flenses the flesh of suffering. To them, however, heaven is full of personalities on permanent vacation; to me, heaven is a permanent vacation from personality.

This inability to face the extinction of personality serves as one of the main reasons for the rejection of evolution by some creationists. They are smart enough to see that, if life evolved, it will continue to do so, and that we (body and soul) may not survive the process. So they seek to preserve their cherished selves by pushing fairy tales—as if, by evangelizing hard enough, they could make it so!

The latest version of the creationist credo delivered to my door is a "textbook" that its makers, the Jehovah's Witnesses, hope to have placed in schools. Entitled *Life—How Did It Get Here? (By Evolution or Creation?)*, this heavily illustrated and simply written tract attempts to convince the reader that evolution is a "lie," claiming: "We should feel even stronger indignation toward the doctrine of evolution and its originator since the intent is to defraud us of eternal life."

The "marvelous new era" that this book promises for believers will have peace and plenty and endless health, youth and life for all. Apparently there will be room for endless population growth, because "mankind will have the enjoyable task of transforming the earth into a paradise." Man's "loving dominion of animals" (*sic*) will be a feature, and "the wilderness and waterless plain will exult." I am struck by the presence, in one of the pretty illustrations of paradise, of a bulldozer. Not my idea of heaven!

Nor is my purpose to make fun. The picture painted *is* a pretty one, and touching, in a way. But were such beliefs to gain many adherents, I would tremble for the stewardship of the earth. What incentive could there possibly be to maintain biological diversity if you didn't believe in its mortality? In the same way, millions trembled on hearing Ronald Reagan speak of biblical Armageddon during the 1984 campaign. If it is inevitable, as foretold, what incentive exists to keep the finger off the button?

Probabilities speak louder than prophecies, but they both speak of annihilation if we carry on the way we have been. I would rather it didn't happen and support peace- and nonnuclear activism for that reason. I see no inevitabilities as regards human behavior. However, should annihilation occur, I console myself that nature will persist.

I call this attitude a cosmic optimism. It simply suggests that nature, *sensu latu*, will carry on, having batted last in its minor-league game with us. We played catch for a brief while, dropped the ball, and threw a tantrum; whereupon nature took her big blue ball and went home to repair the scratches and scuffs we'd inflicted in its soft hide. We lost by default, and there were no more games in the season, for our season was finished. It mattered very much to us, but the rest of nature just didn't care, was rather tired of our company, thought perhaps we'd been a bad recruit to the league of species in the first place, and that she might not try that same experiment again.

That's supposed to make one optimistic? Let me put it another way, dropping the tired metaphor of a ball game like a high fly with the sun in my eyes (which was my first and last act in Little League). Imagine the sun in your eyes—your lizard eyes—through no smog. Imagine the lakes in your fish gills, fresh, pH 7. Conceive the cosmos untroubled by that spot of bother on earth, as all its peaceful, dumb species go back to their business of life and death and evolution, unperturbed by busy-busy men. I like these thoughts.

It would be dishonest to say that I feel no sadness at the prospect of the passing of humanity. Untellable sadness greets the very thought of it. When I consider the moldering of the last lost manuscript of Mozart; the combustion of the libraries when Fahrenheit 451 is reached early in the firestorms; the tumbling of towers and the crumbling of cottages, I could swoon (if I knew how) with earnest, dolorous regret. But think: all of the sadness in the world belongs to us. When we're gone, there will be no sadness, for it is a human conceit. So it would not matter, afterward.

The fact is, nature doesn't care. Only we care. And if we care so much, perhaps we should look for a few good platitudes to guide our critical actions in these days. T-shirts make a good source. "Extinction Is Forever" is a good one; "We All Live Downstream" is another; and "Share the Earth."

"Cosmic discipline," John Hay wrote, "will not allow too much ignorance of what it cherishes." It is that discipline, finally, that lies at the root of my so-called cosmic optimism (just as our mammoth ignorance of what it cherishes makes me dread what comes next). And the "real stature" Hay mentioned, that "comes from an attachment with the unknown," I take to mean a buckling-up of our seat belt for the universal ride. Attaching to the unknown can be acceptance of nature, a faith in the course of natural events, even if they entail our own eventual extinction.

Taking satisfaction from such ideas implies a nonanthropocentric viewpoint. Copernicus saw that we weren't in the middle; why can't we? The natural world does not revolve around us, it merely tolerates us for a spell. We are indulged, yet we continue to indulge our own earthly xenophobia. Biting the land that feeds us, behaving like bulls in nature's china shop, and casting clichés across the littered landscape, we run serious risks. I am told the Finns around here had a saying: "You shouldn't shit in your own house." Taking it literally, they built their johns outside long after others had brought them in. We not only foul our own nest, we do it in the living room.

I can't help but keep on quoting John Hay, who employs never a tired phase and whose phrases never tire: "We have been cutting ourselves off, and we are wise to be alarmed. We have not been meeting the earth, we have only been erasing its opportunities, missing its indefinite, healing associations. Suddenly there is a terrible need for a great cognizance of the unity and interdependence of the world, in the sense of both human and natural communities."

There is another need, among progressive people, to realize that humanism can only take us so far down the agenda of "what," to quote Lenin, "is to be done." Beyond that, speciesism takes over. Liberation doesn't mean a damn in the face of imminent extinction, if we can arrange a rain check on infinity, liberation means everything.

To return to heaven briefly, the popular idea of deferring it till later does nothing for our sense of obligation to the earth. The naturalist knows that heaven is here on earth (for those whose lives are neither too meager nor too glutted, too shackled or too free, to notice). The traditional view holds that there's more, and better, where this came from. I prefer to think that this is all we're gonna get, and it is more than enough, if we take time to experience it and care for it.

Beyond heaven and humanism, an evolutionary view is necessary. Evolution, we find, will adjust in rate and degree to the kinds of stresses imposed upon organisms. Extinctions will occur under stress, but so will resistance evolve, tolerances develop, and tactics adjust. Organic evolution will go on. As the only show in town, it must. Whether it goes without us is another matter entirely. Eventually, it will.

Leading conservation biologists believe that opportunities may have ended already for significant evolution among large mammals, under the management regimes, stresses, and rarity we impose. Otto Frankel and Michael Soule, in *Conservation and Evolution*, argue that this may be the case and that we bear the responsibility of preserving evolutionary potential for as many species as possible. Whether we give them a chance to get back on the world by getting off ourselves, or take them with us one way or the other, will soon be seen. We still have some limited powers to affect the outcome. What stands certain is that we shan't arrest evolution much more without arresting our own.

Whether we choose to remain is our concern, and ours alone. Nature doesn't care. We are but a drip of spittle on the whisker of a beast in a constellation we can't even see. Nature has a right to care, and a sagging sack of grievances against our tenancy, but she doesn't. Nature gets along. Which brings me around at last to Willapa.

In the ravaged land through which we have been rambling, rarities have been lost, common creatures rendered rare, and the productivity of a great forest diminished for ages. The big trees and the bears are nearly gone, and the humans, many of them, are following. But certain species are doing just fine. Natives like the salal and the coyote thrive in the logged-off land, finding opportunities for expansion that they never dreamed of in the old-growth forest. Aliens such as the gaudy foxglove and the dowdy opossum proliferate still more, covering the clearcuts and the roads with their magenta blossoms and gray hides, respectively. These organisms evolved under stress; they know adversity and eat it up.

The weeds do it even better. Farmers with their sprays and archaic weed boards with nefarious powers battle gamely the tansy ragwort, Canadian thistles, and Himalayan blackberry. They make inroads with their powerful poisons. But make no mistake: the weeds will win: nature bats last.

In a sense, all life in the ravaged land is a bunch of weeds—survivors, coping and adapting under adversity. That goes for tenacious families who find something else to do when the creameries go under and the timber companies pull out, as well as for abandoned cats foraging at the local dump, and for huckleberries that clothe the clearcuts as soon as anything can.

Whether or not the weedy, faithful humans choose to remain, nature will not be crowded out, even here. Tonight I watched a possum waddle across the yard and up the slope to the road; every night, it takes what it will from our compost. We look forward to the marsupial's visits and hope it never has a date with a Dodge. I appreciate possums in the same way I admire starlings and cabbage butterflies and reed canary grass—not as native species, but as tough, clever, evolutionarily and ecologically astute organisms—as survivors, against all we dish out.

Last night the coyotes called by the covered bridge: first one tight, metallic yip, then a tentative croon, followed by five minutes of falsetto chorus in many parts, all countertenor and tremolo. "We are here," they say; "we'll eat your apples, your voles, your cats, the afterbirth of your calves; we're here, we set your dogs to barking, we intend to multiply. We are here to stay." That's what they say. Then silence, as they go about their wise, tenacious hunt for whatever there is. No other animal is more systematically or aggressively persecuted across the West. The coyote: evolving, getting better all the time, under heavy pressure.

In his book *Giving Birth to Thunder Sleeping with His Daughter*, Barry Lopez recounts a wide array of North American Indian tales of coyote. Coyote as trickster and in many other incarnations emerges from the pages as from the ancient campfires. I suspect one of those storytellers originated the last aphorism I wish to use. Anyway, whether ancient or modern, it is a good one, and after publication someone will write to say they said it first. Watch for a credit in the second edition, should I be so fortunate. The saying: "When the last man takes to his grave, there will be a coyote on hand to lift his leg over the marker." The image should be struck on a new coin, with Charles Darwin on the other side; not negotiable, but a good-luck coin to remind us of change and evolution, and of creatures that will be happy to adapt if we ourselves cannot.

The land has been hurt. Misuse is not to be excused, and its ill effects will long be felt. But nature will not be eliminated, even here. Rain, moss, and time apply their healing bandage, and the injured land at last recovers.

Nature is evergreen, after all.

DIANE ACKERMAN
b. 1948

Diane Ackerman resembles such other outstanding nature writers as Wendell Berry and Linda Hogan in that she is also a distinguished poet. Her collections of poetry, including Jaguar of Sweet Laughter: New and Selected Poems *(1991), are suffused with the insights and vocabulary of science. Her nature writing, conversely, is often marked by an audacious*

lyrical voice. Among her books of prose are A Natural History of the Senses
(1990), The Moon by Whale Light, and Other Adventures among Bats,
Penguins, Crocodilians, and Whales *(1991), and* A Natural History of
Love *(1994). As two of these titles suggest, she has especially helped to
expand nature writing's range by delving into the physical senses. Where
Emerson reduced the body to a transparent eyeball, she has found in it an
entire landscape, with a biodiversity of meaning that includes the olfactory
and tactile realms as well as the visual.*

WHY LEAVES TURN COLOR IN THE FALL

The stealth of autumn catches one unaware. Was that a goldfinch
perching in the early September woods, or just the first turning leaf? A
red-winged blackbird or a sugar maple closing up shop for the winter?
Keen-eyed as leopards, we stand still and squint hard, looking for signs
of movement. Early-morning frost sits heavily on the grass, and turns
barbed wire into a string of stars. On a distant hill, a small square of yellow appears to be a lighted stage. At last the truth dawns on us: Fall is
staggering in, right on schedule, with its baggage of chilly nights,
macabre holidays, and spectacular, heart-stoppingly beautiful leaves.
Soon the leaves will start cringing on the trees, and roll up in clenched
fists before they actually fall off. Dry seedpods will rattle like tiny
gourds. But first there will be weeks of gushing color so bright, so pas-
tel, so confettilike, that people will travel up and down the East Coast
just to stare at it—a whole season of leaves.

Where do the colors come from? Sunlight rules most living things
with its golden edicts. When the days begin to shorten, soon after the
summer solstice on June 21, a tree reconsiders its leaves. All summer it
feeds them so they can process sunlight, but in the dog days of summer
the tree begins pulling nutrients back into its trunk and roots, pares
down, and gradually chokes off its leaves. A corky layer of cells forms at
the leaves' slender petioles, then scars over. Undernourished, the leaves
stop producing the pigment chlorophyll, and photosynthesis ceases.
Animals can migrate, hibernate, or store food to prepare for winter. But
where can a tree go? It survives by dropping its leaves, and by the end
of autumn only a few fragile threads of fluid-carrying xylem hold leaves
to their stems.

A turning leaf stays partly green at first, then reveals splotches of yel-

A *Natural History of the Senses* (New York: Random House, 1990).

low and red as the chlorophyll gradually breaks down. Dark green seems to stay longest in the veins, outlining and defining them. During the summer, chlorophyll dissolves in the heat and light, but it is also being steadily replaced. In the fall, on the other hand, no new pigment is produced, and so we notice the other colors that were always there, right in the leaf, although chlorophyll's shocking green hid them from view. With their camouflage gone, we see these colors for the first time all year, and marvel, but they were always there, hidden like a vivid secret beneath the hot glowing greens of summer.

The most spectacular range of fall foliage occurs in the northeastern United States and in eastern China, where the leaves are robustly colored, thanks in part to a rich climate. European maples don't achieve the same flaming reds as their American relatives, which thrive on cold nights and sunny days. In Europe, the warm, humid weather turns the leaves brown or mildly yellow. Anthocyanin, the pigment that gives apples their red and turns leaves red or red-violet, is produced by sugars that remain in the leaf after the supply of nutrients dwindles. Unlike the carotenoids, which color carrots, squash, and corn, and turn leaves orange and yellow, anthocyanin varies from year to year, depending on the temperature and amount of sunlight. The fiercest colors occur in years when the fall sunlight is strongest and the nights are cool and dry (a state of grace scientists find vexing to forecast). This is also why leaves appear dizzyingly bright and clear on a sunny fall day: The anthocyanin flashes like a marquee.

Not all leaves turn the same colors. Elms, weeping willows, and the ancient ginkgo all grow radiant yellow, along with hickories, aspens, bottlebrush buckeyes, cottonweeds, and tall, keening poplars. Basswood turns bronze, birches bright gold. Water-loving maples put on a symphonic display of scarlets. Sumacs turn red, too, as do flowering dogwoods, black gums, and sweet gums. Though some oaks yellow, most turn a pinkish brown. The farmlands also change color, as tepees of cornstalks and bales of shredded-wheat-textured hay stand drying in the fields. In some spots, one slope of a hill may be green and the other already in bright color, because the hillside facing south gets more sun and heat than the northern one.

An odd feature of the colors is that they don't seem to have any special purpose. We are predisposed to respond to their beauty, of course. They shimmer with the colors of sunset, spring flowers, the tawny buff of a colt's pretty rump, the shuddering pink of a blush. Animals and flowers color for a reason—adaptation to their environment—but there is no adaptive reason for leaves to color so beautifully in the fall any more than there is for the sky or ocean to be blue. It's just one of the haphazard marvels the planet bestows every year. We find the sizzling colors thrilling, and in a sense they dupe us. Colored like living things, they signal death and disintegration. In time, they will become fragile and, like the body, return to dust. They are as we hope our own fate will be when we die: Not to vanish, just to sublime from one beautiful state into another. Though leaves lose their green life, they bloom with urgent colors, as the woods grow mummified day by day, and Nature becomes more carnal, mute, and radiant.

We call the season "fall," from the Old English *feallan*, to fall, which leads back through time to the Indo-European *phol*, which also means to fall. So the word and the idea are both extremely ancient, and haven't really changed since the first of our kind needed a name for fall's leafy abundance. As we say the word, we're reminded of that other Fall, in the garden of Eden, when fig leaves never withered and scales fell from our eyes. Fall is the time when leaves fall from the trees, just as spring is when flowers spring up, summer is when we simmer, and winter is when we whine from the cold.

Children love to play in piles of leaves, hurling them into the air like confetti, leaping into soft unruly mattresses of them. For children, leaf fall is just one of the odder figments of Nature, like hailstones or snowflakes. Walk down a lane overhung with trees in the never-never land of autumn, and you will forget about time and death, lost in the sheer delicious spill of color. Adam and Eve concealed their nakedness with leaves, remember? Leaves have always hidden our awkward secrets.

But how do the colored leaves fall? As a leaf ages, the growth hormone, auxin, fades, and cells at the base of the petiole divide. Two or three rows of small cells, lying at right angles to the axis of the petiole, react with water, then come apart, leaving the petioles hanging on by only a few threads of xylem. A light breeze, and the leaves are airborne. They glide and swoop, rocking in invisible cradles. They are all wing and may flutter from yard to yard on small whirlwinds or updrafts, swiveling as they go. Firmly tethered to earth, we love to see things rise up and fly—soap bubbles, balloons, birds, fall leaves. They remind us that the end of a season is capricious, as is the end of life. We especially like the way leaves rock, careen, and swoop as they fall. Everyone knows the motion. Pilots sometimes do a maneuver called a "falling leaf," in which the plane loses altitude quickly and on purpose, by slipping first to the right, then to the left. The machine weighs a ton or more, but in one pilot's mind it is a weightless thing, a falling leaf. She has seen the motion before, in the Vermont woods where she played as a child. Below her the trees radiate gold, copper, and red. Leaves are falling, although she can't see them fall, as she falls, swooping down for a closer view.

At last the leaves leave. But first they turn color and thrill us for weeks on end. Then they crunch and crackle underfoot. They *shush*, as children drag their small feet through leaves heaped along the curb. Dark, slimy mats of leaves cling to one's heels after a rain. A damp, stuccolike mortar of semidecayed leaves protects the tender shoots with a roof until spring, and makes a rich humus. An occasional bulge or ripple in the leafy mounds signals a shrew or a field mouse tunneling out of sight. Sometimes one finds in fossil stones the imprint of a leaf, long since disintegrated, whose outlines remind us how detailed, vibrant, and alive are the things of this earth that perish.

JOHN DANIEL
b. 1948

Born in South Carolina and raised outside Washington, D. C., John Daniel is one of a number of transplanted easterners who have found an adopted home in the West, in Daniel's case, Oregon. Daniel was greatly influenced by Wallace Stegner, who was his neighbor for a time in California and who also had an itinerant childhood before rooting himself in a chosen locale. The question of finding and committing oneself to place and family figures largely in both men's work. Daniel's first essay collection, The Trail Home *(1992), examines nature and imagination in the American West.* Looking After *(1996) is a memoir of caring for his Alzheimer's-stricken mother. But one of his merits as a writer is his inclination to examine his own most deeply held convictions. In the following essay, he questions the absolute value of rootedness and suggests the salutary aspects of the outsider and the perpetual wanderer to society as a whole.*

A WORD IN FAVOR OF ROOTLESSNESS

I am one of the converted when it comes to the cultural and economic necessity of finding place. Our rootlessness—our refusal to accept the discipline of living as responsive and responsible members of neighborhoods, communities, landscapes, and ecosystems—is perhaps our most serious and widespread disease. The history of our country, and especially of the American West, is in great part a record of damage done by generations of boomers, both individual and corporate, who have wrested from the land all that a place could give and continually moved on to take from another place. Boomers such as Wallace Stegner's father, who, as we see him in *The Big Rock Candy Mountain*, "wanted to make a killing and end up on Easy Street." Like many Americans, he was obsessed by the fruit of Tantalus: "Why remain in one dull plot of Earth when Heaven was reachable, was touchable, was just over there?"

We don't stand much chance of perpetuating ourselves as a culture, or of restoring and sustaining the health of our land, unless we can outgrow our boomer adolescence and mature into stickers, or nesters—human beings willing to take on the obligations of living in communities rooted in place, conserving nature as we conserve ourselves. And maybe, slowly, we are headed in that direction. The powers and virtues of place are celebrated in a growing body of literature and discussed in conferences across the country. Bioregionalism, small-scale organic farming, urban food co-ops, and other manifestations of the spirit of place seem to be burgeoning, or at least coming along.

That is all to the good. But as we settle into our home places and local communities and bioregional niches, as we become the responsible economic and ecologic citizens we ought to be, I worry a little. I worry, for one thing, that we will settle in place so pervasively that no unsettled places will remain. But I worry about us settlers, too. I feel at least a tinge of concern that we might allow our shared beliefs and practices to harden into orthodoxy, and that the bath water of irresponsibility we are ready to toss out the home door might contain a lively baby or two. These fears may turn out to be groundless, like most of my insomniac broodings. But they are on my mind, so indulge me, if you will, as I address some of the less salutary aspects of living in place and some of the joys and perhaps necessary virtues of rootlessness.

No power of place is more elemental or influential than climate, and I

Originally published in *Orion*, Autumn 1995.

feel compelled at the outset to report that we who live in the wet regions of the Northwest suffer immensely from our climate. Melville's Ishmael experienced a damp, drizzly November in his soul, but only now and again. For us it is eternally so, or it feels like eternity. From October well into June we slouch in our mossy-roofed houses listening to the incessant patter of rain, dark thoughts slowly forming in the dull cloud chambers of our minds. It's been days, weeks, *years*, we believe, since a neighbor knocked or a letter arrived from friend or agent or editor. Those who live where sun and breezes play, engaged in their smiling businesses, have long forgotten us, if they ever cared for us at all. Rain drips from the eaves like poison into our souls. We sit. We sleep. We check the mail.

What but climate could it be that so rots the fiber of the Northwestern psyche? Or if not climate itself, then an epiphenomenon of climate—perhaps the spores of an undiscovered fungus floating out of those decadent forests we environmentalists are so bent on saving. Oh, we try to improve ourselves. We join support groups and twelve-step programs, we drink gallons of cappuccino and café latte, we bathe our pallid bodies in the radiance of full-spectrum light machines. These measures keep us from dissolving outright into the sodden air, and when spring arrives we bestir ourselves outdoors, blinking against the occasional cruel sun and the lurid displays of rhododendrons. By summer we have cured sufficiently to sally forth to the mountains and the coast, where we linger in sunglasses and try to pass for normal.

But it is place we're talking about, the powers of place. As I write this, my thoughts are perhaps unduly influenced by the fact that my right ear has swollen to the size and complexion of a rutabaga. I was working behind the cabin this afternoon, cutting up madrone and Douglas fir slash with the chain saw, when I evidently stepped too close to a yellow jacket nest. I injured none of their tribe, to my knowledge, but one of them sorely injured me. Those good and industrious citizens take place pretty seriously. I started to get out the .22 and shoot every one of them, but thought better of it and drank a tumbler of bourbon instead.

And now, a bit later, a spectacle outside my window only confirms my bitter state of mind. The place in question is the hummingbird feeder, and the chief influence of that place is to inspire in hummingbirds a fiercely intense desire to impale one another on their needlelike beaks. Surely they expend more energy blustering in their buzzy way than they can possibly derive from the feeder. This behavior is not simply a consequence of feeding Kool-Aid to already over-amped birds—they try to kill each other over natural flower patches too. Nor can it be explained as the typically mindless and violent behavior of the male sex in general. Both sexes are represented in the fray, and females predominate. It is merely a demonstration of over-identification with place. Humans do it too. Look at Yosemite Valley on the Fourth of July. Look at any empty parking space in San Francisco. Look at Jerusalem.

When human beings settle in a place for the long run, it may be that good things occur overall. There are dangers, though. Stickers run the substantial risk of becoming sticks-in-the-mud. Consider my own state of Oregon, which was settled by farmers from the Midwest and upper South who had one epic move in them, across the Oregon Trail, and having found paradise resolved not to stir again until the millennium. The more scintillating sorts—writers, murderers, prostitutes, lawyers, other riffraff—tended toward Seattle or San Francisco. And so it happens that we Oregonians harbor behind our bland and agreeable demeanor a serious streak of moralism and conformism. We have some pretty strict notions about the way people should live. It was we who started the nationwide spate of legal attacks on gay and lesbian rights, and it is we who annually rank among the top five states in citizen challenges to morally subversive library books, books such as *Huckleberry Finn*, *The Catcher in the Rye*, and *The Color Purple*.

This pernicious characteristic is strongest, along with some of our best characteristics, where communities are strongest and people live closest to the land—in the small towns. When my girlfriend and I lived in Klamath Falls in the early 1970s, we were frequently accosted by Mrs. Grandquist, our elderly neighbor across the road. She was pointedly eager to lend us a lawn mower, and when she offered it she had the unnerving habit of staring at my hair. Our phone was just inside the front door, and sometimes as we arrived home it rang before we were entirely *through* the door. "You left your lights on," Mrs. Grandquist would say. Or, "You ought to shut your windows when you go out. We've got burglars, you know." Not in that block of Denver Avenue, we didn't. Mrs. Grandquist and other watchful citizens with time on their hands may have kept insurance rates down, but the pressure of all those eyes and inquiring minds was at times intensely uncomfortable. Small towns are hard places in which to be different. Those yellow jackets are wary, and they can sting.

Customs of land use can become as ossified and difficult to budge as social customs. The Amish, among other long-established rural peoples, practice a good and responsible farming economy. But long-term association with a place no more *guarantees* good stewardship than a long-term marriage guarantees a loving and responsible relationship. As Aldo Leopold noted with pain, there are farmers who habitually abuse their land and cannot easily be induced to do otherwise.

Thoreau saw the same thing in Concord—landspeople who, though they must have known their places intimately, mistreated them continually. They whipped the dog every day because the dog was no good, and because that's the way dogs had always been dealt with.

As for us of the green persuasions, settled or on the loose, we too are prone—perhaps more prone than most—to orthodoxy and intolerance. We tend to be overstocked in piety and self-righteousness, deficient in a sense of humor about our values and our causes. Here in the Northwest, where debate in the last decade has focused on logging issues, it's instructive to compare bumper stickers. Ours say, sanctimoniously, "Stumps Don't Lie" or "Love Your Mother." Those who disagree with us, on the other hand, sport sayings such as "Hug a Logger—You'll Never Go Back to Trees," or "Earth First! (We'll Log the Other Planets Later)."

I don't mean to minimize the clear truth that ecological blindness and misconduct are epidemic in our land. I only mean to suggest that rigid ecological correctness may not be the most helpful treatment. All of us, in any place or community or movement, tend to become insiders; we all need the stranger, the outsider, to shake our perspective and keep us honest. Prominent among Edward Abbey's many virtues was his way of puncturing environmentalist pieties (along with every other brand of piety he encountered). What's more, the outsider can sometimes see landscape with a certain clarity unavailable to the longtime resident. It was as a relative newcomer to the Southwest that Abbey took the notes that would become his best book, in which he imagined the canyon country of the Colorado Plateau more deeply than anyone had imagined it before or has imagined it since. His spirit was stirred and his vision sharpened by his outsider's passion. I don't know that he could have written *Desert Solitaire* if he had been raised in Moab or Mexican Hat.

Unlike Thoreau, who was born to his place, or Wendell Berry, who returned to the place he was born to, Edward Abbey came to his place from afar and took hold. More of a lifelong wanderer was John Muir, who we chiefly identify with the Sierra Nevada but who explored and sojourned in and wrote of a multitude of places, from the Gulf of Mexico to the Gulf of Alaska. I think Muir needed continually to see new landscapes and life forms in order to keep his ardent mind ignited. Motion for him was not a pathology but a devotion, an essential joy, a continuous discovery of place and self. Marriage to place is something we need to realize in our culture, but not all of us are the marrying kind. The least happy period of Muir's life was his tenure as a settled fruit farmer in Martinez, California. He was more given to the exhilarated attention and fervent exploration of *wooing*, more given to rapture than to extended fidelity. "Rapture" is related etymologically to "rape," but unlike the boomer, who rapes a place, the authentic wooer allows the place to enrapture him.

Wooing often leads to marriage, of course, but not always. Is a life of wooing place after place less responsible than a life of settled wedlock? It may be less sustainable, but the degree of its responsibility depends on the quality of the wooing. John Muir subjected himself utterly to the places he sought out. He walked from Wisconsin to the Gulf Coast, climbed a tree in a Sierra windstorm, survived a subzero night on the summit of Mount Shasta by scalding himself in a sulfurous volcanic vent. There was nothing macho about it—he loved where he happened to be and refused to miss one lick of it. In his wandering, day to day and minute to minute, he was more placed than most of us ever will be, in a lifetime at home or a life on the move. Rootedness was not his genius and not his need. As the followers of the Grateful Dead like to remind us, quoting J. R. R. Tolkien, "Not all who wander are lost."

Muir's devoted adventuring, of course, was something very different from the random restlessness of many in our culture today. Recently I sat through a dinner party during which the guests, most of them thirty-something, compared notes all evening about their travels through Asia. They were experts on border crossings, train transport, currency exchange, and even local art objects, but nothing I heard that evening indicated an influence of land or native peoples on the traveler's soul. They were travel technicians. Many backpackers are the same, passing through wilderness places encapsulated in maps and objectives and high-tech gear. There *is* a pathology there, a serious one. It infects all of us to one degree or another. We have not yet arrived where we believe—and our color slides show—we have already been.

But if shifting around disconnected from land and community is our national disease, I would argue, perversely perhaps, or perhaps just homeopathically, that it is also an element of our national health. Hank Williams and others in our folk and country traditions stir something in many of us when they sing the delights of the open road, of rambling on the loose by foot or thumb or boxcar through the American countryside. Williams's "Ramblin' Man" believes that God intended him for a life of discovery beyond the horizons. Is this mere immaturity? Irresponsibility? An inability to relate to people or place? Maybe. But maybe also renewal, vitality, a growing of the soul. It makes me very happy to drive the highways and back roads of the West, exchanging talk with people who live where I don't, pulling off somewhere, anywhere, to sleep in the truck and wake to a place I've never seen. I can't defend the cost of that travel in fossil fuel consumption and air befoulment—Williams's rambler at least took the fuel-efficient train—but I do know that it satisfies me as a man and a writer.

Such pleasure in movement—the joy of hitting the trail on a brisk morning, of watching from a train the towns and fields pass by, of riding a skateboard or hang glider or even a 747—must come from a deep and ancient source. All of us are descended from peoples whose way was to roam with the seasons, following game herds and the succession of edible plants, responding to weather and natural calamities and the shifting field of relations with their own kind. And those peoples came, far deeper in the past, from creatures not yet human who crawled and leapt and swung through the crowns of trees for millions of years, evolving prehensile hands and color binocular vision as a consequence, then took to the ground and learned to walk upright and wandered out of Africa (or so it now seems) across the continents of Earth. Along the way we have lost much of the sensory acuity our saga evoked in us, our ability to smell danger or read a landscape or notice nuances of weather, but the old knowing still stirs an alertness, an air of anticipation, when we set out on our various journeys.

Native cultural traditions reflect the value of the traveler's knowing. In Native American stories of the Northwest, I notice that Coyote doesn't seem to have a home. Or if he does, he's never there. "Coyote was traveling upriver," the stories begin. "Coyote came over Neahkanie Mountain," "Coyote was going there. . . ." The stories take place in the early time when the order of the world was still in flux. Coyote, the placeless one, helps people and animals find their proper places. You wouldn't want to base a code of ethics on his character, which is unreliable and frequently ignoble, but he is the agent who introduces human beings to their roles and responsibilities in life. Coyote is the necessary inseminator. (Sometimes literally.) He is the shifty and shiftless traveler who fertilizes the locally rooted bloomings of the world.

Maybe Coyote moves among us as the stranger, often odd or disagreeable, sometimes dangerous, who brings reports from far places. Maybe that stranger is one of the carriers of our wildness, one of the mutant genes that keep our evolution fresh and thriving. It is for that stranger, says Elie Weisel, that an extra place is set at the Seder table. The voyager might arrive, the one who finds his home in the homes of others. He might tell a story, a story no one in the family or local community is capable of telling, and children might hear that story and imagine their lives in a new way.

It could be Hank Williams who stops in, and he'll sing to you half the night (and maybe yours will be the family he needs, and he won't die of whiskey and barbiturates in the back seat of a car). Or Huck Finn might be your stranger, on the run from "sivilization," dressed as a girl and telling stupendous lies. It could be Jack Kerouac and Neal Cassady, on the road with their Beat buddies, hopped-up on speed, and they never *will* stop talking. It might be Gerry Nanapush, the Chippewa power man Louise Erdrich has given us, escaped from jail still again to slip through the mists and snows with his ancient powers. Or it might be Billy Parham or John Grady Cole, Cormac McCarthy's boy drifters. They'll want water for their horses, they'll be ready to eat, and if you're wise you'll feed them. They won't talk much themselves, but you might find yourself telling them the crucial story of your life.

Or yours could be the house where Odysseus calls, a still youngish man returning from war, passionate for his family and the flocks and vineyards of home. Just as likely, though, he could be an old man when he stands in your door. No one's quite sure what became of Odysseus. Homer tells us that he made it to Ithaca and set things in order, but the story leaves off there. Some say he resumed his settled life, living out his days as a placed and prosperous landsman. But others say that after all his adventures he couldn't live his old life again. Alfred, Lord Tennyson writes that he shipped out from Ithaca with his trusted crew. Maybe so. Or maybe the poet got it only half right. Maybe Penelope, island bound for all those years, was stir crazy herself. Maybe they left Telemachus the ranch and set out westward across the sea, two gray spirits "yearning in desire / To follow knowledge like a sinking star, / Beyond the utmost bound of human thought."

DAVID QUAMMEN
b. 1948

"Biology," *asserts David Quammen, "has great potential for vulgar enter-* *tainment."* As *a natural history columnist for* Outside *magazine for* *many years, he not only entertained his readers with his witty and* *provocative style, but also employed biological phenomena as a means of* *understanding human behavior—using, for instance, the structure of a* *chambered nautilus to discuss the nature of memory or the mating pat-* *terns of wild geese to examine marital fidelity. Born in Cincinnati,* *Quammen now lives in Montana, where he studied aquatic entomology at the state university in Missoula. He is not, as he says, a scientist but "a follower of science"—often for the purpose of catching the scientists themselves in biased or untenable positions. Among his books are* Natural Acts *(1985),* Song of the Dodo *(1996), and* Wild Thoughts from Wild Places *(1998), from which the following selection comes.*

STRAWBERRIES UNDER ICE

1. THE GRADIENT OF NET MASS BALANCE

Antarctica is a gently domed continent squashed flat, like a dent in the roof of a Chevy, by the weight of its ice. That burden of ice amounts to seven million cubic miles. Melt it away and the Antarctic interior would bounce upward; Earth itself would change shape. This grand cold fact has, to me, on the tiny and personal scale, a warm appeal. Take away ice and the topography of my own life changes drastically too.

Ice is lighter than water but still heavy. The stuff answers gravity. Ice is a solid but not an absolute solid. The stuff flows. Slowly but inexorably it runs downhill. We think of iciness as a synonym for cold, but cold is relative and ice happens to function well as insulation against heat loss: low thermal conductivity. Also it *releases* heat to immediate surroundings in the final stage of becoming frozen itself. Ice warms. On a certain night, roughly thirteen years ago, it warmed me.

When a tongue of ice flows down a mountain valley, we call it a glacier. When it flows out in all directions from a source point at high elevation, like pancake batter poured on a griddle, we call it a sheet. Out at the Antarctic circumference are glaciers and seaborne shelves, from which icebergs calve off under their own weight. Both sheets and glaciers are supplied with their substance, their impetus, their ice, by snow and other forms of precipitation back uphill at the source. While old ice is continually lost by calving and melting in the lowlands, new ice is deposited in the highlands, and any glacier or sheet receiving more new ice than it loses old, through the course of a year, is a glacier or sheet that is growing. The scientists would say that its net mass balance is positive.

The Antarctic sheet, for instance, has a positive balance. But this is not an essay about Antarctica.

Each point on a great ice body has its own numerical value for mass

Wild Thoughts from Wild Places (New York: Scribner, 1998).

balance. Is the ice right here thicker or thinner than last year? Is the glacier, at this spot, thriving or dying? The collective profile of all those individual soundings—more ice or less? thriving or dying?—is called the gradient of net mass balance. This gradient tells, in broad perspective, what has been lost and what has been gained. On that certain night, thirteen years ago, I happened to be asking myself exactly the same question: *What's been lost and what, if anything, gained?* Because snow gathers most heavily in frigid sky-scraping highlands, the gradient of net mass balance correlates steeply with altitude. Robust glaciers come snaking down out of the Alaskan mountains. Also because snow gathers most heavily in frigid sky-scraping highlands, I had taken myself on that day to a drifted-over pass in the Bitterroot Mountains, all hell-and-gone up on the state border just west of the town of Tarkio, Montana, and started skiing uphill from there.

I needed as much snow as possible. I carried food and a goosedown sleeping bag and a small shovel. The night in question was December 31, 1975.

I hadn't come to measure depths or to calculate gradients. I had come to insert myself into a cold white hole. First, of course, I had to dig it. This elaborately uncomfortable enterprise seems to have been part of a long foggy process of escape and purgation, much of which you can be spared. Suffice to say that my snow cave, to be dug on New Year's Eve into a ten-foot-high cornice on the leeward side of the highest ridge I could ski to, and barely large enough for one person, would be at the aphelion of that long foggy process. At the perihelion was Oxford University.

At Oxford University during one week in late springtime there is a festival of crew races on the river, featuring girls in long dresses, boys in straw hats, an abundance of champagne and strawberries. This event is called Eights Week, for the fact of eight men to a crew. It is innocent. More precisely: It is no more obnoxious, no more steeped in snobbery and dandified xenophobia and intellectual and social complacence than any other aspect of Oxford University. The strawberries are served under heavy cream. Sybaritism is mandatory. For these and other reasons, partly personal, partly political, I had fled the place screaming during Eights Week of 1972, almost exactly coincident (by no coincidence) with Richard Nixon's announcement of the blockade of Haiphong harbor. Nixon's blockade and Oxford's strawberries had nothing logically in common, but they converged to produce in me a drastic reaction to what until then had been just a festering distemper.

It took me another year to arrive in Montana. I had never before set foot in the state. I knew no one there. But I had heard that it was a place where, in the early weeks of September, a person could look up to a looming horizon and see fresh-fallen snow. I had noted certain blue lines on a highway map, knew the lines to be rivers, and imagined those rivers to be dark mountain streams flashing with trout. I arrived during the early weeks of September and lo it was all true.

I took a room in an old-fashioned boardinghouse. I looked for a job. I started work on a recklessly ambitious and doomed novel. I sensed rather soon that I hadn't come to this place temporarily. I began reading the writers—Herodotus, Euripides, Coleridge, Descartes, Rousseau, Thoreau, Raymond Chandler—for whom a conscientious and narrow academic education had left no time. I spent my nest egg and then sold my Volkswagen bus for another. I learned the clownish mortification of addressing strangers with: "Hi, my name is Invisible and I'll be your waiter tonight." I was twenty-six, just old enough to realize that this period was not some sort of prelude to my life but the thing itself. I knew I was spending real currency, hard and finite, on a speculative venture at an unknowable rate of return: the currency of time, energy, stamina. Two more years passed before I arrived, sweaty and chilled, at that high cold cornice in the Bitterroots.

By then I had made a small handful of precious friends in this new place, and a menagerie of acquaintances, and I had learned also to say: "You want that on the rocks or up?" Time was still plentiful but stamina was low. Around Christmas that year, two of the precious friends announced a New Year's Eve party. Tempting, yet it seemed somehow a better idea to spend the occasion alone in a snow cave.

So here I was. There had been no trail up the face of the ridge and lifting my legs through the heavy snow had drenched and exhausted me. My thighs felt as though the Chicago police had worked on them with truncheons. I dug my hole. That done, I changed out of the soaked, freezing clothes. I boiled and ate some noodles, drank some cocoa; if I had been smart enough to encumber my pack with a bottle of wine, I don't remember it. When dark came I felt the nervous exhilaration of utter solitude and, behind that like a metallic aftertaste, loneliness. I gnawed on my thoughts for an hour or two, then retired. The night turned into a clear one and when I crawled out of the cave at 3:00 A.M. of the new year, to empty my bladder, I found the sky rolled out in a stunning pageant of scope and dispassion and cold grace.

It was too good to waste. I went back into the cave for my glasses.

The temperature by now had gone into the teens below zero. I stood there beside the cornice in cotton sweatpants, gaping up. "We never know what we have lost, or what we have found," says America's wisest poet, Robert Penn Warren, in the context of a meditation about John

James Audubon and the transforming power of landscape. We never know what we have lost, or what we have found. All I did know was that the highway maps called it Montana, and that I was here, and that in the course of a life a person could travel widely but could truly open his veins and his soul to just a limited number of places.

After half an hour I crawled back into the cave, where ten feet of snow and a rime of ice would keep me warm.

2. ABLATION

Trace any glacier or ice sheet downhill from its source and eventually you will come to a boundary where the mass balance of ice is zero. Nothing is lost, over the course of time, and nothing is gained. The ice itself constantly flows past this boundary, molecule by molecule, but if any new ice is added here by precipitation, if any old ice is taken away by melting, those additions and subtractions cancel each other exactly. This boundary is called the equilibrium line. Like other forms of equilibrium, it entails a cold imperturbability, a sublime steadiness relative to what's going on all around. Above the equilibrium line is the zone of accumulation. Below is the zone of ablation.

Ablation is the scientists' fancy word for loss. Down here the mass balance is negative. Ice is supplied to this zone mainly by flow from above, little or not at all by direct precipitation, and whatever does come as direct precipitation is less than the amount annually lost. The loss results from several different processes: wind erosion, surface melting, evaporation (ice does evaporate), underside melting of an ice shelf where it rests on the warmer sea water. Calving off of icebergs. *Calving* is the scientists' quaint word for that sort of event when a great hunk of ice—as big as a house or, in some cases, as big as a county—tears away from the leading edge of the sheet or the glacier and falls thunderously into the sea.

Possibly this talk about calving reflects an unspoken sense that the larger ice mass, moving, pulsing, constantly changing its shape, is almost alive. If so, the analogy doesn't go far. Icebergs don't suckle or grow. They float away on the sea, melt, break apart, disappear. Wind erosion and evaporation and most of those other ablative processes work on the ice slowly, incrementally. Calving on the other hand is abrupt. A large piece of the whole is there, and then gone.

The occurrence of a calving event depends on a number of factors—flow rate of the whole ice body, thickness at the edge, temperature, fissures in the ice, stresses from gravity or tides—one of which is the strength of the ice itself. That factor, strength, is hard to measure. You might never know until too late. Certain experiments done on strength-testing machines have yielded certain numbers: a strength of thirty-eight bars (a bar is a unit of pressure equal to 100,000 newtons per square meter) for crushing; fourteen bars for bending; nine bars for tensile. But those numbers offer no absolute guide to the performance of different types of ice under different conditions. They only suggest in a relative way that, though ice may flow majestically under its own weight, though it may stretch like caramel, though it may bend like lead, it gives back rock-like resistance to a force coming down on it suddenly. None of this cold information was available to me on the day now in mind, and if it had been I wouldn't have wanted it.

On the day now in mind I had been off skiing, again, with no thought for the physical properties of ice, other than maybe some vague awareness of the knee strain involved in carving a turn across boilerplate. I came home to find a note in my door.

The note said that a young woman I knew, the great love of a friend of mine, was dead. The note didn't say what had happened. I should call a number in Helena for details. It was not only shocking but ominous. Because I knew that the young woman had lately been working through some uneasy and confusing times, I thought of all the various grim possibilities involving despair. Then I called the Helena number, where a houseful of friends were gathered for communal grieving and food and loud music. I learned that the young woman had died from a fall. A freak accident. In the coldest sense of cold consolation, there was in this information some relief.

She had slipped on a patch of sidewalk ice, the night before, and hit her head. A nasty blow, a moment or two of unconsciousness, but she had apparently been all right. She went home alone and was not all right and died before morning. I suppose she was about twenty-seven. This is exactly why head-trauma cases are normally put under close overnight observation, but I wasn't aware of that at the time, and neither evidently were the folks who had helped her up off that icy sidewalk. She had seemed okay. Even after the fall, her death was preventable. Of course most of us will die preventable deaths; hers was only more vividly so, and earlier.

I had known her, not well, through her sweetheart and the network of friends now assembled in that house in Helena. These friends of hers and mine were mostly a group of ecologists who had worked together, during graduate school, as waiters and bartenders and cooks; I met them in that context and they had nurtured my sanity to no small degree when that context began straining it. They read books, they talked about ideas, they knew a spruce from a hemlock, they slept in snow caves: a balm of good company to me. They made the state of Montana into a place that was not only cold, true, hard, and beautiful, but damn near humanly habitable. The young woman, now dead, was not herself a scientist, but she was one of them in all other senses. She came from a town up on the Hi-Line.

I had worked with her too, and seen her enliven long afternoons that could otherwise be just a tedious and not very lucrative form of self-demeanment. She was one of those rowdy, robust people—robust in good times, just as robust when she was angry or miserable—who are especially hard to imagine dead. She was a rascal of wit. She could be wonderfully crude. We all knew her by her last name, because her first seemed too ladylike and demure. After the phone call to Helena, it took me a long time to make the mental adjustment of tenses. She *had* been a rascal of wit.

The memorial service was scheduled for such-and-such day, in that town up on the Hi-Line.

We drove up together on winter roads, myself and two of the Helena friends, a husband-and-wife pair of plant ecologists. Others had gone ahead. Places available for sleeping, spare rooms and floors; make contact by phone; meet at the church. We met at the church and sat lumpish while a local pastor discoursed with transcendent irrelevance about what we could hardly recognize as her life and death. It wasn't his fault, he didn't know her. There was a reception with the family, followed by a postwake on our own at a local bar, a fervent gathering of young survivors determined not only to cling to her memory but to cling to one another more appreciatively now that such a persuasive warning bell of mortality had been rung, and then sometime after dark as the wind came up and the temperature dropped away as though nothing was under it and a new storm raked in across those wheatlands, the three of us started driving back south. It had been my first trip to the Hi-Line.

Aside from the note in the door, this is the part I remember most clearly. The car's defroster wasn't working. I had about four inches of open windshield. It was a little Honda that responded to wind like a shuttlecock, and on slick pavement the rear end flapped like the tail of a trout. We seemed to be rolling down a long dark tube coated inside with ice, jarred back and forth by the crosswinds, nothing else visible except the short tongue of road ahead and the streaming snow and the trucks blasting by too close in the other lane. How ironic, I thought, if we die on the highway while returning from a funeral. I hunched over the wheel, squinting out through that gap of windshield, until some of the muscles in my right shoulder and neck shortened themselves into a knot. The two plant ecologists kept me awake with talk. One of them, from the backseat, worked at the knot in my neck. We talked about friendship and the message of death as we all three felt we had heard it, which was to cherish the living, while you have them. Seize, hold, appreciate. Pure friendship, uncomplicated by romance or blood, is one of the most nurturing human relationships and one of the most easily taken for granted. This was our consensus, spoken and unspoken.

These two plant ecologists had been my dear friends for a few years, but we were never closer than during that drive. Well after midnight, we reached their house in Helena. I slept on sofa cushions. In the morning they got me to a doctor for the paralytic clench in my neck. That was almost ten years ago and I've hardly seen them since.

The fault is mine, or the fault is nobody's. We got older and busier and trails diverged. They began raising children. I traveled to Helena less and less. Mortgages, serious jobs, deadlines; and the age of sleeping on sofa cushions seemed to have passed. I moved, they moved, opening more geographical distance. Montana is a big place and the roads are often bad. These facts offered in explanation sound even to me like excuses. The ashes of the young woman who slipped on the ice have long since been sprinkled onto a mountaintop or into a river, I'm not sure which. Nothing to be done now either for her or about her. The two plant ecologists I still cherish, in intention anyway, at a regrettable distance, as I do a small handful of other precious friends who seem to have disappeared from my life by wind erosion or melting.

3. LEONTIEV'S AXIOM

The ice mass of a mountain glacier flows down its valley in much the same complicated pattern as a river flowing in its bed. Obviously the glacier is much slower. Glacial ice may move at rates between six inches and six feet per day; river water may move a distance in that range every second. Like the water of a river, though, the ice of a glacier does not all flow at the same rate. There are eddies and tongues and slack zones, currents and swells, differential vectors of mix and surge. The details of the flow pattern depend on variable parameters special to each case: depth of the ice, slope, contour of the bed, temperature. But some generalizations can be made. Like a river, a glacier will tend to register faster flow rates at the surface than at depths, faster flows at mid-channel than along the edges, and faster flows down in the middle reaches than up near the source. One formula scientists use to describe the relations between flow rate and those other factors is:

$$u = k_1 \sin^3 a \, h^4 + k_2 \sin^2 a \, h^2.$$

Everyone stay calm. This formula is not Leontiev's Axiom, and so we aren't going to bother deciphering it.

Turbulent flow is what makes a glacier unfathomable, in the sense of *fathoming* that connotes more than taking an ice-core measurement of depth. Turbulent flow is also what distinguishes a river from, say, a lake. When a river freezes, the complexities of turbulent flow interact with the peculiar physics of ice formation to produce a whole rat's nest of intriguing and sometimes inconvenient surprises. Because of turbulence, the water of a river cools down toward the freezing point uniformly, not in stratified layers as in a lake. Eventually the entire mass of flowing water drops below thirty-two degrees Fahrenheit. Small disks of ice, called frazil ice, then appear. Again because of turbulence, this frazil ice doesn't all float on the surface (despite being lighter than water) but mixes throughout the river's depth. Frazil ice has a tendency toward adhesion, so some of it sticks to riverbed rocks. Some of it gloms onto bridge pilings and culverts, growing thick as a soft cold fur. Some of it aggregates with other frazil ice, forming large dollops of drifting slush. Meanwhile, huge slabs of harder sheet ice, formed along the banks and broken free as the river changed level, may also be floating downstream. The slabs of sheet ice and the dollops of frazil ice go together like bricks and mortar. Stacking up at a channel constriction, they can lock themselves into an ice bridge.

Generally, when such an ice bridge forms, the river will have room to flow underneath. But if the river is very shallow and the slabs of sheet ice are large, possibly not. Short of total blockage, the flow of the river will be slowed where it must pass through that narrowed gap; if it slows to less than some critical value, more ice will collect along the front face of the bridge and the ice cover will expand upstream. The relevant formula here is:

$$V_C = (1 - h/H) \sqrt{2g(p - p_i/p)h}$$

where V_C is the critical flow rate and h is the ice thickness and everything else represents something too. But this also is not Leontiev's Axiom, and so we can ignore it, praise God.

The Madison River where it runs north through Montana happens to be very shallow. Upstream from (that is, south of) the lake that sits five miles north of the small town of Ennis, it's a magnificent stretch of habitat for stoneflies and caddisflies and trout and blue heron and fox and eagles and, half the year anyway, fishermen. The water is warmed at its geothermal source in Yellowstone Park, cooled again by its Montana tributaries such as West Fork, rich in nutrients and oxygen, clear, lambent, unspoiled. Thanks to these graces, it's probably much too famous for its own good, and here I am making it a little more famous still. Upstream from the highway bridge at Ennis, where it can be conveniently floated by fishermen in rafts and guided Mackenzie boats, it gets an untoward amount of attention. This is where the notorious salmonfly hatch happens: boat traffic like the Henley Regatta, during that dizzy two weeks of June while the insects swarm and the fish gluttonize. This is the stretch of the Madison for fishermen who crave trophies but not solitude. Downstream from the Ennis bridge it becomes a different sort of river. It becomes a different sort of place.

Downstream from the Ennis bridge, for that five-mile stretch to the lake, the Madison is a broken-up travesty of a river that offers mediocre fishing and clumsy floating and no trophy trout and not many salmonflies and I promise you fervently you wouldn't like it. This stretch is called the channels. The river braids out into a maze of elbows and sloughs and streams separated by dozens of small and large islands, some covered only with grass and willow, some shaded with buckling old cottonwoods, some holding thickets of water birch and woods rose and raspberry scarcely tramped through by a single fisherman in the course of a summer. The deer love these islands and, in May, so do the nesting geese. Mosquitoes are bad here. The walking is difficult and there are bleached cottonwood deadfalls waiting to tear your waders. At the end of a long day's float, headwinds and choppy waves will come up on the lake just as you try to row your boat across to the ramp. Take my word, you'd hate the whole experience. Don't bother. Give it a miss. I adore that five miles of river more than any other piece of landscape in the world.

Surrounding the braidwork of channels is a zone of bottomland roughly two miles wide, a great flat swatch of subirrigated meadow only barely above the river's springtime high-water level. This low meadow area is an unusual sort of no-man's-land that performs a miraculous service: protecting the immediate riparian vicinity of the channels from the otherwise-inevitable arrival of ranch houses, summer homes, resort lodges, motels, paved roads, development, spoliation, and all other manner of venal doom. Tantalizing and vulnerable as it may appear on a July afternoon, the channels meadowland is an ideal place to raise bluegrass and Herefords and sandhill cranes but, for reasons we'll come to, is not really good for much else.

By late December the out-of-state fishermen are long gone, the duck hunters more recently, and during a good serious stretch of weather the dark river begins to flow gray and woolly with frazil ice. If the big slabs of sheet ice are moving too, a person can stand on the Ennis highway bridge and hear the two kinds of ice rubbing, hissing, whispering to each other as though in conspiracy toward mischief—or maybe revenge. (Through the three years I lived in Ennis myself, I stood on that bridge often, gawking and listening. There aren't too many other forms of legal amusement in a Montana town of a thousand souls during the short days and long weeks of midwinter.) By this time the lake, five miles downstream, will have already frozen over. Then the river water cools still further, the frazil thickens, the slabs bump and tumble into those narrow channels, until somewhere, at a point of constriction down near the lake, mortar meets brick and you begin to get:

$$V_C = (1 - h/H) \sqrt{2g(p - p_i/p)h}.$$

Soon the river is choked with its own ice. All the channels are nearly or totally blocked. But water is still arriving from upstream, and it has to go somewhere. So it flows out across the bottomland. It spills over its banks and, moving quickly, faster than a man can walk, it covers a large part of that meadow area with water. Almost as quickly, the standing floodwater becomes ice.

If you have been stubborn or foolish enough to build your house on that flat, in a pretty spot at the edge of the river, you now have three feet of well-deserved ice in your living room. *Get back away from me,* is what the river has told you. *Show some goddamn respect.* There are memories of this sort of ice-against-man encounter. It hasn't happened often, that a person should come along so mule-minded as to insist on flouting the reality of the ice, but often enough for a few vivid stories. Back in 1863, for instance, a settler named Andrew Odell, who had built his cabin out on the channel meadows, woke up one night in December to find river water already lapping onto his bed. He grabbed his blanket and fled, knee deep, toward higher ground on the far side of a spring creek that runs parallel to the channels a half mile east. That spring creek is now called Odell Creek, and it marks a rough eastern boundary of the zone that gets buried in ice. Nowadays you don't see any cabins or barns in the flat between Odell Creek and the river.

Folks in Ennis call this salutary ice-laying event the Gorge. The Gorge doesn't occur every year, and it isn't uniform or predictable when it does. Two or three winters may go by without serious weather, without a Gorge, without that frozen flood laid down upon hundreds of acres, and then there will come a record year. A rancher named Ralph Paugh remembers one particular Gorge, because it back-flooded all the way up across Odell Creek to fill his barn with a two-foot depth of ice. This was on Christmas Day, 1983. "It come about four o'clock," he recalls. "Never had got to the barn before." His barn has sat on that rise since 1905. He has some snapshots from the 1983 episode, showing vistas and mounds of whiteness. "That pile there, see, we piled that up with the dozer when we cleaned it out." Ralph also remembers talk about the Gorge in 1907, the year he was born. That one took out the old highway bridge, so for the rest of the winter schoolchildren and mailmen and whoever else had urgent reason for crossing the river did so on a trail of planks laid across ice. The present bridge is a new one, the lake north of Ennis is also a relatively recent contrivance (put there for hydroelectric generation about the time, again, when Ralph Paugh was born), but the Gorge of the Madison channels is natural and immemorial.

I used to lace up my Sorels and walk on it. Cold sunny afternoons of January or February, bare willows, bare cottonwoods, exquisite solitude, fox tracks in an inch of fresh snow, and down through three feet of ice below my steps and the fox tracks were spectacular bits of Montana that other folk, outlanders, coveted only in summer.

Mostly I wandered these places alone. Then one year a certain biologist of my recent acquaintance came down for a visit in Ennis. I think this was in late April. I know that the river had gorged that year and that the ice was now melting away from the bottomland, leaving behind its moraine of fertile silt. The channels themselves, by now, were open and running clear. The first geese had arrived. This biologist and I spent that day in the water, walking downriver through the channels. We didn't fish. We didn't collect aquatic insects or study the nesting of *Branta canadensis*. The trees hadn't yet come into leaf and it was no day for a picnic. We just walked in the water, stumbling over boulders, bruising our feet, getting wet over the tops of our waders. We saw the Madison channels, fresh from cold storage, before anyone else did that year. We covered only about three river miles in the course of the afternoon, but that was enough to exhaust us, and then we stumbled out across the muddy fields and walked home on the road. How extraordinary, I thought, to come across a biologist who would share my own demented appreciation of such an arduous, stupid, soggy trek. So I married her.

The channels of the Madison are a synecdoche. They are the part that resonates suggestively with the significance of the whole. To understand how I mean that, it might help to know Leontiev's Axiom. Konstantin Leontiev was a cranky nineteenth-century Russian thinker. He trained as a physician, worked as a diplomat in the Balkans, wrote novels and essays that aren't read much today, became disgusted at the prospect of moral decay in his homeland, and in his last years flirted with becoming a monk. By most of the standards you and I likely share, he was an unsavory character. But even a distempered and retrograde Czarist with monastic leanings can be right about something once in a while.

Leontiev wrote: "To stop Russia from rotting, one would have to put it under ice."

In my mind, in my dreams, that great flat sheet of Madison River whiteness spreads out upon the whole state of Montana. I believe, with Leontiev, in salvation by ice.

4. SOURCES

The biologist whose husband I am sometimes says to me: "All right, so where do we go when Montana's been ruined? Alaska? Norway? Where?" This is a dark joke between us. She grew up in Montana, loves the place the way some women might love an incorrigibly self-destructive man, with pain and fear and pity, and she has no desire to go anywhere else. I grew up in Ohio, discovered home in Montana only fifteen years ago, and I feel the same. But still we play at the dark joke. "Not Norway," I say, "and you know why." We're each half Norwegian and we've actually eaten lutefisk. "How about Antarctica," I say. "Antarctica should be okay for a while yet."

On the desk before me now is a pair of books about Antarctica. Also here are a book on the Arctic, another book titled *The World of Ice*, a book of excerpts from Leontiev, a master's thesis on the subject of goose reproduction and water levels in the Madison channels, an extract from an unpublished fifty-year-old manuscript on the history of the town of Ennis, a cassette tape of a conversation with Ralph Paugh, and a fistful of photocopies of technical and not-so-technical articles. One of the less technical articles is titled "Ice on the World," from a recent issue of *National Geographic*. In this article is a full-page photograph of strawberry plants covered with a thick layer of ice.

These strawberry plants grew in central Florida. They were sprayed with water, says the caption, because subfreezing temperatures had been forecast. The growers knew that a layer of ice, giving insulation, even giving up some heat as the water froze, would save them.

In the foreground is one large strawberry. The photocopy shows it dark gray, but in my memory it's a death-defying red.

LESLIE MARMON SILKO
b. 1948

Leslie Silko addresses the role of ritual and myth in lending order to contemporary life—in helping people both to survive and to grow. This theme is developed in her 1977 novel Ceremony, which tells the story of a World War II veteran trying to make peace with himself and his world on a New Mexico reservation. Her poems and stories, too, (Laguna Woman, 1974; Storyteller, 1981) portray lives within which traditional beliefs and spirits can make sense of a fragmented social world. Silko's essay about naming as a traditional form of storytelling, making the landscape into a sustaining, holy text, brings a crucial element into the American literature of nature. For Indians and non-Indians alike, she suggests that naming may be a form of deep identification, rather than the analytical distancing from nature that other writers about wilderness sometimes assume it to be.

LANDSCAPE, HISTORY, AND THE PUEBLO IMAGINATION

FROM A HIGH ARID PLATEAU IN NEW MEXICO

You see that after a thing is dead, it dries up. It might take weeks or years, but eventually if you touch the thing, it crumbles under your fingers. It goes back to dust. The soul of the thing has long since departed. With the plants and wild game the soul may have already been borne back into bones and blood or thick green stalk and leaves. Nothing is wasted. What cannot be eaten by people or in some way used must then be left where other living creatures may benefit. What domestic animals or wild scavengers can't eat will be fed to the plants. The plants feed on the dust of these few remains.

The ancient Pueblo people buried the dead in vacant rooms or par-

Antaeus, no. 57, Autumn 1986.

tially collapsed rooms adjacent to the main living quarters. Sand and clay used to construct the roof make layers many inches deep once the roof has collapsed. The layers of sand and clay make for easy gravedigging. The vacant room fills with cast-off objects and debris. When a vacant room has filled deep enough, a shallow but adequate grave can be scooped in a far corner. Archaeologists have remarked over formal burials complete with elaborate funerary objects excavated in trash middens of abandoned rooms. But the rocks and adobe mortar of collapsed walls were valued by the ancient people. Because each rock had been carefully selected for size and shape, then chiseled to an even face. Even the pink clay adobe melting with each rainstorm had to be prayed over, then dug and carried some distance. Corn cobs and husks, the rinds and stalks and animal bones were not regarded by the ancient people as filth or garbage. The remains were merely resting at a mid-point in their journey back to dust. Human remains are not so different. They should rest with the bones and rinds where they all may benefit living creatures—small rodents and insects—until their return is completed. The remains of things—animals and plants, the clay and the stones—were treated with respect. Because for the ancient people all these things had spirit and being. The antelope merely consents to return home with the hunter. All phases of the hunt are conducted with love. The love the hunter and the people have for the Antelope People. And the love of the antelope who agree to give up their meat and blood so that human beings will not starve. Waste of meat or even the thoughtless handling of bones cooked bare will offend the antelope spirits. Next year the hunters will vainly search the dry plains for antelope. Thus it is necessary to return carefully the bones and hair, and the stalks and leaves to the earth who first created them. The spirits remain close by. They do not leave us.

The dead become dust, and in this becoming they are once more joined with the Mother. The ancient Pueblo people called the earth the Mother Creator of all things in this world. Her sister, the Corn Mother, occasionally merges with her because all succulent green life rises out of the depths of the earth.

Rocks and clay are part of the Mother. They emerge in various forms, but at some time before, they were smaller particles or great boulders. At a later time they may again become what they once were. Dust.

A rock shares this fate with us and with animals and plants as well. A rock has being or spirit, although we may not understand it. The spirit may differ from the spirit we know in animals or plants or in ourselves. In the end we all originate from the depths of the earth. Perhaps this is how all beings share in the spirit of the Creator. We do not know.

FROM THE EMERGENCE PLACE

Pueblo potters, the creators of petroglyphs and oral narratives, never conceived of removing themselves from the earth and sky. So long as the human consciousness remains *within* the hills, canyons, cliffs, and the plants, clouds, and sky, the term *landscape*, as it has entered the English language, is misleading. "A portion of territory the eye can comprehend in a single view" does not correctly describe the relationship between the human being and his or her surroundings. This assumes the viewer is somehow *outside* or *separate from* the territory he or she surveys. Viewers are as much a part of the landscape as the boulders they stand on. There is no high mesa edge or mountain peak where one can stand and not immediately be part of all that surrounds. Human identity is linked with all the elements of Creation through the clan: you might belong to the Sun Clan or the Lizard Clan or the Corn Clan or the Clay Clan.[1] Standing deep within the natural world, the ancient Pueblo understood the thing as it was—the squash blossom, grasshopper, or rabbit itself could never be created by the human hand. Ancient Pueblos took the modest view that the thing itself (the landscape) could not be improved upon. The ancients did not presume to tamper with what had already been created. Thus *realism*, as we now recognize it in painting and sculpture, did not catch the imaginations of Pueblo people until recently.

The squash blossom itself is *one thing*: itself. So the ancient Pueblo potter abstracted what she saw to be the key elements of the squash blossom—the four symmetrical petals, with four symmetrical stamens in the center. These key elements, while suggesting the squash flower, also link it with the four cardinal directions. By representing only its intrinsic form, the squash flower is released from a limited meaning or restricted identity. Even in the most sophisticated abstract form, a squash flower or a cloud or a lightning bolt became intricately connected with a complex system of relationships which the ancient Pueblo people maintained with each other, and with the populous natural world they lived within. A bolt of lightning is itself, but at the same time it may mean much more. It may be a messenger of good fortune when summer rains are needed. It may deliver death, perhaps the result of manipulations by the Gunnadeyahs, destructive necro-

---

[1] Clan—A *social unit composed of families sharing common ancestors who trace their lineage back to the Emergence where their ancestors allied themselves with certain plants or animals or elements.* [Silko's note]

mancers. Lightning may strike down an evil-doer. Or lightning may strike a person of good will. If the person survives, lightning endows him or her with heightened power.

Pictographs and petroglyphs of constellations or elk or antelope draw their magic in part from the process wherein the focus of all prayer and concentration is upon the thing itself, which, in its turn, guides the hunter's hand. Connection with the spirit dimensions requires a figure or form which is all-inclusive. A "lifelike" rendering of an elk is too restrictive. Only the elk *is* itself. A *realistic* rendering of an elk would be only one particular elk anyway. The purpose of the hunt rituals and magic is to make contact with *all* the spirits of the Elk.

The land, the sky, and all that is within them—the landscape—includes human beings. Interrelationships in the Pueblo landscape are complex and fragile. The unpredictability of the weather, the aridity and harshness of much of the terrain in the high plateau country explain in large part the relentless attention the ancient Pueblo people gave the sky and the earth around them. Survival depended upon harmony and cooperation not only among human beings, but among all things—the animate and the less animate, since rocks and mountains were known to move, to travel occasionally.

The ancient Pueblos believed the Earth and the Sky were sisters (or sister and brother in the post-Christian version). As long as good family relations are maintained, then the Sky will continue to bless her sister, the Earth, with rain, and the Earth's children will continue to survive. But the old stories recall incidents in which troublesome spirits or beings threaten the earth. In one story, a malicious ka'tsina, called the Gambler, seizes the Shiwana, or Rainclouds, the Sun's beloved children.[2] The Shiwana are snared in magical power late one afternoon on a high mountain top. The Gambler takes the Rainclouds to his mountain stronghold where he locks them in the north room of his house. What was his idea? The Shiwana were beyond value. They brought life to all things on earth. The Gambler wanted a big stake to wager in his games of chance. But such greed, even on the part of only one being, had the effect of threatening the survival of all life on earth. Sun Youth, aided by old Grandmother Spider, outsmarts the Gambler and the rigged game, and the Rainclouds are set free. The drought ends, and once more life thrives on earth.

---

[2] Ka'tsina—*Ka'tsinas are spirit beings who roam the earth and who inhabit kachina masks worn in Pueblo ceremonial dances.* [Silko's note]

THROUGH THE STORIES WE HEAR WHO WE ARE

All summer the people watch the west horizon, scanning the sky from south to north for rain clouds. Corn must have moisture at the time the tassels form. Otherwise pollination will be incomplete, and the ears will be stunted and shriveled. An inadequate harvest may bring disaster. Stories told at Hopi, Zuni, and at Acoma and Laguna describe drought and starvation as recently as 1900. Precipitation in west-central New Mexico averages fourteen inches annually. The western pueblos are located at altitudes over 5,600 feet above sea level, where winter temperatures at night fall below freezing. Yet evidence of their presence in the high desert plateau country goes back ten thousand years. The ancient Pueblo people not only survived in this environment, but many years they thrived. In A.D. 1100 the people at Chaco Canyon had built cities with apartment buildings of stone five stories high. Their sophistication as sky-watchers was surpassed only by Mayan and Inca astronomers. Yet this vast complex of knowledge and belief, amassed for thousands of years, was never recorded in writing.

Instead, the ancient Pueblo people depended upon collective memory through successive generations to maintain and transmit an entire culture, a world view complete with proven strategies for survival. The oral narrative, or "story," became the medium in which the complex of Pueblo knowledge and belief was maintained. Whatever the event or the subject, the ancient people perceived the world and themselves within that world as part of an ancient continuous story composed of innumerable bundles of other stories.

The ancient Pueblo vision of the world was inclusive. The impulse was to leave nothing out. Pueblo oral tradition necessarily embraced all levels of human experience. Otherwise, the collective knowledge and beliefs comprising ancient Pueblo culture would have been incomplete. Thus stories about the Creation and Emergence of human beings and animals into this World continue to be retold each year for four days and four nights during the winter solstice. The "humma-hah" stories related events from the time long ago when human beings were still able to communicate with animals and other living things. But, beyond these two preceding categories, the Pueblo oral tradition knew no boundaries. Accounts of the appearance of the first Europeans in Pueblo country or of the tragic encounters between Pueblo people and Apache raiders were no more and no less important than stories about the biggest mule deer ever taken or adulterous couples surprised in cornfields and chicken coops. Whatever happened, the ancient people instinctively sorted events and details into a loose narrative structure. Everything became a story.

Traditionally everyone, from the youngest child to the oldest person, was expected to listen and to be able to recall or tell a portion, if only a small detail, from a narrative account or story. Thus the remembering and retelling were a communal process. Even if a key figure, an elder who knew much more than others, were to die unexpectedly, the system would remain intact. Through the efforts of a great many people, the community was able to piece together valuable accounts and crucial information that might otherwise have died with an individual.

Communal storytelling was a self-correcting process in which listeners were encouraged to speak up if they noted an important fact or detail omitted. The people were happy to listen to two or three different versions of the same event or the same humma-hah story. Even conflicting versions of an incident were welcomed for the entertainment they provided. Defenders of each version might joke and tease one another, but seldom were there any direct confrontations. Implicit in the Pueblo oral tradition was the awareness that loyalties, grudges, and kinship must always influence the narrator's choices as she emphasizes to listeners this is the way *she* has always heard the story told. The ancient Pueblo people sought a communal truth, not an absolute. For them this truth lived somewhere within the web of differing versions, disputes over minor points, outright contradictions tangling with old feuds and village rivalries.

A dinner-table conversation, recalling a deer hunt forty years ago when the largest mule deer ever was taken, inevitably stimulates similar memories in listeners. But hunting stories were not merely after-dinner entertainment. These accounts contained information of critical importance about behavior and migration patterns of mule deer. Hunting stories carefully described key landmarks and locations of fresh water. Thus a deer-hunt story might also serve as a "map." Lost travelers, and lost piñon-nut gatherers have been saved by sighting a rock formation they recognize only because they once heard a hunting story describing this rock formation.

The importance of cliff formations and water holes does not end with hunting stories. As offspring of the Mother Earth, the ancient Pueblo people could not conceive of themselves within a specific landscape. Location, or "place," nearly always plays a central role in the Pueblo oral narratives. Indeed, stories are most frequently recalled as people are passing by a specific geographical feature or the exact place where a story takes place. The precise date of the incident often is less important than the place or location of the happening. "Long, long ago," "a long time ago," "not too long ago," and "recently" are usually how stories are classified in terms of time. But the places where the stories occur are precisely located, and prominent geographical details recalled, even if the landscape is well-known to listeners. Often because the turning point in the narrative involved a peculiarity or special quality of a rock or tree or plant found only at that place. Thus, in the case of many of the Pueblo narratives, it is impossible to determine which came first: the incident or the geographical feature which begs to be brought alive in a story that features some unusual aspect of this location.

There is a giant sandstone boulder about a mile north of Old Laguna, on the road to Paguate. It is ten feet tall and twenty feet in circumference. When I was a child, and we would pass this boulder driving to Paguate village, someone usually made reference to the story about Kochininako, Yellow Woman, and the Estrucuyo, a monstrous giant who nearly ate her. The Twin Hero Brothers saved Kochininako, who had been out hunting rabbits to take home to feed her mother and sisters. The Hero Brothers had heard her cries just in time. The Estrucuyo had cornered her in a cave too small to fit its monstrous head. Kochininako had already thrown to the Estrucuyo all her rabbits, as well as her moccasins and most of her clothing. Still the creature had not been satisfied. After killing the Estrucuyo with their bows and arrows, the Twin Hero Brothers slit open the Estrucuyo and cut out its heart. They threw the heart as far as they could. The monster's heart landed there, beside the old trail to Paguate village, where the sandstone boulder rests now.

It may be argued that the existence of the boulder precipitated the creation of a story to explain it. But sandstone boulders and sandstone formations of strange shapes abound in the Laguna Pueblo area. Yet most of them do not have stories. Often the crucial element in a narrative is the terrain—some specific detail of the setting.

A high dark mesa rises dramatically from a grassy plain fifteen miles southeast of Laguna, in an area known as Swanee. On the grassy plain one hundred and forty years ago, my great-grandmother's uncle and his brother-in-law were grazing their herd of sheep. Because visibility on the plain extends for over twenty miles, it wasn't until the two sheepherders came near the high dark mesa that the Apaches were able to stalk them. Using the mesa to obscure their approach, the raiders swept around from both ends of the mesa. My great-grandmother's relatives were killed, and the herd lost. The high dark mesa played a critical role: the mesa had compromised the safety which the openness of the plains had seemed to assure. Pueblo and Apache alike relied upon the terrain, the very earth herself, to give them protection and aid. Human activities or needs were maneuvered to fit the existing surroundings and conditions. I imagine the last afternoon of my distant ancestors as warm and sunny for late September. They might have been traveling slowly, bringing the sheep closer to Laguna in preparation for the approach of colder weather. The grass was tall and only beginning to change from green to a yellow which matched the late-afternoon sun shining off it. There might have been comfort in the warmth and the sight of the sheep fattening on good pasture which lulled my ancestors into their fatal inattention. They might have had a rifle whereas the Apaches had only bows and arrows. But there would have been four or five Apache raiders, and the surprise attack would have canceled any advantage the rifles gave them.

Survival in any landscape comes down to making the best use of all available resources. On that particular September afternoon, the raiders made better use of the Swanee terrain than my poor ancestors did. Thus the high dark mesa and the story of the two lost Laguna herders became inextricably linked. The memory of them and their story resides in part with the high black mesa. For as long as the mesa stands, people within the family and clan will be reminded of the story of that afternoon long ago. Thus the continuity and accuracy of the oral narratives are reinforced by the landscape—and the Pueblo interpretation of that landscape is *maintained*.

THE MIGRATION STORY: AN INTERIOR JOURNEY

The Laguna Pueblo migration stories refer to specific places—mesas, springs, or cottonwood trees—not only locations which can be visited still, but also locations which lie directly on the state highway route linking Paguate village with Laguna village. In traveling this road as a child with older Laguna people I first heard a few of the stories from that much larger body of stories linked with the Emergence and Migration.[3] It may be coincidental that Laguna people continue to follow the same route which, according to the Migration story, the ancestors followed south from the Emergence Place. It may be that the route is merely the short-

---

[3] The Emergence—*All the human beings, animals, and life which had been created emerged from the four worlds below when the earth became habitable.*

The Migration—*The Pueblo people emerged into the Fifth World, but they had already been warned they would have to travel and search before they found the place they were meant to live.* [Silko's note]

est and best route for car, horse, or foot traffic between Laguna and Paguate villages. But if the stories about boulders, springs, and hills are actually remnants from a ritual that retraces the creation and emergence of the Laguna Pueblo people as a culture, as the people they became, then continued use of that route creates a unique relationship between the ritual-mythic world and the actual, everyday world. A journey from Paguate to Laguna down the long incline of Paguate Hill retraces the original journey from the Emergence Place, which is located slightly north of the Paguate village. Thus the landscape between Paguate and Laguna takes on a deeper significance: the landscape resonates the spiritual or mythic dimension of the Pueblo world even today.

Although each Pueblo culture designates a specific Emergence Place—usually a small natural spring edged with mossy sandstone and full of cattails and wild watercress—it is clear that they do not agree on any single location or natural spring as the one and only true Emergence Place. Each Pueblo group recounts its own stories about Creation, Emergence, and Migration, although they all believe that all human beings, with all the animals and plants, emerged at the same place and at the same time.[4]

Natural springs are crucial sources of water for all life in the high desert plateau country. So the small spring near Paguate village is literally the source and continuance of life for the people in the area. The spring also functions on a spiritual level, recalling the original Emergence Place and linking the people and the spring water to all other people and to that moment when the Pueblo people became aware of themselves as they are even now. The Emergence was an emergence into a precise cultural identity. Thus the Pueblo stories about the Emergence and Migration are not to be taken as literally as the anthropologists might wish. Prominent geographical features and landmarks which are mentioned in the narratives exist for ritual purposes, not because the Laguna people actually journeyed south for hundreds of years from Chaco Canyon or Mesa Verde, as the archaeologists say, or eight miles from the site of the natural springs at Paguate to the sandstone hilltop at Laguna.

The eight miles, marked with boulders, mesas, springs, and river crossings, are actually a ritual circuit or path which marks the interior journey the Laguna people made: a journey of awareness and imagina-

---

[4] Creation—*Tse'itsi'nako, Thought Woman, the Spider, thought about it, and everything she thought came into being. First she thought of three sisters for herself, and they helped her think of the rest of the Universe, including the Fifth World and the four worlds below.* The Fifth World *is the world we are living in today. There are four previous worlds below this world.* [Silko's note]

tion in which they emerged from being within the earth and from everything included in earth to the culture and people they became, differentiating themselves for the first time from all that had surrounded them, always aware that interior distances cannot be reckoned in physical miles or in calendar years.

The narratives linked with prominent features of the landscape between Paguate and Laguna delineate the complexities of the relationship which human beings must maintain with the surrounding natural world if they hope to survive in this place. Thus the journey was an interior process of the imagination, a growing awareness that being human is somehow different from all other life—animal, plant, and inanimate. Yet we are all from the same source: the awareness never deteriorated into Cartesian duality, cutting off the human from the natural world.

The people found the opening into the Fifth World too small to allow them or any of the animals to escape. They had sent a fly out through the small hole to tell them if it was the world which the Mother Creator had promised. It was, but there was the problem of getting out. The antelope tried to butt the opening to enlarge it, but the antelope enlarged it only a little. It was necessary for the badger with her long claws to assist the antelope, and at last the opening was enlarged enough so that all the people and animals were able to emerge up into the Fifth World. The human beings could not have emerged without the aid of antelope and badger. The human beings depended upon the aid and charity of the animals. Only through interdependence could the human beings survive. Families belonged to clans, and it was by clan that the human being joined with the animal and plant world. Life on the high arid plateau became viable when the human beings were able to imagine themselves as sisters and brothers to the badger, antelope, clay, yucca, and sun. Not until they could find a viable relationship to the terrain, the landscape they found themselves in, could they *emerge*. Only at the moment the requisite balance between human and *other* was realized could the Pueblo people become a culture, a distinct group whose population and survival remained stable despite the vicissitudes of climate and terrain.

Landscape thus has similarities with dreams. Both have the power to seize terrifying feelings and deep instincts and translate them into images—visual, aural, tactile—into the concrete where human beings may more readily confront and channel the terrifying instincts or powerful emotions into rituals and narratives which reassure the individual while reaffirming cherished values of the group. The identity of the individual as a part of the group and the greater Whole is strengthened, and the terror of facing the world alone is extinguished.

Even now, the people at Laguna Pueblo spend the greater portion of social occasions recounting recent incidents or events which have occurred in the Laguna area. Nearly always, the discussion will precipitate the retelling of older stories about similar incidents or other stories connected with a specific place. The stories often contain disturbing or provocative material, but are nonetheless told in the presence of children and women. The effect of these inter-family or inter-clan exchanges is the reassurance for each person that she or he will never be separated or apart from the clan, no matter what might happen. Neither the worst blunders or disasters nor the greatest financial prosperity and joy will ever be permitted to isolate anyone from the rest of the group. In the ancient times, cohesiveness was all that stood between extinction and survival, and, while the individual certainly was recognized, it was always as an individual simultaneously bonded to family and clan by a complex bundle of custom and ritual. You are never the first to suffer a grave loss or profound humiliation. You are never the first, and you understand that you will probably not be the last to commit or be victimized by a repugnant act. Your family and clan are able to go on at length about others now passed on, others older or more experienced than you who suffered similar losses.

The wide deep arroyo near the Kings Bar (located across the reservation borderline) has over the years claimed many vehicles. A few years ago, when a Viet Nam veteran's new red Volkswagen rolled backwards into the arroyo while he was inside buying a six-pack of beer, the story of his loss joined the lively and large collection of stories already connected with that big arroyo. I do not know whether the Viet Nam veteran was consoled when he was told the stories about the other cars claimed by the ravenous arroyo. All his savings of combat pay had gone for the red Volkswagen. But this man could not have felt any worse than the man who, some years before, had left his children and mother-in-law in his station wagon with the engine running. When he came out of the liquor store his station wagon was gone. He found it and its passengers upside down in the big arroyo. Broken bones, cuts and bruises, and a total wreck of the car. The big arroyo has a wide mouth. Its existence needs no explanation. People in the area regard the arroyo much as they might regard a living being, which has a certain character and personality. I seldom drive past that wide deep arroyo without feeling a familiarity with and even a strange affection for this arroyo. Because as treacherous as it may be, the arroyo maintains a strong connection between human beings and the earth. The arroyo demands from us the caution and attention that constitute respect. It is this sort of respect the old believers have in mind when they tell us we must respect and love the earth.

Hopi Pueblo elders have said that the austere and, to some eyes, barren plains and hills surrounding their mesa-top villages actually help to nurture the spirituality of the Hopi *way*. The Hopi elders say the Hopi people might have settled in locations far more lush where daily life would not have been so grueling. But there on the high silent sandstone mesas that overlook the sandy arid expanses stretching to all horizons, the Hopi elders say the Hopi people must "live by their prayers" if they are to survive. The Hopi way cherishes the intangible: the riches realized from interaction and interrelationships with all beings above all else. Great abundances of material things, even food, the Hopi elders believe, tend to lure human attention away from what is most valuable and important. The views of the Hopi elders are not much different from those elders in all the Pueblos.

The bare vastness of the Hopi landscape emphasizes the visual impact of every plant, every rock, every arroyo. Nothing is overlooked or taken for granted. Each ant, each lizard, each lark is imbued with great value simply because the creature is there, simply because the creature is alive in a place where any life at all is precious. Stand on the mesa edge at Walpai and look west over the bare distances toward the pale blue outlines of the San Francisco peaks where the ka'tsina spirits reside. So little lies between you and the sky. So little lies between you and the earth. One look and you know that simply to survive is a great triumph, that every possible resource is needed, every possible ally—even the most humble insect or reptile. You realize you will be speaking with all of them if you intend to last out the year. Thus it is that the Hopi elders are grateful to the landscape for aiding them in their quest as spiritual people.

JAMAICA KINCAID
b. 1949

Born in Antigua as Elaine Potter, Jamaica Kincaid adopted her present name on moving to New York as an au pair and staying to become a writer for The New Yorker. *Kincaid made her reputation with a collection of short stories,* At the Bottom of the River *(1983), and a novel,* Annie John *(1985), and has continued to produce fiction. But nonfiction has also been an important part of her work, especially in relation to two themes.* A Small Place *(1988), in which the present selection was published after its original appearance in* The New Yorker, *is one of Kincaid's memoirs of her native island. While resolutely avoiding sentimentality about the history of Antigua and the present conditions of life there, she evokes its vivid, windswept beauty. Gardening, a second focus of her nonfiction, relates both to her girlhood experiences in the West Indies and to Vermont, where she and her family presently live. In both regards she has introduced challenging and controversial questions about the political implications and cultural consequences of human-made landscapes.*

ALIEN SOIL

Whatever it is in the character of the English people that leads them to obsessively order and shape their landscape to such a degree that it looks like a painting (tamed, framed, captured, kind, decent, good, pretty), while a painting never looks like the English landscape, unless it is a bad painting—this quality of character is blissfully lacking in the Antiguan people. I make this unfair comparison (unfair to the Antiguan people? unfair to the English people? I cannot tell but there is an unfairness here somewhere) only because so much of the character of the Antiguan people is influenced by and inherited, through conquest,

The New Yorker, June 21, 1993.

from the English people. The tendency to shower pity and cruelty on the weak is among the traits the Antiguans inherited, and so is a love of gossip. (The latter, I think, is responsible for the fact that England has produced such great novelists, but it has not yet worked to the literary advantage of the Antiguan people.) When the English were a presence in Antigua—they first came to the island as slaveowners, when a man named Thomas Warner established a settlement there in 1632—the places where they lived were surrounded by severely trimmed hedges of plumbago, topiaries of willow (casuarina), and frangipani and hibiscus; their grass was green (odd, because water was scarce; the proper word for the climate is not "sunny" but "drought-ridden") and freshly cut; they kept trellises covered with roses, and beds of marigolds and cannas and chrysanthemums.

Ordinary Antiguans (and by "ordinary Antiguans" I mean the Antiguan people, who are descended from the African slaves brought to this island by Europeans; this turns out to be a not uncommon way to become ordinary), the ones who had some money and could live in houses of more than one room, had gardens in which only flowers were grown. This made it even more apparent that they had some money, in that all their outside space was devoted not to feeding their families but to the sheer beauty of things. I can remember in particular one such family, who lived in a house with many rooms (four, to be exact). They had an indoor kitchen and a place for bathing (no indoor toilet, though); they had a lawn, always neatly cut, and they had beds of flowers, but I can now remember only roses and marigolds. I can remember those because once I was sent there to get a bouquet of roses for my godmother on her birthday. The family also had, in the middle of their small lawn, a willow tree, pruned so that it had the shape of a pine tree—a conical shape—and at Christmastime this tree was decorated with colored lights (which was so unusual and seemed so luxurious to me that when I passed by this house I would beg to be allowed to stop and stare at it for a while). At Christmas, all willow trees would suddenly be called Christmas trees, and for a time, when my family must have had a small amount of money, I, too, had a Christmas tree—a lonely, spindly branch of willow sitting in a bucket of water in our very small house. No one in my family and, I am almost certain, no one in the family of the people with the lighted-up willow tree had any idea of the origins of the Christmas tree and the traditions associated with it. When these people (the Antiguans) lived under the influence of these other people (the English), there was naturally an attempt among some of them to imitate their rulers in this particular way—by rearranging the landscape—and they did it without question. They can't be faulted for not asking what it was they were doing; that is the way these things work. The English left, and most of their landscaping influence went with them. The Americans came, but Americans (I am one now) are not interested in influencing people directly; we instinctively understand the childish principle of monkey see, monkey do. And at the same time we are divided about how we ought to behave in the world. Half of us believe in and support strongly a bad thing our government is doing, while the other half do not believe in and protest strongly against the bad thing. The bad thing succeeds, and everyone, protester and supporter alike, enjoys immensely the results of the bad thing. This ambiguous approach in the many is always startling to observe in the individual. Just look at Thomas Jefferson, a great American gardener and our country's third President, who owned slaves, and strongly supported the idea of an expanded American border, which meant the extinction of the people who already lived on the land to be taken, while at the same time he was passionately devoted to ideas about freedom—ideas that the descendants of the slaves and the people who were defeated and robbed of their land would have to use in defense of themselves. Jefferson, as President, commissioned the formidable trek his former secretary, the adventurer and botany thief Meriwether Lewis, made through the West, sending plant specimens back to the President along the way. The *Lewisia rediviva*, state flower of Montana, which Lewis found in the Bitterroot River valley, is named after him; the clarkia, not a flower of any state as far as I can tell, is named for his co-adventurer and botany thief, William Clark.

What did the botanical life of Antigua consist of at the time another famous adventurer—Christopher Columbus—first saw it? To see a garden in Antigua now will not supply a clue. I made a visit to Antigua this spring and most of the plants I saw there came from somewhere else. The bougainvillea (named for another restless European, the sea adventurer Louis-Antoine de Bougainville, first Frenchman to cross the Pacific) is native to tropical South America; the plumbago is from Southern Africa; the croton (genus *Codiaeum*) is from Malay Peninsula; the *Hibiscus rosa-sinensis* is from Asia and the *Hibiscus schizopetalus* is from East Africa; the allamanda is from Brazil; the poinsettia (named for an American ambassador, Joel Poinsett) is from Mexico; the bird of paradise flower is from Southern Africa; the Bermuda lily is from Japan; the flamboyant tree is from Madagascar; the casuarina is from Australia; the Norfolk pine is from Norfolk Island; the tamarind tree is from Africa; the mango is from Asia. The breadfruit, that most Antiguan (to me) and starchy food, the bane of every Antiguan child's palate, is from the East Indies. This food has been the cause of more disagreement between parents and their children than anything else I can think of. No child has ever liked it. It was sent to the West Indies by Joseph Banks, the English naturalist and world traveller, and the head of Kew Gardens, which was then a clearing house for all the plants stolen from the various parts of the world where the English had been. (One of the climbing roses, *Rosa banksiae*, from China, was named for Banks' wife.) Banks sent tea to India; to the West Indies he sent the breadfruit. It was meant to be a cheap food for feeding slaves. It was the cargo that Captain Bligh was carrying to the West Indies on the ship *Bounty* when his crew so rightly mutinied. It's as though the Antiguan child senses intuitively the part this food has played in the history of injustice and so will not eat it. But, unfortunately for her, it grows readily, bears fruit abundantly, and is impervious to drought. Soon after the English settled in Antigua, they cleared the land of its hardwood forests to make room for the growing of tobacco, sugar, and cotton, and it is this that makes the island drought-ridden to this day. Antigua is also empty of much wildlife natural to it. When snakes proved a problem for the planters, they imported the mongoose from India. As a result there are no snakes at all on the island—nor other reptiles, other than lizards—though I don't know what damage the absence of snakes causes, if any.

What herb of beauty grew in this place then? What tree? And did the people who lived there grow anything beautiful for its own sake? I do not know, I can only make a straightforward deduction: the frangipani, the mahogany tree, and the cedar tree are all native to the West Indies, so these trees are probably indigenous. And some of the botany of Antigua can be learned from medicinal folklore. My mother and I were sitting on the steps in front of her house one day during my recent visit, and I suddenly focussed on a beautiful bush (beautiful to me now; when I was a child I thought it ugly) whose fruit I remembered playing with when I was little. It is an herbaceous plant that has a red stem covered with red thorns, and emerald-green, simple leaves, with the same red thorns running down the leaf from the leafstalk. I cannot remember what its flowers looked like, and it was not in flower when I saw it while I was there with my mother, but its fruit is a small, almost transparent red berry, and it is this I used to play with. We children sometimes called it "chinaberry," because of its transparent, glassy look—it reminded us of china dinnerware, though we were only vaguely familiar with such a thing as china, having seen it no more than once or twice—and sometimes "baby tomato," because of its size, and to signify that it was not real; a baby thing was not a real thing. When I pointed the bush out to my mother, she called it something else; she called it cancanberry bush, and said that in the old days, when people could not afford to see doctors, if a child had thrush they would make a paste of this fruit and rub it inside the child's mouth, and this would make the thrush go away. But, she said, people rarely bother with this remedy anymore. The day before, a friend of hers had come to pay a visit, and when my mother offered her something to eat and drink the friend declined, because, she said, she had some six-sixty-six and maiden-blush tea waiting at home for her. This tea is taken on an empty stomach, and it is used for all sorts of ailments, including to help bring on abortions. I have never seen six-sixty-six in flower, but its leaves are a beautiful ovoid shape and a deep green—qualities that are of value in a garden devoted to shape and color of leaf.

People who do not like the idea that there is a relationship between gardening and wealth are quick to remind me of the cottage gardener, that grim-faced English person. Living on land that is not his own, he has put bits and pieces of things together, things from here and there, and it is a beautiful jumble—but just try duplicating it; it isn't cheap to do. And I have never read a book praising the cottage garden written by a cottage gardener. This person—the cottage gardener—does not exist in a place like Antigua. Nor do casual botanical conversation, knowledge of the Latin names for plants, and discussions of the binomial system. If an atmosphere where these things could flourish exists in this place I am not aware of it. I can remember very well the cruel Englishwoman who was my botany teacher, and that, in spite of her cruelty, botany was one of my two favorite subjects in school. (History was the other.) With this in mind I visited a bookstore (the only bookstore I know of in Antigua) to see what texts are now being used in the schools and to see how their content compares with what was taught to me back then; the botany I had studied was a catalogue of the plants of the British Empire, the very same plants that are now widely cultivated in Antigua and are probably assumed by ordinary Antiguans to be native to their landscape—the mango, for example. But it turns out that botany as a subject is no longer taught in Antiguan schools; the study of plants is now called agriculture. Perhaps that is more realistic, since the awe and poetry of botany cannot be eaten, and the mystery and pleasure in the knowledge of botany cannot be taken to market and sold.

And yet the people of Antigua have a relationship to agriculture that does not please them at all. Their very arrival on this island had to do with the forces of agriculture. When they (we) were brought to this island from Africa a few hundred years ago, it was not for their pottery-making skills or for their way with a loom; it was for the free labor they could provide in the fields. Mary Prince, a nineteenth-century African woman, who was born in Bermuda and spent part of her life as a slave in Antigua, writes about this in an autobiographical account, which I found in "The Classic Slave Narratives," edited by Henry Louis Gates, Jr. She says:

> My master and mistress went on one occasion into the country, to Date Hill, for change of air, and carried me with them to take charge of the children, and to do the work of the house. While I was in the country, I saw how the field negroes are worked in Antigua. They are worked very hard and fed but scantily. They are called out to work before daybreak, and come home after dark; and then each has to heave his bundle of grass for the cattle in the pen. Then, on Sunday morning, each slave has to go out and gather a large bundle of grass, and when they bring it home, they have all to sit at the manager's door and wait till he come out: often they have to wait there till past eleven o'clock, without any breakfast. After that, those that have yams or potatoes, or fire-wood to sell, hasten to market to buy . . . salt fish, or pork, which is a great treat for them.

Perhaps it makes sense that a group of people with such a wretched historical relationship to growing things would need to describe their current relationship to it as dignified and masterly (agriculture), and would not find it poetic (botany) or pleasurable (gardening).

In a book I am looking at (to read it is to look at it: the type is as tall as a doll's teacup), "The Tropical Garden," by William Warren, with photographs by Luca Invernizzi Tettoni, I find statements like "the concept of a private garden planted purely for aesthetic purposes was generally alien to tropical countries" and "there was no such tradition of ornamental horticulture among the inhabitants of most hot-weather places. Around the average home there might be a few specimens chosen especially because of their scented flowers or because they were believed to bring good fortune. . . . Nor would much, if any, attention be paid to attractive landscape design in such gardens: early accounts by travellers in the tropics abound in enthusiastic descriptions of jungle scenery, but a reader will search in vain for one praising the tasteful arrangement of massed ornamental beds and contrasting lawns of well-trimmed grass around the homes of natives." What can I say to that? No doubt it is true. And no doubt contrasting lawns and massed ornamental beds are a sign of something, and that is that someone—someone other than the owner of the lawns—has been humbled. To give just one example: on page 62 of this book is a photograph of eight men, natives of India, pulling a heavy piece of machinery used in the upkeep of lawns. They are without shoes. They are wearing the clothing of schoolboys—khaki shorts and khaki short-sleeved shirts. There is no look of bliss on their faces. The caption for the photograph reads, "Shortage of labour was never a problem in the maintenance of European features in large colonial gardens; here a team of workers is shown rolling a lawn at the Gymkhana Club in Bombay."

And here are a few questions that occur to me: what if the people living in the tropics, the ones whose history isn't tied up with and contaminated by slavery and indenturedness, are contented with their surroundings, are happy to observe an invisible hand at work and from time to time laugh at some of the ugly choices this hand makes; what if they have more important things to do than make a small tree large, a large tree small, or a tree whose blooms are usually yellow bear black blooms; what if these people are not spiritually feverish, restless, and full of envy?

When I was looking at the book of tropical gardens, I realized that the flowers and the trees so familiar to me from my childhood do not now have a hold on me. I do not long to plant and be surrounded by the bougainvillea; I do not like the tropical hibiscus; the corallita (from Mexico), so beautiful when tended, so ugly when left to itself, which makes everything around it look rusty and shabby, is not a plant I like at all. I returned from my visit to Antigua, the place where I was born, to a small village in Vermont, the place where I choose to live. Spring had arrived. The tulips I had planted last autumn were in bloom, and I liked to sit and caress their petals, which felt disgustingly delicious, like scraps of peau de soie. The dizzy-making yellow of dandelions and cowslips was in the fields and riverbanks and marshes. I like these things. (I do not like daffodils, but that's a legacy of the English approach: I was forced to memorize the poem by William Wordsworth when I was a child.) I transplanted to the edge of a grove of pine trees some foxgloves that I grew from seed in late winter. I found some Virginia bluebells in a spot in the woods where I had not expected to find them, and some larches growing grouped together, also in a place I had not expected. On my calendar I marked the day I would go and dig up all the mulleins I could find and re-plant them in a very sunny spot across from the grove of pine trees. This is to be my forest of mulleins, though in truth it will appear a forest only to an ant. I marked the day I would plant the nasturtiums under the fruit trees. I discovered a clump of Dutchman's-breeches in the wildflower bed that I inherited from the man who built and used to own the house in which I now live, Robert Woodworth, the botanist who invented time-lapse photography. I waited for the things I had ordered in the deep cold of winter to come. They started to come. Mr. Pembroke, who represents our village in the Vermont legislature, came and helped me dig some of the holes where some of the things I wanted to put in were to be planted.

Mr. Pembroke is a very nice man. He is never dressed in the clothing of schoolboys. There is not a look of misery on his face; on his face is the complicated look of an ordinary human being. When he works in my garden, we agree on a price; he sends me a bill, and I pay it. The days are growing longer and longer, and then they'll get shorter again. I am now used to that ordered progression, and I love it. But there is no order in my garden. I live in America now. Americans are impatient with memory, which is one of the things order thrives on.

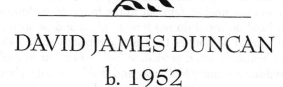

DAVID JAMES DUNCAN
b. 1952

In 1983, David James Duncan's The River Why *became the first novel ever published by the Sierra Club. But in the years after he wrote this passionate celebration of the rivers of Oregon, logging destroyed the watersheds and fisheries surrounding his home, and Duncan eventually moved with his family to Montana. His writing since then, including* River Teeth *(1995), has included both lyrical encounters with the Big Blackfoot and other Montana rivers and forceful protests against the ways in which extractive industries like mining are threatening their survival. Like Rick Bass, he is a writer both drawn personally to the wilderness of his adopted state and impelled by what he has discovered there to enter the public fray.*

Northwest Passage

When I was sixteen and hated high school, one of the things I did to get through a school day was rip pictures out of magazines. I did this in the school library, and the magazines belonged to the school: the ripping was a deliberate act of vandalism. But I only stole photos I loved.

River Teeth: Stories and Writings (New York: Doubleday, 1995).

And I felt that, in taking them home to my bedroom, I was stealing them away to a better life than the one they'd led in the library.

I had festooned my bedroom—floor, walls and ceiling—with blankets, cheap imported tapestries, wooden crates and sheepskins. It looked more like the interior of a desert Bedouin's tent than the generic Sheetrocked cubicle it was. When I'd get my stolen photos into my "tent" I'd prop them like books in front of me, light candles or kerosene lanterns before them, stare at them till they swallowed me, and virtually worship the daydreams, wanderlust and longings these makeshift icons allowed me to feel.

One such photo, plundered from a *National Geographic*, was of the confluence of the Ganges and Jumna rivers in India. It depicted a barren, boulder-strewn plateau beneath a range of mountains. You could see white water churning in the background. But the place was stripped of vegetation, desolate, and would have held no photographic interest at all if the rocky plain along the river hadn't been strewn with huts. *Hundreds* of huts. Maybe a thousand of them. Crude, leaky, diminutive hovels no red-blooded American would dream of keeping a lawnmower in. Yet in these, men were living. "Holy men," the caption called them. In any case long-bearded, huge-eyed, emaciated men, just existing on the rocks by the river there. And the reason they chose to do this, so the caption maintained, was that "the rishis of ancient India considered the confluence of two or more rivers to be a sacred place."

I couldn't have explained why I found this riverside ghetto so appealing. Trying to hone in on my feelings with the help of a dictionary, I looked up the word *holy* and in the margins of the photo wrote "1. hallowed by association with the divine, hence deserving of reverence; 2. inviolable; not to be profaned." But weeks passed and understanding did not deepen. I was still in a kind of love with the confluence-dwelling holy men and still unable to say why, when an old friend named Jered asked me to go fishing.

It was early on a school day, in mid-October. Jered's plan was to try for jack and silver salmon that very afternoon—which we'd lengthen slightly by cutting our last class. Jered and I had been fishing buddies as kids, inseparable for a season or two. In the years between I prided myself on having changed completely, while Jered prided himself on not having changed at all. I'd become a hippie; Jered still had a crew cut; I aspired (with no success) to vegetarianism; Jered was still shooting, cleaning and happily devouring scores of the indigenous mammals and birds of the Northwest; I was antiwar, Jered was ready to kill who he was told; I was trying to piece together some sort of crazy-quilt bhakti/Wishram/Buddhistic/ball-playing mysticism to live by; Jered was your basic working-class, Consciousness One, Huntin'-n'-Fishin' type guy. But I said yes for three reasons. The first was that we shared a respect, despite all differences, based on the fact that we each considered the other to be the best fisherman we knew. The second was that he had enough boraxed salmon roe for both of us, and the fishing had been, in his rudimentary but reliable diction, "hot." And the third reason, the decisive one, was that the place he proposed to fish was one the rishis of ancient India held sacred: a joining of rivers—this one in downtown Camas, Washington, where Lacamas Creek and the Washougal River both met the Columbia, and the Crown Zellerbach papermill met them all.

I knew, before going, that this confluence would be a place I would hate. I'd lived within sight and smell of the Crown Z mill, directly across the Columbia, all my life. My plan for the day, though—having read of the ancient rishis—was to see whether it might be possible to love what I also hate.

To get to our confluence we drove half an hour down the Oregon side of the Columbia, half an hour back up the Washington side, then parked in the Crown Z visitors' lot. Ignoring the NO TRESPASSING, HIGH VOLTAGE and DANGER signs, we passed through a hole in a cyclone fence, detoured round a gigantic mill building, crossed disused railroad tracks and bulldozed fields, reached the Washougal's last long riffle. We clambered downstream through brambles and scrub willow. We eased in alongside another huge mill wall, in front of which lay a kind of bay. The confluence proper—the exact spot where the crystal-clear Washougal blended with the complicated greens of the Columbia—turned out to be a slow, fishy-looking glide directly below the mill wall, just at the neck of the bay.

The riverbank there was interesting: it was made of hard-packed clay; bare rock; spilled oil; logging cable; shards of every kind and color of pop, beer and booze bottle; flood-crushed car and appliance parts; slabs of broken concrete with rebar sticking out of them; driftwood; drift Styrofoam; drift tires and reject mill parts—*huge* reject mill parts.

We found a rusted sprocket the size of a merry-go-round and sat on it.

Our legs fit perfectly between the teeth.

We had come, I felt in my very center, to a joining of everything that created, sustained and warped us. The question was: was it a viable home? Was it still somehow holy? Or was I, with my Bedouin bedroom and stolen Oriental photos, right to long only for escape?

We began setting up the odd tackle we'd together come to prefer: a stiff fiberglass fly rod and light spinning reel (mine open-faced, his closed); monofilament line a third the test of that used by most salmon fishermen (to increase casting distance). There wasn't a holy man in sight. There wasn't any kind of human in sight. The mill rumbled behind us like an insatiable stomach; tugs dragged log rafts up Crown Z's own private slough; a dredge worked the slough a half-mile downriver; trucks and cars whished along the jettied highway across the bay; ships and barges plowed the Columbia beyond. It was the machines you saw and heard, though, not the people inside them. Industrial men, not holy ones. Them's the kind we grew in these parts.

Yet it was still beautiful at the confluence. We could see east to the Cascade foothills, west clear to the Coast Range, and an enormous piece of Oregon lay across the river to the south. There is a Gangian majesty to the lower Columbia, and something awesome, if not holy, in knowing its currents are made of the glaciers, springs, desert seeps and dark industrial secrets of two Canadian provinces and six American states. When I kept my eyes on distant vistas, or on the two waters blending at our feet, I could even imagine some kind of "association with the divine" here. Though I'd no faith in the wisdom of the resident machine operators, I could picture Indians, the American kind, watching these same mountains while awaiting these same fish, some of them maybe knowing what the rishis knew.

But when I turned, as I had known I must, to the third waterway, things immediately began to break down. I knew, from studying maps, that a creek named Lacamas—a genuine little river—entered the Crown Zellerbach mill on the opposite side from us. I knew this creek headed in the mountains north of Camas, and that it'd had its own run of salmon once. But the mill-used fluid that shot from the flume just downstream from our sprocket bore no resemblance to water. It looked like hot pancake batter, gushing forth in a quantity so vast that part of me found it laughable. It looked like Satan's own nostril risen from hell, blowing out an infinite, scalding booger. But it was a steaming, poisonous, killing joke that shot across the Washougal's drought-shriveled mouth in a yellow-gray scythe, curved downstream and coated the Columbia's north shore with what looked like dead human skin for miles. And maybe a rishi could have pondered it and still felt equanimity. All I could feel, though, just as I'd feared, was fury and impotence and sickness. I said nothing to Jered—who'd merely pinched his nose, said, "Pyoo!" and set about fishing. I just squeezed my eyes shut, tried the same with my brain and waited for my friend to take us elsewhere. But I couldn't help but hear the flume's diarrheal gushing.

And then, over the gushing, a strange double splash.

The first half of the splash was a bright coho salmon. The second half was its echo, bouncing so hard off the mill wall behind us that I turned to the sound, half-expecting to see yet another river running through the rubble and broken glass.

"*Look!*" Jered whispered as a second bright coho leapt high in the evening light, fell back in the river—and again the loud, clean echo. Then another one leapt, and another, all amazingly high, all in the same place—a point in midriver, just upstream from the toxic scythe.

My first impulse, lip-service vegetarianism and all, was to grab my rod. A big school of silvers was clearly moving in. But Jered, the unabashed carnivore, had stopped fishing; he just stood watching and listening now. I would not have done the same if he hadn't set the example, but I too grew satisfied to watch: for those salmon leaps were language. They were the salmon people's legend, enacted before our eyes. And though we'd heard the old story a thousand times, thought we knew it by rote, there was a twist, at this confluence, that we two sons of these same troubled waters needed to sit still and hear.

The familiar part of the story, the rote chorus, told how these unlikely creatures had been born way up this mountain river, had grown strong in it, then had left it for the Pacific; yet some impression of their birthplace, some memory or scent touched them years and leagues later in that vastness, brought them schooling in off the Columbia's mouth, forced them to run the gauntlet of nets, hooks and predators, search the big river's murky greens, solve the riddle and enter again the waters of their fatal, yet life-giving, home. Then came the twist—in the shape of a scythe.

Salmon are not stupid. They grow tentative in rivers. They know when to spook, and when to wait quietly; when to leap, when to hide, when to fight for their lives. As these coho entered the confluence and tasted the scythe they must surely have tried everything—must have hesitated, sought another channel, circled back out into the Columbia, come round again and again, waiting for the pain of the thing to diminish, for rain to fall, for rivers to rise, for industries to die. But salmon have no choice: their great speed and long journeys, like ours, create an illusion of freedom, but to live as a race they must finally become as much a part of their river as its water and its stones. So in the end, they entered. With eyes that can't close and breath that can't be held they darted straight into this confounding of the vast, the pure and the insane. And the slashing leaps that now shattered the river's surface were each the coho word for their cold, primordial rage against whatever it was that maimed them—and their equally cold, primordial joy at having reached the waters of their home.

So we never did fish that day, Jered and I. We just sat like a couple of Ganges River hut-jockeys on a reject mill sprocket, watching salmon leap as the day grew dark. Yet after each leap my breath caught as the splash resounded, impossibly loud, against the walls of the mill, and the surface rings tripled, at least for an instant, the two dimensions of the killing scythe. Feeling a frail hope welling, feeling a need to memorize, for life, this same wild leap and passage, I suddenly began to dread my old fishing partner—or to dread, at least, the Consciousness One, Huntin'-n'-Fishin' type summary I expected him to make at any moment, thus crew-cutting the beauty off of all that we were seeing.

But when I finally did turn to my stick-in-the-mud friend, he didn't even notice me, and in his eyes, which were brimming, I saw nothing but that same cold anger, and that same wild joy. Jered was not watching fish jump. He was raging and exulting with the coho as if they were *our* people; as if ours were the ancient instincts that had sorted the Columbia's countless strands, ours the unclosing eyes the scythe betrayed and blinded, ours the bright bodies leaping and falling back into the home waters—falling just to burst them apart; just to force them to receive, even now, our gleaming silver sides.

RAY GONZALEZ
b. 1952

The following essay is a chapter in Ray Gonzalez's 1993 memoir Memory Fever: A Journey Beyond El Paso del Norte. *This was the book in which, after publishing five collections of his own poetry and also editing several important anthologies of contemporary Latino writing, Gonzalez turned for inspiration to the landscape around his boyhood home in El Paso, Texas. The resulting volume is memorable for its evocation of that austere and beautiful region, for the author's exploration of his Chicano heritage and its specific affinities with that terrain, and for the new angles it offers, in both of those connections, on "the sense of place."*

THE THIRD EYE OF THE LIZARD

I killed hundreds of lizards when I was a boy. I shot them with my BB gun because it was a favorite sport of my friends and mine. Shooting lizards helped to relieve the monotony of living in a small desert town. In 1963 my parents' house was built in a newly developed part of northwest El Paso. It sat on the outskirts of town, right in the middle of the desert, which meant that we were surrounded by rattlesnakes and lizards, creatures uprooted from their natural habitat by the new housing developments.

The empty lots around the house were havens for tumbleweeds and dozens of lizards. After a summer rain, I would walk through the lots, kick the weeds out of the way and watch the small gray and white lizards scatter for cover. I carried my loaded BB gun into the desert because I had a fantasy of shooting a huge rattlesnake, something that never happened. I killed many lizards without shame or guilt because I saw them as a threat to my life there, plus it was an ideal way to play out the timeless drama between the hunter and prey, the dance that begins when men are still boys.

Killing lizards became as routine as stepping on ants or swatting flies. There were endless numbers of lizards in the desert and many wandered into our yard, sunned themselves on the sidewalk, or crawled up the brick walls of the house. The most common was the collar lizard, four or five inches long, counting the tail. The creatures were lightning quick, but made good targets. When the lizards paused, they reared up on their hind legs and trembled.

It was the moment to shoot. To be effective, you needed to be at least ten feet close. (BB guns are not very accurate.) I often missed, the pellet springing dirt into the air as the pale lizard leaped away. When I got lucky and hit one, it would bounce into the air and land on its back, twitching, its white belly exposed to the sun. Then it would lie still.

During that moment, at the age of eleven, I felt powerful knowing I was a successful hunter—the conqueror, the proud killer, the triumphant American soldier, a favorite fantasy of young boys who dreamed beyond their big box of plastic toy soldiers. I grew up watching violent cartoons long before anyone cared to make an issue about the carnage kids were exposed to on TV. With my friends, I played "army" and owned a huge pile of toy guns. I even got a G.I. Joe the first

Memory Fever: A Journey Beyond El Paso del Norte (Tucson: University of Arizona Press, 1993).

year they were made in the early sixties. Getting my father to buy me a Daisy BB rifle was the next logical step. I had earned it by having many exciting adventures and fantasies as a neighborhood soldier, defending the long row of new houses against the dangers of living in the desert. By exterminating so many lizards, I kept the neighborhood safe. Dangers were tamed with my BB gun.

One summer I decided to keep track of how many lizards I killed by collecting their tails in a large matchbox. It had never occurred to me to count tails until the first time I encountered lizards shedding them. I cornered a small lizard in a cardboard box one day, and decided not to shoot it. I wanted a live one to exhibit in a glass jar I sometimes carried with me. Unafraid, I reached down to grab the tiny thing. As I touched the tail, it came off. I was so shocked, I dropped it. The tail shook by itself as the lizard ran out of the box. I had never seen a tail with a life of its own.

This reflexive self-dismemberment, or autotomy as it is called, is a widely-known phenomenon among scientists who study such things, but it is a real shocker the first time you discover it. I stopped hunting lizards for several days after that, thinking that some evil desert spirit was punishing me for killing so many. But my friends told me it was part of being a lizard—the tail came off naturally and the lizard would grow a new one. I got over my guilt and proceeded to grab the tails of any lizards I shot. The tails always fell off, to twitch in the matchbox. I don't recall how many I had in the box when it disappeared one day. My mother probably found it. She must have screamed at the decaying little tails and thrown them away.

For awhile, my friends and I wanted to see twitching tails more than dead lizards. Instead of using our BB guns, we started going after the lizards with sticks to cut the tails off. We fell for a lizard's natural defense. If the lizard is frightened enough, touching its body is enough to cause a detachment of the tail. The tail then wiggles strongly and attracts more attention than the lizard. The reptiles in my neighborhood must have realized it was better to leave their tails behind than to be blown away. Casualties went down when my friends and I stood around watching little tails shaking on the ground while the bare-assed lizard took off. We must have eventually caught on though; after the first summer of discovering detachable tails, we went back to shooting the lizards.

By the age of thirteen, after I discovered rock-and-roll music, I lost interest in guns and killing lizards forever. My last summer of killing them is memorable—my greatest challenge as a hunter came with the appearance of the biggest lizard I had ever seen in the neighborhood.

This huge reptile so threatened my territory it made me carry extra BB pellets in the pockets of my torn Levi's.

I first saw it clinging to the back door that led into the garage. It was a fat, dark-brown lizard measuring a good ten inches. It was not a collar lizard. I first thought it was a Gila monster, but knew they didn't exist in Texas, only in the Sonora desert of Arizona. (I had looked that up in the school encyclopedia long ago.)

I grabbed my BB gun and fired the first shot from several yards away. I missed and heard the thunk of the pellet embedding itself in the wooden door. The huge lizard darted into a crevice between the roof of the house and the doorway. I was excited, but knew I should have gotten closer. Usually, I was a very patient hunter, and my friends always kidded me because I was the best marksman among us.

I saw it again a few days later. It lay along the edge of the concrete flower bed in the backyard. I stepped closer and fired from about six feet away. The shot ricocheted off the cement and the lizard flew off. I thought I hit it, but could find no trace of it in the flower box.

I did not see the mysterious, dark creature for over two weeks, until one day I walked past the garage door to find it near the same spot where it had first appeared. I aimed carefully and fired. I missed again and was stunned to see the lizard did not move or run. I quickly recocked the single-shot air rifle and fired a second time. The second pellet buried itself in the door with my other stray shots. The lizard fell with a thud and then ran under the door. It disappeared in a stack of firewood inside the garage.

I counted the BBs stuck in the door and couldn't believe the strange luck of this reptile. I went to the woodpile and kicked logs around until I heard a scraping sound underneath. The lizard was trapped, but I couldn't see it. I came back the next day, poked around and heard it. I wondered why it didn't run out the door. Maybe lizards with their reptilian brains were still not smart enough to find their way out of a room. I waited in the garage, but it didn't come out.

For five days I went into the garage without flushing it out. On the morning of the sixth day, as I crossed the backyard toward the garage, I saw the lizard sitting high on the brick wall of the house. It had finally made its move. I cocked the rifle, aimed, and slowly drew closer. I fired from a few feet away.

The lizard flew off the wall and landed behind two trash cans. It shivered for a few seconds, its large brown feet outstretched, its long, thick tail twitching slowly. Watching it die, I was suddenly afraid. This was the biggest thing I had killed. I felt panic and guilt I had never encountered before as a boy with a nasty BB gun. I was too shaken to bury the lizard or throw it into the trash can. I knew the ants would get it soon. (It was a common sight to walk through the desert and see red army ants picking clean the tiny skeletons of dead lizards.)

The day after I shot the big lizard, I came back to the trash cans to see what the ants had done. The lizard's thick body lay on its stomach. When I hit it, it had landed on its back. I wondered if ants had the power to turn it over.

Hundreds of them crawled over the body. They had eaten half of the right side around its belly. My shot had split its back, offering another entry for the carnivorous ants. What struck me that day was the sudden appearance of a rough, round band on top of the lizard's head. I bent over to study it closer. I had never seen such a design on a lizard. The circle on its head was a lighter brown, almost pale red, and looked like a marking you might find on a rattlesnake. As I gazed at the intricate colorings on the dead animal, the sense of dread and guilt returned. I walked away and did not check on the lizard for several days. When I came back, it was gone. The only trace was a dark spot of blood and the remains of a gnarled foot. Even the skeleton was gone.

Years later, I came across the key to the circle on the lizard's head. In his book *Desert Journal* (University of California, 1977), naturalist Raymond Cowles writes that some species of lizards have a third eye on top of their heads. He calls them parietal eyes and says researchers have found no mechanism for vision in the third eye. He feels the parietal eye in lizards helps regulate daily and seasonal exposure to sunlight, but there are theories some dinosaurs and ancient reptiles may have had a third eye once, a good extra eyeball for defense against approaching enemies.

In his essay "The Three Brains," poet Robert Bly speculates about which parts of ourselves have not truly evolved, and which parts of our brains remain reptilian. He speculates that we all have reptilian brains working to think about the need for food, survival, and security. Perhaps this is what made the lizard so elusive and mysterious, an unexplainable connection to both of our reptilian states. Did the lizard dodge so many of my BBs because it was watching me the whole time? Did I kill lizards because I wanted to survive in the desert without hidden eyes witnessing everything I did? Later in the essay, Bly announces that the desert landscape rarely contains mammal images. He feels that lizards and snakes dominate and influence the way we behave when we live in such arid places.

The lizard I killed had a third eye, and my memory of it on its head is an image that is reptilian, not mammalian. The third eye closes around the mysteries of the desert and how living in it is crucial to learning why we hide ourselves in our three brains. Its third eye made sure I recalled what I killed, why I keep going back to the memory of the huge lizard, why I spent so much time as the great hunter.

The fact I found the lizard's body was turned over the day after I left it on its back, has something to do with the third eye. Perhaps, it made it easier for one last look at me. The third eye stood out from the decaying body the last time I saw it, a detail I missed until the final hunt. No other lizard I killed as a boy revealed a third eye to focus on me.

The big lizard was the last one I killed. Somehow, my need to be a hunter ended. I found my BB gun rusting in the garage the following summer. Eventually, my mother threw it away.

The holes in the garage door are still there. Some of them contain old BB pellets that have found their place in the wood like remnants of an old western shootout. A few years ago, while visiting my family, I walked around the house and spotted the holes in the door. As I counted fourteen holes, I tried to think of a reason why I had loved to shoot lizards. What made me do it? Many of my friends shot birds. A couple of them got into trouble for hurting cats and dogs with their BB guns. I never shot birds or any other animal besides lizards. Were lizards acceptable because they were so abundant in the area? Was it human fear of monsters coming to get me? I rarely see them anymore when I visit the desert.

ABSENCE OF LIZARDS

I haven't seen a lizard
since I left the desert,
though I feel a lizard
behind my eyelids.
It darts in and out,
though I can't see it,
can't really picture it
jumping off a rock
to sit inside my head.

I recall the invasion
of the white lizards,
the season they beat
the desert rain, overflowed
into the arroyos,
sat on the adobe walls
like cut-off fingers,
twitching their tails,

waiting for me to approach
before leaping into
the cactus like torn pieces
of paper I threw away,
white lizards flashing
their mocking dance at me.

The last giant lizard I saw
was shot by a kid with a BB gun.
It was a foot-long,
dark-brown, thick, and fast.
The kid was a good shot
and left it in the dirt lot
across from my house.
I found it on my walk,
ants crawling over the rocks
to get to it, hundreds of them
opening the stiff, pregnant body
to get to the yellow eggs that
spilled out of its belly like
kernels of corn fertilizing
the hot sand.

Writing this poem and finding the twenty-five-year-old holes in the door unlocked the most important detail about the elusive lizard I finally killed. I have several recurring dreams about growing up in the desert. Most of the dreams involve images of old adobe ruins and cliff dwellings—stretches of hot ground crawling with dozens of enormous rattlesnakes, some sleeping entangled in each other, some poised to strike at me. When I dream of snakes, I wake up with no fear or anxiety, but I also wake with the memory of the third eye on the lizard. Somewhere, the lizards I killed as a boy watch me as snakes surround me in my dreams, but the lizards don't show themselves. Perhaps, I tried to erase it from my soul by giving up my ways as a hunter with a BB gun.

What does that say about us as predators? Can we truly forget? The huge lizard may be waiting for me to join it as it scurries across a field of tumbleweeds, fresh ground after a summer rain when the desert opens and hundreds of lizards, large and small, flash their tails before the arrogant hunter.

VICKI HEARNE
1952-2001

Vicki Hearne was a professional animal trainer and a professor of English at Yale. Born in Texas, she studied writing at Stanford University and published two volumes of poetry, Nervous Horses *(1980) and* In the Absence of Horses *(1984).* Adam's Task *(1986) treats a little-explored field of nature writing—the domestication of animals—and explores Hearne's interest in the philosophical and ethical implications of her chosen profession. Further explorations of these themes appear in* Animal Unhappiness, *published in 1994.*

CALLING ANIMALS BY NAME

And Adam gave names to all cattle, and to the fowl of the
air, and to every beast of the field. . . .
                                              Genesis 2:20

In the course of restoring Drummer Girl to herself, I obedience-trained her, and in the course of doing this work with me, she learned what her name was. In fact, although there are often problems even for humans about learning their names, about knowing what one's own or another's name is (as when I don't know whether to call you Freddie or Professor Jones), for us naming the animals is the original emblem of animal responsiveness to and interest in humans, in Genesis, our first text. An apocryphal expansion of the verse that forms the epigraph to this chapter says that not only did Adam name the animals but the moment he did, each recognized his or her name; the cow now knew she was Cow and came when called by name, and so it was like this, as John Hollander describes it:

Adam's Task: Calling Animals by Name (New York: Knopf, 1986).

Every burrower, each flier,
  Came for the name he had to give:
  Gay, first work, ever to be prior,
  Not yet sunk to primitive.

Now it is the case, sadly, that many horses go through their whole lives without even knowing that they have a name, and this misleads some logicians into believing that they can't have names, and therefore can't have the mental faculties that go with knowing one's name. But in fact many horses learn their names, either informally around the barn or stable, just as most humans learn their names, or through formal obedience work of the sort I did with Salty and Drummer Girl.

I would like to take a little time here to consider the general implications of naming and acknowledging naming. I see us—meaning anyone possessed of that particular sort of literacy that makes him/her want to write and read books like this one—as not being in the enviable position Adam was when he named the animals—"not yet sunk to primitive." I don't mean that we are primitive in our consciousness but rather that we have gone on to a further distancing. We did this when we learned to write and thus to add to the possibilities of consciousness conceptions made possible by typography of various sorts. One example of this is the advance in mathematical thinking when numerals were devised and replaced the prose descriptions of arithmetic. It was typography that eventually made statistics possible and all of the errors as well as the epiphanies of statistical thinking.

Typography has also made possible further gaps between us and animals, because we have become able to give them labels without ever calling them by name. The registered names of most horses and dogs are primary examples. Champion Redheath Nimble Gunner, C.D., C.D.X., U.D., for example, is not a name but something halfway between labels (of the sort found on packing lists or in livestock inventories) and titles—not titles such as Sir, Madam or Your Highness, but titles like the titles of books. Such names are bookkeeping.

It is only when I am saying, "Gunner, Come!" that the dog has a name. His name becomes larger when we proceed to "Gunner, Fetch!" and eventually when he and his name become near enough to being the same size, he is as close to having a proper name as anyone ever gets. When Drummer Girl learned her name, one of the things it meant was that she became able to fit into her name properly; when I said in her story that "her soul was several sizes too large for her," I could as accurately have said that her soul was several sizes too large for the truncated version of a "name" she had so far had, not a name she could answer to. Without a name and someone to call her by name, she couldn't enter the moral life.

There are other things at stake. I knew a woman named Shelley Mason, who took a job running an animal shelter in a small desert town. She didn't do this because she thought that the activity of merely housing dogs and feeding them was an especially meaningful activity (especially as one of her duties was destroying unwanted animals) but because she understood the importance of training as a way of increasing the number of animals who were wanted and who would not be abandoned, thus reducing the piles of corpses. She figured that from that small shelter she could insist that anyone who adopted a dog learn at least the rudiments of training a dog to heel and sit before they were allowed to take the dog home, thereby of necessity naming the dog. And she usually had several dogs from the shelter at her house, teaching them more advanced work in order to increase their chances of placement.

One day when I was visiting her, she gestured at the dogs, most of them doomed, in the runs at the shelter and said, "Goddamit! Most of them wouldn't be here if only they knew their names!"

The grammar of the world we imagine when we call creatures by name is not the grammar of the world in which they have no names, is not the same form of life. But our grammar, or maybe I mean punctuation or typography, has given us the possibility of attenuations of naming, of names that are not invocative. Consider for example that

I am involved with a dog

does not indicate a world as fully as when we say

I am involved with a dog called "Annie"

or

I am involved with "Annie."

The last example gives the feel of a more committed and thoughtful relationship than the first two do, but it is still a disturbing (to me) convention of English punctuation to put what philosophers call scare quotes around animal names, to indicate that these aren't real names, in the way Vicki Hearne is, and even many animal lovers conventionally use the pronoun "it" rather than "he" or "she" to refer to an animal. I find this to be extraordinarily weird, evidence of the superstitions that control the institutionalization of thought. It is as weird, to me, as these examples:

I am married to "Robert."
Pass the butter, "Robert."
Kiss me, "Robert."
I wish "Robert" would return.

When I asked my husband, whose name in fact is not "Robert," but Robert, to look at those sentences, he reported feeling a slight jolt of uneasiness, as though what had been a name for a person—his person—had suddenly become something like a label, and the uneasiness—the dis-ease—is the uneasiness of someone the labeler won't and can't talk to.

Obedience-training horses creates a logic that demands not only the use of a call name, since the imperatives demand it, especially for the command "Dobbin, Come!" but also the removal of the quotes from the name, the making of the name into a real name rather than a label for a piece of property, which is what most racehorses' names are.

Which leads me to my final small point about the disciplines of naming, one of which is horse training. I believe that the disciplines come to us in the form they do because deep in human beings is the impulse to perform Adam's task, to name animals and people as well, and to name them in such a way that the grammar is flexible enough to do at least two things. One is to make names that give the soul room for expansion. My talk of the change from utterances such as "Belle, Sit!" to "Belle, Go find!" is an example of names projecting the creature named into more glorious contexts. Our awareness of the importance of this is indicated, at least partially, by the fact that we have occasions to say, "Well, Rosemary has really made a name for herself."

But I think our impulse is also conservative, an impulse to return to Adam's divine condition. I can't imagine how we would do that, or what it would be like, but linguistic anthropology has found out some things about illiterate peoples that suggest at least names that really call, language that is genuinely invocative and uncontaminated by writing and thus by the concept of names as labels rather than genuine invocations.

I once, for example, heard a linguist talking about the days when the interest in learning and especially recording illiterate languages revealed some surprises. One of his stories was about an eager linguist in some culturally remote corner trying to elicit from a peasant the nominative form of "cow" in the peasant's language.

The linguist met with frustrations. When he asked, "What do you call that animal?" pointing to the peasant's cow, he got, instead of the nominative of "cow," the vocative of "Bossie." When he tried again, asking, "Well, what do you call your neighbor's animal that moos and gives milk?" the peasant replied, "Why should I call my neighbor's animal?"

Since I am a creature born to writing, my horses are not born to their names but to their labels, and care and discipline are required. The dog trainer's knowledge of genuine names—"call name," in fact, is the technical term for a true name—is one of the reasons true trainers say, as I reported in my discussion of Salty, "Joe, Sit!" less frequently than most people and to fewer dogs. They know what the peasant in the linguist's story knew—there has to be a reason for a name or else there is no name.

I am not arguing against advances in culture, only pointing out that it is paradoxically the case that some advances create the need for other advances that will take us back to what we call the primitive, even if not all the way back to paradise, to that region of consciousness in which naming is "Gay, first work, ever to be prior, not yet sunk to primitive." But no advance will enable me to call Drummer Girl with anything less than her name, which is why obedience training is centrally a sacred and poetic rather than a philosophical or scientific discipline.

GARY PAUL NABHAN
b. 1952

Gary Nabhan is an ethnobiologist—one who studies the relationship between native cultures and traditionally used plants and animals. He is cofounder of Native Seeds—SEARCH, *a nonprofit organization that maintains seed banks of indigenous southwestern plants. His books,* The Desert Smells Like Rain *(1982) and* Gathering the Desert *(1985), winner of the John Burroughs Medal, came from his work with the Papago Indians of southwestern Arizona. Nabhan's books show more than an ecologist's interest in their desert-adaptive traditions and the potential value of little-known plant varieties. He writes with great appreciation of the Indians' sense of history and ritual interaction with the desert and convincingly of their seemingly paradoxical sense of the desert as a place of great fertility and vitality. Nabhan has more recently published* Songbirds, Truffles, and Wolves *(1993) and* Cultures of Habitat *(1997).*

From THE DESERT SMELLS LIKE RAIN: A NATURALIST IN PAPAGO INDIAN COUNTRY

AN OVERTURE

With many dust storms, with many lightnings, with
   many thunders, with many rainbows, it started to go.
From within wet mountains, more clouds came out
   and joined it.
                              Joseph Pancho, *Mockingbird Speech*

Last Saturday before dusk, the summer's 114-degree heat broke to 79 within an hour. A fury of wind whipped up, pelting houses with dust, debris, and gravel. Then a scatter of rain came, as a froth of purplish clouds charged across the skies. As the last of the sun's light dissipated, we could see Baboquivari Peak silhouetted on a red horizon, lightning dancing around its head.

The rains came that night—they changed the world.

Crusty dry since April, the desert floor softened under the rain's dance. Near the rain-pocked surface, hundreds of thousands of wild sprouts of bloodroot amaranth are popping off their seedcoats and diving toward light. Barren places will soon be shrouded in a veil of green.

Desert arroyos are running again, muddy water swirling after a head of suds, dung, and detritus. Where sheetfloods pool, buried animals awake, or new broods hatch. At dawn, dark egg-shaped clouds of flying ants hover over ground, excited in the early morning light.

In newly filled waterholes, spadefoot toads suddenly congregate. The males bellow. They seek out mates, then latch onto them with their special nuptial pads. The females spew out egg masses into the hot murky water. For two nights, the toad ponds are wild with chanting while the Western spadefoot's burnt-peanut-like smell looms thick in the air.

A yellow mud turtle crawls out of the drenched bottom of an old

The Desert Smells Like Rain: A Naturalist in Papago Indian Country (San Francisco: North Point, 1982).

adobe borrow pit where he had been buried through the hot dry spell.
He plods a hundred yards over to a floodwater reservoir and dives in.
He has no memory of how many days it's been since his last swim, but
the pull of the water—*that* is somehow familiar.

This is the time when the Papago Indians of the Sonoran Desert celebrate the coming of the rainy season moons, the *Jujkiabig Mamsad*,
and the beginning of a new year.

Fields lying fallow since the harvest of the winter crop are now ready
for another planting. If sown within a month after summer solstice,
they can produce a crop quick enough for harvest by the Feast of San
Francisco, October 4.

When I went by the Madrugada home in Little Tucson on Monday,
the family was eagerly talking about planting the flashflood field again.
At the end of June, Julian wasn't even sure if he would plant this
year—no rain yet, too hot to prepare the field, and hardly any water left
in their *charco* catchment basin.

Now, a fortnight later, the pond is nearly filled up to the brim. Runoff
has fed into it through four small washes. Sheetfloods have swept across
the field surface. Julian imagines big yellow squash blossoms in his field,
just another month or so away. It makes his mouth water.

Once I asked a Papago youngster what the desert smelled like to him.
He answered with little hesitation:

"The desert smells like rain."

His reply is a contradiction in the minds of most people. How could
the desert smell like rain, when deserts are, by definition, places which
lack substantial rainfall?

The boy's response was a sort of Papago shorthand. Hearing Papago
can be like tasting a delicious fruit, while sensing that the taste comes
from a tree with roots too deep to fathom.

The question had triggered a scent—creosote bushes after a
storm—their aromatic oils released by the rains. His nose remembered
being out in the desert, overtaken: *the desert smells like rain*.

Most outsiders are struck by the apparent absence of rain in deserts,
feeling that such places lack something vital. Papago, on the other
hand, are intrigued by the unpredictability rather than the paucity of
rainfall—theirs is a dynamic, lively world, responsive to stormy forces
that may come at any time.

A Sonoran Desert village may receive five inches of rain one year
and fifteen the next. A single storm may dump an inch and a half in the

matter of an hour on one field and entirely skip another a few miles away. Dry spells lasting four months may be broken by a single torrential cloudburst, then resume again for several more months. Unseasonal storms, and droughts during the customary rainy seasons, are frequent enough to reduce patterns to chaos.

The Papago have become so finely tuned to this unpredictability that it shapes the way they speak of rain. It has also ingrained itself deeply in the structure of their language.

Linguist William Pilcher has observed that the Papago discuss events in terms of their probability of occurrence, avoiding any assumption that an event will happen for sure:

> . . . it is my impression that the Papago abhor the idea of making definite statements. I am still in doubt as to how close a rain storm must be before one may properly say *t'o tju:* (It is going to rain on us), rather than *tki'o tju:ks* (something like: It looks like it may be going to rain on us).

Since few Papago are willing to confirm that something will happen until it does, an element of surprise becomes part of almost everything. Nothing is ever really cut and dried. When rains do come, they're a gift, a windfall, a lucky break.

Elderly Papago have explained to me that rain is more than just water. There are different ways that water comes to living things, and what it brings with it affects how things grow.

Remedio Cruz was once explaining to me why he plants the old White Sonora variety of wheat when he does. He had waited for some early January rains to gently moisten his field before he planted.

"That Pap'go wheat—it's good to plant just in January or early February. It grows good on just the *rain* water from the sky. It would not do good with water from the *ground*, so that's why we plant it when those soft winter rains come to take care of it."

In the late 1950s, a Sonoran Desert ecologist tried to simulate the gentle winter rains in an attempt to make the desert bloom. Lloyd Tevis used untreated groundwater from a well, sprayed up through a sprinkler, to encourage wildflower germination on an apparently lifeless patch of desert. While Tevis did trigger germination of one kind of desert wildflower with a little less than two inches of fake rain, none germinated with less than an inch. In general, production of other wildflowers required more than three or four inches of fake rain.

Tevis was then surprised to see what happened when less than an inch of real rain fell on his experimental site in January. He noticed in the previously sparse vegetation "a tremendous emergence of seedlings.

Real rain demonstrated an extraordinary superiority over the artificial variety to bring about a high rate of germination." With one particular kind of desert wildflower, seedlings were fifty-six times more numerous after nearly an inch of real rain than they were after the more intense artificial watering.

The stimulating power of rain in the desert is simply more than moisture. Be it the nutrients released in a rainstorm, or the physical force of the water, there are other releasing mechanisms associated with rainwater. But even if someone worked up a better simulation of rain using *fortified* groundwater, would it be very useful in making the desert bloom?

Doubtful. Remedio himself wonders about the value of groundwater pumping for farming, for water is something he *sings* rather than pumps into his field. Every summer, Remedio and a few elderly companions sing to bring the waters from the earth and sky to meet each other. Remedio senses that only with this meeting will his summer beans, corn, and squash grow. A field relying solely on groundwater would not have what it takes. He has heard that well water has some kind of "medicine" (chemical) in it that is no good for crops. In addition, he believes that groundwater pumping as much as twenty miles away adversely affects the availability of moisture to his field.

I joined in a study with other scientists to compare the nutritive value of tepary beans grown in Papago flashflood fields with those grown in modern Anglo-American-style groundwater-irrigated fields nearby. The protein content of the teparies grown in the traditional flashflood environments tended to be higher than that of the same tepary bean varieties grown with water pumped from the ground. Production appeared to be more efficient in the Papago fields—more food energy was gained with less energy in labor and fuel spent. No wonder—it is a way of agriculture that has fine-tuned itself to local conditions over generations.

There they are, Julian and Remedio—growing food in a desert too harsh for most kinds of agriculture—using cues that few of us would ever notice. Their sense of how the desert works comes from decades of day-to-day observations. These perceptions have been filtered through a cultural tradition that has been refined, honed, and handed down over centuries of living in arid places.

If others wish to adapt to the Sonoran Desert's peculiarities, this ancient knowledge can serve as a guide. Yet the best guide will tell you: there are certain things you must learn on your own. The desert is unpredictable, enigmatic. One minute you will be smelling dust. The next, the desert can smell just like rain.

LOUISE ERDRICH
b. 1954

Although she has also published widely in the areas of poetry and nonfiction, Louise Erdrich is best known for a series of novels that include Love Medicine (1984), The Beet Queen (1986), Tracks (1988), and The Bingo Palace (1994). In these works, she relates the intertwined stories of several generations of North Dakota families, while at the same time exploring her own mixed heritage as the daughter of a Chippewa [Turtle Mountain Ojibwa] mother and a German-American father. The way in which lineage and landscape saturate each other in Erdrich's fiction has caused her northern Midwest novels to be compared to William Faulkner's chronicles of Yoknapatawpha County. Though she expresses herself more rarely in the mode of nature writing, Louise Erdrich offers in the following selection a haunting perspective on topography, tradition, and art.

BIG GRASS

My father loves the small and receding wild places in the agribusiness moonscape of North Dakota cropland, and so do I. Throughout my childhood, we hunted and gathered in the sloughs, the sandhills, the brushy shelterbelts and unmowed ditches, on the oxbows and along the banks of mudded rivers of the Red River valley. On the west road that now leads to the new Carmelite monastery just outside of Wahpeton, we picked prairie rosehips in fall and dried them for vitamin C–rich teas in the winter. There was always, in the margins of the cornfield just beyond our yard, in the brushy scraps of abandoned pasture, right-of-ways along the railroad tracks, along the river itself, and in the corners and unseeded lots of the town, a lowly assertion of grass.

Originally published in *Heart of the Land: Essays on the Last Great Places* (New York: Pantheon, 1995).

It was big grass. Original prairie grass—bluestem and Indian grass, side oats grama. The green fringe gave me the comforting assurance that all else planted and tended and set down by humans was somehow temporary. Only grass is eternal. Grass is always waiting in the wings.

Before high-powered rifles and a general dumbing down of hunting attitudes, back when hunters were less well armed, and anxious more than anything to put meat on their tables, my father wore dull green and never blaze orange. He carried a green fiberglass bow with a waxed string, and strapped to his back a quiver of razor-tipped arrows. Predawn on a Saturday in fall he'd take a child or two into the woods near Hankinson, Stack Slough, or the cornfields and box elder and cottonwood scruff along the Wild Rice or Bois de Sioux rivers. Once, on a slim path surrounded by heavy scrub, my father and I heard a distant crack of a rifle shot and soon, crashing toward us, two does and a great gray buck floated. Their bounds carried them so swiftly that they nearly ran us over.

The deer huffed and changed direction midair. They were so close I felt the tang of their panic. My father didn't shoot—perhaps he fumbled for his bow but there wasn't time to aim—more likely, he decided not to kill an animal in front of me. Hunting was an excuse to become intimate with the woods and fields, and on that day, as on most, we came home with bags of wild plums, elmcap mushrooms, more rosehips.

Since my father began visiting the wild places in the Red River valley, he has seen many of them destroyed. Tree cover of the margins of rivers, essential to slow spring runoff and the erosion of topsoil—cut and leveled for planting. Wetlands—drained for planting. Unplowed prairie (five thousand acres in a neighboring Minnesota county)—plowed and planted. From the air, the Great Plains is now a vast earth-toned Mondrian painting, all strict right angles of fields bounded by thin and careful shelterbelts. Only tiny remnants of the tallgrass remain. These pieces in odd cuts and lengths are like the hems of long and sweeping old-fashioned skirts. Taken up, the fabric is torn away, forgotten. And yet, when you come across the original cloth of grass, it is an unfaded and startling experience. Here is a reminder that before this land was a measured product tended by Steiger tractors with air-cooled cabs and hot-red combines, before this valley was wheat and sugar-beet and sunflower country, before the drill seeders and the windbreaks, the section measures and the homesteads, this was the northern tallgrass prairie.

It was a region mysterious for its apparent simplicity.

Grass and sky were two canvases into which rich details painted and destroyed themselves with joyous intensity. As sunlight erases cloud, so fire ate grass and restored grass in a cycle of unrelenting power. A prairie burned over one year blazes out, redeemed in the absolving mist of green the next. On a warm late-winter day, snow slipping down the sides of soft prairie rises, I can feel the grass underfoot collecting its bashful energy. Big bluestem, female and green sage, snakeweed, blue grama, ground cherry, Indian grass, wild onion, purple coneflower, and purple aster all spring to life on a prairie burned the previous year.

To appreciate grass, you must lie down in grass. It's not much from a distance and it doesn't translate well into most photographs or even paint, unless you count Albrecht Dürer's *Grosses Rasenstuck*, 1503. He painted grass while lying on his stomach, with a wondering eye trained into the seed tassles. Just after the snow has melted each spring, it is good to throw oneself on grass. The stems have packed down all winter, in swirls like a sleeper's hair. The grass sighs and crackles faintly, a weighted mat, releasing fine winter dust.

It is that smell of winter dust I love best, rising from the cracked stalk. Tenacious in its cycle, stubborn in its modest refusal to die, the grass embodies the philosopher's myth of eternal return. *All flesh is grass* is not a depressing conceit to me. To see ourselves within our span as creatures capable of quiet and continual renewal gets me through those times when the writing stinks, I've lost my temper, overloaded on wine chocolates, or am simply lost to myself. Snow melts. Grass springs back. Here we are on a quiet rise, finding the first uncanny shoots of green.

My daughters' hair has a scent as undefinable as grass — made up of mood and weather, of curiosity and water. They part the stiff waves of grass, gaze into the sheltered universe. Just to be, just to exist — that is the talent of grass. Fire will pass over. The growth tips are safe underground. The bluestem's still the scorched bronze of late-summer deer pelts. Formaldehyde ants swarm from a warmed nest of black dirt. The ants seem electrified, driven, ridiculous in tiny self-importance. Watching the ants, we can delight in our lucky indolence. They'll follow one another and climb a stem of grass threaded into their nest to the end, until their weight bows it to the earth. There's a clump of crested wheatgrass, a foreigner, invading. The breast feather of a grouse. A low hunker of dried ground cherries. Sage. Still silver, its leaves specks and spindrels, sage is a generous plant, releasing its penetrating scent of freedom long after it is dried and dead. And here, the first green of the year rises in the female sage, showing at the base in the tiny budded lips.

Horned larks spring across the breeze and there, off the rent ice, the first returning flock of Canada geese search out the open water of a local power plant on the Missouri River. In order to recreate as closely as possible the mixture of forces that groomed the subtle prairie, buffalo are included, at Cross Ranch Preserve, for grazing purposes. Along with fire, buffalo were the keepers of the grass and they are coming back now, perhaps because they always made sense. They are easier to raise than cattle, they calve on their own, and find winter shelter in brush and buffalo-berry gullies.

From my own experience of buffalo—a tiny herd lives in Wahpeton and I saw them growing up and still visit them now—I know that they'll eat most anything that grows on the ground. In captivity, though, they relish the rinds of watermelon. The buffalo waited for and seemed to know my parents, who came by every few days in summer with bicycle baskets full of watermelon rinds. The tongue of a buffalo is long, gray, and muscular, a passionate scoop. While they eat watermelon, the buffalo will consent to have their great boulder foreheads scratched but will occasionally, over nothing at all, or perhaps everything, ram themselves into their wire fences. I have been on the other side of a fence charged by a buffalo and I was stunned to a sudden blank-out at the violence.

One winter, in the middle of a great snow, the buffalo walked up and over their fence and wandered from their pen by the river. They took a route through the town. There were reports of people stepping from their trailers into the presence of shaggy monoliths. The buffalo walked through backyards, around garages, took the main thoroughfares at last into the swept-bare scrim of stubble in the vast fields—into their old range, after all.

Grass sings, grass whispers. Ashes to ashes, dust to grass. But real grass, not the stuff that we trim and poison to an acid green mat, not clipped grass never allowed to go to seed, not this humanly engineered lawn substance as synthetic as a carpet. Every city should have a grass park, devoted to grass, long grass, for city children haven't the sense of grass as anything but scarp on a boulevard. To come into the house with needlegrass sewing new seams in your clothes, the awns sharp and clever, is to understand botanical intelligence. Weaving through the toughest boots, through the densest coat, into skin of sheep, needle-grass will seed itself deep in the eardrums of dogs and badgers. And there are other seeds, sharp and eager, diving through my socks, shorter barbs sewn forever into the laces and tongues of my walking boots.

Grass streams out in August, full grown to a hypnotizing silk. The ground begins to run beside the road in waves of green water. A motorist, distracted, pulls over and begins to weep. Grass is emotional, its message a visual music with rills and pauses so profound it is almost dangerous to watch. Tallgrass in motion is a world of legato. Returning from a powwow my daughter and I are slowed and then stopped by the spectacle and we drive side roads, walk old pasture, until we find real grass turned back silver, moving, running before the wind. Our eyes fill with it and on a swale of grass we sink down, chewing the ends of juicy stems.

Soon, so soon.

Your arms reach, dropping across the strings of an air harp. Before long, you want your lover's body in your hands. You don't mind dying quite so much. You don't fear turning into grass. You almost believe that you could continue, from below, to express in its motion your own mesmeric yearning, and yet find cheerful comfort. For grass is a plant of homey endurance, pure fodder after all.

I would be converted to a religion of grass. *Sleep the winter away and rise headlong each spring. Sink deep roots. Conserve water. Respect and nourish your neighbors and never let trees gain the upper hand.* Such are the tenets and dogmas. As for the practice—*grow lush in order to be devoured or caressed, stiffen in sweet elegance, invent startling seeds*—those also make sense. *Bow beneath the arm of fire. Connect underground. Provide. Provide. Be lovely and do no harm.*

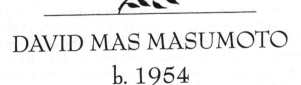

DAVID MAS MASUMOTO
b. 1954

Although he has also written stories and produced a collection of oral history, David Mas Masumoto's reputation as a nature writer is founded on one volume: Epitaph for a Peach *(1995). This lovely and engaging book offers the account of a farmer trying to maintain his livelihood, his integrity, and his joy in the fruitful beauty of the California landscape, even as the tides of agribusiness and the trend toward a monoculture surge around him. His example is an encouraging one, both despite and because of his sober admonitions about the forces that continue to place such family farms in peril. Another engaging aspect of Masumoto's writing is its integration of practices and perspectives from his Japanese-American background that continue to shape his ambitions as a farmer.*

PLANTING SEEDS

In the early spring, the earth lies bare and naked, with not much growing between the grapevines and trees. Cover crops—clovers, vetches,
beans, and barley—can be planted to add nourishment and a green
color of life to the fields.

I grew up playing hide-and-seek and other games, with my brother
and sister in the lush fields of the cover crops. The family dog, Dusty,
would come trotting after me, panting and smiling. I'd shoo her away,
begging her not to disclose my location, but she'd stand next to me,
wagging her tail. It didn't take long to be discovered. (Later I realized
old Dusty followed me because I was the only kid she could keep up
with.) Still, it was a glorious few minutes of hiding, the grasses cool to
the side of my face. I could watch a ladybug crawling up my arm and
feel the goose bumps spreading over my body as I tried not to flinch. I
remember looking up at the pale blue sky and listening to a gentle
breeze rustling through the tall grass. Often the wind would be an
innocent whisper, at other times it could howl. Enveloped by nature, a
child's imagination soars. The cover crops brought another world to
our fields of play.

One season my dad stopped growing cover crops because of the
extra work of planting seeds, irrigating, and battling weeds. Like
many farmers, my dad believed that cover crops were just a cheaper
source of plant nutrients until he could afford synthetic fertilizers. He
explained that work was easier when plant nutrients came in bags,
with guaranteed nitrogen contents. And with synthetic fertilizers, the
kids could help spread the granules. I remember sitting on the back
of a vineyard wagon with Dad while my brother drove the tractor (he
drove because I couldn't reach the clutch pedal). Dad and I would
clench old coffee cans and scoop out the fertilizer with them, tossing
half a can at the base of each peach tree. My old work gloves felt like
oversized baseball mitts, and my little fingers could barely bend the
leather into a curl around the can. After a few hours, though, I could
hold it more easily; something in the fertilizer caused the leather to
stiffen, and the gloves became frozen in a death grip around the
metal can.

I hope to renew the practice of cover cropping. I am part of a generation of grape and tree fruit farmers who never planted clover or beans
or barley. I plant a vine and expect it to last a lifetime; a peach tree

Epitaph for a Peach: Four Seasons on My Family Farm (New York: HarperCollins, 1995).

should last decades. Annual crops feel odd and peculiar—I don't know how to prepare beds and am not used to planting something underground that would be out of sight for weeks. Many of my generation never learned how to sow seeds.

I planted my very first cover crop eight years ago when my first child was born. I didn't do it because cover crops would be good for the soil and build up organic matter. And I didn't do it to provide a habitat for beneficial insects to overwinter and make my land their spring home.

I did it because my wife would be home with a new baby and she was tired of seeing only the gray earth of winter outside our kitchen window. I did it for her dreams of spring walks through the soft clover with the baby in her arms, breathing in the fresh scent of spring growth. I did it for reasons that seemed disconnected with farming at the time.

I planted my cover crops in autumn, motivated by a vision of lush fields by spring. I didn't have a seeder, so at first I tried using an old fertilizer spreader to broadcast the seeds. It sort of worked, but the seeds poured from the outlets at the beginning of a row while near the end the flow was reduced to a trickle. Instead of a nice blend, most of the smaller seeds fell first, leaving the larger ones behind, with some of the largest seeds, like the fava beans, sliced in half by the spreader gears.

Caught up in a pioneer spirit, I then tried sowing a row by hand until my arm ached from carrying the bucketfuls of seed (I was planting at only twenty pounds an acre but I had eighty acres to plant). I ended up on a tractor, throwing seeds over my shoulder with one hand while steering with the other. "Let them fall where they may," I said to myself.

Our daughter was born in early November. By Thanksgiving we had a smiling, cooing child and a germinating cover crop, its green leaves poking through the soil crust. Both child and crop grew through the winter and by spring we took walks through the fields, picking fresh peas and beans and letting ladybugs tickle her soft baby skin.

Later an organic farmer friend introduced me to the real benefits of cover crops, how they improve soils and work as a habitat for insects. "You start with the soil and build from there," he explained. He called it "growing your way back to natural farming."

Cover crops have a multiplier effect. Their lush growth adds organic matter to the earth. Earthworms return to my farm because of the healthy jungle of roots to crawl within. Moisture from the winter rains is held by a dense underground mat of clover roots. Insects breed, prolific in their wild and rank homes.

Planting these cover-crop seeds is my first step to save my Sun Crests. I begin by planting hope—hope that the seeds will germinate, hope that they will add life to the farm and even help save the wonderful taste of my fruits.

I had no training to be a father, I could only hope I'd learn quickly, on the job. As I grew my first cover crop, I had a similar feeling. I hoped an enriching harvest would follow. Babies and planting seeds: they demand that you believe in the magic and mystery of life.

Planting cover crops is more art than work. Subtle differences in each field will affect my seed selections. Some areas of my eighty-acre farm have sandy soils, others have more clay, one area is a low land, another we call "the hill" stands four feet higher than its neighbors. They all add to a diversity and create a patchwork, arranged into small five- or six-acre blocks of vines or peaches.

I like growing a variety of cover crops. The vines on "the hill" are designated for crimson clover, my young peach trees need a healthy start of vetch, the wine grapes enjoy a solid stand of strawberry clover and New Zealand white clover. I feel like Georges Seurat and his "dot, dot, dots," each seed becoming a dot on my farm canvas.

Selecting seeds is simple; planting them continues to be a challenge each year. Hand sowing from a tractor works well for a few acres but grows tedious for eighty acres. Looking for an alternative, I located an old set of Planet Jr. Seeders, a modern implementation of a simple old tool—a hopper with seeds and a roller wheel. With each turn of the wheel, a gear opens a hole and a seed drops from the hopper. The Planet Jr. Company improved the idea, adding gears and a few more moving parts while trying to keep the basic concept simple. The machine works well in smooth, clean fields like the one a vegetable grower may have prior to planting. But vines and orchards are filled with bits of trash—twigs, stems, and sticks—which take time to decompose and love to lodge themselves in Planet Jr. gears.

I discovered this one year in the spring, six months after I had seeded a field. I noticed gaping bare spots in the rows, places where little was growing. At first I thought some disease had ravaged the cover crops, then I imagined birds stealing my precious seeds. But upon closer examination, just where the bald spot ended, I found a tuft of clover or vetch, as if the seeds were piled upon one another. I concluded that the seeds must have jammed in the planter, then poured out all at once when freed.

I've thought of buying a better planter, something adapted to vineyards instead of vegetable beds. But I've become attached to my Planet Jrs. They remind me of a simple age, and I like the name. I also enjoy controlling each individual planter. Unlike an eight-foot-wide, single-hopper machine that uniformly plants an entire field with the same seed mix and in the same pattern, these individual units can be adjusted to create different patterns with a variety of seed combinations. I play artist in my fields, painting with a blend of clover and vetches with a splash of wildflowers. Next to a vine I can plant dense cahaba white vetch that would dominate in the early spring canvas with its white blooms but may begin to wither with the first heat of summer. Along another edge I might weave in some crimson clover with its deep red seed heads or scatter strawberry and red clovers for variety. I would add a combination bur clover and a blanket of yellow flowers with the green hues created by different medics, low-growing but sturdy plants that creep along the surface and replace the wilting vetches and crimson clovers in our valley heat.

My fields have become a crazy quilt of cover crops, a wild blend of patterns, some intended, some a product of nature's whims. The different plants grow to different heights and in different patterns, creating a living appliqué. The casual passerby might not notice my art. From the roadside, it often looks like irregular growth, bald spots, breaks in uniformity. But the farmer walking his fields can feel the changing landscape beneath his boots, he can sense the temperature changes with the different densities of growth and smell the pollen of blooming clover or vetch or wildflowers. He appreciates the precarious character of nature. As if running your fingers over a finely crafted quilt, you can feel pattern upon pattern. Just as a quilter may stitch together emotions with each piece of fabric, I weave the texture of life into my farm.

SHARMAN APT RUSSELL
b. 1954

In addition to her work as an instructor in writing at Western New Mexico University and her writing in the areas of biography, architecture, and archaeology, Sharman Apt Russell has produced three books that contribute in important ways to the sense of place in her region: Songs of the Fluteplayer: Seasons of Life in the Southwest *(1991),* Kill the Cowboy: A Battle of Mythology in the New West *(1993), and* The Humpbacked Fluteplayer *(1994). One of her special gifts as a writer is the professional researcher's accuracy with which she evokes the topography and history of the Southwest. At the same time, she is able to sustain her own voice in a modest, engaging way, interweaving family and what the poet Dave Smith has called "dailiness" with her loftier vistas and speculations. In the following essay, Russell describes a family horse trip through the nation's first "official wilderness" and examines the contradictions inherent in that concept, the relationship between solitude and wilderness, and the way in which the experience of such places is affected by both one's own personal history and the awareness of previous cultures.*

GILA WILDERNESS

Perched high on their horses, completely unafraid, the children are pretending to be English tourists.

"I say, look at that over there, old boy!"

"Old boy, old boy. Look at that!" Maria repeats gleefully. She is not quite five and Eric is seven. They have squabbled off and on all morning over who would ride in front and then who would ride in back and then who would ride in front again. Eric's mother, Lana, is leading the horse, in part because I am nervous when we walk downhill and the big animal crowds behind me. I wonder where these children have picked up the nuances of satire and funny accents. Videos, I suppose. TV. Their ideas, to be sure, are still vague. When we ask Eric where England is, he looks shifty and then says with authority, "In Pennsylvania."

There is plenty of time to talk on this five mile hike and family pack train. My son, nearly two, is fulfilling a dream he has held half his life, for he, too, is high on a horse, his excitement contained by Lana's ten-year-old daughter, his horse led by Lana's husband. At the back, my husband leads two more horses packed with camping gear. At the front, Roberta and her married daughter Carol Greene walk the trail. Carol has left behind her family in Wisconsin to be with her mother now, and while I am freshly creating my children's memories, Carol and Roberta are sorting through their own. A little high in protest, Carol's voice can be heard as they ascend a hill. Roberta's answer is carried away by the wind.

Although I have lived here for eight years—although I can see from the ridge above my home the green edge of the Gila National Forest—

Songs of the Fluteplayer: Seasons of Life in the Southwest (New York: Addison-Wesley, 1991).

the next four days will be my longest trip into the Gila Wilderness, my fourth trip only, the one in which I will best comprehend wilderness, and the one in which I will start scheming, immediately after, as to how I can return.

It's ironic. Like most people, I associate wilderness with being alone. The 1964 National Wilderness Act, which legislated our country's system of wildernesses, is specific about this and includes "outstanding opportunities for solitude" as part of its definition of what a wilderness is. Historically, bred deep in our cultural bones, we think of the solitary explorer and hunter: men with righteous-sounding names like Daniel Boone or Jedediah Smith. Socially, we believe that the point of wilderness is to get away from people. Spiritually, we want to meet Nature stripped of our accoutrements and modern "superficial" selves. We want to be that vulnerable. We want to be that arrogant—the only human being on earth!

Yet here I am, with four children, five adults, and four horses. My husband and Lana's husband are partners in a new outfitting company, and they have condescended to treat their families to what they provide paying customers. Thus we are well-equipped, with therma-rest pads, big tents, and down sleeping bags. We have disposable diapers and storybooks and sunscreen lotion. We have pork tenderloin for tonight's dinner, fresh asparagus, wine glasses, and a cheesecake. As a mother and wife and friend, I have all my usual concerns. We have, I think, left only the dogs, the cats, and the videos behind.

Our trail runs through dry rolling hills dotted with juniper, scrub oak, and pinon pine. There are three colors: the parched yellow of grama grass, the dark of evergreen trees, the pale blue sky. This is peaceful country, and if its vistas are not grand, they call, nonetheless, to those parts of the body which have always yearned to fly. Sometimes the trail crosses an outcropping of bleak rhyolite or a bed of pink, eroded stone. Sometimes we swing around the side of a hill, and Lana's daughter gets nervous, unsure of her horse's footing on the narrow path.

"Davy, Davy, horsie, Davy," she comforts herself by crooning to my son. Ahead of her, Eric and Maria are much alike; three-quarters in an interior world, they squeal so hilariously that Lana, whom I admire for her patience, must tell them to be quiet.

We all stop to watch a red-tailed hawk.

Then we began our descent, down Little Bear Canyon to the Middle Fork of the Gila River.

For my husband, this loss of altitude is a psychic passage. Rather quickly, the canyon begins to narrow and the sky shuts down until it is a swatch of blue in a pattern of pine boughs and fir needles. In a world grown suddenly cool, we are walking on the stream bed with its trickling flow and ledges of rock that rise above us. A grassy bench shelters a stand of yellow columbine; we stop again while the girls exclaim and wax sentimental. All around us, in contrast to the hills above, the murmuring life has been nurtured by water. Insects hatch and burrow in the mud. Emerald green algae swirls in a puddle. The canyon deepens and the rock rises above us, so close we could almost extend our arms and touch each wall. Now there is only rock and water and we are moving darkly to the center of the earth.

Then, like the odd turn in a dream, the stream bed expands to the size of a living room. Above is a small cave which my husband announces as a prehistoric shrine. The children are lifted down from their horses; the older ones scramble up the rock face. My husband tells us that not long ago you could still find prayer sticks and arrowheads here, scattered on the ground. These artifacts were from the Mogollon Culture: weavers, potters, basket-makers, and farmers who by A.D. 900 had upgraded their pithouses into multi-room villages of stone and mud. Slighter and shorter than we are today, the Mogollons had a life expectancy of forty-five years and babies who died rather frequently. At this site, they have left a few faded pictographs—a squiggle I might generously call a lizard, a red hand that is oddly evocative. Carefully, Lana's daughter puts her own hand over the ancient print, covering it completely.

We emerge from Little Bear Canyon, as my husband said we would, into the light. Streaming down from the center of the sky, reflected in the water and trembling cottonwood leaves, bouncing up from banks of white sand, the sun seems to explode around us. It is a drought year, and the Middle Fork of the Gila is not as large as usual. Still, it looks grandly like a river, with riffles and pools and a fiery gleam that disappears around corners left and right. High over the water, red cliffs form the towers of an abandoned city, with tapered ends eroded into strange balancing acts. Here the riparian ecology includes walnut, sycamore, cottonwood, willow, wild grape, and Virginia creeper. Herons and kingfishers hunt the shallows. Trout rise for bugs. These sudden shifts in environment, accomplished in a few hours of walking, are not unusual in the Gila Wilderness. On some day trips, a hiker can move from the Chihuahuan desert to a Subalpine forest. Diversity is the rule, with five life zones and a thousand microclimates, all determined by water or its lack.

Water is our goal as well, and the mothers take the children to the river, letting the men deal with horses and lunch. Maria joins Lana's daughter, who is busy making dams and catching tadpoles. Eric attempts to disappear forever, but Lana knows him well and is prepared for this. She hauls him from the underbrush and informs him firmly: he is only seven, he can not leave her yet. My own son picks up a stick and begins to splash me. Then, who knows why, he is suddenly anxious and wants only to nurse.

A half mile from our planned campsite, we eat in a small grove of ponderosa pine. When it is time to pack up, Maria wants to sit in front on the horse again. It's not your turn, we tell her, it's Eric's turn. Maria has a tantrum. Six parents stand around, cajoling a little, making deals. She can ride in front later; we'll tell her a story. She accepts the good part—she'd love a story—but rejects the bad. Finally, my husband takes a stand. Maria can not ride unless she rides in back.

So I am left amid the ponderosa pine with my screaming child. Ponderosa are beautiful trees. Unlike other local conifers, their long needles extend out and away, giving this pine an oriental delicacy. In many parts of the Gila, ponderosa form vast, parklike forests where the trees rise a hundred feet, the crowns do not touch, and the accumulated needles make a deep carpet free of undergrowth. These are, as well, trees with a secret: breathe deep into their reddish bark and you are suffused with the faint scent of vanilla.

I try to show this to Maria as she cries fiercely, piercingly. I hug her, in a physical effort to contain an emotion of which she has clearly become the victim. Following one school of thought, for a while I simply let her express the emotion. I wait and admire the beautiful trees. My patience for waiting is not long, but in that time I am surprisingly content. This is my job, I think. Socialization. Taking turns. I am grateful for where I am. In a public place, or even a friend's home, I could not endure this blast of feeling.

"Maria," I say, like a fisherman baiting a hook, knowing well she will succumb, "I'll tell you a story while we walk."

The storm subsides slowly; Maria is beached on shore, looking bewildered. Where has she been? We hold hands and follow the trail along the riverbank, through shady groves and then out, once more, into the sun. I tell her about Hansel and Gretel, and when we reach the campsite, everyone is happy to see us. The camp itself is full of miraculous signs. In a pool, deep enough for bathing, a group of fish sway in the shadows. Nearby, three baby birds with wide mouths complain from their nest. A swallowtail butterfly circles the stalk of a purple bull thistle. Two trees for a hammock stand perfectly apart.

As we go about the business of setting up camp, our son opens and empties a bottle of brandy on our clothes. In the rustles inside newly erected tents, I know that other family dramas are going on, and I am beginning to see that older children bring a whole new set of interesting problems. Roberta is hurt by something Carol has said. Lana's daughter is unhappy about the dinner arrangements. Lana is trying to keep Eric from under the horses' feet.

The horses themselves are engaged in complex social arrangements. I had never known that they could be so human, so insistent in their desires. One horse doesn't like the other and won't be picketed beside her. Two of the horses are set free to graze because my husband knows they are too loyal to leave while their partners are tied up. Much later, in the middle of the night, the mare begins to scream with jealousy and rage: another horse has broken loose and is eating grass. This she can not abide.

The afternoon slips away with the sounds of children playing in the river. Around the evening campfire I shelter my son just as these green cottonwoods shelter us. Lana's husband also holds his ten-year-old daughter, the girl's long legs crowding her father's lap. Roberta and Carol are thoughtfully quiet while Maria and Eric can be heard from a tent, whispering secrets.

We are not the first family group to laze under these trees and count our riches and our sorrows. In a wilderness, relatively few humans have come before and it is permissible, I think, to imagine an intersection. I imagine an Indian family, descendents of Asians who crossed the Bering Strait and came to this area after the Mogollons and before the Spanish. The Zunis christened these nomads Apache—the word for enemy. I imagine wicki-ups instead of tents, pine boughs for softness, hides for warmth, *metates, manos,* beads from Mexico.

"In that country which lies around the headwaters of the Gila I was reared," dictated Geronimo when he was old, exiled, and still homesick. "This range was our fatherland; among these mountains our wigwams were hidden; the scattered valleys contained our fields; the boundless prairies, stretching away on every side, were our pastures; the rocky caverns were our burying places. I was fourth in a family of eight children—four boys and four girls. . . . I rolled on the dirt floor of my father's teepee, hung in my cradle at my mother's back, or slept suspended from the bough of a tree. I was warmed by the sun, rocked by the winds, and sheltered as other Indian babes."

As a boy growing up in these forests and mountains, Geronimo's life would have been greatly envied by Lana's son. In the Apache's words, he "played at hide and seek among the rocks and pines" or "loitered in the shade of the cottonwood trees" or worked with his parents in the cornfields. Sometimes, to avoid the latter, he and his friends would sneak out of camp and hide all day in some secret dappled meadow or sunny canyon. If caught, they were subject to ridicule. If not, they could expect to return at twilight, victorious and unpunished. Geronimo's father died when he was small and at seventeen years of age—1846, the year the United States declared war on Mexico—the teenager was admitted into the tribe's council of warriors. Soon after, the young man married his version of a high school sweetheart and together they had three children. This first wife, Alope, was artistic. To beautify their home amid the vanilla-scented pine, she made decorations of beads and drew pictures on buckskin.

Later, as Geronimo tells it, he and his tribe went to Mexico to trade. There Mexican troops attacked the camp while the men were in town. Geronimo's mother, wife, and three children were killed. Stunned, the warrior vowed vengeance and went on to fight both Mexicans and Americans in a guerrilla warfare that was mean and dirty by all accounts and on all sides. "Even babies were killed," one Apache warrior regretted later. "And I love babies!"

Here in the Gila headwaters, local chiefs had long fought the parade of settlers and prospectors. The end was inevitable. In 1886, Geronimo, the last holdout, surrendered and was shipped to Florida along with every other Indian who had ever made the Gila a home. Even the Apache scouts who had helped bring Geronimo in were loaded onto the boxcars. In the 1890s a newspaper reported with nostalgia and some compassion that a "wild and half-starved" Apache family had been seen foraging in the rugged Mogollon Mountains. Desperate and surely lonely, they died or left by the end of the century.

In the morning, my husband gets up early and walks with David and Maria to the nearby hot springs. I follow later and for half an hour, I am, in fact, solitary in the wilderness.

Self-consciously, I look about the scene of a fast-flowing river, lined with leafy trees, against a background of rock. It is conventionally pretty. It is also hard edged and muscular, Southwestern tough. I have the strong feeling that I am not the dominant species here.

This, too, is an echo of the 1964 Wilderness Act, which declares that "a wilderness, in contrast with those areas where man and his own works dominate the landscape, is hereby recognized as a place where the earth and its community of life are untrammeled by man, where man is a visitor who does not remain."

Frankly, I like this lack of power and control. I like being a visitor. Here in the wilderness I can put aside my grievances against humanity. I can exchange, at the very least, one set of complexities for another: the dappled slant of a bank, rustling leaves, straight white trunks, crumbling cliff faces, gravel slopes, turbulent water—all glowing with sunlight, intertwined, patterned; rich with diatoms, moss, algae, caddisflies, dragonflies, damselflies, stoneflies, trout, suckers, bass, minnows, chubs; pinchers, mouthparts, claws, teeth; photosynthesis, decomposition, carbonization. None of it is my doing. I am just a large mammal walking the riverbank. Ahead is my mate.

When I was fifteen, I lied to my mother and hitchhiked from my home in Phoenix to camp out in a sycamore-lined canyon above the desert. The point was to do this alone. The point was to be alone and serene and in touch with beauty. The trip, unsurprisingly for a girl raised in the suburbs, was a disaster, and I ended up leaving a day early. On the way home, the old man who gave me a ride tried once to put his hand on my thigh. The image lingered with me for many years. The stubby white fingers. My revulsion. My ignorance.

When I was eighteen, a girlfriend and I planned a summer-long backpack trip that would take us four hundred miles up the Pacific Crest Trail. The girlfriend dropped out at the last minute, and I went on by myself, determined this time to live alone in the woods. Outside Ashland, Oregon, I watched the dawn beneath layers of a plastic tarp against which mosquitos hammered and whined for my blood. At that time I was still concerned about my alienation from nature, and I perceived a sheet of glass, a terrible wall, between me and life, me and experience. For days, I hiked through a pine forest that never seemed to vary or end, until my thoughts too began to hammer and whine at the bone of my skull. One evening I cried after swimming in a lake and finding my body, my legs and crotch, covered with small red worms. A week later I met a boy my age who was also alone, and we traveled together the rest of the summer, hitchhiking north to mountains that began where timberline ended. We never grew to like each other. We never had the slightest physical contact. Yet we hung on, gamely, blindly, to the comfort and distraction of another human being.

When I was twenty, I set out again, this time bicycling with a college classmate up the East Coast. She ended her tour in Maine, and the next day I started for Nova Scotia. By now I knew what it meant to travel alone as a female: I knew about circumspection, reserve, hiding. In Canada, the ocean exploded against a lushness of farmland, and for me this was exotic, stupendous surrealism. I tried my best to internalize the scenery. But it seemed that I could only turn wheels, pushing my limit, sixty miles up and down the green hills, a hundred miles on the flat inland highway between the tips of the island. By now I knew as well when to recognize misery, and in Halifax I prepared to pack up and head back to school. Instead—a postscript—I met another bicyclist, fell in love with him, and stayed on through a long winter.

Somewhere in all that, I gave up on my ability to conquer solitude. I had tried to be my version of Daniel Boone, brave and self-sufficient, to seek distance and the lonely sound of foreign names. My model could have come straight from the Gila Wilderness. I had tried to be—not Geronimo, who was too much the warrior—but such a man as James Ohio Pattie, a twenty-year-old who trapped beaver on the Gila River five years before Geronimo was ever born. By his own account, Pattie left Missouri in 1824, traveled to Santa Fe where he rescued the governor's daughter from Commanches, managed the copper mines in Santa Rita, escaped massacre by Pimas in Arizona, floated the Colorado River to its salty mouth, starved in a Spanish jail, and crossed Mexico to sell his memoirs to a publisher in St. Louis.

In this case, as I walk beside the Middle Fork, it is not fanciful to imagine that I am following Pattie's footsteps. In his narrative of the 1824 trip, he clearly reaches the hot springs where we will picnic this afternoon. Typically, his description is more dramatic than seems reasonable. He writes of catching a fish and throwing it in the spring's boiling waters where "in six minutes it would be thoroughly cooked." Other tales are equally elongated, and it has become a historian's game to match up Pattie's journey with the rest of history. His account of daily dangers are the most credible: the terror of meeting a grizzly bear or the hunger that forced him and his partners to shoot their dogs. On one sad day, Pattie wrote piteously, "We killed a raven, which we cooked for seven men." By the end of his adventures, he had probably become what he most admired—the quintessential mountain man. Still, it is a lesson to me that James Ohio Pattie, living out the romance, felt so strongly the need to romanticize. For at that time the governor had no daughter, it was another trapper who fought the Pimas, and another man who killed the grizzly.

In 1924, a hundred years after Pattie explored the Middle Fork, three-quarter million acres of the Gila National Forest were designated by the Forest Service as "an area big enough to absorb a two weeks' pack trip and kept devoid of roads, artificial trails, cottages, or other works of man." This was the first official wilderness in the United States, the beginning of our national wilderness system, and the brain-child of a thirty-seven-year-old forester named Aldo Leopold. In my own history, upon returning from Nova Scotia and my first, unsuccessful love affair—upon giving up the idea of becoming a mountain man—I settled instead on becoming Aldo Leopold. I read his famous work *Sand County Almanac* and I changed my college major from drama to natural resources. I took courses in wildlife management, the field that Leopold pioneered, and wrote papers on deer herd reduction.

I even took a course from Leopold's son, Starker Leopold, whom I glamorized on the slightest of proofs. My hero became not the man who lives wilderness, but the one who manages it.

Years later, when I came to live near the Gila Wilderness, my attachment to Leopold increased for an odd reason. I learned more about his mistakes. They were not small. After a boyhood beside the Mississippi River, Aldo Leopold went East to the Yale Forestry School and then Southwest as a greenhorn foreman of a timber crew. At first, this sportsman thought mainly in terms of hunting and fishing. He had no problems with grazing either and eventually had friends and relatives at both of the big ranches in the Gila Forest. With these connections, and in his later role as a game and fish manager, Leopold pushed hard for predator control and vowed to extinguish every killer of deer and cow, "down to the last wolf and lion."

In the Gila area, he hired hunter extraordinaire Ben Lilly, who by 1921 had a lifetime lion kill of five hundred. Today there is a Ben Lilly Monument in the Gila National Forest with a plaque dedicated to the memory of a man who shot more wildlife in the Southwest than anyone else would ever want to. With all his outdoor expertise, Ben Lilly is not a man I would want my children to emulate. Violence was his tie to nature. And when his dogs "betrayed their species" by being poor hunters, he beat them to death.

In the late 1920s an irruption of deer in the Gila and nearby Black Range caused Aldo Leopold to rethink his ideas on predator control. Twenty years after the fact he describes shooting at a wolf and her half-grown cubs from a high rimrock. In seconds, the mother and children were dead or scattered. Leopold rushed down in time to catch a "fierce green fire" dying in the old wolf's eyes.

"I thought that because fewer wolves meant more deer, no wolves would mean a hunter's paradise," the conservationist wrote in his essay "Thinking Like a Mountain." "But after seeing the green fire die, I sensed that neither the wolf nor the mountain agreed with such a view. Since then I have lived to see state after state extirpate its wolves. I have watched the face of many a newly wolfless mountain, and seen the south-facing slopes wrinkle with a maze of new deer trails. I have seen every edible bush and seedling browsed, first to anaemic desuetude, and then to death . . . In the end the starved bones of the hoped-for deer herd, dead of its own too much, bleach with the bones of the dead sage, or molder under the high-lined junipers. I now suspect that just as a deer herd lives in mortal fear of its wolves, so does a mountain live in mortal fear of its deer."

By the time Leopold himself died, in 1949, he saw wilderness in a much richer light than the one that prompted him, in 1924, to push for a "national hunting ground" in the Gila Forest. Wilderness areas were still important as sanctuaries for the primitive arts of canoeing, packing, and hunting. But they were also necessary as part of a larger land ethic and as a laboratory for the study of land health. Culturally, wilderness was a place where Americans could rediscover their history and "organize yet another search for a durable scale of values." Wilderness even had something that Leopold could not name. "The physics of beauty," he noted, "is one department of natural science still in the Dark Ages."

It is, perhaps, not surprising that as the country's first wilderness, the Gila may also be the most mismanaged. In part due to its bloated deer herds, in the late 1920s a road was opened through the heart of the wilderness to allow access to hunters. Another road to the Gila Cliff Dwellings National Monument would later be paved. In 1964, historic grazing leases were granted "in perpetuity," and along certain streams cattle have clearly become the dominant species. The imperfections of the Gila carry their own lessons. To become a visitor, to relinquish control, is not easy.

When I reach the hot springs, I have reached a place like my husband's passage through Little Bear Canyon, a place that conforms to a place inside. Surrounded by ferns and vegetation, the two pools are sheltered against a massive rock upon which the hot water trickles down in a cascade over slick moss and lime green algae. A hand built dam of loose stone creates a four-foot-deep swimming hole in which the older children play and splash. The water temperature is about a hundred degrees. My husband stretches full length and lets his nose touch the tiny yellow wildflowers that bloom at the pool's edge.

I carry my son David against my chest. He is developing his sense of humor and, to amuse him, I simulate disgust when he sticks out his tongue so that it fills his little mouth. "Oooooh!" I make a face. He laughs with power and sticks out his tongue again. "Ooooh!" I say. He grins and sticks out his tongue at everyone. All the girls, excepting his sister, want to hold him in their arms and glide away with him in the warm water. He skims over the surface of the pool and then cries out so beseechingly that they float him back to me. Clinging, he rides my hip like a cowboy in the saddle. We go through cycles with our children, as they do with us, and for now my son, who is twenty-two months, and I, who am thirty-five years, are besotted with each other. I adore his skin and his smell and every stray expression that informs his face with intelligence and personality. This is mutual, unconditional love—an exotic interlude. This has been going on in these hot springs since the first Mogollon mother, since Alope and her children.

The rest of the trip passes in this way. We take turns riding the horses farther up the Middle Fork: here the rock walls loom a thousand feet above a canyon floor that narrows dramatically to the width of its river. Another few miles and the trail runs downhill, faster and faster, as the horses hurry to a grassy bottom land known as The Meadows. The scenery is breathtaking and we claim it as our own. No one has ever seen it, just this way, before. In the cooling twilights, we swim in the water hole. During an afternoon rain, we lie in our tents. We cook. We talk. We clean. Roberta and Carol take long walks together. Spouses, as usual, spar a little, and the children bicker. On our therma-rest pads, we all sleep well.

Later, driving home, I have to wonder why these four days have been such a success. Who was it—my husband, my children, my friends—who helped me to see, just a little more clearly, that I do not need to become more than I am to have a place in the wilderness? I do not need to love solitude more than the company of my own species. I do not need to become a man. Or a manager. The shrine is here already. The graves. The bowls and the baskets and the way we touch a baby or tell stories to children. I need only walk in.

EVELYN WHITE
b. 1954

Evelyn White worked as an editor and writer for the San Francisco Chronicle *from 1986 to 1995. She has also published two influential books relating to African American women's health and well-being. The first edition of* Chain, Chain, Change: For Black Women Dealing with Physical and Emotional Abuse *appeared in 1985. The* Black Women's Health Book: Speaking for Ourselves, *a volume which she edited, came out in 1990. Her essay that appears here, "Black Women and the Wilderness," makes an important and original contribution to the discussion of ways in which racial, sexual, cultural, and political factors affect people's experience of nature.*

BLACK WOMEN AND THE WILDERNESS

> I wanted to sit outside and listen to the roar of the ocean, but I
> was afraid.
> I wanted to walk through the redwoods, but I was afraid.
> I wanted to glide in a kayak and feel the cool water splash in my
> face, but I was afraid.

For me, the fear is like a heartbeat, always present, while at the same time, intangible, elusive, and difficult to define. So pervasive, so much a part of me, that I hardly knew it was there.

In fact, I wasn't fully aware of my troubled feelings about nature until I was invited to teach at a women's writing workshop held each summer on the McKenzie River in the foothills of Oregon's Cascade Mountains. I was invited to Flight of the Mind by a Seattle writer and her friend, a poet who had moved from her native England to Oregon many years before. Both committed feminists, they asked me to teach because they believe, as I do, that language and literature transcend the man-made boundaries that are too often placed upon them. I welcomed and appreciated their interest in me and my work.

Once I got there, I did not welcome the steady stream of invitations to explore the great outdoors. It seemed like the minute I finished my teaching duties, I'd be faced with a student or fellow faculty member clamoring for me to trek to the lava beds, soak in the hot springs, or hike into the mountains that loomed over the site like white-capped security guards. I claimed fatigue, a backlog of classwork, concern about "proper" student/teacher relations; whatever the excuse, I always declined to join the expeditions into the woods. When I wasn't teaching, eating in the dining hall, or attending our evening readings, I stayed holed up in my riverfront cabin with all doors locked and windowshades drawn. While the river's roar gave me a certain comfort and my heart warmed when I gazed at the sun-dappled trees out of a classroom window, I didn't want to get closer. I was certain that if I ventured outside to admire a meadow or to feel the cool ripples in a stream, I'd be taunted, attacked, raped, maybe even murdered because of the color of my skin.

I believe the fear I experience in the outdoors is shared by many African-American women and that it limits the way we move through the

Originally published in *Literature and the Environment: A Reader on Nature and Culture*, edited by Lorraine Anderson et al. (New York: Addison Wesley Longman, 1999).

world and colors the decisions we make about our lives. For instance, for several years now, I've been thinking about moving out of the city to a wooded, vineyard-laden area in Northern California. It is there, among the birds, creeks, and trees that I long to settle down and make a home.

Each house-hunting trip I've made to the countryside has been fraught with two emotions: elation at the prospect of living closer to nature and a sense of absolute doom about what might befall me in the backwoods. My genetic memory of ancestors hunted down and preyed upon in rural settings counters my fervent hopes of finding peace in the wilderness. Instead of the solace and comfort I seek, I imagine myself in the country as my forebears were—exposed, vulnerable, and unprotected—a target of cruelty and hate.

I'm certain that the terror I felt in my Oregon cabin is directly linked to my memories of September 15, 1963. On that day, Denise McNair, Addie Mae Collins, Cynthia Wesley, and Carol Robertson were sitting in their Sunday school class at the Sixteenth Street Church in Birmingham, Alabama. Before the bright-eyed black girls could deliver the speeches they'd prepared for the church's annual Youth Day program, a bomb planted by racists flattened the building, killing them all. In black households throughout the nation, families grieved for the martyred children and expressed their outrage at whites who seemed to have no limits on the depths they would sink in their ultimately futile effort to curtail the civil rights movement.

To protest the Birmingham bombing and to show solidarity with the struggles in the South, my mother bought a spool of black cotton ribbon which she fashioned into armbands for me and my siblings to wear to school the next day. Nine years old at the time, I remember standing in my house in Gary, Indiana, and watching in horror as my mother ironed the black fabric that, in my mind, would align me with the bloody dresses, limbless bodies, and dust-covered patent leather shoes that had been entombed in the blast.

The next morning, I put on my favorite school dress—a V-necked cranberry jumper with a matching cranberry-and-white pin-striped shirt. Motionless, I stared stoically straight ahead, as my mother leaned down and pinned the black ribbon around my right sleeve shortly before I left the house.

As soon as I rounded the corner at the end of our street, I ripped the ribbon off my arm, looking nervously up into the sky for the "evil white people" I'd heard my parents talk about in the aftermath of the bombing. I feared that if I wore the armband, I'd be blown to bits like the black girls who were that moment rotting under the rubble. Thirty years later, I know that another part of my "defense strategy" that day was to wear the outfit that had always garnered me compliments from teachers and friends. "Don't drop a bomb on me," was the message I was desperately trying to convey through my cranberry jumper. "I'm a pretty black girl. Not like the ones at the church."

The sense of vulnerability and exposure that I felt in the wake of the Birmingham bombing was compounded by feelings that I already had about Emmett Till. Emmett was a rambunctious, fourteen-year-old black boy from Chicago, who in 1955 was sent to rural Mississippi to enjoy the pleasures of summer with relatives. Emmett was delivered home in a pine box long before season's end bloated and battered beyond recognition. He had been lynched and dumped in the Tallahatchie River with the rope still dangling around his neck for allegedly whistling at a white woman at a country store.

Those summers in Oregon when I walked past the country store where thick-necked loggers drank beer while leaning on their big rig trucks, it seemed like Emmett's fate had been a part of my identity from birth. Averting my eyes from those of the loggers, I'd remember the ghoulish photos of Emmett I'd seen in *JET* magazine with my childhood friends Tyrone and Lynette Henry. The Henrys subscribed to *JET*, an inexpensive magazine for blacks, and kept each issue neatly filed on the top shelf of a bookcase in their living room. Among black parents, the *JET* with Emmett's story was always carefully handled and treated like one of the most valuable treasures on earth. For within its pages rested an important lesson they felt duty-bound to teach their children: how little white society valued our lives.

Mesmerized by Emmett's monstrous face, Lynette, Tyrone, and I would drag a flower-patterned vinyl chair from the kitchen, take the Emmett *JET* from the bookcase, and spirit it to a back bedroom where we played. Heads together, bellies on the floor as if we were shooting marbles or scribbling in our coloring books, we'd silently gaze at Emmett's photo for what seemed like hours before returning it to its sacred place. As with thousands of black children from that era, Emmett's murder cast a nightmarish pall over my youth. In his pummeled and contorted face, I saw a reflection of myself and the blood-chilling violence that would greet me if I ever dared to venture into the wilderness.

I grew up. I went to college. I traveled abroad. Still, thoughts of Emmett Till could leave me speechless and paralyzed with the heart-stopping fear that swept over me as when I crossed paths with loggers near the McKenzie River or whenever I visited the outdoors. His death seemed to be summed up in the prophetic warning of writer Alice Walker, herself a native of rural Georgia: "Never be the only one, except, possibly, in your own house."

For several Oregon summers, I concealed my pained feelings about the outdoors until I could no longer reconcile my silence with my mandate to my students to face their fears. They found the courage to write openly about incest, poverty, and other ills that had constricted their lives: How could I turn away from my fears about being in nature?

But the one time I'd attempted to be as bold as my students, I'd been faced with an unsettling incident. Legend had it that the source of the McKenzie was a tiny trickle of water that bubbled up from a pocket in a nearby lake. Intrigued by the local lore, two other Flight teachers and a staff person, all white women, invited me to join them on an excursion to the lake. The plan was to rent rowboats and paddle around the lake Sacajawea-style, until we, brave and undaunted women, "discovered the source" of the mighty river. As we approached the lake, we could see dozens of rowboats tied to the dock. We had barely begun our inquiry about renting one when the boathouse man interrupted and tersely announced: "No boats."

We stood shocked and surprised on a sun-drenched dock with a vista of rowboats before us. No matter how much we insisted that our eyes belied his words, the man held fast to his two-note response: "No boats."

Distressed but determined to complete our mission, we set out on foot. As we trampled along the trail that circled the lake we tried to make sense of our "Twilight Zone" encounter. We laughed and joked about the incident and it ultimately drifted out of our thoughts in our jubilation at finding the gurgling bubble that gave birth to the McKenzie. Yet I'd always felt that our triumph was undermined by a searing question that went unvoiced that day: Had we been denied the boat because our group included a black?

In an effort to contain my fears, I forced myself to revisit the encounter and to reexamine my childhood wounds from the Birmingham bombing and the lynching of Emmett Till. I touched the terror of my Ibo and Ashanti ancestors as they were dragged from Africa and enslaved on southern plantations. I conjured bloodhounds, burning crosses, and white-robed Klansmen hunting down people who looked just like me. I imagined myself being captured in a swampy backwater, my back ripped open and bloodied by the whip's lash. I cradled an ancestral mother, broken and keening as her baby was snatched from her arms and sold down the river.

Every year, the Flight of the Mind workshop offers a rafting trip on the McKenzie River. Each day we'd watch as flotillas of rafters, shrieking excitedly and with their oars held aloft, rumbled by the deck where students and teachers routinely gathered. While I always cheered their adventuresome spirit, I never joined the group of Flight women who took the trip. I was always mindful that I had never seen one black person in any of those boats.

Determined to reconnect myself to the comfort my African ancestors felt in the rift valleys of Kenya and on the shores of Sierra Leone, I eventually decided to go on a rafting trip. Familiar with my feelings about nature, Judith, a dear friend and workshop founder, offered to be one of my raftmates.

With her sturdy, gentle and wise body as my anchor, I lowered myself into a raft at the bank of the river. As we pushed off into the current, I felt myself make an unsure but authentic shift from my painful past.

At first the water was calm—nearly hypnotic in its pristine tranquility. Then we met the rapids, sometimes swirling, other times jolting us forward like a runaway roller coaster. The guide roared out commands, "Highside! All forward! All back!" To my amazement, I responded. Periodically, my brown eyes would meet Judith's steady aquamarine gaze and we'd smile at each other as the cool water splashed in our faces and shimmered like diamonds in our hair.

Charging over the river, orange life vest firmly secured, my breathing relaxed and I allowed myself to drink in the stately rocks, soaring birds, towering trees, and affirming anglers who waved their rods as we rushed by in our raft. About an hour into the trip, in a magnificently still moment, I looked up into the heavens and heard the voice of black poet Langston Hughes:

"I've known rivers ancient as the world and older than the flow of human blood in human veins. I bathed in the Euphrates when dawns were young. I built my hut near the Congo and it lulled me to sleep. I looked upon the Nile and raised the pyramids above it. My soul has grown deep like the rivers."

Soaking wet and shivering with emotion, I felt tears welling in my eyes as I stepped out of the raft onto solid ground. Like my African forebears who survived the Middle Passage, I was stronger at journey's end.

Since that voyage, I've stayed at country farms, napped on secluded beaches, and taken wilderness treks all in an effort to find peace in the outdoors. No matter where I travel, I will always carry Emmett Till and the four black girls whose deaths affected me so. But comforted by our tribal ancestors—herders, gatherers, and fishers all—I am less fearful, ready to come home.

BARBARA KINGSOLVER
b. 1955

Barbara Kingsolver is one of America's most acclaimed novelists, the author of such works as The Bean Trees *(1988),* Animal Dreams *(1990), and* The Poisonwood Bible *(1998). But in her book entitled* High Tide in Tucson *she also established herself as a powerful and original voice about landscape and the sense of place. The title essay from that volume reflects on the ways in which we human beings, with all our self-conscious rationality and long-considered agendas, may also be governed by the same larger rhythms that determine the movements and life cycles of other living creatures.*

HIGH TIDE IN TUCSON

A hermit crab lives in my house. Here in the desert he's hiding out from local animal ordinances, at minimum, and maybe even the international laws of native-species transport. For sure, he's an outlaw against nature. So be it.

He arrived as a stowaway two Octobers ago. I had spent a week in the Bahamas, and while I was there, wishing my daughter could see those sparkling blue bays and sandy coves, I did exactly what she would have done: I collected shells. Spiky murexes, smooth purple moon shells, ancient-looking whelks sand-blasted by the tide—I tucked them in the pockets of my shirt and shorts until my lumpy, suspect hemlines gave me away, like a refugee smuggling the family fortune. When it was time to go home, I rinsed my loot in the sink and packed it carefully into a plastic carton, then nested it deep in my suitcase for the journey to Arizona.

I got home in the middle of the night, but couldn't wait till morning to show my hand. I set the carton on the coffee table for my daughter

High Tide in Tucson (New York: HarperCollins, 1995).

to open. In the dark living room her face glowed, in the way of antique stories about children and treasure. With perfect delicacy she laid the shells out on the table, counting, sorting, designating scientific categories like yellow-striped pinky, Barnacle Bill's pocketbook . . . Yeek! She let loose a sudden yelp, dropped her booty, and ran to the far end of the room. The largest, knottiest whelk had begun to move around. First it extended one long red talon of a leg, tap-tap-tapping like a blind man's cane. Then came half a dozen more red legs, plus a pair of eyes on stalks, and a purple claw that snapped open and shut in a way that could not mean We Come in Friendship.

Who could blame this creature? It had fallen asleep to the sound of the Caribbean tide and awakened on a coffee table in Tucson, Arizona, where the nearest standing water source of any real account was the municipal sewage-treatment plant.

With red stiletto legs splayed in all directions, it lunged and jerked its huge shell this way and that, reminding me of the scene I make whenever I'm moved to rearrange the living-room sofa by myself. Then, while we watched in stunned reverence, the strange beast found its bearings and began to reveal a determined, crabby grace. It felt its way to the edge of the table and eased itself over, not falling bang to the floor but hanging suspended underneath within the long grasp of its ice-tong legs, lifting any two or three at a time while many others still held in place. In this remarkable fashion it scrambled around the underside of the table's rim, swift and sure and fearless like a rock climber's dream.

If you ask me, when something extraordinary shows up in your life in the middle of the night, you give it a name and make it the best home you can.

The business of naming involved a grasp of hermit-crab gender that was way out of our league. But our household had a deficit of males, so my daughter and I chose Buster, for balance. We gave him a terrarium with clean gravel and a small cactus plant dug out of the yard and a big cockleshell full of tap water. All this seemed to suit him fine. To my astonishment our local pet store carried a product called Vitaminized Hermit Crab Cakes. Tempting enough (till you read the ingredients) but we passed, since our household leans more toward the recycling ethic. We give him leftovers. Buster's rapture is the day I drag the unidentifiable things in cottage cheese containers out of the back of the fridge.

We've also learned to give him a continually changing assortment of seashells, which he tries on and casts off like Cinderella's stepsisters preening for the ball. He'll sometimes try to squeeze into ludicrous outfits too small to contain him (who can't relate?). In other moods, he will disappear into a conch the size of my two fists and sit for a day, immobilized by the weight of upward mobility. He is in every way the perfect housemate: quiet, entertaining, and willing to eat up the trash. He went to school for first-grade show-and-tell, and was such a hit the principal called up to congratulate me (I think) for being a broad-minded mother.

It was a long time, though, before we began to understand the content of Buster's character. He required more patient observation than we were in the habit of giving to a small, cold-blooded life. As months went by, we would periodically notice with great disappointment that Buster seemed to be dead. Or not entirely dead, but ill, or maybe suffering the crab equivalent of the blues. He would burrow into a gravelly corner, shrink deep into his shell, and not move, for days and days. We'd take him out to play, dunk him in water, offer him a new frock—nothing. He wanted to be still.

Life being what it is, we'd eventually quit prodding our sick friend to cheer up, and would move on to the next stage of a difficult friendship: neglect. We'd ignore him wholesale, only to realize at some point later on that he'd lapsed into hyperactivity. We'd find him ceaselessly patrolling the four corners of his world, turning over rocks, rooting out and dragging around truly disgusting pork-chop bones, digging up his cactus and replanting it on its head. At night when the household fell silent I would lie in bed listening to his methodical pebbly racket from the opposite end of the house. Buster was manic-depressive.

I wondered if he might be responding to the moon. I'm partial to lunar cycles, ever since I learned as a teenager that human females in their natural state—which is to say, sleeping outdoors—arrive at menses in synchrony and ovulate with the full moon. My imagination remains captive to that primordial village: the comradely grumpiness of new-moon days, when the entire world at once would go on PMS alert. And the compensation that would turn up two weeks later on a wild wind, under that great round headlamp, driving both men and women to distraction with the overt prospect of conception. The surface of the land literally rises and falls—as much as fifty centimeters!—as the moon passes over, and we clay-footed mortals fall like dominoes before the swell. It's no surprise at all if a full moon inspires lyricists to corny love songs, or inmates to slamming themselves against barred windows. A hermit crab hardly seems this impetuous, but animals are notoriously responsive to the full moon: wolves howl; roosters announce daybreak all night. Luna moths, Arctic loons, and lunatics have a sole inspiration in common. Buster's insomniac restlessness seemed likely to be a part of the worldwide full-moon fellowship.

But it wasn't, exactly. The full moon didn't shine on either end of his cycle, the high or the low. We tried to keep track, but it soon became clear: Buster marched to his own drum. The cyclic force that moved him remained as mysterious to us as his true gender and the workings of his crustacean soul.

Buster's aquarium occupies a spot on our kitchen counter right next to the coffeepot, and so it became my habit to begin mornings with chin in hands, pondering the oceanic mysteries while awaiting percolation. Finally, I remembered something. Years ago when I was a graduate student of animal behavior, I passed my days reading about the likes of animals' internal clocks. Temperature, photoperiod, the rise and fall of hormones—all these influences have been teased apart like so many threads from the rope that pulls every creature to its regulated destiny. But one story takes the cake. F. A. Brown, a researcher who is more or less the grandfather of the biological clock, set about in 1954 to track the cycles of intertidal oysters. He scooped his subjects from the clammy coast of Connecticut and moved them into the basement of a laboratory in landlocked Illinois. For the first fifteen days in their new aquariums, the oysters kept right up with their normal intertidal behavior: they spent time shut away in their shells, and time with their mouths wide open, siphoning their briny bath for the plankton that sustained them, as the tides ebbed and flowed on the distant Connecticut shore. In the next two weeks, they made a mystifying shift. They still carried out their cycles in unison, and were regular as the tides, but their high-tide behavior didn't coincide with high tide in Connecticut, or for that matter California, or any other tidal charts known to science. It dawned on the researchers after some calculations that the oysters were responding to high tide in Chicago. Never mind that the gentle mollusks lived in glass boxes in the basement of a steel-and-cement building. Nor that Chicago has no ocean. In the circumstances, the oysters were doing their best.

When Buster is running around for all he's worth, I can only presume it's high tide in Tucson. With or without evidence, I'm romantic enough to believe it. This is the lesson of Buster, the poetry that camps outside the halls of science: Jump for joy, hallelujah. Even a desert has tides.

When I was twenty-two, I donned the shell of a tiny yellow Renault and drove with all I owned from Kentucky to Tucson. I was a typical young American, striking out. I had no earthly notion that I was bringing on myself a calamity of the magnitude of the one that befell poor Buster. I am the commonest kind of North American refugee: I believe I like it here, far-flung from my original home. I've come to love the desert that bristles and breathes and sleeps outside my windows. In the course of seventeen years I've embedded myself in a family here—neighbors, colleagues, friends I can't foresee living without, and a child who is native to this ground, with loves of her own. I'm here for good, it seems.

And yet I never cease to long in my bones for what I left behind. I open my eyes on every new day expecting that a creek will run through my backyard under broad-leafed maples, and that my mother will be whistling in the kitchen. Behind the howl of coyotes, I'm listening for meadowlarks. I sometimes ache to be rocked in the bosom of the blood relations and busybodies of my childhood. Particularly in my years as a mother without a mate, I have deeply missed the safety net of extended family.

In a city of half a million I still really look at every face, anticipating recognition, because I grew up in a town where every face meant something to me. I have trouble remembering to lock the doors. Wariness of strangers I learned the hard way. When I was new to the city, I let a man into my house one hot afternoon because he seemed in dire need of a drink of water; when I turned from the kitchen sink I found sharpened steel shoved against my belly. And so I know, I know. But I cultivate suspicion with as much difficulty as I force tomatoes to grow in the drought-stricken hardpan of my strange backyard. No creek runs here, but I'm still listening to secret tides, living as if I belonged to an earlier place: not Kentucky, necessarily, but a welcoming earth and a human family. A forest. A species.

In my life I've had frightening losses and unfathomable gifts: A knife in my stomach. The death of an unborn child. Sunrise in a rain forest. A stupendous column of blue butterflies rising from a Greek monastery. A car that spontaneously caught fire while I was driving it. The end of a marriage, followed by a year in which I could barely understand how to keep living. The discovery, just weeks ago when I rose from my desk and walked into the kitchen, of three strangers industriously relieving my house of its contents.

I persuaded the strangers to put down the things they were holding (what a bizarre tableau of anti-Magi they made, these three unwise men, bearing a camera, an electric guitar, and a Singer sewing machine), and to leave my home, pronto. My daughter asked excitedly when she got home from school, "Mom, did you say bad words?" (I told her this was the very occasion that bad words exist for.) The police said, variously, that I was lucky, foolhardy, and "a brave lady." But it's not good luck to be invaded, and neither foolish nor brave to stand your ground. It's only the way life goes, and I did it, just as years ago I fought off the knife; mourned the lost child; bore witness to the rain

forest; claimed the blue butterflies as Holy Spirit in my private pantheon; got out of the burning car; survived the divorce by putting one
foot in front of the other and taking good care of my child. On most
important occasions, I cannot think how to respond, I simply do. What
does it mean, anyway, to be an animal in human clothing? We carry
around these big brains of ours like the crown jewels, but mostly I find
that millions of years of evolution have prepared me for one thing only:
to follow internal rhythms. To walk upright, to protect my loved ones,
to cooperate with my family group—however broadly I care to define
it—to do whatever will help us thrive. Obviously, some habits that saw
us through the millennia are proving hazardous in a modern context:
for example, the yen to consume carbohydrates and fat whenever they
cross our path, or the proclivity for unchecked reproduction. But it's
surely worth forgiving ourselves these tendencies a little, in light of the
fact that they are what got us here. Like Buster, we are creatures of
inexplicable cravings. Thinking isn't everything. The way I stock my
refrigerator would amuse a level-headed interplanetary observer, who
would see I'm responding not to real necessity but to the dread of
famine honed in the African savannah. I can laugh at my Rhodesian
Ridgeback as she furtively sniffs the houseplants for a place to bury
bones, and circles to beat down the grass before lying on my kitchen
floor. But she and I are exactly the same kind of hairpin.

We humans have to grant the presence of some past adaptations,
even in their unforgivable extremes, if only to admit they are permanent rocks in the stream we're obliged to navigate. It's easy to speculate
and hard to prove, ever, that genes control our behaviors. Yet we are
persistently, excruciatingly adept at many things that seem no more
useful to modern life than the tracking of tides in a desert. At recognizing insider/outsider status, for example, starting with white vs. black
and grading straight into distinctions so fine as to baffle the
bystander—Serb and Bosnian, Hutu and Tutsi, Crip and Blood. We
hold that children learn discrimination from their parents, but they
learn it fiercely and well, world without end. Recite it by rote like a
multiplication table. Take it to heart, though it's neither helpful nor
appropriate, anymore than it is to hire the taller of two men applying
for a position as bank clerk, though statistically we're likely to do that
too. Deference to the physical superlative, a preference for the scent of
our own clan: a thousand anachronisms dance down the strands of our
DNA from a hidebound tribal past, guiding us toward the glories of survival, and some vainglories as well. If we resent being bound by these
ropes, the best hope is to seize them out like snakes, by the throat, look
them in the eye and own up to their venom.

But we rarely do, silly egghead of a species that we are. We invent the most outlandish intellectual grounds to justify discrimination. We tap our toes to chaste love songs about the silvery moon without recognizing them as hymns to copulation. We can dress up our drives, put them in three-piece suits or ballet slippers, but still they drive us. The wonder of it is that our culture attaches almost unequivocal shame to our animal nature, believing brute urges must be hurtful, violent things. But it's no less an animal instinct that leads us to marry (species that benefit from monogamy tend to practice it); to organize a neighborhood cleanup campaign (rare and doomed is the creature that fouls its nest); to improvise and enforce morality (many primates socialize their young to be cooperative and ostracize adults who won't share food).

It's starting to look as if the most shameful tradition of Western civilization is our need to deny we are animals. In just a few centuries of setting ourselves apart as landlords of the Garden of Eden, exempt from the natural order and entitled to hold dominion, we have managed to behave like so-called animals anyway, and on top of it to wreck most of what took three billion years to assemble. Air, water, earth, and fire—so much of our own element so vastly contaminated, we endanger our own future. Apparently we never owned the place after all. Like every other animal, we're locked into our niche: the mercury in the ocean, the pesticides on the soybean fields, all come home to our breastfed babies. In the silent spring we are learning it's easier to escape from a chain gang than a food chain. Possibly we will have the sense to begin a new century by renewing our membership in the Animal Kingdom.

Not long ago I went backpacking in the Eagle Tail Mountains. This range is a trackless wilderness in western Arizona that most people would call Godforsaken, taking for granted God's preference for loamy topsoil and regular precipitation. Whoever created the Eagle Tails had dry heat on the agenda, and a thing for volcanic rock. Also cactus, twisted mesquites, and five-alarm sunsets. The hiker's program in a desert like this is dire and blunt: carry in enough water to keep you alive till you can find a water source; then fill your bottles and head for the next one, or straight back out. Experts warn adventurers in this region, without irony, to drink their water while they're still alive, as it won't help later.

Several canyons looked promising for springs on our topographical map, but turned up dry. Finally, at the top of a narrow, overgrown gorge we found a blessed tinaja, a deep, shaded hollow in the rock about the size of four or five claw-foot tubs, holding water. After we drank our fill, my friends struck out again, but I opted to stay and spend the day in the hospitable place that had slaked our thirst. On either side of the natural water tank, two shallow caves in the canyon wall faced each other, only a few dozen steps apart. By crossing from one to the other at noon, a person could spend the whole day here in shady comfort—or in colder weather, follow the winter sun. Anticipating a morning of reading, I pulled *Angle of Repose* out of my pack and looked for a place to settle on the flat, dusty floor of the west-facing shelter. Instead, my eyes were startled by a smooth corn-grinding stone. It sat in the exact center of its rock bowl, as if the Hohokam woman or man who used this mortar and pestle had walked off and left them there an hour ago. The Hohokam disappeared from the earth in A.D. 1450. It was inconceivable to me that no one had been here since then, but that may have been the case—that is the point of trackless wilderness. I picked up the grinding stone. The size and weight and smooth, balanced perfection of it in my hand filled me at once with a longing to possess it. In its time, this excellent stone was the most treasured thing in a life, a family, maybe the whole neighborhood. To whom it still belonged. I replaced it in the rock depression, which also felt smooth to my touch. Because my eyes now understood how to look at it, the ground under my feet came alive with worked flint chips and pottery shards. I walked across to the other cave and found its floor just as lively with historic debris. Hidden under brittlebush and catclaw I found another grinding stone, this one some distance from the depression in the cave floor that once answered its pressure daily, for the grinding of corn or mesquite beans.

For a whole day I marveled at this place, running my fingers over the knife edges of dark flint chips, trying to fit together thick red pieces of shattered clay jars, biting my lower lip like a child concentrating on a puzzle. I tried to guess the size of whole pots from the curve of the broken pieces: some seemed as small as my two cupped hands, and some maybe as big as a bucket. The sun scorched my neck, reminding me to follow the shade across to the other shelter. Bees hummed at the edge of the water hole, nosing up to the water, their abdomens pulsing like tiny hydraulic pumps; by late afternoon they rimmed the pool completely, a collar of busy lace. Off and on, the lazy hand of a hot breeze shuffled the white leaves of the brittlebush. Once I looked up to see a screaming pair of red-tailed hawks mating in midair, and once a clatter of hooves warned me to hold still. A bighorn ram emerged through the brush, his head bent low under his hefty cornice, and ambled by me with nothing on his mind so much as a cool drink.

How long can a pestle stone lie still in the center of its mortar? That long ago—that recently—people lived here. *Here*, exactly, and not one valley over, or two, or twelve, because this place had all a person needs:

shelter, food, and permanent water. They organized their lives around a catchment basin in a granite boulder, conforming their desires to the earth's charities; they never expected the opposite. The stories I grew up with lauded Moses for striking the rock and bringing forth the bubbling stream. But the stories of the Hohokam—oh, how they must have praised that good rock.

At dusk my friends returned with wonderful tales of the ground they had covered. We camped for the night, refilled our canteens, and hiked back to the land of plumbing and a fair guarantee of longevity. But I treasure my memory of the day I lingered near water and covered no ground. I can't think of a day in my life in which I've had such a clear fix on what it means to be human.

Want is a thing that unfurls unbidden like fungus, opening large upon itself, stopless, filling the sky. But *needs*, from one day to the next, are few enough to fit in a bucket, with room enough left to rattle like brittlebush in a dry wind.

For each of us—furred, feathered, or skinned alive—the whole earth balances on the single precarious point of our own survival. In the best of times, I hold in mind the need to care for things beyond the self: poetry, humanity, grace. In other times, when it seems difficult merely to survive and be happy about it, the condition of my thought tastes as simple as this: let me be a good animal today. I've spent months at a stretch, even years, with that taste in my mouth, and have found that it serves.

But it seems a wide gulf to cross, from the raw, green passion for survival to the dispassionate, considered state of human grace. How does the animal mind construct a poetry for the modern artifice in which we now reside? Often I feel as disoriented as poor Buster, unprepared for the life that zooms headlong past my line of sight. This clutter of human paraphernalia and counterfeit necessities—what does it have to do with the genuine business of life of on earth? It feels strange to me to be living in a box, hiding from the steadying influence of the moon; wearing the hide of a cow, which is supposed to be dyed to match God-knows-what, on my feet; making promises over the telephone about things I will do at a precise hour next *year*. (I always feel the urge to add, as my grandmother does, "Lord willing and the creeks don't rise!") I find it impossible to think, with a straight face, about what colors ought not to be worn after Labor Day. I can become hysterical over the fact that someone, somewhere, invented a thing called the mushroom scrubber, and that many other people undoubtedly feel they *need* to possess one. It's completely usual for me to get up in the morning, take a look around, and laugh out loud.

Strangest of all, I am carrying on with all of this in a desert, two thousand miles from my verdant childhood home. I am disembodied. No one here remembers how I was before I grew to my present height. I'm called upon to reinvent my own childhood time and again; in the process, I wonder how I can ever know the truth about who I am. If someone had told me what I was headed for in that little Renault—that I was stowing away in a shell, bound to wake up to an alien life on a persistently foreign shore—I surely would not have done it. But no one warned me. My culture, as I understand it, values independence above all things—in part to ensure a mobile labor force, grease for the machine of a capitalist economy. Our fairy tale commands: Little Pig, go out and seek your fortune! So I did.

Many years ago I read that the Tohono O'odham, who dwell in the deserts near here, traditionally bury the umbilicus of a newborn son or daughter somewhere close to home and plant a tree over it, to hold the child in place. In a sentimental frame of mind, I did the same when my own baby's cord fell off. I'm staring at the tree right now, as I write—a lovely thing grown huge outside my window, home to woodpeckers, its boughs overarching the house, as dissimilar from the sapling I planted seven years ago as my present life is from the tidy future I'd mapped out for us all when my baby was born. She will roam light-years from the base of that tree. I have no doubt of it. I can only hope she's growing as the tree is, absorbing strength and rhythms and a trust in the seasons, so she will always be able to listen for home.

I feel remorse about Buster's monumental relocation; it's a weighty responsibility to have thrown someone else's life into permanent chaos. But as for my own, I can't be sorry I made the trip. Most of what I learned in the old place seems to suffice for the new: if the seasons like Chicago tides come at ridiculous times and I have to plant in September instead of May, and if I have to make up family from scratch, what matters is that I do have sisters and tomato plants, the essential things. Like Buster, I'm inclined to see the material backdrop of my life as mostly immaterial, compared with what moves inside of me. I hold on to my adopted shore, chanting private vows: wherever I am, let me never forget to distinguish *want* from *need*. Let me be a good animal today. Let me dance in the waves of my private tide, the habits of survival and love.

Every one of us is called upon, probably many times, to start a new life. A frightening diagnosis, a marriage, a move, loss of a job or a limb or a loved one, a graduation, bringing a new baby home: it's impossible to think at first how this all will be possible. Eventually, what moves it all forward is the subterranean ebb and flow of being alive among the living.

In my own worst seasons I've come back from the colorless world of despair by forcing myself to look hard, for a long time, at a single glorious thing: a flame of red geranium outside my bedroom window. And then another: my daughter in a yellow dress. And another: the perfect outline of a full, dark sphere behind the crescent moon. Until I learned to be in love with my life again. Like a stroke victim retraining new parts of the brain to grasp lost skills, I have taught myself joy, over and over again.

It's not such a wide gulf to cross, then, from survival to poetry. We hold fast to the old passions of endurance that buckle and creak beneath us, dovetailed, tight as a good wooden boat to carry us onward. And onward full tilt we go, pitched and wrecked and absurdly resolute, driven in spite of everything to make good on a new shore. To be hopeful, to embrace one possibility after another—that is surely the basic instinct. Baser even than hate, the thing with teeth, which can be stilled with a tone of voice or stunned by beauty. If the whole world of the living has to turn on the single point of remaining alive, that pointed endurance is the poetry of hope. The thing with feathers.

What a stroke of luck. What a singular brute feat of outrageous fortune: to be born to citizenship in the Animal Kingdom. We love and we lose, go back to the start and do it right over again. For every heavy forebrain solemnly cataloging the facts of a harsh landscape, there's a rush of intuition behind it crying out: High tide! Time to move out into the glorious debris. Time to take this life for what it is.

MICHAEL POLLAN
b. 1955

Second Nature: A Gardener's Education (1991) was a timely book, and almost immediately an influential one. At a time when both environmental historians and conservationists were trying to integrate human and natural history more successfully, Michael Pollan called their attention back to the domestic garden, a scene where the two realms inform each other and one with its own rich literature, as well as a landscape central

to many people's sense of place today. Pollan's subsequent books include
A Place of My Own: The Education of an Amateur Builder *(1997), and*
Botany of Desire: A Plant's Eye View of the World *(2001).*

WEEDS ARE US

Ralph Waldo Emerson, who as a lifelong gardener really should
have known better, once said that a weed is simply a plant whose
virtues we haven't yet discovered. "Weed" is not a category of nature
but a human construct, a defect of our perception. This kind of attitude, which comes out of an old American strain of romantic thinking
about wild nature, can get you into trouble. At least it did me. For I had
Emerson's pretty conceit in mind when I planted my first flower bed,
and the result was not a pretty thing.

Having read perhaps too much Emerson, and too many of the sort
of gardening book that advocates "wild gardens" and nails a pair of
knowing quotation marks around the word *weed* (a sure sign of ecological sophistication), I sought to make a flower bed that was as "natural"
as possible. Rejecting all geometry (too artificial!), I cut a more or less
kidney-shaped bed in the lawn, pulled out the sod, and divided the
bare ground into irregular patches that I roughly outlined with a bit of
ground limestone. Then I took packets of annual seeds—bachelor's
buttons, nasturtiums, nicotianas, cosmos, poppies (California and
Shirley both), cleomes, zinnias, and sunflowers—and broadcast a
handful of each into the irregular patches, letting the seeds fall wherever nature dictated. No rows: this bed's arrangement would be
natural. I sprinkled the seeds with loose soil, watered, and waited for
them to sprout.

Pigweed sprouted first, though at the time I was so ignorant that I
figured this vigorous upstart must be zinnia, or sunflower. I had had no
prior acquaintance with pigweed (it grew nowhere else on the property), and did not deduce that it was a weed until I noticed it was coming up in every single one of my irregular patches. Within a week the
entire bed was clothed in tough, hairy pigweeds, and it was clear that I
would have to start pulling them out if I ever expected to see my
intended annuals. The absence of rows or paths made weeding difficult, but I managed to at least thin the lusty pigweeds, and the annuals,
grateful for the intervention on their behalf, finally pushed themselves

Second Nature: A Gardener's Education (New York: Atlantic Monthly, 1991).

up out of the earth. Finding the coast relatively clear, they started to grow in earnest.

That first summer, my little annual meadow thrived, pretty much conforming to the picture I'd had in mind when I planted it. Sky-blue drifts of bachelor's buttons flowed seamlessly into hot spots thick with hunter-orange and fire-engine poppies, behind which rose great sunflower towers. The nasturtiums poured their sand-dollar leaves into neat, low mounds dabbed with crimson and lemon, and the cleomes worked out their intricate architectures high in the air. Weeding this dense tangle was soon all but impossible, but after the pigweed scare I'd adopted a more or less laissez-faire policy toward the uninvited. The weeds that moved in were ones I was willing to try to live with: jewelweed (a gangly orange-flowered relative of impatiens), foxtail grass, clover, shepherd's purse, inconspicuous Galinsoga, and Queen Anne's lace, the sort of weed Emerson must have had in mind, with its ivory lace flowers (as pretty as anything you might plant) and edible, carrotlike root. That first year a pretty vine also crept in, a refugee from the surrounding lawn. It twined its way up the sunflower stalks and in August unfurled white, trumpet-shaped flowers that resembled morning glory. What right had I to oust this delicate vine? To decide that the flowers I planted were more beautiful than ones the wind had sown? I liked how wild my garden was, how peaceably my cultivars seemed to get along with their wild relatives. And I liked how unneurotic I was being about "weeds." Call me Ecology Boy.

"Weeds," I decided that summer, did indeed have a bad rap. I thought back to my grandfather's garden, to his unenlightened, totalitarian approach toward weeds. Each day he patrolled his pristine rows, beheading the merest smudge of green with his vigilant hoe. Hippies, unions, and weeds: all three made him crazy then, an old man in the late sixties, and all three called forth his reactionary wrath. Perhaps because there was little he could do to stop the march of hippies and organized labor, he attacked weeds all the more zealously. He was one of those gardeners who would pull weeds anywhere—not just in his own or other people's gardens, but in parking lots and storefront window boxes too. His world then was under siege, and weeds to him represented the advance guard of the forces of chaos. Had he lived to see it, my little wild garden—this rowless plant be-in, this horticultural Haight-Ashbury—would probably have broken his heart.

My grandfather wasn't the first person to sense a social or political threat in the growth of weeds. Whenever Shakespeare tells us that "darnel, hemlock, and rank fumatory" or "hateful docks, rough thistles, kecksies, burrs" are growing unchecked, we can assume a monarchy is about to fall. Until the romantics, the hierarchy of plants was generally thought to mirror that of human society. Common people, one writer held in 1700, may be "looked upon as trashy weeds or nettles." J. C. Loudon, an early nineteenth-century gardening expert, invited his readers "to compare plants with men, [to] consider aboriginal species as mere savages, and botanical species . . . as civilized beings."

The garden world even today organizes plants into one great hierarchy. At the top stand the hypercivilized hybrids—think of the rose, "queen of the garden"—and at the bottom are the weeds, the plant world's proletariat, furiously reproducing and threatening to usurp the position of their more refined horticultural betters. Where any given plant falls in this green chain of being has a lot to do with fashion, but there are a few abiding rules. In general, the more intensively a plant has been hybridized—the further it's been distanced from its wildflower origins—the higher its station in plant society. Thus a delphinium can lord it over a larkspur, a heavily doubled bourbon rose over a five-petaled rugosa. A corollary of this rule holds that the more "weedy" a plant is—the easier it is to grow—the lower its place: garden phlox, heir to all the fungi, has greater status than indestructible coreopsis.

Color, too, determines rank, and white comes at the top. This is because pure white occurs only rarely in nature, and perhaps also because a taste for the subtleties of white flowers is something that must be acquired. (Gaudy colors have always been associated with the baser elements: gaillardia, a loud, two-toned cousin of the daisy, used to be called "nigger flower.") Just beneath white is blue, a color that has always enjoyed royal and aristocratic connections, and from there it is a descent downscale through the hot and flashy shades, by the all-too-common yellows, past the reds even bulls will take note of, on down to the very bottom: shunned, rebuffed, eschewed, embarrassing, promiscuous magenta. Magenta, the discount pigment with which nature has brushed a thousand weeds, has always been a mark of bad breeding in the garden world. The offspring of hybrid species that have been allowed to set seed will frequently revert to magenta, as base genes reassert themselves.

The nineteenth-century romantics, who looked more kindly on the common man, also looked kindly on the weed. By the time they wrote, the English countryside had been so thoroughly dominated, every acre cleared of trees and bisected by hedgerows, that the idea of a wild landscape acquired a strong appeal, perhaps for the first time in European history. (Nostalgia for wilderness comes easy once it no longer poses any threat.) Ruskin wrote enthusiastically of the wildflower, which had never been "provoked to glare into any gigantic impudence at a flower show." He judged a flower garden as unnatural, "an ugly thing, even when best managed: it is an assembly of unfortunate beings, pampered and bloated above their natural size; . . . corrupted by evil communication into speckled and inharmonious colours; torn from the soil which they loved, and of which they were the spirit and the glory, to glare away their term of tormented life. . . ."

If garden flowers were slaves to men, then weeds were emblems of freedom and wildness—at least among romantic writers who lived at some distance from nature. "Better to me the meanest weed," wrote Tennyson in the early 1830s. "Weed" soon became a standard synecdoche for *wilderness*, as in this stanza of Gerard Manley Hopkins:

> What would the world be, once bereft
> Of wet and wildness? Let them be left,
> O let them be left, wildness and wet;
> Long live the weeds and the wilderness yet.

Predictably, the romance of the weed gained a ready purchase on the American mind, which has always been disposed to regard the works of nature as superior to those of men, and to resist hierarchies wherever they might be found. The weed supplies Emerson, Whitman, Thoreau, and generations of American naturalists with a favorite trope—for unfettered wildness, for the beauty of the unimproved landscape, and of course, when in quotes, for the benightedness of those fellow countrymen who fail to perceive nature as acutely and sympathetically as they do. (We'll leave aside for now the question of how acutely these writers themselves perceived weeds.) Weed worship continues to flower periodically in America, most recently in the sixties. "Weed" became a fond nickname for marijuana, and millions of us consulted our tattered copies of Euell Gibbons's *Stalking the Wild Asparagus*, an improbable best-seller that, essentially, proposed weeds as the basis of a new American cuisine. Whenever history and culture seem stifling, weeds begin to look good.

My own romance of the weed did not survive a second summer. The annuals, which I had allowed to set seed the previous year, did come back, but they proved a poor match for the weeds, who returned heavily reinforced. It was as though news of this sweet deal (this chump gardener!) had spread through the neighborhood over the winter, for the weed population burgeoned, both in number and in kind. Recognizing that what I now tended was a weed garden, and having been taught that a gardener should know the name of every plant in his care, I consulted a few field guides and drew up an inventory of my collection. In addition to the species I've already mentioned, I had milkweed, poke-weed, smartweed, St.-John's-wort, quack grass, crabgrass, plantain, dandelion, bladder campion, fleabane, butter-and-eggs, timothy, mallow, bird's-foot trefoil, lamb's quarters, chickweed, purslane, curly dock, goldenrod, sheep sorrel, burdock, Canada thistle, and stinging nettle. I'm sure I've missed another dozen, and misidentified a few, but this will give you an idea of the various fruits of my romanticism. What had begun as a kind of idealized wildflower meadow now looked like a roadside tangle and, if I let it go another year, would probably pass for a vacant lot.

Since this had not been my aesthetic aim, I set about reclaiming my garden—to at least arrest the process at "country roadside" before it degenerated to "abandoned railroad siding." I would be enlightened about it, though, pardoning the weeds I liked and expelling all the rest. I was prepared to tolerate the fleabane, holding aloft their sunny clouds of tiny asterlike flowers, or milkweed, with its interesting seed-pods, but bully weeds like burdock, Canada thistle, and stinging nettle had to go. Unfortunately, the weeds I liked least proved to be the best armed and most recalcitrant. Burdock, whose giant clubfoot leaves shade out every other plants for yards around, holds the earth in a death grip. Straining to pull out its mile-long taproot, you feel like a boy trying to arm-wrestle a man. Inevitably the root breaks before it yields, with the result that, in a few days' time, you have two tough burdocks where before there had been one. All I seemed able to do was help my burdock reproduce. I felt less like an exterminator of these weeds than their midwife.

That pretty vine with the morning glory blossoms turned out to be another hydra-headed monster. Bindweed, as it's called, grows like kudzu and soon threatened to blanket the entire garden. It can grow only a foot or so high without support, so it casts about like a blind man, lurching this way then that, until it finds a suitable plant to lean on and eventually smother. Here too my efforts at eradication proved counterproductive. Bindweed, whose roots may reach ten feet down, can reproduce either by seed or human-aided cloning. For its root is as brittle as a fresh snapbean; put a hoe to it and it breaks into a dozen pieces, *each of which will sprout an entire new plant.* It is as though the bindweed's evolution took the hoe into account. By attacking it at its root—the approved strategy for eradicating most weeds—I played right into the insidious bindweed's strategy for world domination.

Have I mentioned my annuals? A few managed to hang on gamely. California poppies and Johnny-jump-ups proved adept at finding

niches among the thistles, and a handful of second-generation nicotianas appeared, though these had reverted to the hue of some weedy
ancestor—instead of bright pink, they came back a muddy shade of
pale green. For the most part, my annuals counted themselves lucky to
serve as underplanting for the triumphant weeds. But whatever niches
remained for them the grasses seemed bent on erasing. Stealthy quack
grass moved in, spreading its intrepid rhizomes to every corner of the
bed. Quack grass roots can travel laterally as much as fifty feet, moving
an inch or two beneath the surface and pushing up a blade (or ten)
wherever the opportunity arises. You pull a handful of this grass thinking you've doomed an isolated tuft, only to find you've grabbed hold of
a rope that reaches clear into the next county—where it is no doubt
tied by a very good knot to an oak.

Now what would Emerson have to say? I had given all my weeds the
benefit of the doubt, acknowledged their virtues and allotted them a
place. I had treated them, in other words, as garden plants. But they did
not behave as garden plants. They differed from my cultivated varieties
not merely by a factor of human esteem. No, they seemed truly a different order of being, more versatile, better equipped, swifter, craftier—
simply more adroit at the work of being a plant. What garden plant can
germinate in thirty-six minutes, as a tumbleweed can? What cultivar
can produce four hundred thousand seeds on a single flower stalk, as
the mullein does? Or hitch its seeds to any passing animal, like the burdock? Or travel a foot each day, as kudzu can? ("You keep still enough,
watch close enough," southerners will tell you, "and damn if you can't
see it move.") Or, like the bindweed, clone new editions of itself in
direct proportion to the effort we expend trying to eradicate it? Japanese
knotweed can penetrate four inches of asphalt, no problem. Each summer the roots of a Canada thistle venture another ten feet in every
direction. Lamb's-quarter seeds recovered from an archaeological site
germinated after spending seventeen hundred years in storage,
patiently awaiting their shot. The roots of the witchweed emit a poison
that kills every other plant in its vicinity.

No, it can't just be my lack of imagination that gives the nettle its sting.

So what is a weed? I consulted several field guides and botany books
hoping to find a workable definition. Instead of one, however, I found
dozens, though almost all of them could be divided into two main
camps. "A weed is any plant in the wrong place" fairly summarizes the
first camp, and the second maintains, essentially, that "a weed is an
especially aggressive plant that competes successfully against cultivated
plants." In the first, Emersonian definition, the weed is a human con-

struct; in the second, weeds possess certain inherent traits we did not impose. The metaphysical problem of weeds, I was beginning to think, is not unlike the metaphysical problem of evil: Is it an abiding property of the universe, or an invention of humanity?

Weeds, I'm convinced, are really out there. But I am prepared to concede the existence of a gray area inhabited by Emerson's weeds, plants upon which we have imposed weediness simply because we can find no utility or beauty in them. One man's flowers may indeed be another's weeds. Purple loosestrife, which I planted in my perennial border, has been outlawed as a "noxious weed" in several midwestern states, where it has escaped gardens and now threatens wetland flora. Likewise, certain of my weeds may have value in the eyes of another. Every day I pull easily enough dandelions and purslanes from my vegetable garden to make a tasty salad for Euell Gibbons. What I call weeds he would call lunch.

Not long ago, I had a local excavator over to estimate a job for me. He was one of those venerable old-timers who possess an intimate knowledge of local geography. This fellow knew more about my land than I did: how many gallons a minute my well could pump, and the source of its water; the alkalinity of my soil, due to the limestone ledge it sits on; the fact that the ancient apple trees on my property used to make the best hard cider in town. We walked around the property, looking for a suitable spot for a pond, and he nodded approvingly at the changes I'd made in the landscape: restoring a meadow that had gone to brush, pruning the old apple trees, regrading the hillside to divert spring runoff from the house. But there was one thing I'd done that seemed to bother him, and after a while he spoke up. I'd planted a pair of weeping willows at the edge of a small wetland. The size of phone booths when I put them in, two years later they were already as big as houses. The property has few shade trees—it *looks* hot in the summer—and I planted the willows to create a few cool spots in the landscape; willows seem almost to imply water, and they will amplify the quietest breeze. My visitor tipped his head in the direction of the trees and growled:

"What'n hell did you plant those weeds for?"

"Weeds?! What are you talking about?" The willows were poetry as far as I was concerned.

"Those damn trees. They're good for nothing, they're dirty, and if you don't watch it, their roots are going to crack your foundation one day. You'll see."

I've since discovered that a lot of country people consider willows weeds. Given a good supply of water, their growth is rampant, and their roots have been known to bust through concrete. But my trees were a good fifty yards from the house. The real objection, it seems, is that willows drop a lot of branches in the course of a growing season; in the eyes of a dedicated lawn man, they *are* dirty. They're also faulted for having soft wood, the result of their rapid growth. The world of trees has its own hierarchy, of course, and hardwood stands at the top. Soft, wet woods like willow have no commercial value, either as lumber or firewood.

So a perspective exists from which the weeping willow is a weed. It grows too fast, sullies lawns, damages homes, and burns about as well as celery. It also grows wild around here, another supposedly weedy trait. My excavator's dealings with willows were mainly a matter of economics; he knew you could not sell a cord of willow to save your life, and he'd heard enough stories about cracked concrete. My own dealings were largely aesthetic, and they were informed by another kind of story, one that put the willow in a completely different light.

The weeping willow, I had read, is not a native tree, but an eighteenth-century garden import. It is thought to have first been planted in America not too far from here, in the Stamford, Connecticut, garden of Samuel Johnson, a clergyman and philosopher who was the first president of Kings College (which was renamed Columbia University after independence). Johnson saw his first weeping willow at Twickenham, Alexander Pope's famous garden on the Thames. He was so taken with Pope's ancient tree that he returned to his home on the Housatonic with a cutting. Evidently the banks of that river proved as hospitable to the tree as those of the Thames, because the weeping willow soon escaped his garden and spread north. Today vast, pendulous willows—towering green fountains—line the Housatonic from Stamford to the Berkshires; every one of them, presumably, can trace its lineage back to Pope's great garden. Knowing this, I am more apt to think of Twickenham and the Thames when I see a weeping willow than I am of a cement-busting "weed."

These stories about weeping willows would seem to bolster Emerson's contention that weediness is in the eye of the beholder, that it is a matter of perception. Now ordinarily I am perfectly comfortable with this sort of relativistic thinking, but experience suggests that here it is shallow. And not only my experience: Emerson's own disciple, Henry David Thoreau, runs into some difficulty with his teacher's theory of weeds when he plants his bean field at Walden.

As an observer and naturalist, Thoreau consistently refuses to make "invidious distinctions" between different orders of nature; sworn enemy of hierarchy, the man boasts of the fact that he loves swamps more than gardens. But as soon as he determines to make "the earth say beans instead of grass" he finds he has made enemies in nature: worms, the morning dew, woodchucks, and weeds. The bean field "attached me to the earth," Thoreau felt, giving him positions he must defend if he hopes to prove his experiment in self-reliance a success. And so Thoreau is obliged to wage a long and decidedly uncharacteristic "war, not with cranes, but with weeds, those Trojans who had sun and rain and dews on their side. Daily the beans saw me come to their rescue armed with a hoe, and thin the ranks of their enemies, filling up the trenches with weedy dead." He finds himself "making such invidious distinctions with his hoe, levelling whole ranks of one species, and sedulously cultivating another."

Thoreau is gardening here, of course, and this forces him at least for a time to throw out his romanticism about nature—to drop what naturalists today hail as his precocious "biocentrism" (as opposed to anthropocentrism). But by the end of the chapter, his bean field having achieved its purpose, Thoreau trudges back—lamely, it seems to me—to the Emersonian fold: "The sun looks on our cultivated fields and on the prairies and forests without distinction. . . . Do [these beans] not grow for woodchucks too? . . . How, then, can our harvest fail? Shall I not rejoice also at the abundance of the weeds whose seeds are the granary of the birds?"

Sure, Henry, rejoice. And starve.

My own experience in the garden has convinced me "absolute weediness" does exist—that weeds represent a different order of being, and the fact that Thoreau's beans were no match for his weeds does not mean the weeds have a higher claim to the earth, as Thoreau seems to think. I found support for this hunch in the field guides and botany books I consulted when I was trying to identify my weeds. As I searched these volumes for the *noms de bloom* of my marauders, I jotted down each species' preferred habitat. Here are a few of the most typical: "waste places and roadside"; "open sites"; "old fields, waste places"; "cultivated and waste ground"; "old fields, roadsides, lawns, gardens"; "lawns, gardens, disturbed sites."

What this list suggests is that weeds are not superplants: they don't grow everywhere, which explains why, for all their vigor, they haven't covered the globe entirely. Weeds, as the field guides indicate, are plants particularly well adapted to man-made places. They don't grow in forests or prairies—in "the wild." Weeds thrive in gardens, meadows, lawns, vacant lots, railroad sidings, hard by dumpsters and in the cracks of sidewalks. They grow where we live, in other words, and hardly anywhere else.

Weeds, contrary to what the romantics assumed, are not wild. They are as much a product of cultivation as the hybrid tea rose, or Thoreau's bean plants. They do better than garden plants for the simple reason that they are better adapted to life in a garden. For where garden plants have been bred for a variety of traits (tastiness, nutritiousness, size, aesthetic appeal), weeds have evolved with just one end in view: the ability to thrive in ground that man has disturbed. At this they are very accomplished indeed.

Weeds stand at the forefront of evolution; no doubt they are evolving in my garden at this very moment, their billions of offspring self-selecting for new tactics to outwit my efforts and capitalize on any opening in my garden. Weeds are nature's ambulance chasers, carpetbaggers, and confidence men. Virtually every crop in general cultivation has its weed impostor, a kind of botanical doppelganger that has evolved to mimic the appearance as well as the growth rate of the cultivated crop and so ensure its survival. Some of these impostors, such as wild oats, are so versatile that they can alter their appearance depending on the crop they are imitating, like an insidious agricultural Fifth Column. According to Sara B. Stein's botany, *My Weeds*, wild oats growing in a field of alternating rows of spring and winter barley will mimic the habits of either crop, *depending on the row*. Stein, whose book is a trove of information about weeds, also tells of a rice mimic that became so troublesome that researchers planted a purple variety of rice to expose the weeds once and for all. Within a few years, the weed-rice had turned purple too.

And yet as resourceful and aggressive as weeds may be, they cannot survive without us any more than a garden plant can. Without man to create crop land and lawns and vacant lots, most weeds would soon vanish. Bindweed, which seems so formidable in the field and garden, can grow nowhere else. It lives by the plow as much as we do.

To learn all this was somehow liberating. My weeds were no more natural than my garden plants, had no greater claim to the space they were vying for. Those smug quotes in which naturalists like to coddle weeds were merely a conceit. My battles with weeds did not bespeak alienation from nature, or some irresponsible drive to dominate it. Had Thoreau known this, perhaps he would not have troubled himself so about "what right had I to oust St. Johnswort, and the rest, and to break up their ancient herb garden?"

Thoreau considered his wormwood, pigweed, sorrel, and St.-John's-wort part of nature, his beans part of civilization. He looked to the American landscape, as many of us do, for a path that would lead him out of history and into nature, and this led him to value what grew "naturally" over that which man planted. But as it turns out history is inescapable, even at Walden. Much of the flora in the Walden land-scape is as historical as his beans, his books, even the Mexican battle-field he makes his bean field a foil for. Had Thoreau brought a field guide with him to Walden, he might have noted that most of the weeds that came up in his garden were alien species, brought to America by the colonists. St.-John's-wort, far from being an ancient Walden resi-dent, was brought to America in 1696 by a band of fanatical Rosicrucians who claimed the herb had the power to exorcise evil spir-its. You want to privilege *this* over beans?

It's hard to imagine the American landscape without St.-John's-wort, daisies, dandelions, crabgrass, timothy, clover, pigweed, lamb's-quar-ters, buttercup, mullein, Queen Anne's lace, plantain, or yarrow, but not one of these species grew here before the Puritans landed. America in fact had few indigenous weeds, for the simple reason that it had little disturbed ground. The Indians lived so lightly on the land that they cre-ated few habitats for weeds to take hold in. No plow, no bindweed. But by as early as 1663, when John Josselyn compiled a list "of such plantes as have sprung up since the English planted and kept cattle in New England," he found, among others, couch grass, dandelion, sow thistle, shepherd's purse, groundsel, dock, mullein, plantain, and chickweed.

Some of these weeds were brought over deliberately: the colonists prized dandelion as a salad green, and used plantain (which is millet) to make bread. Other weed seeds, though, came by accident—in for-age, in the earth used for shipboard ballast, even in pants cuffs and cracked boot soles. Once here, the weeds spread like wildfire. According to Alfred W. Crosby, the ecological historian, the Indians considered the Englishman a botanical Midas, able to change the flora with his touch; they called plantain "Englishman's foot" because it seemed to spring up wherever the white man stepped. (Hiawatha claimed that the spread of the plant presaged the doom of the wilder-ness.) Though most weeds traveled with white men, some, like the dandelion, raced west of their own accord (or possibly with the help of the Indians, who quickly discovered the plant's virtues), arriving well ahead of the pioneers. Thus the supposedly virgin landscape upon which the westward settlers gazed had already been marked by their civilization. However, those same pioneers did *not* gaze out on tumble-weed, that familiar emblem of the untamed western landscape. Tumbleweed did not arrive in America until the 1870s, when a group of Russian immigrants settled in Bonhomme County, South Dakota, intending to grow flax. Mixed in with their flax seeds were a few seeds of a weed well known on the steppes of the Ukraine: tumbleweed.

European weeds thrived here, in a matter of years changing the face of the American landscape, helping to create what we now take to be our country's abiding "nature." Why should these species have prospered so? Probably because the Europeans who brought them got busy making the land safe for weeds by razing the forests, plowing fields, burning prairies, and keeping grazing animals. And just as the Europeans helped smooth the way for their weeds, weeds helped smooth the way for Europeans. This is particularly true in the case of the grasses. Native grasses proved poor forage for European livestock, which at first did not fare well in America. Yet colonists noted that after a few years the grasses—and in turn the health of the livestock—seemed to improve. What had happened, according to Crosby, is that Old World livestock had overgrazed the native grasses. Because these species were unaccustomed to such heavy grazing, they had trouble regenerating themselves. This left them vulnerable to the onslaught of European weed grasses which, having co-evolved with the goat and sheep and cow, are better equipped to withstand the grazing pressure of these animals. The European grasses soon conquered American meadows, thereby providing European livestock with their preferred forage once again. Today most of the native grasses have vanished.

Working in concert, European weeds and European humans proved formidable ecological imperialists, rapidly driving out native species and altering the land to suit themselves. The new plant species thrived because they were consummate cosmopolitans, opportunists superbly adapted to travel and change. In a sense, the invading species had less in common with the retiring, provincial plants they ousted than with the Europeans themselves. Or perhaps that should be put the other way around. "If we confine the concept of weeds to species adapted to human disturbance," writes Jack R. Harland in *Crops and Man*, "then man is by definition the first and primary weed under whose influence all other weeds have evolved."

Weeds are not the Other. Weeds are us.

TERRY TEMPEST WILLIAMS
b. 1955

Terry Tempest Williams has declared that she writes "through my biases of gender, geography, and culture, that I am a woman whose ideas have been shaped by the Colorado Plateau and the Great Basin, that these ideas are then sorted out through the prism of my culture—and my culture is Mormon. Those tenets of family and community that I see at the heart of that culture are then articulated through story." Her first book, Pieces of White Shell: A Journey to Navajoland (1984), *based on her experiences as a teacher among the Navajo, is a personal retelling and examination of Native American myths. The strong link between story and landscape is further explored in* Coyote's Canyon (1989), *personal narratives of southern Utah's desert canyon, in collaboration with photographer John Telford.* Refuge (1991), *the book that established Williams as one of our most influential nature writers, is an intense exploration of the links, metaphorical and environmental, between her mother's death from cancer and threats to the ecology of the Great Salt Lake. It was followed by* An Unspoken Hunger (1994), Desert Quartet (1995), *and* Leap (2000).

THE CLAN OF ONE-BREASTED WOMEN

I belong to a Clan of One-Breasted Women. My mother, my grandmothers, and six aunts have all had mastectomies. Seven are dead. The two who survive have just completed rounds of chemotherapy and radiation.

I've had my own problems: two biopsies for breast cancer and a small tumor between my ribs diagnosed as a "borderline malignancy."

This is my family history.

Most statistics tell us breast cancer is genetic, hereditary, with rising

Refuge: An Unnatural History of Family and Place (New York: Random House, 1991).

percentages attached to fatty diets, childlessness, or becoming pregnant after thirty. What they don't say is living in Utah may be the greatest hazard of all.

We are a Mormon family with roots in Utah since 1847. The "word of wisdom" in my family aligned us with good foods—no coffee, no tea, tobacco, or alcohol. For the most part, our women were finished having their babies by the time they were thirty. And only one faced breast cancer prior to 1960. Traditionally, as a group of people, Mormons have a low rate of cancer.

Is our family a cultural anomaly? The truth is, we didn't think about it. Those who did, usually the men, simply said, "bad genes." The women's attitude was stoic. Cancer was part of life. On February 16, 1971, the eve of my mother's surgery, I accidently picked up the telephone and overheard her ask my grandmother what she could expect.

"Diane, it is one of the most spiritual experiences you will ever encounter."

I quietly put down the receiver.

Two days later, my father took my brothers and me to the hospital to visit her. She met us in the lobby in a wheelchair. No bandages were visible. I'll never forget her radiance, the way she held herself in a purple velvet robe, and how she gathered us around her.

"Children, I am fine. I want you to know I felt the arms of God around me."

We believed her. My father cried. Our mother, his wife, was thirty-eight years old.

A little over a year after Mother's death, Dad and I were having dinner together. He had just returned from St. George, where the Tempest Company was completing the gas lines that would service southern Utah. He spoke of his love for the country, the sandstoned landscape, bare-boned and beautiful. He had just finished hiking the Kolob trail in Zion National Park. We got caught up in reminiscing, recalling with fondness our walk up Angel's Landing on his fiftieth birthday and the years our family had vacationed there.

Over dessert, I shared a recurring dream of mine. I told my father that for years, as long as I could remember, I saw this flash of light in the night in the desert—that this image had so permeated my being that I could not venture south without seeing it again, on the horizon, illuminating buttes and mesas.

"You did see it," he said.

"Saw what?"

"The bomb. The cloud. We were driving home from Riverside, California. You were sitting on Diane's lap. She was pregnant. In fact,

I remember the day, September 7, 1957. We had just gotten out of the Service. We were driving north, past Las Vegas. It was an hour or so before dawn, when this explosion went off. We not only heard it, but felt it. I thought the oil tanker in front of us had blown up. We pulled over and suddenly, rising from the desert floor, we saw it, clearly, this golden-stemmed cloud, the mushroom. The sky seemed to vibrate with an eerie pink glow. Within a few minutes, a light ash was raining on the car."

I stared at my father.

"I thought you knew that," he said. "It was a common occurrence in the fifties."

It was at this moment that I realized the deceit I had been living under. Children growing up in the American Southwest, drinking contaminated milk from contaminated cows, even from the contaminated breasts of their mothers, my mother—members, years later, of the Clan of One-Breasted Women.

It is a well-known story in the Desert West, "The Day We Bombed Utah," or more accurately, the years we bombed Utah: above ground atomic testing in Nevada took place from January 27, 1951 through July 11, 1962. Not only were the winds blowing north covering "low-use segments of the population" with fallout and leaving sheep dead in their tracks, but the climate was right. The United States of the 1950s was red, white, and blue. The Korean War was raging. McCarthyism was rampant. Ike was it, and the cold war was hot. If you were against nuclear testing, you were for a communist regime.

Much has been written about this "American nuclear tragedy." Public health was secondary to national security. The Atomic Energy Commissioner, Thomas Murray, said, "Gentlemen, we must not let anything interfere with this series of tests, nothing."

Again and again, the American public was told by its government, in spite of burns, blisters, and nausea, "It has been found that the tests may be conducted with adequate assurance of safety under conditions prevailing at the bombing reservations." Assuaging public fears was simply a matter of public relations. "Your best action," an Atomic Energy Commission booklet read, "is not to be worried about fallout." A news release typical of the times stated, "We find no basis for concluding that harm to any individual has resulted from radioactive fallout."

On August 30, 1979, during Jimmy Carter's presidency, a suit was filed, *Irene Allen v. The United States of America.* Mrs. Allen's case was the first on an alphabetical list of twenty-four test cases, representative of nearly twelve hundred plaintiffs seeking compensation from the United States government for cancers caused by nuclear testing in Nevada.

Irene Allen lived in Hurricane, Utah. She was the mother of five children and had been widowed twice. Her first husband, with their two oldest boys, had watched the tests from the roof of the local high school. He died of leukemia in 1956. Her second husband died of pancreatic cancer in 1978.

In a town meeting conducted by Utah Senator Orrin Hatch, shortly before the suit was filed, Mrs. Allen said, "I am not blaming the government, I want you to know that, Senator Hatch. But I thought if my testimony could help in any way so this wouldn't happen again to any of the generations coming up after us . . . I am happy to be here this day to bear testimony of this."

God-fearing people. This is just one story in an anthology of thousands.

On May 10, 1984, Judge Bruce S. Jenkins handed down his opinion. Ten of the plaintiffs were awarded damages. It was the first time a federal court had determined that nuclear tests had been the cause of cancers. For the remaining fourteen test cases, the proof of causation was not sufficient. In spite of the split decision, it was considered a landmark ruling. It was not to remain so for long.

In April 1987, the Tenth Circuit Court of Appeals overturned Judge Jenkins's ruling on the ground that the United States was protected from suit by the legal doctrine of sovereign immunity, a centuries-old idea from England in the days of absolute monarchs.

In January 1988, the Supreme Court refused to review the Appeals Court decision. To our court system it does not matter whether the United States government was irresponsible, whether it lied to its citizens, or even that citizens died from the fallout of nuclear testing. What matters is that our government is immune: "The King can do no wrong."

In Mormon culture, authority is respected, obedience is revered, and independent thinking is not. I was taught as a young girl not to "make waves" or "rock the boat."

"Just let it go," Mother would say. "You know how you feel, that's what counts."

For many years, I have done just that—listened, observed, and quietly formed my own opinions, in a culture this rarely asks questions because it has all the answers. But one by one, I have watched the women in my family die common, heroic deaths. We sat in waiting rooms hoping for good news, but always receiving the bad. I cared for them, bathed their scarred bodies, and kept their secrets. I watched beautiful women become bald as Cytoxan, cisplatin, and Adriamycin were injected into their veins. I held their foreheads as they vomited green-black bile, and I shot them with morphine when the pain became inhuman. In the end, I witnessed their last peaceful breaths, becoming a midwife to the rebirth of their souls.

The price of obedience has become too high.

The fear and inability to question authority that ultimately killed rural communities in Utah during atmospheric testing of atomic weapons is the same fear I saw in my mother's body. Sheep. Dead sheep. The evidence is buried.

I cannot prove that my mother, Diane Dixon Tempest, or my grandmothers, Lettie Romney Dixon and Kathryn Blackett Tempest, along with my aunts developed cancer from nuclear fallout in Utah. But I can't prove they didn't.

My father's memory was correct. The September blast we drove through in 1957 was part of Operation Plumbbob, one of the most intensive series of bomb tests to be initiated. The flash of light in the night in the desert, which I had always thought was a dream, developed into a family nightmare. It took fourteen years, from 1957 to 1971, for cancer to manifest in my mother—the same time, Howard L. Andrews, an authority in radioactive fallout at the National Institutes of Health, says radiation cancer requires to become evident. The more I learn about what it means to be a "downwinder," the more questions I drown in.

What I do know, however, is that as a Mormon woman of the fifth generation of Latter-day Saints, I must question everything, even if it means losing my faith, even if it means becoming a member of a border tribe among my own people. Tolerating blind obedience in the name of patriotism or religion ultimately takes our lives.

When the Atomic Energy Commission described the country north of the Nevada Test Site as "virtually uninhabited desert terrain," my family and the birds at Great Sale Lake were some of the "virtual uninhabitants."

One night, I dreamed women from all over the world circled a blazing fire in the desert. They spoke of change, how they hold the moon in their bellies and wax and wane with its phases. They mocked the presumption of even-tempered beings and made promises that they would never fear the witch inside themselves. The women danced wildly as sparks broke away from the flames and entered the night sky as stars.

And they sang a song given to them by Shoshone grandmothers:

Ah ne nah, nah	Consider the rabbits
nin nah nah—	How gently they walk on the earth—
ah ne nah, nah	Consider the rabbits

nin nah nah—	How gently they walk on the earth—
Nyaga mutzi	We remember them
oh ne nay—	We can walk gently also—
Nyaga mutzi	We remember them
oh ne nay—	We can walk gently also—

The women danced and drummed and sang for weeks, preparing themselves for what was to come. They would reclaim the desert for the sake of their children, for the sake of the land.

A few miles downwind from the fire circle, bombs were being tested. Rabbits felt the tremors. Their soft leather pads on paws and feet recognized the shaking sands, while the roots of mesquite and sage were smoldering. Rocks were hot from the inside out and dust devils hummed unnaturally. And each time there was another nuclear test, ravens watched the desert heave. Stretch marks appeared. The land was losing its muscle.

The women couldn't bear it any longer. They were mothers. They had suffered labor pains but always under the promise of birth. The red hot pains beneath the desert promised death only, as each bomb became a stillborn. A contract had been made and broken between human beings and the land. A new contract was being drawn by the women, who understood the fate of the earth at their own.

Under the cover of darkness, ten women slipped under a barbed-wire fence and entered the contaminated country. They were trespassing. They walked toward the town of Mercury, in moonlight, taking their cues from coyote, kit fox, antelope squirrel, and quail. They moved quietly and deliberately through the maze of Joshua trees. When a hint of daylight appeared they rested, drinking tea and sharing their rations of food. The women closed their eyes. The time had come to protest with the heart, that to deny one's genealogy with the earth was to commit treason against one's soul.

At dawn, the women draped themselves in mylar, wrapping long streamers of silver plastic around their arms to blow in the breeze. They wore clear masks, that became the faces of humanity. And when they arrived at the edge of Mercury, they carried all the butterflies of a summer day in their wombs. They paused to allow their courage to settle.

The town that forbids pregnant women and children to enter because of radiation risks was asleep. The women moved through the streets as winged messengers, twirling around each other in slow motion, peeking inside homes and watching the easy sleep of men and women. They were astonished by each stillness and periodically would utter a shrill note or low cry just to verify life.

The residents finally awoke to these strange apparitions. Some simply stared. Others called authorities, and in time the women were apprehended by wary soldiers dressed in desert fatigues. They were taken to a white, square building on the other edge of Mercury. When asked who they were and why they were there, the women replied, "We are mothers and we have come to reclaim the desert for our children."

The soldiers arrested them. As the ten women were blindfolded and handcuffed, they began singing:

> You can't forbid us everything
> You can't forbid us to think—
> You can't forbid our tears to flow
> And you can't stop the songs that we sing.

The women continued to sing louder and louder, until they heard the voices of their sisters moving across the mess:

> *Ah ne nah, nah*
> *nin nah nah—*
> *Ah ne nah, nah*
> *nin nah nah—*
> *Nyaga mutzi*
> *oh ne nay—*
> *Nyaga mutzi*
> *oh ne nay—*

"Call for reinforcements," one soldier said.

"We have," interrupted one woman, "we have—and you have no idea of our numbers."

I crossed the line at the Nevada Test Site and was arrested with nine other Utahns for trespassing on military lands. They are still conducting nuclear tests in the desert. Ours was an act of civil disobedience. But as I walked toward the town of Mercury, it was more than a gesture of peace. It was a gesture on behalf of the Clan of One-Breasted Women.

As one officer cinched the handcuffs around my wrists, another frisked my body. She found a pen and a pad of paper tucked inside my left boot.

"And these?" she asked sternly.

"Weapons," I replied.

Our eyes met. I smiled. She pulled the leg of my trousers back over my boot.

"Step forward, please," she said as she took my arm.

We were booked under an afternoon sun and bused to Tonopah, Nevada. It was a two-hour ride. This was familiar country. The Joshua trees standing their ground had been named by my ancestors, who believed they looked like prophets pointing west to the Promised Land. These were the same trees that bloomed each spring, flowers appearing like white flames in the Mojave. And I recalled a full moon in May, when Mother and I had walked among them, flushing out mourning doves and owls.

The bus stopped short of town. We were released.

The officials thought it was a cruel joke to leave us stranded in the desert with no way to get home. What they didn't realize was that we were home, soul-centered and strong, women who recognized the sweet smell of sage as fuel for our spirits.

JANE BROX
b. 1956

In Here and Nowhere Else: Late Seasons of a Farm and Its Family (1995), and Five Thousand Days Like This One (1999), Jane Brox has made her own distinctive contribution to America's literature of farming. From Crèvecoeur to Wendell Berry, such writing has looked at the land through the lens of agriculture and has discovered in a life of cultivating the soil occasions for intimacy, reflectiveness, and political awareness fully equivalent to the insights of Thoreauvian naturalists. Like her predecessors, Brox sometimes sounds an elegiac note—contemplating the disappearance of family farms across her region and the real possibility that her own family's farm in Dracut, Massachusetts, will not long endure. But the concrete details of planting, harvesting, and selling the crops and of maintaining the buildings and machinery also ground her writing and prevent sentimentality. Among her voice's distinctive qualities are trustworthiness and comprehensiveness. She takes pains to locate her family's farm within a larger historical context: both the story of immigrants like her grandfather, who established the farm in 1901 after his arrival in the United States from Lebanon, and the trends of population, economics, and transportation that threaten to uproot it today.

BALDWINS

Apple varieties, like anything else, have their years, and Baldwins—along with Greenings, Russets, Winesaps, Sheep Nose, Ben Davis, Astrakhans—drift now on the edge of sleep. The trees survive as solitaries along what was once a fence or at the far end of old orchards that stretch across gravelly New England hillsides. A heart-lifting surprise to see a hill like that in winter: craggy, evenly spaced trees with staunch, gray-barked trunks and bare branches crazing the air, the contained red life in them glinting in a cold sun. Commercial orchards won't look that way again. There's no economical way to harvest trees that are so big, and most of what hasn't been lost to housing has given way to dwarf or semi-dwarf trees that bear more quickly, are easier to pick, and can be densely planted to give a higher yield per acre.

And so many kinds of apples, fallen to market pressures, have been replaced by more uniform, evenly colored varieties. In 1920, the prominent Massachusetts apples were McIntosh, Baldwin, Wealthy, Red Delicious, and Gravenstein. Of those, only Macs and Red Delicious are commonly known anymore, persisting in larger markets among newcomers such as Empire, Mutsu—developed in Japan—and disease-resistant Liberty. Of course, it is better that some of the old varieties have nearly disappeared. My grandfather had to practically give his Ben Davis away—cottony, tasteless—their virtue, if it is a virtue, being they clung to the branches all winter.

But spicy, juicy Baldwins are another story. By late September the apples have deepened in color to a brownish red with a rusty splay at the stem end. Ripening as the fall itself slopes towards its close, Baldwins—a pie apple, a keeper—taste better after the frost. They're picked in October when the orchard grasses have already turned and morning frost lingers in the shadows. As the winter progresses, the skin of a Baldwin wrinkles in storage, but its flavor and crispness hold, and its wine-dark smell fills closed-in cellars and refrigerators.

Here, the remaining Baldwin tree is framed in my bay window. The late light backs it in all seasons, and I watch its changes as I work, and read, and eat my breakfast and lunch. The man who planted this tree also built my small, white farmhouse—he repaired with scrap, insulated with newspaper, saved string, lived a more frugal life than I could ever imagine. Who knows why, but it's this tree that reminds me of his effort and economy and the rough stone over his grave.

Five Thousand Days Like This One: An American Family History (Boston: Beacon, 1999).

Baldwins bear every other year, and the fall feels different to me in the years the tree is laden. This past September, when I looked up one mild afternoon and noticed how the branches were bowed by the weight of their fruit; I felt my first sting of the coming cold season. It's hard to predict what will fist up your heart—maybe the smell of decay in the garden, or the clearer light, or the leaves whirling in front of the car as you hurry home at night. After that first sting the bright peak weeks follow, then the yellows become brown, the rusts deepen, and the laden branches are what you see every day until the harvest is over and the oaks alone have it. Afterwards, bare branches against a big sky, and the light and forms of the world are too hard to dream into, so you get used to the spare, smaller life, and what once chilled your heart no longer has its old power. What is withheld becomes what is beautiful.

And what of things revived? Of all that is hauled forward so self-consciously? The catalogs that come in my mail are full of old utilitarian things elevated, and as I flip through the pages I'm tugged by the orbit of seamed, thick glass milk bottles, wire egg baskets, and galvanized buckets set apart from their workaday purposes and their strength. Dowdy, plain, taken for granted in their former life, in the warmer rooms of our time such square and solid forms have gained a grace and even seem seductive. And nothing is more seductive than the Baldwins offered for sale in every Williams-Sonoma catalog I've received this fall. Along with Spitzenbergs, Arkansas Blacks, and Winesaps they top off a weathered half-peck basket. Six pounds—about twelve apples—are thirty-two dollars, plus shipping. Available in the catalog only. Each apple, polished to a still life, is more perfect than any Baldwin I've ever seen—larger and redder, no russeting from mites, or sooty mold, or frass at the calyx. Called heirloom or antique apples now, they seem to say, *See how beautiful the old life was?*
And it does take time before I hazard to ask how much of that old life I would want with its aches, its silence and remoteness. The ones who lived it can't afford the price of these goods, nor would they ever pay it. More farmhouses than farms remain. The interiors of the ones my oldest neighbors live in look nearly the way they did eighty years back. A salt box and tea canister by the stove in the kitchen. A plaid wool coat hangs on the coat rack in the hall. The last pansies of the year are set in a milkglass vase on the table. "The younger ones just don't understand," Mrs. Burton says. "They want me to *adjust* since Carl's died. But the world is so different now. I wouldn't know where to begin." Her husband had had a long illness and had left off work

months before his death, though they waited until after the services to sell his herd. She pares away the toughened skin on one of her windfall Baldwins, going a little deeper where there's a bruise from its drop, and nicking out the rough spots in the flesh with the tip of her knife—apple maggot, curculio, codling moth, all the troubles apples are heir to. It's true, the past is a different country.

In this one a late fall storm gathers its strength over the Atlantic and sweeps in a warm rain. It lashes the bay window and mats the mottled leaves on the ground. No moon, no stars. The bark of the Baldwin tree is silver in the wet night, and its resistant branches toss against each other in the gusts.

DAVID ABRAM
b. 1957

David Abram is an ecologist, philosopher, linguist, and professional sleight-of-hand magician—diverse interests and careers that combined to produce his groundbreaking book The Spell of the Sensuous: Perception and Language in a More-than-Human World *(1996). Traveling through Southeast Asia to study the relationship between folk medicine and magic, Abram became increasingly aware of his own culture's alienation from the natural world, its distrust of sensory experience, and the destructive effects of its reliance on abstract thought. In the book he discusses the role of shamans and other traditional healers as intermediaries between human societies and the "more-than-human" world of ambient life. He explores in depth his contentions that the development of the written word was a primary factor in replacing immediate, sensory knowledge with linguistic thought and that many of our abstract concepts and spiritual ideas derive from direct physical experience with the natural world. He does not suggest that we abandon literacy but rather that we learn how "to write our senses back into the land."*

THE ECOLOGY OF MAGIC

Late one evening I stepped out of my little hut in the rice paddies of eastern Bali and found myself falling through space. Over my head the black sky was rippling with stars, densely clustered in some regions, almost blocking out the darkness between them, and more loosely scattered in other areas, pulsing and beckoning to each other. Behind them all streamed the great river of light with its several tributaries. Yet the Milky Way churned beneath me as well, for my hut was set in the middle of a large patchwork of rice paddies, separated from each other by narrow two-foot-high dikes, and these paddies were all filled with water. The surface of these pools, by day, reflected perfectly the blue sky, a reflection broken only by the thin, bright green tips of new rice. But by night the stars themselves glimmered from the surface of the paddies, and the river of light whirled through the darkness underfoot as well as above; there seemed no ground in front of my feet, only the abyss of star-studded space falling away forever.

I was no longer simply beneath the night sky, but also *above* it—the immediate impression was of weightlessness. I might have been able to reorient myself, to regain some sense of ground and gravity, were it not for a fact that confounded my senses entirely: between the constellations below and the constellations above drifted countless fireflies, their lights flickering like the stars, some drifting up to join the clusters of stars overhead, others, like graceful meteors, slipping down from above to join the constellations underfoot, and all these paths of light upward and downward were mirrored, as well, in the still surface of the paddies. I felt myself at times falling through space, at other moments floating and drifting. I simply could not dispel the profound vertigo and giddiness; the paths of the fireflies, and their reflections in the water's surface, held me in a sustained trance. Even after I crawled back to my hut and shut the door on this whirling world, I felt that now the little room in which I lay was itself floating free of the earth.

Fireflies! It was in Indonesia, you see, that I was first introduced to the world of insects, and there that I first learned of the great influence that insects—such diminutive entities—could have upon the human senses. I had traveled to Indonesia on a research grant to study magic—more precisely, to study the relation between magic and medicine, first among the traditional sorcerers, or *dukuns*, of the Indonesian

Adapted from Chapter 1 of *The Spell of the Sensuous: Perception and Language in a More-Than-Human World* (New York: Pantheon, 1996).

archipelago, and later among the *dzankris*, the traditional shamans of Nepal. One aspect of the grant was somewhat unique: I was to journey into rural Asia not outwardly as an anthropologist or academic researcher, but as a magician in my own right, in hopes of gaining a more direct access to the local sorcerers. I had been a professional sleight-of-hand magician for five years back in the United States, helping to put myself through college by performing in clubs and restaurants throughout New England. I had, as well, taken a year off from my studies in the psychology of perception to travel as a street magician through Europe and, toward the end of that journey, had spent some months in London, England, exploring the use of sleight-of-hand magic in psychotherapy, as a means of engendering communication with distressed individuals largely unapproachable by clinical healers. The success of this work suggested to me that sleight-of-hand might lend itself well to the curative arts, and I became, for the first time, interested in the relation, largely forgotten in the West, between folk medicine and magic.

It was this interest that led to the aforementioned grant, and to my sojourn as a magician in rural Asia. There, my sleight-of-hand skills proved invaluable as a means of stirring the curiosity of the local shamans. For magicians—whether modern entertainers or indigenous, tribal sorcerers—have in common the fact that they work with the malleable texture of perception. When the local sorcerers gleaned that I had at least some rudimentary skill in altering the common field of perception, I was invited into their homes, asked to share secrets with them, and eventually encouraged, even urged, to participate in various rituals and ceremonies.

*   *   *

The most sophisticated definition of magic that now circulates through the American counterculture is "the ability or power to alter one's consciousness at will." No mention is made of any *reason* for altering one's consciousness. Yet in tribal cultures that which we call "magic" takes its meaning from the fact that humans, in an indigenous and oral context, experience their own consciousness as simply one form of awareness among many others. The traditional magician cultivates an ability to shift out of his or her common state of consciousness precisely in order to make contact with the other organic forms of sensitivity and awareness with which human existence is entwined. Only by temporarily shedding the accepted perceptual logic of his culture can the sorcerer hope to enter into relation with other species on their own terms; only by altering the common organization of his senses will he be able to enter into a rapport with the multiple nonhuman sensibilities that animate the local landscape. It is this, we might say, that defines a shaman: the ability to readily slip out of the perceptual boundaries that demarcate his or her particular culture—boundaries reinforced by social customs, taboos, and most importantly, the common speech or language—in order to make contact with, and learn from, the other powers in the land. His magic is precisely this heightened receptivity to the meaningful solicitations—songs, cries, gestures—of the larger, more-than-human field.

Magic, then, in its perhaps most primordial sense, is the experience of existing in a world made up of multiple intelligences, the intuition that every form one perceives—from the swallow swooping overhead to the fly on a blade of grass, and indeed the blade of grass itself—is an *experiencing* form, an entity with its own predilections and sensations, albeit sensations that are very different from our own.

*   *   *

It was only gradually that I became aware of this more subtle dimension of the native magician's craft. The first shift in my preconceptions came rather quietly, when I was staying for some days in the home of a young "balian," or magic practitioner, in the interior of Bali. I had been provided with a simple bed in a separate, one-room building in the balian's family compound (most compound homes, in Bali, are comprised of several separate small buildings, for sleeping and for cooking, set on a single enclosed plot of land), and early each morning the balian's wife came to bring me a small but delicious bowl of fruit, which I ate by myself, sitting on the ground outside, leaning against the wall of my hut and watching the sun slowly climb through the rustling palm leaves. I noticed, when she delivered the fruit, that my hostess was also balancing a tray containing many little green plates: actually, they were little boat-shaped platters, each woven simply and neatly from a freshly cut section of palm frond. The platters were two or three inches long, and within each was a little mound of white rice. After handing me my breakfast, the woman and the tray disappeared from view behind the other buildings, and when she came by some minutes later to pick up my empty bowl, the tray in her hands was empty as well.

The second time that I saw the array of tiny rice platters, I asked my hostess what they were for. Patiently, she explained to me that they were offerings for the household spirits. When I inquired about the Balinese term that she used for "spirit," she repeated the same explanation, now in Indonesian, that these were gifts for the spirits of the family compound, and I saw that I had understood her correctly. She handed me a bowl of sliced papaya and mango, and disappeared around the corner. I pondered for a minute, then set down the bowl, stepped to the side of my hut, and peered through the trees. At first unable to see her, I soon caught sight of her crouched low beside the corner of one of the other buildings, carefully setting what I presumed was one of the offerings on the ground at that spot. Then she stood up with the tray, walked to the other visible corner of the same building, and there slowly and carefully set another offering on the ground. I returned to my bowl of fruit and finished my breakfast. That afternoon, when the rest of the household was busy, I walked back behind the building where I had seen her set down the two offerings. There were the little green platters, resting neatly at the two rear corners of the building. But the mounds of rice that had been within them were gone.

The next morning I finished the sliced fruit, waited for my hostess to come by for the empty bowl, then quietly headed back behind the buildings. Two fresh palm-leaf offerings sat at the same spots where the others had been the day before. These were filled with rice. Yet as I gazed at one of these offerings, I abruptly realized, with a start, that one of the rice kernels was actually moving.

Only when I knelt down to look more closely did I notice a line of tiny black ants winding through the dirt to the offering. Peering still closer, I saw that two ants had already climbed onto the offering and were struggling with the uppermost kernel of rice; as I watched, one of them dragged the kernel down and off the leaf, then set off with it back along the line of ants advancing on the offering. The second ant took another kernel and climbed down with it, dragging and pushing, and fell over the edge of the leaf, then a third climbed onto the offering. The line of ants seemed to emerge from a thick clump of grass around a nearby palm tree. I walked over to the other offering and discovered another line of ants dragging away the white kernels. This line emerged from the top of a little mound of dirt, about fifteen feet away from the buildings. There was an offering on the ground by a corner of my building as well, and a nearly identical line of ants. I walked into my room chuckling to myself: the balian and his wife had gone to so much trouble to placate the household spirits with gifts, only to have their offerings stolen by little six-legged thieves. What a waste! But then a strange thought dawned on me: what if the ants were the very "household spirits" to whom the offerings were being made?

I soon began to discern the logic of this. The family compound, like most on this tropical island, had been constructed in the vicinity of several ant colonies. Since a great deal of cooking took place in the compound (which housed, along with the balian and his wife and children, various members of their extended family), and also much preparation of elaborate offerings of foodstuffs for various rituals and festivals in the surrounding villages, the grounds and the buildings at the compound were vulnerable to infestations by the sizable ant population. Such

invasions could range from rare nuisances to a periodic or even constant siege. It became apparent that the daily palm-frond offerings
served to preclude such an attack by the natural forces that surrounded
(and underlay) the family's land. The daily gifts of rice kept the ant
colonies occupied—and, presumably, satisfied. Placed in regular,
repeated locations at the corners of various structures around the compound, the offerings seemed to establish certain boundaries between
the human and ant communities; by honoring this boundary with gifts,
the humans apparently hoped to persuade the insects to respect the
boundary and not enter the buildings.

Yet I remained puzzled by my hostess's assertion that these were gifts
"for the spirits." To be sure, there has always been some confusion
between our Western notion of "spirit" (which so often is defined in
contrast to matter or "flesh"), and the mysterious presences to which
tribal and indigenous cultures pay so much respect. Gross misunder-
standings arose from the circumstance that many of the earliest
Western students of these other customs were Christian missionaries all
too ready to see occult ghosts and immaterial phantoms where the
tribespeople were simply offering their respect to the local winds.
While the notion of "spirit" has come to have, for us in the West, a primarily anthropomorphic or human association, my encounter with the
ants was the first of many experiences suggesting to me that the "spirits"
of an indigenous culture are primarily those modes of intelligence or
awareness that do *not* possess a human form.

As humans, we are well acquainted with the needs and capacities of
the human body—we *live* our own bodies and so know, from within,
the possibilities of our form. We cannot know, with the same familiarity
and intimacy, the lived experience of a grass snake or a snapping turtle;
we cannot readily experience the precise sensations of a hummingbird
sipping nectar from a flower or a rubber tree soaking up sunlight. And
yet we do know how it feels to sip from a fresh pool of water or to bask
and stretch in the sun. Our experience may indeed be a variant of these
other modes of sensitivity; nevertheless, we cannot, as humans, precisely experience the living sensations of another form. We do not
know, with full clarity, their desires or motivations; we cannot know, or
can never be sure that we know, what they know. That the deer does
experience sensations, that it carries knowledge of how to orient in the
land, of where to find food and how to protect its young, that it knows
well how to survive in the forest without the tools upon which we
depend, is readily evident to our human senses. That the mango tree
has the ability to create fruit, or the yarrow plant the power to reduce a
child's fever, is also evident. To humankind, these Others are purveyors
of secrets, carriers of intelligence that we ourselves often need: it is

these Others who can inform us of unseasonable changes in the weather, or warn us of imminent eruptions and earthquakes, who show us, when foraging, where we may find the ripest berries or the best route to follow back home. By watching them build their nests and shelters, we glean clues regarding how to strengthen our own dwellings, and their deaths teach us of our own. We receive from them countless gifts of food, fuel, shelter, and clothing. Yet still they remain Other to us, inhabiting their own cultures and displaying their own rituals, never wholly fathomable.

Moreover, it is not only those entities acknowledged by Western civilizations as "alive," not only the other animals and the plants that speak, as spirits, to the senses of an oral culture, but also the meandering river from which those animals drink, and the torrential monsoon rains, and the stone that fits neatly into the palm of the hand. The mountain, too, has its thoughts. The forest birds whirring and chattering as the sun slips below the horizon are vocal organs of the rain forest itself.

Bali, of course, is hardly an aboriginal culture; the complexity of its temple architecture, the intricacy of its irrigation systems, the resplendence of its colorful festivals and crafts all bespeak the influence of various civilizations, most notably the Hindu complex of India. In Bali, nevertheless, these influences are thoroughly intertwined with the indigenous animism of the Indonesian archipelago; the Hindu gods and goddesses have been appropriated, as it were, by the more volcanic, eruptive spirits of the local terrain.

Yet the underlying animistic cultures of Indonesia, like those of many islands in the Pacific, are steeped as well in beliefs often referred to by ethnologists as "ancestor worship," and some may argue that the ritual reverence paid to one's long-dead human ancestors (and the assumption of their influence in present life), easily invalidates my assertion that the various "powers" or "spirits" that move through the discourse of indigenous, oral peoples are ultimately tied to nonhuman (but nonetheless sentient) forces in the enveloping landscape.

This objection rests upon certain assumptions implicit in Christian civilization, such as the assumption that the "spirits" of dead persons necessarily retain their human form, and that they reside in a domain outside of the physical world to which our senses give us access. However, most indigenous tribal peoples have no such ready recourse to an immaterial realm outside earthly nature. Our strictly human heavens and hells have only recently been abstracted from the sensuous world that surrounds us, from this more-than-human realm that abounds in its own winged intelligences and cloven-hoofed powers. For almost all oral cultures, the enveloping and sensuous earth remains the dwelling place of both the living *and* the dead. The "body"—whether human or otherwise—is not yet a mechanical object in such cultures, but is a magical entity, the mind's own sensuous aspect, and at death the body's decomposition into soil, worms, and dust can only signify the gradual reintegration of one's ancestors and elders into the living landscape, from which all, too, are born.

Each indigenous culture elaborates this recognition of metamorphosis in its own fashion, taking its clues from the particular terrain in which it is situated. Often the invisible atmosphere that animates the visible world—the subtle presence that circulates both within us and between all things—retains within itself the spirit or breath of the dead person until the time when that breath will enter and animate another visible body—a bird, or a deer, or a field of wild grain. Some cultures may burn, or "cremate," the body in order to more completely return the person, as smoke, to the swirling air, while that which departs as flame is offered to the sun and stars, and that which lingers as ash is fed to the dense earth. Still other cultures may dismember the body, leaving certain parts in precise locations where they will likely be found by condors, or where they will be consumed by mountain lions or by wolves, thus hastening the re-incarnation of that person into a particular animal realm within the landscape. Such examples illustrate simply that death, in tribal cultures, initiates a metamorphosis wherein the person's presence does not "vanish" from the sensible world (where would it go?) but rather remains as an animating force within the vastness of the landscape, whether subtly, in the wind, or more visibly, in animal form, or even as the eruptive, ever to be appeased, wrath of the volcano. "Ancestor worship," in its myriad forms, then, is ultimately another mode of attentiveness to nonhuman nature; it signifies not so much an awe or reverence of human powers, but rather a reverence for those forms that awareness takes when it is *not* in human form, when the familiar human embodiment dies and decays to become part of the encompassing cosmos.

This cycling of the human back into the larger world ensures that the other forms of experience that we encounter—whether ants, or willow trees, or clouds—are never absolutely alien to ourselves. Despite the obvious differences in shape, and ability, and style of being, they remain at least distantly familiar, even familial. It is, paradoxically, this perceived kinship or consanguinity that renders the difference, or otherness, so eerily potent.

Several months after my arrival in Bali, I left the village in which I was staying to visit one of the pre-Hindu sites on the island. I arrived on my bicycle early in the afternoon, after the bus carrying tourists from the coast had departed. A flight of steps took me down into a lush, emerald valley, lined by cliffs on either side, awash with the speech of the river and the sighing of the wind through high, unharvested grasses. On a small bridge crossing the river I met an old woman carrying a wide basket on her head and holding the hand of a little, shy child; the woman grinned at me with the red, toothless smile of a beetle nut chewer. On the far side of the river I stood in front of a great moss-covered complex of passageways, rooms, and courtyards carved by hand out of the black volcanic rock.

I noticed, at a bend in the canyon downstream, a further series of caves carved into the cliffs. These appeared more isolated and remote, unattended by any footpath I could discern. I set out through the grasses to explore them. This proved much more difficult than I anticipated, but after getting lost in the tall grasses, and fording the river three times, I at last found myself beneath the caves. A short scramble up the rock wall brought me to the mouth of one of them, and I entered on my hands and knees. It was a wide but low opening, perhaps only four feet high, and the interior receded only about five or six feet into the cliff. The floor and walls were covered with mosses, painting the cave with green patterns and softening the harshness of the rock; the place, despite its small size—or perhaps because of it—had an air of great friendliness. I climbed to two other caves, each about the same size, but then felt drawn back to the first one, to sit cross-legged on the cushioning moss and gaze out across the emerald canyon. It was quiet inside, a kind of intimate sanctuary hewn into the stone. I began to explore the rich resonance of the enclosure, first just humming, then intoning a simple chant taught to me by a balian some days before. I was delighted by the overtones that the cave added to my voice, and sat there singing for a long while. I did not notice the change in the wind outside, or the cloud shadows darkening the valley, until the rains broke—suddenly and with great force. The first storm of the monsoon!

I had experienced only slight rains on the island before then, and was startled by the torrential downpour now sending stones tumbling along the cliffs, building puddles and then ponds in the green landscape below, swelling the river. There was no question of returning home—I would be unable to make my way back through the flood to the valley's entrance. And so, thankful for the shelter, I recrossed my legs to wait out the storm. Before long the rivulets falling along the cliff above gathered themselves into streams, and two small waterfalls cascaded across the cave's mouth. Soon I was looking into a solid curtain of water, thin in some places, where the canyon's image flickered unsteadily, and thickly rushing in others. My senses were all but overcome by the wild beauty of the cascade and by the roar of sound, my body trembling inwardly at the weird sense of being sealed into my hiding place.

And then, in the midst of all this tumult, I noticed a small, delicate activity. Just in front of me, and only an inch or two to my side of the torrent, a spider was climbing a thin thread stretched across the mouth of the cave. As I watched, it anchored another thread to the top of the opening, then slipped back along the first thread and joined the two at a point about midway between the roof and the floor. I lost sight of the spider then, and for a while it seemed that it had vanished, thread and all, until my focus rediscovered it. Two more threads now radiated from the center to the floor, and then another; soon the spider began to swing between these as on a circular trellis, trailing an ever-lengthening thread which it affixed to each radiating rung as it moved from one to the next, spiraling outward. The spider seemed wholly undaunted by the tumult of waters spilling past it, although every now and then it broke off its spiral dance and climbed to the roof or the floor to tug on the radii there, assuring the tautness of the threads, then crawled back to where it left off. Whenever I lost the correct focus, I waited to catch sight of the spinning arachnid, and then let its dancing form gradually draw the lineaments of the web back into visibility, tying my focus into each new knot of silk as it moved, weaving my gaze into the ever-deepening pattern.

And then, abruptly, my vision snagged on a strange incongruity: another thread slanted across the web, neither radiating nor spiraling from the central juncture, violating the symmetry. As I followed it with my eyes, pondering its purpose in the overall pattern, I began to realize that it was on a different plane from the rest of the web, for the web slipped out of focus whenever this new line became clearer. I soon saw that it led to its own center, about twelve inches to the right of the first, another nexus of forces from which several threads stretched to the floor and the ceiling. And then I saw that there was a *different* spider spinning this web, testing its tautness by dancing around it like the first, now setting the silken cross weaves around the nodal point and winding outward. The two spiders spun independently of each other, but to my eyes they wove a single intersecting pattern. This widening of my gaze soon disclosed yet another spider spiraling in the cave's mouth, and suddenly I realized that there were *many* overlapping webs coming into being, radiating out at different rhythms from myriad centers poised—some higher, some lower, some minutely closer to my eyes and some farther—between the stone above and the stone below.

I sat stunned and mesmerized before this ever-complexifying expanse of living patterns upon patterns, my gaze drawn like a breath into one converging group of lines, then breathed out into open space, then drawn down into another convergence. The curtain of water had become utterly silent—I tried at one point to hear it, but could not. My senses were entranced.

I had the distinct impression that I was watching the universe being born, galaxy upon galaxy. . . .

Night filled the cave with darkness. The rain had not stopped. Yet, strangely, I felt neither cold nor hungry—only remarkably peaceful and at home. Stretching out upon the moist, mossy floor near the back of the cave, I slept.

When I awoke, the sun was staring into the canyon, the grasses below rippling with bright blues and greens. I could see no trace of the webs, nor their weavers. Thinking that they were invisible to my eyes without the curtain of water behind them, I felt carefully with my hands around and through the mouth of the cave. But the webs were gone. I climbed down to the river and washed, then hiked across and out of the canyon to where my cycle was drying in the sun, and headed back to my own valley.

I have never, since that time, been able to encounter a spider without feeling a great strangeness and awe. To be sure, insects and spiders are not the only powers, or even central presences, in the Indonesian universe. But they were *my* introduction to the spirits, to the magic afoot in the land. It was from them that I first learned of the intelligence that lurks in nonhuman nature, the ability that an alien form of sentience has to echo one's own, to instill a reverberation in oneself that temporarily shatters habitual ways of seeing and feeling, leaving one open to a world all alive, awake, and aware. It was from such small beings that my senses first learned of the countless worlds within worlds that spin in the depths of this world that we commonly inhabit, and from them that I learned that my body could, with practice, enter sensorially into these dimensions. The precise and minuscule craft of the spiders had so honed and focused my awareness that the very webwork of the universe, of which my own flesh was a part, seemed to be being spun by their arcane art.

* * *

My exposure to traditional magicians and seers was shifting my senses; I became increasingly susceptible to the solicitations of nonhuman things. In the course of struggling to decipher the magicians' odd gestures or to fathom their constant spoken references to powers unseen and unheard, I began to *see* and to *hear* in a manner I never had before. When a magician spoke of a power or "presence" lingering in the corner of his house, I learned to notice the ray of sunlight that was then

pouring through a chink in the roof, illuminating a column of drifting
dust, and to realize that that column of light was indeed a power, influencing the air currents by its warmth, and indeed influencing the whole
mood of the room; although I had not consciously seen it before, it had
already been structuring my experience. My ears began to attend, in a
new way, to the songs of birds—no longer just a melodic background to
human speech, but meaningful speech in its own right, responding to
and commenting on events in the surrounding earth. I became a student of subtle differences: the way a breeze may flutter a single leaf on a
whole tree, leaving the other leaves silent and unmoved (had not that
leaf, then, been brushed by a magic?); or the way the intensity of the
sun's heat expresses itself in the precise rhythm of the crickets. Walking
along the dirt paths, I learned to slow my pace in order to *feel* the difference between one nearby hill and the next, or to taste the presence of a
particular field at a certain time of day when, as I had been told by a
local *dukun*, the place had a special power and proffered unique gifts. It
was a power communicated to my senses by the way the shadows of the
trees fell at that hour, and by smells that only then lingered in the tops
of the grasses without being wafted away by the wind, and other elements I could only isolate after many days of stopping and listening.

And gradually, then, other animals began to intercept me in my
wanderings, as if some quality in my posture or the rhythm of my
breathing had disarmed their wariness; I would find myself face-to- face
with monkeys, and with large lizards that did not slither away when I
spoke, but leaned forward in apparent curiosity. In rural Java, I often
noticed monkeys accompanying me in the branches overhead, and
ravens walked toward me on the road, croaking. While at Pangandaran,
a nature preserve on a peninsula jutting out from the south coast of
Java ("a place of many spirits," I was told by nearby fishermen), I
stepped out from a clutch of trees and found myself looking into the
face of one of the rare and beautiful bison that exist only on that island.
Our eyes locked. When it snorted, I snorted back; when it shifted its
shoulders, I shifted my stance; when I tossed my head, it tossed *its* head
in reply. I found myself caught in a nonverbal conversation with this
Other, a gestural duet with which my conscious awareness had very little to do. It was as if my body in its actions was suddenly being motivated by a wisdom older than my thinking mind, as though it was held
and moved by a logos, deeper than words, spoken by the Other's body,
the trees, and the stony ground on which we stood.

Anthropology's inability to discern the shaman's allegiance to nonhuman nature has led to a curious circumstance in the "developed world"
today, where many persons in search of spiritual understanding are

enrolling in workshops concerned with "shamanic" methods of personal discovery and revelation. Psychotherapists and some physicians have begun to specialize in "shamanic healing techniques." "Shamanism" has thus come to connote an alternative form of therapy; the emphasis, among these new practitioners of popular shamanism, is on personal insight and curing. These are noble aims, to be sure, yet they are secondary to, and derivative from, the primary role of the indigenous shaman, a role that cannot be fulfilled without long and sustained exposure to wild nature, to its patterns and vicissitudes. Mimicking the indigenous shaman's curative methods without his intimate knowledge of the wider natural community cannnot, if I am correct, do anything more than trade certain symptoms for others, or shift the locus of dis-ease from place to place within the *human* community. For the source of stress lies in the relation *between* the human community and the natural landscape.

Western industrial society, of course, with its massive scale and hugely centralized economy, can hardly be seen in relation to any particular landscape or ecosystem; the more-than-human ecology with which it is directly engaged is the biosphere itself. Sadly, our culture's relation to the earthly biosphere can in no way be considered a reciprocal or balanced one: with thousands of acres of nonregenerating forest disappearing every hour, and hundreds of our fellow species becoming extinct each month as a result of our civilization's excesses, we can hardly be surprised by the amount of epidemic illness in our culture, from increasingly severe immune dysfunctions and cancers, to widespread psychological distress, depression, and ever more frequent suicides, to the accelerating number of household killings and mass murders committed for no apparent reason by otherwise coherent individuals.

From an animistic perspective, the clearest source of all this distress, both physical and psychological, lies in the aforementioned violence needlessly perpetrated by our civilization on the ecology of the planet; only by alleviating the latter will we be able to heal the former. While this may sound at first like a simple statement of faith, it makes eminent and obvious sense as soon as we acknowledge our thorough dependence upon the countless other organisms with whom we have evolved. Caught up in a mass of abstractions, our attention hypnotized by a host of human-made technologies that only reflect us back to ourselves, it is all too easy for us to forget our carnal inherence in a more-than-human matrix of sensations and sensibilities. Our bodies have formed themselves in delicate reciprocity with the manifold textures, sounds, and shapes of an animate earth—our eyes have evolved in subtle interaction with *other* eyes, as our ears are attuned by their very structure to the howling of wolves and the honking of geese. To shut ourselves off from these other voices, to continue by our lifestyles to condemn these other sensibilities to the oblivion of extinction, is to rob our own senses of their integrity, and to rob our minds of their coherence. We are human only in contact, and conviviality, with what is not human.

RICK BASS
b. 1958

Like John Muir in Yosemite Valley and Mary Austin in the deserts of the Southwest, Rick Bass is a writer who, on entering the Yaak Valley of Montana as an adult, discovered and fell in love with his true home on Earth. In his ten works of nonfiction and five of fiction, he often focuses on this part of the northern Rockies, where he and his family now live and where he has become a tireless defender of wild beauty against those who would despoil it. In such books as Winter: Notes from Montana *(1991) and* The Ninemile Wolves *(1992), he evokes the drama and integrity of his wilderness home, while in* The Book of Yaak *(1996), he addresses both the disasters looming over the land and the conflicting loyalties he sometimes experiences as an artist who is also the advocate of a particular, embattled community. Readers of Bass appreciate the authenticy and freshness of his voice—conveyed at some moments by a passionate torrent of description and at others by the honesty and humor with which he tracks his own complex emotions.*

From THE NINEMILE WOLVES

CHAPTER ONE

They say not to anthropomorphize—not to think of them as having feelings, not to think of them as being able to think—but late at night I like to imagine that they are killing: that another deer has gone down in a tangle of legs, tackled in deep snow; and that, once again, the wolves are feeding. That they have saved themselves, once again. That the deer or moose calf, or young dumb elk is still warm (steam rising from the belly as that part which contains the entrails is opened first), is now dead, or dying.

They eat everything, when they kill, even the snow that soaks up the blood.

This all goes on usually at night. They catch their prey from behind, often, but also by the nose, the face, the neck—whatever they can dart in and grab without being kicked. When the prey pauses, or buckles, it's over; the prey's hindquarters, or neck, might be torn out, and in that manner, the prey flounders. The wolves swarm it, then. They don't have thumbs. All they've got is teeth, long legs, and—I have to say this—great hearts.

I can say what I want to say. I gave up my science badge a long time ago. I've interviewed maybe a hundred people for or against wolves. The ones who are "for" wolves, they have an agenda: wilderness, and freedom for predators, for prey, for everything. The ones who are "against" wolves have an agenda: they've got vested financial interests. It's about money—more and more money—for them. They perceive the wolves to be an obstacle to frictionless cash flow.

The story's so rich. I can begin anywhere.

I can start with prey, which is what controls wolf numbers (not the other way around), or with history, which is rich in sin, cruelty, sensationalism (poisonings, maimings, torture). You can start with biology, or politics, or you can start with family, with loyalty, and even with the mystic-tinged edges of fate, which is where I choose to begin. It's all going to come together anyway. It has to. We're all following the wolf. To pretend anything else—to pretend that we are protecting the wolf, for instance, or *managing* him—is nonsense of the kind of immense proportions of which only our species is capable.

We're following the wolf. He's returning to Montana after sixty years.

The history of wolves in the West—of wolves in this country—is pretty well documented. Even by the turn of the century they were being diminished, and the wolves were all killed so quickly, and with such essentially religious zeal that we never had time to learn about them, and about their place on the land. And about our place on it. The wolves sure as hell didn't have time to learn about the government's wolf-, buffalo-, and Indian-killing program, and—well, I've got

The Ninemile Wolves: An Essay (Livingston, Mont.: Clark City, 1992).

to say it—cows. Our culture has replaced buffalo with cattle, and wolves kill and eat cows, sometimes.

Peter Matthiessen writes in *Wildlife in America* that "with the slaughter of the bison and other hoofed animals in the late nineteenth century, the wolves . . . turned their attention to . . . livestock, which was already abundant on the grasslands." In the absence of bison, there was the bison's replacement: cattle. The wolves preyed upon these new intruders, without question, but ranchers and the government overreacted just a *tad*. Until very recently, the score stood at Cows, 99,200,000; Wolves, 0.

It took a lot of money to kill every last wolf out of the West. We behaved badly doing it: setting them on fire, feeding them ground-up glass, et cetera. Some people say it was the ranchers who kept the wolves out; others say the government. Other people say this country's lost its wildness, that it's no place for a wolf anymore, that already, wolves are like dinosaurs. There's room for sheep and cows, yes, and deer and elk and other plant eaters—but these woods are not the same woods they were sixty years ago. Even the shadows of trees in the forest are different, the wind's different.

There are a lot of people who believe that last part: that predators are out of date. And that Rocky Mountain gray wolves are the most old-fashioned, out-of-date predator of all. Sometimes I detect, even among wolves' most ardent well-wishers, a little chagrin that wolves succumbed so *easily*: that they wilted before the grazing juggernaut and federal predator-extinction programs of the 1920s. But with the government having been honed into fighting trim by practice on the "Indian problem," the wolves really had no chance. What's interesting (and wise) is that once the West's wolves were exterminated, they *stayed* exterminated. Canada's wolves did not repopulate the gaping wound of our wolf-free country.

Canada's always had the wolves, about 50,000 of them; so many wolves that there's even been a "hunting" (read: *killing*—people don't eat wolves) season on them. Which may not be so bad, in places, because it probably helps keep those wolves wild. With the mild winters of the greenhouse eighties, ungulate populations have increased dramatically. Using Canada as a base, wolves, or rumors of wolves, have historically trickled across the border, making shy, hidden, and often ill-fated sorties into the U.S., hanging out in the deep timber of northwestern Montana, eating deer and trying to stay away from people. The prey that wolves chase is most often found in valley bottoms—along rivers, where ranches and villages are located. I like to think of the wolves hanging back in the woods, up in the mountains, *longing* for the river bottoms, but too wild, too smart, to descend. The deer down along the rivers, among the people, among the barking dogs, among the intricate road systems, are in a way protected from everything but overpopulation, disease and starvation.

The wolves watching those tempting Montana river bottoms, and longing, but then — wild — drifting back into Canada.

But the thing that defines a wolf more than anything — better than DNA, better than fur, teeth, green eyes, better than even the low, mournful howl — is the way it *travels*. The home range of wolves in the northern Rockies averages 200 to 300 square miles, and ranges of 500 square miles are not uncommon. Montana could not be avoided by Canadian wolves. There are too many deer, too many elk: too many for the few predators that still exist.

Glacier National Park, right on the Montana — Canada line, has sometimes harbored a pack, sometimes two, but that's all. In Glacier, the Magic and Wigwam and Camas packs averaged between fifteen and thirty animals in the 1970s and 1980s. The Wigwam pack held the most mystery — eight of their nine members simply disappeared, over the course of a few days in 1989. Poaching is suspected, but they could also have been "assassinated" — wolves will often fight to the death if another pack crosses their boundary. (Deer have learned to seek out these boundaries, the line between two packs' territories acting as a sort of demilitarized zone — a ridge separating two valleys, perhaps, or an uplift between two river forks.)

The Glacier wolves served as a hope, a *longing*, for the rest of Montana, and gave lovely possibility to the occasional rumors that leapt up around the area, in nearby valleys — the Tobacco and Yaak valleys to the west and Swan Valley to the south. But the wolves that left Glacier, if they were indeed the ones being spotted, never (to any scientist's knowledge) mated — never formed what the biologists call the "pair bond" of an alpha male and alpha female — and it was thought that the sightings were unusually large coyotes. There's not as much similarity between the two as you'd think: coyotes often have a lot of red in their coats, and their muzzles are sharper, reminding me of screw-drivers, and their ears are larger and more pointy-looking. Coyotes are 25 percent to 50 percent smaller than long-legged wolves, and wolves have "fur around their face," and a ruffed coat, like a cape, up across their shoulders; coyotes' necks look bare by comparison.

Rumors of wolves outside of Glacier kept drifting in, increasing steadily through the 1980s, which fanned the hopes of those people desiring wolves. A poll taken in 1990 showed that two-thirds of Montanans believed wolves should be allowed to return to the state, while one-third thought wolves should be locked out. In the summer of 1988, in my valley, a mile south of Canada, I saw two big gray wolves—a mated pair, I hoped—lope through the woods, running north, headed to the border; later that winter, friends told me they'd seen a big black wolf and a gray wolf on the other side of their frozen lake, on several mornings, which indicated there might possibly, hopefully, be a pack. They're filtering south, even as you're reading this—moving through the trees, mostly, and eating a deer about every third day; they're coming down out of Canada, and the wild ones are trying to stay out of the populated river bottoms. The less wild ones—well, sometimes people see those.

It's been theorized that in addition to Montana's high deer population, hunting pressure in Canada may be helping send wolves back south, all along the U.S.-Canadian border—into the Cascades, the Olympics, northern Idaho and northwestern Montana, even along Montana's Front Range—but the main reason seems to be game overabundance, a dangerous excess, the simple response of predator-and-prey cycles. Like the tides and the pull of the moon, it's not a thing that we've been able to mess up, yet. It still works, or tries to. The recent warm winters in Montana have led deer and elk populations to all-time highs. There haven't been any dramatic disease die-offs in recent memory, no massive winter kills, no outbreaks of starvation and land-gone-ragged. The deer populations keep climbing as if *desiring* that outcome, though, and each summer you see more and more fawns; each fall you see more and more deer. The deer have the run of the woods, though mountain lion populations are increasing rapidly to join in on the feast. You've got to drive slowly at night. It's hard to find anyone living in the country who hasn't struck a deer at least once, late at night, no matter how carefully he drives. Usually if you hit a deer, you stop and put it in the back of your truck and drive on home and clean it and eat it (illegal, of course, but not immoral)—but sometimes the out-of-town people will keep going after striking a deer, leaving the deer dead on the side of the road or even *in* the road, and in the morning, if you live nearby, you'll hear the ravens, and then later in the afternoon, the little slinky-dog coyotes barking at one another, and by that night, everything's gone: meat, bones, hooves, even the hide.

I should point out that that's the scientists' belief—that Montana's huge deer populations are bringing the wolves back. There is another thought that occurs to me, though I'm sure the scientists would have nothing to do with this idea. I've been reading all the old case histories of wolves in this country and following the new histories, and the species doesn't seem to have changed in a hundred years, a hundred generations. The intricacy of their pack structure—the hierarchy of dominance and submission—is well documented, as is their territoriality, their fierce protection of borders, and their love of travel, of exploring those borders.

But these new wolves—I get the sense that they're a little different. Wiser, of course—even if only bearing wise blood but not knowing it. They seem to be a little *edgier*—pushing for those edges. All wolves travel like crazy, but these new wolves seem a little restless even for their species. They're trying to trickle down, like roots spreading fingers into weathered rock. And this time we'll be able to find out if human nature, and our politics, have changed—metamorphosed, perhaps, into something more advanced—or if at the base our politics are still those of Indian killers.

The wolves intend to find out, too.

Everything travels fast for a wolf. They went from huge buffalo-supported packs of twenty and thirty animals to near-extinction with great speed, and it is in their blood to recover with great speed, given the right conditions. Sixty years is only a blink, and it is a predator's genetic duty to endure hard times—very hard times—along with the sweet times.

The old wolves were natives, residents. These new wolves are foreigners, pioneers: explorers. It could be argued that wolves are never more wolflike than when they're exploring, trying to claim, or reclaim, new territory, rather than holding on and defending old borders. It could be argued that our perverse resistance to wolves helps them *remain* wolves, that they need that great arm's distance to remain always outside of other communities, except perhaps for the community of ravens.

Is the base of our history unchanging, like some *batholith* of sin—are we irretrievable killers?—or can we exist with wolves, this time? I believe we are being given another chance, an opportunity to demonstrate our ability to change. This time, we have a chance to let a swaying balance be struck: not just for wolves, but for humans, too. If I could say any one thing to politicians, and to people with guns and poison (and sheep and cows), it would be this: that to have the balance the majority of people claim to long for, it must be struck by the wolves as much as by the people. The wolves must have some say in defining it, or it will not be valid. And since they do not speak our language, it might be rough for a while: for three years, or five years, or maybe even ten.

*   *   *

BILL McKIBBEN
b. 1960

Bill McKibben's The End of Nature *(1989) has earned a place in the great prophetic tradition of American environmental writing, along with George Perkins Marsh's* Man and Nature *and Rachel Carson's* Silent Spring. *He warns his readers of the grim dangers posed to life on earth by global climate change and documents the relationship between such change and our addiction to the internal-combustion engine. More broadly, McKibben confronts the reader with an assertion that the traditional idea of nature as "a world apart from man" is no longer viable and examines the psychological and ethical consequences of "the end of nature." Like Carson's, his book was immediately criticized by established economic interests; like hers, his ecological conclusions have been vindicated by continuing research in the field. As McKibben continues to call attention to the issue of climate change, he has also explored positive alternatives to our society's wasteful and destructive practices. The Age of Missing Information (1993) contrasts the historically and sensually impoverished realm of television with the good news offered by the natural world. In* Hope Human and Wild *(1995), he reports on three places, in Brazil, India, and the Northeast of the United States, where communities have made positive choices and where there has been an increase of ecological stability.*

From THE END OF NATURE

Almost every day, I hike up the hill out my back door. Within a hundred yards the woods swallows me up, and there is nothing to remind me of human society—no trash, no stumps, no fence, not even a real path. Looking out from the high places, you can't see road or house; it is a world apart from man. But once in a while someone

The End of Nature (New York: Random House, 1989).

will be cutting wood farther down the valley, and the snarl of a chain saw will fill the woods. It is harder on those days to get caught up in the timeless meaning of the forest, for man is nearby. The sound of the chain saw doesn't blot out all the noises of the forest or drive the animals away, but it does drive away the feeling that you are in another, separate, timeless, wild sphere.

Now that we have changed the most basic forces around us, the noise of that chain saw will always be in the woods. We have changed the atmosphere, and that will change the weather. The temperature and rainfall are no longer to be entirely the work of some separate, uncivilizable force, but instead in part a product of our habits, our economies, our ways of life. Even in the most remote wilderness, where the strictest laws forbid the felling of a single tree, the sound of that saw will be clear, and a walk in the woods will be changed—tainted—by its whine. The world outdoors will mean much the same thing as the world indoors, the hill the same thing as the house.

An idea, a relationship, can go extinct, just like an animal or a plant. The idea in this case is "nature," the separate and wild province, the world apart from man to which he adapted, under whose rules he was born and died. In the past, we spoiled and polluted parts of that nature, inflicted environmental "damage." But that was like stabbing a man with toothpicks: though it hurt, annoyed, degraded, it did not touch vital organs, block the path of the lymph or blood. We never thought that we had wrecked nature. Deep down, we never really thought we could: it was too big and too old; its forces—the wind, the rain, the sun—were too strong, too elemental.

But, quite by accident, it turned out that the carbon dioxide and other gases we were producing in our pursuit of a better life—in pursuit of warm houses and eternal economic growth and of agriculture so productive it would free most of us from farming—*could* alter the power of the sun, could increase its heat. And that increase *could* change the patterns of moisture and dryness, breed storms in new places, breed deserts. Those things may or may not have yet begun to happen, but it is too late to altogether prevent them from happening. We have produced the carbon dioxide—we are ending nature.

We have not ended rainfall or sunlight; in fact, rainfall and sunlight may become more important forces in our lives. It is too early to tell exactly how much harder the wind will blow, how much hotter the sun will shine. That is for the future. But the *meaning* of the wind, the sun, the rain—of nature—has already changed. Yes, the wind still blows— but no longer from some other sphere, some inhuman place.

In the summer, my wife and I bike down to the lake nearly every afternoon for a swim. It is a dogleg Adirondack lake, with three beaver lodges, a blue heron, some otter, a family of mergansers, the occasional loon. A few summer houses cluster at one end, but mostly it is surrounded by wild state land. During the week we swim across and back, a trip of maybe forty minutes—plenty of time to forget everything but the feel of the water around your body and the rippling, muscular joy of a hard kick and the pull of your arms.

But on the weekends, more and more often, someone will bring a boat out for waterskiing, and make pass after pass up and down the lake. And then the whole experience changes, changes entirely. Instead of being able to forget everything but yourself, and even yourself except for the muscles and the skin, you must be alert, looking up every dozen strokes to see where the boat is, thinking about what you will do if it comes near. It is not so much the danger—few swimmers, I imagine, ever die by Evinrude. It's not even so much the blue smoke that hangs low over the water. It's that the motorboat gets in your mind. You're forced to think, not feel—to think of human society and of people. The lake is utterly different on these days, just as the planet is utterly different now.

*   *   *

I took a day's hike last fall, walking Mill Creek from the spot where it runs by my door to the place where it crosses the main county road near Wevertown. It's a distance of maybe nine miles as the car flies, but rivers are far less efficient, and endlessly follow pointless, time-wasting, uneconomical meanders and curves. Mill Creek cuts some fancy figures, and so I was able to feel a bit exploratory—a budget Bob Marshall. In a strict sense, it wasn't much of an adventure. I stopped at the store for a liverwurst sandwich at lunchtime, the path was generally downhill, the temperature stuck at an equable 55 degrees, and since it was the week before the hunting season opened I didn't have to sing as I walked to keep from getting shot. On the other hand, I had made an arbitrary plan—to follow the creek—and, as a consequence, I spent hours stumbling through overgrown marsh, batting at ten-foot saplings and vines, emerging only every now and then, scratched and weary, into the steeper wooded sections. When Thoreau was on Katahdin, nature said to him, "I have never made this soil for thy feet, this air for thy breathing, these rocks for they neighbors. I cannot pity nor fondle thee there, but forever relentlessly drive thee hence to where I *am* kind. Why seek me where I have not called thee, and then complain because you find me but a stepmother?" Nature said this to me on Mill Creek, or at least it said, "Go home and tell your wife you walked to Wevertown." I felt I should have carried a machete, or employed a macheteist (The worst thing about battling through brake and bramble of this sort is that it's so anonymous—gray sticks, green stalks with reddish thorns, none of them to be found in any of the many guides and almanacs on my shelf.) And though I started the day with eight dry socks, none saw noon in that pleasant state.

If it was all a little damp and in a minor key, the sky was nonetheless bright blue, and rabbits kept popping out from my path, and pheasants fired up between my legs, and at each turning some new gift appeared: a vein of quartz, or a ridge where the maples still held their leaves, or a pine more than three feet in diameter that beavers had gnawed all the way around and halfway through and then left standing—a forty-foot sculpture. It was October, so there weren't even any bugs. And always the plash of the stream in my ear. It isn't Yosemite, the Mill Creek Valley, but its small beauties are absorbing, and one can say with Muir on his mountaintop, "Up here all the world's prizes seem as nothing."

And so what if it isn't nature primeval? One of our neighbors has left several kitchen chairs along his stretch of the bank, spaced at fifty-yard intervals for comfort in fishing. At one old homestead, a stone chimney stands at either end of a foundation now filled by a graceful birch. Near the one real waterfall, a lot of rusty pipe and collapsed concrete testifies to the old mill that once stood there. But these aren't disturbing sights—they're almost comforting, reminders of the way that nature has endured and outlived and with dignity reclaimed so many schemes and disruptions of man. (A mile or so off the creek, there's a mine where a hundred and fifty years ago a visionary tried to extract pigment for paint and pack it out on mule and sledge. He rebuilt after a fire; finally an avalanche convinced him. The path in is faint now, but his chimney, too, still stands, a small Angkor Wat of free enterprise.) Large sections of the area were once farmed; but the growing season is not much more than a hundred days, and the limits established by that higher authority were stronger than the (powerful) attempts of individual men to circumvent them, and so the farms returned to forest, with only a dump of ancient bottles or a section of stone wall as a memorial. (Last fall, though, my wife and I found, in one abandoned meadow, a hop vine planted at least a century before. It was still flowering, and with its blossoms we brewed beer.) These ruins are humbling sights, reminders of the negotiations with nature that have established the world as we know it.

Changing socks (soaking for merely clammy) in front of the waterfall, I thought back to the spring before last, when a record snowfall melted in only a dozen or so warm April days. A little to the south, an inflamed stream washed out a highway bridge, closing the New York Thruway for months. Mill Creek filled till it was a river, and this waterfall, normally one of those diaphanous-veil affairs, turned into a cataract. It filled me with awe to stand there then, on the shaking ground and think, This is what nature is capable of.

But as I sat there this time, and thought about the dry summer we'd just come through, there was nothing awe-inspiring or instructive, or even lulling, in the fall of the water. It suddenly seemed less like a waterfall than like a spillway to accommodate the overflow of a reservoir. That didn't decrease its beauty, but it changed its meaning. It has begun or will soon begin to rain and snow when the particular mix of chemicals we've injected into the atmosphere adds up to rain or snow—when they make it hot enough over some tropical sea to form a cloud and send it this way. I had no more control, in one sense, over this process than I ever did. But it felt different, and lonelier. Instead of a world where rain had an independent and mysterious existence, the rain had become a subset of human activity: a phenomenon like smog or commerce or the noise from the skidder towing logs on Cleveland Road—all things over which I had no control, either. The rain bore a brand; it was a steer, not a deer. And that was where the loneliness came from. There's nothing there except us. There's no such thing as nature anymore—that other world that isn't business and art and breakfast is now not another world, and there is nothing except us alone.

At the same time that I felt lonely, though, I also felt crowded, without privacy. We go to the woods in part to escape. But now there is nothing except us and so there is no escaping other people. As I walked in the autumn woods I saw a lot of sick trees. With the conifers, I suspected acid rain. (At least I have the luxury of only suspecting; in too many places, they *know*). And so who walked with me in the woods? Well, there were the presidents of the Midwest utilities who kept explaining why they had to burn coal to make electricity (cheaper, fiduciary responsibility, no *proof* it kills trees) and then there were the congressmen who couldn't bring themselves to do anything about it (personally favor but politics the art of compromise, very busy with the war on drugs) and before long the whole human race had arrived to explain its aspirations. We like to drive, they said, air conditioning is a necessity nowadays, let's go to the mall. By this point, the woods were pretty densely populated. As I attempted to escape, I slipped on another rock, and in I went again. Of course, the person I was fleeing most fearfully was myself, for I drive (I drove forty thousand miles one year), and I'm burning a collapsed barn behind the house next week because it is much the cheapest way to deal with it, and I live on about four hundred times what Thoreau conclusively proved was enough, so I've done my share to take this independent, eternal world and turn it into a science-fair project (and not even a good science-fair project but a cloddish one, like pumping poison into an ant farm and "observing the effects").

The walk along Mill Creek, or any stream, or up any hill, or through any woods, is changed forever—changed as profoundly as when it shifted from pristine and untracked wilderness to mapped and deeded and cultivated land. Our local shopping mall now has a club of people who go "mall walking" every day. They circle the shopping center en masse—Caldor to Sears to J. C. Penney; circuit after circuit with an occasional break to shop. This seems less absurd to me now than it did at first. I like to walk in the outdoors not solely because the air is cleaner but because outdoors we venture into a sphere larger than ourselves. Mall walking involves too many other people, and too many purely human sights, ever to be more than good-natured exercise. But now, out in the wild, the sunshine on one's shoulders is a reminder that man has cracked the ozone, that, thanks to us, the atmosphere absorbs where once it released.

The greenhouse effect is a more apt name than those who coined it imagined. The carbon dioxide and trace gases act like the panes of glass on a greenhouse—the analogy is accurate. But it's more than that. We have built a greenhouse, *a human creation*, where once there bloomed a sweet and wild garden.

*   *   *

If nature were about to end, we might muster endless energy to stave it off; but if nature has already ended, what are we fighting for? Before any redwoods had been cloned or genetically improved, one could understand clearly what the fight against such tinkering was about. It was about the idea that a redwood was somehow sacred, that its fundamental identity should remain beyond our control. But once that barrier has been broken, what is the fight about, then? It's not like opposing nuclear reactors or toxic waste dumps, each one of which poses new risks to new areas. This damage is to an idea, the idea of nature, and all the ideas that descend from it. It is not cumulative. Wendell Berry once argued that without a "fascination" with the wonder of the natural world "the energy needed for its preservation will never be developed"—that "there must be a mystique of the rain if we are ever to restore the purity of the rainfall." This makes sense when the problem is transitory—sulfur from a smokestack drifting over the Adirondacks. But how can there be a mystique of the rain now that every drop—even the drops that fall as snow on the Arctic, even the drops that fall deep in the remaining forest primeval—bears the permanent stamp of man? Having lost its separateness, it loses its special power. Instead of being a category like God—something beyond our

control—it is now a category like the defense budget or the minimum
wage, a problem we must work out. This in itself changes its meaning
completely, and changes our reaction to it.

A few weeks ago, on the hill behind my house, I almost kicked the
biggest rabbit I had ever seen. She had nearly finished turning white for
the winter, and we stood there watching each other for a pleasant
while, two creatures linked by curiosity. What will it mean to come
across a rabbit in the woods once genetically engineered "rabbits" are
widespread? Why would we have any more reverence or affection for
such a rabbit than we would for a Coke bottle?

The end of nature probably also makes us reluctant to attach ourselves to its remnants, for the same reason that we usually don't choose
friends from among the terminally ill. I love the mountain outside my
back door—the stream that runs along its flank, and the smaller stream
that slides down a quarter-mile mossy chute, and the place where the
slope flattens into an open plain of birch and oak. But I know that some
part of me resists getting to know it better—for fear, weak-kneed as it
sounds, of getting hurt. If I knew as well as a forester what sick trees
looked like, I fear I would see them everywhere. I find now that I like
the woods best in winter, when it is harder to tell what might be dying.
The winter woods might be perfectly healthy come spring, just as the
sick friend, when she's sleeping peacefully, might wake up without the
wheeze in her lungs.

Writing on a different subject, the bonds between men and women,
Allan Bloom describes the difficulty of maintaining a committed relationship in an age when divorce—the end of that relationship—is so widely
accepted: "The possibility of separation is already the fact of separation,
inasmuch as people today must plan to be whole and self-sufficient and
cannot risk interdependence." Instead of working to strengthen our
attachments, our energies "are exhausted in preparation for independence." How much more so if that possible separation is definite, if that
hurt and confusion is certain. I love winter best now, but I try not to love
it too much, for fear of the January perhaps not so distant when the snow
will fall as warm rain. There is no future in loving nature.

And there may not even be much past. Though Thoreau's writings
grew in value and importance the closer we drew to the end of nature,
the time fast approaches when he will be inexplicable, his notions less
sensible to future men than the cave paintings are to us. Thoreau
writes, on his climb up Katahdin, that the mountain "was vast, Titanic,
and such as man never inhabits. Some part of the beholder, even some
vital part, seems to escape through the loose grating of his ribs. . . .
Nature has got him at a disadvantage, caught him alone, and pilfers

him of some of his divine faculty. She does not smile on him as in the plains. She seems to say sternly, why came ye here before your time. This ground is not prepared for you." This sentiment describes perfectly the last stage of the relationship of man to nature—though we had subdued her in the low places, the peaks, the poles, the jungles still rang with her pure message. But what sense will this passage make in the years to come, when Katahdin, the "cloud factory," is ringed by clouds of man's own making? When the massive pines that ring its base have been genetically improved for straightness of trunk and "proper branch drop," or, more likely, have sprung from the cones of genetically improved trees that began à few miles and a few generations distant on some timber plantation? When the moose that ambles by is part of a herd whose rancher is committed to the enlightened, Gaian notion that "conservation and profit go hand in hand"?

Thoreau describes an afternoon of fishing at the mouth of Murch Brook, a dozen miles from the summit of Katahdin. Speckled trout "swallowed the bait as fast as we could throw in; and the finest specimens . . . that I have ever seen, the largest one weighing three pounds, were heaved upon the shore." He stood there to catch them as "they fell in a perfect shower" around him. "While yet alive, before their tints had faded, they glistened like the fairest flowers, the product of primitive rivers; and he could hardly trust his senses, as he stood over them, that these jewels should have swam away in that Aboljacknagesic water for so long, some many dark ages—these bright fluviatile flowers, seen of Indians only, made beautiful, the Lord only knows why, to swim there!" But through biotechnology we have already synthesized growth hormone for trout. Soon pulling them from the water will mean no more than pulling cars from an assembly line. We won't have to wonder why the Lord made them beautiful and put them there; we will have created them to increase protein supplies or fish-farm profits. If we want to make them pretty, we may. Soon Thoreau will make no sense. And when that happens, the end of nature—which began with our alteration of the atmosphere, and continued with the responses to our precarious situation of the "planetary managers" and the "genetic engineers"—will be final. The loss of memory will be the eternal loss of meaning.

In the end, I understand perfectly well that defiance may mean prosperity and a sort of security—that more dams will help the people of Phoenix, and that genetic engineering will help the sick, and that there is so much progress that can still be made against human misery. And I have no great desire to limit my way of life. If I thought we could put off the decision, foist it on our grandchildren, I'd be willing. As it is, I have no plans to live in a cave, or even an unheated cabin. If it took ten thousand years to get where we are, it will take a few generations to climb back down. But this could be the epoch when people decide at least to go no farther down the path we've been following—when we make not only the necessary technological adjustments to preserve the world from overheating but also the necessary mental adjustments to ensure that we'll never again put our good ahead of everything else's. This is the path I choose, for it offers at least a shred of hope for a living, eternal, meaningful world.

The reasons for my choice are as numerous as the trees on the hill outside my window, but they crystallized in my mind when I read a passage from one of the brave optimists of our managed future. "The existential philosophers—particularly Sartre—used to lament that man lacked an essential purpose," writes Walter Truett Anderson. "We find now that the human predicament is not quite so devoid of inherent purpose after all. To be caretakers of a planet, custodians of all its life forms and shapers of its (and our own) future is certainly purpose enough." This intended rallying cry depresses me more deeply than I can say. That is our destiny? To be "caretakers" of a managed world, "custodians" of all life? For that job security we will trade the mystery of the natural world, the pungent mystery of our own lives and of a world bursting with exuberant creation? Much better, Sartre's neutral purposelessness. But much better than that, another vision, of man actually living up to his potential.

As birds have flight, our special gift is reason. Part of that reason drives the intelligence that allows us, say, to figure out and master DNA, or to build big power plants. But our reason could also keep us from following blindly the biological imperatives toward endless growth in numbers and territory. Our reason allows us to conceive of our species as a species, and to recognize the danger that our growth poses to it, and to feel something for the other species we threaten. Should we so choose, we could exercise our reason to do what no other animal can do: we could limit ourselves voluntarily, *choose* to remain God's creatures instead of making ourselves gods. What a towering achievement that would be, so much more impressive than the largest dam (beavers can build dams) because so much harder. Such restraint—not genetic engineering or planetary management—is the real challenge, the hard thing. Of course we can splice genes. But can we *not* splice genes?

The momentum behind our impulse to control nature may be too strong to stop. But the likelihood of defeat is not an excuse to avoid trying. In one sense it's an aesthetic choice we face, much like Thoreau's, though what is at stake is less the shape of our own lives than the very practical question of the lives of all the other species and the creation they together constitute. But it is, of course, for our benefit, too. Jeffers wrote, "Integrity is wholeness, the greatest beauty is / organic wholeness of life and things, the divine beauty of the universe. Love that, not man / Apart from that, or else you will share man's pitiful confusions, or drown in despair when his days darken." The day has come when we choose between that wholeness and man in it or man apart, between that old clarity or new darkness.

The strongest reason for choosing man apart is, as I have said, the idea that nature has ended. And I think it has. But I cannot stand the clanging finality of the argument I've made, any more than people have ever been able to stand the clanging finality of their own deaths. So I hope against hope. Though not in our time, and not in the time of our children, or their children, if we now, *today,* limited our numbers and our desires and our ambitions, perhaps nature could someday resume its independent working. Perhaps the temperature could someday adjust itself to its own setting, and the rain fall of its own accord.

Time, as I said at the start of this essay, is elusive, odd. Perhaps the ten thousand years of our encroaching, defiant civilization, an eternity to us and a yawn to the rocks around us, could give way to ten thousand years of humble civilization when we choose to pay more for the benefits of nature, when we rebuild the sense of wonder and sanctity that could protect the natural world. At the end of that span we would still be so young, and perhaps ready to revel in the timelessness that surrounds us. I said, much earlier, that one of the possible meanings of the end of nature is that God is dead. But another, if there was or is any such thing as God, is that he has granted us free will and now looks on, with great concern and love, to see how we exercise it: to see if we take the chance offered by this crisis to bow down and humble ourselves, or if we compound original sin with terminal sin.

And if what I fear indeed happens? If the next twenty years sees us pump ever more gas into the sky, and if it sees us take irrevocable steps into the genetically engineered future, what solace then? The only ones in need of consolation will be those of us who were born in the transitional decades, too early to adapt completely to a brave new ethos.

I've never paid more than the usual attention to the night sky, perhaps because I grew up around cities, on suburban blocks lined with street-lights. But last August, on a warm Thursday afternoon, my wife and I hauled sleeping bags high into the mountains and laid them out on a rocky summit and waited for night to fall and the annual Perseid meteor shower to begin. After midnight, it finally started in earnest—every minute, every thirty seconds, another spear of light shot across some corner of the sky, so fast that unless you were looking right at it you had only the sense of a flash. Our bed was literally rock-hard, and when, toward dawn, an unforecast rain soaked our tentless clearing, it was cold—but the night was glorious, and I've since gotten a telescope. When, in *Paradise Lost*, Adam asks about the movements of the heavens, Raphael refuses to answer. "Let it speak," he says, "the Maker's high magnificence, who built / so spacious, and his line stretcht out so far; / That man may know he dwells not in his own; / An edifice too large for him to fill, / Lodg'd in a small partition, and the rest / Ordain'd for uses to his Lord best known." We may be creating microscopic nature; we may have altered the middle nature all around us; but this vast nature above our atmosphere still holds mystery and wonder. The occasional satellite does blip across, but it is almost a self-parody. Someday, man may figure out a method of conquering the stars, but at least for now when we look into the night sky, it is as Burroughs said: "We do not see ourselves reflected there—we are swept away from ourselves, and impressed with our own insignificance."

As I lay on the mountaintop that August night I tried to pick out the few constellations I could identify—Orion's Belt, the Dippers. The ancients, surrounded by wild and even hostile nature, took comfort in seeing the familiar above them—spoons and swords and nets. But we will need to train ourselves not to see those patterns. The comfort we need is inhuman.

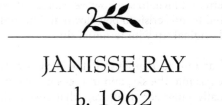

JANISSE RAY
b. 1962

In the introduction to the 1990 Norton Book of Nature Writing, *the editors remarked on the relative scarcity of reflective environmental writing from the southeastern part of the United States. Janisse Ray is one of several new writers making up for that lack. Her 1999 book* Ecology of a